GASEOUS ELECTRONICS

Tables, Atoms, and Molecules

Gorur Govinda Raju

CRC Press is an imprint of the
Taylor & Francis Group, an **informa** business

CRC Press
Taylor & Francis Group
6000 Broken Sound Parkway NW, Suite 300
Boca Raton, FL 33487-2742

First issued in paperback 2017

ISBN-13: 978-1-4398-4894-4 (hbk)
ISBN-13: 978-1-138-07724-9 (pbk)

Library of Congress Cataloging-in-Publication Data

Raju, Gorur G., 1937-
Gaseous electronics : tables, atoms, and molecules / Gorur Govinda Raju.
p. cm.
"A CRC title."
Includes bibliographical references and index.
ISBN 978-1-4398-4894-4 (hardback)
1. Plasma (Ionized gases)--Conductivity. 2. Gases--Electric properties. 3. Electric discharges through gases. 4. Electronic apparatus and appliances--Materials. I. Title.

QC718.5.T7R35 2011
530.4'4--dc23 2011025312

Visit the Taylor & Francis Web site at
http://www.taylorandfrancis.com

and the CRC Press Web site at
http://www.crcpress.com

Dedicated to my grandchildren

Tara and Dhruva

Contents

SECTION IV 4 Atoms

SECTION V 5 Atoms

SECTION VIII 8 Atoms

SECTION IX 9 Atoms

SECTION X 10 Atoms

SECTION XI 11 Atoms

SECTION XII 12 Atoms

SECTION XIII More than 12 Atoms

SECTION XIV Gas Mixtures

SECTION XV Appendices

Preface

> Why keep talking at length? In all the three worlds involving moving and non-moving entities, there is nothing that can be without mathematical sciences.
>
> **Mahavira (c. 850 AD)**
> *Indian Mathematician*

> When you can measure what you are speaking about and express it in numbers you know something about it; but when you cannot measure it, when you cannot express it in numbers, your knowledge is of a meager and unsatisfactory kind; it may be the beginning of knowledge, but you have scarcely, in your thoughts, advanced to the stage of science.
>
> **Lord Kelvin (William Thomson) 1883**

Interactions of energetic electrons and photons with atoms, molecules, excited states, and ions are generally understood to fall in the domain of gaseous electronics. Theoretical and experimental research into several facets of these interactions has continued till now, from the days when the concept of the structure of the atom, composed of electrons, protons and neutrons was revolutionary during the last years of the nineteenth century.

Ingenious methods were developed for the study of the interaction of electron beams with gas molecules, the energy of the beam being controlled and measured to an extraordinary degree of sophistication. Study of electrons undergoing collisions in a swarm with a distribution of energy formed a parallel branch of study. With the increasing complexity of these methods advantage was taken of the enormous storage of data and fast computation of modern computers. Methods were developed to improve the congruence of results obtained from beam studies and swarm measurements.

Theoretical research led to the development of quantum mechanics as a fundamental area of physics. Experimental research has led to the development of x-ray machines, fluorescent bulbs, neon signs and gas lasers without which it is difficult to imagine a modern society functioning. In power engineering applications, equipment such as high voltage cables and circuit breakers insulated with sulfur hexafluoride (SF_6) have contributed to enormous progress in the matter of reducing cost and increasing reliability of electrical power supply.

Innumerable applications have resulted from these studies. Gaseous plasma studies continue in universities, research organizations and industrial laboratories for the purposes of providing more efficient devices and for generating electrical power from controlled fusion processes. In nature, lightning and aurora lights are spectacular examples of electron interactions with neutral and charged atomic entities in the upper atmosphere.

Areas that involve this branch of knowledge include electrical and mechanical engineering, environmental studies, defense applications, agricultural industry just to name a few (Dedrick, 2010). Recent interest is in the areas of dusty plasma (Anderson and Radovanov, 1995) to quality improvement in semiconductor industry, understanding toxicity (Christophorou and Hadjiantoniou, 2006), biomedical applications (Lee, 2011), applications of plasma in medicine (Laroussi, 2009), treatment of skin (Babaeva and Kushner, 2010) and aerodynamics (Kogelshatz, 2004).

Plasma TV has become family décor. Electric arc furnaces, plasma torches for a variety of applications including generation of powders, for spraying, for cutting and welding, and for destruction of hazardous waste materials, miniature circuit breakers, electrical discharge machining, electrostatic precipitators, ozone generators—the list is ever growing—need data covered in this book.

In a previous book (*Gaseous Electronics: Theory and Practice*, Taylor & Francis, LLC, Boca Raton) the present author provided the theoretical background for understanding electron–neutron interactions and the experimental data available for selected atoms and molecules were cited to understand the complexities of various fundamental processes. At the time it was felt the volume of data available in the literature was so extensive, scattered in various journals in different disciplines of scientific study that a separate volume devoted to presentation of most, if not all, data available was desirable as a reference source for researchers in this field.

This book is complementary to the author's previous book (*Gaseous Electronics: Theory and Practice*) and the purposes are

1. To present data on the parameters stated earlier in a properly classified way for most, if not all, molecules that have been studied during the past 100 years.
2. To classify based on electron–neutron interactions, viz., collision cross sections, drift and diffusion parameters, ionization and attachment coefficients, attachment reactions and rates.
3. To supply a resource material for established universities, researchers, industrial laboratories, and research institutions.
4. To obtain data quickly on a variety of target molecules and verify data that are already available or processes that need to be verified. Questions such as, "Is the methane electron attaching?" or "Has the ionization cross section in uranium hexafluoride (UF_6) been measured?" can be checked.

Without violating the boundaries of academic modesty expected from researchers it is safe to claim that no such volume currently exists to meet all the above needs.

The data selected for presentation are

Total collision cross section, electron energy in the range 0–1000 eV
Differential cross section
Elastic and momentum transfer cross section
MERT coefficients
Rotational and vibrational cross section
Excitation cross section
Ionization cross section
Attachment and detachment cross section
Attachment processes and relevant energetics
Attachment rate coefficients
Drift velocity of electrons in a swarm
Mean and characteristic energy
Ionization coefficients
Attachment coefficients
Limited data on Ion mobility

It is appropriate to point out that not all these data are available in the published literature for all molecules included. Total cross section and ionization cross sections are the most frequently available though for theoretical work, simulation and calculation of transport coefficients (drift velocity and characteristic energy) from cross section data one requires momentum transfer cross sections.

In an effort to make the present volume user friendly and access to data rapid, it was decided, after several trial versions, to divide the contents into sections and subdivide each section into chapters. Each section is devoted to the number of atoms in the molecule. For example, Section I consists of atomic gases, Section III of molecules with three atoms, and so on up to Section XII. Section XIII is devoted to molecules with more than twelve atoms. The last section, Section XIV, is devoted to gas mixtures with a single entry, dry and humid air, which is of engineering importance as a natural insulating medium in overhead transmission lines, circuit breakers and so on.

Each section consists of individual chapters arranged in alphabetical order. For example, carbon dioxide (CO_2) takes precedence over water (H_2O) in Chapter 3; methane (CH_4) comes before silane (SiH_4) in Chapter 5.

Because of the volume, not all atoms are included in Section I and a selection has been made. Rare gases (argon, helium, neon etc.) are the obvious choice. In addition, atoms included are mercury (Hg) which is used in the lighting industry, sodium (Na) and cesium (Cs) used for seeding in generation of electricity by plasma techniques. The ionization cross section of common atoms, carbon (C), hydrogen (H), oxygen (O), nitrogen (N), sulfur (S) is required for development of theoretical methods of calculating the ionization cross sections from data of constituent atoms and these have been included in Appendix 6.

For each species electronic polarizability (α_e), dipole moment (μ) and ionization potential (ε_i) are some of the most basic information required for theoretical developments and analysis of experimental results. Of these, α_e and μ are significant factors in low energy electron interactions with neutral species. ε_i is, of course, important in connection with theory and experiments on ionization cross sections. These data (Lide, 2005–2006) are provided in the introductory part of each chapter.

For each species, available data are provided in both tabular and graphical form; the latter includes data from several authors employing a variety of experimental techniques with increasing accuracy as we move forward from early years to the present. Each table consists of the results of at least two different measurements, often at intervals of many years. To obtain data for these tables and to make comparison meaningful it became necessary to convert the available literature in various units into uniform SI units, first by digitizing, then by numerical integration and interpolation over a small range. A number of scientists have assisted the present author in generously providing tabulated data, as individually acknowledged below.

This book has ten appendices. Appendix 1 consists of fundamental constants and values for the most frequently used quantities. It is a fact of life that the study of gaseous electronics as a scientific discipline has resulted in referring to a atomic species either by name or by chemical formula depending upon the molecule's complexity. As an example carbon dioxide (CO_2) is referred to, by name, more frequently; CF_4 (tetrafluoromethane or perfluoromethane) is referred to by formula more frequently. It became necessary to provide cross reference to each species (over 430 of them) by name (Appendix 2) arranged alphabetically and by formula (Appendix 3), similarly arranged.

As stated earlier, not all the parameters listed above have been measured for all species. There are a large number of molecules for which only a single parameter has been reported in the literature. These are only total scattering cross section, ionization cross section, attachment cross section, and attachment rate coefficient. Since an entire chapter could not be devoted to each, in the interest of limiting the volume size, such data are given in appendices: attachment cross sections in Appendix 4, attachment rates in Appendix 5, and ionization cross sections for atoms in Appendix 6, ionization cross sections for molecules in Appendix 7.

Appendix 8 summarizes the relationships between various quantities, and Appendix 9 gives quadrupole moments of selected molecules, which are important in understanding rotational and vibrational excitation of selected molecules. Finally Appendix 10 lists the relative dielectric strengths of gases at atmospheric pressure that are required for engineering purposes.

Acknowledgments

This work would not have been possible without the facilities provided by the University of Windsor in my capacity as emeritus professor. The online facilities of the Leddy Library were an indispensable source that could be utilized from home, including the assistance of the most helpful library professionals for troubleshooting. Professor M.A. Sid-Ahmed very kindly placed all departmental facilities at the author's disposal. Special thanks are due to Alida De Marco for her persistence in tracking and securing a very expensive volume through interlibrary loan.

I am extremely grateful to the Natural Sciences and Engineering Research Council (NSERC) of Canada for their uninterrupted research grants for over 30 years, from 1980 onward. This support has helped in research and almost continuous acquisition and updating of scientific data during this period.

The advice and encouragement of Dr. V. Agarwal and Dr. Lakdawala, United States of America, from the beginning stages till completion, have helped enormously.

Many scientists gracefully provided me with tabulated data and research reprints (along with permission) to include in the volume. Professor Harland from the University of Canterbury, New Zealand kindly provided ionization cross-section data for over 80 molecules. Most of them are the only available data, appearing here for the first time in print. Such generosity to my request is indeed the true hallmark of a perfect gentleman, and humbled me beyond words.

Dr. E.V. Krishnakumar kindly provided tabulated values of attachment cross sections. Dr. J. Rene-Vacher, France kindly provided partial and total ionization cross sections. Dr. V. Lisovskiy, Ukraine kindly provided data on drift velocities, and Dr. V. Lakdawala, United States of America on attachment coefficients. Dr. M. Frechette, Canada kindly provided cross-section data and a large number of reprints. Dr. S. Buckman, Australia, Dr. Lindsay, United States of America, and A. Raju, Canada gave permission to reproduce transport data from research reports. I sincerely acknowledge all this help.

I wish to thank the journals that granted permissions to reproduce materials. The publishing administrator of The Institute of Physics (IOP) in England was extremely prompt and I can't thank IOP enough. The *Journal of American Physical Society, Journal of American Institute of Physics*, and *Journal of Italian Physical Society* are thanked for their permission received without delay. I have made sincere efforts to obtain permission from various publishers and apologize for any omission, which is inadvertent and entirely due to oversight.

I wish to thank many of my graduate students whose demanding queries forced me to be up to date with published literature. Ryan Marchand rendered valuable assistance connected with software in the preparation of the manuscript. Shelby Marchand and Andria Ballo provided efficient support in providing administrative and official help in the departmental office. I gratefully acknowledge the help received from F. Chichello and D. Tersigni for technical support that kept my computer system and associated equipment functioning smoothly.

It is a pleasure to acknowledge my professional association with Dr. Nagu Srinivas, Dr. Ed Cherney, Dr. M. Frichette, Professor S. Jayaram, Professor Ravi Gorur, Professor Reuben Hackam, Dr. Soli Bamji, Dr. Tangli Sudarshan, Professor K. J. Rao, Dr. Sri Hari Gopalakrishna, and Dr. M. Abou-Dakka and recall the many benefits I have derived through technical discussions.

I wish to thank the generosity of Professor S. Ramasesha and his wife at the Indian Institute of Science, Bangalore who made my stay at the institute campus very pleasant. Professor C. N. R. Rao, President, Jawaharlal Nehru Center for Advanced Scientific Research, kindly provided me with opportunity during the initial stages. I received generous hospitality and kindness from Professor N. Rudraiah and his wife, University of Bangalore. Mr Nagaraja Rao and his wife have been extraordinarily generous toward me as personal friends. Lt. Gen. Raghunath (Retd.) and his wife have cheerfully accepted my frequent intrusion into their home, often without prior notice. Professor Mandyam D. Srinivas kindly provided the sanskrit quotation in the front matter.

Words are not enough to express my gratitude to Nora Konopka, Editor, Taylor & Francis, LLC, whose encouragement sustained me until completion. The publication team of Jeniffer Ahringer, Brittany Gilbert, and Joette Lynch has shown admirable patience during the production stage. I am extremely pleased and thankful for the standard of proofreading completed by S.M. Syed and the team at Techset Composition.

Though a recipient of such enormous assistance and kindness, any errors, omissions, and shortfalls are entirely my own responsibility.

My son Anand and his ever welcoming wife Tanny cheerfully reconciled themselves to my preoccupation with completion of this volume. Finally, I thank my wife Padmini who generously forgave all the inconveniences and my inevitable nonparticipation in social and family responsibilities, which resulted from an undertaking such as this. The blessings of my parents from faraway abode have been a source of continuous strength all through my life.

REFERENCES

Anderson, H. M. and S. B. Radovanov, *J. Res. Nat. Inst. Stand. Technol.* 100, 449, 1995.

Babaeva, N. Y. and M. J. Kushner, *J. Phys. D: Appl. Phys.*, 43, 185206, 2010.

Christophorou, L. G. and D. Hadjiantoniou, *Chem. Phys. Lett.*, vol. 419, 405, 2006.

Dedrick, J., R. W. Boswell, and C. Charles, *J. Phys. D: Appl. Phys.*, 43, 342001, 2010.

Raju, G. G., *Gaseous Electronics: Theory and Practice*, Taylor & Francis LLC., Boca Raton, 2005.

Kogelshatz, U., Plasma technology, *Plasma Phys. Control. Fusion,* 46, B63-B75, 2004

Laroussi, M. *IEEE Transactions on Plasma Science,* 37, 714, 2009.

Lee, H. W., G. Y. Park, Y. S. Seo, Y. H. Im, S. B. Shim, and H. J. Lee, *J. Phys. D: Appl. Phys.,* 44, 053001, 2011.

Lide, D. R., *Handbook of Chemistry and Physics*, 86th Ed., 2005–2006, Taylor & Francis, LLC, Boca Raton.

Author

Professor Gorur Govinda Raju obtained a BEng degree from the University of Bangalore (India) and a PhD from the University of Liverpool (England) in 1963. He has held the Leverhulme Fellowship and the Leverhulme Travel Fellowship at the University of Liverpool during graduate studies. He then worked as a research engineer at Associated Electrical Industries (Manchester, England), where he was awarded a research premium for one of his research papers. He joined the Department of High Voltage Engineering, Indian Institute of Science, Bangalore, in 1965 and became professor and chairman during the years 1975–1980. He has held the Commonwealth Fellowship and, concurrently, visiting lecturership at the University of Sheffield (1972–73 and 1973–74). He has been a visiting professor to Bhabha Atomic Research Centre (Trombay) and University of Bangalore.

He joined the University of Windsor (Canada) in 1980 and became professor and head of the Electrical and Computer Engineering Department during 1989–97 and 2000–2002. He has been on the board and program committee of the Conference on Electrical Insulation and Dielectric Phenomena (IEEE) for a number of years. He is currently an emeritus professor at the University of Windsor. He has been a consultant on electrical power and dielectric phenomena to the government of India, Detroit Edison Co., and several other industries. He is a life fellow of the Institution of Engineers (India), a registered professional engineer of Ontario, and a life senior member of the Institution of Electrical and Electronics Engineers (USA) (IEEE) and has been cited by the non-commercial publication American Men and Women of Science. He has published over 140 papers in international journals and conferences and three previous books. His experimental and theoretical contributions to gaseous electronics continue to be cited in research papers on this topic.

Section I

1 ATOM

1

ARGON

Ar

CONTENTS

Argon (Ar) atom has 18 electrons with an electronic polarizability of 1.826×10^{-40} F m^2 and a first ionization potential of 15.76 eV.

1.1 SELECTED REFERENCES FOR DATA

See Table 1.1.

TABLE 1.1
Selected References

Quantity	Range: eV (Td)	Reference
Drift velocity	(1–3000)	**Lisovskiy et al. (2006)**
All cross sections	0–1000	Raju (2005)
Inelastic cross sections	10–1000	Yanguas-Gil et al. (2005)
All cross sections	0–1000	Raju (2004)
Q_i	16–1000	**Kobayashi et al. (2002)**
Q_i	17–1000	**Rejoub et al. (2002)**
Q_i	140–4000	**Sorokin et al. (2000)**
Q_T	0.5–10,000	Zecca et al. (2000)
Q_i, (α/N)	16–1000, (25–2000)	Phelps and Petrović (1999)
Q_{el}, Q_{ex}, Q_i	45–1000	Brusa et al. (1996)
Q_{diff}, Q_{el}, Q_M	1–10	**Gibson et al. (1996)**

continued

Ar

TABLE 1.1 (continued)
Selected References

Quantity	Range: eV (Td)	Reference
Q_T	0.5–220	**Szmytkowski and Maciag (1996)**
Q_i	17–1000	**Straub et al. (1995)**
Q_i	18–5300	**McCallion et al. (1992)**
Q_i	18–660	**Syage (1992)**
Q_M	0.0001–100	**Pack et al. (1992)**
Q_i	20–500	**Ma et al. (1991)**
Q_T	3–20	**Furst et al. (1989)**
Q_i	20–1000	**Krishnakumar and Srivastava (1988)**
Transport coefficients	(0.25–50)	**Nakamura and Kurachi (1988)**
Q_i	20–1000	Lennon et al. (1988)
Diffusion coefficient	56.5–5650	**Al-Amin and Lucas (1987)**
Q_T	0.7–10.0	**Subramanian and Kumar (1987)**
Q_{diff}, Q_{el}, Q_M	40–1000	**Iga et al. (1987)**
Q_{el}	3–300	Nahar and Wadhera (1987)
Q_i	16–200	Wetzel et al. (1987)
Q_T	100–3000	**Zecca et al. (1987)**
Q_T	0.12–20.0	**Buckman and Lohmann (1986)**
Q_T	0.08–20.0	**Ferch et al. (1985)**
Q_T	4–300	**Nickel et al. (1985)**
Q_T	20–100	**Wagenaar and de Heer (1985)**
Q_T	0.05–60	**Jost et al. (1983)**
Q_i	500, 700, 1000	**Nagy et al. (1980)**
Q_i	20–180	**Stephan et al. (1980)**
Q_T	15–800	**Kaupilla et al. (1981)**
Transport coefficients	(2.8–566)	**Kücükarpaci and Lucas (1981)**
Q_{diff}, Q_M	3–100	**Srivastava et al. (1981)**
Q_T, Q_{diff}, Q_{el}, Q_M	15–3000	de Heer et al. (1979)
Q_i	16–500	**Fletcher and Cowling (1973)**
Q_i	16–1000	**Rapp and Englander-Golden (1965)**
α/N	20–2000	**Kruithoff (1940)**

Note: Q_{diff} = differential scattering; Q_T = total scattering; Q_{el} = elastic scattering; Q_M = momentum transfer; α/N = reduced first ionization coefficient; Q_i = ionization; Q_{ex} = excitation. Bold font indicates experimental study.

1.2 TOTAL SCATTERING CROSS SECTION

The highlights of total cross section as a function of electron energy are

1. Increasing cross section as the energy is reduced toward zero, reaching a plateau of 7.5×10^{-20} m^2.
2. Ramsauer–Townsend minimum at ~0.25 eV.
3. A broad resonance at ~20 eV. This feature, attributed to onset of inelastic collisions including ionization, is common to many gases.
4. A monotonic decrease for energy >20 eV. This feature is also common to many gases (see Table 1.2).

1.3 ELASTIC SCATTERING CROSS SECTION

See Table 1.3.

Ar

TABLE 1.2
Recommended Total Scattering Cross Sections for Ar

Energy (eV)	Q_T (10^{-20} m^2)	Energy (eV)	Q_T (10^{-20} m^2)	Energy (eV)	Q_T (10^{-20} m^2)
0.08	1.50	0.9	1.05	30	14.52
0.09	1.36	0.95	1.14	40	12.09
0.1	1.19	1	1.22	50	10.69
0.11	1.10	1.1	1.40	60	9.89
0.12	1.00	1.2	1.57	70	9.32
0.13	0.910	1.3	1.79	80	8.82
0.14	0.820	1.4	1.98	90	8.39
0.15	0.750	1.5	2.13	100	8.00
0.16	0.687	1.6	2.33	125	7.26
0.17	0.625	1.7	2.52	150	6.65
0.2	0.491	1.9	2.88	200	5.83
0.25	0.383	2	2.97	250	5.23
0.3	0.327	3	4.84	300	4.80
0.345	0.311	4	6.45	350	4.51
0.4	0.328	5	8.64	400	4.18
0.45	0.367	6	10.10	450	3.90
0.5	0.416	7	12.10	500	3.66
0.55	0.478	8	14.00	550	3.46
0.6	0.548	10	18.90	600	3.28
0.65	0.624	12	23.10	650	3.12
0.7	0.700	14	24.00	700	2.97
0.75	0.780	16	23.50	800	2.89
0.8	0.870	20	20.20	900	2.70
0.85	0.960	25	16.21	1000	2.51

Note: See Figure 1.1 for graphical presentation.

TABLE 1.3
Elastic Scattering Cross Section for Ar

Energy (eV)	Q_{el} (10^{-20} m^2)	Energy (eV)	Q_{el} (10^{-20} m^2)	Energy (eV)	Q_{el} (10^{-20} m^2)
0.08	1.500	0.8	0.756	75	5.82
0.09	1.360	0.9	0.934	80	5.49
0.1	1.190	1.0	1.07	90	5.00
0.11	1.100	1.5	2.19	100	4.86
0.12	1.037	2.0	3.12	125	4.33
0.13	0.844	3.0	4.96	150	3.79
0.14	0.778	4.0	6.82	200	3.20
0.15	0.680	5.0	8.81	250	2.83
0.16	0.646	6.0	11.13	300	2.47
0.17	0.581	8.0	15.66	350	2.29
0.18	0.553	10	20.06	400	2.12
0.19	0.526	12	21.25	450	2.00
0.20	0.500	14	22.56	500	1.89
0.25	0.334	15	23.22	550	1.81
0.30	0.313	16	22.30	600	1.73
0.34	0.318	18	19.52	650	1.65
0.36	0.318	20	18.60	700	1.57
0.40	0.309	25	16.27	750	1.49
0.45	0.328	30	13.94	800	1.45
0.50	0.374	40	9.51	850	1.41
0.55	0.437	50	7.74	900	1.36
0.60	0.487	60	6.80	950	1.32
0.65	0.555	70	6.15	1000	1.27
0.70	0.629				

Source: Adapted from Raju, G. G. *Gaseous Electronics: Theory and Practice*, Taylor & Francis, New York, NY, 2005.
Note: See Figure 1.2 for graphical presentation.

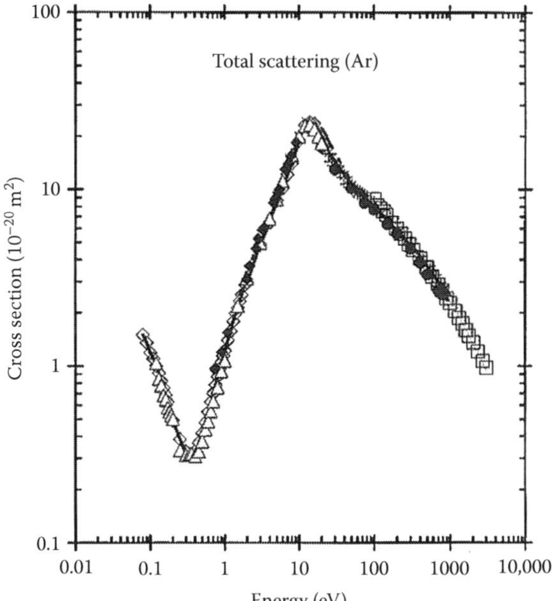

FIGURE 1.1 Total scattering cross section for Ar. (—) Recommended, Raju (2005); (– –) Zecca et al. (2000), empirical formula; (—) Szmytkowski and Maciąg (1996); (◆) Subramanyam and Kumar (1987); (□) Zecca et al. (1987); (△) Buckman and Lohmann (1986); (◊) Ferch et al. (1985); (- — -) Nickel et al. (1985); (+) Wagenaar and de Heer (1985); (×) Jost et al. (1983); (●) Kaupilla et al. (1981); (●—●) de Heer et al. (1979).

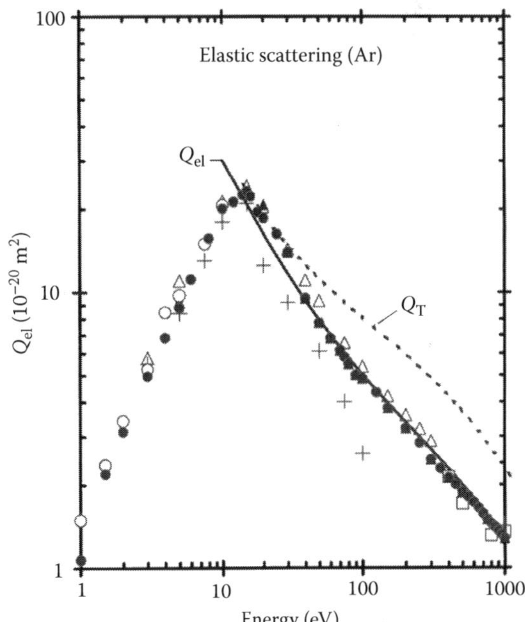

FIGURE 1.2 Elastic scattering cross section for Ar. (•) Raju (2005); (—) Zecca et al. (2000), semiempirical expression; (— —) Q_T, Zecca et al. (2000), semiempirical expression; (○) Gibson et al. (1996); (■) Furst et al. (1989); (□) Iga et al. (1987); (△) Nahar and Wadhera (1987); (+) Srivastava et al. (1981); (▲) de Heer et al. (1979).

Ar

1.4 MOMENTUM TRANSFER CROSS SECTION

See Table 1.4.

TABLE 1.4
Recommended Momentum Transfer Cross Section

Energy (eV)	Q_M (10^{-20} m²)	Energy (eV)	Q_M (10^{-20} m²)	Energy (eV)	Q_M (10^{-20} m²)
0.000	7.5	0.35	0.235	6.0	8.1
0.001	7.5	0.40	0.33	7.0	9.6
0.002	7.1	0.50	0.51	8.0	11.7
0.003	6.7	0.70	0.86	10	15.0
0.005	6.1	1.0	1.38	12	15.2
0.007	5.4	1.2	1.66	15	14.1
0.009	5.05	1.3	1.82	17	13.1
0.010	4.6	1.5	2.1	20	9.5
0.015	3.75	1.7	2.3	25	7.4
0.020	3.25	1.9	2.5	30	6.0
0.030	2.5	2.1	2.8	50	3.5
0.040	2.05	2.2	2.9	75	2.3
0.050	1.73	2.5	3.3	100	1.7
0.070	1.13	2.8	3.8	150	1.1
0.10	0.59	3.0	4.1	200	0.85
0.12	0.4	3.3	4.5	300	0.45
0.15	0.23	3.6	4.9	400	0.28
0.17	0.16	4.0	5.4	500	0.20
0.20	0.103	4.5	6.1	700	0.15
0.25	0.091	5.0	6.7	1000	0.12
0.30	0.153				

Note: See Figure 1.3 for graphical presentation.

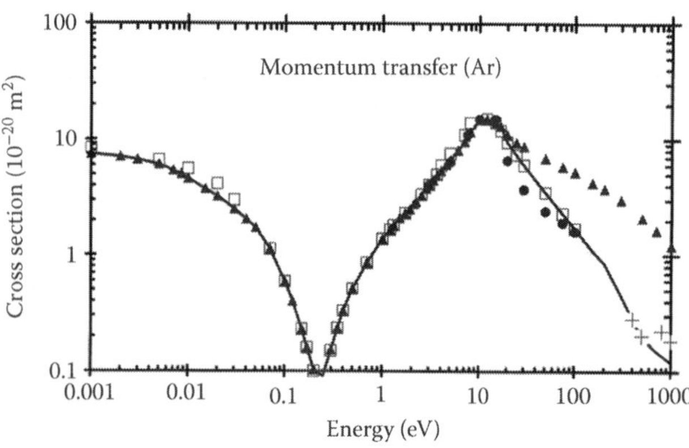

FIGURE 1.3 Momentum transfer cross sections for Ar. (—) Recommended; (▲) Pitchford et al., (1994) courtesy of main author (2003); (□) Pack et al. (1992); (+) Iga et al. (1987); (●) Srivastava et al. (1981).

1.5 SCATTERING LENGTH

See Table 1.5.

TABLE 1.5
Scattering Length (A_0) for Ar

Length (10^{-11} m)	Reference
−7.72	Petrović et al. (1995)
−8.92	Pack et al. (1992)
−7.63	Buckman and Lohmann (1986)
−8.68	Weyhreter et al. (1988)
−7.67	Ferch et al. (1985)
−7.97	McEachran and Stauffer (1983)
−7.87	Haddad and O'Malley (1982)

Note: $4\pi A_0^2$ is the cross section at zero electron energy. Scattering length is assigned a negative value in gases that exhibit the Ramsauer–Townsend effect.

1.6 RAMSAUER–TOWNSEND MINIMUM

Ramsauer–Townsend minimum is observed in elastic scattering, momentum transfer, and total cross sections as shown in Table 1.6.

TABLE 1.6
Ramsauer–Townsend Minimum in Ar

Authors	Method	Energy (eV)	Cross Section (10^{-20} m²)
Pack et al. (1992)	Momentum transfer	0.25	0.091
Buckman and Lohmann (1986)	Total	0.30	0.31
Ferch et al. (1985)	Total	0.34	0.31
Milloy et al. (1977)	Momentum transfer	0.25	0.095
Golden and Bandel (1966)	Total	0.28	0.15

1.7 EXCITATION CROSS SECTION

Only electronic excitation cross sections are considered. Threshold energies for selected excitation are shown in Table 1.7. See Figure 1.4 for graphical presentation of cross sections for individual levels. Table 1.8 and Figure 1.5 show selected excitation cross sections available in the literature.

TABLE 1.7
Selected Excitation Threshold Energy for Ar in the 11–15 eV Energy Range

Level Number	Threshold Energy (eV)	Notation
1	11.55	$4s[3/2]^0_2\ ^3P_2$
2	11.72	$4s'[1/2]^0_0\ ^3P_0$
3	12.91	$4p[1/2]_1$
4	13.27	$4p[1/2]_0$
	13.28	$4p'[3/2]_1$
5	13.48	$4p'[1/2]_0$
6	13.85	$3d[1/2]^0_0$
7	14.15	$3d[3/2]^0_1$
8	14.21	$3d'[3/2]^0_2$
9	14.30	$3d'[3/2]^0_1$
10	14.46	$5p[1/2]^0_1$
11	14.58	$5p[1/2]_0$
12	14.68	$5p'[3/2]_1$
13	14.69	$4d[1/2]^0_0$
14	14.74	$5p'[1/2]_0$
15	14.86	$4d[3/2]^0_1$
16	14.90	$4f[3/2]_2$
17	14.91	$4f[7/2]_3$
18	14.95	$4d'[3/2]^0_1$
19	15.00	$4d'[3/2]^0_1$

Source: Adapted from G. G. Raju, *IEEE Trans. Diel. Elec. Insul.*, 11, 649, 2004.

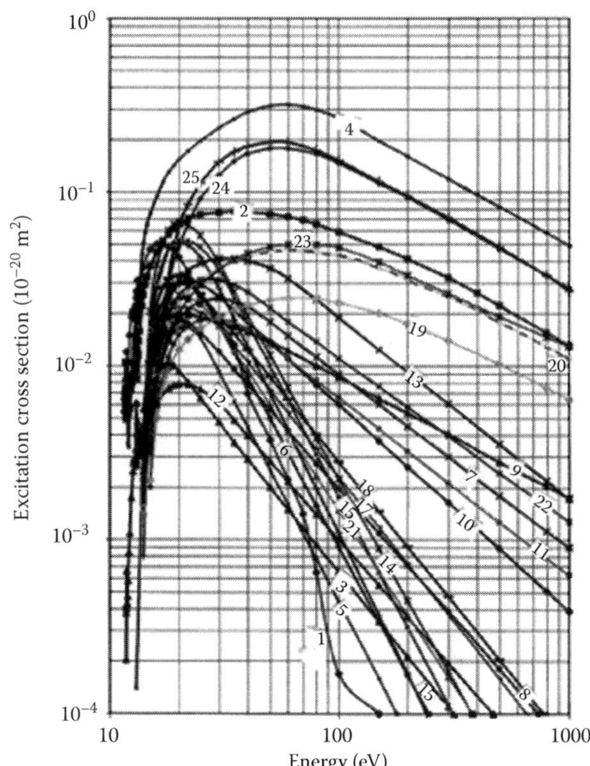

FIGURE 1.4 Excitation cross sections for Ar. Numbers for curves refer to the first column of Table 1.7. (Adapted from M. Hayashi. The Gaseous Electronics Institute, 503 Sakae High Home, 15-14 Sakae 4-chome, Nakaku, Nagoya 460, Japan. See also K. Kosaki and M. Hayashi, Denkigakkai Zenkokutaikai Yokoshu (*Prepr. Natl. Meet. Inst. Electr. Eng. Jpn.*), 1992 (in Japanese).)

1.8 IONIZATION CROSS SECTIONS

Seventeen sets of data are available as shown in Figure 1.6. Recommended values are shown in Table 1.9 and Figure 1.7.

TABLE 1.8
Excitation Cross Section

Energy (eV)	Q_{ex} (10^{-20} m^2)	Energy (eV)	Q_{ex} (10^{-20} m^2)
11.62	0.001	70.0	0.689
12.0	0.003	80.0	0.670
14.0	0.035	90.0	0.649
16.0	0.097	100	0.627
18.0	0.175	200	0.454
20.0	0.256	300	0.357
25.0	0.428	400	0.296
30.0	0.542	500	0.254
35.0	0.615	600	0.223
40.0	0.658	700	0.200
45.0	0.683	900	0.166
50.0	0.696	1000	0.154
60.0	0.700		

Sources: Adapted from Pitchford, L. C., J. P. Bouef, and J. P. Morgan, www.siglo-kinema.com. 1996. Also see Fiala, A., L. C. Pitchford, and J. P. Boeuf, *Phys. Rev. E*, 49, 5607, 1994.

Ar

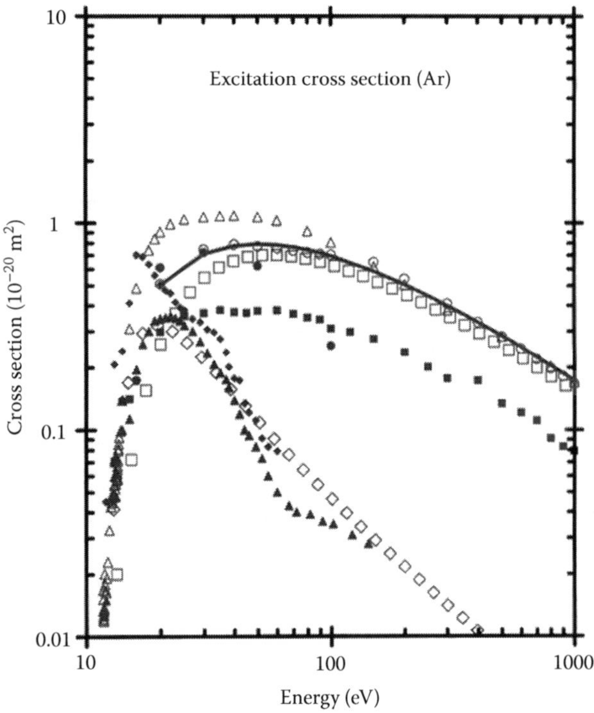

FIGURE 1.5 Selected excitation cross section for Ar. (—) Brusa et al. (1996), semiempirical expression; (◇) Pitchford et al. (1996), total metastables, swarm calculation; (□) resonance levels, swarm calculation; (■) Tsurubuchi et al. (1996), 106.6 and 104.8 nm radiations, measurement of total cascade cross section; (▲) Mason and Newell (1987), total metastable excitation; (◆) Tsurubuchi + Mason and Newell, derived; (●) Chutjian and Cartwright (1981); (△) Hayashi (1992), sum of 25 states; (○) de Heer et al. (1979).

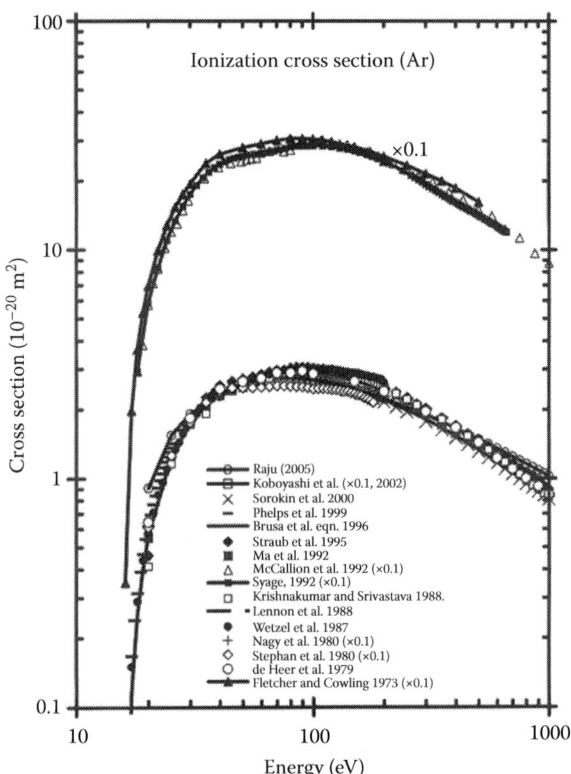

FIGURE 1.6 Absolute total ionization cross sections in Ar. The data are presented by two curves for more clarity. The values shown by the top curve should be multiplied by 0.1. Figure compiled by Raju (2005). (•—•) Raju (2005); (—□—) Kobayashi et al. (2002); (×) Sorokin et al. (2000); (–) Phelps and Petrović (1999); (——) Brusa et al. (1996); (□) Straub et al. (1995); (○) McCallion et al. (1992); (—■—) Syage (1992); (▲) Ma et al. (1991); (△) Krishnakumar and Srivastava (1988). Semiempirical values: (— - —) Lennon et al. (1988); (– –) Wetzel et al. (1987); (◇) Nagy et al. (1980); (◇) Stephan et al. (1980); (●) de Heer et al. (1979); (♦) Fletcher and Cowling, (1973).

1.9 DRIFT VELOCITY OF ELECTRONS

See Table 1.10.

1.10 DIFFUSION COEFFICIENT

Radial diffusion coefficients expressed as a ratio to mobility (D_r/μ), also termed as characteristic energy, at 293 K are shown in Table 1.11 and Figure 1.10 (Raju, 2005).

1.11 MEAN ENERGY

See Table 1.12.

TABLE 1.9
Ionization Cross Sections for Ar

Rejoub et al. (2002)		Kobayashi et al. (2002)		Rapp and Englander-Golden (1965)	
Energy (eV)	**Q_i (10^{-20} m^2)**	**Energy (eV)**	**Q_i (10^{-20} m^2)**	**Energy (eV)**	**Q_i (10^{-20} m^2)**
17	0.159	17	0.128	17	0.134
18.5	0.419	18	0.281	18.5	0.377
20	0.604	19	0.440	20	0.627
21	0.769	20	0.600	21	0.787
22.5	1.000	22	0.893	22.5	0.994
25	1.250	25	1.246	25	1.302
27.5	1.580	26	1.348	28	1.601
30	1.750	30	1.727	30	1.803
32.5	2.070	32	1.878	32	1.962
35	2.210	36	2.148	36	2.243
40	2.410	40	2.291	40	2.393
45	2.491	45	2.380	45	2.489
50	2.554	50	2.428	50	2.533
55	2.608	55	2.489	55	2.595
60	2.671	60	2.544	60	2.656
65	2.738	65	2.614	65	2.727
70	2.794	70	2.648	70	2.771
75	2.804	75	2.697	75	2.815
80	2.860	80	2.722	80	2.841
90	2.898	90	2.737	90	2.859
100	2.873	100	2.726	100	2.850
110	2.848	110	2.708	110	2.832
120	2.785	120	2.690	120	2.806
140	2.680	140	2.612	140	2.727
160	2.582	160	2.524	160	2.621
180	2.488	180	2.413	180	2.516
200	2.382	200	2.334	200	2.393
225	2.277	225	2.219		
250	2.167	250	2.111	250	2.173
275	2.033				
300	1.955	300	1.917	300	1.979
350	1.783	350	1.855	350	1.812
400	1.659	400	1.797	400	1.680
500	1.427	500	1.643	500	1.460
600	1.269	600	1.592	600	1.302
700	1.139	700	1.527	700	1.161
800	1.030	800	1.450	800	1.064
900	0.957	900	1.410	900	0.985
1000	0.881	1000	1.364	1000	0.915

Note: See Figures 1.7 and 1.8 for graphical presentation.

Sources: Adapted from Rejoub, R., B. G. Lindsay, and R. F. Stebbings, *Phys. Rev. A*, 65, 042713, 2002; Courtesy of Dr. Kobayashi for tabulated data in columns 3 and 4.

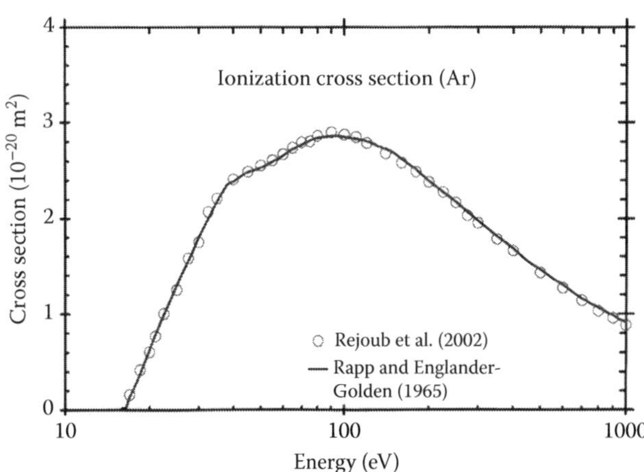

FIGURE 1.7 Total ionization cross sections for Ar. (○) Rejoub et al. (2002), recommended; (—) Rapp and Englander-Golden (1965).

TABLE 1.10
Recommended Drift Velocity of Electrons

E/N (Td)	*W* (10^3 m/s)	*E/N* (Td)	*W* (10^3 m/s)	*E/N* (Td)	*W* (10^3 m/s)
1×10^{-2}	0.935	0.4	2.39	40	34.4
1.2	0.972	0.5	2.52	50	41.2
1.4	1.005	0.60	2.63	60	49.1
1.7	1.047	0.70	2.73	70	58.4
2.0	1.084	0.80	2.81	80	68.0
2.5	1.144	1.0	2.95	100	85.5
3.0	1.205	2.0	3.44	200	149
3.5	1.252	3.0	3.81	300	217
4.0	1.294	4.0	4.12	400	282
5.0	1.368	5.0	4.75	500	351
6.0	1.437	6.0	5.64	600	424
7.0	1.500	7.0	6.54	700	497
8.0	1.556	8.0	7.61	800	568
0.1	1.654	10	9.56	1000	704
0.12	1.741	12	11.7	2000	1190
0.14	1.820	14	13.4	3000	1600
0.17	1.918	17	16.3	4000	1770
0.20	2.00	20	18.8	5000	1860
0.25	2.13	25	22.4	6000	2060
0.30	2.23	30	27.1		
0.35	2.31	35	31.1		

Source: Adapted from Raju, G. G., *Gaseous Electronics: Theory and Practice*, Taylor & Francis, New York, NY, 2005.

Note: See Figure 1.9 for graphical presentation.

Ar

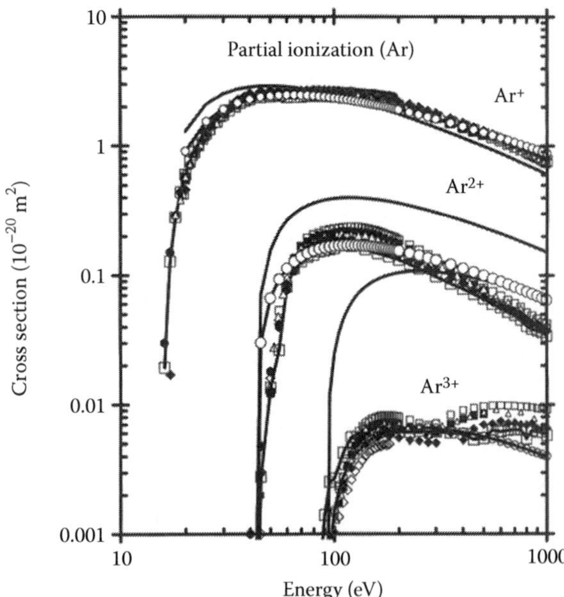

FIGURE 1.8 Partial ionization cross sections in Ar. (○—○) Raju (2005); (—□—) Kobayashi et al. (2002); (♦) Straub et al. (1995); (■) Ma et al. (1992); (△) McCallion et al. (1992); (□) Krishnakumar and Srivastava (1988); (—) Lennon et al. (1988); (•) Wetzel et al. (1987); (◇) Stephan et al. (1980). The results of Syage (1992) and Rejoub et al. (2002) are not shown.

TABLE 1.11
Characteristic Energy for Ar; $0.01 \le E/N \le 50$[a], $56.5 \le E/N \le 5650$[b]

E/N (Td)	D_r/μ (V)	E/N (Td)	D_r/μ (V)	E/N (Td)	D_r/μ (V)
0.01	0.45	10	7.70	254	7.38
0.02	0.8	12	7.42	424	7.45
0.04	1.05	14	7.37	565	8.05
0.07	1.45	20	7.36	678	8.55
0.10	1.85	25	7.50	848	8.86
0.35	2.8	30	7.61	1130	11.2
1.0	4.78	35	7.60	1413	11.8
2.0	6.35	40	7.46	1695	13.0
3.0	7.36	50	6.91	1978	14.6
4.0	8.04	56.5	6.40	2260	15.1
5.0	8.24	84.8	7.66	2825	16.4
6.0	8.29	113	7.35	3390	15.2
7.0	8.54	141	7.79	4238	14.0
8.0	8.18	198	7.11	5650	12.8

Sources: [a] Adapted from Raju, G. G. *Gaseous Electronics: Theory and Practice*, Taylor & Francis, New York, NY, 2005.
[b] Adapted from Al-Amin, S. A. J. and J. Lucas, *J. Phys. D: Appl. Phys.*, 20, 1590, 1987.

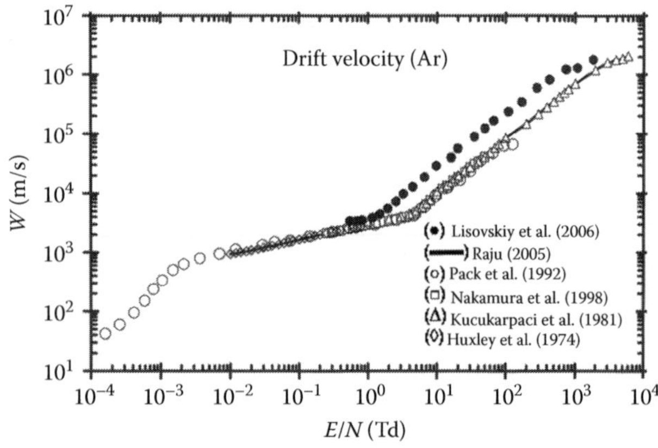

FIGURE 1.9 Drift velocity of electrons for Ar. (•) Lisovskiy et al. (2006), digitized; (—) recommended, Raju (2005); (○) Pack et al. (1992); (□) Nakamura and Kurachi (1988); (△) Kücükarpaci and Lucas (1981); (◊) Huxley and Crompton (1974).

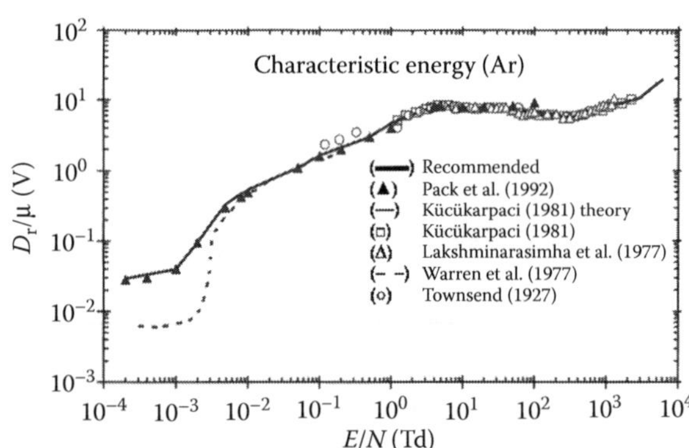

FIGURE 1.10 Characteristic energy for Ar at 293 K. Unless otherwise specified, values are experimental. (—) Recommended; (▲) Pack et al. (1992); (—) Kücükarpaci and Lucas (1981), theory; (□) Kücükarpaci and Lucas (1981); (△) Lakshminarasimha and Lucas (1977); Warren and Parker (1962), 77 K; (○) Townsend and Bailey (1922).

TABLE 1.12
Mean Energy of Electrons for Ar

E/N (Td)	$\bar{\varepsilon}$ (eV)	*E/N* (Td)	$\bar{\varepsilon}$ (eV)	*E/N* (Td)	$\bar{\varepsilon}$ (eV)
1	5.90	90	7.74	700	12.47
2	6.00	100	7.91	800	13.12
5	6.47	150	8.46	900	13.61
10	6.38	200	8.75	1000	13.94
20	7.14	250	9.38	1500	15.28
30	7.47	300	10.09	2000	18.09
40	7.44	350	10.59	3000	23.62
50	7.40	400	10.89	4000	31.42
60	7.40	450	11.09	5000	37.07
70	7.47	500	11.27		
80	7.58	600	11.80		

Note: See Figure 1.11 for graphical presentation.

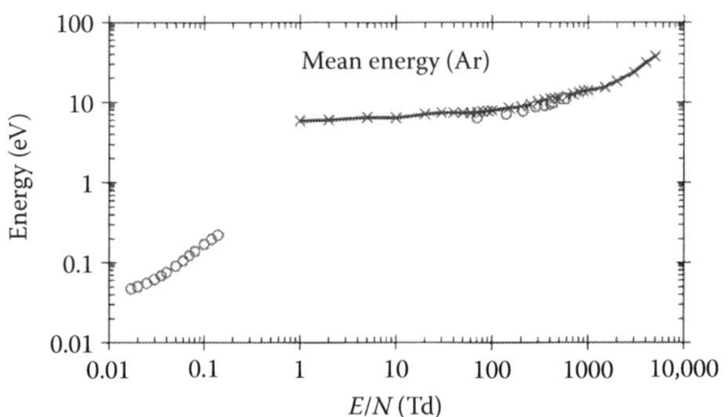

FIGURE 1.11 Mean energy of electron swarm in Ar as a function of *E/N*. (—×—) Recommended, Raju (2005). Experimental values unless otherwise mentioned. (□) Makabe et al. (1977); (○) Losee and Burch (1972), theory; (- - -) Kücükarpaci and Lucas (1981), overlapping the recommended values; (○) mean energy in Ar + 5.44% H_2 (Petrović et al., 1995) showing the dramatic reduction of $\bar{\varepsilon}$.

1.12 REDUCED FIRST IONIZATION COEFFICIENT

See Table 1.13.

1.13 GAS CONSTANTS

Gas constants evaluated according to expression (see inset of Figure 1.12)

$$\frac{\alpha}{N} = F \exp\left(\frac{GN}{E}\right) \tag{1.1}$$

TABLE 1.13
Reduced Ionization Coefficient for Ar

E/N (Td)	Phelps et al. (1999) α/N (10^{-20} m²)	Kruithoff (1940) α/N (10^{-20} m²)	Recommended α/N (10^{-20} m²)
20		3.00E–04	3.00E–04
25	9.28E–04	9.00E–04	9.14E–04
30	0.0021	0.0022	0.0022
35	0.0040	0.0042	0.0041
40	0.0069	0.0071	0.0070
50	0.0157	0.0151	0.0154
60	0.0277	0.0257	0.0267
70	0.0421	0.0389	0.0405
80	0.0581	0.0550	0.0566
90	0.0751	0.0734	0.0743
100	0.0930	0.0923	0.0927
125	0.141	0.151	0.146
150	0.195	0.213	0.204
175	0.255	0.277	0.266
200	0.320	0.342	0.331
225	0.390	0.406	0.398
250	0.463	0.472	0.468
300	0.614	0.606	0.610
350	0.764	0.735	0.750
400	0.910	0.860	0.885
450	1.05	0.981	1.02
500	1.18	1.10	1.14
550	1.30	1.22	1.26
600	1.41	1.33	1.37
650	1.51	1.44	1.48
700	1.61	1.54	1.58
750	1.70	1.63	1.67
800	1.78	1.72	1.75
850	1.86	1.81	1.84
900	1.93	1.90	1.92
950	1.99	1.97	1.98
1000	2.06	2.05	2.06
1500	2.50	2.65	2.58
1750	2.64	2.88	2.76
2000	2.76	3.07	2.92
2500	2.91	3.36	3.14
4000	3.10	3.76	3.43
4500	3.11	3.86	3.49

Note: See Figure 1.12 for graphical presentation.

are $F = 7.28 \times 10^{-21}$ m² and $G = 179.8$ Td for the range $25 \le E/N \le 350$ Td.

1.14 PENNING IONIZATION CROSS SECTION

Penning effect occurs according to

$$Ar^* + B \rightarrow Ar + B^+ + e \tag{1.2}$$

Ar

where Ar* is an excited atom and B is a second atom with ionization potential less than the Ar excitation potential (see Table 1.14).

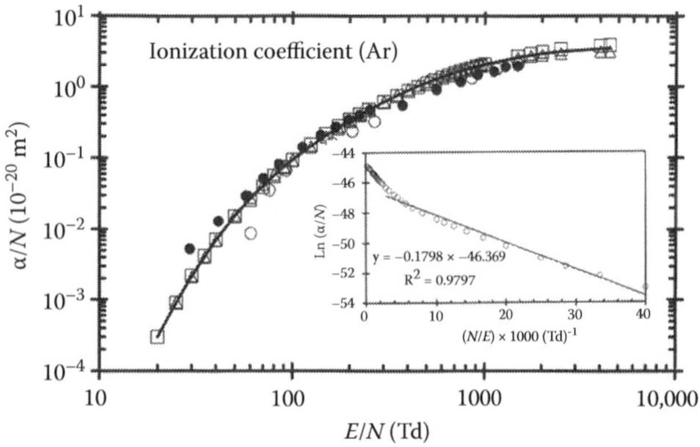

FIGURE 1.12 Reduced first ionization coefficient for Ar as a function of reduced electric field. (—) Recommended, Raju (2005); (△) Phelps and Petrović (1999); (●) Lakshminarasimha and Lucas (1977); (○) Milne and Davis (1959); (□) Kruithoff (1940). Inset shows plot according to equation.

TABLE 1.14
Penning Ionization Cross Section (in Units of 10^{-20} m^2) for Ar with Other Gases

Excitation Level (Ar*)	Ar* Energy (eV)	B: Kr	B: O_2
$^3P_2$	11.55	1	1.2
$^1P_1$	11.83		1.8
$^3P_0$	11.72		1.8
$^3P_1$	11.623		1.6

Source: Adapted from Eletskii, A. V. in I. S. Grigoriev and Z. Meilikhov (Eds.), *Handbook of Physical Quantities*, CRC Press, Boca Raton, FL, 1999, Chapter 20.

1.15 POSITIVE ION MOBILITY

See Table 1.15.

TABLE 1.15
Reduced Ion Mobility and Drift Velocity for Ar^+

E/N (Td)	μ_0 (10^{-4} m²/V s)	W^+ (10^2 m/s)	*E/N* (Td)	μ_0 (10^{-4} m²/V s)	W^+ (10^2 m/s)
8.00	1.53	0.329	150	1.16	4.68
10.0	1.53	0.411	200	1.06	5.70
12.0	1.53	0.493	250	0.99	6.65
15.0	1.52	0.613	300	0.95	7.66
20.0	1.51	0.811	400	0.85	9.14
25.0	1.49	1.00	500	0.78	10.5
30.0	1.47	1.18	600	0.72	11.6
40.0	1.44	1.55	800	0.63	13.5
50.0	1.41	1.89	1000	0.56	15.0
60.0	1.38	2.22	1200	0.51	16.4
80.0	1.32	2.84	1500	0.46	18.5
100	1.27	3.41	2000	0.40	21.5
120	1.22	3.93			

Source: Adapted from Akridge, G. R. et al., *J. Chem. Phys.*, 62, 4578, 1975.
Note: See Figure 1.13 for graphical presentation.

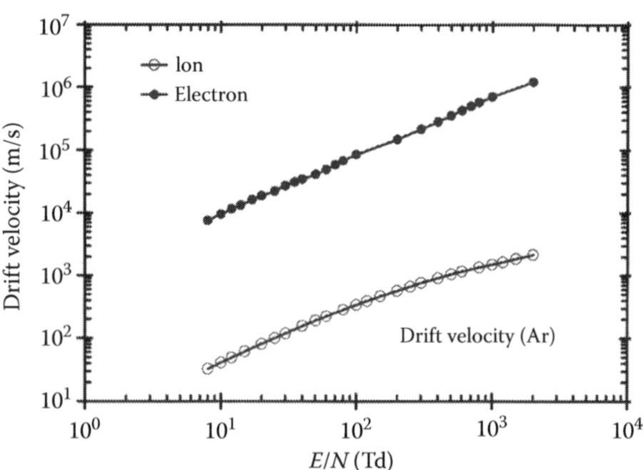

FIGURE 1.13 Positive ion drift velocity for Ar. Electron drift velocity added for comparison. (Adapted from Akridge, G. R. et al., *J. Chem. Phys.*, 62, 4578, 1975.)

REFERENCES

Akridge, G. R., H. W. Ellis, R. Y. Pai, and E. W. McDaniel, *J. Chem. Phys.*, 62, 4578, 1975.

Al-Amin, S. A. J. and J. Lucas, *J. Phys. D: Appl. Phys.*, 20, 1590, 1987.

Brusa, R. S., G. P. Karwasz, and A. Zecca, *Z. Phys. D*, 38, 279, 1996.

Buckman, S. J. and B. Lohmann, *J. Phys. B: At. Mol. Phys.*, 19, 2547, 1986.

de Heer, F. J., R. H. J. Jansen, and W. van der Kaay, *J. Phys. B: At. Mol. Phys.*, 12, 979, 1979.

Eletskii, A. V., Transport phenomena in weakly ionized plasma in I. S. Grigoriev and Z. Meilikhov (Eds.), *Handbook of Physical Quantities*, CRC Press, Boca Raton, FL, 1999, Chapter 20.

Ferch, J., B. Granitza, C. Masche, and W. Raith, *J. Phys. B: At. Mol. Phys.*, 18, 967, 1985.

Fletcher, J. and I. R. Cowling, *J. Phys. B: At. Mol. Phys.*, 6, L258, 1973.

Furst, J. E., D. E. Golden, M. Mahgerefteh, J. Zhou, and D. Mueller, *Phys. Rev. A*, 40, 5592, 1989.

Gibson, J. C., R. J. Gulley, J. P. Sullivan, S. J. Buckman, V. Chan, and P. D. Burrow, *J. Phys. B: At. Mol. Opt. Phys.*, 29, 3177, 1996.

Golden, D. E. and H. W. Bandel, *Phys. Rev.*, 138, A14, 1965; 149, 58, 1966.

Haddad, G. N. and T. F. O'Malley, *Austr. J. Phys.*, 35, 35, 1982.

Hayashi, M. The Gaseous Electronics Institute, 503 Sakae High Home, 15-14 Sakae 4-chome, Nakaku, Nagoya 460, Japan. See also K. Kosaki and M. Hayashi, Denkigakkai Zenkokutaikai Yokoshu (*Prepr. Natl. Meet. Inst. Electr. Eng. Jpn.*), 1992 (in Japanese).

Huxley, L. G. H. and R. W. Crompton, *The Diffusion and Drift of Electrons in Gases*, John Wiley and Sons, New York, NY, 1974.

Iga, I., L. Mu-Tao, J. C. Nogueira, and R. S. Barbieri, *J. Phys. B: At. Mol. Phys.*, 20, 1095, 1987.

Jost, K., P. G. F. Bisling, F. Eschen, M. Felsmann, and L. Walther, in J. Eichler et al. (Eds.), *Proceedings of the 13th International Conference on the Physics of Electronic Atomic Collisions*, Amsterdam, North-Holland, 1983, p. 91, cited by R. W. Wagenaar and F. J. de Heer (1985). Tabulated values above 7.5 eV are given in the latter reference. The lower range of 0.05 eV is obtained from Nickel et al. (1985).

Kaupilla, W. E., T. S. Stein, J. H. Smart, M. S. Dababneh, Y. K. Ho, J. P. Downing, and V. Pol, *Phys. Rev. A*, 24, 725, 1981.

Kobayashi, A., G. Fujiki, A. Okaji, and T. Masuoka, *J. Phys. B: At. Mol. Opt. Phys.*, 35, 2087, 2002. The author thanks Dr. Kobayashi for sending tabulated results of cross sections (January, 2003).

Krishnakumar, E. and S. K. Srivastava, *J. Phys. B: At. Mol. Phys.*, 21, 1055, 1988.

Kruithof, A. A., *Physica*, 7, 519, 1940.

Kücükarpaci, H. N. and J. Lucas, *J. Phys. D: Appl. Phys.*, 14, 2001, 1981.

Lennon, M. A., K. L. Bell, H. B. Gilbody, J. G. Hughes, A. E. Kingston, M. J. Murray, and F. J. Smith, *J. Phys. Chem. Ref. Data*, 17, 1285, 1988.

Lisovskiy, V., J.-P. Booth, K. Landry, D. Douai, V. Cassagne, and V. Yegorenkov, *J. Phys. D: Appl. Phys.*, 39, 660, 2006.

Losee, J. R. and D. S. Burch, *Phys. Rev.*, A6, 1652, 1972.

Ma, C., C. R. Sporeleder, and R. A. Bonham, *Rev. Sci. Instr.*, 62, 909, 1991.

McEachran, R. P. and A. D. Stauffer, *J. Phys. B., At. Mol. Phys.*, 16, 4023, 1983.

Makabe, T., T. Goto, and T. Mori, *J. Phys. B: At. Mol. Phys.*, 10, 1781, 1977.

Mason, N. J. and W. R. Newell, *J. Phys. B: At. Mol. Opt. Phys.*, 20, 1357, 1987.

McCallion, P., M. B. Shah, and H. B. Gilbody, *J. Phys. B: At. Mol. Phys.*, 25, 1061, 1992.

Milloy, H. B., R. W. Crompton, J. A. Rees, and A. G. Robertson, *Aust. J. Phys.*, 30, 61, 1977.

Nagy, P., A. Skutzlart, and V. Schmidt, *J. Phys. B: At. Mol. Phys.*, 13, 1249, 1980.

Nahar, S. N. and J. M. Wadhera, *Phys. Rev.*, 35, 2051, 1987.

Nakamura, Y. and M. Kurachi, *J. Appl. Phys.*, 21, 718, 1988.

Nickel, J. C., K. Imre, D. F. Register, and S. Trajmar, *J. Phys. B: At. Mol. Phys.*, 18, 125, 1985.

Pack, J. L., R. E. Voshall, and A. V. Phelps, *J. Appl. Phys.*, 71, 5363, 1992.

Petrović, Z. Lj., T. F. O'Mallory, and R. W. Crompton, *J. Phys. B: At. Mol. Opt. Phys.*, 28, 3309, 1995.

Phelps, A. V. and Z. Lj. Petrović, *Plasma Sources Sci. Tech.*, 8, R21–R44, 1999.

Pitchford, L. C., J. P. Bouef, and J. P. Morgan, www.siglo-kinema.com. 1996. Also see Fiala, A., L. C. Pitchford, and J. P. Boeuf, *Phys. Rev. E*, 49, 5607, 1994.

Raju, G. G., *Gaseous Electronics: Theory and Practice*, Taylor & Francis, New York, NY, 2005.

Raju, G. G., *IEEE Trans. Diel. Elec. Insul.*, 11, 649, 2004.

Rapp, D. and P. Englander-Golden, *J. Chem. Phys.*, 43, 1464, 1965.

Rejoub, R., B. G. Lindsay, and R. F. Stebbings, *Phys. Rev. A*, 65, 042713, 2002.

Sorokin, A. A., L. A. Schmaenok, S. B. Bobashev, B. Möbus, M. Richter, and G. Ulm, *Phys. Rev. A*, 61, 022723–1, 2000. Cross sections are measured for Ar, Kr, Xe.

Srivastava, S. K., H. Tanaka, A. Chutjian, and S. Trajmar, *Phys. Rev. A*, 23, 2156, 1981.

Stephan, K., H. Helm, and T. D. Märk, *J. Chem. Phys.*, 73, 3763, 1980. Gases studied are He, Ne, Ar, Kr; *J. Chem. Phys.*, 81, 3116. Gas studied is Xe.

Straub, H. C., P. Renault, B. G. Lindsay, K. A. Smith, and R. F. Stubbings, *Phys. Rev.*, A52, 1115, 1995.

Subramanian, K. P. and V. Kumar, *J. Phys. B: At. Mol. Phys.*, 20, 5505, 1987.

Syage, J. A., *Phys. Rev. A*, 46, 5666, 1992.

Szmytkowski, C. and K. Maciąg, *Phys. Scripta*, 54, 271, 1996.

Townsend, J. S. and V. A. Bailey, *Philos. Mag.*, 43, 593, 1922.

Tsurubuchi, S., T. Miyazaki, and K. Motohashi, *J. Phys. B: At. Mol. Opt. Phys.*, 29, 1785, 1996.

Wagenaar, R. W. and F. J. de Heer, *J. Phys. B: At. Mol. Phys.*, 18, 2021, 1985.

Warren, R. W. and J. H. Parker, *Phys. Rev.*, 128, 2661, 1962.

Wetzel, R. C., F. A. Baiocchi, T. R. Hayes, and R. S. Freund, *Phys. Rev. A*, 35, 559, 1987.

Weyhreter, M., B. Barzick, A. Mann, and F. Linder, *Z. Phys. D*, 7, 333, 1988.

Yanguas-Gil, A., J. Cotrino, and L. Alves, *J. Phys. D: Appl. Phys.*, 38, 1588, 2005.

Zecca, A., G. P. Karwasz, and R. S. Brusa, *J. Phys. B: At. Mol. Opt. Phys.*, 33, 843, 2000.

Zecca, A., S. Oss, G. Karwasz, R. Grisenti, and R. S. Brusa, *J. Phys. B: At. Mol. Phys.*, 20, 5157, 1987.

2

CESIUM

CONTENTS

Cesium (Cs) atom has 55 electrons with electron configuration $5p^6 6s\,(^2S_{1/2})$. The electronic polarizability is 66.11×10^{-40} F m^2 and the ionization potential is 3.894 eV.

2.1 SELECTED REFERENCES FOR DATA

See Table 2.1.

TABLE 2.1
Selected References for Data

Quantity	Range: eV, (Td)	Reference
Compilation	0–1000	Zecca et al. (1996)
Elastic scattering cross section	0–4	Bartschat (1993)
Elastic scattering cross section	0–2	Crown and Russek (1992)
Total cross section	2–18	**Jaduszliwer and Chan (1992)**
Momentum transfer cross section	0.05–2.0	Stefanov (1980)[a]
Drift velocity	2–100	**Saelee and Lucas (1979)**
Excitation cross section	2–1500	**Chen and Gallagher (1978)**
Excitation rates	0.25	Deutsch (1973)
Ionization coefficient	(150–1800)	**Garamoon and Surplice (1973)**
Momentum transfer cross section	0.01–10	**Nighan and Postma (1972)**[b]
Total scattering cross section	0.3–9	**Visconti et al. (1971)**[c]
Excitation cross section	2–30	Moiseiwitsch and Smith (1968)
Ionization cross section	5–100	**Nygaard (1968)**
Ionization cross section	5–500	**McFarland and Kinney (1965)**
Ionization cross section	300, 500	**Brink (1964)**
Drift velocity	(2.8–60)	**Chanin and Steen (1964)**
Ionization cross section	0–700	**Tate and Smith (1934)**
Total cross section	0–400	**Brode (1929)**

Note: Bold font indicates experimental study.

[a] An interesting approach was adopted to derive the momentum transfer cross sections in the energy range 0.05–2 eV. Six quantities, all of which involve the momentum transfer cross sections, were considered. The six quantities are (a) width of the electron cyclotron resonance, (b) attenuation of microwaves, (c) electrical conductivity in both equilibrium and nonequilibrium pure Cs or Ar–Cs plasmas, (d) electron thermal conductivity in a strong magnetic field, (e) perpendicular electrical conductivity in a strong magnetic field, and (f) electron drift velocity. An algorithm was generated to obtain the best fit with the experimental data. The Ramsauer–Townsend minimum was observed at 0.278 eV and the minimum cross section was 72×10^{-20} m^2. Two maxima are observed at 0.15 and 0.45 eV. Tabulated results are given in terms of the velocity of the electron. The conversion factor is ε (eV) = $[W(10^7)$ cm s$^{-1}/5.93]^2$.

[b] Electron drift velocity was analyzed to obtain the momentum transfer cross sections in the energy range 0–1 eV. The maximum cross section of 2500×10^{-20} m^2 is observed at 0.25 eV.

[c] Total cross sections are provided for cesium, rubidium, and potassium in the energy range 0–10 eV by the crossed beam atom recoil technique which comprises of measuring the energy of scattered atoms.

Cs

2.2 CROSS SECTIONS

Figure 2.1 presents available cross sections for Cs atom. Ramsauer–Townsend minimum occurs at 0.08 eV (Crown and Russek, 1965). The excitation cross section to levels $6^2P_{1/2}$ (threshold 1.39 eV) (Saelee and Lucas, 1979) and $6^2P_{3/2}$ (threshold 1.45 eV) are the major contributions and according to Chen and Gallagher (1978) the total is 1.5 times the latter.

Figure 2.2 shows the optical excitation cross sections for resonance for K, Rb, and Cs up to 1500 eV.

Tables 2.2, 2.3 and Figure 2.3 show the ionization cross section. Figure 2.3 also includes excitation cross section.

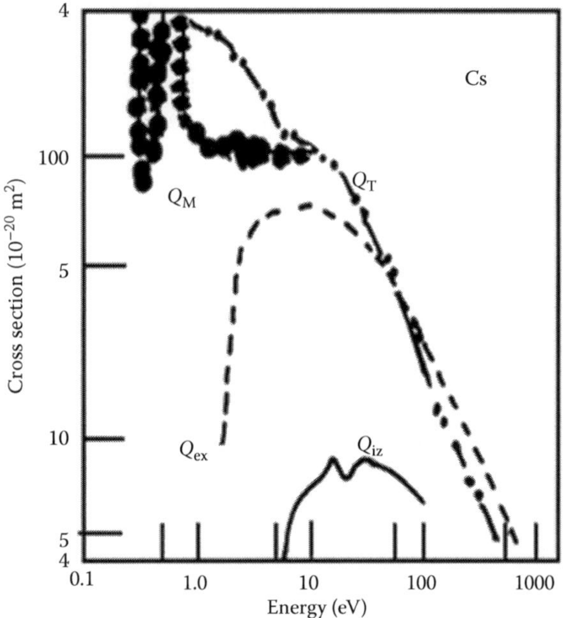

FIGURE 2.1 Cross sections for Cs atom. Q_{ex} = excitation cross section; Q_{iz} = ionization; Q_M = momentum transfer; Q_T = total. (Reproduced from Zecca, A., G. P. Karwasz, and R. S. Brusa, *Riv. Nuovo Cimento*, 19, 1, 1996. With kind permission from Italian Physical Society.)

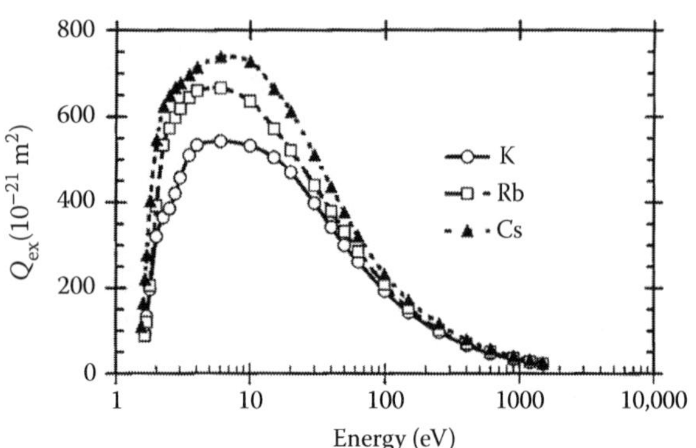

FIGURE 2.2 Excitation cross sections for selected atoms for optical resonance levels. (—○—) K (4^2P); (—□—) Rb (5^2P); (—▲—) Cs ($6^2P_{3/2}$ ×3/2), Chen and Gallagher (1978).

TABLE 2.2
Ionization Cross Section for Cs Atom

Energy (eV)	50	100	200	300	400	500
Q_i (10^{-20} m^2)	8.8	7.0	5.45	4.0	3.4	2.8

Source: Adapted from McFarland, R. H. and J. D. Kinney, *Phys. Rev.*, 137, A1058, 1965.

TABLE 2.3
Ionization Cross Sections for Cs Atom, Digitized

Energy (eV)	Q_i (10^{-20} m^2)	Energy (eV)	Q_i (10^{-20} m^2)
4.25	0.04	28.14	9.48
5.08	1.79	34.30	9.38
5.49	2.74	40.24	9.16
5.91	3.81	46.19	8.93
6.32	4.67	51.29	8.71
6.95	5.70	57.24	8.40
8.00	6.85	63.18	8.09
8.63	7.62	71.68	7.69
9.68	8.35	78.05	7.33
11.37	8.95	85.28	6.98
13.49	9.29	90.37	6.71
15.19	9.42	94.41	6.62
17.95	9.28	97.17	6.44
22.62	9.45	99.72	6.39

Source: Adapted from Nygaard, K. J., *J. Chem. Phys.*, 49, 1995, 1968.

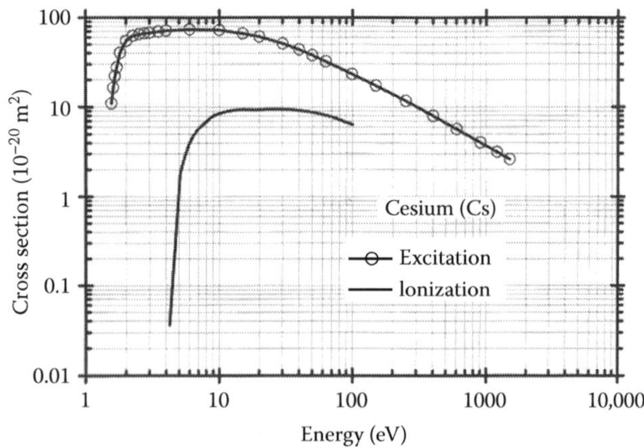

FIGURE 2.3 Total ionization and excitation cross sections for Cs atom. (—○—) Total excitation (Chen and Gallagher, 1978); (—) total ionization (Nygaard, 1968).

A recent review (Borovik and Kupliauskiene, 2009) shows that Nygaard's data are still the most recent one available.

2.3 TRANSPORT PARAMETERS

Figure 2.4 presents transport parameters for the Cs atom.

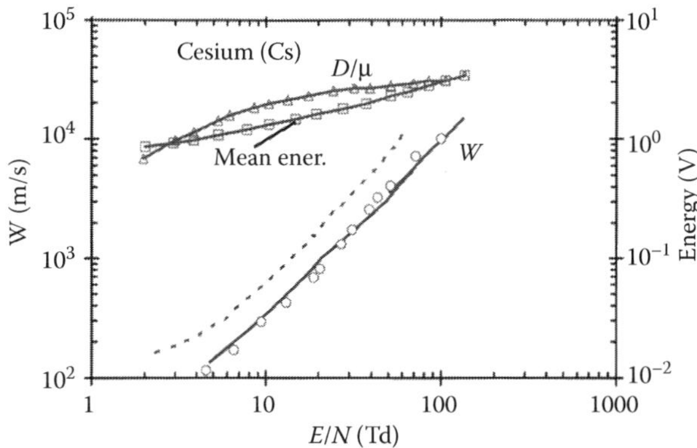

FIGURE 2.4 Transport parameters for the Cs atom. *W*: Saelee and Lucas (1979), experiment; (– –) Chanin and Steen (1964), experiment; (—) Saelee and Lucas (1979); characteristic energy: (—△—) Saelee and Lucas (1979), theory; mean energy: (—□—) Saelee and Lucas (1979).

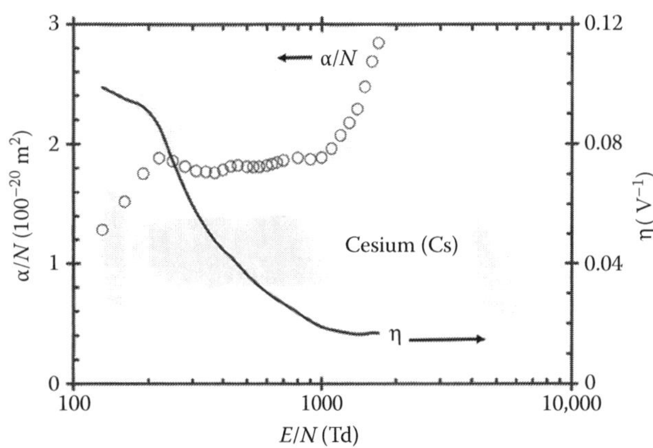

FIGURE 2.5 Reduced first ionization coefficient for Cs atom. (Adapted from Garamoon, A.A.W.M. and N.A. Surplice, *J. Phys. D: Appl. Phys.*, 6, 206, 1973.)

2.4 IONIZATION COEFFICIENT

Table 2.4 and Figure 2.5 show the results of a single study (Garamoon and Surplice, 1973).

TABLE 2.4
Reduced Ionization Coefficient for Cs Atom

E/N (Td)	α/N (10^{-20} m^2)	*E/N* (Td)	α/N (10^{-20} m^2)
130	1.285	600	1.821
160	1.520	630	1.832
190	1.754	660	1.845
220	1.884	700	1.863
250	1.858	800	1.885
280	1.812	900	1.873
310	1.778	1000	1.887
340	1.770	1100	1.962
370	1.762	1200	2.070
400	1.784	1300	2.173
430	1.814	1400	2.286
460	1.820	1500	2.471
500	1.811	1600	2.685
530	1.808	1700	2.842
560	1.810		

Source: Adapted from Garamoon, A. A. W. M. and N. A. Surplice, *J. Phys. D: Appl. Phys.*, 6, 206, 1973.

REFERENCES

Bartschat, K. J., *J. Phys. B: At. Mol. Opt. Phys.*, 26, 3595, 1993.

Borovik, A. and A. Kupliauskiene, *J. Phys. B: At. Mol. Opt. Phys.*, 42, 165202, 1–5, 2009.

Brink, G. O., *Phys. Rev.*, 134, A345, 1964.

Brode, R. B., *Phys. Rev.*, 34, 673, 1929.

Chanin, L. M. and R. D. Steen, *Phys. Rev.*, 136, A138, 1964.

Chen, S. T. and A. C. Gallagher, *Phys. Rev.*, A17, 551, 1978.

Crown, J. C. and A. Russek, *Phys. Rev,*, 138, A669, 1965.

Deutsch, C. *J. Appl. Phys.*, 44, 1142, 1973.

Garamoon, A. A. W. M. and N. A. Surplice, *J. Phys. D: Appl. Phys.*, 6, 206, 1973.

Jaduszliwer, B. and Y. C. Chan, *Phys. Rev.*, A45, 197, 1992. The atom recoil technique was adopted and the energy range covered is 2–18 eV.

McFarland, R. H. and J. D. Kinney, *Phys. Rev.*, 137, A1058, 1965.

Moiseiwitsch, B. L. and S. J. Smith, *Rev. Mod. Phys.*, 40, 238, 1968.

Nighan, W. L. and A. J. Postma, *Phys. Rev.*, A6, 2109, 1972.

Nygaard, K. J., *J. Chem. Phys.*, 49, 1995, 1968.

Saelee, H. T. and J. Lucas, *J. Phys. D: Appl. Phys.*, 12, 1275, 1979.

Stefanov, B., *Phys. Rev.*, A22, 427, 1980.

Tate, J. T. and P. T. Smith, *Phys. Rev.*, 46, 773, 1934.

Visconti, P. J., J. A. Slevin, and K. Rubin, *Phys. Rev. A*, 3, 1310, 1971.

Zecca, A., G. P. Karwasz, and R. S. Brusa, *Riv. Nuovo Cimento*, 19, 1, 1996.

3

HELIUM

CONTENTS

Helium (He) is an atomic gas, with atomic number 2 and electronic polarizability of 0.228×10^{-40} F m^2 and ionization potential of 24.587 eV.

3.1 SELECTED REFERENCES FOR DATA

See Table 3.1.

3.2 TOTAL SCATTERING CROSS SECTION

Highlights of the cross section are

1. The cross section as a function of electron energy remains constant up to about 3 eV.
2. For energy greater than about 10 eV, the cross section decreases monotonically.

Due to the lowest number of electrons, the features of the cross section curve are different, particularly for energy <5 eV when compared with other target atoms or molecules. See Table 3.2.

3.3 DIFFERENTIAL SCATTERING CROSS SECTION

Figure 3.2 shows the differential scattering cross sections as a function of energy. Figure 3.3 shows the differential cross

TABLE 3.1
Selected References for Data

Quantity	Range: eV, (Td)	Reference
Ionization cross section	25–1000	**Rejoub et al. (2002)**
Total scattering cross section	10–3000	Zecca et al. (2000)
Several cross sections	20–1000	Brusa et al. (1996)
Total scattering cross section	0.5–220	**SzmytKowski and Maciąg (1996)**
Excitation cross sections	25–500	Cartwright et al. (1992)
Differential scattering	1.5–50	**Brunger et al. (1992)**
Momentum transfer	0–700	**Pack et al. (1992)**
Transport parameters	(0.001–100)	**Pack et al. (1992)**
Excitation cross section	30, 50, 100	**Trajmar et al. (1992)**
Ionization cross section	25–1000	**Krishnakumar and Srivastava, (1988)**
Ionization cross section	25–1000	Lennon et al. (1988)
Diffusion coeficient	(56.5–5650)	**Al-Amin and Lucas (1987)**
Excitation cross section	20–140	**Mason and Newell (1987)**
Ionization cross section	26.6–10000	**Shah et al. (1987)**
Ionization cross section	20–200	**Wetzel et al. (1987)**
Total scattering cross section	0.1–20.0	**Buckman and Lohmann (1986)**
Total scattering cross section	4–300	**Nickel et al. (1985)**

continued

He

TABLE 3.1 (continued)
Selected References for Data

Quantity	Range: eV, (Td)	Reference
Ionization cross section	26–750	**Montague et al. (1984)**
Total scattering cross section	30–600	**Kaupilla et al. (1981)**
Transport coefficients	(1–850)	**Kücükarpaci et al. (1981)**
Total scattering cross section	16–700	**Blaauw et al. (1980)**
Ionization cross section	500, 700, 1000	**Nagy et al. (1980)**
Differential cross section	5–200	**Register and Trajmar (1980)**
Elastic cross section	2–400	**Shyn (1980)**
Ionization cross section	25–180	**Stephan et al. (1980)**
Total scattering cross section	100–1400	**Dalba et al. (1979)**
Total scattering cross section	0.5–50	**Kennerly and Bonham (1978)**
Total scattering cross section	30–1000	de Heer and Jansen (1977)
Ionization coefficients	(15–1000)	**Lakshminarasimha et al. (1975)**
Excitation cross sections	19.82–1000	**Alkhazov (1970)**
Ionization cross section	25–1000	**Rapp and Englander-Golden (1965)**
Ionization cross section	25–100	**Asundi and Kurepa (1963)**
Ionization cross section	100–1000	**Schram et al. (1965)**
Ionization coefficient	(15–300)	**Davies et al. (1962)**

Note: Bold font indicates experimental study.

TABLE 3.2
Recommended Total Scattering Cross Sections for He

Energy (eV)	Q_T (10^{-20} m^2)	Energy (eV)	Q_T (10^{-20} m^2)	Energy (eV)	Q_T (10^{-20} m^2)
0.10	5.52	1.25	5.99	29.0	2.422
0.12	5.65	1.50	6.02	30.0	2.372
0.14	5.73	2.00	5.88	32.5	2.248
0.16	5.73	3.00	5.70	35.0	2.156
0.18	5.80	4.00	5.47	37.5	2.055
0.20	5.83	5.00	5.26	40.0	1.952
0.25	5.87	6.00	5.06	42.5	1.896
0.30	5.99	8.00	4.67	45.0	1.826
0.35	5.95	10.0	4.30	47.5	1.817
0.40	5.98	12.0	3.96	50	1.728
0.45	6.03	14.0	3.67	55	1.627
0.50	6.03	16.0	3.41	60	1.529
0.55	6.03	18.0	3.17	65	1.448
0.60	6.02	20.0	2.99	70	1.378
0.65	6.04	21.0	2.954	75	1.310
0.70	6.03	22.0	2.867	80	1.254
0.75	6.05	23.0	2.786	85	1.215
0.80	6.07	24.0	2.713	90	1.176
0.85	6.07	25.0	2.640	95	1.142
0.90	6.07	26.0	2.576	100	1.112
0.95	6.08	27.0	2.517	150	0.871

TABLE 3.2 (continued)
Recommended Total Scattering Cross Sections for He

Energy (eV)	Q_T (10^{-20} m^2)	Energy (eV)	Q_T (10^{-20} m^2)	Energy (eV)	Q_T (10^{-20} m^2)
1.00	6.08	28.0	2.456	200	0.722
250	0.624	800	0.256	1150	0.186
300	0.554	850	0.242	1200	0.177
400	0.459	900	0.227	1250	0.170
500	0.378	950	0.217	1300	0.161
600	0.333	1000	0.207	1350	0.157
700	0.291	1050	0.198	1400	0.148
750	0.268	1100	0.193		

Note: See Figure 3.1 for graphical presentation.

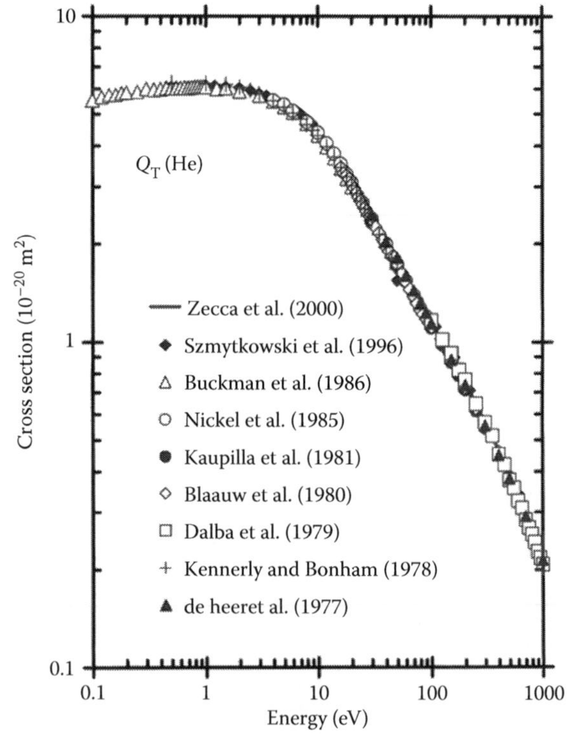

FIGURE 3.1 Absolute total cross sections in He. (—) Brusa et al. (17, 1996) with the correction applied according to Zecca et al. (18, 2000), (– –) empirical formula; (Δ) Buckman and Lohmann (1986); (○) Nickel et al. (1985); (●) Kauppila et al. (1981); (◊) Blaauw et al. (1980); (□) Dalba et al. (1979); (+) Kennerly and Bonham (1978); (▲) de Heer and Jansen (1977), semiempirical.

sections as a function of scattering angle at selected electron energy.

3.4 ELASTIC SCATTERING CROSS SECTION

See Table 3.3.

3.5 MOMENTUM TRANSFER CROSS SECTION

See Table 3.4.

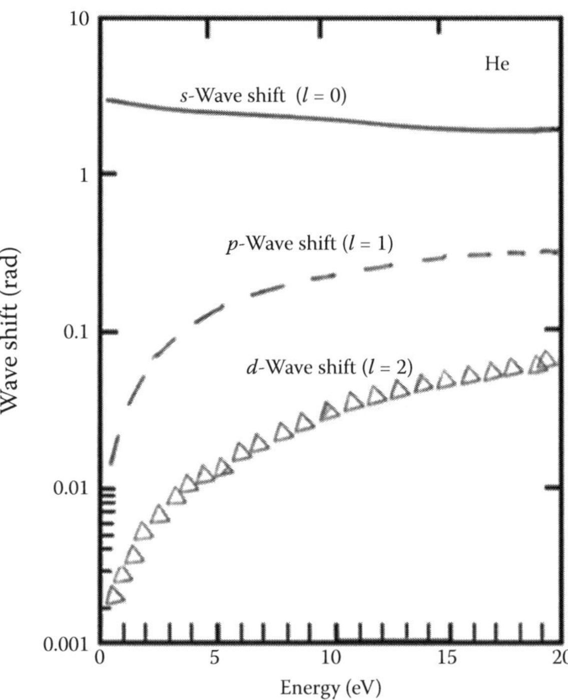

FIGURE 3.2 Phase shifts for electron scattering from He as a function of energy. (Drawn from the tabulated data of Williams, J. F., *J. Phys. B: At. Mol. Phys.*, 12, 265, 1979.)

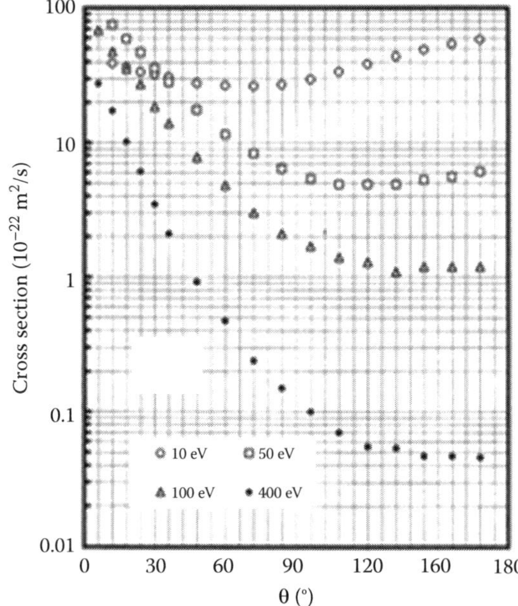

FIGURE 3.3 Differential cross sections for elastic scattering of electrons in He at selected electron energies. (Adapted from Raju, G. G., *Gaseous Electronics: Theory and Practice*, Taylor & Francis, London, 2005.)

3.6 ENERGY LEVELS

See Table 3.5 and Figure 3.6.

TABLE 3.3
Recommended Elastic Scattering Cross Section for He

Energy (eV)	Q_{el} (10^{-20} m^2)	Energy (eV)	Q_{el} (10^{-20} m^2)	Energy (eV)	Q_{el} (10^{-20} m^2)
1	6.23	30	2.24	100	0.610
2	5.88	35	1.96	150	0.381
3	5.74	40	1.73	200	0.274
5	5.26	45	1.60	300	0.165
10	4.29	50	1.37	400	0.120
12	4.00	60	1.19	500	0.09
15	3.58	70	0.989	600	0.077
18	3.23	75	0.898	700	0.063
20	3.00	80	0.816	800	0.060
25	2.51	90	0.709	1000	0.041

Note: See Figure 3.4 for graphical presentation.

He

FIGURE 3.4 Elastic scattering cross sections in He. (—) Raju (2005); (●) Brunger et al. (1992); (○) Register et al. (1980); (Δ) Shyn (1980); (□) de Heer and Jansen (1977).

TABLE 3.4
Recommended Momentum Transfer Cross Section for He

Energy (eV)	Q_M (10^{-20} m^2)	Energy (eV)	Q_M (10^{-20} m^2)	Energy (eV)	Q_M (10^{-20} m^2)
0.000	4.95	8.0	5.50	100	0.22
0.0025	5.00	14	3.60	150	0.12
0.0036	5.10	18	2.90	200	0.07
0.010	5.27	20	2.69	250	0.05
0.032	5.52	25	2.00	300	0.036
0.20	6.20	35	1.26	500	0.016
0.60	6.66	40	1.00	700	0.009
1.4	6.98	50	0.7	100	0.22
3.0	6.93	75	0.36	150	0.12

Source: Adapted from Pack, J. L., R. E. Voshall, and A. V. Phelps, *J. Appl. Phys.*, 71, 5363, 1992.
Note: See Figure 3.5 for graphical presentation.

He

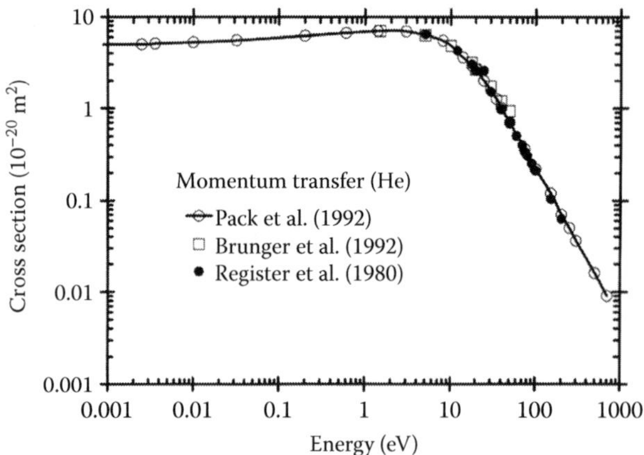

FIGURE 3.5 Momentum transfer cross section for He. (—○—) Pack et al. (1992); (□) Brunger et al. (1992); (●) Register et al. (1980).

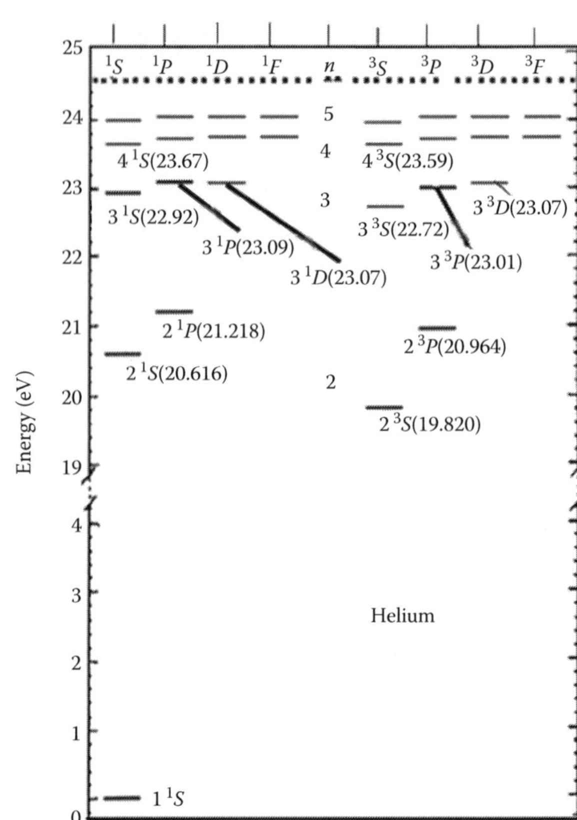

FIGURE 3.6 Energy levels for He atom.

TABLE 3.5
Selected Energy Levels for He

Notation	Energy (eV)	Notation	Energy
$1\,^1S$	0	$2\,^3S$	19.82
$2\,^1S$	20.616	$2\,^3P$	20.964
$2\,^1P$	21.218	$3\,^3S$	22.72
$3\,^1S$	22.92	$3\,^3P$	23.01
$3\,^1P$	23.09	$3\,^3D$	23.07
$3\,^1D$	23.07	$4\,^3S$	23.59
$4\,^1S$	23.67	$4\,^3P$	23.71
$4\,^1P$	23.74	$4\,^3D$	23.74
$4\,^1D$	23.74	$4\,^3F$	23.74
$4\,^1F$	23.74	$5\,^3S$	23.97

Source: Adapted from Grigoriev, I. S. and E. Z. Meilikhov, *Handbook of Physical Quantities*, CRC Press, Boca Raton, FL, 1999, Chapter 32.
Note: See Figure 3.6 also.

3.7 EXCITATION CROSS SECTIONS

Figure 3.7 shows excitation cross sections for selected levels (Alkhazov, 1970). Table 3.6 and Figure 3.8 show the total excitation cross section.

3.8 IONIZATION CROSS SECTIONS

See Table 3.7 and Figure 3.9.

3.9 DRIFT VELOCITY OF ELECTRONS

Recommended drift velocity for electrons is shown in Table 3.8 and graphical presentation in Figure 3.10.

3.10 MEAN ENERGY OF ELECTRONS

Figure 3.11 shows the mean energy of electrons as a function of E/N. Table 3.9 shows the characteristic energy (D_r/μ).

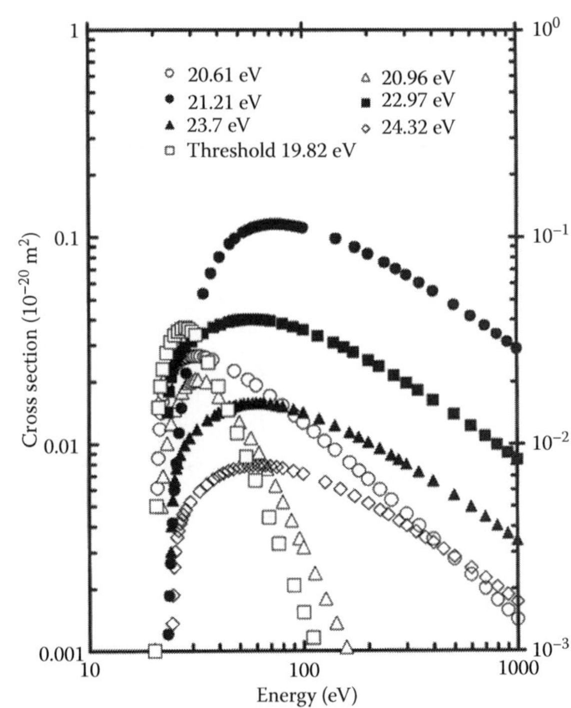

FIGURE 3.7 Excitation cross sections for selected levels for He atom. (Adapted From Alkhazov, G. D., *Sov. Phys. Tech. Phys.*, 15, 66, 1970.)

TABLE 3.6
Total Excitation Cross Sections

Energy (eV)	Q_{ex} (10^{-20} m²)	Energy (eV)	Q_{ex} (10^{-20} m²)	Energy (eV)	Q_{ex} (10^{-20} m²)
19.82	0.0000	31	0.1670	180	0.1383
20	0.0010	35	0.1944	190	0.1348
20.5	0.0050	40	0.2023	200	0.1310
20.61	0.0072	45	0.2061	225	0.1236
20.78	0.0167	50	0.2096	250	0.1134
20.94	0.0224	55	0.2030	275	0.1082
21	0.0241	60	0.2049	300	0.1023
21.5	0.0374	70	0.2062	350	0.0934
22	0.0462	80	0.1999	400	0.0845
23	0.0592	90	0.1894	500	0.0725
24	0.0949	100	0.1847	600	0.0637
25	0.1153	110	0.1786	700	0.0570
26	0.1262	125	0.1695	800	0.0517
27	0.1380	140	0.1582	900	0.0474
28	0.1480	155	0.1496	1000	0.0438
29	0.1550	165	0.1458		
30	0.1617	175	0.1424		

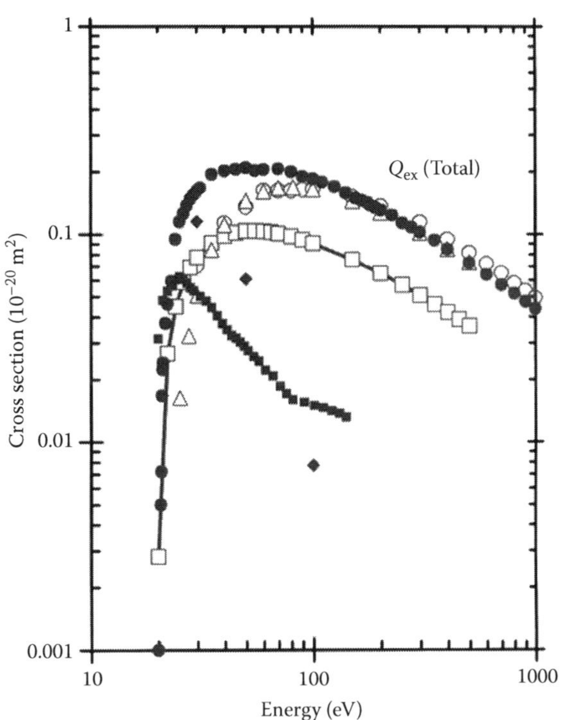

FIGURE 3.8 Total excitation cross sections for He atom. (—□—) Brusa et al. (1996); (Δ) Cartwright et al. (1992); (◆) Trajmar et al. (1992); (■) Mason and Newell (1987); (○) de Heer and Jansen (1977); (▲) sum of first six levels, Alkhazov (1970); (●) Alkhazov (1970), sum of seven levels shown in Figure 3.7.

3.11 REDUCED FIRST IONIZATION COEFFICIENTS

See Table 3.10 and Figure 3.12.

TABLE 3.7
Total Ionization Cross Section for He

Rejoub et al. (2002)		Rapp and Englander-Golden (1965)	
Energy (eV)	Q_i, Total (10^{-21} m²)	Energy (eV)	Q_i, Total (10^{-20} m²)
25	0.037	25	0.0052
		26	0.0175
27.5	0.290	27	0.0303
		28	0.0425
		29	0.0552
30	0.568	30	0.0668
32.5	0.851	32	0.0924
		34	0.114
35	1.09	36	0.135
		38	0.155
40	1.52	40	0.172
45	1.90	45	0.202
50	2.24	50	0.243
55	2.47	55	0.271
60	2.69	60.0	0.308
		65	0.308
70	2.96	70	0.321
		75	0.334
80	3.16	80	0.344
		85	0.351
90	3.28	90	0.357
		95	0.362
100	3.37	100	0.366
110	3.41	110	0.370
120	3.46	120	0.373
130	3.48	130	0.374
140	3.45	140	0.372
150	3.39	150	0.369
160	3.37		
180	3.26	175	0.359
200	3.16	200	0.347
225	3.05		
250	2.92	250	0.321
300	2.72	300	0.296
350	2.49	350	0.275
400	2.34	400	0.257
		450	0.239
500	2.05	500	0.224
600	1.81	600	0.200
700	1.64	700	0.180
800	1.48	800	0.164_5
900	1.37	900	0.150
1000	1.25	1000	0.141

Note: Gross cross sections of Rejoub et al. (2002) have been calculated from partial cross sections. See Figure 3.9 for graphical presentation.

He

He

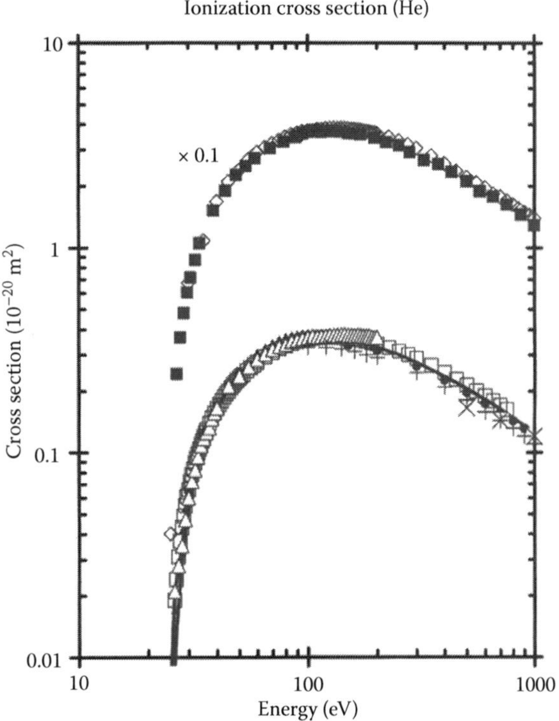

FIGURE 3.9 Ionization cross sections in He. The data are divided into two groups for clarity. The cross sections of the upper curve should be multiplied by 0.1. Legends are for both curves. (Δ, ■) Shah et al. (1988); (◊) Krishnakumar et al. (1988), (— —) Lennon et al. (1988); Wetzel et al. (1987); (□) Montague et al. (1984); (×) Nagy et al. (41, 1980); (○) Stephan et al. (1980); (●) de Heer and Jansen (1977); (◆, +) Schram et al. (1966B); (—) Rapp and Englander-Golden (1965); (▲) Asundi and Kurepa (1963).

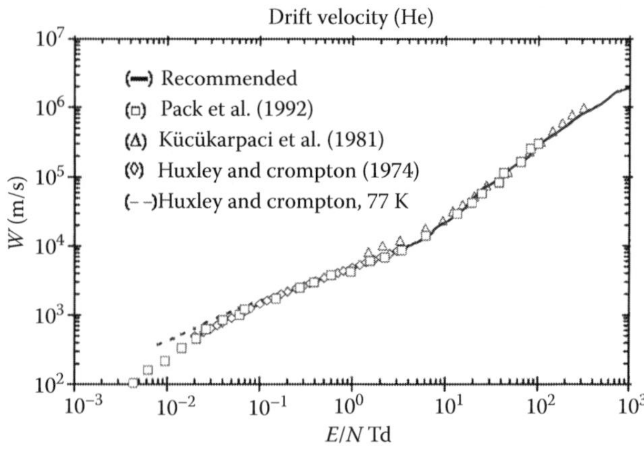

FIGURE 3.10 Drift velocity of electrons in He. Temperature 293 K unless otherwise mentioned. (—) Raju (2005); (□) Pack et al. (1992); (Δ) Kücükarpaci et al. (1981); (◊) Huxley and Crompton (1974); (– –) Huxley and Crompton (1974), 77 K.

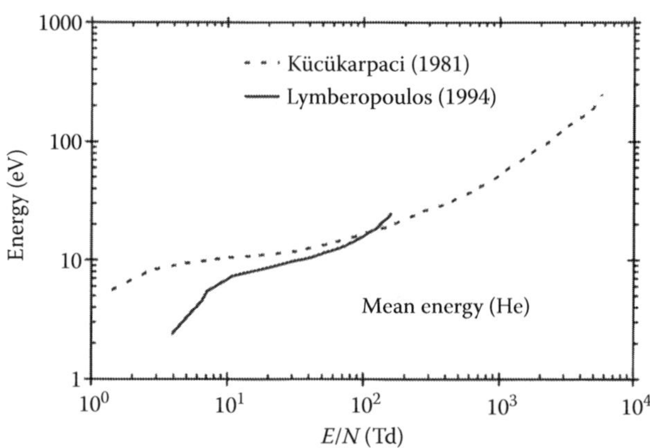

FIGURE 3.11 Mean energy of electron swarm. (—) Lymberopoulos and Schieber (1994); (– –) Kücükarpaci et al. (1981).

TABLE 3.8
Recommended Drift Velocity of Electrons

E/N (Td)	*W* (10^3 m/s)	*E/N* (Td)	*W* (10^3 m/s)	*E/N* (Td)	*W* (10^3 m/s)
0.02	0.453	1	4.85	40	99
0.03	0.631	2	6.86	50	134
0.04	0.786	3	8.51	60	159
0.05	0.922	4	10.07	70	176
0.06	1.044	5	11.52	80	203
0.07	1.156	6	13.13	100	285
0.08	1.259	7	15.03	200	534
0.1	1.446	8	18.77	300	797
0.2	2.14	10	22.2	400	965
0.3	2.65	15	32.6	500	1147
0.4	3.08	20	49.2	600	1380
0.5	3.44	25	68.1	700	1642
0.6	3.77	30	77.8	800	1729
0.8	4.35	35	87.5	1000	1915

Source: Adapted from Raju, G. G., *Gaseous Electronics: Theory and Practice*, Taylor & Francis, London, 2005.

TABLE 3.9
Characteristic Energy for He

E/N (Td)	D_r/μ (V)	*E/N* (Td)	D_r/μ (V)
15	3.91	565	21.7
20	4.72	706	18.9
25	5.39	848	17.4
30	5.93	1130	16.4
40	6.72	1413	13.4
56.5	6.52	1695	12.0
70.6	7.06	1978	10.0
84.6	9.52	2260	9.18
113	10.1	2825	8.62
141	11.1	3390	6.59
282	18.3	4238	6.29
339	19.8	5650	5.14
423	20.9		

Note: Values in the range $56.5 \le E/N \le 5650$ are from Al-Amin and Lucas (1987).

He

TABLE 3.10
Recommended Ionization Coefficient for He

E/N (Td)	α/N (10^{-20} m²)	E/N (Td)	α/N (10^{-20} m²)	E/N (Td)	α/N (10^{-20} m²)
10	0.0005	80	0.090	500	0.564
15	0.0020	90	0.105	550	0.583
20	0.0048	100	0.119	600	0.598
25	0.010	150	0.191	650	0.610
30	0.017	200	0.263	700	0.622
35	0.023	250	0.330	750	0.637
40	0.029	300	0.392	800	0.659
50	0.043	350	0.449	850	0.689
60	0.059	400	0.498	900	0.731
70	0.075	450	0.536		

Note: See Figure 3.12 for graphical presentation.

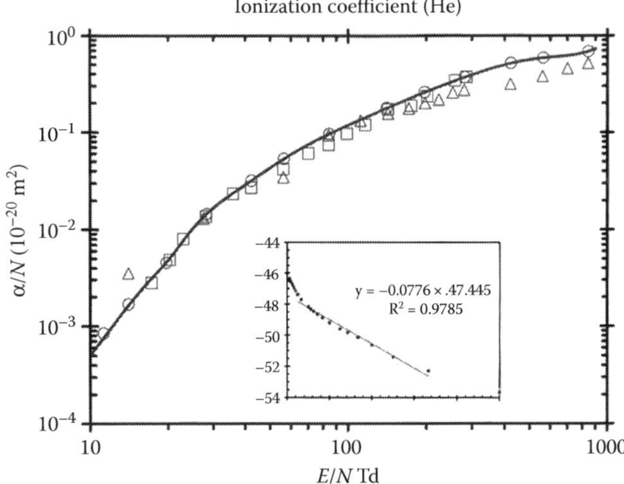

FIGURE 3.12 Reduced ionization coefficients for He. (—) Recommended, Table 3.10; (Δ) Lakshminarasimha et al. (1975); (○) Chanin and Rork (1964); (□) Davies et al. (1962). Inset shows plot according to Equation 3.1.

3.12 GAS CONSTANTS

Gas constants evaluated according to expression (see inset of Figure 3.12)

$$\frac{\alpha}{N} = F \exp\left(-\frac{GN}{E}\right) \tag{3.1}$$

are $F = 2.48 \times 10^{-21}$ m² and $G = 77.6$ Td for the range $15 \le E/N \le 108$ (Td).

3.13 ION MOBILITY

See Table 3.8.

TABLE 3.11
Reduced Mobility ($N = 2.69 \times 10^{25}$ m^{-3}; $T = 273.16$ K) and Drift Velocity of He⁺ Ions Drifting in He Gas

E/N (Td)	μ_0 ($\times 10^{-4}$ m²/V s)	W^+ (m/s)
6.00	10.3	166
8.00	10.2	219
10.0	10.2	274
12.0	10.1	326
15.0	10.0	403
20.0	9.90	532
25.0	9.74	654
30.0	9.60	774
40.0	9.28	997
50.0	8.97	12.1
60.0	8.67	14.0
80.0	8.12	17.5
100	7.67	20.6
120	7.25	23.4
150	6.78	27.3
200	6.12	32.9
250	5.60	37.6
300	5.19	41.8
400	4.58	49.2
500	4.17	56.0
600	3.81	61.4
700	3.57	67.1

Sources: Adapted from Ellis, H. W. et al., *Atom. Data Nucl. Data Tab.*, 17, 177, 1976; Beaty, E. C. and P. L. Paterson, *Phys. Rev.*, 137, A346, 1965.

Note: See Figure 3.13 for graphical presentation. Drift velocity of electrons shown for comparison.

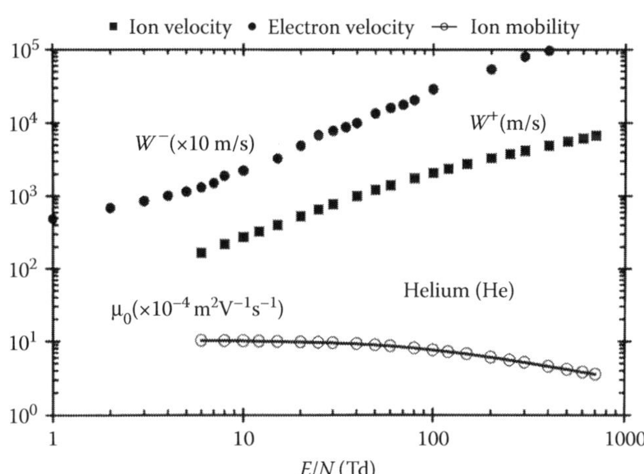

FIGURE 3.13 Reduced mobility ($N = 2.69 \times 10^{25}$ m^{-3}; $T = 273.16$ K) and drift velocity of He⁺ ions drifting in He gas (○) reduced mobility of He⁺ ions (×10^{-4} m²/V s) (Ellis et al., 1976; Beaty and Paterson, 1965); (■) drift velocity of He⁺ ions (m/s); drift velocity of electrons (×10 m/s) shown for comparison.

REFERENCES

Al-Amin, S. A. J. and J. Lucas, *J. Phys. D: Appl. Phys.*, 20, 1590, 1987.

Alkhazov, G. D., *Sov. Phys. Tech. Phys.*, 15, 66, 1970. Tabulated values are taken from L. C. Pitchford, J. P. Bouef, and J. P. Morgan, www.siglo-kinema.com (1996). Also see A. Fiala, L. C. Pitchford, and J. P. Boeuf, *Phys. Rev. E*, 49, 5607, 1994.

Asundi, R. K. and M. V. Kurepa, *J. Electron. Control*, 15, 41, 1963.

Beaty, E. C. and P. L. Paterson, *Phys. Rev.*, 137, A346, 1965.

Blaauw, H. J., R. W. Wagenaar, D. H. Barends, and F. J. de Heer, *J. Phys. B: At. Mol. Phys.*, 13, 359, 1980.

Brunger, M. J., S. J. Buckman, L. J. Allen, I. E. McCarthy, and Nd K. Rathnavelu, *J. Phys. B: At. Mol. Op. Phys.*, 25, 1823, 1992.

Brusa, R. S., G. P. Karwasz, and A. Zecca, *Z. Phys. D*, 38, 279, 1996. Zecca et al. (2000) is erratum for this reference.

Buckman, S. J. and B. Lohmann, *J. Phys. B: At. Mol. Phys.*, 19, 2547, 1986.

Cartwright, D. C., G. Csanak, S. Trajmar, and D.F. Register, *Phys. Rev. A*, 45, 1602, 1992.

Chanin, L. M. and G. D. Rork, *Phys. Rev.*, 133, A1005, 1964.

Dalba, G., P. Fornasini, I. Lazzizzera, G. Ranieri, and A. Zecca, *J. Phys. B: At. Mol. Phys.*, **12**, 3787, 1979; G. Dalba, P. Fornasini, R. Grisenti, I. Lazzizzera, G. Ranieri, and A. Zecca, *Rev. Sci. Instr.*, 52, 979, 1981.

Davies, D. K., F. Llewellyn-Jones, and C. G. Morgan, *Proc. Phys. Soc. London*, 80, 898, 1962.

de Heer, F. J. and R. H. J. Jansen, *J. Phys. B: At. Mol. Phys.*, 10, 3741, 1977.

Ellis, H. W., R. Y. Pai, E. W. McDaniel, E. A. Mason and L. A. Viehland, *Atom. Data Nucl. Data Tab.*, 17, 177, 1976.

Grigoriev, I. S. and E. Z. Meilikhov, *Handbook of Physical Quantities*, CRC Press, Boca Raton, FL, 1999, Chapter 32.

Huxley, L. G. H. and R. W. Crompton, *Diffusion and Drift of Electrons in Gases*, John Wiley and Sons Inc., New York, NY, 1974.

Kauppila, W. E., T. S. Stein, J. H. Smart, M. S. Dababneh, Y. K. Ho, J. P. Downing, and V. Pol, *Phys. Rev. A*, 24, 725, 1981.

Kennerly, R. E. and R. A. Bonham, *Phys. Rev. A*, 17, 1844, 1978.

Krishnakumar, E. and S. K. Srivastava, *J. Phys. B: At. Mol. Phys.*, 21, 1055, 1988.

Kücükarpaci, H. N., H. T. Saelee, and J. Lucas, *J. Phys. D: Appl. Phys.*, 14, 9, 1981.

Lakshminarasimha, C. S., J. Lucas, and R. A. Snelson, *Proc. IEE*, 122, 1162, 1975.

Lennon, M. A., K. L. Bell, H. B. Gilbody, J. G. Hughes, A. E. Kingston, M. J. Murray, and F. J. Smith, *J. Phys. Chem. Ref. Data*, 17, 1285, 1988.

Lymberopoulos, D. P. and J. D. Schieber, *Phys. Rev. E*, 50, 4911, 1994.

Mason, N. J. and W. R. Newell, *J. Phys. B: At. Mol. Phys.*, 20, 1357, 1987.

Montague, R. G., M. F. A. Harrison, and A. C. H. Smith, *J. Phys. B: At. Mol. Opt. Phys.*, 17, 3295, 1984.

Nagy, P., A. Skutzlart, and V. Schmidt, *J. Phys. B: At. Mol. Phys.*, 13, 1249, 1980.

Nickel, J. C., K. Imre, D. F. Register, and S. Trajmar, *J. Phys. B: At. Mol. Phys.*, 18, 125, 1985.

Pack, J. L., R. E. Voshall, and A. V. Phelps, *J. Appl. Phys.*, 71, 5363, 1992.

Raju, G. G., *Gaseous Electronics: Theory and Practice*, Taylor & Francis, London, 2005.

Rapp, D. and P. Englander-Golden, *J. Chem. Phys.*, 43, 1464, 1965.

Register, D. F. and S. Trajmar, *Phys. Rev. A*, 21, 1134, 1980.

Rejoub, R., B. G. Lindsay, and R. F. Stebbings, *Phys. Rev. A*, 65, 042713, 2002.

Schram, B. L., F. J. deHeer, F. J. van Der Wieland, and J. Kistemaker, *Physica*, 31, 94, 1965; B. L. Schram, H. R. Moustafa, H. R. Schutten, and F. J. de Heer, *Physica*, 32, 734, 1966a; B. L. Schram, J. H. Boereboom, and J. Kistemaker, *Physica*, 32, 85, 1966b; B. L. Schram, *Physica*, 32, 197, 1966.

Shah, M. B., D. S. Elliot, and H. B. Gilbody, *J. Phys. B: At. Mol. Phys.*, 20, 3501, 1987; 21, 2751, 1988.

Shyn, T. W., *Phys. Rev. A*, 22, 916, 1980.

Stephan, K., H. Helm, and T. D. Märk, *J. Chem. Phys.*, 73, 3763, 1980. Gases studied are He, Ne, Ar, Kr; *J. Chem. Phys.*, 81, 3116, 1984. Gas studied is Xe.

SzmytKowski, C. and K. Maciag, *Physica Scripta*, 54, 271, 1996.

Trajmar, S., D. F. Register, D. C. Cartwright, and G. Csanak, *J. Phys. B: At. Mol. Opt. Phys.*, 25, 4889, 1992.

Wetzel, R. C., F. A. Baiocchi, T. R. Hayes, and R. S. Freund, *Phys. Rev. A*, 35, 559, 1987.

Williams, J. F., *J. Phys. B: At. Mol. Phys.*, 12, 265, 1979.

Zecca, A., G. P. Karwasz, and R. S. Brusa, *J. Phys. B: At. Mol. Opt. Phys.*, 33, 843, 2000.

4

KRYPTON

CONTENTS

Krypton (Kr) atom has 36 electrons with configuration $1s^2\ 2s^2p^6\ 3s^2p^6\ 3d^{10}\ 4s^2p^6$. The electronic polarizability is 2.764×10^{-40} F m^2 and the first ionization potential is 14.00 eV.

4.1 SELECTED REFERENCES FOR DATA

See Table 4.1.

4.2 TOTAL SCATTERING CROSS SECTION

Table 4.2 and Figure 4.1 show the total scattering cross section for Kr. The highlights are

1. As the energy is decreased toward zero the total cross section increases, due to the relatively large electronic polarizability of the Kr atom.
2. Ramsauer–Townsend minimum at ~1 eV.
3. The total cross section increases to the right of the Ramsauer–Townsend minimum, reaching a peak at ~10 eV. This feature is common to many gases.
4. For electron energy higher than about 20 eV, the cross section decreases monotonically up to 3000 eV.

4.3 DIFFERENTIAL CROSS SECTION

The differential cross sections as a function of scattering angle at selected electron energy exhibit the following general characteristics (Srivastava et al., 1981):

1. A sharp forward peak arising mainly from high partial waves. The peak becomes more pronounced as the energy increases (Raju, 2005).
2. A peak in the backward cross section, reflecting the difficulty of the electron penetrating the core. As the target atom becomes heavier the probability of electron being reflected also increases.
3. A formation of double minima at energies below 70 eV.
4. A third minimum slowly developing above 70 eV. This is attributed to the higher partial waves taking over, from *d*- to *f*-wave.

Figure 4.2 shows phase shifts for elastic scattering of electrons for Kr as function of electron energy *s*- and *p*-waves pass through zero at low electron energy whereas *d*-waves show a peak at ~27 eV.

Kr

TABLE 4.1
Selected References for Data

Quantity	Range: eV (Td)	Reference
Ionization cross section	14.5–1000	**Kobayashi et al. (2002)**
Ionization cross section	15–1000	**Rejoub et al. (2002)**
Excitation cross section	10–250	**Chilton et al. (2000)**
Excitation cross section	12–20	**Guo et al. (2000)**
Cross section	12–3000	Zecca et al. (2000)
Excitation cross sections	15–100	Kaur et al. (1998)
Ionization cross section	140–4000	**Sorokin et al. (2000)**
Cross sections	12–3000	Brusa et al. (1996)
Total cross section	0.5–220	**Szmytkowski and Maciąg (1996)**
Momentum transfer cross section	0.01–40	Mimnagh et al. (1993)
Total cross section	5–300	**Kanik et al. (1992)**
Transport coefficients	0–50 (10^{-3}–10^2)	**Pack et al. (1992)**
Ionization cross section	18–466	**Syage (1992)**
Total cross section	81–4000	**Zecca et al. (1991)**
Momentum transfer cross section	0–25	**Mitroy (1990)**
Excitation cross section	20–80	**Danjo (1989)**
Transport coefficients	(0.001–10)	Suzuki et al. (1989)
Elastic scattering cross section	5–200	**Danjo (1988)**
Drift velocity	(0.002–3.0)	**Hunter et al. (1988)**
Ionization cross section	15–1000	**Krishnakumar and Srivastava (1988)**
Transport coefficients	(1.41–5650)	**Al-Amin and Lucas (1987)**
Total cross section	0.18–20.0	**Buckman and Lohmann (1987)**
Excitation cross section	20–140	**Mason and Newell (1987)**
Total cross section	0.73–9.14	**Subramanian and Kumar (1987)**
Ionization cross section	14–200	**Wetzel et al. (1987)**
Total cross section	700–1000	**Garcia et al. (1986)**
Characteristic energy	(0.005–0.90)	**Koizumi et al. (1986)**
Differential cross section	0.1–120	Fon et al. (1984)
Elastic scattering cross section	3–50	McEachran and Stauffer (1984)
Total cross section	7.5–60.0	**Jost et al. (1983)**
Total cross section	20–750	**Dababneh et al. (1982)**
Elastic scattering cross section	0.01–30	**Sin Fai Lam (1982)**
Excitation cross section	9.91–18.00	**Specht et al. (1981)**
Differential cross section	3–100	**Srivastava et al. (1981)**
Excitation cross sections	15–100	**Trajmar et al. (1981)**
Transport coefficients	(14–849)	**Kücükarpaci and Lucas (1981)**
Ionization cross section	500–1000	**Nagy et al. (1980)**
Ionization cross section	15–180	**Stephan et al. (1980)**
Total cross section	22.5–750	**Wagenaar and de Heer (1980)**
Total cross section	1–9–99.4	**Dababneh et al. (1980)**
Cross sections	20–4000	de Heer et al. (1979)
Differential scattering	3–10.5	**Heindörff et al. (1976)**
Differential scattering	20–400	**Williams and Crowe (1975)**
Ionization cross section	100–600	**Schram et al. (1966)**
Ionization cross section	600–20000	**Schram et al. (1965)**
Ionization cross section	14.5–1000	**Rapp and Englander-Golden (1965)**
Drift velocity	(0.001–10)	**Frost and Phelps (1964)**
Drift velocity	(6×10^{-4}–3.0)	**Pack et al. (1962)**
Drift velocity	(0.3–7.0)	**Bowe (1960)**

Note: Bold font indicates experimental study.

TABLE 4.2
Recommended Total Scattering Cross Section

Energy (eV)	Q_T (10^{-20} m^2)	Energy (eV)	Q_T (10^{-20} m^2)	Energy (eV)	Q_T (10^{-20} m^2)
0.18	5.030	1.40	1.540	30	19.05
0.20	4.380	1.50	1.840	40	16.53
0.23	3.770	1.75	2.230	50	14.91
0.25	3.220	2.00	3.020	60	13.87
0.28	2.670	2.25	3.830	70	13.00
0.30	2.310	2.50	4.680	80	12.09
0.35	1.750	3.00	6.360	90	11.32
0.40	1.300	3.50	8.240	100	10.59
0.45	1.040	4.00	10.100	125	9.69
0.50	0.828	4.50	12.040	150	8.73
0.55	0.649	5.00	14.080	200	7.44
0.60	0.570	6.00	18.030	250	6.82
0.65	0.495	7.00	21.670	300	6.00
0.70	0.455	8.00	24.650	324	5.932
0.72	0.450	9.00	26.200	361	5.589
0.74	0.441	10.0	27.040	400	5.251
0.76	0.451	11.0	27.270	441	4.946
0.80	0.458	12.0	27.290	500	4.628
0.85	0.482	13.0	26.570	576	4.328
0.90	0.530	14.0	26.140	676	3.896
0.95	0.600	15.0	25.400	700	3.81
1.00	0.672	16.0	24.590	784	3.519
1.10	0.853	18.0	23.340	900	3.25
1.20	1.070	20.0	22.290	1000	3.088
1.30	1.320	25	21.04		

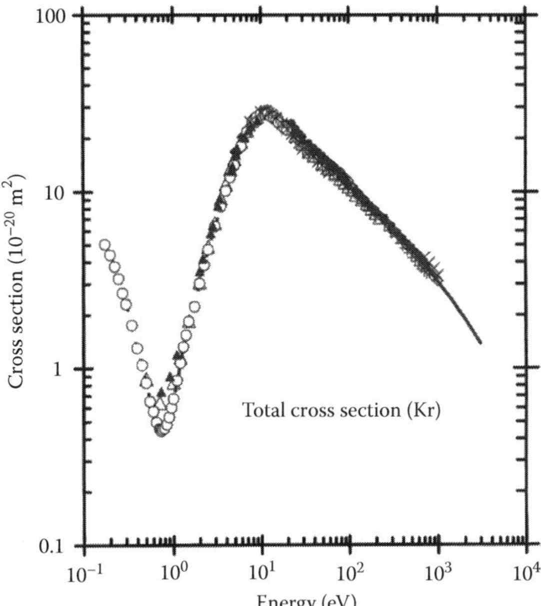

FIGURE 4.1 Absolute total scattering cross sections for electrons on Kr. (●) Wagenaar et al. (44, 1980); (—) Zecca et al. (2000); (△) Szmytkowski and Maciąg (1996); (◆) Kanik et al. (1992); (◊) Zecca et al. (86, 1991); (—○—) Buckman and Lohmann (1987); (▲) Subramanian and Kumar (1987); (×) Garcia et al. (1986); (□) Wagenaar et al. (1985); (— —) Jost et al. (1983); (+) Dababneh et al. (1982).

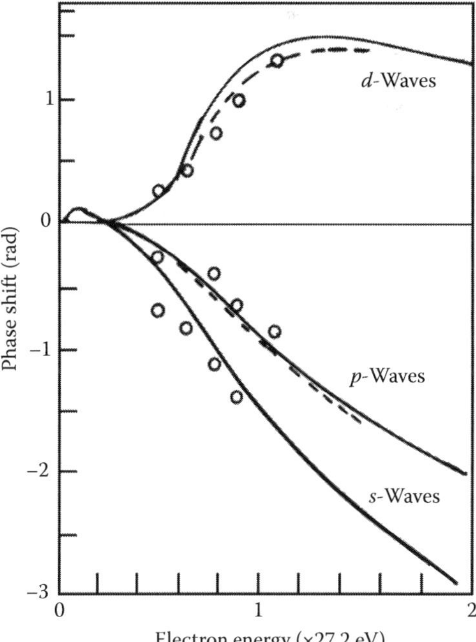

FIGURE 4.2 Phase shifts for elastic scattering of electrons for Kr as function of electron energy. (——) McEachran and Stauffer (1984); (–––) spin up and spin down results of Sin Fai Lam (1982); (○) Srivastava et al. (1981). *s*-Waves and *p*-waves pass through zero at low electron energy whereas *d*-waves show a peak at ~27 eV. (Reproduced from McEachran, R. P. and A. D. Stauffer, *J. Phys. B: At. Mol. Phys*, 17, 2507, 1984 with kind permission of Institute of Physics, England.)

4.4 ELASTIC SCATTERING CROSS SECTION

See Table 4.3.

TABLE 4.3
Elastic Scattering Cross Section for Kr

Buckman and Lohmann (1987)				Srivastava et al. (1981)	
Energy (eV)	Q_{el} (10^{-20} m^2)	Energy (eV)	Q_{el} (10^{-20} m^2)	Energy (eV)	Q_{el} (10^{-20} m^2)
0.175	5.03	1.00	0.67	3	10.0
0.200	4.38	1.10	0.85	5	14.4
0.225	3.77	1.20	1.07	7.5	18.5
0.250	3.22	1.30	1.32	10	19.4
0.275	2.67	1.40	1.54	15	22.1
0.30	2.31	1.50	1.84	20	16.2
0.35	1.75	1.75	2.23	30	10.4
0.40	1.30	2.00	3.02	50	8.9
0.45	1.04	2.25	3.83	75	5.2
0.50	0.83	2.50	4.68	100	4.0
0.55	0.65	3.00	6.36		
0.60	0.57	3.50	8.24		
0.65	0.50	4.00	10.10		
0.70	0.46	4.50	12.04		
0.72	0.45	5.00	14.08		
0.74	0.44	6.00	18.03		
0.76	0.45	7.00	21.67		
0.80	0.46	8.00	24.65		
0.85	0.48	9.00	26.20		
0.90	0.53	10.00	27.04		
0.95	0.60				

Note: See Figure 4.3 for graphical presentation.

FIGURE 4.3 Elastic scattering cross sections in Kr. (- - -) (——) Zecca et al. (2000); (●) Danjo (1988); Buckman and Lohmann (1987); (△) Fon et al. (1984); (◆) McEachran and Stauffer (1984); (□) Sin Fai Lam (1982); (○) Srivastava et al. (1981); (■) Dababneh et al. (1980); (▲) de Heer et al. (1979).

4.5 MOMENTUM TRANSFER CROSS SECTION

See Table 4.4.

4.6 RAMSAUER–TOWNSEND MINIMUM

See Table 4.5.

TABLE 4.4
Recommended Momentum Transfer Cross Section for Kr

Energy (eV)	Q_M (10^{-20} m^2)	Energy (eV)	Q_M (10^{-20} m^2)	Energy (eV)	Q_M (10^{-20} m^2)
0.000	39.70	0.25	1.30	2.50	4.40
0.001	39.70	0.300	0.86	3.00	6.00
0.003	35.00	0.350	0.55	4.00	10.00
0.005	30.00	0.400	0.26	5.00	14.00
0.0085	27.00	0.500	0.10	6.00	16.00
0.010	26.20	0.540	0.11	7.00	17.00
0.020	21.40	0.600	0.15	8.00	16.50
0.040	15.32	0.700	0.27	10.00	15.50
0.060	11.68	0.800	0.42	12.00	13.50
0.080	9.13	1.00	0.80	20.00	6.00
0.100	7.23	1.20	1.30	50.00	1.55
0.16	4.00	1.60	2.00		
0.200	2.37	2.00	3.00		

Source: Adapted from Pack, J. L., R. E. Voshall, and A. V. Phelps, *Phys. Rev.*, 127, 2084, 1962.

Note: See Figure 4.4 for graphical presentation.

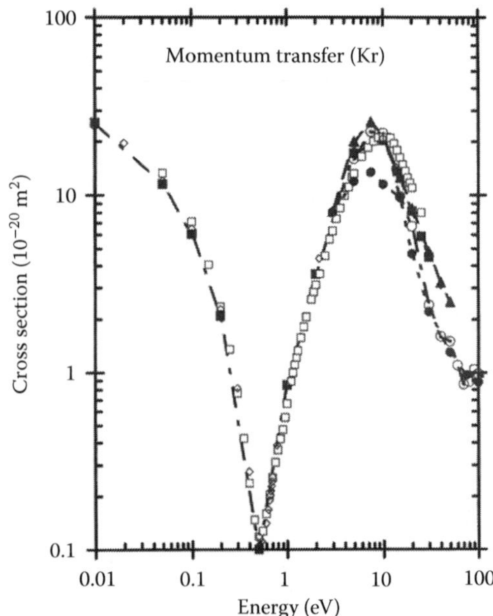

FIGURE 4.4 Momentum transfer cross sections in Kr. (—△—) Pack et al. (1992); (□) Mitroy (1990); (—○—) Danjo (1988); (—▲—) McEachran and Stauffer (1984); (—■—) Sin Fai Lam (1982); (—●—) Srivastava et al. (1981).

4.7 SCATTERING LENGTH

See Table 4.6.

4.8 EXCITATION LEVELS

See Table 4.7. Total excitation cross sections are shown in Table 4.8 and Figure 4.5.

TABLE 4.5
Comparison of Ramsauer–Townsend Minimum

Method	Minimum (eV)	Cross Section (10^{-20} m^2)	Reference
Elastic cross section	0.5	0.74	Sin Fai Lam (1982)
Momentum transfer	0.5	0.10	
Momentum transfer	0.5	0.10	Koizumi et al. (1986)
Total	0.74	0.441	Buckman et al. (1987)
Total	0.5	0.1182	Mitroy (1990)

TABLE 4.6
Comparison of Scattering Length

Scattering Length (nm)	Method	Reference
~–0.143	Theory	Mimnagh et al. (1993)
–0.178	Momentum transfer cross section	Pack et al. (1992)
–0.179	Momentum transfer	Mitroy (1990)
–0.178	Momentum transfer	Hunter et al. (1988)
–0.169	Total cross section	Buckman et al. (1987)
–0.129	Total cross section	Jost et al. (1983)
–0.176	Momentum transfer	Frost and Phelps (1964)

Note: Negative sign denotes the presence of distinct Ramsauer–Townsend minimum.

Kr

TABLE 4.7
Energy Levels for Excitation States in Kr in 9–13 eV Range

Designation	Designation[a]		J	Designation[b]	Energy (eV)
Ground state	$4p^6\text{–}{}^1S_0$		0	0	0.0
1	$4p^5\,5s[3/2]^0_2$		2	A	9.915
2	$5s[3/2]^0_1$		1	B	10.033
3	$5s'[1/2]^0_0$		0	C	10.563
4	$5s'[1/2]^0_1$		1	D	10.644
5	$5p[1/2]_1$		1	E	11.304
6	$5p[5/2]$	§[c]	3	F	11.443
	$5p[5/2]$	§	2		11.445
8	$5p[3/2]^2$			G	11.546
9	$5p[1/2]_0$		0	H	11.666
10	$4d[1/2]_0$		0		11.998
11	$4d[1/2]_1$		1	I	12.037
12	$5p'[3/2]_1$		1		12.101
13	$4d[1\,1/2]$	§	2		12.112
	$4d[1\,1/2]$	§	4		12.126
14	$5p'[0\,1/2]$	§	1	J	12.141
	$5p[1\,1/2]$	§	2		12.144
15	$4d[3\,1/2]$		3		12.179
16	$5p'[1/2]$	§	0	K	12.257
	$4d[2\,1/2]$	§	2		12.258
17	$4d[2\,1/2]$		3		12.284
18	$6s[1\,1/2]$	#	2	L	12.352
	$4d[1\,1/2]$	#	1		12.355
19	$6s[1\,1/2]$	#	1		12.386
	$6p[2\,1/2]$	§	2,3	M	12.785
	$6p[1/2]$	#	0	N	12.856
	$4d'[2\,1/2]$	#	2		12.860
	$4d'[1\,1/2]$	§	1	P	13.005

[a] J–L coupling notation. See Delâge and Carette (1976) and Guo et al. (2000).
[b] Notation of Delâge and Carette (1976).
[c] § and # mean that the levels are not distinguished.

Kr

TABLE 4.8
Total Excitation Cross Section for Kr

Energy (eV)	Q_{ex} (10^{-20} m^2)	Energy (eV)	Q_{ex} (10^{-20} m^2)
20	0.728	200	0.622
30	0.983	300	0.479
40	1.000	400	0.398
50	0.977	500	0.330
60	0.949	600	0.291
70	0.927	700	0.255
80	0.907	800	0.230
90	0.900	900	0.213
100	0.882	1000	0.199
150	0.756		

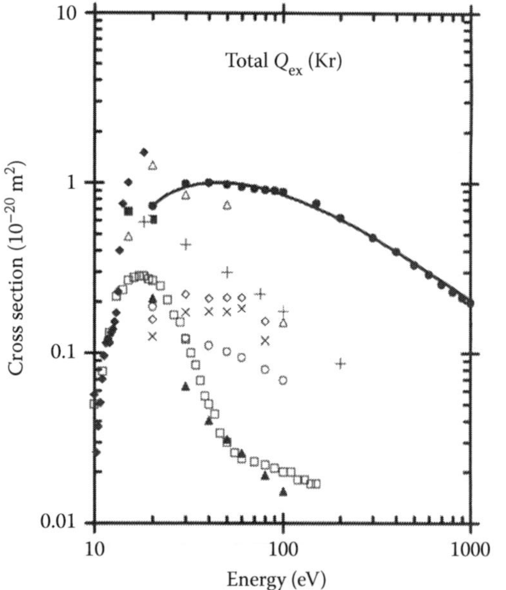

FIGURE 4.5 Excitation cross sections in Kr. (■) Guo et al. (2000a,b), sum of $4p^5$ $5s$ levels; experimental; (+) Chilton et al. (2000), experimental; (▲) Kaur et al. (1998), total of $4p^5$ $5p$ states, theory (method 1); (○) Kaur et al. (1998), total of $4p^5$ $5p$ states, theory (method 2); (——) Brusa et al. (1996), semiempirical; (◊) Danjo (1989), level 2, experimental; (×) Danjo (1989), level 4, experimental; (□) Mason et al. (1987), total metastable, experimental; (△) Trajmar et al. (1981), total of 19 levels, experimental; (◆) Specht et al. (1981), total excitation, from ionization coefficients; (●) de Heer et al. (1979), semiempirical.

4.9 IONIZATION CROSS SECTION

See Table 4.9.

4.10 DRIFT VELOCITY OF ELECTRONS

See Table 4.10.

TABLE 4.9
Total Ionization Cross Sections for Kr Computed from Partial Cross Section Data of Rejoub et al.

Rejoub et al. (2002)		Kobayashi et al. (2002)		Rapp et al. (1965)	
Energy (eV)	Total (10^{-20} m^2)	Energy (eV)	Total (10^{-20} m^2)	Energy (eV)	Total (10^{-20} m^2)
		14.5	0.070	14.5	0.078
15	0.102	15	0.142	15	0.160
16	0.367	16	0.318	16	0.358
17	0.529	17	0.512	17	0.400
18	0.732	18	0.710	18	0.799
20	1.170	20	1.086	20	1.223
22	1.480	22	1.407	22	1.583
24	1.760	24	1.712	24	1.926
26	2.140	26	1.993	26	2.243
28	2.340	28	2.244	28	2.524
30	2.550	30	2.462	30	2.771
35	2.990	34	2.783	36	3.263
40	3.240	40	3.103	40	3.492
45	3.429	45	3.263	45	3.668
50	3.615	50	3.413	50	3.835
55	3.754	55	3.525	55	3.967
60	3.824	60	3.639	60	4.090
65	3.942	65	3.705	65	4.169
70	3.956	70	3.745	70	4.213
75	4.010	75	3.780	75	4.248
80	4.026	80	3.788	80	4.257
90	4.025	90	3.763	90	4.231
100	3.967	100	3.732	100	4.196
110	3.894	110	3.680	110	4.143
120	3.853	120	3.632	120	4.081
130	3.754	135	3.515	130	3.993
140	3.659	140	3.497	140	3.914
150	3.582			150	3.826
160	3.495	160	3.342	160	3.747
170	3.438				
180	3.335	180	3.188	180	3.606
190	3.266				
200	3.218	200	3.058	200	3.457
225	3.055				
250	2.912	250	2.758	250	3.131
275	2.764				
300	2.646	300	2.526	300	2.867
350	2.438	350	2.327	350	2.656
400	2.268	400	2.160	400	2.463
500	1.979	500	1.886	500	2.164
600	1.783	600	1.681	600	1.944
700	1.610	700	1.511	700	1.759
800	1.464	800	1.375	800	1.601
900	1.359	900	1.266	900	1.487
1000	1.249	1000	1.175	1000	1.390

Note: Counting cross section data of Kobayashi et al. (2002) have been converted to gross cross sections by data kindly provided by Dr. Kobayashi. See Figures 4.6 and 4.7 for graphical presentation.

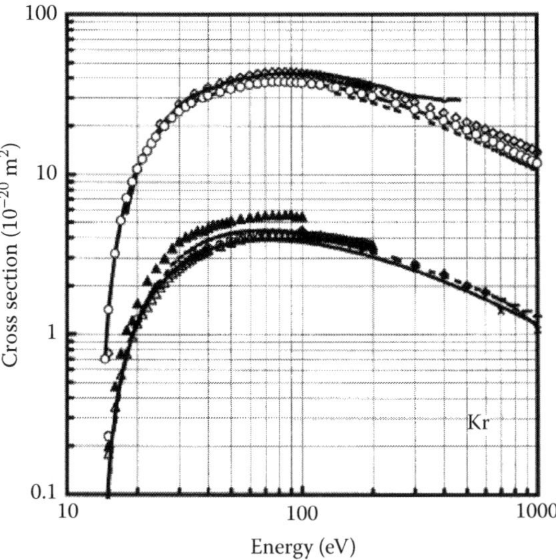

FIGURE 4.6 Total ionization cross sections in Kr. (○) Kobayashi et al. (2002); (▲) Rejoub et al. (2002); (——) Brusa et al. (1996), analytical equation; (— —) Krishnakumar and Srivastava (1988), analytical equation; (△) Wetzel et al. (1987); (×) Nagy et al. (1980); (○) Stephan et al. (1980); (◆) Schram et al. (1966); (■) Schram et al. (1965). Top curve, multiply the ordinates by 0.1. (— —) Sorokin et al. (2000); (——) Syage (1992); (◊) Krishnakumar and Srivastava (1988), experimental.

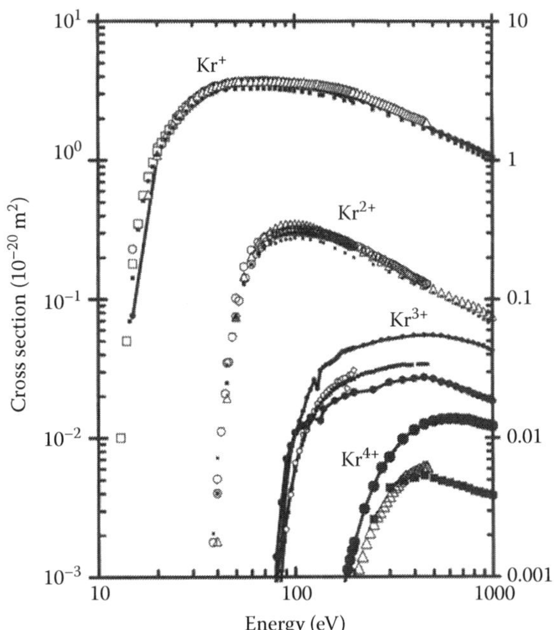

FIGURE 4.7 Partial absolute ionization cross sections in Kr. (■) Kobayashi (2002); (△) Syage (1992); (—●—) Krishnakumar et al. (1988); (□) Wetzel et al. (1987); (○) Stephan et al. (1980).

TABLE 4.10
Experimental and Theoretical Drift Velocity in Kr

E/N (Td)	*W* (m/s)	*E/N* (Td)	*W* (m/s)	*E/N* (Td)	*W* (m/s)
0.001	12.00	0.4	1.57 $(\times 10^3)$	100	7.74 $(\times 10^4)$
0.002	23.70	0.7	1.78 $(\times 10^3)$	200	1.41 $(\times 10^5)$
0.004	48.30	1	1.94 $(\times 10^3)$	300	1.92 $(\times 10^5)$
0.007	95.00	2	2.30 $(\times 10^3)$	400	2.45 $(\times 10^5)$
0.01	166.8	4	3.38 $(\times 10^3)$	500	3.29 $(\times 10^5)$
0.02	625.00	7	5.75 $(\times 10^3)$	750	4.95 $(\times 10^5)$
0.04	992.00	10	8.75 $(\times 10^3)$	1000	7.34 $(\times 10^5)$
0.07	1.12 $(\times 10^3)$	20	1.69 $(\times 10^4)$	2000	1.05 $(\times 10^6)$
0.1	1.20 $(\times 10^3)$	40	3.16 $(\times 10^4)$	3000	1.35 $(\times 10^6)$
0.2	1.37 $(\times 10^3)$	70	4.85 $(\times 10^4)$	4000	1.52 $(\times 10^6)$

Note: $0.002 \le E/N \le 2.0$ Td (Hunter et al., 1988), $2 < E/N \le 100$ Td averaged from Figure 4.8; $E/N > 100$ Td, digitized from Kücükarpaci and Lucas (1981). *E/N* in units of Td and *W* in units of m/s.

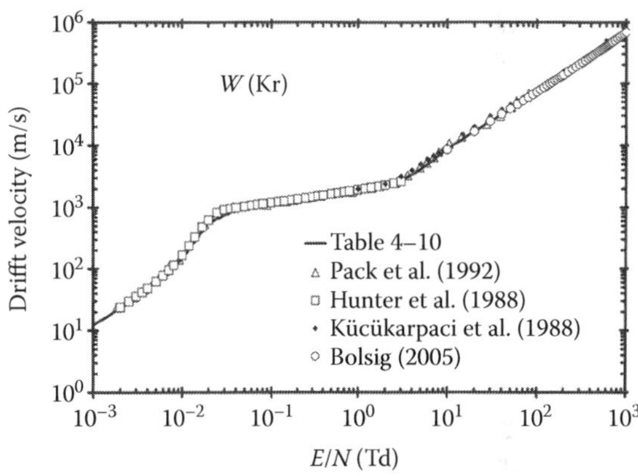

FIGURE 4.8 Experimental and theoretical drift velocity for electrons in Kr. (△) Pack et al. (1992); (○) Hunter et al. (1988); (◆) Kücükarpaci and Lucas (1981); (○) computed by the author using Bolsig software; (——) Table 4.10.

4.11 CHARACTERISTIC ENERGY

See Table 4.11.

4.12 REDUCED IONIZATION COEFFICIENT

See Table 4.12.

4.13 GAS CONSTANTS

Gas constants evaluated according to the expression (see inset of Figure 4.10)

$$\frac{\alpha}{N} = F \exp\left(-\frac{GN}{E}\right) \tag{4.1}$$

are $F = 3.83 \times 10^{-21}$ m^2 and $G = 159.1$ Td for the range $20 \le E/N \le 150$ Td.

TABLE 4.11
Characteristic Energy (D_r/μ) for Kr

Al-Amin and Lucas (1987)		Koizumi et al. (1988)	
E/N (Td)	D_r/μ (V)	*E/N* (Td)	D_r/μ (V)
1.41	5.24	0.005	0.0337
1.98	5.96	0.006	0.0396
2.82	6.67	0.007	0.0497
4.24	7.45	0.008	0.0715
5.65	6.92	0.009	0.0989
14.1	6.47	0.01	0.144
28.2	6.4	0.012	0.269
56.5	4.75	0.014	0.418
113	5.57	0.018	0.67
141	5.69	0.08	2.67
198	5.51	0.09	2.89
282	4.46	0.1	2.94
565	5.79	0.12	3.25
848	7	0.14	3.5
1130	8.86	0.16	3.61
1413	10.4	0.18	3.85
1695	12	0.2	3.99
1978	11.4	0.22	4.18
2260	11.8	0.25	4.47
2825	13.3	0.28	4.71
4238	11.9	0.3	4.76
5650	8.92	0.35	5.12
		0.4	5.31
		0.5	5.66
		0.7	6.56
		0.9	6.92

Note: See Figure 4.9 for graphical presentation.

4.14 PENNING IONIZATION CROSS SECTION

Penning effect occurs according to

$$Kr^* + B \rightarrow Kr + B^+ + e \tag{4.2}$$

where Kr* is an excited atom and B is a second atom with ionization potential less than the Kr excitation potential (Table 4.13).

4.15 POSITIVE ION MOBILITY

See Table 4.14 and Figure 4.11.

TABLE 4.12
Reduced Ionization Coefficients for Kr

E/N (Td)	α/N (10^{-20} m²)	*E/N* (Td)	α/N (10^{-20} m²)	*E/N* (Td)	α/N (10^{-20} m²)
20	0.0002	150	0.209	700	1.69
25	0.0007	200	0.344	750	1.80
30	0.0016	250	0.483	800	1.90
35	0.0031	300	0.626	850	2.01
40	0.0055	350	0.776	900	2.11
50	0.0129	400	0.918	1000	2.30
60	0.0236	450	1.06	1250	2.71
70	0.0367	500	1.20	1500	3.07
80	0.0524	550	1.33	1750	3.37
90	0.0700	600	1.45	2000	3.61
100	0.0897	650	1.57		

Note: See Figure 4.10 for graphical presentation.

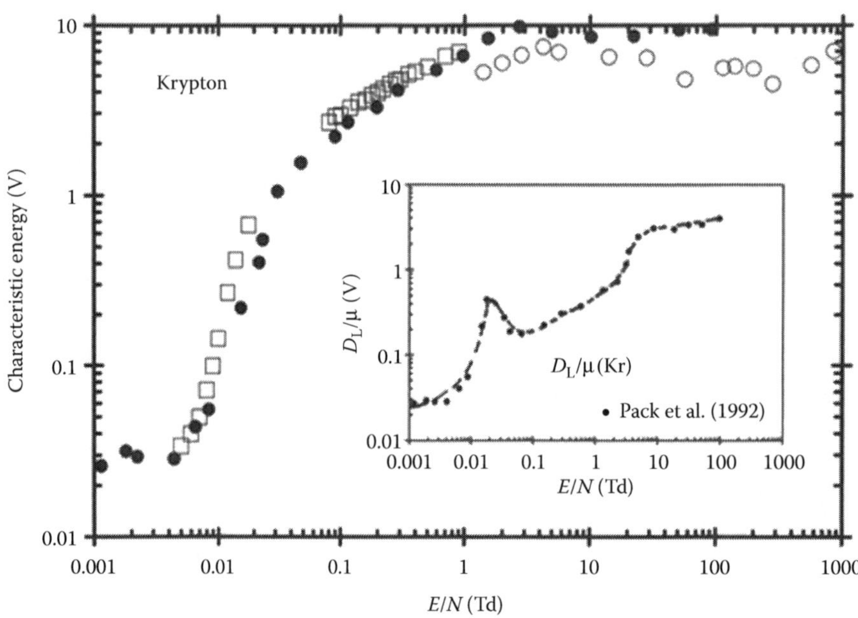

FIGURE 4.9 Characteristic energy for Kr. (●) Pack et al. (1992); (○) Al-Amin and Lucas (1987); (□) Kaizumi et al. (1986). Inset shows the ratio (D_L/μ), Phelps et al. (1992).

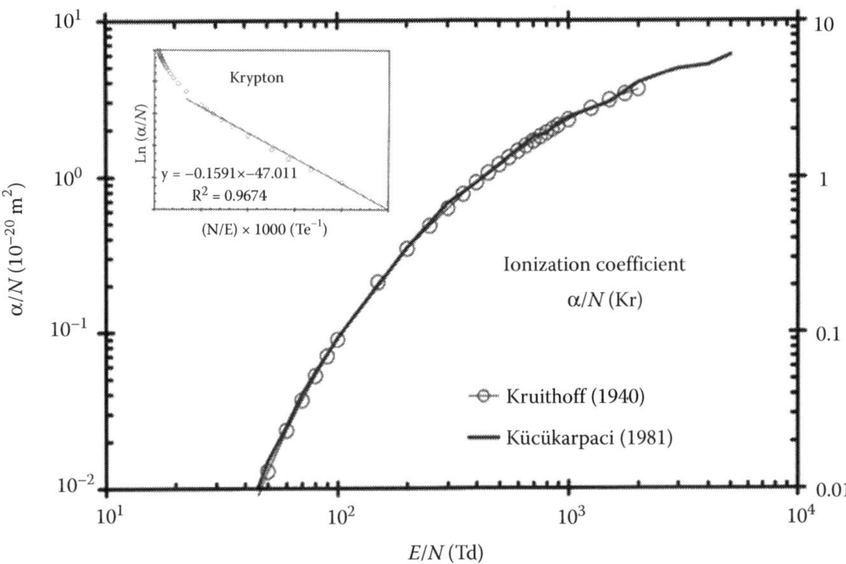

FIGURE 4.10 Reduced ionization coefficients for Kr. Inset shows the plot according to Equation 4.1 for evaluation of gas constants. (Adapted from Kruithoff, A. A., *Physica*, 7, 519, 1940.)

Kr

TABLE 4.13
Penning Ionization Cross Section (in Units of 10^{-20} m²) for Argon with Other Gases

	Krypton			
B	$^3P_2$ (9.915 eV)	$^1P_1$ (10.644 eV)	$^3P_0$ (10.563 eV)	$^3P_1$ (10.032 eV)
O_2	1.9	1.6	1.8	2.2
Zn			93	
Cd			110	

Source: Adapted from Eletskii, A. V. in I. S. Grigoriev and Z. Meilikhov (Ed.), *Handbook of Physical Quantities*, CRC Press, Boca Raton, FL, 1999, Chapter 20.

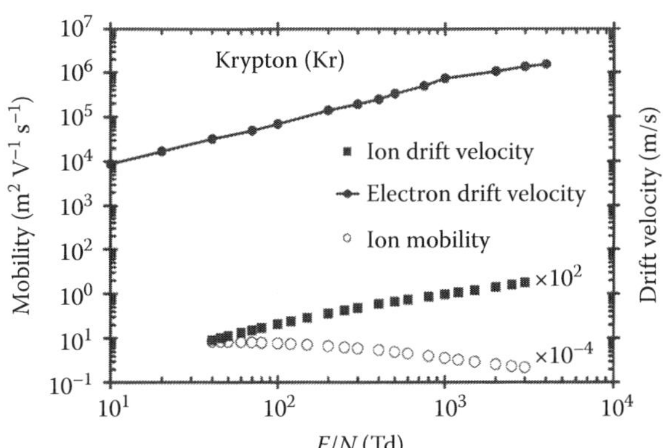

FIGURE 4.11 Reduced positive ion mobility for Kr. The corresponding drift velocity and electron drift velocities are also shown for comparison. (Adapted from Ellis, H. W. et al., *Atomic and Nuclear Data Tables*, 17, 177, 1976.)

TABLE 4.14
Reduced Positive Ion Mobility ($N = 2.69 \times 1025$ m⁻³; $T = 273.16$ K) for Kr and Drift Velocity

E/N (Td)	μ_0 ($\times 10^{\times 4}$ m²/V s)	W^+ ($\times 10^2$ m/s)
40	0.838	0.901
45	0.833	1.01
50	0.828	1.11
60	0.816	1.32
70	0.803	1.51
80	0.791	1.7
100	0.767	2.06
120	0.743	2.4
150	0.711	2.87
200	0.666	3.58
250	0.627	4.21
300	0.592	4.77
400	0.546	5.87
500	0.491	6.6
600	0.453	7.3
800	0.398	8.56
1000	0.359	9.65
1200	0.329	10.6
1500	0.294	11.8
2000	0.259	13.9
2500	0.234	15.7
3000	0.22	17.7

Source: Adapted from Ellis, H. W. et al., *Atomic and Nuclear Data Tables*, 17, 177, 1976.

Note: See Figure 4.11 for graphical presentation.

Kr

REFERENCES

Al-Amin, S. A. J. and J. Lucas, *J. Phys. D: Appl. Phys.*, 20, 1590, 1987.

Bowe, J. C., *Phys. Rev.*, 117, 1411, 1960.

Brusa, R. S., G. P. Karwasz, and A. Zecca, *Z. Phys. D*, 38, 279, 1996.

Buckman, S. J. and B. Lohmann, *J. Phys. B: At. Mol. Phys.*, 20, 5807, 1987.

Chilton, J. E., M. D. Stewart, Jr., and C. C. Lin, *Phys. Rev. A*, 62, 032714–1, 2000.

Dababneh, M. S., W. E. Kauppila, J. P. Downing, F. Laperriere, V. Pol, J. H. Smart, and T. S. Stein, *Phys. Rev*, 22, 1872, 1980.

Dababneh, M. S., Y. F. Hsieh, W. E. Kauppila, V. Pol, and T. S. Stein, *Phys. Rev.*, 26, 1252, 1982.

Danjo, A., *J. Phys. B: At. Mol. Opt. Phys.*, 21, 3759, 1988.

Danjo, A., *J. Phys. B: At. Mol. Opt. Phys.*, 22, 951, 1989.

de Heer, F. J., R. H. J. Jansen, and W. van der Kaay, *J. Phys. B: At. Mol. Phys.*, 12, 979, 1979.

Delâge, A. and J.-D. Carette, *J. Phys. B: At. Mol. Phys.*, 9, 2399, 1976.

Eletskii, A. V., in I. S. Grigoriev and Z. Meilikhov (Eds.), *Handbook of Physical Quantities*, CRC Press, Boca Raton, FL, 1999.

Ellis, H. W., R. Y. Pai, E. W. McDaniel. E. A. Mason, and L. A. Viehland, *Atomic and Nuclear Data Tables*, 17, 177, 1976.

Fon, W. C., K. A. Berrington, P. G., and A. Hibbert, *J. Phys. B: At. Mol. Phys.*, 17, 3279, 1984.

Frost, L. S. and A. V. Phelps, *Phys. Rev.*, 136, A1538, 1964.

Garcia, G., F. Arqueros, and J. Campos, *J. Phys. B: At. Mol. Phys.*, 19, 3777, 1986.

Guo, X., D. F. Mathews, G. Mikaelien, M. A. Khakoo, A. Crowe, I. Kanik, S. Trajmar, V. Zeman, K. Bartschat, and C. J. Fontes, *J. Phys. B: At. Mol. Opt. Phys.*, 33, 1895, 2000a; *J. Phys. B: At. Mol. Opt. Phys.*, 33, 1921, 2000b.

Heindörff, T., J. Hofft, and P. Dabkiewicz, *J. Phys. B: At. Mol. Phys.*, 9, 89, 1976.

Hunter, S. R., J. G. Carter, and L. G. Christophorou, *Phys. Rev. A*, 38, 5539, 1988.

Jost, K., P. G. F. Bisling, F. Eschen, M. Felsmann, and L. Walther, in J. Eichler et al. (Ed.), *Proceedings of the 13th International Conference on the Physics of Electronic Atomic Collisions*, Berlin, North-Holland, 1983. p. 91, cited by Wagenaar et al. (1985). Tabulated values above 7.5 eV are given in the latter reference. (The lower range of 0.05 eV is obtained from Nickel et al. (1985)).

Kanik, I., J. C. Nickel, and S. Trajmar, *J. Phys. B: At. Mol. Opt. Phy.*, 25, 2189, 1992.

Kaur, S., R. Srivastava, R. P. McEachran, and A. D. Stauffer, *J. Phys. B: At. Mol. Opt. Phys.*, 31, 4833, 1998.

Kobayashi, A., G. Fujiki, A. Okaji, and T. Masuoka, *J. Phys. B: At. Mol. Opt. Phys.*, 35, 2087, 2002.

Koizumi, T., E. Shirakawa, and I. Ogawa, *J. Phys. B: At. Mol. Phys.*, 19, 2331, 1986.

Krishnakumar, E. and S. K. Srivastava, *J. Phys. B: At. Mol. Phys.*, 21, 1055, 1988.

Kruithoff, A. A., *Physica*, 7, 519, 1940.

Kücükarpaci, H. N. and J. Lucas, *J. Phys. D: Appl. Phys.*, 14, 2001, 1981.

Mason, N. J. and W. R. Newell, *J. Phys. B: At. Mol. Phys.*, 20, 1357, 1987.

McEachran, R. P. and A. D. Stauffer, *J. Phys. B: At. Mol. Phys*, 17, 2507, 1984.

Mimnagh, D. J. R., R. P. McEachran, and A. D. Stauffer, *J. Phys. B: At. Mol. Opt. Phys.*, 26, 1727, 1993.

Mitroy, J., *Aust. J. Phys.*, 43, 19, 1990.

Nagy, P., A. Skutzlart, and V. Schmidt, *J. Phys. B: At. Mol. Phys.*, 13, 1249, 1980.

Pack, J. L., R. E. Voshall, and A. V. Phelps, *Phys. Rev.*, 127, 2084, 1962.

Pack, J. L., R. E. Voshall, and A. V. Phelps, *J. Appl. Phys.*, 71, 5363, 1992

Rapp, D. and P. Englander-Golden, *J. Chem. Phys.*, 13, 1464, 1965.

Rejoub, R., B. G. Lindsay, and R. F. Stebbings, *Phys. Rev. A*, 65, 042713-1, 2002.

Raju, G. G., *Gaseous Electronics: Theory and practice*, Taylor & Francis LLC., Boca Rotan, 2005.

Schram, B. L., F. J. deHeer, F. J. van Der Wieland, and J. Kistemaker, *Physica*, 31, 94, 1965.

Schram, B. L., *Physica*, 32, 197, 1966.

Sin Fai Lam, L. T., *J. Phys. B: At. Mol. Phys*, 15, 119, 1982.

Sorokin, A. A., L. A. Schmaenok, S. B. Bobashev, B. Möbus, M. Richter, and G. Ulm, *Phys. Rev. A*, 61, 022723–1, 2000. (Cross sections are measured in Ar, Kr, and Xe.)

Specht, L. T., S. A. Lawton, and T. A. de Temple, *J. Appl. Phys.*, 51, 166, 1981

Srivastava, S. K., H. Tanaka, A. Chutjian, and S. Trajmar, *Phys. Rev. A*, 23, 2156, 1981.

Stephan, K., H. Helm, and T. D. Märk, *J. Chem. Phys.*, 73, 3763, 1980.

Subramanian, K. P. and V. Kumar, *J. Phys. B: At. Mol. Phys.*, 20, 5505, 1987.

Suzuki, M., T. Taniguchi, and H. Tagashira, *J. Phys. D: Appl. Phys.*, 22, 1848, 1989.

Syage, J. A., *Phys. Rev. A*, 46, 5666, 1992.

Szmytkowski, C. and K. Maciąg, *Physica Scripta*, 54, 271, 1996.

Trajmar, S., S. K. Srivastava, H. Tanaka, and H. Nishimura, *Phys. Rev.*, 23, 2167, 1981.

Wagenaar, R. W. and F. J. de Heer, *J. Phys. B: At. Mol. Phys.*, 13, 3855, 1980.

Wetzel, R. C., F. A. Baiocchi, T. R. Hayes, and R. S. Freund, *Phys. Rev. A*, 35, 559, 1987.

Williams, J. F. and A. Crowe, *J. Phys. B: At. Mol. Phys.*, 8, 2233, 1975.

Zecca, A., G. Karwasz, R.S. Brusa, and R. Grisenti, *J. Phys. B. At. Mol. Opt. Phys.*, 24, 2737, 1991.

Zecca, A., G. P. Karwasz, and R. S. Brusa, *J. Phys. B: At. Mol. Opt. Phys.*, 33, 843, 2000. (This is an erratum to Brusa et al. (1996).)

5

MERCURY

CONTENTS

Mercury (Hg) has 80 electrons. The electronic polarizability is 5.585×10^{-40} F m^2 and the ionization potential is 10.437 eV.

5.1 SELECTED REFERENCES FOR DATA

See Table 5.1.

5.2 TOTAL SCATTERING CROSS SECTION

The highlights of the total cross section are (Jost and Ohnemus, 1979)

1. A pronounced structure at 0.4 eV attributed to shape resonance
2. Resonance structure at ~6 eV and ~9–11 eV range
3. A monotonic decrease for energy >~40 eV, which is a common feature for many gases (see Table 5.2)

TABLE 5.1
Selected References for Data on Hg Vapor

Quantity	Range: eV, (Td)	Reference
Momentum transfer cross section	0.01–10.0	McEachran and Elford (2003)
Elastic scattering cross section	9–25	**Zubek et al. (1995)**
Excitation cross section	15–100	**Panajotović et al. (1993)**
Excitation cross section	15–100	Srivastava et al. (1993)
Transport coefficients	(10–2000)	Liu and Raju (1992)
Momentum transfer cross section	0.08–3.0	**England and Elford (1991)**
Excitation cross section	15, 60, 100	**Pietzmann and Kessler (1990)**
Differential scattering	1–180	Sienkiewicz (1990)
Excitation cross sections	80–400	Heddle and Gallagher (1989)
Transport coefficients	(20–1200)	Sakai et al. (1989)
Excitation cross section	25–300	**Holtkamp et al. (1987)**
Elastic scattering cross section	1–50	McEachran and Stauffer (1987)
Transport coefficients	(150–2100)	Garamoon and Abdelhaleem (1979)
Total cross section	0.1–500	Jost and Ohnemus (1979)
Drift velocity	(0.07–14)	**Nakamura and Lucas (1978)**
Ionization processes	–	Vriens et al. (1978)

continued

Hg

TABLE 5.1 (continued)
Selected References for Data on Hg Vapor

Quantity	Range: eV, (Td)	Reference
Excitation cross section	0–8	**Burrow et al. (1976)**
Swarm coefficients	(0.1–1000)	Rockwood (1973)
Excitation cross section	4–15	**Borst (1969)**
Ionization coefficient	(≤2500)	**Overton and Davies (1968)**
Ionization cross section	10–10000	Kieffer and Dunn (1966)
Elastic cross section	100–2000	Bunyan and Schonfelder (1965)
Ionization coefficient	(≤2500)	**Davies and Smith (1965)**
Drift velocity	(0.75–10)	**McCutchen (1958)**
Drift velocity	(100–600)	**Klarfeld (1938)**
Total cross section	8–800	**Arnot (1931)**
Ionization cross section	15–800	**Smith (1931)**
Ionization cross section	12–315	**Bleakney (1930)**

Note: Bold font indicates experimental study.

TABLE 5.2
Total Scattering Cross Section for Hg

Energy (eV)	Q_T (10^{-20} m^2)	Energy (eV)	Q_T (10^{-20} m^2)	Energy (eV)	Q_T (10^{-20} m^2)
0.10	153.44	2.60	92.62	5.30	53.23
0.13	157.86	2.80	87.25	5.35	53.68
0.16	184.16	3.00	81.12	5.40	54.32
0.20	188.52	3.20	76.75	5.45	54.82
0.23	221.42	3.30	72.80	5.47	54.99
0.26	236.77	3.60	69.27	5.50	54.85
0.30	245.56	3.80	66.42	5.52	54.40
0.32	254.32	4.00	63.59	5.55	52.78
0.36	263.09	4.20	60.51	5.60	51.07
0.40	265.27	4.30	58.86	5.70	48.02
0.43	263.09	4.40	57.88	5.80	45.95
0.46	263.09	4.50	57.34	5.90	44.55
0.50	260.90	4.54	57.34	6.00	43.85
0.53	256.51	4.58	57.26	6.20	41.69
0.56	252.14	4.62	57.12	6.4	39.90
0.60	247.74	4.66	56.95	6.6	38.28
0.63	238.98	4.70	56.64	6.7	37.35
0.66	234.58	4.74	56.42	6.8	36.71
0.70	228.00	4.78	56.39	6.9	36.12
0.80	212.66	4.82	56.34	7.0	35.62
0.90	197.32	4.84	56.48	7.2	34.61
1.00	184.16	4.86	56.64	7.4	33.77
1.10	173.21	4.88	57.01	7.6	32.62
1.20	164.44	4.90	57.06	7.8	31.84
1.30	157.02	4.92	56.84	8.0	30.88
1.40	149.52	4.94	55.92	8.2	29.76
1.50	140.76	4.96	54.71	8.3	29.37
1.60	135.49	5.00	53.62	8.4	29.18
1.70	128.91	5.02	53.31	8.5	29.06
1.80	124.10	5.04	53.23	8.55	28.95
1.90	117.96	5.10	53.06	8.6	28.81
2.00	113.12	5.15	52.98	8.7	28.76
2.20	106.54	5.20	52.98	8.75	28.98
2.40	99.96	5.25	53.06	8.8	29.06

TABLE 5.2 (continued)
Total Scattering Cross Section for Hg

Energy (eV)	Q_T (10^{-20} m^2)	Energy (eV)	Q_T (10^{-20} m^2)	Energy (eV)	Q_T (10^{-20} m^2)
8.85	29.04	10.3	25.31	15.00	19.52
8.9	28.84	10.35	25.31	18.00	18.82
8.95	28.45	10.40	25.03	20.00	18.84
9.0	28.11	10.45	24.61	25.00	19.68
9.05	27.89	10.50	24.47	30.00	21.92
9.1	27.58	10.55	24.47	40.00	21.48
9.2	27.36	10.60	24.61	50.00	21.06
9.3	27.41	10.70	24.61	60.00	19.77
9.4	27.44	10.75	24.70	70.00	17.89
9.5	27.66	10.80	24.78	80.00	17.53
9.6	27.72	10.85	24.86	100.00	16.52
9.65	27.72	10.90	24.81	150.00	12.80
9.7	27.10	10.95	24.64	200.00	11.09
9.75	26.21	11.00	24.33	250.00	9.13
9.8	25.79	11.10	23.86	300.00	8.76
9.85	25.48	11.20	23.80	350.00	7.11
9.9	25.17	11.30	23.60	400.00	6.75
10.0	25.17	11.40	23.24	450.00	6.27
10.1	25.12	12.00	22.15	500.00	5.80
10.2	25.23	14.00	20.38		

Source: Reprinted with permission from Jost, K. and B. Ohnemus, *Phys. Rev. A*, 19, 641, 1979. Copyright (1979) by the American Physical Society.
Note: See Figure 5.1 for graphical presentation.

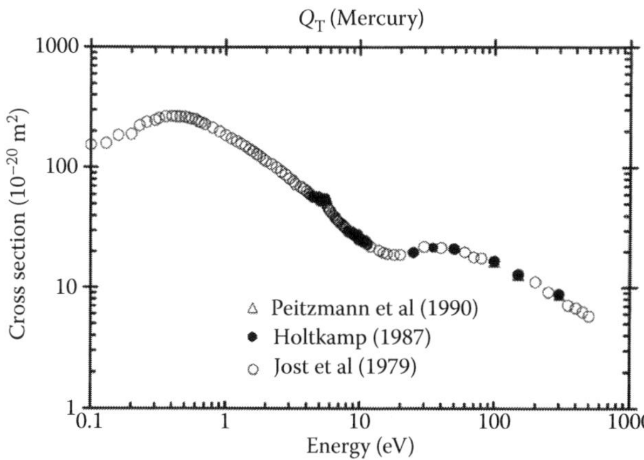

FIGURE 5.1 Total scattering cross sections for Hg. (Δ) Peitzmann et al. (1990); (•) Holtkamp (1987); (○) Jost and Ohnemus (1979).

5.3 DIFFERENTIAL SCATTERING CROSS SECTION

Figure 5.2 shows the differential scattering cross sections for elastic scattering (Holtkamp et al., 1987). Points to note are

1. A pronounced forward scattering decreasing to <1% at angles of ~35° for 25 eV, ~90° for 50 eV, ~60° for 100 eV, and ~35° for 300 eV

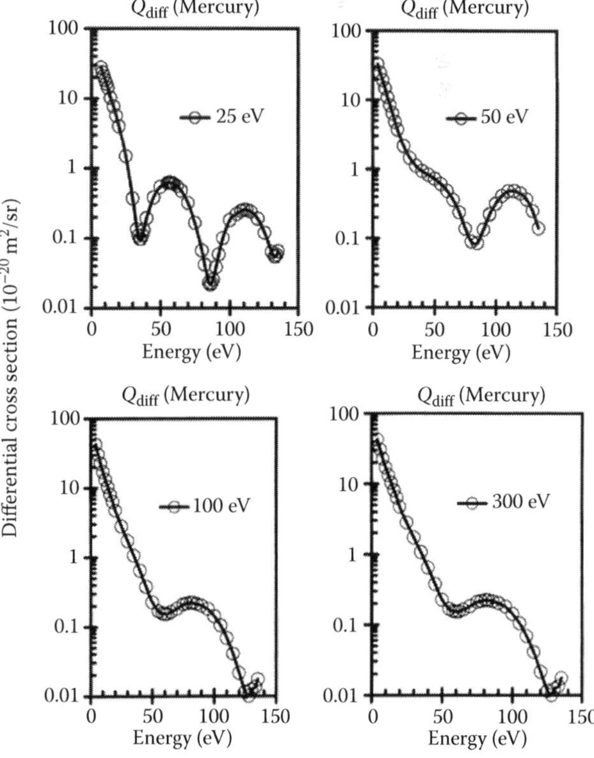

FIGURE 5.2 Differential cross section for elastic scattering for Hg at selected electron energy. (Adapted from Holtkamp, G. et al., *J. Phys. B: At. Mol. Phys.*, 20, 4543, 1987.)

2. Several peaks due to diffraction effects between the incoming and reflected waves, characteristic of atoms having large number of electrons (e.g., see xenon)

5.4 ELASTIC SCATTERING CROSS SECTION

See Table 5.3.

Hg

5.5 MOMENTUM TRANSFER CROSS SECTION

Momentum transfer cross section for Hg is shown in Figure 5.3. Data from Nakamura and Lucas[15] and McEachran and Stauffer[12] are not shown as they agree with Rockwood (1973).

TABLE 5.3
Elastic Scattering Cross Section for Hg

Zubek et al. (1995)		Holtkamp et al. (1987)	
Energy (eV)	Q_{el} (10^{-20} m^2)	Energy (eV)	Q_{el} (10^{-20} m^2)
9	13.6	25	9.97
12.2	12	35	10.14
15	12.1	50	10.05
17.5	10.2	100	8.99
20	8.85	150	7.95
25	9	300	4.93

Note: Below 4.667 eV, which is the lowest excitation potential, the elastic scattering cross section is the same as total cross section.

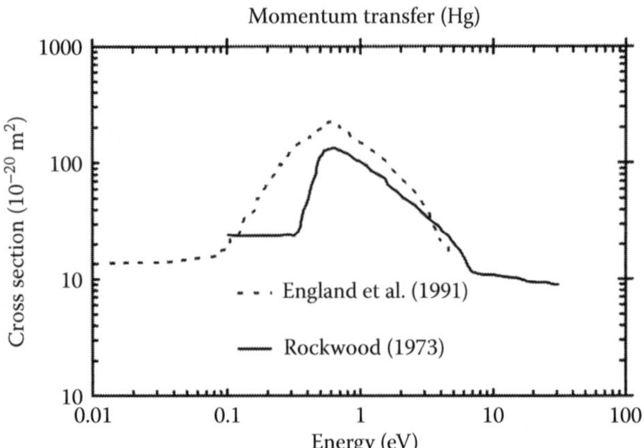

FIGURE 5.3 Momentum transfer cross section for Hg. (– –) England and Elford (1991); (—) Rockwood (1973). Data from Nakamura and Lucas (1978) and McEachran and Elford (2003) are not shown as they agree with Rockwood (1973).

5.6 EXCITATION CROSS SECTION

See Tables 5.4 and 5.5.

TABLE 5.4
Lower Excitation Levels for Hg

Designation	Energy (eV)
6^3P_0	4.667
6^3P_1	4.887
6^3P_2	5.461
6^1P_1	6.704
7^3S_1	7.731
7^1S_0	7.926
7^3P_0	8.619
7^3P_1	8.637
7^3P_2	8.829
7^1P_1	8.840

Note: Ground state designation $5d^{10}6s^2$ ($^1S_0$).

TABLE 5.5
Excitation Cross Sections for Hg Recommended by Liu and Raju (1992)

Energy (eV)	Cross Section (10^{-20} m^2)
6.00	0.25
7.00	9.25
10.00	10.22
12.50	15.10
17.78	14.25
25.00	12.20
30.00	10.30
50.00	8.40
100.00	6.50
300.00	2.82

Note: See Figure 5.4 for graphical presentation.

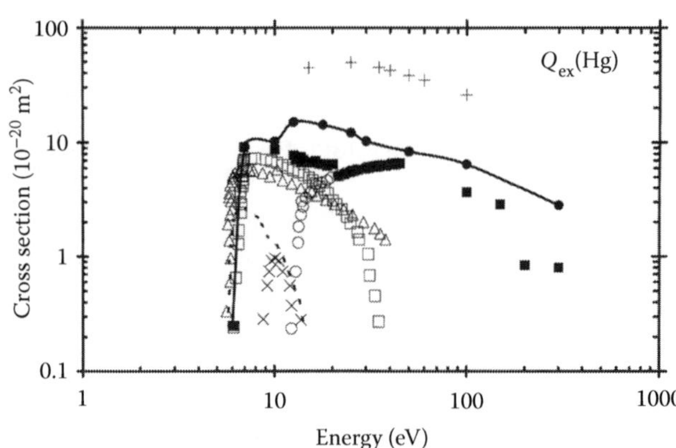

FIGURE 5.4 Excitation cross section for Hg. (+) Srivastava et al. (1993); (– –) 6^3P_0; (Δ) 6^3P_1; (□) 6^3P_2; (○) 6^1P_1; (×) 7^3S_1; (■) sum, Liu and Raju (1992); (—) recommended.

TABLE 5.6
Ionization Cross Section for Hg

Energy (eV)	Q_i (10^{-20} m^2)	Energy (eV)	Q_i (10^{-20} m^2)	Energy (eV)	Q_i (10^{-20} m^2)
12	0.75	55	6.24	350	4.13
15	1.76	60	6.37	400	3.78
18	2.60	70	6.46	450	3.52
20	3.08	80	6.42	500	3.33
25	4.08	90	6.34	600	3.08
30	4.77	100	6.24	700	2.86
35	5.18	150	5.79	800	2.63
40	5.46	200	5.40	900	2.40
45	5.75	250	4.98	1000	2.14
50	6.03	300	4.54		

Note: See Figure 5.5 for graphical presentation.
Source: Digitized from L. J. Kieffer and G. H. Dunn, *Rev. Mod. Phys.*, 38, 1, 1966.

FIGURE 5.5 Ionization cross section for Hg. (○) Liu and Raju (1992); (■) Pietzmann and Kessler (1990); (—△—) Holtkamp et al. (1987); (—•—) Kieffer and Dunn (1966); (□) Bleakney (1930). The larger values of Liu and Raju (1992) are possibly due to ionization of excited states (Raju, 2005).

5.7 IONIZATION CROSS SECTION

See Table 5.6.

5.8 DRIFT VELOCITY

See Table 5.7.

5.9 REDUCED IONIZATION COEFFICIENTS

See Table 5.8.

TABLE 5.7
Suggested Drift Velocity of Electrons in Hg

E/N (Td)	*W* (10^3 m/s)	*E/N* (Td)	*W* (10^3 m/s)	*E/N* (Td)	*W* (10^3 m/s)
0.08	0.112	3.00	0.934	120	48.1
0.10	0.123	4.0	1.12	130	52.1
0.12	0.132	5.0	1.50	150	62.3
0.14	0.140	6.0	1.89	200	84.7
0.17	0.152	8.0	2.58	250	109.6
0.20	0.162	10.0	3.20	300	141.6
0.25	0.178	15	5.02	350	178.7
0.30	0.192	20	8.00	400	204.6
0.35	0.205	25	10.43	450	232.5
0.40	0.216	30	12.4	500	261.2
0.50	0.237	35	14.4	550	295.8
0.60	0.256	40	16.3	600	323.4
0.70	0.273	45	18.0	700	378.2
0.80	0.290	50	19.8	800	433.6
1.00	0.321	60	24.1	900	482.7
1.20	0.352	70	28.0	1000	536.2
1.40	0.384	80	31.6	1200	632.9
1.70	0.440	90	36.2	1400	719.0
2.00	0.512	100	41.3	1500	725.6
2.50	0.692	110	44.9		

Note: See Figure 5.6 for graphical presentation.

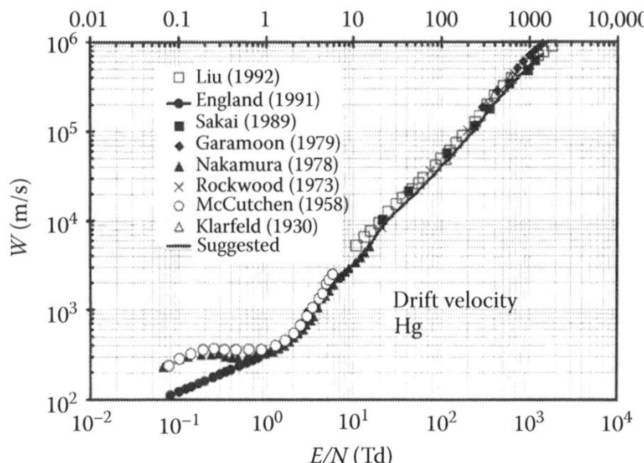

FIGURE 5.6 Drift velocity of electrons for Hg. (□) Liu and Raju (1992); (—•—) England and Elford (1991); (■) Sakai (1989); (♦) Garamoon and Abdelhaleem (1979); (▲) Nakamura (1978); (×) Rockwood (1973); (○) McCutchen (1958); (Δ) Klarfeld (1938); (—) suggested.

TABLE 5.8
Reduced Ionization Coefficient

E/N (Td)	α/N (10^{-20} m^2)	E/N (Td)	α/N (10^{-20} m^2)
208	0.0483	1750	4.103
280	0.113	2000	4.448
390	0.305	2500	5.100
500	0.615	3000	5.602
600	0.947	3500	6.116
700	1.170	4000	6.537
800	1.527	4500	6.919
900	1.888	5000	7.182
1000	2.152	6000	7.577
1250	2.825	7000	7.727
1500	3.482		

Note: See Figure 5.7 for graphical presentation.

Hg

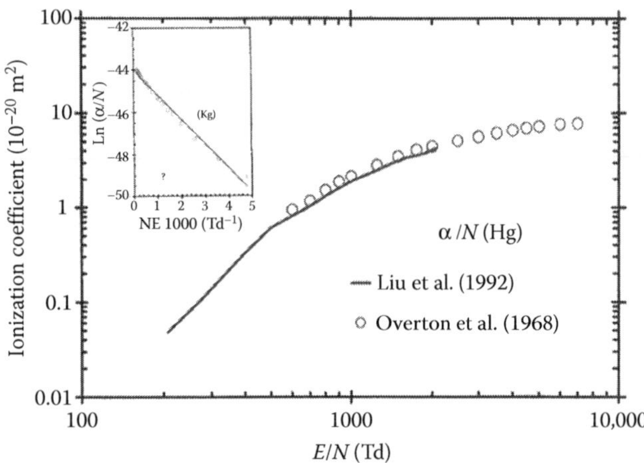

FIGURE 5.7 Reduced ionization coefficients for Hg. (—) Liu and Raju (1992); (○) Overton and Davies (1968). Inset shows plot according to Equation 5.1.

5.10 GAS CONSTANTS

Gas constants evaluated according to expression (see inset of Figure 5.7)

$$\frac{\alpha}{N} = F \exp\left(-\frac{GN}{E}\right) \quad (5.1)$$

are $F = 7.78 \times 10^{-20}$ m^2 and $G = 1157$ Td for the range $280 \leq E/N \leq 5000$ Td.

REFERENCES

Arnot, F. L., *Proc. Roy. Soc.*, 130, 655, 1931.

Bleakney, W., *Phys. Rev.*, 35, 139, 1930.

Borst, W. L., *Phys. Rev.*, 181, 257, 1969.

Bunyan, P. J. and J. L. Schonfelder, *Proc. Phys. Soc.*, 85, 455, 1965.

Burrow, P. D., J. A. Michejda, and J. Comer, *J. Phys. B: Atom. Mol. Phys.*, 9, 3225, 1976.

Davies, D. E. and D. Smith, *Brit. J. Appl. Phys.*, 16, 697, 1965.

England, J. P. and M. T. Elford, *Aust. J. Phys.*, 44, 647, 1991.

Garamoon, A. A. and A. S. Abdelhaleem, *J. Phys. D: Appl. Phys.*, 12, 2181, 1979.

Heddle, D. W. O. and J. W. Gallagher, *Rev. Mod. Phys.*, 61, 221, 1989.

Holtkamp, G., K. Jost, F. J. Peitzmann, and J. Kessler, *J. Phys. B: At. Mol. Phys.*, 20, 4543, 1987.

Jost, K. and B. Ohnemus, *Phys. Rev. A*, 19, 641, 1979.

Kieffer, L. J. and G. H. Dunn, *Rev. Mod. Phys.*, 38, 1, 1966.

Klarfeld, B., *Tech. Phys. USSR*, 5, 913, 1938.

Liu, J. and G. R. Govinda Raju, *J. Phys. D: Appl. Phys.*, 25, 167, 1992.

McCutchen, C. W., *Phys. Rev.*, 112, 1848, 1958.

McEachran, R. P. and A. D. Stauffer, *J. Phys. B: At. Mol. Phys.*, 20, 5517, 1987.

McEachran, R. P. and M. T. Elford, *J. Phys. B: At. Mol. Opt. Phys.*, 36, 427, 2003.

Nakamura, Y. and J. Lucas, *J. Phys. D: Appl. Phys.*, 11, 325, 1978.

Overton, G. D. N. and D. E. Davies, *Brit. J. Appl. Phys.*, 1, 881, 1968.

Panajotović, R., V. Pejčev, M. Konstantinović, D. Filipović, V. Bočvarski, and B. Marinković, *J. Phys. B: At. Mol. Opt. Phys.*, 26, 1005, 1993.

Peitzmann, F. J. and J. Kessler, *J. Phys. B: At. Mol. Opt. Phys.*, 23, 2629, 1990.

Rockwood, S. D., *Phys. Rev. A*, 8, 2348, 1973.

Sakai, Y., S. Sawada, and H. Tagashira, *J. Phys. D: Appl. Phys.*, 22, 276, 1989.

Sienkiewicz, J. E., *J. Phys. B: At. Mol. Opt. Phys.*, 23, 1869, 1990.

Smith, P. T., *Phys. Rev.*, 37, 808, 1931.

Srivastava, R., T. Zuo, R. P. McEachran, and A. D. Stauffer, *J. Phys. B: At. Mol. Opt. Phys.*, 26, 1025, 1993.

Vriens, L., A. J. Keijser, and F. A. S. Ligthart, *J. Appl. Phys.*, 49, 3807, 1978.

Zubek, M., A. Danjo, and G. C. King, *J. Phys. B: At. Mol. Opt. Phys.*, 28, 4117, 1995.

6

NEON

CONTENTS

Neon (Ne) is a rare gas that has 10 electrons with configuration $1s^2 2s^2 2p^6$. The electronic polarizability is 4.40×10^{-41} F m^2 and ionization potential 21.564 eV.

6.1 SELECTED REFERENCES FOR DATA

See Table 6.1.

6.2 TOTAL SCATTERING CROSS SECTION

Points to note are

1. The cross section increases monotonically with increasing electron energy up to 40 eV.
2. The cross section decreases monotonically for electron energy >50 eV.

TABLE 6.1
Selected References for Data

Quantity	Range: eV, (Td)	Reference
Ionization cross section	22–1000	**Kobayashi et al.** (2002)
Ionization cross section	22.5–1000	**Rejoub et al. (2002)**
Excitation cross section	25–200	**Chilton et al. (2000)**
Total scattering cross section	60–1000	Zecca et al. (2000)
Ionization cross section	140–1000	**Sorokin et al. (1998)**
Excitation cross section	30–1000	Brusa et al. (1996)
Total scattering cross section	0.5–220	**Szmytkowski and Maciąg (1996)**
Excitation cross section	20–400	**Kanik et al. (1996)**
Ionization cross section	140–1000	**Almeida et al. (1995)**

continued

Ne

TABLE 6.1 (continued)
Selected References for Data

Quantity	Range: eV, (Td)	Reference
Total scattering cross section	0.1–5.0	**Gulley et al. (1994)**
Transport parameters	(0.001–1000)	Puech and Mizzi (1991)
Differential cross section	0.1–70	Saha (1989)
Ionization cross section	25–1000	Krishnakumar and Srivastava (1988)
Ionization cross section	22–1000	Lennon et al. (1988)
Diffusion coefficients	1.41–5650	**Al-Amin and Lucas (1987)**
Total scattering cross section	0.7–10.0	**Kumar et al. (1987)**
Excitation cross section	17–147	**Mason and Newell (1987)**
Ionization cross section	21–200	**Wetzel et al. (1987)**
Total scattering cross section	700–1000	**Garcia et al. (1986)**
Differential scattering cross section	0–50	McEachran and Stauffer (1985)
Total scattering cross section	4–300	**Nickel et al. (1985)**
Excitation cross section	25–300	**Phillips et al. (1985)**
Excitation cross section	17–500	**Teubner et al. (1985)**
Excitation cross section	25–100	**Register et al. (1984)**
Differential cross section	5–200	Fon and Berrington (1981)
Total scattering cross section	20–700	**Kaupilla et al. (1981)**
Transport coefficients	(1.4–339)	**Kücükarpaci et al. (1981)**
Total scattering cross section	0–25	**O'Malley and Crompton (1980)**
Ionization cross section	500–1000	**Nagy et al. (1980)**
Ionization cross section	25–180	**Stephan et al. (1980)**
Total scattering cross section	25–750	**Wagenaar and de Heer (1980)**
Total scattering cross section	20–3000	de Heer et al. (1979)
Differential cross section	20–3000	McCarthy et al. (1977)
Ion mobility	(6.00–1500)	Ellis et al. (1976)
Differential cross section	100–500	Gupta and Rees (1975)
Differential cross section	20–400	Williams and Crowe (1975)
Ionization cross section	22–500	**Fletcher and Cowling (1973)**
Drift velocity	(0.03–7.0)	**Robertson (1972)**
Differential cross section	0–14	Thompson (1971)
Excitation cross section	22–200	Sharpton et al. (1970)
Ionization coefficients	(28–110)	**Dutton et al. (1969)**
Ionization cross section	22–1000	Rapp and Englander-Golden (1965)
Ionization cross section	100–20000	**Schram et al. (1965, 1966)**
Ionization coefficients	(5–1100)	**Chanin and Rork (1963)**
Ionization coefficients	(6–1100)	**Kruithoff (1940)**

Note: Bold font indicates experimental study.

3. Most measurements shown in Figure 6.1 fall within a band of 5% (Raju, 2005). See Table 6.2.

6.3 DIFFERENTIAL SCATTERING CROSS SECTION

Figure 6.2 shows the differential cross section for Ne at selected electron energy. Points to note are

1. A sharp forward peak at all energies except the lowest, attributed to the contribution of higher partial waves to the differential cross section. At low energies only *s*- and *p*-waves contribute, not resulting in a forward peak (Figure 6.2a).
2. Formation of a deep minimum at all energies higher than 20 eV.

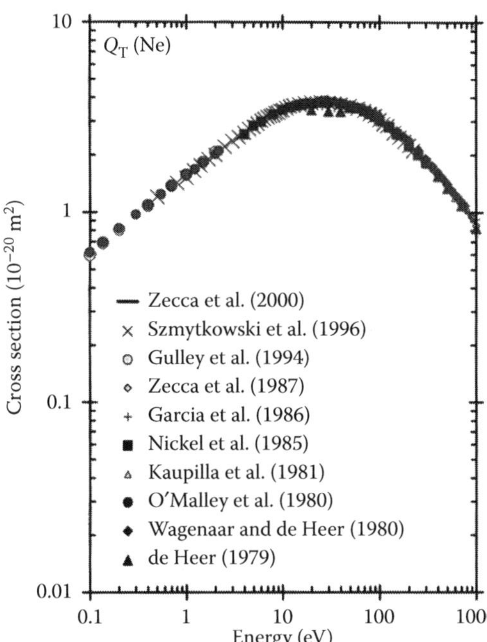

FIGURE 6.1 Total cross section for Ne. (——) Zecca et al. (2000), analytical expression; (×) Szmytkowski and Maciąg (1996); (○) Gulley et al. (1994); (◇) Zecca et al. (1987); (+) Garcia et al. (1986); (■) Nickel et al. (1985); (△) Kauppila et al. (1981); (●) O'Malley and Crompton (1980); (◆) Wagenaar and de Heer (1980)); (▲) de Heer et al. (1979). Not shown are the results of Kumar et al. (1987).

TABLE 6.2
Total Scattering Cross Sections in Ne in the Energy Range $0.1 \leq \varepsilon \leq 300$ eV

Gulley et al. (1994)		Gulley et al. (1994)		Nickel et al. (1985)	
Energy (eV)	Q_T (10^{-20} m^2)	Energy (eV)	Q_T (10^{-20} m^2)	Energy (eV)	Q_T (10^{-20} m^2)
0.100	0.595	0.760	1.415	4	2.565
0.120	0.658	0.780	1.425	5	2.843
0.140	0.690	0.800	1.446	6	2.984
0.160	0.734	0.820	1.457	8	3.26
0.180	0.766	0.840	1.471	10	3.443
0.200	0.808	0.860	1.490	12	3.555
0.220	0.845	0.880	1.502	14	3.625
0.240	0.878	0.900	1.506	16	3.668
0.250	0.884	0.920	1.511	18	3.706
0.360	1.042	0.940	1.520	20	3.727
0.380	1.057	0.960	1.544	25	3.766
0.400	1.084	0.980	1.559	30	3.78
0.420	1.105	1.000	1.569	40	3.709
0.440	1.133	1.250	1.708	50	3.613
0.460	1.153	1.500	1.827	60	3.509
0.480	1.177	1.750	1.944	70	3.398
0.500	1.200	2.000	2.060	80	3.283
0.520	1.227	2.250	2.160	90	3.172
0.540	1.238	2.500	2.260	100	3.041
0.560	1.262	2.750	2.339	125	2.803
0.580	1.276	3.000	2.425	150	2.58
0.600	1.291	3.250	2.488	200	2.25
0.620	1.311	3.500	2.569	250	2.008
0.640	1.326	3.750	2.617	300	1.827
0.660	1.341	4.000	2.691	400	1.573
0.680	1.354	4.250	2.752	500	1.376
0.700	1.371	4.500	2.812	600	1.226
0.720	1.360	4.750	2.852	800	1.012
0.740	1.413	5.000	2.902	1000	0.865

Note: See Figure 6.1 for graphical presentation.

Ne

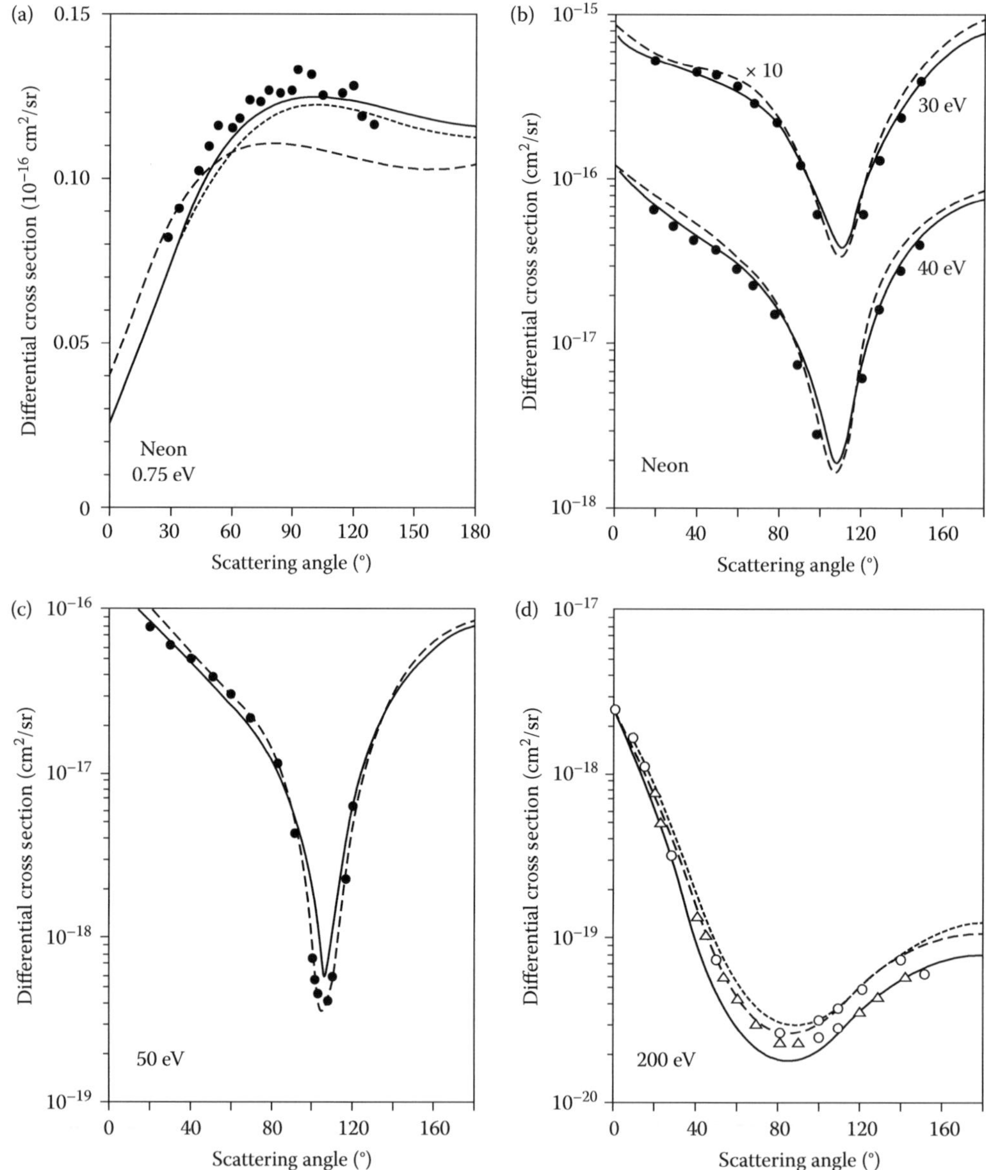

FIGURE 6.2 Experimental and theoretical differential scattering cross sections in Ne. (a): (●) Gulley et al. (1994); (—) Saha (1989); (– –) McEachran and Stauffer (1985); (- - -) O'Malley and Crompton (1980); (b), (c), and (d): (●) Williams and Crowe (1975); (—) Fon and Berrington (1981); (– –) Thompson (1971); (–●–) McCarthy et al. (1977); (○) Jansen et al. (quoted by Williams and Crowe, 1975); (△) Gupta and Rees (1975).

6.4 ELASTIC SCATTERING CROSS SECTION

See Table 6.3.

6.5 MOMENTUM TRANSFER CROSS SECTION

See Table 6.4.

6.6 SCATTERING LENGTH

See Table 6.5.

6.7 EXCITATION CROSS SECTIONS

See Table 6.6. Suggested total excitation cross section for Ne in the range $30 \le \varepsilon \le 1000$ eV are given by the analytical expression of Brusa et al. (1996):

$$Q_{ex} = \frac{1}{85.97(0.0317 + 0.001\varepsilon)} \times \ln\left(\frac{\varepsilon}{16.619}\right) \quad (6.1)$$

where ε is the energy in eV and the cross section is 10^{-20} m^2. Figure 6.5 summarizes the selected data.

TABLE 6.3
Recommended Elastic Scattering Cross Section

Energy (eV)	Q_{el} (10^{-20} m^2)	Energy (eV)	Q_{el} (10^{-20} m^2)	Energy (eV)	Q_{el} (10^{-20} m^2)
0.02	0.338	20.0	3.77	450	0.837
0.04	0.402	25.0	3.71	475	0.806
0.06	0.454	30.0	3.66	500	0.777
0.08	0.502	40	3.34	525	0.751
0.1	0.550	50	3.04	550	0.727
0.2	0.773	60	2.80	575	0.704
0.4	1.01	70	2.60	600	0.683
0.6	1.20	80	2.43	625	0.663
0.8	1.35	90	2.29	650	0.644
1.0	1.49	100	2.16	675	0.627
2.0	1.92	125	1.90	700	0.611
3.0	2.23	150	1.71	725	0.595
4.0	2.45	175	1.56	750	0.580
5.0	2.59	200	1.43	775	0.566
6.0	2.76	225	1.33	800	0.553
7.0	2.94	250	1.24	825	0.541
8.0	3.08	275	1.17	850	0.529
9.0	3.17	300	1.10	875	0.517
10.0	3.24	325	1.05	900	0.507
12.0	3.36	350	0.994	925	0.496
14.0	3.46	375	0.949	950	0.486
16.0	3.53	400	0.908	975	0.477
18.0	3.56	425	0.871	1000	0.468

Note: See Figure 6.3 for graphical presentation.

TABLE 6.4
Suggested Momentum Transfer Cross Sections for Ne

Energy (eV)	Q_M (10^{-20} m^2)	Energy (eV)	Q_M (10^{-20} m^2)	Energy (eV)	Q_M (10^{-20} m^2)
0.000	0.142	0.500	1.31	1.400	1.75
0.025	0.386	0.544	1.35	1.450	1.76
0.050	0.502	0.550	1.35	1.500	1.77
0.075	0.593	0.600	1.39	1.600	1.79
0.100	0.670	0.650	1.43	1.700	1.81
0.125	0.738	0.700	1.46	1.800	1.83
0.136	0.766	0.750	1.50	1.900	1.84
0.150	0.799	0.800	1.53	1.965	1.85
0.175	0.854	0.850	1.55	2.000	1.86
0.200	0.904	0.900	1.58	5.000	2.07
0.225	0.951	0.950	1.60	10.0	2.42
0.250	0.994	1.000	1.62	20.0	2.96
0.275	1.03	1.050	1.64	50.0	2.80
0.300	1.07	1.100	1.66	60	2.27
0.325	1.11	1.150	1.68	70	2.08
0.350	1.14	1.200	1.70	100	1.55
0.375	1.17	1.225	1.70	130	1.21
0.400	1.20	1.250	1.71	150	1.04
0.425	1.23	1.300	1.72	170	0.91
0.450	1.26	1.350	1.74	200	0.76
0.475	1.28	0.500	1.31		

Note: See Figure 6.4 for graphical presentation.

Ne

FIGURE 6.3 Integral elastic cross sections for Ne. (——) Zecca et al. (2000); (■) Saha (1989); (△) McEachran and Stauffer (1985); (□) Register and Trajmar (1984), Method A; (×) Register and Trajmar (1984), Method B (see Raju 2005); (▲) Fon and Berrington (1981); (○) de Heer et al. (1979); (●) Gupta and Rees (1975).

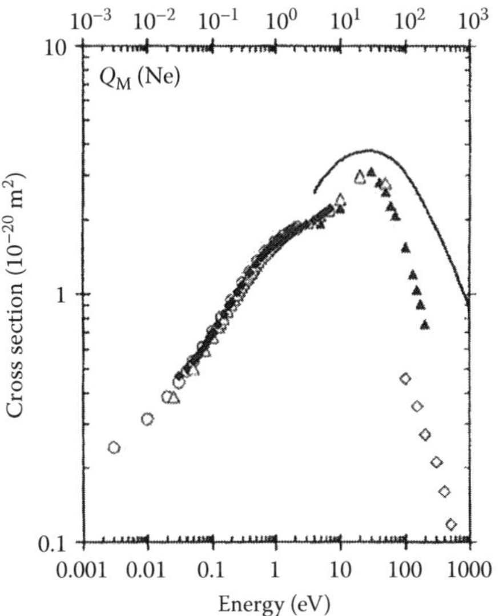

FIGURE 6.4 Momentum transfer cross sections in Ne. (□) Gulley et al. (1994); (△) McEachran and Stauffer (1985); (▲) Fon and Berrington (1981); (○) O'Malley and Crompton (1980); (◇) Gupta and Rees (1975); (◆) Robertson (1972); (—). Total cross sections are shown for comparison (from Figure 6.1).

TABLE 6.5
Scattering Length (A_0) for Ne

Scattering Length (10^{-11} m)	Method	Reference
1.093	Total cross section	Gulley et al. (1994)
1.063	Momentum transfer	McEachran and Stauffer (1985)
1.114	Momentum transfer	O'Malley and Crompton (1980)
1.27	Momentum transfer	Robertson (1972)

Note: $4\pi A_0^2$ is the scattering cross section at zero electron energy.

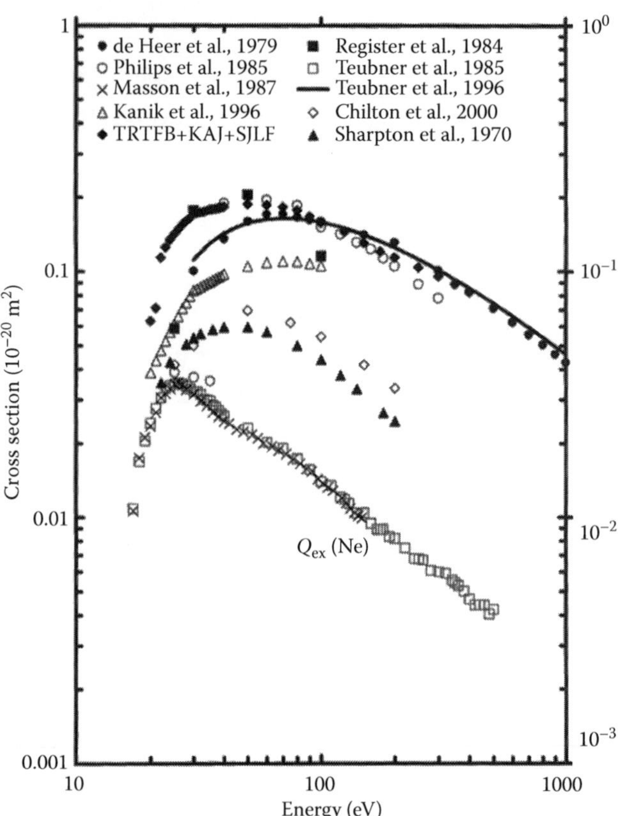

FIGURE 6.5 Selected excitation cross sections in Ne. (▲) Sharpton et al. (1970); (●) de Heer et al. (1979); (■) Register et al. (1984); (○) Phillips et al. (1985); (□) Teubner et al. (1985); (×) Mason and Newell (1987); (—) Brusa et al. (1996); (△) Kanik et al. (1996); (◇) Chilton et al. (2000); (◆) Teubner et al. (1985); (+) Kanik et al. (1996); Sharpton et al. (1970); sum of three separate measurements.

TABLE 6.6
Excitation Levels and Corresponding Energies in Ne

Paschen Notation	Designation $2p^6\ ^1S_0$	J 0	Energy (eV) 0 (Ground State)	L–S Coupling Notation $1S_0$
$1s_5$	$3s[3/2]_2$	2	16.619	$^3P_2$
$1s_4$	$3s[3/2]_1$	1	16.671	$^3P_1$
$1s_3$	$3s'[1/2]_0$	0	16.716	$^3P_0$
$1s_2$	$3s'[1/2]_1$	1	16.848	$^1P_1$
$2p_{10}$	$3p[1/2]_1$	1	18.382	$^3S_1$
$2p_9$, $2p_8$	$3p[5/2]_3$, $3p[5/2]_2$	3, 2	18.556, 18.576	$^3D_3$, $^3D_2$
$2p_7$, $2p_6$	$3p[3/2]_1$, $3p[3/2]_2$	1, 2	18.613, 18.637	$^3D_1$, $^1D_2$
$2p_3$, $2p_5$	$3p[1/2]_0$, $3p'[3/2]_1$,	0,1,	18.712,18.694,	$^3P_0$, $^1P_1$,
$2p_4$, $2p_2$	$3p'[3/2]_2$, $3p'[1/2]_1$	2, 1	18.704, 18.727	$^3P_2$, $^3P_1$
$2p_1$	$3p'[1/2]_0$	0	18.966	$^1S_0$
$2s_5$, $2s_4$	$4s[3/2]_2$, $4s[3/2]_1$	2, 1	19.664, 19.689	—
$2s_3$, $2s_2$	$4s'[1/2]_0$, $4s'[1/2]_1$	0, 1	19.761, 19.780	—
$3d_6$, $3d_{5.}$	$3d[1/2]_0$, $3d[1/2]_1$,	0,1,	20.025, 20.027,	
$3d'_4$, $3d_{4,}$	$3d[7/2]_{4,}$ $3d[7/2]_3$,	4,3,	20.035, 20.035,	
$3d_3$, $3d_2$,	$3d[3/2]_2$, $3d[3/2]_1$,	2,1,	20.037, 20.041,	—
$3d_1''$, $3d_1'$	$3d[5/2]_2$, $3d[5/2]_3$,	2,3	20.049, 20.049	
$3s_1''''$, $3s_1'''$	$3d'[5/2]_2$,$3d'[5/2]_3$,	2,3,	20.137, 20.137, 20.138,	
$3s_1''$, $3s_1'$	$3d'[3/2]_2$, $3d'[3/2]_{1,}$	2,1,	20.140,	—
$3p_{10}$	$4p[1/2]_1$	1	20.150	—
$3p_{9,}$ $3p_{8,}$	$4p[5/2]_3$, $4p[5/2]_2$,	3, 2,	20.189, 20.197,	
$3p_{7,}$ $3p_{6,}$	$4p[3/2]_1$, $4p[3/2]_2$,	1, 2	20.211, 20.215,	—
$3p_{3,}$	$4p[1/2]_{0,}$	0, 1,	20.260, 20.291,	
$3p_5$, $3p_4$,	$4p'[3/2]_1$, $4p'[3/2]_2$,	2, 1,	20.298, 20.298,	
$3p_2$, $3p_1$	$4p'[1/2]_1$, $4p'[1/2]_0$	0	20.369	—

Source: Adapted from Menses, G. D. *J. Phys. B: At. Mol. Op. Phys.*, 35, 3119, 2002.
Note: Energy data are from Register et al. (1984).

TABLE 6.7
Recommended Ionization Cross Section for Ne

Energy (eV)	Q_i (10^{-20} m^2)	Energy (eV)	Q_i (10^{-20} m^2)	Energy (eV)	Q_i (10^{-20} m^2)
22.5	0.006	90	0.576	225	0.684
25	0.025	100	0.614	250	0.670
27.5	0.051	110	0.649	275	0.650
30	0.081	120	0.661	300	0.631
32.5	0.109	130	0.676	350	0.598
35	0.141	140	0.685	400	0.560
40	0.202	150	0.696	500	0.507
45	0.250	160	0.700	600	0.462
50	0.305	170	0.700	700	0.422
60	0.396	180	0.699	800	0.388
70	0.475	190	0.696	900	0.358
80	0.536	200	0.694	1000	0.335

6.8 IONIZATION CROSS SECTION

Table 6.7 shows the recommended cross sections for Ne. See Figures 6.6 and 6.7 for graphical presentation.

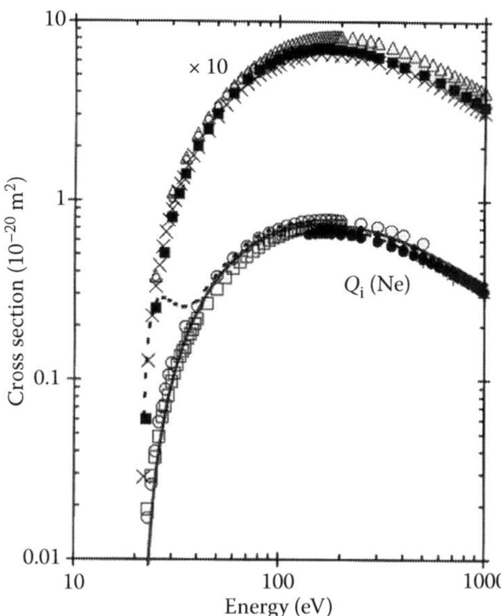

FIGURE 6.6 Absolute total ionization cross sections in Ne. Ordinates of the top curve should be multiplied by 0.1. (×) Kobayashi et al. (2002), (× 0.1); (△) Rejoub et al. (2002), (×0.1); (△) Krishnakumar and Srivastava (1988), (×0.1); (●) Sorokin et al. (105, 1998), (×0.1); (◆) Almeida et al. (1995); (—— ——) Lennon et al. (1987); (□) Wetzel et al. (1987); (+) Nagy et al. (1980); (○) Stephan et al. (1980), (× 0.1); Fletcher and Cowling (1973); (–) Schram et al. (1966 A); (×) Schram et al. (1966 B); (◇) Schram et al. (1965).

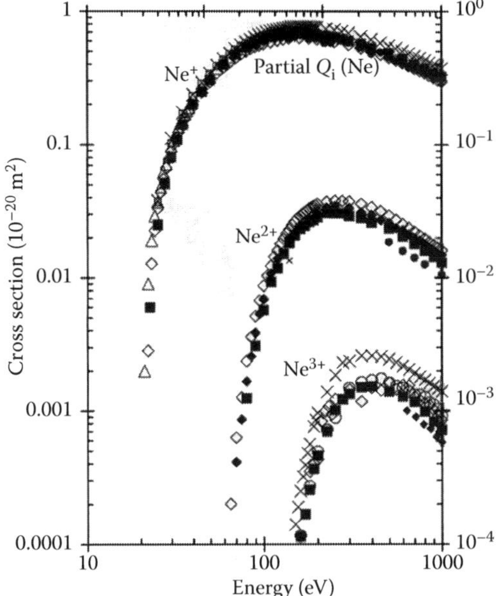

FIGURE 6.7 Partial ionization cross sections in neon. (■) Rejoub et al. (2002); (○) Kobayashi et al. (2002); (●) Almeida et al. (177, 1995); (×) Krishnakumar et al. (1988); (△) Wetzel et al. (1987); (□) Stephan et al. (1980); (◆) Schram et al. (1966b).

6.9 DRIFT VELOCITY

See Table 6.8.

TABLE 6.8
Recommended Drift Velocity for Ne

E/N (Td)	*W* (10^3 m/s)	*E/N* (Td)	*W* (10^3 m/s)	*E/N* (Td)	*W* (10^3 m/s)
(0.01518)	(0.976)				
(0.01821)	(1.052)	(0.4250)	(4.004)		
0.02	1.09	(0.4553)	(4.130)	40	120
(0.02125)	(1.122)	(0.4857)	(4.255)		
(0.02428)	(1.185)	0.5	4.31		
(0.02732)	(1.243)	(0.5464)	(4.478)		
0.03	1.29	0.6	4.67	50	146
(0.03035)	(1.300)	(0.6071)	(4.699)		
0.04	1.46	0.7	5.00	60	175
(0.04553)	(1.532)	0.7589	(5.170)		
0.05	1.59	0.8	5.28	70	201
0.06	1.72	(0.9106)	(5.576)	80	222
(0.06071)	(1.723)	1	5.80		
0.07	1.83	(1.062)	(5.945)	90	239
0.08	1.93	(1.214)	(6.280)	100	255
(0.09106)	(2.040)	(1.336)	(6.542)		
0.1	2.12	(1.821)	(7.813)	200	427
(0.1214)	(2.300)	2	8.48		
(0.1518)	(2.532)	(2.003)	(8.491)		
(0.1821)	(2.741)	4	16.10		
0.2	2.86	7	27.60	300	552
(0.2125)	(2.938)	10	35.00		
(0.2428)	(3.115)				
(0.2732)	(3.284)				
0.3	3.43	20	65.90	400	646
(0.3036)	(3.445)				
(0.3643)	(3.738)				
0.4	3.90	30	96.0	600	829

Note: Numbers in brackets are from Robertson (1972). See Figure 6.8 for graphical presentation.

TABLE 6.9
Ratio of Radial Diffusion Coefficient to Mobility

E/N (Td)	D_r/μ (V)	*E/N* (Td)	D_r/μ (V)	*E/N* (Td)	D_r/μ (V)
1.41	1.23	141	11.6	1695	23.6
2.82	2.36	198	13.8	1978	20.8
5.65	3.94	282	15.8	2260	17.5
14.1	5.67	424	19.0	2825	16.7
28.2	7.30	565	20.3	4238	12.3
56.5	8.25	848	28.0	5650	11.2
84.8	9.23	1130	27.2		
113	10.5	1413	24.6		

Source: Adapted from Al-Amin, S. A. J. and J. Lucas, *J. Phys. D: Appl. Phys.*, 20, 1590, 1987.
Note: See Figure 6.8 for graphical presentation.

6.10 DIFFUSION COEFFICIENT

See Table 6.9.

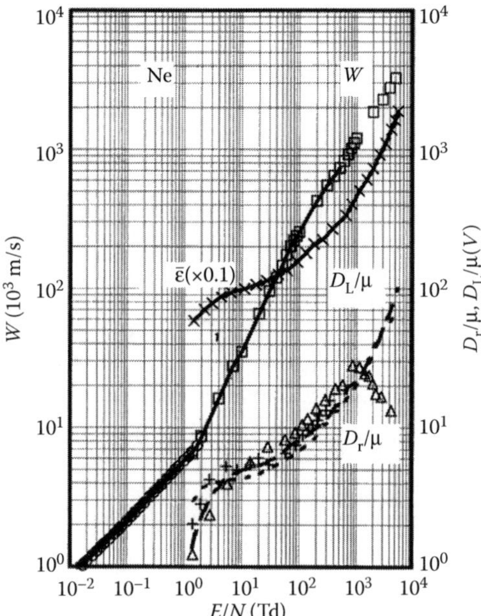

FIGURE 6.8 Transport parameters in Ne as a function of *E/N*. Unless otherwise specified the temperature of the gas is 300 K. W: (— —) recommended, Raju (2005); (□) Kücükarpaci et al. (1981); (●) Huxley and Crompton (1974); 77K; (○) Huxley and Crompton (1974);. D_r/μ: (— — —). Kücükarpaci et al. (1981) theory; (△) Al-Amin and Lucas (1987), experiment. D_L/μ: (+) Kücükarpaci et al. (1981) experimental; (— —) Kücükarpaci et al. (1981) theory. Mean energy ($\bar{\varepsilon}$): (—×—) Kücükarpaci et al. (1981) theory.

TABLE 6.10
Reduced Ionization Coefficient for Ne

E/N (Td)	α/*N* (10^{-20} m²)	*E/N* (Td)	α/*N* (10^{-20} m²)	*E/N* (Td)	α/*N* (10^{-20} m²)
6	0.0002	60	0.060	500	0.67
7	0.0003	70	0.077	550	0.72
8	0.0005	80	0.094	600	0.76
9	0.0007	90	0.11	650	0.80
10	0.0008	100	0.13	700	0.83
15	0.003	150	0.22	750	0.86
20	0.006	200	0.30	800	0.89
25	0.011	250	0.37	850	0.91
30	0.016	300	0.44	900	0.93
35	0.023	350	0.51	1000	0.97
40	0.029	400	0.57	1100	1.00
50	0.044	450	0.62		

Note: See Figure 6.9 for graphical presentation.

6.11 REDUCED IONIZATION COEFFICIENT

See Table 6.10.

6.12 GAS CONSTANTS

Gas constants evaluated according to the expression (see inset of Figure 6.9)

$$\frac{\alpha}{N} = C\exp\left[-D\left(\frac{N}{E}\right)^{1/2}\right] \quad (6.2)$$

are $C = 2.547 \times 10^{-20}$ m²; $D = 29.95$ (Td)$^{1/2}$.

6.13 ION MOBILITY

See Table 6.11.

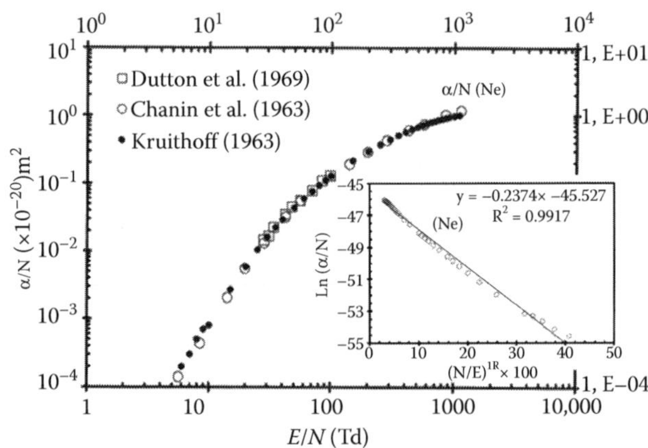

FIGURE 6.9 Reduced ionization coefficient for Ne. (□) Dutton et al. (1969); (○) Chanin and Rork (1963); (●) Kruithoff (1940). Inset shows plot according to Equation 6.2.

TABLE 6.11
Reduced Ion Mobility ($N = 2.69 \times 10^{25}$ m⁻³ and $T = 273.16$ K) and Drift Velocity

E/N (Td)	Mobility (10^{-4} m²/V s)	Drift Velocity (10^2 m/s)
6	4.07	0.656
8	4.05	0.871
10	4.04	1.09
12	4.02	1.3
15	3.98	1.6
20	3.91	2.1
25	3.84	2.58
30	3.76	3.03
40	3.61	3.88
50	3.48	4.68
60	3.35	5.4
80	3.13	6.73
100	2.96	7.95
120	2.81	9.06
150	2.61	10.5

TABLE 6.11 (continued)
Reduced Ion Mobility ($N = 2.69 \times 10^{25}$ m^{-3} and $T = 273.16$ K) and Drift Velocity

E/N (Td)	Mobility (10^{-4} m^2/V s)	Drift Velocity (10^2 m/s)
200	2.36	12.7
250	2.17	14.6
300	2.02	16.3
400	1.8	19.3
500	1.63	21.9
600	1.51	24.3
800	1.32	28.4
1000	1.19	32
1200	1.09	35.1
1500	0.99	39.9

Source: Adapted from Ellis, H. W., R. Y. Pai, and E. W. McDaniel, *Atomic Nuclear Data Tables*, 17, 177, 1976.
Note: See Figure 6.10 for graphical presentation.

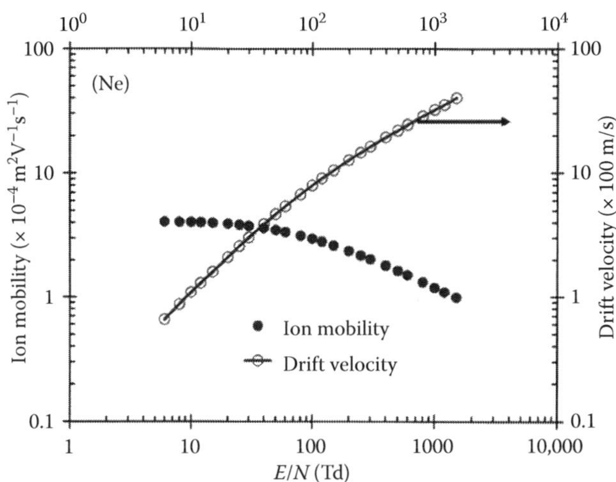

FIGURE 6.10 Reduced ion mobility and drift velocity for Ne. (●) Mobility; (—○—) drift velocity.

REFERENCES

Al-Amin, S. A. J. and J. Lucas, *J. Phys. D: Appl. Phys.*, 20, 1590, 1987.

Almeida, D. P., A. C. Fontes, and C. F. L. Godinho, *J. Phys. B: At. Mol. Op. Phys.*, 28, 3335, 1995.

Brusa, R. S., G. P. Karwasz, and A. Zecca, *Z. Phys. D*, 38, 279, 1996.

Chanin, L. M. and G. D. Rork, *Phys. Rev.*, 132, 2547, 1963.

Chilton, J. E., M. D. Stewart, Jr., and C. C. Lin, *Phys. Rev. A*, 61, 052708-1, 2000.

de Heer, F. J., R. H. J. Jansen, and W. van der Kaay, *J. Phys. B: At. Mol. Phys.*, 12, 979, 1979.

Dutton, J., M. H. Hughes, and B. C. Tan, *J. Phys. B: At. Mol. Phys.*, 2, 890, 1969.

Ellis, H. W., R. Y. Pai, and E. W. McDaniel, *Atomic Nuclear Data Tables*, 17, 177, 1976.

Fletcher, J. and I. R. Cowling, *J. Phys. B: At. Mol. Phys.*, 6, L258, 1973.

Fon, W. C. and K. A. Berrington, *J. Phys. B: At. Mol. Phys.*, 14, 323, 1981.

Garcia, G., F. Arqueros, and J. Campos, *J. Phys. B: At. Mol. Phys.*, 19, 3777, 1986.

Gulley, R. J., D. T. Alle, M. J. Brennan, M. J. Brunger, and S. J. Buckman, *J. Phys. B: At. Mol. Opt. Phys.*, 27, 2593, 1994.

Gupta, S. C. and J. A. Rees, *J. Phys. B: At. Mol. Phys.*, 8, 417, 1975.

Huxley, L. G. H. and R. W. Crompton, *Diffusion and Drift of Electrons in Gases*, John Wiley and Sons Inc., New York, NY, 1974.

Janson et al., quoted by Williams and Crowe (1975) as private Communication.

Kanik, I., J. M. Ajello, and G. K. James, *J. Phys. B: At. Mol. Phys.*, 29, 2355, 1996.

Kauppila, W. E., T. S. Stein, J. H. Smart, M. S. Dababneh, Y. K. Ho, J. P. Downing, and V. Pol, *Phys. Rev. A*, 24, 725, 1981.

Kobayashi, A., G. Fujiki, A. Okaji, and T. Masuoka, *J. Phys. B: At. Mol. Opt. Phys.*, 35, 2087, 2002. The author thanks Dr. Kobayashi for sending tabulated results of cross sections.

Krishnakumar, E. and S. K. Srivastava, *J. Phys. B: At. Mol. Phys.*, 21, 1055, 1988.

Kruithoff, A. A., *Physica*, VII, 519, 1940.

Kumar, V., E. Krishnakumar, and K. P. Subramanian, *J. Phys. B: At. Mol. Phys.*, 20, 2899, 1987.

Kücükarpaci, H. N., H. T. Saelee, and J. Lucas, *J. Phys. D: Appl. Phys.*, 14, 9, 1981.

Lennon, M. A., K. L. Bell, H. B. Gilbody, J. G. Hughes, A. E. Kingston, M. J. Murray, and F. J. Smith, *J. Phys. Chem. Ref. Data*, 17, 1285, 1988.

Mason, N. J. and W. R. Newell, *J. Phys. B: At. Mol. Phys.*, 20, 1357, 1987.

Menses, G. D. *J. Phys. B: At. Mol. Op. Phys.*, 35, 3119, 2002.

McCarthy, J. E., C. J. Noble, B. A. Phillips, and A. D. Turnbull, *Phys. Rev. A*, 15, 2173, 1977.

McEachran, R. P. and A. D. Stauffer, *Phys. Lett.*, 107A, 397, 1985.

Nagy, P., A. Skutzlart, and V. Schmidt, *J. Phys. B: At. Mol. Phys.*, 13, 1249, 1980.

Nickel, J. C., K. Imre, D. F. Register, S. Trajmar, *J. Phys. B: At. Mol. Phys.*, 18, 125, 1985.

O'Malley, T. F. and R. W. Crompton, *J. Phys. B.*, 13, 3451, 1980.

Phillips, M. H., L. W. Anderson, and C. C. Lin, *Phys. Rev. A*, 32, 2117, 1985.

Puech, V. and S. Mizzi, *J. Phys. D: Appl. Phys.*, 24, 1974, 1991.

Raju, G. G., *Gaseous Electronics: Theory and Practice*, Taylor & Francis, London, 2005.

Rapp, D. and P. Englander-Golden, *J. Chem. Phys.*, 43, 1464, 1965.

Register, D. F., S. Trajmar, D. F. Register, S. Trajmar, G. Steffensen, and D. C. Cartwright, *Phys. Rev. A*, 29, 1793, 1984.

Register, D. F. and S. Trajmar, *Phys. Rev. A.*, 29, 1785, 1984.

Rejoub, R., B. G. Lindsay, and R. F. Stebbings, *Phys. Rev. A*, 65, 042713-1, 2002.

Robertson, A. G., *J. Phys. B: At. Mol. Phys.*, 5, 648, 1972.

Saha, H. P., *Phys. Rev.*, 39, 5048, 1989.

Schram, B. L., F. J. deHeer, F. J. van Der Wieland, and J. Kistemaker, *Physica*, 31, 94, 1965; B. L. Schram, H. R. Moustafa, H. R. Schutten, and F. J. de Heer, *Physica*, 32, 734, 1966 A; B. L. Schram, J. H. Boereboom, and J. Kistemaker, *Physica*, 32, 1966 B, 185; B. L. Schram, *Physica*, 32, 197, 1966.

Sharpton, F. A., R. M. St. John, C. C. Lin, and E. Fajen, *Phys. Rev. A*, 2, 1305, 1970.

Sorokin, A. A., L. A. Shmaenok, S. B. Bobashev, B. Möbus, and G. Ulm, *Phys. Rev. A*, 58, 2900, 1998. Cross sections are measured in neon.

Stephan, K., H. Helm, and T. D. Märk, *J. Chem. Phys.*, 73, 3763, 1980. Gases studied are He, Ne, Ar, K; *J. Chem. Phys.*, 81, 1984, 3116. Gas studied is Xe.

Szmytkowski, C. and K. Maciag, *Phys. Scrip.*, 54, 271, 1996.

Teubner, P. J. O., J. L. Riley, M. C. Tonkin, J. E. Furst, and S. J. Buckman, *J. Phys. B: At. Mol. Phys.*, 18, 3641, 1985.

Thompson, D. G., *J. Phys. B: At. Mol. Phys.*, 4, 468, 1971.

Wagenaar, R. W. and F. J. de Heer, *J. Phys. B: At. Mol. Phys.*, 13, 3855, 1980.

Wetzel, R. C., F.A. Baiocchi, T. R. Hayes, and R. S. Freund, *Phys. Rev. A*, 35, 559, 1987.

Williams, J. F. and A. Crowe, *J. Phys. B: At. Mol. Phys.*, 8, 2233, 1975.

Zecca, A., G. P. Karwasz, and R. S. Brusa, *J. Phys. B: At. Mol. Opt. Phys.*, 33, 843, 2000.

Zecca, A. S., S. Oss, G. Karwasz, R. Grisewti, and R. S. Brusa, *J. Phys. B: At. Mol. Phys.*, 20, 5157, 1987.

7

POTASSIUM

CONTENTS

Potassium (K) atom has 19 electrons with configuration $3p^6 4s\ ^2S_{1/2}$. The electronic polarizability is 48.29×10^{-40} F m^2 and the ionization potential is 4.341 eV.

7.1 SELECTED REFERENCES FOR DATA

See Table 7.1.

7.2 TOTAL CROSS SECTION

See Table 7.2.

TABLE 7.1
Selected References for Data

Quantity	Range: eV, (Td)	Reference
Total cross section	4.4–101.9	**Kwan et al. (1991)**
Ionization cross section	3.5–1000	Lennon et al. (1988)
Total cross section	7–100	**Vušković and Srivastava (1980)**
Total cross section	0.5–500	**Kasdan and Miller (1973)**
Total cross section	0.3–9.0	**Visconti et al. (1971)**
Ionization cross section	50–500	**McFarland and Kinney (1965)**
Ionization cross section	200, 300, 500	**Brink (1964)**
Total cross section	1–10	**Perel et al. (1962)**
Ionization cross section	0–700	**Tate and Smith (1934)**
Total cross section	0.25–400	**Brode (1929)**

Note: Bold font indicates experimental study.

7.3 IONIZATION CROSS SECTION

See Table 7.3.

TABLE 7.2
Total Cross Section for K Atom

Energy (eV)	Q_T (10^{-20} m^2)	Energy (eV)	Q_T (10^{-20} m^2)
Kwan et al. (1991)		Kasdan and Miller (1973)	
4.4	90.3	0.5	304
6.2	89.6	1	234
11.0	77.7	1.6	205
21.2	65.1	2	185
31.3	51.6	3	152
41.4	43.9	5	122
51.4	42.1	8	109
76.8	37.5	10	105
101.9	31.5	20	93
		50	75

Note: See Figure 7.1 for graphical presentation.

TABLE 7.3
Ionization Cross Section

Energy (eV)	Q_i (10^{-20} m^2)	Energy (eV)	Q_i (10^{-20} m^2)
40	6.7	300	4.22
50	7.3	400	3.52
100	6.82	500	3.11
200	5.46		

Source: Adapted from McFarland, R. H. and J. D. Kinney, *Phys. Rev.*, 137, A1058, 1965.
Note: See Figure 7.2 for graphical presentation.

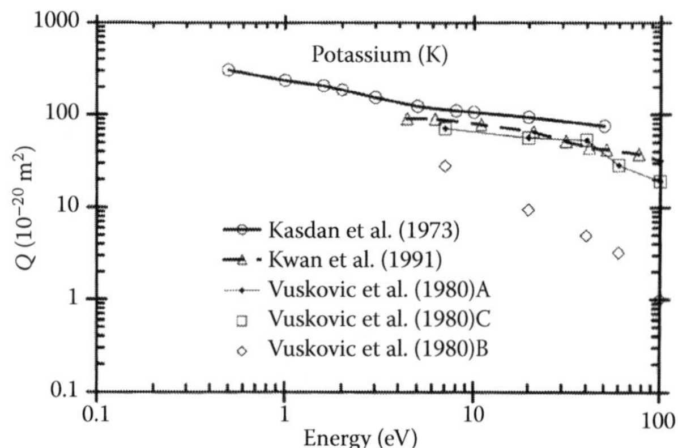

FIGURE 7.1 Cross sections for K atom. Total: (—△—) Kwan et al. (1991). Elastic: (—◆—) Vušković and Srivastava (1980)A. Momentum transfer: (◊) Vušković and Srivastava (1980)B. Excitation: (◇) Vušković and Srivastava (1980)C Total.

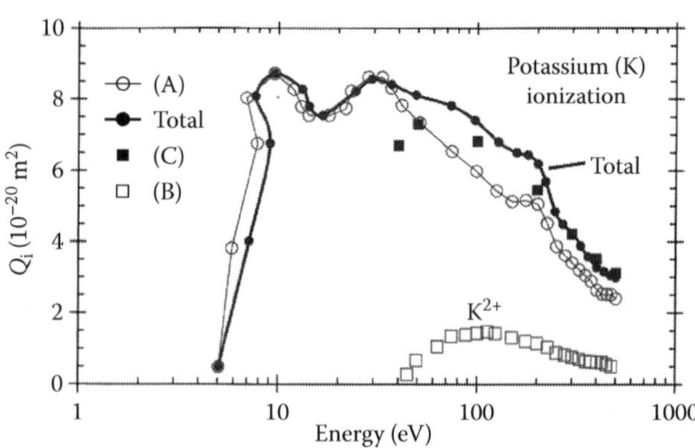

FIGURE 7.2 Ionization cross section for K atom. (—○—) Tate and Smith (1934), K^+; (□) Tate and Smith (1934), K^{2+}; (—●—) Tate and Smith (1934), total; (■) McFarland and Kinney (1965). The relative cross sections of Tate and Smith (1934) are normalized with McFarland data at 50 eV and 7.3×10^{-20} m^2. A, B, and C refer to K^+, K^{2+} and total cross section respectively.

REFERENCES

Brink, G. O., *Phys. Rev.*, 134, A345, 1964.

Brode, R. B., *Phys. Rev.*, 34, 673, 1929.

Kasdan, A. and T. M. Miller, *Phys. Rev. A*, 8, 1562, 1973.

Kwan, C. K., W. E. Kauppilla, R. A. Lukaszew, S. P. Parikh, T. S. Stein, Y. J. Wan, and M. S. Dababneh, *Phys. Rev.*, 44, 1620, 1991.

Lennon, M. A., K. L. Bell, H. B. Gilbody, J. G. Hughes, A. E. Kingston, M. J. Murray, and F. J. Smith, *J. Phys. Chem. Ref. Data*, 17, 1285, 1988.

McFarland, R. H. and J. D. Kinney, *Phys. Rev.*, 137, A1058, 1965.

Perel, J., P. Englander, and B. Bederson, *Phys. Rev.*, 128, 1148, 1962.

Tate, J. T. and P. T. Smith, *Phys. Rev.*, 46, 773, 1934.

Visconti, P. J., J. A. Slevin, and K. Rubin, *Phys. Rev A.*, 3, 1310, 1971.

Vušković, L. and S. K. Srivastava, *J. Phys. B: At. Mol. Phys.*, 13, 4849, 1980.

8

SODIUM

CONTENTS

Sodium (Na) atom has 11 electrons with configuration $2p^63s$ $^2S_{1/2}$. The electronic polarizability is 26.825×10^{-40} F m^2 and the ionization potential is 5.139 eV.

8.1 SELECTED REFERENCES FOR DATA

See Table 8.1.

8.2 TOTAL CROSS SECTIONS

See Table 8.2.

TABLE 8.1
Selected References for Data

Quantity	Range	Reference
Total cross section	3–102	**Kwan et al. (1991)**
Ionization cross section	0–1000	Lennon et al. (1988)
Elastic scattering	10–54.4	**Srivastava and Vušković (1980)**
Drift velocity	(1–100)	**Nakamura and Lucas (1978a)**
Momentum transfer cross section	0.05–10	Nakamura and Lucas (1978b)
Total cross section	0.5–50	**Kasdan and Miller (1973)**
Ionization cross section	50–500	**McFarland and Kinney (1965)**
Ionization cross section	200, 300, 500	**Brink (1964)**
Total cross section	1–10	**Perel et al. (1962)**
Ionization cross section	0–700	**Tate and Smith (1934)**
Total cross section	0–400	**Brode (1929)**

Note: Bold font indicates experimental study.

TABLE 8.2
Total Scattering Cross Section

Energy (eV)	Q_T (10^{-20}m^2)	Energy (eV)	Q_T (10^{-20} m^2)
Kwan et al. (1991)		Kasdan and Miller (1973)	
4.1	67.1	0.5	352
5.9	66.5	0.75	246
10.8	55.9	1	197
20.7	43.3	1.5	152
30.7	32.6	2	129
40.8	30.0	2.5	113
50.8	26.2	3	102
60.9	22.9	3.5	97
76.1	22.0	4	93
		4.5	88
		5	82
		5.5	79
		6	72
		7	73
		8	66
		9	69
		10	75
		20	65
		50	57

Note: See Figure 8.1 for graphical presentation.

8.3 IONIZATION CROSS SECTION

See Table 8.3.

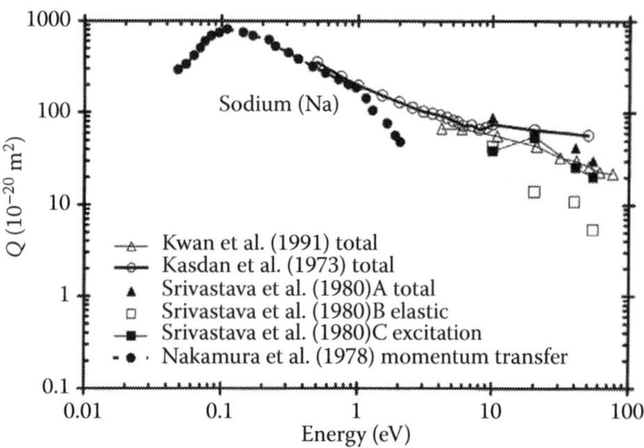

FIGURE 8.1 Cross sections for Na atom. (—△—) Kwan et al. (1991) total; (▲) Srivastava and Vušković (1980) total; (□) Srivastava and Vušković (1980) elastic; (—■—) Srivastava and Vušković (1980) excitation; (– –●– –) Nakamura and Lucas (1978b) momentum transfer; (—○—) Kasdan and Miller (1973) total.

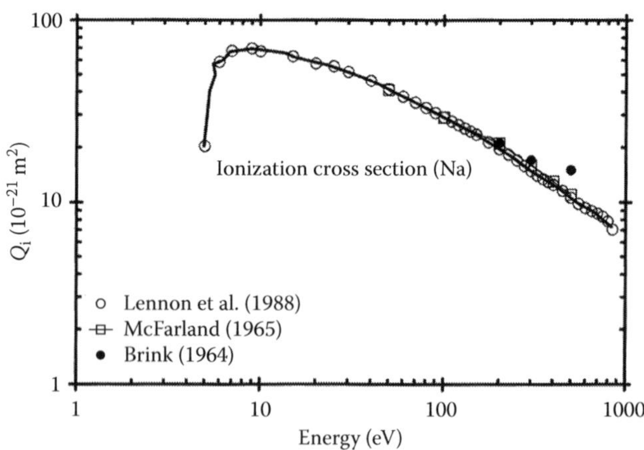

FIGURE 8.2 Ionization cross section for Na. (—○—) Lennon et al. (1988), recommended. (-□-) McFarland and Kinney (1965); (●) Brink (1964).

TABLE 8.3
Ionization Cross Section, Digitized and Interpolated

Energy (eV)	Q_i (10^{-21} m²)	Energy (eV)	Q_i (10^{-21} m²)	Energy (eV)	Q_i (10^{-21} m²)
5	20.30	90	30.75	350	13.41
6	58.72	100	29.10	375	12.91
7	67.54	110	27.69	400	12.46
9	69.71	120	26.46	450	11.55
10	67.14	130	25.37	500	10.60
15	63.04	140	24.39	550	9.81
20	57.59	150	23.47	600	9.29
25	55.56	175	21.33	650	8.95
30	51.75	200	19.60	700	8.68
40	46.19	225	18.23	750	8.35
50	41.44	250	16.99	800	7.85
60	37.77	275	15.83	850	7.05
70	35.09	300	14.82		
80	32.73	325	14.03		

Source: Adapted from Lennon, M. A. et al., *J. Phys. Chem. Ref. Data*, 17, 1285, 1988

Note: See Figure 8.2 for graphical presentation.

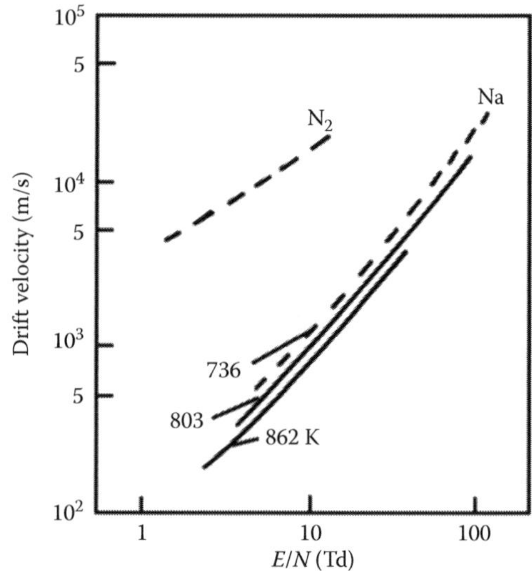

FIGURE 8.3 Drift velocity of electrons in Na vapor. The mean values are shown by dotted curve. Data for N_2 added for comparison. (Adapted from Nakamura, Y. and J. Lucas, *J. Phys. D: Appl. Phys.*, 11, 325, 1978a; Nakamura, Y. and J. Lucas, *J. Phys. D: Appl. Phys.*, 11, 337, 1978b.)

8.4 DRIFT VELOCITY OF ELECTRONS

Figure 8.3 shows the experimental drift velocity of electrons (Nakamura and Lucas, 1978a), the only results available. The drift velocity depends on the vapor pressure, at constant *E/N*, and this observation is attributed to the formation of dimers.

8.5 SCALING LAW

Several scaling laws have been proposed in the literature for ionization cross sections and the law proposed by Szłuińska et al. (2002) for atoms is relatively simple to use. A scaling law attempts to represent the cross sections of a large number of gases by a single formula and a single curve. The scaling law is

$$\frac{Q_i(\varepsilon)}{Q_{i\max}} = f\left(\frac{\varepsilon}{\varepsilon_i}\right) \tag{8.1}$$

where ε is the electron energy, ε_i is the ionization potential, Q_i is the ionization cross section at ε, and $Q_{i\max}$ is the maximum ionization cross section.

Figure 8.4 shows the cross sections of 17 atoms plotted according to Equation 8.1. Except for xenon (Xe), there is reasonable correspondence, specially for $(\varepsilon/\varepsilon_i) \le 50$. Szłuińska et al. (2002) have demonstrated the applicability of Equation 8.1 for 27 species of target particles, including noble gases, halogens, and gases in columns V and VI in the periodic table,

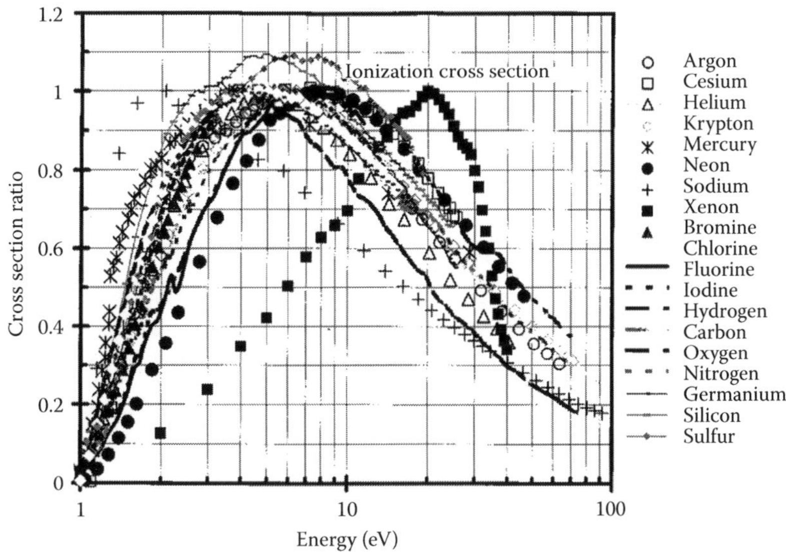

FIGURE 8.4 Ionization cross section for 17 atomic target particles plotted according to the law proposed by Szłuińska et al. (2002).

by suggesting small correction factors though Xe is not included by them.

REFERENCES

Brink, G. O., *Phys. Rev.*, 134, A345, 1964.

Brode, R. B., *Phys. Rev.*, 34, 673, 1929.

Kasdan, A. and T. M. Miller, *Phys. Rev. A*, 8, 1562, 1973.

Kwan, C. K., W. E. Kauppilla, R. A. Lukaszew, S. P. Parikh, T. S. Stein, Y. J. Wan, and M. S. Dababneh, *Phys. Rev.*, 44, 1620, 1991.

Lennon, M. A., K. L. Bell, H. B. Gilbody, J. G. Hughes, A. E. Kingston, M. J. Murray, and F. J. Smith, *J. Phys. Chem. Ref. Data*, 17, 1285, 1988.

McFarland, R. H. and J. D. Kinney, *Phys. Rev.*, 137, A1058, 1965.

Nakamura, Y. and J. Lucas, *J. Phys. D: Appl. Phys.*, 11, 325, 1978a.

Nakamura, Y. and J. Lucas, *J. Phys. D: Appl. Phys.*, 11, 337, 1978b.

Perel, J., P. Englander, and B. Bederson, *Phys. Rev.*, 128, 1148, 1962.

Srivastava, S. K. and L. Vušković, *J. Phys. B: At. Mol. Phys.*, 13, 2633, 1980.

Szłuińska, M., P. Van Reeth, and G. Laricchia, *J. Phys. B: At. Mol. Opt. Phys.*, 35, 4059, 2002.

Tate, J. T. and P. T. Smith, *Phys. Rev.*, 46, 773, 1934.

9

XENON

CONTENTS

Xenon (Xe) is a heavier rare gas with 54 electrons and electronic polarizability of 4.50×10^{-40} F m^2. The ground-state configuration is $5p^6\ {}^1S_0$ and the ionization potential is 12.130 eV. The lowest metastable energy is 8.315 eV and the lowest radiative level is 8.437 eV (Grigoriev and Meilikhov, 1999).

9.1 SELECTED REFERENCES FOR DATA

See Table 9.1.

9.2 TOTAL SCATTERING CROSS SECTION

Table 9.2 presents the total scattering cross sections in Xe. See Figure 9.1 for graphical presentation.

TABLE 9.1
Selected References for Data

Quantity	Range: eV, (Td)	Reference
Q_i	12–1000	Rejoub et al. (2002)
Q_i	12.5–1000	Kobayashi et al. (2002)
Q_M, Q_{el}, Q_i, Q_T	1–1000	Zecca et al. (2000)
W	(50–600)	Xiao et al. (2000)
Q_{diff}	5–200	Danjo (1998)
Q_{ex}	Onset–150	Fons and Lin (1998)
Q_M, Q_{el}	0.67–50	Gibson et al. (1998)
Q_{ex}	10–80	Khakoo et al. (1996b)
Q_{ex}	10–30	Khakoo et al. (1996a)

continued

Xe

TABLE 9.1 (continued)
Selected References for Data

Quantity	Range: eV, (Td)	Reference
Q_T	0.5–220	Szmytkowski et al. (1996)
Q_{ex}	20, 30	Khakoo et al. (1994)
D_L	(0.002–25)	Pack et al. (1992)
Q_i	18–466	Syage (1992)
W	(0.01–500)	Puech and Mizzi (1991)
Q_{ex}	100, 400, 500	Suzuki et al. (1991)
Q_i	15–1000	Krishnakumar and Srivastava (1988)
Q_i	Onset–1000	Nishimura et al. (1987)
Q_{el}	40–100	Ester and Kessler (1994)
W	(2–16)	Santos et al. (1994)
Q_T	81–1000	Zecca et al. (1991)
Q_{ex}	15–80	Filipović et al. (1988)
W	(0.004–3.0)	Hunter et al. (1988)
Q_{diff}	0.05–2.0	Weyherter et al. (1988)
Q_{diff}	1–100	Register et al. (1986)
Q_M	0.010–6.00	Koizumi et al. (1986)
Q_T	4–300	Nickel et al. (1985)
Q_T	20–750	Wagenaar and de Heer (1985)
Q_M	3–50	McEachran and Stauffer (1984)
Q_M, Q_{ex}	0–10000	Hayashi (1983)
W	(1–100)	Brooks et al. (1982)
Q_{diff}	2–300	Klewer and van der Wiel (1980)
Q_i	13–200	Wetzel et al. (1980)
Q_T	20–3000	De Heer et al. (1979)
W	(0.005–5)	Huang and Freeman (1978)
α/N	(75–600)	Makabe and Mori (1978)
Q_{diff}	20–400	Williams et al. (1975)
Q_i	12.5–1000	Rapp and Englander-Golden (1965)
W	(0.002–80)	Pack et al. (1962)
W	(0–12)	Bowe (1960)
α/N	(20–7200)	Kruithoff (1940)

Note: Q_{diff} = differential; Q_{el} = elastic; Q_{ex} = excitation; Q_M = momentum transfer; Q_i = ionization; Q_T = total; D_L = longitudinal diffusion coefficient; W = drift velocity; α/N = reduced ionization coefficient.

TABLE 9.2
Total Scattering Cross Section in Xe

Szmytkowski et al. (1996)		Zecca et al. (1991)		Wagenaar et al. (1985)	
Energy (eV)	Q_T (10^{-20} m^2)	Energy (eV)	Q_T (10^{-20} m^2)	Energy (eV)	Q_T (10^{-20} m^2)
0.5	2.73	81	12.45	20	35.48
0.8	1.4	90.25	12.39	22.5	32.17
1	1.79	100	12.24	25	28.73
1.2	3.55	110.25	11.95	27.5	24.53
1.5	6.21	121	11.53	30	20.85
2.5	14.9	132.25	11.15	35	17.10
3	20.4	144	10.81	40	15.67
3.5	24.9	169	10.39	45	14.92
4	30.1	196	9.476	50	14.23
4.5	34.8	225	9.069	55	13.76

TABLE 9.2 (continued)
Total Scattering Cross Section in Xe

Szmytkowski et al. (1996)		Zecca et al. (1991)		Wagenaar et al. (1985)	
Energy (eV)	Q_T (10^{-20} m^2)	Energy (eV)	Q_T (10^{-20} m^2)	Energy (eV)	Q_T (10^{-20} m^2)
5	37.8	256	8.612	60	13.31
5.5	41.4	289	8.273	65	13.13
6	42.4	324	7.717	70	12.93
6.5	43.4	361	7.376	75	12.86
7	43.3	400	7.244	80	12.78
7.5	43.0	484	6.599	85	12.64
8	42.6	576	5.89	90	12.50
8.5	42.1	676	5.245	95	12.34
9	41.5	784	4.956	100	12.18
9.5	41.0	900	4.516	125	11.60
10	40.5	1000	4.171	150	11.03
12	38.6			175	10.53
15	36.7			200	10.09
17	36.2			250	9.32
20	34.7			300	8.67
22	31.9			350	8.10
25	27.2			400	7.65
30	20.0			450	7.23
35	17.5			500	6.87
38	16.4			550	6.55
40	15.6			600	6.27
42	14.9			650	6.01
50	13.7			700	5.76
60	12.8			750	5.51
70	12.5				
80	12.3				
90	12.0				
100	11.7				
110	11.5				
120	11.1				
140	10.7				
160	10.4				
180	10.1				
200	9.51				
220	8.97				

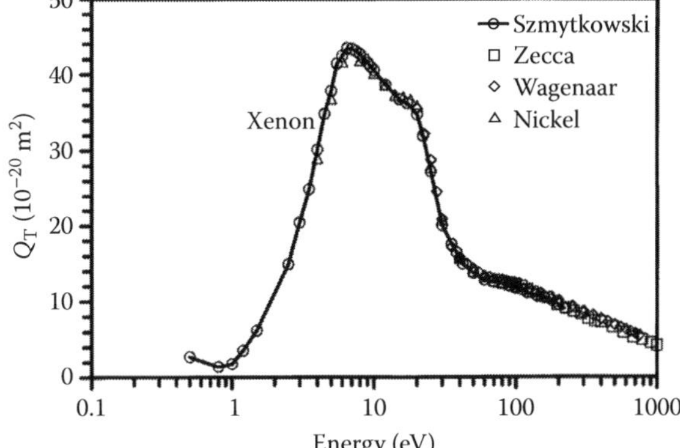

FIGURE 9.1 Total scattering cross section in Xe. (–⊖–) Szmytkowsky et al. (1996); (□) Zecca et al. (1991); (◊) Wagenaar et al. (1985); (Δ) Nickel et al. (1985).

9.3 DIFFERENTIAL SCATTERING CROSS SECTION

The differential scattering cross sections in Xe are shown in Table 9.3. See Figure 9.2 for graphical presentation. The dominant feature of the differential cross section is the diffraction pattern peaks and valleys. At low energy (~1 eV), two valleys are observed; with increasing energy (~20 eV), three shallow minima are observed. A further increase in energy (~60 eV) increases the minima to four, clearly demonstrating the dominance of the l = 4 partial wave. Some of the minima are not evident or become less pronounced at higher energies and obviously many partial waves contribute to the observed pattern (Raju, 2005).

Xe

TABLE 9.3
Differential Cross Section for Elastic Scattering in Xe in Units of 10^{-20} m²/sr at Selected Electron Energies

	Energy (eV)			Energy (eV)		
Angle (°)	**1.0**	**2.0**	**Angle (°)**	**25**	**40**	**50**
15	1.121		10	18.26	14.03	11.72
20	0.482	0.490	15	23.5	7.57	4.77
25	0.289	0.245	20	13.69	3.67	1.94
30	0.128	0.149	25	9.42	1.83	0.699
35	0.0406	0.170	30	5.01	0.869	0.238
40	0.0102	0.277	35	2.54	0.405	0.085
45	0.0126	0.425	40	0.948	0.172	0.030
50	0.0359	0.622	45	0.223	0.066	0.0088
55	0.0772	0.815	47.5	0.079	0.045	0.0037
			50	0.074	0.040	0.002
60	0.125	0.991	52.5	0.152	0.046	0.0032
65	0.175	1.142	55	0.258	0.058	0.0074
70	0.208	1.221	60	0.532	0.092	0.020
75	0.238	1.278	65	0.745	0.120	0.034
80	0.254	1.257	70	0.848	0.127	0.043
85	0.255	1.200	75	0.784	0.119	0.039
90	0.254	1.057	80	0.583	0.099	0.032
95	0.234	0.925	85	0.362	0.085	0.029
100	0.197	0.731	90	0.173	0.097	0.048
105	0.159	0.562	92.5	0.088	—	0.066
110	0.124	0.384	95	0.065	0.146	0.095
115	0.0723	0.238	97.5	0.045	—	0.131
120	0.0515	0.117	100	0.056	0.223	0.173
122.5		0.080	105	0.120	0.299	0.269
125	0.027	0.0479	110	0.210	0.400	0.377
130	0.0177	0.027	115	0.273	0.458	0.449
135		0.047	120	0.277	0.457	0.464
			125	0.209	0.391	0.423
			130	0.100	0.278	0.311

Source: Reproduced from Gibson, J. C. et al., *J. Phys. B: At. Mol. Opt. Phys.*, 31, 3949, 1998. With kind permission of Institute of Physics, England.

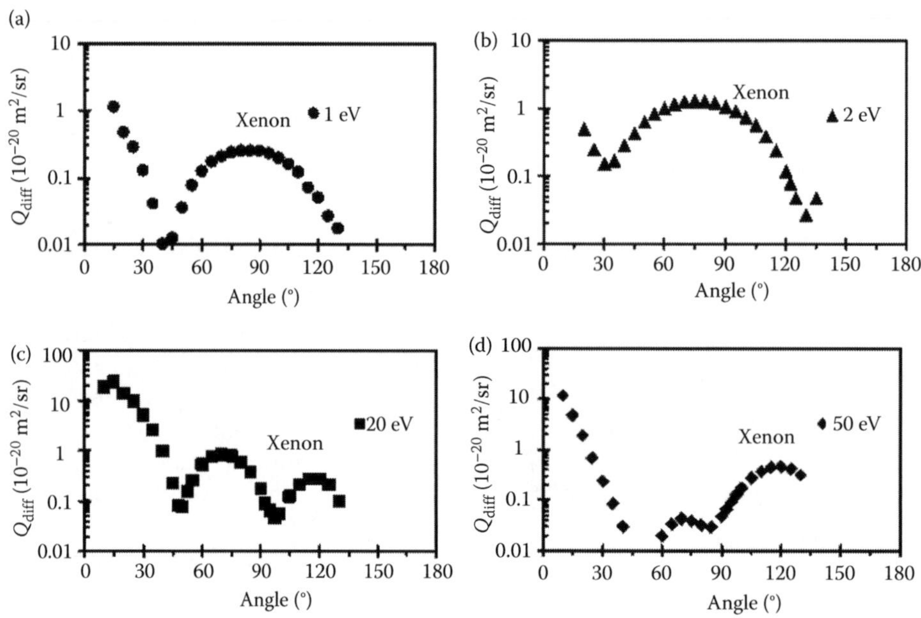

FIGURE 9.2 Elastic differential scattering cross sections in Xe. (a) (●) 1.0 eV; (b) (▲) 2.0 eV; (c) (■) 20 eV; (d) 50 eV. (Adapted from Gibson, J. C. et al. *J. Phys. B: At. Mol. Opt. Phys.*, 31, 3949, 1998.)

9.4 MOMENTUM TRANSFER CROSS SECTION

The momentum transfer cross sections in Xe are shown in Table 9.4. See Figure 9.3 for graphical presentation.

TABLE 9.4
Momentum Transfer Cross Section

Pack et al. (1992)[a]		Register et al. (1986)[b]		Hayashi (1983)[c]			
Energy (eV)	Q_M (10^{-20} m^2)	Energy (eV)	Q_M (10^{-20} m^2)	Energy (eV)	Q_M (10^{-20} m^2)	Energy (eV)	Q_M (10^{-20} m^2)
0.000	176	1.00	1.24	1.00E-05	176	12	13.5
0.002	160	1.75	6.32	0.01	116	15	9.5
0.005	139	2.75	15.20	0.02	80	20	7
0.007	127	3.75	23.19	0.03	61.3	25	5.9
0.010	116	4.75	27.37	0.04	48	30	5.1
0.030	61.3	5.75	28.04	0.05	39.5	40	4.2
0.050	39.5	9.75	14.62	0.06	33.5	50	3.6
0.080	25.6	14.75	8.33	0.08	25.6	60	3.2
0.100	20.4	19.75	5.44	0.1	20.4	80	2.7
0.160	12.0	29.75	5.33	0.13	15.1	100	2.4
0.200	8.4	49.75	3.95	0.16	12	120	2.15
0.250	5.9	59.75	2.12	0.2	8.4	150	1.9
0.300	3.3	63.00	1.92	0.25	5.35	200	1.65
0.400	1.6	80.00	1.24	0.3	3.3	250	1.45
0.500	0.5	100.00	1.36	0.4	1.6	300	1.3
0.600	0.4			0.5	0.955	400	1.08
0.700	0.4			0.6	0.8	500	0.94
0.800	0.6			0.7	0.82	600	0.83
0.900	0.9			0.8	1.05	800	0.7
1.000	1.2			1	1.7	1000	0.58
1.200	1.9			1.2	2.55	1200	0.48
1.500	3.5			1.5	4	1500	0.37
2.000	7.5			2	7.5	2000	0.255
2.500	11.5			2.5	11.5	2500	0.19
3.000	16.5			3	16	3000	0.15
4.000	24.5			4	24.5	4000	0.100
5.000	28.0			5	28	5000	0.075
6.000	26.0			6	28	6000	0.057
7.000	27.0			8	26	8000	0.038
8.000	26.0			10	20	10000	0.027
10.000	20.0						
12.000	13.5						
15.000	9.5						
20.000	7.0						
30.000	5.1						
50.000	3.6						
100.000	2.4						

[a] Derived from longitudinal diffusion coefficients.
[b] From measured differential scattering cross section.
[c] From swarm coefficients.

Xe

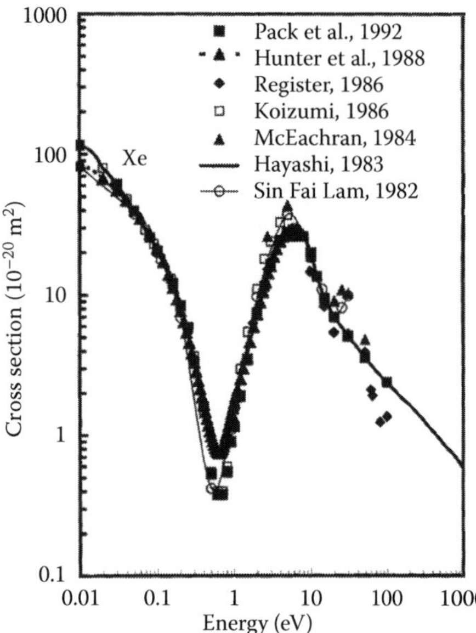

FIGURE 9.3 Momentum transfer cross sections in Xe. (—⊖—) Sin Fai Lam (1982); (–) Hayashi (1983); (▲) McEachran and Stauffer (1985); (◆) Register et al. (1986); (●) Koizumi et al. (1986); (-▲-) Hunter et al. (1988); (■) Pack et al. (1992). The cross sections from Frost and Phelps (1964) are not shown.

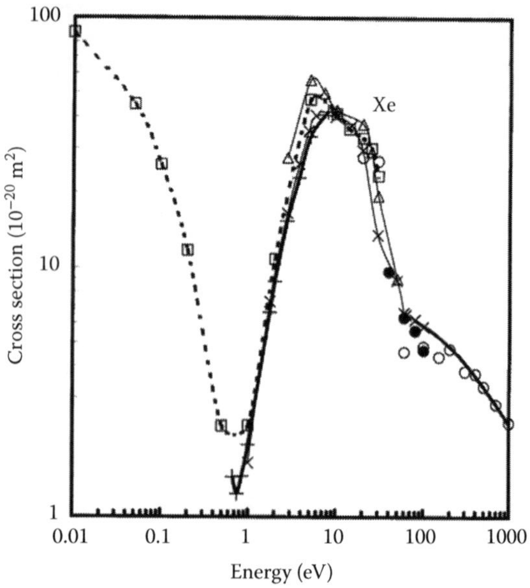

FIGURE 9.4 Total elastic scattering cross section in Xe. (○) de Heer et al. (1979); (—□—) Sin Fai Lam (1982); (—Δ—) McEachran and Stauffer (1985); (×) Register et al. (1986); (●) Ester and Kessler (1994); (—+—) Gibson et al. (1998); (——) Zecca et al. (2000). The analytical formula of Zecca et al. (2000) shows poorer agreement with experimental results in contrast with the quality of agreement in other rare gases.

9.5 ELASTIC SCATTERING CROSS SECTION

Elastic scattering cross sections in Xe are shown in Table 9.5. See Figure 9.4 for graphical presentation.

TABLE 9.5
Elastic Scattering Cross Section in Xe

Zecca et al. (2000)[a]				Gibson et al. (1998)[b]		Register et al. (1986)[c]	
Energy (eV)	Q_{el} (10^{-20} m^2)	Energy (eV)	Q_{el} (10^{-20} m^2)	Energy (eV)	Q_{el} (10^{-20} m^2)	Energy (eV)	Q_{el} (10^{-20} m^2)
80	6.169	400	3.678	0.67	1.45	1	1.65
90	6.020	450	3.490	0.75	1.24	1.75	7.3
100	5.880	500	3.325	0.85	1.46	2.75	16.5
125	5.563	550	3.180	1	1.95	3.75	26
150	5.286	600	3.050	1.75	6.62	4.75	35.3
175	5.042	650	2.934	2	8.8	5.75	40.7
200	4.825	700	2.829	2.75	15.3	9.75	40.4
250	4.455	750	2.733	3.75	22.9	14.75	37.2
300	4.151	800	2.646	5	33.6	19.75	29.6
350	3.896	900	2.492	7.9	42.2	29.75	13.5
		1000	2.361			49.75	8.89
						59.75	6.6
						63	6.49
						80	6.21
						100	5.82

[a] Semianalytical.
[b] Experimental.
[c] Experimental.

9.6 RAMSAUER–TOWNSEND MINIMUM

Ramsauer–Townsend minimum cross section and the corresponding energy are shown in Table 9.6.

9.7 MERT COEFFICIENTS

The phase shifts for electron–atom scatterings in the MERT formulation are (Raju, 2005)

$$\tan \eta_0 + -Ak = \left[\frac{\pi\alpha}{3a_0}\right]k^2 - Ak\left[\frac{4\alpha}{3a_0}\right]k^2 \ln(ka_0) + Dk^3 + Fk^4 \quad (9.1)$$

where

$$\varepsilon = 13.605\,(ka_0)^2 \quad (9.2)$$

The coefficients are shown in Table 9.7.

TABLE 9.6
Ramsauer–Townsend Minimum in Xe

	Momentum Transfer		Integral Elastic	
Authors	**Energy (eV)**	**Q_M (10^{-20} m²)**	**Energy (eV)**	**Q_{el} (10^{-20} m²)**
Sin Fai Lam (1982)	0.5	0.42	0.5	2.32
Hayashi (1983)	0.6	0.80	—	—
Koizumi et al. (1986)	0.7	0.4	—	—
Subramanian and Kumar (1987)	—	—	0.73	1.10
Hunter et al. (1988)	0.64	0.7530	—	—
Pack et al. (1992)	0.7	0.38	—	—
Gibson et al. (1998)	0.67	0.41	0.75	1.24

TABLE 9.7
MERT Coefficients in Xe

Authors	**A/a_0**	**D/a_0^3**	**F/a_0^4**	**A_1/a_0^3**	**Energy Range (eV)**
Swarm Method (Drift Velocity, D/μ)					
Hunter et al. (1988)	−6.09	490.2	−627.5	22.0	0.01–0.75
O'Malley (1963)[a]	−6.0				
Pack et al. (1992)	−7.072				
Electron Beam Method					
O'Malley (1963)[b]	−6.50	388.0		23.2	
Jost et al. (1983)	−5.83	490.0	−708.0	22.8	0.1–0.5
Weyherter et al. (1988)	−6.527	517.0	−717.8	21.65	0.05–0.5
Theory					
Sin Fai Lam (1982)	−6.04				
McEachran and Stauffer (1985)	−5.232				

[a] O'Malley (1963) drift velocities from Frost and Phelps (1964) were used.
[b] Beam studies from Ramsauer and Kollath (1932).

9.8 EXCITATION CROSS SECTION

Designation of selected lower levels of excitation in Xe is shown in Table 9.8 (Khakoo et al., 1996a). Figure 9.5 shows

TABLE 9.8
Designation of Lower Excitation Levels and Threshold Energy for Each Level

Level Number	Level Designation	Energy (eV)
0	$5p^6\ ^1S_0$	0.0
1	$6s[3/2]_2$	8.315
2	$6s[3/2]_1$	8.437
3	$6s'\,[1/2]_0$	9.447
4	$6s'\,[1/2]_1$	9.570
5	$6p[1/2]_1$	9.580
6	$6p[5/2]_2$	9.686
7	$6p[5/2]_3$	9.721
8	$6p[3/2]_1$	9.789
9	$6p[3/2]_2$	9.821
10	$5d[1/2]_0$	9.891
11	$5d[1/2]_1$	9.917
12	$6p[1/2]_0$	9.934
13	$5d[7/2]_4$	9.643
14	$5d[3/2]_2$	9.959
15	$5d[7/2]_3$	10.039
16	$5d[5/2]_2$	10.159
17	$5d[5/2]_3$	10.220
18	$5d[3/2]_1$	10.401
19	$7s[3/2]_2$	10.562
20	$7s[3/2]_1$	10.593

Sources: Adapted from Khakoo, M. A. et al., *J. Phys. B: At. Mol. Opt. Phys.*, 29, 3455, 1996a; Khakoo, M. A. et al., *J. Phys. B: At. Mol. Opt. Phys.*, 29, 3477,1996b.

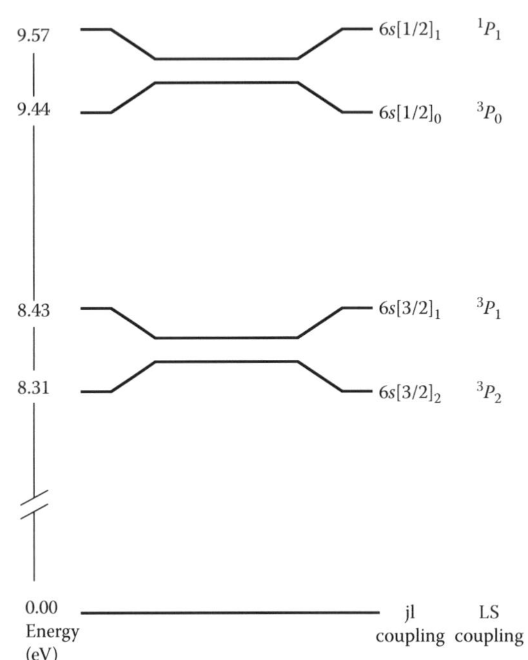

FIGURE 9.5 A simplified diagram of the first four excitation levels of Xe with corresponding energy threshold. (Adapted from Ester, T. and J. Kessler, *J. Phys. B: At. Mol. Opt. Phys.*, 27, 4295, 1994.)

Xe

the first four excitation levels of Xe. Table 9.9 shows the excitation cross sections. See Figure 9.6 for graphical presentation.

9.9 IONIZATION CROSS SECTIONS

Absolute total ionization cross sections are shown in Table 9.10. See Figure 9.7 for graphical presentation.

TABLE 9.9
Excitation Cross Sections in Xe

Kaur et al. (1998)		Ester and Kessler (1994)		Mason et al. (1987)		Hayashi (1983)	
Energy (eV)	Q_{ex} (10^{-20} m^2)	Energy (eV)	Q_{ex} (10^{-20} m^2)	Energy (eV)	Q_{ex} (10^{-20} m^2)	Energy (eV)	Q_{ex} (10^{-20} m^2)
20	0.183	15	0.4312	9	0.246	8.32	0
30	0.221	30	0.3528	10	0.318	8.5	0.026
40	0.0522	40	0.3472	11	0.691	9	0.126
50	0.0422	80	0.3528	12	0.996	9.5	0.131
60	0.0415	100	0.3444	13	1.077	10	0.18
80	0.0272			14	1.14	10.5	0.24
100	0.0185			15	1.169	11	0.42
				16	1.132	11.5	0.62
				17	1.092	12	0.84
				18	1.033	12.5	1.05
				19	0.989	13	1.28
				20	0.923	14	1.7
				21	0.82	15	2.14
				22	0.724	16	2.55
				23	0.607	18	3.35
				25	0.46	20	3.73
				27	0.074	25	3.85
				29	0.061	30	3.57
				31	0.049	40	2.85
				33	0.04	50	2.4
				35	0.034	60	2.1
				37	0.031	70	1.85
				39	0.026	80	1.66
				41	0.026	90	1.52
				44	0.024	100	1.38
				47	0.022	150	1.0
				51	0.021	200	0.8
				56	0.02	300	0.568
				60	0.02	400	0.465
				70	0.019	500	0.395
				80	0.018	600	0.344
				90	0.018	700	0.302
				100	0.017	800	0.277
				110	0.016	900	0.252
				120	0.015	1000	0.231
				130	0.015		
				140	0.014		
				150	0.014		

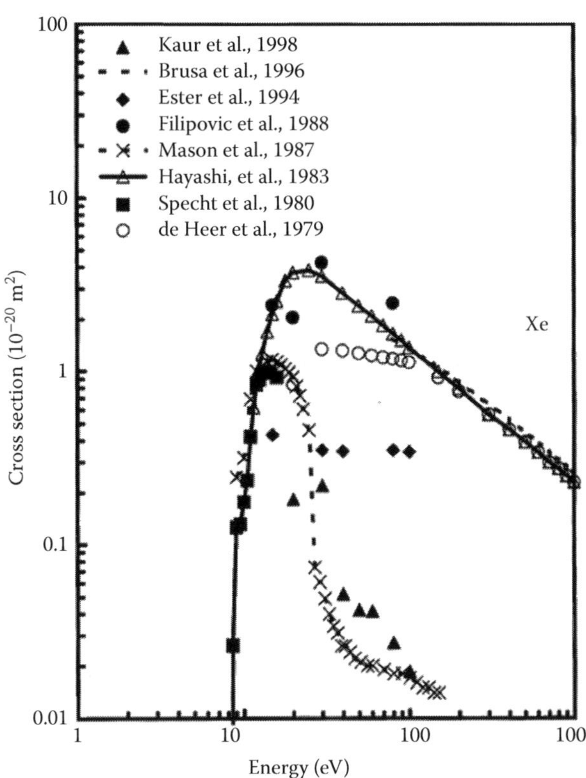

FIGURE 9.6 Excitation cross sections in Xe. (○) Total, de Heer et al. (1979), semiempirical; (■) total, Specht et al. (1980), swarm coefficients; (-△-) Hayashi (1983), swarm coefficients; (—×—) Mason and Newell (1987), measurement; (●) Filipović et al. (1988); (♦) Ester and Kessler (1994); (——) Brusa et al. (1996), semianalytical; (▲) Kaur et al. (1998).

Xe

TABLE 9.10
Absolute Total Ionization Cross Sections in Xe

Rejoub et al. (2002)		Kobayashi et al. (2002)		Krishnakumar and Srivastava (1988)		Rapp and Englander-Golden (1965)	
Energy (eV)	Q_i (10^{-20} m^2)	Energy (eV)	Q_i (10^{-20} m^2)	Energy (eV)	Q_i (10^{-20} m^2)	Energy (eV)	Q_i (10^{-20} m^2)
12	0.170	12.5	0.108	15	0.553	12.5	0.110
14	0.730	13	0.253	20	2.074	13	0.256
16	1.370	14	0.566	25	3.203	13.5	0.412
18	2.010	15	0.897	30	3.853	14	0.572
20	2.430	16	1.220	35	4.348	14.5	0.742
22	2.900	17	1.516	40	4.662	15	0.906
24	3.330	18	1.787	45	4.908	15.5	1.073
26	3.620	19	2.031	50	5.114	16	1.231
28	3.800	20	2.257	55	5.251	16.5	1.381
30	4.010	21	2.466	60	5.324	17	1.530
35	4.480	22	2.711	65	5.371	17.5	1.671
40	4.728	23	2.902	70	5.420	18	1.803
45	4.948	24	3.068	75	5.446	18.5	1.926
50	5.094	26	3.338	80	5.482	19	2.050
60	5.346	28	3.582	85	5.555	19.5	2.164
70	5.442	30	3.818	90	5.610	20	2.278
80	5.466	32	4.009	95	5.688	20.5	2.384
90	5.549	34	4.149	100	5.746	21	2.489
100	5.675	40	4.437	105	5.810	21.5	2.621
110	5.764	45	4.637	110	5.853	22	2.736
120	5.724	50	4.792	115	5.900	22.5	2.832
130	5.637	55	4.897	120	5.903	23	2.929
140	5.529	60	4.984	125	5.906	24	3.096

continued

TABLE 9.10 (continued)
Absolute Total Ionization Cross Sections in Xe

Rejoub et al. (2002)		Kobayashi et al. (2002)		Krishnakumar and Srivastava (1988)		Rapp and Englander-Golden (1965)	
Energy (eV)	Q_i (10^{-20} m^2)	Energy (eV)	Q_i (10^{-20} m^2)	Energy (eV)	Q_i (10^{-20} m^2)	Energy (eV)	Q_i (10^{-20} m^2)
150	5.422	65	5.037	130	5.856	26	3.3690
160	5.251	70	5.072	135	5.793	28	3.615
170	5.096	75	5.089	140	5.717	30	3.853
180	4.982	80	5.133	145	5.647	32	4.046
190	4.934	85	5.203	150	5.560	34	4.187
200	4.839	90	5.221	155	5.453	36	4.292
225	4.601	95	5.281	160	5.379	38	4.389
250	4.360	100	5.333	165	5.307	40	4.477
300	4.021	105	5.368	170	5.245	45	4.680
350	3.710	110	5.403	175	5.184	50	4.838
400	3.455	120	5.404	180	5.128	55	4.943
500	3.056	140	5.270	185	5.076	60	5.0313
600	2.718	160	5.001	190	5.036	65	5.0841
700	2.489	180	4.764	195	4.992	70	5.119
800	2.262	200	4.562	200	4.961	75	5.137
900	2.099	225	4.348	225	4.726	80	5.181
1000	1.961	250	4.173	250	4.512	85	5.251
		300	3.817	275	4.326	90	5.269
		350	3.529	300	4.164	95	5.330
		400	3.286	350	3.866	100	5.383
		450	3.081	400	3.632	105	5.418
		500	2.884	450	3.417	110	5.454
		550	2.730	500	3.239	115	5.462
		600	2.581	550	3.062	120	5.454
		650	2.453	600	2.926	125	5.410
		700	2.326	650	2.796	130	5.418
		750	2.216	700	2.697	135	5.348
		800	2.118	750	2.589	140	5.286
		850	2.029	800	2.511	145	5.251
		900	1.945	850	2.432	150	5.190
		950	1.877	900	2.339	160	5.066
		1000	1.807	950	2.262	180	4.838
				1000	2.199	200	4.583
						250	4.222
						300	3.897
						350	3.598
						400	3.351
						450	2.956
						500	2.948
						550	2.762
						600	2.621
						650	2.489
						700	2.384
						750	2.278
						800	2.181
						850	2.093
						900	2.014
						950	1.944
						1000	1.882

9.10 PARTIAL IONIZATION CROSS SECTIONS

Table 9.11 shows the partial ionization cross sections in Xe. See Figure 9.8 for graphical presentation.

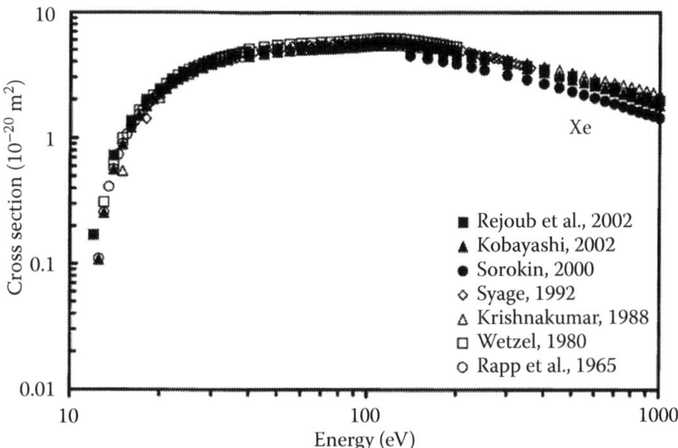

FIGURE 9.7 Absolute total scattering cross section in Xe. (■) Rejoub et al. (2002); (▲) Kobayashi et al. (2002); (●) Sorokin et al. (2000); (◊) Syage (1992); (Δ) Krishnakumar and Srivastava (1988); (□) Wetzel et al. (1980); (○) Rapp and Englander-Golden (1965).

TABLE 9.11
Partial Ionization Cross Section in Xe, in Units of 10^{-20} m^2

Energy (eV)	Xe^{+}	Xe^{2+}	Xe^{3+}	Xe^{4+}	Xe^{5+}	Xe^{6+}
12.5	0.108					
13	0.253					
14	0.566					
15	0.897					
16	1.220					
17	1.516					
18	1.787					
19	2.031					
20	2.257					
21	2.466					
22	2.711					
23	2.902					
24	3.068					
26	3.338					
28	3.582					
30	3.818					
32	4.009	0.0002				
34	4.147	0.0008				
40	4.345	0.0459				
45	4.463	0.0868				
50	4.368	0.212				
55	4.367	0.265				
60	4.332	0.326				
65	4.346	0.345	2.95E-04			
70	4.366	0.352	7.29E-04			
75	4.367	0.357	0.0027			
80	4.346	0.382	0.0076			
85	4.341	0.406	0.0166			
90	4.286	0.428	0.0262			
95	4.236	0.455	0.0449			
100	4.196	0.474	0.0630			
105	4.163	0.490	0.0748	3.31E-05		

continued

Xe

TABLE 9.11 (continued)
Partial Ionization Cross Section in Xe, in Units of 10^{-20} m²

Energy (eV)	Xe^{+}	Xe^{2+}	Xe^{3+}	Xe^{4+}	Xe^{5+}	Xe^{6+}
110	4.098	0.508	0.0963	0.0001		
120	3.993	0.509	0.130	0.0009		
140	3.810	0.476	0.163	0.0047		
160	3.627	0.429	0.156	0.0119		
180	3.458	0.394	0.144	0.0208	4.80E-04	
200	3.292	0.370	0.136	0.0287	0.0013	
225	3.092	0.350	0.127	0.0379	0.0047	
250	2.913	0.336	0.128	0.0397	0.0090	8.28E-05
300	2.639	0.303	0.124	0.0360	0.0106	5.54E-04
350	2.425	0.276	0.118	0.0334	0.0117	9.80E-04
400	2.242	0.254	0.115	0.0320	0.0110	0.0013
450	2.085	0.241	0.108	0.0317	0.0104	0.0018
500	1.941	0.2230	0.103	0.0314	0.0102	0.0019
550	1.823	0.2130	0.100	0.0297	0.0097	0.0022
600	1.722	0.1980	0.0969	0.0281	0.0093	0.0021
650	1.627	0.1890	0.0941	0.0270	0.0090	0.0021
700	1.543	0.1820	0.0863	0.0260	0.0086	0.0021
750	1.469	0.1710	0.0838	0.0249	0.0083	0.0021
800	1.402	0.1650	0.0793	0.0242	0.0079	0.0020
850	1.344	0.1580	0.0754	0.0233	0.0076	0.0019
900	1.295	0.1520	0.0699	0.0222	0.0072	0.0019
950	1.245	0.1470	0.0692	0.0213	0.0069	0.0018
1000	1.200	0.1410	0.0659	0.0207	0.0067	0.0018

Source: Adapted from Kobayashi, A. et al., *J. Phys. B: At. Mol. Opt. Phys.*, 35, 2087, 2002.

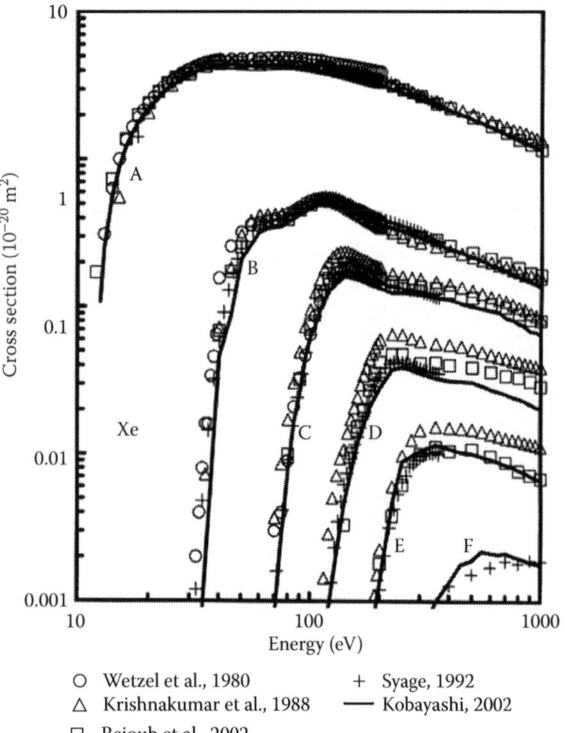

FIGURE 9.8 Partial ionization cross sections in Xe. (○) Wetzel et al. (1980); (Δ) Krishnakumar (1988); (+) Syage (1992); (——) Kobayashi et al. (2002); (□) Rejoub et al. (2002). A – Xe^{+}; B – Xe^{2+}; C – Xe^{3+}; D – Xe^{4+}; E – Xe^{5+}; F – Xe^{6+}.

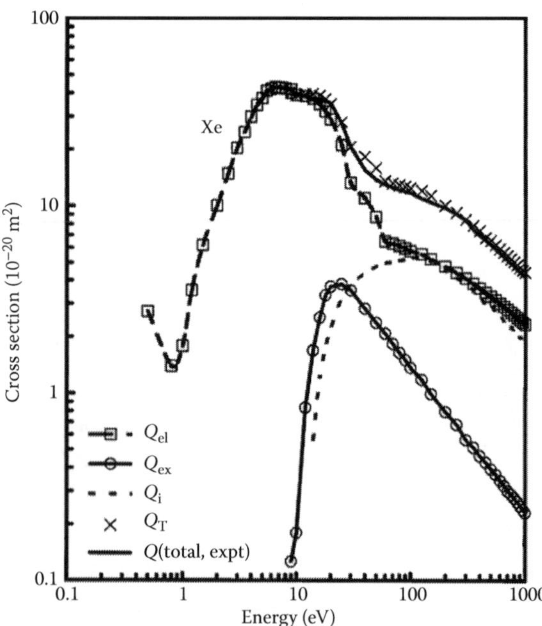

FIGURE 9.9 Consolidated presentation of all cross sections in Xe. (—□—) Q_{el}; (—○—) Q_{ex}; (– – – –) Q_i; (×) Q_T obtained by adding all cross sections; (——) Q_T (experimental).

9.11 CONSOLIDATED PRESENTATION

Figure 9.9 presents the consolidated view of cross sections in the previous sections.

9.12 DRIFT VELOCITY OF ELECTRONS

Table 9.12 shows the drift velocity of electrons in Xe. See Figure 9.10 for graphical presentation.

TABLE 9.12
Drift Velocity of Electrons in Xe at 293 K

Pack et al. (1992) Digitized		Peuch and Mizzi (1991) Digitized		Hunter et al. (1988) Tabulated	
E/N (Td)	*W* (m/s)	*E/N* (Td)	*W* (m/s)	*E/N* (Td)	*W* (m/s)
2.00E-03	6.00	1.00E-02	40.00	0.004	15.20
4.00E-03	19.00	2.00E-02	155.40	0.005	19.20
7.00E-03	28.00	4.00E-02	380.90	0.006	23.00
1.00E-02	42.00	7.00E-02	697.30	0.007	26.80
2.00E-02	83.00	0.1	880.60	0.008	31.10
4.00E-02	323.00	0.2	1.12E + 03	0.010	38.70
7.00E-02	732.00	0.4	1.19E + 03	0.012	46.60
1.00E-01	964.00	0.7	1.23E + 03	0.014	54.40
0.20	927.00	1	1.23E + 03	0.017	67.30
0.40	1.10E + 03	2	1.60E + 03	0.020	82.0
0.70	1.20E + 03	4	2.25E + 03	0.025	114.7
1	1.25E + 03	7	4.00E + 03	0.030	165.4
2	1.49E + 03	10	7.27E + 03	0.035	247.0
4	2.38E + 03	20	1.36E + 04	0.040	336.0
7	3.99E + 03	40	2.13E + 04	0.050	526.0
10	6.93E + 03	70	3.85E + 04	0.060	662.0
20	1.42E + 04	100	5.48E + 04	0.070	743.0
30	1.81E + 04	200	1.01E + 05	0.080	805.0
40	2.15E + 04	300	1.49E + 05	0.10	858.0
50	2.63E + 04	400	2.04E + 05	0.12	900.0
60	3.24E + 04	500	2.33E + 05	0.14	924.0
70	3.84E + 04			0.17	956.0
80	4.34E + 04			0.20	981.0
				0.25	1.02E + 03
				0.30	1.05E + 03
				0.40	1.11E + 03
				0.50	1.15E + 03
				0.60	1.19E + 03
				0.80	1.25E + 03
				1.00	1.31E + 03
				1.20	1.37E + 03
				1.40	1.41E + 03
				1.70	1.49E + 03
				2.00	1.55E + 03
				2.50	1.65E + 03
				3.00	1.79E + 03

Note: See Figure 9.10 for graphical presentation.

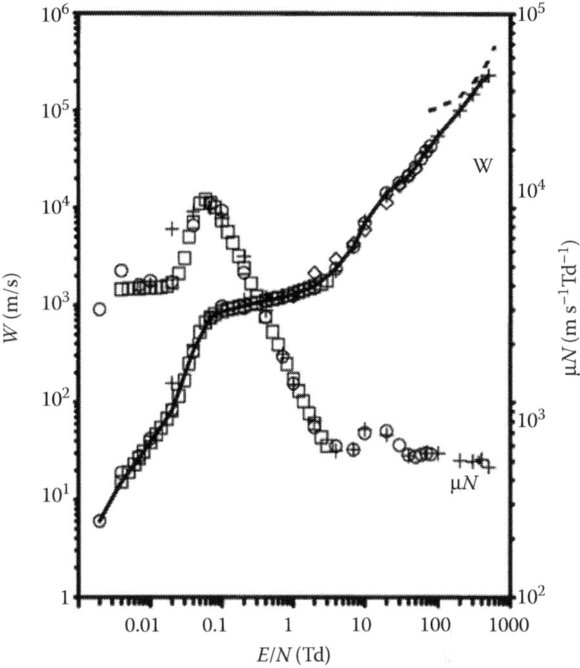

FIGURE 9.10 Drift velocity and density-normalized mobility of electrons in Xe at 293 K. (•) Recommended, Raju (2005), compilation; (+) Xiao et al. (2000), experimental; (—Δ—) Peuch et al. (1991), theory; (□) Hunter et al. (1988), experimental; (○) Brooks et al. (1982); (◊) Pack et al. (1962), experimental.

Xe

9.13 CHARACTERISTIC ENERGY (D/μ)

The diffusion coefficients expressed as characteristic energy D_r/μ and D_L/μ are shown in Tables 9.13 and 9.14. See Figure 9.11 for graphical presentation.

TABLE 9.13
Ratio of Diffusion Coefficient to Mobility. D_r = Radial Diffusion, D_L = Longitudinal Diffusion

E/N (Td)	D_r/μ (V)	E/N (Td)	D_L/μ (V)
0.001	0.026	0.001	0.017
0.002	0.025	0.002	0.014
0.004	0.024	0.004	0.013
0.007	0.026	0.007	0.018
0.01	0.028	0.01	0.023
0.02	0.10	0.02	0.030
0.04	0.44	0.04	0.61
0.07	1.44	0.07	0.26
0.1	1.86	0.1	0.18
0.2	3.06	0.2	0.18
0.4	4.12	0.4	0.24
0.7	5.82	0.7	0.29
1	6.76	1	0.36
2	7.18	2	0.53
4	7.36	4	1.52
7	7.39	7	2.37
10	7.07	10	2.19
20	7.10	20	2.13
30	7.32	30	2.28
40	7.03	40	2.19
50	6.49	50	2.17
60	5.95	60	2.22
70	5.66	70	2.32
80	5.88	80	2.46
90	6.86	90	2.60

Source: Adapted from Pack, J. L. et al., *J. Appl. Phys.*, 71, 5363, 1992.

9.14 MEAN ENERGY OF ELECTRONS

Mean energy of electrons in the swarm as a function of *E/N*, deduced by Hayashi, (1983) is shown in Table 9.15.

9.15 REDUCED IONIZATION COEFFICIENTS

Reduced ionization coefficients (α/N) in Xe are presented in Table 9.16 (Kruithoff et al., 1940). See Figure 9.12 for graphical presentation.

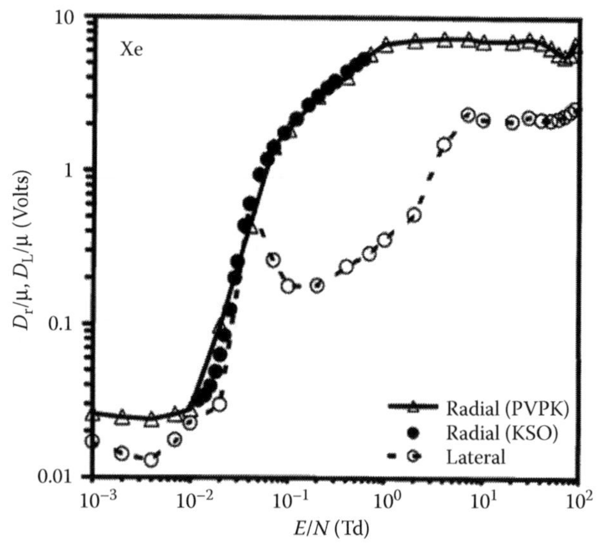

FIGURE 9.11 Radial and lateral diffusion coefficients as ratios of mobility in Xe. (—Δ—) Radial, Pack et al. (1992); (●) radial, Koizumi et al. (1986); (—○—) lateral, Pack et al. (1992).

TABLE 9.14
Characteristic Energy in Xe

E/N (Td)	D_r/μ (V)	E/N (Td)	D_r/μ (V)	E/N (Td)	D_r/μ (V)	E/N (Td)	D_r/μ (V)
0.012	0.0320	0.025	0.125	0.060	1.19	0.25	3.49
0.014	0.0343	0.028	0.200	0.070	1.43	0.30	3.83
0.016	0.0395	0.030	0.255	0.090	1.77	0.40	4.43
0.018	0.0490	0.035	0.439	0.12	2.18	0.50	4.93
0.020	0.0632	0.040	0.607	0.16	2.69	0.60	5.38
0.022	0.0848	0.050	0.943	0.20	3.07		

Source: Adapted from Koizumi, T. Shirakawa, F., and Ogawa, I. *J. Phys. B: At. Mol. Phys.*, 19, 2331, 1986.

Xe

TABLE 9.15
Mean Energy of Electrons

E/N (Td)	$\bar{\varepsilon}$ (eV)	*E/N* (Td)	$\bar{\varepsilon}$ (eV)	*E/N* (Td)	$\bar{\varepsilon}$ (eV)
100	5.8	600	8.1	4000	31
150	5.9	800	9.2	5000	39
200	6.2	1000	10.3	6000	48
300	6.7	1500	13.4	8000	66
400	7.2	2000	16.5	10000	85
500	7.6	3000	23.5		

TABLE 9.16
Density Reduced Ionization Coefficients for Xe

E/N (Td)	α/N (10^{-20} m^2)	*E/N* (Td)	α/N (10^{-20} m^2)	*E/N* (Td)	α/N (10^{-20} m^2)
25	0.0001	200	0.265	750	1.964
30	0.0003	250	0.411	800	2.110
35	0.0006	300	0.563	850	2.249
40	0.001	350	0.711	900	2.381
50	0.004	400	0.867	1000	2.625
60	0.008	450	1.025	1250	3.216
70	0.014	500	1.186	1500	3.713
80	0.024	550	1.347	1750	4.194
90	0.036	600	1.506	2000	4.576
100	0.049	650	1.661	2400	7.370
150	0.144	700	1.814		

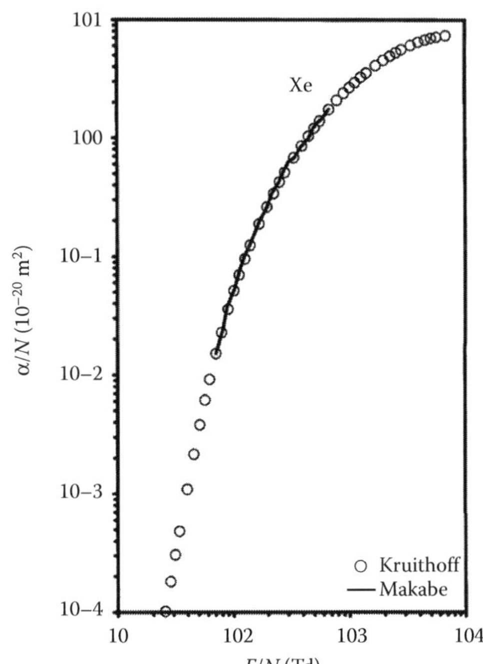

FIGURE 9.12 Reduced ionization coefficients in Xe. (○) Kruithoff (1940), experimental; (——) Makabe and Mori (1978), theory.

TABLE 9.17
Reduced Mobility

E/N (Td)	μ_{red} (10^{-4} m^2/Vs)	*E/N* (Td)	μ_{red} (10^{-4} m^2/Vs)
40	0.5195	140	0.4530
50	0.5132	150	0.4472
60	0.5061	160	0.4411
70	0.4990	170	0.4351
80	0.4926	180	0.4294
100	0.4794	190	0.4241
120	0.4662	220	0.4085

Xe

9.16 GAS CONSTANTS

The gas constants for Xe according to equation

$$\frac{\alpha}{N} = C_{\mathrm{T}} \exp\left[-\frac{D_{\mathrm{T}}}{(E/N)^{1/2}}\right] \quad (9.2)$$

are $C_T = 18.12 \times 10^{-20}$ m^2 and $D_T = 60.33$ (Td)$^{1/2}$ (see Raju, 2005).

9.17 REDUCED MOBILITY OF POSITIVE IONS

The reduced mobility of ground-state positive ions in Xe is shown in Table 9.17 (Helm, 1976).

REFERENCES

Bowe, J. C., *Phys. Rev.*, 117, 1411, 1960.

Brooks, H. L., M. C. Cornell, J. Fletcher, L. M. Littlewood, and K. J. Nygaard, *J. Phys. D: Appl. Phys.*, 15, L51, 1982.

Brusa, R. S., G. P. Karwasz, and A. Zecca, *Z. Phys. D*, 38, 279, 1996.

Danjo, A., *J. Phys. B: At. Mol. Opt. Phys.*, 21, 3759, 1998.

de Heer, F. J., R. H. J. Jansen, and W. van der Kaay, *J. Phys. B: At. Mol. Phys.*, 12, 979, 1979.

Ester, T. and J. Kessler, *J. Phys. B: At. Mol. Opt. Phys.*, 27, 4295, 1994.

Filipović, D., B. Marinković, V. Pejćev, and L. Vusković, *Phys. Rev. A*, 37, 356, 1988.

Fons, J. T. and C. C. Lin, *Phys. Rev. A*, 58, 4603, 1998.

Frost, L. S. and A. V. Phelps, *Phys. Rev. A*, 136, 1538, 1964.

Gibson, J. C., D. R. Lunt, L. J. Allen, R. P. McEachran, L. A. Parcell, and S. J. Buckman, *J. Phys. B: At. Mol. Opt. Phys.*, 31, 3949, 1998.

Grigoriev, I. S. and E. Z. Meilikhov (Eds.), *Handbook of Physical Quantities*, CRC Press, Boca Raton, FL, 1999.

Hayashi, M. J., *J. Phys. D: Appl. Phys.*, 15, 1411, 1982; 16, 581, 1983.

Helm, H., *J. Phys. B: At. Mol. Phys.*, 9, 2931, 1976.

Huang, S. S. S. and G. R. Freeman, *J. Chem. Phys.*, 68, 1355, 1978.

Hunter, S. R., J. C. Carter, and L. G. Christophorou, *Phys. Rev. A*, 38, 5539, 1988.

Jost, K., P. G. F. Bisling, F. Eschen, M. Felsmann, and L. Walther, in J. Eichler et al. (Eds.), *Proceedings of the 13th International Conference on the Physics of Electronic Atomic Collisions*, Amsterdam, Berlin, North-Holland, 1983, p. 91, cited by R. W. Wagenaar and F. J. de Heer (1985). Tabulated values above 7.5 eV are given in the latter reference. The lower range of 0.05 eV is obtained from Nickel et al. (1985).

Kaur, S., R. Srivastava, P. McEachran, and A. D. Stauffer, *J. Phys. B: At. Mol. Opt. Phys.*, 31, 4833, 1998.
Khakoo, M. A., C. E. Beckmann, S. Trajmar, and G. Csanak, *J. Phys. B: At. Mol. Opt. Phys.*, 27, 3159, 1994.
Khakoo, M. A., S. Trajmar, R. LeClair, I. Kanik, G. Csanak, and C. J. Fontes, *J. Phys. B: At. Mol. Opt. Phys.*, 29, 3455, 1996a.
Khakoo, M. A., S. Trajmar, R. LeClair, I. Kanik, G. Csanak, and C. J. Fontes, *J. Phys. B: At. Mol. Opt. Phys.*, 29, 3477, 1996b.
Klewer, M. and M. J. M. van der Wiel, *J. Phys. B: At. Mol. Opt. Phys.*, 13, 571, 1980.
Kobayashi, A., G. Fujiki, A. Okaji, and T. Masuoka, *J. Phys. B: At. Mol. Opt. Phys.*, 35, 2087, 2002.
Koizumi, T., F. Shirakawa, and I. Ogawa, *J. Phys. B: At. Mol. Phys.*, 19, 2331, 1986.
Krishnakumar, E. and S. K. Srivastava, *J. Phys. B: At. Mol. Opt. Phys.*, 21, 1055, 1988.
Kruithoff, A. A., *Physica*, 7, 519, 1940.
Makabe, T. and T. Mori, *J. Phys. B: At. Mol. Phys.*, 11, 3785, 1978.
McEachran, R. P. and A. D. Stauffer, *J. Phys. B: At. Mol. Phys.*, 17, 2507, 1984.
McEachran, R. P. and A. D. Stauffer, *Phys. Lett.*, 107A, 397, 1985.
Mason, N. J., and W. R. Newell, *J. Phys. B: At Mol. Phys.*, 20, 1357, 1987.
Nickel, J. C., K. Imre, D. F. Register, and S. Trajmar, *J. Phys. B: At. Mol. Phys.*, 18, 125, 1985.
Nishimura, H., T. Matsuda, and A. Danjo, *J. Phys. Soc. Jpn.*, 56, 70, 1987.
O'Malley, T. F. *Phys. Rev.*, 130, 1020, 1963; T. F. O'Malley and R. W. Crompton, 13, 3451, 1980.
Pack, J. L., R. E. Voshall, A. V. Phelps, and L. E. Kline, *J. Appl. Phys.*, 71, 5363, 1992.
Pack, J. L., R. E. Voshall, and A. V. Phelps, *J. Appl. Phys.*, 71, 5363, 1992.
Pack, J. L., R. E. Voshall, and A. V. Phelps, *Phys. Rev.*, 127, 2084, 1962.
Puech, V. and S. Mizzi, *J. Phys. D: Appl. Phys.*, 24, 1974, 1991.
Puech, V., and S. Mizzi, *J. Phys. D: Appl. Phys.* 24, 1974, 1991.
Raju, G. G., *Gaseous Electronics: Theory and Practice*, Taylor & Francis, Boca Raton, FL, 2005.
Ramsauer, C., and R. Kollath, *Ann. Phys. LPZ.*, 12, 529, 1932.
Rapp, D. and P. Englander-Golden, *J. Chem. Phys.*, 13, 1464, 1965.
Register, D. F., L. Viscovic, and S. Trajmar, *J. Phys. B: At. Mol. Phys.*, 19, 1685, 1986.
Rejoub, R., B. G. Lindsay, and R. F. Stebbings, *Phys. Rev. A*, 65, 042713-01, 2002.
Santos, F. P., T. H. V. T. Dias, A. D. Stauffer, and C. A. N. Conde, *J. Phys. D: Appl. Phys.*, 27, 42, 1994.
Sin Fai Lam, L. T. *J. Phys. B: At. Mol. Phys.*, 15, 119, 1982.
Specht, L. T., S. A. Lawton, and T. A. De Temple, *J. Appl. Phys.*, 51, 166, 1980.
Subramanian, K. P. and V. Kumar, *J. Phys. B: At. Mol. Phys.*, 20, 5505, 1987.
Suzuki, T. Y., Y. Sakai, B. S. Min, T. Takayanagi, K. Wakiya, H. Suzuki, T. Inaba, and H. Takuma, *Phys. Rev. A*, 43, 5867, 1991.
Syage, J. A., *Phys. Rev. A*, 46, 5666, 1992.
Szmytkowsky, C., C. K. Mąciag, and G. Karwasz, *Phys. Scrip.*, 54, 271, 1996.
Wagenaar, R. W. and F. J. de Heer, *J. Phys. B: At. Mol. Phys.*, 18, 2021, 1985.
Wetzel, R. C., F. A. Balochhi, T. R. Hayes, and R. S. Freund, *Phys. Rev. A*, 35, 339, 1980.
Weyherter, M., B. Barzick, A. Mann, and F. Linder, *Z. Phys. D.*, 7, 333, 1988.
Williams, W., S. Trajmar, and A. Kuperman, *J. Chem. Phys.*, 62, 3031, 1975.
Xiao, D. M., L. L. Zhu, and X. G. Li., *J. Phys. D: Appl. Phys.*, 33, L145, 2000.
Zecca, A., G. Karwasz, R. S. Brusa, and R. Grisenti, *J. Phys. B: At. Mol. Opt. Phys.*, 24, 2737, 1991.
Zecca, A., G. P. Karwasz, and R. S. Brusa, *J. Phys. B: At. Mol. Opt. Phys.*, 33, 843, 2000.

Section II

2 ATOMS

10

BROMINE

Br$_2$

CONTENTS

Molecular bromine (Br_2) with 70 electrons belongs to the class of halogens. It has electronic polarizability of 7.81×10^{-40} F m^2 and ionization potential of 10.516 eV. The dissociation (Br–Br) energy is 2.01 eV and the electron affinity is 2.55 eV. The (Raman) vibrational levels are 39.3, 78.2, and 117.1 meV for $\Delta v = 1, 2, 3$, respectively (Stammerich, 1950). The total vibrational excitation cross section at 0.5 eV is $\sim 2.5 \times 10^{-20}$ m^2 (see Hall and Nighan, 1977). There has been a single study of scattering cross sections for the molecule by Arnot (1934).

10.1 TOTAL IONIZATION CROSS SECTION

Total ionization cross section consists of formation of Br_2^+ ions and Br^+ ions by dissociative ionization. Total ionization cross sections are shown in Table 10.1 and Figure 10.1 (Kurepa et al., 1981).

10.2 DISSOCIATION RATE CONSTANT

Dissociation rate constants for Br_2 at high temperatures behind shock waves have been measured by Boyd et al. (1968).

10.3 ATTACHMENT PROCESSES

Negative ions are formed by three processes (Kurepa et al., 1981):

$$Br_2 + e \rightarrow Br_2^{-*} \rightarrow Br + Br^- \quad (10.1)$$

$$Br_2 + e + M \rightarrow Br^- \quad (10.2)$$

$$Br_2 + e \rightarrow Br^+ + Br^- + e \quad (10.3)$$

$$Br^- + Br_2 \rightarrow Br + Br_2^- \quad (10.4)$$

$$Br_2 + e \rightarrow Br_2^{-*}; \quad Br_2^{-*} + M \rightarrow Br_2^- + M \quad (10.5)$$

Reaction 10.1 is dissociative attachment, Reaction 10.2 is a three-body process, and Reaction 10.3 is ion pair formation. Reaction 10.4 is known as charge transfer (Massey, 1950) and is least probable on energetic considerations. Reaction 10.5 is attachment to parent molecule followed by collisional stabilization. This process shows pressure dependence of the attachment coefficient as in Br_2 (see below) and unlike in Cl_2 (Božin and Goodyear, 1967).

At energies below 10 eV, only dissociative attachment occurs. The threshold energy for ion pair formation is 10.41 eV.

10.4 ATTACHMENT CROSS SECTIONS

The potential energy curves for the dissociative electron attachment to Br_2 molecule are shown in Figure 10.2 (Kurepa et al., 1981). Molecular Br_2, like Cl_2 and F_2, has a peak in the attachment cross section at zero electron energy. The peak positions of the attachment cross sections are shown in Table 10.2.

Total attachment cross sections are shown in Table 10.3 and Figure 10.3. Figures 10.4 and 10.5 show the dissociative attachment and ion pair formation cross sections, respectively.

Br$_2$

TABLE 10.1
Total Ionization Cross Sections

Energy (eV)	Q_i (10^{-20} m^2)	Energy (eV)	Q_i (10^{-20} m^2)	Energy (eV)	Q_i (10^{-20} m^2)
10.6	0.04	17.0	2.99	44	8.56
10.8	0.10	17.2	3.07	46	8.71
11.0	0.20	17.4	3.20	48	8.84
11.2	0.30	17.6	3.27	50	8.97
11.4	0.40	17.8	3.37	52	9.17
11.6	0.47	18.0	3.44	54	9.33
11.8	0.57	18.2	3.54	56	9.46
12.0	0.67	18.4	3.61	58	9.69
12.2	0.74	18.6	3.71	60	9.84
12.4	0.84	18.8	3.77	62	9.94
12.6	0.94	19.0	3.84	64	10.05
12.8	1.04	19.2	3.94	66	10.12
13.0	1.11	19.4	4.01	68	10.20
13.2	1.21	19.6	4.04	70	10.22
13.4	1.34	19.8	4.16	72	10.23
13.6	1.41	20	4.21	74	10.24
13.8	1.48	21	4.54	76	10.24
14.0	1.61	22	4.87	78	10.25
14.2	1.68	23	5.16	80	10.25
14.4	1.81	24	5.43	82	10.25
14.6	1.92	25	5.73	84	10.24
14.8	2.02	26	5.96	86	10.29
15.0	2.08	27	6.20	88	10.29
15.2	2.19	28	6.37	90	10.20
15.4	2.32	29	6.57	92	10.23
15.6	2.38	30	6.74	94	10.21
15.8	2.49	32	7.18	96	10.22
16.0	2.59	34	7.48	98	10.17
16.2	2.66	36	7.74	100	10.21
16.4	2.76	38	7.99		
16.6	2.83	40	8.17		
16.8	2.90	42	8.35		

Source: Reproduced from M. V. Kurepa, D. S. Babić, and D. S. Belić, *J. Phys. B: At. Mol. Phys.*, 14, 375, 1981. With kind permission of Institute of Physics, England.

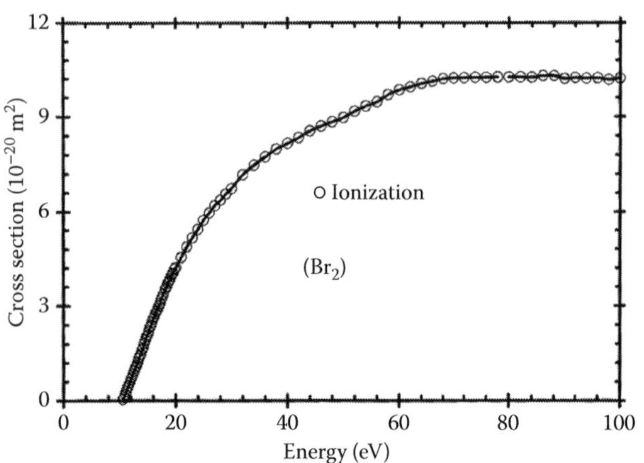

FIGURE 10.1 Total ionization cross section for Br$_2$. (Adapted from M. V. Kurepa, D. S. Babić, and D. S. Belić, *J. Phys. B: At. Mol. Phys.*, 14, 375, 1981.)

10.5 ATTACHMENT RATE CONSTANT

Table 10.4 shows the attachment rate constants for Br$_2$.

10.6 TRANSPORT PARAMETERS

The data available for transport parameters for Br$_2$ are from Bailey et al. (1937). The drift velocities are shown in Figure 10.6 and mean energy (1.5 D/μ) in Figure 10.7. The diffusion coefficients measured at 1 atm buffer of NO and Ar gas are 0.11×10^{-4} and 0.21×10^{-4} m^2/s, respectively (Zittel and Little, 1979).

10.7 IONIZATION AND ATTACHMENT COEFFICIENTS

The data available are from Razzak and Goodyear (1969). Reduced ionization coefficients (α/N) are shown in Figure 10.8 and the values do not show significant dependence on gas

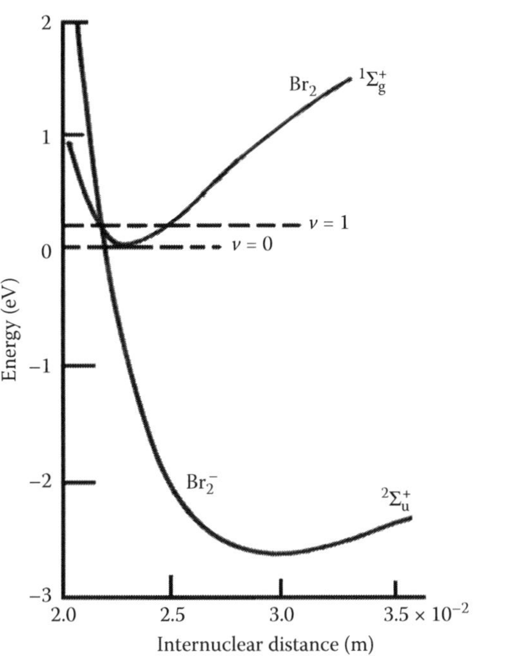

FIGURE 10.2 Potential energy curves for the Br_2 molecule and parent negative ion (Br_2^-). For clarity, only the lowest energy state of the ion is shown. (Adapted from M. V. Kurepa, D. S. Babić, and D. S. Belić, *J. Phys. B: At. Mol. Phys.*, 14, 375, 1981.)

TABLE 10.2
Peak Energy in the Attachment Cross Section

Authors	Peak Energy (eV)
Kurepa et al. (1981)	0.0
	0.50
	1.45
	3.75
	5.30
	8.50
Tam and Wong (1978)	0.07
	1.4
	3.7

TABLE 10.3
Total Attachment Cross Sections for Br_2

Energy (eV)	Q_a (10^{-20} m^2)	Energy (eV)	Q_a (10^{-20} m^2)	Energy (eV)	Q_a (10^{-20} m^2)
0.0	17.74	6.0	0.49	13.0	2.89
0.1	17.00	6.2	0.34	13.5	3.41
0.2	12.0	6.4	0.255	14.0	3.82
0.3	5.40	6.6	0.08	14.5	4.08
0.4	4.20	6.8	0.025	15.0	4.34
0.5	2.80	7.0	0.038	15.5	4.50
0.6	2.60	7.2	0.072	16.0	4.70
0.7	2.50	7.4	0.108	16.5	4.90
0.8	2.35	7.6	0.152	17.0	5.07

TABLE 10.3 (continued)
Total Attachment Cross Sections for Br_2

Energy (eV)	Q_a (10^{-20} m^2)	Energy (eV)	Q_a (10^{-20} m^2)	Energy (eV)	Q_a (10^{-20} m^2)
0.9	2.38	7.8	0.196	18.5	5.25
1.0	2.50	8.0	0.250	19.0	5.25
1.2	2.60	8.2	0.300	19.5	5.17
1.4	2.78	8.4	0.320	20	5.12
1.6	2.90	8.6	0.330	22	4.79
1.8	2.75	8.8	0.320	24	4.50
2.0	2.50	9.0	0.280	26	4.27
2.2	2.05	9.2	0.250	28	4.10
2.4	1.90	9.4	0.200	30	3.88
2.6	1.95	9.6	0.190	32	3.71
2.8	2.42	9.8	0.180	34	3.55
3.0	3.23	10.0	0.180	36	3.48
3.2	3.84	10.6	0.21	38	3.43
3.4	4.43	10.8	0.42	40	3.44
3.6	4.93	11.0	0.62		
3.8	5.05	11.2	0.81		
4.0	4.90	11.4	0.98		
4.2	4.53	11.6	1.22		
4.4	4.04	11.8	1.44		
4.6	3.50	12.0	1.67		
4.8	2.76	12.2	1.92		
5.0	2.23	12.4	2.16		
5.2	1.65	12.6	2.41		
5.4	1.22	12.8	2.66		
5.6	0.89	17.5	5.21		
5.8	0.70	18.0	5.26		

Source: Reproduced from M. V. Kurepa, D. S. Babić, and D. S. Belić, *J. Phys. B: At. Mol. Phys.*, 14, 375, 1981. With kind permission of Institute of Physics, England.

Note: For energy ≥10.6 eV, ion pair formation is the dominant cross section.

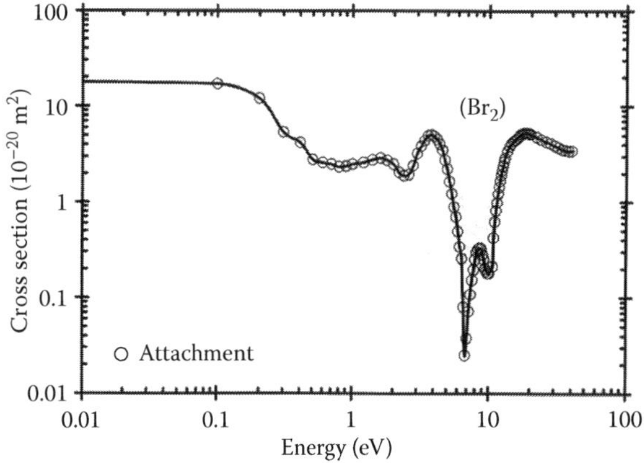

FIGURE 10.3 Total attachment cross sections for Br_2. Above 10 eV energy, the process is ion pair formation. (Adapted from M. V. Kurepa, D. S. Babić, and D. S. Belić, *J. Phys. B: At. Mol. Phys.*, 14, 375, 1981.)

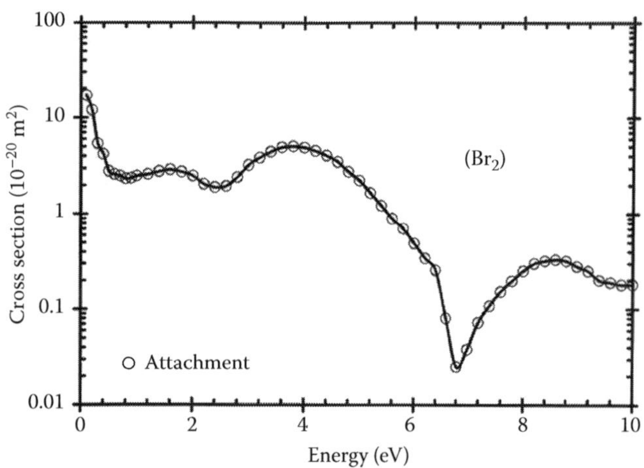

FIGURE 10.4 Total attachment cross sections for Br_2 (low energy details). Note the peak at zero energy. (Adapted from M. V. Kurepa, D. S. Babić, and D. S. Belić, *J. Phys. B: At. Mol. Phys.*, 14, 375, 1981.)

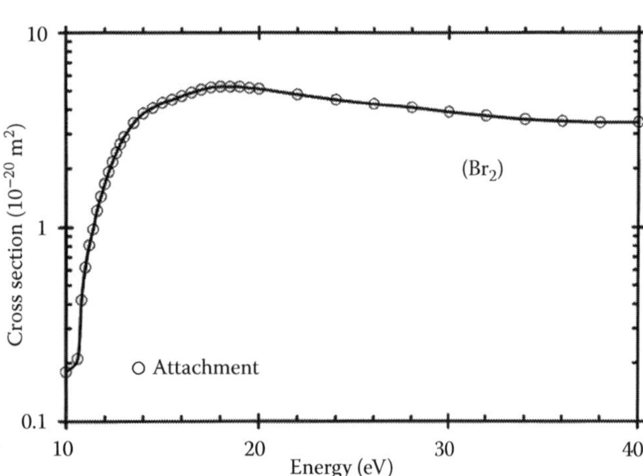

FIGURE 10.5 Total attachment cross section for ion pair formation process. (Adapted from M. V. Kurepa, D. S. Babić, and D. S. Belić, *J. Phys. B: At. Mol. Phys.*, 14, 375, 1981.)

TABLE 10.4
Attachment Rate Constants for Br_2

Temperature, K (eV)	Rate Constant ($\times 10^{-19}$ m^3/s)	Method	Reference
(0.3)	800	Beam	Trainor, D. W. and M. J. M. Boness (1978)
(0.45)	1200		
(0.75)	1250		
(1.0)	1000		
350	100	FA	Sides et al. (1976)
296	8.2	MWC	Truby (1971)

Note: Temperature is thermal unless otherwise mentioned. FA = flowing afterglow; MWC = microwave cavity method.

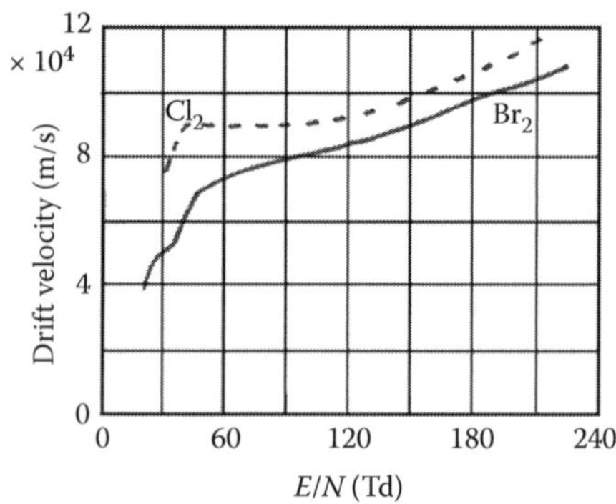

FIGURE 10.6 Drift velocity of electrons for Br_2. (Adapted from J. E. Bailey, R. E. B. Makinson, and J. M. Somerville, *Philos. Mag.*, 24, 177, 1937.)

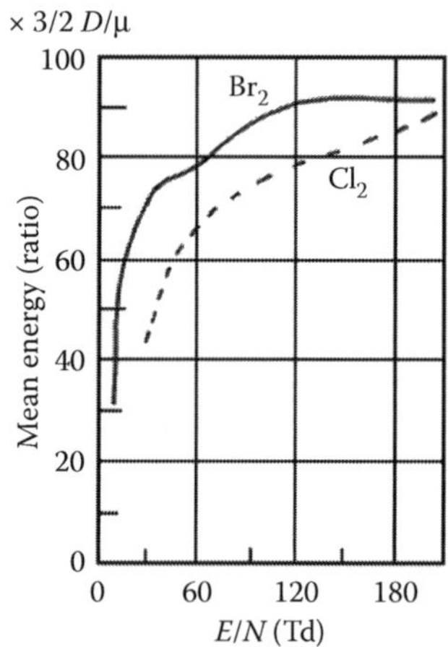

FIGURE 10.7 Mean energy of electrons for Br_2. (Adapted from J. E. Bailey, R. E. B. Makinson, and J. M. Somerville, *Philos. Mag.*, 24, 177, 1937.)

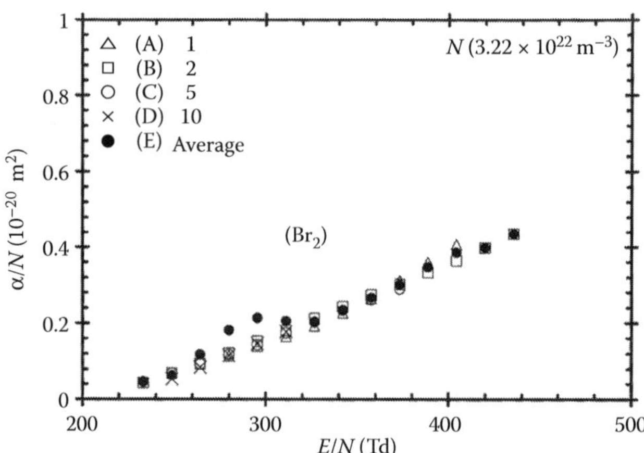

FIGURE 10.8 Reduced ionization coefficients as a function of reduced electric field. Gas number density as shown. (e) (●) Average values. (Adapted from Razzak, S. A. A. and C. C. Goodyear, *Brit. J. Appl. Phys.*, 2, 1577, 1969.)

TABLE 10.5
Reduced Ionization, Attachment, and Effective Ionization Coefficients for Br_2

	Coefficients (10^{-20} m^2)		
E/N (Td)	α/N	η/N	$(\alpha-\eta)/N$
235	0.044	0.169	−0.125
250	0.066	0.166	−0.100
265	0.121	0.157	−0.036
280	0.181	0.152	−0.029
295	0.213	0.137	0.076
310	0.206	0.124	0.082
325	0.201	0.104	0.097
340	0.229	0.092	0.137
355	0.261	0.079	0.182
370	0.293	0.065	0.228
385	0.336	0.062	0.274
400	0.377	0.061	0.316
415	0.398	0.048	0.350
430			0.419
445			0.476
460			0.543
475			0.607
490			0.660
505			0.707
520			0.752
535			0.798
550			0.846
565			0.896
580			0.948
595			1.000
610			1.052
625			1.103
640			1.151

Source: Adapted from Razzak, S. A. A. and C. C. Goodyear, *Brit. J. Appl. Phys.*, 2, 1577, 1969.

number density. Table 10.5 shows the average values of α/N, η/N, and $(\alpha-\eta)/N$ as a function of *E/N*. Figure 10.9 shows that the attachment coefficients (η/N) are dependent on gas number densities, suggesting that Reaction 10.5 is the most likely process. Figure 10.10 shows that the density-reduced effective ionization coefficients ($\alpha-\eta/N$) are marginally density dependent.

10.8 LIMITING *E/N*

Limiting values of *E/N* ($\alpha/N = \eta/N$) are density dependent as shown in Table 10.6 (Razzak and Goodyear, 1969).

10.9 PASCHEN MINIMUM

Paschen minimum voltage for Br_2 is 380 V at ~2.9×10^{20} m^{-2} (see Razzak and Goodyear, 1969).

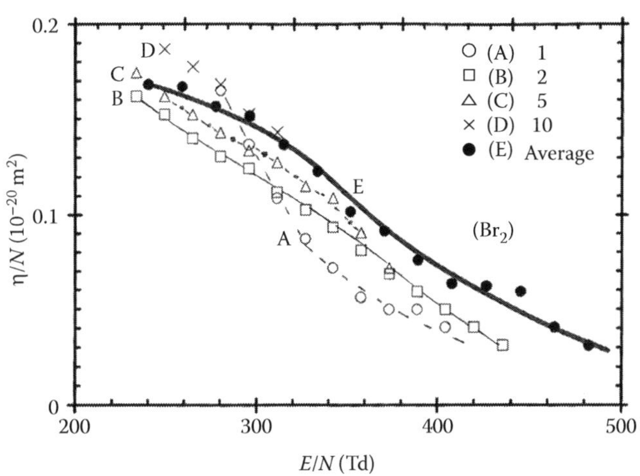

FIGURE 10.9 Density-reduced attachment coefficients for Br_2. Gas number densities in units of 3.22×10^{22} m^{-3}. (A) 1; (B) 2; (C) 5; (D) 10; (E) average. (Adapted from Razzak, S. A. A. and C. C. Goodyear, *Brit. J. Appl. Phys.*, 2, 1577, 1969.)

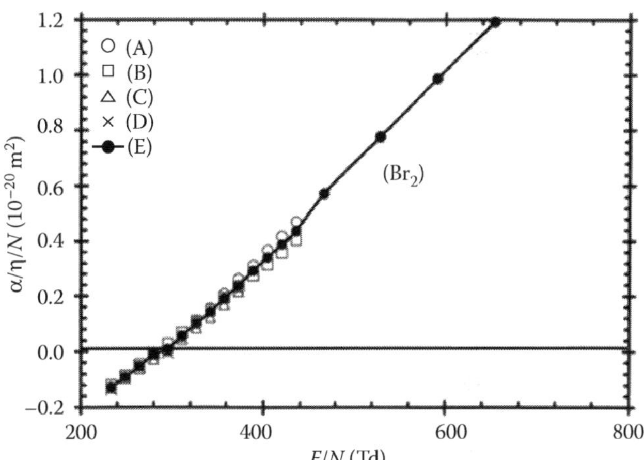

FIGURE 10.10 Density-reduced net ionization coefficients for Br_2. Gas number densities in units of 3.22×10^{22} m^{-3}. (A) 1; (B) 2; (C) 5; (D) 10; (E) average. (Adapted from Razzak, S. A. A. and C. C. Goodyear, *Brit. J. Appl. Phys.*, 2, 1577, 1969.)

TABLE 10.6
Limiting Value of *E/N*

N (10^{22} m^{-3})	*E/N* (Td)
6.44	282
16.1	291
32.2	298

REFERENCES

Arnot, F. L., *Proc. Roy. Soc.*, A144, 360, 1934.

Bailey, J. E., R. E. B. Makinson, and J. M. Somerville, *Philos. Mag.*, 24, 177, 1937.

Boyd, R. K., J. D. Brown, G. Burns, and J. H. Lippiatt, *J. Chem. Phys.*, 49, 3822, 1968.

Božin, S. E. and C. C. Goodyear, *Br. J. Appl. Phys.*, 18, 49, 1967.
Hall, R. B. and W. L. Nighan, *Abstracts of the 30th Annual Gaseous Electronics Conference*, Palo Alto, CA October 18–21, 1977, cited by W.-C. Tam and S. F. Wong, *J. Chem. Phys.*, 68, 5626, 1978.
Kurepa, M. V., D. S. Babić, and D. S. Belić, *J. Phys. B: At. Mol. Phys.*, 14, 375, 1981.
Massey, H. S. W. *Negative Ions*, Cambridge University Press, London, 1950.
Razzak, S. A. A. and C. C. Goodyear, *Brit. J. Appl. Phys.*, 2, 1577, 1969.
Sides, G. D., T. O. Tiernan, and R. J. Hanrahan, *J. Chem. Phys.*, 65, 1966, 1976.
Stammerich, H., *Phys. Rev.*, 78, 79, 1950.
Tam, W.-C. and S. F. Wong, *J. Chem. Phys.*, 68, 5626, 1978.
Trainor, D. W. and M. J. M. Boness, *Appl. Phys. Lett.*, 32, 604, 1978.
Truby, F. K., *Phys. Rev. A*, 4, 613, 1971.
Zittel, P. F. and D. D. Little, *J. Chem. Phys.*,71, 723, 1979.

11

CARBON MONOXIDE

CONTENTS

Carbon monoxide (CO) is a weakly polar, moderately electron-attaching molecule that has 14 electrons. The electronic polarizability is 2.17×10^{-40} F m^2, the dipole moment is 0.11 D, and the ionization potential is 14.014 eV.

11.1 SELECTED REFERENCES FOR DATA

See Table 11.1.

TABLE 11.1
Selected References for Data on CO

Quantity	Range: eV, (Td)	Reference
Excitation cross section	10–15	Poparić et al. (2001)
Ionization cross section	15–1000	Mangan et al. (2000)
Attachment process	0–16	Denifl et al. (1998)
Elastic cross section	300–1300	Maji et al. (1998)
Ionization cross section	17.5–600	Tian and Vidal (1998)
Elastic scattering	1.0–30	Gibson et al. (1996)
Rotational cross section	0.002–0.160	Randell et al. (1996)
Total cross section	0.4–250	Szmytkowski et al. (1996)
Excitation cross section	0–3.7	Zobel et al. (1996)
Rotational cross section	10–200	Gote and Ehrhardt (1995)
Excitation cross section	0–3.7	Zobel et al. (1995)
Total cross section	400–2600	Xing et al. (1995)
Excitation cross section	13.5–198.5	Cosby (1993)

continued

TABLE 11.1 (continued)
Selected References for Data on CO

Quantity	Range: eV, (Td)	Reference
Total cross section	**80–4000**	**Karwasz et al. (1993)**
Excitation cross section	6–18	Morgan and Tennyson (1993)
Total cross section	10–5000	Jain and Norcross (1992)
Total cross section	**5–300**	**Kanik et al. (1992)**
Ionization cross section	**16–200**	**Freund et al. (1990)**
Total cross section	**380–5200**	**Garcia et al. (1990)**
Total cross section	150–1000	Garcia et al. (1990)
Excitation cross section	**10–60**	**Mason and Newell (1988)**
Elastic scattering	**20–100**	**Nickel et al. (1988)**
Transport coefficients	**(0.3–300)**	**Nakamura (1987)**
Ionization cross section	**20–1000**	**Orient and Srivastava (1987)**
Total cross section	**0.5–4.90**	**Buckman and Lohman (1986)**
Transport coefficients	**(50–6000)**	**Al-Amin et al. (1985)**
Total cross section	0.1–5.0	Jain and Norcross (1985)
Vibrational cross section	**0.37–1.26**	**Sohn et al. (1985)**
Drift velocity	**(50–150)**	**Roznerski et al. (1984)**
Total cross section	**1.0–500**	**Kwan et al. (1983)**
Vibrational excitation	**3–100**	**Chutjian and Tanaka (1980)**
Total cross section	0.005–10	Chandra (1977)
Swarm parameters	**(30–3000)**	Saelee and Lucas (1977)
Drift velocity	**(0.04–25)**	**Huxley et al. (1974)**
Swarm parameters	**(4–1500)**	**Lakshminarasimha et al. (1974)**
Diffusion coefficients	**(0.1–95)**	**Lowke and Parker (1969)**
Diffusion coefficients	**(0.5–12)**	**Wagner et al. (1967)**
Drift velocity	**0.01–1000**	**Hake and Phelps (1967)**
Ionization cross section	**14.5–1000**	**Rapp and Englander-Golden (1965)**
Diffusion coefficient	**(0.001–5.00)**	**Warren and Parker (1962)**
Ionization coefficients	**(108–600)**	**Bhalla and Craggs (1961)**
Total cross section	**1.0–5.0**	**Bruche (1927)**

Note: Bold font indicates experimental study.

11.2 TOTAL SCATTERING CROSS SECTION

Table 11.2 shows the recommended cross section for CO. Figures 11.1 and 11.2 show the data available in the literature. The highlights of the cross section are

1. A trend toward lower cross section as the energy is decreased to zero from about 1 eV. The very low dipole moment and the weak attachment properties do not seem to influence the cross section in this energy range.
2. A shape resonance in the energy range 1.80–1.93 eV (Raju, 2005). This characteristic appears common to many gases.
3. A broad peak in the 10–30 eV range, attributed to resonance due to inelastic processes. This feature is also common to several gases.
4. A monotonic decrease in the high energy range 50–4000 eV. This feature is also common to several gases.

11.3 DIFFERENTIAL SCATTERING CROSS SECTION

Figure 11.3 shows the differential cross sections at two angles, 20° and 60° (Gibson et al., 1996), at low electron energy. The peak in the cross section at resonance is clearly discernible.

TABLE 11.2
Recommended Total Scattering Cross Section for CO

Energy (eV)	Q_T (10^{-20} m^2)	Energy (eV)	Q_T (10^{-20} m^2)	Energy (eV)	Q_T (10^{-20} m^2)
0.500	10.69	2.10	42.30	60	11.19
0.533	11.03	2.20	40.38	70	10.62
0.575	11.45	2.30	38.24	80	10.11
0.625	11.73	2.40	36.02	90	9.69
0.670	11.96	2.50	33.56	100	9.29
0.717	12.19	2.60	31.45	125	8.41
0.763	12.46	2.70	29.01	150	7.65
0.805	12.74	2.80	27.07	200	6.60
0.854	13.06	2.90	25.62	250	5.63
0.905	13.44	3.15	22.07	300	4.98
0.945	13.77	3.40	19.62	350	4.50
0.995	14.19	3.90	16.90	400	4.14
1.064	14.90	4.40	15.12	500	3.56
1.158	16.45	4.90	14.44	600	3.16
1.253	17.89	5.0	14.73	700	2.80
1.346	19.78	6.0	13.63	800	2.52
1.405	24.02	7.0	13.61	900	2.29
1.50	29.25	8.0	13.88	1000	2.10
1.60	34.54	9.0	13.62	1100	1.91
1.70	37.95	10	13.27	1250	1.71
1.80	41.61	11	13.18	1500	1.46
1.82	41.94	12	13.34	1750	1.26
1.84	42.32	14	13.75	2000	1.11
1.86	42.73	16	14.07	2250	0.992
1.88	43.02	18	14.26	2500	0.890
1.90	43.52	20	14.46	2750	0.819
1.92	43.03	25	14.06	3000	0.755
1.94	43.44	30	13.54	3250	0.706
1.96	43.17	40	12.67	3500	0.654
1.98	43.24	50	11.90	4000	0.565
2.00	43.19				

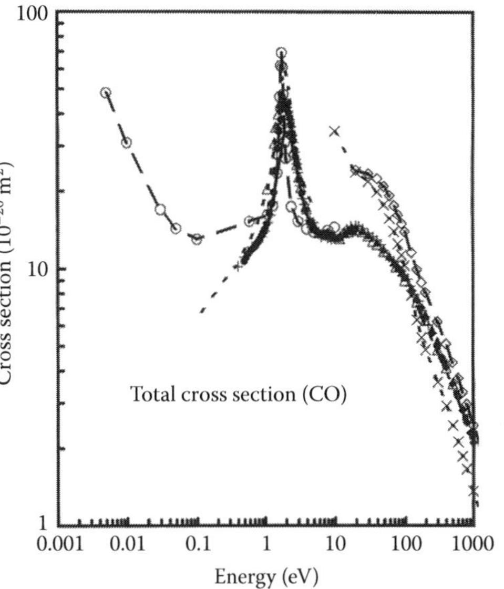

FIGURE 11.1 Total scattering cross section for CO. (+) Szmytkowski et al. (1996); (◆) Karwasz et al. (1993); (—×—) Jain and Norcross (1992), theory, without anisotropic term; (—◆—) Jain and Norcross (1992), with anisotropic term; (▲) Kanik et al. (1992); (■) Garcia et al. (1990); (—) Garcia et al. (1990), theory; (●) Buckman and Lohmann (1986); (— —) Jain and Norcross (1985), theory; (△) Kwan et al. (1983); (—) Bruche (1927); (—○—) Chandra (1977), theory.

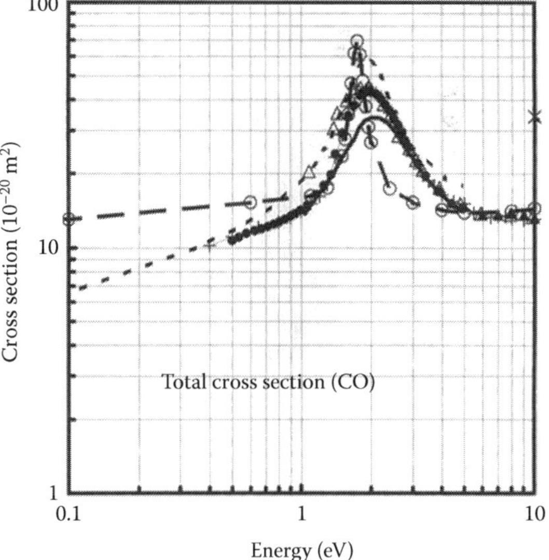

FIGURE 11.2 Details of total cross section for CO in the low energy range. (+) Szmytkowski et al. (1996); (◆) Karwasz et al. (1993); (—×—) Jain and Norcross (1992), theory, without anisotropic term; (—◆—) Jain and Norcross (1992), with anisotropic term; (▲) Kanik et al. (1992); (■) Garcia et al. (1990); (—) Garcia et al. (1990), theory; (●) Buckman and Lohmann (1986); (— —) Jain and Norcross (1985), theory; (△) Kwan et al. (1983); (—) Bruche (1927); (—○—) Chandra (1977), theory.

CO

11.4 ELASTIC SCATTERING CROSS SECTION

See Table 11.3.

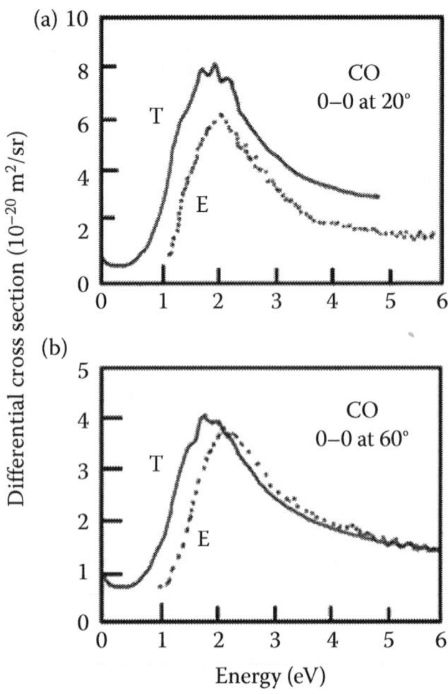

FIGURE 11.3 Differential cross section for elastic scattering for electrons in CO at (a) 20° and (b) 60° angles. (. . . . .) and (●) experimental; (–––) theory. (Adapted from Gibson, J. C. et al., *J. Phys, B: At. Mol. Opt. Phys.*, 29, 3197, 1996.)

TABLE 11.3
Elastic Scattering Cross Sections for CO

Kanik et al. (1992)				Gibson et al. (1996)	
Energy (eV)	Q_{el} (10^{-20} m²)	Energy (eV)	Q_{el} (10^{-20} m²)	Energy (eV)	Q_{el} (10^{-20} m²)
1.0	13.56	60	5.50	1.0	15.4
2.0	37.60	70	5.10	1.3	20.4
3.0	23.20	80	4.70	1.5	26.8
4.0	16.84	90	4.40	1.9	35.8
5.0	14.90	100	4.20	2.5	30.2
6.0	13.00	150	3.30	3.0	22.7
7.0	12.20	200	2.80	5.0	14
8.0	12.00	300	2.21	6.0	12.9
9.0	11.70	400	1.90	7.5	12
10	11.50	500	1.65	9.9	11.4
12	11.20	600	1.50	20	11
15	11.00	700	1.38	30	10.3
20	9.40	800	1.28		
30	7.80	900	1.20		
40	6.80	1000	1.10		
50	6.00				

Note: See Figure 11.4 for graphical presentation.

11.5 MOMENTUM TRANSFER CROSS SECTION

See Table 11.4.

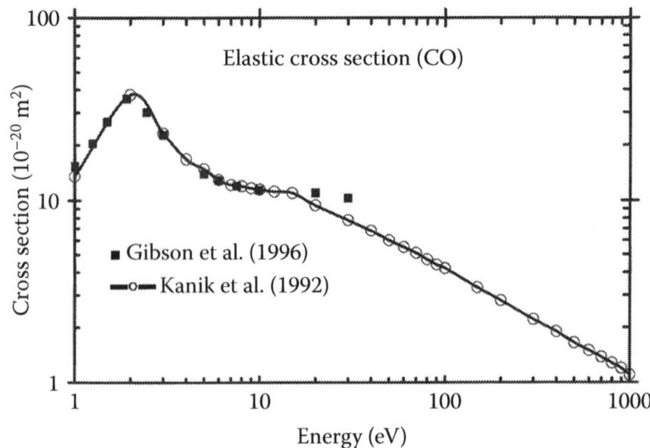

FIGURE 11.4 Elastic scattering cross section for CO. (■) Gibson et al. (1996); (—○—) Kanik et al. (1992).

TABLE 11.4
Momentum Transfer Cross Sections for CO

Gibson et al. (1996)		Land (1978)			
Energy (eV)	Q_M (10^{-20} m²)	Energy (eV)	Q_M (10^{-20} m²)	Energy (eV)	Q_M (10^{-20} m²)
1.00	19.1	0.000	60	1.5	42
1.25	25.4	0.001	40	1.7	40
1.50	27.8	0.002	25	1.9	32
1.91	30.1	0.003	17.7	2.1	23.5
2.45	21.4	0.005	12.3	2.2	21.5
3.00	15.8	0.007	9.8	2.5	17.5
5.00	10.3	0.0085	8.6	2.8	16.0
6.00	9.8	0.010	7.8	3.0	15.4
7.50	8.9	0.015	6.5	3.3	14.6
9.90	8.3	0.02	5.9	3.6	14.2
20.00	6.6	0.03	5.4	4.0	13.8
30.00	5.6	0.04	5.2	4.5	13.3
		0.05	5.4	5	12.9
		0.07	6.1	6	12.3
		0.10	7.3	7	11.8
		0.15	8.8	8	11.3
		0.20	10	10	10.6
		0.25	11.2	12	10.4
		0.30	12.1	15	10.2
		0.35	13	17	10.1
		0.40	13.8	20	9.8
		0.5	15.4	25	9.1
		0.7	16.5	30	8.6
		1	18.5	50	7.1
		1.2	28	75	6.1
		1.3	37	100	5.5

Note: See Figure 11.5 for graphical presentation.

11.6 RO-VIBRATIONAL CROSS SECTION

Rotational energy for CO molecule is quite low at 0.24 meV and vibrational excitation energy is 0.266 eV (Hake and Phelps, 1967). Figure 11.6 shows rotational excitation cross section at low energies.

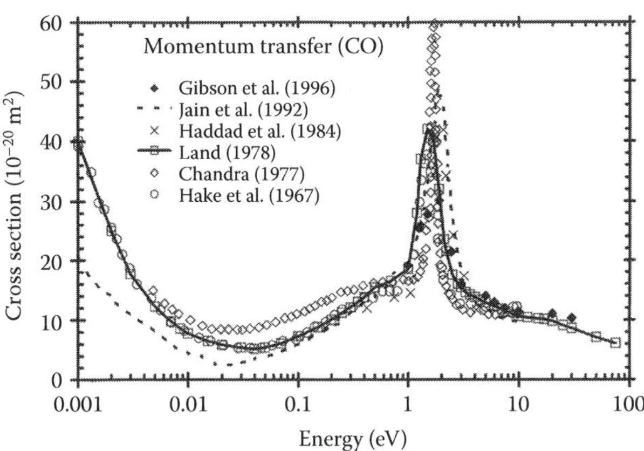

FIGURE 11.5 Momentum transfer cross section for CO. (◆) Gibson et al. (1996); (– – –) Jain and Norcross (1992); (×) Haddad and Milloy (1984); (—□—) Land[46] (1978); (◊) Chandra (1977); (○) Hake and Phelps (1967).

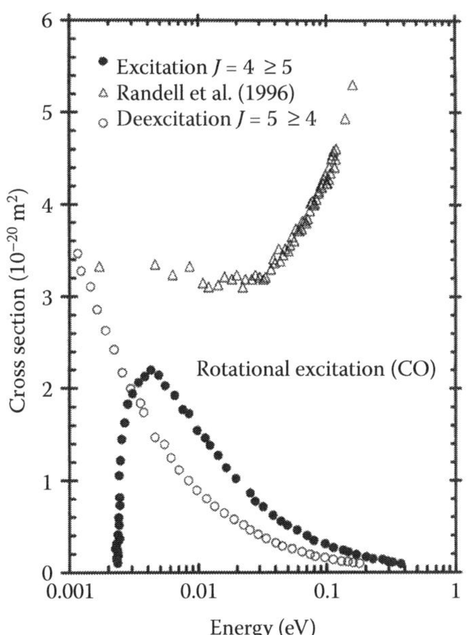

FIGURE 11.6 Rotational excitation cross section for CO. (△) Randell et al. (1996); (●) excitation, $J = 4 \rightarrow 5$; (○) de-excitation, $J = 5 \rightarrow 4$, Hake and Phelps (1967).

Tables 11.5 and 11.6 show selected differential cross section for rotational excitation at higher energies. Cross sections integrated by the present author are shown in Table 11.7 and Figure 11.7.

TABLE 11.5
Differential Cross Section for Rotational Excitation

	Energy (eV)				
Angle (°)	5	10	25	50	75
10	1.808	2.542	6.915	9.718	8.649
20	2.045	2.117	4.247	4.195	3.064
30	2.133	1.787	2.171	1.444	0.995
40	2.057	1.464	1.136	0.545	0.303
50	1.927	1.180	0.649	0.244	0.141
60	1.693	0.969	0.396	0.150	0.092
70	1.417	0.674	0.264	0.096	0.068
80	1.145	0.488	0.188	0.067	0.055
90	0.930	0.384	0.162	0.053	0.050
100	0.809	0.391	0.183	0.058	0.053
110	0.708	0.459	0.222	0.094	0.075
120	0.683	0.550	0.302	0.149	0.104
130	0.705	0.619	0.390	0.243	0.135
140	0.750	0.712	0.496	0.321	0.167
150	0.851	0.791	0.634	0.425	0.208
160	0.955	0.806	0.751	0.529	0.240

Source: Adapted from Gote, M. and H. Ehrhardt, *J. Phys. B: At. Mol.Opt. Phys.*, 28, 3957, 1995.

TABLE 11.6
Differential Cross Section for Rotational Excitation

	Energy (eV)				
Angle (°)	100	125	150	175	200
10	7.849	7.200	6.303	6.361	5.646
20	2.389	2.360	2.781	2.893	2.430
30	0.721	0.645	0.807	0.885	0.662
40	0.202	0.212	0.262	0.318	0.241
50	0.093	0.130	0.129	0.170	0.141
60	0.095	0.082	0.090	0.117	0.089
70	0.050	0.060	0.070	0.087	0.059
80	0.048	0.049	0.052	0.066	0.041
90	0.055	0.450	0.041	0.054	0.033
100	0.053	0.044	0.038	0.052	0.031
110	0.061	0.045	0.040	0.056	0.032
120	0.071	0.052	0.046	0.060	0.035
130	0.091	0.061	0.053	0.062	0.036
140	0.106	0.074	0.058	0.062	0.037
150	0.129	0.094	0.062	0.062	0.041
160	0.160	0.104	0.075	0.069	0.046

Table 11.8 and Figure 11.8 show the vibrational excitation cross sections.

11.7 ELECTRONIC CONFIGURATION

Figure 11.9 shows the potential energy of electronic states and Table 11.9 lists the important properties (Zobel et al., 1995).

TABLE 11.7
Integral Cross Sections for Rotational Excitation and Momentum Transfer

Energy (eV)	Q_{rot} (10^{-20} m^2)	$Q_{m(rot)}$ (10^{-20} m^2)
5	11.52	14.96
10	10.15	7.69
25	8.52	4.73
50	6.80	2.62
75	5.44	1.53
100	4.72	1.18
125	4.54	1.26
150	3.69	0.88
175	3.95	1.02
200	3.18	0.63

Source: Adapted from Raju, G. G. unpublished data, 2008. These cross sections were obtained by integration of the differential cross section of Gote and Ehrhatdt (1995).

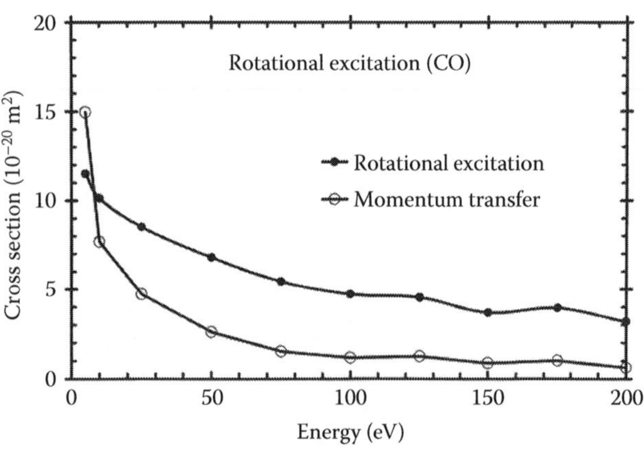

FIGURE 11.7 Integral cross sections for rotational excitation and associated momentum transfer for CO. (Adapted from From Raju G. G., unpublished data, 2008. These cross sections were obtained by integration of the differential cross section of Gote and Ehrhatdt (1995).)

11.8 ELECTRONIC EXCITATION

Details of the electronic excitation cross sections are discussed by Raju (2005). Table 11.10 and Figure 11.10 show the recommended cross section (Raju, 2005) and selected previous results.

TABLE 11.8
Vibrational Excitation Cross Section for CO

Land (1978)				Chutjian et al. (1980)	
Energy (eV)	Q_V (10^{-20} m^2)	Energy (eV)	Q_V (10^{-20} m^2)	Energy (eV)	Q_V (10^{-20} m^2)
0.266	0.0	1.514	5.377	3	0.365
0.290	0.095	1.645	5.054	5	0.104
0.320	0.125	1.740	5.934	9	0.024
0.350	0.144	1.821	6.578	20	0.097
0.400	0.156	1.902	5.843	30	0.021
0.500	0.159	1.982	5.216	50	0.007
0.600	0.157	2.086	5.683	75	0.005
0.700	0.154	2.170	4.965	100	0.007
0.800	0.165	2.281	4.176		
0.850	0.224	2.316	4.285		
0.900	0.300	2.404	3.747		
0.950	0.397	2.508	3.120		
1.000	0.513	2.688	2.455		
1.031	0.604	2.872	1.828		
1.130	0.924	3.072	1.290		
1.215	1.350	3.294	0.861		
1.307	2.137	3.528	0.555		
1.410	3.602	3.816	0.287		

Note: See Figure 11.8 for graphical presentation.

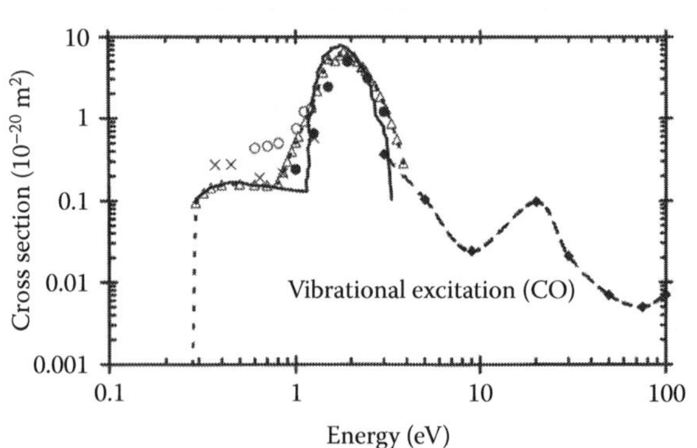

FIGURE 11.8 Vibrational excitation cross section for CO. (●) Gibson et al. (1996); (×) Sohn et al. (1985); (—◆—) Chutjian and Tanaka (1980); (--△--) Land (1978); (—) Hake and Phelps (1967); (○) Schulz (1959).

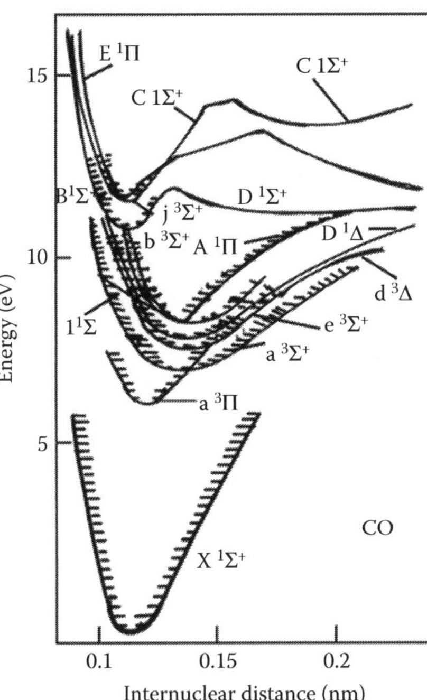

FIGURE 11.9 Potential energy diagrams of states of CO. Threshold energies are tabulated in Table 4.4. (Adapted from Zobel, J. et al., *J. Phys. B: At. Mol. Opt. Phys.*, 29, 813, 1996.)

TABLE 11.10
Electronic Excitation Cross Section for CO

	Q_{ex} (10^{-22} m²)		
Energy (eV)	**Recommended**	**Kanik et al. (1992)**	**Sawada (1972)**
7.0	5.75	0.00	11.5
8.0	79.6	101	58
9.0	137	143	131
10.0	162	143	181
12	194	187	201
15	239	288	190
20	291	415	168
30	300	452	148
40	284	430	138
50	243	360	127
60	226	331	121
70	200	284	116
80	185	261	109
90	184	263	106
100	157	214	100
150	138	192	83
200	108	142	74
300	91.9	119	65
400	76.8	96	58
500	60.3	72	49
600	57.5	71	44
700	54.7	69	40
800	44.5	53	36
900	40.6	48	33
1000	39.5	48	31

CO

TABLE 11.9
Electronic Configuration and the Important Properties of the Vibrational Level ($V = 0$) of the Electronic States of CO

	Electronic Structure								
	Valence Orbitals			Rydberg Orbitals					
Electronic State	**1π**	**5σ**	**2π**	**6σ**	**3σ**	**3pπ**	**3pπ**	**Energy (eV)**	r_e **(10^{-10} m)**
X 1Σ⁺	4	2						0	1.128
a ³Π	4	1	1					6.006 (6.034)	1.206
a′ ³Σ⁺	3	2	1					6.863 (6.928)	1.352
D ³Δ	3	2	1					7.513	1.369
e ³Σ⁻	3	2	1					7.898	1.384
A ¹Π	4	1	1					8.024 (8.065)	1.235
I ¹Σ⁻	3	2	1	—	—	—	—	8.137	1.391
D ¹Δ	3	2	1	—	—	—	—	8.241	1.399
b ³Σ⁺	4	1	—	—	1	—	—	10.399 (10.387)	1.113
B ¹Σ⁺	4	1	—	—	1	—	—	10.777 (10.776)	1.119
D′ ¹Σ⁺	3	2	1	—	—	—	—	10.995	
j ³Σ⁺	4	1	—	—	—	1	—	11.269	1.144
C ¹Σ⁺	4	1	—	—	—	1	—	11.396 (11.393)	1.122
c ³Π	4	1	—	—	—	—	1	11.414	1.127
E ¹Π	4	1	—	—	—	—	1	11.524	1.115

Note: Core orbitals are 1σ², 2σ², 3σ², and 4σ². Energy values given in parentheses are spectroscopic values from Herzberg (1957). r_e = internuclear distance at equilibrium (Zobel et al., 1996).

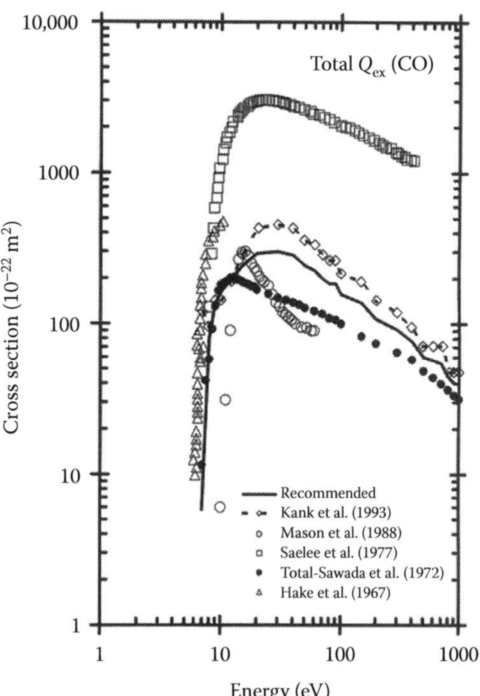

FIGURE 11.10 Total excitation cross section for CO. (—) Recommended, Raju (2005); (—◇—) Kanik et al. (1992); (○) Mason and Newell (1988); (□) Saelee and Lucas (1977); (●) Sawada et al. (1972); (◇) Hake and Phelps (1967). (With kind permission of Institute of Physics, England.)

TABLE 11.11
Total Ionization Cross Sections for CO

Mangan et al. (2000)		Rapp and Englander-Golden (1965)		Recommended (Raju, 2005)	
Energy (eV)	Q_i (10^{-20} m^2)	Energy (eV)	Q_i (10^{-20} m^2)	Energy (eV)	Q_i (10^{-20} m^2)
		14.5	0.027		
15	0.06	15	0.051	15	0.051
20	0.44	20	0.428	20	0.428
25	0.76	25	0.85	30	1.24
30	1.21	30	1.24	40	1.78
35	1.54	36	1.60	50	2.12
40	1.79	40	1.79	60	2.34
50	2.13	50	2.12	70	2.5
60	2.35	60	2.34	80	2.58
70	2.53	70	2.50	90	2.63
80	2.59	80	2.56	100	2.65
90	2.64	90	2.63	200	2.36
100	2.65	100	2.65	300	1.99
125	2.61	125	2.63	400	1.72
150	2.52	150	2.57	500	1.49
200	2.30	200	2.37	600	1.35
250	2.09	250	2.16	700	1.21
300	1.90	300	1.99	800	1.11
400	1.61	400	1.72	900	1.03
500	1.39	500	1.49_5	1000	1.01
700	1.12	600	1.35		
1000	0.87	700	1.21		
		800	1.11		
		900	1.03		
		1000	0.96		

Source: Adapted from Mangan, M. A. et al., *J. Phys. B: At. Mol. Opt. Phys.*, 33, 3225, 2000.

Note: See Figure 11.11 for graphical presentation.

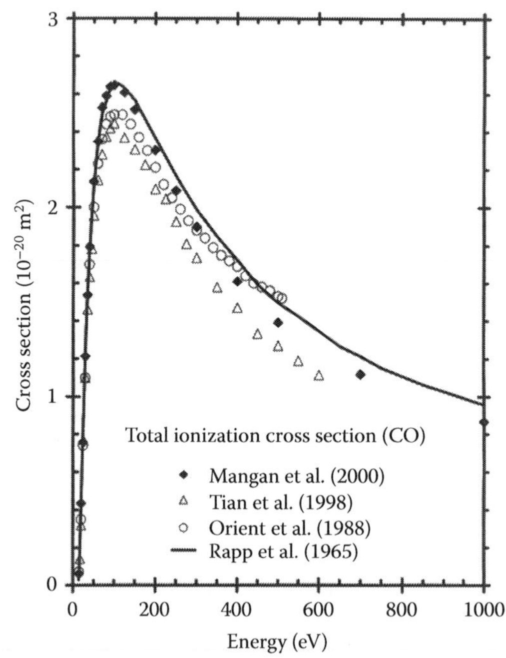

FIGURE 11.11 Total ionization cross sections for CO. (◆) Mangan et al. (2000); (△) Tian and Vidal (1998); (○) Orient and Srivastava (1987); (—) Rapp and Englander-Golden (1965).

11.9 IONIZATION CROSS SECTION

Table 11.11 and Figure 11.11 show the total ionization cross sections for CO. A comparative analysis is presented in Raju (2005). Figure 11.12 shows the partial ionization cross sections.

11.10 ATTACHMENT CROSS SECTION

Attachment of electrons occurs according to Raju (2005):

$$CO + e \rightarrow C + O^- \quad (11.1)$$

$$CO + e \rightarrow C^+ + O^- + e \quad (11.2)$$

$$CO + e \rightarrow C^- + O^+ + e \quad (11.3)$$

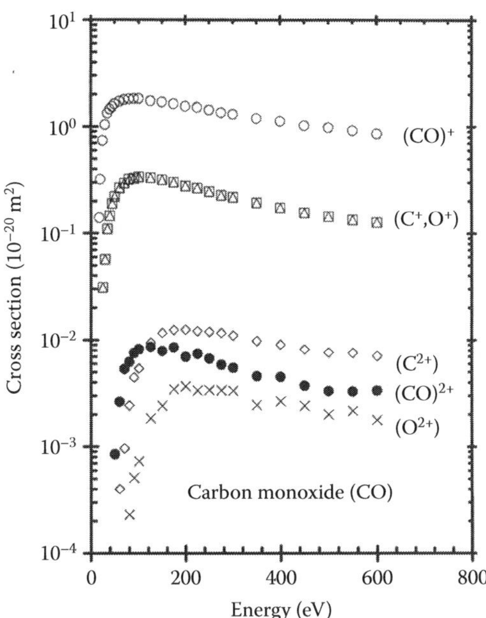

FIGURE 11.12 Partial ionization cross sections for CO. (○) CO^+; (□) C^+; (◇) O^+; (◊) C^{2+}; (●) CO^{2+}; (×) O^{2+}. (Adapted from Tian, C. and C. R. Vidal, *J. Phys. B: At. Mol. Opt. Phys.*, 31, 895, 1998.)

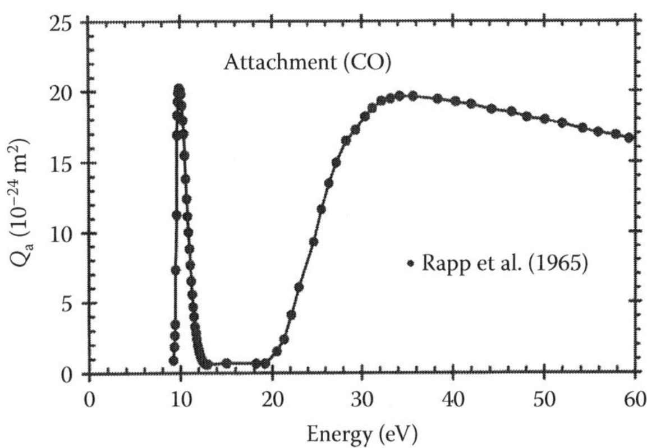

FIGURE 11.13 Attachment cross sections for CO. (Adapted from Rapp, D. and D. D. Briglia, *J. Chem. Phys.*, 43, 1480, 1965.)

TABLE 11.12
Attachment Cross Sections for CO

Energy (eV)	Q_a (10^{-24} m^2)	Energy (eV)	Q_a (10^{-24} m^2)	Energy (eV)	Q_a (10^{-24} m^2)
9.20	0.88	11.2	6.51	24.7	9.30
9.30	1.85	11.3	5.54	25.6	11.61
9.35	2.64	11.4	4.66	26.4	13.49
9.40	3.43	11.5	3.96	27.2	14.95
9.45	7.30	11.6	3.25	28.3	16.48
9.60	11.26	11.7	2.81	29.4	17.25
9.65	16.89	11.8	2.37	30.4	18.18
9.70	18.30	11.9	1.94	31.2	18.78
9.75	19.26	12.0	1.67	32.2	19.29
9.80	19.88	12.1	1.41	33.2	19.45
9.85	19.97	12.2	1.14	34.2	19.62
9.90	20.23	12.3	0.97	35.7	19.61
10.0	20.14	12.4	0.88	38.3	19.42
10.1	19.79	12.5	0.79	40.3	19.23
10.2	19.00	12.6	0.70	42.0	19.05
10.3	17.94	12.8	0.62	44.2	18.69
10.4	16.98	13.0	0.62	46.4	18.50
10.5	15.48	15.0	0.70	48.1	18.14
10.6	13.81	18.2	0.68	50.1	17.96
10.7	12.40	19.2	0.67	52.0	17.69
10.8	11.17	20.5	1.52	54.2	17.33
10.9	10.03	21.3	2.37	55.9	17.06
11.0	8.80	22.1	4.08	57.9	16.87
11.1	7.65	23.0	6.05	59.3	16.61

Source: Digitized and interpolated from Rapp, D. and D. D. Briglia, *J. Chem. Phys.*, 43, 1480, 1965.

Note: See Figure 11.13 for graphical presentation.

TABLE 11.13
Recommended Drift Velocity of Electrons for CO

E/N (Td)	*W* (10^3 m/s)	*E/N* (Td)	*W* (10^3 m/s)	*E/N* (Td)	*W* (10^3 m/s)
0.04	0.744	0.8	7.522	40	42.61
0.05	1.003	1	8.620	50	46.68
0.06	1.280	2	12.301	60	54.56
0.07	1.520	3	14.778	70	64.81
0.08	1.693	4	16.778	80	65.74
0.09	1.873	5	18.329	90	74.73
0.1	2.125	6	19.265	100	87.22
0.2	4.026	8	19.977	150	139.42
0.3	5.039	10	20.164	200	211.86
0.4	5.572	20	26.134	250	240.60
0.5	5.693	25	31.97	300	303.27
0.6	6.035	30	34.59	400	436.62
0.7	6.725	35	38.77	500	583.57

Note: See Figure 11.14 for graphical presentation.

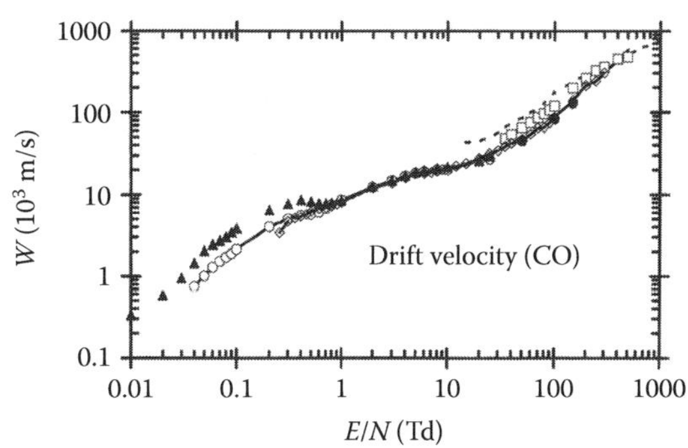

FIGURE 11.14 Drift velocity of electrons for CO. Temperature 293 K unless otherwise mentioned. (—) Recommended, Raju (2005); (◊) Nakamura (1987); (●) Roznerski and Leja (1984); (□) Saelee and Lucas (1977), theory; (— —) Saelee and Lucas (1977), experimental; (◊) Huxley et al. (1974); (▲) Huxley et al. (1974), 77 K.

CO

$$CO + e \rightarrow (CO^-)^* \quad (11.4)$$

Reaction 11.1 is dissociative attachment with an appearance potential of 9.65 eV. Reactions 11.2 and 11.3 are ion pair formation at 21 and 23 eV, respectively. Reaction 11.4 is the excited state of the negative ion formed by direct attachment. The electron affinity of O^- in CO is 1.6 eV.

See Table 11.12 and Figure 11.13 for attachment cross section.

TABLE 11.14
Recommended Ratio D_r/μ for CO

E/N (Td)	D_r/μ (V)	E/N (Td)	D_r/μ (V)	E/N (Td)	D_r/μ (V)
0.01	0.007	2	0.251	300	5.42
0.02	0.008	3	0.332	400	6.10
0.03	0.009	4	0.387	500	6.54
0.04	0.009	5	0.443	600	6.87
0.05	0.009	7	0.611	700	7.23
0.07	0.01	10	0.881	800	7.72
0.1	0.013	20	1.48	900	8.25
0.2	0.027	30	2.19	1000	8.69
0.3	0.062	40	2.58	2000	10.52
0.4	0.075	50	2.78	3000	12.67
0.5	0.101	70	3.14	4000	13.82
0.7	0.133	100	3.82	5000	14.78
1	0.169	200	4.75	6000	16.55

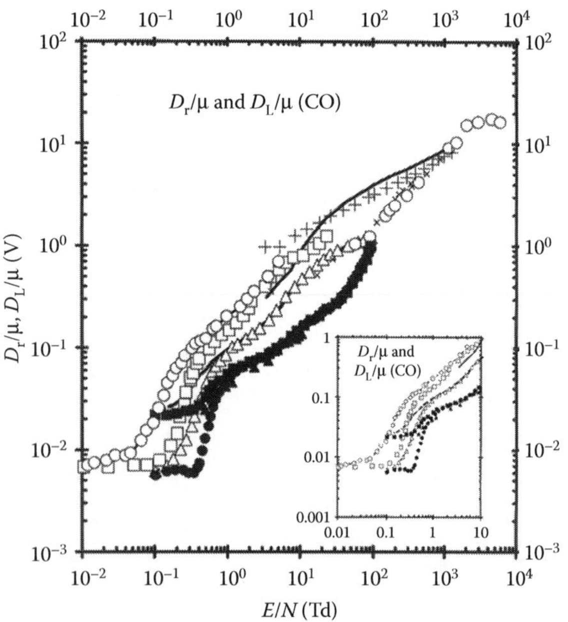

FIGURE 11.15 Radial and longitudinal diffusion coefficients of electrons in CO. Unless otherwise mentioned the temperature is 293 K. Open symbols for radial and closed symbols for longitudinal diffusion. The letters E and T mean experimental and theoretical, respectively. Radial diffusion: (○) Al-Amin (1985), E; Lakshminarasimha et al. (1974), E; (— —) Lowke and Parker (1969), T; (—△—) Lowke and Parker (1969), 77 K, T; (□) Hake and Phelps (1967), T; (—○—) Warren and Parker (1962) 77 K, E; Longitudinal diffusion: (—) Saelee and Lucas (1977), T. (■) Lowke and Parker (1969), T; (●) Lowke and Parker (1969), 77 K, T; (▲) Wagner et al. (1967), E. The inset shows details at low *E/N*.

11.11 DRIFT VELOCITY OF ELECTRONS

Table 11.13 shows the recommended drift velocity (Raju, 2005). Figure 11.14 shows selected data.

11.12 DIFFUSION COEFFICIENT OF ELECTRONS

Recommended radial diffusion coefficients expressed as the ratio D_r/μ (μ = mobility) at 293 K are shown in Table 11.14. Figure 11.15 shows selected data at 293 and 77 K for radial and longitudinal diffusion.

11.13 REDUCED IONIZATION COEFFICIENTS

See Table 11.15.

TABLE 11.15
Reduced Ionization Coefficients for CO

E/N (Td)	α/N (10^{-20} m²)	E/N (Td)	α/N (10^{-20} m²)	E/N (Td)	α/N (10^{-20} m²)
125	0.0023	400	0.260	900	1.32
150	0.0069	450	0.376	1000	1.48
175	0.016	500	0.489	1100	1.64
200	0.022	550	0.597	1200	1.81
250	0.054	600	0.701	1300	1.96
300	0.101	700	0.906	1400	2.08
350	0.172	800	1.12	1500	2.16

Note: See Figure 11.16 for graphical presentation.

TABLE 11.16
Attachment Coefficients for CO

E/N (Td)	η/N (10^{-20} m²) $N = 8.05 \times 10^{23}$ m⁻³
120	4.98×10^{-3}
126	5.60×10^{-3}
132	6.22×10^{-3}
138	8.71×10^{-3}
144	1.12×10^{-2}
150	1.31×10^{-2}
156	1.55×10^{-2}
162	1.74×10^{-2}
168	1.87×10^{-2}
	$N = 8.05 \times 10^{23}$ m⁻³

Source: Adapted from Bhalla, M. S. and J. D. Craggs, *Proc. Phys. Soc.*, London, 78, 438, 1961.

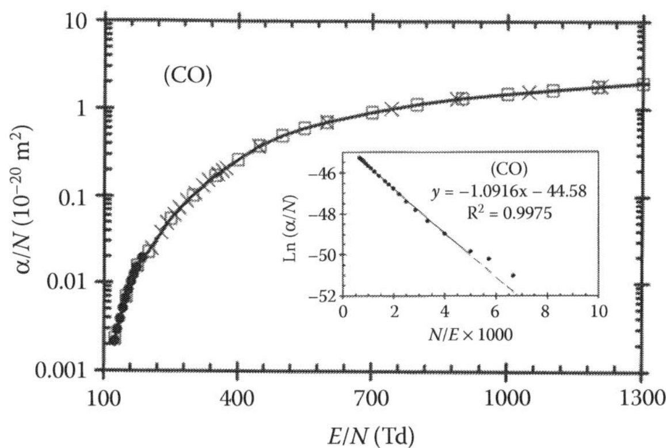

FIGURE 11.16 Density-reduced ionization coefficients for CO. (—) Recommended, Raju (2005); (×) Lakshminarasimha et al. (1974); (●) Bhalla and Craggs (1961). The inset shows the plot according to the equation.

11.14 ATTACHMENT COEFFICIENTS

See Table 11.16.

11.15 GAS CONSTANTS

Gas constants evaluated according to the equation

$$\frac{\alpha}{N} = F\exp\left(-\frac{GN}{E}\right) \quad (11.5)$$

are $F = 4.36 \times 10^{-20}$ m²; $G = 1092$ Td for the range $200 \leq E/N \leq 1500$ Td (see inset of Figure 11.16).

REFERENCES

Al-Amin, S. A. J., J. Lucas, and H. N. Kücükarpaci, *J. Phys. D: Appl. Phys.*, 18, 2007, 1985.

Bhalla, M. S. and J. D. Craggs, *Proc. Phys. Soc. London*, 78, 438, 1961.

Brüche, E., *Ann. Phys.*, 83, 1065, 1927.

Buckman, S. J. and B. Lohman, *Phys. Rev. A*, 34, 1561, 1986.

Chandra, N., *Phys. Rev. A*, 16, 80, 1977.

Chutjian, A. and H. Tanaka, *J. Phys. B: At. Mol. Phys.*, 13, 1901, 1980.

Cosby, P. C., *J. Chem. Phys.*, 98, 7804, 1993.

Denifl, G., D. Muigg, A. Stamatovic, and T. D. Märk, *Chem. Phys. Lett.* 288, 105, 1998.

Freund, R. S., R. C. Wetzel, and R. J. Shul, *Phys. Rev. A*, 41, 5861, 1990.

Garcia, G., C. Aragon, and J. Campos, *Phys. Rev. A*, 42, 4400, 1990.

Gibson, J. C., L. A. Morgan, R. J. Gulley, M. J. Brunger, C. T. Bundschu, and S. J. Buckman, *J. Phys, B: At. Mol. Opt. Phys.*, 29, 3197, 1996.

Gote, M. and H. Ehrhardt, *J. Phys. B: At. Mol. Opt. Phys.*, 28, 3957, 1995.

Haddad, G. N. and H. B. Milloy, *Austr. J. Phys.*, 36, 473, 1984.

Hake, R. D. and A. V. Phelps, *Phys. Rev.*, 158, 70, 1967.

Herzberg, G., *Molecular Spectra and Molecular Structure*, Van Nostrand, Princeton, NJ, 1957.

Huxley, L. G. H. and R. W. Crompton, *The Diffusion and Drift of Electrons in Gases*, John Wiley and Sons, New York, NY, 1974.

Jain, A. and D. W. Norcross, in Abstracts, *Proceedings of the XIV International Conference on the Physics of Electronic Atomic Collisions*, Palo Alto, CA, 1985, M. J. Coggiola, D. L. Huestis, and R. P. Saxon, (Eds.) p. 214.

Jain, A. and D. W. Norcross, *Phys. Rev. A*, 45, 1644, 1992.

Kanik, I., J. C. Nickel, and S. Trajmar, *J. Phys. B: At. Mol. Opt. Phys.*, 25, 2189, 1992.

Karwasz, G., R. S. Brusa, A. Gasparoli, and A. Zecca, *Chem. Phys. Lett.*, 211, 529, 1993.

Kwan, Ch. K., Y.-F. Hsieh, W. E. Kauppila, S. J. Smith, T. S. Stein, M. N. Uddin, and M. S. Dababneh, *Phys. Rev. A*, 27, 1328, 1983.

Lakshminarasimha, C. S., J. Lucas, and N. Kontoleon, *J. Phys. D: Appl. Phys.*, 7, 2545, 1974.

Land, J. E., *J. Appl. Phys.*, 49, 5716, 1978.

Lowke, J. J. and J. H. Parker (Jr.), *Phys. Rev.*, 181, 302, 1969.

Maji, S., G. Basavaraju, S. M. Bharathi, K. G. Bhushan, and S. P. Khare, *J. Phys. B: At. Mol. Opt. Phys.*, 31, 4975, 1998.

Mangan, M. A., B. G. Lindsay, and R. F. Stebbings, *J. Phys. B: At. Mol. Opt. Phys.*, 33, 3225, 2000.

Mason, N. J. and W. R. Newell, *J. Phys. B: At. Mol. Opt. Phys.*, 21, 1293, 1988.

Morgan, L. A. and J. Tennyson, *J. Phys. B: At. Mol. Opt. Phys.*, 26, 2429, 1993.

Nakamura, Y., *J. Phys. D: Appl. Phys.*, 20, 933, 1987.

Nickel, J. C., C. Mott, I. Kanik, and D. C. McCollum, *J. Phys. B: At. Mol. Opt. Phys.*, 21, 1867, 1988.

Orient, O. J. and S. K. Srivastava, *J. Phys. B: At. Mol. Phys.*, 20, 3923, 1987.

Poparić, G., M. Vićić, and D. S. Belić, *J. Phys. B: At. Mol. Opt. Phys.*, 34, 381, 2001.

Raju, G. G., *Gaseous Electronics: Theory and Practice*, Taylor & Francis, London, 2005.

Randell, J., R. J. Gulley, S. L. Lunt, J.-P. Ziesel, and D. Field, *J. Phys. B: At. Mol. Opt. Phys.*, 29, 2049, 1996.

Rapp, D. and D. D. Briglia, *J. Chem. Phys.*, 43, 1480, 1965.

Rapp, D. and P. Englander-Golden, *J. Chem. Phys.*, 43, 1464, 1965.

Roznerski, W. and K. Leja, *J. Phys. D: Appl. Phys.*, 17, 279, 1984.

Saelee, H. T. and J. Lucas. *J. Phys. D: Appl. Phys.*, 10, 343, 1977.

Sawada, T., D. L. Sellin, and A. E. S. Green, *J. Geophys. Res.*, 77, 4819, 1972.

Schulz, G. J., *Phys. Rev.*, 116, 1141, 1959; also see *Phys. Rev.*, 135, 988, 1964.

Sohn, W., K.H. Kochem, K. Jung, H. Ehrhardt, and E. S. Chang, *J. Phys. B: At. Mol. Phys.*, 18, 2049, 1985.

Szmytkowski, C., K. Maciag, and G. Karwasz, *Physica Scripta*, 54, 271, 1996.

Tian, C. and C. R. Vidal, *J. Phys. B: At. Mol. Opt. Phys.*, 31, 895, 1998.

Wagner, E. B., F. J. Davis, and G. S. Hurst, *J. Chem. Phys.*, 47, 3138, 1967.

Warren, R. W. and J. H. Parker, *Phys. Rev.*, 128, 2661, 1962.

Xing, S. L., Q. C. Shi, X. J. Chen, K. Z. Xu, B. X. Yang, S. L. Wu, and R. J. Feng, *Phys. Rev. A*, 51, 414, 1995.

Zobel, J., U. Mayer, K. Jung, and H. Ehrhardt, *J. Phys. B: At. Mol. Opt. Phys.*, 29, 813, 1996.

Zobel, J., U. Mayer, K. Jung, H. Ehrhardt, H. Pritchard, C. Winstead, and V. McKoy, *J. Phys. B: At. Mol. Opt. Phys.*, 28, 839, 1995.

12

CHLORINE

CONTENTS

Chlorine (Cl_2) is an electron-attaching molecule with 34 electrons. Its electronic polarizability is 5.13×10^{-40} F m^2 and its ionization potential is 11.48 eV (molecular) and 12.968 eV (atomic). The dissociation energy (Cl–Cl) is 2.513 eV. The electron affinity of the atom is 3.61 eV and of the molecule 2.38 eV.

12.1 SELECTED REFERENCES FOR DATA

Table 12.1 shows selected references for data on Cl_2.

TABLE 12.1
Selected References

Parameter	Range: eV, (Td)	Reference
Q_i	12–900	**Basner and Becker (2004)**
Review	0–1000	Christophorou and Olthoff (2004)
Q_T	0.8–600	**Makochekanwa et al. (2003)**
Q_a	0–9	**Feketeova et al. (2003)**
Q_a	0–0.2	**Barsotti et al. (2002)**
Q_T	0.3–23.0	**Cooper et al. (1999)**
Q_T	0.02–9.5	**Gulley et al. (1998)**
Q_{rot}	0.01–1000	Kutz and Meyer (1995)
Q_{rot}	10–200	**Gote and Erhardt (1995)**
Q_a	0–1.0	**McCorkle et al. (1986)**
Q_i	13.5–102	**Stevie and Vasile (1981)**
Q_i, Q_a	0–100	**Kurepa and Belić (1978)**
Q_a	0–8	**Tam and Wong (1978)**
k_a	(350)	**Sides et al. (1976)**
α/N, η/N	(210–450)	**Božin et al. (1967)**

Note: Q_a = attachment cross section; Q_i: = ionization cross section; Q_{rot} = rotational excitation; Q_T: = total cross section; α/N = density-reduced ionization coefficient; η/N = density-reduced attachment coefficient. Bold font indicates experimental study.

12.2 TOTAL SCATTERING CROSS SECTION

Total scattering cross sections suggested by Christophorou and Olthoff (2004), on the basis of measurements of Gulley et al. (1998) and Cooper et al. (1999) are shown in Table 12.2 and Figure 12.1. The essential features of the total scattering cross section are

1. The cross section increases toward zero energy, attributed to electron attachment. Low-energy attachment takes place most probably through *p*-wave attachment, unlike *s*-wave attachment in other species such as SF_6. This explains the relatively lower cross section at low energies for Cl_2 (Gulley et al. 1998).

Cl$_2$

2. A deep Ramsauer–Townsend minimum at ~0.4 eV.
3. A strong peak at ~7 eV, a feature common to many gases (see Table 12.2).

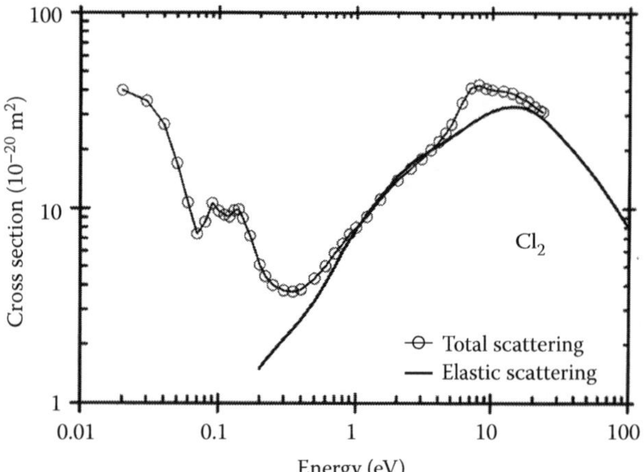

FIGURE 12.1 Total scattering cross sections (—○—) and elastic scattering (—) for Cl$_2$. (Adapted from Christophorou, L. G. and J. K. Olthoff, *Fundamental Electron Interactions with Plasma Processing Gases*, Kluwer Academic/Plenum Publishers, New York, NY, 2004.)

TABLE 12.2
Total Scattering Cross Sections for Cl$_2$

Energy (eV)	Q_T (10^{-20} m^2)	Energy (eV)	Q_T (10^{-20} m^2)
0.02	40.0	0.80	6.55
0.03	35.2	0.90	7.36
0.04	26.8	1.00	7.97
0.05	17.0	1.20	9.06
0.06	10.7	1.50	11.1
0.07	7.36	2.00	13.9
0.08	8.50	2.50	16.0
0.09	10.6	3.00	17.9
0.10	9.68	3.5	19.9
0.11	9.26	4.0	21.9
0.12	9.06	4.5	24.2
0.13	9.76	5.0	26.8
0.14	9.89	6.0	34.5
0.15	8.90	7.0	41.2
0.17	7.19	8.0	42.8
0.20	5.09	9.0	41.0
0.22	4.44	10.0	40.3
0.25	4.00	12.0	39.7
0.30	3.75	14.0	38.6
0.35	3.70	16.0	36.7
0.40	3.80	18.0	35.1
0.50	4.32	20.0	33.0
0.60	5.00	22.0	31.5
0.70	5.83	23.0	31.0

Source: Adapted from Christophorou, L. G. and J. K. Olthoff, *Fundamental Electron Interactions with Plasma Processing Gases*, Kluwer Academic/Plenum Publishers, New York, NY, 2004.

Note: See Table 12.11 also.

12.3 ELASTIC SCATTERING CROSS SECTION

Elastic scattering cross sections suggested by Christophorou and Olthoff (2004) are shown in Table 12.3 and included in Figure 12.1.

12.4 ROTATIONAL CROSS SECTION

Rotational cross section is visualized as comprising of two components: rotational scattering and rotational excitation. Rotational scattering designates $j = 0 \rightarrow 0$ and it is an elastic process. Rotational excitation designates $j = 0 \rightarrow 2, 4, 6, \ldots$ and it is an inelastic process (Kutz and Meyer, 1995). The essential features of rotational scattering are (Cooper et al., 1999)

1. Except for electron energies near the minimum (~0.4 eV), rotational scattering is the largest contribution of the total scattering cross section for energy up to ~20 eV.
2. Near the peak (~8 eV) rotational excitation makes a significant contribution to the total scattering cross section.

Gote and Erhardt (1995) have measured the differential rotational excitation cross sections for electrons impacting with energy from 2 to 200 eV. These results are shown in

TABLE 12.3
Elastic Scattering Cross Sections for Cl$_2$

Energy (eV)	Q_{el} (10^{-20} m^2)	Energy (eV)	Q_{el} (10^{-20} m^2)
0.20	1.50	7.0	27.1
0.22	1.64	8.0	28.8
0.25	1.82	9.0	30.2
0.30	2.11	10.0	31.3
0.35	2.38	12.0	32.7
0.40	2.66	14.0	33.1
0.50	3.30	16.0	32.9
0.60	4.10	18.0	32.1
0.70	4.98	20	30.9
0.80	5.99	22	29.5
0.90	6.89	23	28.8
1.0	7.77	25	27.3
1.2	9.34	30	24.0
1.5	11.4	40	19.4
2.0	14.6	50	16.1
2.5	16.9	60	13.6
3.0	18.6	70	11.6
3.5	19.9	80	10.1
4.0	21.1	90	8.87
4.5	22.1	100	7.99
5.0	23.2	150	6.31
6.0	25.2	200	6.16

Source: Adapted from Christophorou, L. G. and J. K. Olthoff, *Fundamental Electron Interactions with Plasma Processing Gases*, Kluwer Academic/Plenum Publishers, New York, NY, 2004.

TABLE 12.4
Differential Rotational Excitation Scattering Cross Sections for Cl_2 in Units of 10^{-20} m²/sr

	Energy (eV)									
Angle (°)	**2**	**5**	**10**	**20**	**30**	**50**	**70**	**100**	**150**	**200**
10	1.591	5.819	21.283	31.605	31.838	28.722	22.238	15.695	14.123	12.915
20	1.129	4.584	14.387	19.320	15.896	8.607	5.434	3.498	2.397	2.518
30	0.856	3.349	7.491	9.406	5.860	2.409	1.457	0.912	0.778	0.831
40	1.017	2.984	4.615	3.970	2.263	1.012	0.592	0.415	0.345	0.357
50	1.318	2.542	1.740	1.834	1.180	0.505	0.262	0.188	0.185	0.230
60	1.528	2.023	1.411	1.316	0.687	0.196	0.131	0.133	0.154	0.190
70	1.615	1.720	1.082	0.961	0.406	0.158	0.152	0.148	0.154	0.132
80	1.577	1.506	0.963	0.754	0.431	0.304	0.219	0.201	0.126	0.094
90	1.453	1.381	0.843	0.799	0.621	0.440	0.295	0.192	0.086	0.064
100	1.291	1.177	0.858	0.886	0.712	0.485	0.320	0.148	0.054	0.035
110	1.029	1.115	0.874	0.893	0.593	0.435	0.237	0.076	0.026	0.026
120	0.723	1.196	0.977	0.831	0.393	0.288	0.117	0.036	0.030	0.048
130	0.540	1.146	1.080	0.651	0.218	0.179	0.093	0.073	0.084	0.110
140	0.510	1.156	1.569	0.548	0.111	0.154	0.177	0.153	0.154	0.239
150	0.600	1.225	2.058	0.571	0.107	0.327	0.435	0.590	0.328	0.339
160	0.744	1.294	3.124	0.886	0.198	0.688	1.049	0.909	0.547	0.580

Source: Adapted from Gote, M. and H. Erhardt, *J. Phys. B: At. Mol. Opt. Phys.*, 28, 3957, 1995.

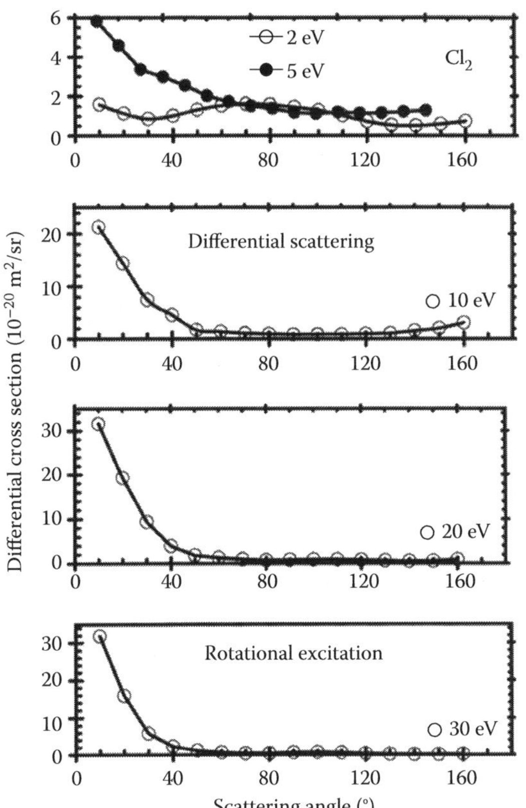

FIGURE 12.2 Differential scattering cross section for rotational excitation for Cl_2 molecule. Measurements of Gote and Erhardt (1995). Energy of impacting electron as shown. For integrated values, see Table 12.5.

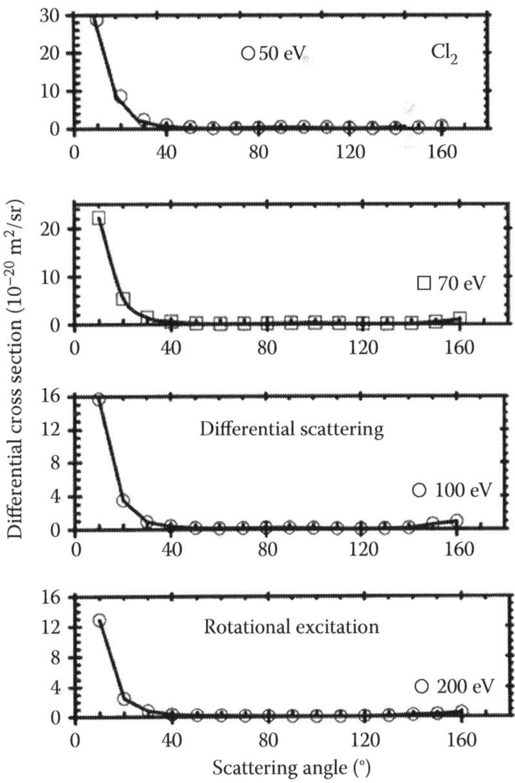

FIGURE 12.3 Differential scattering cross section for rotational excitation for Cl_2 molecule. Measurements of Gote and Erhardt (1995). Energy of impacting electron as shown. For integrated values, see Table 12.5.

Table 12.4. Figures 12.2 and 12.3 are graphical presentations. Cross sections for rotational excitation integrated by Gorur Govinda Raju (unpublished data, 2007) are shown in Table 12.5. Theoretical cross sections up to 1000 eV are given by Kutz and Meyer (1995).

Cl_2

12.5 VIBRATIONAL EXCITATION CROSS SECTION

The vibrational excitation energy for Cl_2 is 69.4 meV (Christophorou and Olthoff, 2004). Further experimental data on vibrational cross sections are not available. Theoretical or derived values are given by Christophorou and Olthoff (2004).

12.6 IONIZATION CROSS SECTION

Total ionization cross sections for Cl_2 are shown in Table 12.6 and Figure 12.4. Partial ionization cross sections have been measured by Calandra et al. (2000).

TABLE 12.5
Cross Sections for Rotational Excitation for Cl_2

Energy (eV)	Q_{rot} (10^{-20} m^2)	Energy (eV)	Q_{rot} (10^{-20} m^2)
2	14.22	50	15.57
5	22.61	70	11.41
10	31.25	100	7.89
20	32.0	150	6.41
30	23.12	200	6.34

Source: Adapted from Integrated by Gorur Govinda Raju, unpublished data, 2007.

TABLE 12.6
Total Ionization Cross Sections for Cl_2

	Total Ionization Cross Section (10^{-20} m^2)	
Energy (eV)	Stevie and Vasile (1981)	Kurepa and Belić (1978)
11.0		0
11.2		0.016
11.4		0.028
11.6		0.050
11.8		0.068
12.0		0.0927
12.2		0.122
12.4		0.146
12.6		0.158
12.8		0.192
13.0		0.231
13.2		0.283
13.4		0.317
13.5	0.17	
13.6		0.371
13.8		0.414

TABLE 12.6 (continued)
Total Ionization Cross Sections for Cl_2

	Total Ionization Cross Section (10^{-20} m^2)	
Energy (eV)	Stevie and Vasile (1981)	Kurepa and Belić (1978)
14.0	0.34	0.463
14.2		0.521
14.4		0.581
14.6		0.631
14.8		0.682
15.0	0.68	0.747
15.2		0.797
15.4		0.859
15.6		0.913
15.8		0.979
16.0	0.85	1.03
16.2		1.09
16.4		1.15
16.6		1.21
16.8		1.26
17.0	1.24	1.32
17.2		1.38
17.4		1.43
17.6		1.48
17.8		1.54
18.0	1.75	1.59
18.2		1.64
18.4		1.70
18.6		1.75
18.8		1.80
19.0	2.20	1.86
19.2		1.91
19.4		1.95
19.6		2.00
19.8		2.07
20	2.65	2.12
21	3.44	2.38
22	4.00	2.56
23	4.28	2.79
24		2.97
25		3.18
26	4.90	3.30
27		3.51
28	5.35	3.66
29		3.79
30		3.90
31	5.80	
32		4.09
34		4.27
36	6.25	4.42
38		4.53
40		4.61
42	6.31	4.68
44		4.80
46		4.87
47	6.54	

continued

TABLE 12.6 (continued)
Total Ionization Cross Sections for Cl_2

Energy (eV)	Total Ionization Cross Section (10^{-20} m^2) Stevie and Vasile (1981)	Kurepa and Belić (1978)
48		4.98
50		5.06
52		5.13
53	6.71	
54		5.20
56		5.26
58	6.87	5.32
60		5.39
62		5.43
64	6.87	5.468
66		5.48
68		5.52
69	7.10	
70		5.53
72		5.54
74		5.56
76		5.57
78		5.58
80	1.17	5.58
82		5.58
84		5.58
86		5.58
88		5.572
90		5.568
91	6.82	
92		5.56
94		5.55
96		5.54
98		5.535
100		5.524
102	6.82	

Note: Also see Table 12.12.

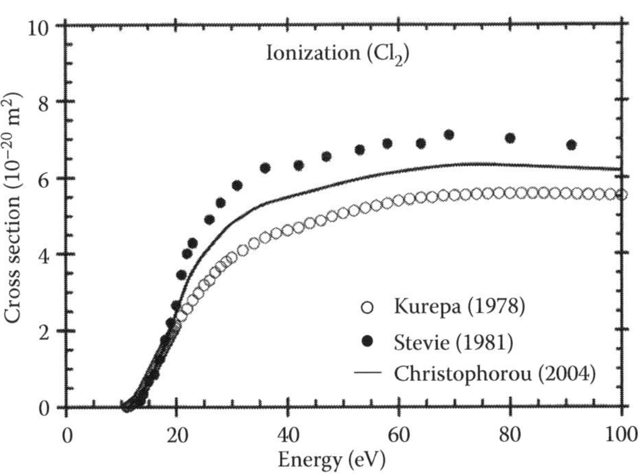

FIGURE 12.4 Total ionization cross sections for Cl_2. (—) Suggested by Christophorou and Olthoff (2004); (●) Stevie and Vasile (1981); (○) Kurepa and Belić (1978).

12.7 ATTACHMENT PROCESSES

Negative ions are formed due to dissociative attachment

$$Cl_2 + e \rightarrow Cl + Cl^- \quad (12.1)$$

and attachment

$$Cl_2 + e \rightarrow Cl_2 \quad (12.2)$$

In addition, at energies >10 eV, ion pair production is possible. The threshold energy for Cl^+ ion appearance by this process is 11.87 eV (Kurepa and Belić, 1978). The potential energy diagram of the Cl_2 molecule and the lowest of the four states of negative ion Cl_2^- are shown in Figure 12.5 (Barsotti et al., 2002).

12.8 ATTACHMENT CROSS SECTION

Appearance potentials of negative ions and positions of peaks from electron attachment to the Cl_2 molecule are shown in Table 12.7.

The total attachment cross sections for Cl_2 molecule are shown in Table 12.8 and Figure 12.6 (Kurepa and Belić, 1978).

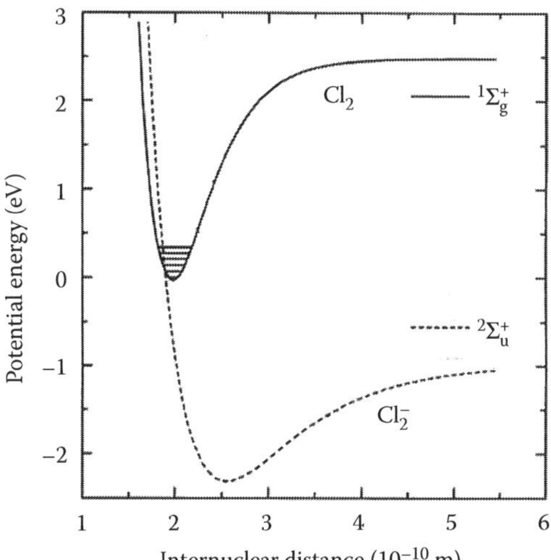

FIGURE 12.5 Potential energy curves for the dissociative electron attachment to Cl_2 molecules. For clarity only one of the four states of the negative ion is shown. (Adapted from S. Barsotti, M.-W. Ruf, and H. Hotop, *Phys. Rev. Lett.*, 89, 083201, 2002.)

TABLE 12.7
Appearance Potentials and Peak Positions

Process	Appearance Potential (eV)	Peak (eV)	Reference
$Cl_2 + e \rightarrow Cl_2^{-*}$	0.0	0.0	Kurepa and Belić (1978)
$Cl_2 + e \rightarrow Cl + Cl^-$	1.0	2.5	
	3.4	5.75	
	8.2	9.7	
$Cl_2 + e \rightarrow Cl + Cl^-$		0.03	Tam and Wong (1978)
		2.5	
		5.5	

Cl$_2$

TABLE 12.8
Total Attachment Cross Sections for Cl$_2$

Energy (eV)	Q_{att} (10^{-22} m^2)	Energy (eV)	Q_{att} (10^{-22} m^2)	Energy (eV)	Q_{att} (10^{-22} m^2)
0.0	201.6	6.8	3.55	28	1.35
0.1	80.0	7.0	3.01	30	1.08
0.2	24.5	7.2	2.31	32	0.89
0.3	6.2	7.4	1.78	34	0.74
0.4	2.0	7.6	1.38	36	0.65
0.5	0.98	7.8	1.00	38	0.62
0.6	0.68	8.0	0.699	40	0.60
0.7	0.63	8.2	0.51	42	0.40
0.8	0.50	8.4	0.45	44	0.62
0.9	0.44	8.6	0.41	46	0.64
1.0	0.42	8.8	0.395	48	0.69
1.2	0.48	9.0	0.395	50	0.76
1.4	0.59	9.2	0.380	52	0.88
1.6	0.84	9.4	0.386	54	1.05
1.8	1.32	9.6	0.392	56	1.28
2.0	1.81	9.8	0.390	58	1.52
2.2	2.42	10.0	0.380	60	1.76
2.4	2.77	10.2	0.372	62	1.96
2.6	2.79	10.4	0.365	64	2.18
2.8	2.49	10.6	0.355	66	2.34
3.0	1.92	10.8	0.348	68	2.52
3.2	1.36	11.0	0.344	70	2.72
3.4	1.06	11.2	0.325	72	3.18
3.6	0.95	11.4	0.316	74	3.73
3.8	1.05	11.6	0.318	76	4.06
4.0	1.29	11.8	0.328	78	4.24
4.2	1.69	12.0	0.365	79	4.28
4.4	2.09	12.2	0.440	80	4.25
4.6	2.57	12.4	0.530	82	4.20
4.8	3.08	12.6	0.625	84	4.13
5.0	3.63	12.8	0.720	86	4.04
5.2	4.08	13.0	0.510	88	3.94
5.4	4.46	14	0.900	90	3.82
5.6	4.74	16	1.80	92	3.70
5.8	4.84	18	2.40	94	3.57
6.0	4.78	20	2.58	96	3.40
6.2	4.62	22	2.40	98	3.22
6.4	4.38	24	2.07	100	3.07
6.6	3.99	26	1.66		

Note: For electron energy ≥14 eV, cross section for ion pair production is also included.

More recent measurements have confirmed that the cross section at zero energy peak is 2.5×10^{-20} m^2 (see Barsotti et al., 2002). Figure 12.7 shows the details at low energy. The total cross section at zero energy is much higher at 40×10^{-20} m^2 (see Gulley et al., 1998) and the discrepancy has not been satisfactorily resolved.

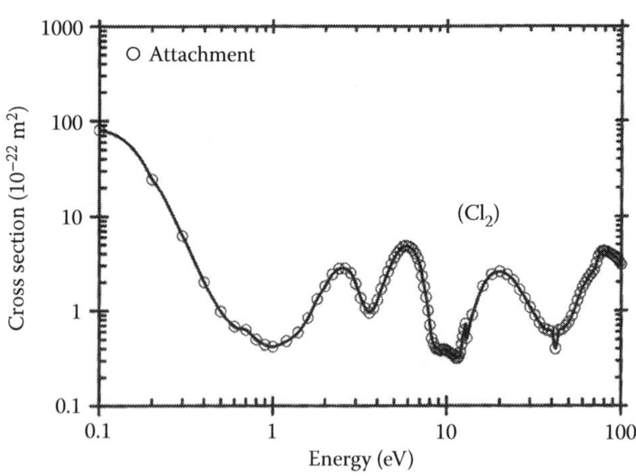

FIGURE 12.6 Total attachment cross section for Cl$_2$. Note the rise toward zero energy due to attachment. (Adapted from M. V. Kurepa and D. Belić, *J. Phys. B: At. Mol. Phys.*, 11, 3719, 1978.)

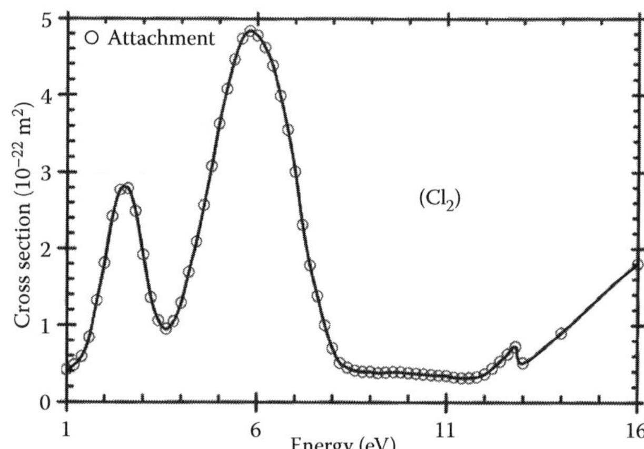

FIGURE 12.7 Details of total electron attachment cross section for Cl$_2$ at low energy. (Adapted from M. V. Kurepa and D. Belić, *J. Phys. B: At. Mol. Phys.*, 11, 3719, 1978.)

12.9 ATTACHMENT RATE CONSTANTS

Christophorou and Olthoff (2004) summarize the data on attachment rates at various temperatures, reduced electric field *E*/*N*, and derived mean energy of the swarm. Table 12.9 shows the rates at various temperatures obtained by several researchers.

12.10 ELECTRON TRANSPORT

Figure 12.8 (Bailey and Healy, 1935) shows the drift velocity and characteristic energy (D/μ) at low values of *E*/*N*. Both sets are the only data available.

12.11 IONIZATION AND ATTACHMENT COEFFICIENTS

Table 12.10 and Figure 12.9 show the density-reduced ionization coefficients (α/N), density-reduced attachment coefficients (η/N), and density-reduced effective ionization coefficients $(\alpha-\eta)/N$, deduced by Christophorou and Olthoff (2004) based on the measurements of Božin and Goodyear (1967). The attachment coefficients are not found to depend on pressure, unlike for Br_2 (see Razzak and Goodyear (1969)) (Table 12.10).

TABLE 12.9
Attachment Rate Constants

Temperature (K)	Rate (10^{-16} m³/s)	Method	Reference
213	12.2	Swarm	McCorkle et al. (1984)
233	13.5		
253	15.1		
273	16.7		
298	18.6		
323	21.4		
203	<10	FA/LP	Smith et al. (1984)
300	20		
455	33		
590	48		
350	37	FA/ECR	Sides et al. (1976)
293	3.1	Swarm	Christodoulides et al. (1975)

Note: ECR = electron cyclotron resonance; FA = flowing afterglow; LP = Langmuir probe.

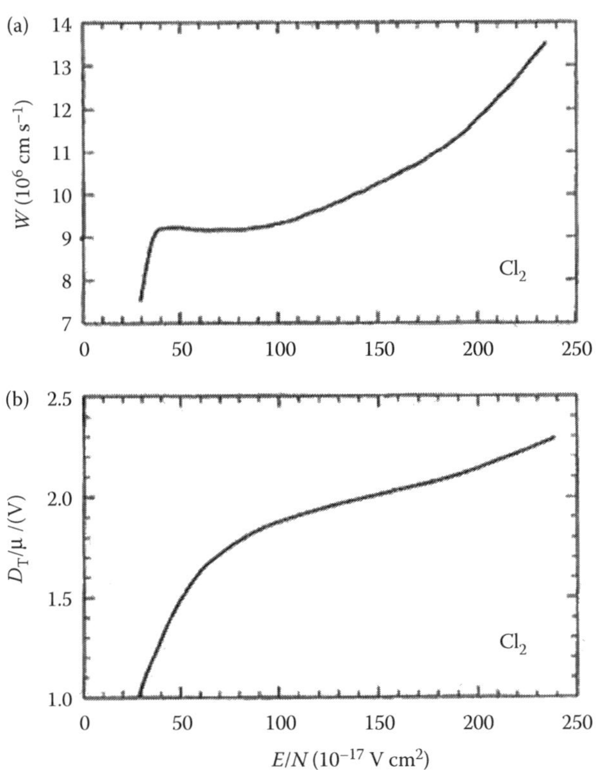

FIGURE 12.8 Drift velocity and characteristic energy for Cl_2 at 288 K. Both figures are the only data available. (Adapted from V. A. Bailey and R. H. Healy, *Philos. Mag.*, 19, 725, 1935.)

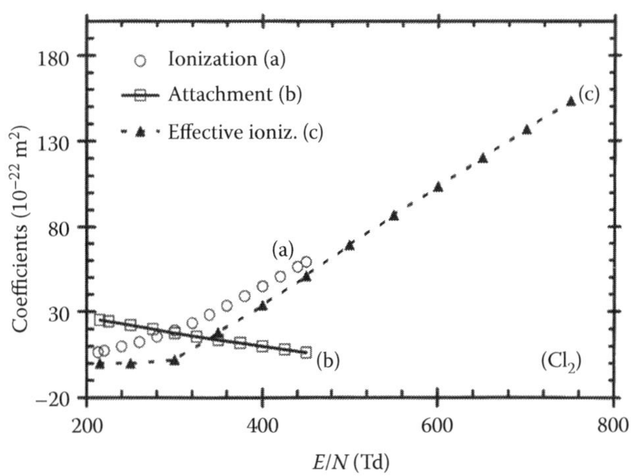

FIGURE 12.9 Reduced ionization coefficients (a), attachment coefficients (b), and effective ionization coefficients for Cl_2(c). (Adapted from S. E. Božin and C. C. Goodyear, *Br. J. Appl. Phys.*, 18, 49, 1967.)

TABLE 12.10
Ionization and Attachment Coefficients for Cl_2

E/N (Td)	α/N (10^{-22} m²)	η/N (10^{-22} m²)	$(\alpha-\eta)/N$ (10^{-22} m²)
213	6.45		
215		25.3	−18.5
220	7.34		
225		24.4	
240	9.82		
250		22.3	−10.7
260	12.4		
275		20.0	
280	15.5		
300	19.2	17.6	2.13
320	23.4		
325		15.6	
340	28.2		
350		13.7	18.0
360	33.4		
375		11.9	
380	39.0		
400	44.7	10.0	33.7
420	50.4		
425		8.14	
440	56.2		
450	59.1	6.26	50.9
500			69.1
550			86.8
600			103.7
650			120.3
700			136.9
750			153.4

Cl_2

TABLE 12.11
Total Scattering Cross Sections for Cl_2

Energy (eV)	Q_T (10^{-20} m^2)	Energy (eV)	Q_T (10^{-20} m^2)	Energy (eV)	Q_T (10^{-20} m^2)
0.8	5.45	6.5	25.84	28.0	22.39
1.0	4.96	7.1	27.99	30.0	21.55
1.2	5.95	7.3	30.33	35.0	20.19
1.4	6.77	7.7	32.12	40.0	19.90
1.6	7.38	8.1	31.28	45.0	19.26
1.8	7.97	8.5	30.71	50.0	18.20
2.0	7.58	9.0	30.71	60.0	16.49
2.2	10.04	9.5	30.60	70.0	16.04
2.5	12.17	10.0	29.29	80.0	14.64
2.9	13.37	11.0	29.66	90.0	13.85
3.2	13.81	12.0	30.20	100	12.67
3.5	14.50	14.0	30.46	150	10.42
3.8	15.32	16.0	28.34	200	8.96
4.1	16.37	18.0	26.81	300	6.59
4.7	18.18	20.0	26.56	400	6.24
5.1	19.93	22.0	24.10	500	5.50
5.6	22.22	24.0	23.09	600	4.57
5.9	23.89	26.0	22.98		

Source: Digitized and interpolated from C. Makochekanwa et al. *J. Phys. B: At. Mol. Opt. Phys.*, 36, 1673, 2003.

12.12 ADDENDUM

12.12.1 Total Scattering Cross Section

Makochekanwa et al. (2003) have measured the total scattering cross sections as shown in Table 12.11 and Figure 12.10.

12.12.2 Ionization Cross Section

See Table 12.12.

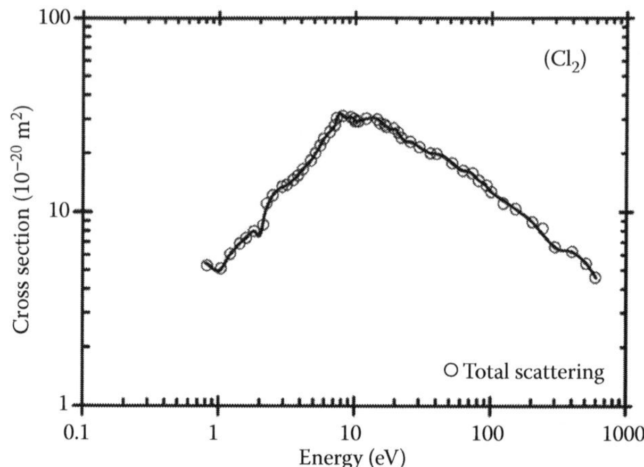

FIGURE 12.10 Total scattering cross sections for Cl_2. (Adapted from C. Makochekanwa et al., *J. Phys. B: At. Mol. Opt. Phys.*, 36, 1673, 2003.)

TABLE 12.12
Total Ionization Cross Sections for Cl_2

Energy (eV)	Q_i (10^{-20} m^2)	Energy (eV)	Q_i (10^{-20} m^2)	Energy (eV)	Q_i (10^{-20} m^2)	Energy (eV)	Q_i (10^{-20} m^2)
12	0.100	28	6.34	52.5	8.56	180	7.56
13	0.346	30	6.75	55	8.67	200	7.18
14	0.778	32	7.26	57.5	8.76	300	5.73
15	1.19	34	7.47	60	8.86	400	4.84
16	1.66	36	7.58	65	9.00	500	4.14
17	2.13	38	7.68	70	9.11	600	3.59
18	2.60	40	7.75	80	9.20	700	3.26
19	3.02	42	7.89	90	9.08	800	2.92
20	3.57	44	7.99	100	9.03	900	2.67
22	4.37	46	8.10	120	8.73		
24	5.08	48	8.26	140	8.12		
26	5.73	50	8.40	160	7.89		

Source: Adapted from R. Basner and K. Becker, *New J. Phys.*, 6, 118, 2004.
Note: For partial ionization cross sections (Cl_2^+, Cl^+, Cl_2^{2+}, and Cl_2^+ ions), see original paper.

REFERENCES

Bailey, V. A. and R. H. Healy, *Philos. Mag.*, 19, 725, 1935.
Barsotti, S., M.-W. Ruf, and H. Hotop, *Phys. Rev. Lett.*, 89, 083201, 2002.
Basner, R. and K. Becker, *New J. Phys.*, 6, 118, 2004.
Božin, S. E. and C. C. Goodyear, *Br. J. Appl. Phys.*, 18, 49, 1967.
Calandra, P., C. S. S. O'Connor, and S. D. Price, *J. Chem. Phys.*, 112, 10821, 2000.
Christophorou, L. G. and J. K. Olthoff, *Fundamental Electron Interactions with Plasma Processing Gases*, Kluwer Academic/Plenum Publishers, New York, NY, 2004.
Cooper, G. D., J. E. Sanabia, J. H. Moore, J. K. Olthoff, and L. G. Christophorou, *J. Chem. Phys.*, 110, 682, 1999.
Feketeova, L., D. J. Skalny, G. Hanel, B. Gstir, M. Francis, and T. D. Mark, *Int. J. Mass Spectrom.*, 223–224, 661, 2003.
Gote, M. and H. Erhardt, *J. Phys. B: At. Mol. Opt. Phys.*, 28, 3957, 1995.
Gulley, R. J., T. A. Field, W. A. Steer, N. J. Mason, S. L. Lunt, J.-P. Ziesel, and D. Field, *J. Phys. B: At. Mol. Opt. Phys.*, 31, 2971, 1998.
Kurepa, M. V. and D. Belić, *J. Phys. B: At. Mol. Phys.*, 11, 3719, 1978.
Kutz, H. and H.-D. Meyer, *Phys. Rev.*, 51, 3819, 1995.
Makochekanwa, C., H. Kawate, O. Sueoka, and M. Kimura, *J. Phys. B: At. Mol. Opt. Phys.*, 36, 1673, 2003.
McCorkle, D. L., A. A. Christodoulides, and L. G. Christophorou, *Chem. Phys. Lett.*, 109, 276, 1984.
McCorkle, D. L., L. G. Christophorou, and A. A. Christodoulides, *J. Chem. Phys.*, 85, 1966, 1986.
Razzak, S. A. A. and C. C. Goodyear, *Br. J. Appl. Phys.* (*J. Phys. D*), 2, 1577, 1969.
Sides, G. D., T. O. Tiernan, and R. J. Hanrahan, *J. Chem. Phys.*, 65, 1966, 1976.
Smith, D., N. G. Adams, and E. Alge, *J. Phys. B: At. Mol. Opt. Phys.*, 17, 461, 1984.
Stevie, F. A. and M. J. Vasile, *J. Chem. Phys.*, 74, 5106, 1981.
Tam, W. and S. F. Wong, *J. Chem. Phys.*, 68, 5626, 1978.

13

DEUTERIUM

CONTENTS

Deuterium (D_2) is an isotope of hydrogen. Its electronic polarizability is 8.81×10^{-41} F m^2 and the ionization potential is 15.426 eV.

13.1 SELECTED REFERENCES FOR DATA

See Table 13.1.

13.2 IONIZATION CROSS SECTION

See Table 13.2.

TABLE 13.1
Selected References for Data

Quantity	Range: eV, (Td)	Reference
Drift velocity	(3–125)	**Roznerski et al. (1994)**
Ionization coefficients	(100–850)	**Cowling and Fletcher (1973)**
Diffusion coefficients	(0.03–10)	**Lowke and Parker (1969)**
Ionization cross section	16–1000	**Rapp and Englander-Golden (1965)**
Ionization coefficient	(50–300)	**Barna et al. (1964)**
Ionization coefficient	(85–800)	**Rork and Chanin (1964)**
Transport coefficients	(0.001–100)	Engelhardt and Phelps (1963)
Drift velocity	(0.00075–125)	**Pack et al. (1962)**
Transport coefficients	(0.0003–10)	**Warren and Parker (1962)**
Ionization coefficient	(50–1700)	**Rose (1956)**

Note: Bold font indicates experimental study.

13.3 DRIFT VELOCITY OF ELECTRONS

See Table 13.3.

TABLE 13.2
Ionization Cross Sections for D_2

	Q_i			Q_i	
Energy	Cowling (1973)	Rapp (1965)	Energy	Cowling (1973)	Rapp (1965)
16	0.034	0.034	125	0.890	0.877
17	0.097	0.104	150	0.833	0.813
18	0.159	0.173	175	0.781	
19	0.218	0.239	200	0.735	0.716
20	0.273	0.300	250	0.661	0.638
21	0.325	0.355	300	0.604	0.576
22	0.378	0.404	350	0.553	0.523
23	0.423	0.454	400	0.510	0.482
24	0.470	0.498	450	0.469	0.446
25	0.515	0.537	500	0.431	0.414
30	0.700	0.699	550	0.387	0.387
35	0.809	—	600		0.366
40	0.876	0.876	650		0.344
45	0.917	—	700		0.325
50	0.951	0.950	750		0.310
60	0.977	0.977	800		0.295
70	0.986	0.981	850		0.281
80	0.981	0.974	900		0.271
90	0.964	0.958	950		0.257
100	0.946	0.939	1000		0.247

Note: See Figure 13.1 for graphical presentation.

TABLE 13.3
Recommended Drift Velocity of Electrons for D_2

E/N (Td)	*W* (m/s)	*E/N* (Td)	*W* (m/s)	*E/N* (Td)	*W* (m/s)
0.002	37.49	0.20	2310	6.0	1.27(4)
0.003	47.67	0.25	2700	7.0	1.38(4)
0.004	77.08	0.30	3030	8.0	1.48(4)
0.006	98.02	0.35	3340	10.0	1.67(4)
0.009	158.5	0.40	3595	12.0	1.85(4)
0.012	193.0	0.50	4017	14.0	2.02(4)
0.02	308.0	0.60	4350	17.0	2.28(4)
0.025	383.0	0.70	4650	20.0	2.51(4)
0.03	457.0	0.80	4920	25.0	2.90(4)
0.035	530.0	1.0	5.38(3)	30.0	3.38(4)
0.04	601.0	1.2	5.79(3)	35.0	3.72(4)
0.05	740.0	1.4	6.19(3)	40.0	4.17(4)
0.06	873.0	1.7	6.74(3)	50	4.96(4)
0.07	1002	2.0	7.27(3)	60	5.85(4)
0.08	1130	2.5	8.10(3)	70	6.54(4)
0.10	1335	3.0	8.88(3)	80	7.47(4)
0.12	1553	3.5	9.61(3)	100	9.20(4)
0.14	1759	4.0	1.02(4)	120	1.05(5)
0.17	2046	5.0	1.12(4)		

Note: a (b) means a × 10^b. See Figure 13.2 for graphical presentation.

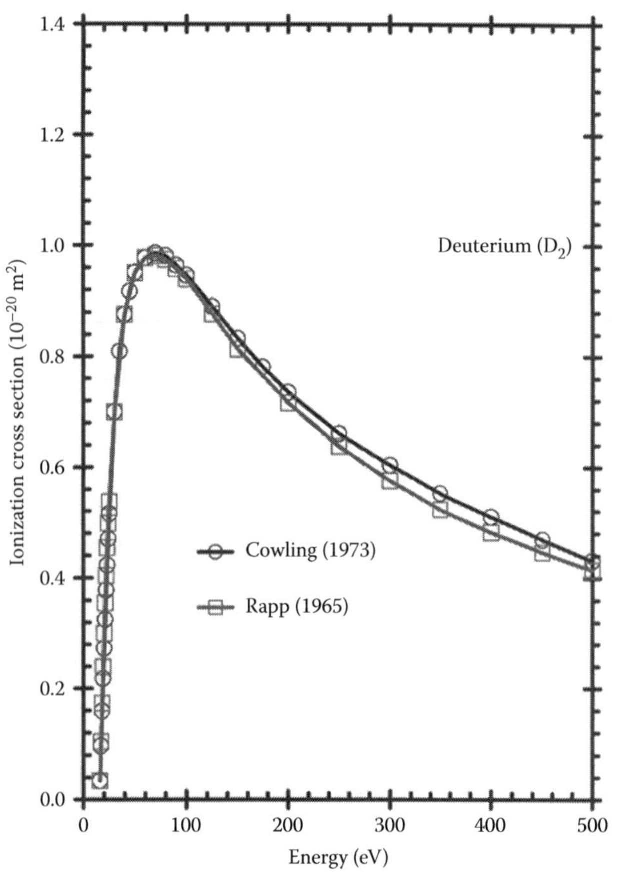

FIGURE 13.1 Ionization cross section for D_2. (–○–) Cowling and Fletcher (1973); (–□–) Rapp and Englander-Golden (1965).

13.4 DIFFUSION COEFFICIENTS

Figure 13.3 shows the longitudinal and radial diffusion coefficients expressed as ratio of mobility at 77 K and 293 K.

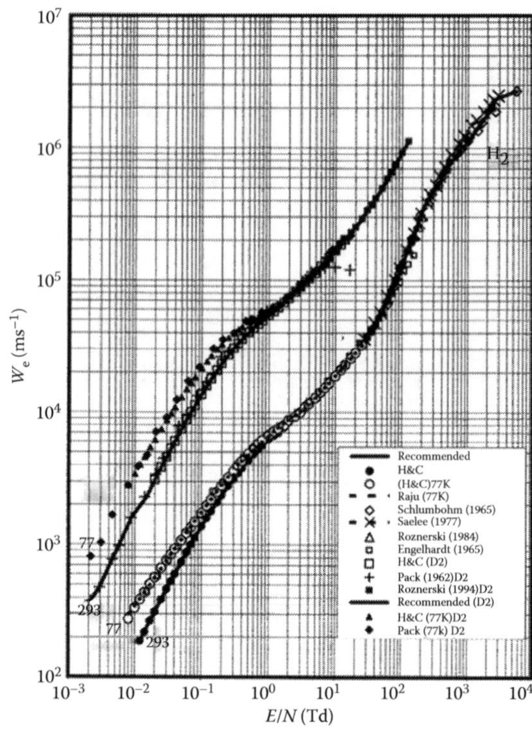

FIGURE 13.2 Drift velocity as a function of *E/N* in hydrogen and deuterium. Unless otherwise specified, temperature is 293 K. Hydrogen (293 K): (Δ) Roznerski and Leja (1984); (–×–) Saelee and Lucas (1977); (●) Huxley and Crompton (1974); (◊) Schlumbohm (1965); (– –) recommended. Hydrogen (77 K): (–●–) Raju (1985); (○) Huxley and Crompton (1974); deuterium (293 K): (■) Roznerski et al. (1994); (□) Huxley and Crompton (1974); (+) Pack et al. (1962); (– –) recommended. Deuterium (77 K): (▲) Huxley and Crompton (1974); (♦) Pack et al. (1962). Ordinates of D_2 points are multiplied by a factor of 10 for the sake of clarity.

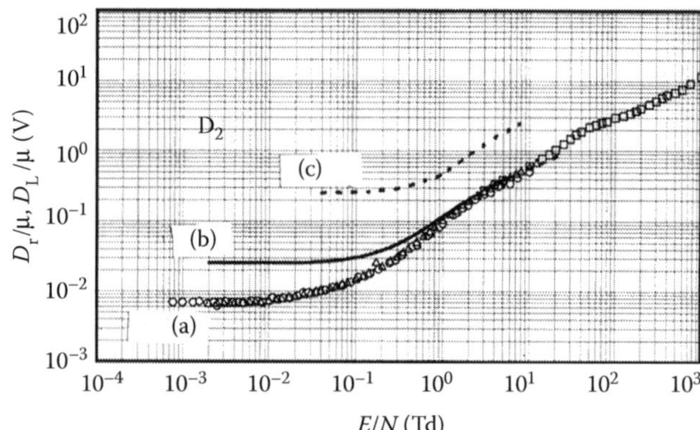

FIGURE 13.3 Ratios D_r/μ and D_L/μ as a function of *E/N* in D_2. (a): D_r/μ, 77 K: (Δ) Huxley and Crompton (1974); (- - -) Engelhardt and Phelps (1963); (○) Warren and Parker (1962). (b): D_r/μ, 293 K: (□) Roznerski et al. (1994); (—) Huxley and Crompton (1974); (c): D_L/μ, 77 K: (- - -) Lowke and Parker (1969). Ordinates of curve (c) have been multiplied by a factor of 10 for better presentation.

TABLE 13.4
Ionization Coefficients for D_2

	α/N (10^{-20} m^2)	
E/N (Td)	**Rose (1956)**	**Cowling and Fletcher (1973)**
50	0.0012	
60	0.0048	
70	0.0111	
80	0.0199	
85	0.0281	
100	0.061	
125	0.0846	
150	0.135	0.106
175	0.190	0.158
200	0.245	0.218
225	0.297	0.281
250	0.352	0.335
300	0.448	0.414
350	0.530	0.480
400	0.616	0.535
450	0.688	0.580
500	0.743	0.620
600	0.843	0.679
700	0.928	0.710
800	0.988	0.727
900	1.03	
1000	1.07	
1250	1.16	
1500	1.24	
1700	1.27	

Note: See Figure 13.4 for graphical presentation.

13.5 IONIZATION COEFFICIENTS

See Table 13.4.

13.6 GAS CONSTANTS

Gas constants evaluated according to the expression (see inset of Figure 13.4)

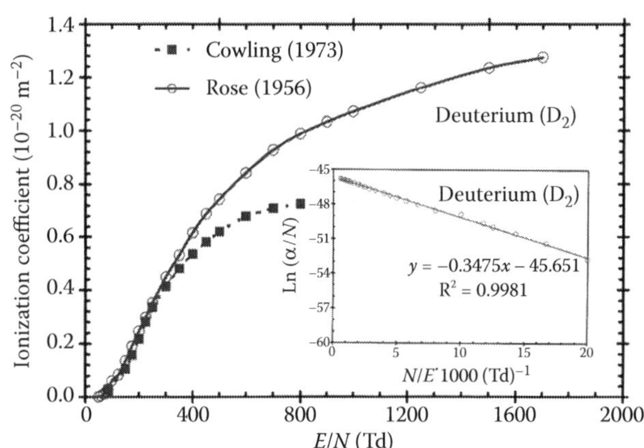

FIGURE 13.4 Reduced ionization coefficient (α/N) for D_2. Inset shows plot according to Equation 13.1. (■) Cowling and Fletcher (1973); (—○—) Rose (1956).

$$\frac{\alpha}{N} = F \exp\left(-\frac{GN}{E}\right) \tag{13.1}$$

are $F = 1.49 \times 10^{-20}$ m^2 and $G = 347.6$ Td.

REFERENCES

Barna, S. F., D. Edelson, and K. B. McAffee, *J. Appl. Phys.*, 35, 2781, 1964.

Cowling, I. R. and J. Fletcher, *J. Phys. B: Atom. Mol. Phys.*, 6, 665, 1973. Both ionization cross sections and reduced ionization coefficients in H_2 and D_2 are measured.

Engelhardt, A. G. and A. V. Phelps, *Phys. Rev.*, 131, 2115, 1963.

Lowke, J. J. and J. H. Parker Jr., *Phys. Rev.*, 181, 302, 1969. D_L/μ and D_r/μ for several gases are calculated and compared with measured ones.

Pack, J. L., R. E. Voshall, and A. V. Phelps, *Phys. Rev.*, 127, 2084, 1962.

Rapp, D. and P. Englander-Golden, *J. Chem. Phys.*, 43, 1464, 1965.

Rork, G. D. and L. M. Chanin, *J. Appl. Phys.*, 35, 2801, 1964.

Rose, D. J. *Phys. Rev.*, 104, 273, 1956.

Roznerski, W., J. Mechlińska-Drewco, K. Leja, and Z. Lj. Petrović, *J. Phys. D: Appl. Phys.*, 27, 2060, 1994.

Warren, R. W. and J. H. Parker, *Phys. Rev.*, 128, 2661, 1962.

14

DEUTERIUM BROMIDE

DBr

CONTENTS

Deuterium bromide (DBr) has properties very similar to hydrogen bromide (HBr) as expected. The attachment rate constant for the molecule is 3.1×10^{-17} m^3/s (Christophorou et al., 1968). The attachment cross sections are very similar to that for HBr, the ion formed being Br^- due to dissociative attachment. Table 14.1 shows the electron attachment data for DBr and HBr.

TABLE 14.1
Electron Attachment Data for DBr and HBr

Molecule	Appearance Potential	Peak (eV)	Q_a (peak) m^2
DBr	0.11	0.28	18.7×10^{-21}
HBr	0.11	0.28	27×10^{-21}
		0.5	5.8×10^{-21}

Source: Adapted from Christophorou, L. G., R. N. Compton, and H. W. Dickson, *J. Chem. Phys.*, 48, 1949, 1968.

TABLE 14.2
Dissociative Attachment Cross Sections for DBr

Energy (eV)	Q_a (10^{-20} m^2)	Energy (eV)	Q_a (10^{-20} m^2)
0.08	0.02	0.50	0.41
0.10	0.09	0.55	0.31
0.15	0.26	0.60	0.23
0.20	1.14	0.65	0.17
0.25	1.40	0.70	0.15
0.30	1.41	0.75	0.14
0.35	1.14	0.80	0.13
0.40	0.84	0.85	0.09
0.45	0.61	0.90	0.04

Source: Digitized from Christophorou, L. G., R. N. Compton, and H. W. Dickson, *J. Chem. Phys.*, 48, 1949, 1968.

14.1 ATTACHMENT CROSS SECTION

Table 14.2 and Figure 14.1 show the attachment cross sections for DBr.

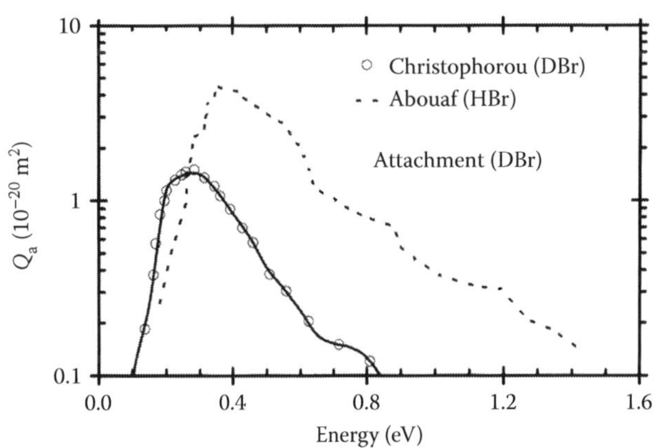

FIGURE 14.1 Dissociative attachment cross sections for DBr. The attachment cross sections for HBr measured by Abouaf and Teillet-Billy (1980) are also shown as they were not included in the HBr section. (Adapted from Christophorou, L. G., R. N. Compton, and H. W. Dickson, *J. Chem. Phys.*, 48, 1949, 1968.)

REFERENCES

Abouaf, R. and D. Teillet-Billy, *Chem. Phys. Lett.*, 73, 106, 1980.
Christophorou, L. G., R. N. Compton, and H. W. Dickson, *J. Chem. Phys.*, 48, 1949, 1968.

15

DEUTERIUM CHLORIDE

DCl

CONTENTS

Deuterium chloride (DCl) has properties very similar to hydrogen chloride (HCl) as expected. The attachment rate constant is 1.2×10^{-16} m^3/s (Christophorou et al., 1968). The attachment cross sections are very similar to those for HBr, the ion formed being Br^- due to dissociative attachment. Table 15.1 shows the electron attachment data for DCl and HCl. Table 15.2 and Figure 15.1 show the attachment cross sections for DCl. The step-like structure in the attachment cross section curve is attributed to the vibrational modes $v = 4$ to 6 (Abouaf and Tillet-Billy, 1977). The energy for vibrational excitation of DCl molecule is 266 meV.

15.1 ATTACHMENT DATA

See Figure 15.1.

TABLE 15.1
Electron Attachment Data for DCl and HCl

Molecule	Appearance Potential	Peak Energy (eV)	Q_a (peak, m^2)
DCl	0.63	0.75	14×10^{-22}
HCl	0.63	0.78	19.5×10^{-22}

Source: Adapted from Christophorou, L. G., R. N. Compton, and H. W. Dickson, *J. Chem. Phys.*, 48, 1949, 1968.

TABLE 15.2
Dissociative Attachment Cross Sections for DCl

Energy (eV)	Q_a (10^{-22} m^2)	Energy (eV)	Q_a (10^{-22} m^2)
0.65	4.42	1.35	2.44
0.70	8.18	1.40	2.19
0.75	11.46	1.45	1.94
0.80	14.80	1.50	1.69
0.85	15.86	1.55	1.45
0.90	13.42	1.60	1.23
0.95	10.42	1.70	0.89
1.00	7.42	1.75	0.75
1.05	6.09	1.80	0.63
1.10	4.78	1.85	0.51
1.15	4.01	1.90	0.38
1.20	3.38	1.95	0.24
1.25	2.96	2.00	0.07
1.30	2.68		

Source: Digitized from Christophorou, L. G., R. N. Compton, and H. W. Dickson, *J. Chem. Phys.*, 48, 1949, 1968.

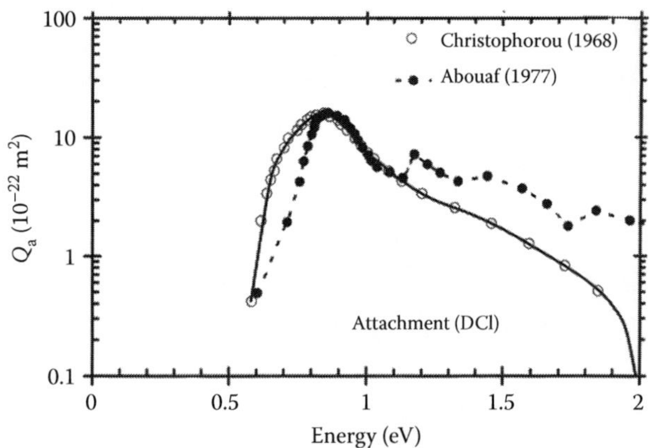

FIGURE 15.1 Dissociative attachment cross sections for DCl. (—○—) Christophorou et al. (1968); (—●—) Abouaf and Tillet-Billy (1977), normalized to the peak cross section of Christophorou et al. (1968). The step-like structures are attributed to the vibrational excitation (ν = 4 to 6).

REFERENCES

Abouaf, R. and D. Tillet-Billy, *J. Phys. B: Atom. Mol. Phys.*, 10, 2261, 1977.

Christophorou, L. G., R. N. Compton, and H. W. Dickson, *J. Chem. Phys.*, 48, 1949, 1968.

16

DEUTERIUM IODIDE

CONTENTS

Deuterium iodide (DI) has properties similar to hydrogen iodide (HI) as expected, though the cross sections for the deuterated isotope tend to be smaller. The attachment rate constant is ~8 × 10^{-14} m^3/s (Christophorou et al., 1968), the ion formed being I^- due to dissociative attachment. Table 16.1 shows the electron attachment data for DI and HI. Table 16.2 and Figure 16.1 show the attachment cross sections for DI.

TABLE 16.1
Electron Attachment Data for DI

Molecule	Appearance Potential	Peak Energy (eV)	Q_a (peak, m^2)
DI	~0	~0	1.38 × 10^{-18}
HI	~0	~0	2.56 × 10^{-18}

Source: Adapted from L. G. Christophorou, R. N. Compton, and H. W. Dickson, *J. Chem. Phys.*, 48, 1949, 1968.

TABLE 16.2
Dissociative Attachment Cross Sections for DI

Energy (eV)	Q_a (10^{-18} m^2)	Energy (eV)	Q_a (10^{-18} m^2)
0.0	1.39	0.25	0.145
0.02	1.16	0.30	0.1
0.05	0.88	0.35	0.08
0.07	0.67	0.40	0.06
0.1	0.48	0.45	0.05
0.15	0.3	0.50	0.04
0.2	0.19		

Source: Adapted from L. G. Christophorou, R. N. Compton, and H. W. Dickson, *J. Chem. Phys.*, 48, 1949, 1968.

16.1 ATTACHMENT DATA

See Figure 16.1.

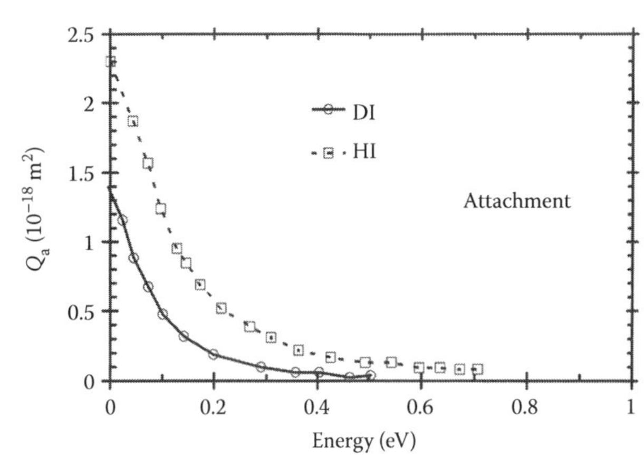

FIGURE 16.1 Attachment cross sections for DI. Cross sections for hydrogen iodide are also given for comparison. (Adapted from L. G. Christophorou, R. N. Compton, and H. W. Dickson, *J. Chem. Phys.*, 48, 1949, 1968.)

REFERENCE

Christophorou, L. G., R. N. Compton, and H. W. Dickson, *J. Chem. Phys.*, 48, 1949, 1968.

17

FLUORINE

F_2

CONTENTS

Fluorine (F_2) is an electron-attaching diatomic molecule with 18 electrons. It has electronic polarizability of 1.54×10^{-40} F m^2, ionization potential of 15.697 eV, bond dissociation energy of 1.64 eV, and electron affinity of 3.08 eV. The lowest excitation level is 3.16 eV (Hayashi and Nimura, 1983).

17.1 SELECTED REFERENCES

Selected references for data are shown in Table 17.1. In view of the fragmentary data available, a set of cross sections derived by Hayashi and Nimura (1983) is given.

TABLE 17.1
Selected References for Data for Molecular F_2

Quantity	Range: eV, (Td)	Reference
Q_a, Q_v	0–2.5 eV	Brems et al. (2002)
k_a	0.4–2.95	**McCorkle et al. (1986)**
α/N, η/N	(100–3000)	Hayashi and Nimura (1983)
Q_a	0–4	**Chantry (1982)**
Q_{ex}	5–100	Hazi (1981)
Q_i	12–100	**Stevie and Vasile (1981)**
Q_{el}	5–40	Fliplet et al. (1980)
k_a	—	Nygaard et al. (1978)
Q_a, Q_v	0–5	Hall (1978)
k_a	0.9–4.0	Schneider and Brau (1978)
k_a	0.9–4	Chen et al. (1977)
Q_{el}	0–5	Schneider and Hay (1976a)
Q_{el}	0–5	Schneider and Hay (1976b)
k_a	0–0.1	**Sides et al. (1976)**
Q_{ex}	0–13.6	Rescigno et al. (1976)
Q_{diss}	0–32	**Decorpo et al. (1970)**

Note: Bold font indicates experimental study. k_a = Attachment rate; Q_a = dissociative attachment cross section; Q_{el} = elastic; Q_{ex} = excitation; Q_V = vibrational excitation; α/N = reduced ionization coefficient; η/N = reduced attachment coefficient.

17.2 TOTAL SCATTERING CROSS SECTIONS

The total scattering cross sections obtained by adding the known cross sections are shown in Table 17.2. Figure 17.1 also shows these data. The essential features are

1. A very large cross section at near zero energy due to electron attachment
2. A resonance at 2.2 eV energy
3. A monotonic decrease of the cross section with increasing energy for energies >10 eV

F$_2$

TABLE 17.2
Total Cross Sections for F_2

Energy (eV)	Q_T (10^{-20} m^2)	Energy (eV)	Q_T (10^{-20} m^2)	Energy (eV)	Q_T (10^{-20} m^2)
0.0001	80.00	3	20.57	60	5.14
0.01	54.71	4	15.56	80	4.67
0.02	34.96	6	12.78	100	4.24
0.04	18.40	8	12.55	150	3.23
0.06	18.40	10	10.40	200	2.66
0.08	17.02	15	9.56	300	2.33
0.1	16.77	20	7.98	400	1.65
0.2	16.93	25	7.70	500	1.19
0.4	14.11	30	6.94	600	1.04
0.6	19.13	35	6.40	800	0.89
0.8	16.69	40	6.03	1000	0.71
1	18.15	45	5.71		
2	28.21	50	5.47		

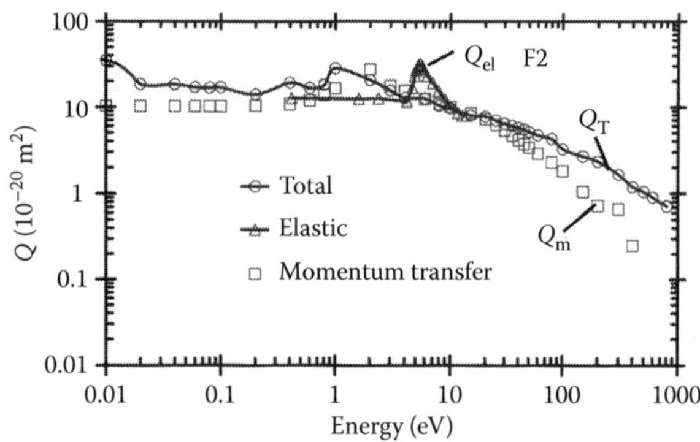

FIGURE 17.1 Total, elastic, and momentum transfer cross sections for F$_2$. (—○—) Total, Hayashi and Nimura (1983); (—△—) elastic, Rescigno et al. (1976); (□) momentum transfer cross section, Hayashi and Nimura (1983). The total cross section has been derived by adding all the cross sections as shown in the following sections (17.3 through 17.8).

17.3 MOMENTUM TRANSFER CROSS SECTIONS

Table 17.3 and Figure 17.1 show the momentum transfer cross sections.

17.4 ELASTIC SCATTERING CROSS SECTIONS

Theoretically derived elastic scattering cross sections (Rescigno et al., 1976) are shown in Figure 17.1.

17.5 VIBRATIONAL EXCITATION CROSS SECTIONS

Vibrational excitation threshold is 110 meV (Hiyashi and Nimura, 1983). Theoretically derived total vibrational excitation cross sections are shown in Table 17.4.

TABLE 17.3
Derived Momentum Transfer Cross Sections for F_2

Energy (eV)	Q_M (10^{-20} m^2)	Energy (eV)	Q_M (10^{-20} m^2)
0.01	10.31	10	10.07
0.02	10.26	15	8.50
0.04	10.20	20	7.18
0.06	10.20	25	6.14
0.08	10.22	30	5.32
0.10	10.26	35	4.65
0.20	10.42	40	4.12
0.40	10.81	45	3.70
0.60	11.92	50	3.36
0.80	13.87	60	2.87
1	16.48	80	2.28
2	27.33	100	1.81
3	17.72	150	1.04
4	15.43	200	0.72
6	12.43	300	0.66
8	10.85	400	0.06

Source: Digitized from Hayashi, M. and T. Nimura, *J. Appl. Phys.*, 54, 4879, 1983.

TABLE 17.4
Vibrational Excitation Cross Sections

Energy (eV)	Q_v (10^{-20} m^2)
0.1	0.010
0.2	0.652
0.4	2.410
0.6	2.164
0.8	1.344
1	0.859
2	0.151
3	0.051
4	0.034
6	0.015

Source: Adapted from Hayashi, M. and T. Nimura, *J. Appl. Phys.*, 54, 4879, 1983.

17.6 DISSOCIATION CROSS SECTIONS

Dissociation of F$_2$ molecule occurs according to the following reactions:

$$F_2 + e \rightarrow F + F^- \quad (17.1)$$

$$F_2 + e \rightarrow F^- + F^+ + e \quad (17.2)$$

$$F_2 + e \rightarrow F^+ + F + 2e \quad (17.3)$$

Reaction 17.1 is dissociative attachment with appearance potential of 0.0 eV, Reaction 17.2 is attachment by ion pair

TABLE 17.5
Dissociation Cross Sections Measured by DeCorpo et al. (1970) Normalized at 1.0 eV

Energy (eV)	Q_{diss} (10^{-22} m^2)	Energy (eV)	Q_{diss} (10^{-22} m^2)
0.2	1.749	6	0.611
0.4	3.501	8	0.352
0.6	3.980	10	0.077
0.8	6.472	15	0.279
1	10.00	20	1.689
2	3.546	25	5.200
3	1.186	30	8.814
4	0.591		

production with appearance potential of 15.8 eV, and Reaction 17.3 is dissociative ionization with appearance potential of 19.2 eV (DeCorpo et al., 1970).

DeCorpo et al. (1970) have measured the relative intensity for the dissociation cross section, and the cross sections normalized to 1.0×10^{-21} m^2 at 1.0 eV energy are shown in Table 17.5 and Figure 17.2.

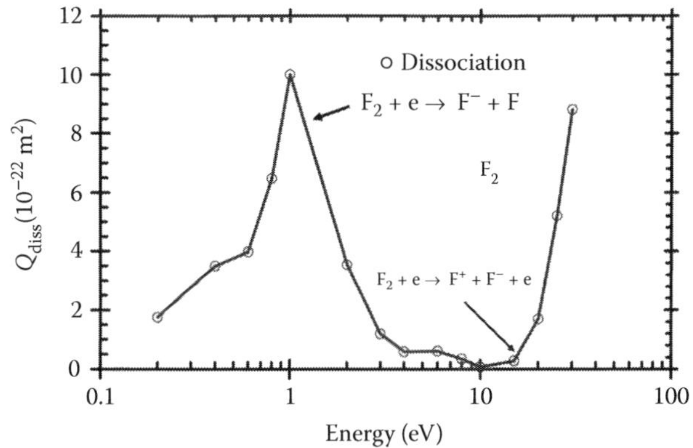

FIGURE 17.2 Dissociation cross section for F_2. Measurements of DeCorpo et al. (1970) are normalized to 1.0×10^{-21} m^2.

17.7 IONIZATION CROSS SECTIONS

Ionization cross sections have been measured by Stevie and Vasile (1981) in the energy range 17–102 eV as shown in Table 17.6 and Figure 17.3. The cross sections derived by Hayashi and Nimura (1983) for higher energy range are also shown in the same figure.

17.8 ATTACHMENT CROSS SECTIONS

Attachment processes are shown by Reactions 17.1 and 17.2. Attachment cross sections measured by Chantry et al. (1982) are shown in Table 17.7 and Figures 17.4 and 17.5.

TABLE 17.6
Ionization Cross Sections Measured by Stevie and Vasile (1981)

Energy (eV)	Q_i (10^{-20} m^2)	Energy (eV)	Q_i (10^{-20} m^2)
17	0.02	38	0.64
18	0.04	41	0.69
19	0.08	43	0.74
20	0.10	45	0.79
21	0.14	47	0.82
22	0.15	49	0.84
23	0.20	53	0.91
26	0.28	58	1.00
28	0.35	64	1.05
30	0.43	69	1.10
32	0.49	80	1.17
34	0.56	91	1.21
36	0.59	102	1.25

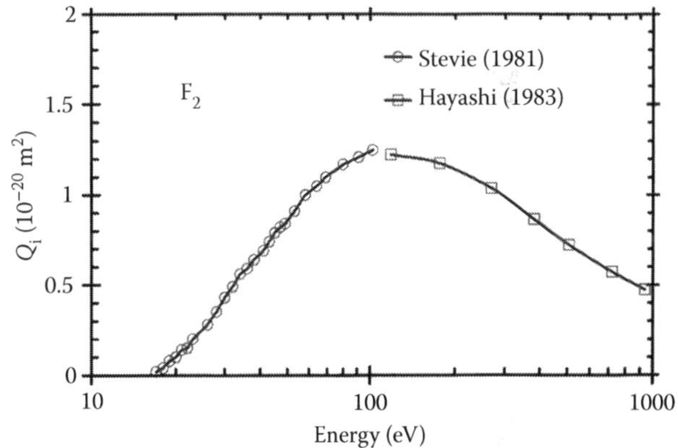

FIGURE 17.3 Ionization cross sections for F_2. (—○—) Stevie and Vasile (1981), measured; (—□—) Hayashi and Nimura (1983), derived.

TABLE 17.7
Attachment Cross Sections for F_2

Energy (eV)	Q_i (10^{-20} m^2)	Energy (eV)	Q_i (10^{-20} m^2)
1.00E–04	80	0.4	2.65
0.01	44.4	0.45	2.25
0.02	24.7	0.5	1.92
0.03	13.7	0.6	1.34
0.04	8.2	0.7	0.94
0.05	7.4	0.8	0.655
0.07	7.1	0.9	0.455
0.1	6.5	1	0.32
0.15	5.45	1.2	0.153
0.2	4.8	1.4	0.075
0.25	4.25	1.6	0.037
0.3	3.65	1.8	0.022
0.35	3.1	2	0.014

Source: Adapted from Chantry, P. J. in H. S. W. Massey, E. W. McDaniel, and B. Bederson (Eds.), *Applied Collision Physics Vol. 3: Gas Lasers*, Academic, New York, NY, 1982, p. 35.

F2

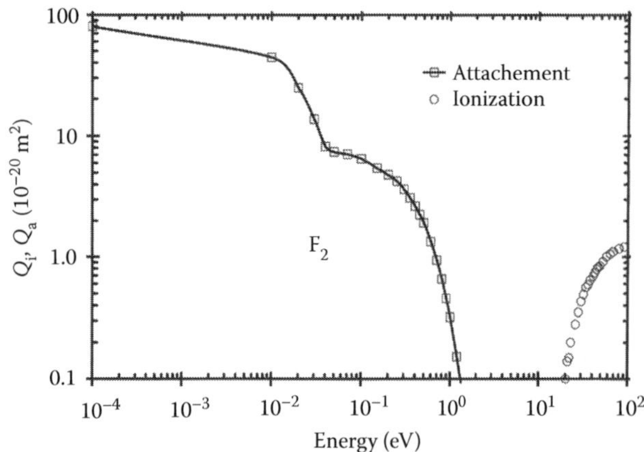

FIGURE 17.4 Low-energy attachment cross sections for F_2. (—□—) Chantry (1982), measurement. The measured ionization cross sections (○) from Stevie and Vasile (1981) are shown for comparison.

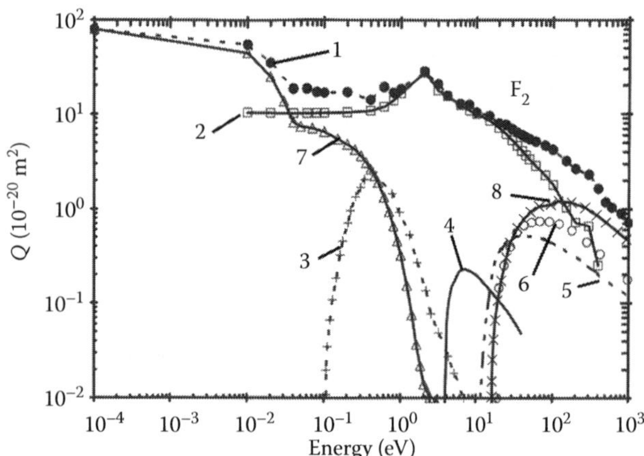

FIGURE 17.5 Relative magnitudes of all cross sections for F_2. 1—total; 2—momentum transfer; 3—vibrational excitation; 4—excitation (3.16 eV threshold); 5—excitation (11.57 eV threshold); 6—excitation (13.08 eV threshold); 7—electron attachment; 8—ionization.

TABLE 17.8
Density-Reduced Ionization Coefficients (α/N), Attachment Coefficients (η/N), and Effective Ionization Coefficients (α–η)/N

E/N (Td)	α/N (10^{-20} m²)	η/N (10^{-20} m²)	(α–η)/N (10^{-20} m²)
100	0.020	0.522	− 0.502
125	0.039	0.421	− 0.382
150	0.081	0.331	− 0.251
175	0.126	0.263	− 0.138
200	0.181	0.224	− 0.043
225	0.225	0.195	0.029
250	0.275	0.165	0.110
275	0.346	0.139	0.208
300	0.411	0.124	0.287
325	0.462	0.114	0.348
350	0.508	0.103	0.405
375	0.557	0.091	0.467
400	0.617	0.081	0.536
425	0.682	0.074	0.609
450	0.748	0.068	0.679
475	0.805	0.064	0.742
500	0.852	0.059	0.793
550	0.920	0.050	0.871
600	0.975	0.042	0.934
650	1.035	0.035	1.000
700	1.103	0.031	1.072
750	1.174	0.028	1.146
800	1.247	0.026	1.221
850	1.318	0.023	1.295
900	1.387	0.020	1.367
950	1.453	0.018	1.435
1000	1.517	0.017	1.500
1250	1.785	0.011	1.774
1500	1.970		1.970
1750	2.101		2.101
2000	2.230		2.230
2250	2.409		2.409

Source: Adapted from Hayashi, M. and T. Nimura, *J. Appl. Phys.*, 54, 4879, 1983.

17.9 REDUCED IONIZATION COEFFICIENTS

Density-reduced ionization coefficients (α/N) theoretically derived by Hayashi and Nimura (1983) are shown in Table 17.8 and Figure 17.6. Figure 17.7 shows the details at lower E/N.

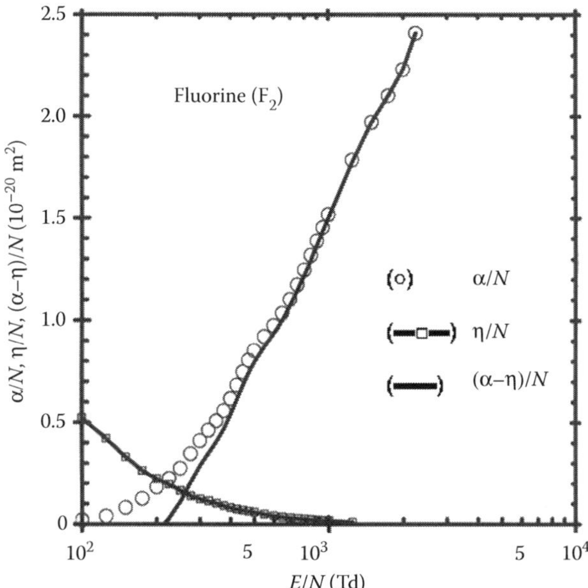

FIGURE 17.6 Density-reduced ionization and attachment coefficients for F_2 theoretically derived by Hayashi and Nimura (1983). (○) α/N; (—□—) η/N; (—) $(\alpha-\eta)/N$.

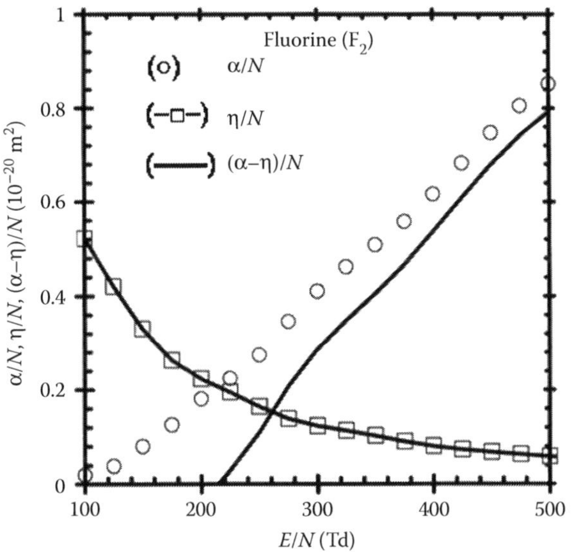

FIGURE 17.7 Details of density-reduced ionization and attachment coefficients for F_2 theoretically derived by Hayashi and Nimura (1983). (○) α/N; (—□—) η/N; (—) $(\alpha-\eta)/N$.

17.10 REDUCED ATTACHMENT COEFFICIENTS

Low-energy attachment coefficients determined by Nygaard et al. (1978) using the current pulse method are shown in Table 17.9. Density-reduced attachment coefficients (η/N) theoretically derived by Hayashi and Nimura (1983) are shown in Table 17.8 and Figure 17.6. Figure 17.7 shows the details at lower E/N.

17.11 ATTACHMENT RATES

Table 17.10 and Figure 17.8 show the attachment rates for F_2. The essential feature is that the rate decreases with increasing energy in the range 0.04–3.0 eV (McCorkle et al., 1983).

TABLE 17.9
Attachment Coefficients at Low E/N for F_2

E/N (Td)	η/N (10^{-20} m²)	E/N (Td)	η/N (10^{-20} m²)
3.0	6.29	9.0	2.16
4.0	5.64	10.0	2.28
4.5	5.09	12.0	1.92
5.0	4.56	13.0	1.66
5.5	4.28	14.0	1.64
6.0	3.92	15.0	1.85
7.0	2.60	16.0	1.90
8.0	2.86		

Source: Adapted from Nygaard, K. J. et al., *Appl. Phys. Lett.*, 32, 351, 1978.

F₂

TABLE 17.10
Attachment Rates for F_2

Temperature (K); Energy (eV)	Attachment Rate (m³/s)	Method	Reference
233	1.2×10^{-14}	Swarm	McCorkle et al. (1986)
298	1.7×10^{-14}		
373	1.9×10^{-14}		
(1)	7.0×10^{-15}	Beam	Schneider and Brau (1978)
(5)	7.5×10^{-16}	Current pulse	Nygaard et al. (1978)
(1)	2.9×10^{-15}	Beam	Chen et al. (1977)
350	3.1×10^{-15}	FA	Sides et al. (1976)
600	4.6×10^{-15}		

Note: Unless otherwise mentioned the temperature (energy) is thermal. FA = flowing afterglow.

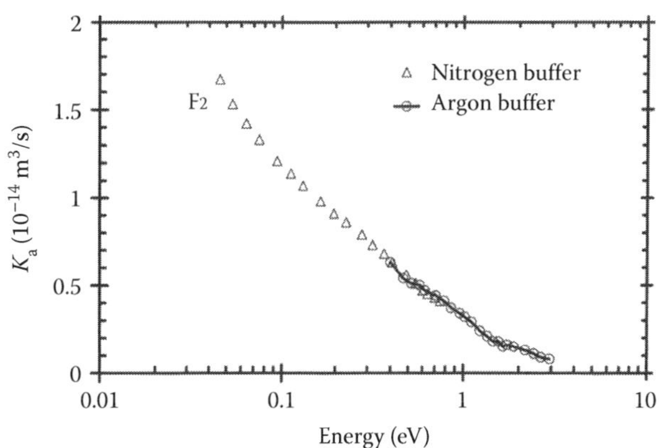

FIGURE 17.8 Attachment rate coefficients for F_2 as a function of mean energy of the swarm. (Adapted from McCorkle, D. L. et al., *J. Chem. Phys.*, 85, 1966, 1986.)

REFERENCES

Brems, V., T. Beyer, B. M. Nestman, H. D. Meyer, and L. S. Cederbaum, *J. Chem. Phys.*, 117, 10635, 2002.

Chantry, P. J., Negative ion formation in gas lasers, in H. S. W. Massey, E. W. McDaniel, and B. Bederson (Eds.), *Applied Collision Physics Vol. 3: Gas Lasers*, Academic, New York, NY, 1982, p. 35, cited by Brems et al. (2002) as their reference [54].

Chen, H. L., R. E. Center, D. W. Trainor, and W. I. Fyfe, *J. Appl. Phys.*, 48, 2297, 1977.

Decorpo, J. J., R. P. Steiger, J. L. Franklin, and J. L. Margrave, *J. Chem. Phys.*, 53, 936, 1970.

Fliplet, A. W., V. McCoy, and T. N. Rescigno, *Phys. Rev. A*, 21, 788, 1980.

Hall, R. J., *J. Chem. Phys.*, 68, 1803, 1978.

Hayashi, M. and T. Nimura, *J. Appl. Phys.*, 54, 4879, 1983.

Hazi, A. U., *Phys. Rev. A*, 23, 2232, 1981.

McCorkle, D. L., L. G. Christophorou, A. A. Christodoulides, and L. Pichiarella, *J. Chem. Phys.*, 85, 1966, 1986.

Nygaard, K. J., S. R. Hunter, J. Fletcher, and S. R. Foltyn, *Appl. Phys. Lett.*, 32, 351, 1978.

Rescigno, T. N., C. F. Binder, C. W. McCurdy, and V. McKoy, *J. Phys. B: At. Mol. Phys.*, 9, 2141, 1976.

Schneider, B. I. and C. A. Brau, *Applied Phys. Lett.*, 33, 569, 1978.

Schneider, B. I. and P. J. Hay, *J. Phys. B: At. Mol. Phys.*, 9, L165, 1976a.

Schneider, B. I. and P. J. Hay, *J. Phys. Phys. Rev. A*, 13, 2049, 1976b.

Sides, G. D., T. O. Tiernan, and R. J. Hanrahan, *J. Chem. Phys.*, 65, 1966, 1976.

Stevie, F. A. and M. J. Vasile, *J. Chem. Phys.*, 74, 5106, 1981.

18

HYDROGEN

H$_2$

CONTENTS

Molecular hydrogen (H_2) has two electrons, with electronic polarizability of 8.944×10^{-41} F m^2 and ionization potential of 15.426 eV.

18.1 SELECTED REFERENCES FOR DATA

See Table 18.1.

TABLE 18.1
Selected References for Data

Quantity	Range: eV (Td)	Reference
Drift velocity	90–1700	**Lisovskiy et al. (2006)**
Analysis	0–1000	Raju (2005)
Excitation cross section	17.5, 20, 30	**Wrkich et al. (2002)**
Excitation cross section	8–15	Trevison and Tennyson (2001)
Momentum transfer cross section	0–1000	Morgan et al. (1999)
Ionization cross section	17–1000	**Straub et al. (1996)**
Total scattering cross section	0.4–250	**Szmytkowski et al. (1996)**
Analysis	0–1000	**Zecca et al. (1996)**
Ionization cross section	400–2000	**Jacobsen et al. (1995)**
Ionization cross section	20–1000	**Krishnakumar and Srivastava (1994)**
Total scattering cross section	0.1–0.175	**Randell et al. (1994)**
Excitation cross section	9.2–20.2	**Khakoo and Segura (1994)**
Ionization cross section	16–1000	van Zyl and Stephan (1994)
Total scattering cross section	10–5000	Jain and Baluja (1992)
Total scattering cross section	4–300	**Nickel et al. (1992)**
Differential scattering cross section	1–5	**Brunger et al. (1991)**

continued

H_2

TABLE 18.1 (continued)
Selected References for Data

Quantity	Range: eV (Td)	Reference
Ionization cross section	100–1000	**Kossmann et al. (1990)**
Ionization cross section	16–1000	**Rudd (1991)**
Total scattering cross section	0.21–9.4	**Subramanian and Kumar (1989)**
Ionization cross section	408–1906	**Edwards et al. (1988)**
Momentum transfer cross section	0–25	**England et al. (1988)**
Total scattering cross section	100–2000	Liu (1987)
Ro-vibrational cross section	0.0–10.0	**Morrison et al. (1987)**
Momentum transfer cross section	0.1–14	**Nesbet et al. (1986)**
Diffusion coefficient	28–5650	**Al-Amin et al. (1985)**
Total scattering cross section	1–50	**Jones (1985)**
Differential scattering cross section	25–500	**Nishimura et al. (1985)**
Analysis	(0.01–150)	**Raju (1985)**
Elastic scattering cross section	1–19	**Furst et al. (1984)**
Drift velocity	(50–250)	**Roznerski and Leja (1984)**
Total scattering cross section	6–400	**Deuring et al. (1983)**
Total scattering cross section	2–500	**Hoffman et al. (1982)**
Excitation cross section	10–50	**Mu-Tao et al. (1982)**
Differential scattering cross section	2.0–200	**Shyn and Sharp (1981)**
Total scattering cross section	0.2–100	**Dalba et al. (1980)**
Total scattering cross section	0.02–2.0	**Ferch et al. (1980)**
Total scattering cross section	25–750	**van Wingerden et al. (1980)**
Ionization coefficient	50–170	**Blevin et al. (1978)**
Transport coefficients	25–3000	Saelee and Lucas (1977)
Ionization coefficient	76–576	**Shimozuma et al. (1977)**
Differential scattering cross section	2–20	**Srivastava et al. (1975)**
Analysis	(0.01–150)	Huxley and Crompton (1974)
Ionization coefficient	100–700	**Cowling and Fletcher (1973)**
Elastic scattering cross section	0.5–10	**Ehrhardt et al. (1968)**
Transport coefficients	90–1050	**Kontoleon et al. (1972)**
Diffusion coefficients	15–90	**Virr et al. (1972)**
Ionization coefficient	175–3000	**Shallal and Harrison (1971)**
Ro-vibrational cross section	0–1.5	Crompton et al. (1970)
Excitation cross section	10–1000	Phelps (1968)
Diffusion coefficients	0.03–12.0	Lowke and Parker (1969)
Analysis	10–3000	Vroom and de Heer (1969)
Ionization cross section	50–900	**Adamczyk et al. (1966)**
Total scattering cross section	0.25–15.0	**Golden et al. (1966)**
Ionization coefficient	50–1200	**Fletcher and Haydon (1966)**
Dissociation cross section	9–90	**Corrigan (1965)**
Ionization cross section	16–1000	**Rapp and Englander-Golden (1965)**
Drift velocity	(60–5500)	**Schlumbohm (1965)**
Vibrational cross section	1–5	**Schulz and Asundi (1967)**
Ionization coefficient	60–1500	**Chanin and Rork (1963)**
Transport coefficients	0.01–200	Engelhardt and Phelps (1963)
Diffusion coefficient	0.0005–10.0	**Warren and Parker (1962)**
Ionization coefficients	150–1500	**Haydon and Robertson (1961)**
Ionization coefficient	40–2800	**Rose (1956)**

Note: Bold font denotes experimental study.

18.2 TOTAL SCATTERING CROSS SECTION

See Table 18.2.

18.3 ELASTIC SCATTERING CROSS SECTION

See Table 18.3.

TABLE 18.2
Recommended Total Scattering Cross Section

Energy (eV)	Q_T (10^{-20} m²)	Energy (eV)	Q_T (10^{-20} m²)	Energy (eV)	Q_T (10^{-20} m²)
0.015	7.24	0.50	11.50	50	3.63
0.020	7.50	0.60	12.00	60	3.36
0.030	7.90	0.80	12.80	70	3.15
0.040	7.89	1.0	13.50	80	2.93
0.050	8.30	2.0	17.00	90	2.80
0.060	8.53	3.0	17.40	100	2.61
0.070	8.70	4.0	17.65	110	2.43
0.090	8.92	5.0	15.60	120	2.25
0.100	9.20	10.0	10.80	140	2.12
0.125	9.44	12	9.51	160	1.96
0.150	9.70	14	8.73	180	1.89
0.180	10.00	16	7.82	200	1.71
0.190	10.09	18	7.16	300	1.27
0.22	10.34	20	6.66	400	1.04
0.25	10.60	25	5.82	500	0.870
0.30	10.86	30	4.88	600	0.736
0.35	11.12	35	4.55	700	0.648
0.40	11.25	40	4.17	800	0.580
0.45	11.38	45	3.99	1000	0.481

Note: See Figure 18.1 for graphical presentation.

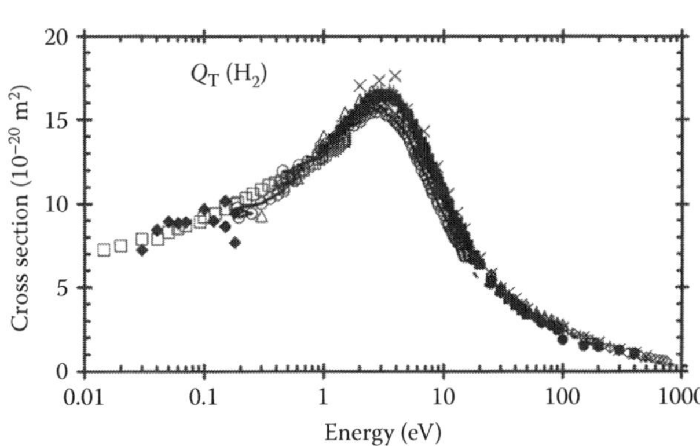

FIGURE 18.1 Absolute total scattering cross sections in H_2. (+) Szmytkowski et al. (1996); (◆) Randell et al. (1994); (— —) Jain and Baluja (1992); (▲) Nickel et al. (1992); (■) Jones (1985); (●) Deuring et al. (1983); (×) Hoffman et al. (1982); (△) Dalba et al. (1980); (□) Ferch et al. (1980); (◆) van Wingerden et al. (1980); (○) Golden et al. (1966).

TABLE 18.3
Recommended Elastic Scattering Cross Section

Energy (eV)	Q_{el} (10^{-20} m²)	Energy (eV)	Q_{el} (10^{-20} m²)	Energy (eV)	Q_{el} (10^{-20} m²)
0.20	11.47	4.80	15.22	80.0	1.06
0.40	11.28	5.00	15.28	90.0	0.958
0.60	11.54	5.25	14.85	100.0	0.874
0.75	12.39	5.50	14.57	125	0.713
0.90	12.66	6.50	13.67	150	0.599
1.10	13.67	7.20	13.05	175	0.514
1.15	13.64	7.80	12.22	200	0.450
2.00	15.59	9.10	11.23	300	0.296
2.20	16.03	10.0	8.55	400	0.220
2.70	16.18	20.0	5.33	500	0.174
2.85	16.46	30.0	3.00	600	0.144
3.20	16.30	40.0	2.18	700	0.123
3.40	16.34	50.0	1.71	800	0.107
4.40	15.65	60.0	1.40	1000	0.086
4.60	15.75	70.0	1.19		

Note: See Figure 18.2 for graphical presentation.

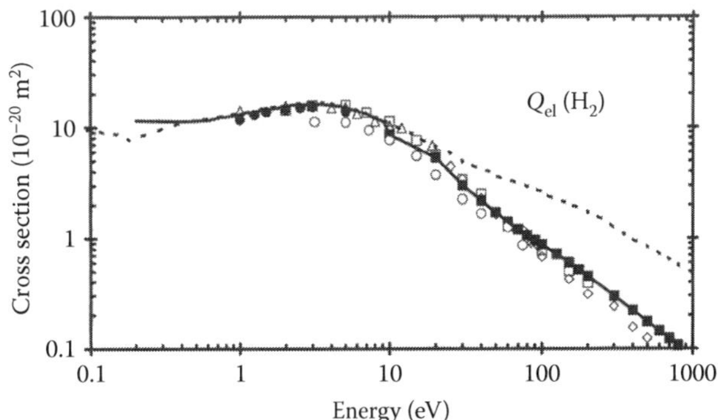

FIGURE 18.2 Elastic scattering cross sections in H_2. (○) Srivastava et al. (76, 1975); (□) Shyn and Sharp (1981); (Δ) Furst et al. (1984); (◊) Nishimura et al. (1985); (●) Brunger et al. (1991); (■) Jain and Baluja (1992); (—) recommended. Total cross sections (— —) have also been plotted for the sake of comparison (Jain and Baluja, 1992); cross sections multiplied by 1.45 for smooth joining. Note the linear scale of the ordinate for Figure 18.1 which explains the apparent change of the shape of Q_T curve.

18.4 MOMENTUM TRANSFER CROSS SECTION

See Table 18.4.

18.5 RO-VIBRATIONAL EXCITATION

See Tables 18.5 through 18.7.

TABLE 18.4
Recommended Momentum Transfer Cross Sections

Energy (eV)	Q_M (10^{-20} m^2)	Energy (eV)	Q_M (10^{-20} m^2)	Energy (eV)	Q_M (10^{-20} m^2)
0.0001	6.35	0.35	13.45	7	8.90
0.001	6.40	0.4	13.90	8	7.85
0.002	6.50	0.5	14.75	10	6.80
0.003	6.60	0.7	16.30	12	6.20
0.005	6.80	1.0	17.40	15	5.50
0.007	7.10	1.2	17.80	17	5.20
0.0085	7.20	1.3	18.05	20	3.80
0.01	7.30	1.5	18.25	30	2.70
0.015	7.65	1.7	18.15	40	2.10
0.02	8.05	1.9	18.10	50	1.70
0.03	8.50	2.1	17.90	70	1.20
0.04	8.96	2.2	17.70	80	1.00
0.05	9.28	2.5	17.20	100	0.80
0.07	9.85	2.8	16.90	200	0.40
0.10	10.50	3	16.30	300	0.29
0.12	10.85	3.3	15.60	400	0.21
0.15	11.40	3.6	15.05	500	0.18
0.17	11.60	4.0	14.75	600	0.16
0.20	12.05	4.5	13.90	800	0.15
0.25	12.50	5.0	13.10	1000	0.12
0.30	13.00	6.0	11.50		

Note: See Figure 18.3 for graphical presentation.

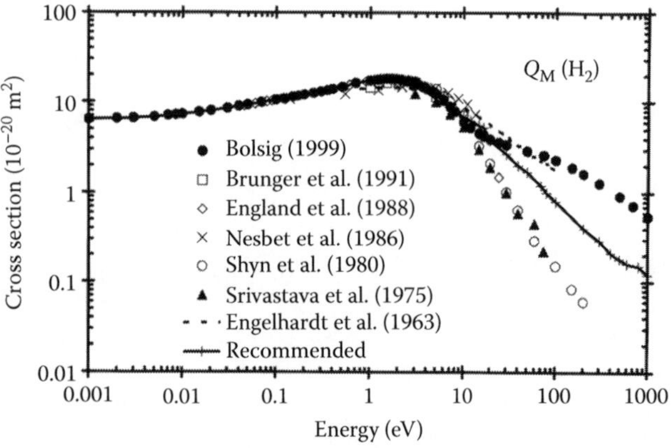

FIGURE 18.3 Momentum transfer cross sections in H_2. (●) Bolsig, Morgan et al. (1999); (□) Brunger et al. (1991); (◊) England et al. (1988); (×) Nesbet et al. (1986), theory; (○) Shyn and Sharp (1981); (▲) Srivastava et al. (1975); (— —) Engelhardt and Phelps (1963); (—+—) recommended.

TABLE 18.5
Threshold Energy and Terminology

Transition	Energy (eV)	Q (Maximum) (10^{-20} m^2)	Terminology
$J = 0 \to 2$	0.046	1.758	Rotational
$J = 1 \to 3$	0.073	1.050	Rotational
$J = 2 \to 4$	0.101	0.802	Rotational
$J = 3 \to 5$	0.128	0.828	Rotational
$v = 0 \to 1, \Delta J = 0$	0.516	0.201	Vibrational, rotationally elastic
$v = 0 \to 1, \Delta J = 0$	0.558	0.267	Vibrational, rotationally elastic
$v = 1 \to 2$	1.03	7.42×10^{-3}	Vibrational
$v = 2 \to 3$	1.55	6.2×10^{-4}	Vibrational

Source: Adapted from Raju, G. G., *Gaseous Electronics: Theory and Practice*, Taylor & Francis, London, 2005.

Note: The cross sections in the last two rows are at threshold and not maximum.

TABLE 18.6
Rotational Excitation Cross Section (10^{-20} m^2)

Energy (eV)	Q_{rot} $J = 0 \to 2$	Energy (eV)	Q_{rot} $J = 1 \to 3$	Energy (eV)	Q_{rot} $J = 2 \to 4$	Q_{rot} $J = 3 \to 5$
0.0439	0.0	0.0727	0.0	0.1008	0.0	
0.047	0.019	0.075	0.0070	0.1280		0.0
0.050	0.027	0.08	0.0140	0.15	0.0272	0.016
0.055	0.035	0.085	0.0198	0.20	0.0474	0.037
0.060	0.042	0.09	0.0237	0.25	0.0557	0.051
0.065	0.048	0.095	0.0265	0.30	0.0663	0.062
0.070	0.053	0.1	0.0280	0.35	0.0766	0.072
0.080	0.062	0.11	0.0330	0.40	0.0872	0.082
0.090	0.068	0.12	0.0364	0.45	0.0955	0.093
0.10	0.074	0.13	0.0394	0.50	0.1054	0.104
0.11	0.079	0.15	0.0450	0.60	0.132	0.129
0.13	0.088	0.2	0.0580	0.70	0.162	0.160
0.15	0.097	0.25	0.0719	0.80	0.193	0.194
0.20	0.115	0.3	0.0860	0.90	0.227	0.233
0.25	0.132	0.35	0.1000	1.0	0.266	0.271
0.30	0.152	0.4	0.1140	1.5	0.463	0.478
0.35	0.175	0.45	0.1285	2.0	0.619	0.637
0.40	0.200	0.5	0.1439	2.5	0.719	0.742
0.45	0.228	0.56	0.1633	3.0	0.774	0.799
0.50	0.260	0.6	0.1776	3.5	0.799	0.825
0.60	0.323	0.66	0.1996	4.0	0.802	0.828
0.70	0.394	0.7	0.2135	4.5	0.790	0.818
0.80	0.469	0.8	0.2518	5.0	0.771	0.797
0.90	0.555	0.9	0.2919	5.5	0.748	0.774
1.0	0.636	1.01	0.3338	6.0	0.721	0.747
1.2	0.796	1.2	0.420	7.0	0.669	0.692
1.5	1.036	1.4	0.510	8.0	0.617	0.640
2.0	1.370	1.6	0.610	10.0	0.529	0.548
2.5	1.585	1.8	0.700			

TABLE 18.6 (continued)
Rotational Excitation Cross Section (10^{-20} m²)

Rotational Excitation Cross Section						
Energy (eV)	Q_{rot} $J = 0 \to 2$	Energy (eV)	Q_{rot} $J = 1 \to 3$	Energy (eV)	Q_{rot} $J = 2 \to 4$	Q_{rot} $J = 3 \to 5$
3.0	1.704	2.0	0.786			
3.5	1.755	2.5	0.937			
4.0	1.758	3.0	1.014			
4.5	1.732	3.5	1.046			
5.0	1.689	4.0	1.050			
6.0	1.579	4.5	1.036			
7.0	1.462	5.0	1.011			
8.0	1.350	6.0	0.946			
9.0	1.248	7.0	0.876			
10.0	1.156	8.0	0.809			
15.0	0.730	9.0	0.748			
		10.0	0.694			
		15.0	0.450			

Source: Adapted from England, J. P., M. T. Elford, and R. W. Crompton, *Aust. J. Phys.*, 41, 573, 1988.
Note: See Figure 18.4 for graphical presentation.

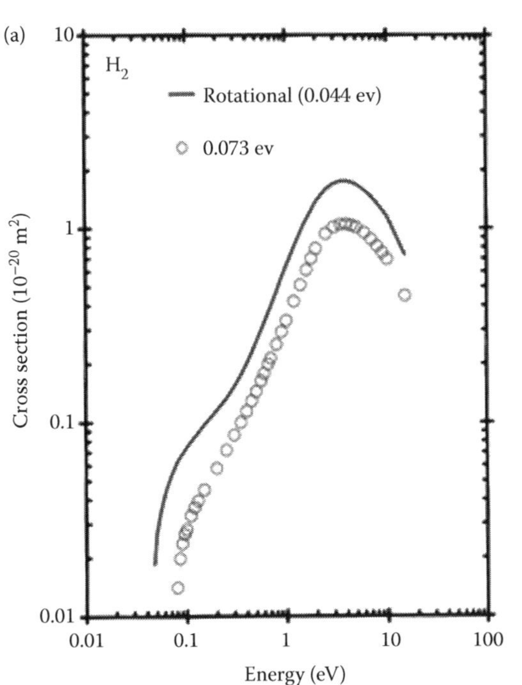

FIGURE 18.4 Rotational and vibrational cross sections in H_2 derived from swarm parameters. Rotational excitation cross sections with threshold energies: (a) (—) 0.044 eV; (○) 0.073 eV. (b) (—) 0.101 eV; (○) 0.126 eV. (Adapted from J. P. England, M. T. Elford, and R. W. Crompton, *Aust. J. Phys.*, 41, 573, 1988.)

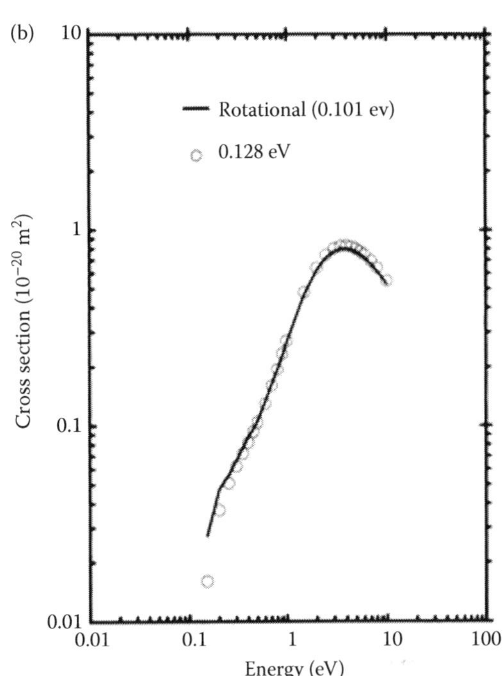

FIGURE 18.4 Continued.

TABLE 18.7
Vibrational Excitation Cross Section

Energy (eV)	Q_V (10^{-20} m²) $v = 0 \to 1, \Delta J = 0$	Q_V (10^{-20} m²) $v = 0 \to 1, \Delta J = 2$
0.516	0.0	
0.558		0.0
0.56	0.0028	
0.575		0.0005
0.60	0.0053	0.0013
0.65	0.0082	0.0032
0.75	0.0143	0.0078
0.85	0.0206	0.0134
0.95	0.0280	0.0205
1.00	0.0322	0.0248
1.05	0.0363	0.0287
1.10	0.0407	0.0333
1.15	0.0450	0.0380
1.20	0.0499	0.0437
1.30	0.0594	0.0549
1.40	0.0688	0.0653
1.60	0.0865	0.0892
1.80	0.1038	0.1139
2.2	0.1394	0.1639
2.4	0.1561	0.1869
2.6	0.1709	0.2121
3.0	0.1916	0.2494
3.5	0.2008	0.2672
4.0	0.1860	0.2540
4.5	0.1630	0.2270
5.0	0.1460	0.2040
6.0	0.1160	0.1640
7.0	0.0876	0.1224
8.0	0.0637	0.0879

continued

TABLE 18.7 (continued)
Vibrational Excitation Cross Section

Energy (eV)	Q_V (10^{-20} m^2) $v = 0 \rightarrow 1, \Delta J = 0$	Q_V (10^{-20} m^2) $v = 0 \rightarrow 1, \Delta J = 2$
9.0	0.0506	0.0684
10.0	0.0376	0.0498
11.0	0.0292	0.0388
12.0	0.0215	0.0285
13.0	0.0180	0.0200
14.0	0.0170	0.0150
15.0	0.0150	0.0100

Source: Adapted from England, J. P., M. T. Elford, and R. W. Crompton, *Aust. J. Phys.*, 41, 573, 1988.
Note: See Figures 18.5 and 18.6 for graphical presentation.

H_2

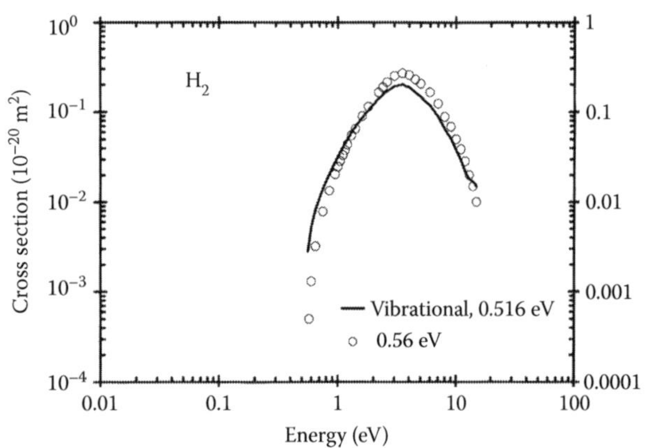

FIGURE 18.5 Vibrational cross sections in H_2 derived from swarm parameters. Rotational excitation cross sections with threshold energies: (—) 0.516 eV; (○) 0.56 eV. (Adapted from England, J. P., M. T. Elford, and R. W. Crompton, *Aust. J. Phys.*, 41, 573, 1988.)

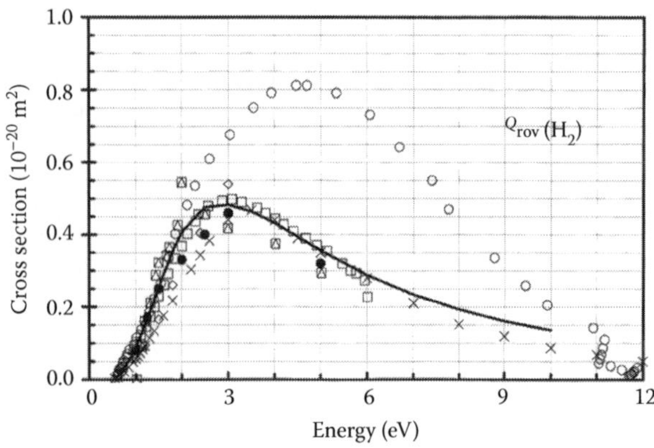

FIGURE 18.6 Vibrational cross sections ($Q_{v=0\rightarrow1}$) in H_2 determined by several methods. Swarm method: (×) England et al. (1988); (◊) Crompton et al. (1970); (○) Engelhardt and Phelps (1963). Beam method: (●) Brunger et al. (1991); (□) Ehrhardt et al. (1968); (△) Schulz and Asundi (1964). The cross sections of England et al. have been obtained by adding the rotationally elastic and inelastic $Q_{v=0\rightarrow1}$ cross sections. The curve is from Morrison et al. (1987).

18.6 ELECTRONIC EXCITATION CROSS SECTION

See Tables 18.8 through 18.10.

18.7 IONIZATION CROSS SECTION

See Table 18.11.

TABLE 18.8
Excitation Levels of H_2 Molecule from Ground State

State	Onset (eV)	Q_{ex} (Maximum) (10^{-20} m^2)	Remarks
b $^3\Sigma_u^+$	6.9[KS]	0.83	Unstable lower state of continuous spectrum; leads to dissociation
B $^1\Sigma_u^+$	11.183[W], 11.36[H]	0.55	Optically allowed transition (Lyman bands)
c $^3\Pi_u$	11.789[W], 11.87[H]	0.20	Optically forbidden transition to triplet state
a $^3\Sigma_g^+$	11.793[W], 11.89[H]	0.25	Optically forbidden transition to triplet state
C $^1\Pi_u$	12.295[W]	4.25	Optically allowed transition (Werner bands)
B′ $^1\Sigma_u^+$	14.8[MLM]	0.05	Optically allowed transition[Z]

Source: Adapted from Raju, G. G., *Gaseous Electronics: Theory and Practice*, Taylor & Francis, London, 2005.
Note: H = Herzberg (1950); KS = Khakoo and Segura (1994); Mu-Tao et al., 1982; Wrkich et al. (2002); Z = Zecca et al. (1996).

TABLE 18.9
Cross Section Derived from Monte Carlo Simulation

Level	Threshold (eV)	Peak Energy	X-Section at Peak (10^{-20} m^2)
b $^3\Sigma_u^+$	8.8	16	0.28
B $^1\Sigma_u^+$	11.37	40	0.48
c $^3\pi_u$	11.87	15	0.56
a $^3\Sigma_g^+$	11.89	15	0.09
C $^1\pi_u$	12.40	35	0.24
E $^1\Sigma_g^+$	12.40	50	0.076
e $^3\Sigma_u^+$	13.36	16.5	0.068
B′ $^1\Sigma_u^+$	13.70	40	0.038
d $^3\pi_u$	13.97	18.5	0.034
D $^1\pi_u$	14.12	50	0.012
Ionization	15.425	70	0.972
r_{02}	0.0439	4	1.84
r_{13}	0.0727	4	1.07
v_{01}	0.516	3	0.50
v_{02}	1.003	4.3	0.040

Source: Adapted from Hayashi, M., *J. Phys.*, Colloque c7, 45, 1979.

TABLE 18.10
Recommended Total Excitation Cross Section

Energy (eV)	Q_{ex} (10^{-20} m^2)	Energy (eV)	Q_{ex} (10^{-20} m^2)	Energy (eV)	Q_{ex} (10^{-20} m^2)
8.9	9.00E–05	19	1.11	150	0.91
9.0	9.00E–05	20	1.18	200	0.88
10	0.12	25	1.28	300	0.63
11	0.21	30	1.35	400	0.55
12	0.49	40	1.33	500	0.47
13	0.69	50	1.29	600	0.39
14	0.84	60	1.23	700	0.36
15	0.95	70	1.17	800	0.33
16	1.04	80	1.10	900	0.64
17	1.08	90	1.04	1000	0.28
18	1.10	100	0.99		

Note: See Figure 18.7 for graphical presentation.

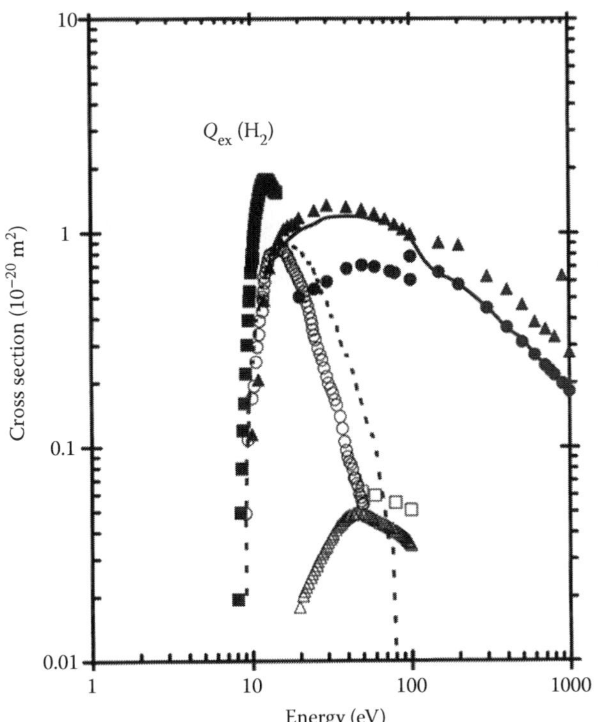

FIGURE 18.7 Excitation and dissociation cross sections in H_2 in units of 10^{-20} m^2. (◆) Trevisan and Tennyson (2001); (▲) Morgan et al. (1999), sum of all states; (○) Khakoo and Segura (1994), b $^3\Sigma_u^+$ continuum; (△) Mu-Tao et al. (1982), B′ $^1\Sigma_u^+$ state only; (□) Vroom and de Heer (1969); (——) Corrigan (1965), dissociation cross section; (●) excitation cross sections of van Wingerden et al. (1980) for the Lyman and Werner bands are shown for comparison.

18.8 DRIFT VELOCITY OF ELECTRONS

See Table 18.12.

18.9 DIFFUSION COEFFICIENT

See Table 18.13 and Figure 18.11.

18.10 IONIZATION COEFFICIENT

See Table 18.14.

TABLE 18.11
Recommended Ionization Cross Section for H_2

Energy (eV)	H_2^+ (10^{-20} m^2)	H^+ (10^{-20} m^2)	Total (10^{-20} m^2)
17	0.0611		0.0611
20	0.279		0.2790
25	0.513	0.0046	0.5176
30	0.690	0.0092	0.6992
35	0.795	0.0188	0.8138
40	0.868	0.0307	0.8987
45	0.896	0.0436	0.9396
50	0.916	0.0514	0.9674
55	0.930	0.0608	0.9908
60	0.937	0.0664	1.0034
65	0.934	0.0724	1.0064
70	0.931	0.0748	1.0058
75	0.922	0.0781	1.0001
80	0.913	0.0782	0.9912
85	0.902	0.0794	0.9814
90	0.890	0.0797	0.9697
95	0.881	0.0803	0.9613
100	0.868	0.0800	0.9480
110	0.839	0.0784	0.9174
120	0.820	0.0761	0.8961
140	0.774	0.0704	0.8444
160	0.732	0.0668	0.7988
180	0.684	0.0618	0.7458
200	0.648	0.0569	0.7049
225	0.609	0.0526	0.6616
250	0.572	0.0467	0.6187
275	0.534	0.0428	0.5768
300	0.507	0.0406	0.5476
350	0.458	0.0350	0.4930
400	0.419	0.0303	0.4493
450	0.383	0.0267	0.4097
500	0.358	0.0248	0.3828
550	0.326	0.0217	0.3477
600	0.305	0.0201	0.3251
650	0.290	0.0185	0.3085
700	0.272	0.0175	0.2895
750	0.261	0.0162	0.2772
800	0.247	0.0152	0.2622
850	0.238	0.0140	0.2520
900	0.226	0.0137	0.2397
950	0.213	0.0127	0.2257
1000	0.202	0.0119	0.2139

Note: See Figure 18.8 for graphical presentation.

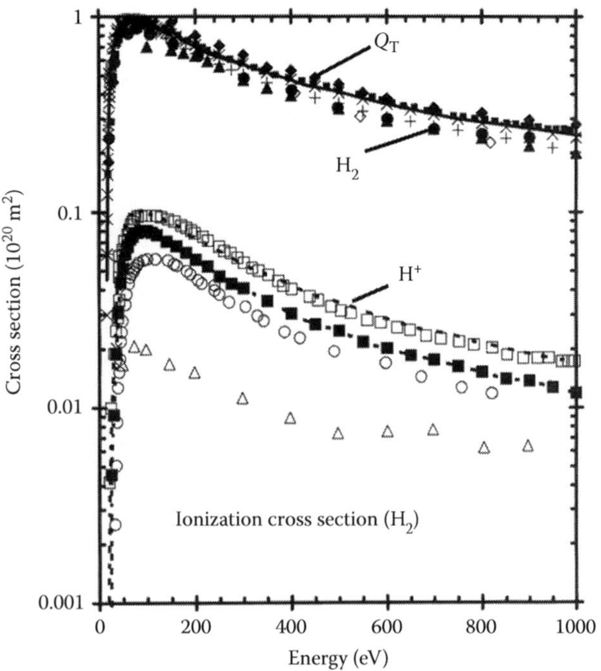

FIGURE 18.8 Dissociative and nondissociative ionization cross sections in H_2. Dissociative: (—■—) Straub et al. (1996); (□) Krishnakumar and Srivastava (1994); (— —) van Zyl and Stephan (1994); (△) Adamczyk et al. (1966); (○) Rapp and Englander-Golden (1965). Nondissociative: (■) Krishnakumar and Srivastava (1994); (◇) Kossmann et al. (1990); (◇) Edwards et al. (1988). Total: (+) Straub et al. (1996), calculated by adding the two cross sections ($H_2^+ + H^+$); (◆) Krishnakumar and Srivastava (1994); (—) Rudd (1991); (×) Rapp and Englander-Golden (1965).

TABLE 18.12
Recommended Drift Velocity for Electrons in H_2

E/N (Td)	*W* (m/s)	*E/N* (Td)	*W* (m/s)	*E/N* (Td)	*W* (m/s)
0.002		0.25	2760	8.0	1.65(4)
0.003		0.30	3130	10.0	1.87(4)
0.004		0.35	3470	12.0	2.07(4)
0.006		0.40	3790	14.0	2.27(4)
0.009		0.50	4330	17.0	2.55(4)
0.012	186.2	0.60	4820	20.0	2.81(4)
0.02	311.0	0.70	5240	25.0	3.22(4)
0.025	385.0	0.80	5610	30.0	3.66(4)
0.03	459.0	1.0	6.23(3)	35.0	4.08(4)
0.035	530.0	1.2	6.62(3)	40.0	4.54(4)
0.04	600.0	1.4	6.92(3)	50	5.71(4)
0.05	737.0	1.7	7.64(3)	60	6.95(4)
0.06	870.0	2.0	8.37(3)	70	8.30(4)
0.07	998.0	2.5	9.13(3)	80	9.80(4)
0.08	1120	3.0	9.82(3)	100	1.28(5)
0.10	1370	3.5	1.08(4)	120	1.66(5)
0.12	1580	4.0	1.15(4)	140	1.94(5)
0.14	1780	5.0	1.29(4)	150	2.13(5)
0.17	2070	6.0	1.42(4)	200	3.36(5)
0.20	2350	7.0	1.54(4)	250	4.19(5)
300	4.68(5)	700	9.23(5)	2300	2.00(6)
350	5.22(5)	800	9.94(5)	2650	2.35(6)

TABLE 18.12 (continued)
Recommended Drift Velocity for Electrons in H_2

E/N (Td)	*W* (m/s)	*E/N* (Td)	*W* (m/s)	*E/N* (Td)	*W* (m/s)
400	6.02(5)	900	1.04(6)	5050	2.69(6)
500	7.25(5)	1000	1.13(6)		
600	8.30(5)	2000	1.80(6)		

Note: a(b) means a × 10^b. See Figure 18.9 for graphical presentation.

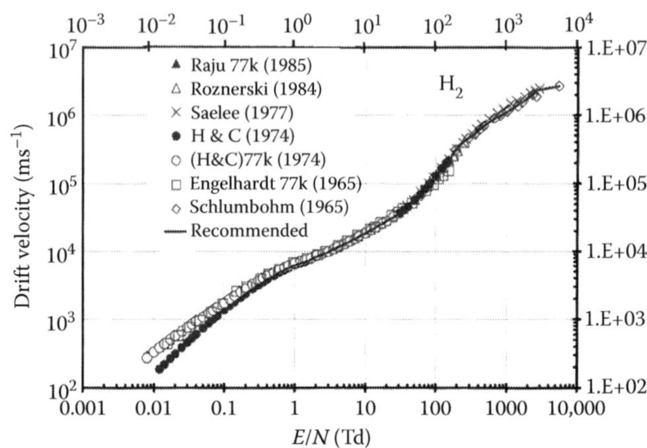

FIGURE 18.9 Drift velocity as a function of *E/N* in hydrogen. Unless otherwise specified, temperature is 293 K. (△) Roznerski and Leja (1984); (—×—) Saelee and Lucas (1977); (●) Huxley and Crompton (1974); (◊) Schlumbohm (1965); (—) recommended. 77 K: (—●—) Raju (1985); (○) Huxley and Crompton (3, 1974); (□) Engelhardt and Phelps (1963).

TABLE 18.13
Recommended Characteristic Energy for H_2

E/N (Td)	D_r/μ (V)	*E/N* (Td)	D_r/μ (V)	*E/N* (Td)	D_r/μ (V)
0.02	0.0259	0.400	0.0524	8.00	0.447
0.025	0.0262	0.500	0.0593	10.0	0.511
0.03	0.0265	0.600	0.0663	12.0	0.573
0.035	0.0268	0.700	0.0735	14.0	0.630
0.04	0.0271	0.800	0.0809	17.0	0.710
0.05	0.0278	1.000	0.0953	20.0	0.787
0.06	0.0285	1.20	0.110	25.0	0.916
0.07	0.0292	1.40	0.125	30.0	1.05
0.08	0.0300	1.70	0.147	35.0	1.18
0.100	0.0315	2.00	0.168	40.0	1.33
0.120	0.0329	2.50	0.201	50.0	1.72
0.140	0.0344	3.00	0.231	60.0	1.98
0.170	0.0366	3.50	0.259	70.0	2.19
0.200	0.0387	4.00	0.285	80.0	2.37
0.250	0.0423	5.00	0.330	100	2.67
0.300	0.0457	6.00	0.371	120	2.93
0.350	0.0490	7.00	0.410	140	3.19
170	3.49	420	7.42	1970	10.54
200	3.82	570	9.71	2850	9.51
300	5.12	880	11.45		

TABLE 18.13 (continued)
Recommended Characteristic Energy for H_2

E/N (Td)	D_r/μ (V)	*E/N* (Td)	D_r/μ (V)	*E/N* (Td)	D_r/μ (V)
360	6.29	1450	11.22		

Note: See Figure 18.10 for graphical presentation.

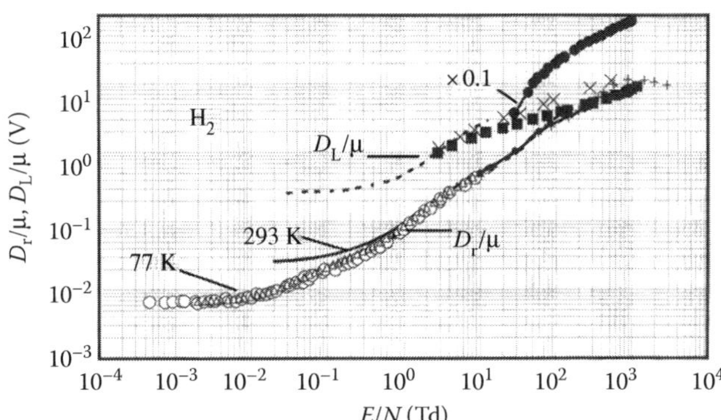

FIGURE 18.10 Ratios of radial diffusion coefficient to mobility (D_r/μ) and longitudinal diffusion coefficient to mobility (D_L/μ) in H_2 as a function of *E/N*. Unless otherwise specified, temperature is 293 K. D_r/μ: (+) Al-Amin et al. (1985); (■) Saelee and Lucas (1977); (△) Huxley and Crompton (1974) 77 K; (●) Kontoleon et al. (1972); (×) Virr et al. (1972); (— — —) Engelhardt and Phelps (1963) 77 K; (○) Warren and Parker (1962) 77 K. D_L/μ: (●) Saelee and Lucas (1977) ; (———) Lowke and Parker (1969). The experimental results of Wagner et al. (1967) (D_L/μ) have not been shown because they are in excellent agreement with those of Lowke and Parker (1969). Note that the ratio D_L/μ is multiplied by a factor of 10 for better presentation.

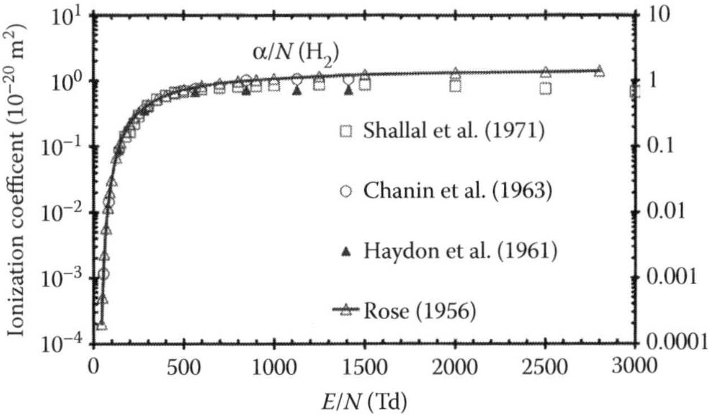

FIGURE 18.11 Reduced first ionization coefficient for H_2. (□) Shallal and Harrison (1971); (○) Chanin and Rork (1963); (▲) Haydon and Robertson (1961); (Δ) Rose (1956); (—) recommended.

18.11 GAS CONSTANTS

Gas constants evaluated according to expression

$$\frac{\alpha}{N} = F \exp\left(-\frac{GN}{E}\right) \quad (18.1)$$

are $F = 1.44 \times 10^{-20}$ m^2 and $G = 391.4$ Td.

18.12 ION MOBILITY (H^+/H_2)

See Table 18.15.

Most recent measurements of dissociative attachment cross sections in H_2 and D_2 have revealed three peaks, due to processes in 4 eV, 7–13 eV and 14 eV energies. The cross sections are two orders of magnitude lower when compared with dissociative attachment cross sections in O_2 (Krishnakumar et al., 2011).

TABLE 18.14
Recommended Reduced Ionization Coefficient

E/N (Td)	α/N (10^{-20} m^2)	*E/N* (Td)	α/N (10^{-20} m^2)	*E/N* (Td)	α/N (10^{-20} m^2)
45	0.0002	200	0.220	800	0.974
50	0.0005	225	0.275	900	1.022
60	0.0023	250	0.328	1000	1.07
70	0.0058	300	0.426	1250	1.16
80	0.0119	350	0.514	1500	1.23
90	0.0203	400	0.594	2000	1.31
100	0.0314	450	0.665	2500	1.35
125	0.0687	500	0.728	2800	1.39
150	0.115	600	0.826		
175	0.167	700	0.912		

TABLE 18.15
Reduced H^+ Ion Mobility and Drift Velocity for H_2

E/N (Td)	μ_0 (10^{-4} m^2/V s)	W^+ (10^2 m/s)
4.00	16.0	1.72
5.00	16.0	2.15
6.00	16.0	2.58
8.00	16.0	3.44
10.0	16.0	4.30
12.0	16.0	5.16
15.0	15.9	6.41
20.0	15.8	8.49
25.0	15.7	10.5
30.0	15.5	12.5
40.0	15.2	16.3
50.0	14.9	20.0
60.0	14.5	23.4
80.0	13.9	29.9
100	13.4	36.0
120	13.2	42.6
150	13.1	52.8
200	13.1	70.4
250	13.2	88.7
300	13.3	107
400	13.7	147

Source: Adapted from Ellis, H. W. et al., *Atom. Nucl. Data Tables*, 17, 177–210, 1981.

REFERENCES

Adamczyk, B., A. J. H. Boerboom, B. L. Schram, and J. Kistemaker, *J. Chem. Phys.*, 44, 4640, 1966.

Al-Amin, S. A. J., J. Lucas, and H. N. Kücükarpaci, *J. Phys. D: Appl. Phys.*, 18, 2007, 1985. (D_r/μ in H_2, N_2, and CO are measured.)

Blevin, H. A., J. Fletcher, and S. R. Hunter, *J. Phys. D: Appl. Phys.*, 11, 2295, 1978.

Brunger, M. J., S. J. Buckman, D. S. Newman, and D. T. Alee, *J. Phys. B: At. Mol. Opt. Phys.*, 24, 1435, 1991.

Chanin, L. M. and G. D. Rork, *Phys. Rev.*, 132, 2547, 1963.

Corrigan, S. J. B., *J. Chem. Phys.*, 43, 4381, 1965.

Cowling, I. R. and J. Fletcher, *J. Phys. B: At. Mol. Phys.*, 6, 665, 1973.

Crompton, R. W., D. K. Gibson, and A. G. Robertson, *Phys. Rev. A*, 2, 1386, 1970.

Dalba, G., P. Fornasini, I. Lazzizzera, G. Ranieri, and A. Zecca, *J. Phys. B: At. Mol. Phys.*, 13, 2839, 1980.

Deuring, A., K. Floeder, D. Fromme, W. Raith, A. Schwab, G. Sinapius, P. W. Zitzewitz, and J. Krug, *J. Phys. B: At. Mol. Phys.*, 16, 1633, 1983.

Edwards, A. K., R. M. Wood, A. S. Beard, and R. L. Ezell, *Phys. Rev. A*, 37, 3697, 1988.

Ehrhardt, H., L. Langhans, F. Linder and H. S. Taylor, *Phys. Rev.*, 173, 222, 1968.

Ellis, H. W., R. Y. Pai, E. W. McDaniel, E. A. Mason, and L. A. Viehland, *Atom. Nucl. Data Tables*, 17, 177–210, 1981.

Engelhardt, A. G. and A. V. Phelps, *Phys. Rev.*, 131, 2115, 1963.

England, J. P., M. T. Elford, and R. W. Crompton, *Aust. J. Phys.*, 41, 573, 1988.

Ferch, J., W. Raith, and K. Schröder, *J. Phys. B: At. Mol. Phys.*, 13, 1481, 1980.

Fletcher, J. and S. C. Haydon, *Aust. J. Phys.*, 19, 615, 1966.

Furst, J., M. Mahgerefteh, and D. Golden, *Phys. Rev. A*, 30, 2256, 1984.

Golden, D. E., H. W. Bandel, and J. A. Salerno, *Phys. Rev.*, 146, 40, 1966.

Hayashi, M., *J. Phys.*, Colloque c7, 45, 1979.

Haydon, S. C. and A. G. Robertson, *Proc. Phys. Soc. London*, 78, 92, 1961.

Herzberg, G., *Molecular Spectra and Molecular Structure*, Van Nostrand and Co., Princeton, NJ, 1950.

Hoffman, K. R., M. S. Dababneh, Y. F. Hsieh, W. E. Kaippila, V. Pol, J. H. Smart, and T. S. Stein, *Phys. Rev. A*, 25, 1393, 1982.

Huxley, L. G. H. and R. W. Crompton, *Diffusion and Drift of Electrons in Gases*, John Wiley and Sons Inc., New York, NY, 1974.

Jacobsen, F. M., N. P. Frandsen, H. Knudsen, and U. Mikkelsen, *J. Phys. B: At. Mol. Opt. Phys.*, 28, 4675, 1995.

Jain, A. and K. L. Baluja, *Phys. Rev. A.*, 45, 202, 1992.

Jones, R. K., *Phys. Rev. A*, 31, 2898, 1985.

Khakoo, M. A. and J. Segura, *J. Phys. B: At. Mol. Opt. Phys.*, 27, 2355, 1994.

Kontoleon, N., J. Lucas, and L. E. Virr, *J. Phys. D: Appl. Phys.*, 5, 956, 1972.

Kossmann, H., O. Schwarzkopf, and V. Schmidt, *J. Phys. B: At. Mol. Opt. Phys.*, 23, 301, 1990.

Krishnakumar, E. and S. K. Srivastava, *J. Phys. B: At. Mol. Opt. Phys.*, 27, L251–L258, 1994.

Krishnakumar, E.V., S. Deniff, I. Cadez, S. Markeli, and N.J. Mason, *Phys. Rev. Letters*, 106, 243201, June 2011.

Lisovskiy, V., J.-P. Booth, K. Landry, D. Douai, V. Cassagne, and V. Yegorenkov, *J. Phys. D: Appl. Phys.*, 39, 660, 2006.

Liu, J. W., *Phys. Rev. A*, 35, 591, 1987.

Lowke, J. J. and J. H. Parker, Jr., *Phys. Rev.*, 181, 302, 1969. D_L/μ and D_r/μ for several gases are calculated and compared with measured ones.

Morgan, W. L., J. P. Bouef, and L. C. Pitchford, www.siglo-kinema.com, 1999.

Morrison, M. A., R. W. Crompton, B. C. Saha, and Z. Lj. Petrović, *Aust. J. Phys.*, 40, 239, 1987.

Mu-Tao, L., R. R. Lucchese, and V. McKoy, *Phys. Rev. A*, 26, 3240, 1982.

Nesbet, R. K., C. J. Noble, and L. A. Morgan, *Phys. Rev. A*, 34, 2798, 1986.

Nickel, J. C., I. Kanik, S. Trajmar, and K. Imre, *J. Phys. B: At. Mol. Phys.*, 25, 2427, 1992.

Nishimura, H., A. Danjo, and H. Sugahara, *J. Phys. Soc. Japan*, 54, 1757, 1985.

Phelps, A. V., *Rev. Mod. Phys.*, 40, 399, 1968.

Raju, A., *Basic Data on Gas Discharges*, internal report, University of Windsor, 1985.

Raju, G. G., *Gaseous Electronics: Theory and Practice*, Taylor & Francis, London, 2005.

Randell, J., S. L. Hunt, G. Mrotzek, J.-P. Ziesel, and D. Field, *J. Phys. B: At. Mol. Opt. Phys.*, 27, 2369, 1994.

Rapp, D. and P. Englander-Golden, *J. Chem. Phys.*, 43, 1464, 1965.

Rose, D. J., *Phys. Rev.*, 104, 273, 1956.

Roznerski, W. and K. Leja, *J. Phys. D: Appl. Phys.*, 17, 279, 1984.

Rudd, M. E., *Phys. Rev. A*, 44, 1644, 1991.

Saelee, H. T. and J. Lucas, *J. Phys. D: Appl. Phys.*, 10, 343, 1977.

Schlumbohm, H., *Z. Physik*, 182, 317, 1965.

Schulz, G. J. and R. K. Asundi, *Phys. Rev.*, 158, 25, 1967. Also see 135, A988, 1964. (In this paper, Schulz gives data for all the three gases, CO, H_2, and N_2.)

Shallal, M. A. and J. A. Harrison, *J. Phys. D: Appl. Phys.*, 4, 1550, 1971. (Ionization currents were measured in both uniform and non-uniform electric fields between 300 and 3300 Td.)

Shimozuma, M., Y. Sakai, H. Tagashira, and S. Sakamoto, *J. Phys. D: Appl. Phys.*, 10, 1671, 1977.

Shyn, T. W. and W. E. Sharp, *Phys. Rev. A*, 24, 1734, 1981.

Srivastava, S. K., A. Chutjian, and S. Trajmar, *J. Chem. Phys.*, 63, 2659, 1975.

Straub, H. C., P. Renault, B. G. Lindsay, K. A. Smith, and R. F. Stebbings, *Phys. Rev. A*, 54, 2146, 1996.

Subramanian, K. P. and V. Kumar, *J. Phys. B: At. Mol. Opt. Phys.* 22, 2387, 1989.

Szmytkowski, C., K. Maciąg, and G. Karwasz, *Phys. Scripta*, 54, 271, 1996.

Trevisan, C. S. and J. Tennyson, *J. Phys. B: At. Mol. Opt. Phys.*, 34, 2935, 2001.

van Wingerden, B., R. W. Wagenaar, and F. J. de Heer, *J. Phys. B: At. Mol. Phys.*, 13, 3481, 1980.

van Zyl, B. and T. M. Stephan, *Phys. Rev. A*, 40, 3164, 1994.

Virr, L. E., J. Lucas, and N. Kontoleon, *J. Phys. D: Appl. Phys.*, 5, 542, 1972.

Vroom, D. A. and F. J. de Heer, *J. Chem. Phys.*, 50, 580, 1969.

Wagner, E. B., F. J. Davis, and G. S. Hurst, *J. Chem. Phys.* 47, 3138, 1967.

Warren, R. W. and J. H. Parker, *Phys. Rev.*, 128, 2661, 1962

Wrkich, J., D. Mathews, I. Kanik, S. Trajmar, and M. A. Khakoo, *J. Phys. B: At. Mol. Opt. Phys.*, 35, 4695, 2002.

Zecca, A., G. P. Karwasz, and R. S. Brusa, *Rivista del Nu. Ci.*, 19, 1, 1996.

19

HYDROGEN BROMIDE

CONTENTS

Hydrogen bromide (HBr) is an electron-attaching, polar molecule with 36 electrons. It has electronic polarizability of 4.017×10^{-40} F m^2, dipole moment of 0.827 D, and ionization potential of 11.66 eV. The dissociation energy is 3.795 eV and the vibrational energy is 328.4 meV. The electron affinity of the molecule is 0.11 eV (Christophorou et al., 1968), and that of the attaching atom, Br, is 3.364 eV.

19.1 SELECTED REFERENCES FOR DATA

Table 19.1 shows selected references for data on HBr.

TABLE 19.1
Selected References for Data on HBr

Parameter	Range: eV, (Td), [K]	Reference
Q_v	0–10	**Rohr (1978)**
k_a	[48–170]	**Speck et al. (2001)**
k_a	0–2	**Wang and Lee (1988)**
k_a	[300, 515]	**Smith and Adams (1987)**
k_a	[300, 515]	**Adams et al. (1986)**
AP	0–1	**Ziesel et al. (1975)**
k_a	0–2.0	**Christophorou et al. (1968)**
AP	—	**Frost and McDowell (1958)**

Note: AP = attachment process; k_a = attachment rate constant; Q_v = vibrational cross section. Bold font indicates experimental study.

19.2 VIBRATIONAL EXCITATION CROSS SECTION

The threshold energy for vibrational excitation ($\nu = 1$) is 0.328 eV and shape resonance is observed at 0.9 eV. Table 19.2 and Figure 19.1 (Rohr, 1978) show these data.

19.3 ATTACHMENT PROCESSES

Dissociative attachment according to reaction 19.1 is the dominant process.

$$HBr + e \rightarrow H + Br^- \quad (19.1)$$

TABLE 19.2
Vibrational Excitation for HBr

	Threshold Resonance ($\nu = 1$)			Shape Resonance ($\nu = 1$)		
	Position (eV)	Width (meV)	Cross Section (m^2)	Position (eV)	Width (eV)	Cross Section (m^2)
HBr	0.41	<70	4.0×10^{-19}	0.9	1	2.7×10^{-19}
HCl	0.51	180	2.0×10^{-19}	2.5	2	3.0×10^{-19}
HF	0.51	<30	7.0×10^{-20}			1.0×10^{-20}

Note: Data for hydrogen chloride (HCl) and hydrogen fluoride (HF) are also given for comparison purposes.

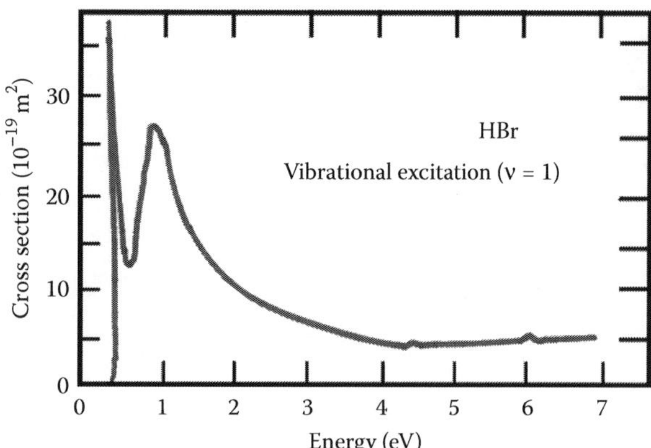

FIGURE 19.1 Integral cross section for vibrational excitation of HBr molecule for ν = 1 eV. The excitation cross section has large value near threshold and the shape resonance at 0.9 eV. (Adapted from Rohr, K., *J. Phys. B: At. Mol. Phys.*, 11, 1849, 1978.)

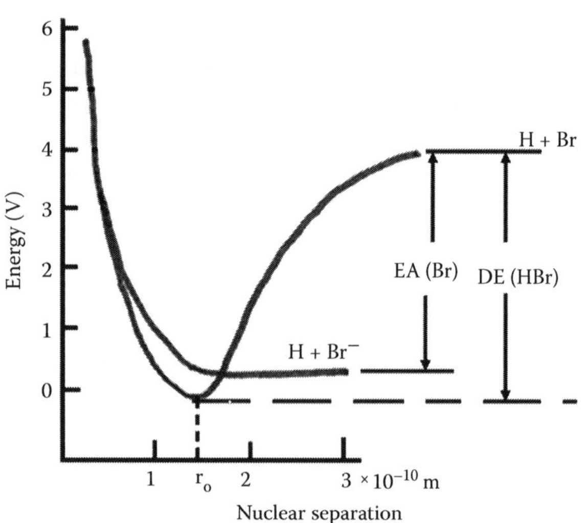

FIGURE 19.2 Potential energy curves for the formation of Br^- by dissociative attachment. DE = dissociation energy of the molecule, EA = electron affinity of the bromine atom. Appearance potential (DE – EA) is positive.

The potential energy diagram for the formation of Br^- ion is shown in Figure 19.2. Table 19.3 shows the electron impact data for HBr, and data for other hydrogen halides are included for comparison (Frost and McDowell, 1958).

TABLE 19.3
Electron Impact Data for Hydrogen Halides

Molecule/				Experimental		
Ion	DE	EA	AP	Onset	Peak	Reference
HBr/Br^-			0.396		0.39	Ziesel et al. (1975)
			0.11		0.28	Christophorou et al. (1968)
	3.75	3.54	0.21	0.10	0.21	Frost and McDowell (1958)
HCl/Cl^-			0.819		0.84	Ziesel et al. (1975)
			0.64		0.78	Christophorou et al. (1968)
	4.43	3.78	0.65	0.62	0.77	Frost and McDowell (1958)
					0.84	
HF/F^-	5.83	3.63	2.20	1.88	4.0	Frost and McDowel (1958)
HI/I^-			~0		~0	Christophorou et al. (1968)
	3.06	3.24	–0.18	0.03	0.05	Frost and McDowell (1958)

Note: AP = appearance potential; DE = dissociation energy of the molecule; EA = electron affinity of the atom. Energy in eV.

TABLE 19.4
Dissociative Attachment Cross Sections

Energy (eV)	Q_a (10^{-20} m^2)	Energy (eV)	Q_a (10^{-20} m^2)
0.08	0.027	0.32	2.565
0.10	0.146	0.34	2.329
0.12	0.283	0.36	2.056
0.14	0.391	0.38	1.824
0.16	0.619	0.40	1.643
0.18	1.255	0.45	1.151
0.20	1.913	0.50	0.787
0.22	2.366	0.55	0.537
0.24	2.568	0.60	0.429
0.26	2.622	0.65	0.377
0.28	2.704	0.70	0.319
0.30	2.701	0.72	0.288

Source: Adapted from Christophorou, L. G., R. N. Compton, and H. W. Dickson, *J. Chem. Phys.*, 48, 1949, 1968.

19.4 ATTACHMENT CROSS SECTION

Attachment cross sections for HBr, digitized from the measurements of Christophorou et al. (1968) are shown in Table 19.4 and Figure 19.3.

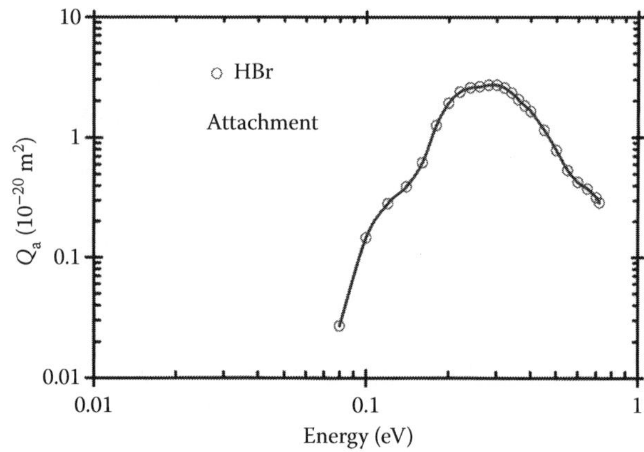

FIGURE 19.3 Dissociative attachment cross sections for HBr. (Adapted from Christophorou, L. G., R. N. Compton, and H. W. Dickson, *J. Chem. Phys.*, 48, 1949, 1968.)

TABLE 19.5
Attachment Rate Constants for HBr

Temperature, K, (eV)	Method	Rate Constant ($\times10^{-18}$ m³/s)	Reference
169.7	P/LP	<1.0	Speck et al. (2001)
145.4		2.5	
84.6		9.4	
74.5		18.0	
71.5		11.0	
47.7		120[a]	
	Swarm	100	Wang and Lee (1988)
300	FA/LP	≤3	Smith and Adams (1987)
515		300	
300	FA/LP	3.3	Adams et al. (1986)
515		280	
	Swarm	100	Christophorou et al. (1968)

Note: Temperature is thermal unless otherwise mentioned. FA/LP = flowing afterglow/Langmuir probe; P/LP= plasma/Langmuir probe.
[a] Possibly clusters of HBr molecules.

19.5 ATTACHMENT RATE CONSTANTS

Table 19.5 shows the attachment rate constants for HBr. The attachment rate constants as a function of *E/N* and mean energy are shown in Figure 19.4 (Wang and Lee, 1988).

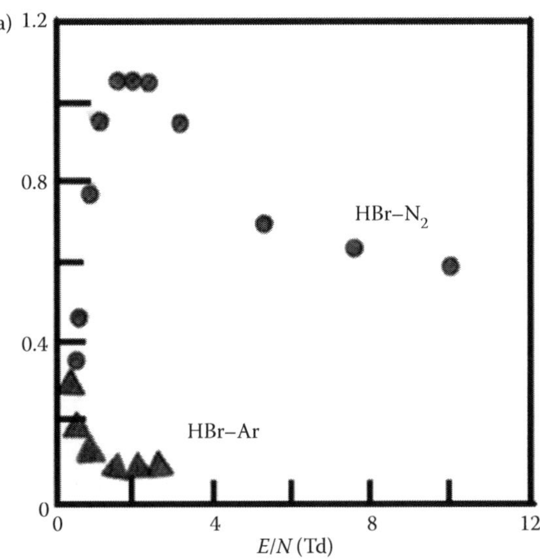

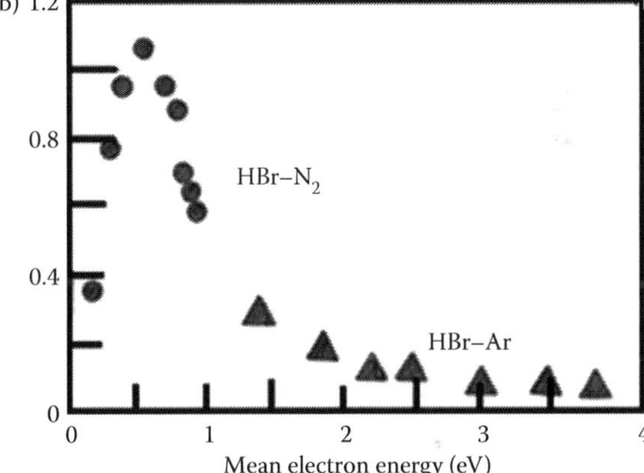

FIGURE 19.4 Attachment rate constants for HBr as function of (a) *E/N* and (b) mean energy. (Adapted from Wang, W. C. and L. C. Lee, *J. Appl. Phys.*, 63, 4905, 1988.)

REFERENCES

Adams, N. G., D. Smith, A. A. Viggiano, J. F. Paulson, and M. J. Henchman, *J. Chem. Phys.*, 84, 6728, 1986.

Christophorou, L. G., R. N. Compton, and H. W. Dickson, *J. Chem. Phys.*, 48, 1949, 1968.

Frost, D. C. and M. A. McDowell, *J. Chem. Phys.*, 29, 503, 1958.

Rohr, K., *J. Phys. B: At. Mol. Phys.*, 11, 1849, 1978.

Smith, D. and N. G. Adams, *J. Phys. B: At. Mol. Phys.*, 20, 4903, 1987.

Speck, T., J.-L. Le Garrec, S. Le Picard, A. Canosa, J. B. A. Mitchell, and B. R. Rowe, *J. Chem. Phys.*, 144, 8303, 2001.

Wang, W. C. and L. C. Lee, *J. Appl. Phys.*, 63, 4905, 1988.

Ziesel, J. P., I. Nenner, and G. J. Schulz, *J. Chem. Phys.*, 63, 1943, 1975.

HBr

20

HYDROGEN CHLORIDE

CONTENTS

Hydrogen chloride (HCl) is a polar, electron-attaching molecule with 18 electrons. It has electronic polarizability of 2.93×10^{-40} F m^2, dipole moment of 1.109 D, and ionization potential of 12.749 eV. The energy for the fundamental vibrational mode is 370.8 meV.

20.1 SELECTED REFERENCES FOR DATA

Table 20.1 shows the selected references for data on HCl.

TABLE 20.1
Selected References for Data

Parameter	Range: eV, (Td), [K]	Reference
k_a	<0.04	**Speck et al. (2001)**
Q_T	0.8–400	**Hamada and Sueoka (1994)**
Q_T	10–5000	Jain and Baluja (1992)
Q_v	0–2	**Schafer and Allen (1991)**
Q_v	0–10	**Knoth et al. (1989)**
Q_v	0–10	**Knoth et al. (1989)**
Q_{el}	0–10	**Rädle et al. (1989)**
Q_a	0.5–2.0	**Petrović et al. (1988)**
k_a	[300, 515]	**Adams et al. (1986)**
Q_a	5.0–12.0	**Orient and Srivastava (1985)**
k_a, W	(20–200)	Penetrante and Bardsley (1983)
k_a	0.149–0.213	**Miller and Gould (1978)**
Q_v	0–6	**Rohr and Linder (1976)**
k_a	0.04	**Davis et al. (1973)**
Q_a	0–2.0	**Christophorou et al. (1968)**
Q_{ex}	0–14	**Compton et al. (1968)**
Q_a	0–2	**Frost and McDowell (1958)**
η	(0–30)	**Bradbury (1934)**

Note: k_a = Attachment rate; Q_a = attachment cross section; Q_{el} = elastic scattering cross section; Q_T = total scattering; Q_v = vibrational excitation cross section; η = attachment coefficient. Bold font indicates experimental study.

20.2 TOTAL SCATTERING CROSS SECTIONS

Table 20.2 and Figure 20.1 present total scattering cross sections (Hamada and Sueoka, 1994).

20.3 RO-VIBRATIONAL EXCITATION CROSS SECTIONS

The rotational constant for HCl molecule is very low, 1.31 meV, and the energy for the first rotational excitation ($\Delta j = 1$) is 2.62 eV Knoth et al. (1989) have measured the differential scattering cross sections for the first vibrational level, simultaneous with strong rotational levels for HCl as shown in Table 20.3. Vibrational excitation is the dominant energy loss mechanism in a discharge for values of *E*/*N* from 30 to 120 Td (Penetrante and Bardsley, 1983).

20.4 ELASTIC SCATTERING CROSS SECTIONS

Rädle et al. (1989) have measured the integral elastic scattering cross sections as shown in Table 20.4 and Figure 20.2 which also include the theoretical calculations of Padial and Norcross (1984). The minimum at 1.5 eV is identified as Ramsauer–Townsend minimum (Rädle et al., 1989).

HCl

TABLE 20.2
Total Scattering Cross Sections

Energy (eV)	Q_T (10^{-20} m²)	Energy (eV)	Q_T (10^{-20} m²)	Energy (eV)	Q_T (10^{-20} m²)
0.8	35.39	6.5	24.15	22	20.26
1.0	32.55	7.0	24.84	25	19.20
1.2	28.53	7.5	25.34	30	16.95
1.4	26.23	8.0	27.44	35	15.67
1.6	24.27	8.5	27.80	40	14.71
1.8	23.75	9.0	28.07	50	12.80
2.0	22.97	9.5	27.66	60	11.94
2.2	22.03	10.0	28.31	70	10.90
2.5	21.72	11.0	28.12	90	9.90
2.8	22.55	12.0	27.63	100	9.43
3.1	23.26	13.0	26.92	120	8.49
3.4	23.39	14.0	26.91	150	7.65
3.7	23.54	15.0	25.72	200	6.68
4.0	23.43	16.0	24.82	250	5.88
4.5	24.04	17.0	23.81	300	5.27
5.0	23.10	18.0	22.62	350	4.89
5.5	24.18	19.0	22.01	400	4.43
6.0	23.44	20.0	21.67		

Source: Adapted from Hamada, A. and O. Sueoka, *J. Phys. B: At. Mol. Opt. Phys.*, 27, 5055, 1994.

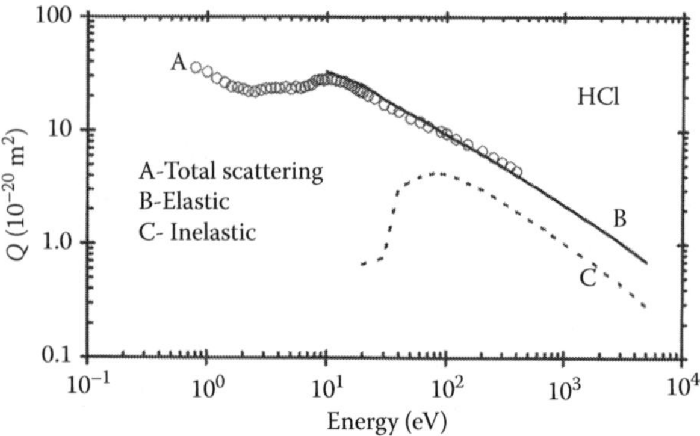

FIGURE 20.1 Scattering cross sections for HCl. A: Hamada and Sueoka (1994), total scattering, experimental; B: Jain and Baluja (1992), elastic, theoretical; C: Jain and Baluja (1992), inelastic, theoretical.

TABLE 20.3
Differential Cross Sections for the First Vibrational Level for HCl

Energy (eV)	Differential Cross Section (10^{-20} m²/sr)	
Angle (°)	0.5	1.5
15	0.34	0.13
30	0.57	0.18
45	0.64	0.17
60	0.89	0.15
75	0.93	0.16
90	0.91	0.18
105	0.85	0.20
120	0.87	0.20
135	0.90	0.21

Source: Adapted from Knoth, G. et al., *J. Phys. B: At. Mol. Opt. Phys.*, 22, 299, 1989.

TABLE 20.4
Integral Elastic Scattering Cross Sections for HCl

Energy (eV)	Q_{el} (10^{-20} m²)	Energy (eV)	Q_{el} (10^{-20} m²)
0.5	19.88	4.5	12.93
1.0	9.43	5.0	13.52
1.5	6.35	6.0	17.80
2.0	6.33	7.0	19.62
2.5	6.30	8.0	21.43
3.0	8.15	9.0	24.43
3.5	10.26	10.0	27.46
4.0	11.87		

Source: Adapted from Rädle, M. et al., *J. Phys. B: At. Mol. Opt. Phys.*, 22, 1455, 1989.

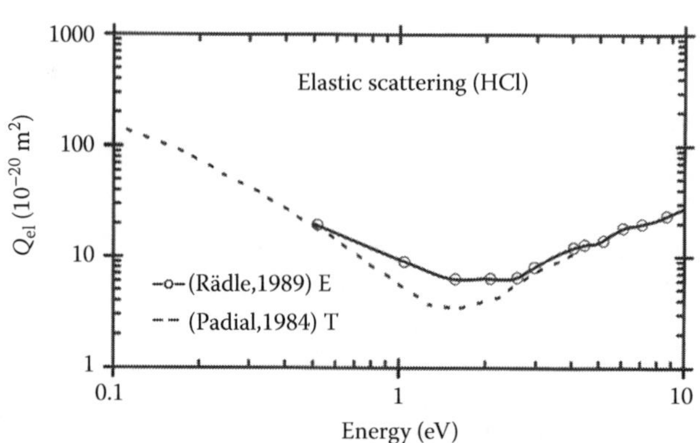

FIGURE 20.2 Integral elastic scattering cross sections for HCl.

20.5 ATTACHMENT PROCESSES

Attachment of electrons to HCl molecule occurs through two processes (Petrović et al., 1988):

$$HCl + e \rightarrow H + Cl^- \quad (20.1)$$

$$HCl + e \rightarrow H^- + Cl \quad (20.2)$$

Appearance potentials and peak cross sections measured for the process (Reaction 20.1) are shown in Table 20.5.

20.6 ATTACHMENT CROSS SECTIONS

The attachment cross sections tabulated by Itikawa (2003) on the basis of measurements of Petrović et al. (1988) for Cl^- ion and Orient and Srivastava (1985) for H^- ion are shown in Table 20.6 and Figure 20.3.

TABLE 20.5
Attachment Process

Process	Appearance Potential (eV)	Peak Energy (eV)	Peak Cross Section (10^{-22} m²)	Reference
Reaction 20.1		0.85	26.59	Orient and Srivastava (1985)
	0.64	0.81	19.8	Christophorou et al. (1968)
	0.62	0.77		Frost and McDowell (1958)
Reaction 20.2	~5.0	7.1	2.07	Orient and Srivastava (1985)
		9.05	0.93	

TABLE 20.6
Attachment Cross Sections for HCl

Cl^- Ion		H^- Ion	
Energy (eV)	Q_a (10^{-20} m²)	Energy (eV)	Q_a (10^{-20} m²)
0.63	0.0024	5.0	4.5×10^{-5}
0.678	0.004	5.5	2.7×10^{-3}
0.70	0.008	6.0	8.44×10^{-3}
0.715	0.0141	6.5	1.68×10^{-2}
0.747	0.0396	7.0	2.05×10^{-2}
0.772	0.0496	7.08	2.067×10^{-2}
0.80	0.0707	7.5	1.69×10^{-2}
0.8212	0.0892	8.0	1.16×10^{-2}
0.854	0.113	8.5	9.6×10^{-3}
0.877	0.124	9.0	9×10^{-3}
0.891	0.127	9.5	7×10^{-3}
0.90	0.127	10.0	4.62×10^{-3}
0.914	0.125	10.5	2.76×10^{-3}
0.9263	0.119	11.0	1.24×10^{-3}
0.9463	0.1128	11.5	4×10^{-4}
0.9679	0.103	12.0	7×10^{-5}
1.0	0.097		
1.042	0.0848		
1.0672	0.074		
0.1403	0.0672		
1.2	0.056		
1.244	0.045		
1.3	0.033		
1.35	0.028		
1.4	0.023		
1.5	0.018		
1.7	0.0116		
1.9	0.0048		

Source: Adapted from Itikawa, Y., in *Interactions of Photons and Electrons with Molecules*, Springer-Verlag, New York, NY, 2003, Chapter 5-2, pp. 5–87.

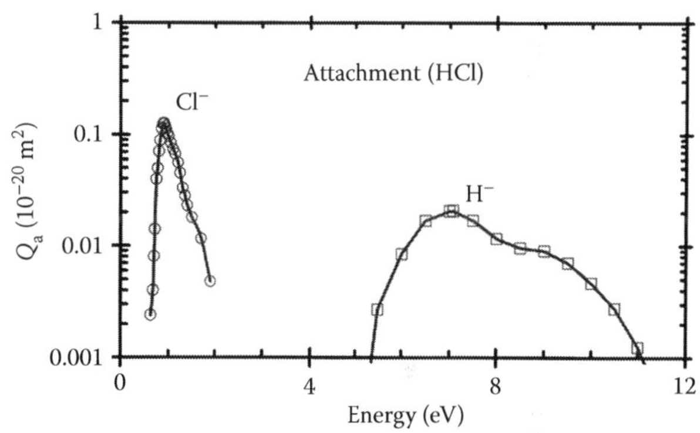

FIGURE 20.3 Electron attachment cross sections for HCl. (–○–) Cl^- ion; (–□–) H^- ion.

TABLE 20.7
Attachment Rate Constants

Temperature (K)	Rate Constant (10^{-18} m³/s)	Method	Reference
170	<1.0	P/LP	Speck et al. (2001)
300	<10	FA/LP	Adams et al. (1986)
515	<10		
	~100	e-Beam	Kliger et al. (1981)
2475	1.1	Flame/MWC	Miller and Gould (1978)
2200	0.6		
1950	0.34		
1730	0.03		

Note: Temperature is thermal unless otherwise mentioned. MWC = microwave cavity; P/LP = plasma/Langmuir probe; FA= flowing afterglow.

20.7 ATTACHMENT RATE CONSTANTS

Table 20.7 provides the attachment rate constants for HCl. The attachment rate constants as a function of E/N are shown in Figure 20.4.

HCl

HCl

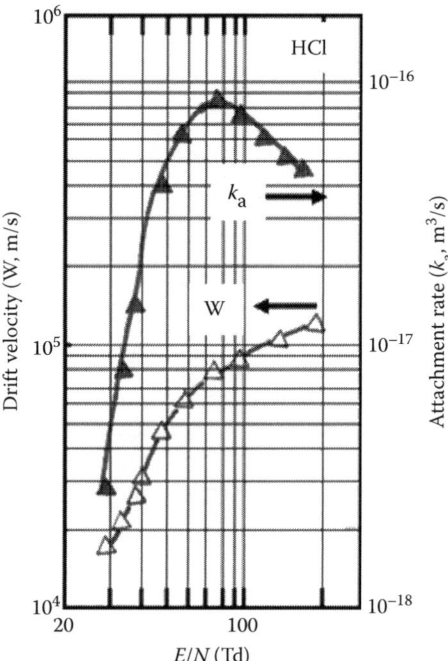

FIGURE 20.4 Drift velocity and attachment rate constant for HCl. Open triangles, Davies (1983), measurements; closed triangles, Penetrante and Bardsley (1983), simulation.

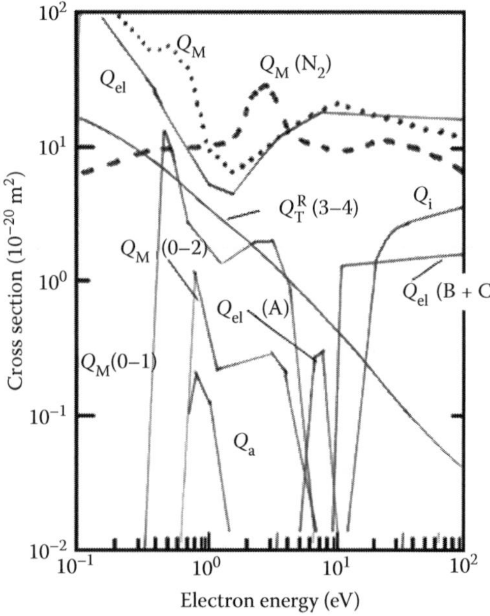

FIGURE 20.5 Consolidated cross sections used for simulation. Nitrogen curve is drawn for comparison. Q_a = attachment cross section; Q_{el} = elastic scattering; Q_{ex} = excitation; Q_i = ionization; Q_M = rotational excitation (momentum transfer); Q_T^R = rotational excitation (integrated); Q_v = vibrational. (Adapted from B. M. Penetrante and J. N. Bardsley, *J. Appl. Phys.*, 54, 6150, 1983.) With permission from American Institute of Physics.

20.8 DRIFT VELOCITY OF ELECTRONS

Figure 20.4 also shows the drift velocity as a function of *E/N* obtained from simulation studies (Penetrante and Bardsley, 1983).

20.9 CONSOLIDATED CROSS SECTIONS

A set of cross sections used for Monte Carlo simulation is available in Penetrante and Bardsley (1983). See Figure 20.5.

REFERENCES

Adams, N. G., D. Smith, A. A. Viggiano, J. F. Paulson, and M. J. Henchman, *J. Chem. Phys.*, 84, 6728, 1986.

Bradbury, N. E., *J. Chem. Phys.*, 2, 827, 1934.

Christophorou, L. G., R. N. Compton, and H. W. Dickson, *J. Chem. Phys.*, 48, 1949, 1968.

Compton, R. N., R. H. Huebner, P. W. Reinhardt, and L. G. Christophorou, *J. Chem. Phys.*, 48, 901, 1968.

Davies, D. K., *Bull. Am. Phys. Soc.*, 26, 726, 1981, cited by Penetrante and Bardsley, 1983.

Davis, F. J., R. N. Compton, and D. R. Nelson, *J. Chem. Phys.*, 59, 2324, 1973.

Frost, D. C. and C. A. McDowell, *J. Chem. Phys.*, 29, 503, 1958.

Hamada, A. and O. Sueoka, *J. Phys. B: At. Mol. Opt. Phys.*, 27, 5055, 1994.

Itikawa, Y., Electron attachment, in *Interactions of Photons and Electrons with Molecules*, Springer-Verlag, New York, NY, 2003, Chapter 5-2, pp. 5–87.

Jain, A. and K. L. Baluja, *Phys. Rev.*, A45, 202, 1992.

Kliger, D., Z. Rosenberg, and M. Rokni, *Appl. Phys. Lett.*, 39, 319, 1981.

Knoth, G., M. Gote, M. Rädle, F. Leber, K. Jung, and H. Erhardt, *J. Phys. B: At. Mol. Opt. Phys.*, 22, 2797, 1989.

Knoth, G., M. Rädle, M. Gote, H. Erhardt, and K. Jung, *J. Phys. B: At. Mol. Opt. Phys.*, 22, 299, 1989.

Miller, W. J. and R. K. Gould, *J. Chem. Phys.*, 68, 3542, 1978.

Orient, O. J. and S. K. Srivastava, *Phys. Rev.*, A32, 2678, 1985.

Padial, N. T. and D. W. Norcross, *Phys. Rev.*, A29, 1990, 1984, cited by Rädle et al. 1989.

Penetrante, B. M. and J. N. Bardsley, *J. Appl. Phys.*, 54, 6150, 1983.

Petrović, Z. L., W. C. Wang, and L. C. Lee, *J. Appl. Phys.*, 64, 1625, 1988.

Rohr, K. and F. Linder, *J. Phys. B: At. Mol. Opt. Phys.*, 9, 2521, 1976.

Rädle, M., G. Knoth, K. Jung, and H. Erhardt, *J. Phys. B: At. Mol. Opt. Phys.*, 22, 1455, 1989.

Schafer, O. and M. Allen, *J. Phys. B: At. Mol. Opt. Phys.*, 24, 3069, 1991.

Speck, T., J.-L. Le Garrec, S. Le Picard, A. Canosa, J. B. A. Mitchell, and B. R. Rowe, *J. Chem. Phys.*, 114, 8303, 2001.

21

HYDROGEN FLUORIDE

CONTENTS

Hydrogen fluoride (HF) is a polar, electron-attaching molecule with 10 electrons. It has electronic polarizability of 0.89×10^{-40} F m^2, dipole moment of 1.826 D, and ionization potential of 16.04 eV. The energy for the fundamental vibrational mode is 513.1 meV.

21.1 SELECTED REFERENCES FOR DATA

Table 21.1 shows the selected references for data on HF.

21.2 RO-VIBRATIONAL EXCITATION CROSS SECTION

The rotational constant for HF molecule is very low, 2.6 meV, and the energy for the first rotational excitation is 5.2 meV. Knoth et al. (1989) have measured the differential scattering cross sections for the first vibrational level, $v = 0 \rightarrow 1$, simultaneous with rotational excitation as shown in Table 21.2.

TABLE 21.1
Selected References for Data

Parameter	Range: eV, (K)	Reference
Q_v	0.5–10	**Knoth et al. (1989)**
Q_{el}	0.5–10	**Rädle et al. (1989)**
k_a	(300, 510)	**Adams et al. (1986)**
Q_a	0–4	**Allan and Wong (1981)**
Q_v	0–6	**Rohr and Lindner (1976)**

Note: Q_a = attachment cross section; Q_{el} = elastic scattering cross section; Q_v = vibrational excitation cross section. Bold face indicates experimental study.

TABLE 21.2
Differential Cross Sections for the First Vibrational Level for HF

	Differential Cross Section (10^{-20} m^2/sr)		
	Energy (eV)		
Angle (°)	**0.63**	**0.9**	**1.2**
15	0.40		
30	0.47	0.12	0.067
45	0.54	0.083	0.045
60	0.60	0.078	0.043
75	0.61	0.086	0.044
90	0.61	0.10	0.048
105	0.60	0.10	0.054
120	0.58	0.11	0.060
135	0.60	0.12	0.064

Source: Adapted from Knoth, G. et al., *J. Phys. B: At. Mol. Opt. Phys.*, 22, 2797, 1989.

HF, like HCl, shows a strong threshold peak of 10^{-19} m^2 in vibrational excitation cross section, with scattering becoming isotropic (Allan and Wong, 1981).

21.3 ATTACHMENT PROCESSES

The appearance potential of F^- ion is given by the following equation (Forst and McDowell, 1958):

$$AP(F^-) = DE(HF) - EA(F) + KE \tag{21.1}$$

where

AP is the appearance potential of F^- ion
DE(HF) is the dissociation energy of the HF molecule
EA(F) is the electron affinity of the F atom
KE is the kinetic energy of the products

Table 21.3 shows these quantities along with experimentally observed values for the HF molecule. For comparison, data for other hydrogen halides are also included.

The dissociation energy of the hydrogen iodide molecule is smaller than the electron affinity of the iodine atom, rendering the electron affinity negative, according to Equation 21.1. The potential energy diagrams for the molecules in the first and last row of Table 21.3 are shown in Figures 21.1 and 21.2.

TABLE 21.3
Attachment Processes in Hydrogen Halides

Ion	DE, eV (Molecule)	EA (Atom)	AP, Equation 21.1	Experimental Onset (eV)	Experimental Peak (eV)
F^-	5.83 (HF)	3.63 (F)	2.20	1.88	4.0
Cl^-	4.43 (HCl)	3.78 (Cl)	0.65	0.62	0.77
Br^-	3.75 (HBr)	3.54 (Br)	0.21	0.10	0.21
I^-	3.06 (HI)	3.24 (I)[a]	−0.18	0.03	0.05

Source: Adapted from Frost, D. C. and C. A. McDowell, *J. Chem. Phys.*, 29, 503, 1958.

[a] More recent determination of electron affinity of I is 3.059 eV (Klar et al., 2001).

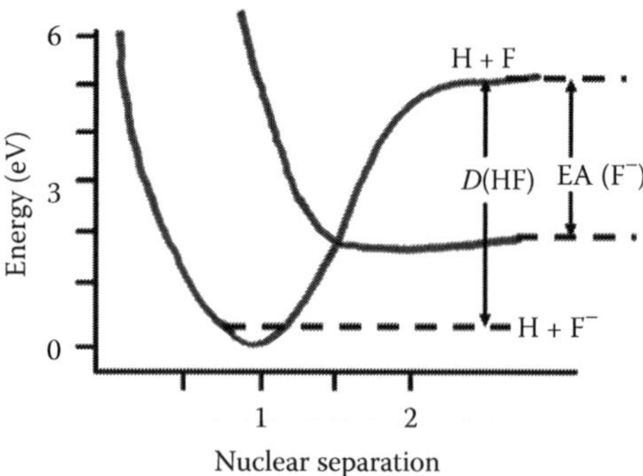

FIGURE 21.1 Potential energy diagram (schematic) for the reaction HF + e → H + F^-. Note the relative values of dissociation energy of the molecule (*D*) and the electron affinity (EA) of the fluorine atom.

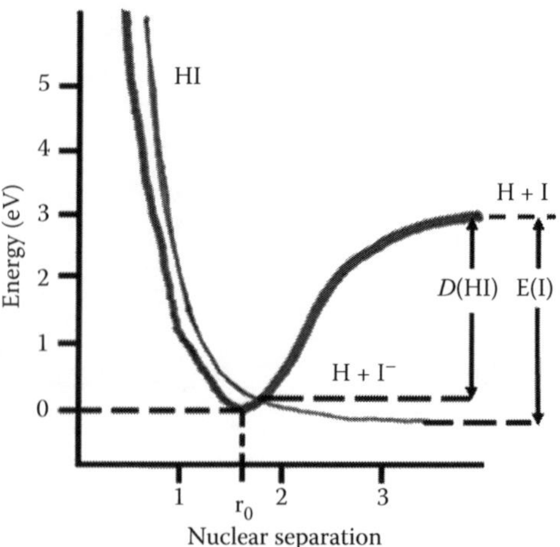

FIGURE 21.2 Potential energy diagram (schematic) for the reaction HI + e → H + I^-. Note the relative values of dissociation energy of the molecule (*D*) and the electron affinity (E(I)) of the iodine atom.

21.4 ATTACHMENT RATE CONSTANT

Adams et al. (1986) have measured the attachment rate constant for HF, using flowing afterglow/Langmuir probe technique, as $<1 \times 10^{-17}$ m^3/s.

REFERENCES

Adams, N. G., D. Smith, A. A. Viggiano, J. F. Paulson, and M. J. Henchman, *J. Chem. Phys.*, 84, 6728, 1986.

Allan, M. and S. F. Wong, *J. Chem. Phys.*, 74, 1687, 1981.

Frost, D. C. and C. A. McDowell, *J. Chem. Phys.*, 29, 503, 1958.

Klar, D., M.-W. Ruf, I. I. Fabrikant, and H. Hotop, *J. Phys. B: At. Mol. Opt. Phys.*, 34, 3855, 2001.

Knoth, G., M. Gote, M. Rädle, F. Leber, K. Jung, and H. Ehrhardt, *J. Phys. B: At. Mol. Opt. Phys.*, 22, 2797, 1989.

Rohr, K. and F. Linder, *J. Phys. B: At. Mol. Phys.*, 9, 2521, 1976.

Rädle, M., G. Knoth, K. Jung, and H. Ehrhardt, *J. Phys. B: At. Mol. Opt. Phys.*, 22, 1455, 1989.

22

HYDROGEN IODIDE

HI

CONTENTS

Hydrogen iodide (HI) is an electron-attaching, polar molecule that has 54 electrons. The electronic polarizability is 6.053×10^{-40} F m^2, the dipole moment is 0.448 D, and the ionization potential is 10.386 eV. The dissociation energy of the molecule is 3.09 eV and the electron affinity of the iodine atom is 3.06 eV. The vibrational excitation energy is 286.3 meV and the rotational excitation energy ($\Delta j = 1$) is 1.6 meV.

22.1 SELECTED REFERENCES FOR DATA

Table 22.1 shows selected references for data on HI.

22.2 ATTACHMENT PROCESSES

Common with other hydrogen halides, the predominant attachment process in HI is dissociative attachment, according to Reaction 22.1

$$\mathrm{HI} + \mathrm{e} \rightarrow \mathrm{H} + \mathrm{I}^- \quad (22.1)$$

TABLE 22.1
Selected References for Data on HBr

Parameter	Range: eV, (Td), [K]	Reference
k_a, Q_a	0–3	**Klar et al. (2001)**
k_a	[300, 510]	**Smith and Adams (1987)**
k_a	[300, 510]	**Adams et al. (1986)**
k_a, Q_a	0–1.5, (0–2)	**Christophorou et al. (1968)**
AP	0–1	**Frost and McDowell (1958)**

Note: Bold font indicates experimental study. AP = attachment process; k_a = attachment rate constant, Q_a = attachment cross section.

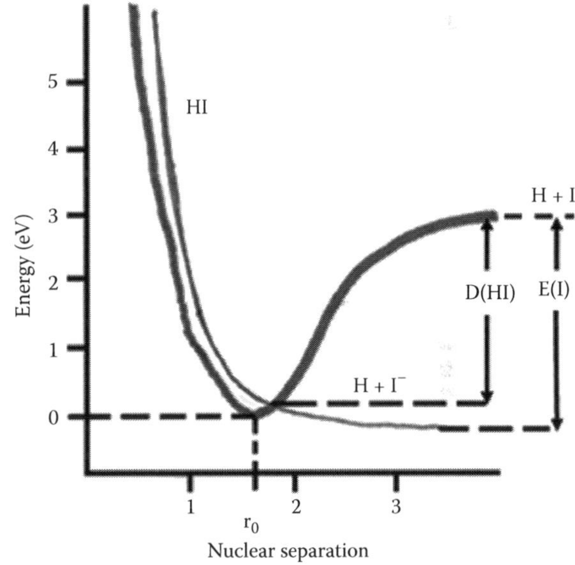

FIGURE 22.1 Potential energy curves for the neutral molecule and the formation of I⁻ ion.

The potential energy diagram for the formation of I⁻ ion is shown in Figure 22.1 (Klar et al., 2001). The significant feature is that the dissociation energy of the molecule is almost equal to the electron affinity of the iodine atom with a peak in the attachment cross section occurring at 0 eV.

Table 22.2 shows the electron impact data at low energies for HI. Data for other halogen halides are also included for comparison.

22.3 ATTACHMENT CROSS SECTION

Attachment cross sections for the dissociative process are shown in Table 22.3 and Figure 22.2.

HI

TABLE 22.2
Electron Impact Data for Hydrogen Halides

Molecule /Ion	DE	EA	AP	Experimental Onset	Experimental Peak	Reference
HI/I^-			~0		~0	Christophorou et al. (1968)
	3.06	3.24	−0.18	0.03	0.05	Frost and McDowell (1958)
HBr/Br^-			0.396		0.39	Ziesel et al. (1975)
			0.11		0.28	Christophorou et al. (1968)
	3.75	3.54	0.21	0.10	0.21	Frost and McDowell (1958)
HCl/Cl^-			0.819		0.84	Ziesel et al. (1975)
			0.64		0.78	Christophorou et al. (1968)
	4.43	3.78	0.65	0.62	0.77	Frost and McDowell (1958)
					0.84	
HF/F^-	5.83	3.63	2.20	1.88	4.0	Frost and McDowell (1958)

Note: AP = appearance potential; DE = dissociation energy of the molecule; EA = electron affinity of the atom. Energy in eV.

TABLE 22.3
Dissociative Attachment Cross Sections for HI

Klar et al. (2001)		Christophorou et al. (1968)	
Energy (meV)	Q_a (10^{-20} m²)	Energy (meV)	Q_a (10^{-20} m²)
0.2	18,256	10	225.4
1.0	5595	20	223.1
2.0	3311	30	211.5
3.0	2485	40	197.4
4.0	1802	50	185.9
5.0	1395	60	173.7
6.0	1194	70	154.3
7.0	1080	80	147.4
8.0	974	90	134.1
9.0	871	100	121.2
10.0	779	120	97.89
12.5	603	140	82.86
15.0	487	160	72.96
17.5	407	180	62.22
20.0	354	200	52.87
25	290	220	47.84
30	244	240	43.67
35	206	260	38.79
40	174	280	35.88
45	151	300	33.72
50	133	320	30.72
55	121	340	26.87
60	113	360	23.97
65	107	380	21.11
70	102	400	18.56
80	90	420	17.12
90	79	440	15.66
100	68	460	13.86
110	61	480	12.23
120	57	500	11.13

TABLE 22.3 (continued)
Dissociative Attachment Cross Sections for HI

Klar et al. (2001)		Christophorou et al. (1968)	
Energy (meV)	Q_a (10^{-20} m²)	Energy (meV)	Q_a (10^{-20} m²)
130	55	550	10.64
140	54	600	9.15
150	52	650	7.04
160	47	700	6.41
170	40	800	3.16
—	—	900	2.03

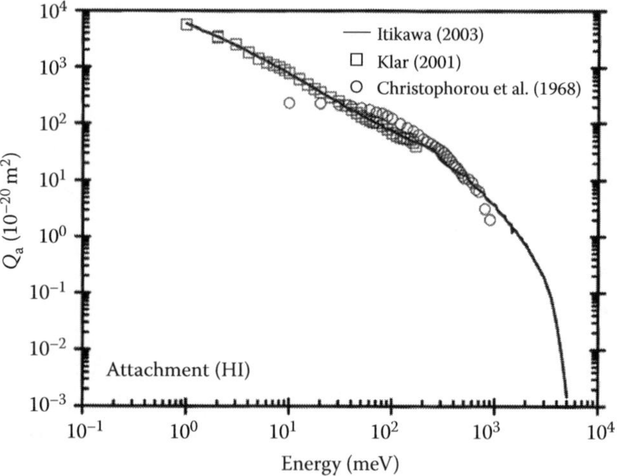

FIGURE 22.2 Attachment cross sections for HI. (——) Recommended by Itikawa (2003); (□) Klar et al. (2001); (○) Christophorou et al. (1968).

TABLE 22.4
Attachment Rate Constants for HI

Temperature, K, (Td)	Method	Rate Constant (10^{-13} m^3/s)	Reference
	LPA	3.0	Klar et al. (2001)
300	FA/LP	3.0	Smith and Adams (1987)
300	FA/LP	3.5	Adams et al. (1986)
510		3.5	
(0.6)	Swarm	1.86	Christophorou et al. (1968)

Note: Temperature is thermal unless otherwise mentioned. FA/LP = flowing afterglow/Langmuir probe; LPA = laser photoelectron attachment method.

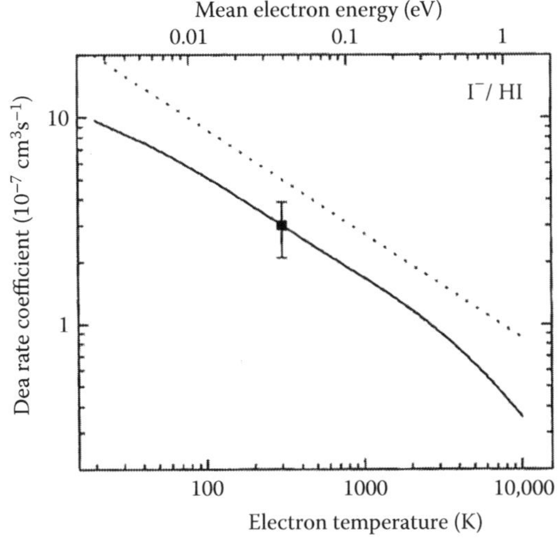

FIGURE 22.3 Attachment rate constants for HI. (——) Klar et al. (2001), from measured cross sections; (- - -) calculated from maximum s-wave capture cross sections; (■) Smith and Adams (1987) measurements. (Reproduced with kind permission of Institute of Physics, England.)

22.4 ATTACHMENT RATES

Table 22.4 shows the attachment rate constants for HI and Figure 22.3 is a graphical presentation.

REFERENCES

Adams, N. G., D. Smith, A. A. Viggiano, J. F. Paulson, and M. J. Henchman, *J. Chem. Phys.*, 84, 6728, 1986.

Christophorou, L. G., R. N. Compton, and H. W. Dickson, *J. Chem. Phys.*, 48, 1949, 1968.

Frost, D. C. and C. A. McDowell, *J. Chem. Phys.*, 29, 503, 1958.

Itikawa, Y. Electron attachment, in *Interactions of Photons and Electrons with Molecules*, Vol. 17c, Springer-Verlag, New York, NY, 2003, Chapter 5.

Klar, D., M.-W. Ruf, I. I. Fabrikant, and H. Hotop, *J. Phys. B: At. Mol. Opt. Phys.*, 34, 3855, 2001.

Smith, D. and N. G. Adams, *J. Phys. B: At. Mol. Phys.*, 20, 4903, 1987.

Ziesel, J. P., I. Nenner, and G. J. Schulz, *J. Chem. Phys.*, 63, 1943, 1975.

HI

23

IODINE

I_2

CONTENTS

Iodine (I_2) molecule has 106 electrons, electronic polarizability of 5.95×10^{-40} F m^2, and ionization potential of 9.31 eV. The electron affinities of the atom and molecule are given in Table 23.1 along with those for other halogen molecules added for comparison.

The vibrational states and corresponding energies for I_2 are shown in Table 23.2 (Brooks et al., 1979).

23.1 SELECTED REFERENCES FOR DATA

Table 23.3 provides selected references for data for I_2 molecule.

TABLE 23.1
Electron Affinities of Halogen Atoms and Molecules

	Electron Affinity (eV)	
	Atom	**Molecule**
Bromine	3.364	2.55
Chlorine	3.613	2.38
Fluorine	3.401	3.08
Iodine	3.059	2.55

TABLE 23.2
Vibrational States and Energies

Vibrational State	0	1	2	3	4	5
Energy (meV)	13.3	39.9	66.5	93.1	119.8	146.3

Note: Energy for the fundamental mode is 26.6 meV.

TABLE 23.3
Selected References for Data

Parameter	Range: eV, (Td)	Reference
Q_a	0–2 (1–50)	**Brooks et al. (1979)**
Q_a	0.1–0.3	Birtwistle and Modinos (1978)
AP	0–8	**Tam and Wong (1978)**
k_a	0.04–0.27	**Truby (1969)**
Q_a	0–10	**Biondi and Fox (1958)**
Q_a	0.039	**Biondi (1958)**
AP	0–3	**Fox (1958)**
AP	0–7	**Buchdal (1941)**
$W, \bar{\varepsilon}$	0–200	**Healy (1938)**

Note: AP = Attachment processes; k_a = attachment rate; Q_a = attachment cross section; W = drift velocity, $\bar{\varepsilon}$ = mean energy. Bold font indicates experimental study.

23.2 ATTACHMENT PROCESS

The dominant process is due to dissociative attachment

$$I_2 + e \rightarrow I + I^- \quad (23.1)$$

Negative ion peaks due to dissociative attachment are observed as shown in Table 23.4.

23.3 ATTACHMENT CROSS SECTION

Figure 23.1 shows the attachment cross sections for I_2 in the low-energy region. At 300 K the cross section measured is 3.9×10^{-20} m^2 (see Biondi, 1958).

I2

TABLE 23.4
Peak Energy for Negative Ion Formation

Peak Energy (eV)	Peak Cross Section (m^2)	Reference
0.05		Tam and Wong (1978)
0.9		
2.5		
0		Fox (1958)
	3.2×10^{-19}	Biondi and Fox (1958)
0.4		Buchdal (1941)

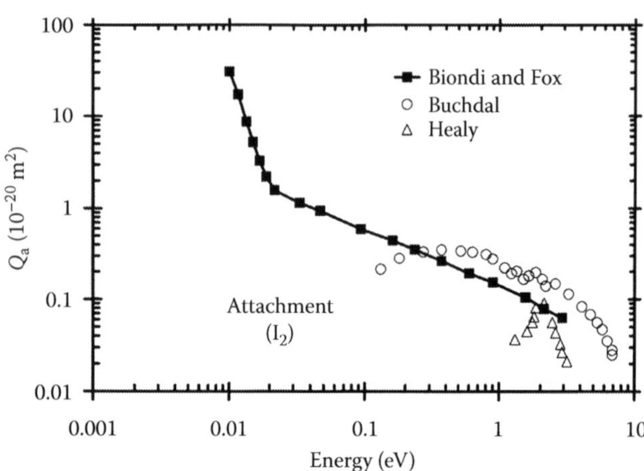

FIGURE 23.1 Attachment cross sections for I_2. (—■—) Biondi and Fox (1958); (○) Buchdal (1941); (Δ) Healy (1938).

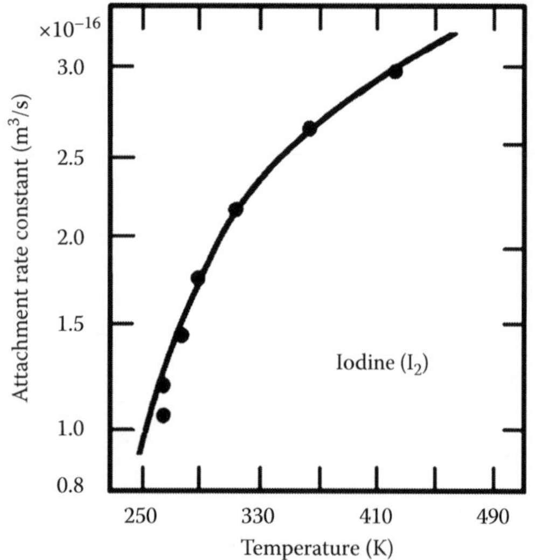

FIGURE 23.2 Attachment rate constants as a function of temperature for I_2. (Adapted from Truby, F. K., *Phys. Rev.*, 188, 508, 1969.)

23.4 ATTACHMENT RATE CONSTANTS

Figure 23.2 shows the attachment rate constants as a function of temperature (see Truby, 1969). Thermal (0.04 eV) attachment rate constant for I_2 is 1.8×10^{-16} m^3/s increasing to 2.8×10^{-16} m^3/s at 0.27 eV energy (Truby, 1969).

23.5 ATTACHMENT COEFFICIENTS

Figure 23.3 shows the density-reduced attachment coefficients in N_2 as buffer gas at number density 1.6×10^{22} m^{-3} (Brooks et al. 1979). The attachment coefficient increases with temperature in the range from 35°C to 90°C. The earlier results of Healy (1938) for I_2 are also shown (Figure 23.4).

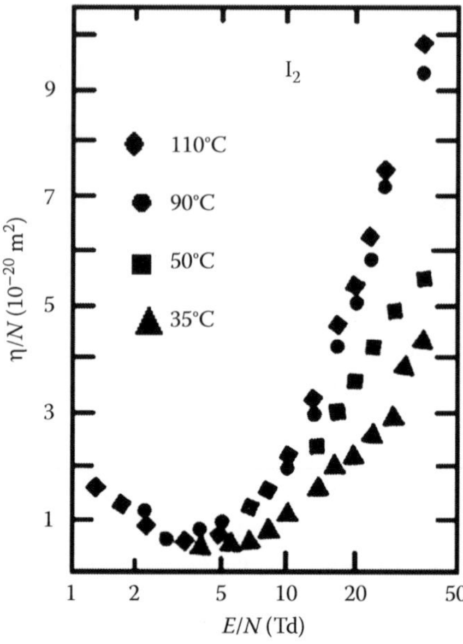

FIGURE 23.3 Density-reduced attachment coefficients as a function of *E*/*N* at various constant temperatures. With $N' = 1\%I_2$. N' is the partial gas number density in buffergas(*N*). (Adapted from Brooks, H. L., S. R. Hunter, and K. J. Nygaard, *J. Chem. Phys.*, 71, 1870, 1979.)

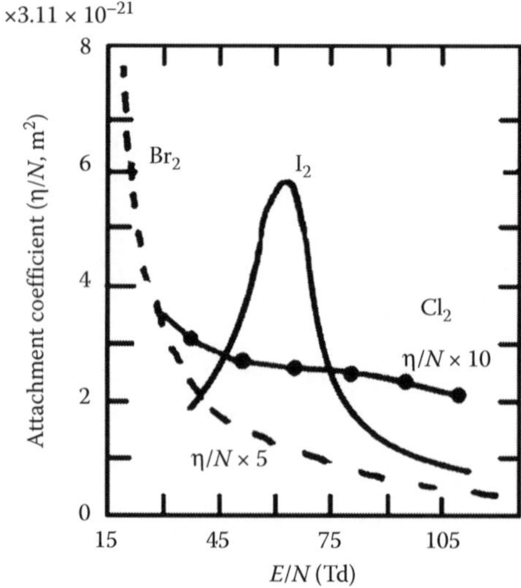

FIGURE 23.4 Density-reduced attachment coefficients for I_2 as a function of *E*/*N* in N_2 buffer gas. Data for Br_2 and Cl_2 are added for comparsion. (Adapted from Healy, R. H., *Philos. Mag.*, 26, 940, 1938.)

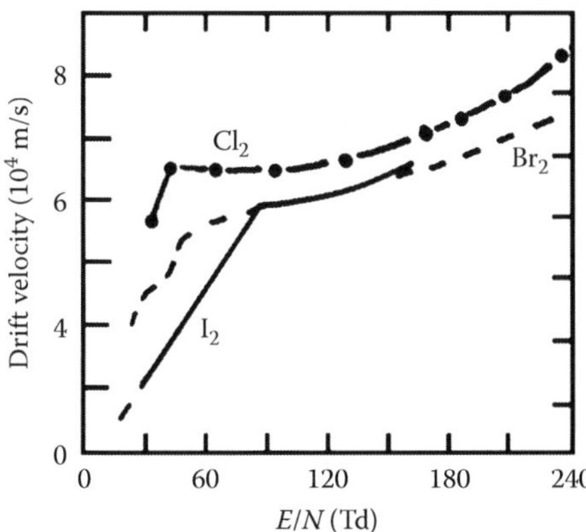

FIGURE 23.5 Drift velocity of electrons in I_2 as a function of *E/N*. Data for bromine and chlorine added for comparison. (Adapted from Healy, R. H., *Philos. Mag.*, 26, 940, 1938.)

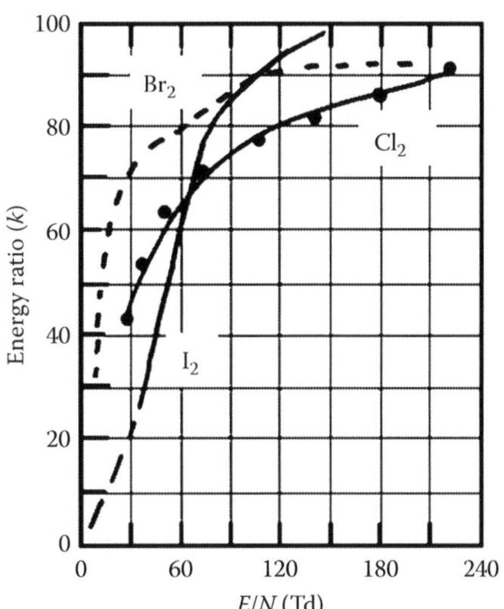

FIGURE 23.6 Mean energy of electrons as function of *E/N* in I_2. Data for bromine and chlorine are added for comparison. (Adapted from Healy, R. H., *Philos. Mag.*, 26, 940, 1938.)

23.6 TRANSPORT PARAMETERS

Experiments of Healy (1938) are the only available data for drift velocities and mean energies, as shown in Figures 23.5 and 23.6.

REFERENCES

Biondi, M. A., *Phys. Rev.*, 109, 2005, 1958.
Biondi, M. A. and R. E. Fox, *Phys. Rev.*, 109, 2012, 1958.
Birtwistle, D. T. and A. Modinos, *J. Phys. B: At. Mol. Phys.*, 11, 2949, 1978.
Brooks, H. L., S. R. Hunter, and K. J. Nygaard, *J. Chem. Phys.*, 71, 1870, 1979.
Buchdal, R. *J., Chem. Phys.*, 9, 146, 1941.
Fox, R. E., *Phys. Rev.*, 109, 2008, 1958.
Healy, R. H., *Philos. Mag.*, 26, 940, 1938.
Tam, W. and S. F. Wong, *J. Chem. Phys.*, 68, 5626, 1978.
Truby, F. K., *Phys. Rev.*, 188, 508, 1969.

24

NITRIC OXIDE

NO

CONTENTS

Nitric oxide (NO) is a polar, electron-attaching molecule that has 15 electrons. Its electronic polarizability is 1.89×10^{-40} F m^2, dipole moment 0.159 D, and ionization potential 9.264 eV.

24.1 SELECTED REFERENCES FOR DATA

See Table 24.1.

TABLE 24.1
Selected References for Data on NO

Quantity	Range: eV, (Td)	Reference
Total cross section	10–2000	Joshipura et al. (2007)
Ionization cross section	9.5–200	**Lopez et al. (2003)**
Excitation cross section	15–50	**Brunger et al. (2000)**
Ionization cross section	12.5–1000	**Lindsay et al. (2000)**
Diffusion coefficients	(12–300)	**Mechlińska-Drewko et al. (1999)**
Total cross section	0.2–5	**Alle et al. (1996)**
Ionization cross section	10–1000	Hwang et al. (1996)
Total cross section	0.4–250	**Szmytkowski and Maciąg (1996)**
Elastic cross section	1.5–40	**Mojarrabi et al. (1995)**
Elastic cross section	5–500	**Lee et al. (1992)**
Total cross section	0.5–160	**Szmytkowski and Maciąg (1991)**
Attachment cross section	0–55	**Krishnakumar and Srivastava (1988)**
Total cross section	100–1600	**Dalba et al. (1980)**
Characteristic energy	(56–1412)	**Lakshminarasimha and Lucas (1977)**
Vibrational excitation	0.1–3.0	**Tronc et al. (1975)**
Total cross section	0–10	**Zecca et al. (1974)**
Swarm parameters	0.1–10	**Parkes and Sugden (1972)**
Attachment processes	0–3	**Spence and Schulz (1971)**
Ionization cross section	9.5–1000	**Rapp and Englander-Golden (1965)**
Attachment cross section	6.5–13	**Rapp and Briglia (1965)**
Attachment coefficient	(0–15)	**Bradbury (1934)**

Note: Bold font indicates experimental study.

24.2 TOTAL SCATTERING CROSS SECTIONS

Total scattering cross sections for NO are shown in Tables 24.2 and 24.3. Figures 24.1 and 24.2 show graphical presentation. The highlights of the cross sections are

1. As the energy decreases toward zero the cross section increases. This is attributed to the formation of negative ion, NO^-.
2. A series of sharp peaks (about 10) attributed to vibrational excitation of the NO^-, with a spacing of ~165 meV. Table 24.4 shows the energy corresponding to the peaks.

TABLE 24.2
Total Scattering Cross Sections for NO in the Low Energy Region

Energy (eV)	Q_T (10^{-20} m²)	Energy (eV)	Q_T (10^{-20} m²)	Energy (eV)	Q_T (10^{-20} m²)
0.037	17.51	0.400	10.125	0.780	10.650
0.039	13.41	0.410	10.375	0.790	10.620
0.040	12.65	0.420	10.400	0.800	10.501
0.050	12.13	0.430	10.400	0.810	10.501
0.060	10.68	0.440	10.170	0.820	10.625
0.070	9.5783	0.450	9.875	0.830	10.875
0.080	9.0410	0.460	9.625	0.840	11.125
0.090	8.3253	0.470	9.250	0.850	11.370
0.100	7.9364	0.480	8.875	0.860	11.625
0.110	7.6997	0.490	8.834	0.870	11.759
0.120	7.5913	0.500	8.800	0.880	11.882
0.130	7.4725	0.510	8.810	0.890	11.995
0.140	7.3367	0.520	9.000	0.900	12.000
0.150	7.1673	0.530	9.375	0.910	11.995
0.160	7.1410	0.540	9.600	0.920	11.750
0.170	7.1250	0.550	9.875	0.930	11.375
0.180	7.2000	0.560	10.250	0.940	11.250
0.190	7.3750	0.570	10.500	0.950	10.750
0.200	7.750	0.580	10.700	0.960	10.620
0.210	8.125	0.590	10.750	0.970	10.620
0.220	8.675	0.600	10.800	0.980	10.750
0.230	9.125	0.610	10.750	0.990	11.000
0.240	9.600	0.620	10.605	1.000	11.125
0.250	9.750	0.630	10.375	1.010	11.250
0.260	9.750	0.640	10.245	1.020	11.500
0.270	9.700	0.650	10.138	1.030	11.750
0.280	9.500	0.660	10.250	1.040	11.875
0.290	9.125	0.670	10.375	1.050	11.895
0.300	8.875	0.680	10.625	1.060	11.914
0.310	8.500	0.690	10.825	1.070	11.915
0.320	8.250	0.700	11.000	1.080	11.875
0.330	8.119	0.710	11.161	1.090	11.750
0.340	8.250	0.720	11.250	1.100	11.625
0.350	8.400	0.730	11.262	1.110	11.416
0.360	8.750	0.740	11.250	1.120	11.250
0.370	9.125	0.750	11.161	1.130	11.114
0.380	9.400	0.760	11.000	1.140	11.103
0.390	9.800	0.770	10.875	1.150	11.114

TABLE 24.2 (continued)
Total Scattering Cross Sections for NO in the Low Energy Region

Energy (eV)	Q_T (10^{-20} m²)	Energy (eV)	Q_T (10^{-20} m²)	Energy (eV)	Q_T (10^{-20} m²)
1.160	11.125	1.470	10.827	1.900	10.115
1.170	11.250	1.480	10.801	1.920	10.050
1.180	11.416	1.490	10.773	1.940	10.000
1.190	11.563	1.500	10.765	1.960	9.951
1.200	11.630	1.520	10.801	1.980	9.916
1.220	11.700	1.540	10.875	2.000	9.840
1.240	11.625	1.560	10.889	2.050	9.750
1.250	11.500	1.580	10.855	2.20	9.570
1.260	11.375	1.600	10.750	2.50	9.375
1.270	11.250	1.620	10.625	3.00	9.134
1.280	11.130	1.640	10.510	3.50	9.063
1.290	11.120	1.650	10.500	4.00	9.000
1.300	11.110	1.660	10.450	4.50	9.000
1.320	11.110	1.680	10.382	5.00	8.950
1.340	11.130	1.700	10.382	5.50	9.000
1.360	11.325	1.720	10.382	6.00	9.000
1.380	11.375	1.740	10.383	6.50	9.100
1.390	11.370	1.760	10.375	7.00	9.130
1.400	11.360	1.780	10.313	7.50	9.200
1.410	11.265	1.800	10.250	8.00	9.250
1.420	11.249	1.820	10.200	8.50	9.344
1.430	11.130	1.840	10.120	9.00	9.437
1.440	11.005	1.850	10.115	9.50	9.500
1.450	10.887	1.860	10.110		
1.460	10.875	1.880	10.110		

Source: With kind permission from Professor Buckman.
Note: Data in tabulated form (Brunger and Buckman, 2002).

TABLE 24.3
Total Scattering Cross Sections for NO in the High Energy Range

Serial Number	Energy (eV)	Q_T (10^{-20} m²)	Serial Number	Energy (eV)	Q_T (10^{-20} m²)
1	0.4	9.1	15	12	10.9
2	0.5	9.25	16	15	11.6
3	0.6	10.2	17	20	11.3
4	0.7	10.8	18	25	11.2
5	0.8	10.6	19	35	10.4
6	1	11.4	20	50	9.65
7	1.2	12.3	21	75	8.91
8	1.5	12.1	22	100	8.41
9	2	10.5	23	125	7.95
10	2.5	9.68	24	150	7.49
11	3.5	9.25	25	175	7.05
12	5	9.3	26	200	6.81
13	7.5	9.58	27	225	6.38
14	10	10.3	28	250	6.02

TABLE 24.3 (continued)
Total Scattering Cross Sections for NO in the High Energy Range

Serial Number	Energy (eV)	Q_T (10^{-20} m^2)	Serial Number	Energy (eV)	Q_T (10^{-20} m^2)
29	121	8.47	38	484	3.78
30	144	8.22	39	576	3.37
31	169	7.23	40	676	2.89
32	196	6.72	41	784	2.65
33	225	6.10	42	900	2.34
34	256	5.83	43	1024	2.12
35	289	5.29	44	1156	1.93
36	324	5.10	45	1296	1.69
37	400	4.25	46	1444	1.58

Source: Reproduced with kind permission of Institute of Physics, England.
Note: Serial numbers (1–28) are from Szmytkowski and Maciąg (1996) and (29–46) are from Dalba et al. (1980).

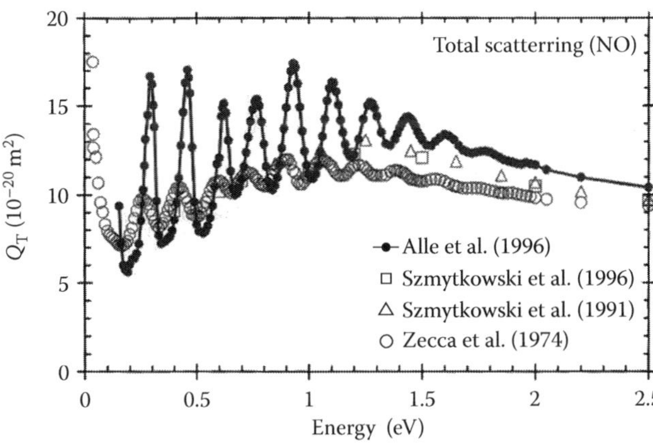

FIGURE 24.1 Total scattering cross section for NO in the low energy region. (—●—) Alle et al. (1996); (□) Szmytkowski and Maciąg (1996); (△) Szmytkowski and Maciąg (1991); (○) Zecca et al (1974).

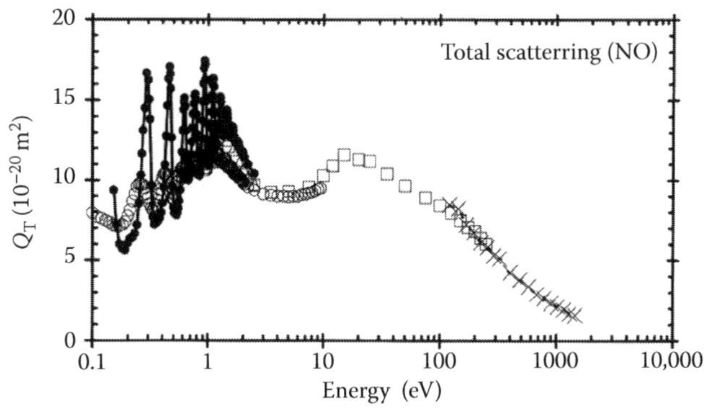

FIGURE 24.2 Total scattering cross section for NO. (—●—) Alle et al. (1996); (□) Szmytkowski and Maciąg (1996); (△) Szmytkowski and Maciąg (1991); (—×—) Dalba et al. (1980).

3. A broad peak centered at 10–12 eV; this feature is common to many gases.
4. A monotonic decrease for energy >15 eV; this feature is also common to many gases.

24.3 DIFFERENTIAL CROSS SECTIONS

Tables 24.5 and 24.6 and Figures 24.3 and 24.4 show the differential cross sections for elastic scattering for NO. Points to note are (Brunger and Buckman, 2002):

1. In the low energy range, 1.5–5.0 eV, the differential cross section shows a peak at about 60° angle, falling off in magnitude toward both smaller and larger angles.

TABLE 24.4
Experimental Positions of Peaks in Total Scattering Cross Section

Peak	Energy (eV)	Separation (meV)
1	0.293	
2	0.460	164
3	0.624	161
4	0.768	149
5	0.933	165
6	1.104	171
7	1.272	168
8	1.441	169
9	1.608	167
10	1.790	182

Source: Adapted from Alle, D. T., M. J. Brennan, and S. J. Buckman, *J. Phys. B: At. Mol. Opt. Phys.*, 29, L277, 1996.

TABLE 24.5
Differential Cross Sections for Elastic Scattering for NO

	Differential Cross Section (10^{-20} m^2/sr)				
	Energy (eV)				
Angle (°)	1.5	3.0	5.0	7.5	10.0
15		0.812	0.941	1.33	1.87
20	0.799	0.85	0.949	1.234	1.629
25		0.908	1.01	1.186	1.475
30	0.843	0.942	1.095	1.202	1.375
35		0.993	1.104	1.209	1.333
40	0.91	1.028	1.121	1.221	1.292
45		1.056	1.174	1.193	1.23
50	0.955	1.081	1.148	1.207	1.172
55		1.064		1.13	1.086
60	0.96	1.074	1.114	1.067	0.996
65		1.054			0.92
70	0.947	1.004	0.976	0.873	0.752
75		0.928			0.631
80	0.895	0.864	0.802	0.647	0.517

continued

NO

NO

TABLE 24.5 (continued)
Differential Cross Sections for Elastic Scattering for NO

	Differential Cross Section (10^{-20} m²/sr)				
	Energy (eV)				
Angle (°)	**1.5**	**3.0**	**5.0**	**7.5**	**10.0**
85		0.81			0.439
90	0.838	0.742	0.642	0.496	0.373
95		0.696			0.345
100	0.805	0.622	0.534	0.409	0.333
105		0.625			0.355
110	0.774	0.597	0.472	0.395	0.366
115	0.761	0.568			0.404
120		0.538	0.443	0.421	0.448
125	0.75	0.526			0.473
130		0.517	0.441	0.47	0.519

Source: Adapted from Mojarrabi, B. et al., *J. Phys. B: At. Mol. Opt. Phys.*, 28, 487, 1995.

TABLE 24.6
Differential Cross Section for NO at Higher Energy

	Differential Cross Section (10^{-20} m²/sr)			
	Energy (eV)			
Angle (°)	**15**	**20**	**30**	**40**
10				9.874
15	3.6	4.236	6.484	6.594
20	2.907	3.444	4.49	4.364
25	2.424	2.968	3.387	3.042
30	1.922	2.132	2.29	1.958
35	1.639	1.764	1.671	1.275
40	1.409	1.502	1.323	0.885
45	1.264	1.231	0.987	0.578
50	1.109	1.036	0.739	0.410
55	0.910	0.737	0.576	0.318
60	0.739	0.567	0.417	0.245
65	0.622	0.481	0.296	0.189
70	0.529	0.361	0.249	0.149
75	0.428	0.285	0.193	0.121
80	0.343	0.233	0.157	0.101
85	0.284	0.202	0.148	0.08
90	0.278	0.186	0.138	0.07
95	0.274	0.183	0.137	0.066
100	0.305	0.197	0.136	0.068
105	0.326	0.213	0.166	0.082
110	0.382	0.246	0.206	0.110
115	0.436	0.294	0.262	0.152
120	0.472	0.326	0.315	0.206
125	0.511	0.382	0.417	0.277
130	0.536	0.454	0.498	0.345

Source: Adapted from Mojarrabi, B. et al., *J. Phys. B: At. Mol. Opt. Phys.*, 28, 487, 1995.

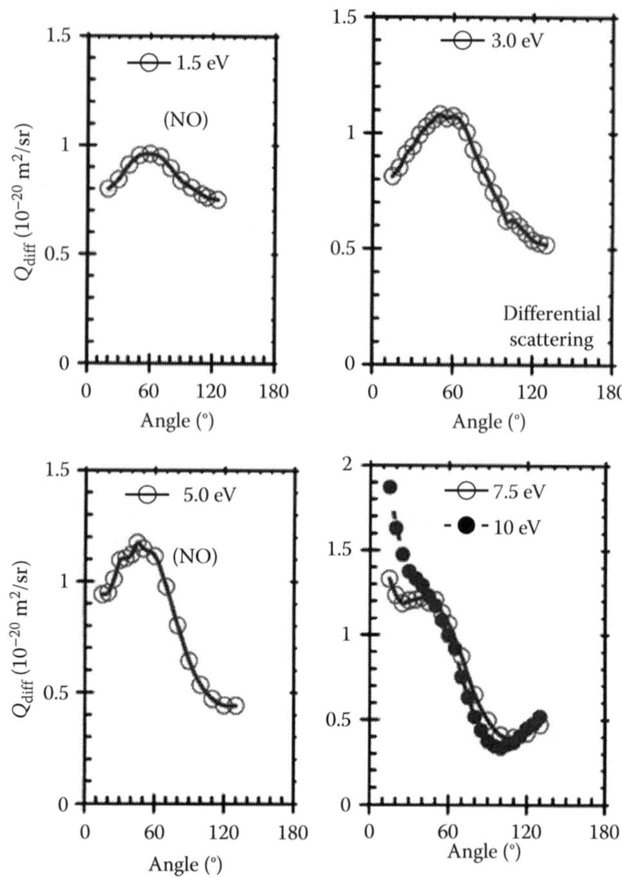

FIGURE 24.3 Differential cross section for elastic scattering for NO as a function of scattering angle and electron energy. (Adapted from Mojarrabi, B. et al., *J. Phys. B: At. Mol. Opt. Phys.*, 28, 487, 1995.)

2. At energies in the range from 7.5 to 40 eV the differential cross section decreases with increasing angle and this characteristic is attributed to the electronic polarization of the molecule.

24.4 ELASTIC AND MOMENTUM TRANSFER CROSS SECTIONS

See Tables 24.7, 24.8 and Figure 24.5.

24.5 VIBRATIONAL EXCITATION CROSS SECTIONS

The ro-vibrational excitation cross sections for $v = 0 \rightarrow 1$ and $v = 0 \rightarrow 2$ are shown in Table 24.9. See Table 24.4 for individual energy levels.

24.6 ELECTRONIC EXCITATION CROSS SECTIONS

Figure 24.6 shows the lower excitation states of NO (Raju, 2005). Figure 24.7 shows the integral excitation cross sections (Brunger et al., 2000).

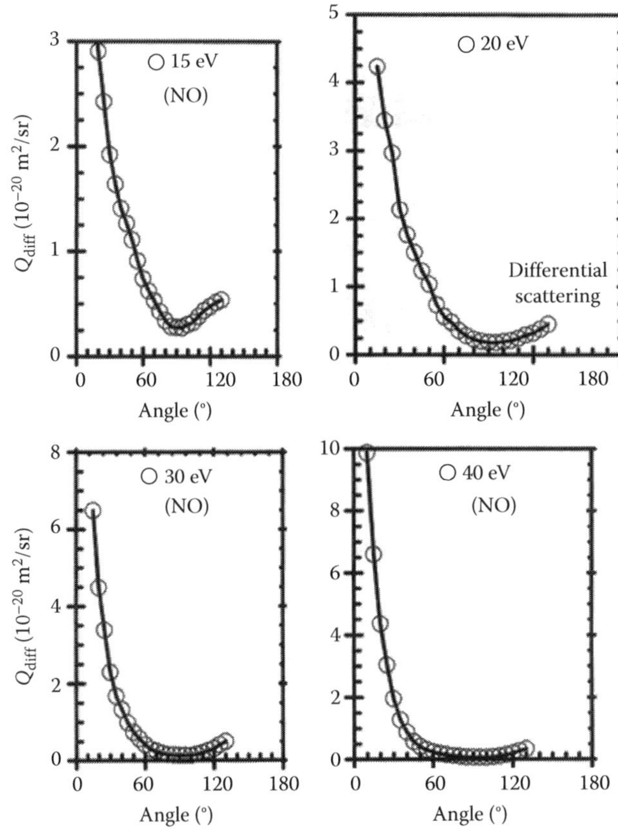

FIGURE 24.4 Differential cross section for elastic scattering for NO as a function of scattering angle and electron energy in the higher energy range. (Adapted from Mojarrabi, B. et al., *J. Phys. B: At. Mol. Opt. Phys.*, 28, 487, 1995.)

TABLE 24.8
Theoretical Cross Sections for NO in the High Energy Range

Energy (eV)	Q_{el} (10^{-20} m^2)	Q_M (10^{-20} m^2)
5.0	9.45	5.21
10.0	12.14	8.64
20.0	14.33	14.33
50	7.66	7.66
100	4.33	4.33
200	3.22	3.22
500	1.04	1.04

Note: Differential cross sections of Lee et al. (1992) integrated by Raju (2005).

NO

TABLE 24.9
Cross Section for Ro-Vibrational Excitation

Energy (eV)	$Q_{0\to1}$ (10^{-20} m^2)	$Q_{0\to2}$ (10^{-20} m^2)
7.5	0.028	
10.0	0.074	0.014
15.0	0.270	0.073
20.0	0.097	0.022
30.0	0.022	
40.0	0.014	

Source: Adapted from Brunger, M. J. et al., *J. Phys. B: At. Mol. Opt. Phys.*, 33, 809, 2000.

TABLE 24.7
Integral Elastic and Momentum Transfer Cross Sections for NO

Energy (eV)	Q_{el} (10^{-20} m^2)	Q_M (10^{-20} m^2)
1.5	10.473	8.415
3	9.604	7.044
5	9.239	6.296
7.5	9.095	5.797
10	9.241	5.539
15	9.714	5.116
20	9.707	4.232
30	9.314	3.547
40	8.214	2.546
50	6.444	

Note: See Figure 24.5 for graphical presentation (Brunger and Buckman 2002).

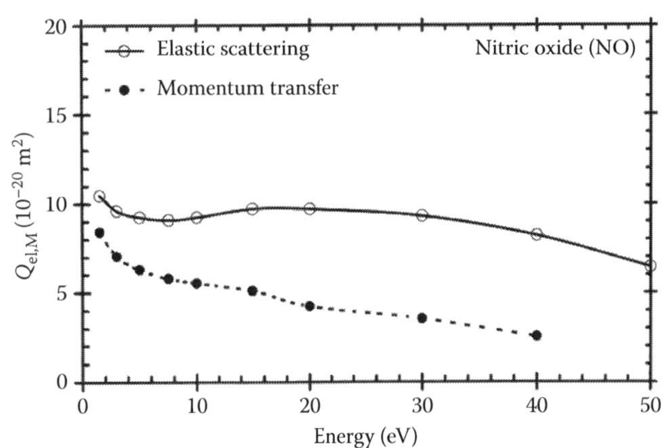

FIGURE 24.5 Integral elastic and momentum transfer cross sections for NO. (Adapted from Brunger, M. J. and S. J. Buckman, *Phys. Rep.*, 357, 215, 2002.)

NO

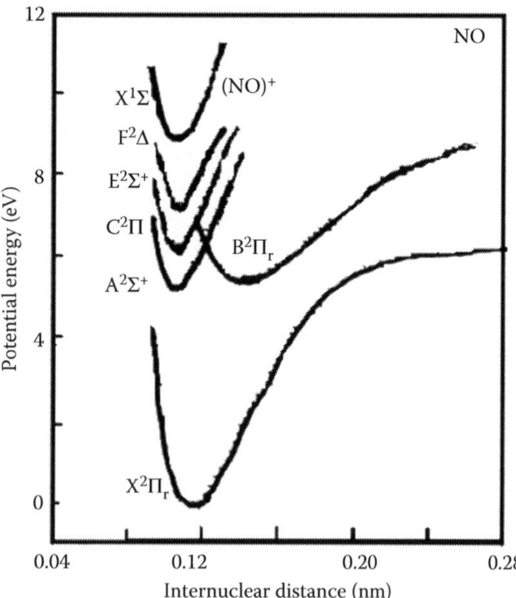

FIGURE 24.6 Lower excitation states of NO. The threshold energies are as follows with spectroscopic values shown in brackets: A $^2\Sigma^+$ = 5.484 eV (5.449 eV), B $^2\Pi_r$ = 5.769 eV (5.691), C $^2\Pi$ = 6.499 eV, E $^2\Sigma^+$ = (7.514 eV), F $^2\Delta$ = 7.722 eV, X $^1\Sigma$ = (9.4 eV). The energy for states is from Brunger et al. (2000). Spectroscopic values from Herzberg (1950).

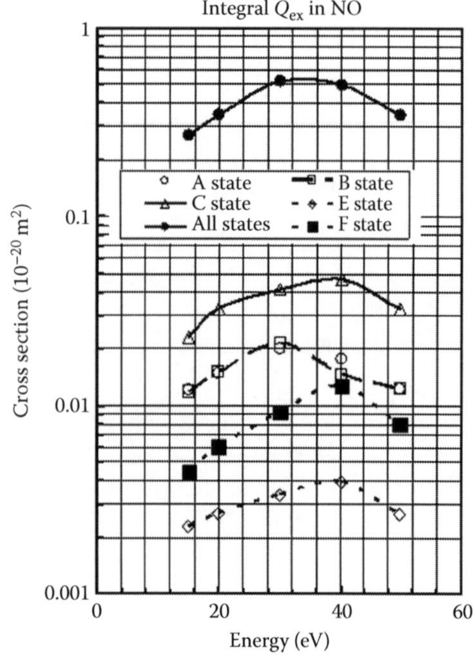

FIGURE 24.7 Integral excitation cross sections in NO. (—○—) A state; (—□—) B state; (—△—) C state; (—◊—) E state; (—■—) F state; (—●—) sum of all states. (Adapted from Brunger, M. J. et al., *J. Phys. B: At. Mol. Opt. Phys.*, 33, 809, 2000.)

24.7 IONIZATION CROSS SECTIONS

See Table 24.10.

24.8 ATTACHMENT CROSS SECTIONS

Electron attachment occurs by different processes according to

$$e + NO \rightarrow (^2\Pi, v = 0) \rightarrow NO^-(^3\Sigma^-, v')$$
$$\rightarrow NO\,(^2\Pi, v = 0) + e \qquad (24.1)$$

TABLE 24.10
Partial and Total Ionization Cross Sections for NO

Lindsay et al. (2000)		Rapp and Englander-Golden (1965)			
Energy (eV)	Q_i (10^{-20} m^2)	Energy (eV)	Q_i (10^{-20} m^2)	Energy (eV)	Q_i (10^{-20} m^2)
12.5	0.048	9.5	0.011	60	2.74
15	0.21	10	0.018	65	2.82
17.5	0.48	10.5	0.031	70	2.91
20	0.59	11	0.047	75	2.97
22.5	0.75	11.5	0.064	80	3.03_5
25	1.02	12	0.092	85	3.07
30	1.33	12.5	0.13	90	3.11
35	1.68	13	0.18	95	3.13
40	1.99	13.5	0.24	100	3.140
45	2.32	14	0.30_5	105	3.15
50	2.58	14.5	0.36	110	3.15
55	2.76	15	0.42	115	3.15
60	2.97	15.5	0.46	120	3.14
70	3.18	16	0.50	125	3.14
80	3.36	16.5	0.54_5	130	3.12
90	3.51	17	0.585	135	3.10_5
100	3.55	17.5	0.625	140	3.10
125	3.56	18	0.663	145	3.09
150	3.45	18.5	0.701	150	3.08
200	3.16	19	0.736	160	3.04
250	2.91	17	0.58_5	180	2.96
300	2.60	17.5	0.62_5	200	2.86
400	2.24	18	0.66	250	2.65
500	1.96	18.5	0.70	300	2.44_5
600	1.74	19	0.74	350	2.27
800	1.41	19.5	0.77	400	2.11
1000	1.20	20	0.81	450	1.98
		22	0.96	500	1.86
		24	1.11	550	1.76
		26	1.26	600	1.67
		28	1.40	650	1.59
		30	1.52	700	1.51
		32	1.65	750	1.45
		34	1.77	800	1.39
		36	1.87	850	1.33
		38	1.99	900	1.21_5
		40	2.08	950	1.24
		45	2.30_5	1000	1.20_5
		50	2.48		
		55	2.61		

Note: See Figures 24.8 and 24.9 for graphical presentation.

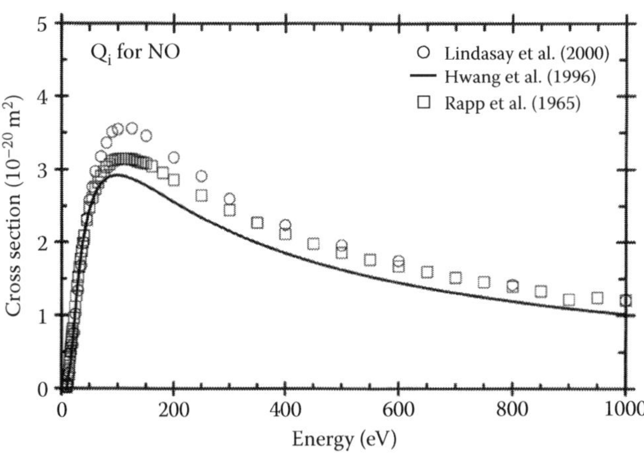

FIGURE 24.8 Total ionization cross section for NO. (○) Lindsay et al. (2000); (——) Hwang et al. (1996); (□) Rapp and Englander-Golden (1965). The agreement between measurements is very good. In a previous publication it was erroneously stated that the agreement was less than satisfactory. (Adapted from Raju, G. G., *Gaseous Electronics: Theory and Practice*, Taylor & Francis, London, 2005, p. 254.)

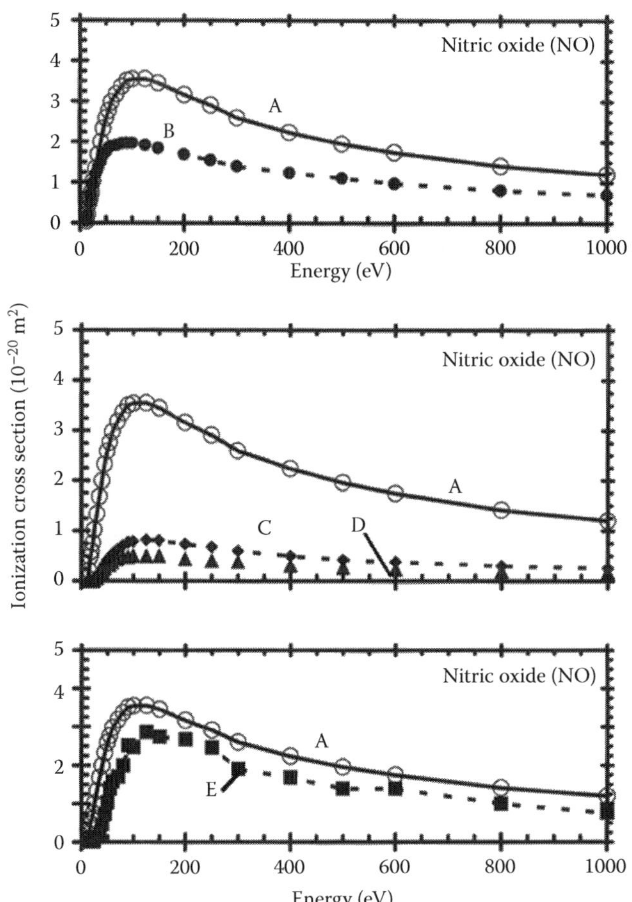

FIGURE 24.9 Partial ionization cross sections for NO. Total cross sections are also included to show the relative contribution. (A) Total; (B) NO^+; (C) ($N^+ + O^+ + NO_2^+$); (D) N^+; (E) (O^+)-multiplication factor 10. (Adapted from Lindsay, B. G. et al., *J. Chem. Phys.*, 112, 9404, 2000.)

$$e + NO \rightarrow (^2\Pi, v = 0) \rightarrow NO^-(^1\Delta, v') \rightarrow NO(^2\Pi, v = 0) + e \quad (24.2)$$

$$e + NO \rightarrow (^2\Pi, v = 0) \rightarrow NO^-(^1\Sigma^+, v') \rightarrow NO(^2\Sigma^+, v = 0) + e \quad (24.3)$$

$$e + NO \rightarrow N + O^- \quad (24.4)$$

$$e + NO \rightarrow N^* + O^- \quad (24.5)$$

The first step in Reactions 24.1 through 24.3 is electron attachment, the state of the NO^- ion as shown in parenthesis, and the second step is auto detachment of the electron with the negative ion reverting to the ground state. All three reactions occur below 2.0 eV. Reaction 24.4 is dissociative attachment with onset energy of ~5.074 eV (Denifl et al., 1998). Reaction 24.5 yields excited nitrogen atom, with two channels and corresponding threshold energies of 7.457 (excitation energy = 2.383 eV) and 8.650 eV (excitation energy = 3.576 eV) (Denifl et al., 1998). See Table 24.11.

TABLE 24.11
Dissociative Attachment Cross Sections for NO

Energy (eV)	Q_a (10^{-22} m²)	Energy (eV)	Q_a (10^{-22} m²)
6.7	0.009	9.1	1.003
6.8	0.018	9.2	0.950
6.9	0.044	9.3	0.888
7.0	0.079	9.4	0.827
7.1	0.150	9.5	0.748
7.2	0.334	9.6	0.651
7.3	0.537	9.7	0.581
7.4	0.712	9.8	0.510
7.5	0.862	9.9	0.440
7.6	0.959	10.0	0.378
7.7	1.038	10.1	0.317
7.8	1.076	10.2	0.264
7.9	1.103	10.3	0.220
8.0	1.114	10.4	0.176
8.1	1.117	10.5	0.141
8.2	1.116	10.6	0.114
8.3	1.109	10.7	0.092
8.4	1.100	10.8	0.079
8.5	1.103	10.9	0.070
8.6	1.106	11.0	0.062
8.7	1.100	11.5	0.044
8.8	1.088	12.0	0.044
8.9	1.069	12.5	0.035
9.0	1.038	13.0	0.035

Source: Adapted from Rapp, D. and D. D Briglia, *J. Chem. Phys.*, 43, 1480, 1965.
Note: See Figure 24.10 for graphical presentation.

NO

24.9 DRIFT AND DIFFUSION COEFFICIENT

Figure 24.11 shows the drift velocity of electrons for NO at low values of *E*/*N* (Parkes and Sugden, 1972). Slight dependence on gas number density is observed and attributed to the three-body processes at low values of *E*/*N*.

Table 24.12 shows the characteristic energy D_r/μ and D_L/μ for NO. See Figure 24.12 for graphical presentation.

NO

24.10 IONIZATION COEFFICIENT

See Table 24.13.

24.11 GAS CONSTANTS

Gas constants evaluated according to the expression

$$\frac{\alpha}{N} = F\exp\left(-\frac{GN}{E}\right) \tag{24.6}$$

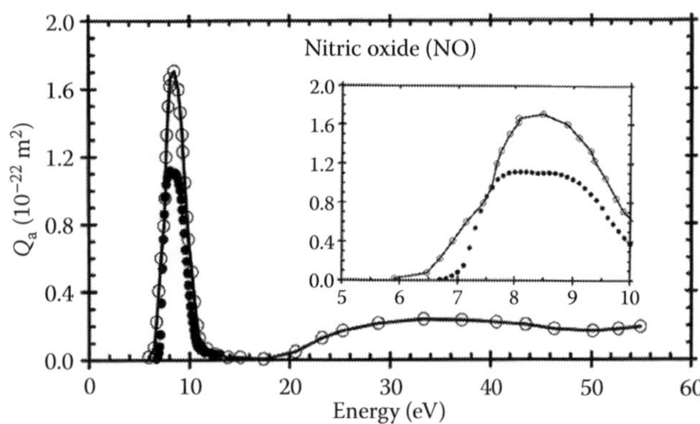

FIGURE 24.10 Attachment cross sections for NO. (—○—) Krishnakumar and Srivastava (1988); (●) Rapp and Briglia (1965). Inset shows details in the 5–10 eV region.

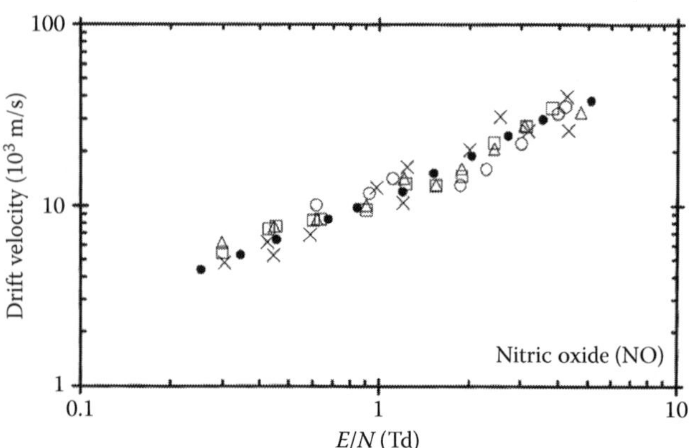

FIGURE 24.11 Drift velocity of electrons in NO. (○) $N = 21 \times 10^{22}/m^3$; (□) $32 \times 10^{22}/m^3$; (Δ) $52 \times 10^{22}/m^3$; (●) $60 \times 10^{22}/m^3$; (×) $71 \times 10^{22}/m^3$. (Adapted from Parkes, D. A. and T. M. Sugden, *J. Chem. Soc. Farad. Trans. Part II*, 68, 600–614, 1972.)

TABLE 24.12
Characteristic Energy for NO

Mechlińska-Drewko et al. (1999)			Lakshminarasimha and Lucas (1977)	
E/*N* (Td)	D_r/μ (V)	D_L/μ (V)	*E*/*N* (Td)	D_r/μ (V)
12	0.2578	0.094	285	3.92
20	0.315	0.14	300	4.08
25	0.432		350	4.44
30	0.594		400	4.68
40	0.978	0.48	450	4.95
50	1.46	0.68	500	5.31
60	1.69	0.81	600	5.97
70	1.91		700	6.20
100	2.49	1.52	800	6.11
150	3.21	2.29	900	6.34
200	3.91	3.12	1000	6.89
250	4.19	3.84	1100	7.35
300	4.35	4.33	1200	7.95

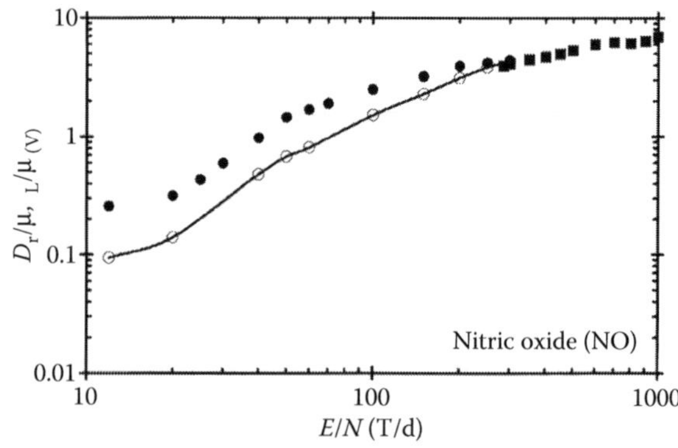

FIGURE 24.12 D_r/μ and D_L/μ for electrons in NO. D_r/μ: (●) Mechlińska-Drewko et al. (1999); (■) Lakshminarasimha and Lucas (1977). D_L/μ: (—○—) Mechlińska-Drewko et al. (1999).

TABLE 24.13
Density-Reduced Ionization Coefficients for NO

E/*N* (Td)	α/*N* (10^{-21} m^2)	*E*/*N* (Td)	α/*N* (10^{-21} m^2)
50	0.0479	350	5.90
60	0.0823	400	7.17
70	0.151	450	8.28
80	0.230	500	9.34
90	0.308	600	11.64
100	0.412	700	14.05
125	0.718	800	15.60
150	1.03	900	17.17
175	1.47	1000	19.45
200	2.03	1100	21.66
250	3.37	1200	22.93
300	4.46		

Source: Adapted from Lakshminarasimha, C. S. and J. Lucas, *J. Phys. D: Appl. Phys.*, 10, 313, 1977.

Note: See Figure 24.13 for graphical presentation.

are $F = 8 \times 10^{-21}$ m² and $G = 274.5$ Td⁻¹ for the range $50 \leq E/N \leq 200$ Td. See inset of Figure 24.13.

24.12 ATTACHMENT COEFFICIENTS

Attachment coefficients for NO measured by Parkes and Sugden (1972) are found to depend on N^2, as shown in Figure 24.14.

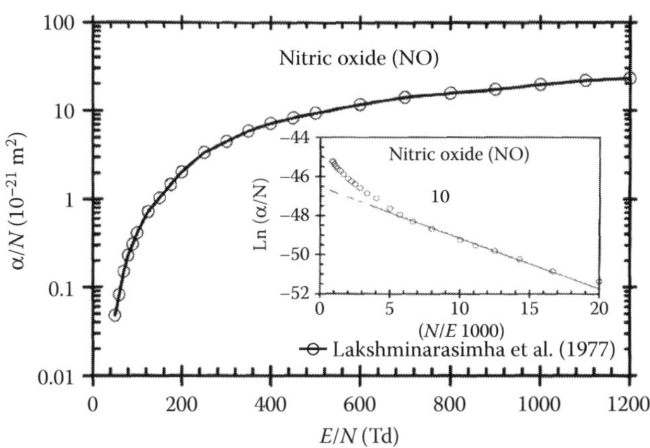

FIGURE 24.13 Reduced Townsend's first ionization coefficient as a function of E/N for NO. Inset shows the evaluation of gas constants according to Equation 24.6. (Adapted from Lakshminarasimha, C. S. and J. Lucas, *J. Phys. D: Appl. Phys.*, 10, 313, 1977.)

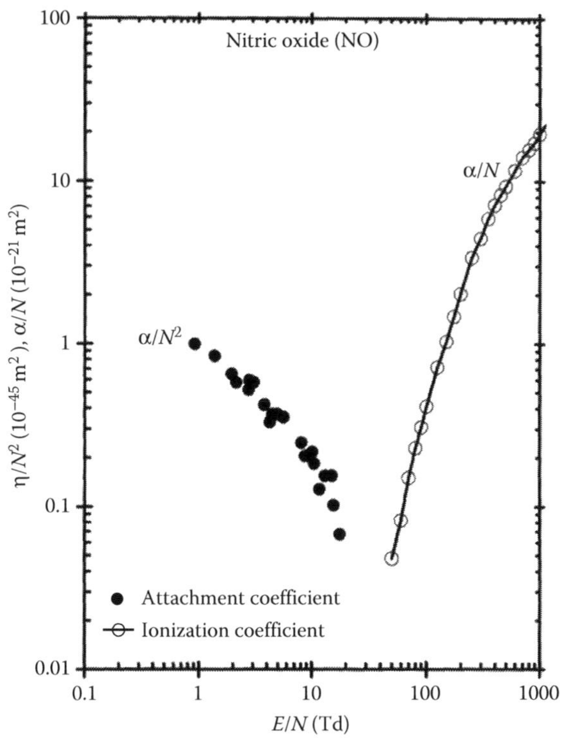

FIGURE 24.14 Attachment coefficients for NO. Ionization coefficients are replotted for comparison. (Adapted from Parkes, D. A. and T. M. Sugden, *J. Chem. Soc. Farad. Trans. Part II*, 68, 600–614, 1972.)

24.13 ATTACHMENT RATE CONSTANTS

See Table 24.14.

24.14 POSITIVE ION MOBILITY

The measured mobility of NO^+ referred to 273 K and $N = 2.69 \times 10^{25}$/m³ is 1.9×10^{-4} m²/s V (Gunton and Shaw, 1965).

24.15 CONSOLIDATED CROSS SECTIONS

Table 24.15 shows the consolidated cross sections for NO. See Figure 24.15 for graphical presentation.

TABLE 24.14
Attachment Rate Constants for NO

Rate	Temperature (eV)	Method	Reference
5.80×10^{-43} (m⁶/s)	300 K	FA	Mcfarland et al. (1972)
1.4×10^{-17} (m³/s)[a]	296 K	SA	Puckett et al. (1971)
6.80×10^{-44} (m⁶/s)	296 K		
6.80×10^{-43} (m⁶/s)	196 K	MWCW	Gunton and Shaw (1965)
2.2×10^{-43} (m⁶/s)	298 K		
1.1×10^{-43} (m⁶/s)	358 K		

Note: Three-body attachment unless otherwise mentioned. FA = flowing afterglow; MWCM = microwave cavity method; SA = stationary afterglow.

[a] Two-body attachment processes.

TABLE 24.15
Consolidated Cross Section for NO (1.5–500 eV)

Energy (eV)	Cross Section (10^{-20} m²) Q_{el}	Q_V	Q_{ex}	Q_i	Q_T (sum)	Q_T (expt)	Difference (%)
1.5	10.473	—	—	—	10.473	12.3	14.85
3.0	9.604	—	—	—	9.604	9.45	−1.56
5.0	9.239	—	—	—	9.239	9.22	−0.22
7.5	9.095	0.028			9.123	9.52	4.17
10.0	9.241	0.088	—	0.018	9.347	10.1	7.45
15.0	9.714	0.343	0.270	0.418	10.745	11.50	6.56
20.0	9.707	0.119	0.346	0.813	10.985	11.40	3.60
30.0	9.314	0.022	0.521	1.522	11.379	10.80	−5.36
40.0	8.214	0.014	0.497	2.086	10.811	10.3	−4.96
50.0	6.444	—	0.344	2.482	9.270	9.6	3.44
100	4.33		0.3	3.14	7.78	8.48	8.25
200	3.22		0.3	2.86	6.38	6.81	6.31
500	1.03		0.3	1.86	3.19	3.55	10.14

Source: Adapted from Raju, G. G., *Gaseous Electronics: Theory and Practice*, Taylor & Francis, London, 2005, p. 254.

NO

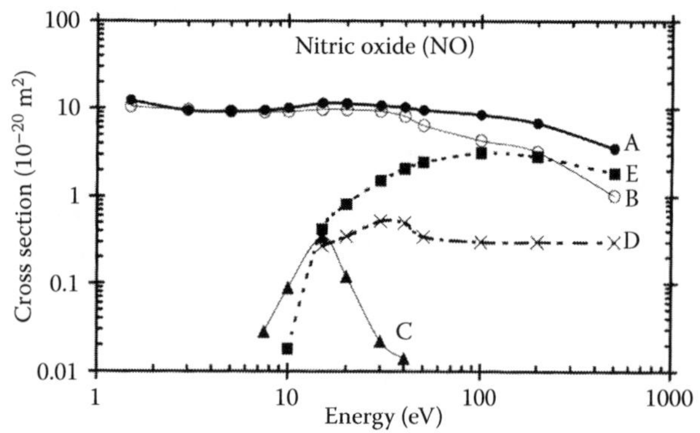

FIGURE 24.15 Consolidated cross section data for NO. (A) total; (B) elastic; (C) vibrational excitation; (D) electronic excitation; (E) ionization cross section.

NO

REFERENCES

Alle, D. T., M. J. Brennan, and S. J. Buckman, *J. Phys. B: At. Mol. Opt. Phys.*, 29, L277, 1996.

Bradbury, N. E., *J. Chem. Phys.*, 2, 827, 1934.

Brunger, M. J. and S. J. Buckman, *Phys. Rep.*, 357, 215, 2002.

Brunger, M. J., L. Campbell, D. C. Cartwright, A. G. Middleton, B. Mojarrabi, and P. J. O. Teubner, *J. Phys. B: At. Mol. Opt. Phys.*, 33, 809, 2000.

Dalba, G., P. Fornasini, R. Grisenti, G. Ranieri, and A. Zecca, *J. Phys. B: At. Mol. Phys.*, 13, 4695, 1980.

Denifl, G., D. Muigg, A. Stamatovic, and T. D. Märk, *Chem. Phys. Lett.*, 288, 105, 1998.

Gunton, R. C. and T. M. Shaw, *Phys. Rev.*, 140, A748, 1965.

Herzberg, G., *Molecular Spectra and Molecular Structure*, Van Nostrand, Princeton, NJ, 1950.

Hwang, W., Y.-K. Kim, and M. E. Rudd, *J. Chem. Phys.*, 104, 2956, 1996.

Joshipura, K. N., S. Gangopadhyay, and B. G. Vaishnav, *J. Phys. B: At. Mol. Opt. Phys.*, 40, 199, 2007.

Krishnakumar, E. and S. K. Srivastava, *J. Phys. B: At. Mol. Opt. Phys.*, 21, L607, 1988.

Lakshminarasimha, C. S. and J. Lucas, *J. Phys. D: Appl. Phys.*, 10, 313, 1977.

Lee, M.-T., M. J. Fujimato, S. E. Michelin, L. E. Machado, and L. M. Brescansin, *J. Phys. B: At. Mol. Opt. Phys.*, 25, L505, 1992.

Lindsay, B. G., M. A. Mangan, H. C. Straub, and R. F. Stebbings, *J. Chem. Phys.*, 112, 9404, 2000.

Lopez, J., V. Tarnovsky, M. Gutkin, and K. Becker, *Int. J. Mass Spectrom.*, 225, 25, 2003.

McFarland, M., D. B. Dunkin, F. C. Fehsenfeld, A. L. Schmeltekopf, and F. E. Ferguson, *J. Chem. Phys.*, 56, 2358, 1972.

Mechlińska-Drewko, J., W. Roznerski, Z. J. Petrović, and G. P. Karwasz, *J. Phys. D: Appl. Phys.*, 32, 2746, 1999.

Mojarrabi, B., R. J. Gulley, A. G. Middleton, D. C. Cartwright, P. J. O. Teubner, S. J. Buckman, and M. J. Brunger, *J. Phys. B: At. Mol. Opt. Phys.*, 28, 487, 1995.

Parkes, D. A. and T. M. Sugden, *J. Chem. Soc. Farad. Trans. Part II*, 68, 600–614, 1972.

Puckett, L. J., M. D. Kregel, and M. W. Teague, *Phys. Rev. A*, 4, 1659, 1971.

Raju, G. G., *Gaseous Electronics: Theory and Practice*, Taylor & Francis, London, 2005, p. 254.

Rapp, D. and D. D., Briglia, *J. Chem. Phys.*, 43, 1480, 1965.

Rapp, D. and P. Englander-Golden, *J. Chem. Phys.*, 43, 1464, 1965.

Spence, D. and G. J. Schulz, *Phys. Rev. A*, 3, 1968, 1971.

Szmytkowski, C. and K. Maciąg, *J. Phys. B: At. Mol. Opt. Phys.*, 24, 4273, 1991.

Szmytkowski, C. and K. Maciąg, *Phys. Scrip.*, 54, 271, 1996.

Tronc, M., A. Huetz, M. Landau, F. Pichou, and J. Reinhardt, *J. Phys. B: Atom. Mol. Phys.*, 8, 1160, 1975.

Zecca, A., I. Lazzizzera, M. Krauss, and C. E. Kuyatt, *J. Chem. Phys.*, 61, 4560, 1974.

25

NITROGEN

CONTENTS

Nitrogen (N_2) molecule possesses 14 electrons with electronic polarizability of 1.936×10^{-40} F m^2 and ionization potential of 15.581 eV.

25.1 SELECTED REFERENCES FOR DATA

See Table 25.1.

TABLE 25.1
Selected References for Data for N_2

Quantity	Range: eV, (Td)	Reference
Drift velocity	(1.5–6000)	**Lisovskiy et al. (2006)**
Swarm parameters	(50–250)	Tanaka (2004)
Excitation cross section	15–50	**Campbell et al. (2001)**
Ionization cross section	20–600	**Tian and Vidal (1998)**
Transport coefficients	(20–1000)	**Hasegawa et al. (1996)**
Diffusion coefficient	50–1150	**Roznerski (1996)**
Ionization cross section	17–1000	**Straub et al. (1996)**
Total cross section	0.4–250	**Szmytkowski and Maciag (1996)**
Vibrational cross section	0.55–50.0	Zecca et al. (1996)
Ionization cross section	16–1000	Van Zyl and Stephan (1994)
Excitation cross section	18.5–148.5	**Cosby (1993)**
Total cross section	250–1000	**Karwasz et al. (1993)**
Transport coefficients	(20–2000)	Liu and Raju (1993)
Elastic cross section	1.5–5.0	**Brennan et al. (1992)**
Swarm parameters	(100–600)	Liu and Raju (1992)
Total cross section	4–300	**Nickel et al. (1992)**
Excitation cross section	15–50	**Brunger and Teubner (1990)**
Ionization cross section	16–200	**Freund et al. (1990)**
Ionization cross section	20–1000	**Krishnakumar and Srivastava (1990)**
Ionization coefficient	(210–1500)	**Watts and Heylen (1989)**
Total cross section	600–1000	García et al. (1988)

continued

TABLE 25.1 (continued)
Selected References for Data for N_2

Quantity	Range: eV, (Td)	Reference
	0.5–300	**Nakamura and Kurachi (1988)**
Ro-vibrational cross section	0.01–14.0	**Morrison et al. (1987)**
Drift velocity	(0.5–300)	**Nakamura (1987)**
Vibrational excitation	1.95–4.66	**Allan (1985)**
Diffusion coefficient	10–6000	**Al-Amin et al. (1985)**
Excitation cross section	0–1000	Phelps and Pitchford (1985)
Transport coefficients	(10–1500)	Phelps and Pitchford (1985)
Swarm parameters	(140–905)	**Wedding et al. (1985)**
Ionization coefficient	(90–420)	Novak and Frechette (1984)
Drift velocity	(0.1–250)	**Roznerski and Leja (1984)**
Total cross section	2.2–700	**Hoffman et al. (1982)**
Ionization coefficients	(300–600)	**Raju and Dincer (1982)**
Ionization cross section	10.1–240	**Armentrout et al. (1981)**
Ionization coefficients	(100–700)	**Raju and Hackam (1981)**
Total cross section	11.5–770	**Blaauw et al. (1980)**
Total cross section	121–1024	**Dalba et al. (1980)**
Transport parameters	(50–500)	**Fletcher and Reid (1980)**
Total cross section	0.519–10.0	**Kennerly (1980)**
Elastic cross section	1.5–400	**Shyn and Carignan (1980)**
Swarm parameters	(14–3000)	Kücükarpaci and Lucas (1979)
Ionization coefficients	(60–600)	**Raju and Gurumurthy (1978)**
Swarm parameters	(60–1200)	Taniguchi et al. (1978)
Total cross section	15–750	**Blaauw et al. (1977)**
Excitation cross section	10–50	**Cartwright et al. (1977)**
Total cross section	1–4	**Mathur and Hasted (1977)**
Elastic cross section	20–800	**DuBois and Rudd (1976)**
Ionization coefficients	(140–3400)	**Haydon and Williams (1976)**
Elastic cross section	85–1000	**Herrmann et al. (1976)**
Ionization coefficients	(3–30)	**Raju and Gurumurthy (1976)**
Ionization coefficients	(300–1100)	**Raju and Rajapandian (1976)**
Elastic cross section	5–75	Srivastava et al. (1976)
Ionization cross section	16–170	**Märk (1975)**
Drift velocity	(55–710)	**Saelee et al. (1977)**
Drift velocity	(20–60)	**Snelson and Lucas (1975)**
Total cross section	0.30–1.6	**Baldwin (1974)**
Ionization coefficient	(300–3000)	**Maller and Naidu (1974)**
Elastic cross section	40–1000	**Wedde and Strand (1974)**
Ionization cross section	60–300	**Crowe and McConkey (1973)**
Ionization coefficients	(180–6000)	**Folkard and Haydon (1973)**
Ionization coefficients	90–3000	**Heylen and Dargan (1973)**
Swarm parameters	(30–1350)	**Kontoleon et al. (1973)**
Elastic cross section	5–90	**Shyn et al. (1980)**
Ion mobility	(110–150)	**Allen and Prew (1970)**
Ionization coefficient	(100–150)	**Daniel et al. (1969)**
Diffusion coefficient	(0.01–10)	**Lowke and Parker (1969)**
Diffusion coefficient	(0.01–6.0)	**Wagner et al. (1967)**
Ionization coefficient	(300–1800)	**Fletcher and Haydon (1966)**
Total cross section	0.3–5.0	**Golden (1966)**
Ionization cross section	16–1000	**Rapp and Englander-Golden (1965)**
Ionization cross section	16–1000	**Rapp et al. (1965)**
Ionization coefficients	(140–10000)	**Schlumbohm (1965)**
Transport coefficients	(0.001–300)	Engelhardt et al. (1964)
Transport coefficients	(0.0003–30.0)	Pack and Phelps (1961)
Drift velocity	(0.3–6.0)	**Bowe (1960)**
Diffusion coefficient	(0.0006–30.0)	**Warren and Parker (1962)**

Note: Bold font indicates experimental study.

25.2 TOTAL SCATTERING CROSS SECTION

See Tables 25.2 and 25.3.

TABLE 25.2
Total Cross Section for N_2 in the Low Energy Region

Energy (eV)	Q_T (10^{-20} m^2)	Energy (eV)	Q_T (10^{-20} m^2)	Energy (eV)	Q_T (10^{-20} m^2)	Energy (eV)	Q_T (10^{-20} m^2)
0.519	8.79	1.258	10.32	2.07	23.69	3.089	19.17
0.535	8.74	1.272	10.43	2.086	23.83	3.118	19.24
0.552	8.91	1.287	10.42	2.102	24.13	3.147	18.69
0.57	8.94	1.302	10.31	2.118	25.26	3.176	18.22
0.588	9.14	1.317	10.44	2.134	27.18	3.206	17.46
0.607	8.9	1.333	10.36	2.151	29.13	3.236	16.95
0.628	9.43	1.349	10.61	2.167	30.77	3.267	16.75
0.649	9.07	1.365	10.66	2.184	32.78	3.298	16.69
0.672	9.3	1.381	10.55	2.201	33.22	3.328	16.60
0.695	9.38	1.398	10.75	2.219	33.58	3.36	16.51
0.72	9.21	1.415	10.76	2.236	32.62	3.391	16.02
0.747	9.14	1.432	10.8	2.254	31.11	3.423	15.90
0.774	9.53	1.45	10.96	2.272	29.75	3.456	15.37
0.804	9.38	1.468	10.97	2.29	28.32	3.489	15.16
0.835	9.55	1.487	10.98	2.308	27.14	3.523	15.05
0.868	9.73	1.505	11.24	2.326	26.19	3.557	15.05
0.898	9.89	1.524	11.26	2.345	26.36	3.591	14.55
0.907	9.91	1.544	11.4	2.364	27.64	3.626	14.51
0.916	9.81	1.563	11.58	2.383	28.82	3.662	14.16
0.925	9.97	1.584	11.79	2.403	30.79	3.698	14.14
0.934	9.75	1.604	11.86	2.422	32.06	3.734	13.86
0.943	9.85	1.625	12.04	2.442	33.03	3.772	13.71
0.953	9.84	1.647	12.3	2.463	31.8	3.809	13.63
0.962	9.93	1.668	12.66	2.483	30.85	3.848	13.34
0.972	9.96	1.691	13	2.504	29.19	3.886	13.50
0.982	9.87	1.713	13.51	2.525	27.25	3.926	13.31
0.992	9.97	1.736	13.79	2.546	25.83	3.966	13.09
1.002	9.91	1.76	14.44	2.567	25.18	4.007	13.21
1.012	9.88	1.784	15.28	2.589	25.19	4.048	12.92
1.023	9.86	1.809	16.36	2.611	25.63	4.090	12.89
1.033	9.96	1.834	17.46	2.633	27.03	4.132	12.88
1.044	10.05	1.853	18.61	2.656	27.9	4.176	12.48
1.055	10.04	1.866	19.12	2.679	27.78	4.220	12.44
1.066	10.01	1.879	20.23	2.702	26.74	4.264	12.54
1.078	10.21	1.892	21.19	2.725	25.9	4.310	12.34
1.089	10.1	1.906	22.26	2.749	24.33	4.356	12.31
1.101	10.17	1.922	23.23	2.773	22.95	4.402	12.25
1.113	10.07	1.936	24.26	2.798	22.19	4.450	12.19
1.125	10.06	1.951	25.4	2.822	22.32	4.498	12.15
1.137	9.95	1.965	25.95	2.848	22.49	4.547	11.95
1.15	10.03	1.98	25.88	2.873	22.84	4.597	12.00
1.162	10.22	1.994	25.84	2.899	23.28	4.648	11.99
1.175	10.31	2.009	25.61	2.925	22.39	4.700	11.89
1.188	10.15	2.024	25.37	2.951	21.64	4.752	11.90
1.202	10.14	2.039	24.52	2.978	20.51	4.806	11.79
1.215	10.34	2.055	23.81	3.005	19.73	4.860	11.74
1.229	10.27	2.039	24.52	3.033	19.29	4.915	11.64
1.243	10.33	2.055	23.81	3.061	19.30	4.971	11.63
						5.028	11.56

Source: Reprinted with permission from Kennerly, R. E., *Phys. Rev. A* 21, 1876, 1980. Copyright (1980) by the American Physical Society.

Note: See Figure 25.1 for graphical presentation.

N_2

N$_2$

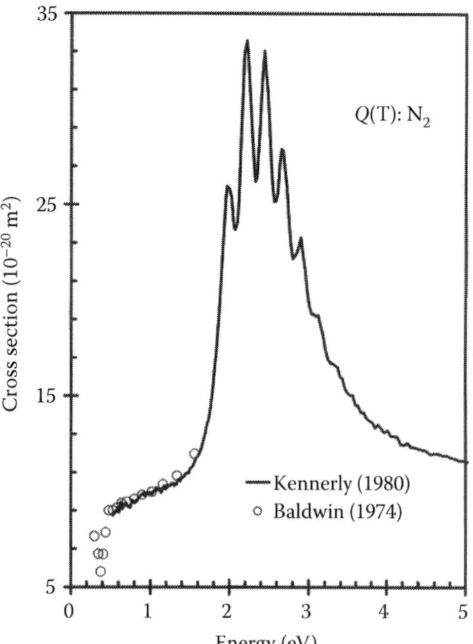

FIGURE 25.1 Total scattering cross section for N_2 in the low energy region. (—) Kennerly (1980); (○) Baldwin (1974).

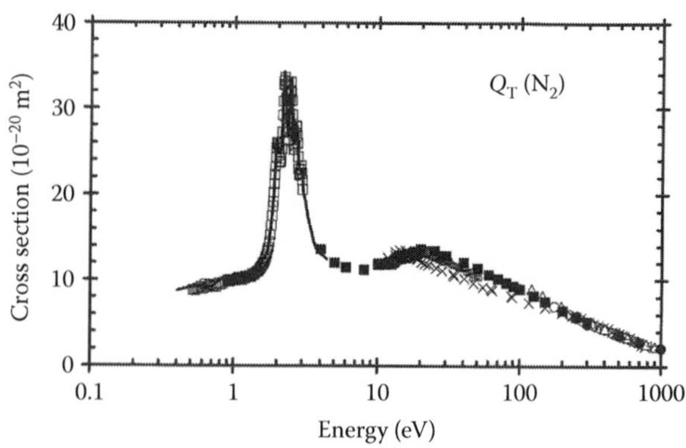

FIGURE 25.2 Total scattering cross sections in N_2. (—) Szmytkowski and Maciag (1996); (●) Karwasz et al. (1993); (■) Nickel et al. (1992); (+) García et al. (1988); (×) Blaauw et al. (1980); (△) Dalba et al. (1980); (□) Kennerly (1980); (○) Blaauw et al. (1977).

TABLE 25.3
Recommended Total Scattering Cross Section for N_2

Energy (eV)	Q_T (10^{-20}m^2)	Energy (eV)	Q_T (10^{-20}m^2)	Energy (eV)	Q_T (10^{-20}m^2)
5	11.5	100	8.85	350	4.53
10	12.5	110	8.46	400	4.215
20	13.8	120	8.09	450	3.90
30	13.1	140	7.67	500	3.58
40	12.4	160	7.22	600	3.185
50	11.2	180	6.81	700	2.79
60	10.7	200	6.41	800	2.55
70	10.0	250	5.67	900	2.31
80	9.45	300	4.85	1000	2.08
90	9.21				

Note: See Figure 25.2 for graphical presentation.

TABLE 25.4
Recommended Elastic Scattering Cross Section for N_2

Energy (eV)	Q_{el} (10^{-20} m^2)	Energy (eV)	Q_{el} (10^{-20} m^2)	Energy (eV)	Q_{el} (10^{-20} m^2)
0.10	5.82	15	11.30	500	1.93
0.50	9.84	20	12.10	550	1.75
1.0	10.00	30	10.30	600	1.66
1.5	9.60	40	9.40	650	1.60
1.9	16.73	50	8.50	700	1.44
2.0	18.38	70	7.30	750	1.40
3.0	14.80	100	5.60	800	1.32
4.0	11.10	150	4.50	850	1.27
5.0	11.20	200	4.50	900	1.21
7.0	12.50	300	3.70	950	1.17
10	11.70	400	2.60	1000	1.11

Note: See Figure 25.3 for graphical presentation.

25.3 ELASTIC SCATTERING CROSS SECTION

See Table 25.4.

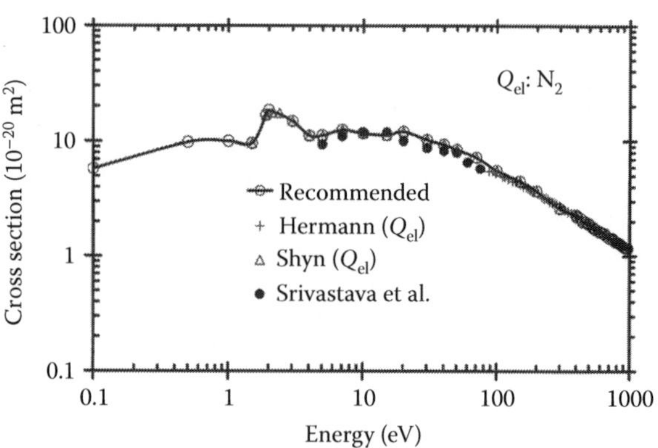

FIGURE 25.3 Elastic scattering cross section for N_2. (—○—) Recommended (Raju, 2005). (Δ) Shyn and Carignan (1980); (●) Srivastava et al. (1976); (+) Herrmann et al. (1976).

25.4 MOMENTUM TRANSFER CROSS SECTION

See Table 25.5.

25.5 RO-VIBRATIONAL EXCITATION

See Tables 25.6 through 25.8.

TABLE 25.5
Momentum Transfer Cross Section for N_2

Energy (eV)	Q_M (10^{-20} m^2)	Energy (eV)	Q_M (10^{-20} m^2)
0.0006	0.90	1.5	11.87
0.0008	0.90	1.7	13.47
0.0009	0.91	1.9	16.41
0.0010	0.91	2.1	16.85
0.0012	0.96	2.2	18.02
0.0014	0.97	2.5	17.92
0.0017	1.02	2.8	21.00
0.002	1.05	3.0	17.20
0.0025	1.21	3.30	15.30
0.003	1.23	3.60	13.96
0.004	1.37	4.0	12.42
0.0041	1.44	4.5	11.19
0.005	1.50	5.0	10.86
0.006	1.80	6.0	10.36
0.007	1.70	7.0	10.00
0.008	1.82	8.0	10.20
0.01	2.0	10.0	9.9
0.02	2.08	12.0	9.5
0.03	3.48	15.0	8.7
0.04	3.82	17.0	8.26
0.05	4.23	20	7.6
0.06	4.76	25	6.70
0.08	5.25	30	5.9
0.100	5.93	50	3.8
0.2	7.82	75	2.56
0.3	9.04	100	1.80
0.4	9.52	150	1.13
0.50	9.84	200	0.8
0.60	9.93	300	0.48
0.70	10.07	500	0.23
1.0	9.96	700	0.14
1.2	10.34	1000	0.01
1.3	10.92		

Note: Momentum transfer cross sections in N_2 based on Ramanan and Freeman (1990) for the range $0 < \varepsilon \leq 1$ eV; Morgan, Bouef and Pitchford (1999) for the range $1 \leq \varepsilon \leq 1000$ eV. It is important to realize that the total Q_M is higher than that for elastic scattering only, particularly at high energies. See Figure 25.4 for graphical presentation.

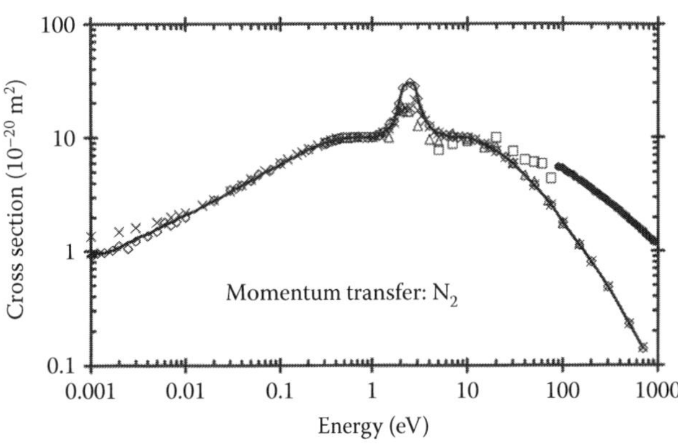

FIGURE 25.4 Momentum transfer cross sections for N_2. (—) Recommended cross sections (Raju 2005); (◊) Ramanan and Freeman (1990); (×) Phelps and Pitchford (1985); (△) Shyn et al. (1980); (□) Srivastava et al. (1976); (○) Baldwin (1974).

TABLE 25.6
Rotational Excitation, $J = 0$

Energy (eV)	Q_{rot} (10^{-20} m^2)	Energy (eV)	Q_{rot} (10^{-20} m^2)	Energy (eV)	Q_{rot} (10^{-20} m^2)
0.010	1.711	1.520	10.765	3.508	9.784
0.020	2.265	1.572	11.140	3.616	9.535
0.030	2.704	1.625	11.516	3.885	9.288
0.040	3.070	1.677	12.016	4.153	9.165
0.050	3.393	1.726	13.142	4.422	9.043
0.060	3.684	1.779	13.517	4.798	8.922
0.070	3.950	1.829	14.643	5.228	8.676
0.080	4.195	1.881	15.018	5.550	8.554
0.100	4.633	1.927	17.269	6.194	8.560
0.136	5.287	1.978	18.269	6.838	8.317
0.150	5.508	2.022	21.145	7.321	8.321
0.272	6.923	2.072	22.270	7.964	8.327
0.350	7.527	2.121	23.521	8.662	8.334
0.550	8.524	2.275	25.897	9.305	8.340
0.680	8.928	2.385	25.024	10.110	8.223
1.000	9.565	2.440	24.524	10.861	8.231
1.040	9.760	2.517	17.275	11.666	8.113
1.094	9.761	2.639	13.026	12.524	8.122
1.200	10.012	2.748	12.527	13.007	8.251
1.307	10.138	2.857	12.028	13.543	8.257
1.413	10.514	2.966	11.279		
		3.292	10.032		

Source: Adapted from Morrison, M. A., B. C. Saha, and T. L. Gibson, *Phys. Rev.*, 36, 3682, 1987.

Note: Digitized and interpolated from graphical data. See Figure 25.5 for graphical presentation.

N$_2$

TABLE 25.7
Ro-Vibrational Cross Section for N_2, $J = 0 \rightarrow 2$

Energy (eV)	Q_{rot} (10^{-20} m^2)	Energy (eV)	Q_{rot} (10^{-20} m^2)	Energy (eV)	Q_{rot} (10^{-20} m^2)
0.010	0.121	2.506	4.524	8.410	3.207
0.020	0.199	2.561	4.025	8.625	3.209
0.030	0.219	2.615	3.775	8.893	3.336
0.040	0.222	2.671	3.276	9.107	3.338
0.050	0.222	2.832	3.028	9.322	3.341
0.060	0.225	2.887	2.903	9.429	3.342
0.070	0.229	3.049	2.530	9.644	3.344
0.080	0.232	3.317	2.407	9.751	3.345
0.100	0.236	3.639	2.285	10.019	3.347
0.136	0.240	3.854	2.287	10.234	3.349
0.150	0.242	4.069	2.290	10.394	3.351
0.272	0.260	4.283	2.417	10.609	3.478
0.350	0.274	4.604	2.545	10.824	3.355
0.550	0.311	4.872	2.547	10.931	3.356
0.680	0.337	5.033	2.549	11.145	3.358
1.000	0.407	5.462	2.553	11.360	3.360
1.125	0.261	5.676	2.680	11.521	3.362
1.340	0.263	5.891	2.807	11.574	3.362
1.554	0.390	6.105	2.934	11.735	3.364
1.660	0.766	6.373	2.937	11.896	3.491
1.713	1.017	6.534	2.939	12.057	3.367
1.926	1.269	6.802	3.066	12.218	3.369
1.978	1.769	7.284	3.071	12.486	3.371
2.030	2.270	7.445	3.072	12.701	3.373
2.184	4.521	7.660	3.074	12.862	3.375
2.234	5.522	7.874	3.077	13.130	3.378
2.339	6.398	8.088	3.204	13.291	3.379
2.393	6.148	8.196	3.205	13.505	3.381

Source: Adapted from Morrison, M. A., B. C. Saha, and T. L. Gibson, *Phys. Rev.*, 36, 3682, 1987.
Note: Digitized and interpolated from graphical data. See Figure 25.5 for graphical presentation.

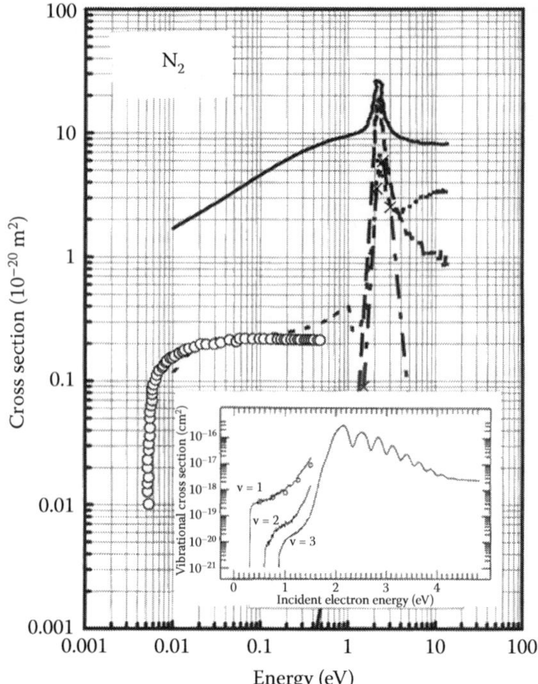

FIGURE 25.5 Rotational excitation cross sections in N_2. (- - -, short broken line) Morrison et al. (156, 1987), $J = 0 \rightarrow 2$; (— —, long broken line) Morrison et al. (1987), $J = 0 \rightarrow 4$; (—, full line) Morrison et al. (156, 1987), elastic scattering cross section included for comparison; (○) Engelhardt et al. (1964), rotational excitation $J = 4 \rightarrow 6$. Untabulated values are digitized from the original publication and replotted. The inset shows the vibrational cross sections measured by Allan (1985) for comparison. Open circles in the inset are Engelhardt et al. (1964).

25.6 ELECTRONIC EXCITATION

Onset Energies and total excitation cross sections as a function of energy are shown in Tables 25.9 and 25.10 (Leclair and Trajmar, 1996). See Figure 25.6 for graphical presentation.

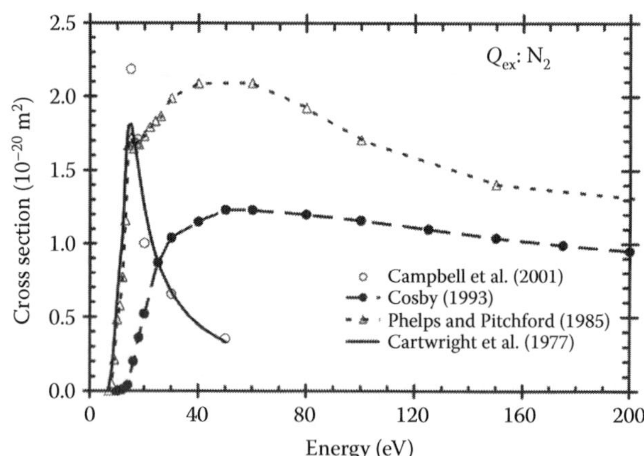

FIGURE 25.6 Total excitation cross sections in N_2. (○) Campbell et al. (2001); (—●—) Cosby (1993), dissociation cross sections; (—△—) Phelps and Pitchford (1985); (—) Cartwright et al. (1977).

TABLE 25.8
Ro-Vibrational Excitation Cross Section for N2, $J = 0 \rightarrow 4$

Energy (eV)	Q_{rot} (10^{-20} m²)	Energy (eV)	Q_{rot} (10^{-20} m²)	Energy (eV)	Q_{rot} (10^{-20} m²)
0.010	0.001	2.30	18.273	6.75	1.316
0.020	0.001	2.37	12.023	7.24	1.070
0.030	0.001	2.48	11.149	7.72	1.200
0.040	0.001	2.60	7.525	8.20	1.080
0.050	0.001	2.66	7.026	8.69	1.084
0.060	0.001	2.72	5.651	9.11	1.089
0.070	0.001	2.83	5.153	9.38	1.091
0.080	0.001	2.94	4.404	9.65	1.094
0.100	0.001	3.21	3.156	10.03	0.972
0.136	0.001	3.37	2.783	10.40	0.976
0.150	0.001	3.53	2.409	10.67	0.979
0.272	0.001	3.69	2.286	11.15	0.983
0.350	0.001	3.91	2.038	11.42	1.111
0.550	0.002	4.18	1.916	11.74	0.864
0.680	0.006	4.39	1.668	12.23	0.869
1.000	0.003	4.61	1.545	12.44	0.871
1.9615	6.894	4.93	1.548	12.71	0.873
2.011	8.270	5.36	1.427	13.03	1.002
2.10	13.521	5.79	1.306	13.25	0.879
2.20	17.147	6.16	1.310	13.46	0.881
				13.57	0.882

Source: Adapted from Morrison, M. A., B. C. Saha, and T. L. Gibson, *Phys. Rev.*, 36, 3682, 1987.

Note: Digitized and interpolated from graphical data. See Figure 25.5 for graphical presentation.

TABLE 25.9
Onset Energies of Selected States of N_2 Molecule

State	Onset (eV)	State	Onset (eV)	State	Onset (eV)	State	Onset (eV)
$A\,^3\Sigma_u^+$	6.17	$a\,^1\Pi_g$	8.55	$b\,^1\Pi_u$	12.58	$F\,^3\Pi_u$	12.98
$B\,^3\Pi_g$	7.35	$w\,^1\Delta_u$	8.89	$D\,^3\Sigma_u^+$	12.81	$o\,^1\Pi_u$	13.10
$W\,^3\Delta_u$	7.36	$C\,^3\Pi_u$	11.03	$G\,^3\Pi_u$	12.84	$b'\,^1\Sigma_u^+$	13.22
$B'\,^3\Sigma_u^-$	8.16	$E\,^3\Sigma_g^+$	11.88	$c\,^1\Pi_u$	12.91		
$a'\,^1\Sigma_u^-$	8.40	$a''\,^1\Sigma_g^+$	12.25	$c'\,^1\Sigma_u^+$	12.94		

Source: Adapted from Leclair, L. R. and S. Trajmar, *J. Phys. B: At. Mol. Opt. Phys.*, 29, 5543, 1996.

TABLE 25.10
Excitation Cross Section

Energy (eV)	Q_{ex} (10^{-20} m²)	Energy (eV)	Q_{ex} (10^{-20} m²)	Energy (eV)	Q_{ex} (10^{-20} m²)
7	0.001	15	1.717	30	1.988
8	0.050	16	1.647	40	2.088
9	0.213	17	1.686	60	2.091
10	0.484	18	1.674	80	1.923
11	0.576	20	1.730	100	1.706
12	0.769	22	1.792	150	1.405
13	1.159	24	1.831	500	0.777
14	1.670	26	1.866	1000	0.540

25.7 IONIZATION CROSS SECTION

See Table 25.11.

25.8 DRIFT VELOCITY OF ELECTRONS

See Table 25.12.

TABLE 25.11
Total and Partial Ionization Cross Section for N_2

	Q_i (10^{-20} m²)			
Energy (eV)	N_2^+	$N + N_2^{2+}$	N^{2+}	Total
17	0.0242			0.0242
20	0.218			0.218
25	0.571			0.571
30	0.998	0.0349		1.033
35	1.24	0.0969		1.337
40	1.47	0.178		1.648
45	1.63	0.262		1.892
50	1.70	0.340		2.040
55	1.77	0.415		2.185
60	1.83	0.466		2.296
65	1.85	0.511		2.361
70	1.88	0.554	0.000181	2.434
75	1.90	0.594	0.000697	2.495
80	1.92	0.621	0.00129	2.542
85	1.92	0.639	0.00215	2.561
90	1.94	0.667	0.00346	2.611
95	1.95	0.681	0.00463	2.636
100	1.94	0.692	0.00522	2.637
110	1.93	0.695	0.00763	2.633
120	1.91	0.695	0.00974	2.615
140	1.87	0.684	0.0128	2.567
160	1.80	0.662	0.0144	2.476
180	1.75	0.622	0.0161	2.388
200	1.68	0.590	0.016	2.286
225	1.61	0.537	0.016	2.163
250	1.53	0.512	0.0148	2.057
275	1.47	0.475	0.0147	1.960
300	1.41	0.453	0.0132	1.876
350	1.32	0.406	0.0121	1.738
400	1.24	0.362	0.0106	1.613
450	1.15	0.334	0.00967	1.494
500	1.08	0.307	0.0083	1.395
550	1.02	0.281	0.00816	1.309
600	0.966	0.254	0.00778	1.228
650	0.900	0.239	0.00717	1.146
700	0.862	0.221	0.00663	1.090
750	0.812	0.209	0.00599	1.027
800	0.780	0.203	0.00605	0.989
850	0.752	0.195	0.00552	0.952
900	0.731	0.186	0.00531	0.922
950	0.709	0.179	0.00514	0.893
1000	0.686	0.169	0.00492	0.860

Source: Adapted from Straub, H. C. et al., *Phys. Rev. A*, 54, 2146, 1996.

Note: See Figure 25.7 for graphical presentation.

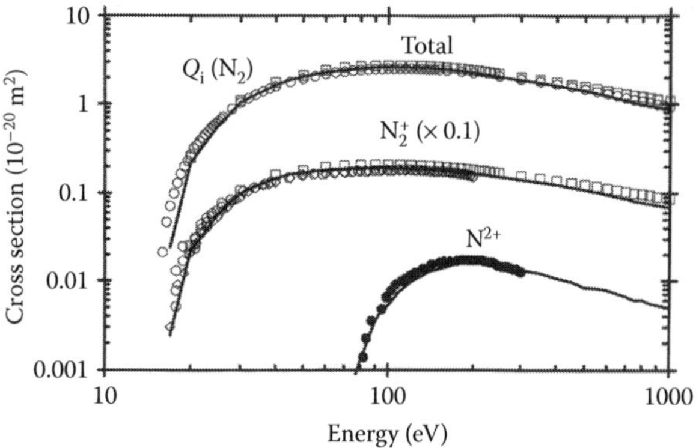

FIGURE 25.7 Ionization cross sections for N_2. The cross sections for N_2^+ are shown reduced for clarity purposes. The ordinates of this curve should be multiplied by 10 to obtain the cross sections. The symbols for all the curves are (——) Straub et al. (1996); (◊) Freund et al. (1990); (□) Krishnakumar and Srivastava (144, 1990); (●) Crowe and McConkey (1973); (○) Rapp and Englander-Golden (1965).

TABLE 25.12
Recommended Drift Velocity of Electrons

E/N (Td)	*W* (m/s)	*E/N* (Td)	*W* (m/s)	*E/N* (Td)	*W* (m/s)
0.014	5.23 (2)	1.00	4.43 (3)	70.0	7.90 (4)
0.017	6.30 (2)	1.2	4.74 (3)	80.0	8.74 (4)
0.02	7.33 (2)	1.4	5.03 (3)	100.0	1.05 (5)
0.025	9.02 (2)	1.7	5.49 (3)	120.0	1.28 (5)
0.03	1.06 (3)	2.0	5.98 (3)	140.0	1.48 (5)
0.035	1.21 (3)	2.5	6.79 (3)	170.0	1.81 (5)
0.04	1.35 (3)	3.0	7.67 (3)	200.0	2.12 (5)
0.05	1.60 (3)	3.5	8.51 (3)	300	3.21 (5)
0.06	1.81 (3)	4.0	9.33 (3)	400	4.12 (5)
0.07	1.99 (3)	5.0	1.10 (4)	500	5.00 (5)
0.08	2.14 (3)	6.0	1.26 (4)	600	5.65 (5)
0.10	2.38 (3)	7.0	1.41 (4)	700	6.21 (5)
0.12	2.54 (3)	8.0	1.55 (4)	800	6.59 (5)
0.14	2.66 (3)	10.0	1.84 (4)	900	7.26 (5)
0.17	2.77 (3)	12.0	2.09 (4)	1000	8.06 (5)
0.20	2.87 (3)	14.0	2.35 (4)	1500	1.10 (6)
0.25	2.98 (3)	17.0	2.73(4)	2000	1.25 (6)
0.30	3.09 (3)	20.0	3.09 (4)	2500	1.35 (6)
0.35	3.19 (3)	25.0	3.65 (4)	3000	1.53 (6)
0.40	3.30 (3)	30.0	4.17 (4)		
0.50	3.53 (3)	35.0	4.68 (4)		
0.60	3.75 (3)	40.0	5.18 (4)		
0.70	3.93 (3)	50.0	6.09 (4)		
0.80	4.09 (3)	60.0	7.03 (4)		

Note: a (b) means a × 10^b. See Figure 25.8 for graphical presentation.

25.9 DIFFUSION COEFFICIENT

See Table 25.13.

25.10 IONIZATION COEFFICIENT

See Table 25.14.

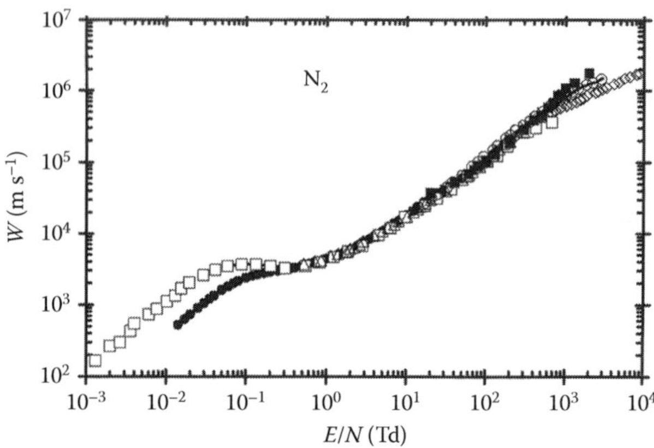

FIGURE 25.8 Drift velocity of electrons in N_2. Unless otherwise specified the temperature is 293 K. W: (×) Hasegawa et al. (249, 1996); (■) Liu and Raju (1993); (- - -) Liu and Raju (1992); Nakamura (245, 1987); (○) Kücükarpaci and Lucas (1979); (+) Saelee et al., (238, 1977); (■) Rajapandian and Raju (1976); (▲) Snelson and Lucas (1975); (●) Huxley and Crompton[79] (1974); (◊) Schlumbohm (1965); (□) Engelhardt et al. (1964) 77 K; (—) recommended.

TABLE 25.13
Recommended Radial Diffusion Coefficients Expressed as Characteristic Energy for N_2

E/N (Td)	D_r/μ (V)	*E/N* (Td)	D_r/μ (V)	*E/N* (Td)	D_r/μ (V)
0.020	0.0265	0.6	0.180	30.0	1.16
0.025	0.0271	0.7	0.204	40.0	1.27
0.030	0.0277	0.8	0.229	50.0	1.31
0.035	0.0283	1.0	0.277	60.0	1.38
0.040	0.0291	1.2	0.233	70.0	1.52
0.050	0.0308	1.4	0.368	100.0	1.90
0.060	0.0327	1.7	0.428	150.0	2.66
0.070	0.0348	2.0	0.48	200.0	3.33
0.080	0.037	2.5	0.55	300	4.17
0.10	0.0418	3.0	0.606	400	4.64
0.12	0.0471	3.5	0.653	500	5.11
0.14	0.0528	4.0	0.691	600	5.56
0.17	0.0618	5.0	0.744	700	6.03
0.20	0.0713	6.0	0.797	800	6.45
0.25	0.0869	7.0	0.835	900	6.95
0.30	0.102	8.0	0.873	1000	7.35
0.35	0.116	10.0	0.932	1100	7.84
0.40	0.130	12	0.979		
0.50	0.156	14.0	1.019		

Note: See Figure 25.9 for graphical presentation of D_r/μ and D_L/μ ratios.

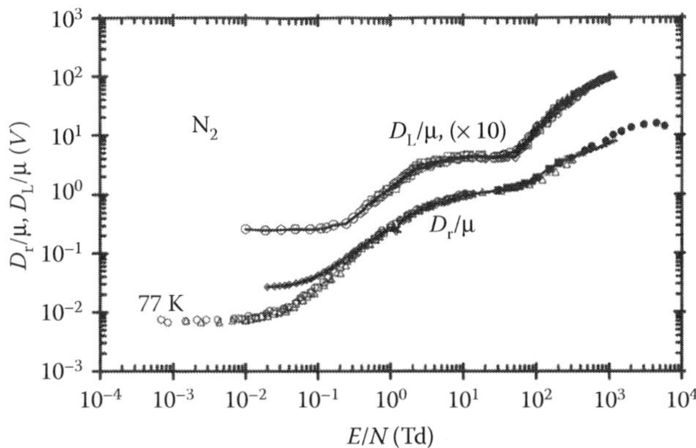

FIGURE 25.9 Radial diffusion coefficient (D_r) and longitudinal diffusion coefficient (D_L) as ratio of mobility (μ) for N_2. Temperature 293 K unless otherwise mentioned. Hasegawa et al. (1996) D_L/μ; (■) Roznerski and Leja (1996) D_r/μ; (▲) Roznerski and Leja (1996) D_L/μ; (—□—) Nakamura (1987) D_L/μ; (●) Al-Amin et al. (1985) D_r/μ; (*) Fletcher and Reid (1980) D_r/μ; (□) Huxley and Crompton (1974) 77 K; (◊) Huxley and Crompton (1974); (—○—) Lowke and Parker (1969) D_L/μ; (Δ) Engelhardt et al. (1964) D_r/μ, 77 K; (○) Warren and Parker (1962) D_r/μ, 77 K; (—) recommended, D_r/μ. D_L/μ values are multiplied by factor of 10 for improved presentation.

TABLE 25.14
Recommended Ionization Coefficients

E/N (Td)	α/N (10^{-20} m²)	E/N (Td)	α/N (10^{-20} m²)
85	2.00(−4)		
90	3.00 (−4)	600	0.668
100	7.00 (−4)	700	0.851
125	3.30 (−3)	800	1.03
150	9.50 (−3)	900	1.21
175	2.02 (−2)	1000	1.37
200	3.60 (−2)	1250	1.72
225	5.70 (−2)	1500	2.00
250	8.26 (−2)	1750	2.23
300	0.145	2000	2.43
350	0.220	2500	2.70
400	0.302	3000	2.84
450	0.391	3300	2.87
500	0.483		

Note: See Figure 25.10 for graphical presentation.

25.11 GAS CONSTANTS

Gas constants evaluated according to the expression

$$\frac{\alpha}{N} = F\exp\left(-\frac{GN}{E}\right) \tag{25.1}$$

are shown in Table 25.15. See inset of Figure 25.10.

TABLE 25.15
Gas Constants

F (10^{-20} m²)	G (Td)	E/N Range	Reference
4.37	1099	600–3000	Raju and Gurumurthy (1977)
3.39	1038.5	280–1690	Heylen and Dargan (1973)

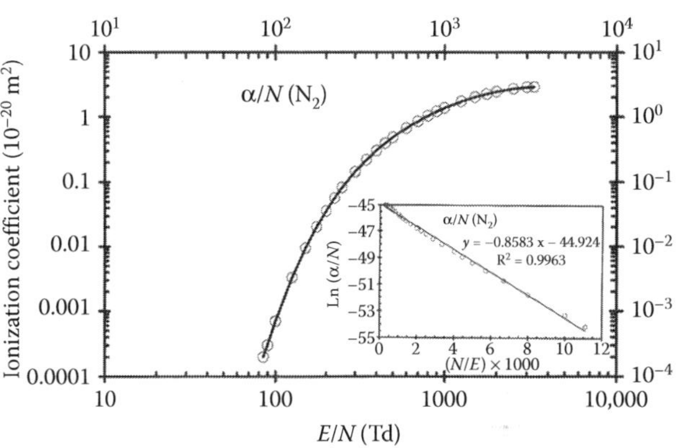

FIGURE 25.10 Density-reduced ionization coefficients for N_2. (—) Raju (2005); (○) Haydon and Williams (1976) tabulated. See Raju (2005) for previous data.

25.12 POSITIVE ION MOBILITY

Reduced positive ion mobility (Allen and Prew, 1970) is 2.9×10^{-4} m²/V s.

REFERENCES

Al-Amin, S. A. J., J. Lucas, and H. N. Kücükarpaci, *J. Phys. D: Appl. Phys.*, 18, 2007, 1985.

Allan, M., *J. Phys. B: At. Mol. Phys.*, 18, 4511, 1985.

Allen, N. L. and B. A. Prew, *J. Phys. B: At. Mol. Phys.*, 3, 1127, 1970.

Armentrout, P. B., S. M. Tarr, A. Sori, and R. S. Freund, *J. Chem. Phys.*, 75, 2786, 1981.

Baldwin, G. C., *Phys. Rev. A*, 9, 1225, 1974.

Blaauw, H. J., F. J. de Heer, R. W. Wagenaar, and D. H. Barends, *J. Phys. B: At. Mol. Phys.*, 10, L299, 1977.

Blaauw, H. J., R. W. Wagenaar, D. H. Barends, and F. J. de Heer, *J. Phys. B: At. Mol. Phys.*, 13, 359, 1980.

Bowe, J. C., *Phys. Rev.*, 117, 1411, 1960.

Brennan, M. J., D. T. Alle, P. Euripides, S. J. Buckman, and M. J. Brunger, *J. Phys. B: At. Mol. Opt. Phys.*, 25, 2669, 1992.

Brunger, M. J. and P. J. O. Teubner, *Phys. Rev. A*, 41, 1413, 1990.

Campbell, L., M. J. Brunger, A. M. Nolan, L. J. Kelly, A. B. Wedding, J. Harrison, P. J. O. Teubner, D. C. Cartwright, and B. Mclaughlin, *J. Phys. B: At. Mol. Opt. Phys.*, 34, 1185, 2001.

Cartwright, D. C., S. Trajmar, A. Chutjian, and W. Williams, *Phys. Rev. A*, 16, 1013, 1977a; *Phys. Rev. A*, 16, 1041, 1977b.

Cosby, P. C., *J. Chem. Phys.*, 98, 9560, 1993.

Crowe, A. and J. W. McConkey, *J. Phys. B: At. Mol. Phys.*, 6, 2108, 1973.

Dalba, G., P. Foransini, R. Grisenti, G. Ranieri, and A. Zecca, *J. Phys. B: At. Mol. Phys.*, 13, 4695, 1980.

Daniel, T. N., J. Dutton, and F. M. Harris, *J. Phys. D: Appl. Phys.*, 2, 1559, 1969. Accurate sparking potentials in nitrogen at high values of Nd (1-6 × 10^{24} m^{-2}) are also reported in this paper.

DuBois, R. D. and M. E. Rudd, *J. Phys. B: At. Mol. Phys.*, 9, 2657, 1976.

Engelhardt, A. G., A. V. Phelps, and C. G. Risk, *Phys. Rev.*, 135, A1566, 1964.

Fletcher, J. and I. D. Reid, *J. Phys. D: Appl. Phys.*, 13, 2275, 1980.

Fletcher, J. and S. C. Haydon, *Aust. J. Phys.*, 19, 615, 1966.

Folkard, M. A. and S. C. Haydon, *J. Phys. B: At. Mol. Phys.*, 6, 214, 1973.

Freund, R. S., R. C. Wetzel, R. J. Shul, *Phys. Rev. A*, 41, 5861, 1990.

García, G., A. Pérez, and J. Campos, *Phys. Rev. A*, 38, 654, 1988.

Golden, D. E., *Phys. Rev. Lett.*, 17, 847, 1966.

Hasegawa, H., H. Date, M. Shimozuma, K. Yoshida, and H. Tagashira, *J. Phys. D: Appl. Phys.*, 29, 2664, 1996.

Haydon, S. C. and O. M. Williams, *J. Phys. D: Appl. Phys.*, 9, 523, 1976. Both spatial and temporal current measurements are reported in nitrogen with tabulated values of α/N (cm^2) in the range of *E*/*N* from 85 to 3400 Td.

Herrmann, D., K. Jost, and J. Kessler, *J. Chem. Phys.*, 64, 1, 1976.

Heylen, A. E. D. and C. L. Dargan, *Int. J. Electron*, 35, 433, 1973.

Hoffman, K. R., M. S. Dababneh, Y. F. Hsieh, W. E. Kaippila, V. Pol, J. H. Smart, and T. S. Stein, *Phys. Rev. A*, 25, 1393, 1982.

Huxley, L. G. H., and R. W. Crompton, *The Diffusion and Drift of Electrons in Gases*, John Wiley and Sons, New York, NY, 1974.

Karwasz, G., R. S. Brusa, A. Gasparoli, and A. Zecca, *Chem. Phys. Lett.*, 211, 529, 1993.

Kennerly, R. E., *Phys. Rev. A*, 21, 1876, 1980.

Kontoleon, N., J. Lucas, and L. E. Virr, *J. Phys. D: Appl. Phys.*, 6, 1237, 1973.

Krishnakumar, E. and S. K. Srivastava, *J. Phys. B: At. Mol. Phys.*, 23, 1893, 1990.

Kücükarpaci, H. N. and J. Lucas, *J. Phys. D: Appl. Phys.*, 12, 2123, 1979.

Leclair, L. R. and S. Trajmar, *J. Phys. B: At. Mol. Opt. Phys.*, 29, 5543, 1996.

Lisovskiy, V., J.-P. Booth, K. Landry, D. Douai, V. Cassagne, and V. Yegorenkov, *J. Phys. D: Appl. Phys.*, 39, 660, 2006.

Liu, J. and G. R. G. Raju, *J. Franklin. Inst.*, 329, 181, 1992.

Liu, J. and G. R. G. Raju, *IEEE Trans. Elec. Insul.*, 28, 154, 1993.

Lowke, J. J. and J. H. Parker, *Phys. Rev.*, 181, 302, 1969.

Maller, V. N. and M. S. Naidu, *J. Phys. D: Appl. Phys.*, 7, 1406, 1974.

Märk, T. D., *J. Chem. Phys.*, 63, 3731, 1975.

Mathur, D. and J. B. Hasted, *J. Phys. B: At. Mol. Phys.*, 10, L265, 1977.

Morgan, W. L., J. P. Bouef, and L. C. Pitchford, www. siglo-kinema. com, 1999.

Morrison, M. A., B. C. Saha, and T. L. Gibson, *Phys. Rev.*, 36, 3682, 1987.

Nakamura, Y., *J. Phys. D: Appl. Phys.*, 20, 933, 1987.

Nakamura, Y. and M. Kurachi, *J. Phys. D: Appl. Phys.*, 21, 718, 1988.

Nickel, J. C., I. Kanik, S. Trajmar, and K. Imre, *J. Phys. B: At. Mol. Opt. Phys.*, 25, 2427, 1992.

Novak, J. P. and M. Frechette, *J. Appl. Phys.*, 55, 107, 1984.

Pack, J. L. and A. V. Phelps, *Phys. Rev.*, 121, 798, 1961.

Phelps, A. V. and L. C. Pitchford, *Phys. Rev. A*, 31, 2932, 1985. In this paper, a set of cross sections for N_2 is derived and a detailed discussion of transport coefficients available upto 1985 is provided. Also see ftp://jila.colorado.edu/collision_data/electronneutral/electron.txt.

Raju, G. G. and G. R. Gurumurthy, *IEEE Trans. Plasma Sci.*, 4, 241, 1976.

Raju, G. G. and G.R. Gurumurthy, *IEEE Tran. Dielectrics*, EI-12, 325, 1977.

Raju, G. G. and G. R. Gurumurthy, *Int. J. Electron.*, 44, 714, 1978.

Raju, G. G., *Gaseous Electronics: Theory and Practice*, Taylor & Francis, London, 2005.

Raju, G. R. G. and R. Hackam, *J. Appl. Phys.*, 52, 3912, 1981.

Raju, G. R. G. and M. S. Dincer, *J. Appl. Phys.*, 53, 8562, 1982.

Raju, G. R. G. and S. Rajapandian, *Int. J. Electron.*, 40, 65, 1976.

Ramanan, G. and G. R. Freeman, *J. Chem. Phys.*, 93, 3120, 1990.

Rapp, D., P. Englander-Golden, and D. D. Briglia, *J. Chem. Phys.*, 42, 4081, 1965. Dissociative cross sections are measured in H_2 expressed as a fraction of the total ionization cross sections for ions of energy greater than 0.25 eV.

Rapp, D. and P. Englander-Golden, *J. Chem. Phys.*, 43, 1464, 1965. The covered energy range is from threshold to 1000 eV in closely spaced intervals. Tabulated values are given in units of $\pi a_0^2 = 0.88 \times 10^{-20}$ m^2.

Roznerski, W., *J. Phys. D: Appl. Phys.*, 29, 614, 1996.

Roznerski, W. and K. Leja, *J. Phys. D: Appl. Phys.*, 17, 279, 1984.

Saelee, H. T., J. Lucas, and J. W. Limbeck, *IEE Solid State Electron Devices*, 1, 111, 1977.

Schlumbohm, H., *Z. Phys.*, 182, 317, 1965.

Shyn, T. W., R. S. Stolarski, and G. R. Corignan, *Phys. Rev. A*, 6, 1009, 1980.

Shyn, T. W. and G. R. Carignan, *Phys. Rev. A*, 22, 923, 1980.

Snelson, R. A. and J. Lucas, *Proc. IEE*, 122, 107, 1975.

Srivastava, S. K., A. Chutjian, and S. Trajmar, *J. Chem. Phys.*, 64, 1340, 1976.

Straub, H. C., P. Renault, B. G. Lindsay, K. A. Smith, and R. F. Stebbings, *Phys. Rev. A*, 54, 2146, 1996.

Szmytkowski, C. and K. Maciąg, *Phys. Scripta*, 54, 271, 1996.

Tanaka, Y., *J. Phys. D: Appl. Phys.*, 37, 851, 2004. Boltzmann energy distribution is calculated for N_2, O_2, and mixtures thereof in the temperature range of 300–3500 K.

Taniguchi, T., H. Tagashira, and Y. Sakai., *J. Phys. D: Appl. Phys.*, 11, 1757, 1978.

Tian, C. and C. R. Vidal, *J. Phys. B: At. Mol. Opt. Phys.*, 31, 5369, 1998.

Van Zyl, B. and T. M. Stephan, *Phys. Rev. A*, 40, 3164, 1994. Corrections to dissociative ionization cross section for H_2, N_2, and O_2 measured by Rapp et al. (1965) are given.

Wagner, E. B., F. J. Davis, and G. S. Hurst, *J. Chem. Phys.*, 47, 3138, 1967.

Warren, R. W. and J. H. Parker, *Phys. Rev.*, 128, 2661, 1962.

Watts, M. P. and A. E. D. Heylen, *Int. J. Elect.*, 67, 661, 1989.

Wedde, T. and T. G. Strand, *J. Phys. B: At. Mol. Phys.*, 7, 1091, 1974.

Wedding, A. B., H. A. Blevin, and J. Fletcher, *J. Phys. D: Appl. Phys.*, 18, 2361, 1985.

Zecca, A., G. P. Karwasz, and R. S. Brusa, *Riv. Nu. Ci.*, 19, 1, 1996.

26

OXYGEN

CONTENTS

Oxygen (O_2) molecule is an electron-attaching molecule with 16 electrons, electronic polarizability of 1.759×10^{-40} F m^2, and ionization potential of 12.07 eV.

26.1 SELECTED REFERENCES FOR DATA

See Table 26.1.

TABLE 26.1
Selected References for Data

Quantity	Range: eV, (Td)	Reference
Drift velocity	(200–4800)	**Lisovskiy et al. (2006)**
Transport coefficients	(1.7–350)	**Jeon and Nakamura (1998)**
Ionization cross section	13–1000	Hwang et al. (1996)
Ionization cross section	13–998	**Straub et al. (1996)**
Total scattering cross section	0.4–250	**Szmytkowski et al. (1996)**
Elastic scattering cross section	1–30	**Sullivan et al. (1995)**
Total scattering cross section	0.001–0.175	**Randell et al. (1994)**
Total scattering cross section	1–1000	Kanik et al. (1993)
Ionization cross section	50–400	**Evans et al. (1982)**

TABLE 26.1 (continued)
Selected References for Data

Quantity	Range: eV, (Td)	Reference
Total scattering cross section	5–300	**Kanik et al. (1992)**
Ionization cross section	13–1000	**Krishnakumar et al. (1992)**
Swarm parameters	(30–5000)	Liu and Raju (1992)
Total scattering cross section	0.15–9.14	**Subramanian and Kumar (1990)**
Elastic scattering cross section	0.5–1000	Itikawa et al. (1989)
Total scattering cross section	5–500	**Dabbneh et al. (1988)**
Elastic scattering cross section	300–1000	**Iga et al. (1987)**
Attachment coefficients	(0.5–50)	**Hunter et al. (1986)**
Total scattering cross section	0.2–100	**Zecca et al. (1986)**
Transport coefficients	(25–850)	**Al-Amin et al. (1985)**
Elastic scattering cross section	2–200	**Shyn and Sharp (1982)**
Total scattering cross section	100–1024	**Dalba et al. (1980)**
Attachment coefficients	(0.1–20)	**Taniguchi et al. (1978)**

continued

TABLE 26.1 (continued)
Selected References for Data

Quantity	Range: eV, (Td)	Reference
Elastic scattering cross section	20–500	**Wakiya (1978a)**
Attachment coefficient	(20–160)	**Masek et al. (1977)**
Swarm coefficients	(240–1120)	**Gurumurthy and Raju (1975)**
Ionization cross section	15–180	**Märk (1975)**
Elastic scattering cross section	40–1000	Wedde and Strand (1974)
Ionization coefficient	(100–350)	**Price and Moruzzi (1973)**
Diffusion coefficients	(1.0–16.0)	**Fleming et al. (1972)**
Diffusion coefficients	(0.1–2.25)	**Nelson and Davis (1972)**
Attachment coefficients	(1–20)	**Chatterton and Craggs (1971)**
Elastic scattering cross section	3.5–45	**Trajmar et al. (1971)**
Attachment coefficients	(15–90)	**Grünberg (1969)**
Transport coefficient	(15–150)	**Naidu and Prasad (1970)**
Diffusion coefficient	(0.1–150)	**Lowke and Parker (1969)**
Transport coefficients	(0.01–180)	Hake and Phelps (1967)
Attachment cross section	4.2–55.0	**Rapp and Briglia (1965)**
Ionization cross secction	12.5–1000	**Rapp and Englander-Golden (1965)**
Attachment coefficients	(0.1–20)	**Rees (1965)**
Swarm parameters	(200–900)	**Schlumbohm (1965)**
Attachment coefficient	(40–150)	**Freely and Fischer (1964)**
Attachment coefficient	(100–150)	**Frommhold (1964)**
Attachment coefficient	(100–120)	**Dutton et al. (1963)**
Attachment coefficients	(10–80)	**Chanin et al. (1962)**
Attachment coefficient	(80–160)	**Prasad and Craggs (1961)**
Attachment coefficient	(1–20)	**Kuffel (1959)**
Ionization coefficient	(200–9000)	**Schlumbohm (1959)**
Ionization coefficient	(25–150)	**Geballe and Harrison (1952)**
Attachment coefficient	(0.1–20)	**Herreng (1952)**
Total scattering cross section	0.1–30	**Fisk (1936)**

Note: Bold font denotes experimental study.

26.2 TOTAL SCATTERING CROSS SECTION

See Table 26.2.

26.3 ELASTIC SCATTERING CROSS SECTION

See Table 26.3.

TABLE 26.2
Recommended Total Scattering Cross Section

Energy (eV)	Q_T (10^{-20} m^2)	Energy (eV)	Q_T (10^{-20} m^2)	Energy (eV)	Q_T (10^{-20} m^2)
0.15	4.69	30	11.2	300	4.85
0.23	4.27	40	10.0	400	4.00
0.49	4.95	50	9.8	500	3.5
1	6.12	60	9.5	600	3.1
2	6.27	70	8.9	700	2.8
3	6.46	80	8.55	800	2.55
5	7.18	90	8.3	900	2.35
10	10.4	100	8.1	1000	2.2
15	10.8	150	7.33	—	—
20	10.7	200	5.85	—	—

Note: See Figure 26.1 for graphical presentation.

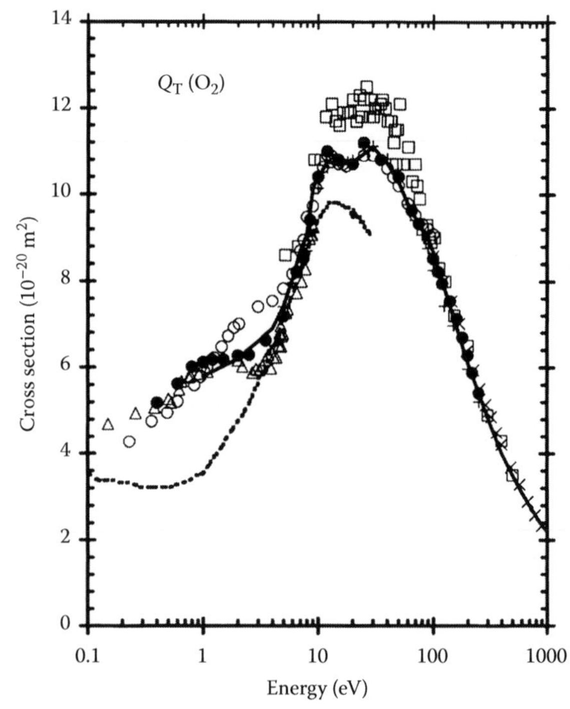

FIGURE 26.1 Total scattering cross sections in O_2. (●) Szmytkowski et al. (1996); (——) Kanik et al. (1993), recommended; (+) Kanik et al. (1992); (△) Subramanian and Kumar (1990); (□) Dababneh et al. (1988); (○) Zecca et al. (1986); (×) Dalba et al. (1980); (——) theoretical values, Fisk (1936).

26.4 MOMENTUM TRANSFER CROSS SECTION

See Table 26.4.

26.5 RO-VIBRATIONAL EXCITATION CROSS SECTION

Table 26.5 shows the rotational vibrational cross section for selected transitions as a function of electron energy.

TABLE 26.3
Elastic Scattering Cross Sections for O_2

Energy (eV)	Q_{el} (10^{-20} m^2)	Energy (eV)	Q_{el} (10^{-20} m^2)	Energy (eV)	Q_{el} (10^{-20} m^2)
0.15	4.7	30	8.8	300	2.4
0.23	4.6	40	7.1	400	2
0.49	4.5	50	6.5	500	1.72
1	6.1	60	6.1	600	1.53
2	6.7	70	5.7	700	1.37
3	6.9	80	5.4	800	1.27
5	7.1	90	5	900	1.18
10	8.6	100	4.8	1000	1.12
15	8.8	150	3.98	—	—
20	8.7	200	3.15	—	—

Note: See Figure 26.2 for graphical presentation.

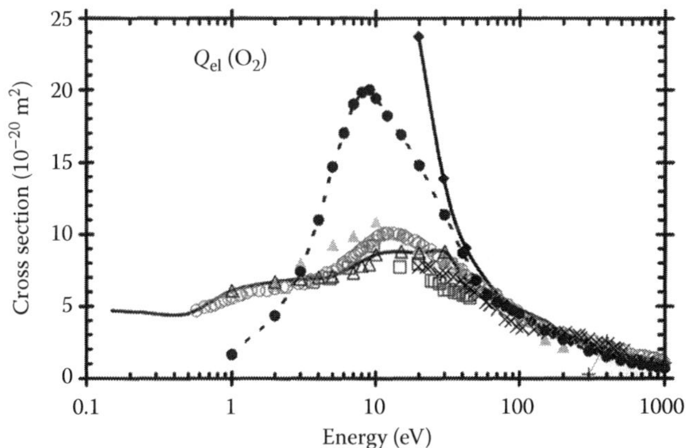

FIGURE 26.2 Elastic scattering cross sections in O_2. (——) Recommended, Raju (2005); (△) Sullivan et al. (1995); (—●—) Kanik et al. (1993); (○) Itikawa et al. (1989); (– + –) Iga et al. (1987); (▲) Shyn and Sharp (1982); (×) Wakiya (1978a); (–◆–) Wedde and Strand (1974); (□) Trajmar et al. (1971).

TABLE 26.4
Recommended Momentum Transfer Cross Sections for O_2

Energy (eV)	Q_m (10^{-20} m^2)	Energy (eV)	Q_m (10^{-20} m^2)	Energy (eV)	Q_m (10^{-20} m^2)
0.01	0.74	5	6.28	100	1.55
0.02	1.06	7	6.95	150	1.22
0.04	1.52	10	7.65	200	0.92
0.07	2.00	15	6.90	300	0.55
0.1	2.50	20	6.10	400	0.40
0.25	3.90	25	5.30	500	0.35
0.4	5.00	30	4.80	600	0.25
0.6	5.65	40	4.02	700	0.15
0.8	6.00	50	3.50	800	0.08
1	6.30	60	2.85	1000	0.07
2	6.50	70	2.40		
3	6.10	80	2.10		
4	6.00	90	1.75		

Note: See Figure 26.3 for graphical presentation.

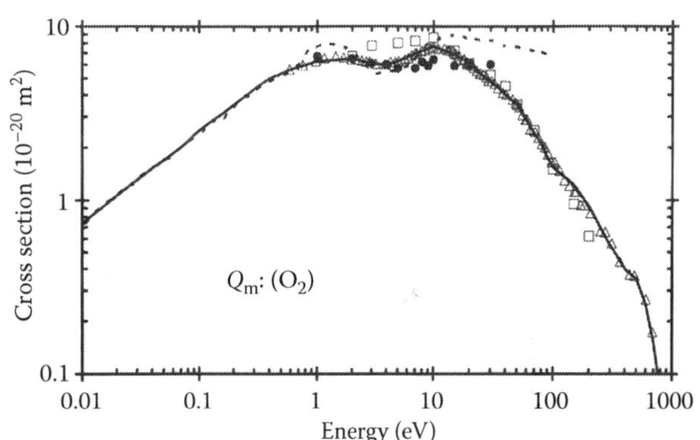

FIGURE 26.3 Momentum transfer cross sections for O_2. (——) Recommended, Raju (2005); (•) Sullivan et al. (1995); (△) Itikawa et al. (1989); (□) Shyn and Sharp (1982); (□□) Hake and Phelps (1967).

TABLE 26.5
Integral Ro-Vibrational Cross Sections in O_2, in Units of 10^{-22} m^2

Transition	Energy (eV) 5	7	10	15
0→1	9.5	30.5	31.2	5.7
0→2	3.4	11.4	16.5	1.5
0→3	—	4.5	7.5	0.65
0→4	—	3.5	4.0	0.31

	Energy (eV) 5	6	7	8	9	9.5	10	11	15	20
0→1	5.21	11.55	20.24	25.62	40.26	46.36	53.63	40.56	11.57	1.75
0→2	—	2.26	6.04	9.60	16.32	20.11	25.52	17.47	3.89	—
0→3	—	—	2.64	4.07	8.11	10.12	11.45	9.02	2.15	—
0→4	—	—	1.15	2.28	4.73	5.82	6.53	5.30	1.20	—

Note: See Figure 26.4 for graphical presentation.

O_2

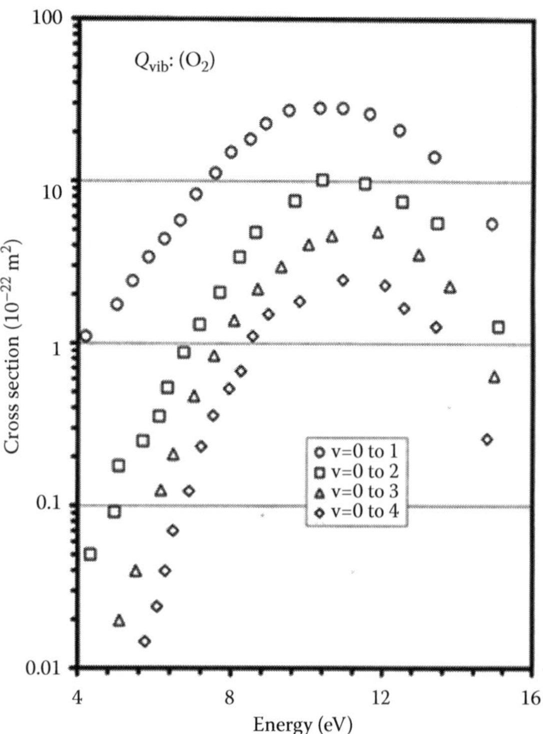

FIGURE 26.4 Recommended integral vibrational cross sections in units of 10^{-22} m^2 for levels as shown. (Adapted from Raju, G. G., *Gaseous Electronics: Theory and Practice*, Taylor & Francis, London, 2005.)

26.6 ELECTRONIC EXCITATION CROSS SECTION

See Tables 26.6 and 26.7.

26.7 IONIZATION CROSS SECTION

See Table 26.8.

26.8 DRIFT VELOCITY

See Table 26.9.

26.9 DIFFUSION COEFFICIENT

See Tables 26.10 and 26.11.

TABLE 26.6
Threshold Energy for Excitation Processes

State	Energy (eV)	State	Energy (eV)	State	Energy (eV)
$X^3\Sigma_g^-$	0	$A^3\Sigma_u^+$(F)	4.340	$e(^1\Delta_{2u})$	9.346
$a^1\Delta_g$(F)	0.977	$B^3\Sigma_u^-$	6.120	$\beta^3\Sigma_u^+$	9.355
$b^1\Sigma_g^+$(F)	1.627	$^1\Pi_g$	8.141	$\alpha^1\Sigma_u$	9.455
$c^1\Sigma_u^-$(F)	4.050	$d(^1\Pi_g)$	8.595		
$C^3\Delta_u$(F)	4.262	$e'(^1\Delta_{2u})$	9.318		

TABLE 26.7
Excitation Cross Sections for $c^1\sum_u^- + A'^3\Delta_u + A^3\sum_u^+$ States

Energy (eV)	Q_{ex} (10^{-20} m^2)	Energy (eV)	Q_{ex} (10^{-20} m^2)
4.23	0.000	15	0.107
9	0.092	20	0.094
10	0.132	45	0.038
12	0.120		

Source: Adapted from Brunger, M., S. Buckman, and M. Elford, in *Interactions of Photons and Electrons with Molecules*, Vol. 17c, Springer-Verlag, New York, NY, 2003.

Note: See Figure 26.5 for additional data.

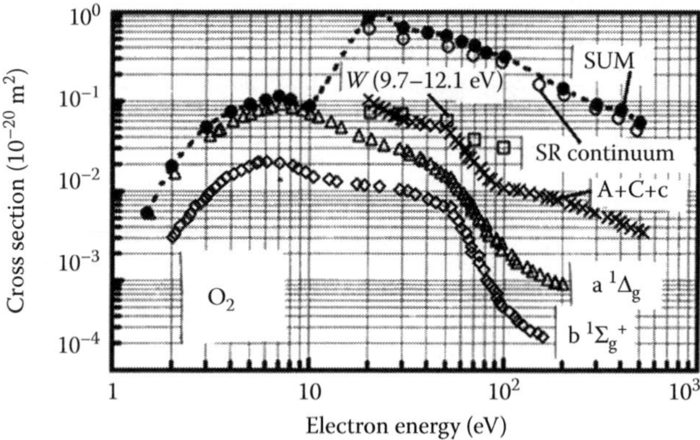

FIGURE 26.5 Excitation cross sections in O_2 showing the relative magnitudes of the selected states. SR means Schumann–Runge; W means the optical emissions in the energy range shown; A + C + c is the abbreviation for ($A\,^3\Sigma_u^+ + C\,^3\Delta_u + c\,^1\Sigma_u^-$) states. Sum is the total for all states.

TABLE 26.8
Partial and Total Ionization Cross Sections

Energy (eV)	Q_i (O_2^+) (10^{-20} m^2)	Q_i ($O^+ + O_2^{2+}$) (10^{-21} m^2)	Q_i (O^{2+}) (10^{-22} m^2)	Total (10^{-20} m^2)
13	0.0127			0.0127
15.5	0.0792			0.0792
18	0.178			0.178
23	0.395	0.180		0.413
28	0.606	0.840		0.690
33	0.813	1.82		1.00
38	0.994	2.76		1.27
43	1.16	3.56		1.52
48	1.27	4.47		1.72
53	1.37	5.22		1.89
58	1.45	5.88		2.04
63	1.51	6.59		2.17
68	1.56	7.21		2.28
73	1.59	7.60	0.125	2.35
78	1.6	7.95	0.200	2.40

TABLE 26.8 (continued)
Partial and Total Ionization Cross Sections

Energy (eV)	$Q_i(O_2^+)$ (10^{-20} m²)	$Q_i(O^+ + O_2^{2+})$ (10^{-21} m²)	$Q_i(O^{2+})$ (10^{-22} m²)	Total (10^{-20} m²)
83	1.62	8.47	0.254	2.47
88	1.64	8.73	0.372	2.52
93	1.64	9.03	0.462	2.55
98	1.64	9.18	0.644	2.56
108	1.62	9.47	0.850	2.58
118	1.61	9.57	1.01	2.58
138	1.57	9.57	1.44	2.54
158	1.54	9.47	1.88	2.51
178	1.49	9.31	2.09	2.44
198	1.45	9.01	2.20	2.37
223	1.39	8.64	2.39	2.28
248	1.36	8.25	2.35	2.21
273	1.28	7.83	2.21	2.09
298	1.24	7.46	2.14	2.01
348	1.16	6.81	1.96	1.86
398	1.08	6.30	1.76	1.73
448	1.01	5.78	1.58	1.60
498	0.948	5.41	1.40	1.50
548	0.904	4.99	1.26	1.42
598	0.847	4.68	1.13	1.33
648	0.818	4.42	1.10	1.27
698	0.777	4.24	1.01	1.21
748	0.735	3.96	1.00	1.14
798	0.7	3.76	0.85	1.08
848	0.684	3.62	0.81	1.05
898	0.654	3.42	0.78	1.00
948	0.627	3.31	0.75	0.97
998	0.607	3.22	0.76	0.94

Source: Reprinted with permission from Straub, H. C. et al., *Phys. Rev.*, 1996. Copyright (1996) by the American Physical Society.
Note: See Figures 26.6 and 26.7 for graphical presentation.

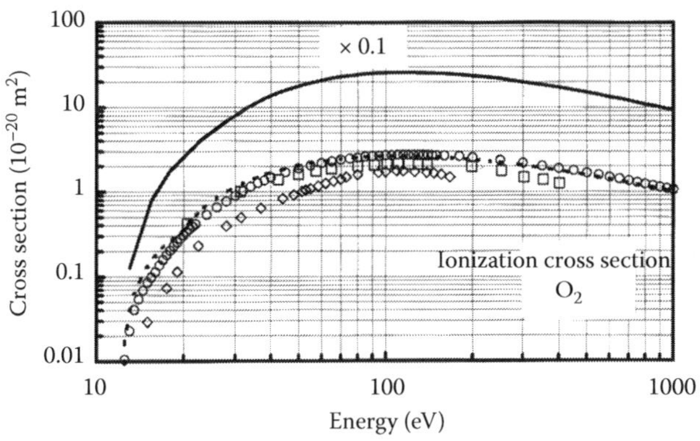

FIGURE 26.6 Ionization cross sections for O_2. (——) Hwang et al. (200, 1996), theory; (——), Straub et al. (1996), recommended; (□) Evans et al. (1988); (◊) Märk (1975); (○) Rapp and Englander-Golden (1965). The recommended cross sections are shown after multiplying with the factor 10 for easy readability. The agreement between measurements is very good. The lower values of Evans et al. (1988) are due to vibrationally excited O_2 at 5300 K.

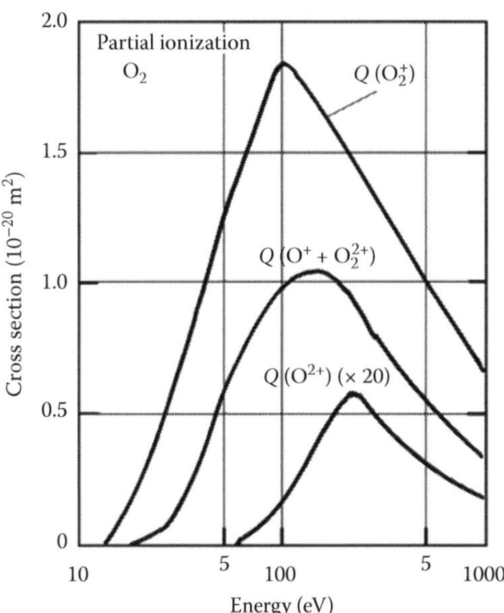

FIGURE 26.7 Averaged partial cross sections for O_2. References: Straub et al. (1996); Krishnakumar and Srivastava (1992); Märk (1975); Rapp et al. (1965). The data of Rapp et al. (1965) are for the production of ions with kinetic energies >0.25 eV.

TABLE 26.9
Recommended Drift Velocity in O_2

E/N	*W*	*E/N*	*W*
0.009	417.8	1	1.14 (04)
0.014	590.3	2	1.78 (04)
0.020	787.5	4	2.44 (04)
0.03	1.32 (03)	5	2.57 (04)
0.04	1.66 (03)	6	2.66 (04)
0.05	2.10 (03)	7	2.76 (04)
0.08	3.13 (03)	8	2.87 (04)
0.10	3.52 (03)	10	3.14 (04)
0.20	4.71 (03)	12	3.46 (04)
0.40	5.94 (03)	14	3.82 (04)
0.6	7.20 (03)	17	4.36 (04)
0.8	9.40 (03)	20	4.94 (04)
25	5.59 (04)	500	4.76 (05)
30	7.08 (04)	600	5.17 (05)
40	9.14 (04)	700	5.94 (05)
50	1.06 (05)	800	7.04 (05)
60	1.18 (05)	900	7.17 (05)
70	1.29 (05)	1000	7.71 (05)
80	1.44 (05)	1300	8.96 (05)
90	1.58 (05)	2000	1.19 (06)
100	1.72 (05)	3000	1.48 (06)
200	2.78 (05)	4000	1.70 (06)
300	3.55 (05)	5000	1.91 (06)
400	4.25 (05)		

Note: *E/N* in units of Td and *W* in units of m/s. a (b) means $a \times 10b$. See Figure 26.8 for graphical presentation.

O_2

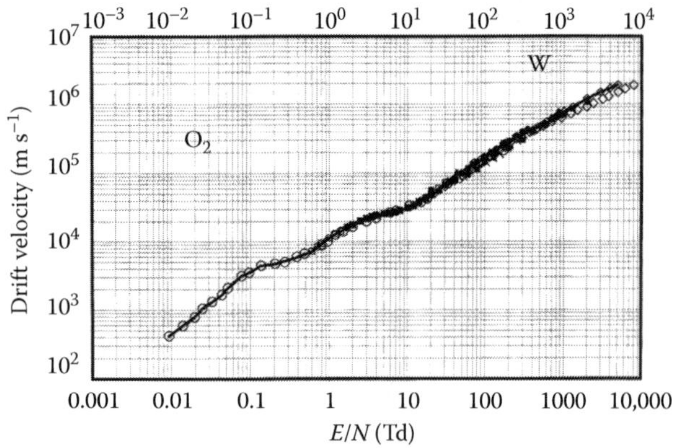

FIGURE 26.8 Drift velocity of electrons in O_2 as a function of *E/N*. Letter E or T after citation means experimental or theoretical, respectively. (——) Recommended, Raju (2005); (•) Jeon and Nakamura (1998) E; (■) Liu and Raju (1993) T; (♦) Liu and Raju (1992) T; (×) Roznerski (1984) E; (□) Huxley and Crompton (1974) E; (○) Hake and Phelps (1967) T; (◊) Schlumbohm (1965) E.

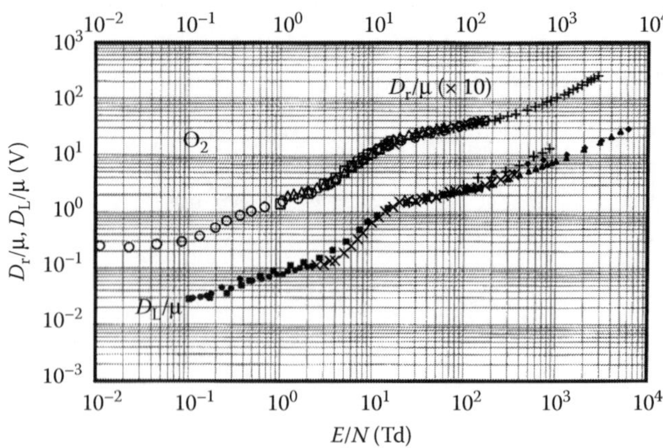

FIGURE 26.9 Ratios of radial and longitudinal diffusion coefficients to mobility as a function of *E/N*. Letter E or T after citation indicates experimental or theoretical, respectively. D_r/μ: (+); A l-Amin et al. (1985) T; (△) Huxley and Crompton (1974) E; (□) Fleming et al. (1972) E; (×) Naidu and Prasad (1970) E; (○) Hake and Phelps (1967) T. D_r/μ values are multiplied by a factor of 10 for clarity of presentation. D_L/μ: (×) Jeon and Nakamura (1998) E (+) Al-Amin et al. (1985) E; (•) Nelson and Davis (1972) E; (■) Lowke and Parker (1969) T; Schlumbohm (1965) E.

TABLE 26.10
Lateral Diffusion Coefficient Expressed as Ratio of Mobility (D_L/μ)

E/N (Td)	D_L/μ (V)	ND_L (10^{24} 1/ms)	*E/N* (Td)	D_L/μ (V)
Jeon and Nakamura (1998)			Liu and Raju (1992)	
2.5	0.113	0.914	30.0	1.55
3.0	0.116	0.854	40.0	1.74
3.5	0.135	0.910	50.0	1.87
4.0	0.137	0.849	60.0	1.99
5.0	0.187	0.959	70.0	2.08
6.0	0.247	1.098	80.0	2.21
7.0	0.332	1.318	90.0	2.24
8.0	0.423	1.530	100.0	2.33
10.0	0.647	2.024	130	2.54
12.0	0.813	2.257	200	3.05
14.0	1.038	2.760	300	3.84
17.0	1.241	3.087	400	4.56
20.0	1.501	3.557	500	5.27
25.0	1.601	3.539	600	5.83
30.0	1.568	3.533	700	6.40
35.0	1.688	3.560	800	6.90
40.0	1.768	3.748	900	7.40
50.0	1.817	3.613	1000	7.95
60.0	1.982	3.839	1300	8.98
70.0	2.214	4.168	2000	12.65
80.0	2.333	4.243	3000	16.78
100.0	2.45	4.260	4000	21.29
120.0	2.601	4.288	5000	26.18
140.0	2.689	4.251		
170.0	2.938	4.583		
200.0	3.361	4.820		
250.0	3.747	5.113		
300.0	4.195	5.412		
350.0	4.565	5.470		

Note: See Figure 26.9 for graphical presentation.

TABLE 26.11
Recommended Ratio of Radial Diffusion Coefficient to Mobility (D_r/μ)

E/N (Td)	D_r/μ (V)	*E/N* (Td)	D_r/μ (V)
0.01	0.025	20	2.060
0.02	0.024	40	2.700
0.04	0.025	70	2.985
0.07	0.029	100	3.391
0.1	0.033	120	3.576
0.2	0.053	140	3.846
0.4	0.094	200	4.047
0.7	0.127	400	6.500
1	0.152	700	7.850
2	0.204	800	8.700
4	0.360	1000	10.270
7	0.728	2000	17.800
10	1.090	3000	25.000

Note: See Figure 26.9 for graphical presentation.

26.10 IONIZATION COEFFICIENT

See Table 26.12.

26.11 ATTACHMENT CROSS SECTION

Table 26.13 shows the attachment cross section for production of O^- ions due to dissociative attachment. See Figure 26.11 for graphical presentation.

TABLE 26.12
Recommended Density-Reduced Ionization Coefficients (α/N): (O_2)

E/N (Td)	α/N (10^{-22} m^2)	E/N (Td)	α/N (10^{-22} m^2)
70	0.10	700	114.2
80	0.30	800	133.1
90	0.50	900	151.1
100	0.70	1000	167.7
130	2.20	1300	210.3
200	9.2	2000	287.4
300	27.0	3000	358.2
400	48.9	4000	392.2
500	71.5	5000	396.0
600	93.2		

Note: See Figure 26.10 for graphical presentation.

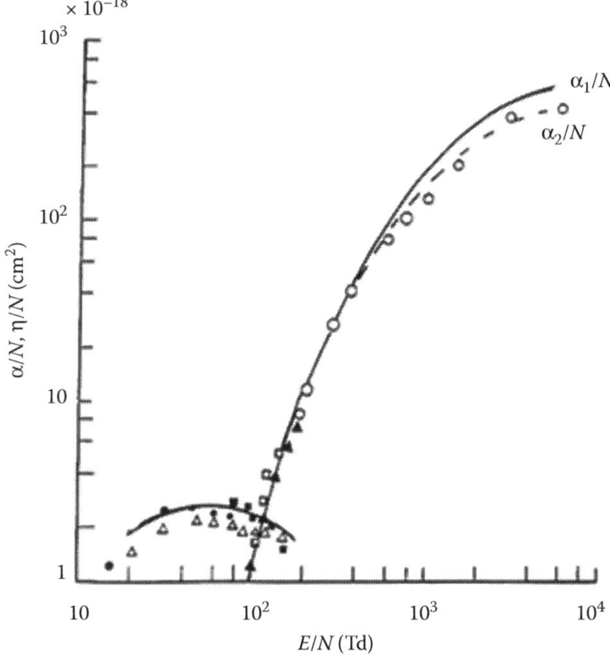

FIGURE 26.10 Reduced ionization and attachment coefficients as a function of E/N for O_2. α/N: (——) Liu and Raju (1992) corrected for diffusion; (– –) Liu and Raju (1992) uncorrected for diffusion; (▲) Price et al. (1973); (■) Naidu and Prasad (1970); (○) Schlumbohm (1959). η/N: (△) Geballe and Harrison (1952); (•) Grünberg (1969); (■) Naidu and Prasad (1970).

26.12 ATTACHMENT COEFFICIENT

See Table 26.14.

26.13 GAS CONSTANTS

See Table 26.15.

TABLE 26.13
Attachment Cross Section for O_2

Energy (eV)	Q_a (10^{-23} m^2)	Energy (eV)	Q_a (10^{-23} m^2)	Energy (eV)	Q_a (10^{-23} m^2)
4.2	0.00	6.1	12.31	8.0	4.49
4.3	0.09	6.2	13.11	8.1	3.87
4.4	0.26	6.3	13.63	8.2	3.34
4.5	0.70	6.4	13.11	8.3	2.81
4.6	0.70	6.5	14.07	8.4	2.37
4.7	0.97	6.6	13.99	8.5	2.02
4.8	1.32	6.7	13.72	8.6	1.67
4.9	1.76	6.8	13.37	8.7	1.41
5.0	2.20	6.9	12.84	8.8	1.23
5.1	2.90	7.0	12.23	8.9	1.06
5.2	3.61	7.1	11.43	9.0	0.88
5.3	4.49	7.2	10.64	9.1	0.70
5.4	5.37	7.3	9.85	9.2	0.70
5.5	6.33	7.4	8.97	9.3	0.62
5.6	7.48	7.5	8.18	9.4	0.53
5.7	8.53	7.6	7.39	9.5	0.44
5.8	9.59	7.7	6.42	9.6	0.44
5.9	10.47	7.8	5.72	9.8	0.35
6.0	11.43	7.9	5.01	9.9	0.35

Source: Adapted from Rapp, D. and D. D. Briglia, *J. Chem. Phys.*, 43, 1480, 1965.

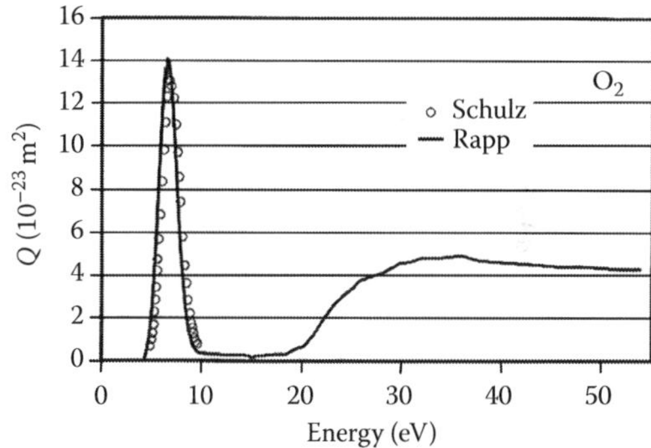

FIGURE 26.11 Cross section for electron attachment in O_2. The 6.7 eV peak is due to dissociative attachment and the high energy continuum due to ion pair formation. (Adapted from Rapp, D. and D. D. Briglia, *J. Chem. Phys.*, 43, 1480, 1965.)

TABLE 26.14
Recommended Attachment Coefficients for O_2

E/N (Td)	η/N (10^{-22} m^2)	E/N (Td)	η/N (10^{-22} m^2)
20	1.8	80	2.5
30	2.4	90	2.5
40	2.6	100	2.3
50	2.7	130	2.1
60	2.7	200	2.1
70	2.5		

Source: Adapted from Raju, G. G., *Gaseous Electronics: Theory and Practice*, Taylor & Francis, London, 2005.
Note: See Figures 26.10, 26.12, and 26.13 for graphical presentation.

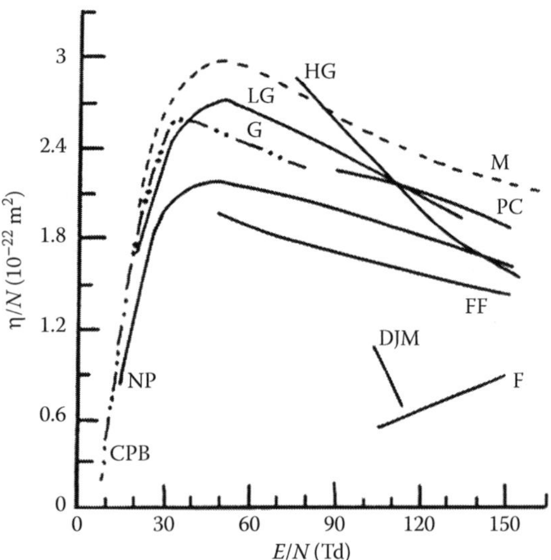

FIGURE 26.12 Density-reduced attachment coefficient as a function of E/N for O_2. The letters refer to (LG) Liu and Raju (162, 1992); (M) Masek et al., (157, 1977); (NP) Naidu and Prasad (154, 1970); (G) Grünberg (145, 1969); (FF) Freely and Fischer (152, 1964); (F) Frommhold (153, 1964); (DJM) Dutton et al. (151, 1963); (CPB) Chanin et al. (143, 1962); (PC) Prasad and Craggs (150, 1961); (HG) Geballe and Harrison (148, 1952).

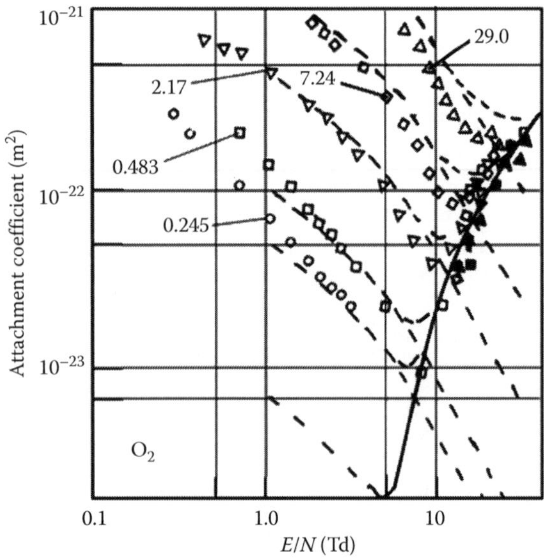

FIGURE 26.13 Electron attachment coefficient (η/N) as a function of E/N at low values of E/N. The left side of the curves shows three-body attachment processes below $E/N = 10$ Td and the density-reduced attachment coefficient is dependent on N. N is as shown ($\times 10^{24}$ m^{-3}). Note the increase of η/N with increasing N. (-□-□-) both branches, Hunter et al. (1986); (– – –) Taniguchi et al. (1978), computations; (▲) Chatterton and Craggs (1971); (▼) Rees (1965) agreeing with Herreng (1952); left branch open symbols (Chanin et al., 1962); ∇, right branch (Kuffel, 1959).

TABLE 26.15
Gas Constants

E/N range (Td)	F (10^{-20} m^2)	G (Td)	Deviation
257–642	1.97	646	±2.5%
642–1026	3.87	1066	±2.5%

Source: Adapted from Gurumurthy, G. R. and G. R. Govinda Raju, *IEEE Trans. Plas. Sci.*, 3, 131, 1975.
Note: Deviation is calculated from measured values.

REFERENCES

Al-Amin, S. A. J., H. N. Kücükarpaci, and J. Lucas, *J. Phys. D: Appl. Phys.*, 18, 1781, 1985.

Brunger, M., S. Buckman, and M. Elford, Cross sections for scattering- and excitation-processes in electron-molecule collisions in *Interactions of Photons and Electrons with Molecules*, Vol. 17c, Springer-Verlag, New York, NY, 2003.

Chanin, L. M., A. V. Phelps, and M. A. Biondi, *Phys. Rev.*, 128, 219, 1962. Also, L. M. Chanin, A. V. Phelps, and M. A. Biondi, *Phys. Rev. Lett.*, 2, 344, 1959.

Chatterton, P. A. and J. D. Craggs, *J. Electron. Control.*, 11, 425, 1971.

Dababneh, M. S., Y.-F. Hsieh, W. E. Kauppila, C. K. Kwan, S. J. Smith, T. S. Stein, and M. N. Uddin, *Phys. Rev. A*, 38, 1207, 1988.

Dalba, G., P. Fornasini, R. Grisenti, G. Ranieri, and A. Zecca, *J. Phys. B: At. Mol. Phys.* 13, 4695, 1980.

Dutton, J., F. Llewelyn-Jones, and G. B. Morgan, *Nature*, 198, 680, 1963.

Evans, B., S. Ono, R. M. Hobson, A. W. Yau, S. Teii, and J. S. Chang, *Proceedings of the 13th International Symposium on Shock Tubes and Waves*, SUNY Press, Albany, NY, 1982, pp. 535–542.

Evans, B., J. S. Chang, A. W. Yau, R. W. Nicholls, and R. M. Hobson, *Phys. Rev.*, 38, 2782, 1988.

Fisk, J. B., *Phys. Rev.*, 49, 167, 1936.

Fleming, I. A., D. R. Gray, and J. A. Rees, *J. Phys. D: Appl. Phys.*, 5, 291, 1972.

Freely, J. B. and L. H. Fischer, *Phys. Rev.*, 133, A304, 1964.

Frommhold, I., *Fortschr. Physik*, 12, 597, 1964.

Geballe, R. and M. A. Harrison, *Phys. Rev.*, 85, 372, 1952.

Grünberg, R., *Z. Naturforsch*, 24 a, 1039, 1969.

Gurumurthy, G. R. and G. R. Raju, *IEEE Trans. Plas. Sci.*, 3, 131, 1975.

Hake, R. D. Jr. and A. V. Phelps, *Phys. Rev.*, 158, 70, 1967.

Herreng, P., *Cahiers Phys.*, 38, 1, 1952.

Hunter, S. R., J. G. Carter, and L. G. Christophorou, *J. Appl. Phys.*, 60, 24, 1986.

Huxley, L. G. H. and R. W. Crompton, *The Diffusion and Drift of Electrons in Gases*, John Wiley & Sons, New York, NY 1974.

Hwang, M., Y. K. Kim, and M. E. Rudd, *J. Chem. Phys.*, 104, 2956, 1996. For tabulated values see http://physics.nist.gov/PhysRefData/Ionization.

Iga, I., L. Mu-Tao, J. C. Nogueira, and R. S. Barbieri, *J. Phys. B: At. Mol. Phys.*, 20, 1095, 1987.

Itikawa, Y., A. Ichimura, K. Onda, K. Sakimoto, K. Takayanagi, Y. Hatano, M. Hayashi, H. Nishimura, and S. Tsurubuchi, *J. Phys. Chem. Ref. Data*, 18, 23, 1989.

Jeon, B. and Y. Nakamura, *J. Phys. D: Appl. Phys.*, 31, 2145, 1998.

Kanik, I., S. Trajmar, and J. C. Nickel, *J. Geophys. Res.*, 98, 7447, 1993.
Kanik, J. C. Nickel, S. Trajmar, *J. Phys. B: At. Mol. Opt. Phys.*, 25, 2189, 1992.
Krishnakumar, E., S. K. Srivastava et al., *Int. J. Mass Spectr. Ion. Proc.*, 113, 1, 1992.
Kuffel, E., *Proc. Phys. Soc.*, 74, 297, 1959.
Lisovskiy, V., J.-P. Booth, K. Landry, D. Douai, V. Cassagne, and V. Yegorenkov, *J. Phys. D: Appl. Phys.*, 39, 660, 2006.
Liu, J. and G. R. G. Raju, *Can. J. Phys.*, 70, 216, 1992.
Lowke, J. J. and J. H. Parker, *Phys. Rev.*, 181, 302, 1969.
Mark, T. D., *J. Chem. Phys.*, 63, 3731, 1975.
Masek, K., L. Laska, and T. Ruzicka, *J. Phys. D: Appl. Phys.*, 10, L125, 1977.
Naidu, M. S., and A. N. Prasad, *J. Phys. D: Appl. Phys.*, 3, 957, 1970.
Nelson, D. R. and F. J. Davis, *J. Chem. Phys.*, 57, 4079, 1972.
Prasad, A. N. and J. D. Craggs, *Proc. Phys. Soc. London*, 77, 385, 1961.
Price, D. A. and J. L. Moruzzi, *J. Phys. D: Appl. Phys.*, 6, L17, 1973.
Raju, G. G., *Gaseous Electronics: Theory and Practice*, Taylor & Francis, London, 2005.
Randell, J., S. L. Hunt, G. Mrotzek, J.-P. Ziesel, and D. Field, *J. Phys. B: At. Mol. Opt. Phys.*, 27, 2369, 1994.
Rapp, D. and D. D. Briglia, *J. Chem. Phys.*, 43, 1480, 1965.
Rapp, D. and P. Englander-Golden, *J. Chem. Phys.*, 43, 1464, 1965.
Rapp, D., P. Englander-Golden, and D. D. Briglia, *J. Chem. Phys.*, 42, 4081, 1965.
Rees, J. A., *Austr. J. Phys.*, 18, 41, 1965.
Schlumbohm, H., *Z. Angew. Phys.*, 11, 156, 1959.
Schlumbohm, H., *Z. Phys.*, 182, 317, 1965.
Shyn, T. W. and W. E. Sharp, *Phys. Rev. A*, 26, 1369, 1982.
Straub, H. C. et al., *Phys. Rev.*, 54, 2146, 1996.
Subramanian, K. P. and V. J. Kumar, *J. Phys. B: At. Mol. Phys.*, 23, 745, 1990.
Sullivan, J. P., J. C. Gibson, R. J. Gulley, and S. J. Buckman, *J. Phys. B: At. Mol. Opt. Phys.*, 28, 4319, 1995.
Szmytkowski, C., K. Maciąg, and G. Karwasz, *Phys. Scripta*, 54, 271, 1996.
Taniguchi, T., H. Tagashira, I. Okada, and Y. Sakai, *J. Phys. D: Appl. Phys.*, 11, 2281, 1978.
Trajmar, S., D. C. Cartwright, and W. Williams, *Phys. Rev. A*, 41, 1482, 1971.
Wakiya, K., *J. Phys. B: At. Mol. Phys.*, 11, 3913, 1978(a); *J. Phys. B: At. Mol. Phys.*, 11, 3931, 1978(b).
Wedde, J. and T. G. Strand, *J. Phys. B: Atom. Molec. Phys.*, 7, 1091, 1974.
Zecca, A., R. S. Brusa, R. Grisenti, S. Oss, and C. Szmytkowski, *J. Phys. B: At. Mol. Phys.*, 19, 3353, 1986.

Section III

3 ATOMS

27

CARBON DIOXIDE

CONTENTS

Carbon dioxide (CO_2) is a nonpolar, slightly attaching gas that has 22 electrons. The electronic polarizability is 3.24×10^{-40} F m^2 and the ionization potential is 13.773 eV. The molecule possesses three fundamental vibrational modes: the symmetric stretch mode (100) which does not change the molecular symmetry with excitation energy of 172 meV, the bending mode (010) (energy 83 meV), and the asymmetrical stretching mode (001) with energy of 291 meV (Kochem et al., 1985).

27.1 SELECTED REFERENCES FOR DATA

See Table 27.1.

TABLE 27.1
Selected References for Data

Quantity	Range: eV, (Td)	Reference
Cross sections and swarm data	0–1000 (0.1–1000)	Raju (2005)
Transport parameters	**(2–400)**	**Hernández-Ávila et al. (2002)**
Differential cross section	**1.5–30**	**Kitajima et al. (2001)**
Elastic cross section	0.1–100	Lee et al. (1999)
Elastic cross section	**1–50**	**Gibson et al. (1999)**
Vibrational cross section	4–50	Takekawa and Itikawa (1998)
Elastic cross section	1.5–100	Tanaka et al. (1998)
Total and elastic cross section	0.3–100	Kimura et al. (1997)
Total cross sections	**400–5000**	**Garcia and Manero (1996)**
Transport parameters	**20–1000**	**Hasegawa et al. (1996)**

continued

CO_2

TABLE 27.1 (continued)
Selected References for Data

Quantity	Range: eV, (Td)	Reference
Ionization cross section		**Hwang et al. (1996)**
Ionization cross section	**15–1000**	**Straub et al. (1996)**
Drift velocity	**(1–500)**	**Nakamura (1995)**
Excitation cross section	**10–1000**	**LeClair and McConkey (1994)**
Total cross section	**1–500**	**Sueoka and Hamada (1993)**
Total cross section	10–5000	Jain and Baluja (1992)
Ionization cross section	**14–200**	**Freund et al. (1990)**
Total cross section	**0.5–9.0**	**Ferch et al. (1989)**
Total cross section	**0.12–2.0**	**Buckman et al. (1987)**
Ionization cross section	**15–510**	**Orient and Srivastava (1987)**
Total cross section	**0.5–3000**	**Szmytkowski et al. (1987)**
Elastic cross section	**500–1000**	**Iga et al. (1984)**
Drift velocity	**(50–250)**	**Roznerski and Leja (1984)**
Total cross section	**2–50**	**Hoffman et al. (1982)**
Total cross section	**0.07–5.0**	**Ferch et al. (1981)**
Rotational cross section	0.07–10	Morrison and Lane (1977)
Drift velocity	**0.1–10**	**Haddad and Elford (1979)**
Swarm parameters	**(14–3000)**	**Kucukarpaci and Lucas (1979)**
Total cross section	**1.5–8.0**	**Szmytkowski and Zubek (1978)**
Elastic scattering	**3–90**	**Shyn et al. (1978)**
Ionization coefficient	**(61–152)**	**Alger and Rees (1976)**
Vibrational excitation	**3–5**	**Boness and Schulz (1974)**
Total cross section	**3–5**	**Čadež et al. (1974)**
Swarm parameters	**(30–1650)**	**Lakshminarasimha et al. (1974)**
Vibrational excitation	**0–6**	**Sanche and Schulz (1973)**
Drift velocity	**(0.01–200)**	**Hake and Phelps (1965)**
Ionization cross section	**14.5–1000**	**Rapp and Englander-Golden (1965)**
Ionization coefficients	**(80–3600)**	**Bhalla and Craggs (1960)**

Note: Bold font indicates experimental study.

27.2 TOTAL SCATTERING CROSS SECTION

Table 27.2 and Figure 27.1 show the total scattering cross section in the low-energy range, $0.05 \leq \varepsilon \leq 9.0$ eV. Table 27.3 and Figure 27.2 show the total cross section in the energy range, $0.05 \leq \varepsilon \leq 1000$ eV. Points to note are

1. The total scattering cross section increases as the energy decreases toward zero; the effect is attributed to the electron attachment process, referred to as virtual state in the literature (Raju, 2005). The influence of the dipole moment of the vibrationally excited molecule may also contribute. It is noted that CO_2 is a nonpolar molecule in the ground state, but possesses a dipole moment in the vibrationally excited state in bending and asymmetrical stretching mode (Kitajima et al., 2001).
2. A shape resonance at 3.8 eV attributed to vibrational excitation (Boness and Schulz, 1974).
3. A broad maximum near 30 eV possibly due to inelastic processes, including electronic excitation and ionization. This feature is common to many gases.
4. A monotonic decrease beyond 50 eV which is also common to many gases.

TABLE 27.2
Total Scattering Cross Sections for CO_2

Energy (eV)	Q_T (10^{-20} m^2)	Energy (eV)	Q_T (10^{-20} m^2)
0.05	67.94	2.75	6.26
0.06	63.82	3.00	7.32
0.07	60.12	3.20	9.05
0.09	56.41	3.30	10.21
0.09	51.47	3.35	10.81
0.11	47.35	3.40	11.42
0.12	43.65	3.45	12.03
0.14	39.53	3.50	12.64
0.16	36.24	3.55	13.24
0.18	34.18	3.60	13.84
0.19	30.06	3.65	14.38
0.24	27.18	3.70	14.80
0.27	25.12	3.75	15.09
0.30	22.24	3.80	15.34
0.40	19.35	3.85	15.51
0.48	16.88	3.90	15.47
0.50	15.56	3.95	15.29
0.60	12.92	4.0	15.06
0.7	11.46	4.1	14.48
0.8	10.12	4.2	13.82
0.9	9.02	4.3	13.15
1.0	8.17	4.5	11.71
1.2	7.06		
1.4	6.33		
1.6	5.90		
1.8	5.67		
2.0	5.56		
2.25	5.65		
2.50	5.79		

Source: Adapted from Ferch, J., C. Masche, and W. Raith, *J. Phys. B: At. Mol. Phys.*, 14, L97, 1981.
Note: $0.05 \le \varepsilon \le 4.5$ eV. Digitized from graphical presentation.

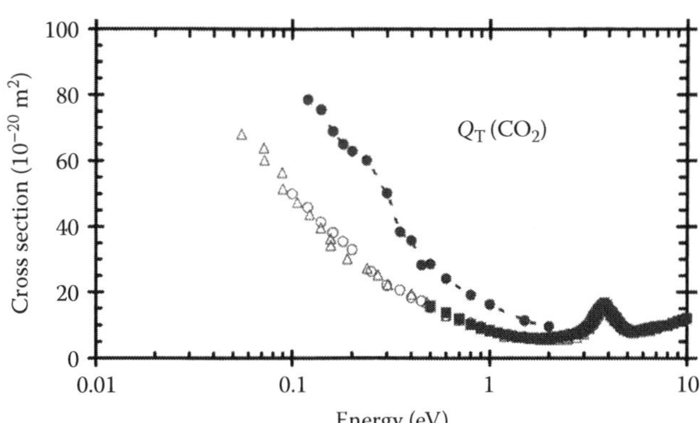

FIGURE 27.1 Total scattering cross section for CO_2 in the low-energy region. (△) Ferch et al. (1981, 1989); (○) Buckman et al. (1987), 310 K; (—●—) Buckman et al. (1987), 510 K; (■) Szmytkowski et al. (1987); (□) Szmytkowski and Zubek (1978).

TABLE 27.3
Total Scattering Cross Sections for CO_2

Szmytkowski et al. (1987)[a]				Szmytkowski et al. (1987)[a]	
Energy (eV)	Q_T (10^{-20} m^2)	Energy (eV)	Q_T (10^{-20} m^2)	Energy (eV)	Q_T (10^{-20} m^2)
0.5	15.80	4.6	9.6	72.2	14.4
0.6	13.70	4.7	9.3	81	13.81
0.7	11.90	4.8	8.66	90.2	13.39
0.8	10.20	5.0	8.14	100	12.54
0.9	9.15	5.2	7.78	110.2	12.14
1.0	8.35	5.4	7.70	121	11.65
1.1	7.63	5.6	8.10	132.2	11.17
1.2	7.10	5.8	8.10	144	11.03
1.3	6.72	5.9	8.10	169	9.93
1.4	6.34	6.0	8.30	196	9.25
1.5	6.12	6.5	8.42	225	8.71
1.6	6.07	7.0	9.01	256	8.31
1.8	5.89	7.5	9.74	289	7.84
2.0	5.90	8.0	9.83	324	7.31
2.2	6.11	8.5	10.5	361	6.79
2.4	6.73	9.0	11.1	400	6.40
2.6	7.23	9.5	11.6	441	6.12
2.8	7.60	10	12.0	484	5.58
3.0	8.60	12	13.2	576	4.93
3.1	9.10	13.5	13.6	676	4.41
3.2	9.60	15	13.9	784	3.94
3.3	11.10	17.5	14.9	900	3.53
3.4	12.80	20	15.8	1024	3.28
3.5	13.60	22.5	15.8	1156	2.95
3.6	14.80	25	16.0	1296	2.69
3.65	15.40	27.5	16.2	1444	2.38
3.7	16.00	30	16.2	1600	2.13
3.8	16.7	35	16.1	1764	1.92
3.9	16.1	40	15.9	1936	1.79
3.95	15.2	45	15.8	2116	1.62
4.0	14.2	50	15.1	2304	1.5
4.1	13.0	60	14.9	2500	1.29
4.2	12.4	65	14.4	2704	1.27
4.3	11.4	70	13.9	2916	1.15
4.4	11.1	80	13.4		
4.5	10.4				

Source: Szmytkowski, C. et al., *J. Phys. B: At. Mol. Phys.*, 20, 5817, 1987. With kind permission of Institute of Physics, England.
[a] Measurements in another laboratory.

27.3 DIFFERENTIAL SCATTERING CROSS SECTION

Figure 27.3 shows the differential scattering cross sections for the bending modes (010) and (020). General comments for differential scattering cross section are

1. The rotational and vibrational excitation produces a change in the shape of the angular distribution of the differential scattering cross section (Figure 27.4).
2. The differential cross section for the bending mode is approximately symmetric about 90°, but not so for symmetrical stretching.
3. The cross section for asymmetric mode is about 10% compared to the two other modes.

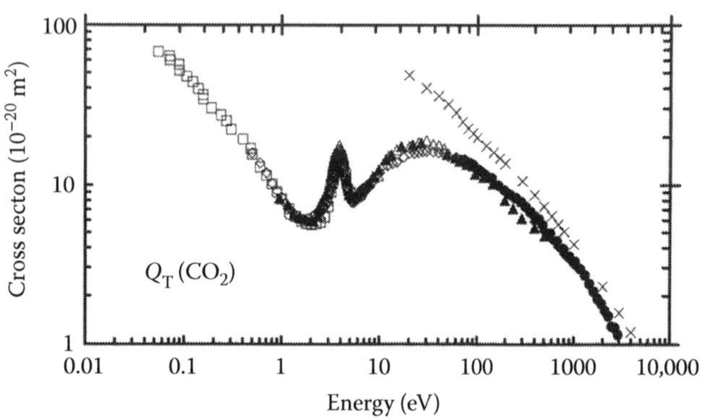

FIGURE 27.2 Total scattering cross sections for the entire energy range for CO_2. (◆) Garcia and Manero (1996); (□) Ferch et al. (1981, 1989); (△) Sueoka and Hamada; (×) Jain and Baluja (1992), theory; (◊) Szmytkowski et al. (1987); (△) Hoffmann et al. (1982).

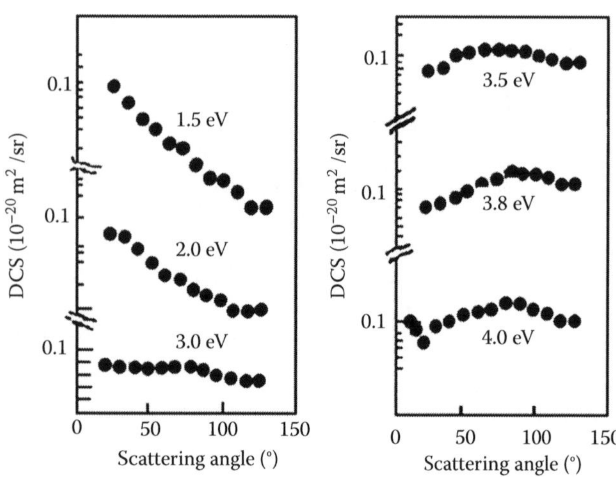

FIGURE 27.3 Differential scattering cross sections for CO_2 for (010) and (020) bending modes. (Adapted from Kitajima, M. et al., *J. Phys. B: At. Mol. Opt. Phys*., 34, 1929, 2001.)

27.4 ELASTIC SCATTERING CROSS SECTION

See Table 27.4.

27.5 MOMENTUM TRANSFER CROSS SECTION

Table 27.5 and Figure 27.6 show the momentum transfer cross section for the energy range 0–1000 eV.

27.6 INELASTIC PROCESSES

Inelastic processes are listed in Table 27.6 (Lowke et al., 1973). Table 27.7 gives detailed vibrational energy levels (Boness and Schulz, 1974). Energy and vibrational spacing of CO_2^- ion are shown in Table 27.8 (Sanche and Schulz, 1973).

27.7 VIBRATIONAL EXCITATION CROSS SECTION

See Table 27.9.

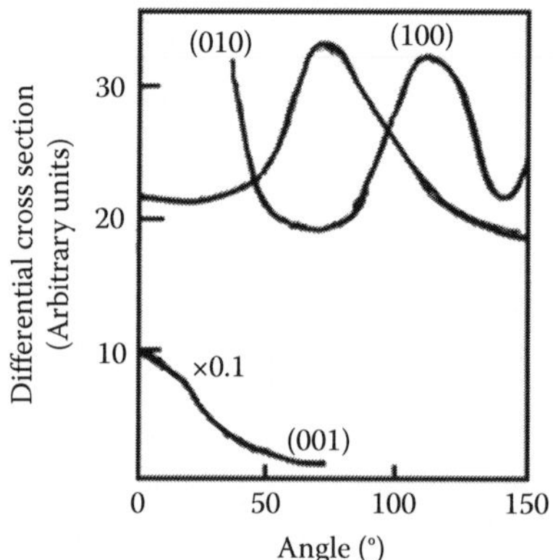

FIGURE 27.4 Relative magnitude of differential cross sections for vibrational excitation for CO_2. (100) is symmetrical stretch, (010) is bending, and (001) is asymmetrical stretching mode. (Adapted from Andrick, D. and F. H. Read, *J. Phys. B: At. Mol. Phys*., 4, 389, 1971.)

TABLE 27.4
Elastic Scattering Cross Sections for CO_2

Energy (eV)	Q_{el} (10^{-20} m^2)	Energy (eV)	Q_{el} (10^{-20} m^2)
0.155	25.0	60.0	11.0
1.05	5.8	90.0	7.5
2.0	4.62	100	6.8
4.0	11.0	500	2.98
10.0	10.61	800	2.31
20.0	14.59	1000	1.92
50.0	11.7		

Source: Adapted from Raju, G. G., *Gaseous Electronics: Theory and Practice*, Taylor & Francis, London, 2005.
Note: See Figure 27.5 for graphical presentation.

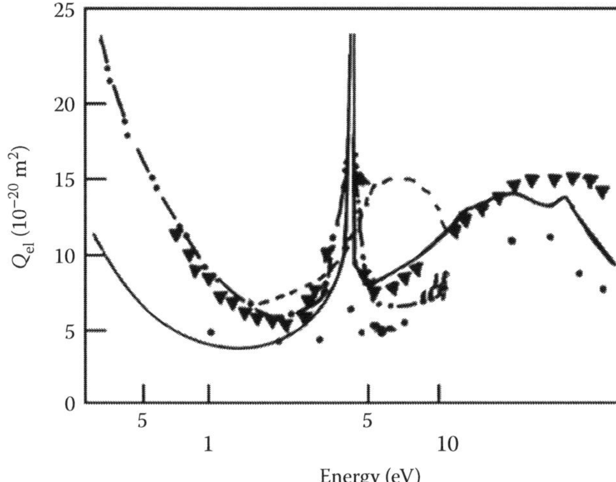

FIGURE 27.5 Integral elastic cross sections for CO_2. (●) Gibson et al. (1999), ANU data; (◆) Gibson et al. (1999), Flinders data; (— · · —) Morrison and Lane (1977); (- - -) Takekawa and Itikawa (1996); (––) Lee et al. (1999). Also shown for comparison is the total scattering cross section of (▼) Szmytkowski et al. (1987). (With kind permission of Institute of Physics, England.)

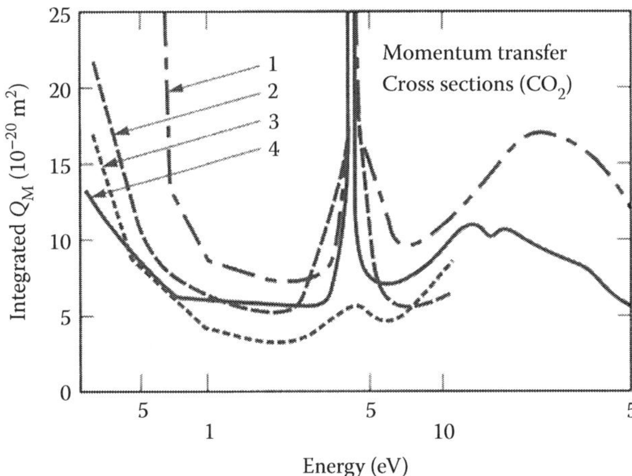

FIGURE 27.6 Momentum transfer cross sections in CO_2. Experimental: (●) Gibson et al. (1999). Theory: (1) Morrison and Lane (1977); (2) Nakamura (1995); (3) Swarm derived: Lowke et al. (1973); (4) Lee et al. (40, 1999). (With kind permission of Institute of Physics, England.)

TABLE 27.5
Momentum Transfer Cross Sections in CO_2

Energy (eV)	Q_M (10^{-20} m^2)	Energy (eV)	Q_M (10^{-20} m^2)	Energy (eV)	Q_M (10^{-20} m^2)
0.000	600.0	1.000	5.55	8.0	8.07
0.001	540.0	1.2	5.02	9.0	9.24
0.002	387.0	1.3	4.90	10.0	9.94
0.010	170.0	1.5	4.83	15.0	11.19
0.020	119.0	2.0	4.53	20.0	10.17
0.080	58.0	3.0	5.96	30.0	7.51
0.100	52.0	3.8	7.69	60.0	4.15
0.150	40.0	4.0	7.22	100.0	2.65
0.200	31.0	5.0	5.66	200.0	1.08
0.300	20.30	6.0	6.69	300.0	0.66
0.400	14.30	6.5	6.56	500.0	0.36
0.500	10.90	7.0	6.56	1000.0	0.14

Note: 0–1.5 eV and 100–1000 eV, L. Pitchford, personal communication, 2003, see Raju (2005); 1.5–100 eV, Tanaka et al. (37, 1998).

TABLE 27.6
Selected Inelastic Processes and Threshold Energies in CO_2

Energy loss	Threshold (eV)	Process	Remarks
0.083	0.083	000 → 010	I Bending mode
0.167	0.167	000 → 020 + 100	Bending and symmetrical stretching
0.291	0.291	000 → 001	I Asymmetrical stretching
0.252	2.5	000 → 0n0 + n00	Bending and symmetrical stretching
0.339	1.5	000 → 0n0 + n00	Bending and symmetrical stretching
0.422	2.5	000 → 0n0 + n00	Bending and symmetrical stretching
0.505	2.5	000 → 0n0 + n00	Bending and symmetrical stretching
2.5	2.5	000 → 0n0 + n00	Bending and symmetrical stretching
3.85	3.85	$e + CO_2 \rightarrow CO + O^-$	Dissociative attachment
7.0	7.0		Electronic excitation
10.5	10.5		Electronic excitation
13.3	13.3	$e + CO_2 \rightarrow CO_2^+ + 2e$	Ionization
20.9	20.9	$e + CO_2 \rightarrow CO^+ + O + 2e$	Dissociative ionization[a]
22.6	22.6	$e + CO_2 \rightarrow CO + O^+ + 2e$	Dissociative ionization[a]
24.6	24.6	$e + CO_2 \rightarrow C^+ + O_2 + 2e$	Dissociative ionization[a]

Sources: Adapted from Lowke, J. J., A. V. Phelps, and B. W. Irwin, *J. Appl. Phys.*, 44, 4664, 1973; Crowe, A. and J. W. McConkey, *J. Phys. B: At. Mol. Phys.*, 7, 349, 1974.

Note: A more detailed energy list is given in Table 27.7.

[a] Reference for the last three rows

CO₂

TABLE 27.7
Vibrational Energy Levels in CO_2

	Calculated Energy Level (eV)		
Designation	***n*10**	***n*00**	**Experimental Energy Loss Peak (eV)**
01^10	0.083		0.083
10^00		0.167	0.168
11^10	0.250		0.250
20^00		0.333	0.335
21^10	0.416		0.415
30^00		0.498	0.500
310	0.582		0.580
400		0.662	0.665
410	0.746		0.740
500		0.825	0.820
510	0.910		0.900
600		0.967	0.985
610	1.072		1.06
700		1.149	1.14
710	1.234		1.23
800		1.309	1.30
810	1.395		1.39
900		1.469	1.47
910	1.555		1.55
10,00		1.627	1.61
10, 10	1.714		1.70
11, 00		1.785	
11, 10	1.872		1.87
12, 00		1.942	
12, 10	2.030		2.02
13, 00		2.098	
13, 10	2.186		2.17
14, 00		2.253	
14, 10	2.342		2.35
15, 00		2.407	
15, 10	2.496		2.50
16, 00		2.561	
16, 10	2.650		2.65
17, 00		2.713	
17, 10	2.803		2.78
18, 00		2.864	
18, 10	2.955		2.93
19, 00		3.015	
19, 10	3.106		3.10
20, 00		3.165	
20, 10	3.256		3.23
21, 00		3.313	
21, 10	3.405		

Source: Adapted from Boness, M. J. W. and G. J. Schulz, *Phys. Rev. A*, 9, 1969, 1974.

TABLE 27.8
Vibrational Excitation Levels of CO_2^-

Vibrational Transition: $v \rightarrow v + 1$	Spacing (meV)	Energy of Vibrational State (eV)
$0 \rightarrow 1$	138	3.14
$1 \rightarrow 2$	136	3.28
$2 \rightarrow 3$	134	3.41
$3 \rightarrow 4$	132	3.54
$4 \rightarrow 5$	130	3.67
$5 \rightarrow 6$	128	3.80
$6 \rightarrow 7$	127	3.93
$7 \rightarrow 8$	125	4.06
$8 \rightarrow 9$	123	4.18
$9 \rightarrow 10$	122	4.31
$10 \rightarrow 11$	121	4.42
$11 \rightarrow 12$	...	4.54

Source: Adapted from Sanche, L. and G. J. Schulz, *J. Chem. Phys.*, 58, 479, 1973.
Note: See Figure 27.7 for graphical presentation.

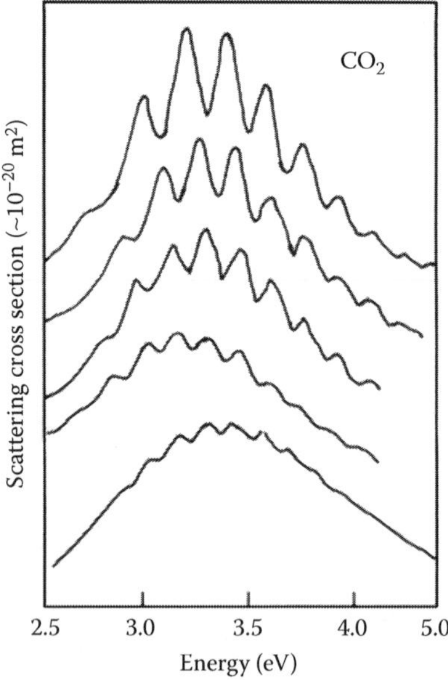

FIGURE 27.7 Schematic diagram of oscillations in vibrational excitation scattering cross sections in the resonance region. The approximate energy range is as shown. Energy levels for individual peaks in excitation mode are given by Čadež et al. (1974).

TABLE 27.9
Vibrational Excitation Cross Section for CO_2

Energy (eV)	Mode (100) Q_{vib} (10^{-20} m^2)	Mode (010) Q_{vib} (10^{-20} m^2)	Mode (001) Q_{vib} (10^{-20} m^2)
0.1		2.224	
0.2		1.882	
0.3		1.236	1.054
0.4		0.967	1.695
0.5	0.050	0.835	1.628
0.8		0.627	1.216
1.0	0.075	0.498	0.997
1.5	0.225	0.411	0.761
2.0	0.562	0.453	0.646
2.5	2.946	0.678	0.558
3.0	2.737	1.122	0.490
3.5	2.158	2.945	0.466
4.0	1.291	3.400	0.450
4.5	0.290	2.661	0.434
5.0	0.874	1.320	0.427
6.0	0.405	0.564	0.408
7.0	0.215	0.460	0.452
8.0	0.225	0.452	0.467
9.0	0.193	0.321	0.370
10.0	0.165	0.211	0.243
12.0	0.059	0.053	0.231
14.0	0.020	0.013	0.199
15.0		0.007	
16.0	0.007		0.183
18.0	0.002		0.156
20.0	0.002		0.152
30			0.100
40			0.08
50			0.065
60			0.057
70			0.050
80			0.046
90			0.043
100			0.040

Source: Adapted from Nakamura, Y., *Aust. J. Phys.*, 48, 357, 1995.
Note: Digitized and interpolated from original presentation. See Figure 27.8 for graphical presentation.

27.8 ELECTRONIC EXCITATION CROSS SECTION

A discussion of excitation levels of the molecule is given by Raju (2005). Table 27.10 shows excitation cross sections recommended by Raju (2005) and Figure 27.8 shows selected available data. Also see Figure 27.9.

TABLE 27.10
Excitation Cross Sections for CO_2 Recommended

Energy (eV)	Q_{ex} (10^{-20} m^2)	Energy (eV)	Q_{ex} (10^{-20} m^2)
12	0.02	65	2.60
13	0.17	70	2.60
14	0.39	75	2.60
15	0.60	80	2.63
16	0.88	85	2.62
17	1.16	90	2.60
18	1.35	95	2.60
19	1.48	100	2.61
20	1.61	125	2.60
25	2.01	150	2.58
30	2.20	175	2.54
35	2.36	200	2.48
40	2.45	225	2.42
45	2.50	250	2.36
50	2.56	275	2.26
55	2.57	300	2.22
60	2.59		

Source: Adapted from Raju, G. G., *Gaseous Electronics: Theory and Practice*, Taylor & Francis, London, 2005.

27.9 IONIZATION CROSS SECTION

See Table 27.11.

27.10 ATTACHMENT CROSS SECTION

Attachment processes according to Spence and Schulz (1974) are as follows

$$CO_2 + e \rightarrow CO + O^- \quad (27.1)$$

$$CO_2 + e \rightarrow C + O_2^- \quad (27.2)$$

$$O^- + CO_2 \rightarrow CO + O_2^- \quad (27.3)$$

$$CO_2 + e \rightarrow C^- + 2O \quad (27.4)$$

Following Reaction 27.1, three-body process may also occur at higher gas number densities according to

$$O^- + CO_2 + CO_2 \rightarrow O_3^- + CO_2 \quad (27.5)$$

Reactions 27.1, 27.2, and 27.4 are dissociative attachments and Reaction 27.3 is an ion–molecule reaction. Reaction 27.5 exhibits dependence on gas number density (Table 27.17). The observed peak energy and peak cross section are shown in Table 27.12.

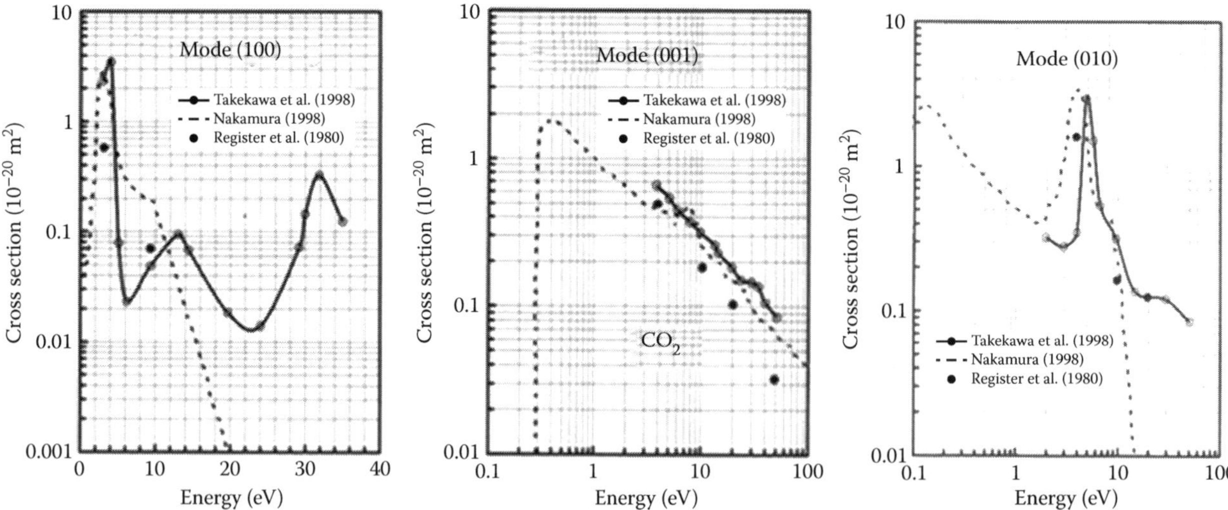

FIGURE 27.8 Selected vibrational excitation cross sections in CO_2. (—) Takekawa and Itikawa (1998) for (100) and (001) modes, Takekawa and Itikawa (1999) for (010) mode; (— —) Nakamura (31b, 1995); (●) Register et al. (1980).

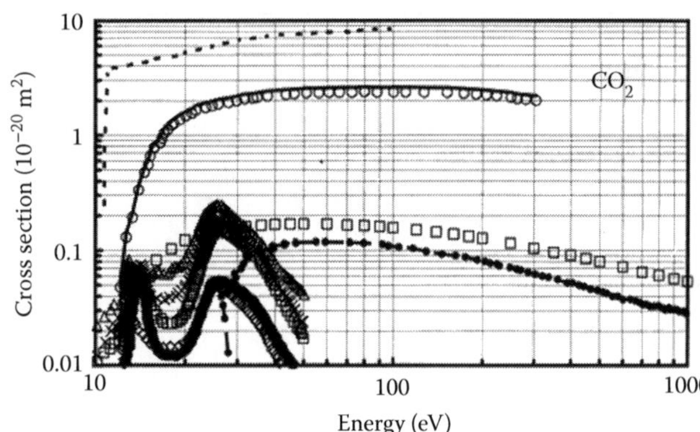

FIGURE 27.9 Excitation and dissociation cross sections in CO_2. (—●—) Strickland and Green (1969)–$^1\Sigma u^+$; (○) Ajello (1971); (— —) Lowke et al. (1973); (□) Leclair and McConkey (1994); (△) Lee et al. (1999)–X $^1\Sigma g^+ \rightarrow 3\Sigma u^+$; (◊) ibid.–X $^1\Sigma g^+ \rightarrow 1\Delta u$; (×) ibid.–X $^1\Sigma g^+ \rightarrow {}^3\Delta u$; (●) ibid.–X $^1\Sigma g^+ \rightarrow {}^1\Sigma u^-$; (■) ibid.–X $^1\Sigma g^+ \rightarrow {}^3\Sigma u^-$; (——) recommended. Untabulated values are digitized from the original publications and replotted.

Table 27.13 and Figure 27.13 present the attachment cross sections for CO_2 (Rapp and Briglia, 1965).

TABLE 27.11
Recommended Total Ionization Cross Section for CO_2

Straub et al. (1996)		Rapp and Englander-Golden (1965)	
Energy (eV)	Q_i (10^{-20} m²)	Energy (eV)	Q_i (10^{-20} m²)
		14.5	0.055
15	0.143	15	0.097
20	0.564	15.5	0.135
25	1.115	16	0.174
30	1.698	16.5	0.215
35	2.095	17	0.255
40	2.414	18	0.333
45	2.673	18.5	0.373
50	2.881	19	0.427
55	3.062	19.5	0.452
60	3.239	21	0.577
65	3.356	22	0.676
70	3.454	23	0.777
75	3.533	24	0.880
80	3.623	26	1.117
85	3.682	28	1.337
90	3.739	30	1.513
95	3.777	32	1.654
100	3.810	34	1.777
110	3.823	36	1.891
120	3.813	38	1.997
140	3.774	40	2.111
160	3.643	45	2.366
180	3.546	50	2.586
200	3.426	55	2.762
225	3.300	60	2.929
250	3.139	65	3.070
275	3.046	70	3.175
300	2.887	75	3.272
350	2.632	80	3.351
400	2.480	85	3.413

TABLE 27.11 (continued)
Recommended Total Ionization Cross Section for CO_2

Straub et al. (1996)		Rapp and Englander-Golden (1965)	
Energy (eV)	Q_i (10^{-20} m^2)	Energy (eV)	Q_i (10^{-20} m^2)
450	2.273	90	3.457
500	2.122	100	3.518
550	1.988	105	3.527
600	1.879	110	3.545
650	1.785	115	3.554
700	1.702	120	3.554
750	1.622	125	3.548
800	1.541	130	3.545
850	1.464	135	3.536
900	1.419	140	3.518
950	1.371	145	3.510
1000	1.312	150	3.483
		160	3.439
		180	3.360
		200	3.255
		250	3.017
		300	2.780
		350	2.595
		400	2.419
		450	2.269
		500	2.137
		600	1.909
		700	1.733
		800	1.574
		900	1.469
		1000	1.399

Note: The total cross sections from Straub et al. (1996) are calculated from partial ionization cross sections. See Figures 27.10 through 27.12 for graphical presentation.

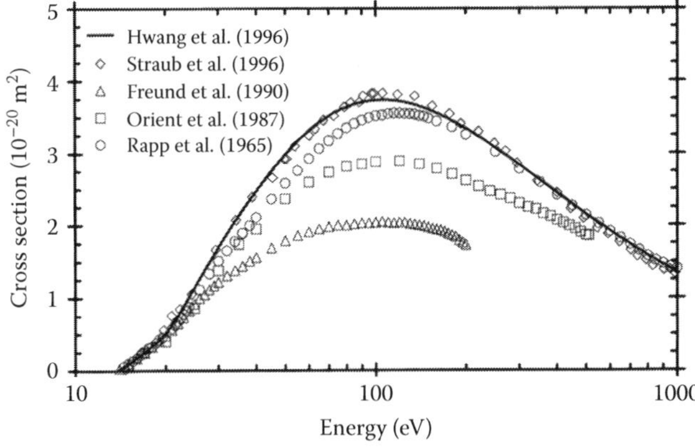

FIGURE 27.10 Total ionization cross sections for CO_2. (——) Hwang et al. (1996); (◊) Straub et al. (1996); (△) Freund et al. (1990); (□) Orient et al. (1987); (○) Rapp and Englander-Golden (1965).

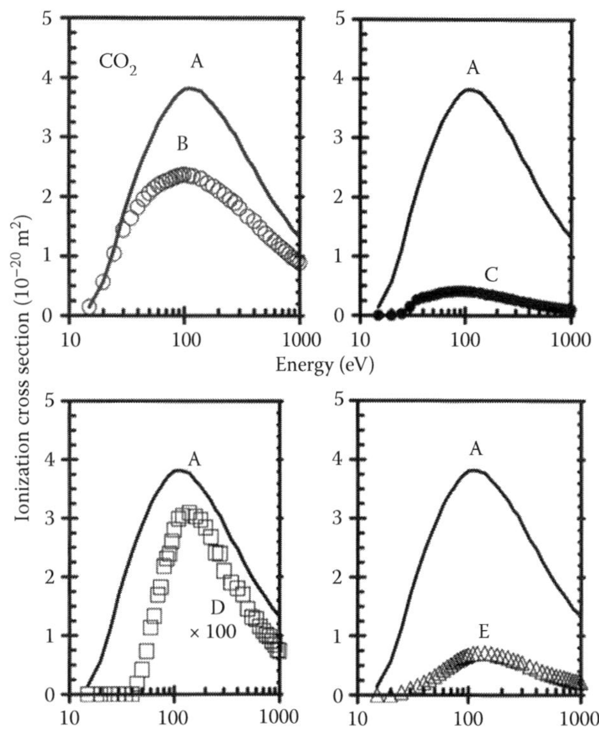

FIGURE 27.11 Partial ionization cross sections for CO_2. The total cross section is plotted to show the relative contribution. (A) total; (B) CO_2^+; (C) CO^+; (D) CO_2^{2+}, note the multiplication factor; (E) O^+. (Adapted from Straub, H. C. et al., *J. Chem. Phys.*, 105, 4015, 1996.)

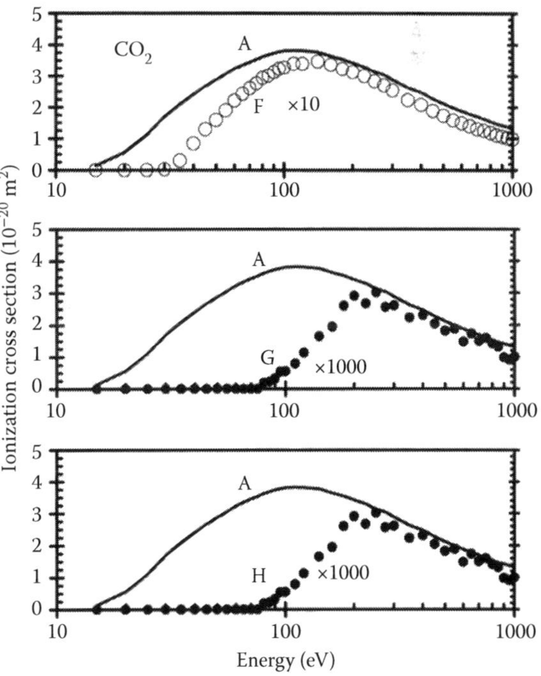

FIGURE 27.12 Further partial ionization cross sections for CO_2. The total cross section is plotted to show the relative contribution. (A) total; (F) C^+, ×10; (G) O^{2+}, ×1000; (H) CO_2^{2+}, ×1000; note the multiplication factors. (Adapted from Straub, H. C. et al., *J. Chem. Phys.*, 105, 4015, 1996.)

CO$_2$

TABLE 27.12
Attachment Processes

Ion Species	Peak Energy (eV)	Peak Cross Section (10^{-24} m^2)	Process
O^-	4.3	1.68	27.1
	8.1	4.87	27.1
O_2^-	8.2		27.1 followed by 27.3
	11.3	~1 × 10^{-4}	27.3
	12.9	~1 × 10^{-4}	27.3
C^-	16.0		27.4
	17.0		27.4
	18.7	~0.20	27.4

Source: Adapted from Spence, D. and G. J. Schulz, *J. Chem. Phys.*, 60, 216, 1974.

TABLE 27.13
Total Attachment Cross Sections for CO$_2$

Energy (eV)	Q_i (10^{-24} m^2)	Energy (eV)	Q_i (10^{-24} m^2)
3.3	0.0	6.7	2.90
3.4	0.176	6.8	3.87
3.5	0.616	6.9	5.28
3.6	1.41	7.0	6.86
3.7	2.73	7.1	8.98
3.8	5.28	7.2	11.44
3.9	8.18	7.3	14.52
4.0	10.65	7.4	17.78
4.1	12.76	7.5	21.65
4.2	14.08	7.6	26.66
4.3	14.78	7.7	31.24
4.4	13.64	7.8	35.73
4.5	12.06	7.9	39.60
4.6	9.77	8.0	42.42
4.7	7.74	8.1	42.86
4.8	5.98	8.2	41.36
4.9	4.40	8.3	38.02
5.0	2.82	8.4	33.62
5.1	1.94	8.5	28.34
5.2	1.32	8.6	21.47
5.3	0.97	8.7	17.25
5.4	0.616	8.8	13.64
5.5	0.264	8.9	10.21
5.6	0.176	9.0	7.83
5.7	0.088	9.1	6.16
5.8	0.000	9.2	4.84
5.9	0.088	9.3	3.70
6.0	0.176	9.4	2.90
6.1	0.264	9.5	2.29
6.2	0.440	9.6	1.76
6.3	0.616	9.7	1.32
6.4	1.06	9.8	1.06
6.5	1.41	9.9	0.792
6.6	2.02	10.0	0.616

Note: See Figure 27.13 for graphical presentation.

27.11 DRIFT VELOCITY

See Table 27.14.

27.12 CHARACTERISTIC ENERGY

See Table 27.15.

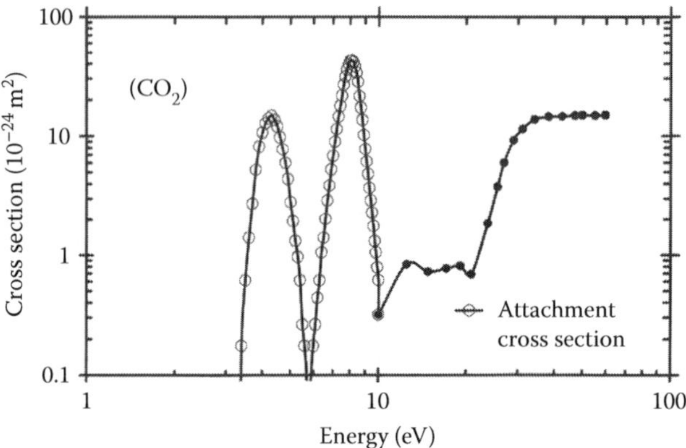

FIGURE 27.13 Attachment cross sections for CO$_2$. (—○—) Tabulated results; (—●—) digitized results. (Adapted from Rapp and Briglia, 1965.)

TABLE 27.14
Drift Velocity of Electrons

E/N (Td)	*W* (10^3 m/s)	*E/N* (Td)	*W* (10^3 m/s)
0.3	0.536	80	124.4
0.4	0.714	90	133.8
0.5	0.89	100	142.6
0.6	1.068	150	178.7
0.7	1.246	200	219.3
0.8	1.424	250	266.0
1.00	1.781	300	294.1
2.00	3.56	350	305.7
3.00	5.37	400	334.5
4.00	7.2	450	373.4
5.00	9.12	500	410.3
6.00	11.12	600	450.1
7.00	13.24	700	490.4
8.00	15.51	800	546.2
10.00	20.6	900	572.0
12.00	26.8	1000	663.0
14.00	34.6	2000	1411
17.00	48.7	3000	1793
20.00	63.2	4000	2125
60	115.9	5000	2425
70	124.5		

Source: Adapted from Raju, G. G., *Gaseous Electronics: Theory and Practice*, Taylor & Francis, London, 2005.

Note: See Figure 27.14 for graphical presentation. Figure 27.15 shows the influence of gas pressure on drift velocity.

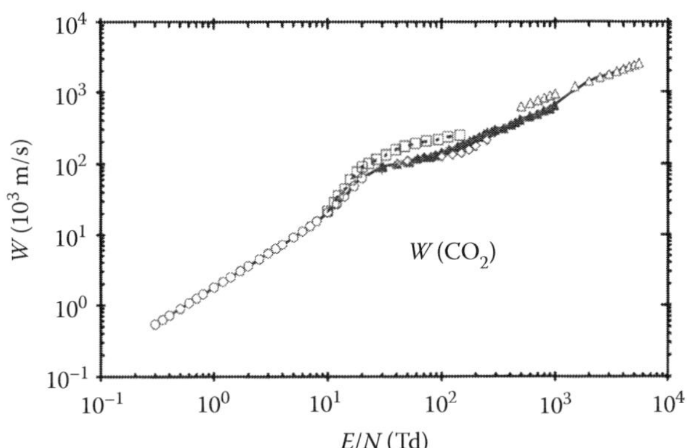

FIGURE 27.14 Drift velocity of electrons for CO_2. (—●—) Raju (2005); (▲) Hasegawa et al. (1996); (◊) Roznerski and Leja (1984); (+) Kucukarpaci and Lucas (1979); (∗) Saelee et al. (1977); (○) Huxley and Crompton (1974); (×) Elford (1966); (△) Schlumbohm (1965); (–□–) Hake and Phelps (1965).

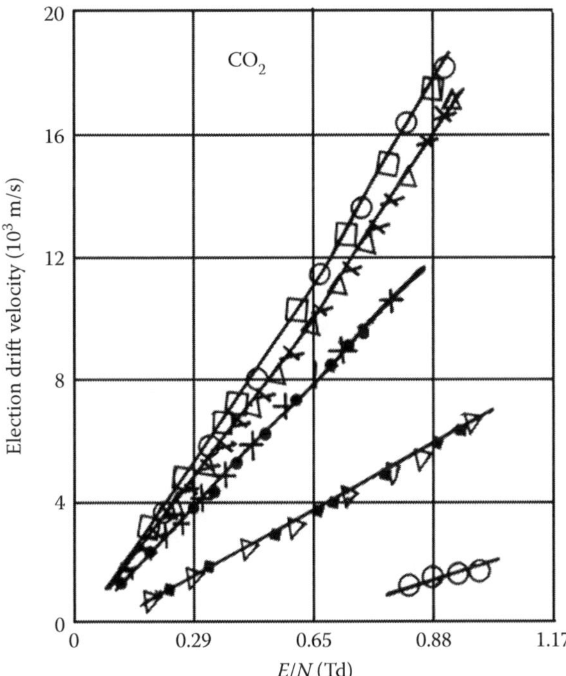

FIGURE 27.15 Drift velocity of electrons in CO_2 as a function of *E*/*N* at various gas number densities. Gas number densities and pressure are (○, □) $N = 5.9 \times 10^{25}$ m^{-3}, $P = 237.9$ kPa; (×, △) $N = 19.5 \times 10^{25}$ m^{-3}, $P = 757$ kPa; (●, +) $N = 26.4 \times 10^{25}$ m^{-3}, $P = 1.01$ MPa; (■, ∇) $N = 35.7 \times 10^{25}$ m^{-3}, $P = 1.33$ MPa; (○) $N = 47.0 \times 10^{25}$ m^{-3}, $P = 1.73$ MPa. The drift velocity decreases with increasing gas pressure at constant *E*/*N*. (Adapted from Allen, N. L. and B. A. Prew, *J. Phys B: At. Mol. Phys.*, 3, 1113, 1970.)

27.13 LONGITUDINAL DIFFUSION COEFFICIENT

Figure 27.17 shows the longitudinal diffusion coefficient (D_L) as a ratio of mobility (D_L/μ) and density-normalized product (ND_L).

TABLE 27.15
Recommended Characteristic Energy (D_r/μ)

E/*N* (Td)	D_r/μ (V)	*E*/*N* (Td)	D_r/μ (V)
0.07	0.024	30	1.11
0.10	0.024	40	1.50
0.3	0.024	50	1.75
0.50	0.025	70	2.23
0.80	0.028	100	2.82
1.0	0.029	200	5.34
2.0	0.032	300	4.84
3.0	0.027	400	5.58
4.0	0.026	500	6.11
5.0	0.030	700	6.88
7.0	0.039	1000	8.11
10.0	0.053	1500	10.50
20	0.47		

Source: Adapted from Raju, G. G., *Gaseous Electronics: Theory and Practice*, Taylor & Francis, London, 2005.
Note: See Figure 27.16 for graphical presentation.

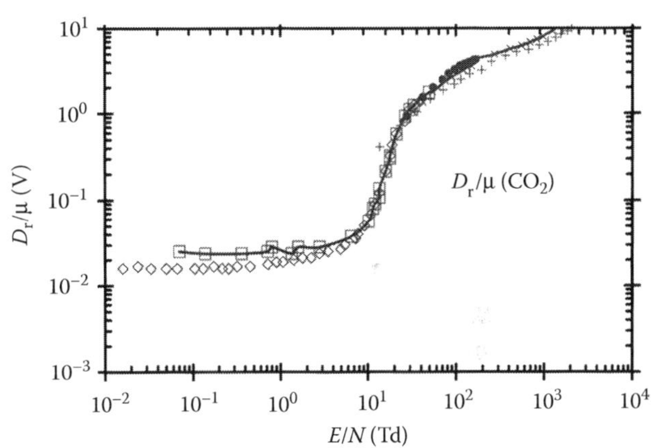

FIGURE 27.16 Characteristic energy for CO_2. (——) Raju (2005); (+) Kucukarpaci and Lucas (1979); (×) Lakshminarasimha et al. (1974); (●) Rees (1964); (□) Warren and Parker (1962), 300 K; (◆) Warren and Parker (1962), 195 K.

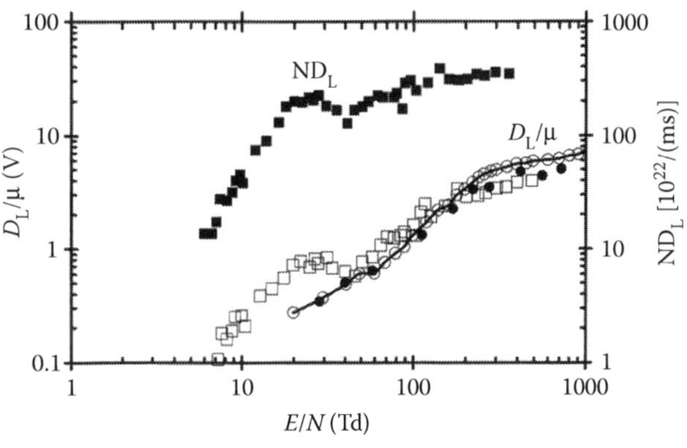

FIGURE 27.17 Longitudinal diffusion coefficient for CO_2. D_L/μ: (□) Hernández-Ávila (2002); (—○—) Hasegawa et al. (1996); (●) Saelee et al. (1977). ND_L: (■) Hernández-Ávila (2002).

CO$_2$

CO$_2$

TABLE 27.16
Density-Reduced Effective Ionization Coefficient

E/N (Td)	(α–η)/*N* (10^{-20} m^2)	*E/N* (Td)	(α–η)/*N* (10^{-20} m^2)
80	0.0022	900	1.64
90	0.0040	1000	1.82
100	0.0075	1100	1.98
125	0.020	1200	2.12
150	0.036	1300	2.28
175	0.059	1400	2.46
200	0.094	1500	2.66
250	0.170	1600	2.85
300	0.242	1700	3.00
350	0.409	1800	3.14
400	0.525	1900	3.27
450	0.626	2000	3.39
500	0.744	2250	3.73
550	0.864	2500	4.06
600	0.974	3000	4.63
700	1.19	3500	4.88
800	1.43	3600	4.88

Source: Adapted from Bhalla, M. S. and J. D. Craggs, *Proc. Phys. Soc. London*, 76, 369, 1960.
Note: See Figure 27.18 for graphical presentation.

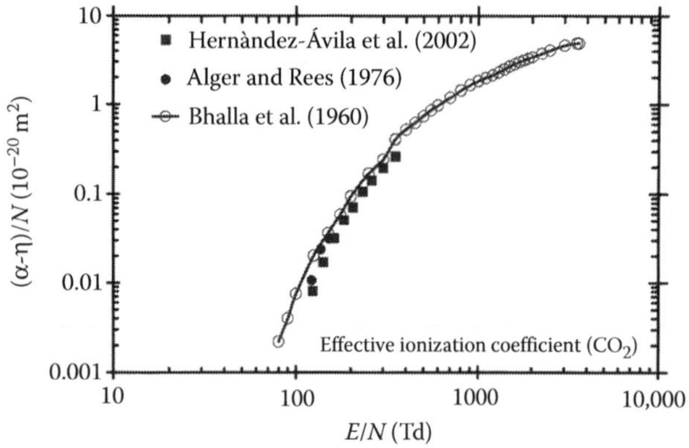

FIGURE 27.18 Density-reduced effective ionization coefficient for CO$_2$. (■) Hernández-Ávila et al. (2002); (●) Alger and Rees (1976); (—○—) Bhalla and Craggs (1960).

27.14 EFFECTIVE IONIZATION COEFFICIENT

See Table 27.16.

27.15 ATTACHMENT COEFFICIENT

Density-reduced attachment coefficients are shown in Figure 27.19. Table 27.17 shows the gas number density dependence of attachment coefficients for three-body process.

27.16 GAS CONSTANTS

The gas constants evaluated from Table 27.16 according to the expression

$$\frac{\alpha}{N} = F\exp\left(-\frac{GN}{E}\right) \tag{27.6}$$

are $F = 2.97 \times 10^{-20}$ m^2 and $G = 616$ Td for the range $80 \le E/N \le 1700$ Td. Schlumbohm (1965) obtains $F = 1.48 \times 10^{-20}$ m^2 and $G = 568$ Td^{-1} for the range $225 \le E/N \le 450$ Td.

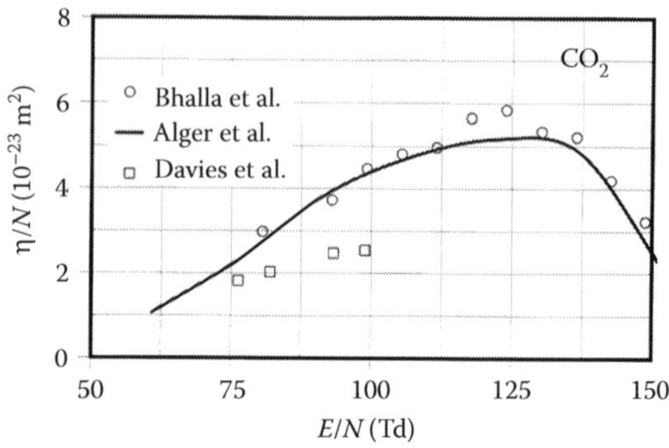

FIGURE 27.19 Density-reduced attachment coefficients for CO$_2$. (□) Davies (1978); (—) Alger and Rees (1976); (○) Bhalla and Craggs (1960).

TABLE 27.17
Density-Reduced Swarm Coefficient for CO$_2$

E/N (Td)	*N* (10^{23}/m^3)	(α–η)/*N* (10^{-22} m^2)	η/*N* (10^{-22} m^2)	α/*N* (10^{-22} m^2)
84.8	192	0.025	0.15	0.18
96.7	4.52	0.32	0.12	0.44
96.7	7.22	0.24	0.48	0.72
96.7	14.0	0.26	0.33	0.59
96.7	35.9	0.25	0.36	0.62
96.7	63.1	0.29	0.45	0.75
96.7	101	0.30	0.18	0.48
96.7	192	0.27	—	0.27
96.7	220	0.31	—	0.31
109	4.52	0.83	0.09	0.92
109	7.22	0.78	0.39	1.17
109	14.0	0.67	0.52	1.20
109	35.9	0.68	0.27	0.95
109	63.1	0.78	0.12	0.90
109	78.7	0.78	—	0.78
109	101	0.72	—	0.72
109	220	0.80	—	0.80
121	4.52	1.2	0.39	1.6
121	7.22	1.4	0.42	1.8
121	14.0	1.3	0.06	1.4
121	35.9	1.3	0.17	1.5
121	63.1	1.4	—	1.4
121	78.7	1.5	0.04	1.5
121	101	1.3	—	1.3
136	25.9	2.3	0.23	2.5
152	25.9	3.5	—	3.5
167	25.9	5.0	—	5.0

Source: Adapted from Conti, V. J. and A. W. Williams, *J. Phys. D: Appl. Phys.*, 8, 2198, 1975.

REFERENCES

Ajello, J. M., *J. Chem. Phys.*, 55, 3169, 1971.

Alger, S. R. and J. A. Rees, *J. Phys. D: Appl. Phys.*, 9, 2359, 1976.

Allen, N. L. and B. A. Prew, *J. Phys B: At. Mol. Phys.*, 3, 1113, 1970.

Andrick, D. and F. H. Read, *J. Phys. B: At. Mol. Phys.*, 4, 389, 1971. Also *J. Phys. B: At. Mol. Phys.*, 4, 911, 1971.

Bhalla, M. S. and J. D. Craggs, *Proc. Phys. Soc. London*, 76, 369, 1960.

Boness, M. J. W. and G. J. Schulz, *Phys. Rev. A*, 9, 1969, 1974.

Buckman, S. J., M. T. Elford, and D. S. Newman, *J. Phys. B: At. Mol. Phys.*, 20, 5175, 1987.

Čadež, I., M. Tronc, and R. I. Hall, *J. Phys. B: At. Mol. Opt. Phys.*, 7, L132, 1974.

Conti, V. J. and A. W. Williams, *J. Phys. D: Appl. Phys.*, 8, 2198, 1975.

Crowe, A. and J. W. McConkey, *J. Phys. B: At. Mol. Phys.*, 7, 349, 1974.

Davies, D. K., *J. Appl. Phys.*, 49, 127, 1978.

Elford, M. T., *Aust. J. Phys.*, 19, 629, 1966.

Ferch, J., C. Masche, and W. Raith, *J. Phys. B: At. Mol. Phys.*, 14, L97, 1981.

Ferch, J., C. Masche, W. Raith, and L. Wiemann, *Phys. Rev. A*, 40, 5407, 1989.

Freund, R. S., R. C. Wetzel, and R. J. Shul, *Phys. Rev. A*, 41, 5861, 1990.

Garcia, G. and F. Manero, *Phys. Rev. A*, 53, 250, 1996.

Gibson, J. C., M. A. Green, K. W. Trantham, S. J. Buckman, P. J. O. Teubner, and M. J. Brunger, *J. Phys B: At. Mol. Opt. Phys.*, 32, 213, 1999.

Haddad, G. N. and M. T. Elford, *J. Phys. B: At. Mol. Opt. Phys.*, 12, L743, 1979.

Hake, R. D., Jr. and A. V. Phelps, *Phys. Rev.*, 158, 70, 1965.

Hasegawa, H., H. Date, M. Shimozuma, K. Yoshida, and H. Tagashira, *J. Phys. D: Appl. Phys.*, 29, 2664, 1996.

Hernández-Ávila, J. L., E. Basurto, and J. de Urquiho, *J. Phys. D: Appl. Phys.*, 35, 2264, 2002.

Hoffman, K. R., M. S. Dababneh, Y.-F. Hseih, W. E. Kauppila, V. Pol, J. H. Smart, and T. S. Stein, *Phys. Rev.*, 25, 1393, 1982.

Huxley, L. G. H. and R. W. Crompton, *The Diffusion and Drift of Electrons in Gases*, John Wiley and Sons, New York, NY, 1974.

Hwang, W., Y.-K. Kim, and M. E. Rudd, *J. Chem. Phys.*, 104, 2956, 1996.

Iga, I., J. C. Nogueira, and L. Mu-Tao, *J. Phys. B: At. Mol. Phys.*, 17, L185, 1984.

Jain, A. and K. L. Baluja, *Phys. Rev. A*, 45, 202, 1992.

Kimura, M., O. Sueoka, A. Hamada, M. Takekawa, Y. Itikawa, H. Tanaka, and L. Boesten, *J. Chem. Phys.*, 107, 6616, 1997.

Kitajima, M., S. Watanabe, H. Tanaka, M. Takekawa, M. Kimura, and Y. Itikawa, *J. Phys. B: At. Mol. Opt. Phys.*, 34, 1929, 2001.

Kochem, K.-H., W. Sohn, N. Hebel, K. Jung, and H. Ehrhardt, *J. Phys. B: At. Mol. Phys.*,18, 4455, 1985.

Kucukarpaci, H. N. and J. Lucas, *J. Phys D: Appl. Phys.*, 12, 2123, 1979.

Lakshminarasimha, C. S., J. Lucas, and N. Kontoleon, *J. Phys. D: Appl. Phys.*, 7, 2545, 1974.

Leclair, L. R. and J. W. McConkey, *J. Phys. B: At. Mol. Opt. Phys.*, 27, 4039, 1994.

Lee, C., C. Winstead, and V. McKoy, *J. Chem. Phys.*, 111, 5056, 1999.

Lowke, J. J., A. V. Phelps, and B. W. Irwin, *J. Appl. Phys.*, 44, 4664, 1973.

Morrison, M. A. and N. F. Lane, *Phys. Rev. A*, 16, 975, 1977.

Nakamura, Y., *Aust. J. Phys.*, 48, 357, 1995.

Orient, O. J. and S. K. Srivastava, *J. Phys. B: At. Mol. Phys.*, 20, 3923, 1987.

Raju, G. G., *Gaseous Electronics: Theory and Practice*, Taylor & Francis, London, 2005.

Rapp, D. and D. D. Briglia, *J. Chem. Phys.* 43, 1480, 1965.

Rapp, D. and P. Englander-Golden, *J. Chem. Phys.*, 43, 1464, 1965.

Rees, J. A., *Aust. J. Phys.*, 17, 462, 1964.

Register, D. F., H. Nishimura, and S. Trajamar, *J. Phys. B: At. Mol. Opt. Phys.*, 13, 1651, 1980.

Roznerski, W. and K. Leja, *J. Phys. D: Appl. Phys.*, 17, 279, 1984.

Saelee, H. T., J. Lucas, and J. W. Limbeeck, *Solid State Electron Devices*, 1, 111, 1977.

Sanche, L. and G. J. Schulz, *J. Chem. Phys.*, 58, 479, 1973.

Schlumbohm, H., *Z. Phys.*, 182, 317, 1965.

Schlumbohm, H., *Z. Phys.*, 184, 492, 1965.

Shyn, T. W., W. E. Sharp, and G. R. Carignan, *Phys. Rev. A*, 17, 1855, 1978.

Spence, D. and G. J. Schulz, *J. Chem. Phys.*, 60, 216, 1974.

Straub, H. C., B. G. Lindsay, K. A. Smith, and R. F. Stebbings, *J. Chem. Phys.*, 105, 4015, 1996.

Strickland, D. J. and A. E. S. Green, *J. Geophys. Res.*, 74, 6415, 1969.

Sueoka, O. and A. Hamada, *J. Phys. Soc. Jpn.*, 62, 2669, 1993. The cross sections are digitized from Kimura et al., *J. Chem. Phys.*, 107, 6616, 1997.

Szmytkowski, C., A. Zecca, G. Karwasz, S. Oss, K. Maciąg, B. Marinković, R. S. Brusa, and R. Grisenti, *J. Phys. B: At. Mol. Phys.*, 20, 5817, 1987.

Szmytkowski, C. and M. Zubek, *Chem. Phys. Lett.*, 57, 105, 1978.

Takekawa, M. and Y. Itikawa, *J. Phys. B: At. Mol. Opt. Phys.*, 29, 4227, 1996. Elastic scattering cross sections are calculated theoretically in the 3–60 eV range.

Takekawa, M. and Y. Itikawa, *J. Phys. B: At. Mol. Opt. Phys.*, 31, 3245, 1998.

Takekawa, M. and Y. Itikawa, *J. Phys. B: At. Mol. Opt. Phys.*, 32, 4209, 1999.

Tanaka, H., T. Ishikawa, T. Masai, T. Sagara, L. Boesten, M. Takekawa, Y. Itikawa, and M. Kimura, *Phys. Rev. A*, 57, 1798, 1998.

Warren, R. W. and J. H. Parker, *Phys. Rev.*, 128, 2661, 1962.

28

CARBON DISULFIDE

CS_2

CONTENTS

Carbon disulfide (CS_2) is an electron-attaching molecule that has 38 electrons. The electronic polarizability is 9.7×10^{-40} F m^2 and the ionization potential is 10.068 eV. The molecule has three modes of vibrational excitation (Shimanouchi, 1972) as shown in Table 28.1.

28.1 SELECTED REFERENCES FOR DATA

See Table 28.2.

TABLE 28.2
Selected References for Data

Quantity	Range (eV)	Reference
Ionization cross section	11–1000	Lindsay et al. (2003)
Attachment cross section	0–50	Krishnakumar and Nagesha (1992)
Attachment processes	0–10	Dressler et al. (1987)
Total scattering cross section	0.4–80	Szmytkowski (1987)
Attachment cross section	0–5	Ziesel et al. (1975)
Attachment processes	0–10	Dorman (1966)

28.2 TOTAL SCATTERING CROSS SECTION

Table 28.2 and Figure 28.1 show the total scattering cross section for CS_2 (Szmytkowski, 1987). The following points are noted

1. A trend toward increase of scattering cross section toward lower electron energies, <0.8 eV. The reason is not specified.
2. A weak resonance at 1.8 eV, showing up in change of slope of the cross section–energy plot.
3. A broad peak in the range 5–10 eV common to many other gases.

TABLE 28.1
Vibrational Modes and Energies for CS_2

Designation	Mode	Energy (meV)
v_1	Symmetrical stretch	84.9
v_2	Bend	49.2
v_3	Antisymmetrical stretch	190.3

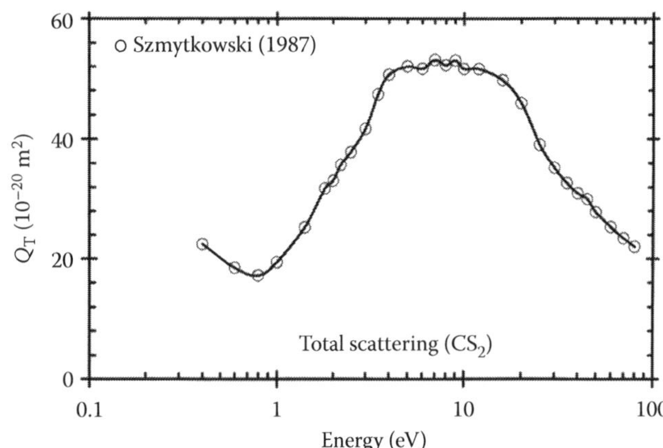

FIGURE 28.1 Total scattering cross section for CS_2. (Adapted from Szmytkowski, Cz., *J. Phys. B: At. Mol. Phys.*, 20, 6613, 1987.)

4. A monotonic decrease of the cross section above ionization threshold, which is also a common feature of several gases.

A sharp peak at 1.8 eV obtained by theory by Lynch et al. (1979) is not observed in the experiment. See Table 28.3.

TABLE28.3
Total Scattering Cross Section for CS_2

Energy (eV)	Q_T (10^{-20} m^2)	Energy (eV)	Q_T (10^{-20} m^2)
0.4	22.4	8.0	52.2
0.6	18.5	9.0	53.0
0.8	17.2	10.0	51.6
1.0	19.4	12.0	51.6
1.4	25.2	16.0	49.7
1.8	31.8	20	45.9
2.0	33.1	25	39.0
2.2	35.7	30	35.2
2.5	37.8	35	32.7
3.0	41.6	40	31.0
3.5	47.3	45	30.0
4.0	50.6	50	27.8
5.0	52.0	60	25.3
6.0	51.6	70	23.4
7.0	53.1	80	22.0

Source: Adapted from Szmytkowski, Cz., *J. Phys. B: At. Mol. Phys.*, 20, 6613, 1987.

28.3 IONIZATION CROSS SECTION

See Table 28.4.

28.4 ATTACHMENT CROSS SECTION

Dissociative attachment yields ions of species S^-, CS^-, and C^-. The first of these, S^-, is produced according to (Doman, 1966)

$$CS_2 + e \rightarrow CS + S^- \quad (28.1)$$

There are two pathways for production of C^- according to (Krishnakumar and Nagesha, 1992)

$$CS_2 + e \rightarrow C^- + S_2 \quad (28.2)$$

$$CS_2 + e \rightarrow C^- + 2S \quad (28.3)$$

TABLE 28.4
Total and Partial Cross Sections for CS_2

Energy (eV)	CS_2^+ (10^{-20} m^2)	S_2^+ (10^{-20} m^2)	CS^+ (10^{-20} m^2)	S^+ (10^{-20} m^2)	C^+ (10^{-20} m^2)	CS_2^{2+} (10^{-20} m^2)	Total (10^{-20} m^2)
11	0.58						0.580
12	1.29						1.290
13	1.84						1.840
14	2.36						2.360
16	3.11						3.110
18	3.68		0.251	0.231			4.162
20	4.14		0.507	0.354			5.001
22.5	3.95		0.890	0.590			5.430
25	4.00	0.100	1.120	0.654			5.874
30	4.04	0.120	1.410	1.190	0.106		6.866
35	4.16	0.110	1.590	1.630	0.139	0.106	7.735
40	4.25	0.110	1.640	1.850	0.168	0.136	8.154
50	4.36	0.081	1.660	2.120	0.209	0.172	8.602
60	4.47	0.073	1.610	2.250	0.252	0.178	8.833
80	4.51	0.079	1.490	2.290	0.301	0.181	8.851
100	4.44	0.076	1.370	2.320	0.324	0.177	8.707
125	4.23	0.066	1.200	2.200	0.302	0.171	8.169
150	4.00	0.056	1.130	2.020	0.271	0.153	7.630
200	3.51	0.055	0.999	1.630	0.245	0.112	6.551
300	2.99	0.037	0.800	1.160	0.193	0.089	5.269
400	2.65	0.033	0.670	0.937	0.148	0.072	4.510
500	2.37	0.026	0.584	0.813	0.142	0.054	3.989
600	2.18	0.029	0.513	0.726	0.113	0.050	3.611
800	1.73	0.026	0.445	0.641	0.102	0.058	3.002
1000	1.5	0.028	0.413	0.615	0.095	0.046	2.697

Source: Adapted from Lindsay, B. G., R. Rejoub, and R. F. Stebbings, *J. Chem. Phys.*, 118, 5894, 2003.
Note: See Figures 28.2 and 28.3 for graphical presentation.

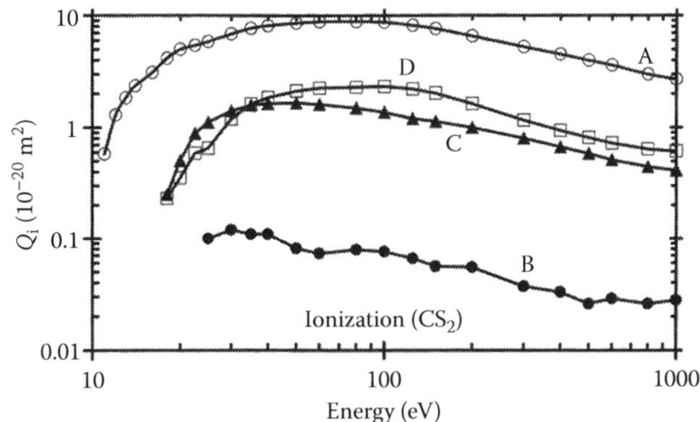

FIGURE 28.2 Total and partial ionization cross sections for CS_2. (A) Total; (B) CS_2^+; (C) S_2^+; (D) CS^+. (Adapted from Lindsay, B. G., R. Rejoub, and R. F. Stebbings, *J. Chem. Phys.*, 118, 5894, 2003.)

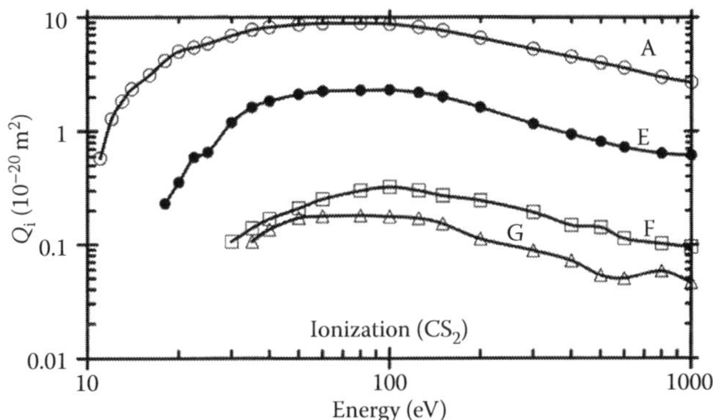

FIGURE 28.3 Total and partial ionization cross sections for CS_2. (A) Total, repeated for comparison; (E) S^+; (F) C^+; (G) CS_2^{2+}. (Adapted from Lindsay, B. G., R. Rejoub, and R. F. Stebbings, *J. Chem. Phys.*, 118, 5894, 2003.)

Ion species S_2^- is produced by dissociative attachment (Krishnakumar and Nagesha, 1992) and ion–molecule reaction according to

$$S^- + CS_2 \rightarrow S_2^- + CS \quad (28.4)$$

Dorman (1966) identified three fragment ions, S^- at 3.7, 6.4, and 8.1 eV, and S_2^- and CS^- peaks at 6.4 eV. Ziesel et al. (1975) obtained a peak of S^- cross section at 3.35 eV with a peak cross section of 3.7×10^{-23} m^2. Although CS_2^- ion was observed, it is attributed to the presence of dimers in the molecular beam (Krishnakumar and Nagesha, 1992).

Tables 28.5 through 28.8 show the attachment cross sections tabulated by Itikawa (2003) on the basis of measurements made by Krishnakumar and Nagesha (1992). Graphical presentation is given in Figures 28.4 and 28.5 drawn to the same scale for comparison.

28.5 ADDENDUM

28.5.1 Vibrational Excitation

Sohn et al. (1987) have measured the elastic scattering and vibrational excitation cross section as shown in Table 28.9.

TABLE 28.5
Attachment Cross Section for CS_2 for Ion Species S^-

Energy (eV)	Q_a (10^{-24} m^2)	Energy (eV)	Q_a (10^{-24} m^2)	Energy (eV)	Q_a (10^{-24} m^2)
1	0.20	9.25	0.70	25.0	13.70
1.25	1.20	9.50	0.50	25.5	14.20
1.5	1.40	9.75	0.50	26.0	14.70
1.75	2.30	10.0	0.30	26.5	15.00
2.00	3.30	10.5	0.10	27.0	15.10
2.25	4.00	11.0	0.10	27.5	14.90
2.50	7.90	11.5	0.20	28.0	14.70
2.75	14.20	12.0	0.30	28.5	14.20
3.00	25.60	12.5	0.50	29.0	14.00
3.25	33.00	13.0	0.60	29.5	13.70
3.50	35.10	13.5	0.80	30.0	13.50
3.75	27.40	14.0	1.00	30.5	13.30
4.00	22.30	14.5	1.50	31.0	13.00
4.25	12.10	15.0	1.20	31.5	12.80
4.50	6.50	15.5	2.80	32.0	12.60
4.75	4.20	16.0	3.40	32.5	12.30
5.00	6.00	16.5	3.70	33.0	12.40
5.25	8.40	17.0	4.10	33.5	12.10
5.50	12.30	17.5	4.80	34.0	11.70
5.75	20.20	18.0	5.10	34.5	11.60
6.00	22.10	18.5	5.60	35.0	11.50
6.25	23.00	19.0	5.90	35.5	11.40
6.50	16.50	19.5	6.40	36.0	11.30
6.75	12.60	20.0	6.70	36.5	11.30
7.00	6.00	20.5	7.20	37.0	11.30
7.25	4.90	21.0	7.90	37.5	11.20
7.50	4.80	21.5	8.60	38.0	11.20
7.75	4.90	22.0	9.30	38.5	11.20
8.00	4.50	22.5	10.10	39.0	11.20
8.25	3.40	23.0	10.70	39.5	11.20
8.50	2.30	23.5	11.40	40.0	11.20
8.75	1.40	24.0	12.20		
9.00	1.00	24.5	13.40		

Source: Adapted from Itikawa, Y., *Interactions of Photons and Electrons with Molecules*, Vol. 17c, Springer-Verlag, New York, NY, 2003, Chapter 5.

TABLE 28.6
Attachment Cross Section for CS_2 for Ion Species CS^-

Energy (eV)	Q_a (10^{-24} m^2)	Energy (eV)	Q_a (10^{-24} m^2)	Energy (eV)	Q_a (10^{-24} m^2)
2.50	0	3.75	0.31	5.00	1.30
2.75	0.02	4.00	0.39	5.25	2.69
3.00	0.02	4.25	0.45	5.50	4.03
3.25	0.04	4.50	0.49	5.75	5.74
3.50	0.18	4.75	0.81	6.00	6.50

continued

CS$_2$

TABLE 28.6 (continued)
Attachment Cross Section for CS_2 for Ion Species CS^-

Energy (eV)	Q_a (10^{-24} m^2)	Energy (eV)	Q_a (10^{-24} m^2)	Energy (eV)	Q_a (10^{-24} m^2)
6.25	6.70	15.5	0.47	28.5	1.22
6.50	4.92	16.0	0.61	29.0	1.20
6.75	4.23	16.5	0.79	29.5	1.16
7.00	2.12	17.0	1.00	30.0	1.12
7.25	0.88	17.5	1.14	30.5	1.08
7.50	0.39	18.0	1.30	31.0	1.04
7.75	0.22	18.5	1.38	31.5	1.00
8.00	0.14	19.0	1.49	32.0	0.98
8.25	0.12	19.5	1.53	32.5	0.92
8.50	0.12	20.0	1.57	33.0	0.90
8.75	0.10	20.5	1.61	33.5	0.88
9.00	0.08	21.0	1.63	34.0	0.81
9.25	0.08	21.5	1.65	34.5	0.79
9.50	0.06	22.0	1.63	35.0	0.77
9.75	0.06	22.5	1.61	35.5	0.75
10.0	0.05	23.0	1.59	36.0	0.69
10.5	0.06	23.5	1.57	36.5	0.69
11.0	0.06	24.0	1.55	37.0	0.67
11.5	0.07	24.5	1.53	37.5	0.67
12.0	0.04	25.0	1.49	38.0	0.65
12.5	0.03	25.5	1.44	38.5	0.63
13.0	0.02	26.0	1.42	39.0	0.61
13.5	0.02	26.5	1.38	39.5	0.59
14.0	0.04	27.0	1.34	40.0	0.59
14.5	0.16	27.5	1.30		
15.0	0.28	28.0	1.26		

Source: Adapted from Itikawa, Y., *Interactions of Photons and Electrons with Molecules*, Vol. 17c, Springer-Verlag, New York, NY, 2003, Chapter 5.

TABLE 28.7
Attachment Cross Section for CS_2 for Ion Species C^-

Energy (eV)	Q_a (10^{-24} m^2)	Energy (eV)	Q_a (10^{-24} m^2)	Energy (eV)	Q_a (10^{-24} m^2)
3.00	0.01	9.00	0.05	15.00	0.07
3.50	0.02	9.50	0.04	15.50	0.08
4.00	0.05	10.00	0.04	16.00	0.09
4.50	0.07	10.50	0.04	16.50	0.10
5.00	0.04	11.00	0.05	17.00	0.12
5.50	0.09	11.50	0.05	17.50	0.13
6.00	0.14	12.00	0.06	18.00	0.14
6.50	0.22	12.50	0.08	18.50	0.15
7.00	0.15	13.00	0.09	19.00	0.16
7.50	0.17	13.50	0.08	19.50	0.17
8.00	0.14	14.00	0.06	20.00	0.18
8.50	0.09	14.50	0.06	20.50	0.19

TABLE 28.7 (continued)
Attachment Cross Section for CS_2 for Ion Species C^-

Energy (eV)	Q_a (10^{-24} m^2)	Energy (eV)	Q_a (10^{-24} m^2)	Energy (eV)	Q_a (10^{-24} m^2)
21.00	0.21	31.00	1.56	41.00	1.50
21.50	0.24	31.50	1.54	41.50	1.53
22.00	0.29	32.00	1.52	42.00	1.55
22.50	0.33	32.50	1.51	42.50	1.58
23.00	0.36	33.00	1.49	43.00	1.62
23.50	0.42	33.50	1.48	43.50	1.64
24.00	0.46	34.00	1.47	44.00	1.67
24.50	0.58	34.50	1.45	44.50	1.70
25.00	0.68	35.00	1.44	45.00	1.72
25.50	0.82	35.50	1.43	45.50	1.74
26.00	0.91	36.00	1.43	46.00	1.76
26.50	1.02	36.50	1.42	46.50	1.77
27.00	1.14	37.00	1.42	47.00	1.79
27.50	1.23	37.50	1.42	47.50	1.79
28.00	1.29	38.00	1.42	48.00	1.80
28.50	1.37	38.50	1.45	48.50	1.80
29.00	1.41	39.00	1.45	49.00	1.81
29.50	1.44	39.50	1.46	49.50	1.82
30.00	1.53	40.00	1.47	50.00	1.82
30.50	1.57	40.50	1.49		

Source: Adapted from Itikawa, Y., *Interactions of Photons and Electrons with Molecules*, Vol. 17c, Springer-Verlag, New York, NY, 2003, Chapter 5.

TABLE 28.8
Attachment Cross Section for CS_2 for Ion Species S_2^-

Energy (eV)	Q_a (10^{-24} m^2)	Energy (eV)	Q_a (10^{-24} m^2)	Energy (eV)	Q_a (10^{-24} m^2)
3.00	0.004	5.00	0.26	7.00	0.97
3.25	0.01	5.25	0.43	7.25	0.36
3.50	0.03	5.50	1.16	7.50	0.26
3.75	0.04	5.75	1.61	7.75	0.23
4.00	0.08	6.00	2.10	8.00	0.16
4.25	0.10	6.25	2.48	8.25	0.08
4.50	0.12	6.50	2.30	8.50	0.04
4.75	0.14	6.75	2.20	8.75	0.02

Source: Adapted from Itikawa, Y., *Interactions of Photons and Electrons with Molecules*, Vol. 17c, Springer-Verlag, New York, NY, 2003, Chapter 5.

28.5.2 Attachment Peaks

Attachment peak energy and cross sections measured by Krishnakumar and Nagesha (1992) are summarized in Table 28.10.

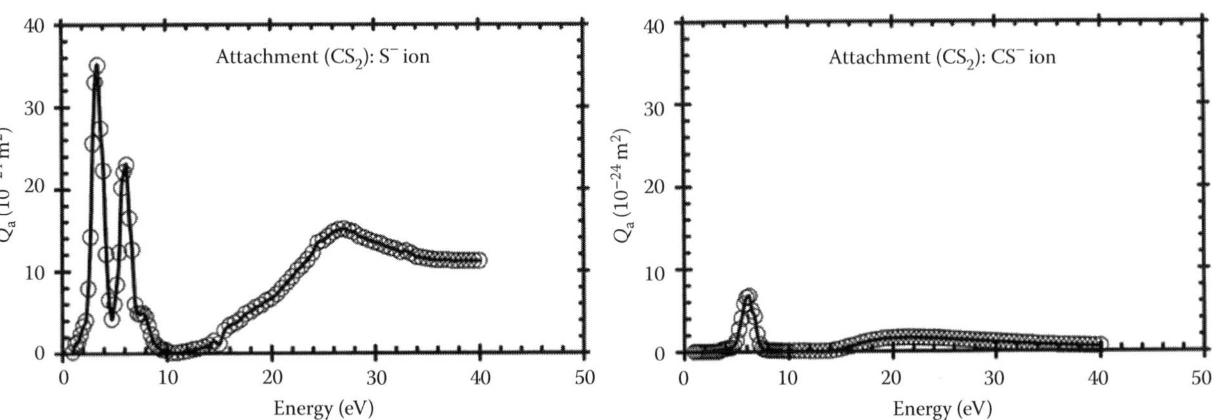

FIGURE 28.4 Attachment cross section for CS_2. Tabulated values are from Itikawa (2003) and measurements from Krishnakumar and Nagesh (1992). Left frame: S^- ion; right frame: CS^- ion. Both frames are drawn for the same scales for easy comparison.

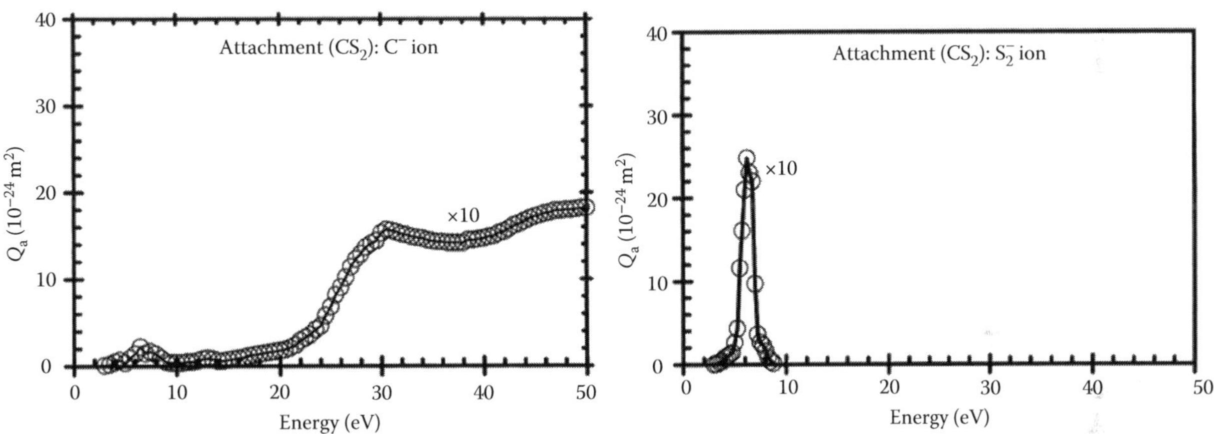

FIGURE 28.5 Attachment cross section for CS_2. Tabulated values are from Itikawa (2003) and measurements from Krishnakumar and Nagesh (1992). Left frame: C^- ion; right frame: S_2^- ion. Both frames are drawn for the same scales for easy comparison.

TABLE 28.9
Vibrational Excitation Cross Section in Units of 10^{-20} m²

Energy (eV)	Elastic	ν_2 (49 meV)	ν_1 (81 meV)	ν_3 (190 meV)	Total
0.3	18.7	1.06	2.02	3.26	25.04
0.5	12.8	0.40	0.79	2.62	16.61
0.8	9.5	0.14	0.61	1.67	11.92
1.0	12.3	0.11	0.55	1.73	14.68
1.2	14.9	0.09	0.49	1.54	17.02
1.5	17.3	0.11	0.37	1.28	19.06
1.8	22.4	0.12	0.24	1.07	23.83
2.2	25.4	0.26	0.27	1.01	29.95
3.0	31.6	0.41	0.23	0.84	33.08
3.5	35.5	0.54	0.29	0.83	37.16
5.0	35.1	0.23	0.35	0.70	36.38

Source: Adapted from Sohn W. et al. *J. Phys. B.: At. Mol. Phys.* 20, 3217,1987.

TABLE 28.10
Attachment Peaks and Corresponding Cross Sections

Ion	Energy	Cross Section (10^{-24} m²)
S^-	3.6	35
	6.2	23
	7.7	4.9
	9.2	1.1
CS^-	6.2	6.7
	21.5	1.65
S_2^-	6.2	2.5
C^-	6.5	0.22
	7.7	0.19
	12.8	0.09

REFERENCES

Dorman, F. H., *J. Chem. Phys.,* 44, 3856, 1966.

Dressler, R., Allan, M., and Tronc, M., *J. Phys. B: At. Mol. Phys.*, 20, 393, 1987.

Itikawa, Y., Electron attachment, in *Interactions of Photons and Electrons with Molecules*, Vol. 17c, Springer-Verlag, New York, NY, 2003, Chapter 5.

Krishnakumar, E. and Nagesha, K., *J. Phys. B: At. Mol. Opt. Phys.,* 25, 1645, 1992.

Lindsay, B. G., Rejoub, R., and Stebbings, R. F., *J. Chem. Phys.*, 118, 5894, 2003.

Lynch, M. G., Dill, D., Siegel, J., and Dehmer, J. L., *J. Chem. Phys.*, 71, 4249, 1979.

Shimanouchi, T., *Tables of Molecular Vibrational Frequencies Consolidated Volume I*, NSRDS-NBS 39, US Department of Commerce, Washington (DC), 1972.

Sohn W, K.-H. Kochem, K. M. Scheuerlein, K. Jung, and H. Ehrgardt, *J. Phys. B: At. Mol. Phys.* 20, 3217,1987.

Szmytkowski, Cz., *J. Phys. B: At. Mol. Phys.,* 20, 6613, 1987.

Ziesel, J. P., Schulz, G. J., and Milhaud, J., *J. Chem. Phys.*, 62, 1936, 1975.

29

CARBON OXYSULFIDE

CONTENTS

Carbon oxysulfide (COS) (synonym carbonyl sulfide) is an inorganic polar molecule that has 30 electrons. The electronic polarizability is 6.35×10^{-40} F m^2, the dipole moment is 0.715 D, and the ionization potential is 11.18 eV. The vibrational excitation energies and modes are shown in Table 29.1 (Shimanouchi, 1972).

29.1 SELECTED REFERENCES FOR DATA

See Table 29.2.

29.2 TOTAL SCATTERING CROSS SECTION

Table 29.3 and Figure 29.1 show the total scattering cross section for COS. The data are composed of two contributions: Karwasz et al. (2003) in the energy range ($0.2 \le \varepsilon \le 1000$ eV) and Zecca et al. (1995) in the energy range ($1000 \le \varepsilon \le 4000$ eV). The highlights are (Szmytkowski et al., 1989):

1. As the energy decreases toward zero, there is a trend for increasing total cross section, possibly due to the dipole moment of the target particle.
2. A shape resonance at 1.15 eV with a maximum cross section of 53×10^{-20} m^2.
3. A weak enhancement of the total scattering cross section at 4.1 eV.
4. A broad maximum of 35×10^{-20} m^2 at 11 eV; this feature is common for many target particles.
5. A monotonic decrease for energy >15 eV, which is also a common feature for many target particles.
6. The total cross sections are, in general, noticeably higher than N_2, CO_2, N_2O, and SF_6, probably due to the dipole moment of the COS molecule.

TABLE 29.1
Vibrational Energies and Modes for COS

Designation	Type of Motion	Energy (meV)
ν_1	CO stretch	255.7
ν_2	Bend	64.5
ν_3	CS stretch	106.5

Source: Adapted from Shimanouchi, T., *Tables of Molecular Vibrational Frequencies Consolidated Volume I*, NSRDS-NBS, 39, US Department of Commerce, 1972.

29.3 ELASTIC SCATTERING CROSS SECTION

The integral elastic scattering cross sections and vibrational excitation cross sections measured by Sohn et al. (1987) are shown in Table 29.4.

TABLE 29.2
Selected References for Data on COS

Quantity	Range: eV, (Td)	Reference
Ionization cross section	11.5–5000	Kim et al. (2008)
Total cross section	90–4000	**Zecca et al. (1995)**
Elastic scattering cross section	0.4–5.0	**Sohn et al. (1987)**
Total cross section	40–100	**Szmytkowski et al. (1989)**
Total cross section	0.4–40	**Szmytkowski et al. (1984)**
Attachment cross section	0.8–2.4	**Ziesel et al. (1975)**

Note: Bold font indicates experimental study.

TABLE 29.3
Total Scattering Cross Sections for COS

Energy (eV)	Q_T (10^{-20} m^2)	Energy (eV)	Q_T (10^{-20} m^2)
0.20	51.6	60	22.3
0.25	48.0	70	21.1
0.30	44.0	80	20.2
0.35	39.9	90	19.2
0.40	36.0	100	18.4
0.45	32.7	120	16.9
0.50	29.8	150	15.1
0.60	25.8	170	14.2
0.70	25.7	200	13.0
0.80	29.0	250	11.5
0.90	34.3	300	10.3
1.0	42.5	350	9.39
1.2	51.3	400	8.63
1.5	31.8	450	7.98
1.7	24.4	500	7.43
2.0	20.3	600	6.58
2.5	19.4	700	6.08
3.0	22.5	800	5.50
3.5	26.3	900	5.02
4.0	27.5	1000	4.63
4.5	27.5	1025	4.57
5.0	27.5	1150	4.14
6.0	28.2	1300	3.68
7.0	29.4	1450	3.27
8.0	30.6	1600	3.05
9.0	31.7	1770	2.80
10	32.3	1940	2.60
12	32.8	2000	2.44
15	32.2	2150	2.35
17	31.2	2300	2.21
20	30.0	2500	1.97
25	28.5	2750	1.87
30	27.4	3000	1.73
35	26.5	3250	1.55
40	25.7	3500	1.47
45	24.8	4000	1.29
50	23.9		

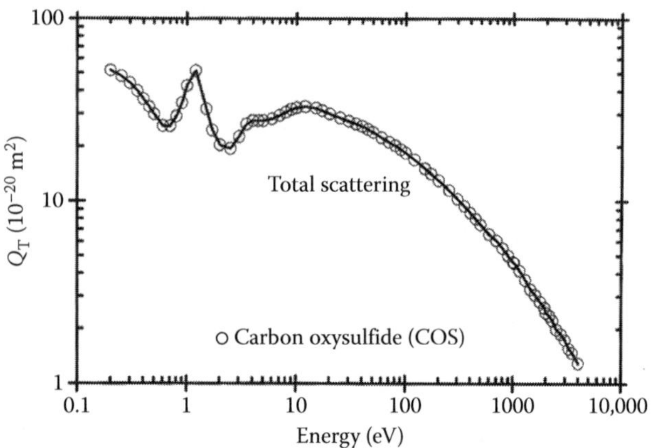

FIGURE 29.1 Total scattering cross sections for COS. Data are from Karwasz et al. (2003) in the energy range ($0.2 \le \varepsilon \le 1000$ eV) and Zecca et al. (1995) in the energy range ($1000 \le \varepsilon \le 4000$ eV).

29.4 IONIZATION CROSS SECTION

See Table 29.5.

29.5 ATTACHMENT CROSS SECTION

Dissociative attachment yields S^- ions, with an onset energy of 1.07 eV (Ziesel et al. 1975). The attachment cross section rises steeply, reaching a peak at 1.35 eV and peak attachment cross section of 2.9×10^{-21} m^2. The attachment cross sections are shown in Table 29.6 and Figure 29.3.

TABLE 29.4
Integral Elastic Scattering Cross Sections and Vibrational Excitation Cross Sections

Energy (eV)	Q_{el} (10^{-20} m^2)	Q_v (10^{-20} m^2)
0.4	57.2	—
0.6	39.4	6.65
1.15	36.1	15.4
1.7	18.9	—
2	16.9	—
2.5	15.3	1.7
3	19.6	1.3
4	26.3	1.8
5	25.8	—

Source: Adapted from Sohn, W. et al. *J. Phys. B: At. Mol. Phys.*, 20, 3217, 1987.

TABLE 29.5
Total Ionization Cross Section for COS

	Q_i (10^{-20} m^2)	
Energy (eV)	Kim et al. (2008)	S. K. Srivastava (1997, unpublished)[a]
11.17	0.000	
11.5	0.050	
12.0	0.128	
12.5	0.208	
13.0	0.289	
13.5	0.370	
14.0	0.451	
14.5	0.530	
15.0	0.607	0.57
15.5	0.682	
16.0	0.756	
16.5	0.827	
17.0	0.896	
17.5	0.969	
18.0	1.046	
18.5	1.121	
19.0	1.216	

TABLE 29.5 (continued)
Total Ionization Cross Section for COS

	Q_i (10^{-20} m^2)	
Energy (eV)	**Kim et al. (2008)**	**S. K. Srivastava (1997, unpublished)[a]**
19.5	1.309	
20.0	1.401	1.78
20.5	1.491	
21.0	1.579	
21.5	1.665	
22.0	1.755	
22.5	1.843	
23.0	1.930	
23.5	2.014	
24	2.096	
25		2.46
26	2.404	
28	2.680	
30	2.929	3.00
32	3.161	
34	3.370	
35		3.40
36	3.556	
38	3.723	
40	3.873	3.76
45	4.187	
50	4.430	4.38
55	4.614	
60	4.752	4.68
65	4.853	
70	4.925	4.91
75	4.975	
80	5.005	5.07
85	5.021	
90	5.025	5.19
95	5.019	
100	5.006	5.29
105	4.986	
110	4.961	
115	4.931	
120	4.898	5.38
125	4.863	
130	4.826	
135	4.786	
140	4.746	5.40
145	4.704	
150	4.662	
160	4.577	5.32
170	4.491	
180	4.405	5.14
190	4.321	
200	4.238	4.91
210	4.158	
220	4.080	
230	4.003	
240	3.929	
250	3.857	4.31
300	3.531	3.84
350	3.254	
400	3.018	3.27
450	2.814	
500	2.637	2.97
550	2.482	
600	2.346	2.74
650	2.224	
700	2.115	2.59
750	2.017	
800	1.928	2.46
850	1.848	
900	1.774	2.35
950	1.706	
1000	1.644	2.26

Source: Calculated by Kim, Y.-K. et al., *Electron-Impact Ionization Cross Section for Ionization and Excitation Database (Version 3.0).* [Online], 2004, available at http://physics.nist.gov/ionxsec (2008, February 5). National Institute of Standards and Technology, Gaithersberg, MD.

Note: Numbers are from S. K. Srivastava (1997, unpublished, quoted by Kim et al.) See Figure 29.2 for graphical presentation.

[a] Values are digitized.

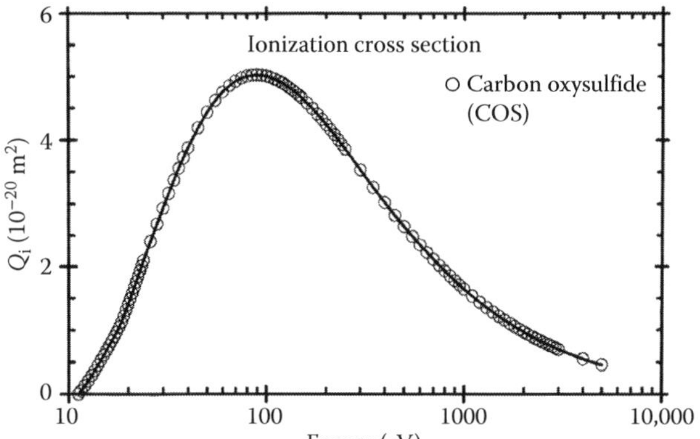

FIGURE 29.2 Ionization cross section for COS. (Adapted from Kim, Y.-K. et al., *Electron-Impact Ionization Cross Section for Ionization and Excitation Database (Version 3.0).* [Online], 2004, available at http://physics.nist.gov/ionxsec (2008, February 5). National Institute of Standards and Technology, Gaithersberg, MD.)

COS

TABLE 29.6
Attachment Cross Section for COS for Production of S^- Ion

Energy (eV)	Q_a (10^{-20} m^2)	Energy (eV)	Q_a (10^{-20} m^2)
0.8	0.003	1.49	0.2654
0.846	0.0043	1.524	0.2585
0.9	0.006	1.526	0.259
0.952	0.009	1.6	0.217
1.0	0.0184	1.644	0.191
1.042	0.035	1.743	0.1267
1.078	0.049	1.8	0.1
1.08	0.05	1.89	0.067
1.152	0.111	1.937	0.0543
1.2	0.172	2.0	0.042
1.293	0.2756	2.2	0.016
1.347	0.29	2.4	0.006
1.4	0.281		

Note: Data tabulated by Itikawa (2003) from measurements of Ziesel et al. (1975).

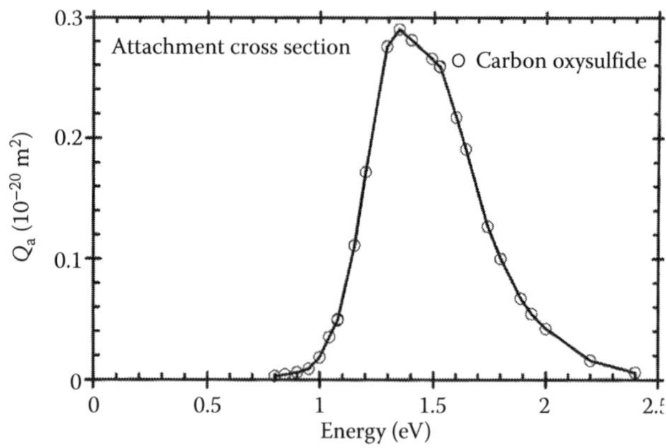

FIGURE 29.3 Attachment cross section for production of S^- ions from COS. (Adapted from Ziesel et al., *J. Chem. Phys.*, 62, 1936, 1975.)

REFERENCES

Itikawa, Y., Cross sections for ion production by electron collosions with molecules, in B. G. Lindsay and M. A. Mangan (Eds.), *Interactions of Photons and Electrons with Molecules*, Vol. 17c, Springer-Verlag, New York, NY, 2003, Chapter 5.1.

Karwasz, G., R. Brusa, and A. Zecca, Cross sections for scattering- and exicitation-processes in electron-molecule collisions, in Y. Etikawa (Ed.), *Interactions of Photons and Electrons with Molecules*, Vol. 17c, Springer-Verlag, New York, NY, 2003, Chapter 6.1.

Kim, Y.-K., K. K. Irikura, M. E. Rudd, M. A. Ali, P. M. Stone, J. Chang, J. S. Coursey et al., *Electron-Impact Ionization Cross Section for Ionization and Excitation Database (Version 3.0).* [Online], 2004, available at http://physics.nist.gov/ionxsec (2008, February 5). National Institute of Standards and Technology, Gaithersberg, MD.

Shimanouchi, T. *Tables of Molecular Vibrational Frequencies Consolidated Volume I*, NSRDS-NBS, 39, US Department of Commerce, Washington (DC), 1972.

Sohn, W., H.-H. Kochem. K. M. Scheuerlein, K. Jung, and H. Ehrhardt, *J. Phys. B: At. Mol. Phys.*, 20, 3217, 1987.

Srivastava, S. K., 1997, unpublished, quoted by Kim et al.

Szmytkowski, C., G. Karwasz, and K. Maciąg, *Chem. Phys. Lett.*, 107, 481, 1984.

Szmytkowski, C., K. Maciąg, G. Karwasz, and D. Filipović, *J. Phys. B: At. Mol. Opt. Phys.*, 22, 525, 1989.

Zecca, A., J. C. Nogueira, G. P. Karwasz, and R. S. Brusa, *J. Phys. B: At. Mol. Opt. Phys.*, 28, 477, 1995.

Ziesel, J. P., G. J. Schulz, and J. J. Milhaud, *J. Chem. Phys.*, 62, 1936, 1975.

30

CHLORINE DIOXIDE

ClO_2

CONTENTS

Chlorine dioxide (ClO_2 or OClO) is a molecule with both the species of atoms electronegative, chlorine being more electronegative than oxygen. It has 33 electrons, it is nonpolar, and it has ionization potentials of 10.55 eV ($OClO^+$ fragment) and 13.37 eV (ClO^+ fragment) (Probst et al., 2002).

30.1 TOTAL SCATTERING CROSS SECTION

Field et al. (2000) have measured the total scattering cross sections in the range 20–500 meV, improving upon the earlier measurements of Gulley et al. (1998) subsequently found to have had impurities. The measurements of the latter authors up to 10 eV are much lower than the corrected values and they have been tabulated by Karwasz et al. (2003) To remove the discontinuity at 0.5 eV, the values of Gulley et al. (1998) have been adjusted by multiplying with a factor of 2.5, as shown in Table 30.1 and Figure 30.1.

TABLE 30.1
Total Scattering Cross Sections for ClO_2

Energy (eV)	Q_T (10^{-20} m^2)	Energy (eV)	Q_T (10^{-20} m^2)
0.022	1146.10	0.049	390.18
0.023	1066.84	0.053	437.67
0.025	918.87	0.057	474.59
0.028	797.33	0.062	495.64
0.030	628.21	0.072	463.78
0.032	543.68	0.075	442.61
0.034	411.55	0.080	442.51
0.036	295.50	0.090	415.94
0.038	179.04	0.10	399.97
0.043	147.26	0.11	346.90
0.047	305.69	0.12	315.02

TABLE 30.1 (continued)
Total Scattering Cross Sections for ClO_2

Energy (eV)	Q_T (10^{-20} m^2)	Energy (eV)	Q_T (10^{-20} m^2)
0.13	288.47	0.43	82.99
0.14	272.50	0.44	77.53
0.15	251.19	0.45	77.37
0.16	235.16	0.46	77.21
0.17	224.46	0.47	77.05
0.18	218.99	0.48	76.91
0.19	203.01	0.49	76.73
0.20	192.28	0.50	76.59
0.21	186.84	0.50	77.00
0.22	176.09	0.60	64.50
0.23	165.39	0.70	55.75
0.24	154.66	0.80	50.25
0.25	149.20	0.90	46.25
0.26	143.78	1.0	43.75
0.27	143.64	1.2	40.00
0.28	138.17	1.5	37.00
0.29	127.47	1.7	35.50
0.30	122.03	2.0	34.00
0.31	121.87	2.2	33.50
0.32	105.86	2.5	32.25
0.33	110.98	3.0	30.00
0.34	105.54	3.5	27.50
0.35	105.38	4.0	24.85
0.36	99.94	4.5	22.43
0.37	99.76	5.0	20.28
0.38	89.05	6.0	18.75
0.39	94.20	7.0	19.63
0.40	88.73	8.0	20.13
0.41	88.59	9.0	20.48
0.42	88.42	10.0	20.80

Note: For energies >0.5 eV, the values of Gulley et al. (1998) have been increased by a factor of 2.5. Digitized by the author.

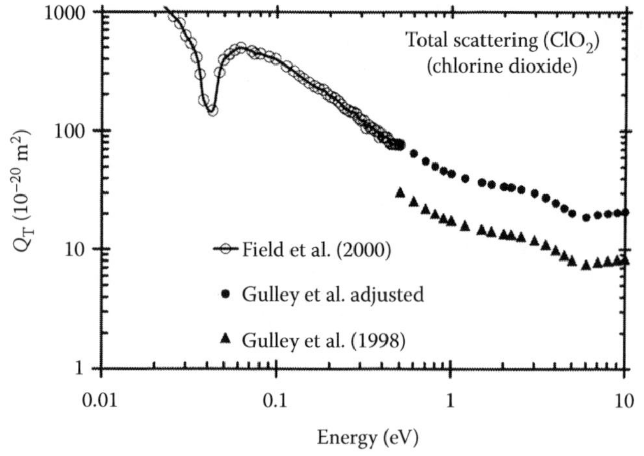

FIGURE 30.1 Total scattering cross sections for ClO_2. (—○—) Field et al. (2000); (▲) Gulley et al. (1998); (●) adjusted by a factor of 2.5.

ClO_2

REFERENCES

Field, D., N. C. Jones, J. M. Gingell, N. J. Mason, S. L. Hunt, and J-P. Ziesel, *J. Phys. B: At. Mol. Opt. Phys.*, 33, 1039, 2000.

Gulley, R. J., T. A. Field, W. A. Steer, N. J. Mason, S. L. Hunt, J. P. Ziesel, and D. Field, *J. Phys. B: At. Mol. Opt. Phys.*, 31, 5197, 1998.

Karwasz, G., R. Brusa, and A. Zecca, Cross sections for scattering- and excitations-processes in electron–molecule collisions, in *Interactions of Photons and Electrons with Molecules*, Vol. 17c, Springer-Verlag, New York, NY, 2003, Chapter 6.

Probst, M., K. Hermansson, J. Urban, P. Mach, D. Muigg, G. Denifl, T. Fiegele, N. J. Mason, A. Stamatovic, and T. D. Märk, *J. Chem. Phys.*, 116, 984, 2002.

31

HEAVY WATER

D_2O

CONTENTS

Heavy water (D_2O), also named as deuterated water, has 10 electrons. The electronic polarizability is 1.40×10^{-40} F m^2. The molecule has three modes of vibrational excitation as shown in Table 31.1 (Shimanouchi, 1972).

31.1 SELECTED REFERENCES FOR DATA

See Table 31.2.

31.2 IONIZATION CROSS SECTIONS

Table 31.3 shows the total and partial ionization cross sections for selected fragments from Straub et al. (1998) See Figure 31.1 for graphical presentation.

31.3 NEGATIVE ION APPEARANCE POTENTIALS

See Table 31.4.

TABLE 31.1
Vibrational Excitation Modes for D_2O

Designation	Type of Motion	Energy (meV)
ν_1	Symmetrical stretching	331.2
ν_2	Bending	146.1
ν_3	Antisymmetrical stretching	345.7

Source: Adapted from Shimanouchi, T., *Tables of Molecular Vibrational Frequencies Consolidated Volume I*, NSRDS-NBS 39, US Department of Commerce, 1972.

TABLE 31.2
Selected References for Data on D_2O

Quantity	Range: eV, (Td)	Reference
Ionization cross section	13.5–1000	**Straub et al. (1998)**
Vibrational excitation cross section	4–10	**Cvejanović et al. (1993)**
Transport coefficients	(1.5–75.0)	**Wilson et al. (1975)**
Attachment rate	(1.2–3.0)	**Compton and Christophorou (1967)**

Note: Bold font denotes experimental study.

D$_2$O

TABLE 31.3
Recommended Ionization Cross Sections for D$_2$O

Energy (eV)	D_2O Q_i (10^{-20} m^2)	H_2O Q_i (10^{-20} m^2)
13.5	0.028	0.0340
15	0.136	0.133
17.5	0.296	0.292
20	0.462	0.460
22.5	0.657	0.638
25	0.815	0.801
30	1.094	1.087
35	1.344	1.322
40	1.515	1.529
45	1.667	1.677
50	1.791	1.802
60	1.947	1.992
70	2.049	2.097
80	2.140	2.187
90	2.164	2.224
100	2.189	2.252
110	2.174	2.252
125	2.144	2.216
150	2.054	2.152
175	1.972	2.076
200	1.897	2.008
250	1.710	1.821
300	1.554	1.647
400	1.316	1.417
500	1.145	1.205
600	1.005	1.062
700	0.903	0.932
800	0.814	0.848
900	0.750	0.776
1000	0.694	0.707

Source: Adapted from Straub, H. C. et al., *J. Chem. Phys.*, 108, 109, 1998.
Note: Cross sections for H$_2$O are given for comparison.

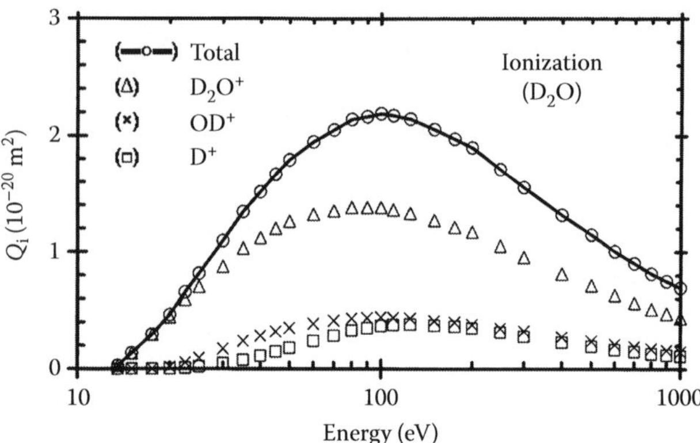

FIGURE 31.1 Total and partial ionization cross sections for D$_2$O. (—○—) Total; (Δ) D_2O^+; (×) OD^+; (□) D^+. (Adapted from Straub, H. C. et al., *J. Chem. Phys.*, 108, 109, 1998.)

31.4 ELECTRON TRANSPORT DATA

See Table 31.5 and Figure 31.2.

31.5 ATTACHMENT RATE CONSTANT

See Table 31.6.

TABLE 31.4
Negative Ion Appearance Potentials

Process	Onset (eV)	Peak Position (eV)	Peak (m^2)
$D^- + OD$	5.7	6.5	5.2×10^{-22}
$D^- + OD$	—	8.6	6.0×10^{-23}
$O^- + D_2$	5.0	7.0	—
$O^- + \cdots$	7.7	9.0	—
$O^- + \cdots$	—	11.8	—

Source: Adapted from Compton, R. N. and L. G. Christophorou, *Phys. Rev.*, 154, 110, 1967.

TABLE 31.5
Electron Transport Data for D$_2$O

E/N (Td)	*W* (m/s)	D_L/μ (V)
6.0	1.35×10^3	0.0278
9.0	2.05×10^3	0.0271
12.0	2.76×10^3	0.0282
15.0	3.52×10^3	0.0304
18.0	4.26×10^3	0.0331
21.0	5.10×10^3	0.0392
24.0	5.87×10^3	0.0503
27.0	6.84×10^3	0.0927
30.0	7.82×10^3	0.225
33.0	9.54×10^3	0.678
36.0	1.16×10^4	1.24
39.0	1.49×10^4	2.08
42.0	1.99×10^4	2.93
45.0	2.59×10^4	3.50
48.0	3.33×10^4	4.01
51.0	4.08×10^4	4.24
54.0	4.98×10^4	4.21
57.0	5.88×10^4	4.13
60.0	6.74×10^4	3.62
63.0	7.75×10^4	3.00
66.0	8.40×10^4	2.83
69.0	8.96×10^4	2.90
72.0	9.52×10^4	2.97
75.0	1.01×10^5	2.75
78.0	1.06×10^5	2.82
81.0	1.11×10^5	2.84

Source: Adapted from Wilson, J. F. et al., *J. Chem. Phys.*, 62, 4204, 1975.
Note: See Figure 31.2 for graphical presentation.

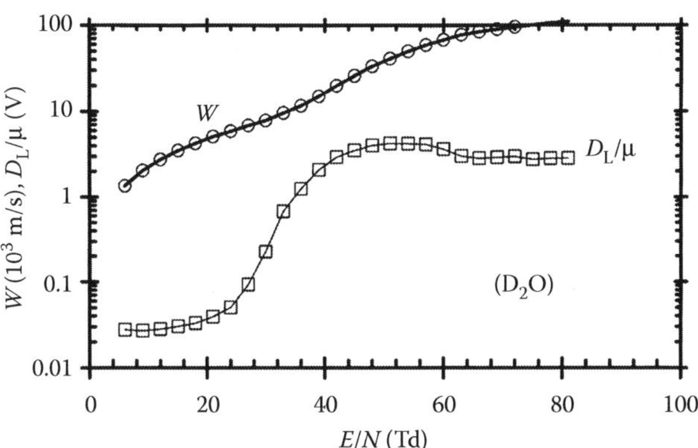

FIGURE 31.2 Electron transport data for D_2O. (—○—) W; (—□—) D_L/μ. (Adapted from Wilson, J. F. et al., *J. Chem. Phys.*, 62, 4204, 1975.)

TABLE 31.6
Attachment Rate Constants for D_2O

E/N (Td)	k_a (m^3/s)
1.35	2.88×10^{-18}
0.50	5.86×10^{-18}
0.55	1.05×10^{-17}
0.60	1.70×10^{-17}
0.65	2.40×10^{-17}
0.70	3.10×10^{-17}
0.74	3.83×10^{-17}

Source: Adapted from Compton, R. N. and L. G. Christophorou, *Phys. Rev.*, 154, 110, 1967.

REFERENCES

Compton, R. N. and L. G. Christophorou, *Phys. Rev.*, 154, 110, 1967.

Cvejanović, D., L. Andrić, and R. I. Hall, *J. Phys. B: At. Mol. Opt. Phys.*, 26, 2899, 1993.

Shimanouchi, T., *Tables of Molecular Vibrational Frequencies Consolidated Volume I*, NSRDS-NBS 39, US Department of Commerce, Washington (DC), 1972.

Straub, H. C., B. G. Lindsay, K. A. Smith, and R. F. Stebbings, *J. Chem. Phys.*, 108, 109, 1998.

Wilson, J. F., F. J. Davis, D. R. Nelson, and R. N. Compton, *J. Chem. Phys.*, 62, 4204, 1975.

32 HYDROGEN SULFIDE

CONTENTS

Hydrogen sulfide (H_2S) is a polar molecule with 18 electrons. The electronic polarizability is 4.21×10^{-40} F m^2, the dipole moment is 0.978 D, and the ionization potential is 10.457 eV. The dissociation energy into H–HS fragments is 3.949 eV. The vibrational modes and energies are shown in Table 32.1.

32.1 SELECTED REFERENCES FOR DATA

Selected references are shown in Table 32.2.

32.2 TOTAL SCATTERING CROSS SECTIONS

Table 32.3 and Figure 32.1 show the total scattering cross sections. The highlights are

1. The total scattering cross section shows a trend to increase for electron impact energies <1 eV due to dipole moment of the molecule.
2. A peak is observed at 2.5 eV due to resonance scattering (Karwasz et al., 2003).
3. A stronger peak is observed at 8 eV, also due to resonance scattering.
4. The cross section decreases monotonically in the energy range from 10 to 1000 eV. This feature is common to several gases.

TABLE 32.1
Vibrational Modes and Energies of H_2S Molecule

Mode	Type of Motion	Energy (meV)
v_1	Symmetrical stretch	324.2
v_2	Bend	146.7
v_2	Antisymmetrical stretch	325.6

Source: Adapted from Shimanouchi, T., *Tables of Molecular Vibrational Frequencies Consolidated Volume I*, NSRDS-NBS 39, US Department of Commerce, 1972.

TABLE 32.2
Selected References for Data

Quantity	Range: eV, (Td)	Reference
Total ionization C. S.	16–1000	**Lindsay et al. (2003)**
Total scattering C. S.	0.5–370	**Szmytkowski et al. (2003)**
Elastic and momentum transfer C. S.	1–30	**Gulley et al. (1993)**
Attachment C. S.	0–13	**Rao and Srivastava (1993)**
Total scattering C. S.	10–5000	Jain and Baluja (1992)
Total scattering C. S.	75–4000	**Zecca et al. (1992)**
Characteristic energy	(5–200)	**Millican and Walker (1987)**
Total scattering C. S.	1.3–70	**Szmytkowski and Maciąg (1986)**
Differential scattering C. S.	0–4	**Rohr (1978)**
Attachment C. S.	0–12	**Azria et al. (1972)**
Attachment energy	0–5	**Fiquet-Fayard et al. (1972)**
Attachment coefficient	(15–66)	**Bradbury and Tatel (1934)**

Note: Bold font indicates experimental study. C. S.: cross section.

H$_2$S

TABLE 32.3
Total Scattering Cross Sections

Energy (eV)	Q_T (10^{-20} m^2)	Energy (eV)	Q_T (10^{-20} m^2)
1.0	32.0	40	20.4
1.2	28.4	45	19.1
1.5	25.1	50	18.2
1.7	25.9	60	17.3
2.0	29.1	70	16.1
2.5	31.8	80	15.0
3.0	31.7	90	14.0
3.5	31.3	100	13.1
4.0	31.2	120	11.7
4.5	31.6	150	10.3
5.0	32.5	170	9.57
6.0	35.5	200	8.71
7.0	38.7	250	7.61
8.0	39.9	300	6.79
9.0	39.8	350	6.14
10	39.0	400	5.61
12	36.4	450	5.17
15	32.1	500	4.80
17	30.9	600	4.20
20	30.0	700	3.74
25	28.0	800	3.37
30	25.0	900	3.06
35	22.1	1000	2.81

Source: Adapted from Karwasz, G., R. Brusa, and A. Zecca, *Interactions of Photons and Electrons with Molecules*, Vol. 17c, Springer-Verlag, New York, NY, 2003, Chapter 6.

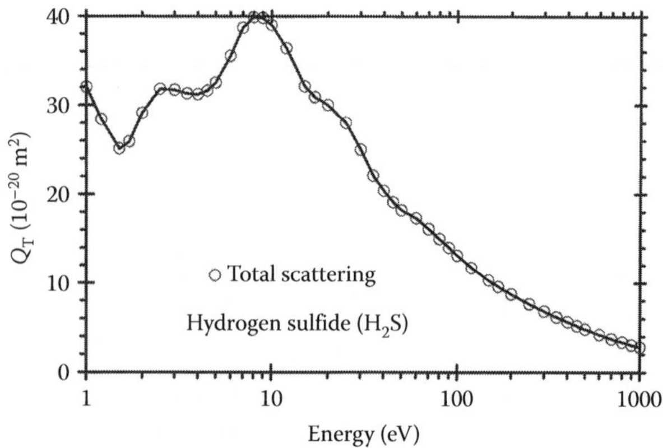

FIGURE 32.1 Total scattering cross sections for H$_2$S. Tabulated values are from Karwasz et al. (2003).

32.3 ELASTIC SCATTERING AND MOMENTUM TRANSFER CROSS SECTIONS

Gulley et al. (1993) have measured the cross sections in the range from 1 to 30 eV as shown in Table 32.4 and Figure 32.2.

32.4 TOTAL IONIZATION CROSS SECTION

Partial and total ionization cross sections are measured by Lindsay et al. (2003) as shown in Table 32.5 and Figure 32.3. The fragments of ionization are H_2S^+, HS^+, S^+, and H^+.

TABLE 32.4
Elastic Scattering and Momentum Transfer Cross Sections for H$_2$S

Energy (eV)	Q_{el} (10^{-20} m^2)	Q_{MT} (10^{-20} m^2)	Energy (eV)	Q_{el} (10^{-20} m^2)	Q_{MT} (10^{-20} m^2)
1.0	26.3	8.1	10.0	26.2	12.8
2.0	26.0	18.9	15.0	22.9	8.3
3.0	25.1	17.0	20.0	18.0	5.2
5.0	31.1	21.8	30.0	14.8	3.3

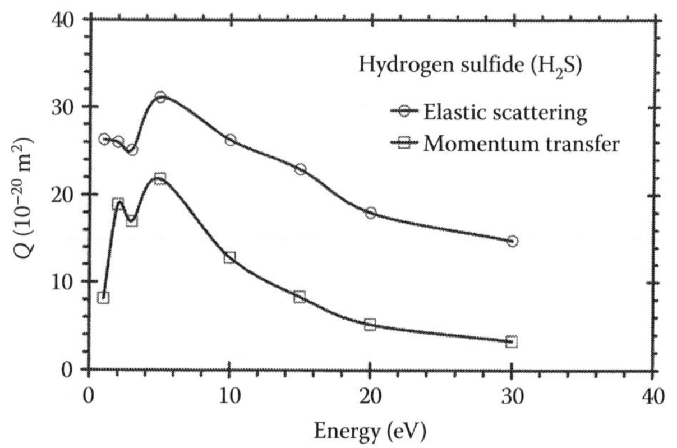

FIGURE 32.2 Elastic integral and momentum transfer cross sections for H$_2$S. (Adapted from Gulley, R. J., M. J. Brunger, and S. J. Buckman, *J. Phys. B: At. Mol. Opt. Phys.*, 26, 2913, 1993.)

TABLE 32.5
Partial and Total Ionization Cross Sections for H$_2$S

Energy (eV)	($H_2S^+ + HS^+ + S^+$) (10^{-20} m^2)	H^+ (10^{-20} m^2)	Total (10^{-20} m^2)
16	2.11		2.11
18	2.92		2.92
20	3.87		3.87
25	4.51	0.48	4.99
30	4.88	1.93	6.81
40	5.03	4.51	9.54
50	4.92	5.56	10.48
80	4.92	6.05	10.97
80	4.69	6.28	10.97
100	4.54	6.07	10.61
150	3.91	4.37	8.28
200	3.36	3.69	7.05
300	2.77	2.66	5.43
400	2.42	2.19	4.61
500	2.19	1.89	4.08
600	1.87	1.44	3.31
800	1.52	1.36	2.88
1000	1.29	0.93	2.22

32.5 ATTACHMENT CROSS SECTION

Dissociative attachment occurs according to the reactions:

$$H_2S + e \rightarrow H + HS^- \quad (32.1)$$

$$H_2S + e \rightarrow H_2 + S^- \quad (32.2)$$

with an onset potential of 1.55 eV for Reaction 32.1 (Fiquet-Fayard). S^- ion has also been observed (Azria et al., 1972). The cross sections for attachment have been tabulated by Itikawa (2003). See Table 32.6 and Figure 32.4.

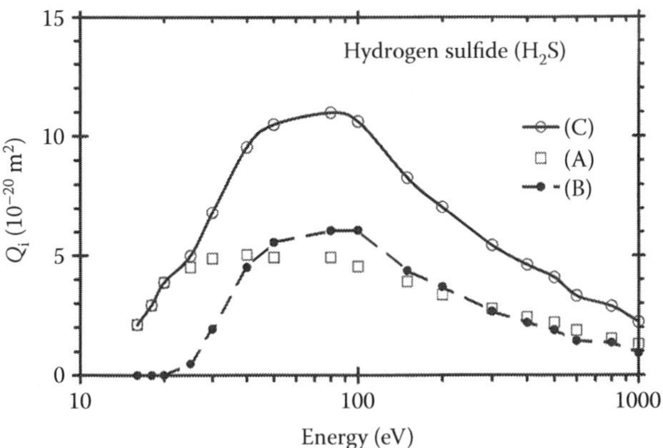

FIGURE 32.3 Partial and total ionization cross section for H_2S. (□) H_nS^+; (—●—) H^+; (—○—) total. (Adapted from Lindsay, B. G., R. Rejoub, and R. F. Stebbings, *J. Chem. Phys.*, 118, 5894, 2003.)

TABLE 32.6
Attachment Cross Sections for H_2S

Energy (eV)	S^- (10^{-20} m^2)	HS^- (10^{-20} m^2)	Total (10^{-20} m^2)
0	3×10^{-5}	2×10^{-4}	
1.0	3×10^{-5}	4.6×10^{-4}	
1.5	1×10^{-4}	2.57×10^{-3}	
1.75		6.57×10^{-3}	
2.0	2.98×10^{-4}	0.014	0.012
2.25	3.4×10^{-4}	0.018	0.0194
2.35			0.0183
2.45			0.0186
2.5	3.2×10^{-4}	0.0152	
2.55			0.0181
2.75		5.06×10^{-3}	
3.0	2.5×10^{-4}	2.3×10^{-3}	2.4×10^{-3}
3.3			0
3.5		0	
4.0	9×10^{-5}		0
4.75			0
5.0	1.1×10^{-4}		1.5×10^{-3}
5.5	2.3×10^{-4}		0.012
5.6			0.0106
5.7			0.0115
5.83	3.12×10^{-4}		

TABLE 32.6 (continued)
Attachment Cross Sections for H_2S

Energy (eV)	S^- (10^{-20} m^2)	HS^- (10^{-20} m^2)	Total (10^{-20} m^2)
6.0			0.004
6.5	2×10^{-4}		0
7.0			5×10^{-4}
7.2	1.1×10^{-4}		
7.5			0.002
8.0	2×10^{-4}		0.003
8.5	1.16×10^{-3}		0.002
9.0	2.97×10^{-3}		0.0022
9.5	4.8×10^{-3}		0.0035
9.6	4.85×10^{-3}		
9.7			0.004
10.0	3.92×10^{-3}		0.004
10.5	1.78×10^{-3}		0.0015
11.0	7.2×10^{-4}		4×10^{-4}
11.5	3×10^{-4}		0
12.0	8.7×10^{-5}		
13.0	2×10^{-5}		

Source: Adapted from Itikawa, Y., Electron attachment, in *Interactions of Photons and Electrons with Molecules*, Vol. 17c, Springer-Verlag, New York, NY, 2003, Chapter 5.2.

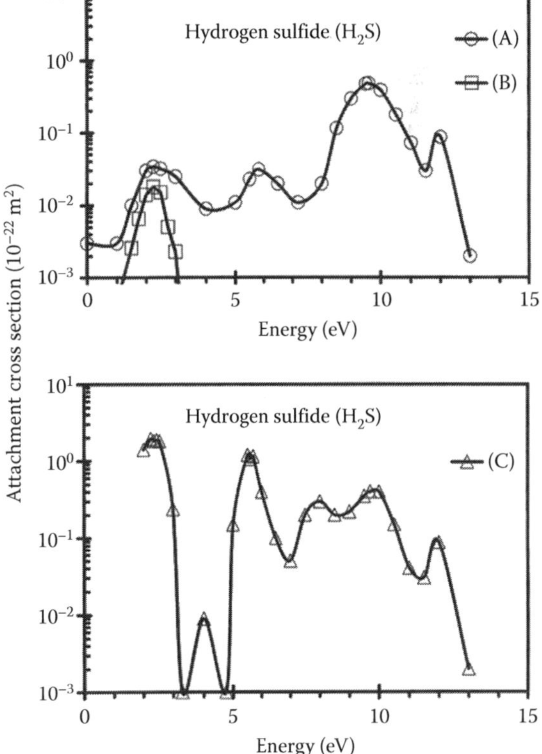

FIGURE 32.4 Attachment cross section for H_2S. (A) S^-; (B) HS^-; (C) total. (Adapted from Itikawa, Y., Electron attachment, in *Interactions of Photons and Electrons with Molecules*, Vol. 17c, Springer-Verlag, New York, NY, 2003, Chapter 5.2.)

32.6 REDUCED ATTACHMENT COEFFICIENT

Only a single study of reduced attachment coefficient as a function of E/N is available (Bradbury and Tatel, 1934), as shown in Table 32.7 and Figure 32.5.

32.7 ADDENDUM

32.7.1 Characteristic Energy

H_2S

See Table 32.8.

32.7.2 Attachment Rate Coefficient

Attachment rate constant obtained from Rydberg state study is 5×10^{-13} m^3/s (Dunning, 1995).

TABLE 32.7
Reduced Attachment Coefficients for H_2S

E/N (Td)	η/N (10^{-20} m^2)	E/N (Td)	η/N (10^{-20} m^2)
18	0.0028	40	0.25
22	0.025	45	0.34
25	0.056	53	0.49
27	0.082	58	0.58
30	0.12	63	0.68
34	0.16	66	0.75
37	0.21	68	0.81

Source: Adapted from Bradbury, N. E. and E. Tatel, *J. Chem. Phys.*, 2, 835, 1934.

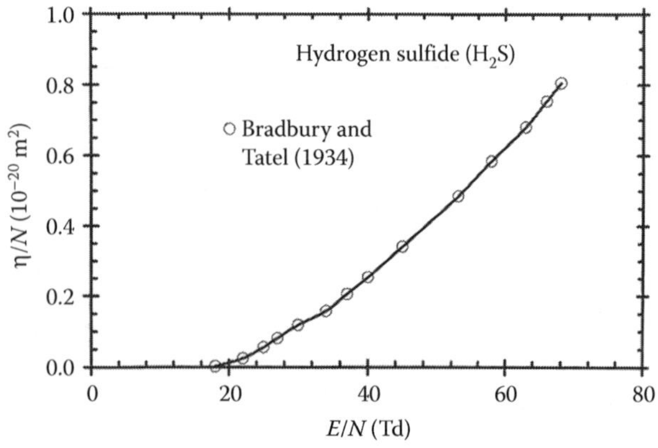

FIGURE 32.5 Reduced attachment coefficient as a function of reduced electric field for H_2S. (Adapted from Bradbury, N. E. and E. Tatel, *J. Chem. Phys.*, 2, 835, 1934.)

TABLE 32.8
Characteristic Energy

E/N (Td)	D_r/μ (V)	E/N (Td)	D_r/μ (V)
5	0.0262	35	0.346
6	0.0273	40	0.382
7	0.0284	50	0.412
8	0.030	60	0.410
10	0.035	70	0.437
12	0.041	80	0.451
14	0.054	100	0.492
17	0.085	120	0.558
20	0.120	140	0.625
25	0.193	170	0.814
30	0.296	200	1.14

Source: Adapted from Millican, P. G. and I. C. Walker, *J. Phys. D: Appl. Phys.*, 20, 193, 1987.

REFERENCES

Azria, R., M. Tronc, and S. Goursaud, *J. Chem. Phys.*, 56, 4234, 1972.

Bradbury, N. E. and E. Tatel, *J. Chem. Phys.*, 2, 835, 1934.

Dunning, F. B., *J. Phys. B: At. Mol. Opt. Phys.*, 28, 1645, 1995.

Fiquet-Fayard, F., J. P. Ziesel, R. Azria, M. Tronc, and J. Chiari, *J. Chem. Phys.*, 56, 2540, 1972.

Gulley, R. J., M. J. Brunger, and S. J. Buckman, *J. Phys. B: At. Mol. Opt. Phys.*, 26, 2913, 1993.

Itikawa, Y., Electron attachment, in *Interactions of Photons and Electrons with Molecules*, Vol. 17c, Springer-Verlag, New York, NY, 2003, Chapter 6.

Jain, A. and K. L. Baluja, *Phys. Rev.*, 45, 202, 1992.

Karwasz, G., R. Brusa, and A. Zecca, Cross sections for scattering and excitation processes in electron–molecule collisions, in *Interactions of Photons and Electrons with Molecules*, Vol. 17c, Springer-Verlag, New York, NY, 2003, Chapter 6.

Lindsay, B. G., R. Rejoub, and R. F. Stebbings, *J. Chem. Phys.*, 118, 5894, 2003.

Millican P. G. and I. C. Walker, *J. Phys. D: Appl. Phys.*, 20, 193, 1987.

Rao, M. V. V. S. and S. K. Srivastava, *J. Geophysical Res.*, 98, 13137, 1993.

Rohr, K., *J. Phys. B: At. Mol. Phys.*, 11, 4109, 1978.

Shimanouchi, T., *Tables of Molecular Vibrational Frequencies Consolidated Volume I*, NSRDS-NBS 39, US Department of Commerce, Washington (DC), 1972.

Szmytkowski Cz. and K. Maciąg, *Chem. Phys. Lett.*, 129, 321, 1986.

Szmytkowski, Cz., P. Możejko, and A. Krzysztofowicz, *Rad. Phys. Chem.*, 68, 307, 2003.

Zecca, A., G. P. Karwasz, and R. S. Brusa, *Phys. Rev.*, 45, 2777, 1992.

33

NITROGEN DIOXIDE

CONTENTS

Nitrogen dioxide (NO_2) is a polar, electron-attaching, symmetrically bent molecule that has 23 electrons. Its electronic polarizability is 3.36×10^{-40} F m^2, dipole moment is 0.316 D, and ionization potential is 9.586 eV. Selected references for data are shown in Table 33.1. The vibrational excitation energies are (Benoit and Abouaf, 1991): $\nu_1 = 163$ meV (symmetrical stretch), $\nu_2 = 93$ meV (bending), and $\nu_3 = 20$ meV (asymmetrical stretch).

33.1 SELECTED REFERENCES FOR DATA

See Table 33.1.

TABLE 33.1
Selected References for Data

Quantity	Range: eV, (Td)	Reference
Ionization cross section	10–2000	Joshipura et al. (2007)
Ionization cross section	10–200	**Lopez et al. (2003)**
Ionization cross section	10–200	**Jiao et al. (2002)**
Ionization cross section	10–1000	**Lukić et al. (2001)**
Ionization cross section	13.5–1000	**Lindsay et al. (2000)**
Total cross section	90–4000	**Zecca et al. (1995)**
Total cross section	0.6–220	**Szmytkowsky et al. (1992)**
Vibrational excitation	0–2	**Benoit and Abouaf (1991)**
Ionization cross section	11–180	**Stephan et al. (1980)**
Attachment rate constant	0.025	**Puckett et al. (1971)**
Attachment cross sections	0–2.4	**Sanche and Schulz (1973)**
Attachment rate constant	0.025	**Hasted and Beg (1965)**

Note: Bold font indicates experimental study.

33.2 TOTAL SCATTERING CROSS SECTION

Table 33.2 and Figure 33.1 show the total scattering cross sections for NO_2. The highlights are

1. At low energies below 1 eV, there is a trend toward increasing cross sections, due to the dipole moment of the molecule. Dissociative attachment also occurs in the energy range.
2. At ~11 eV, there is a maximum in the cross section at ~15×10^{-20} m^2. This feature is common to many gases.
3. Beyond the maximum, the total cross section decreases monotonically, possibly due to electronic polarization of the molecule (Szmytkowski et al., 1992).

TABLE 33.2
Recommended Total Scattering Cross Section for NO_2

Energy (eV)	Q_T (10^{-20} m^2)	Energy (eV)	Q_T (10^{-20} m^2)
0.6	16.9	3.5	12.6
0.7	16.4	4	12.6
0.8	16	4.5	12.7
0.9	15.6	5	12.8
1	15.3	6	13.2
1.2	14.7	7	13.7
1.5	14.1	7.5	14.0
1.7	13.7	8.5	14.5
2	13.3	9.5	14.8
2.5	12.9	10.5	14.9
3	12.7	11.5	14.9

continued

NO$_2$

TABLE 33.2 (continued)
Recommended Total Scattering Cross Section for NO$_2$

Energy (eV)	Q_T (10^{-20} m^2)	Energy (eV)	Q_T (10^{-20} m^2)
12.5	14.8	225	7.7
13.5	14.6	256	7.49
14.5	14.5	289	7.22
15.5	14.6	324	6.7
16.5	14.7	361	6.69
18.5	14.7	400	6.0
20.5	14.6	484	5.28
24.2	14.5	576	4.72
25	14.5	676	4.15
30	14.1	785	3.77
35	13.8	900	3.34
40	13.8	1000	3.11
45	13.5	1025	3.0
50	13.5	1150	2.75
55	13.3	1300	2.48
60	13.1	1450	2.22
65	12.9	1600	2.08
70	12.6	1770	1.88
80	12.4	1940	1.73
90	12.1	2000	1.68
100	11.8	2120	1.62
110	11.4	2300	1.48
120	11.1	2500	1.37
140	10.3	2750	1.29
160	9.8	3000	1.17
180	9.06	3250	1.1
200	8.44	3500	1.03
220	7.96	4000	0.915

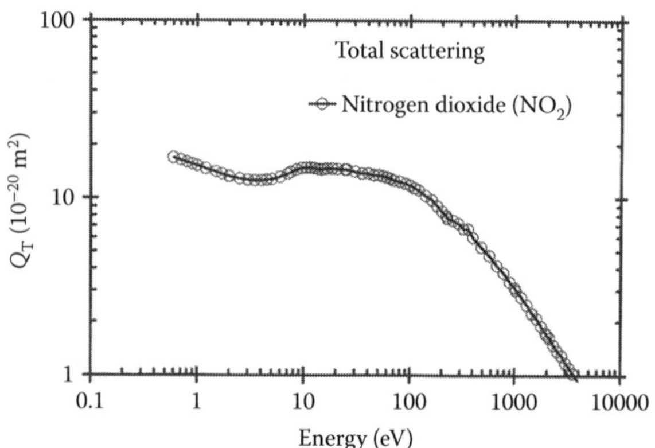

FIGURE 33.1 Total scattering cross section for NO$_2$. Data from Szmytkowski et al. (1992), energy range 0.60–220 eV; Zecca et al. (1995), energy range 220–4000 eV.

33.3 VIBRATIONAL CROSS SECTION

Figure 33.2 shows the relative change in vibrational excitation cross sections with electron energy for NO$_2$. The structure below 1.1 eV is attributed to the formation of NO$_2^-$ (Sanche and Schulz, 1973). Above 1.1 eV the symmetric stretch mode is active. These authors quote an average spacing of 65 and 128 meV for bending and symmetric stretch modes, respectively. Absolute cross sections are not available. A point to note: when a molecule in the ground state forms stable parent negative ions with zero electron energy (NO$_2$, O$_2$, and NO) the structure is evident at zero energy, as shown in Figure 33.2. When a molecule does not have a stable parent negative ion at zero electron energy (CO, N$_2$, and CO$_2$), the structure is observed in the 2–4 eV energy range (Sanche and Schulz, 1973).

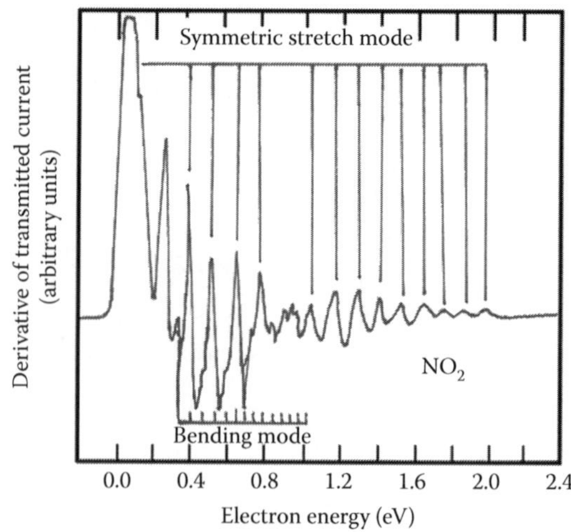

FIGURE 33.2 Relative change in vibrational excitation cross sections with electron energy for NO$_2$. The peaks and valleys are too weak to appear in the total scattering cross section. (Adapted from Sanche, L. and G. J. Schulz, *J. Chem. Phys.*, 58, 479, 1973.)

33.4 IONIZATION CROSS SECTION

See Tables 33.3 and 33.4.

TABLE 33.3
Total Ionization Cross Sections for NO$_2$

Energy (eV)	Lukić et al. (2001) (10^{-20} m^2)	Lindsay et al. (2000) (10^{-20} m^2)
10	0.009	
12	0.035	
13.5		0.146
14	0.088	
16	0.175	0.307
20	0.440	0.689
25		1.051
26	0.976	
30		1.605
35		2.120
36	1.91	
40	2.22	2.610
50	2.83	3.379
60	3.25	4.014
70	3.51	
80	3.67	4.686
90	3.76	
100	3.806	5.010

TABLE 33.3 (continued)
Total Ionization Cross Sections for NO_2

Energy (eV)	Lukić et al. (2001) (10^{-20} m^2)	Lindsay et al. (2000) (10^{-20} m^2)
110	3.80	
120	3.78	5.190
130	3.74	
140	3.70	
150	3.65	
160	3.59	4.994
170	3.53	
180	3.47	
190	3.40	
200	3.34	4.716
225	3.19	
250	3.05	4.344
300	2.79	3.952
350	2.57	
400	2.39	3.377
450	2.23	
500	2.09	2.924
550	1.96	
600	1.86	2.604
700	1.70	
800	1.56	2.156
900	1.46	
1000	1.35	1.803

Note: See Figure 33.3 for graphical presentation.

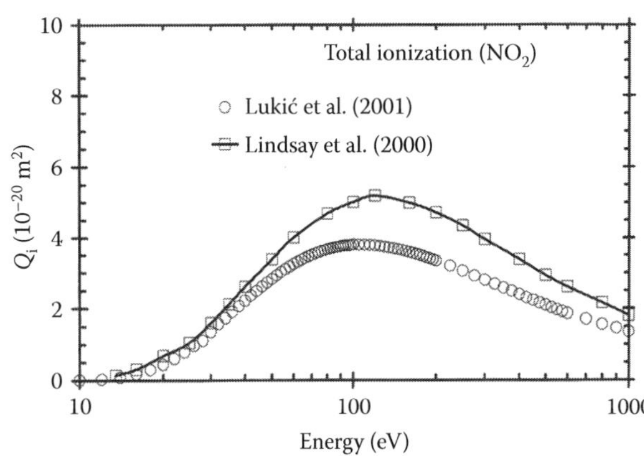

FIGURE 33.3 Total ionization cross section for NO_2. (○) Lukić et al. (2001); (—□—) Lindsay et al. (2000). No explanation is available for the differences between the two sets.

TABLE 33.4
Partial Ionization Cross Section for NO_2

Energy (eV)	(NO_2^+) (10^{-20} m^2)	(NO^+) (10^{-20} m^2)	($N^+ + O^+$) (10^{-20} m^2)	(N^+) (10^{-20} m^2)	(O^+) (10^{-20} m^2)	($N_2^+ + O^{2+}$) (10^{-20} m^2)
13.5	0.0909	0.055				
16	0.159	0.148				
20	0.244	0.401	0.022		0.022	
25	0.317	0.612	0.061		0.061	
30	0.389	0.906	0.155	0.015	0.14	
35	0.436	1.14	0.272	0.055	0.217	
40	0.454	1.3	0.428	0.114	0.314	
50	0.525	1.57	0.642	0.189	0.453	
60	0.546	1.77	0.849	0.234	0.615	
80	0.561	1.89	1.12	0.337	0.778	
100	0.542	1.95	1.26	0.365	0.892	0.00140
120	0.543	1.95	1.35	0.407	0.938	0.00238
160	0.507	1.86	1.31	0.386	0.927	0.00430
200	0.47	1.76	1.24	0.397	0.843	0.00614
250	0.432	1.63	1.14	0.36	0.776	0.00640
300	0.399	1.51	1.02	0.327	0.69	0.00615
400	0.353	1.32	0.85	0.275	0.574	0.00507
500	0.311	1.17	0.719	0.221	0.498	0.00452
600	0.278	1.06	0.631	0.202	0.429	0.00413
800	0.235	0.886	0.516	0.136	0.38	0.00330
1000	0.201	0.756	0.422	0.117	0.305	0.00211

Source: Adapted from Lindsay, B. G. et al., *J. Chem. Phys.*, 112, 9404, 2000.
Note: See Figure 33.4 for graphical presentation.

NO$_2$

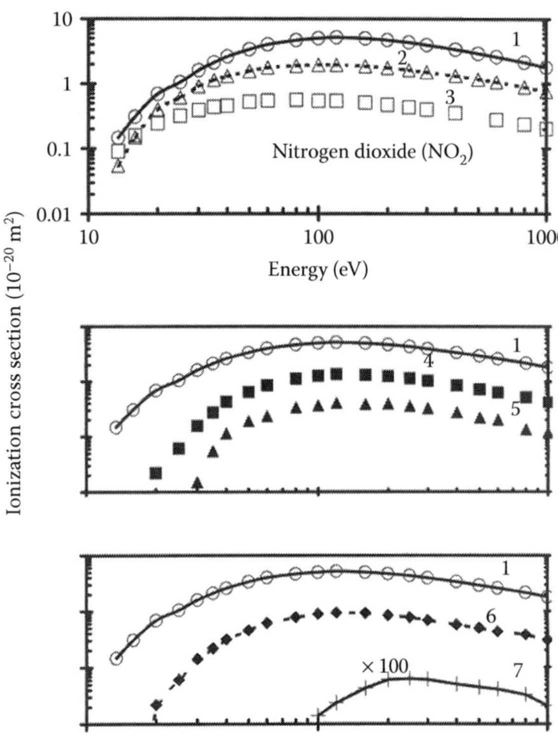

FIGURE 33.4 Total and partial ionization cross sections for NO_2. Total cross section plotted for comparison. (1) Total; (2) NO^+; (3) NO_2^+; (4) $N^+ + O^+$; (5) N^+; (6) O^+; (7) $N^{2+} + O^{2+}$. Note the multiplication factor for this fraction. Note that the cross section for the parent ion is smaller than that for NO^+ (top frame), which is unusual. (Adapted from Lindsay, B. G. et al., *J. Chem. Phys.*, 112, 9404, 2000.)

33.5 ATTACHMENT RATE COEFFICIENT

See Table 33.5.

TABLE 33.5
Attachment Rate Coefficient

Rate Coefficient (10^{-17} m^3/s)	Method	Reference
1.4	Stationary afterglow	Puckett et al. (1971)
0.11	Microwave afterglow	Hasted and Beg (1965)

REFERENCES

Benoit, C. and R. Abouaf, *Chem. Phys. Lett.*, 177, 573, 1991.

Hasted, J. B. and S. Beg, *Br. J. Appl. Phys.*, 16, 1779, 1965.

Jiao, C. Q., C. D. Joseph, J., and A. Garscadden, *J. Chem. Phys.*, 117, 161, 2002.

Joshipura, K. N., S. Gangopadhyay, and B. G. Vaishnav, *J. Phys. B: At. Mol. Opt. Phys.*, 40, 199, 2007.

Lindsay, B. G., M. A. Mangan, H. C. Straub, and R. F. Stebbings, *J. Chem. Phys.*, 112, 9404, 2000.

Lopez, J., V. Tarnovsky, M. Gutkin, and K. Becker, *Int. J. Mass Spectrom.*, 225, 25, 2003.

Lukić, D., G. Josifov, and M. V. Kurepa, *Int. J. Mass Spectrom*, 205, 1, 2001.

Puckett, L. J., M. D. Kregel, and M. W. Teague, *Phys. Rev. A*, 4, 1659, 1971.

Sanche, L. and G. J. Schulz, *J. Chem. Phys.*, 58, 479, 1973.

Stephan, K., H. Helm, Y. B. Kim, G. Seykara, J. Ramler, M. Grössl, E. Märk, and T. Märk, *J. Chem. Phys.*, 73, 303, 1980.

Szmytkowsky, C., K. Maciąg, and M. Krzystofowicz, *Chem. Phys. Lett.*, 190, 141, 1992.

Zecca, A., J. C. Nogueira, G. P. Karwasz, and R. S. Brusa, *J. Phys. B: At. Mol. Opt. Phys.*, 28, 477, 1995.

34

NITROUS OXIDE

CONTENTS

Nitrous oxide (N_2O) is a polar, electron-attaching molecule that has 22 electrons. It is a linear asymmetric molecule with the structure N–N–O with bond lengths 0.113 and 0.119 nm for the N–N bond and N–O bond, respectively. Its electronic polarizability is 3.37×10^{-40} F m^2, dipole moment is 0.161 D, and ionization potential is 12.886 eV. The vibrational modes are shown in Table 34.1 (Shimanouchi, 1972).

TABLE 34.1
Vibrational Modes for N_2O

Designation	Type of Motion	Energy (meV)
v_1	NN stretch	267.2
v_2	Bend	70.9
v_3	NO stretch	156.8

Source: Adapted from Shimanouchi, T., *Tables of Molecular Vibrational Frequencies Consolidated Volume I*. NSRDA-NBS 39, US Department of Commerce, Washington (DC), 1972.

34.1 SELECTED REFERENCES FOR DATA

See Table 34.2.

34.2 TOTAL SCATTERING CROSS SECTION

See Table 34.3. Highlights of Figure 34.1 are

1. As the energy decreases toward zero, there is a trend toward an increase of cross section, possibly due to the attachment of electrons and a smaller contribution from the dipole moment, which is also relatively low, 0.167 D.
2. A shape resonance at 2.5 eV.
3. A second shape resonance at 8.9 eV (Kitajima et al., 1999), though this resonance is not seen in the total scattering cross section.
4. A broad peak in the 10–30 eV range due to inelastic scattering including ionization. This feature is common to many gases.
5. A monotonic decrease beyond 30 eV which is also a common feature observed in many gases.

N_2O

TABLE 34.2
Selected References for Data on N_2O

Quantity	Range: eV, (Td)	Reference
Drift velocity	(250–2500)	**Lisovskiy et al. (2006)**
Ionization cross section	13–1000	**Lindsay et al. (2003)**
Ionization cross section	13–200	**Lopez et al. (2003)**
Swarm parameters	(10–300)	**Mechlińska-Drewko et al. (2003)**
Differential cross section	1.5–100	**Kitajima et al. (1999)**
Swarm parameters	(7–5000)	**Yoshida et al. (1999)**
Attachment cross section	0–4	**Brüning et al. (1998)**
Elastic cross section	0.01–50	Winstead and McKoy (1998)
Ionization cross section	13–1000	Kim et al. (1997)
Total cross section	600–4250	**Shilin et al. (1997)**
Ionization cross section	20–1000	**Iga et al. (1996)**
Total cross section	90–4000	**Zecca et al. (1995)**
Elastic cross section	5–80	**Johnstone and Newell (1993)**
Attachment cross section	0–50	**Krishnakumar and Srivastava (1990)**
Total cross section	40–100	**Szmytkowski et al. (1989)**
Elastic cross section	10–80	**Marinković et al. (1986)**
Total cross section	1–500	**Kwan et al. (1984)**
Total cross section	0.8–40	**Szmytkowski et al. (1984)**
Mean energy	(10–900)	Raju and Hackam (1981)
Attachment rate constant	0.02–0.03	**Shimamori and Fessenden (1979)**
Ionization coefficient	(140–160)	**Dutton et al. (1975a)** **Dutton et al. (1975b)**
Total cross section	0–10	**Zecca et al. (1974)**
Excitation coefficient	(30)	**Austin and Smith (1973)**
Attachment rate constant	0–4	**Christophorou et al. (1971)**
Attachment rate constant	(0–1.5)	**Chaney and Christophorou (1969)**
Attachment coefficients	(0.1–200)	**Phelps and Voshall (1968)**
Ionization cross section	13–1000	**Rapp and Englander-Golden (1965)** **Rapp et al. (1965)**
Attachment cross section	0.4–5.0	**Rapp and Briglia (1965)**
Drift velocity	(0.1–30)	**Pack et al. (1962)**
Total cross section	0.15–1.5	**Ramsauer and Kollath (1930)**

Note: Bold font indicates experimental study.

TABLE 34.3
Total Scattering Cross Sections for N_2O

Energy (eV)	Q_T (10^{-20} m²)	Energy (eV)	Q_T (10^{-20} m²)
0.15	10.3	60	15.0
0.17	9.69	70	14.4
0.2	8.96	80	13.8
0.25	8.08	90	13.2
0.3	7.48	100	12.6
0.35	7.06	120	11.6
0.40	6.77	150	10.4
0.45	6.58	170	9.78
0.50	6.48	200	8.87
0.60	6.57	250	7.80
0.70	6.7	300	7.12
0.80	6.98	350	6.46
0.90	7.37	400	5.92
1.0	8.08	450	5.47
1.2	10.2	500	5.08
1.5	14.2	600	4.44
1.7	18.1	700	3.94
2.0	25.6	800	3.56
2.5	26.9	900	3.26
3.0	18.6	1000	3.01
3.5	13.6	1100	2.80
4.0	11.0	1200	2.71
4.5	9.73	1300	2.54
5.0	9.24	1400	2.35
6.0	9.37	1600	2.12
7.0	10.4	1800	1.91
8.0	11.9	2000	1.77
9.0	13.4	2200	1.62
10.0	14.5	2400	1.52
12	15.9	2600	1.40
15	16.4	2800	1.36
17	16.6	3000	1.27
20	17.4	3250	1.19
25	17.7	3500	1.12
30	17.6	3750	1.11
35	17.2	4000	1.05
40	16.7	4250	1.01
45	16.3		
50	15.8		

Note: Recommended by Karwasz et al. (2003) for the energy range $0.15 \le \varepsilon \le 1000$ eV and Shilin et al. (1997) for the energy range $1000 \le \varepsilon \le 4250$ eV. Figure 34.1 shows these and selected other data.

34.3 DIFFERENTIAL SCATTERING CROSS SECTION

Figure 34.2 shows the differential scattering cross sections for N_2O (Kitajima et al., 1999).

34.4 ELASTIC AND MOMENTUM TRANSFER CROSS SECTIONS

See Table 34.4.

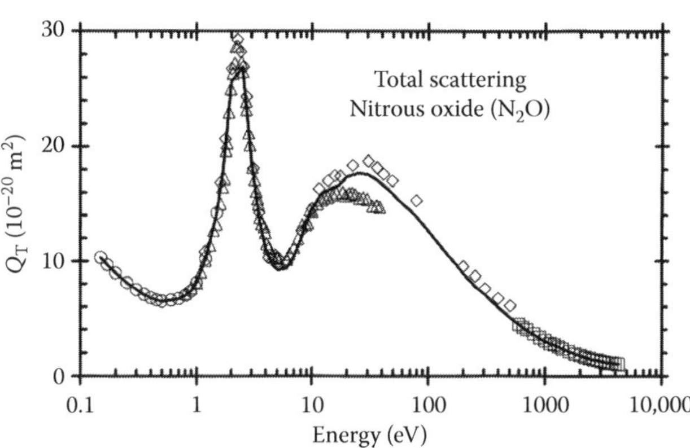

FIGURE 34.1 Total scattering cross sections for N_2O. (——) Recommended by Karwasz et al. (2003); (○) Ramsauer and Kollath (1930); (△) Szmytkowski et al. (1984); (◊) Kwan et al. (1984); (□) Shilin et al. (1997).

TABLE 34.4
Elastic and Momentum Transfer Cross Sections for N_2O

Karwasz et al. (2003)		Johnstone and Newell (1993)		
Energy (eV)	Q_{el} (10^{-20} m^2)	Energy (eV)	Q_{el} (10^{-20} m^2)	Q_M (10^{-20} m^2)
2.0	11.4			
2.2	12.1			
2.4	13.2			
2.6	10.3			
4.0	7.9			
5.0	7.8	5.0	9.7	7.8
6.0	8.5	7.5	9.1	8.9
8.0	10.8	8.0	10.8	11.6
10	13.9	10.0	13.7	11.7
12	15.6	12	17.0	14.2
15	15.0	15	16.2	12.3
20	13.2	20	14.8	8.6
30	9.8	30	12.3	6.4
50	5.6	50	9.3	4.8
75	3.4			
80	4.5	80	4.5	2.1
100	2.9			

Note: See Figures 34.3 and 34.4 for graphical presentation.

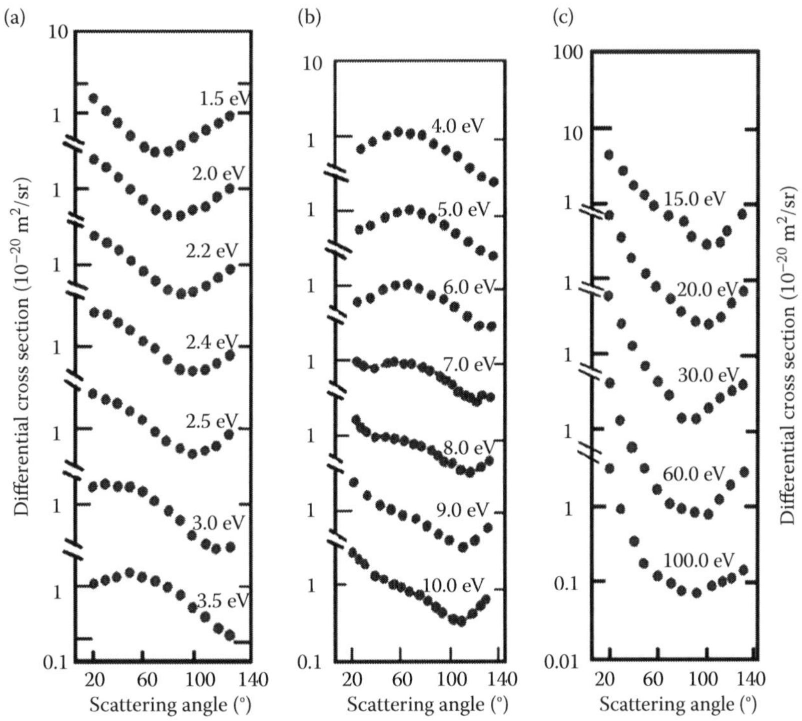

FIGURE 34.2 Differential scattering cross section for elastic collisions for N_2O with electron energy. (Adapted from Kitajima, M. et al., *Chem. Phys. Lett.*, 309, 414, 1999.)

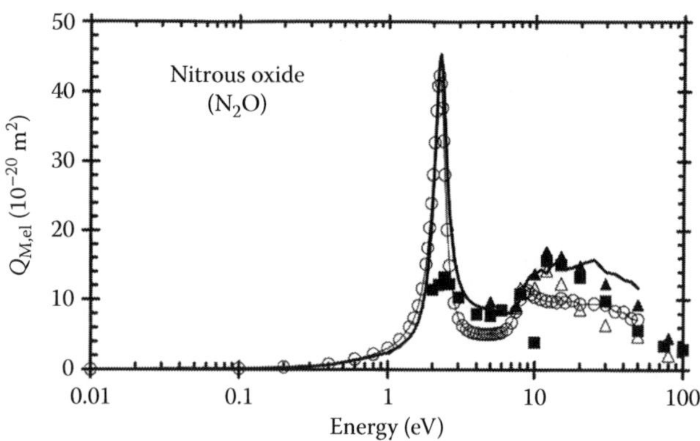

FIGURE 34.3 Integrated elastic and momentum transfer cross sections for N_2O. (——) Elastic, (■) momentum transfer, Winstead and McKoy (1998); (■) elastic, Karwasz et al. (2003); (▲) elastic, (△) momentum transfer cross section. (Adapted from Johnstone, W. M. and W. R. Newell, *J. Phys. B: At. Mol. Phys.*, 26, 129, 1993.)

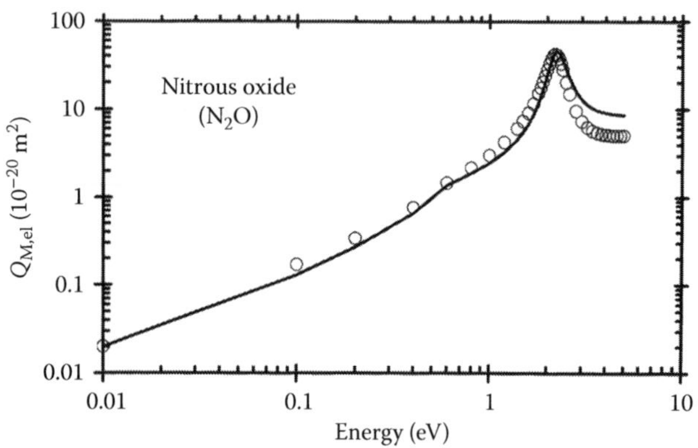

FIGURE 34.4 Low-energy details of elastic and momentum transfer cross sections for N_2O. (Adapted from Winstead, C. and V. McKoy, *Phys. Rev. A*, 57, 3589, 1998.)

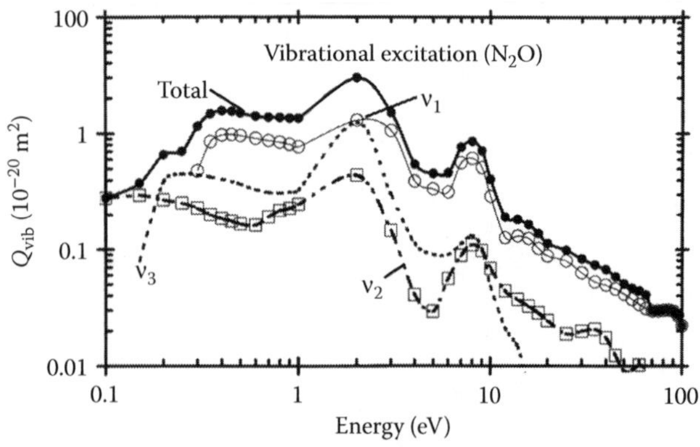

FIGURE 34.5 Vibrational excitation cross section for N_2O recommended by Hayashi (1992).

34.5 VIBRATIONAL EXCITATION CROSS SECTIONS

A set of vibrational cross sections recommended by Hayashi (1992) is digitized and shown in Table 34.5. See Figure 34.5 for graphical presentation.

TABLE 34.5
Vibrational Excitation Cross Section for N_2O

	Q_{vib} (10^{-20} m^2)			
Energy (eV)	ν_1	ν_2	ν_3	Total
0.10		0.28		0.28
0.15		0.30	0.080	0.38
0.20		0.27	0.39	0.66
0.25		0.25	0.45	0.70
0.30	0.48	0.23	0.44	1.15
0.35	0.85	0.20	0.43	1.48
0.40	0.97	0.19	0.41	1.57
0.45	0.98	0.18	0.39	1.55
0.50	0.96	0.17	0.37	1.50
0.60	0.91	0.16	0.33	1.41
0.70	0.87	0.19	0.31	1.38
0.80	0.85	0.22	0.31	1.37
0.90	0.81	0.23	0.32	1.35
1.0	0.77	0.25	0.33	1.35
2.0	1.31	0.44	1.25	2.99
3.0	1.06	0.15	0.30	1.51
4.0	0.39	0.041	0.11	0.54
5.0	0.34	0.029	0.09	0.46
6.0	0.32	0.056	0.09	0.46
7.0	0.56	0.088	0.11	0.76
8.0	0.61	0.110	0.13	0.85
9.0	0.51	0.097	0.10	0.71
10	0.29	0.069	0.048	0.41
12	0.13	0.044	0.021	0.19
14	0.13	0.037	0.015	0.18
16	0.12	0.033	0.007	0.16
18	0.10	0.028	0.007	0.14
20	0.088	0.024		0.11
25	0.079	0.019		0.10
30	0.063	0.020		0.083
35	0.052	0.021		0.073
40	0.049	0.018		0.067
45	0.046	0.012		0.058
50	0.041	0.009		0.050
55	0.037	0.009		0.046
60	0.034	0.010		0.044
65	0.031	0.0095		0.041
70	0.030			0.030
75	0.030			0.030
80	0.030			0.030
85	0.030			0.030
90	0.030			0.030
95	0.027			0.027
100	0.022			0.022

Source: Adapted from Hayashi, M., in Ohm-Sha (Ed.), *Handbook on Plasma Material Science*, Vol. 4(9), Committee No. 153 on Plasma Material Science, The Japan Society for Promotion of Science, 1992.

34.6 ELECTRONIC EXCITATION CROSS SECTIONS

See Table 34.6 (Brunger et al., 2003).

34.7 IONIZATION CROSS SECTION

See Table 34.7.

34.8 PARTIAL IONIZATION CROSS SECTIONS

See Figure 34.7.

TABLE 34.6
Electronic Excitation Cross Section

Energy (eV)	Q_{ex} (10^{-20} m²): $2^1\Sigma^+$	Q_{ex} (10^{-20} m²): $^1\Pi$
20	0.42	0.047
30	0.93	0.064
50	0.55	0.057
80	0.67	0.036

Source: Adapted from Brunger, M. S., Buckman, and M. Elford, in *Interactions of Photons and Electrons with Molecules*, Vol. 17c, Springer-Verlag, New York, NY, 2003, Chapter 6.4.

TABLE 34.7
Total Ionization Cross Sections for N_2O

Lindsay et al. (2003)		Rapp and Englander-Golden (1965)			
Energy (eV)	Q_i (10^{-20} m²)	Energy (eV)	Q_i (10^{-20} m²)	Energy (eV)	Q_i (10^{-20} m²)
14	0.07	14	0.054	23.5	1.01
16	0.27	14.5	0.121	24	1.07
18	0.42	15	0.158	26	1.28
20	0.60	15.5	0.199	28	1.47
22.5	0.86	16	0.238	30	1.64
25	1.15	16.5	0.280	32	1.80
27.5	1.38	17	0.319	34	1.94
30	1.58	17.5	0.362	36	2.07
35	2.05	18	0.404	38	2.19
40	2.37	18.5	0.451	40	2.31
50	2.84	19	0.496	45	2.59
60	3.19	19.5	0.551	50	2.81
70	3.49	21	0.723	55	3.01
80	3.65	21.5	0.781	60	3.18
90	3.73	22	0.840	65	3.32
100	3.77	22.5	0.898	70	3.44
120	3.76	23	0.959	75	3.53

TABLE 34.7 (contiuned)
Total Ionization Cross Sections for N_2O

Lindsay et al. (2003)		Rapp and Englander-Golden (1965)			
Energy (eV)	Q_i (10^{-20} m²)	Energy (eV)	Q_i (10^{-20} m²)	Energy (eV)	Q_i (10^{-20} m²)
140	3.74	80	3.61	200	3.45
160	3.62	85	3.66	250	3.18
200	3.43	90	3.70	300	2.94
250	3.16	95	3.73	350	2.72
300	2.94	100	3.75	400	2.53
400	2.49	105	3.76	450	2.37
500	2.21	110	3.77	500	2.23
600	1.94	115	3.78	550	2.09
800	1.61	120	3.78	600	1.97
1000	1.42	125	3.77	650	1.87
		130	3.76	700	1.78
		135	3.75	750	1.70
		140	3.73	800	1.63
		145	3.71	850	1.55
		150	3.70	900	1.50
		160	3.64	950	1.45
		180	3.55	1000	1.42

Note: See Figure 34.6 for graphical presentation.

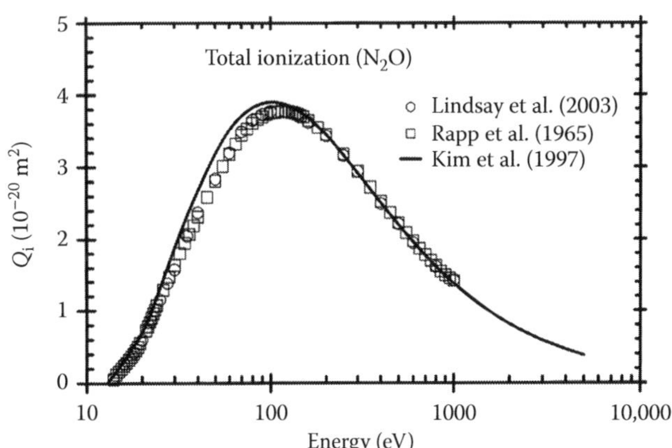

FIGURE 34.6 Total ionization cross section for NO_2. (○) Lindsay et al. (2003); (—) Kim et al. (1997); (□) Rapp and Englander-Golden (1965).

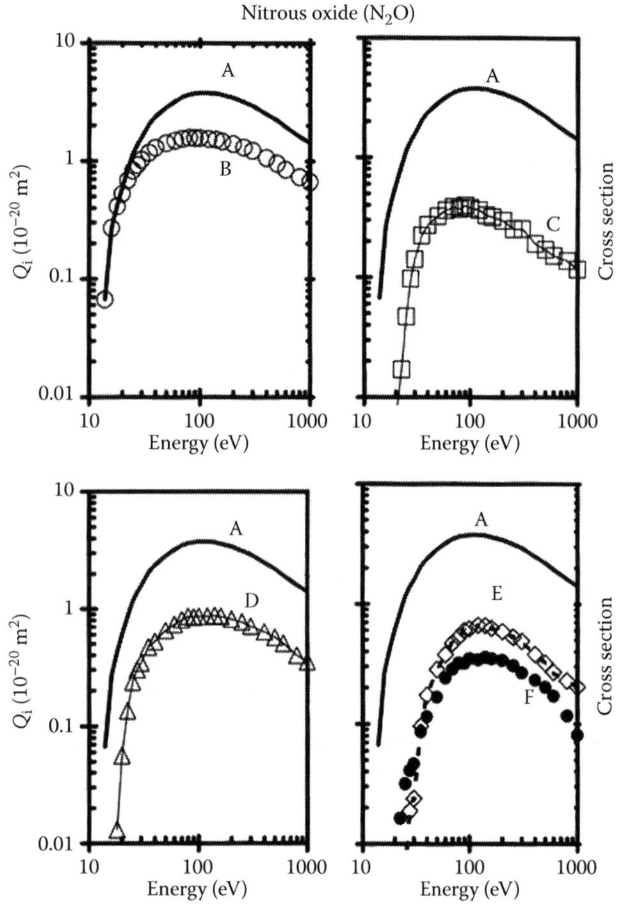

FIGURE 34.7 Partial ionization cross sections for N_2O. Total cross section is also shown to indicate the relative contribution. (A) Total; (B) N_2O^+; (C) N_2^+; (D) NO^+; (E) N^+; (F) O^+. (Adapted from Lindsay, B. G., R. Rejoub, and R. F. Stebbings, *J. Chem. Phys.*, 118, 5894, 2003.)

TABLE 34.8
Partial Ionization Cross Sections

	Ionization Cross Section (10^{-20} m²)				
Energy (eV)	N_2O^+	N_2^+	NO^+	N^+	O^+
14	0.067				
16	0.269				
18	0.411		0.0131		
20	0.527	0.0066	0.0564		0.0066
22.5	0.684	0.017	0.136	0.0023	0.0163
25	0.827	0.047	0.237	0.0059	0.0315
27.5	0.922	0.097	0.303	0.0187	0.0407
30	1.02	0.141	0.344	0.0237	0.0466
35	1.17	0.223	0.471	0.0958	0.086
40	1.29	0.272	0.523	0.173	0.115
50	1.41	0.323	0.654	0.282	0.167
60	1.48	0.363	0.735	0.375	0.24
70	1.54	0.387	0.809	0.467	0.285
80	1.57	0.372	0.864	0.526	0.319
90	1.56	0.398	0.854	0.596	0.32
100	1.56	0.379	0.865	0.618	0.347
120	1.53	0.365	0.865	0.663	0.341
140	1.51	0.33	0.882	0.657	0.357
160	1.46	0.318	0.866	0.627	0.347
200	1.39	0.297	0.819	0.585	0.337
250	1.3	0.252	0.775	0.525	0.305
300	1.22	0.253	0.712	0.486	0.265
400	1.06	0.191	0.634	0.377	0.232
500	0.95	0.171	0.569	0.32	0.202
600	0.844	0.152	0.507	0.269	0.171
800	0.731	0.138	0.4	0.228	0.117
1000	0.666	0.117	0.353	0.203	0.0807

Source: Adapted from Lindsay, B. G., Rejoub, R., and R. F. Stebbings, *J. Chem. Phys.*, 118, 5894, 2003.

Note: See Figure 34.7 for graphical presentation.

34.9 REACTION SCHEMES

1. Primary ionization:

$$N_2O + e \rightarrow N_2O^+ + 2e \quad (34.1)$$

2. Direct attachment:

$$N_2O + e \rightarrow N_2O^- \quad (34.2)$$

While this process is observed at higher temperatures, there is some disagreement toward its occurrence at room temperature (Brüning et al., 1998).

3. Dissociative ionization:

$$N_2O + e \rightarrow X^+ + Y + 2e \quad (34.3)$$

where X^+ is a fragment as shown in Table 34.8, and Y is the corresponding neutral fragment.

4. Dissociative attachment:

$$N_2O + e \rightarrow N_2 + O^- \quad (34.4)$$

This is the only negative ion observed in mass spectrometric studies (Krishnakumar and Srivastava, 1990).

5. Negative ion–molecule reactions (Dutton et al., 1975b):

$$O^- + N_2O \rightarrow NO + NO^- \quad (34.5)$$

$$O^- + 2N_2O \rightarrow N_2O + N_2O_2^- \quad (34.6)$$

$$NO^- + 2N_2O \rightarrow N_2O + N_3O_2^- \quad (34.7)$$

6. Collisional detachment from NO^-, Reaction 34.5:

$$NO^- + N_2O \rightarrow N_2O + NO + e \quad (34.8)$$

34.10 ATTACHMENT CROSS SECTIONS

The energy and peak cross section observed are shown in Table 34.9.

34.11 ATTACHMENT RATE CONSTANT

See Tables 34.10 through 34.12.

TABLE 34.9
Attachment Energy and Peak Cross Section for N_2O for Production of O^- Ion

Peak Energy (eV)	Peak Cross Section (10^{-22} m^2)	Reference
2.25	9.0	Krishnakumar and Srivastava (1990)
5.40	0.007	
8.10	0.014	
35.0 (continuum)	0.18	
2.4	10.0	Christophorou et al. (1971)
2.42 (323 K)	7.07	Chaney and Christophorou (1969)
2.32 (373 K)	7.21	
2.12 (473 K)	7.58	
2.2	8.6	Rapp and Briglia (1965)

Note: Temperature 293 K, unless otherwise mentioned.

TABLE 34.10
Attachment Cross Sections for N_2O

	Attachment Cross Section (10^{-22} m^2)	
Energy (eV)	Krishnakumar and Srivastava (1990)	Rapp et al. (1965)
0.4	1.53	0.46
0.5	1.81	1.33
0.6	2.09	1.73
0.7	2.28	1.93
0.8	2.49	2.04
0.9	2.67	2.08
1.0	2.79	2.16
1.1	2.93	2.23
1.2	3.14	2.33
1.3	3.41	2.49
1.4	3.74	2.79
1.5	4.19	3.29
1.6	4.81	3.92
1.7	5.37	4.94
1.8	6.13	5.95
1.9	6.97	6.63
2.0	8.07	7.58
2.1	8.72	8.28
2.2	8.91	8.60
2.3	8.89	8.57
2.4	8.43	8.04
2.5	7.55	7.10
2.6	6.39	5.98
2.7	5.13	4.84
2.8	4.28	3.57
2.9	3.22	2.60
3.0	2.36	1.92
3.1	1.77	1.39
3.2	1.13	0.97
3.3	0.77	0.63
3.4	0.58	0.47

TABLE 34.10 (continued)
Attachment Cross Sections for N_2O

	Attachment Cross Section (10^{-22} m^2)	
Energy (eV)	Krishnakumar and Srivastava (1990)	Rapp et al. (1965)
3.5	0.30	0.35
3.6	0.15	0.28
3.7	0.09	0.23
3.8	0.05	0.19
3.9	0.03	0.17
4.0	0.02	0.13
4.1		0.12
4.2		0.11
4.3		0.10
4.5		0.10
5.0		0.10

Note: See Figure 34.8 for graphical presentation.

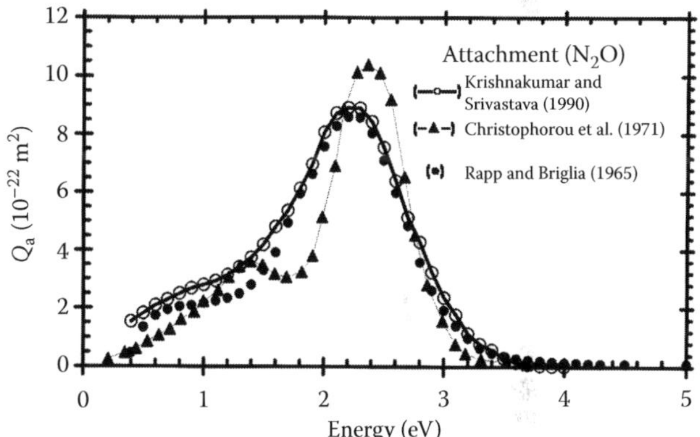

FIGURE 34.8 Attachment cross sections for N_2O. (—○—) Krishnakumar and Srivastava (1990), beam method; (–▲–) Christophorou et al. (1971), swarm method; (●) Rapp and Briglia (1965), beam method.

TABLE 34.11
Attachment Rate Constant for N_2O for Two-Body Process

K (eV)	Method	Rate (m^3/s)	Reference
261	M. C.	1.7×10^{-19}	Shimamori and Fessenden (1979)
275		3.2×10^{-19}	
295		5.0×10^{-19}	
313		9.2×10^{-19}	
335		17×10^{-19}	
323	Swarm	2.34×10^{-17}	Chaney and Christophorou (1969)
373		4.25×10^{-17}	
473		1.71×10^{-16}	

continued

N₂O

TABLE 34.11 (continued)
Attachment Rate Constant for N_2O for Two-Body Process

K (eV)	Method	Rate (m³/s)	Reference
(0.5)	Beam	2.82×10^{-17}	
(1.0)		1.10×10^{-16}	
(1.5)		2.25×10^{-16}	
(2.0)		3.65×10^{-16}	
(2.5)		9.05×10^{-16}	
(3.0)		1.70×10^{-16}	
(0.025)	Swarm	$<3.0 \times 10^{-21}$	Phelps and Voshall (1968)

Note: M. C. = microwave conductivity.

TABLE 34.12
Attachment Rate Constant for N_2O for Three-Body Process with N_2 by the Swarm Method

Energy (eV)	Rate (m⁶/s) × 10^{-43}
0.025	0.03
0.2	0.30
0.3	1.27
0.4	1.85
0.5	2.81
0.6	3.68
0.7	5.07

Source: Adapted from Chaney, E. L., and L. G. Christophorou, *J. Chem. Phys.*, 51, 883, 1969.

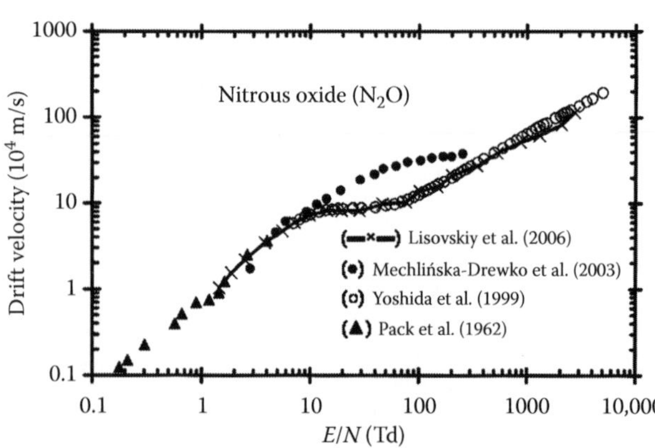

FIGURE 34.9 Drift velocity of electrons for N_2O. (—×—) Lisovskiy et al. (2006); (●) Mechlińska-Drewko et al. (2003); (○) Yoshida et al. (1999); (▲) Pack et al. (1962).

34.12 DRIFT VELOCITY OF ELECTRONS

See Figure 34.9 and Table 34.13.

34.13 LONGITUDINAL DIFFUSION COEFFICIENT (D_L/M)

See Table 34.14.

TABLE 34.13
Drift Velocity of Electrons for N_2O

E/N (Td)	*W* (10⁴ m/s)	*E/N* (Td)	*W* (10⁴ m/s)
7	5.85	260	24.8
8	6.48	280	25.8
9	7.02	300	27.4
10	7.39	350	30.3
12	7.98	400	33.5
14	8.34	500	39.4
16	8.47	600	44.7
18	8.64	700	49.9
20	8.60	800	55.2
25	8.74	900	60.0
30	8.78	1000	64.7
40	9.02	1100	69.8
50	9.46	1200	75.0
60	9.77	1300	78.8
70	10.5	1400	83.50
80	11.1	1500	85.9
90	11.9	1800	98.6
100	12.6	2000	104
110	13.6	2100	109
120	14.3	2200	113
130	15.2	2300	115
140	16.0	2500	121
150	17.0	3000	135
170	18.00	3500	153
200	20.1	4000	165
220	21.7	5000	193
240	23.2		

Source: Adapted from Yoshida, K. et al., *J. Phys. D: Appl. Phys.*, 32, 862, 1999.

TABLE 34.14
Longitudinal Diffusion Coefficient (ND_L) and Ratio D_L/μ for N_2O

E/N (Td)	ND_L (10^{24} 1/ms)	D_L/μ (eV)
7	0.746	0.0893
8	0.919	0.113
9	0.832	0.107
10	0.791	0.107
12	0.774	0.116
14	0.723	0.121
16	0.684	0.129
18	0.676	0.141
20	0.733	0.170
25	0.618	0.177
30	0.588	0.201
40	0.536	0.238
50	0.583	3.07
60	0.482	2.96
70	0.795	5.28
80	0.831	5.98
90	1.17	8.79
100	1.27	1.00

TABLE 34.14 (continued)
Longitudinal Diffusion Coefficient (ND_L) and Ratio D_L/μ for N_2O

E/N (Td)	ND_L (10^{24} m/s)	D_L/μ (eV)
110	1.53	1.24
120	1.67	1.40
130	1.84	1.58
140	1.87	1.64
150	2.13	1.88
170	2.17	2.04
200	2.97	2.96
220	2.94	2.99
240	2.67	2.76
260	2.57	2.70
280	2.94	3.19
300	2.65	2.91
350	2.95	3.41
400	2.85	3.41
500	2.93	3.72
600	3.10	4.16
700	3.50	4.90
800	4.03	5.85
900	4.44	6.66
1000	4.87	7.53
1100	5.13	8.09
1200	5.41	8.64
1300	5.77	9.52
1400	7.17	12.0
1500	6.84	11.9
1800	6.53	11.9
2000	7.08	13.6
2100	8.51	16.4
2200	8.3	16.2
2300	8.10	16.2
2500	7.75	16.0
3000	10.7	23.9
3500		
4000		
5000	8.55	22.1

Source: Adapted from Yoshida, K. et al., *J. Phys. D: Appl. Phys.*, 32, 862, 1999.

Note: See Figure 34.10 for graphical presentation.

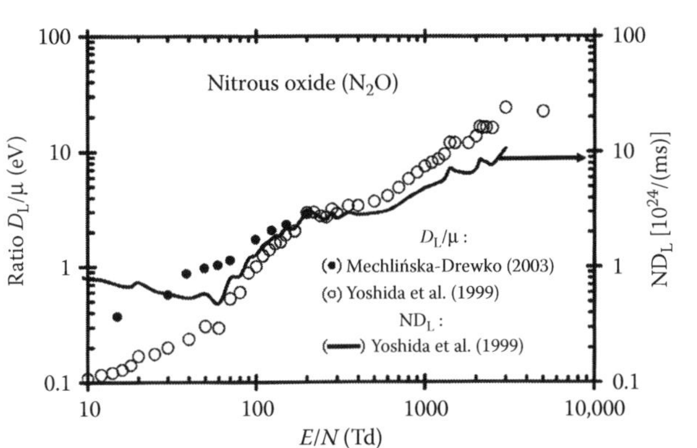

FIGURE 34.10 Density-normalized longitudinal diffusion coefficient (ND_L) and ratio of longitudinal diffusion coefficient to mobility (D_L/μ) for N_2O. D_L/μ: (●) Mechlińska-Drewko et al. (2003); (○) Yoshida et al. (1999). ND_L: (—) Yoshida et al. (1999).

34.14 ELECTRON MEAN ENERGY

Figure 34.11 shows the mean energy of electrons as a function of *E/N*. Note that the data of Mechlińska-Drewko et al. (2003) are characteristic energy and should be multiplied by 3/2 for Maxwellian energy distribution (Raju, 2005).

34.15 IONIZATION AND ATTACHMENT COEFFICIENTS

See Table 34.15.

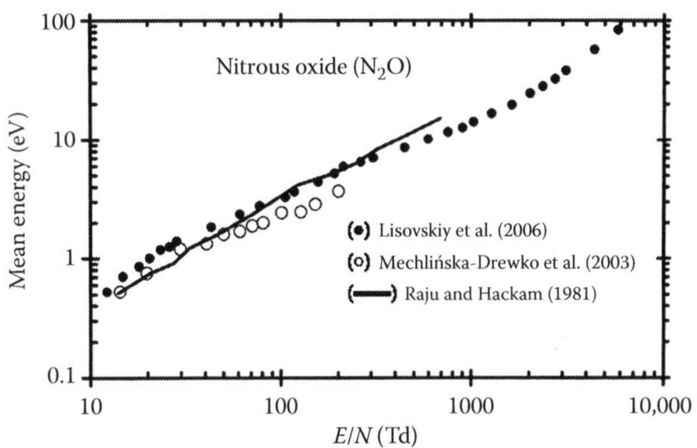

FIGURE 34.11 Mean energy of electrons as a function of *E/N*. (●) Lisovskiy et al. (2006); (○) Mechlińska-Drewko et al. (2003), characteristic energy and should be multiplied by 3/2 for Maxwellian energy distribution (Raju, 2003); (—) Raju and Hackam (1981).

TABLE 34.15
Density-Reduced Ionization and Attachment Coefficients

E/N (Td)	α/N (10^{-21} m²)	η/N (10^{-21} m²)
7	−0.193	0.193
8	−0.11	0.11
9	−0.178	0.178
10	−0.194	0.193
12	−0.241	0.241
14	−0.287	0.287
16	−0.342	0.342
18	−0.374	0.374
20	−0.394	0.394
25	−0.55	0.55
30	−0.883	0.833
40	−0.837	0.837
50	−1.71	1.71
60	−1.84	1.84
70	−1.82	1.82
80	−1.79	1.79
90	−1.35	1.35
100	−1.23	1.23
110	−1.16	1.16
120	−1.03	1.03

continued

N$_2$O

TABLE 34.15 (continued)
Density-Reduced Ionization and Attachment Coefficients

E/N (Td)	α/*N* (10^{-21} m^2)	η/*N* (10^{-21} m^2)
130	−0.951	0.951
140	−0.845	0.933
150	−0.705	
170	−0.534	
200	−0.0546	0.514
220	0.196	0.436
240	0.437	0.417
260	0.687	0.372
280	0.882	0.438
300	1.22	0.302
350	2.02	0.196
400	2.88	0.121
500	4.77	0.0877
600	6.68	
700	8.66	
800	10.6	
1000	14.4	
1500	22.1	
2000	28	
2500	33.1	
3000	35.9	
3500	37.3	
4000	39.9	
5000	42.7	

Source: Adapted from Yoshida, K. et al., *J. Phys. D: Appl. Phys.*, 32, 862, 1999.
Note: See Figure 34.12 for graphical presentation.

34.16 GAS CONSTANTS

Gas constants according to the expression

$$\frac{\alpha}{N} = F\exp\left(-\frac{GN}{E}\right) \quad (34.9)$$

where α/*N* expressed in m^2 and *E*/*N* in Td (Figure 34.13) are obtained from the data of Lisovskiy et al. (2006): $F = 4.49 \times 10^{-20}$ m^2 and $G = 719.2$ Td^{-1}.

34.17 CONSOLIDATED CROSS SECTIONS

Figure 34.14 shows the relative magnitudes of various cross sections presented.

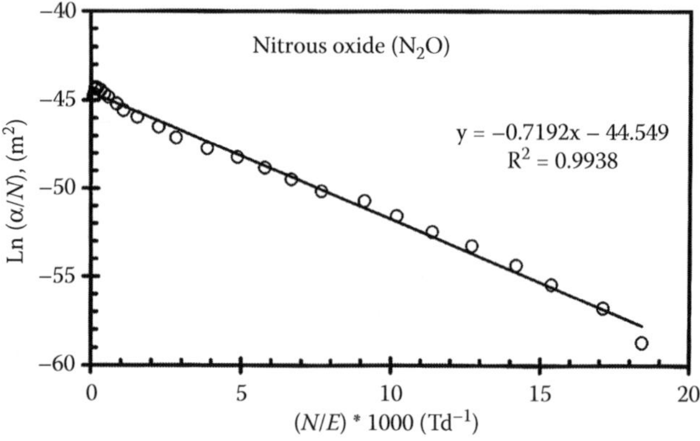

FIGURE 34.13 Evaluation of gas constants, *F* and *G*, from the data of Lisovskiy et al. (2006).

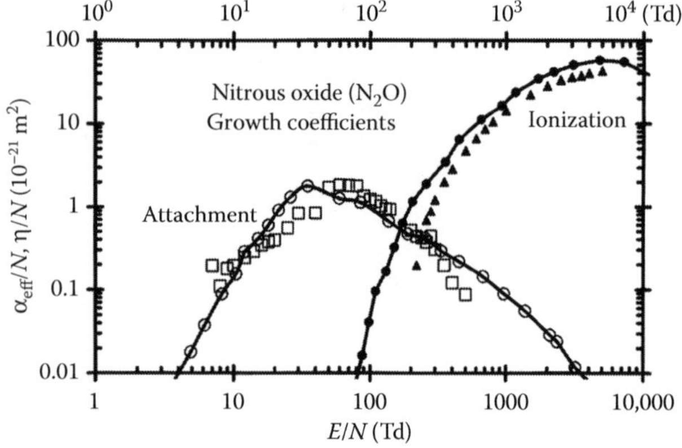

FIGURE 34.12 Density-normalized effective ionization and attachment coefficients for N$_2$O. α_{eff}/N: (—●—) Lisovskiy et al. (2006); (▲) Yoshida et al. (1999). η/*N*: (—○—) Lisovskiy et al. (2006); (□) Yoshida et al. (1999).

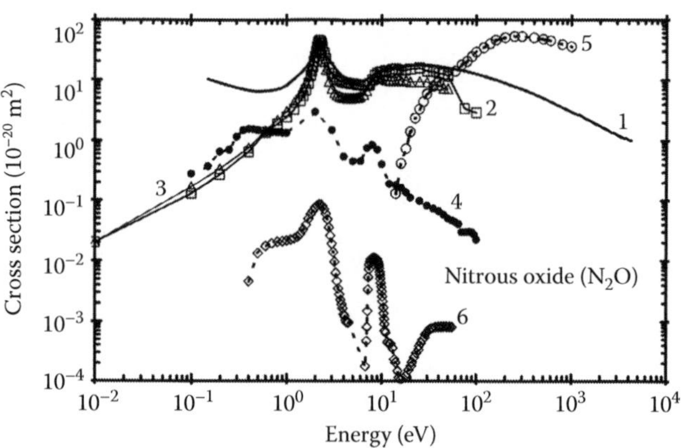

FIGURE 34.14 Consolidated cross section data for N$_2$O. 1—Total; 2—elastic scattering; 3—momentum transfer cross section; 4—vibrational excitation; 5—ionization; 6—attachment.

REFERENCES

Austin, J. M. and A. L. S. Smith, *J. Phys. D: Appl. Phys.*, 6, 2236, 1973.

Brunger, M., S. Buckman, and M. Elford, in *Interactions of Photons and Electrons with Molecules*, Vol. 17c, Springer-Verlag, New York, NY, 2003, Chapter 6.4.

Brüning, F., S. Matejcik, E. Illenberger, Y. Chu, G. Senn, D. Muigg, G. Denifl, and T. D. Märk, *Chem. Phys. Lett.*, 292, 177, 1998.

Chaney, E., L. and L. G. Christophorou, *J. Chem. Phys.*, 51, 883, 1969.

Christophorou, L. G., D. L. McCorkle, and V. E. Anderson, *J. Phys. B: At. Mol. Phys.*, 4, 1163, 1971.

Dutton, J., F. M. Harris, and D. B. Hughes, *J. Phys. B: At. Mol. Phys.*, 8, 313, 1975b.

Dutton, J., F. M. Harris, and D. B. Hughes, *J. Phys. D: Appl. Phys.*, 8, 1640, 1975a.

Govinda Raju, G. R. and R. Hackam, *J. Appl. Phys.*, 52, 3912, 1981.

Hayashi, M. in Ohm-Sha (Ed.), *Electron Collision Cross Sections. Handbook on Plasma Material Science*, Vol. 4(9), Committee No. 153 on Plasma Material Science The Japan Society for Promotion of Science (1992), Cited by Mechlińska-Drewko et al. (2003).

Iga, I., M. V. V. S. Rao, and S. K. Srivastava, *J. Geophys. Res.*, 101, 9261, 1996.

Johnstone, W. M. and W. R. Newell, *J. Phys. B: At. Mol. Phys.*, 26, 129, 1993.

Karwasz, G. R., Brusa, and A. Zecca, in Cross sections for scattering- and excitation-processes in electron–molecule collisions, *Interactions of Photons and Electrons with Molecules*, Vol. 17c, Springer-Verlag, New York, NY, 2003, Chapter 6.1.

Kim, Y.-K., W. Hwang, N. M. Weinberger, M. A. Ali, and M. E. Rudd, *J. Chem. Phys.*, 106, 1026, 1997.

Kitajima, M., Y. Sakamoto, S. Watanabe, T. Suzuki, T. Ishikawa, H. Tanaka, and M. Kimura, *Chem. Phys. Lett.*, 309, 414, 1999.

Krishnakumar, E. and S. K. Srivastava, *Phys. Rev.*, 41, 2445, 1990.

Kwan, Ch. K., Y.-F. Hsieh, W. E. Kaupila, S. J. Smith, T. S. Stein, M. N. Uddin, and M. S. Dababneh, *Phys. Rev. Lett.*, 52, 1417, 1984.

Lindsay, B. G., R. Rejoub, and R. F. Stebbings, *J. Chem. Phys.*, 118, 5894, 2003.

Lisovskiy, V., J.-P. Booth, K. Landry, D. Douai, and V. Cassagne, *J. Phys. D: Appl. Phys.*, 39, 1866, 2006.

Lopez, J., V. Tarnovsky, M. Gutkin, and K. Becker, *Int. J. Mass Spectrom.*, 225, 25, 2003.

Marinković, B., Cz. Szmytkowski, V. Pejčev, D. Filipović, and L. Vušković, *J. Phys. B: At. Mol. Phys.*, 19, 2365, 1986.

Mechlińska-Drewko, J., T. Wróblewski, Z. Lj. Petrović, V. Novaković, and G. P. Karwasz, *Rad. Phys. Chem.*, 68, 205, 2003.

Pack, J. L., R. E. Voshall, and A. V. Phelps, *Phys. Rev.*, 127, 2084, 1962.

Phelps, A. V. and R. E. Voshall, *J. Chem. Phys.*, 49, 3246, 1968.

Raju, G. G., *Gaseous Electronics: Theory and Practice*, Taylor & Francis, London, 2005.

Ramsauer, C., and R. Kollath, *Ann. Phys. Leipzig*, 7, 176, 1930.

Rapp, D. and D. D. Briglia, *J. Chem. Phys.*, 43, 1480, 1965.

Rapp, D. and P. Englander-Golden, *J. Chem. Phys.*, 43, 1464, 1965.

Rapp, D., P. Englander-Golden, and D. D. Briglia, *J. Chem. Phys.*, 42, 4081, 1965.

Shilin, X., Z. Fang, Y. Liqiang, Y. Changking, and X. Kezun, *J. Phys. B: At. Mol. Opt. Phys.*, 30, 2867, 1997.

Shimamori, H. and R. W. Fessenden, *J. Chem. Phys.*, 70, 1137, 1979.

Shimanouchi, T., *Tables of Molecular Vibrational Frequencies Consolidated Volume I*. NSRDA-NBS 39, US Department of Commerce, Washington (DC), 1972.

Szmytkowski, C., G. Karwasz, and K. Maciąg, *Chem. Phys. Lett.*, 107, 481, 1984.

Szmytkowski, C., K. Maciąg, G. Karwasz, and D. Filipović, *J. Phys. B: At. Mol. Opt. Phys.*, 22, 525, 1989.

Winstead, C. and V. McKoy, *Phys. Rev. A*, 57, 3589, 1998.

Yoshida, K., N. Sasaki, H. Ohuchi, H. Hasegawa, M. Shimozuma, and H. Tagashira, *J. Phys. D: Appl. Phys.*, 32, 862, 1999.

Zecca, A., I. Lazzizzera, M. Krauss, and C. E. Kuyatt, *J. Chem. Phys.*, 61, 4560, 1974.

Zecca, A. J. C., Nogueira, G. P. Karwasz, and R. S. Brusa, *J. Phys. B: At. Mol. Opt. Phys.*, 28, 477, 1995.

35

OZONE

CONTENTS

Ozone (O_3) is a polar, electron-attaching molecule that has 24 electrons. Its electronic polarizability is 3.57×10^{-40} F m^2, the dipole moment is 0.534 D, and the ionization potential is 12.43 eV.

35.1 SELECTED REFERENCES FOR DATA

See Table 35.1.

TABLE 35.1
Selected References Data on O_3

Quantity	Range: eV, (Td)	Reference
Ionization cross section	12.75–5000	Kim et al. (2004)
Total cross section	50–2000	Joshipura et al. (2002)
Total cross section	350–5000	de Pablos et al. (2002)
Attachment cross sections	0–10	**Rangwala et al. (1999)**
Attachment cross sections	0–10	**Senn et al. (1999)**
Total cross section	0.009–10	**Gulley et al. (1998)**
Excitation cross section	0–12	**Allan et al. (1996)**
Excitation cross section	3.0–11.0	**Mason et al. (1996)**
Attachment cross section	0–9	**Skalny et al. (1996)**
Excitation cross section	7, 10, 15, 20	**Sweeny and Shyn (1996)**
Attachment cross sections	0–10	**Walker et al. (1996)**
Ionization cross section	40–500	**Newson et al. (1995)**
Vibrational excitation cross sections	3–7	**Davies et al. (1993)**
Elastic scattering cross sections	3.0–20.0	**Shyn and Sweeny (1993)**
Excitation cross sections	3.75–12.5	**Johnstone et al. (1992)**
Ionization cross sections	12.5–100	**Siegel (1982)**
Attachment cross section	0–1	**Stelman et al. (1972)**
Attachment processes	0–3	**Curran (1961)**

Note: Bold font indicates experimental study.

35.2 TOTAL SCATTERING CROSS SECTION

Total scattering cross sections measured in a single study in the range 0.1–1000 eV are not available. Also, experimental values in the range 10–350 eV are not available. Tables 35.2 and 35.3 show the data in the low-energy range and high-energy range. The essential features of Figure 35.1 which shows the cross sections are

1. The cross section increases sharply for electron energy <30 meV due to the dipole moment of the O_3 molecule.
2. A weak structure between 40–60 eV, attributed to dissociative attachment.
3. A shape resonance due to formation of temporary negative ion at ~40 eV; the increase of Q_T at shape

resonance is not as dominant as in several other gases.

4. A monotonic decrease of Q_T at higher energy which is a common feature to most gases.

TABLE 35.2
Total Scattering Cross Sections in the Low-Energy Range

	Q_T (10^{-20} m²)		Energy	Q_T (10^{-20} m²)	
Energy (eV)	O_3	ClO_2	(eV)	O_3	ClO_2
0.010	117		0.40	17.8	88.73
0.012	108		0.45	16.7	77.37
0.015	99		0.5	15.8	76.59
0.017	93.9		0.6	14.8	64.50
0.020	87.6		0.7	13.7	55.75
0.022	84.0	1146.10	0.8	13.1	50.25
0.025	79.4	918.87	0.9	12.8	46.25
0.030	73.1	628.1	1.0	12.6	43.75
0.035	68.1	353.52	1.2	12.2	40.00
0.040	63.9		1.5	12.0	37.00
0.045	60.3	226.50	1.7	12.1	35.50
0.050	57.3	392.5	2.0	12.3	34.0
0.06	52.3	495.65	2.2	12.6	33.50
0.07	48.2	463.80	2.5	13.0	32.25
0.08	45.0	442.50	3.0	13.8	30.00
0.09	42.2	415.95	3.5	14.8	27.50
0.10	39.8	400.0	4.0	15.4	24.85
0.12	35.9	315.0	4.5	14.6	22.45
0.15	31.6	251.2	5	14.1	20.30
0.17	29.4	224.5	6	14.2	18.75
0.20	26.7	192.3	7	14.7	19.65
0.22	25.2	176.1	8	15.3	20.1
0.25	23.4	149.2	9	15.9	20.5
0.30	21.0	122.1	10	16.5	20.8
0.35	19.2	105.4			

Source: Adapted from Gulley, R. J. et al., *J. Phys. B: At. Mol. Opt. Phys.*, 31, 5197, 1998.

Note: Data for ClO_2 are given for comparison. Also see Karwasz et al. (2003).

TABLE 35.3
Total Scattering Cross Sections for O_3

Energy (eV)	Q_T (10^{-20} m²)	Energy (eV)	Q_T (10^{-20} m²)
350	27.7	1500	8.91
400	24.6	1750	7.90
450	22.5	2000	7.12
500	20.7	2500	6.00
600	18.1	3000	5.18
700	15.9	3500	4.61
800	14.3	4000	4.15
900	13.2	4500	3.75
1000	12.1	5000	3.46
1250	10.3		

Source: Adapted from de Pablos, J. L. et al., *J. Phys. B: At. Mol. Opt. Phys.*, 35, 865, 2002.

35.3 DIFFERENTIAL SCATTERING CROSS SECTION

Figure 35.2 shows experimental differential cross sections in the low-energy range, from 3 to 20 eV (Shyn and Sweeny, 1993). Figure 35.3 shows the theoretical differential cross sections in the intermediate and high-energy range, from 30 to 10,000 eV (de Pablos et al., 2002).

35.4 ELASTIC SCATTERING AND MOMENTUM TRANSFER CROSS SECTIONS

A single study is available for the low-energy range as shown in Table 35.4 (Shyn and Sweeny, 1993).

35.5 VIBRATIONAL EXCITATION CROSS SECTION

See Table 35.5.

The excitation cross sections for the Hartley band lying between 4–6 eV (240–749 nm more specifically) are shown in Table 35.6 and Figure 35.4 (Sweeny and Shyn, 1996).

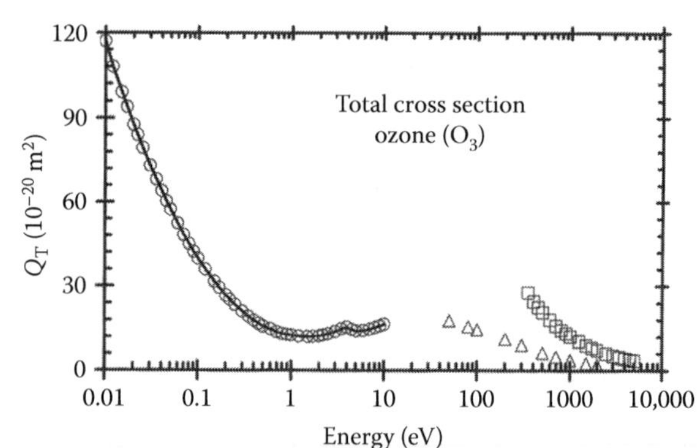

FIGURE 35.1 Total scattering cross sections for O_3. (Δ) Joshipura et al. (2002), theory; (□) de Pablos et al. (2002); (—□—) Gulley et al. (1998).

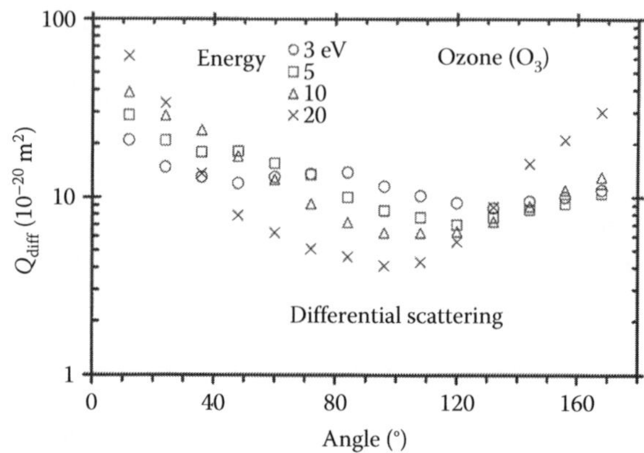

FIGURE 35.2 Differential scattering cross sections for O_3. (Adapted from Shyn, T. W. and C. J. Sweeny, *Phys. Rev.*, 47, 2919, 1993.)

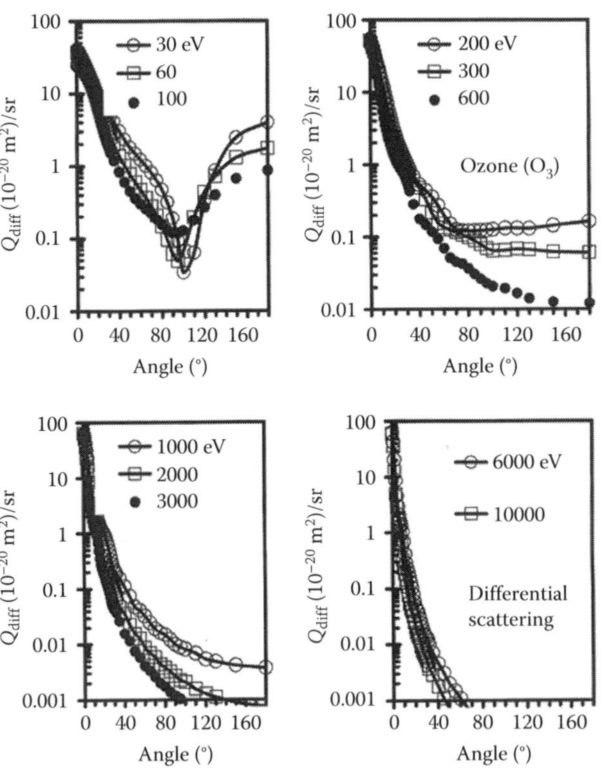

FIGURE 35.3 Theoretical differential scattering cross sections for O_3in the medium and high-energy range. (Adapted from de Pablos, J. L. et al., *J. Phys. B: At. Mol. Opt. Phys.*, 35, 865, 2002.)

TABLE 35.4
Elastic and Momentum Transfer Cross Sections for O_3

Energy (eV)	Q_{el} (10^{-20} m²)	Q_M(10^{-20} m²)
3.0	14.8	13.4
5.0	14.9	11.8
7.0	14.1	10.2
10	14.4	10.7
15	14.1	10.6
20	13.0	11.6

Source: Adapted from Shyn, T. W. and C. J. Sweeny, *Phys. Rev.*, 47, 2919, 1993.

35.6 ENERGY LOSS SPECTRUM

See Table 35.7.

35.7 IONIZATION CROSS SECTION

Table 35.8 and Figure 35.5 show the total ionization cross section. Figure 35.6 shows the partial ionization cross sections.

TABLE 35.5
Properties of the First Three Vibrational Peaks of O_3

Property	Elastic Peak	$\nu = 1$ (meV)	$\nu = 2$ (meV)	$\nu = 3$ (meV)
Threshold		137	256	378
Peak energy	0	130	260	380
Composition of successive peaks		I peak	II peak	III peak
		60% ν_1	25% $2\nu_1$	60% $2\nu_1$
		15% ν_2	20% $\nu_1 + \nu_3$	15% $\nu_1 + \nu_3$
		30% ν_3	30% $2\nu_3$	30% $2\nu_3$
Differential cross section (10^{-22} m²/sr)	66	10.0	3.8	1.6

Source: Adapted from Davies, J. A. et al., *J. Phys. B: At. Mol. Opt. Phys.*, 26, L767, 1993.

TABLE 35.6
Excitation Cross Sections

Energy (eV)	7	10	15	20
Q_{ex} (10^{-22} m²)	69	88	95	60

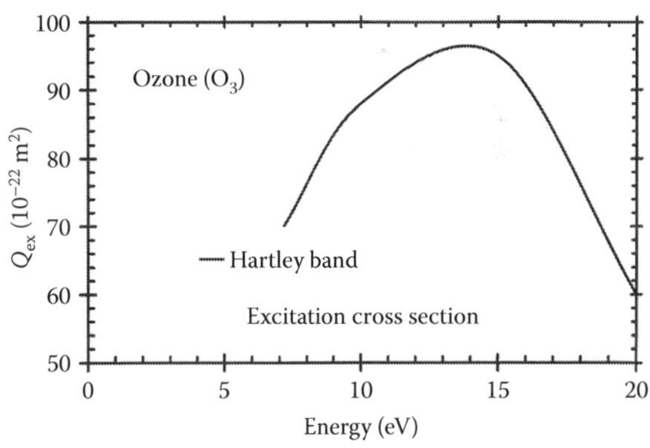

FIGURE 35.4 Excitation cross sections for O_3 for the Hartley band. (Adapted from Sweeny, C. J. and T. W. Shyn, *Phys. Rev. A*, 53, 1576, 1996.)

35.8 ATTACHMENT CROSS SECTION

Attachment occurs according to the following reactions (Walker et al., 1996):

$$O_3 + e \rightarrow [O_3^-] \rightarrow \begin{cases} O^- + O_2 & (35.1a) \\ O_2^- + O & (35.1b) \\ O^- + 2O & (35.1c) \\ O_3^* + e & (35.1d) \end{cases}$$

TABLE 35.7
Energy Loss Spectrum

State	Energy (eV)	Reference
	4.86	Mason et al. (1996)
$^1B_2$	4.89	Johnstone et al. (1992)
	4.85	Molina and Molina[20] (1986)
	7.25	Mason et al. (1996)
$^1B_1$	7.04	Johnstone et al. (1992)
	8.80	Mason et al. (1996)
$^3A_1$	8.76	Johnstone et al. (1992)
	9.31	Mason et al. (1996)
$^1B_2$	9.21	Johnstone et al. (1992)
$^1A_1$	9.39	Johnstone et al. (1992)
	10.21	Mason et al. (1996)
$^1B_2$	10.17	Johnstone et al. (1992)
	10.59	Mason et al. (1996)
$^1B_1$	10.59	Johnstone et al. (1992)
	11.1	Mason et al. (1996)
$^1A_1$	11.07	Johnstone et al. (1992)
	11.42	Johnstone et al. (1992)
	12.26	Johnstone et al. (1992)

TABLE 35.8
Total and Partial Ionization Cross Sections for O_3

	Newson et al. (1995)				Siegel (1982)
Energy (eV)	$(O_3)^+$ (10^{-21} m^2)	$(O_2)^+$ (10^{-21} m^2)	$(O)^+$ (10^{-21} m^2)	Total (10^{-21} m^2)	Total (10^{-21} m^2)
22					1.96
24					2.13
26					3.55
28					10.17
30					12.17
32					13.86
34					14.82
36					15.84
38					17.02
40	8.08	6.39	1.00	15.47	17.95
45	8.12	6.53	1.41	16.06	19.34
50	8.72	7.42	1.84	17.98	20.68
55	9.17	7.81	2.15	19.13	21.69
60	9.82	8.79	2.59	21.2	22.46
65	9.97	8.71	2.76	21.44	23.06
70	10.23	9.63	3.05	22.91	23.48
75	10.25	9.34	3.24	22.83	23.76
80	10.58	9.92	3.49	23.99	23.94
85	10.5	9.67	3.56	23.73	24.04
90	10.57	9.8	3.73	24.1	24.07
95	10.58	9.68	3.77	24.03	23.93
100	10.6	9.9	3.78	24.28	23.55
152	10.07	8.66	3.65	22.38	
200	9.59	8.44	3.34	21.37	
259	9.11	8.27	3.14	20.52	
306	8.54	7.61	2.81	18.96	
355	7.97	7.00	2.6	17.57	
413	7.57	6.79	2.5	16.86	
462	6.82	5.65	2.21	14.68	
509	6.9	6.49	2.31	15.7	

Note: Data from Siegel (1982) are digitized and interpolated.

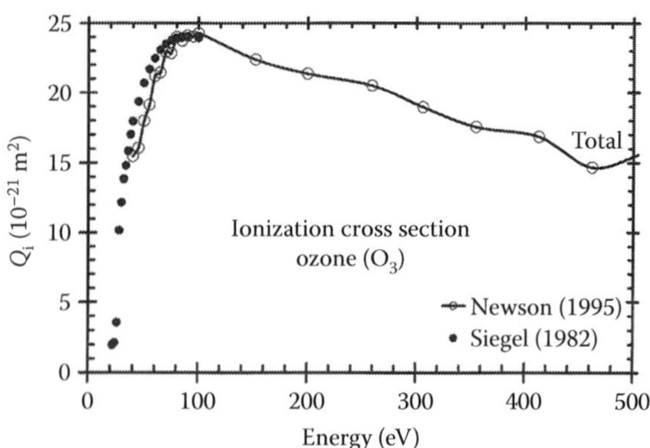

FIGURE 35.5 Total ionization cross sections for O_3. (—○—) Newson et al. (1995); (●) Siegel (1982).

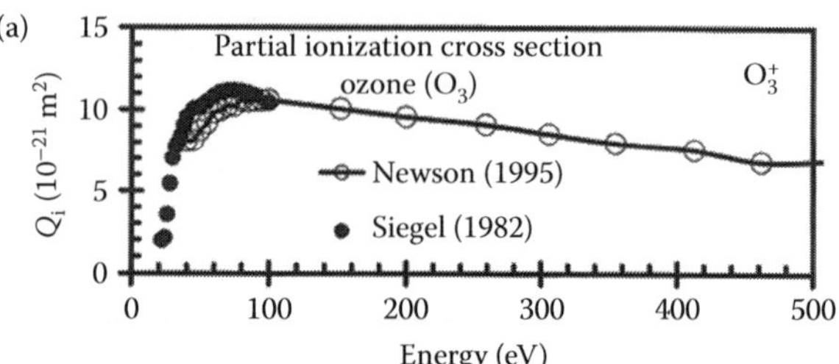

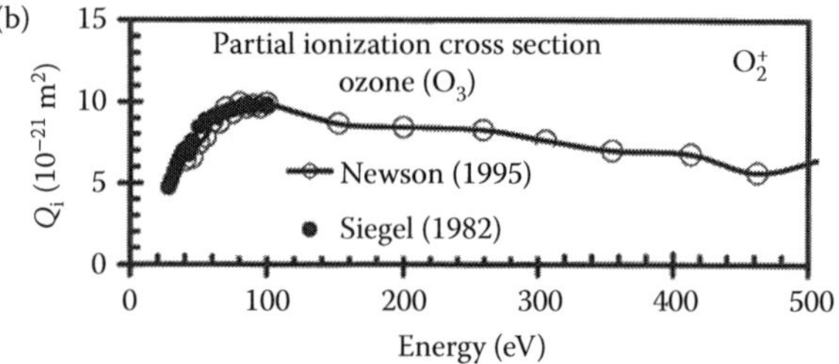

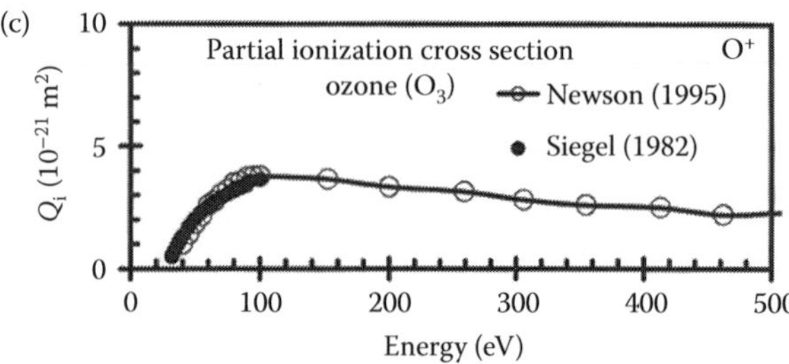

FIGURE 35.6 Partial ionization cross sections for O_3. (a) O_3^+; (b) O_2^+; (O^+). (—○—) Newson et al. (1995); (●) Siegel (1982).

Reactions 35.1a through 35.1c are dssociative attachment and Reaction 35.1d is autodetachment. Electron affinity of O_3 is 2.108 eV.

Three resonant regions are observed centered around 1.4, 3.5, and 7.5 eV. The principal features are shown in Table 35.9 (Rangwala et al., 1999).

Table 35.10 and Figures 35.7 and 35.8 show the attachment cross sections for the formation of O^- and O_2^- ions, tabulated by Itikawa (2003) on the basis of measurements of Senn et al. (1999).

TABLE 35.9
Principal Features of Attachment Cross Sections for O_3

Energy (eV)	Feature	O^- Cross Section (10^{-22} m^2)	O_2^- Cross Section (10^{-22} m^2)
0	Peak	140	
1.2	Peak	—	16.8
1.4	Peak	37.0	—
3.0	Shoulder	8.7	—
6.5	Gradual slope	3.0	—
7.3	Peak	—	0.4
7.5	Peak	6.0	

Source: Adapted from Rangwala, S. A. et al., *J. Phys. B: At. Mol. Opt. Phys.*, 32, 3795, 1999.
Note: Values at zero energy is from Senn et al. (1999).

35.9 ATTACHMENT RATE COEFFICIENT

TABLE 35.10
Attachment Cross Sections for O_3

Energy (eV)	O^- (10^{-22} m^2)	O_2^- (10^{-22} m^2)	Total (10^{-22} m^2)
0.0	8.06	2.00	10.06
0.1	7.50	2.22	9.72
0.2	6.92	1.80	8.72
0.3	8.20	2.00	10.20
0.4	10.01	2.16	12.17
0.5	11.94	3.06	15.00
0.6	14.69	4.02	18.71
0.7	18.70	6.24	24.94
0.8	20.87	8.64	29.51
0.9	24.88	11.88	36.76
1.0	28.24	13.26	41.50
1.1	31.58	15.00	46.58
1.2	36.21	16.80	53.01
1.3	37.03	15.48	52.51
1.4	37.21	14.16	51.37
1.5	36.47	11.52	47.99
1.6	34.67	9.30	43.97
1.7	31.29	6.96	38.25
1.8	27.97	4.68	32.65
1.9	23.78	3.96	27.74
2.0	21.11	2.28	23.39
2.1	19.05	1.86	20.91
2.2	15.99	1.56	17.55
2.3	13.30	1.20	14.50
2.4	11.51	0.60	12.11
2.5	11.25	0.69	11.94
2.6	9.59	0.30	9.89
2.7	8.93	0.24	9.17
2.8	8.63	0.24	8.87
2.9	9.18	0.18	9.36
3.0	8.47	0.18	8.65
3.1	8.94	0.18	9.12
3.2	8.25	0.19	8.44
3.3	8.20	0.20	8.40

TABLE 35.10 (continued)
Attachment Cross Sections for O_3

Energy (eV)	O^- (10^{-22} m^2)	O_2^- (10^{-22} m^2)	Total (10^{-22} m^2)
3.4	7.85	0.21	8.06
3.5	7.66	0.20	7.86
3.6	6.86	0.20	7.06
3.7	6.62	0.19	6.81
3.8	6.12	0.18	6.30
3.9	6.29	0.18	6.47
4.0	5.53	0.17	5.70
4.1	4.62	0.16	4.78
4.2	4.00	0.16	4.16
4.3	3.85	0.16	4.01
4.4	2.84	0.16	3.00
4.5	2.77	0.15	2.92
4.6	2.34	0.14	2.48
4.7	2.10	0.14	2.24
4.8	2.00	0.14	2.14
4.9	1.77	0.14	1.91
5.0	1.08	0.14	1.22
5.1	1.31	0.14	1.45
5.2	1.03	0.14	1.17
5.3	0.72	0.15	0.87
5.4	0.59	0.16	0.75
5.5	0.76	0.17	0.93
5.6	1.06	0.17	1.23
5.7	1.27	0.19	1.46
5.8	1.39	0.20	1.59
5.9	1.37	0.21	1.58
6.0	1.67	0.22	1.89
6.1	1.82	0.24	2.06
6.2	2.12	0.25	2.37
6.3	2.31	0.27	2.58
6.4	2.56	0.28	2.84
6.5	2.79	0.30	3.09
6.6	3.07	0.31	3.38
6.7	3.92	0.32	4.24
6.8	3.72	0.32	4.04
6.9	4.36	0.33	4.69
7.0	4.98	0.34	5.32
7.1	5.56	0.34	5.90
7.2	5.71	0.34	6.05
7.3	6.01	0.33	6.34
7.4	6.31	0.33	6.64
7.5	5.71	0.32	6.03
7.6	6.08	0.31	6.39
7.7	5.86	0.30	6.16
7.8	5.03	0.29	5.32
7.9	5.35	0.29	5.64
8.0	5.30	0.26	5.56
8.1	4.07	0.25	4.32
8.2	3.89	0.23	4.12
8.3	3.58	0.22	3.80
8.4	2.58	0.20	2.78
8.5	2.46	0.18	2.64
8.6	1.81	0.16	1.97
8.7	1.37	0.14	1.51
8.8	1.35	0.13	1.48
8.9	1.01	0.11	1.12

continued

TABLE 35.10 (continued)
Attachment Cross Sections for O_3

Energy (eV)	O^- (10^{-22} m^2)	O_2^- (10^{-22} m^2)	Total (10^{-22} m^2)
9.0	0.79	0.10	0.89
9.1	0.60	0.08	0.68
9.2	0.43	0.07	0.50
9.3	0.38	0.06	0.44
9.4	0.31	0.06	0.37
9.5	0.19	0.05	0.24
9.6	0.42	0.05	0.47
9.7	0.31	0.05	0.36
9.8	0.19	0.05	0.24
9.9	0.16	0.04	0.20
10.0	0.17	0.04	0.21

Note: Tabulated values are from Itikawa (2003) based on the measurements of Senn et al. (1999).

O_3

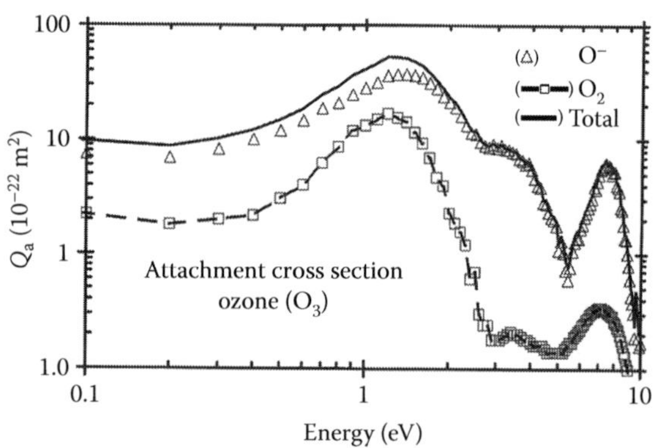

FIGURE 35.7 Attachment cross sections for O_3. (Δ) O^-; (—□—) O_2^-; (—■—) total. Note the logarithmic scales for the axes. (Adapted from Senn, G. et al., *Phys. Rev. Lett.*, 82, 5028, 1999.)

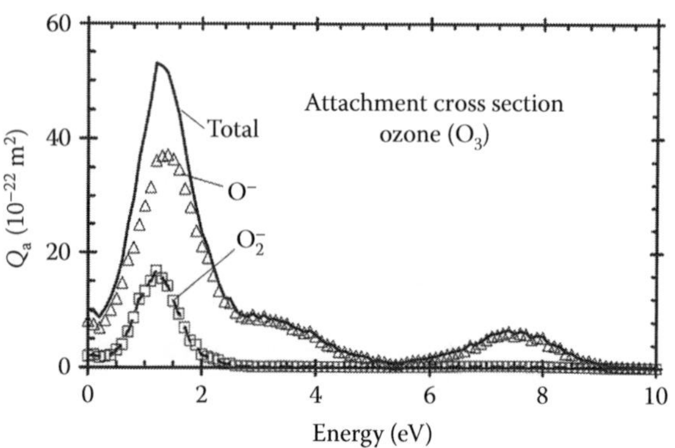

FIGURE 35.8 Attachment cross sections for O_3. (Δ) O^-; (—□—) O_2^-; (——) total. Note the linear scales for the axes. (Adapted from Senn, G. et al., Märk, *Phys. Rev. Lett.*, 82, 5028, 1999.)

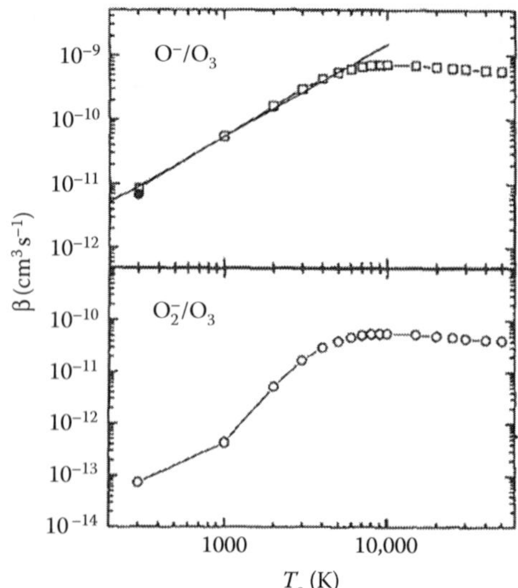

FIGURE 35.9 Attachment rate coefficients for O_3 determined by the beam technique. (Adapted from Skalny, J. D. et al., *Chem. Phys. Lett.*, 255, 112, 1996.)

Thermal rate coefficient at 300 K for two-body electron attachment is 9×10^{-18} m^3/s (Stelman et al., 1972). Attachment rate coefficients measured by the beam technique are shown in Figure 35.9 (Skalny et al., 1996).

REFERENCES

Allan, M., N. J. Mason, and J. A. Davies, *J. Chem. Phys.*, 105, 5665, 1996.

Curran, R. K., *J. Chem. Phys.*, 35, 1849, 1961.

Davies, J. A., W. M. Johnstone, N. J. Mason, P. Biggs, and R. P. Wayne, *J. Phys. B: At. Mol. Opt. Phys.*, 26, L767, 1993.

de Pablos, J. L., P. A. Kendall, P. Tegeder, A. Williart, F. Blanco, G. Garcia, and N. J. Mason, *J. Phys. B: At. Mol. Opt. Phys.*, 35, 865, 2002.

Gulley, R. J., T. A. Field, W. A. Steer, N. J. Mason, S. L. Hunt, J. P. Ziesel, and D. Field, *J. Phys. B: At. Mol. Opt. Phys.*, 31, 5197, 1998.

Itikawa, Y., in *Interactions of Photons and Electrons with Molecules*, Vol. 17C, Springer-Verlag, New York, NY, 2003, Chapter 5.2, pp. 78–114.

Johnstone, W. M., N. J. Mason, W. R. Newell, P. Biggs, G. Marston, and R. P. Wayne, *J. Phys. B: At. Mol. Opt. Phys.*, 25, 3873, 1992.

Joshipura, K. N., B. K. Antony, and M. Vinodkumar, *J. Phys. B: At. Mol. Opt. Phys.* 35, 4211, 2002.

Karwasz, G., R. Brusa, and A. Zecca, in *Interactions of Photons and Electrons with Molecules*, Vol. 17C, Springer-Verlag, New York, NY, 2003, Chapter 6-1.

Kim, Y.-K., K. K. Irikura, M. E. Rudd, M. A. Ali, P. M. Stone, J. Chang, J. S. Coursey, et al., *Electron-Impact Ionization Cross Section for Ionization and Excitation Database (Version 3.0)*. (Online). National Institute of Standards and Technology, Gaithersburg, MD, 2004. Available at http://physics.nist.gov/ionxsec. Accessed August 30, 2007.

Mason, N. J., J. M. Gingell, J. A. Davies, H. Zhao, I. C. Walker, and M. R. F. Siggel, *J. Phys. B: At. Mol. Opt. Phys.*, 29, 3075, 1996.

Molina, L. T. and M. J. Molina, *J. Geophys. Res.*, 91, 14501, 1986.

Newson, K. A., S. M. Luc, S. D. Price, and N. J. Mason, *Int. J. Mass Spectrom. Ion Proc.*, 148, 203, 1995.

Rangwala, S. A., S. V. K. Kumar, E. Krishnakumar, and N. J. Mason, *J. Phys. B: At. Mol. Opt. Phys.*, 32, 3795, 1999.

Senn, G., J. D. Skalny, A. Stamatovic, N. J. Mason, P. Schier, and T. D. Märk, *Phys. Rev. Lett.*, 82, 5028, 1999.

Shyn, T. W. and C. J. Sweeny, *Phys. Rev.*, 47, 2919, 1993.

Siegel, M. W., *Int. J. Mass Spectrom. Ion Phy.*, 44, 19, 1982.

Skalny, J. D., S. Matejcik, A. Kiendler, A. Stamatovic, and T. D. Märk, *Chem. Phys. Lett.*, 255, 112, 1996.

Stelman, D., J. L. Moruzzi, and A. V. Phelps, *J. Chem. Phys.*, 56, 4183, 1972.

Sweeny, C. J. and T. W. Shyn, *Phys. Rev. A*, 53, 1576, 1996.

Walker, I. C., J. M. Gingell, N. J. Mason, and G. Marston, *J. Phys. B: At. Mol. Opt. Phys.*, 29, 4749, 1996.

36

SULFUR DIOXIDE

CONTENTS

Sulfur dioxide (SO_2) is a polar, electron-attaching gas that has 32 electrons. The electronic polarizability is 4.14×10^{-40} F m^2, the dipole moment is 1.633 D, and the ionization potential is 12.349 eV. The vibrational modes and energies are shown in Table 36.1 (Shimanouchi, 1972). Eleven vibrational levels between 2.8 and 3.8 eV have been observed, attributed to the symmetric stretch (see below) belonging to a temporary state of SO_2^- (Sanche and Schulz, 1973).

36.1 SELECTED REFERENCES FOR DATA

See Table 36.2.

TABLE 36.1
Vibrational Modes and Energies for SO_2

Designation	Type of Motion	Energy (meV)
ν_1	Symmetrical stretch	142.7
ν_2	Bend	64.2
ν_3	Antisymmetrical stretch	168.9

Source: Adapted from Shimanouchi, T., *Tables of Molecular Vibrational Frequencies Consolidated Volume I*, NSRDS-NBS 37, US Department of Commerce, Washington (DC), 1972.

TABLE 36.2
Selected References for Data

Quantity	Range: eV, (Td)	Reference
Attachment cross section	2.6–10.2	**Krishnakumar et al. (1997)**
Ionization cross section	15–1000	**Lindsay et al. (1996)**
Ionization cross section	13–200	**Basner et al. (1995)**
Total cross section	90–4000	**Zecca et al. (1995)**
Momentum transfer cross section	1–30	**Gulley and Buckman (1994)**
Elastic scattering cross section	5–50	**Trajmar and Shyn (1989)**
Attachment rate coefficient	1.92–4.81	**Spyrou et al. (1986)**
Total cross section	1.5–15.7	**Szmytkowski and Maciąg (1986)**
Ionization cross section	12–200	**Orient and Srivastava (1984)**
Attachment cross section	0–9.5	**Čadež et al. (1983)**
Attachment cross section	2.5–9.5	**Orient and Srivastava (1983)**

continued

TABLE 36.2 (continued)
Selected References for Data

Quantity	Range: eV, (Td)	Reference
Attachment coefficients	(3–300)	**Lakdawala and Moruzzi (1981)**
Attachment coefficient	(3–300)	**Moruzzi and Lakdawala (1979)**
Vibrational excitation	0–17	**Sanche and Schulz (1973)**
Ionization coefficient	(300–360)	**Schlumbohm (1962)**
Attachment energy	≤7.0	**Reese et al. (1958)**

Note: Bold font denotes experimental study.

SO_2

TABLE 36.3
Total Scattering Cross Section for SO_2

Energy (eV)	Q_T (10^{-20} m^2)	Energy (eV)	Q_T (10^{-20} m^2)
1.5	36.4	225	11.6
2	37.2	256	11.0
3	36.4	289	10.1
4	35.4	324	9.54
5	36.4	361	8.96
6	34.7	400	8.55
7	33.8	484	7.50
8	31.8	576	6.66
9	31.6	676	5.94
10	31.2	785	5.56
12	30.9	900	4.90
15	26.3	1025	4.37
20	21.0	1150	4.05
25	18.3	1300	3.72
30	18.7	1450	3.27
35	18.9	1600	3.17
40	19.0	1770	2.82
50	18.5	1940	2.55
60	16.9	2120	2.36
70	15.7	2300	2.17
90	16.0	2500	2.01
100	15.2	2750	1.81
110	15.0	3000	1.71
121	14.7	3250	1.65
144	14.3	3500	1.48
169	12.9	4000	1.32
200	12.4		

Note: Energy range: $1.5 \le \varepsilon \le 70$ eV (Szmytkowski and Maciąg, 1986); $90 \le \varepsilon \le 4000$ eV (Zecca et al., 1995). See Figure 36.1 for graphical presentation.

36.2 TOTAL SCATTERING CROSS SECTIONS

See Table 36.3.

36.3 DIFFERENTIAL SCATTERING CROSS SECTION

Figure 36.2 shows the differential cross sections for elastic scattering as a function of electron energy and scattering angle (Gulley and Buckman, 1994).

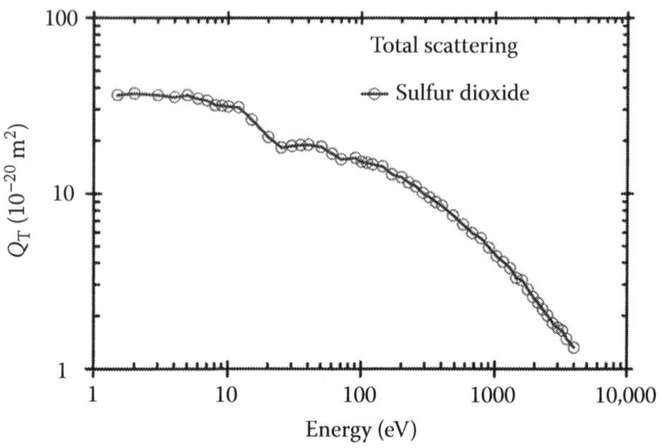

FIGURE 36.1 Total scattering cross sections for SO_2. Energy range: $1.5 \le \varepsilon \le 70$ eV (Szmytkowski and Maciąg, 1986); $90 \le \varepsilon \le 4000$ eV (Zecca et al., 1995).

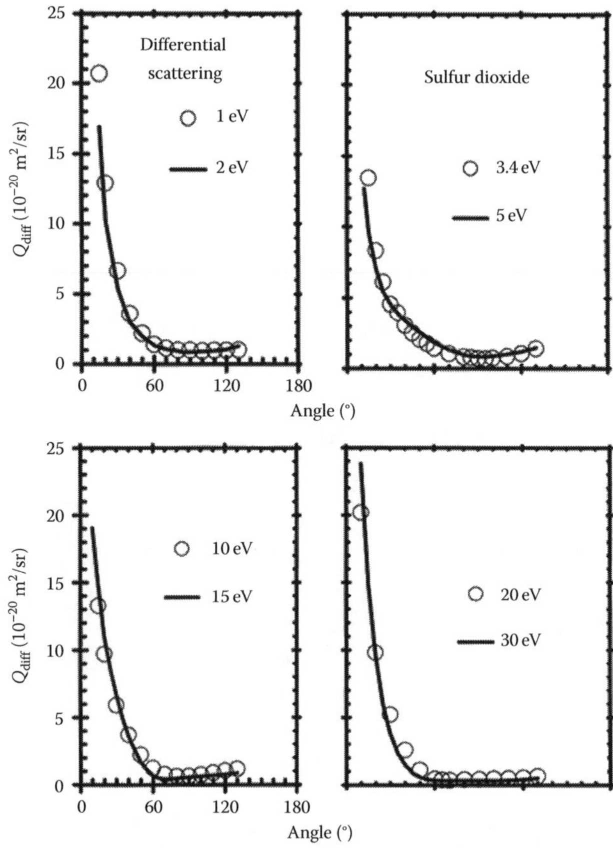

FIGURE 36.2 Differential scattering cross section for elastic scattering for SO_2(a–d). (Adapted from Gulley, R. J. and S. Buckman, *J. Phys. B: At. Mol. Opt. Phys.*, 27, 1833, 1994.)

36.4 ELASTIC SCATTERING AND MOMENTUM TRANSFER CROSS SECTIONS

See Table 36.4.

36.5 ION APPEARANCE POTENTIALS

See Table 36.5.

36.6 IONIZATION CROSS SECTION

Products of ionization processes are (Čadež et al., 1983)

$$SO_2 + e \rightarrow \begin{cases} SO_2^+ + 2e \\ SO + O^+ + 2e \\ S + O_2^+ + 2e \\ O_2 + S^+ + 2e \\ O + SO^+ + 2e \end{cases} \quad (36.1)$$

See Table 36.6.

TABLE 36.4
Elastic and Momentum Transfer Cross Sections for SO_2

Energy (eV)	Q_{el} (10^{-20} m²)	Q_{MT} (10^{-20} m²)
1.0	37.9	15.1
2.0	36.3	16.7
3.4	32.1	18.5
5.0	29.4	19.7
10	28.1	14.5
12	26.4	13.7
15	24.2	11.4
20	19.9	8.6
30	17.0	6.8

Source: Adapted from Gulley, R. J. and S. Buckman, *J. Phys. B: At. Mol. Opt. Phys.*, 27, 1833, 1994.
Note: See Figure 36.3 for graphical presentation.

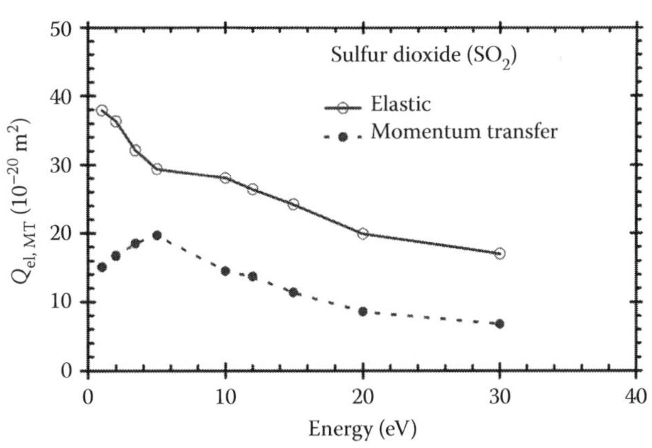

FIGURE 36.3 Integral elastic and momentum transfer cross sections for SO_2. (Adapted from Gulley, R. J. and S. Buckman, *J. Phys. B: At. Mol. Opt. Phys.*, 27, 1833, 1994.)

TABLE 36.5
Ion Appearance Potentials for SO_2

Ion Species	Process	Appearance Potential (eV)	Relative Abundance (%)
	Positive Ions		
SO_2^+	$SO_2 + e \rightarrow SO_2^+ + 2e$	12.4	100.0
SO^+	$\rightarrow SO^+ + O + 2e$	16.2	72.0
S^+	$\rightarrow S^+ + O_2 + 2e$	17.5	14.0
O_2^+	$\rightarrow O_2^+ + S + 2e$	17.5	—
S^+	$\rightarrow S^+ + 2O + 2e$	22.6	—
O^+	$\rightarrow O^+ + SO + 2e$	20.6	1.6
	Negative Ions		
SO^-	$SO_2 + e \rightarrow SO^- + O$	4.6[a]	0.4
SO^-	$\rightarrow SO^- + O$	7.2[a]	—
O^-	$\rightarrow O^- + SO$	4.3[a]	0.3
O^-	$\rightarrow O^- + SO$	7.0[a]	—

Source: Adapted from Reese, R. M., V. H. Diebler, and J. L. Franklin, *J. Chem. Phys.*, 29, 880, 1958.
Note: See Table 36.7 for additional data.
[a] Indicates peak energy.

TABLE 36.6
Total Ionization Cross Sections for SO_2

Lindsay et al. (1996)		Čadež et al. (1983)		Recommended	
Energy (eV)	Q_i (10^{-20} m²)	Energy (eV)	Q_i (10^{-20} m²)	Energy (eV)	Q_i (10^{-20} m²)
15	0.26	13	0.262	15	0.45
20	0.99	14	0.481	20	1.31
25	1.88	15	0.645	25	2.18
30	2.73	16	0.848	30	3.00
35	3.20	17	1.05	35	3.52
40	3.62	18	1.23	40	3.98
50	4.18	19	1.45	50	4.55
60	4.61	20	1.63	60	4.96
80	4.97	30	3.31	80	5.42
100	5.16	40	4.34	100	5.58
120	5.05	50	4.92	120	5.48
160	4.75	60	5.38	150	5.15
200	4.51	70	5.67	200	4.83
250	4.07	80	5.87	250	4.30
300	3.79	90	5.96	300	4.00
400	3.24	100	5.99	400	3.40
500	2.80	120	5.90	500	3.00
600	2.45	150	5.68	600	2.65
800	2.02	200	5.15	800	2.20
1000	1.72	250	4.52	1000	1.95

Note: See Figure 36.4 for graphical presentation.

SO$_2$

TABLE 36.7
Peak Energy and Peak Attachment Cross Sections for SO$_2$

Ion Species	Appearance Potential (eV)	Peak Energy (eV)	Peak Cross Section (10^{-22} m^2)
	Krishnakumar et al. (1997)		
O^-		4.6	5.9
		7.2	3.7
S^-		4.2	0.3
		7.4	0.08
		9.0	0.035
SO^-		4.8	4.3
		7.3	0.5
Total		4.6	10.2
		7.2	4.3
	Spyrou et al. (1986) Beam Data		
O^-	3.9	4.55	2.46
	6.0	7.3	1.27
S^-	3.2	4.2	0.17
	6.1	7.4	0.044
SO^-	4.2	4.85	4.11
		7.0	0.41
Total		4.7	6.44
		7.3	1.72
	Spyrou et al. (1986) Swarm Data		
Total		4.7	4.6
	Čadež et al. (1983)		
O^-	3.5, 4.2	4.8	4.44
		7.4	2.44
S^-	3.6	4.4	0.136
SO^-	4.36	4.8	1.4
		7.4	0.072
Total		4.7	6.21
		7.4	2.53
	Orient and Srivastava (1983)		
O^-	3.5	4.3	8.08
	5.8	7.1	2.68
S^-	3.2	4.0	0.313
		7.5	0.036
		8.9	0.03
SO^-	3.75	4.7	10.98
		7.3	0.51
Total		4.5	17.55
		7.1	3.11

Source: Adapted from Krishnakumar, E. et al., *Phys. Rev. A*, 56, 1945, 1997.

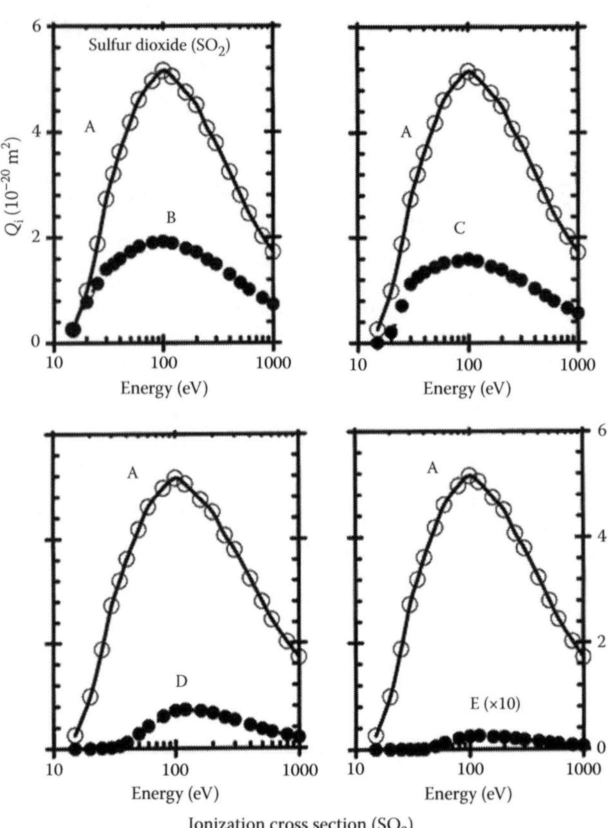

FIGURE 36.4 Partial and total ionization cross sections for SO$_2$. Total ionization cross sections have been drawn in each frame for comparison. The fragment ions are (A) total; (B) SO_2^+; (C) SO^+; (D) O^+; (E) SO^{2+}; note the multiplication factor. (Adapted from Lindsay, B. G. et al., *J. Geophys. Res.*, 101, 26155, 1996.)

36.7 ATTACHMENT PROCESSES

The molecule has stable negative ion with an electron affinity of 1.12 eV and the formation of SO_2^- ion (Sanche and Schulz, 1973). Dissociative attachment produces ions by processes (Čadež et al., 1983):

$$SO_2 + e \rightarrow \begin{cases} SO + O^- \\ S + O + O^- \\ O_2 + S^- \\ O + O + S^- \\ O + SO^- \\ S + O_2^- \end{cases} \tag{36.2}$$

Ion pair formation processes are

$$SO_2 + e \rightarrow \begin{cases} SO^+ + O^- + e \\ SO^- + O^+ + e \\ S^+ + O_2^- + e \end{cases} \tag{36.3}$$

The three-body attachment process has been observed by Lakdawala and Moruzzi (1981) according to the reaction

$$e + SO_2 + X \rightarrow SO_2^- + X + \text{energy} \tag{36.4}$$

The appearance potentials and peak cross section for selected ion formation are shown in Table 36.7.

36.8 ATTACHMENT CROSS SECTIONS

Recommended attachment cross sections for SO_2 (Itikawa, 2003) based on the measurements of Krishnakumar et al. (1997) are given in Table 36.8. See Figure 36.5 for graphical presentation.

TABLE 36.8
Attachment Cross Sections for SO_2

	Recommended				Čadež et al. (1983)
Energy (eV)	O^- (10^{-22} m^2)	S^- (10^{-22} m^2)	SO^- (10^{-22} m^2)	Total (10^{-22} m^2)	Total (10^{-22} m^2)
2.6	0.02	0.00	0.02	0.04	
2.7	0.04	0.01	0.02	0.07	
2.8	0.09	0.01	0.03	0.13	
2.9	0.07	0.02	0.03	0.12	
3.0	0.11	0.03	0.07	0.21	
3.1	0.14	0.04	0.05	0.23	
3.2	0.23	0.05	0.05	0.33	
3.3	0.43	0.08	0.09	0.60	
3.4	0.53	0.10	0.13	0.76	
3.5	0.82	0.12	0.19	1.13	
3.6	1.22	0.16	0.36	1.74	0.092
3.7	1.64	0.18	0.45	2.27	0.157
3.8	2.18	0.21	0.72	3.11	0.325
3.9	2.85	0.24	0.92	4.01	0.636
4.0	3.65	0.26	1.26	5.17	1.13
4.1	4.00	0.28	1.67	5.95	1.87
4.2	5.03	0.28	2.14	7.45	2.71
4.3	5.53	0.27	2.60	8.40	3.61
4.4	5.69	0.27	3.18	9.14	4.65
4.5	5.85	0.24	3.68	9.77	5.46
4.6	5.90	0.22	4.04	10.16	6.04
4.7	5.66	0.21	4.30	10.17	6.21
4.8	4.89	0.19	4.16	9.24	6.05
4.9	4.82	0.17	4.04	9.03	5.58
5.0	4.03	0.14	3.48	7.65	5.02
5.1	3.43	0.12	3.09	6.64	4.34
5.2	2.91	0.10	2.46	5.47	3.70
5.3	2.53	0.09	2.02	4.64	3.09
5.4	2.02	0.08	1.56	3.66	2.47
5.5	1.71	0.06	1.27	3.04	1.98
5.6	1.35	0.03	0.94	2.32	1.54
5.7	1.18	0.04	0.73	1.95	1.19
5.8	0.94	0.04	0.57	1.55	0.987
5.9	0.89	0.03	0.48	1.40	0.820
6.0	0.92	0.03	0.37	1.32	0.737
6.1	1.06	0.03	0.44	1.53	0.700
6.2	1.18	0.03	0.38	1.59	0.719
6.3	1.43	0.03	0.32	1.78	0.804
6.4	1.65	0.04	0.48	2.17	0.907
6.5	1.92	0.04	0.41	2.37	1.03
6.6	2.19	0.05	0.49	2.73	1.22
6.7	2.67	0.05	0.52	3.24	1.44
6.8	3.04	0.05	0.50	3.59	1.65
6.9	3.24	0.05	0.48	3.77	1.90
7.0	3.49	0.06	0.52	4.07	2.10
7.1	3.70	0.06	0.46	4.22	2.25
7.2	3.81	0.07	0.47	4.35	2.41
7.3	3.64	0.07	0.50	4.21	2.48
7.4	3.45	0.08	0.51	4.04	2.53
7.5	3.39	0.08	0.47	3.94	2.45
7.6	2.90	0.07	0.48	3.45	2.31
7.7	2.74	0.07	0.38	3.19	2.14
7.8	2.26	0.07	0.42	2.75	1.91
7.9	1.84	0.07	0.30	2.21	1.67
8.0	1.61	0.06	0.29	1.96	1.44
8.1	1.20	0.06	0.22	1.48	1.24
8.2	0.97	0.05	0.18	1.20	1.03
8.3	0.83	0.04	0.15	1.02	0.843
8.4	0.65	0.04	0.14	0.83	0.691
8.5	0.55	0.03	0.13	0.71	0.576
8.6	0.49	0.03	0.09	0.61	0.462
8.7	0.34	0.03	0.10	0.47	0.372
8.8	0.25	0.03	0.08	0.36	0.248
8.9	0.20	0.03	0.08	0.31	0.208
9.0	0.16	0.03	0.08	0.27	0.165
9.1	0.09	0.03	0.08	0.20	0.125
9.2	0.08	0.03	0.07	0.18	0.103
9.3	0.06	0.02	0.06	0.14	0.082
9.4	0.05	0.02	0.07	0.14	0.060
9.5	0.05	0.01	0.08	0.14	
9.6	0.03	0.01	0.05	0.09	
9.7	0.03	0.01	0.06	0.10	
9.8	0.04	0.00	0.05	0.09	
9.9	0.03	0.00	0.07	0.10	
10.0	0.04	0.00	0.09	0.13	
10.1	0.03	0.00	0.07	0.10	
10.2	0.06	0.00	0.07	0.13	

The influence of temperature on the total dissociative attachment cross section has been investigated by Spyrou et al. (1986). The findings are

1. The peak attachment cross section increases with temperature from 300 to 700 K.
2. The onset energy and the peak energy remain relatively independent of temperature.

Figure 36.6 shows the temperature dependence. Peak cross sections at various temperatures are given in Appendix 4.

SO2

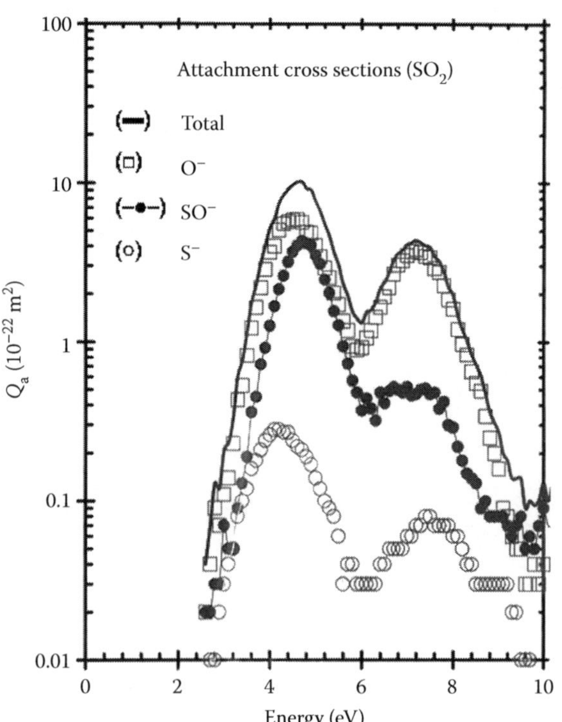

FIGURE 36.5 Attachment cross sections for SO_2. (——) Total; (□) O^-; (-•-) SO^-; (○) S^-. (Adapted from Krishnakumar, E. et al., *Phys. Rev. A*, 56, 1945, 1997.)

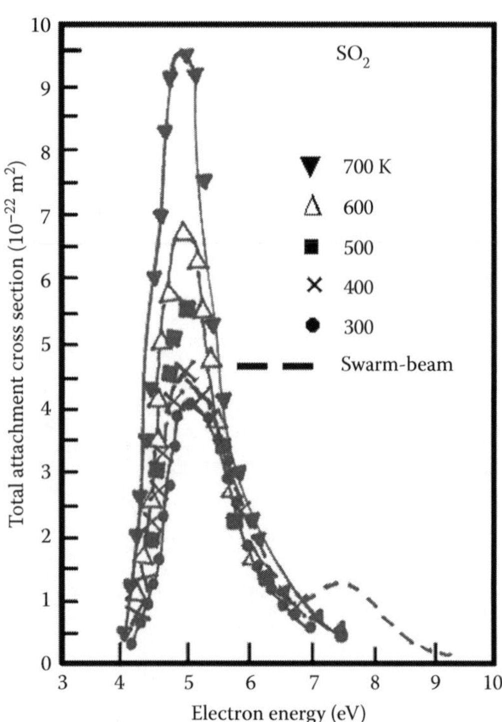

FIGURE 36.6 Total electron attachment cross sections as a function of electron energy for SO_2 in the temperature range 300–700 K. The curve (. . . .) is the cross section at 300 K obtained by the swarm–beam analysis and attachment rate constants as function of mean energy at 300 K. (Adapted from Spyrou, S. M., I. Sauers, and L. G. Christophorou, *J. Chem. Phys.*, 84, 239, 1986.)

36.9 ATTACHMENT RATE CONSTANT

Figure 36.7 shows the attachment rate constant for SO_2 obtained from swarm studies (Spyrou et al., 1986).

36.10 REDUCED ATTACHMENT COEFFICIENTS

At low values of *E*/*N* (*E*/*N* ≤ 30 Td) three-body attachment, Reaction 36.3 occurs with dependence on *N*; at higher value (30 ≤ *E*/*N* ≤ 250 Td) dissociative attachment, Reactions 36.2 occur, dependent on *E*/*N* only. Table 36.9 shows the reduced attachment coefficients (Moruzzi and Lakdawala, 1979). See Figure 36.8.

36.11 REDUCED ION MOBILITY

Reduced negative ion mobilities of 0.69, 0.62, 0.55, and 0.39 in units of 10^{-4} m²/V s have been reported by Lakdawala and Moruzzi (1981) in the *E*/*N* range 5–300 Td. Schlumbohm

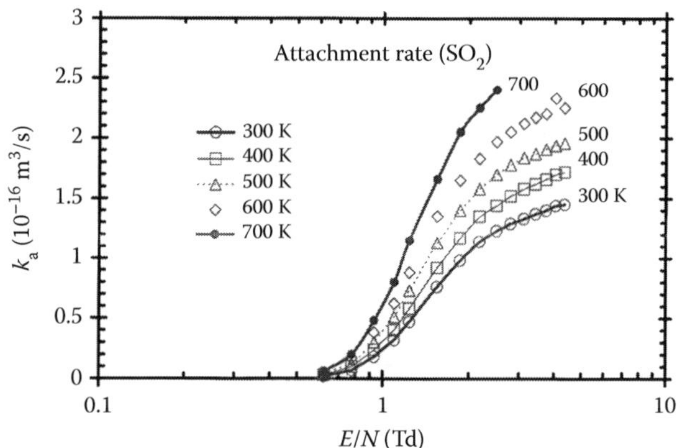

FIGURE 36.7 Attachment rate constant for SO_2 at various constant temperatures. (Adapted from Spyrou, S. M., I. Sauers, and L. G. Christophorou, *J. Chem. Phys.*, 84, 239, 1986.)

TABLE 36.9
Reduced Attachment Coefficients for SO_2

E/*N* (Td)	η/*N* (10^{-22} m²)	*E*/*N* (Td)	η/*N* (10^{-22} m²)
3.0	3.16	60	0.91
6.0	3.60	70	1.15
9.0	3.44	80	1.44
12	2.73	90	1.76
15	2.08	100	2.01
18	1.60	125	2.31
21	1.24	150	2.48
24	0.98	200	2.38
27	0.79	250	2.20
30	0.63		

Note: See Figure 36.7 for graphical presentation.

(1962) reports mobility of 3.6×10^{-5} m²/V s in the range $330 \leq E/N \leq 360$ Td.

36.12 ADDENDUM

36.12.1 Vibrational Excitation Cross Sections

The relative change in the vibrational cross section with electron energy has been given by Sanche and Schulz (1973) and is shown in Figure 36.9. Two bands are observed in the energy range from 2.8 to 3.8 eV, and above 3.8 eV. In the lower-energy band, 11 vibrational levels with an average spacing of 88 meV are observed, attributed to the symmetric stretch mode. In the high-energy range, attachment with the formation of SO_2^- becomes evident. The oscillations are too weak to be observed in the total scattering cross section. A point to note: when a molecule in the ground state forms stable parent negative ions with zero electron energy (NO_2, O_2, NO), the structure is evident at zero energy. When a molecule does not have a stable parent negative ion at zero electron energy (CO, N_2, CO_2), the structure is observed in the 2–4 eV energy range. SO_2 exhibits a structure that deviates from this observation.

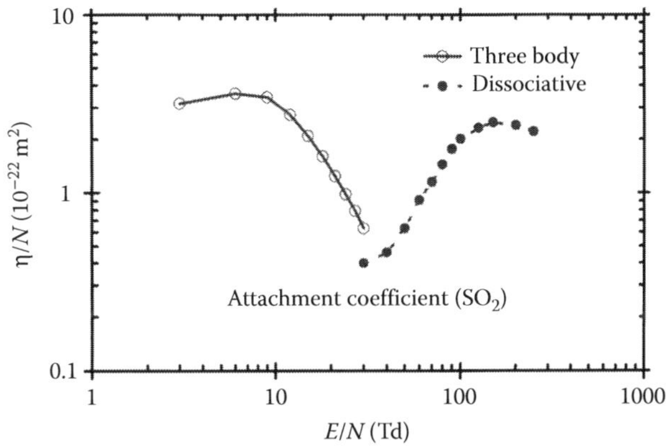

FIGURE 36.8 Density-reduced attachment coefficients for SO_2. At low E/N the attachment coefficients shown are at a number density of 6.4×10^{22} m^{-3} and decrease with increasing N at the same E/N. At higher E/N, the attachment coefficients are dependent on E/N only. (Adapted from Moruzzi, J. L. and V. K. Lakdawala, *J. Physique*, Colloque C7, Supplément au n°7, 40, c7–11, 1979.)

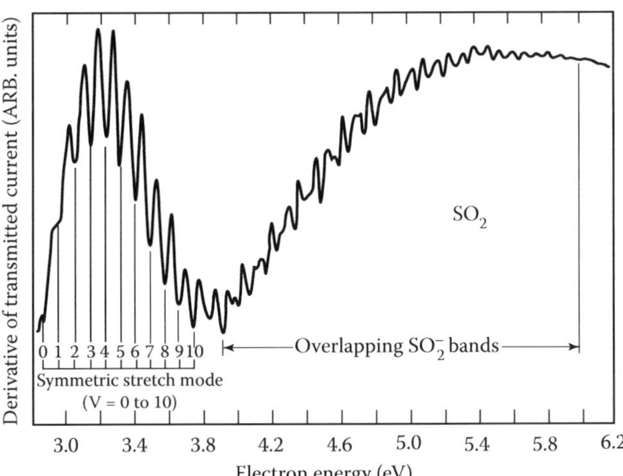

FIGURE 36.9 Relative variation of vibrational cross sections with electron energy for SO_2. The structures are too weak to appear in the total scattering cross section. (Adapted from Sanche and Schulz, *J. Chem. Phys.*, 58, 479, 1973.)

TABLE 36.10
Reduced Ionization Coefficients for SO_2

E/N (Td)	α/*N* (10^{-22} m²)
330	4.05
337.5	5.05
345	5.90
352.5	7.0
360	7.92

Source: Adapted from Schlumbohm, H., *Z. Phys.*, 166, 192, 1962.

36.13 REDUCED IONIZATION COEFFICIENTS

Reduced ionization coefficients (α/*N*) over a limited range of *E/N* have been measured by Schlumbohm (1962) as shown in Table 36.10.

REFERENCES

Basner, R., M. Schmidt, H. Deutsch, V. Tarnovsky, A. Levin, and K. Becker, *J. Chem. Phys.*, 103, 211, 1995.

Čadež, I. M., V. M. Pejčev, and M. V. Kurepa, *J. Phys. D: Appl. Phys.*, 16, 305, 1983.

Gulley, R. J. and S. Buckman, *J. Phys. B: At. Mol. Opt. Phys.*, 27, 1833, 1994.

Itikawa, Y., Interactions of photons and electrons with molecules, in *Electron Attachment*, Springer-Verlag, 2003, Chapter 5.2.

Krishnakumar, E., S. V. K. Kumar, S. A. Rangwala, and S. K. Mitra, *Phys. Rev. A*, 56, 1945, 1997.

Lakdawala, V. K. and J. L. Moruzzi, *J. Phys D: Appl. Phys.*, 14, 2015, 1981.

Lindsay, B. G., H. C. Straub, K. A. Smith, and R. F. Stebbings, *J. Geophys. Res.*, 101, 26155, 1996.

Moruzzi, J. L. and V. K. Lakdawala, *J. Phys. Coll.*, C7, Supplément au n°7, 40, c7–11, 1979.

Orient, O. J. and S. K. Srivastava, *J. Chem. Phys.*, 78, 2949, 1983.

Orient, O. J. and S. K. Srivastava, *J. Chem. Phys.*, 80, 140, 1984.

Reese, R. M., V. H. Diebler, and J. L. Franklin, *J. Chem. Phys.*, 29, 880, 1958.

Sanche, L. and G. J. Schulz, *J. Chem. Phys.*, 58, 479, 1973.

Schlumbohm, H., *Z. Phys.*, 166, 192, 1962.

Shimanouchi, T., *Tables of Molecular Vibrational Frequencies Consolidated Volume I*, NSRDS-NBS 37, US Department of Commerce, Washington (DC), 1972.

Spyrou, S. M., I. Sauers, and L. G. Christophorou, *J. Chem. Phys.*, 84, 239, 1986.

Szmytkowski, Cz. and K. Maciąg, *Chem. Phys. Lett.*, 124, 463, 1986.

Trajmar, S. and T. W. Shyn, *J. Phys. B: At. Mol. Opt. Phys.*, 22, 2911, 1989.

Zecca, A., J. C. Nogueira, G. Karwasz, and R. S. Brusa, *J. Phys. B: At. Mol. Opt. Phys.*, 28, 477, 1995.

37

WATER VAPOR

CONTENTS

Water (H_2O) molecule is polar and electron-attaching, and it has 10 electrons. The electronic polarizability is 1.61×10^{-40} F m^2, the dipole moment is 1.845 D, and the ionization potential is 12.621 eV. The molecule has three vibrational modes as shown in Table 37.1 (Shimanouchi, 1972).

37.1 SELECTED REFERENCES FOR DATA

See Table 37.2.

TABLE 37.1
Vibrational Modes and Energies

Designation	Motion	Energy (meV)
ν_1	Symmetrical stretching	453.4
ν_2	Bending	197.8
ν_3	Asymmetical stretching	465.7

Source: Adapted from Shimanouchi, T., *Tables of Molecular Vibrational Frequencies Consolidated Volume I*, NSRDS-NBS 39, U.S. Department of Commerce, Washington (DC), 1972.

37.2 TOTAL SCATTERING CROSS SECTIONS

Table 37.3 shows the recommended total scattering cross sections for water vapor. Figure 37.1 shows these and measured data. The main points to note are

1. A minimum in the total cross section at ~4 eV and a maximum at ~10 eV, the latter phenomenon being attributed to attachment processes.
2. A decrease in the total cross section for energies >10 eV, a feature common to many gases.
3. A trend toward increasing cross section as the energy decreases below 1 eV. It is noted that the dipole moment of water molecule exceeds the minimum value of 1.67 D for the formation of parent negative ion (Crawford, 1967).

37.3 DIFFERENTIAL SCATTERING CROSS SECTIONS

Figure 37.2 shows the differential scattering cross sections for elastic scattering in the low-energy range 2.2–20 eV

H_2O

TABLE 37.2
Selected References for Data on H_2O

Quantity	Range: eV, (Td)	Reference
Total scattering cross section	15–10,000	Champion et al. (2002)
Dissociation cross section	10–300	**Harb et al. (2001)**
Vibrational excitation cross section	6–20	**El-Zein et al. (2000)**
Ionization cross section	13.5–1000	**Straub et al. (1998)**
Ionization cross section	12.6–5000	Hwang et al. (1996)
Vibrational excitation cross section	4–10	Cvejanović et. al. (1993)
Elastic scattering	6–50	**Johnstone and Newell (1991)**
Total scattering cross section	25–300	**Sağlam and Aktekin (1990)**
Ionization cross section	15–400	**Orient and Srivastava (1987)**
Elastic scattering cross section	2.2–20	**Shyn and Cho (1987)**
Total scattering cross section	81.3–3000	**Zecca et al. (1987)**
Ionization cross section	50–2000	**Bolorizadeh and Rudd (1986)**
Elastic scattering cross section	100–1000	**Katase et al. (1986)**
Total scattering cross section	1–400	**Sueoka et al. (1986)**
Elastic scattering cross section	3–200	**Nishimura (1985)**
Ro-vibrational cross section	0.5–6.0	**Jung et al. (1982)**
Ionization coefficients	(30–3000)	**Risbud and Naidu (1979)**
Attachment coefficients	(40–180)	**Risbud and Naidu (1979)**
Vibrational excitation	0.5–3.0	**Rohr (1977)**
Vibrational excitation cross section	0.5–10	Seng and Linder (1976)
Transport coefficients	(1–25)	**Wilson et al. (1975)**
Attachment cross sections	0–13	**Melton (1972)**
Cross sections	Compilation	Olivero et al. (1972)
Attachment coefficients	(40–80)	**Parr and Moruzzi (1972)**
Attachment cross section	0–15	**Compton and Christophorou (1967)**
Ionization cross section	100–20,000	**Schutten et al. (1966)**
Drift velocity	(170–265)	**Ryżko (1965)**
Drift velocity	(0.06–60)	**Pack et al. (1962)**
Ionization and attachment coefficients	(75–150)	**Prasad and Craggs (1960)**
Attachment coefficients	(50–80)	**Kuffel (1959)**
Attachment processes	–	**Mann et al. (1940)**

Note: Bold font indicates experimental study.

TABLE 37.3
Recommended Total Scattering Cross Sections for Water Vapor

Energy (eV)	Q_T (10^{-20} m^2)	Energy (eV)	Q_T (10^{-20} m^2)
1.0	29.2	20.0	11.3
1.2	24.3	22.0	11.0
1.4	20.7	25.0	10.2
1.6	17.9	30.0	9.5
1.8	17.1	35.0	9.0
2.0	15.8	40.0	8.5
2.2	15.0	50.0	7.5
2.5	13.9	60.0	7.1
2.8	13.0	70.0	6.7
3.1	12.2	80.0	6.3
3.4	12.0	90.0	6.0
3.7	11.7	100	5.6
4.0	12.2	120	5.2
4.5	11.7	150	4.6
5.0	11.9	200	4.0
5.5	12.0	250	3.5
6.0	12.5	300	3.2
6.5	12.3	350	2.9
7.0	12.6	400	2.6
7.5	13.0	500	2.5
8.0	13.0	600	2.1
8.5	13.3	700	1.9
9.0	13.2	800	1.7
9.5	12.7	900	1.6
10.0	12.7	1000	1.4
11.0	13.0	1250	1.2
12.0	12.8	1500	1.00
13.0	12.7	1750	0.87
14.0	12.5	2000	0.75
15.0	12.3	2250	0.69
16.0	12.0	2500	0.60
17.0	11.8	2750	0.54
18.0	11.7	3000	0.49
19.0	11.6		

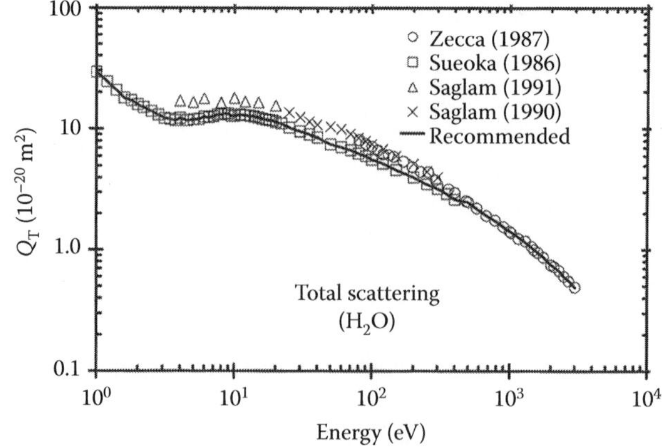

FIGURE 37.1 Total scattering cross sections. (Δ) Sağlam and Aktekin (1991); (×) Sağlam and Aktekin (1990); (○) Zecca et al. (1987); (□) Sueoka et al. (1986); (—) recommended.

(Shyn and Cho, 1987). A minimum is observed at ~150° at low energy (~2 eV) which shifts to lower angles (~100°) as the electron energy is increased to 20 eV. The same trend continues up to ~50 eV (not shown) (Johnstone and Newell, 1991). The minimum tends to flatten out at energies >100 eV as shown in Figure 37.3 (Katase et al., 1986). Note the increased forward scattering at higher energies.

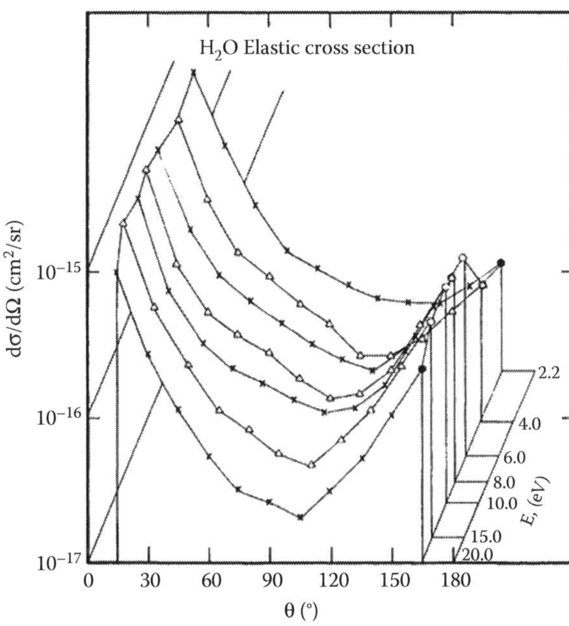

FIGURE 37.2 Differential scattering cross sections as function of electron energy and scattering angle. (Adapted from Shyn, T. W. and S. Y. Cho, *Phys. Rev. A*, 36, 5138, 1987.)

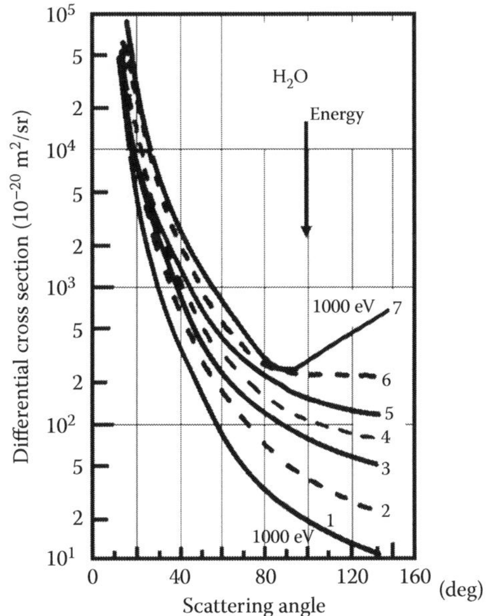

FIGURE 37.3 Differential scattering for water molecule. Energy: 1–1000 eV; 2–700 eV; 3–500 eV; 4–400 eV; 5–300 eV; 6–200 eV; 7–100 eV. Note the increased forward scattering. (Adapted from Katase, A. et al., *J. Phys. B: At. Mol. Phys.*, 19, 2715, 1986.)

37.4 ELASTIC AND MOMENTUM TRANSFER CROSS SECTIONS

Table 37.4 and Figures 37.4 and 37.5 show the suggested integral elastic and momentum transfer cross sections.

TABLE 37.4
Suggested Integral Elastic and Momentum Transfer Cross Sections

Energy (eV)	Q_{el} (10^{-20} m²)	Q_M (10^{-20} m²)
2	11.70	6.23
3	12.83	6.87
5	14.05	7.99
7	14.45	8.74
10	15.32	8.82
15	14.54	9.12
20	13.40	8.04
30	11.04	6.64
40	9.53	5.08
50	8.14	4.22
60	6.84	3.25
70	5.99	2.27
80	5.14	1.84
90	4.67	1.46
100	4.27	1.16
125	3.35	0.81
150	2.88	0.60
175	2.49	0.51
200	2.17	0.41
300	1.69	0.26
400	1.25	0.19
500	1.06	0.15
600	0.91	0.11
700	0.84	0.10
800	0.76	0.07
900	0.71	0.07
1000	0.59	0.06

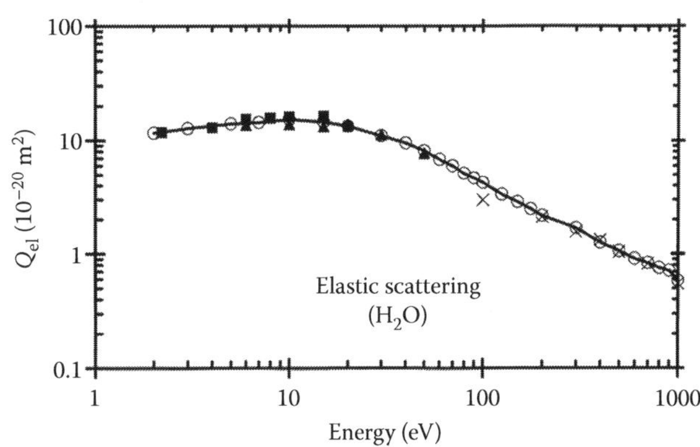

FIGURE 37.4 Integral elastic scattering cross sections. (—○—) Suggested; (▲) Johnstone and Newell (1991); (■) Shyn and Cho (1987); (×) Katase et al. (1986).

37.5 VIBRATIONAL EXCITATION CROSS SECTIONS

The bending mode is designated as (010), symmetric stretch as (100), and asymmetric stretch as (001). In view of the proximity of the quanta of energies for the latter two excitations modes (Table 37.1), excitation cross sections for (100 + 001) are measured together. These are shown in Table 37.5 (Brunger et al., 2003) and Figure 37.6.

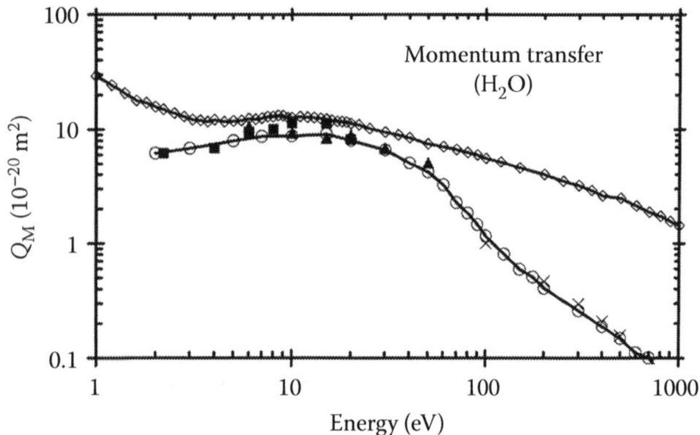

FIGURE 37.5 Momentum transfer cross sections. (—○—) Suggested; (▲) Johnstone and Newell (1991); (■) Shyn and Cho (1987); (×) Katase et al. (1986); (—◆—) total scattering, plotted for comparison.

TABLE 37.5
Vibrational Excitation Cross Sections

	Q_{ex} (10^{-20} m^2)	
Energy (eV)	**Mode: (000 → 001)**	**Mode: (000 → 100 + 001)**
1	0.370	0.500
2	0.200	
2.1		0.320
2.2	0.190	
3	0.160	0.310
4	0.150	0.385
5	0.157	0.430
6	0.163	0.489
7		0.520
7.5	0.317	0.529
8		0.495
8.75	0.227	
8.875		0.413
10	0.180	0.325
15	0.130	0.190
20	0.100	0.080

Source: Adapted from Brunger, M., S. Buckman, and M. Elford, Excitation cross sections, in *Interactions of Photons and Electrons with Molecules*, Vol. 17, Springer-Verlag, New York, NY, 2003, Chapter 6.4, pp. 118–201.

37.6 ION APPEARANCE POTENTIALS

Table 37.6 shows the ion appearance potentials and probable processes for ion formation (Mann et al., 1940). Process 1 is direct ionization, 2–6 are dissociative ionization, 7–9 dissociative attachment, and 10–11 ion pair formation.

37.7 IONIZATION CROSS SECTIONS

Table 37.7 shows the total ionization cross sections (Straub et al., 1998; Schutten et al., 1966) with good agreement between the two sets up to 1000 eV. For higher energies the cross sections of Schutten et al. (1966) are the only available data. Figure 37.7 shows comparison with selected data. Figure 37.8 shows partial ionization cross sections of selected fragments.

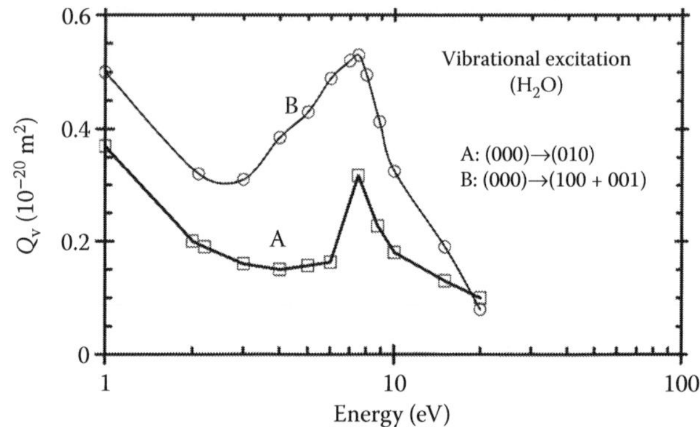

FIGURE 37.6 Vibrational excitation cross sections for water molecule. (Adapted from Brunger, M., S. Buckman, and M. Elford, Excitation cross sections, in *Interactions of Photons and Electrons with Molecules*, Vol. 17, Springer-Verlag, New York, NY, 2003, Chapter 6.4, pp. 118–201.)

TABLE 37.6
Ion Appearance Potentials and Probable Processes

	Ion	Appearance Potential (eV)	Process
1	H_2O^+	13.0	$H_2O + e \rightarrow H_2O^+$
2	OH^+	18.7	$H_2O + e \rightarrow H + OH^+$
3	O^+	18.8	$H_2O + e \rightarrow H_2 + O^+$
4		28.1	$H_2O + e \rightarrow 2H + O^+$
5	H^+	19.5	$H_2O + e \rightarrow OH + H^+$
6	H_2^+	23.0	$H_2O + e \rightarrow O + H_2^+$
7	H^-	5.6	$H_2O + e \rightarrow OH + H^-$
8			$H_2O + e \rightarrow O + H + H^-$
9	O^-	0.15	$H_2O + e \rightarrow 2H + O^-$
10			$H_2O + e \rightarrow H + H^+ + O^-$
11			$H_2O + e \rightarrow H^+ + H^+ + O^-$
12	OH^-	4.3[a]	$H_2O + H^- \rightarrow OH^- + H_2$
		6.9[a]	
		8.8[a]	

Source: Adapted from Mann, M. V., A. Hustrulid, and J. T. Tate, *Phys. Rev.*, 58, 340, 1940.

[a] Indicates Melton (1972).

TABLE 37.7
Total Ionization Cross Sections

Straub et al. (1998)		Schutten et al. (1966)	
Energy (eV)	Q_i (10^{-20} m²)	Energy (eV)	Q_i (10^{-20} m²)
13.5	0.0340	20	0.46
15	0.133	30	0.97
17.5	0.292	50	1.57
20	0.460	80	1.98
22.5	0.638	90	2.01
25	0.801	100	1.98
30	1.087	120	1.96
35	1.322	160	1.83
40	1.529	200	1.71
45	1.677	300	1.46
50	1.802	400	1.26
60	1.992	500	1.12
70	2.097	600	1.00
80	2.187	1000	0.622
90	2.224	2000	0.375
100	2.252	3000	0.261
110	2.252	4000	0.220
125	2.216	5000	0.183
150	2.152	6000	0.160
175	2.076	8000	0.123
200	2.008	10,000	0.101
250	1.821	12,000	0.088
300	1.647	14,000	0.078
400	1.417	16,000	0.064
500	1.205	18,000	0.058
600	1.062	20,000	0.054
700	0.932		
800	0.848		
900	0.776		
1000	0.707		

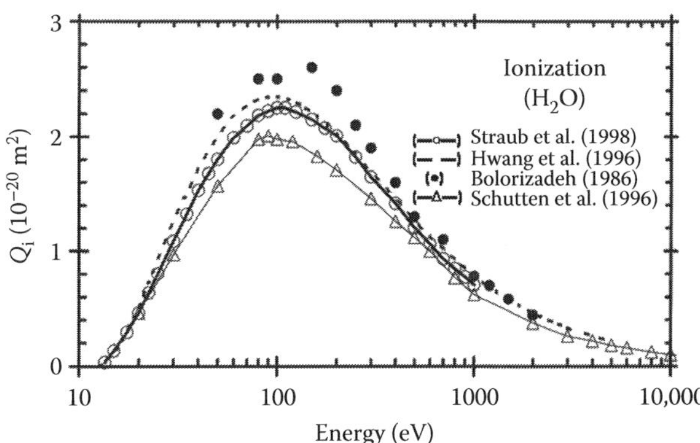

FIGURE 37.7 Total ionization cross sections for H_2O molecule. (—○—) Straub et al. (1998); (— —) Hwang et al. (1996); (●) Bolorizadeh and Rudd (1986); (—Δ—) Schutten et al. (1966).

37.8 ATTACHMENT CROSS SECTIONS

The negative ion species detected are H^-, O^-, and OH^- with decreasing abundance in that order (Itikawa, 2003). Data for the production of H^- and O^- ions are shown in Table 37.8.

Tables 37.9 through 37.12 show the attachment cross sections for the formation of negative ions from H_2O (Itikawa,

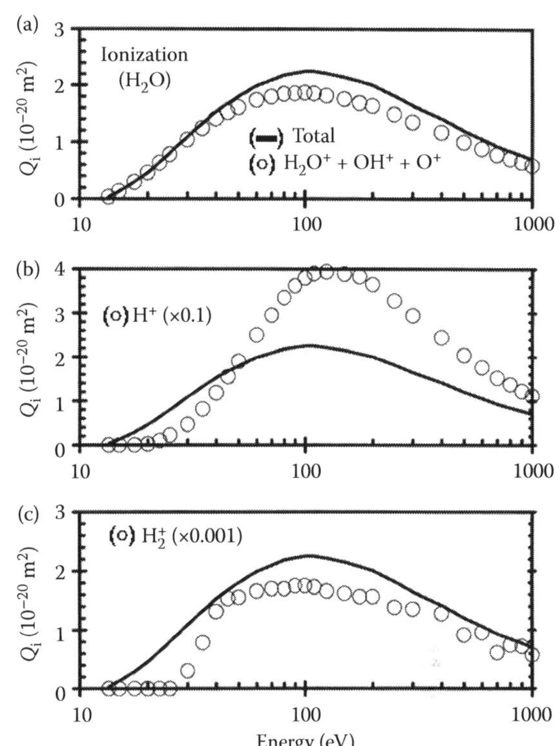

FIGURE 37.8 Partial ionization cross sections for H_2O molecule for selected fragments (Straub et al., 1998). (a) (——) Total; (○) $H_2O^+ + OH^+ + O^+$; (b) (○) H^+ (×0.1); (c) (○) H_2^+ (×0.001). Note the multiplication factors for (b) and (c).

TABLE 37.8
Data for Dissociative Attachment Cross Section

Ion	Onset Energy (eV)	Peak Energy (eV)	Peak Cross Section (m²)	Reference
H^-	5.5	6.4	6.6×10^{-22}	Melton (1972)
	7.3	8.6	1.6×10^{-22}	
O^-	4.4	6.6	1.3×10^{-23}	
	7.0	8.4	3.2×10^{-23}	
	8.5	11.2	5.7×10^{-23}	
OH^-	4.3	6.4	1.2×10^{-25}	
	6.9	8.4	8.5×10^{-26}	
	8.8	11.2	8.5×10^{-26}	
H^-	5.7	6.5	6.9×10^{-22}	Compton and Christophorou (1967)
	7.3	8.6	1.3×10^{-22}	
O^-	4.9	6.9		
	7.8	8.9		
		11.4		
H^-		6.7		Dorman (1966)
		8.8		
O^-		6.6		
		8.9		
		11.4		
H^-	5.6	6.5		Schulz (1960)
		8.5		

continued

H_2O

TABLE 37.8 (continued)
Data for Dissociative Attachment Cross Section

Ion	Onset Energy (eV)	Peak Energy (eV)	Peak Cross Section (m²)	Reference
O^-		~12		
H^-	4.8	6.0		Cottin (1959)
		8.0		
O^-	7.4	9.15		
		11.25		
H^-	5.6	7.1		Mann et al. (1940)
		8.9		
O^-	7.5	8.2		
		11.1		
		12.8		
H^-		6.8		Lozier (1930)
		8.8		

TABLE 37.9
Attachment Cross Sections for Formation of H^-/H_2O

Energy (eV)	Q_a (10^{-22} m²)	Energy (eV)	Q_a (10^{-22} m²)
5.50	0.020	7.69	0.877
5.74	0.160	7.89	0.740
5.90	0.985	8.00	0.790
6.01	4.300	8.09	0.995
6.17	6.220	8.14	1.090
6.29	6.317	8.54	1.166
6.40	6.370	8.40	1.040
6.52	6.250	8.79	0.760
6.65	5.790	9.01	0.620
6.81	4.890	9.57	0.280
7.00	3.560	9.80	0.170
7.47	1.290	10.00	0.098

Source: Adapted from Itikawa, Y., *Interactions of Photons and Electrons with Molecules*, Vol. 17, Springer-Verlag, New York, NY, 2003, Chapter 5.2, pp. 78–114.

TABLE 37.10
Attachment Cross Sections for Formation of O^-/H_2O

Energy (eV)	Q_a (10^{-22} m²)	Energy (eV)	Q_a (10^{-22} m²)
4.59	0.004	6.64	0.100
4.71	0.011	7.00	0.079
5.00	0.025	7.19	0.031
5.29	0.050	7.30	0.029
5.72	0.091	7.43	0.036
6.00	0.116	7.56	0.062
6.19	0.128	7.70	0.133
6.32	0.133	8.00	0.230
6.45	0.122	8.21	0.286

TABLE 37.10 (continued)
Attachment Cross Sections for Formation of O^-/H_2O

Energy (eV)	Q_a (10^{-22} m²)	Energy (eV)	Q_a (10^{-22} m²)
8.35	0.310	10.66	0.550
8.44	0.316	10.80	0.570
8.60	0.310	10.90	0.576
8.76	0.285	11.00	0.576
9.00	0.244	11.18	0.553
9.22	0.213	11.50	0.466
9.40	0.208	12.00	0.327
9.56	0.214	12.50	0.200
9.67	0.226	13.00	0.108
9.89	0.256	13.28	0.069
10.00	0.285	13.63	0.038
10.14	0.337	13.80	0.026
10.46	0.493	14.00	0.023

Source: Adapted from Itikawa, Y., *Interactions of Photons and Electrons with Molecules*, Vol. 17, Springer-Verlag, New York, NY, 2003, Chapter 5.2, pp. 78–114.

TABLE 37.11
Attachment Cross Sections for Formation of OH^-/H_2O

Energy (eV)	Q_a (10^{-22} m²)	Energy (eV)	Q_a (10^{-22} m²)
4.51	0.0009	8.19	0.0780
4.75	0.0019	8.31	0.0820
5.00	0.0040	8.39	0.0830
5.21	0.0066	8.53	0.0810
5.39	0.0109	8.64	0.0756
5.56	0.0165	8.85	0.0570
5.69	0.0246	9.00	0.0436
5.84	0.0379	9.23	0.0304
6.00	0.0537	9.36	0.0244
6.10	0.0757	9.49	0.0201
6.27	0.1048	9.57	0.0194
6.36	0.1140	9.65	0.0202
6.44	0.1160	9.78	0.0229
6.54	0.1154	10.01	0.0358
6.63	0.1105	10.26	0.0530
6.77	0.0950	10.52	0.0669
6.87	0.0763	10.83	0.0775
7.02	0.0624	11.00	0.0824
7.15	0.0489	11.13	0.0847
7.32	0.0376	11.30	0.0830
7.41	0.0356	11.45	0.0795
7.49	0.0345	11.60	0.0699
7.60	0.0360	11.87	0.0482
7.73	0.0417	12.00	0.0402
7.83	0.0480	12.19	0.0311
8.02	0.0670	12.47	0.0184

Source: Adapted from Itikawa, Y., *Interactions of Photons and Electrons with Molecules*, Vol. 17, Springer-Verlag, New York, NY, 2003, Chapter 5.2, pp. 78–114.

2003). Figure 37.9 shows graphical presentation. Note that the total cross section has been obtained by the author by interpolating the component contributions at the same electron energy, a process involving considerable computation.

37.9 DISSOCIATION CROSS SECTION

Dissociation cross sections due to electron impact with production of ground-state OH molecule have been measured by Harb et al. (2001) as shown in Table 37.13 and Figure 37.10.

TABLE 37.12
Total Attachment Cross Sections for H_2O

Energy (eV)	Q_a (10^{-22} m^2)	Energy (eV)	Q_a (10^{-22} m^2)
4.5	0.0009	9.00	0.910
4.75	0.015	9.25	0.738
5.00	0.029	9.50	0.554
5.25	0.054	9.75	0.447
5.50	0.104	10.00	0.418
5.75	0.223	10.25	0.443
6.00	4.191	10.50	0.574
6.25	6.548	10.75	0.640
6.50	6.522	11.00	0.658
6.75	5.459	11.25	0.621
7.00	3.703	11.50	0.543
7.25	2.170	11.75	0.452
7.50	1.280	12.00	0.367
8.00	1.085	12.50	0.217
8.25	1.479	13.00	0.108
8.50	1.545	13.50	0.048
8.75	1.186	14.00	0.023

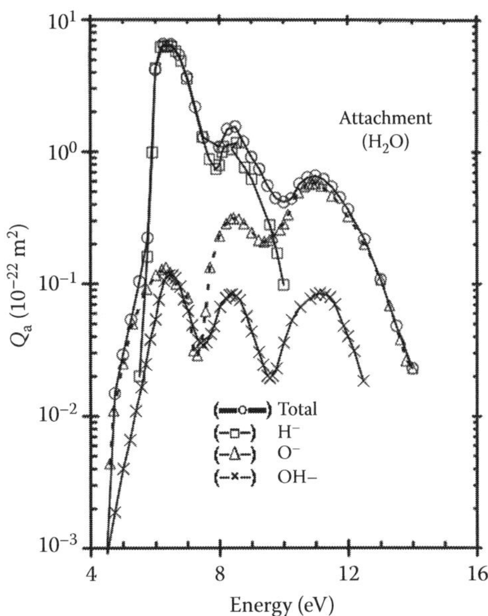

FIGURE 37.9 Electron attachment cross sections for H_2O molecule. Measurements of component cross sections are from Melton (1972). (—○—) Total; (-□-) H^-; (-Δ-) O^-; (- × -) OH^-.

TABLE 37.13
Dissociation Cross Section

Energy (eV)	Q_a (10^{-22} m^2)	Energy (eV)	Q_a (10^{-22} m^2)
11	0.01	80	2.06
13	0.03	90	2.04
15	0.38	100	2.03
17	0.51	120	2.02
20	0.76	140	2.01
25	1.17	160	1.94
30	1.28	180	1.83
35	1.56	200	1.78
40	1.66	220	1.67
45	1.70	240	1.60
50	1.81	260	1.54
60	2.02	280	1.50
70	2.07	300	1.50

Source: Digitized and interpolated from Harb, T., W. Kedzierski, and J. W. McConkey, *J. Chem. Phys.*, 115, 5507, 2001.

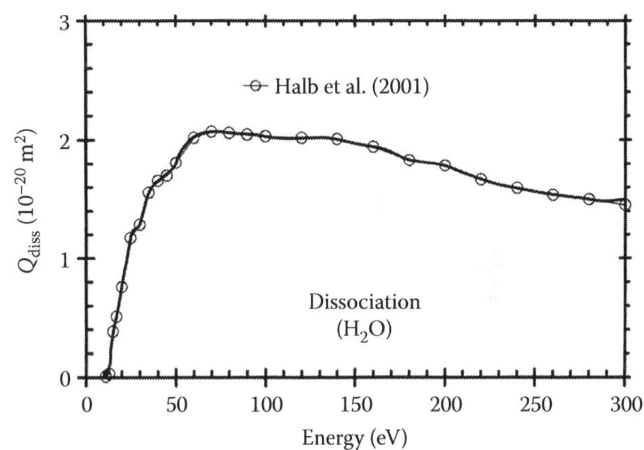

FIGURE 37.10 Dissociation cross section for production of ground-state OH from H_2O. (Adapted from Harb, T., W. Kedzierski, and J. W. McConkey, *J. Chem. Phys.*, 115, 5507, 2001.)

37.10 ELECTRON DRIFT VELOCITY

See Tables 39.14 and 39.15.

37.11 DIFFUSION COEFFICIENTS

See Tables 39.16 (Crompton, 1965) and 39.17.

Figure 37.13 shows the number density-normalized diffusion coefficient (ND) for water vapor.

37.12 IONIZATION COEFFICIENTS

Density-reduced ionization coefficients have been measured by Risbud and Naidu (1979) as shown in Table 37.18 and Figure 37.14.

37.13 GAS CONSTANTS

Gas constants evaluated according to the expression

TABLE 37.14
Drift Velocities of Electron in Water Vapor

E/N (Td)	*W* (m/s)	*E/N* (Td)	*W* (m/s)
1.50	2.77×10^2	33.0	8.06×10^3
2.25	5.02×10^2	36.0	8.84×10^3
3.00	6.53×10^2	39.0	1.03×10^4
4.50	9.90×10^2	42.0	1.18×10^4
6.00	1.33×10^3	45.0	1.39×10^4
7.50	1.67×10^3	48.0	1.68×10^4
10.0	2.22×10^3	51.0	2.10×10^4
12.0	2.67×10^3	54.0	2.34×10^4
13.5	3.02×10^3	57.0	2.95×10^4
15.0	3.37×10^3	60.0	3.56×10^4
16.5	3.70×10^3	63.0	4.25×10^4
18.0	4.04×10^3	66.0	5.18×10^4
21.0	4.79×10^3	69.0	5.91×10^4
24.0	5.59×10^3	72.0	6.95×10^4
27.0	6.35×10^3	75.0	7.70×10^4

Source: Adapted from Wilson, J. F. et al., *J. Chem. Phys.*, 62, 4204, 1975.
Note: See Figure 37.11.

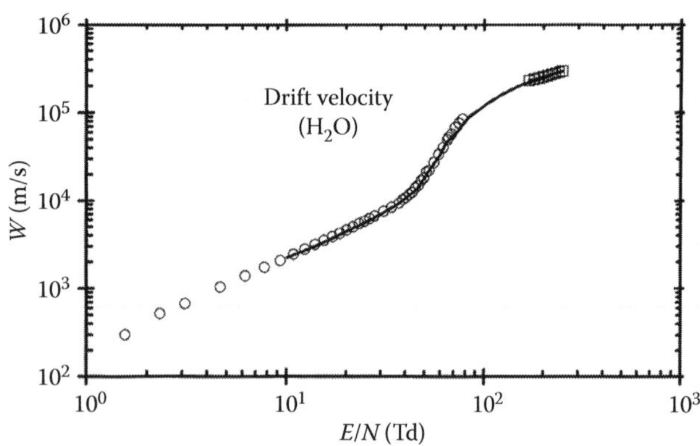

FIGURE 37.11 Electron drift velocities in water vapor. (○) Wilson et al. (1975); (—) Lowke and Parker (1969), theoretical values cited by Wilson et al. (1975); (□) Ryżko (1965).

TABLE 37.15
Drift Velocities of Electron in Water Vapor

E/N (Td)	*W* (10^5 m/s)	*E/N* (Td)	*W* (10^5 m/s)
170	2.29	220	2.70
180	2.37	230	2.78
190	2.45	240	2.87
200	2.54	250	2.95
210	2.62	220	2.70

Source: Adapted from Ryżko, H., *J. Phys. B: At. Mol. Phys.*, 85, 1283, 1965.

TABLE 37.16
Ratio of Radial Diffusion Coefficients to Mobility (D/μ)

E/N (Td)	D_r/μ (V)	*E/N* (Td)	D_r/μ (V)
60	1.8	82.5	2.85
67.75	2.03	90	2.96
75	2.6		

Source: Adapted from Crompton, R. W., J. A. Rees, and R. L. Jory, *Aust. J. Phys.*, 18, 541, 1965.
Note: See Figure 37.12 for graphical presentation.

TABLE 37.17
Ratio of Longitudinal Diffusion Coefficient to Mobility

E/N (Td)	D_L/μ (V)	*E/N* (Td)	D_L/μ (V)
4.5	0.0260	35	0.0264
6	0.0255	37.5	0.13
7.5	0.0261	40	0.37
9	0.0280	45	1.02
10.5	0.0269	50	2.20
12	0.0284	55	3.32
15	0.0292	60	4.19
18	0.0287	65	4.34
21	0.0311	70	4.07
24	0.0350	75	3.35
27	0.0414	77.5	3.64
30	0.0425		

Source: Adapted from Wilson, J. F. et al., *J. Chem. Phys.*, 62, 4204, 1975.
Note: See Figure 37.12 for graphical presentation.

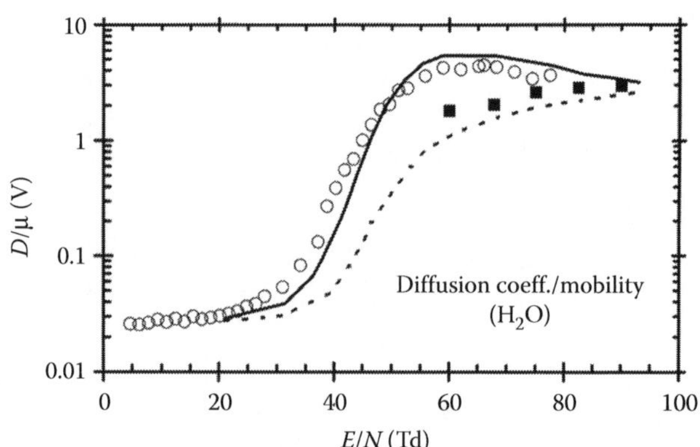

FIGURE 37.12 Ratio of diffusion coefficient to mobility for water vapor. Radial diffusion: (■) Crompton et al. (1965); (— —) Lowke and Parker (1969), theory. Longitudinal diffusion: (○) Wilson et al. (1975); (—) Lowke and Parker (1969).

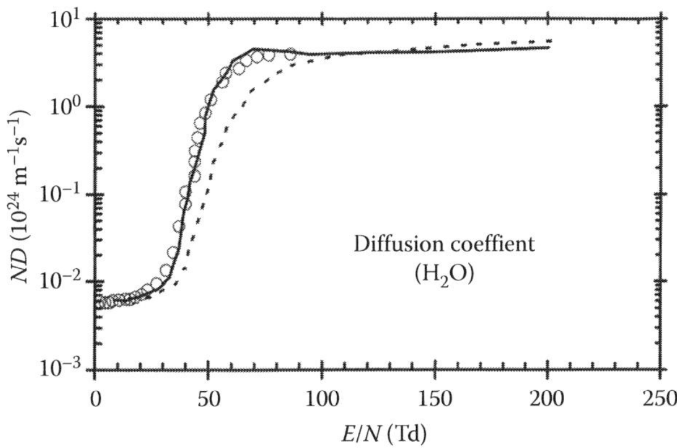

FIGURE 37.13 Density-normalized diffusion coefficients. Longitudinal diffusion: (○) Wilson et al. (1975); (—) Lowke and Parker (1969), theory. Radial diffusion: (— —) Lowke and Parker (1969), theory.

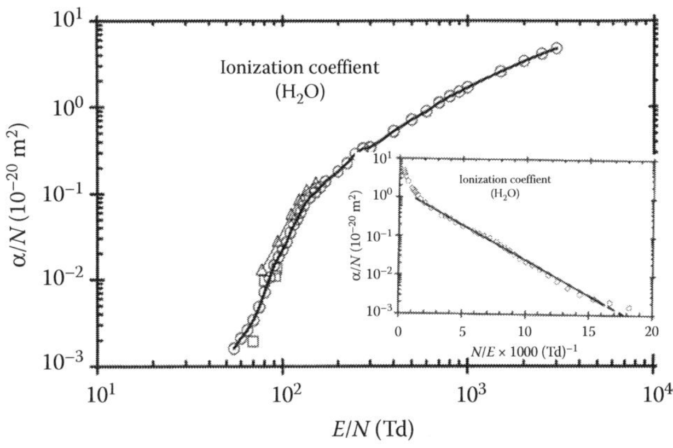

FIGURE 37.14 Reduced first ionization coefficients as function of reduced electric field. (—○—) Risbud and Naidu (1979); (□) Ryżko (1965); (Δ) Prasad and Craggs (1960). Inset shows the plot according to Equation 37.1. (Adapted from Risbud, A. V. and M. S. Naidu, *J. Physique Colloque*, c7, Supplément au n°7, 40, c7–77, 1979.)

TABLE 37.18
Density-Reduced Ionization Coefficients for Water Vapor

E/N (Td)	α/*N* (10^{-20} m²)	*E/N* (Td)	α/*N* (10^{-20} m²)
50.0	0.0007	160	0.12
55.0	0.002	170	0.14
60.0	0.002	200	0.18
65.0	0.003	225	0.23
70.0	0.003	250	0.29
75.0	0.005	275	0.33
80.0	0.007	300	0.35
85.0	0.011	400	0.53
90.0	0.015	500	0.71
95.0	0.018	600	0.89
100	0.022	700	1.12
105	0.027	800	1.32
110	0.035	900	1.49
115	0.043	1000	1.67
120	0.051	1500	2.57
125	0.061	2000	3.40
130	0.073	2500	4.16
140	0.091	3000	4.80
150	0.10		

Source: Adapted from Risbud, A. V. and M. S. Naidu, *J. Physique Colloque*, c7, Supplément au n°7, 40, c7–77, 1979.

$$\frac{\alpha}{N} = F\exp\left(-\frac{GN}{E}\right) \tag{37.1}$$

are $F = 1.25 \times 10^{-20}$ m² and $G = 385.4$ Td⁻¹. See inset of Figure 37.14.

TABLE 37.19
Density-Reduced Attachment Coefficients for Water Vapor

E/N (Td)	η/*N* (10^{-22} m²)	*E/N* (Td)	η/*N* (10^{-22} m²)
70	3.90	115	6.16
75	4.54	120	6.13
80	4.99	125	6.00
85	5.38	130	5.82
90	5.71	140	5.60
95	5.89	150	5.17
100	5.99	160	4.76
105	6.06	170	4.28
110	6.13		

Source: Adapted from Risbud, A. V. and M. S. Naidu, *J. Physique Colloque*, c7, Supplément au n°7, 40, c7–77, 1979.

37.14 ATTACHMENT COEFFICIENTS

Table 37.19 and Figure 37.15 show density-reduced attachment coefficients.

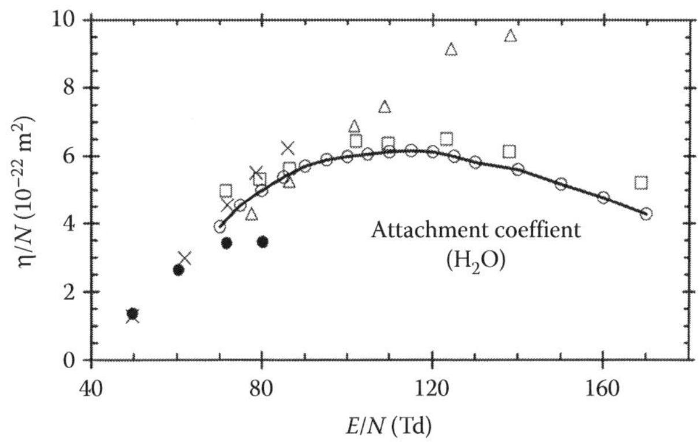

FIGURE 37.15 Density-reduced attachment coefficients for water vapor. (—○—) Risbud and Naidu (1979); (×) Ryżko (1966); (□) Crompton et al. (1965); (Δ) Prasad and Craggs (1960); (●) Kuffel (1959).

TABLE 37.20
Attachment Rate Constants for Water Vapor

E/N (Td)	k_a (m^3/s)	E/N (Td)	k_a (m^3/s)
1.20	2.3×10^{-18}	1.95	4.39×10^{-17}
1.35	5.7×10^{-18}	2.10	5.83×10^{-17}
1.50	1.23×10^{-17}	2.25	7.26×10^{-17}
1.65	2.11×10^{-17}	2.4	8.79×10^{-17}
1.80	3.21×10^{-17}		

Source: Adapted from Compton, R. N. and L. G. Christophorou, *Phys. Rev.*, 154, 110, 1967.

H_2O

37.15 ATTACHMENT RATE CONSTANT

Table 37.20 shows the attachment rate constants as function of E/N under swarm conditions (Compton and Christophorou, 1967).

37.16 POSITIVE ION MOBILITY

The mobility of positive ions, 0.061 m^2/V s is constant over the range of 150–270 Td at $N = 3.22 \times 10^{22}$ m^{-3} (Ryżko, 1965).

37.17 ADDENDUM

A more complete set of vibrational excitation cross sections has been provided by Olivero et al. (1972).

REFERENCES

Bolorizadeh, M. A. and M. A. Rudd, *Phys. Rev. A*, 33, 882, 1986.

Brunger, M., S. Buckman, and M. Elford, Excitation cross sections, in *Interactions of Photons and Electrons with Molecules*, Vol. 17, Springer-Verlag, New York, NY, 2003, Chapter 6.4, pp. 118–201.

Champion, C., J. Hanssen, and P. A. Hervieux, *J. Chem. Phys.*, 117, 197, 2002.

Compton, R. N. and L. G. Christophorou, *Phys. Rev.*, 154, 110, 1967.

Cottin, M., *J. Chem. Phys.*, 56, 1024, 1959.

Crawford, O. H., *Proc. Phys. Soc.*, 91, 279, 1967.

Crompton, R. W., J. A. Rees, and R. L. Jory, *Aust. J. Phys.*, 18, 541, 1965.

Cvejanović, D., L. Andrić, and R. I. Hall, *J. Phys. B: At. Mol. Opt. Phys.*, 26, 2899, 1993.

Dorman, F. H., *J. Chem. Phys.*, 44, 3856, 1966.

El-Zein, A. A. A., M. J. Brunger, and W. R. Newell, *J. Phys. B: At. Mol. Opt. Phys.*, 33, 5033, 2000.

Harb, T., W. Kedzierski, and J. W. McConkey, *J. Chem. Phys.*, 115, 5507, 2001.

Hwang, W., Y.-K. Kim, and M. E. Rudd, *J. Chem. Phys.*, 104, 2956, 1996.

Itikawa, Y., Excitation cross sections, in *Interactions of Photons and Electrons with Molecules*, Vol. 17, Springer-Verlag, New York, NY, 2003, Chapter 5.2, pp. 78–114.

Johnstone, W. M. and W. R. Newell, *J. Phys. B: At. Mol. Phys.*, 24, 3633, 1991.

Jung, K., Th. Antoni, R. Müller, K.-H. Kochem, and H. Ehrhardt, *J. Phys. B: At. Mol. Phys.*, 15, 3535, 1982.

Katase, A., K. Ishibashi, Y. Matsumoto, T. Sakae, S. Maezono, E. Murakami, K. Watanabe, and H. Maki, *J. Phys. B: At. Mol. Phys.*, 19, 2715, 1986.

Kuffel, E., *Proc. Phys. Soc.*, 74, 297, 1959.

Lowke, J. J. and J. H. Parker, Jr., *Phys. Rev.*, 181, 302, 1969.

Lozier, W. W., *Phys. Rev.*, 36, 1417, 1930.

Mann, M. V., A. Hustrulid, and J. T. Tate, *Phys. Rev.*, 58, 340, 1940.

Melton, C. E., *J. Chem. Phys.*, 57, 4218, 1972.

Nishimura, H., *J. Phys. Soc. Japan*, 54, 1227, 1985.

Olivero, J. J., R. W. Stagat, and A. E. S. Green, *J. Geophys. Res.*, 77, 4797, 1972.

Orient, O. J. and S. K. Srivastava, *J. Phys. B: At. Mol. Phys.*, 20, 3923, 1987.

Pack, J. L., R. E. Voshall, and A. V. Phelps, *Phys. Rev.*, 127, 2084, 1962.

Parr, J. E. and J. L. Moruzzi, *J. Phys. D: Appl. Phys.*, 5, 514, 1972.

Prasad, A. N. and J. D. Craggs, *Proc. Phys. Soc.*, 76, 223, 1960.

Risbud, A. V. and M. S. Naidu, *J. Physique Colloque*, c7, Supplément au n°7, 40, c7–77, 1979.

Rohr, K., *J. Phys. B: At. Mol. Phys.*, 10, L735, 1977.

Ryżko, A., *Arkiv. Fys.*, 32, 1, 1966.

Ryżko, H., *J. Phys. B: At. Mol. Phys.*, 85, 1283, 1965.

Sağlam, Z. and N. Aktekin, *J. Phys. B: At. Mol. Opt. Phys.*, 23, 1529, 1990.

Sağlam, Z. and N. Aktekin, *J. Phys. B: At. Mol. Opt. Phys.*, 24, 3491, 1991

Schulz, G. J., *J. Chem. Phys.*, 33, 1661, 1960.

Schutten, J., F. J. de Heer, H. R. Moustafa, A. J. H. Boerboom, and J. Kistemaker, *J. Chem. Phys.*, 44, 3924, 1966.

Seng, G. and F. Linder, *J. Phys. B: At. Mol. Phys.*, 9, 2539, 1976.

Shimanouchi, T., *Tables of Molecular Vibrational Frequencies Consolidated Volume I*, NSRDS-NBS 39, U.S. Department of Commerce, Washington (DC), 1972.

Shyn, T. W. and S. Y. Cho, *Phys. Rev. A*, 36, 5138, 1987.

Straub, H. C., B. G. Lindsay, K. A, Smith, and R. F. Stebbings, *J. Chem. Phys.*, 108, 109, 1998.

Sueoka, O., S. Mori, and Y. Katayama, *J. Phys. B: At. Mol. Phys.*, 19, L373–L378, 1986.

Wilson, J. F., F. J. Davis, D. R. Nelson, and R. N. Compton, *J. Chem. Phys.*, 62, 4204, 1975.

Zecca, A., G. Karwasz, S. Oss, R. Grisenti, and R. S. Brusa, *J. Phys. B: At. Mol. Phys.*, 20, L133, 1987.

Section IV

4 ATOMS

38 ACETYLENE

CONTENTS

Acetylene (C_2H_2) is a simple hydrocarbon with a triple C–C bond and it is isoelectronic with N_2 and CO with 14 electrons. The ground-state electronic configuration is given by Mu-Tao et al. (1990).

The length of the C–C bond is 0.121 nm. In the ground state, the molecule is nonpolar, but acquires a dipole moment when certain modes of vibration are excited, as shown in Table 38.1. The bond dissociation energy (H–CCH) is 5.78 eV, the ionization energy is 11.40 eV, and the electron affinity is 0.49 eV.

The molecule has five fundamental modes of vibration classified as shown in Table 38.1 (Kochem et al., 1985).

The ν_4 mode is unobservable and the cross sections for other modes are presented in Table 38.5.

TABLE 38.1
Details of Vibration Modes of C_2H_2 Molecule

Designation	Description	ε_v (eV)	μ (D)	Representation
ν_1	Symmetric C–H stretching	418	0	H–C–C–H
ν_2	Symmetric C≡C stretching	245	0	H–C—C–H
ν_3	Asymmetric stretching	409	0.07	H–C–C—H
ν_4	Symmetric bending	73	0	
ν_5	Asymmetric bending	91	0.243	

Note: ε_v = vibrational energy (meV); μ = dipole moment (D).

38.1 SELECTED REFERENCES FOR DATA

See Table 38.2.

TABLE 38.2
Selected References for Data on C_2H_2

Quantity	Energy Range (eV)	Reference
	Cross Sections	
Q_i	10–600	**Tian and Vidal (1998)**
Q_i	10–800	**Zheng and Srivastava (1996)**
Q_T	400–2600	**Xing et al. (1995)**
Q_{diff}	5–100	**Khakoo et al. (1993)**
Q_{el}, Q_{inel}	10–1000	Jain and Baluja (1992)
Q_{diff} (T)	10–200	Mu-Tao et al. (1990)
Q_T	1–400	**Sueoka and Mori (1989)**
Q_a	0–12	**Dressler and Allan (1987)**
Q_V	0–3.6	**Kochem et al. (1985)**
Q_a	0–15	**Rutkowsky et al. (1980)**
Q_{diff}	100–1000	**Fink et al. (1975)**
Q_M	0.01–0.06	**Bowman and Gordon (1967)**
Q_{diff}	10–100	**Hughes and McMillen (1933)**
	Swarm Properties	
W_e	0.6–10	**Bowman and Gordon (1967)**
α/N	125–9000	**Heylen (1963)**

Note: Bold font denotes experimental study. Q_a = Attachment; Q_{diff} = differential; Q_{el} = elastic; Q_i = ionization; Q_{inel} = inelastic; Q_T = total; T = theory; W_e = electron drift velocity; E/N = reduced electric field; Td = Townsend.

TABLE 38.3
Total Scattering Cross Sections of C_2H_2

Energy (eV)	Q_T (10^{-20} m²)	Energy (eV)	Q_T (10^{-20} m²)	Energy (eV)	Q_T (10^{-20} m²)	Energy (eV)	Q_T (10^{-20} m²)
Dressler and Allan (1987)		Sueoka and Mori (1989)				Xing et al. (1994)	
0.05	24.10	1.0	18.1	12	20.9	400	5.38
0.10	23.61	1.2	20.6	13	20.8	500	4.59
0.15	23.27	1.4	22.9	14	20.1	600	4.06
0.25	24.38	1.6	26.3	15	18.9	700	3.67
0.35	25.38	1.8	29.4	16	18.9	800	3.19
0.45	25.83	2.0	32.7	17	19.0	900	2.86
0.60	26.30	2.2	34.0	18	18.2	1000	2.61
0.75	26.71	2.5	35.8	19	18.6	1100	2.41
0.90	27.06	2.8	34.4	20	17.7	1200	2.22
1.00	27.36	3.1	32.6	22	16.6	1300	2.11
1.20	27.95	3.4	29.6	25	15.9	1400	1.95
1.40	28.84	3.7	28.3	30	14.9	1500	1.84
1.60	30.26	4.0	26.9	35	14.6	1600	1.74
1.80	31.85	4.5	25.5	40	13.3	1800	1.57
2.00	33.62	5.0	23.7	50	12.5	2000	1.44
2.50	35.72	5.5	23.8	60	12.0	2200	1.36
3.00	35.29	6.0	24.1	70	11.3	2400	1.32
4.00	32.73	6.5	24.6	80	10.8	2600	1.2
4.50	31.90	7.0	24.2	100	9.3		
5.00	32.70	7.5	24.4	150	7.8		
		8.0	24.0	200	6.9		
		8.5	23.1	250	6.0		
		9.0	22.8	300	5.3		
		9.5	22.5	350	4.9		
		10	22.1	400	4.9		

Note: The energy ranges are (1) $0.05 \le \varepsilon \le 5$ eV (Dressler and Allan, 1987); (2) $1 \le \varepsilon \le 400$ eV (Sueoka and Mori, 1989); (3) $400 \le \varepsilon \le 2600$ eV (Xing et al., 1994). The relative cross sections of Dressler and Allan are normalized to the resonance peak measured by Sueoka and Mori. See Figure 38.1 for graphical presentation.

38.2 TOTAL SCATTERING CROSS SECTION

See Table 38.3. The dominant features of the total scattering cross section are

1. Evidence of Ramsauer–Townsend minimum at 0.15 eV.
2. A shape resonance at 2.6 eV. C_2H_2 exhibits low-energy shape resonance with designation $^2\Pi_g$, common with N_2 and CO. The resonance is associated with predissociation of the molecule into C_2H^- and H fragments (Dressler and Allan, 1987).
3. A second peak at about 8 eV possibly due to dissociation and other inelastic collisions.
4. A monotonic decrease for energy >10 eV.

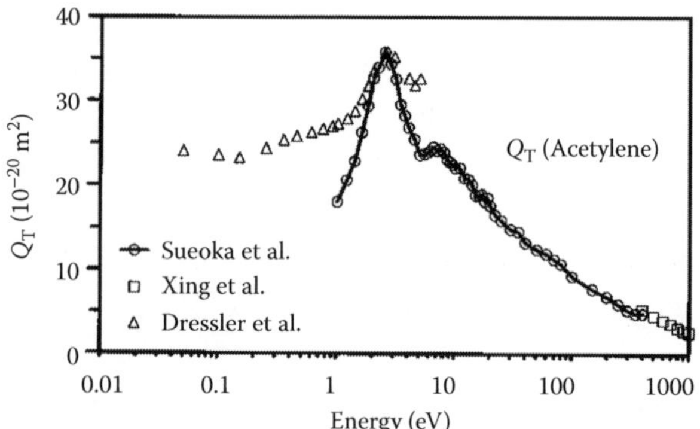

FIGURE 38.1 Total cross sections in C_2H_2. The measurements of Dressler and Allan (1987) are relative cross sections normalized to the shape resonance peak of Sueoka and Mori (1989) at 2.6 eV.

38.3 DIFFERENTIAL SCATTERING CROSS SECTIONS

Peaks and valleys in the differential cross sections show the presence of interference effects between waves scattered from different atoms of the molecule. Figures 38.2 and 38.3 show the differential cross sections in the low- (≤100 eV) and high-energy ranges, respectively. The highlights of differential

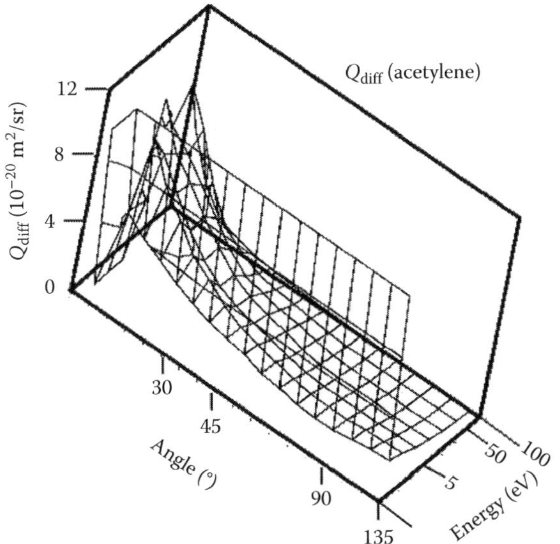

FIGURE 38.2 Three-dimensional view of differential scattering cross section of C_2H_2 as a function of scattering angle and electron energy. The cross sections have a peak at low angles and energy in the range from 5 to 10 eV. (Adapted from Khakoo, M. A. et al., *J. Phys. B: At. Mol. Opt. Phys.*, 26, 4845, 1993.)

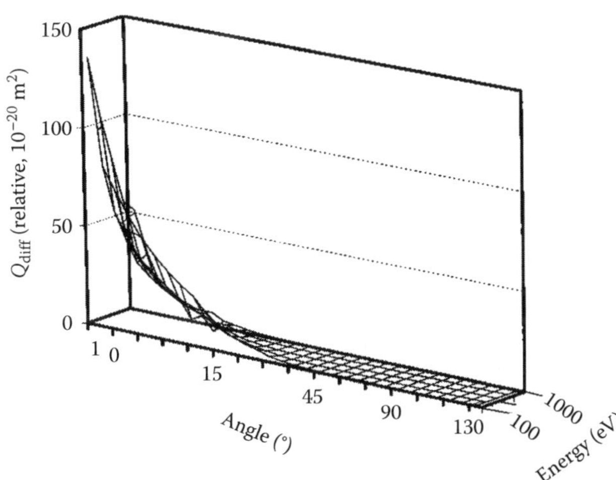

FIGURE 38.3 Three-dimensional view of differential scattering cross section of C_2H_2 as a function of scattering angle and electron energy in the range 100–1000 eV. Note that the differential cross sections shown are relative. (Adapted from Fink, M., K. Jost, and D. Hermann, *J. Chem. Phys.*, 63, 1985, 1975.)

cross sections as a function of energy and angle of scattering are (Khakoo et al., 1993)

1. The differential scattering cross section decreases monotonically with increasing angle of scattering in the range of 10°–90°. For angles >125°, there is increased backward scattering (Hughes and McMillen, 1933).
2. The differential scattering cross section shows a peak in the energy range from 5 to 10 eV.
3. For electron energy >10 eV, the contribution of vibrational excitation is less than 5%.

38.4 ELASTIC SCATTERING AND MOMENTUM TRANSFER CROSS SECTIONS

The measured integral elastic cross sections (Table 38.4) agree well as shown in the last column, except at 100 eV. However, the theoretical values are much higher, particularly at energies below 100 eV and the differences have not been resolved. The integrated and normalized values of elastic scattering cross section from Fink et al. (1975) are taken from Karwasz et al. (2001).

See Table 38.5.

TABLE 38.4
Integral Elastic Scattering Cross Sections of C_2H_2

	Q_{el}				
Energy (eV)	**Hughes et al.**	**Fink et al.**	**Mu-Tao et al. (T)**	**Jain et al. (T)**	**Khakoo et al.**
5					
10	14.5		63.65	41.736	20.3
15	14.44				15.1
20				32.660	12.8
25	14.44				
30				26.900	8.5
50	4.15		20.00	18.692	5.6
100	0.58	11.3	12.42	11.55	3.7
200		6.1	7.12	5.184	
300				3.822	
400		3.4		3.040	
500				2.522	
600		1.0		2.155	
700				1.880	
800				1.667	
1000		0.68		1.359	

Note: Energy in units of eV and Q_{el} in units of 10^{-20} m². T = Theoretical. See Figure 38.4 for graphical presentation.

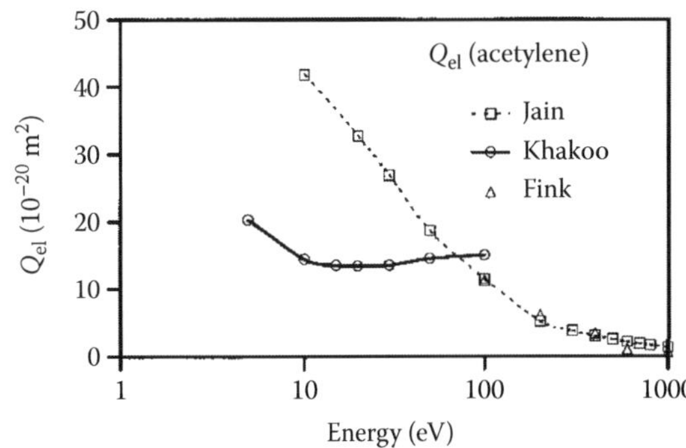

FIGURE 38.4 Elastic scattering cross sections of C_2H_2. The disagreement between theory and experiment has not been resolved.

TABLE 38.5
Momentum Transfer Cross Section

Energy (eV)	Q_M (10^{-20} m²)	Energy (eV)	Q_M (10^{-20} m²)
0.01	191.14	0.04	32.71
0.02	81.14	0.05	29.88
0.03	51.88	0.06	23.13

Source: Adapted from Bowman, C. R. and D. E. Gordon, *J. Chem. Phys.*, 46, 1878, 1967.
Note: Method: drift velocity measurements. See Figure 38.5 for graphical presentation.

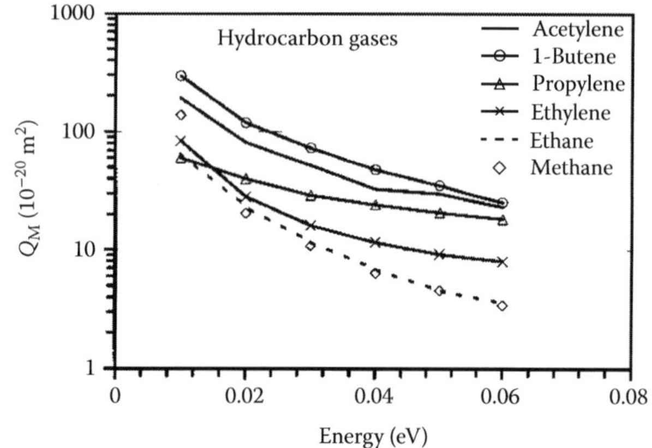

FIGURE 38.5 Momentum transfer cross section of C_2H_2 and selected hydrocarbon gases derived from drift velocity measurements. Note that the energy is too low for Ramsauer–Townsend minimum to show. (Adapted from Bowman, C. R. and D. E. Gordon, *J. Chem. Phys.*, 46, 1878, 1967.)

38.5 RO-VIBRATIONAL SCATTERING CROSS SECTION

Molecular vibrations in the ground state can be excited by three mechanisms, operating individually or in combination (Kochem et al., 1985). The three modes are

1. Scattering of low-energy electrons through dipole, quadrupole, and polarization potentials. This is a nonresonance phenomenon.
2. Formation of a short-lived (~10^{-14} s) negative ion, and subsequent autodetachment of the electron. This process is frequently referred to as shape resonance (Schulz, 1973).
3. The vibrational excitation proceeds via a virtual state, which is a special type of resonance and has the largest cross section at zero energy. This phenomenon was experimentally discovered by Rohr and Linder (1976) and theoretically explained by Dubé and Herzenberg (1977). This phenomenon is observed in CO_2.

All three mechanisms can be operative in principle, and lead to interference effects. Vibrational excitation in C_2H_2 occurs via the first dominant mechanism; the shape resonance has already been referred to.

See Table 38.6.

38.6 ION APPEARANCE POTENTIALS

See Table 38.7.

38.7 IONIZATION CROSS SECTIONS OF SELECTED HYDROCARBONS

See Table 38.8.

38.8 DISSOCIATIVE IONIZATION CROSS SECTIONS

See Table 38.9.

TABLE 38.6
Vibrational Cross Sections (10^{-20} m²) in the Fundamental Modes

	Modes			
Energy (eV)	v_1	v_2	v_3	v_5
0.235	—	—	—	2.6
1.0		—	0.21	1.74
1.6	0.012		0.18	1.0
2.0	0.027	0.40	0.26	1.34
2.6	0.143	0.84	0.24	1.15
3.6	0.062	0.37	0.11	0.69

Source: Adapted from Kochem, K.-H. et al., *J. Phys. B: At. Mol. Opt. Phys.*, 18, 1253, 1985.
Note: The v_4 cross sections are estimated to be no larger than the v_1 mode. The branching ratio for v_1: v_2: v_3: v_4: v_5 modes are 1: 6: 2: 1: 8. See Table 38.1.

TABLE 38.7
Appearance Potentials of Ions in C_2H_2

Ion	Reaction (Products)	Energy (eV)	Reference
C_2H^-	$C_2H_2 + e \rightarrow C_2H^- + H$	2.81	Dressler and Allan (1987)
$C^2H_2^+$	$C_2H_2 + e \rightarrow C_2H_2^+ + 2e$	11.4	Zheng and Srivastava (1996)
C_2H^+	$C_2H^+ + H$	17.2	Zheng and Srivastava (1996)
H^+	$H^+ + C_2H$	19.2	Zheng and Srivastava (1996)
C_2^+	$C_2^+ + 2H$	22.33	Zheng and Srivastava (1996)
CH^+	$CH^+ + H + C$	23.94	Zheng and Srivastava (1996)
C^+	$C^+ + H + CH$	24.9	Zheng and Srivastava (1996)

TABLE 38.8
Ionization Cross Section

	Q_i (10^{-20} m^2)						
	CH_4	C_2H_2	C_2H_4	C_2H_6	$C\text{–}C_3H_6$	C_3H_6	C_3H_8
Energy (eV)	Raju (2005)	Zheng and Srivastava (1996)	Nishimura and Tawara (1994)				
10					0.034	0.093	
12			0.097	0.074			0.206
12.5					0.312	0.498	
15	0.235		0.581	0.618	1.04	1.26	1.14
17.5	0.687		1.15	1.39	1.97	2.15	2.30
20	1.279	1.650	1.71	2.24	2.61	3.07	3.31
22.5	1.879						
25	2.042	2.930	3.01	3.48	4.27	4.54	5.21
30	2.544	3.656	3.52	4.45	5.36	5.54	6.47
35	2.766	4.033	4.17	4.94	6.13	6.42	7.37
40	2.964	4.336	4.52	5.41	6.71	7.18	8.00
45	3.177	4.534	4.82	5.84	7.42	7.54	8.54
50	3.279	4.685	5.11	6.04	7.84	8.00	9.22
55		4.803					
60	3.435	4.906	5.48	6.67	8.27	8.42	9.79
65		4.973					
70	3.524	503.5	5.74	6.93	8.48	8.82	10.09
75		5.068					
80	3.528	5.099	5.76	6.86	8.83	9.04	10.20
85		5.120					
90	3.505	5.121	5.79	6.84	8.87	9.17	10.24
95		5.121					
100	3.461	5.100	5.70	6.89	8.27	9.02	10.23
105		5.056					
110	3.396	5.019					
115		4.986					
120		4.946					
125	3.283	4.902	5.58	6.53	8.30	8.62	9.90
130		4.861					
135		4.815					
140		4.765					
145		4.702					
150	3.086	4.638	5.20	6.32	8.10	8.14	9.36
155		4.576					
160		4.517					
165		4.458					
170		4.410					
175	2.921	4.362	4.80	5.98	7.29	7.83	8.84
180		4.308					
185		4.257					
190		4.212					
195		4.163					
200	2.774	4.121	4.58	5.68	7.25	7.34	8.35
225		3.890					
250	2.445	3.683	3.92	5.01	6.59	6.78	7.80
275		3.475					
300	2.211	3.296	3.56	4.60	5.88	5.99	6.84
350		2.983	3.18	4.18	5.34	5.48	6.25
400	1.841	2.717	2.87	3.86	4.88	4.95	5.78
450		2.506	2.64	3.47	4.56	4.66	5.26
500	1.595	2.316	2.45	3.33	4.33	4.50	4.93
550		2.142					
600	1.391	2.006	2.19	3.03	3.77	3.96	4.33
650		1.875					
700	1.247	1.741	1.96	2.71	3.51	3.61	3.99
750		1.649					
800	1.103	1.544	1.75	2.38	3.12	3.18	3.67
900	1.019		1.63	2.25	2.77	2.98	3.27
1000	1.006		1.52	2.03	2.58	2.79	3.05

Note: CH_4 = methane; C_2H_2 = acetylene (ethyne); C_2H_4 = ethene (ethylene); C_2H_6 = ethane; $C\text{–}C_3H_6$ = cyclopropane; C_3H_6 = propene (propylene); C_3H_8 = propane. See Figures 38.6 and 38.7 for graphical presentation.

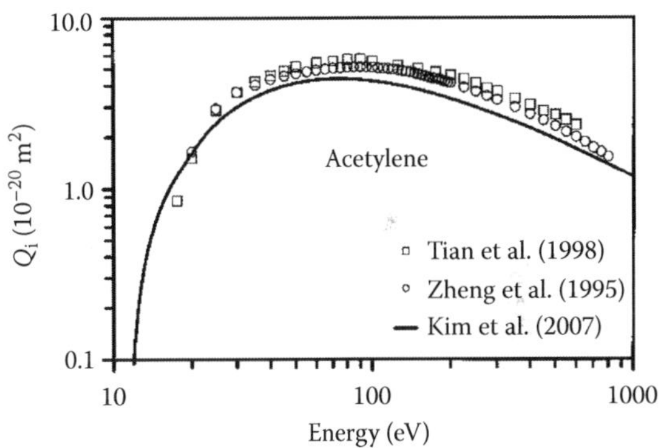

FIGURE 38.6 Total ionization cross section in C_2H_2. The experimental results of Tian and Vidal (1998) and Zheng and Srivastava (1996) agree very well with the theoretical computations of Kim et al. (2001).

38.9 ATTACHMENT CROSS SECTIONS

Both direct and dissociative attachments occur in the gas. The mechanisms are (Rutkowsky et al., 1980)

$$e(>6\,\text{eV}) + C_2H_2 \rightarrow C_2H_2^{-*} \rightarrow \begin{cases} H^- + C_2H \\ C_2^- + H_2 \end{cases} \quad (38.1)$$

Dressler and Allan (1987) suggest the mechanism:

$$e + C_2H_2 \rightarrow H + C_2H^- \quad (38.2)$$

with the energy consideration: dissociation energy = 5.75 eV, electron affinity of C_2H = 2.94 eV, and dissociative attachment

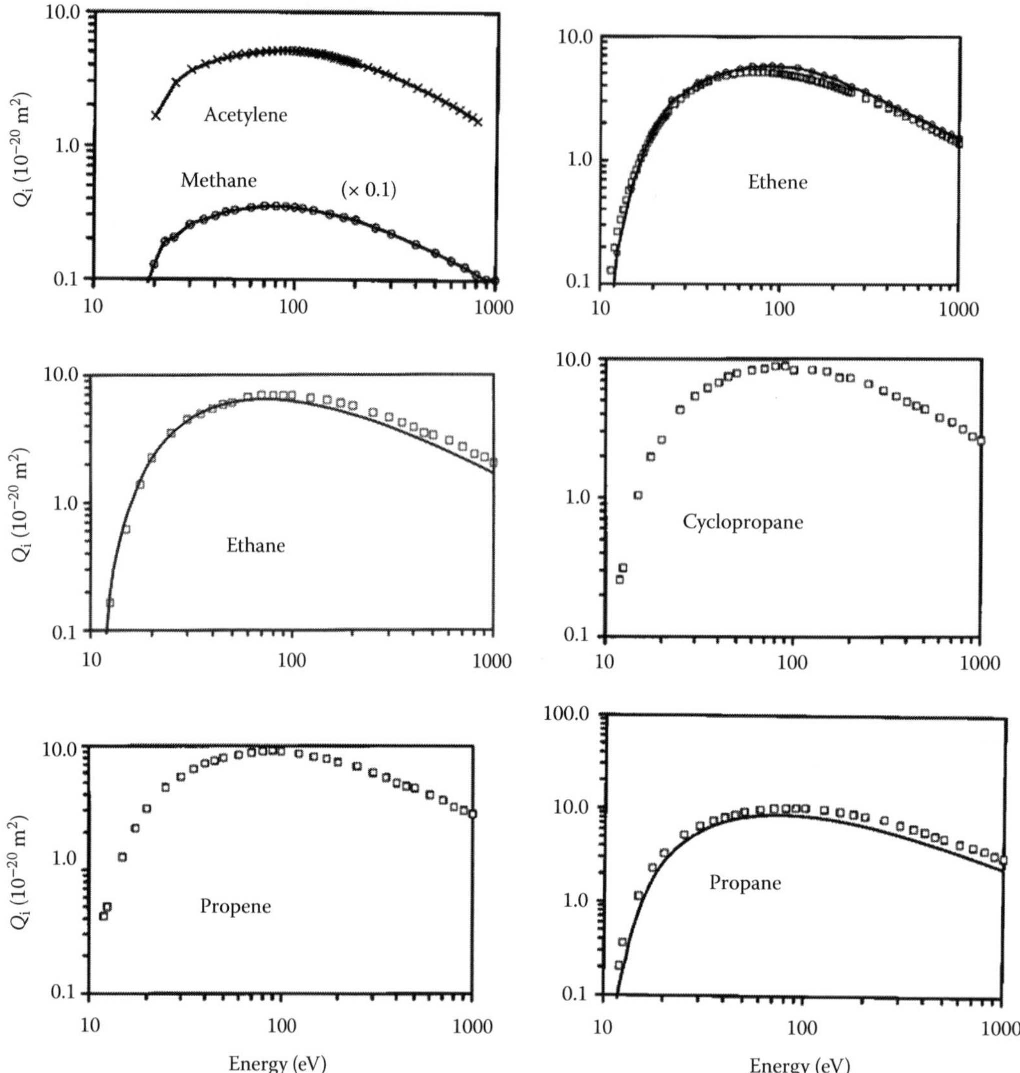

FIGURE 38.7 Ionization cross sections in hydrocarbon gases. Methane (CH_4): (–○–) Raju (2005); acetylene: (×) Zheng and Srivastava (1996); (——) Kim et al. (2001). The cross sections for methane have been multiplied by a factor 0.1 for better presentation. Remaining gases: (——) Kim et al. (2001); (□) Nishimura and Tawara (1994).

threshold = 2.81 eV. The attachment cross section as a function of energy shows the following characteristics (Dressler and Allan, 1987)

1. An intense band centered around 3 eV
2. A weak structureless C_2H^- band centered around 7.4 eV
3. A group of structured C_2^- bands in the 7.6–9 eV region

See Table 38.10.

38.10 MOBILITY AND DRIFT VELOCITY

Experimental values of the reduced mobility, μN, as a function of temperature and with selected gases are shown in Table 38.11 (Bowman and Gordon, 1967).

See Table 38.12.

TABLE 38.9
Cross Section for Dissociative Ionization of C_2H_2 (Units of 10^{-22} m^2) by Electron Impact

Energy	Ion Species					
(eV)	$C_2H_2^+$	C_2H^+	C_2^+	CH^+	C^+	H^+
20	158.9	6.07				
25	248.2	39.5	1.54	0.799	2.92	293.0
30	287.0	56.0	10.1	3.43	1.03	8.02
35	304.0	62.1	14.8	5.56	2.29	14.6
40	314.3	66.8	18.0	8.64	4.00	21.9
45	320.1	69.3	19.1	12.4	4.73	27.8
50	325.2	70.4	19.6	15.2	5.21	32.8
55	329.0	70.9	19.9	18.4	5.57	36.5
60	333.9	71.2	20.1	20.7	5.78	38.9
65	336.2	71.5	20.3	22.5	5.98	40.7
70	339.0	71.7	20.5	24.6	6.10	41.6

TABLE 38.9 (continued)
Cross Section for Dissociative Ionization of C_2H_2 (Units of 10^{-22} m^2) by Electron Impact

Energy (eV)	$C_2H_2^+$	C_2H^+	C_2^+	CH^+	C^+	H^+
75	340.3	71.7	20.6	26.0	6.23	42.1
80	342.4	71.6	20.5	26.7	6.33	42.3
85	343.8	71.4	20.5	27.3	6.42	42.5
90	343.6	71.3	20.4	27.8	6.51	42.5
95	343.5	71.1	20.3	28.2	6.54	42.5
100	341.6	71.0	20.2	28.4	6.62	42.2
105	337.9	70.7	20.0	28.5	6.66	41.8
110	335.0	70.6	19.9	28.5	6.72	41.1
115	332.9	70.4	19.7	28.5	6.75	40.4
120	330.1	70.3	19.5	28.4	6.77	39.6
125	327.0	70.0	19.3	28.1	6.78	39.0
130	324.3	69.6	19.0	27.9	6.78	38.4
135	321.6	69.2	18.7	27.5	6.77	37.8
140	318.7	68.4	18.4	27.1	6.73	37.1
145	314.2	67.8	18.2	26.8	6.69	36.5
150	310.1	66.9	17.9	26.4	6.64	35.8
155	306.0	66.2	17.7	26.0	6.55	35.2
160	302.1	65.6	17.4	25.7	6.48	34.5
165	298.4	64.8	17.2	25.2	6.42	33.9
170	295.6	63.9	17.1	24.9	6.37	33.2
175	292.8	63.1	16.9	24.6	6.30	32.5
180	289.5	62.4	16.7	24.2	6.22	31.8

TABLE 38.9 (continued)
Cross Section for Dissociative Ionization of C_2H_2 (Units of 10^{-22} m^2) by Electron Impact

Energy (eV)	$C_2H_2^+$	C_2H^+	C_2^+	CH^+	C^+	H^+
185	286.1	61.8	16.5	24.0	6.16	31.2
190	283.2	61.2	16.3	23.8	6.10	30.6
195	280.2	60.5	16.1	23.5	6.04	29.8
200	277.7	60.0	15.9	23.3	5.98	29.2
225	263.2	56.9	15.1	21.9	5.67	26.3
250	250.3	54.1	14.1	20.5	5.39	23.8
275	236.8	51.2	13.3	19.3	5.09	21.9
300	225.0	48.9	12.4	18.1	4.78	20.4
350	204.5	44.8	11.0	15.9	4.18	18.0
400	187.1	41.2	9.83	14.1	3.70	15.8
450	173.8	38.0	8.75	12.4	3.32	14.3
500	161.3	35.4	7.93	11.3	3.00	12.8
550	149.7	33.2	7.27	10.1	2.75	11.2
600	140.4	31.3	6.76	9.36	2.55	10.2
650	131.3	29.3	6.22	8.54	2.38	9.72
700	122.0	27.2	5.69	7.75	2.24	9.18
750	115.9	25.9	5.31	7.18	2.15	8.51
800	108.7	24.3	4.89	6.56	2.04	8.02

Source: Adapted from Zheng, S.-H. and S. K. Srivastava, *J. Phys. B: At. Mol. Opt. Phys.*, 29, 3235, 1996.
Note: See Figure 38.8 for graphical presentation.

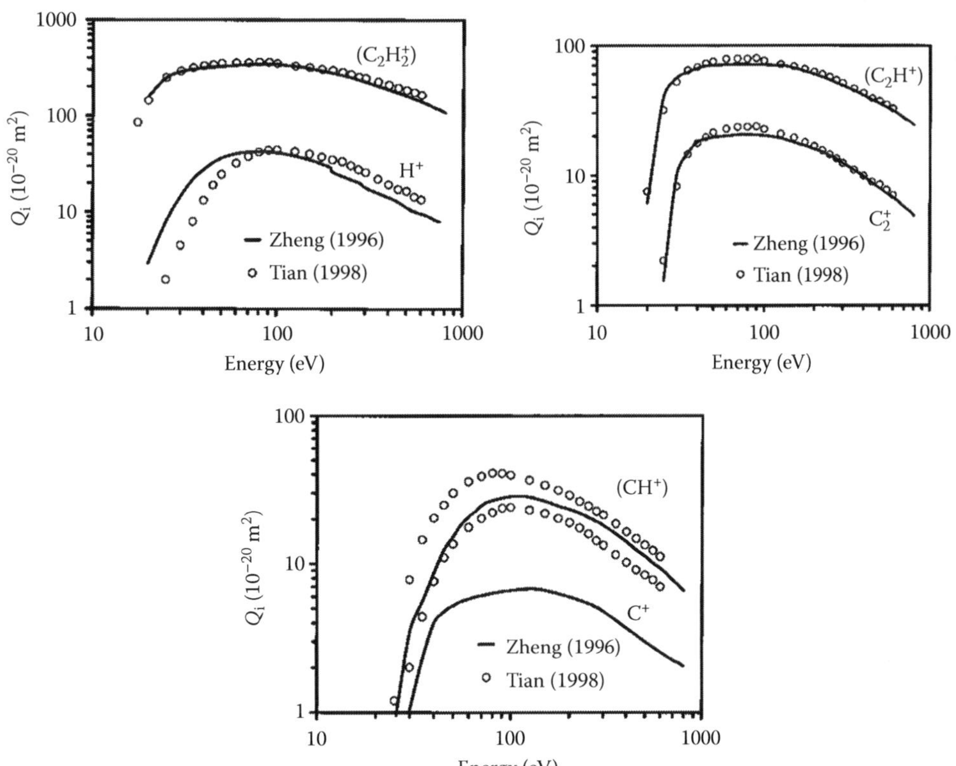

FIGURE 38.8 Dissociative ionization cross sections of C_2H_2. Except for C^+, the agreement between the experimental data of Zheng and Srivastava (1996) and Tian and Vidal (1998) is excellent.

C$_2$H$_2$

38.11 IONIZATION COEFFICIENTS

See Table 38.13.

TABLE 38.10
Attachment Cross Sections of C_2H_2

Energy (eV)	Q_a (10^{-24} m^2)	Energy (eV)	Q_a (10^{-24} m^2)
0	0	4	0.39
1.7	0.16	5	0.07
2	0.99	6	0.80
2.5	6.90	7	3.76
2.65	7.58	7.4	7.48
3	3.21	8	5.16
3.5	1.54	9	0.15

Source: Adapted from Rutkowsky, J., H. Drost, and H. J. Spangenberg, *Ann. Phys. Leipzig*, 37, 259, 1980.
Note: See Figure 38.9 for graphical presentation. Note the low magnitude of the cross section.

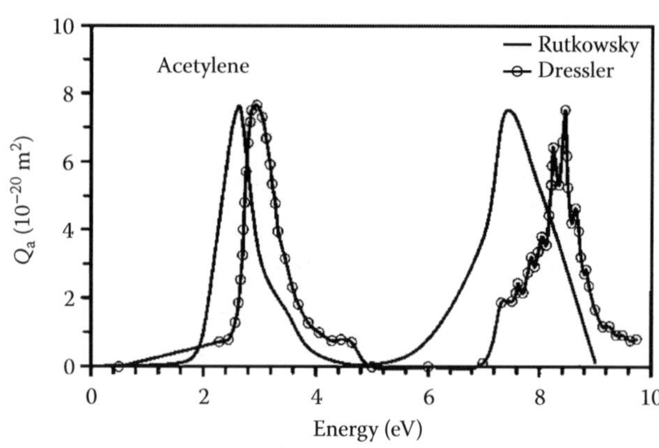

FIGURE 38.9 Electron attachment cross section in C_2H_2. (——) Rutkowsky et al. (1980); (—o—) Dressler and Allan (1987). The left branch is due to the formation of C_2H^-. The relative cross sections of Dressler and Allan have been normalized to the peak cross sections of Rutkowsky et al.

TABLE 38.11
μN as a Function of Temperature and Comparison with Selected Gases

	μN (10^{24} m/V s)					
T (K)	Methane	Ethane	Ethylene	Acetylene	Propylene	1-Butene
225		29.0				
229	29.0		21.0			
298	29.2	39.2	29.3	5.41	6.62	4.84
367			32.2			
370	52.2	56.2		6.44		

Source: Adapted from Bowman, C. R. and D. E. Gordon, *J. Chem. Phys.*, 46, 1878, 1967.

TABLE 38.12
Drift Velocity of Electrons in C_2H_2

298 K		370 K	
E/N (Td)	W_e (10^3 m/s)	*E/N* (Td)	W_e (10^3 m/s)
1	4.41	1	5.83
2	10.17	2	14.47
3	16.30	3	21.52
4	22.24		
5	26.61		
6	30.05		
7	32.46		

Source: Adapted from Bowman, C. R. and D. E. Gordon, *J. Chem. Phys.*, 46, 1878, 1967.
Note: See Figure 38.10 for graphical presentation.

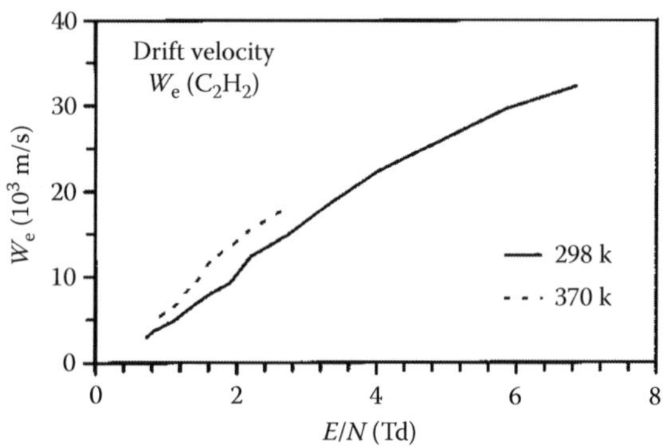

FIGURE 38.10 Drift velocity of electrons in C_2H_2. (——) 298 K; (– – – –) 370 K. At low values of *E/N* the electrons are in thermal equilibrium with gas molecules and therefore have a Maxwellian distribution. Under this condition the applied field does not influence the average energy of the electrons and the collision rate will be independent of the applied field. This will lead to a drift velocity that increases linearly with *E/N*. (Adapted from Bowman, C. R. and D. E. Gordon, *J. Chem. Phys.*, 46, 1878, 1967.)

TABLE 38.13
Reduced Ionization Coefficients for C_2H_2

E/N (Td)	α/*N* (10^{-20} m^2)	*E/N* (Td)	α/*N* (10^{-20} m^2)	*E/N* (Td)	α/*N* (10^{-20} m^2)
125	0.0019	450	0.524	2000	3.83
150	0.0078	500	0.648	2500	4.45
175	0.020	600	0.940	3000	4.98
200	0.040	700	1.20	4000	5.74
225	0.067	800	1.46	5000	6.26
250	0.100	900	1.71	6000	6.59
300	0.190	1000	1.94	7000	6.77
350	0.292	1250	2.52	8000	6.94
400	0.410	1500	3.06	9000	7.26

Source: Adapted from Heylen, A. E. D., *J. Chem. Phys.*, 38, 765, 1963.
Note: Figure 38.11 shows data for several hydrocarbons.

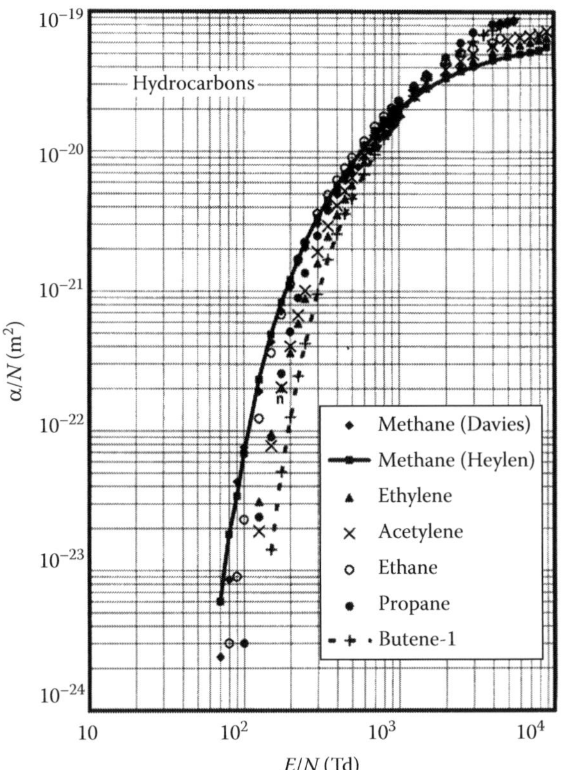

FIGURE 38.11 Reduced ionization coefficients for C_2H_2 and selected hydrocarbon gases. Except for methane (◆, Davies et al., 1989), all other measurements are from Heylen: methane, ethylene, acetylene (Heylen, 1963); ethane and propane (Heylen, 1975); butane-1 (Heylen, 1978).

REFERENCES

Bowman, C. R. and D. E. Gordon, *J. Chem. Phys.*, 46, 1878, 1967.

Davies, D. K., L. E. Kline, and W. E. Bies, *J. Appl. Phys.*, 65, 3311, 1989.

Dressler, R. and M. Allan, *J. Chem. Phys.*, 87, 4510, 1987.

Dubé, L. and A. Herzenberg, *Phys. Rev. Lett.*, 38, 820, 1977.

Fink, M., K. Jost, and D. Hermann, *J. Chem. Phys.*, 63, 1985, 1975. Differential cross sections for elastic scattering of electrons in C_2H_2, C_2H_4, and C_2H_6 are measured,

Heylen, A. E. D., *J. Chem. Phys.*, 38, 765, 1963.

Heylen, A. E. D., *Int. J. Electron.*, 39, 653, 1975.

Heylen, A. E. D., *Int. J. Electron.*, 11, 367, 1978.

Hughes, L. and J. H. McMillen, *Phys. Rev.*, 44, 876, 1933.

Jain, A. and K. I. Baluja, *Phys. Rev. A*, 45, 202, 1992.

Karwasz, G. P., R. S. Brusa, and A. Zecca, *La Rivista del Nuovo Cimento*, 24(3), 1, 2001.

Khakoo, M. A., T. Jayaweera, S. Wang, and S. Trajmar, *J. Phys. B: At. Mol. Opt. Phys.*, 26, 4845, 1993.

Kim, Y.-K., K. K. Inkuro, M. E. Rudd, D. S. Zucker, M. A. Zucker, J. S. Coursey, K. I. Olsen, and G. G. Wiersma 2001. Available at: http://physics.nist.gov/PhysRefData/Contents.html.

Kochem, K.-H., W. Sohn, K. Jung, H. Ehrhardt, and E. S. Chang, *J. Phys. B: At. Mol. Opt. Phys.*, 18, 1253, 1985.

Mu-Tao, L., L. Brescansin, M. A. P. Lima, L. E. Machado, and E. P. Leal, *J. Phys. B: At. Mol. Opt. Phys.*, 23, 4331, 1990.

Nishimura, H. and H. Tawara, *J. Phys. B: At. Mol. Opt. Phys.*, 27, 2063,1994. Total ionization cross sections are measured in CH_4, C_2H_4, C_3H_6, and C_3H_8.

Raju, G. G., *Gaseous Electronics*, Taylor & Francis, Boca Raton, FL 2005.

Rohr, K. and F. Linder, *J. Phys. B: At. Mol. Phys.*, 9, 2521, 1976.

Rutkowsky, J., H. Drost, and H. J. Spangenberg, *Ann. Phys. Leipzig*, 37, 259, 1980. Direct and dissociative attachment cross sections are measured in a number of hydrocarbon gases.

Schulz, G. H., *Rev. Mod. Phys.*, 5, 473, 1973.

Sueoka, O. and S. Mori, *J. Phys. B: At. Mol. Opt. Phys.*, 22, 963, 1989.

Tian, C. and C. Vidal, *J. Phys. B: At. Mol. Opt. Phys.*, 31, 895, 1998.

Xing, S. L., Q. C. Shi, X. J. Chen, K. Z. Xu, B. X. Yang, S. L. Wu, and R. F. Feng, *Phys. Rev. A*, 51, 414, 1995.

Zheng, S.-H. and S. K. Srivastava, *J. Phys. B: At. Mol. Opt. Phys.*, 29, 3235, 1996.

39

AMMONIA

CONTENTS

Ammonia (NH_3) is a polar, electron-attaching gas that has 10 electrons. The electronic polarizability is 3.12×10^{-40} F m^2, the dipole moment is 1.47 D, and the ionization potential is 10.07 eV. The molecule has four vibrational modes as shown in Table 39.1 (Shimanouchi, 1972).

39.1 SELECTED REFERENCES

See Table 39.2.

TABLE 39.1
Vibrational Modes and Energies

Designation	Type of Motion	Energy (meV)
v_1	Symmetrical stretch	413.7
v_2	Symmetrical deformation	117.8
v_3	Degenerate stretch	427.0
v_4	Degenerate deformation	201.7

TABLE 39.2
Selected References for Data

Quantity	Range: eV, (Td)	Reference
Drift velocity	1–1000	**Lisovskiy et al. (2005)**
Compilation	—	Itikawa (2003)
Ionization cross section	11.5–1000	**Rejoub et al. (2001)**
Differential scattering cross section	2–30	**Boesten et al. (1996)**
Transport coefficients	0.1–1000	Yousfi and Benabdessadok (1996)
Differential scattering	2–30	**Alle et al. (1992)**
Vibrational excitation	5, 7.5, 15	**Gulley et al. (1992)**
Ionization cross section	15–1000	**Rao and Srivastava (1992)**
Ionization cross section	10–270	**Syage (1992)**
Total scattering cross section	75–4000	**Zecca et al. (1992)**
Total scattering cross section	1–100	**Szmytkowski et al. (1989)**
Total scattering cross section	1–400	**Sueoka et al. (1987)**
Drift velocity	(1–10)	**Christophorou et al. (1982)**
Ionization coefficients	(30–3000)	**Risbud and Naidu (1979)**
Attachment coefficients	(60–120)	**Risbud and Naidu (1979)**
Attachment coefficients	(45–90)	**Parr and Moruzzi (1972)**
Attachment cross section	4–13.5	**Sharp and Dowell (1969)**
Drift velocity	(0.01–20)	**Pack et al. (1962)**
Attachment coefficients	(45–60)	**Bradbury (1934)**
Attachment coefficients	(45–90)	**Bailey and Duncanson (1930)**

Note: Bold font indicates experimental study.

39.2 TOTAL SCATTERING CROSS SECTIONS

Total scattering cross sections are shown in Table 39.3 and Figure 39.1. The highlights are

1. A trend of increasing cross section toward lower energy below 1 eV.
2. A shallow Ramsauer–Townsend minimum at ~2.5 eV.
3. A broad shape resonance at ~10 eV attributed to the temporary attachment of electron.
4. For energy >10 eV, the cross section decreases approximately according to $\varepsilon^{-1/2}$.

TABLE 39.3
Total Scattering Cross Sections for NH_3

Energy (eV)	Q_T (10^{-20} m^2)	Energy (eV)	Q_T (10^{-20} m^2)	Energy (eV)	Q_T (10^{-20} m^2)
Itikawa (2003)		Zecca et al. (1992)		Sueoka et al. (1987)	
1.0	14.5	75	10.0	1.0	14.7
1.2	13.3	80	9.63	1.2	13.4
1.5	12.0	90	9.07	1.4	12.4
1.7	11.4	100	8.54	1.6	11.6
2.0	10.8	110	7.87	1.8	11.1
2.5	10.5	125	7.37	2.0	10.9
3.0	10.7	150	6.86	2.2	10.4
3.5	11.4	175	6.18	2.5	10.1
4.0	12.6	200	5.64	2.8	10.8
4.5	14.0	225	5.11	3.1	11.2
5.0	15.6	250	4.81	3.4	11.3
6.0	18.0	300	4.25	3.7	11.3
7.0	20.3	350	3.86	4.0	12.0
8.0	21.9	400	3.47	4.5	13.4
9.0	22.8	450	3.15	5.0	14.5
10	22.9	500	2.94	5.5	15.2
12	21.8	600	2.51	6.0	15.5
15	20.1	700	2.20	6.5	15.7
17	19.1	800	1.99	7.0	16.9
20	17.8	900	1.78	7.5	17.5
25	16.1	1000	1.61	8.0	17.1
30	14.8	1100	1.47	8.5	17.1
35	13.8	1250	1.31	9.0	17.5
40	13.0	1500	1.10	9.5	17.5
45	12.3	1750	0.939	10.0	17.5
50	11.8	2000	0.854	11.0	17.5
60	10.9	2250	0.757	12.0	16.9
70	10.3	2500	0.68	13.0	16.9
80	9.68	2750	0.622	14.0	16.4
90	9.10	3000	0.554	15.0	16.0
100	8.57	3250	0.505	16.0	15.5
120	7.64	3500	0.466	17.0	15.0
150	6.67	4000	0.413	18.0	14.9
170	6.17			19.0	14.5
200	5.55			20.0	14.2
250	4.76			22.0	14.0
300	44.21			25.0	12.9
350	3.75			30.0	12.1
400	3.38			35.0	11.0

TABLE 39.3 (continued)
Total Scattering Cross Sections for NH_3

Energy (eV)	Q_T (10^{-20} m^2)	Energy (eV)	Q_T (10^{-20} m^2)	Energy (eV)	Q_T (10^{-20} m^2)
Itikawa (2003)		Zecca et al. (1992)		Sueoka et al. (1987)	
450	3.13			40.0	10.5
500	2.88			50.0	9.3
600	2.50			60.0	8.8
700	2.21			70.0	8.3
800	1.99			80.0	7.9
900	1.81			90.0	7.3
1000	1.66			100	7.0
				120	6.3
				150	5.6
				200	5.0
				250	4.2
				300	3.9
				350	3.5
				400	3.2

Source: Adapted from ItikawaY., (Ed.), *Interactions of Photons and Electrons with Molecules*, Vol. 17c, Landolt-Börnstein, Springer, 2003.

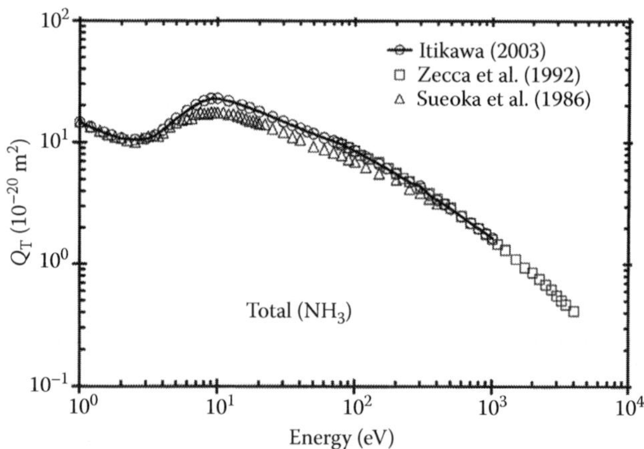

FIGURE 39.1 Total scattering cross sections for NH_3. (—○—) Itikawa (2003); (□) Zecca et al. (1992); (△) Sueoka et al. (1987).

39.3 DIFFERENTIAL SCATTERING CROSS SECTIONS

Figure 39.2 shows the differential scattering cross sections for NH_3 (Boesten et al., 1996). The highlights are

1. A steep decline in differential cross section as the angle increases, signifying forward scattering. This behavior is attributed to the dipole moment of the molecule, though molecules having a smaller dipole moment, such as (N_2O, 0.167 D), also demonstrate the same behavior.
2. The forward scattering is observed at all energies in the range 2–30 eV.

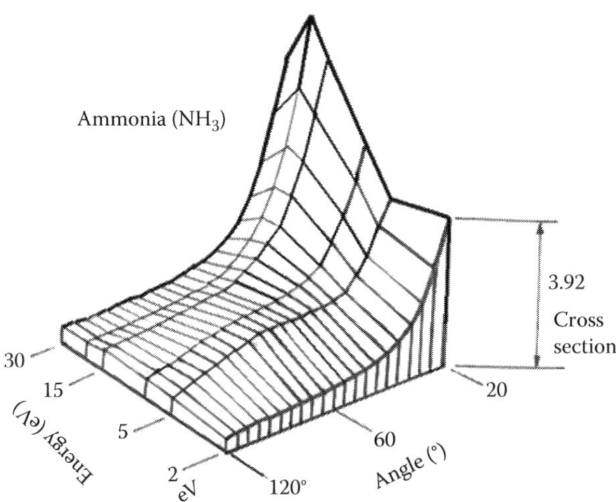

FIGURE 39.2 Differential scattering cross sections for NH_3. (Adapted from Boesten, L. et al., *J. Phys. B: At. Mol. Phys.*, 29, 5475, 1996.)

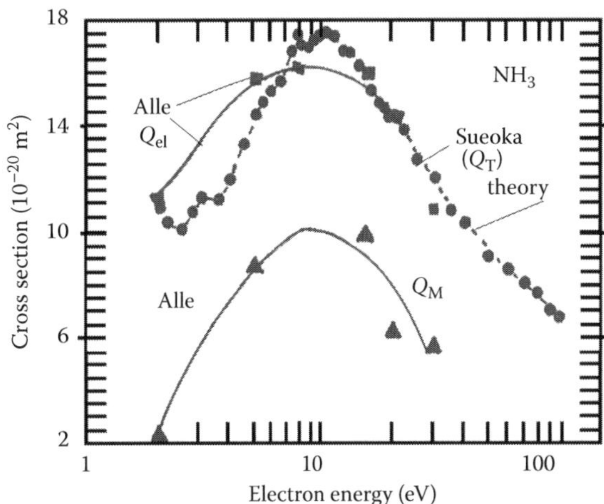

FIGURE 39.3 Momentum transfer and elastic scattering cross sections for NH_3. (□) and (△) from Table 39.4; (○), Sueoka et al. (1987) total cross section; broken curve, theory, Rescigno et al. (1995). (Adapted from Boesten, L., *J. Phys. B: At. Mol. Phys.*, 29, 5475, 1996.)

TABLE 39.4
Integral Elastic and Momentum Transfer Cross Sections

Energy (eV)	Q_{el} (10^{-20} m^2)	Q_M (10^{-20} m^2)
2	11.22	2.24
5	15.83	8.75
7.5	16.21	9.15
15	16.16	10.06
20	14.43	6.40
30	10.97	5.73

39.4 ELASTIC AND MOMENTUM TRANSFER CROSS SECTIONS

Pack et al. (1962) have derived the momentum transfer cross section at very low energies ($0.1 \le \varepsilon \le 1.0$ eV) according to

$$\frac{1}{Q_M} = 4.93 \times 10^{17} \varepsilon^{1/2} + 1.51 \times 10^{19} \varepsilon^{3/2} \text{ m}^{-2} \quad (39.1)$$

Table 39.4 shows the integral elastic and momentum transfer cross sections for NH_3 (Alle et al., 1992). Figure 39.3 shows the cross sections (Boesten et al., 1996).

39.5 VIBRATIONAL EXCITATION CROSS SECTIONS

A single value of vibrational excitation cross section for ($\nu_1 + \nu_3$) mode of 0.67×10^{-20} m^2 at 7.5 eV has been reported by Gulley et al. (1992).

39.6 ION APPEARANCE POTENTIALS

Table 39.5 shows the ion appearance potentials for NH_3 (Mann et al., 1940).

TABLE 39.5
Ion Appearance Potentials

Ion	Energy (eV)	Probable Process	Description
NH_3^+	10.5	$NH_3 + e \to NH_3^+$	Ionization
NH_2^+	15.7	$NH_3 + e \to NH_2^+ + H$	Dissociative ionization
NH^+	19.4	$NH_3 + e \to NH^+ + H_2$	Dissociative ionization
	23.7	$NH_3 + e \to NH^+ + 2H$	Dissociative ionization
N^+	24.9	$NH_3 + e \to N^+ + 3H$	Dissociative ionization
	28.0	$NH_3 + e \to N^{+*} + 3H$	Dissociative ionization
H^+	23.3	$NH_3 + e \to NH + H + H^+$	Dissociative ionization
	26.9	$NH_3 + e \to N^* + 2H + H^+$	Dissociative ionization
H_2^+	15.5	$H_2 \to H_2^+$	Dissociative ionization
		$NH_3 + e \to NH + H_2^+$	Dissociative ionization
NH_3^{2+}	42	$NH_3 + e \to NH_3^{2+}$	Double ionization
H^-	3.76	$NH_3 + e \to NH_2 + H^-$	Dissociative attachment[a]
	5.03	$NH_3 + e \to NH_2^* + H^-$	Dissociative attachment[a]
	5.8	$NH_3 + e \to N + H_2^+ H^-$	Dissociative attachment
	23.0	$NH_3 + e \to NH^+ + H + H^-$	Ion pair production
NH_2^-	3.30	$NH_3 + e \to H + NH_2^-$	Dissociative attachment[a]
	5.78	$NH_3 + e \to H + NH_2^{-*}$	Dissociative attachment[a]
	6.0	$NH_3 + e \to H + NH_2^-$	Dissociative attachment
	13.50	$NH_3 + e \to H^* + NH_2^-$	Dissociative attachment[a]

[a] Data are from Sharp and Dowell (1969).

NH$_3$

TABLE 39.6
Total Ionization Cross Sections for NH$_3$

	Q_i (10^{-20} m^2)	
Energy (eV)	**Rejoub et al. (2001)**	**Rao and Srivastava (1992)**
11.5	0.080	
12.5	0.162	
13.5	0.216	
15	0.331	0.005
17.5	0.628	
20	0.946	0.457
22.5	1.200	
25	1.420	1.108
27.5	1.667	
30	1.821	1.569
35	2.162	1.970
40	2.485	2.347
50	2.808	2.774
60	3.025	2.931
70	3.118	3.010
80	3.177	3.035
90	3.196	3.034
100	3.186	3.007
120	3.063	2.924
140	2.934	2.823
160	2.803	2.725
200	2.585	2.524
250	2.350	2.285
300	2.103	2.081
400	1.748	1.760
500	1.543	1.513
600	1.341	1.341
800	1.089	1.120
1000	0.947	0.991

Note: Intermediate values of Rao and Srivastava are omitted for space economy.

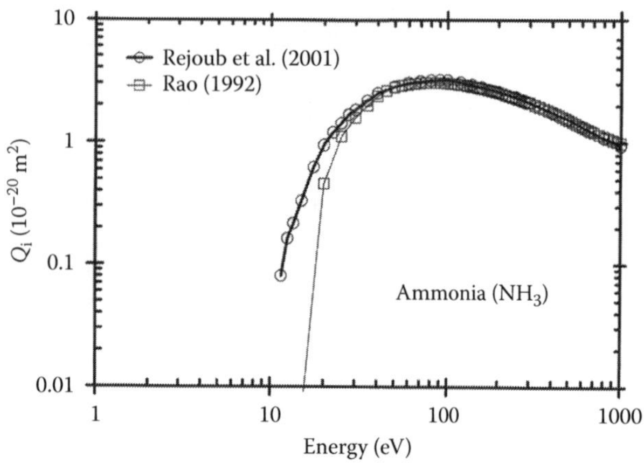

FIGURE 39.4 Total ionization cross sections for NH$_3$. (—○—) Rejoub et al. (2001); (—□—) Rao and Srivastava (1992).

39.7 TOTAL AND PARTIAL IONIZATION CROSS SECTIONS

Table 39.6 and Figures 39.4 and 39.5 show the ionization cross sections for NH$_3$. Figure 39.4 shows the partial ionization cross sections from Rejoub et al. (2001).

39.8 ATTACHMENT CROSS SECTIONS

Attachment cross sections measured by Sharp and Dowell (1969) have been renormalized by Itikawa (2003) as shown in Table 39.7 and Figure 39.6.

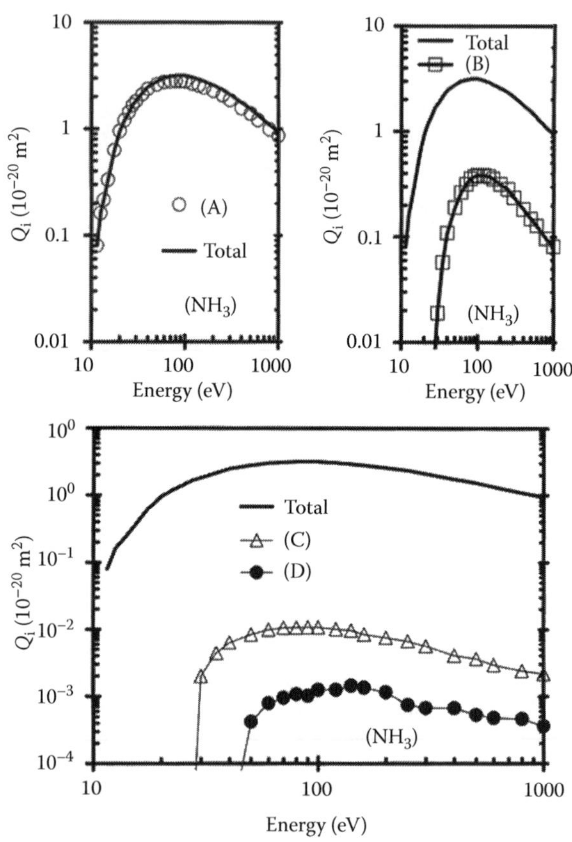

FIGURE 39.5 Partial ionization cross sections for NH$_3$. Total cross sections (——) are plotted to show the relative contributions of the partial ionization cross sections. (○, A) NH$_n^+$ (NH$_3^+$ + NH$_2^+$ + NH$^+$ + N$^+$); (—□—, B) H$^+$; (—Δ—, C) H$_2^+$; (—●—, D) NH$_3^{2+}$. (Adapted from Rejoub, R., B. G. Lindsay, and R. F. Stebbings, *J. Chem. Phys.*, 115, 5053, 2001.)

TABLE 39.7
Attachment Cross Sections for NH$_3$

NH$_2^-$		Total	
Energy (eV)	**Q_a (10^{-22} m^2)**	**Energy (eV)**	**Q_a (10^{-22} m^2)**
4.0	0.000	4.5	0.018
4.4	0.013	4.61	0.064
4.5	0.031	4.77	0.178
4.62	0.091	5.0	0.756
4.72	0.146	5.25	2.056

TABLE 39.7 (continued)
Attachment Cross Sections for NH_3

NH_2^-		Total	
Energy (eV)	Q_a (10^{-22} m^2)	Energy (eV)	Q_a (10^{-22} m^2)
5.0	0.578	5.5	3.386
5.186	1.033	5.55	3.555
5.22	1.109	5.59	3.649
5.5	1.784	5.68	3.697
5.593	1.956	5.76	3.649
5.68	1.976	5.86	3.446
5.81	1.966	6.0	2.902
5.893	1.861	6.2	2.056
6.0	1.612	6.38	1.270
6.5	0.509	6.5	0.868
6.813	0.187	6.7	0.411
7.0	0.091	7.0	0.130
7.193	0.048	7.2	0.064
7.5	0.014	8.0	0.000
8.0	0.002	9.0	0.076
8.5	0.002	9.28	0.168
9.0	0.035	9.5	0.269
9.7	0.006	9.752	0.408
10.0	0.055	10.0	0.484
10.5	0.042	10.268	0.550
11.0	0.022	10.5	0.568
11.5	0.019	10.714	0.555
12.0	0.000	11.0	0.505
		11.255	0.438
		11.5	0.360
		11.76	0.290
		12.0	0.218
		12.25	0.145
		12.5	0.106
		12.737	0.067
		13.0	0.045
		13.5	0.030

Source: Adapted from Itikawa Y., (Ed.), *Interactions of Photons and Electrons with Molecules*, Vol. 17c, Landolt-Börnstein, Springer, 2003.

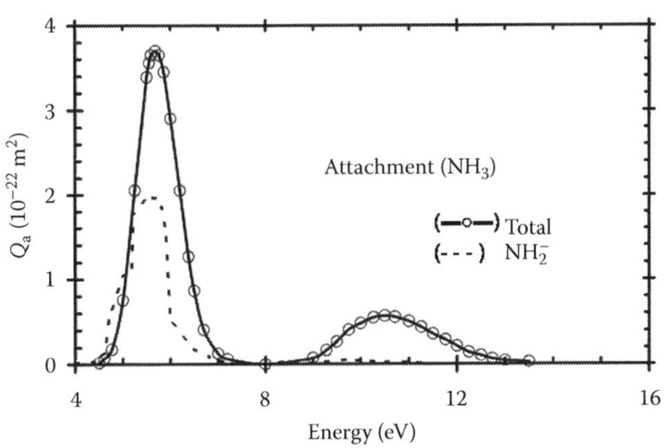

FIGURE 39.6 Attachment cross sections for NH_3. (—○—) Total; (- - -) NH_2^-. (Adapted from Itikawa, Y. (Ed.), *Interactions of Photons and Electrons with Molecules*, Vol. 17c, Landolt-Börnstein, Springer, 2003.)

39.9 DRIFT VELOCITY OF ELECTRONS

Pack et al. (1962) have measured the drift velocity of electrons as shown in Table 39.8 and Figure 39.7. Reduced mobilities (μN) as function of temperature at $E/N \to 0$ are shown in Table 39.9.

39.10 REDUCED IONIZATION COEFFICIENTS

Risbud and Naidu (1979) have measured the reduced ionization coefficients for NH_3 as shown in Table 39.10 and Figure 39.8. Townsend's semiempirical expression

$$\frac{\alpha}{N} = F \exp\left(-\frac{GN}{E}\right) \quad (39.2)$$

TABLE 39.8
Drift Velocity of Electrons

E/N (Td)	*W* (m/s)	*E/N* (Td)	*W* (m/s)
0.03	14.19	5.0	2234
0.06	27.44	10	4441
0.09	44.33	15	7369
0.12	57.50	20	1.21×10^4
0.15	67.43	25	1.90×10^4
0.18	82.44	30	2.73×10^4
0.21	99.65	35	4.61×10^4
0.24	118.4	40	6.22×10^4
0.27	136.2	45	6.92×10^4
0.3	152.1	50	7.18×10^4
0.6	278.4	55	7.46×10^4
0.9	426.3	60	8.24×10^4
1.2	534.1	65	9.99×10^4

Source: Digitized and interpolated from Pack, J. L., R. E. Voshall, and A. V. Phelps, *Phys. Rev.*, 127, 2084, 1962.

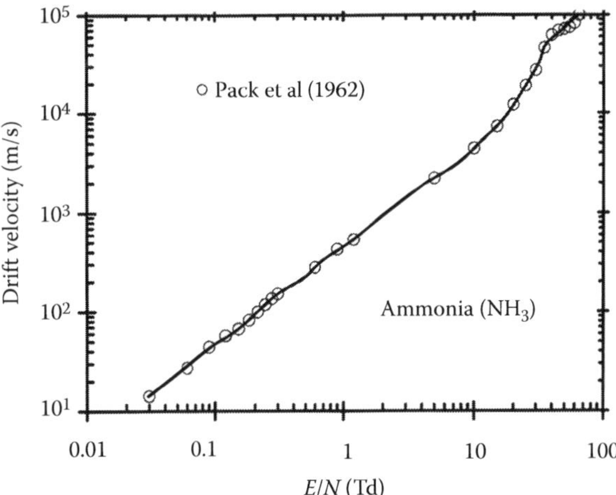

FIGURE 39.7 Drift velocity of electrons for NH_3. (Adapted from Pack, J. L., R. E. Voshall, and A. V. Phelps, *Phys. Rev.*, 127, 2084, 1962.)

NH$_3$

TABLE 39.9
Reduced Mobilities of Electrons

T (K)	μN ($m^{-1} V^{1} s^{-1}$)
195	3.35×10^{23}
300	4.34×10^{23}
381	5.15×10^{23}

Source: Digitized and interpolated from Pack, J. L., R. E. Voshall, and A. V. Phelps, *Phys. Rev.*, 127, 2084, 1962.

TABLE 39.10
Reduced Ionization Coefficients for NH$_3$

E/N (Td)	α/N (10^{-20} m^2)	*E/N* (Td)	α/N (10^{-20} m^2)
115	0.002	300	0.265
120	0.005	400	0.411
125	0.010	500	0.562
130	0.015	600	0.732
140	0.030	700	0.899
150	0.053	800	1.05
160	0.073	900	1.21
170	0.086	1000	1.37
200	0.123	1500	1.87
225	0.163	2000	2.22
250	0.204	2500	2.51
275	0.237	3000	2.73

Source: Adapted from Risbud A. V. and M. S. Naidu, *J. Physique Colloque*, supplement c7, 40, 77, 1979.

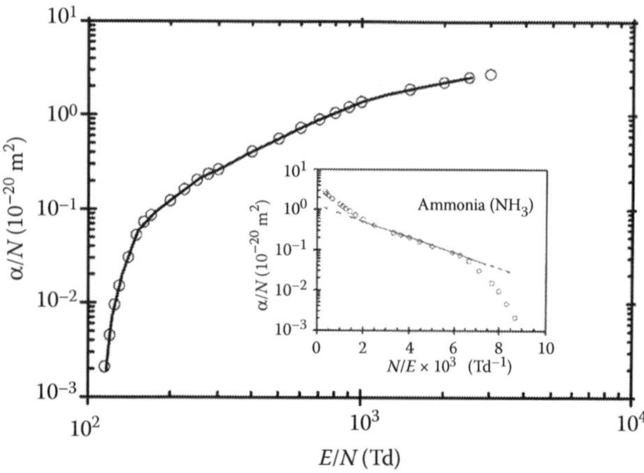

FIGURE 39.8 Reduced ionization coefficients as a function of reduced electric field for NH$_3$. Inset shows range of applicability of Equation 39.2. (Adapted from Risbud A. V. and M. S. Naidu, *J. Physique Colloque*, supplement c7, 40, 77, 1979.)

where *F* and *G* are gas constants applicable in the range $160 \le E/N \le 400$ Td (see inset of Figure 39.8) and the constants evaluated are $F = 1.4 \times 10^{-20}$ m^2 and $G = 488$ (Td)$^{-1}$.

TABLE 39.11
Reduced Attachment Coefficients

E/N (Td)	η/N (10^{-22} m^2)	*E/N* (Td)	η/N (10^{-22} m^2)
60	1.40	90	1.56
65	1.45	95	1.49
70	1.57	100	1.51
75	1.59	110	1.40
80	1.50	120	1.26
85	1.60		

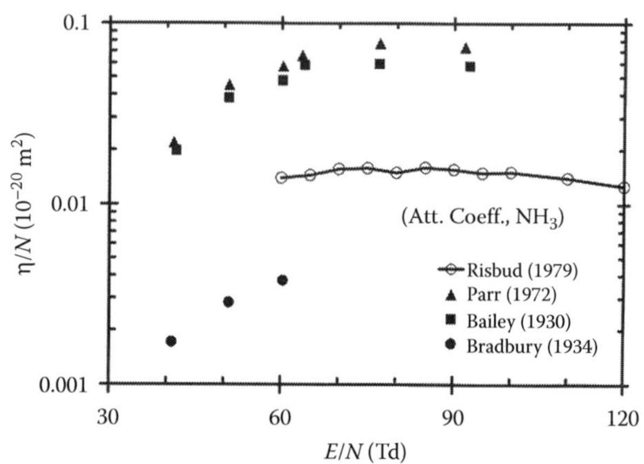

FIGURE 39.9 Reduced attachment coefficients for NH$_3$.

39.11 REDUCED ATTACHMENT COEFFICIENTS

Table 39.11 and Figure 39.9 show the reduced attachment coefficients for NH$_3$.

REFERENCES

Alle, D. T., R. J. Gulley, S. J. Buckman, and M. J. Brunger, *J. Phys. B: At. Mol. Opt. Phys.*, 25, 1533, 1992.

Bailey, V. A., and W. E. Duncanson, *Philos. Mag.*, 2, 145, 1934.

Boesten, L., Y. Tachibana, Y. Nakano, T. Shinohara, H. Tanaka, and M. A. Dillon, *J. Phys. B: At. Mol. Phys.*, 29, 5475, 1996.

Bradbury, N. E., *J. Chem. Phys.*, 2, 827, 1934.

Christophhorou, L. G., J. G. Carter, and D. V. Maxey, *J. Chem. Phys.*, 76, 2653, 1982.

Gulley, R. J., M. J. Brunger, and S. J. Buckman, *J. Phys. B: At. Mol. Opt. Phys.*, 25, 2435, 1992.

Itikawa Y., (Ed.), *Interactions of Photons and Electrons with Molecules*, Vol. 17c, Landolt-Börnstein, Springer, New York, 2003.

Lisovskiy, V., S. Martins, K. Landry, D. Douai, J.-P. Booth, and V. Cassagne, *J. Phys. D: Appl. Phys.*, 38, 872, 2005.

Mann, M. M., A. Hustrulid, and J. T. Tate, *Phys. Rev.*, 58, 340, 1940.

Pack, J. L., R. E. Voshall, and A. V. Phelps, *Phys. Rev.*, 127, 2084, 1962.

Parr, J. E. and J. L. Moruzzi, *J. Phys. D: Appl. Phys.*, 5, 514, 1972.

Rao, M. V. V. S., and S. K. Srivastava, *J. Phys. B: At. Mol. Opt. Phys.*, 25, 2175, 1992.

Rejoub, R., B. G. Lindsay, and R. F. Stebbings, *J. Chem. Phys.*, 115, 5053, 2001.

Rescigno, T. N. 1995, *Phys. Rev.*, A52, 329, 1995.

Risbud, A. V. and M. S. Naidu, *J. Physique Colloque*, supplement c7, 40, 77, 1979.
Sharp, T. E. and J. T. Dowell, *J. Chem. Phys.*, 50, 3024, 1969.
Shimanouchi, T., *Tables of Molecular Vibrational Frequencies Consolidated Volume I*, NSRDS-NBS 39, Washington (DC), 1972.
Sueoka, O., S. Mori, and Y. Katayama, *J. Phys. B: At. Mol. Phys.*, 20, 3237, 1987.
Syage, J. A., *J. Chem. Phys.*, 97, 6085, 1992.
Szmytkowski, Cz., K. Maciąg, G. Karwasz, and D. Filipović, *J. Phys. B: At. Mol. Opt. Phys.*, 22, 525, 1989.
Yousfi, M. and M. D. Benabdessadok, *J. Appl. Phys.*, 80, 6619, 1996.
Zecca, A., G. P. Karwasz, and R. S. Brusa, *Phys. Rev.*, 45, 2777, 1992.

40

BORON TRICHLORIDE

BCl$_3$

CONTENTS

Boron trichloride (BCl_3) is an electron-attaching gas that has 56 electrons. The electronic polarizability is 1.04×10^{-40} F m^2 and the ionization potential is 11.60 eV. The vibrational modes and energies are shown in Table 40.1 (Herzberg, 1945).

TABLE 40.1
Vibrational Modes and Energies

Designation	Mode	Energy (meV)
v_1	Symmetric stretching	58.4
v_2	Out-of-plane bending	57.3
v_3	Asymmetric stretching	118.8
v_4	Asymmetric bending	30.0

Source: Adapted from Herzberg, G., *Molecular Spectra and Molecular Structure II: Infrared and Raman Spectra of Polyatomic Molecules*, Van Nostrand Reinhold Co., New York, NY, p. 178, 1945.

40.1 TOTAL AND PARTIAL IONIZATION CROSS SECTIONS

Jiao et al. (1997) have measured the total and partial ionization cross sections as shown in Table 40.2 and Figure 40.1. Cross sections suggested by Christophorou and Olthoff (2004) in the low-energy region are also shown.

40.2 ATTACHMENT CROSS SECTIONS

Electron attachment processes are (Christophorou and Olthoff, 2004)

$$BCl_3 + e \rightarrow BCl_3^{-\bullet} \rightarrow BCl_2 + Cl^- \quad (40.1)$$

$$BCl_3 + e \rightarrow BCl_3^{-\bullet} \rightarrow BCl + Cl_2^- \quad (40.2)$$

$$BCl_3 + e \rightarrow BCl_3^{-\bullet} \text{ (long lived)} \quad (40.3)$$

$$BCl_3 + e \rightarrow BCl_3^{-\bullet} \rightarrow BCl_2^+ \, Cl^- + e \quad (40.4)$$

Reactions 40.1 and 40.2 are dissociative attachment processes. The dissociation energy for Reaction 40.1 is 1.59 eV. Reaction 40.3 is parent negative ion formation with a threshold energy of 0 eV (Tav et al., 1998). Reaction 40.4 is ion pair formation with energy >9 eV.

Low-energy total attachment cross sections derived by Tav et al. (1998) are shown in Figure 40.2. The peak of the

TABLE 40.2
Total Ionization Cross Sections for BCl_3

Ionization Cross Section (10^{-20} m^2)			
Christophorou Olthoff (2004)		Jiao et al. (1997)	
Energy (eV)	Cross Section	Energy (eV)	Cross Section
13	1.2	12.5	0.20
15	2.6	14	2.0
17	4.4	15	2.8
19	6.0	17.5	4.4
21	7.0	20	5.2
23	7.3	22.5	8.0
25	8.0	25	8.2
27	8.3	27.5	8.3
29	8.3	30	8.4
		32.5	8.6
		35	8.8
		40	8.9
		42.5	8.9
		45	8.9_5
		50	9.0
		60	9.2

Note: Data of Jiao et al. (1997) are digitized.

BCl$_3$

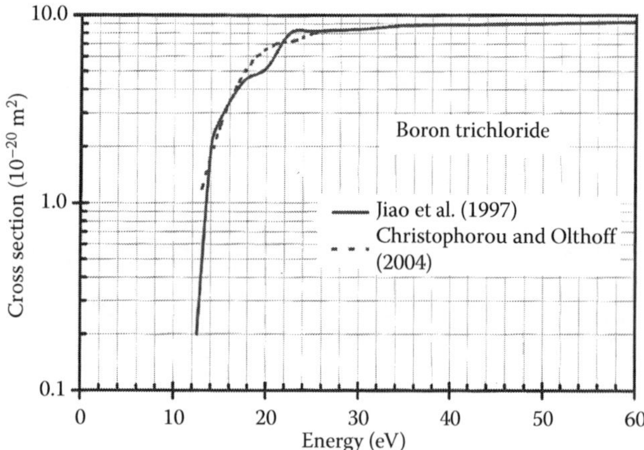

FIGURE 40.1 Total and partial ionization cross sections of BCl$_3$. (Adapted from Jiao, C. Q., R. Nagpal, and P. Haaland, *Chem. Phys. Lett.*, 265, 239, 1997.)

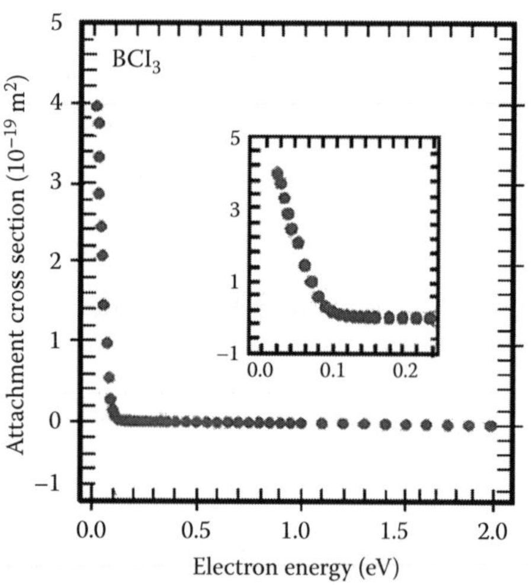

FIGURE 40.2 Swarm-derived attachment cross sections for BCl$_3$. (From Tav, C., P. G. Datskos, and L. A. Pinnaduwage, *J. Appl. Phys.*, 84, 5805, 1998. with kind permission of American Institute of Physics.)

total attachment cross section obtained using the beam technique is 2.8×10^{-21} m^2 at 0.4 eV (Buchełnikova, 1959).

40.3 ATTACHMENT RATE COEFFICIENTS

Table 40.3 shows the thermal attachment rate constants for BCl$_3$. Figure 40.3 shows the attachment rate constants as a function of *E*/*N*.

TABLE 40.3
Thermal Attachment Rate Constants for BCl$_3$

Energy, K (eV)	Rate Constant (10^{-14} m^3/s)	Reference
(~0)	1.8	Tav et al. (1998)
0.4	0.34	Petrovic et al. (1990)
(0.025)	0.28	Stockdale et al. (1972)

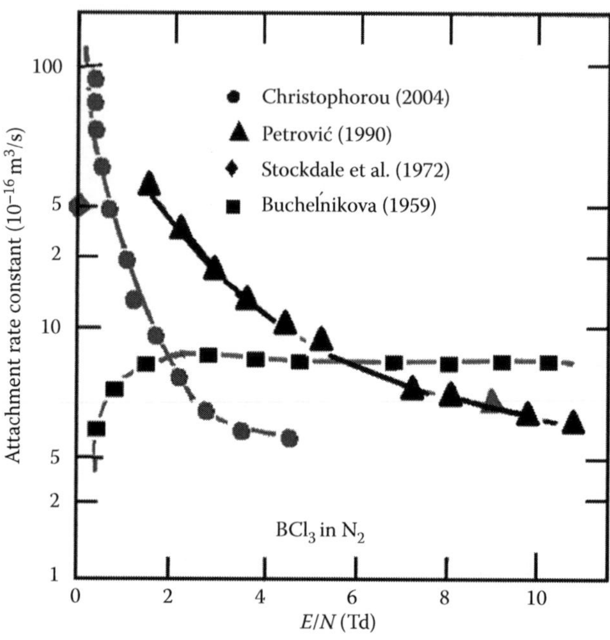

FIGURE 40.3 Attachment rate constants for BCl$_3$. (Adapted from Tav, C., P. G. Datskos, and L. A. Pinnaduwage, *J. Appl. Phys.*, 84, 5805, 1998.)

REFERENCES

Buchełnikova, I. S., *Sov. Phys., JETP*, 35, 783, 1959.

Christophorou, L. G. and J. K. Olthoff, *Fundamental Electron Interactions with Plasma Processing Gases*, Kluwer/Plenum Publishers, New York, NY, 2004.

Herzberg, G. *Molecular Spectra and Molecular Structure II: Infrared and Raman Spectra of Polyatomic Molecules*, Van Nostrand Reinhold Co., New York, NY, p. 178, 1945.

Jiao, C. Q., R. Nagpal, and P. Haaland, *Chem. Phys. Lett.*, 265, 239, 1997.

Petrovic, Z. L., W. C. Wang, M. Suto, J. C. Han, and L. C. Lee, *J. Appl. Phys.*, 67, 675, 1990.

Stockdale, J. A., D. R. Nelson, F. J. Davis, and R. N. Compton, *J. Chem. Phys.*, 56, 3336, 1972.

Tav, C., P. G. Datskos, and L. A. Pinnaduwage, *J. Appl. Phys.*, 84, 5805, 1998.

41

BORON TRIFLUORIDE

CONTENTS

Boron trifluoride (BF_3) is a nonpolar, electron-attaching molecule that has 32 electrons. Its electronic polarizability is 3.68×10^{-40} F m^2 and its ionization potential is 15.7 eV.

41.1 SELECTED REFERENCES FOR DATA

See Table 41.1.

41.2 TOTAL SCATTERING CROSS SECTION

The highlights of the cross section are (see Table 41.2)

1. A shape resonance at ~3.6 eV resulting in a vibrationally excited molecule.
2. A broad hump in the region 20–40 eV, common for many perfluorinated molecules (see Figure 41.1 for NF_3, for example), is possibly due to elastic scattering.
3. A monotonic decrease for higher energy which is common to many gases.

TABLE 41.1
Selected References for Data

Quantity	Range: eV, (Td)	Reference
Ionization cross section	16–5000	Kim et al. (2005)
Total scattering cross section	0.6–370	**Szmytkowski et al. (2004)**
Swarm parameters	(0.5–300)	**Hunter et al. (1989)**
Dissociative ionization	16–70	**Farber and Srivastava (1984)**
Ionization coefficients	(107–170)	**Davies (1976)**
Ionization and attachment cross sections	10–250	**Kurepa et al. (1976)**
Attachment processes	0–15	**Stockdale et al. (1972)**
Ionization potentials	15.5–30.1	**Law and Margrave (1956)**

Note: Bold font denotes experimental study.

TABLE 41.2
Total Scattering Cross Section for BF_3

Energy (eV)	Q_T (10^{-20} m^2)	Energy (eV)	Q_T (10^{-20} m^2)	Energy (eV)	Q_T (10^{-20} m^2)
0.6	10.92	4.5	15.32	60	20.71
0.8	10.59	5.0	14.21	70	20.20
1.0	10.43	6.0	13.67	80	19.57
1.2	10.10	8.0	14.28	90	18.70
1.4	9.89	10.0	15.15	100	17.81
1.6	10.12	12	16.13	125	16.46
1.8	10.21	14	18.63	150	14.94
2.0	10.72	16	20.31	175	14.04
2.2	11.26	18	20.80	200	13.09
2.4	12.16	20	20.71	250	11.66
2.6	13.69	25	20.57	300	11.05
2.8	15.38	30	20.78	350	10.13
3.0	17.74	35	21.11	380	9.96
3.5	21.87	40	21.36		
4.0	18.41	50	21.24		

Source: Digitized from Szmytkowski, C. et al., *J. Chem. Phys.*, 121, 1790, 2004.

Note: See Figure 41.1 for graphical presentation.

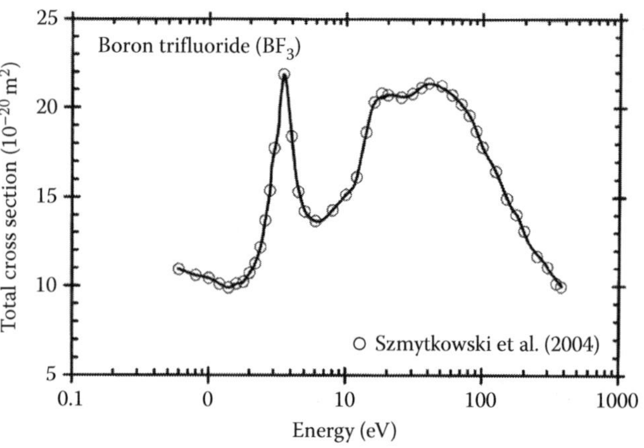

FIGURE 41.1 Total scattering cross section for BF_3. (Adapted from Szmytkowski, C. et al., *J. Chem. Phys.*, 121, 1790, 2004.)

TABLE 41.3
Ion Appearance Potentials

Ion	Energy (eV)	Reference
BF_3^+	15.5	Law and Margrave (1956)
BF_2^+	16.2	
BF^+	27.2	
B^+	30.1	
F^-	10.4	Harland and Franklin (1974)
F_2^-	10.2	Decorpo and Franklin (1971)
BF_2^-		

41.3 ION APPEARANCE POTENTIALS

See Table 41.3.

41.4 IONIZATION AND ATTACHMENT CROSS SECTIONS

See Tables 41.4 and 41.5.

41.5 DRIFT VELOCITY OF ELECTRONS

See Table 41.6.

41.6 IONIZATION AND ATTACHMENT COEFFICIENTS

See Table 41.7.

TABLE 41.4
Ionization and Attachment Cross Sections

Energy (eV)	Q_i (10^{-20} m^2)	Energy (eV)	Q_a (10^{-22} m^2)
Ionization		Attachment	
16	0.037	10.4	0.20
17	0.075	10.5	0.32
18	0.16	10.6	0.46
19	0.25	10.7	0.63
20	0.34	10.8	0.84
22	0.56	10.9	1.0
24	0.81	11.0	1.4
26	1.0	11.1	1.7
28	1.3	11.2	2.0
30	1.5	11.3	2.2
32	1.8	11.4	2.4
34	2.0	11.5	2.5
36	2.2	11.6	2.5
38	2.4	11.7	2.3
40	2.7	11.8	2.1
42	2.8	11.9	1.8
44	3.0	12.0	1.5
46	3.2	12.0	1.2
48	3.3	12.2	0.97
50	3.5	12.3	0.69
55	3.8	12.4	0.48
60	4.1	12.5	0.33
65	4.3	12.6	0.22
70	4.5	12.7	0.15
75	4.7	12.8	0.092
80	4.9	12.9	0.061
85	5.0	13.0	0.024
90	5.1		
95	5.2		
100	5.3		
110	5.4		
120	5.5		
130	5.5		
140	5.5		
150	5.5		
160	5.5		
170	5.5		
180	5.4		
190	5.4		
200	5.3		
210	5.2		
220	5.1		
230	5.1		
240	5.1		
250	5.0		

Source: Adapted from Kurepa, M. V., V. M. Pejčev, and I. M. Cadež, *J. Phys. D: Appl. Phys.*, 9, 481, 1976.

Note: See Figure 41.2 for graphical presentation.

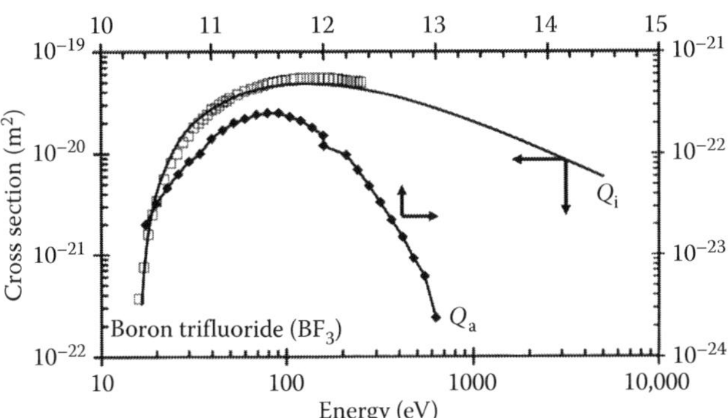

FIGURE 41.2 Ionization and attachment cross sections for BF_3. Note the linear scale for abscissa for attachment curve. (Adapted from Kurepa, M. V., V. M. Pejčev, and I. M. Cadež, *J. Phys. D: Appl. Phys.*, 9, 481, 1976.)

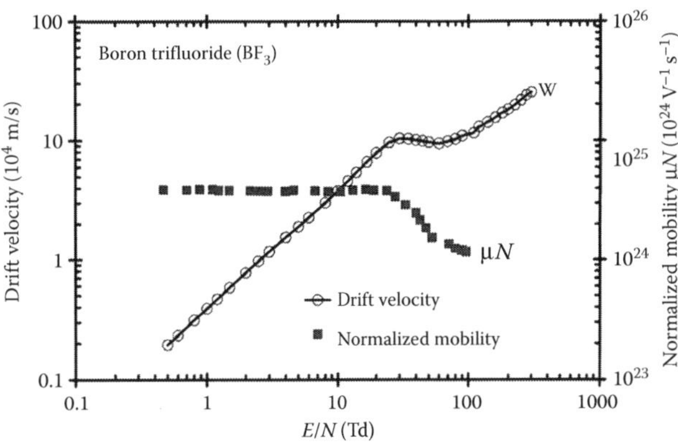

FIGURE 41.3 Electron drift velocity and density-normalized mobility. The normalized mobility below 20 Td is constant, 3.88×10^{24} $(V\ ms)^{-1}$. (Adapted from Hunter, S. R., J. G. Carter, and L. G. Christophorou, *J. Appl. Phys.*, 65, 1858, 1989.)

TABLE 41.5
Relative Abundance of Dissociative Ions

	Relative Abundance (a.u.)			
Energy (eV)	BF_3^+	BF_2^+	BF^+	F^+
15	...	...	...	...
16	1	1	...	...
17	2	2	...	...
18	4	4	...	...
20	6	12	...	...
30	16	60	4	...
40	40	400	30	8
70	100	1000	80	30

Source: Adapted from Farber M. and R. D. Srivastava, *J. Chem. Phys.*, 81, 241, 1984.

TABLE 41.6
Suggested Drift Velocity of Electrons

E/N (Td)	*W* (10^3 m/s)	*E/N* (Td)	*W* (10^3 m/s)	*E/N* (Td)	*W* (10^3 m/s)
0.5	1.95	10.0	38.0	80.0	102.7
0.6	2.33	12.0	46.0	90.0	109.5
0.8	3.12	14.0	54.0	11.0	116.0
1.0	3.88	17.0	66.3	120.0	130.0
1.2	4.66	20.0	78.5	140.0	142.0
1.5	5.82	25.0	96.0	160.0	155.0
2.0	7.73	30.0	103.5	180.0	169.0
2.5	9.69	35.0	102.8	200.0	181.0
3.0	11.6_5	40.0	100.5	225.0	197.0
4.0	15.5	45.0	98.8	250.0	216
5.0	19.0	50.0	96.6	275.0	237
6.0	22.7	60.0	94.1	300.0	252
8.0	30.2	70.0	98.2		

Note: See Figure 41.3 for graphical presentation. Note the negative differential conductivity, which is also observed in CH_4, CF_4, and SiF_4.

TABLE 41.7
Ionization and Attachment Coefficients for BF_3

E/N (Td)	α/N (10^{-22} m^2)	η/N (10^{-22} m^2)	α_{eff}/N (10^{-22} m^2)
30		0.0019	−0.0019
35		0.0236	−0.0236
40		0.085	−0.085
45		0.207	−0.207
50		0.382	−0.382
60		0.722	−0.722
70		0.947	−0.947
80	0.045	1.08	−1.04
90	0.232	1.19	−0.958
100	0.514	1.26	−0.746
110	0.87	1.31	−0.44
119			(0.231)
120	1.26	1.39	−0.13
124			(0.528)
130	1.75	1.45	0.30
130			(0.879)
136			(1.27)
140	2.32	1.54	0.78
141			(1.72)
155			(2.92)
160	3.67	1.68	1.99
170			(4.34)
180	5.34	1.86	3.48
198			(7.69)
200	7.40	2.00	5.40
220	9.56	2.24	7.32
226			(11.80)
240	11.9	2.51	9.39
254			(16.20)
260	14.5	2.68	11.8
280	17.2	2.91	14.3
283			(21.7)
300	20.3	3.22	17.1

continued

BF$_3$

TABLE 41.7 (continued)
Ionization and Attachment Coefficients for BF_3

E/N (Td)	α/N (10^{-22} m^2)	η/N (10^{-22} m^2)	α_{eff}/N (10^{-22} m^2)
424			(47.8)
565			(76.1)
707			(106)
848			(133)
989			(159)
1130			(182)
1270			(206)
1410			(227)
1550			(248)
1700			(263)

Source: Adapted from Hunter, S. R., J. G. Carter, and L. G. Christophorou, *J. Appl. Phys.*, 65, 1858, 1989.

Note: Values in parentheses are from Davies (1976). See Figures 41.4 and 41.5 for graphical presentation.

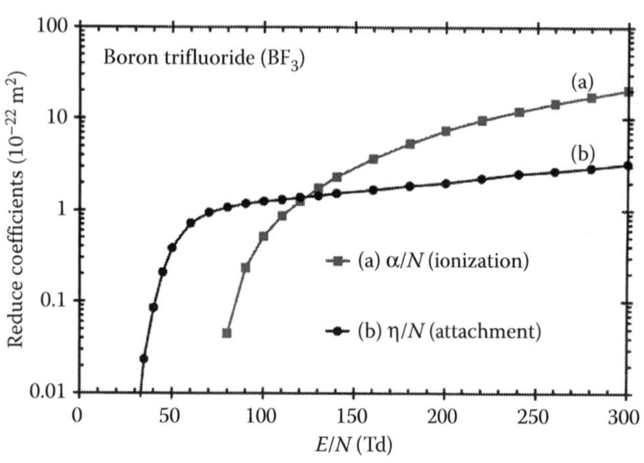

FIGURE 41.4 Density-reduced ionization and attachment coefficients for BF_3. (Adapted from Hunter, S. R., J. G. Carter, and L. G. Christophorou, *J. Appl. Phys.*, 65, 1858, 1989.)

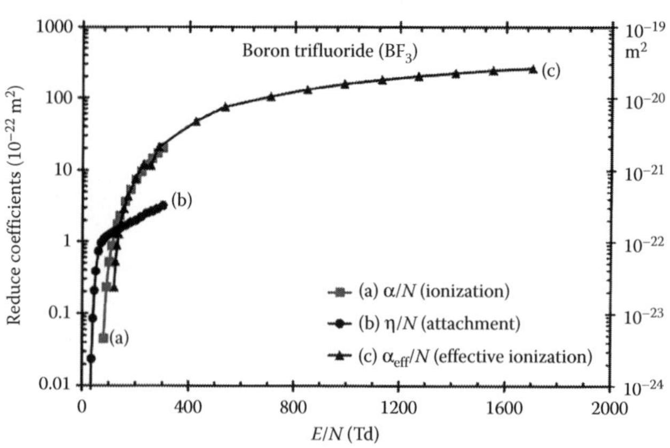

FIGURE 41.5 Density-reduced ionization and attachment coefficients for BF_3. (a) Ionization, Hunter et al. (1989); (b) attachment, Hunter et al. (1989); and (c) effective ionization (Davies, 1976).

REFERENCES

Davies, D. K., *J. Appl. Phys.*, 47, 1920, 1976.

DeCorpo, J. J. and J. L. Franklin, *J. Chem. Phys.*, 54, 1885, 1971.

Farber, M. and R. D. Srivastava, *J. Chem. Phys.*, 81, 241, 1984.

Harland, P. W. and J. L. Franklin, *J. Chem. Phys.*, 61, 1621, 1974.

Hunter, S. R., J. G. Carter, and L. G. Christophorou, *J. Appl. Phys.*, 65, 1858, 1989.

Kim, Y.-K., K. K. Irikura, M. E. Rudd, M. A. Ali, P. M. Stone, J. Chang, J. S. Coursey, http://physics.nist.gov/PhysRefData/Ionization/index.html, 2005.

Kurepa, M. V., V. M. Pejčev, and I. M. Cadež, *J. Phys. D: Appl. Phys.*, 9, 481, 1976.

Law, R. W. and J. L. Margrave, *J. Chem. Phys.*, 25, 1086, 1956.

Stockdale, J. A., D. R. Nelson, F. J. Davis, and R. N. Compton, *J. Chem. Phys.*, 56, 3336, 1972.

Szmytkowski, C., M. Piotrowicz, A. Domaracka, and L. Kłosowski. *J. Chem. Phys.*, 121, 1790, 2004.

42

DEUTERATED AMMONIA

CONTENTS

Deuterated ammonia (ND_3) is a molecule with all the three hydrogen atoms of ammonia (NH_3) replaced with deuterium atoms. It has electronic polarizability of 1.89×10^{-40} F m^2. The vibrational modes and energies of the molecule are shown in Table 42.1 (Shimanouchi, 1972).

TABLE 42.1
Vibrational Modes and Energies of ND_3

Designation	Type of Motion	Energy (meV)
v_1	Symmetrical stretch	300.0
v_2	Symmetrical deformation	92.7
v_3	Degenerate stretch	317.9
v_4	Degenerate deformation	147.7

TABLE 42.2
Total Ionization Cross Sections

		Cross Section for Ion Pair Production		
Energy (eV)	Total (10^{-20} m^2)	Energy (eV)	Q_i (ND^+, D^+) (10^{-22} m^2)	Q_i (N^+, D^+) (10^{-23} m^2)
11.5	0.065	60	0.81	2.8
12.5	0.159	70	1.18	3.6
13.5	0.202	80	1.39	5.3
15	0.345	90	1.64	5.9
17.5	0.604	100	1.67	7.7
20	0.910	120	1.73	7.1

TABLE 42.2 (continued)
Total Ionization Cross Sections

		Cross Section for Ion Pair Production		
Energy (eV)	Total (10^{-20} m^2)	Energy (eV)	Q_i (ND^+, D^+) (10^{-22} m^2)	Q_i (N^+, D^+) (10^{-23} m^2)
22.5	1.157	140	1.70	6.7
25	1.456	160	1.58	8.2
27.5	1.651	200	1.49	6.3
30	1.850	250	1.37	6.3
35	2.181	300	1.13	5.3
40	2.441	400	0.82	3.8
50	2.772	500	0.75	3.5
60	2.968	600	0.63	3.0
70	3.079	800	0.48	1.4
80	3.132	1000	0.32	1.0
90	3.143			
100	3.128			
120	3.060			
140	2.935			
160	2.786			
200	2.562			
250	2.315			
300	2.101			
400	1.765			
500	1.511			
600	1.340			
800	1.070			
1000	0.920			

Source: Adapted from Rejoub, R., B. G. Lindsay, and R. F. Stebbings, *J. Chem. Phys.*, 115, 5053, 2001.

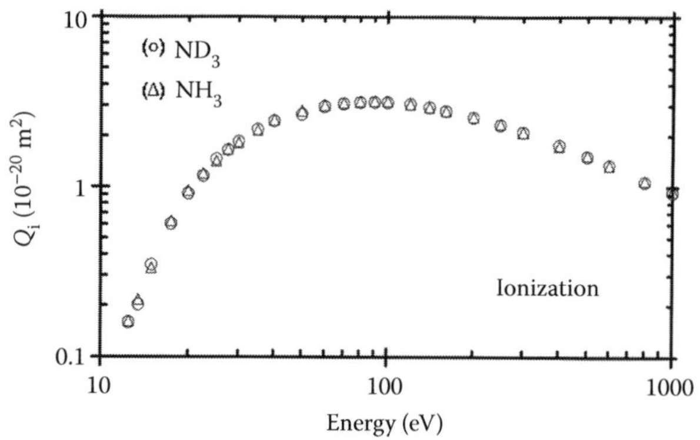

FIGURE 42.1 Total ionization cross sections for ND_3. Total ionization cross sections for NH_3 are also plotted for comparison. No isotope effect is present. (Adapted from Rejoub, R., B. G. Lindsay, and R. F. Stebbings, *J. Chem. Phys.*, 115, 5053, 2001.)

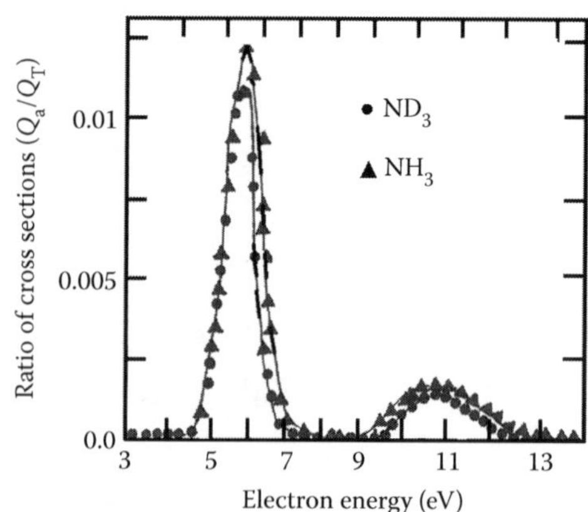

FIGURE 42.3 Total attachment cross section ratios (Q_a/Q_T) for ND_3 and NH_3. (Reprinted with permission from Sharp, T. E. and J. T. Dowell, *J. Chem. Phys.*, 50, 3024, 1969. Copyright (1969), American Institute of Physics.)

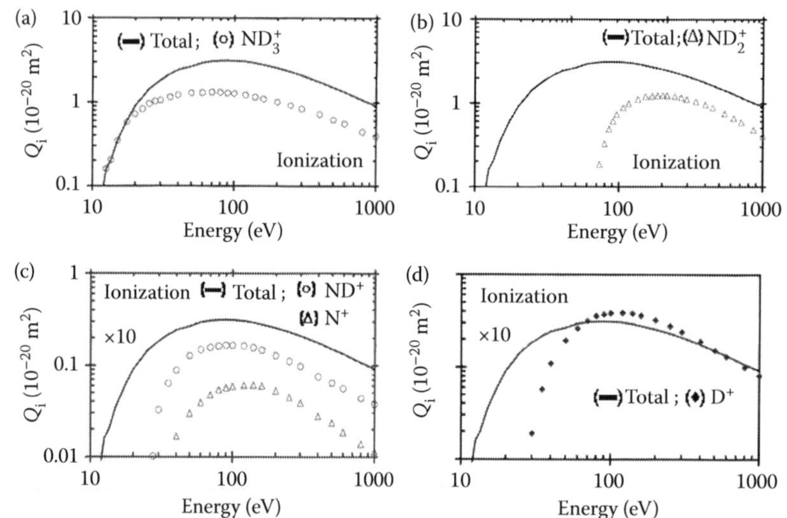

FIGURE 42.2 Partial ionization cross sections for ND_3. Total cross sections have also been plotted to show the relative contributions. (a) ND_3^+; (b) ND_2^+; (c) ND^+ and N^+; (d) D^+. Note the multiplying factor for (c) and (d). (Adapted from Rejoub, R., B. G. Lindsay, and R. F. Stebbings, *J. Chem. Phys.*, 115, 5053, 2001.)

42.1 TOTAL AND PARTIAL IONIZATION CROSS SECTIONS

Rejoub et al. (2001) have measured the absolute ionization cross sections as shown in Table 42.2 and Figures 42.1 and 42.2.

42.2 ATTACHMENT CROSS SECTIONS

The attachment cross sections for ND_3 have the following principal features (Sharp and Dowell, 1969)

1. No negative ions were observed below 5.3 eV.
2. Two peaks are observed in the attachment cross sections at 5.65 and at 10.5 eV.
3. Both peaks contain D^- and ND_2^- ions.
4. The cross sections show a small isotope effect (Figure 42.3), ND_3 having a smaller cross section. See chapter on NH_3 for tabulated values.

REFERENCES

Rejoub, R., B. G. Lindsay, and R. F. Stebbings, *J. Chem. Phys.*, 115, 5053, 2001.

Sharp, T. E. and J. T. Dowell, *J. Chem. Phys.*, 50, 3024, 1969.

Shimanouchi, T. *Tables of Molecular Vibrational Frequencies*, NSRDS-NBS 39, U.S. Department of Commerce, Washington (DC), 1972.

43

NITROGEN TRIFLUORIDE

CONTENTS

Nitrogen trifluoride (NF_3) is a polar, electron-attaching gas that has 34 electrons. The electronic polarizability is 4.05×10^{-40} F m^2, dipole moment 0.235 D (see Szmytkowski et al., 2004), and ionization potential 13.00 eV (see Szmytkowski et al., 2004). The molecule has four modes of vibration as shown in Table 43.1 (Shimanouchi, 1972).

43.1 SELECTED REFERENCES FOR DATA

Table 43.2 shows the selected references for data for NF_3.

TABLE 43.1
Vibrational Modes and Energies for NF_3

Designation	Type of Motion	Energy (meV)
ν_1	Symmetrical stretch	128
ν_2	Symmetrical deformation	80.2
ν_3	Degenerate stretch	112.5
ν_4	Degenerate deformation	61.0

TABLE 43.2
Selected References for Data

Parameter	Range: eV, (Td)	Reference
Total scattering cross section	0.5–370	**Szmytkowski et al. (2004)**
Attachment cross section	0–6	**Nandi et al. (2001)**
Ionization cross section	13–220	**Haaland et al. (2001)**
Elastic scattering cross section	1.5–100	**Boesten et al. (1996)**
Attachment rate	0.02–0.05	**Miller et al. (1995)**
Total scattering cross section	0–10	Rescigno (1995)
Ionization cross section	13–220	**Tarnovsky et al. (1994)**
Attachment coefficients	0.1–10	Ushiroda et al. (1990)
Attachment coefficients	0.3–30.0	**Lakdawala and Moruzzi (1980)**
Attachment coefficients	1–16	**Nygaard et al. (1979)**
Attachment rate	0.025	**Sides and Tiernan (1977)**
Attachment rate	0.025	**Mothes et al. (1972)**
Ionization energy	11–18	**Reese and Diebler (1956)**

Note: Bold font indicates experimental study.

NF$_3$

43.2 TOTAL SCATTERING CROSS SECTIONS

Table 43.3 and Figure 43.1 show the total scattering cross sections. The highlights of the cross sections are

1. The total scattering cross section increases from 0.5 eV to a peak at 2.7 eV, attributed to shape resonance, that is, the electron attaches to the neutral molecule for a resident time longer than the time taken by the electron to transit through a region of molecular dimension.
2. The total scattering cross section has a minimum at ~10 eV and shows a second, broad peak in the region of 30–45 eV. This peak is attributed mostly to elastic scattering with contributions from ionization and dissociative excitation.
3. Above the second peak, the cross section decreases, up to ~370 eV according to $Q_T \sim \varepsilon^{-1/2}$.

TABLE 43.3
Total Scattering Cross Sections for NF$_3$

Energy (eV)	Q_T (10^{-20} m^2)	Energy (eV)	Q_T (10^{-20} m^2)	Energy (eV)	Q_T (10^{-20} m^2)
0.5	15.9	3.5	26.5	27	18.9
0.6	16.6	3.7	25.8	30	19.2
0.8	18.1	4.0	24.9	35	19.4
1.0	19.6	4.5	22.7	40	19.4
1.2	21.2	5.0	20.9	45	19.4
1.4	22.7	5.5	19.5	50	19.3
1.5	23.3	6.0	18.5	60	18.8
1.6	24.0	6.5	17.6	70	18.3
1.7	24.8	7.0	17.2	80	17.3
1.8	25.2	7.5	17.0	90	16.6
1.9	25.4	8.0	16.8	100	16.1
2.0	26.3	8.5	16.7	110	15.6
2.1	26.9	9	16.7	120	15.1
2.2	27.4	9.5	16.7	140	14.1
2.3	27.5	10	16.7	160	13.1
2.4	27.5	11	16.6	180	12.3
2.5	27.7	12	16.7	200	11.5
2.6	27.8	14	16.9	220	11.0
2.7	28.0	16	17.2	250	10.4
2.8	27.9	18	17.6	275	9.94
2.9	27.8	20	17.9	300	9.63
3.0	27.7	22	18.2	350	9.18
3.2	27.2	25	18.6	370	9.07

Source: Adapted from Szmytkowski, C. et al., *Phys. Rev. A*, 70, 032707, 2004.

43.3 DIFFERENTIAL SCATTERING CROSS SECTIONS

Figure 43.2 shows the three-dimensional variation of the differential scattering cross section as a function of angle of scattering and electron energy in the 1.5–100 eV range (Boesten et al., 1996). At low energies, the dependence of the differential scattering cross section on the angle of scattering is relatively small, with a small evidence of forward scattering. At higher energies, a minimum is observed in the range of 120° to 90° as the energy is increased from ~5 to 100 eV. The many ridges and valleys observed (Figure 43.2) are not satisfactorily explained so far.

43.4 ELASTIC AND MOMENTUM TRANSFER CROSS SECTIONS

Table 43.4 and Figure 43.3 show the integral elastic and momentum transfer cross sections for NF$_3$ (Boesten et al., 1996).

43.5 ION APPEARANCE POTENTIALS

Ion appearance potentials and relative abundances are shown in Table 43.5 (Reese and Dibeler, 1956). The probable products of electron impact are also shown.

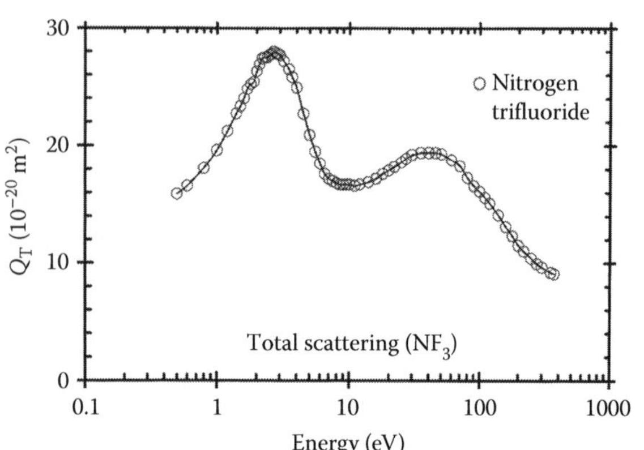

FIGURE 43.1 Total scattering cross sections for NF$_3$. (Adapted from Szmytkowski, C. et al., *Phys. Rev. A*, 70, 032707, 2004.)

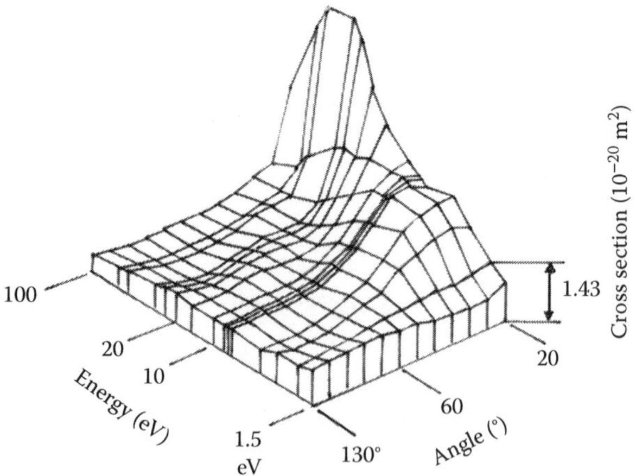

FIGURE 43.2 Differential scattering cross section for NF$_3$. (Adapted from Boesten, L. et al., *J. Phys. B*, 29, 5475, 1996.)

TABLE 43.4
Integral Elastic and Momentum Transfer Cross Sections

Energy (eV)	Q_{el} (10^{-20} m^2)	Q_M (10^{-20} m^2)
1.5	11.90	12.46
2.0	12.98	12.39
3.0	17.24	14.24
4.0	18.41	14.92
7.0	17.35	14.24
7.5	17.47	14.17
8.0	17.89	12.82
10.0	16.91	13.53
15	14.60	10.41
20	14.48	9.87
25	14.05	8.54
30	13.33	7.63
50	12.32	6.62
60	11.03	5.81
100	9.72	5.42

Source: Adapted from Boesten, L. et al., *J. Phys. B*, 29, 5475, 1996.

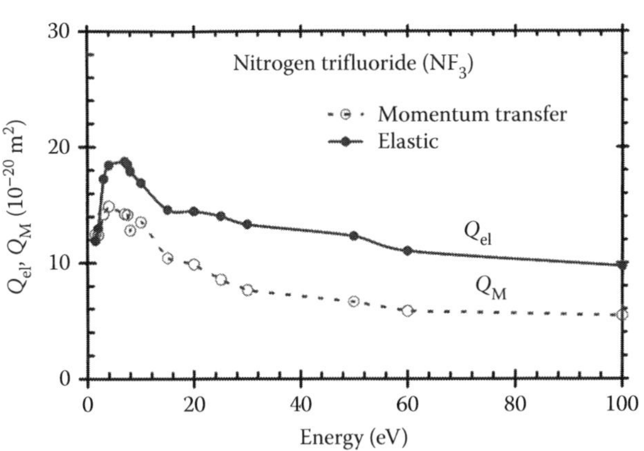

FIGURE 43.3 Integral elastic and momentum transfer cross sections for NF_3. (Adapted from Boesten, L. et al., *J. Phys. B*, 29, 5475, 1996.)

43.6 IONIZATION CROSS SECTIONS

Ionization cross sections measured by Haaland et al. (2004) are shown in Table 43.6 and Figure 43.4.

43.7 ATTACHMENT PROCESSES

Reactions that occur at thermal energy are (sides and Tiernan, 1977)

$$e + NF_3 \rightarrow NF_2 + F^- \quad (43.1)$$

$$\rightarrow NF + F + F^- \quad (43.2)$$

$$\rightarrow N + F + F + F^- \quad (43.3)$$

$$\rightarrow F_2 + NF^- \quad (43.4)$$

$$\rightarrow F + F + NF^- \quad (43.5)$$

$$\rightarrow NF + F_2^- \quad (43.6)$$

TABLE 43.5
Ion Appearance Potentials

Ion Species	Appearance Potential (eV)	Relative Abundance at 70 eV	Products
F^-	~0	—	$F^- + NF_2$
F_2^-	~0	—	?
NF_3^+	13.2	59.7	NF_3^+
NF_2^+	14.2	100	$NF_2^+ + F$
NF^+	17.9	39.1	$NF^+ + 2F$
F^-	22	—	$N + F^+ + F^- + F$
N^+	22.2	5.4	$N^+ + 3F$
F^+	25	4.8	$F^+ + 2F + N$

Source: Adapted from Reese, R. M. and V. H. Dibeler, *J. Chem. Phys.*, 24, 1175, 1956.

TABLE 43.6
Ionization Cross Sections for NF_3

	Q_i (10^{-20} m^2)					
Energy (eV)	NF_2^+	NF^+	NF_3^+	F^+	N^+	Total
16	0.02		0.037			0.057
18	0.07	0.0025	0.043			0.114
20	0.29	0.0078	0.072			0.369
22	0.27	0.020	0.085			0.374
25	0.40	0.055	0.112			0.562
30	0.63	0.106	0.173			0.912
35	0.86	0.173	0.233		0.0006	1.266
40	1.10	0.258	0.265	0.003	0.0011	1.628
45	1.24	0.317	0.305	0.007	0.002	1.874
50	1.36	0.356	0.342	0.010	0.0027	2.075
55	1.47	0.384	0.362	0.014	0.0031	2.230
60	1.50	0.408	0.369	0.018	0.0034	2.299
65	1.55	0.430	0.371	0.021	0.0035	2.377
70	1.62	0.451	0.374	0.026	0.0037	2.480
75	1.65	0.472	0.381	0.028	0.0039	2.539
80	1.69	0.491	0.392	0.031	0.0042	2.605
85	1.76	0.509	0.403	0.035	0.0044	2.708
90	1.82	0.523	0.410	0.037	0.0045	2.792
100	1.83	0.534	0.410	0.039	0.0045	2.814
125	1.82	0.542	0.410	0.039	0.0045	2.816
150	1.77	0.551	0.434	0.045	0.0048	2.803
175	1.90	0.550	0.420	0.049	0.0048	2.921
200	1.95	0.588	0.415	0.052	0.0048	3.007

Source: Digitized from Haaland, P. D., C. Q. Jiao, and A. Garscadden, *Chem. Phys. Lett.*, 340, 479, 2001.

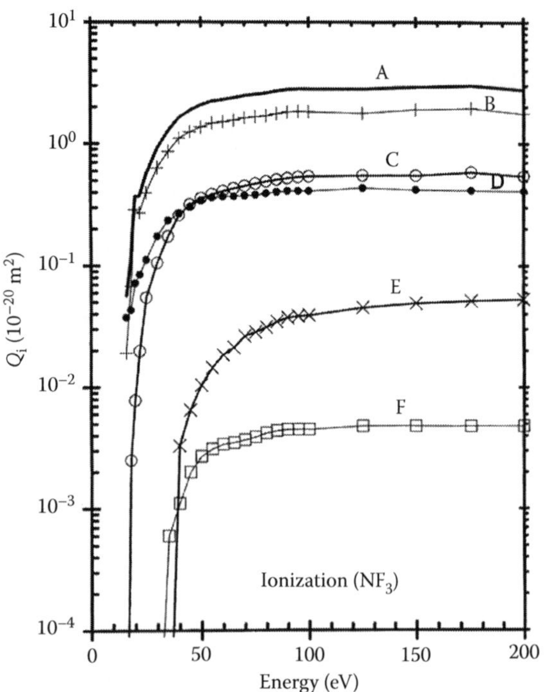

FIGURE 43.4 Total and partial ionization cross sections for NF_3 measured by Haaland et al. (2001). (A) Total; (B) NF_2; (C) NF^+; (D) NF_3^+; (E) F^+; (F) N^+.

TABLE 43.7
Attachment Rate Constants

Rate (10^{-17} m^3/s)	Method	Temperature (K)	Reference
0.5–1.1	FA/LP	300	Miller et al. (1995)
0.6–1.8		343	
1.4–3.4		363	
2.1–5.7		402	
4.1–9.5		434	
11.3–22.7		462	
25.2–50.8		504	
29.4–54.6		551	
<39.0		293	Lakdawala and Moruzzi (1980)
2.1	FA/LP		Sides and Tiernan (1977)
2.4	ECR		Mothes et al. (1972)

Note: ECR = electron cyclotron resonance; FA/LP = flowing afterglow/Langmuir probe.

$$\rightarrow N + F + F_2^- \quad (43.7)$$

$$\rightarrow F + F_2^- \quad (43.8)$$

$$e + NF_3 + Ar \rightarrow Ar + NF_3^- \quad (43.9)$$

Formation of NF_3^- by two-body attachment process has not been observed and Reaction 43.1 is the only significant attachment process at thermal energy. NF_2^- and F_2^- ions appear only above 1 eV (Nandi et al., 2001).

43.8 ATTACHMENT RATES

Table 43.7 shows the thermal attachment rate constants for NF_3. Attachment rate constants as a function of reduced electric field (*E*/*N*) have been measured by Lakdawala and Moruzzi (1980) by the drift tube method, as shown in Table 43.8 and Figure 43.5.

TABLE 43.8
Attachment Rate Constants as Function of *E*/*N*

	Attachment Rate (10^{-15} m^3/s)		
***E*/*N* (Td)**	**$NF_3 + N_2$**	**$NF_3 + Ar$**	**$NF_3 + He$**
0.3	0.39		0.91
0.4	0.45		1.16
0.5	0.48		1.39
0.6	0.53	0.38	1.61
0.8	0.62	0.43	2.16
1	0.69	0.48	2.91
2	1.07	0.67	6.14
4	1.86	1.04	10.03
6	2.59	1.50	12.66
8	3.49	2.04	14.06
10	4.44	2.51	14.93
20	7.65	3.95	
30	8.01	4.90	

Source: Digitized from Lakdawala, V. K. and J. L. Moruzzi, *J. Phys. D: Appl. Phys.*, 13, 377, 1980.

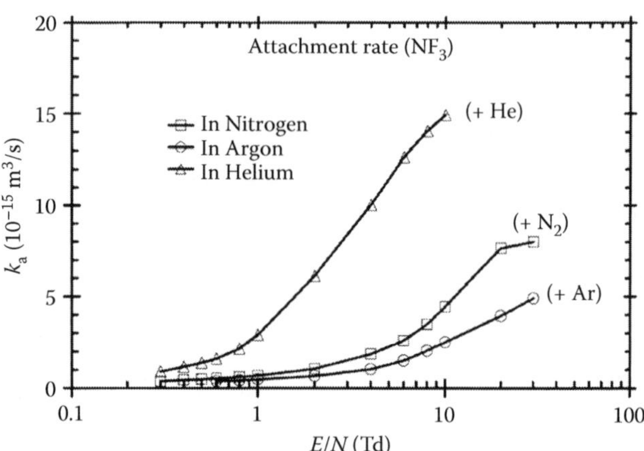

FIGURE 43.5 Attachment rate coefficients as function of reduced electric field. (Adapted from Lakdawala, V. K. and J. L. Moruzzi, *J. Phys. D: Appl. Phys.*, 13, 377, 1980.)

43.9 ATTACHMENT CROSS SECTIONS

Attachment cross sections for selected fragment ions are shown in Table 43.9 and Figure 43.6.

TABLE 43.9
Attachment Cross Sections for Selected Fragment Ions

Energy (eV)	Q_a (F^-) (10^{-20} m²)	Q_a (F_2^-) (10^{-23} m²)	Q_a (NF_2^-) (10^{-24} m²)	Q_a (Total) (10^{-20} m²)
0.00	0.53	0.18	0.48	0.530
0.10	0.58	0.23	0.34	0.580
0.20	0.64	0.30	0.38	0.640
0.30	0.72	0.34	0.41	0.720
0.40	0.79	0.35	0.46	0.790
0.50	0.92	0.38	0.50	0.920
0.60	1.04	0.48	0.47	1.041
0.70	1.21	0.51	0.67	1.211
0.80	1.33	0.63	0.71	1.331
0.90	1.52	0.71	0.93	1.521
1.00	1.68	0.82	1.17	1.681
1.10	1.95	1.03	1.78	1.951
1.30	2.03	1.15	2.20	2.031
1.40	2.11	1.25	2.43	2.111
1.50	2.16	1.40	3.34	2.162
1.60	2.18	1.51	3.94	2.182
1.70	2.20	1.60	4.50	2.202
1.80	2.17	1.64	4.46	2.172
1.90	2.14	1.59	4.93	2.142
2.00	2.08	1.65	4.62	2.082
2.10	2.03	1.59	4.01	2.032
2.20	1.93	1.52	3.73	1.932
2.30	1.83	1.40	3.20	1.832
2.40	1.74	1.31	2.68	1.742
2.50	1.63	1.15	2.13	1.631
2.60	1.50	1.07	1.55	1.501
2.70	1.40	0.99	1.41	1.401
2.80	1.28	0.87	1.06	1.281
2.90	1.16	0.78	1.01	1.161
3.00	1.05	0.65	0.63	1.051
3.10	0.94	0.58	0.54	0.941
3.20	0.84	0.50	0.50	0.841
3.30	0.75	0.45	0.33	0.750
3.40	0.66	0.38	0.35	0.660
3.50	0.59	0.31	0.21	0.590
3.60	0.50	0.28	0.29	0.500
3.70	0.44	0.24	0.17	0.440
3.80	0.38	0.20	0.15	0.380
3.90	0.32	0.15	0.20	0.320
4.00	0.28	0.15	0.11	0.280
4.10	0.23	0.20	0.08	0.230
4.20	0.20	0.10	0.11	0.200
4.30	0.16	0.09	0.06	0.160
4.40	0.14	0.07	0.09	0.140
4.50	0.12	0.06	0.16	0.120
4.60	0.10	0.04	0.09	0.100
4.70	0.08	0.04	0.04	0.080
4.80	0.07	0.03	0.03	0.070
4.90	0.05	0.04	0.06	0.050

TABLE 43.9 (continued)
Attachment Cross Sections for Selected Fragment Ions

Energy (eV)	Q_a (F^-) (10^{-20} m²)	Q_a (F_2^-) (10^{-23} m²)	Q_a (NF_2^-) (10^{-24} m²)	Q_a (Total) (10^{-20} m²)
5.00	0.04	0.02	0.01	0.040
5.10	0.04	0.02	0.03	0.040
5.20	0.03	0.02	0.06	0.030
5.30	0.02	0.01	0.05	0.020
5.40	0.02	0.01	0.05	0.020
5.50	0.02	0.01	0.06	0.020
5.60	0.01	0.01	0.07	0.010
5.70	0.01	0.01	0.06	0.010
5.80	0.01	0.01	0.06	0.010
5.90	0.01	0.01	0.07	0.010
6.00	0.01	0.01	0.03	0.010

Source: Adapted from Nandi, D. et al., *Int. J. Mass Spectrom.*, 205, 111, 2001.

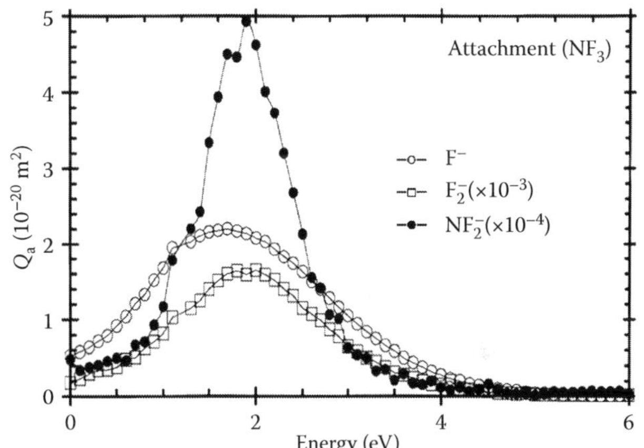

FIGURE 43.6 Electron attachment cross sections for NF_3 for selected ions formation. (Adapted from Nandi, D. et al., *Int. J. Mass Spectrom.*, 205, 111, 2001.)

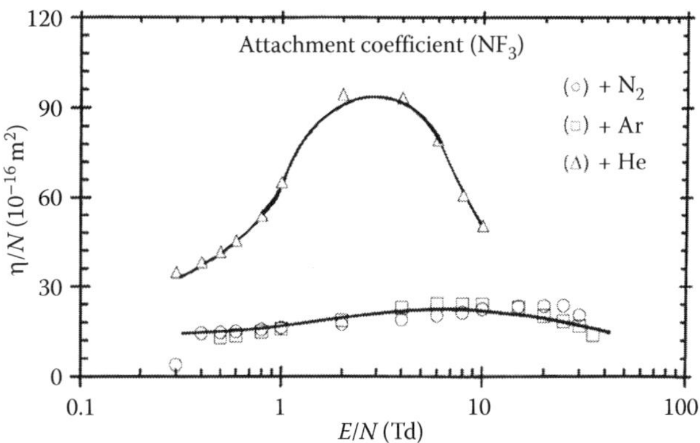

FIGURE 43.7 Reduced attachment coefficients for NF_3 with several buffer gases. (Adapted from Lakdawala, V. K. and J. L. Moruzzi, *J. Phys. D: Appl. Phys.*, 13, 377, 1980.)

NF_3

NF$_3$

43.10 ATTACHMENT COEFFICIENTS

See Table 43.10.

TABLE 43.10
Attachment Coefficients as a Function of Reduced Electric Field

	Reduced Attachment Coefficient (10^{-20} m^2)		
E/N (Td)	$NF_3 + N_2$	$NF_3 + Ar$	$NF_3 + He$
0.3	3.92		34.80
0.4	14.23		38.20
0.5	14.57	12.95	41.67
0.6	14.93	13.64	45.32
0.8	15.62	14.83	53.82
1	16.20	15.80	65.13
2	17.30	18.69	94.42
4	18.96	22.95	93.16
6	20.32	24.17	79.39
8	21.22	24.06	60.68
10	22.34	23.90	50.30
15	23.27	23.02	
20	23.50	20.22	
25	23.63	18.38	
30	20.41	16.70	
35		13.81	

Source: Digitized from Lakdawala, V. K. and J. L., Moruzzi, *J. Phys. D: Appl. Phys.*, 13, 377, 1980.
Note: See Figure 43.7 for graphical presentation.

REFERENCES

Boesten, L., Y. Tachibana, Y. Nakano, T. Shinohara, H. Tanaka, and M. A. Dillon, *J. Phys. B*, 29, 5475, 1996.

Haaland, P. D., C. Q. Jiao, and A. Garscadden, *Chem. Phys. Lett.*, 340, 479, 2001.

Lakdawala, V. K., and J. L. Moruzzi, *J. Phys. D: Appl. Phys.*, 13, 377, 1980.

Miller, T. M., J. F. Friedman, A. E. S. Miller, and J. F. Paulson, *Int. J. Mass Spectrom. Ion Processes*, 149/150, 111, 1995.

Mothes, K. G., E. Schultes, and R. N. Schindler, *J. Phys. Chem.*, 76, 3758, 1972.

Nandi, D., S. A. Rangwala, S. V. K. Kumar, and E. Krishnakumar, *Int. J. Mass Spectrom.*, 205, 111, 2001.

Nygaard, K. J., H. L. Brooks, and S. R. Hunter, *IEEE J. Quantum Electron.*, QE-15, 1216, 1979.

Reese, R. M., and V. H. Dibeler, *J. Chem. Phys.*, 24, 1175, 1956.

Rescigno, T. N., *Phys. Rev. A*, 52, 329, 1995.

Shimanouchi, T. *Tables of Molecular Vibrational Frequencies Consolidated Volume I*, NSRDS-NBS 39, Washington (DC), 1972.

Sides, G. D., and T. O. Tiernan, *J. Chem. Phys.*, 67, 2382, 1977.

Szmytkowski, C., A. Domaracka, P. Możejko, E. Ptasińska-Denga, and L. Kłosowski, *Phys. Rev. A*, 70, 032707, 2004.

Tarnovsky, V., A. Levia, K. Becker, R. Masner, and M. Schmidt, *Int. J. Mass Spectrom. Ion Processes*, 133, 175, 1994.

Ushiroda, S., S. Kajita, and Y. Kondo, *J. Phys. D: Appl. Phys.*, 23, 47, 1990.

44

PHOSPHINE

CONTENTS

Phosphine (PH_3) is a polar, weakly attaching molecule with 18 electrons, electronic polarizability of 5.39×10^{-40} F m^2, dipole moment of 0.574 D, and ionization potential of 9.869 eV.

44.1 SELECTED REFERENCES FOR DATA

See Table 44.1.

44.2 TOTAL SCATTERING CROSS SECTION

See Table 44.2 and Figure 44.1.

44.3 IONIZATION CROSS SECTION

See Table 44.3.

44.4 CHARACTERISTIC ENERGY

See Table 44.4.

TABLE 44.1
Selected References for Data

Quantity	Range: eV, (Td)	Reference
Total scattering cross section	0.5–370	Szmytkowski et al. (2004)
Total scattering cross section	90–3500	Ariyasinghe et al. (2003)
Transport coefficients	(2–80)	Millican and Walker (1987)
Ionization cross section	10.3–183.0	Märk and Egger (1977)

TABLE 44.2
Total Scattering Cross Section for PH_3

Energy (eV)	Q_T (10^{-20} m^2)	Energy (eV)	Q_T (10^{-20} m^2)	Energy (eV)	Q_T (10^{-20} m^2)
0.5	12.1	5.5	45.3	45	22.7
0.6	11.5	6.0	45.3	50	21.2
0.8	11.5	6.5	45.0	60	19.5
1.0	12.8	7.0	44.5	70	17.9
1.2	16.2	7.5	44.3	80	16.4
1.4	23.9	8.0	43.8	90	15.5
1.6	31.2	8.5	43.4	100	14.5
1.8	39.2	9.0	43.2	110	13.8
2.0	45.1	9.5	42.5	120	13.0
2.2	47.5	10.0	41.7	140	11.9
2.4	48.3	11	40.2	160	10.8
2.6	47.7	12	38.1	180	10.2
2.8	46.8	14	35.6	200	9.42
3.0	46.0	16	34.2	220	8.88
3.2	45.0	18	32.6	250	8.30
3.5	43.8	20	31.7	275	7.88
3.7	43.3	25	29.0	300	7.42
4.0	43.1	30	26.8	350	6.83
4.5	43.9	35	25.1	370	6.57
5.0	45.0	40	23.8		

Source: Szmytkowski, C. et al., *J. Phys. B: At. Mol. Opt. Phys.*, 37, 1833, 2004. With kind permission of Institute of Physis, England.

PH$_3$

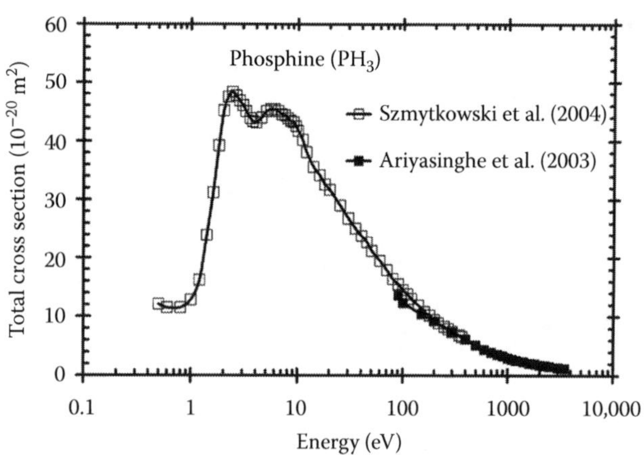

FIGURE 44.1 Total scattering cross sections. (—□—) Szmytkowski et al. (2004); (—■—) Ariyasinghe et al. (2003).

TABLE 44.3
Ionization Cross Section for PH$_3$

Energy (eV)	Q_i (10^{-20} m^2)	Energy (eV)	Q_i (10^{-20} m^2)	Energy (eV)	Q_i (10^{-20} m^2)
10.3	0.018	54.9	4.12	118.5	4.07
11.2	0.098	62.1	4.22	123.5	4.02
12.3	0.265	69.5	4.29	129.5	3.97
15.1	0.707	76.5	4.28	135.5	3.90
19.0	1.75	83.5	4.25	142.5	3.81
22.5	2.45	90.5	4.20	150.5	3.69
26.1	2.91	98	4.18	159.5	3.54
33.3	3.40	103.5	4.14	170.5	3.34
40.5	3.73	108.5	4.12	183	3.12
47.7	3.98	112.5	4.11		

Source: Adapted from. Märk, T. D. and F. Egger, *J. Chem. Phys.*, 67, 2629, 1977.

TABLE 44.4
Characteristic Energy for PH$_3$

E/N (Td)	D_r/μ (V)	*E/N* (Td)	D_r/μ (V)	*E/N* (Td)	D_r/μ (V)
2	0.0257	8	0.065	30	0.719
2.5	0.0259	10	0.106	35	0.796
3	0.0259	12	0.169	40	0.901
4	0.0289	14	0.238	50	1.08
5	0.032	17	0.338	60	1.21
6	0.037	20	0.439	70	1.31
7	0.045	25	0.603	80	1.33

Source: Adapted from Millican, P. G. and I. C. Walker, *J. Phys. D: Appl. Phys.*, 20, 193, 1987.

REFERENCES

Ariyasinghe, W. M., T. Wijerathna, and T. Powers, *Phys. Rev. A*, 68, 032708, 2003.

Märk, T. D. and F. Egger, *J. Chem. Phys.*, 67, 2629, 1977.

Millican, P. G. and I. C. Walker, *J. Phys. D: Appl. Phys.*, 20, 193, 1987.

Szmytkowski, C., Ł. Kłowsowski, A. Domaracka, M. Piotrowicz, and E. Ptasińska-Denga, *J. Phys. B: At. Mol. Opt. Phys.*, 37, 1833, 2004.

45

PHOSPHORUS TRIFLUORIDE

PF_3

CONTENTS

Phosphorus trifluoride (PF_3) is a polar, electron-attaching molecule with 42 electrons. The electronic polarizability is 4.93×10^{-40} F m^2, dipole moment 1.03 D, and ionization potential (PF_3^+ ion) 11.60 eV.

45.1 SELECTED REFERENCES FOR DATA

See Table 45.1.

45.2 TOTAL SCATTERING CROSS SECTION

See Table 45.2.

45.3 ATTACHMENT PROCESSES

Negative ion formation in PF_3 occurs due to dissociative attachment at high energy (>9 eV). The appearance potential and cross section details are shown in Table 45.3 (MacNeil and Thynne, 1970).

TABLE 45.1
Selected References for Data

Quantity	Range: eV, (Td)	Reference
Total scattering cross section	0.5–370	Szmytkowski et al. (2004)
Attachment processes	0–15	Harland et al. (1974)

TABLE 45.2
Total Scattering Cross Section for PF_3

Energy (eV)	Q_T (10^{-20} m^2)	Energy (eV)	Q_T (10^{-20} m^2)	Energy (eV)	Q_T (10^{-20} m^2)
0.5	69.62	4.5	30.36	60	25.53
0.8	65.36	5	29.46	70	24.06
1	61.80	6	29.49	80	22.64
1.2	59.29	8	31.51	90	21.37
1.4	55.50	10	32.98	100	20.34
1.6	52.85	12	32.55	125	18.50
1.8	50.38	14	30.49	150	16.86
2	46.27	16	28.45	175	15.40
2.2	44.33	18	27.04	200	14.37
2.4	42.02	20	26.48	250	12.96
2.6	39.80	25	26.83	300	11.64
2.8	38.02	30	27.55	350	10.54
3	36.66	35	27.62	380	10.08
3.5	34.37	40	27.46		
4	32.25	50	26.74		

Source: Digitized from Szmytkowski, C. et al., *J. Chem. Phys.*, 121, 1790, 2004.
Note: See Figure 45.1 for graphical presentation.

PF$_3$

TABLE 45.3
Ion Appearance Potentials

Ion Species	Appearance Potential (eV)	Peak (eV)
F^-	9.0	11.3
F_2^-	10.9	
PF^-	11.4	
PF_2^-	10.3	

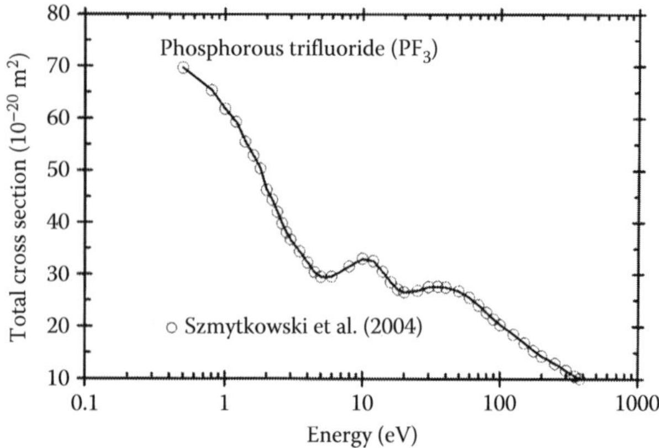

FIGURE 45.1 Total scattering cross sections for PF$_3$. (Adapted from Szmytkowski, C. et al., *J. Chem. Phys.*, 121, 1790, 2004.)

REFERENCES

Harland, P. W., D. W. H. Rankin, and J. C. J. Thynne, *Int. J. Mass Spectrom. Ion Phys.*, 13, 395, 1974.

MacNeil, K. A. G., and J. C. J. Thynne, *J. Phys. Chem.*, 74, 2257, 1970.

Szmytkowski, C., M. Piotrowicz, A. Domaracka, and Ł. Kłosowski, *J. Chem. Phys.*, 121, 1790, 2004.

Section V

5 ATOMS

46

BROMOCHLORO-METHANE

CONTENTS

Bromochloromethane (CH_2BrCl), synonym Halon 1011, is a polar, electron-attaching gas that has 60 electrons. Its electronic polarizability is quoted as 8.34×10^{-40} F m^2 (see Beran and Keran, 1969), dipole moment 1.66 D, and ionization potential 10.77 eV. The vibrational modes and energies are shown in Table 46.1.

46.1 IONIZATION CROSS SECTION

A single value of total ionization cross section of 14.7×10^{-20} m^2 at 70 eV electron energy is reported by Beran and Kevan (1969).

TABLE 46.1
Vibrational Modes and Energies for CH_2BrCl

Mode	Type of Motion	Energy (meV)
ν_1	CH_2 s-stretch	372.3
ν_2	CH_2 scissor	183.7
ν_3	CH_2 wag	152.6
ν_4	CCl stretch	92.2
ν_5	CBr stretch	76.1
ν_6	CBrCl scissor	28.4
ν_7	CH_2 a-stretch	380.1
ν_8	CH_2 twist	139.9
ν_9	CH_2 rock	105.6

Source: Adapted from Shimanouchi, T., *Tables of Molecular Vibrational Frequencies Consolidated Volume 1*, NSRDS-NBS 39, Washington (DC), 1972.

Note: a = Antisymmetrical; s = symmetrical.

46.2 ATTACHMENT RATE COEFFICIENT

The dissociative attachment processes are (Matejcik et al., 2003)

$$CH_2ClBr + e(0\ eV) \rightarrow CH_2Cl + Br^- + 0.41\ eV \quad (46.1)$$

$$CH_2ClBr + e(0\ eV) \rightarrow CH_2Br + Cl^- + 0.18\ eV \quad (46.2)$$

The relative abundances of Br^- and Cl^- are approximately the same, though the temperature affects the abundance in opposite ways; increase of temperature increases the abundance of Br^- ions and decreases that of the Cl^- ions (Matejcik et al., 2003). The energetics for the two processes are shown in Table 46.2.

The molecule possesses a fairly large dipole moment (1.66 D) and the lower critical dipole moment for the formation of the parent negative ion is 1.67 D (Crawford, 1967). Though the formation of CH_2BrCl^- has not been reported, the possibility of the formation of this species has not been completely ruled out (Matejcik et al., 2003).

The thermal attachment rate coefficient for CH_2BrCl is 7.1×10^{-15} m^3/s (Sunagawa and Shimamori, 1997). Figure 46.1

TABLE 46.2
Energetics for Attachment Processes

Ion	Dissociation Energy (eV)	Electron Affinity (eV)	Appearance Potential (eV)
Br^-	2.95 (Br–CH_2Cl)	3.36 (Br)	–0.41
Cl^-	3.43 (Cl–CH_2Br)	3.61 (Cl)	–0.18

CH_2BrCl

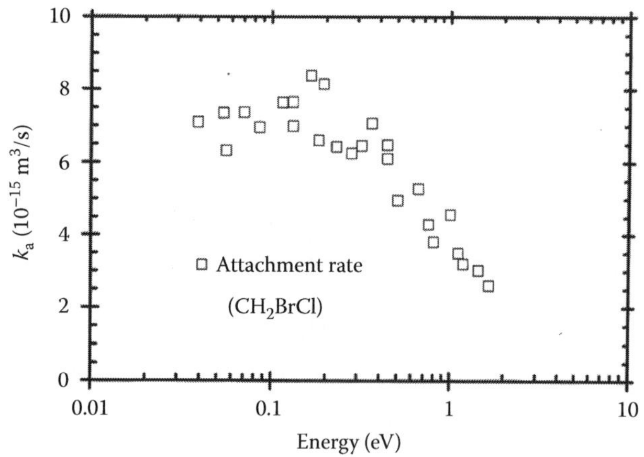

FIGURE 46.1 Attachment rate constants for CH_2BrCl. (Adapted From Sunagawa T. and H. Shimamori, *J. Chem. Phys.*, 107, 7876, 1997.)

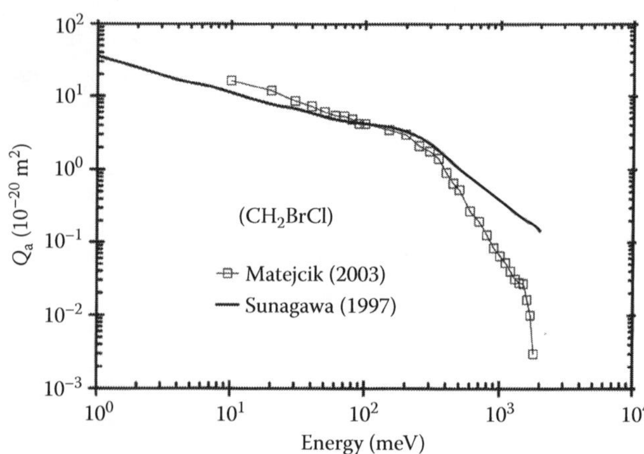

FIGURE 46.2 Attachment cross sections for CH_2BrCl. (—□—) Matejcik et al. (2003), beam method; (——) Sunagawa and Shimamori (1997), deconvoluted from swarm measurements.

TABLE 46.3
Attachment Cross Sections for CH_2BrCl Digitized

Energy (meV)	Matejcik et al. (2003) Q_a (10^{-20} m^2)	Sunagawa and Shimamori (1997) Q_a (10^{-20} m^2)
1		36.01
2		25.03
4		16.99
7		13.63
10	16.34	11.24
20	11.95	7.75
30	8.57	6.78
40	7.25	5.86
50	6.09	5.22
60	5.43	4.84
70	5.30	4.62
80	4.81	4.47
90	4.19	4.34
100	4.19	4.24
150	3.47	3.79
200	3.00	3.30
250	2.10	2.78
300	1.78	2.28
350	1.41	1.84
400	0.91	1.48
450	0.65	1.21
500	0.53	1.02
600	0.27	0.78
700	0.19	0.63
800	0.13	0.53
900	0.083	0.45
1000	0.065	0.38
1100	0.052	0.33
1200	0.040	0.29
1300	0.031	0.26
1400	0.028	0.23
1500	0.027	0.21
1600	0.016	0.20
1700	0.010	0.19
1800	0.003	0.18
2000		0.14

shows the attachment rate coefficients as function of mean electron energy.

46.3 ATTACHMENT CROSS SECTIONS

Attachment cross sections measured by cross beam technique (Matejcik et al., 2003) and deconvoluted by Sunagawa and Shimamori (1997) are shown in Table 46.3 and Figure 46.2.

REFERENCES

Beran, J. A. and L. Kevan, *J. Phys. Chem.*, 73, 3873, 1969.

Crawford, O. H., *Proc. Phys. Soc.*, 91, 279, 1967.

Matejcik, S., I. Ipolyi, and E. Illenberger, *Chem. Phys. Lett.*, 375, 660, 2003.

Shimanouchi, T., *Tables of Molecular Vibrational Frequencies Consolidated Volume 1*, NSRDS-NBS 39, Washington (DC), 1972.

Sunagawa, T. and H. Shimamori, *J. Chem. Phys.*, 107, 7876, 1997.

47

BROMOMETHANE

CH_3Br

CONTENTS

Bromomethane (CH_3Br) (also called methyl bromide) is a molecule that has a hydrogen atom replaced with a bromine atom in methane. Selected properties of halogen-substituted methanes are shown in Table 47.1. The highlights of the table are the larger polarizability with increasing number of electrons and essentially the same dipole moment except for iodine substitution.

The molecule of CH_3Br has six modes of vibration as shown in Table 47.2. The vibrational modes of deuterated molecule (CD_3Br) are also shown for comparison. Vibrations of similar type tend to have similar frequencies (Allan and Andric, 1996).

47.1 SELECTED REFERENCES FOR DATA

See Table 47.3.

TABLE 47.1
Selected Properties of Halogen-Substituted Methanes

Property	CH_3Br	CH_3Cl	CH_3F	CH_3I
z	44	26	18	62
α_e (10^{-40} F m^2)	6.18, 6.53, 6.71	5.25, 5.95	3.30	8.87
μ (D)	1.820	1.896	1.858	1.641
ε_a (eV)				0.11
ε_d (eV)	4.43	4.34	4.39	4.47
ε_i (eV)	10.541	11.22	12.47	9.538

Note: z = Number of electrons; α_e = electronic polarizability; μ = dipole moment; ε_a = electron affinity; ε_d = dissociation energy; ε_i = ionization potential.

TABLE 47.2
Vibrational Modes and Energies

	Energy (meV)	
Mode Designation	**CH_3Br**	**CD_3Br**
ν_1	368.5	267.4
ν_2	161.8	123.2
ν_3	75.5	69.8
ν_4	379.0	284.3
ν_5	179.0	131.0
ν_6	118.3	88.2

Source: Adapted from L. M. Sverdlov, M. A. Kovner, and E. P. Krainov, *Vibrational Spectra of Polyatomic Molecules*, John Wiley & Sons, New York, NY, 1974, p. 381.

47.2 TOTAL SCATTERING CROSS SECTION

Table 47.4 shows the total scattering cross section in CH_3Br and Figure 47.1 is a graphical presentation. The highlights of the cross section variation are

1. The cross section toward zero energy shows an increasing trend, common to most polar molecules.
2. A small peak at 2.5 eV attributed to the formation of negative ions (Guerra et al., 1991).
3. A much larger peak at 10 eV also attributed to the attachment process.
4. A monotonic decrease for energies >10 eV, up to 600 eV.

5. The total cross section increases with the size of the molecule within the family of halomethanes, $CH_3I > CH_3Br > CH_3Cl$.

47.3 TOTAL IONIZATION CROSS SECTION

Table 47.5 shows the total ionization cross section in CH_3Br (Vallance et al., 1997) and Figure 47.2 includes the graphical presentation.

CH_3Br

TABLE 47.3
Selected References for Data

Quantity	Range: eV, (Td), (K)	Reference
Q_a	0–0.2	**Braun et al. (2007)**
Q_i	13–1000	**Rejoub et al. (2002)**
Q_T	0.8–600	**Kimura et al. (2001)**
Q_a	0–10	Wilde et al. (2000)
Q_i	10.5–220	**Vallance et al. (1997)**
k_a	—	**Burns et al. (1996)**
Q_T	0.5–200	**Krzysztofowicz and Szmytkowski (1994)**
k_a, W	≤1	**Datskos et al. (1992)**
Q_T	0.5–8	**Benitez et al. (1988)**
k_a	0–1.2	**Wang and Lee (1988)**
k_a	(293–500 K)	**Petrović and Crompton (1987)**
k_a	(200–600 K)	**Alge et al. (1984)**
Q_a	<0.2, (300–1200 K)	**Spence and Schulz (1973)**
k_a	(0.3–6 Td)	**Christodoulides and Christophorou (1971)**

Note: k_a = Attachment rate; W = drift velocity; Q_a = attachment cross section; Q_i = ionization; Q_T = total scattering; $\bar{\varepsilon}$ = mean energy. Bold font denotes experimental study.

TABLE 47.4
Total Scattering Cross Section in CH_3Br

Energy (eV)	Q_T (10^{-20} m²)	Energy (eV)	Q_T (10^{-20} m²)	Energy (eV)	Q_T (10^{-20} m²)
0.8	51.8	6.5	40.4	22	34.6
1.0	44.8	7.0	41.0	25	32.5
1.2	43.1	7.5	42.7	30	30.0
1.4	41.5	8.0	42.3	40	25.7
1.6	29.4	8.5	42.1	50	23.2
1.8	38.3	9.0	42.2	60	21.1
2.0	38.1	9.5	42.7	70	20.0
2.2	38.4	10	42.9	80	18.2
2.5	38.7	11	42.1	90	17.7
2.8	37.1	12	41.7	100	16.3
3.1	36.5	13	40.6	120	15.4
3.4	35.6	14	39.8	150	13.8
3.7	35.6	15	39.0	200	11.7
4.0	34.9	16	38.7	250	10.9
4.5	37.2	17	37.7	300	9.1
5.0	37.4	18	37.3	400	8.2
5.5	38.0	19	36.7	500	6.9
6.0	38.3	20	35.2	600	6.0

Source: Adapted from Kimura, M. et al., *J. Chem. Phys.*, 115, 7442, 2001.

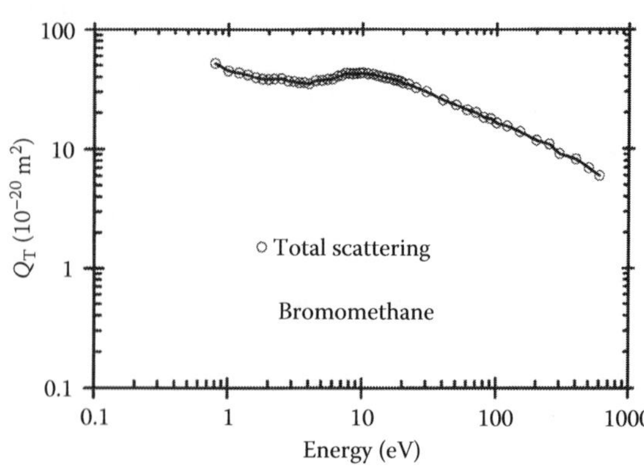

FIGURE 47.1 Total scattering cross sections in CH_3Br. (Adapted from Kimura, M. et al., *J. Chem. Phys.*, 115, 7442, 2001.)

TABLE 47.5
Ionization Cross Section in CH_3Br

Energy (eV)	Q_T (10^{-20} m²)	Energy (eV)	Q_T (10^{-20} m²)	Energy (eV)	Q_T (10^{-20} m²)
12	0.83	82	7.22	153	6.20
16	1.43	86	7.20	156	6.13
20	2.79	90	7.14	160	6.05
23	3.99	94	7.06	164	6.00
27	4.90	97	6.97	167	5.94
31	5.29	101	6.94	171	5.88
34	5.72	105	6.90	175	5.80
38	6.07	108	6.84	179	5.76
42	6.39	112	6.76	182	5.76
45	6.65	116	6.68	186	5.76
49	6.85	119	6.65	190	5.74
53	7.00	123	6.63	193	5.67
57	7.09	127	6.58	197	5.59
60	7.18	130	6.49	201	5.51
64	7.22	134	6.39	204	5.46
68	7.25	138	6.32	208	5.42
71	7.28	142	6.27	212	5.40
75	7.28	145	6.25	215	5.38
79	7.25	149	6.24		

Source: Courtesy of Professor P. Harland (2006).

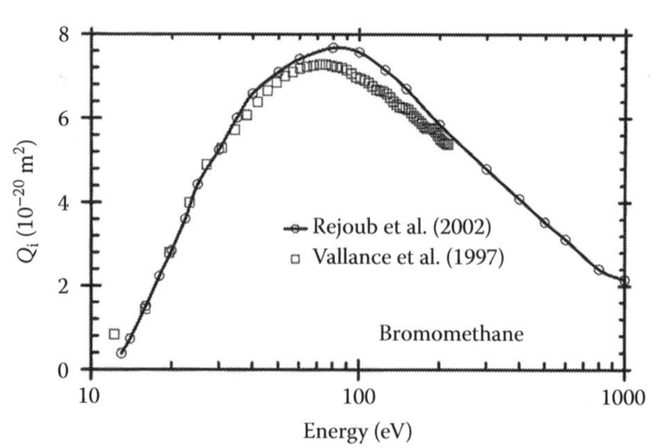

FIGURE 47.2 Total ionization cross sections for CH_3Br. (—○—) Rejoub et al. (2002); (□) Vallance et al. (1997).

47.4 POSITIVE ION APPEARANCE POTENTIALS

See Table 47.6.

47.5 PARTIAL IONIZATION CROSS SECTIONS

Rejoub et al. (2002) have measured the partial ionization cross sections for the fragments CH_nBr^+, CH_n^+, Br^+, and H^+, as shown in Table 47.7 and Figure 47.3.

TABLE 47.6
Appearance Potentials of Positive Ions of Selected Species

Process	Appearance Potential (eV) Photoionization[a]	Electron Ionization[b]
$CH_3^+ + Br^-$	<10.2	10.7
CH_3Br^+	10.75	10.5
$CH_3^+ + Br$	12.75	13.0
$CH_2Br^+ + H$	13.75	
$CHBr^+ + H_2$	15.25	
$CH_2 + HBr$	15.75	
$CH_3 + Br^+$	18.25	14.7
$CBr^+ + H_2 + H$	20.75	
$CH + HBr + H^+$	24.25	

[a] Olney et al. (1997).
[b] Tsuda et al. (1964).

TABLE 47.7
Selected Partial Ionization Cross Sections

Energy (eV)	Q_i (CH_nBr^+) (10^{-20} m²)	Q_i (CH_n^+) (10^{-20} m²)	Q (Total) (10^{-20} m²)
13	0.382		0.382
14	0.729		0.729
16	1.35	0.158	1.508
18	1.82	0.415	2.235
20	2.16	0.684	2.844
22.5	2.57	0.958	3.608
25	3.01	1.25	4.427
30	3.24	1.66	5.249
35	3.43	2.02	6.007
40	3.53	2.29	6.581
50	3.59	2.49	7.091
60	3.58	2.60	7.411
80	3.57	2.69	7.679
100	3.52	2.62	7.571
125	3.27	2.54	7.145
150	3.10	2.38	6.699
200	2.82	2.07	5.850
300	2.36	1.72	4.797
400	2.02	1.51	4.085
500	1.77	1.29	3.531
600	1.58	1.14	3.117
800	1.22	0.876	2.407
1000	1.09	0.799	2.149

Source: Adapted from Rejoub, R., B. G. Lindsay, and R. F. Stebbings, *J. Chem. Phys.*, 117, 6450, 2002.

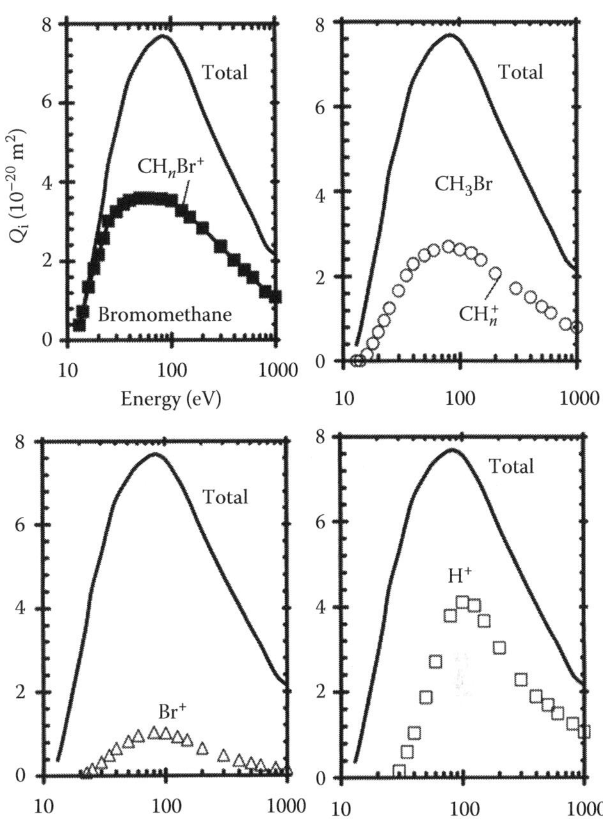

FIGURE 47.3 Partial ionization cross sections for CH_3Br fragments. The total ionization cross section has also been shown to indicate the relative contributions. (Adapted from Rejoub, R., B. G. Lindsay, and R. F. Stebbings, *J. Chem. Phys.*, 117, 6450, 2002.)

47.6 ATTACHMENT RATE AND CROSS SECTION

Pshenichnyuk et al. (2006) observed peaks in negative ion formation at 0.0, 0.6, 6.8, and 9.0 eV in the ratio 3:9:0.8:0.4.

Dissociative attachment of electrons below ~1 eV energy occurs via reaction

$$e + CH_3Br \rightarrow CH_3 + Br^- \qquad (47.1)$$

The electron affinity (ε_a) of bromine is 3.37 eV and the dissociation energy (D) is 3.078 eV so that the negative ion state has the energy (D–ε_a) −0.292 eV. This is below the ground-state vibrational level of CH_3–Br (Datskos et al., 1992). Christophorou and Hadjiantoniou (2006) quote that the electron affinity is >–0.46 eV.

The thermal attachment rate coefficient as a function of E/N (reduced electric field, Td) is shown in Table 47.9. Datskos et al. (1992) have measured the attachment rates as a function of temperature as shown in Table 47.10 and Figure 47.4.

Table 47.8 also shows the attachment rate coefficient at higher temperatures measured by several methods.

The swarm-derived attachment cross section in CH_3Br as a function of electron energy and temperature is shown in Figure 47.5 (Tsuda et al., 1964).

CH$_3$Br

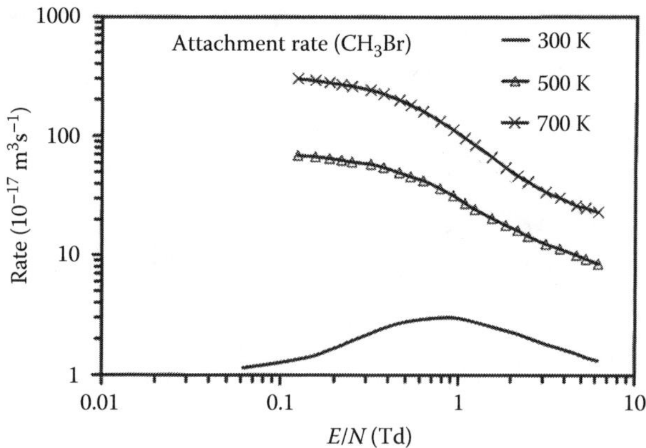

FIGURE 47.4 Attachment rate for CH$_3$Br as a function of reduced electric field and temperature. (Adapted from Datskos, P. G., L. G. Christophorou, and J. G. Carter, *J. Chem. Phys.*, 97, 9031, 1992.)

TABLE 47.8
Thermal Attachment Rates in CH$_3$Br

Temperature (K)	Method	Attachment Rate (10^{-18} m^3/s)	Reference
293	FA/ECR	6.0	Burns et al. (1996)
300	Swarm	10.8	Datskos et al. (1992)
~298	Swarm	6.0	Wang and Lee (1988)
293	CM	6.7	Petrović and Crompton (1987)
300	FA/LP	6.0	Alge et al. (1984)
298	MC	7.0	Bansal and Fessenden (1972)

Source: Adapted from Christophorou, L. G. and D. Hadjiantoniou, *J. Chem. Phys. Lett.*, 419, 405, 2006.

Note: CM = Cavalleri method; FA/ECR = flowing afterglow/electron cyclotron resonance; MC = microwave conductivity. Recommended value = 6.5 × 10^{-18} m^3/s.

TABLE 47.9
Selected Attachment Rates as a Function of Temperature and *E/N*

E/N (Td)	K_a (10^{-17} m^3/s) *T* = 300 K	K_a (10^{-17} m^3/s) *T* = 400 K
0.062	1.15	17.5
0.124	1.35	18.2
0.248	1.91	17.7
0.528	2.81	15.4
1.087	2.91	11.5
1.240	2.77	10.7
1.550	2.53	9.2
1.860	2.35	8.2
2.170	2.21	7.3
2.480	2.05	6.6
3.100	1.82	5.8
3.730	1.68	5.2
4.660	1.52	4.6

TABLE 47.9 (continued)
Selected Attachment Rates as a Function of Temperature and *E/N*

E/N (Td)	K_a (10^{-17} m^3/s) *T* = 300 K	K_a (10^{-17} m^3/s) *T* = 400 K
5.280	1.42	4.4
6.210	1.33	4.2

Source: Adapted from Datskos, P. G., L. G. Christophorou, and J. G. Carter, *J. Chem. Phys.*, 97, 9031, 1992.

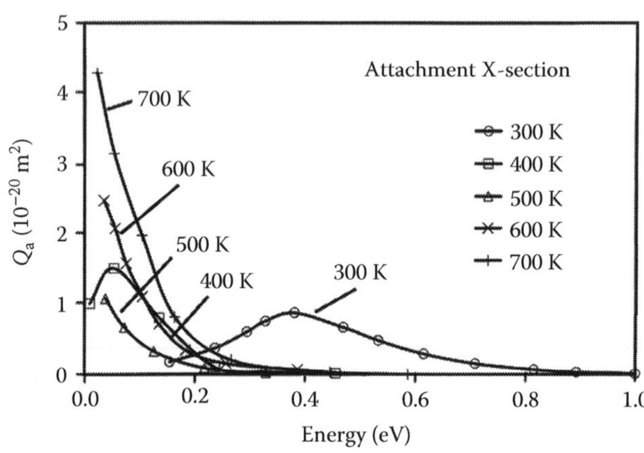

FIGURE 47.5 Attachment cross sections for CH$_3$Br. The rate at 300 K is multiplied by 50 and at 400 K by 10 for better presentation. (Adapted from Datskos, P. G., L. G. Christophorou, and J. G. Carter, *J. Chem. Phys.*, 97, 9031, 1992.)

TABLE 47.10
Attachment Rate as a Function of Temperature

Temperature	Rate (10^{-18} m^3/s)	Reference
293	6.0	Burns et al. (1996)
615	830	
777	9500	
300	11.0	Datskos et al. (1992)
400	170	
500	710	
600	1600	
700	3300	
293	6.7	Petrović and Crompton (1987)
445	180	
499	440	
300	6.0	Alge et al. (1984)
452	230	
585	2500	

47.7 DRIFT VELOCITY OF ELECTRONS

See Table 47.11.

47.8 MEAN ENERGY OF ELECTRONS

See Table 47.12.

TABLE 47.11
Selected Drift Velocity for ($N_2 + CH_3Br$) as a Function of Temperature and *E/N*

E/N (Td)	Drift Velocity (10^4 m/s) *T* = 300 K	*T* = 500 K	*T* = 700 K
0.062	0.185	0.132	0.108
0.124	0.257	0.210	0.181
0.248	0.299	0.277	0.261
0.528	0.363	0.359	0.358
1.087	0.462	0.462	0.470
2.480	0.692	0.707	0.728
5.280	1.142	1.177	1.205

Source: Adapted from Datskos, P. G., L. G. Christophorou, and J. G. Carter, *J. Chem. Phys.*, 97, 9031, 1992.
Note: See Figure 47.6 for graphical presentation.

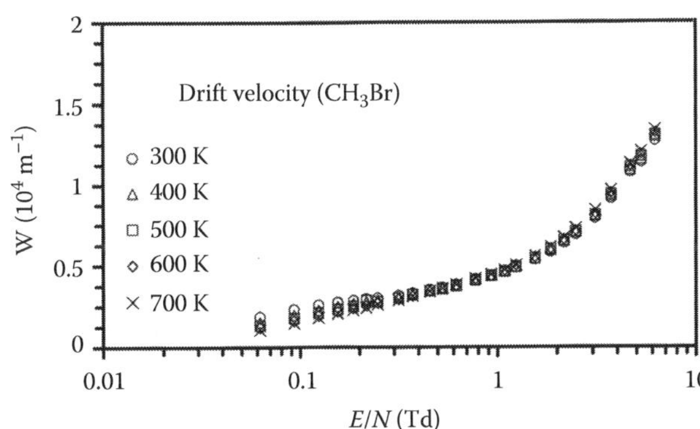

FIGURE 47.6 Drift velocity for ($N_2 + CH_3Br$) as a function of reduced electric field and temperature. (Adapted from Datskos, P. G., L. G. Christophorou, and J. G. Carter, *J. Chem. Phys.*, 97, 9031, 1992.)

TABLE 47.12
Selected Mean Energy for ($N_2 + CH_3Br$) as a Function of Temperature and *E/N*

E/N (Td)	Mean Energy (eV) *T* = 300 K	*T* = 500 K	*T* = 700 K
0.062	0.046	0.070	0.092
0.124	0.065	0.085	0.104
0.248	0.111	0.125	0.139
0.528	0.204	0.217	0.221
1.087	0.374	0.378	0.388
2.170	0.601	0.601	0.601
4.660	0.812	0.812	0.812
5.280	0.839	0.839	0.839
6.210	0.872	0.872	0.872

Source: Adapted from Datskos, P. G., L. G. Christophorou, and J. G. Carter, *J. Chem. Phys.*, 97, 9031, 1992.
Note: See Figure 47.7 for graphical presentation.

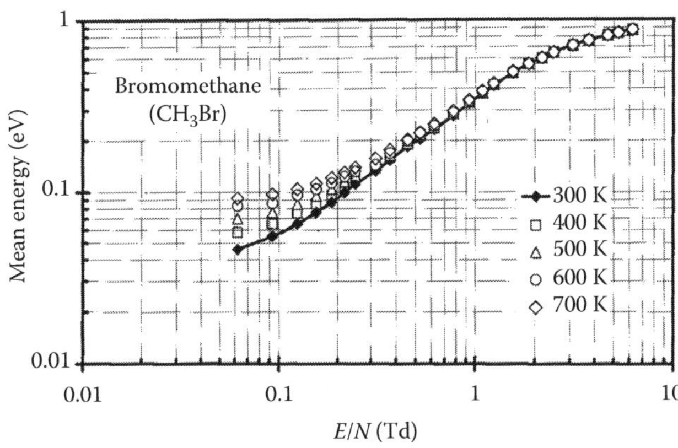

FIGURE 47.7 Mean energy for ($N_2 + CH_3Br$) as a function of reduced electric field and temperature. (Adapted from Datskos, P. G., L. G. Christophorou, and J. G. Carter, *J. Chem. Phys.*, 97, 9031, 1992.)

CH_3Br

REFERENCES

Alge, E., N. G. Adams, and D. Smith, *J. Phys. B: At. Mol. Phys.*, 17, 3827, 1984.

Allan, M. and L. Andric, *J. Chem. Phys.*, 105, 3559, 1996.

Bansal, K. M. and R. W. Fessenden, *Chem. Phys. Lett.*, 15, 21, 1972.

Benitez, A., J. H. Moore, and J. A. Tossell, *J. Chem. Phys.*, 88, 6691, 1988.

Braun, M., I. I. Fabrikant, M. W. Ruf, and H. Hotop, *J. Phys. B: At. Mol. Opt. Phys.*, 40, 659, 2007.

Burns, S. J., J. M. Matthews, and D. L. McFadden, *J. Phys. Chem.* 100, 19437, 1996.

Christodoulides, A. A. and L. G. Christophorou, *J. Chem. Phys.*, 54, 4691, 1971.

Christophorou, L. G. and D. Hadjiantoniou, *J. Chem. Phys. Lett.*, 419, 405, 2006.

Datskos, P. G., L. G. Christophorou, and J. G. Carter, *J. Chem. Phys.*, 97, 9031, 1992.

Guerra, M., D. Jones, G. Distefano, F. Scagnolari, and A. Modelli, *J. Chem. Phys.*, 94, 484, 1991.

Kimura, M., O. Sueoka, C. Makochekanwa, H. Kawatw, and M. Kawada, *J. Chem. Phys.*, 115, 7442, *2001.*

Krzysztofowicz, A. M. and C. Szmytkowski, *Chem. Phys. Lett.*, 219, 86, 1994.

Olney, T. N., G. Cooper, W. F. Chan, G. R. Burton, C. E. Brion, and K. H. Tan, *Chem. Phys.*, 218, 127, 1997.

Petrović, Z. Lj. and R. W. Crompton, *J. Phys. B: At. Mol. Phys.*, 20, 5557, 1987.

Pshenichnyuk, S. A., I. A. Pshenichnyuk, E. P. Nafikova, and N. L. Asfandiarov, *Rapid. Comm. Mass Spectrom.*, 20, 1097, 2006.

Rejoub, R., B. G. Lindsay, and R. F. Steb*bings, J. Chem. Phys.*, 117, 6450, 2002.

Spence, D. and G. J. Schulz, *J. Chem. Phys.*, 58, 1800, 1973.

Sverdlov, L. M., M. A. Kovner, and E. P. Krainov, *Vibrational Spectra of Polyatomic Molecules*, John Wiley & Sons, New York, NY, 1974, p. 381.

Tsuda, S., C. E. Melton, and W. H. Hamill, *J. Chem. Phys.*, 41, 689, 1964.

Vallance, C., S. A. Harris, J. E. Hudson, and P. W. Harland, *J. Phys. B: At. Mol. Opt. Phys., 30,* 2465, 1997.

Wang, W. C. and L. C. Lee, *J. Appl. Phys.*, 63, 4905, *1988.*

Wilde, R. S., G. A. Gallup, and I. I. Fabrikant, *J. Phys. B: At. Mol. Opt. Phys.,* 33, 5479, 2000.

48

BROMOTRICHLOROMETHANE

CONTENTS

Bromotrichloromethane ($CBrCl_3$) is a polar, electron-attaching gas that has 92 electrons. The ionization potential is 10.6 eV.

48.1 ATTACHMENT RATES

Attachment occurs through the formation of Br^- and Cl^- ions according to reactions (Spanel et al., 1997)

$$CCl_3Br + e \rightarrow CCl_3 + Br^- + 0.4\ \text{eV} \quad (48.1)$$

$$CCl_3Br + e \rightarrow CCl_2Br + Cl^- + 0.2\ \text{eV} \quad (48.2)$$

At 300 K the abundance of Cl^- ion is about 20% of the total ($Cl^- + Br^-$) increasing to 80% at 540 K. The fraction of Cl^- ions increases with increasing temperature till the ratio becomes $Cl^-:Br^- = 3:1$ (Spanel et al., 1997).

The attachment rate constants are shown in Table 48.1. Figure 48.1 shows the variation of attachment rate coefficient as a function of mean energy of electrons (Sunagawa and Shimamori, 1997).

TABLE 48.1
Attachment Rate Constants for $CBrCl_3$

Method	Temperature (K)	Rate (10^{-14} m³/s)	Reference
FA/LP	300	6.2	Spanel et al. (1997)
	540	13.0	
PR/MW	293	5.7	Sunagawa and Shimamori (1997)
FA/LP	293	8.2	Adams et al. (1988)
TPI/LS	293	4.9 (Br^-)	Alajajian et al. (1988)

Note: FA/LP = flowing afterglow/Langmuir probe; PR/MW = pulse radiolysis/microwave heating; TPI/LS = threshold photoionization/line shape.

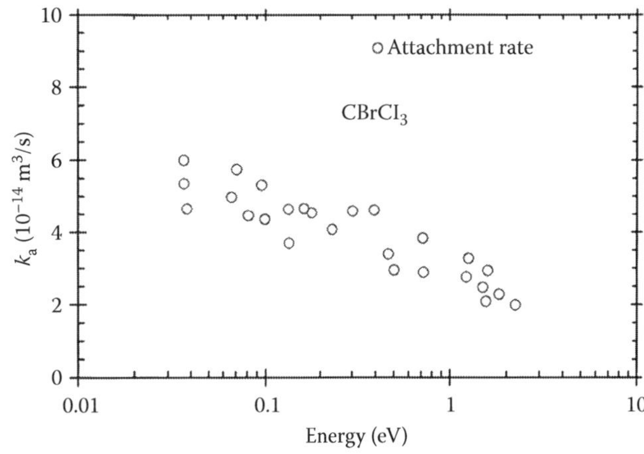

FIGURE 48.1 Attachment rate for $CBrCl_3$. (Adapted from Sunagawa, T. and H. Shimamori, *J. Chem. Phys.*, 7876, 107, 1997.)

48.2 ATTACHMENT CROSS SECTIONS

Deconvoluted attachment cross sections and those measured by beam technique are shown in Table 48.2 and Figure 48.2.

CBrCl$_3$

TABLE 48.2
Attachment Cross Sections for $CBrCl_3$ (Digitized and Interpolated)

	Q_a (10^{-20} m^2)	
Energy (meV)	**Sunagawa and Shimamori (1997)**	**Spanel et al. (1997)**
1.00	344	
2.00	239	
3.00	203	
4.00	181	
6.00	138	
8.00	115	
10.00	105	184
15.0	85.3	143
20.0	69.1	126
25.0	59.8	117
30.0	54.3	104
35.0	50.0	88.5
40.0	46.2	76.9
45.0	42.9	68.6
50.0	40.0	61.6
55.0	37.4	55.2
60.0	35.1	49.8
65.0	33.0	45.7
70.0	31.2	42.6
75.0	29.6	40.1
80.0	28.2	38.0
85.0	27.0	35.8
90.0	26.0	33.3
95.0	25.1	30.1
100	24.3	26.6
150	19.4	7.25
200	15.8	5.48
250	13.4	4.74
300	12.0	3.84
350	11.0	3.29
400	10.1	3.26
450	9.29	3.38
500	8.56	3.70
550	7.93	4.26
600	7.38	4.97
650	6.89	5.69
700	6.46	6.17
750	6.06	6.16
800	5.68	5.71
850	5.32	5.02
900	4.98	4.29
1000	4.41	3.13
1500	2.46	1.24
2000	1.88	1.00

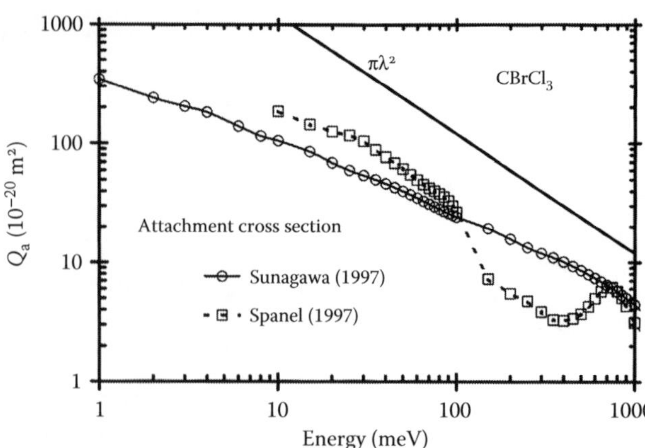

FIGURE 48.2 Attachment cross sections for $CBrCl_3$. (—○—) Sunagawa and Shimamori (1997); (---□---) Spanel et al. (1997). *s*-Wave maximum cross section is also shown for comparison.

REFERENCES

Adams, N. G., D. Smith, and C. R. Herd, *Int. J. Mass Spectrom. Ion Proc.*, 243, 84, 1988.

Alajajian, S. H., M. T. Bernius, and A. Chutjian, *J. Phys. B: At. Mol. Opt. Phys.*, 21, 4021, 1988.

Spanel, P., D. Smith, S. Matejcik, A. Kiendler, and T. D. Märk, *Int. J. Mass Spectrom. Ion Proc.*, 167/168, 1, 1997.

Sunagawa, T. and H. Shimamori, *J. Chem. Phys.*, 7876, 107, 1997.

49

BROMOTRIFLUORO-METHANE

CONTENTS

Bromotrifluoromethane ($CBrF_3$) is a weakly polar molecule that has 58 electrons. The electronic polarizability is 6.45×10^{-40} F m^2 (see Beran and Kevan, 1969), the dipole moment is 0.65 D, and the ionization potential is 11.40 eV. The vibrational modes of the molecule are shown in Table 49.1. The vibrational polarizability is 1.2×10^{-40} F m^2 (see Bishop and Cheung, 1982).

49.1 SELECTED REFERENCES FOR DATA

See Table 49.2.

TABLE 49.1
Vibrational Modes and Energies for $CBrF_3$

Mode	Energy (meV)
ν_1	134.5
ν_2	94.4
ν_3	43.4
ν_4	149.9
ν_5	68.2
ν_6	37.8

Source: Adapted from Bishop, D. M. and L. M. Cheung, *J. Phys. Chem. Ref. Data*, 11, 119, 1982.

TABLE 49.2
Selected References for $CBrF_3$

Quantity	Range, Energy (Temperature) eV (K)	Reference
k_a	—	Christophorou and Hadjiantoniou (2006)
k_a, Q_a	0–2	**Marienfeld et al. (2006)**
k_a	Thermal	**Barszczewska et al. (2004)**
Q_i	12–215	**Bart et al. (2001)**
k_a	Thermal	**Le Garrec et al. (1997)**
k_a	0.03–2.0	**Sunagawa and Shimamori (1997)**
k_a	(300–800)	**Burns et al. (1996)**
Q_a	0–6	**Underwood-Lemons et al. (1995)**
Q_T	0–12	**Underwood-Lemons et al. (1994)**
k_a	0.03–2.0	**Shimamori et al. (1992)**
k_a	0–0.1	**Kalamarides et al. (1990)**
k_a	Thermal	**Marotta et al. (1989)**
Q_a	0–0.2	**Alajajian et al. (1988)**
k_a	Thermal	McCorkle et al. (1987)
Q_a	(300–1200)	**Spence and Schulz (1973)**
Q_i	70	**Beran and Kevan (1969)**
k_a	0–0.7	**Blaunstein et al. (1968)**

Note: k_a = Attachment rate; Q_a = attachment cross section; Q_i = ionization cross section, Q_T = total scattering cross section. Bold font indicates experimental study.

CBrF3

49.2 TOTAL SCATTERING CROSS SECTIONS

Figure 49.1 shows the total scattering cross sections measured by Underwood-Lemons et al. (1994). Note the rise toward zero energy attributed to the attachment of electrons with possible contribution due to the dipole moment of the molecule.

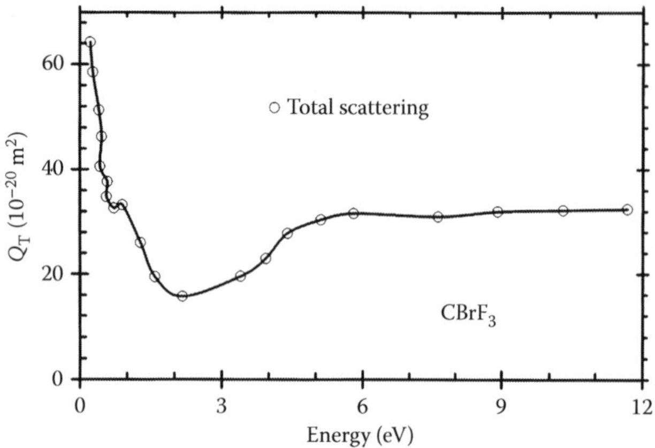

FIGURE 49.1 Total scattering cross sections for $CBrF_3$. (Adapted from Underwood-Lemons, T. et al., *J. Chem. Phys.*, 100, 9117, 1994.)

TABLE 49.3
Ionization Cross Sections for $CBrF_3$

Energy (eV)	Q_i (10^{-20} m^2)	Energy (eV)	Q_i (10^{-20} m^2)
12	0.41	116	7.71
16	0.96	119	7.70
20	1.83	123	7.68
23	2.81	127	7.64
27	3.69	130	7.58
31	4.41	134	7.52
34	5.02	138	7.46
38	5.53	142	7.43
42	5.97	145	7.41
45	6.35	149	7.39
49	6.63	153	7.35
53	6.92	156	7.30
57	7.13	160	7.23
60	7.36	164	7.15
64	7.48	167	7.08
68	7.60	171	7.02
71	7.70	175	6.98
75	7.78	179	6.95
79	7.81	182	6.93
82	7.84	186	6.91
86	7.90	190	6.87
90	7.87	193	6.81
94	7.87	197	6.74
97	7.85	201	6.67
101	7.82	204	6.61
105	7.78	208	6.56
108	7.76	212	6.54
112	7.73	215	6.53

Sources: Adapted from Bart, M. et al., *Phys. Chem. Chem. Phys.*, 3, 800, 2001; tabulated values courtesy of Professor P. Harland (2006).

49.3 IONIZATION CROSS SECTIONS

Ionization cross sections measured by Bart et al. (2001) are shown in Table 49.3 and Figure 49.2.

49.4 ATTACHMENT CROSS SECTIONS

Dissociative attachment of electrons occurs according to the reaction

$$CBrF_3 + e \rightarrow CF_3 + B_r^- \quad (49.1)$$

The electron affinity of the molecule is 0.91 eV (Christophorou and Hadjiantoniou, 2006). The attachment

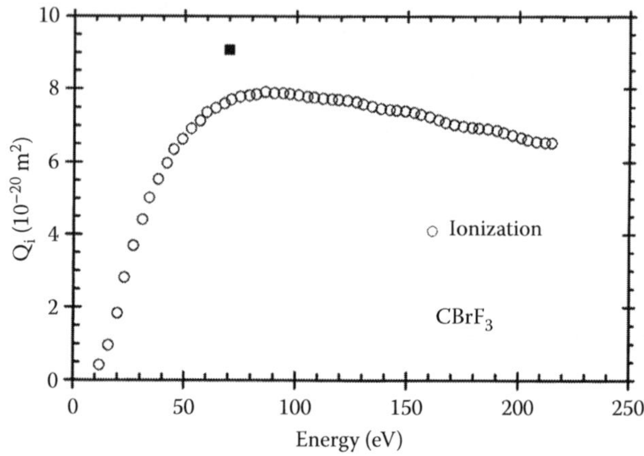

FIGURE 49.2 Total ionization cross sections for $CBrF_3$. (○) Bart et al. (2001); (■) Beran and Kevan (1969).

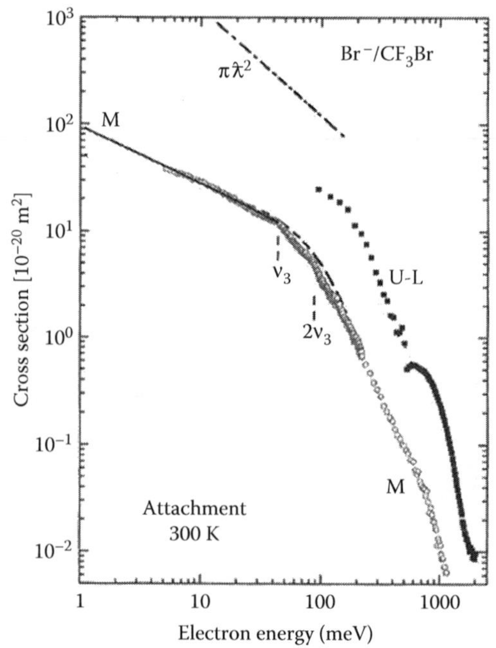

FIGURE 49.3 Dissociative attachment cross sections for $CBrF_3$. Full line and open circles, M = Marienfeld et al. (2006); full squares, U-L = Underwood-Lemons et al. (1995); vibrational energy for mode ν_3 and maximum *s*-wave capture cross section ($\pi ƛ^2$) are also shown. (With kind permission of American Institute of Physics.)

cross sections measured by various techniques have been summarized by Marienfeld et al. (2006) as shown in Figure 49.3. Note the increase of the attachment cross sections toward zero energy.

TABLE 49.4
Attachment Rate Constants for $CBrF_3$

Rate Constant	Method	Temperature	Reference
1.4 (−14)			Christophorou and Hadjiantoniou (2006)
1.3 (−14)	Swarm		Barszczewska et al. (2004)
8.6 (−15)	MCM		Sunagawa and Shimamori (1997)
1.2 (−14)	FA/ECR	293	Burns et al. (1996)
3.9 (−14)		615	
1.2 (−13)		777	
8.6 (−15)	MCM		Shimamori et al. (1992)
1.5 (−14)	ECR		Marotta et al. (1989)
5.3 (−15)	FA/LP	205	Alge et al. (1984)
1.6 (−14)		300	
4.9 (−14)		452	
7.7 (−14)		585	
1.36 (−14)	Swarm		Blaunstein and Christophorou (1968)

Note: Thermal rates unless otherwise mentioned. MCM = microwave cavity method; FA/ECR = flowing afterglow/electron cyclotron resonance; ECR = electron cyclotron resonance; FA/LP = flowing afterglow/Langmuir probe.

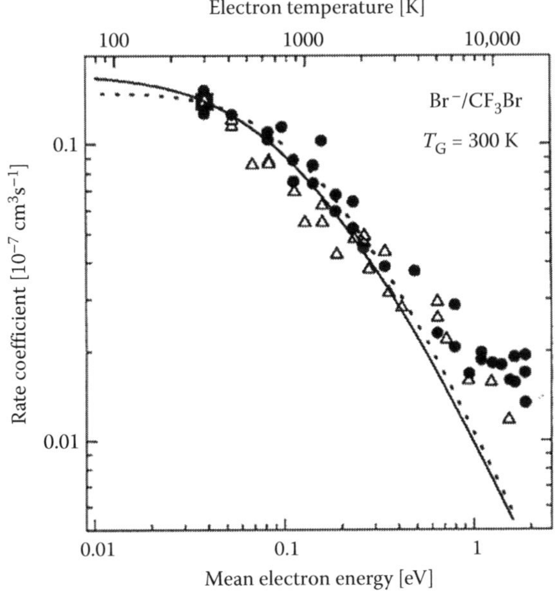

FIGURE 49.4 Attachment rate constants for $CBrF_3$. Full circles, Shimamori et al. (1992); open triangles, Marienfeld et al. (2006); dashed and full curves, theoretical, Marienfeld et al. (2006). (Adapted from Marienfeld, S. et al., *J. Chem. Phys.*, 124, 154316, 2006.)

The attachment cross sections as a function of gas temperature show an increasing trend up to 1400 K and the effect is attributed to increasing vibrational excitation of the molecule (Spence and Schulz, 1973).

49.5 ATTACHMENT RATES

Table 49.4 shows the attachment rate constants obtained by various methods. Figure 49.4 is a graphical presentation as a function of temperature (Marienfeld et al., 2006).

REFERENCES

Alajajian, S. H., M. T. Bernius, and A. Chutjian, *J. Phys. B: At. Mol. Opt. Phys.*, 21, 4021, 1988.

Alge, E., N. G. Adams, and D. Smith *J. Phys. B: At. Mol. Phys.* 17, 3827, 1984.

Barszczewska, W., J. Kopyra, J. Wnorowska, and I. Szamrej, *Int. J. Mass. Spectrom.*, 233, 199, 2004.

Bart, M., P. W. Harland, J. E. Hudson, and C. Vallance, *Phys. Chem. Chem. Phys.*, 3, 800, 2001.

Beran, J. A. and L. Kevan, *J. Phys. Chem.*, 73, 3866, 1969.

Bishop, D. M. and L. M. Cheung, *J. Phys. Chem. Ref. Data*, 11, 119, 1982.

Blaunstein, R. P. and L. G. Christophorou, *J. Chem. Phys.*, 49, 1526, 1968.

Burns, S. J., J. M. Mathews, and D. L. McFadden, *J. Phys. Chem.*, 100, 19436, 1996.

Christophorou, L. G. and D. Hadjiantoniou, *Chem. Phys. Lett.*, 419, 405, 2006.

Kalamarides, A., R. W. Marawar, X. Ling, C. W. Walter, B. G. Lindsay, and K. A. Smith, *J. Chem. Phys.*, 92, 1672, 1990.

Le Garrec, J. L., O. Sidko, J. L. Queffelec, S. Hamon, J. B. A. Mitchell, and B. R. Rowe, *J. Chem. Phys.*, 107, 54, 1997.

Marienfeld, S., T. Sunagawa, I. I. Fabrikant, M. Braun, M.-W. Ruf, and H. Hotop, *J. Chem. Phys.*, 124, 154316, 2006.

Marotta, C. J., C-p. Tsai, and D. L. McFadden, *J. Chem. Phys.*, 91, 2194, 1989.

McCorkle, D. L. 1987, cited by Alajajian et al. 1988, as private communication.

Shimamori, H., Y. Tatsumi, Y. Ogawa, and T. Sunagawa, *J. Chem. Phys.*, 97, 6335, 1992.

Spence, D. and G. J. Schulz, *J. Chem. Phys.*, 58, 1800, 1973.

Sunagawa, T. and H. Shimamori, *J. Chem. Phys.*, 107, 7876, 1997.

Underwood-Lemons, T., T. J. Gergei, and J. H. Moore, *J. Chem. Phys.*, 102, 119, 1995.

Underwood-Lemons, T., D. C. Winkler, J. A. Tossell, and J. H. Moore, *J. Chem. Phys.*, 100, 9117, 1994.

50

CARBON TETRACHLORIDE

CONTENTS

Carbon tetrachloride (CCl_4), synonym tetrachloromethane, has 74 electrons. It is nonpolar and possesses electronic polarizability of 12.46×10^{-40} F m^2. The ionization potential is 11.47 eV. It has four vibrational modes as shown in Table 50.1 (Shimanouchi, 1972). The molecule is electron-attaching with a high electron affinity of 2.12 eV (see Sierra, 2005) and the lowest electronic excitation potential is below 7 eV (Jones, 1986).

TABLE 50.1
Vibrational Modes of CCl_4

Designation	Energy (meV)	Mode
ν_1	49.9	Symmetric stretch
ν_2	26.9	Degenerate deformation
ν	96.2	Degenerate stretch
ν	38.9	Degenerate deformation

50.1 SELECTED REFERENCES FOR DATA

See Table 50.2.

50.2 TOTAL SCATTERING CROSS SECTIONS

Total scattering cross sections in CCl_4 are shown in Tables 50.3 and 50.4. See Figure 50.1 for graphical presentation. The highlights of the total cross section are

1. At low electron energy (<1 eV) the cross section increases enormously to 860×10^{-20} m^2 at ~0.01 eV due to dissociative attachment (Randell et al., 1993).
2. A Ramsauer–Townsend minimum occurs at very low energy of ~40–60 meV (Randell et al., 1993).
3. There is a resonance peak at ~1.2 eV due to dissociative attachment.
4. A second peak at ~8 eV which is also due to resonance in dissociative attachment (Jones, 1986). This

TABLE 50.2
Cross Sections

Quantity	Energy Range (eV)	Reference
Q_i	15–85	Sierra et al. (2005)
Q_i	15–1000	Lindsay et al. (2004)
k_a	0–2	Matejčik et al. (2003)
k_a	0–0.173	Klar et al. (2001)
Q_i	12–215	Hudson et al. (2001)
k_a	—	Burns et al. (1996)
Q_a	10^{-6}–0.1	Dunning (1995)
Q_T	0.8–400	Hamada and Sueoka (1995)
Q_a	0.001–1.5	Matejcik et al. (1995)
Q_{el}	0–40	Natalense et al. (1995)
k_a	0.026–0.35	Spanel et al. (1995)
Q_T	0.01–1.0	Randell et al. (1993)
k_a	0.005–2.5	Shimamori et al. (1992)

continued

CCl$_4$

TABLE 50.2 (continued)
Cross Sections

Quantity	Energy Range (eV)	Reference
Q_a	0–100	Ling et al. (1992)
Q_T	75–4000	Zecca et al. (1992)
Q_T, Q_a	0–12	Wan et al. (1991)
k_a	0.01–1.0	Harth et al. (1989)
k_a	—	Harth et al. (1989)
Q_T	0.5–50	Jones (1986)
Q_a	0–0.08	Chutjian and Alajajian (1985)
Q_i	15–180	Leiter et al. (1984)
k_a	—	Smith et al. (1984)
Q_{el}	10–400	Daimon et al. (1983)
k_a	0–2	Ayala et al. (1981)
Q_a	0–2	Christophorou et al. (1981)
k_a	0.008–0.025	Boltz et al. (1977)
k_a	Thermal	Davis et al. (1973)
k_a	—	Mothes et al. (1972)
k_a	0.05–0.8	Christodoulides and Christophorou (1971)
Q_a	0.1–2.5	Christophorou and McCorkle (1971)
Q_a	0–2	Blaunstein and Christophorou (1968)

Note: k_a = Attachment rate; Q_a = attachment; Q_i = ionization; Q_T = total cross section.

TABLE 50.3
Total Scattering Cross Section in CCl$_4$

Randell et al. (1993)		Hamada and Sueoka (1995)		Zecca et al. (1992)	
Energy (eV)	Q_T (10^{-20} m^2)	Energy	Q_T (10^{-20} m^2)	Energy	Q_T (10^{-20} m^2)
		0.8	56.4	75	38.2
0.01	864.84	1.0	58.2	80	37.0
0.02	650.39	1.2	57.0	90	36.1
0.03	400.78	1.4	55.7	100	36.6
0.05	221.48	1.6	54.5	110	35.0
0.10	112.50	1.8	50.7	125	34.2
0.20	66.80	2.0	47.6	150	31.0
0.40	49.22	2.2	44.4	175	28.9
0.60	35.16	2.5	41.1	200	28.0
0.80	49.22	2.8	40.7	225	25.6
1.0	52.74	3.1	39.5	250	24.4
		3.4	38.7	275	23.5
		3.7	39.3	300	22.6
		4.0	40.2	350	20.8
		4.5	42.3	400	19.4
		5.0	43.7	450	17.9
		5.5	48.7	500	16.4
		6.0	49.9	600	14.7
		6.5	54.2	700	13.3
		7.0	57.9	800	12.5
		7.5	58.3	900	11.1
		8.0	62.0	1000	10.3
		8.5	60.0	1100	9.4

TABLE 50.3 (continued)
Total Scattering Cross Section in CCl$_4$

Randell et al. (1993)		Hamada and Sueoka (1995)		Zecca et al. (1992)	
Energy (eV)	Q_T (10^{-20} m^2)	Energy	Q_T (10^{-20} m^2)	Energy	Q_T (10^{-20} m^2)
		9.0	59.5	1250	8.7
		9.5	58.7	1500	7.6
		10.0	59.7	1750	6.8
		11.0	57.5	2000	6.0
		12.0	55.0	2250	5.5
		13.0	55.0	2500	5.0
		14.0	54.4	2750	4.6
		15.0	54.0	3000	4.3
		16.0	53.0	3250	4.0
		17.0	50.7	3500	3.7
		18.0	50.3	4000	3.4
		19.0	50.5		
		20.0	50.3		
		22.0	47.8		
		25.0	45.6		
		30.0	41.9		
		35.0	40.5		
		40.0	38.4		
		50.0	36.5		
		60.0	35.2		
		70.0	33.1		
		80.0	32.5		
		90.0	31.5		
		100.0	31.1		
		120.0	29.5		
		150.0	28.6		
		200.0	27.5		
		250.0	24.7		
		300.0	22.3		
		350.0	21.0		
		400.0	19.2		

TABLE 50.4
Total Scattering Cross Section in CCl$_4$ at Selected Intervals of Energy

Energy	Q_T (10^{-20} m^2)
0.6	41.76
1.0	66.36
1.5	66.44
2.0	51.98
2.5	42.28
3.0	39.94
3.5	41.35
4.0	44.20
5.0	51.28
6.0	60.14
7.0	69.49
8.0	69.17
10.0	65.70

continued

TABLE 50.4 (continued)
Total Scattering Cross Section in CCl_4 at Selected Intervals of Energy

Energy	Q_T (10^{-20} m^2)
12.0	64.65
14.0	61.53
16.0	58.36
18.0	55.80
20.0	53.38
25.0	48.99
30.0	46.04
35.0	43.51

Source: Adapted from Jones, R. K., *J. Chem. Phys.*, 84, 813, 1986.

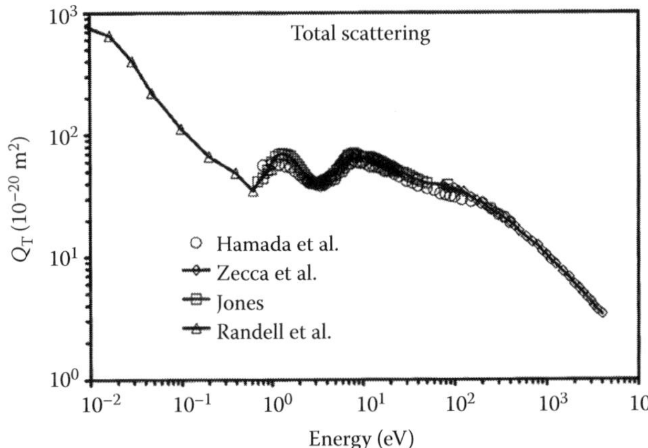

FIGURE 50.1 Total scattering cross sections in CCl_4. (○) Hamada and Sueoka (1995); (—△—) Randell et al. (1993); (—◊—) Zecca et al. (1992); (—□—) Jones (1986).

TABLE 50.5
Elastic Scattering Cross Section in CCl_4

Energy	Q_{el} (10^{-20} m^2)	Energy	Q_{el} (10^{-20} m^2)	Energy	Q_{el} (10^{-20} m^2)
1.00	37.58	4.50	65.47	12.5	52.00
1.50	50.48	5.00	60.65	15	52.21
2.00	59.38	5.50	57.07	20	50.10
2.50	48.04	6.00	54.00	25	46.93
3.00	61.68	6.50	52.00	30	42.72
3.25	71.86	7.75	51.60	35	38.42
3.60	76.92	8.00	58.19	40	35.19
4.00	72.79	8.10	46.50		
4.25	68.80	10.00	53.30		

Source: Digitized and interpolated at shown energy intervals from Natalense, A. P. P. et al., *Phys. Rev. A*, 52, R1, 1995.

peak, though not the process from which it results, is common to many gases.

5. A monotonic decrease beyond this peak, also common to many gases.

50.3 ELASTIC SCATTERING CROSS SECTIONS

Only theoretical cross sections calculated by Natalense et al. (1995) are available as shown in Table 50.5 and Figure 50.2.

50.4 POSITIVE ION APPEARANCE POTENTIALS

Table 50.6 shows the positive ion appearance potentials for CCl_4 (Sierra et al., 2005).

50.5 IONIZATION CROSS SECTIONS

Selected experimental ionization cross sections are shown in Figure 50.3. Tabulated values of Lindsay et al. (2004) and Sierra et al. (2005) are shown in Table 50.7.

Partial ionization cross sections are shown in Tables 50.8 and 50.9, with a graphical presentation in Figure 50.4. A further table of ionization cross sections from Hudson et al. (2001) is shown in Table 58.2 in the section on dichloromethane (CH_2Cl_2).

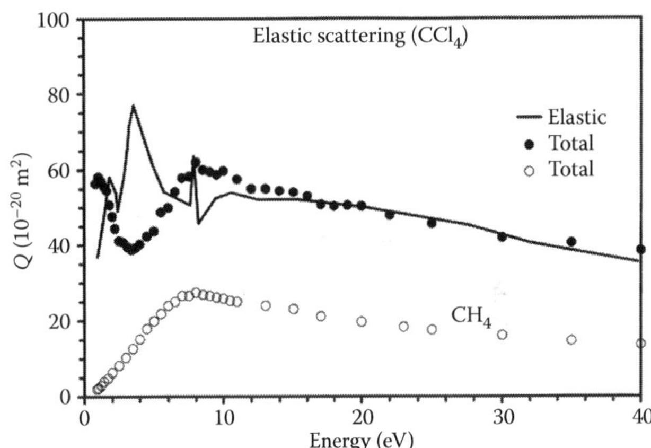

FIGURE 50.2 Elastic scattering cross section in CCl_4 calculated by Natalense et al. (1995). The total cross sections in CCl_4 (Hamada and Sueoka 1995) and CH_4 (Zecca et al., 1991) are also shown for comparison.

TABLE 50.6
Positive Ion Appearance Potentials

Ion Species	Reaction	Energy (eV)
CCl_4^+	$CCl_4 + e \rightarrow CCl_4^+$	Not observed
CCl_3^+	$CCl_4 + e \rightarrow CCl_3^+ + Cl$	13.01
CCl_2^+	$CCl_4 + e \rightarrow CCl_2^+ + 2Cl$	18.03
CCl^+	$CCl_4 + e \rightarrow CCl^+ + Cl_2 + Cl$	21.17
CCl_3^{2+}	$CCl_4 + e \rightarrow CCl_3^{2+} + Cl$	32.3
Cl_2^+	$CCl_4 + e \rightarrow Cl^{2+} + C + 2Cl$	28.05
C^+	$CCl_4 + e \rightarrow C^+ + 4Cl$	27.36
Cl^+, CCl_3^+	$CCl_4 + e \rightarrow Cl^+ + CCl_3^+$	26.09

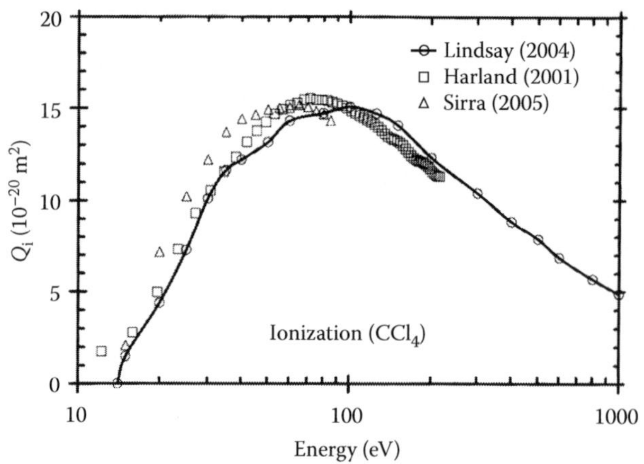

FIGURE 50.3 Total ionization cross sections in CCl_4. (Δ) Sierra et al. (2005); (—○—) Lindsay et al. (2004); (□) Hudson et al. (2001). (Tabulated values for their data are from courtesy of Professor P. Harland.)

TABLE 50.7
Ionization Cross Sections for CCl_4

	Q_i (10^{-20} m^2)	
Energy (eV)	**Lindsay et al. (2004)**	**Sierra et al. (2005)**
15	1.50	2.10
20	4.38	7.17
25	7.28	10.23
30	10.11	12.23
35	11.65	13.70
40	12.23	14.42
45		14.63
50	13.16	14.92
55		15.02
60	14.31	15.03
65		15.18
70		15.06
75		14.83
80	14.73	14.66
85		14.32
100	15.07	
125	14.70	
150	14.07	
200	12.33	
300	10.40	
400	8.83	
500	7.86	
600	6.85	
800	5.68	
1000	4.86	

TABLE 50.8
Partial Ionization Cross Sections for CCl_4 for Selected Species

	Q_i (10^{-20} m^2)			
Energy (eV)	**CCl_3^+ (A)**	**Cl^+ (B)**	**CCl^+ (C)**	**CCl_2^+ (D)**
15	1.5			
20	4.03			0.35
25	5.38	0.085	0.62	1.19
30	6.14	0.415	1.72	1.77
35	6.59	0.979	2.15	1.7
40	6.36	1.660	2.23	1.64
50	6.19	2.760	2.18	1.62
60	6.53	3.430	2.2	1.69
80	6.66	3.740	2.17	1.64
100	6.85	3.970	2.01	1.7
125	6.85	3.750	1.89	1.69
150	6.63	3.570	1.76	1.62
200	6.12	2.880	1.47	1.46
300	5.47	2.150	1.17	1.31
400	4.75	1.770	0.954	1.11
500	4.25	1.530	0.867	0.998
600	3.89	1.180	0.715	0.88
800	3.18	1.040	0.574	0.724
1000	2.72	0.888	0.52	0.622

Source: Adapted from Lindsay, B. G. et al., *J. Chem. Phys.*, 121, 1350, 2004.

Note: Labels A, B, C, and so on, correspond to those in Figure 50.4.

TABLE 50.9
Cross Sections for Ion Pair Generation

Energy (eV)	**Q (CCl_2^+, Cl^+) (10^{-21} m^2)**	**Q (CCl^+, Cl^+) (10^{-21} m^2)**
50	2.6	3.7
60	3.1	4.7
80	3.0	5.3
100	2.8	5.5
125	2.9	5.9
150	2.5	5.3
200	1.9	4.0
300	1.5	2.8
400	1.4	2.6
500	1.0	2.2
600	0.79	2.3
800		1.0
1000		1.4

50.6 ION PAIR PRODUCTION

Ion pairs are produced according to the following reactions (Lindsay et al., 2004):

$$e + CCl_4 \rightarrow CCl_2^+ + Cl^+ \quad (50.1)$$

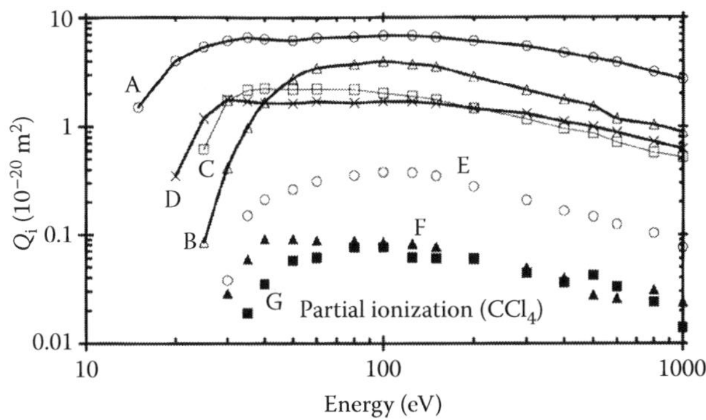

FIGURE 50.4 Partial ionization cross sections for CCl_4. A-CCl_3F; B-Cl^+; C-CCl^+; D-CCl_2^+; E-C^+; F-Cl_2^+. Note the absence of parent ion CCl_4^+. (Adapted from Lindsay, B. G. et al., *J. Chem. Phys.*, 121, 1350, 2004.)

TABLE 50.10
Attachment Rates for CCl_4 at Room Temperature

k_a (m^3/s)	T (K)	Method	Reference
3.79×10^{-13}	300	LPA	Klar et al. (2001)
3.6×10^{-13}	293	EPR spectrometer	Burns et al. (1996)
4.0×10^{-13}	~300	Microwave heating	Shimamori et al. (1992)
3.7×10^{-13}	~300	ECR	Marotta et al. (1989)
3.79×10^{-13}	294	Cavalleri sampling	Orient et al. (1989)
4.0×10^{-13}	300	FA/LP	Adams et al. (1988)
3.9×10^{-13}	300	FA/LP	Smith et al. (1984)
4.4×10^{-13}	293	PS	Ayala et al. (1981)
2.5×10^{-13}	293	Electron swarm	Christophorou et al. (1981)
3.5×10^{-13}	~300	Electron swarm	Davis et al. (1973)
4.1×10^{-13}	293	ECR	Mothes et al. (1972)
2.8×10^{-13}	293	Electron swarm	Christodoulides and Christophorou (1971)
3.6×10^{-13}	293	Electron swarm	Davis et al. (1971)
2.77×10^{-13}	~300	Microwave conductivity	Fessenden and Bansal (1970)
2.8×10^{-13}	293	Electron swarm	Blaunstein and Christophorou (1968)

Note: ECR = electron cyclotron resonance; EPR = electron paramagnetic resonance; FA/LP = flowing afterglow/Langmuir probe; LPA = laser photoelectron attachment; PS = pulse sampling.

$$e + CCl_4 \rightarrow CCl^+ + Cl^+ \quad (50.2)$$

The cross sections for generation of ion pairs are given by Lindsay et al. (2004) as shown in Table 50.9.

50.7 ATTACHMENT RATES

Dissociative attachment in CCl_4 occurs in two steps: formation of the temporary negative ion followed by the dissociation of the temporary negative ion, according to the reaction (Matejcik et al., 1995)

$$CCl_4 + e \rightarrow CCl_4^{-*} \rightarrow CCl_3 + Cl^- + 0.45\,eV \quad (50.3)$$

Table 50.10 shows the attachment rates obtained by various methods at room temperature. There is reasonable agreement between beam and swarm studies. Table 50.11 shows the attachment rates as a function of the mean energy of the electron swarm. Figure 50.5 is a graphical presentation of these data and other selected studies. The highlight of the variation of the attachment rate is that it increases with

TABLE 50.11
Attachment Rates in CCl_4 as a Function of Mean Energy at Room Temperature

Mean Energy (eV)	Rate (10^{-15} m^3/s)	Mean Energy (eV)	Rate (10^{-15} m^3/s)
0.04	400	0.5	52.7
0.05	304	0.6	31.6
0.06	282	0.7	26.0
0.08	289	0.8	20.2
0.1	212	0.9	17.6
0.15	143	1.0	17.8
0.2	97.9	1.1	19.2
0.25	84.9	1.2	19.6
0.3	66.5	1.3	17.4
0.4	59.8		

Source: Adapted from Shimamori, H., et al., *J. Chem. Phys.*, 97, 6335, 1992.

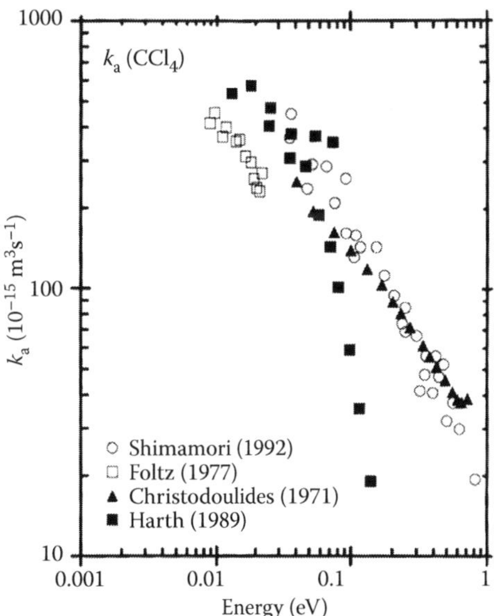

FIGURE 50.5 Selected attachment rate cross sections as a function of electron mean energy for CCl_4. (○) Shimamori et al. (1992); (■) Harth et al. (1989); (□) Foltz et al. (1977); (▲) Christodoulides and Christophorou (1971).

CCl$_4$

TABLE 50.12
Attachment Rates as Function of Temperature for CCl$_4$

Temperature (K)	Attachment Rate (m^3/s)	Reference
293	3.6×10^{-13}	Burns et al. (1996)
467	2.1×10^{-13}	
579	1.4×10^{-13}	
777	1.2×10^{-13}	
294	3.79×10^{-13}	Orient et al. (1989)
400	2.96×10^{-13}	
500	2.33×10^{-13}	
205	4.1×10^{-13}	Smith et al. (1984)
300	3.9×10^{-13}	
455	3.7×10^{-13}	
590	3.5×10^{-13}	

TABLE 50.13
Attachment Cross Sections in CCl$_4$ at Low Energies

Energy (μeV)	Q_a (10^{-15} m^2)	Energy (μeV)	Q_a (10^{-15} m^2)	Energy (μeV)	Q_a (10^{-15} m^2)
1	1.73	50	0.24	2×10^3	0.03
2	1.33	70	0.20	4×10^3	0.02
4	0.86	100	0.17	6×10^3	0.014
5	0.74	200	0.12	7×10^3	0.010
7	0.63	400	0.08	1×10^4	0.007
10	0.56	500	0.07	1×10^5	0.0006
20	0.41	700	0.06		
40	0.27	10^3	0.04		

Source: Adapted from Dunning, F. B., *J. Phys. B: At. Mol. Opt. Phys.*, 28, 1645, 1995.

Note: Digitized and interpolated at energy intervals as shown. See Figure 50.7.

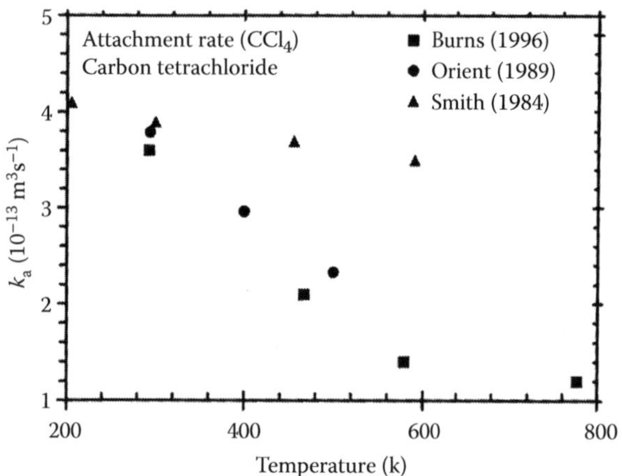

FIGURE 50.6 Attachment rates as a function of temperature in CCl$_4$. (■) Burns et al. (1996); (●) Orient et al. (1989); (▲) Smith et al. (1984).

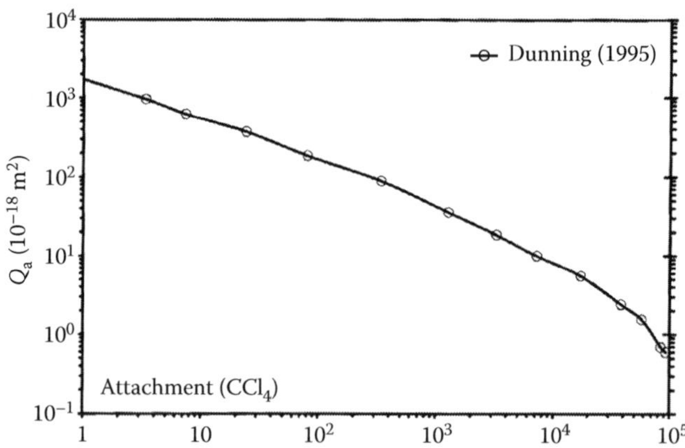

FIGURE 50.7 Attachment cross sections in CCl$_4$, digitized and interpolated at energy intervals as shown from Dunning (1995). Tabulated values are shown in Table 50.13.

decreasing mean energy, essentially due to increase in dissociative attachment cross section as the energy tends to zero (see Section 50.8).

Table 50.12 and Figure 50.6 show selected attachment rates as a function of temperature in CCl$_4$.

50.8 ATTACHMENT CROSS SECTIONS

See Table 50.13. Dissociative attachment cross sections up to 2 eV are shown in Table 50.14 and Figure 50.8 (Klar et al., 2001). The highlights of the figure are

1. There are two maxima of the dissociative attachment cross section: a major peak at 0.3 meV and a much smaller one at 0.8 eV.
2. According to Klar et al. (2001) the apparent peak position of zero energy peak, the apparent width of zero energy peak, and the apparent cross section at

TABLE 50.14
Attachment Cross Sections in CCl$_4$

Energy (meV)	Q_a (10^{-20} m^2)	Energy (meV)	Q_a (10^{-20} m^2)	Energy (meV)	Q_a (10^{-20} m^2)
0.1	18695	6.0	1297	200	12.00
0.2	13072	8.0	1061	400	2.00
0.4	8723	10	948.0	600	4.00
0.6	6690	20	463.0	800	5.00
0.8	5420	40	256.0	1000	3.00
1.0	4748	60	122.0	1500	0.20
2.0	3050	80	89.00	2000	0.01
4.0	1832	100	55.00		

Source: Digitized from Klar, D., M.-W. Ruf, and H. Hotop, *Int. J. Mass Spectrom.*, 205, 93, 2001.

Note: See Figure 50.8.

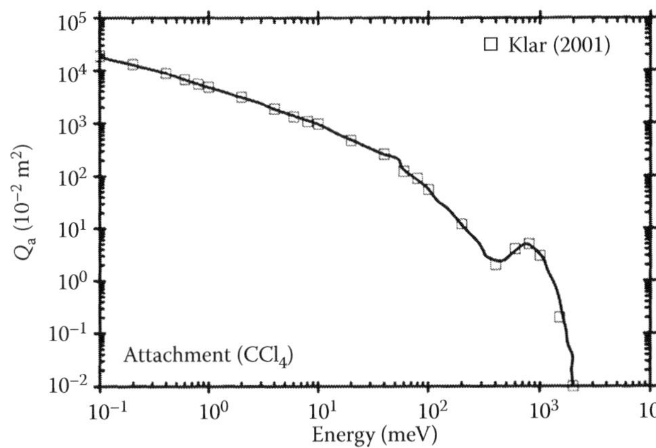

FIGURE 50.8 Dissociative attachment cross sections in CCl_4. Tabulated values are given in Table 50.14. (Adapted from Klar, D., M.-W. Ruf, and H. Hotop, *Int. J. Mass Spectrom.*, 205, 93, 2001.)

TABLE 50.15
Effects of Energy Resolution on the True Cross Section for Attachment for CCl_4

Experimental Energy Resolution (meV)	Apparent Peak Position of Zero Energy Peak (meV)	Apparent Width of Zero Energy Peak (meV)	Apparent Cross Section of Zero Energy Peak (10^{-20} m^2)
1	0.3	1.7	8800
6	1.5	8.5	3106
10	2.3	13.7	2240
20	4.1	26	1413
30	5.6	38	1058
50	8.5	61	730
70	11.0	82	567
100	13.5	113	425
150	17	163	302
200	19	213	236

zero energy depend on the energy width. Table 50.15 shows the variations in the energy width range 1–200 meV, partly explaining the differences in the zero energy maximum.

Attachment cross sections for CCl_4, as in SF_6, are nearly independent of the gas temperature.

REFERENCES

Adams, N. G., D. Smith, and C. R. Herd, *Int. J. Mass Spectrom. Ion Proc.*, 84, 243, 1988.

Ayala, J. A., W. E. Wentworth, and E. C. M. Chen, *J. Phys. Chem.*, 85, 3989, 1981.

Blaunstein, R. P. and L. G. Christophorou, *J. Chem. Phys.*, 49, 1526, 1968.

Boltz, G. W., C. J. Latimer, G. F. Hilderbrand, F. G. Kellert, K. A. Smith, W. P. West, F. B. Dunning, and R. F. Stebbings, *J. Chem. Phys.*, 67, 1352, 1977.

Burns, S. J., J. M. Matthews, and D. L. McFadden, *J. Phys. Chem.*, 100, 19437, 1996.

Christodoulides, A. A. and L. G. Christophorou, *J. Chem. Phys.*, 54, 4691, 1971.

Christophorou, L. G., D. L. McCorkle, and V. E. Anderson, *J. Phys. B: At. Mol. Phys.*, 4, 1163, 1971.

Christophorou, L. G., R. A. Mathis, D. R. James, and L. McCorkle, *J. Phys. D: Appl. Phys.*, 14, 1889, 1981.

Chutjian, A. and S. H. Alajajian, *Phys. Rev. A*, 31, 2885, 1985.

Daimon, H., T. Kondow, and K. Kuchitsu, *J. Phys. Soc. Jpn.*, 52, 84, 1983, cited by Zecca et al., *Phys. Rev. A*, 46, 3877, 1992.

Davis, F. J., R. N. Compton, and D. R. Nelson, *J. Chem. Phys.*, 59, 2324, 1971.

Davis, F. J., R. N. Compton, and D. R. Nelson, *J. Chem. Phys.*, 59, 2324, 1973.

Dunning, F. B., *J. Phys. B: At. Mol. Opt. Phys.*, 28, 1645, 1995.

Fessenden, R. W. and K. M. Bansal, *J. Chem. Phys.*, 53, 3468, 1970.

Hamada, A. and O. Sueoka, *Appl. Sur. Sci.*, 85, 64, 1995.

Harth, K., M.-W. Ruf, and H. Hotop, *Z. Phys. D.*, 14, 149, 1989, cited by Shimamori et al., 1992.

Hudson, J. E., C. Vallance, M. Bart, and P. W. Harland, *J. Phys. B: At. Mol. Opt. Phys.*, 34, 3025, 2001.

Jones, R. K., *J. Chem. Phys.*, 84, 813, 1986.

Klar, D., M.-W. Ruf, and H. Hotop, *Int. J. Mass Spectrom.*, 205, 93, 2001.

Leiter, K., K. Stephan, E. Märk, and T. D. Märk, *Plasma Chem. Plasma Proc.*, 4, 235, 1984.

Lindsay, B. G., K. F. Mcdonald, W. S. Yu, and R. F. Stebbings, *J. Chem. Phys.*, 121, 1350, 2004.

Ling, X., B. G. Lindsay, K. A. Smith, and Dunning, F. B., *Phys. Rev. A*, 45, 242, 1992.

Marotta, C. J., C-p. Tsai, and D. L. M. McFadden, *J. Chem. Phys.*, 91, 2194, 1989.

Matejčik, Š., V. Fotlin, M. Stano, and J. D. Skalný, *Int. J. Mass Spectromet.*, 223–224, 9, 2003.

Matejcik, S., A. Kiendler, A. Stamatovic, and T. D. Märk, *Int. J. Mass Spectr. Ion Proc.*, 149/150, 311, 1995.

Mothes, K. G., E. Schultes, and R. N. Schlinder, *J. Phys. Chem.*, 76, 3758, 1972.

Natalense, A. P. P., M. H. F. Bettega, L. G. Ferriera, and M. A. P. Lima, *Phys. Rev. A*, 52, R1, 1995.

Orient, O. J., A. Chutjian, R. W. Crompton, and B. Cheung, *Phys. Rev.*, 39, 4494, 1989.

Randell, J., J.-P. Ziesel, S. L. Lunt, G. Mrotzek, and D. Field, *J. Phys. B: At. Mol. Opt. Phys.*, 26, 3423, 1993.

Shimamori, H., Y. Tatsumi, Y. Ogawa, and T. Sunagawa, *J. Chem. Phys.*, 97, 6335, 1992.

Shimanouchi, T., *Tables of Molecular Vibrational Frequencies Consolidated Volume I*, NSRDS-NBS 39, Washington (DC), 1972.

Sierra, B., R. Martínez, C. Redondo, and F. Castaño, *Int. J. Mass Spectromet.*, 246, 105, 2005.

Smith, D., N. G. Adams, and E. J. Alge, *J. Phys. B: At. Mol. Phys.*, 17, 461, 1984.

Spanel, P., S. Matejcik, and D. Smith, *J. Phys B: At. Mol. Opt. Phys.*, 28, 2941, 1995.

Wan, H., J. H. Moore, and J. A. Tossell, *J. Chem. Phys.*, *94*, 1868, 1991.

Zecca, A., G. Karwasz, R. S. Brusa, and C. Szmytkowski, *J. Phys. B: At. Mol. Phys.*, 24, 2747, 1991.

Zecca, A., G. P. Karwasz, and R. S. Brusa, *Phys. Rev.*, 46, 3877, 1992.

51

CHLORODIBROMO-METHANE

CONTENTS

Chlorodibromomethane ($CHBr_2Cl$) is a polar, electron-attaching gas that has 95 electrons. The ionization potential is 10.59 eV.

51.1 ATTACHMENT RATE COEFFICIENTS

Dissociative attachment occurs through the formation of Br^- and Cl^- ions according to the following reactions (Ipolyi et al., 2004):

$$CHBr_2Cl + e\ (0\ eV) \rightarrow CHBrCl + Br + 0.68\ eV \quad (51.1)$$

$$CHBr_2Cl + e\ (0\ eV) \rightarrow CHBr_2 + Cl^- + 0.27\ eV \quad (51.2)$$

Br^- ions are relatively more abundant than Cl^- ions at room temperature. In addition, the reaction observed in swarm experiments only is

$$CHBr_2Cl + e\ (0\ eV) \rightarrow CHBr + CHBr^- \quad (51.3)$$

Attachment rate coefficients for $CHBr_2Cl$ are shown in Table 51.1. The attachment rates as a function of electron mean energy are shown in Figure 51.1.

TABLE 51.1
Attachment Rate Coefficients for $CHBr_2Cl$

Method	Temperature, K, (eV)	Rate (10^{-14} m^3/s)	Reference
Beam	321	2.7	Ipolyi et al. (2004)
FA/LP	300	3.0 (total)	Španěl and Smith (2001)
		2.8 (Br^-)	
		0.14 (Cl^-)	
	540	4.9 (total)	
		2.8 (Br^-)	
		1.3 (Cl^-)	
PR/MW	293	12.0	Sunagawa and Shimamori (1997)

Note: FA/LP = flowing afterglow/Langmuir probe; PR/MW = pulse radiolysis/microwave heating.

51.2 ATTACHMENT CROSS SECTIONS

Figure 51.2 shows the cross sections measured in beam experiments (Ipolyi et al., 2004) for ion formation. The resonance peaks that occur at ~0.4 and 4.7 eV are also shown.

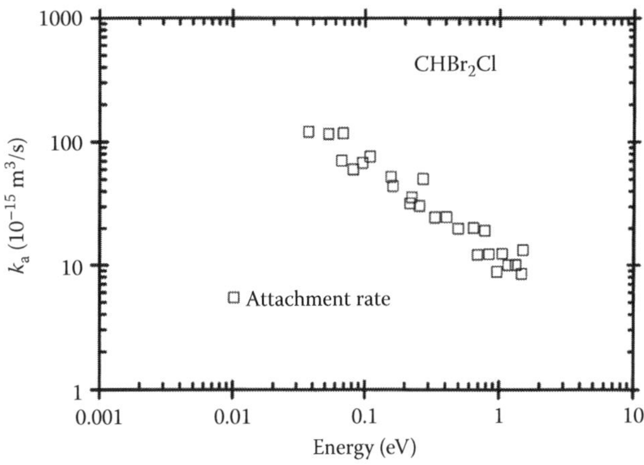

FIGURE 51.1 Attachment rate constants for $CHBr_2Cl$. (Adapted from Sunagawa, T. and H. Shimamori, *J. Chem. Phys.*, 107, 7876, 1997.)

$CHBr_2Cl$

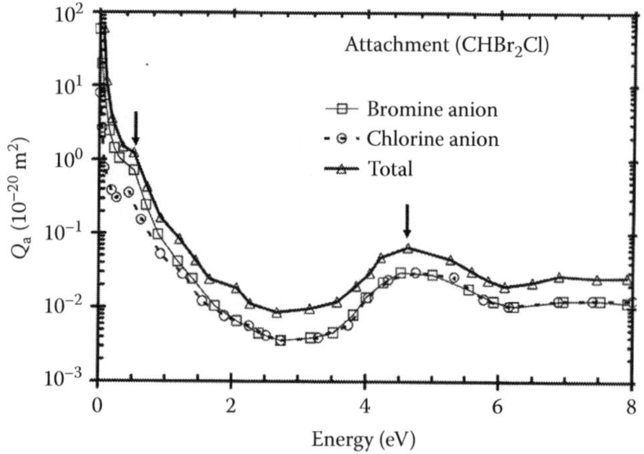

FIGURE 51.2 Attachment cross sections for $CHBr_2Cl$. The arrows indicate resonances at ~0.4 and ~4.7 eV. (Adapted from Ipolyi, I. et al., *Int. J. Mass. Spectrom.*, 233, 193, 2004.)

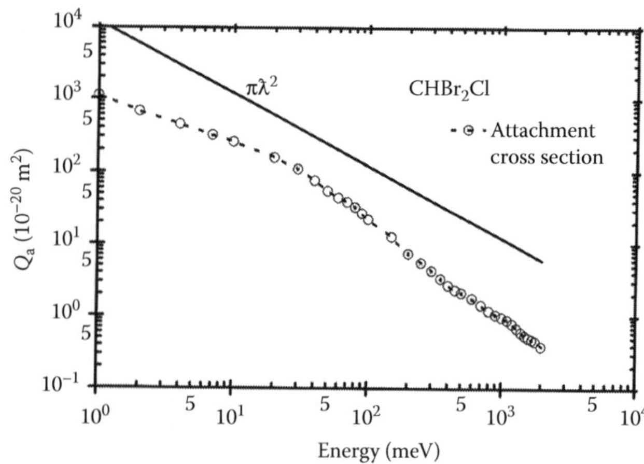

FIGURE 51.3 Attachment cross sections for $CHBr_2Cl$. *s*-Wave maximum cross section is also shown for comparison. (Adapted from T. Sunagawa and H. Shimamori, *J. Chem. Phys.*, 107, 7876, 1997.)

TABLE 51.2
Attachment Cross Sections for $CHBr_2Cl$

Energy (meV)	Q_a (10^{-20} m^2)	Energy (meV)	Q_a (10^{-20} m^2)
1.00	1100	350	3.30
2.00	667	400	2.66
4.00	448	450	2.30
7.00	320	500	2.10
10.0	257	600	1.77
20.0	156	700	1.43
30.0	109	800	1.19
40.0	75.9	900	1.05
50.0	54.5	1000	0.96
60.0	44.1	1100	0.89
70.0	38.4	1200	0.79
80.0	32.6	1300	0.68
90.0	26.7	1400	0.60
100	22.1	1500	0.54
150	12.5	1600	0.51
200	7.31	1700	0.48
250	5.46	1800	0.45
300	4.26	2000	0.39

Source: Digitized from Sunagawa, T. and H. Shimamori, *J. Chem. Phys.*, 107, 7876, 1997.

Table 51.2 and Figure 51.3 show the attachment cross sections deconvoluted from measured attachment rates. The *s*-wave maximum cross section is also shown for comparison. The attachment cross section, in common with other bromine containing targets, shows a peak toward zero energy. Data on species of ions produced and their relative abundance are not available.

REFERENCES

Ipolyi, I., S. Matejcik, P. Lukac, J. D. Skalny, P. Mach, and J. Urban, *Int. J. Mass. Spectrom.*, 233, 193, 2004.

Španěl, P. and D. Smith, *Int. J. Mass Spectrom.*, 205, 243, 2001.

Sunagawa, T. and H. Shimamori, *J. Chem. Phys.*, 107, 7876, 1997.

52

CHLOROMETHANE

CONTENTS

Chloromethane (CH_3Cl), synonym methyl chloride, with 26 electrons is a molecule with one of the hydrogen atoms in methane replaced with a chlorine atom. The substitution renders the molecule both dipolar (dipole moment = 1.896 D; electronic polarizability = 5.25, 5.95×10^{-40} F m^2) and electron attaching. The electron affinity (or vertical-attachment energy) of mono-substituted methanes are shown in Table 52.1.

The six vibrational modes and energies are shown in Table 52.2 (Shimanouchi, 1972). The vibrational modes of deuterated CH_3Cl (see Shimanouchi, 1972) are also included for comparison sake.

TABLE 52.1
Attachment Energy in Mono-Substituted Methanes

Gas	CH_3Cl	CH_3Br	CH_3I
Energy (eV)	−3.4	−0.46	0.31

Source: Adapted from Christophorou, L. G. and D. Hadjiantoniou, *Chem. Phys. Lett.*, 419, 405, 2006.

TABLE 52.2
Vibrational Modes and Energies of CH_3Cl

Chloromethane (CH_3Cl)			Chloromethane-d$_3$ (CD_3Cl)		
Designation	**Deformation**	**Energy (meV)**	**Designation**	**Deformation**	**Energy (meV)**
ν_1	CH_3 s-stretch	364.1	ν_1	CD_3 s-stretch	267.8
ν_2	CH_3 s-deform	168.0	ν_2	CD_3 s-deform	127.6
ν_3	C-Cl stretch	90.8	ν_3	C-Cl stretch	86.9
ν_4	CH_3 d-stretch	376.8	ν_4	CD_3 d-stretch	283.1
ν_5	CH_3 d-deform	180.7	ν_5	CD_3 d-deform	131.5
ν_6	CH_3 rock	126.1	ν_6	CD_3 rock	95.2

Source: Adapted from Shimanouchi, T., *Tables of Vibrational Frequencies Consolidated Volume I*, NSRDS-NBS 39, Washington (DC) 1972.
Note: d = degenerate; s = symmetrical.

CH$_3$Cl

52.1 SELECTED REFERENCES FOR DATA

See Table 52.3.

52.2 TOTAL SCATTERING CROSS SECTION

The total scattering cross sections are shown in Tables 52.4 through 52.6 and Figure 52.1. The highlights of the total cross section are

1. A trend of increase toward zero energy due to the dipole moment.
2. A broad resonance peak centered at 3.5 eV attributed to vibrational excitation (Shi et al., 1992).
3. A less-pronounced peak at 11.0 eV that is common to most gases.
4. A monotonic decrease with increase of energy in the range $11.0 < \varepsilon < 600$ eV, that is also common with most gases.
5. The total cross section increases with the size of the molecule within the family of halomethanes, $CH_3I > CH_3Br > CH_3Cl$.

TABLE 52.3
Selected References (CH_3Cl)

Quantity	Energy Range (eV)	Reference
k_a	—	Christophorou and Hadjiantoniou (2006)
k_a	0–12	Pshenichnyuk et al. (2006)
k_a	—	Barszczewska et al. (2003)
Q_i	14–1000	Rejoub et al. (2002)
Q_a	0–5	Aflatooni and Burrow (2001)
Q_T	0.8–600	Kimura et al. (2001)
Q_T	75–4000	Karwasz et al. (1999)
Q_i	10.5–220	Vallance et al. (1997)
Q_v	0.5–8.0	Shi et al. (1996)
Q_T	0.5–200	Krzysztofowicz and Szmytkowski (1994)
Q_a	0–4	Pearl et al. (1995)
Q_v	1.5–5.0	Fabrikant (1994)
Q_a	0–9	Pearl and Burrow (1994)
Q_v	0.5–8.0	Shi et al. (1992)
Q_a, Q_v	0–20	Fabrikant (1991)
Q_a	0.2–12.0	Wan et al. (1991)
k_a	0–8	Datskos et al. (1990)
Q_a	0–10	Chu and Burrow (1990)
k_a	—	Petrović et al. (1989)
Q_T	0.5–8	Benitez et al. (1988)
—	—	Scheunemann et al. (1980)
Q_a	Thermal	Blaunstein and Christophorou (1968)

Note: k_a = attachment rate; Q_a = attachment cross section; Q_T = total; Q_i = ionization; Q_v = vibrational.

TABLE 52.4
Total Scattering Cross Section in CH_3Cl

Energy (eV)	Q_T (10^{-20} m^2)	Energy (eV)	Q_T (10^{-20} m^2)	Energy (eV)	Q_T (10^{-20} m^2)
0.8	36.0	6.5	35.6	22	31.2
1.0	33.4	7.0	36.6	25	31.2
1.2	31.9	7.5	36.7	30	27.1
1.4	30.5	8.0	37.5	40	23.4
1.6	31.1	8.5	37.7	50	21.8
1.8	30.6	9.0	38.4	60	19.8
2.0	29.8	9.5	38.0	70	18.0
2.2	29.6	10	38.1	80	17.1
2.5	29.8	11	39.6	90	16.7
2.8	29.7	12	37.8	100	15.2
3.1	32.3	13	37.4	120	14.3
3.4	33.8	14	36.4	150	12.2
3.7	34.4	15	35.7	200	10.9
4.0	34.9	16	35.0	250	9.5
4.5	34.9	17	34.1	300	8.9
5.0	35.2	18	33.5	400	7.4
5.5	34.9	19	32.7	500	6.4
6.0	35.5	20	32.2	600	5.3

Source: Adapted from Kimura, M. et al., *J. Chem. Phys.*, 115, 7442, 2001.

TABLE 52.5
Total Scattering Cross Section in CH_3Cl at High Energies

Energy (eV)	Q_T (10^{-20} m^2)	Energy (eV)	Q_T (10^{-20} m^2)	Energy (eV)	Q_T (10^{-20} m^2)
75	21.7	300	9.76	1500	2.73
80	20.7	350	8.8	1750	2.42
90	19.3	400	7.92	2000	2.19
100	18.3	450	7.28	2250	1.87
110	17.1	500	6.72	2500	1.71
125	16.3	600	6	2750	1.6
150	14.4	700	5.19	3000	1.48
175	13.3	800	4.62	3250	1.38
200	12.3	900	4.22	3480	1.28
225	11.7	1000	3.87	4000	1.16
250	11	1100	3.67		
275	10.3	1250	3.28		

Source: Adapted from Karwasz, G. P. et al., *Phys. Rev. A*, 59, 1341, 1999.

52.3 ELASTIC SCATTERING CROSS SECTION

The differential scattering cross sections depend on the angle of scattering and electron impact energy whose behavior is common to most gases. Table 52.7 and Figure 52.2 show the energy dependence at selected angles (Shi et al., 1996). Note

TABLE 52.6
Total Scattering Cross Section in CH_3Cl at Selected Intervals of Energy

Energy (eV)	Q_T (10^{-20} m^2)	Energy (eV)	Q_T (10^{-20} m^2)	Energy (eV)	Q_T (10^{-20} m^2)
0.3	58.2	10.5	32.4	45	22.0
0.5	46.0	12	32.0	50	20.6
0.9	35.4	14	30.9	60	19.3
1.5	31.7	16	39.7	70	18.1
2.0	31.7	18	28.6	80	17.0
2.5	32.6	20	28.0	90	16.3
3.0	33.4	25	26.1	100	15.5
4.0	34.7	30	24.4	140	12.8
5.0	34.5	35	23.3	180	11.0
7.5	35.4	40	22.5	200	10.2
				250	8.45

Source: Adapted from Krzysztofowicz, A. M. and C. Szmytkowski, *J. Phys. B: At. Mol. Phys.*, 28, 1593, 2005.

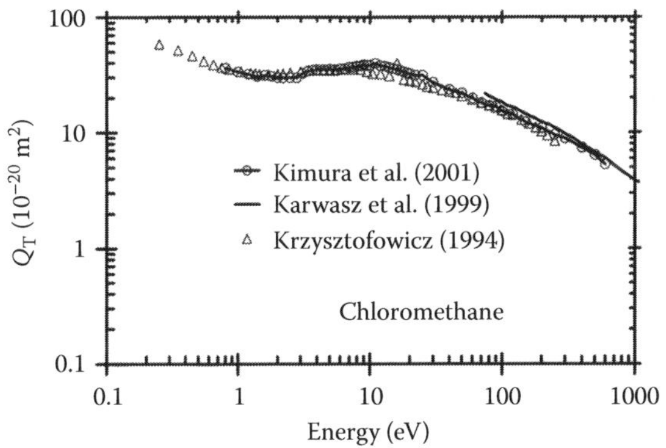

FIGURE 52.1 Total scattering cross sections for CH_3Cl. (—○—) KSMKK (Kimura et al., 2001); (—) KBPZ (Karwasz et al., 1999); (Δ) KS (Krzysztofowicz and Szmytkowski, 2005).

the peak at 3.6 eV for 100° scattering, attributed to the formation of the anion. This feature is nonexistent at 30° due to sharply rising contribution from the dipole interaction at lower energies and steadily increasing potential scattering, other than the dipole, at energies above resonance.

52.4 MOMENTUM TRANSFER CROSS SECTIONS

Theoretical values for the momentum transfer cross sections are given by Natalense et al. (1999) as shown in Table 52.8.

52.5 VIBRATIONAL EXCITATION CROSS SECTIONS

Fabrikant (1994) has calculated the vibrational excitation cross sections for the levels ν = 1–3, see Table 52.1. Figure 52.3 shows these data.

TABLE 52.7
Differential Scattering Cross Sections for CH_3Cl

Energy (eV)	Q_{diff} (10^{-20} m^2/sr) 30°	100°	Energy (eV)	Q_{diff} (10^{-20} m^2/sr) 30°	100°
0.5	19.32	1.89	4.5	4.70	1.79
0.8	12.25	1.43	5.0	5.11	1.62
1.0	9.51	1.31	5.2	5.14	1.54
1.5	7.02	1.38	5.4	5.27	1.49
2.0	5.56	1.52	5.6	5.39	1.44
2.5	4.87	1.64	5.8	5.40	1.37
3.0	4.72	1.85	6.0	5.50	1.34
3.2	4.53	1.97	6.5	5.74	1.19
3.4	4.63	1.92	7.0	5.86	1.16
3.6	4.83	1.93	8.0	6.10	1.08
4.0	4.73	1.87	9.0	6.57	1.05

Source: Adapted from Shi, X. et al., *J. Chem. Phys.*, 104, 1855, 1996.

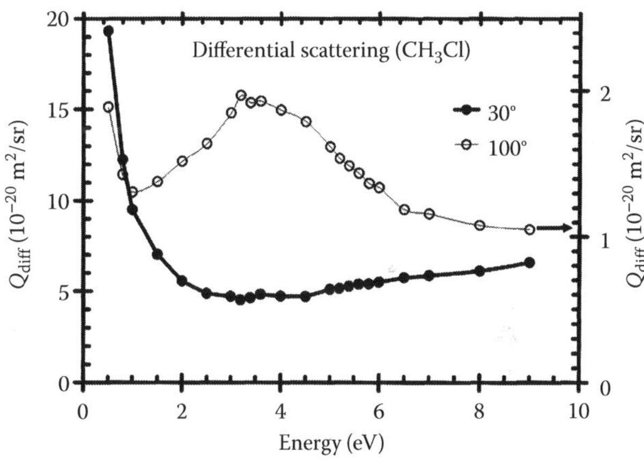

FIGURE 52.2 Differential scattering cross sections for CH_3Cl at 30° and 100° scattering angle. Note the resonance at ~3.6 eV and the secondary ordinate for 100° scattering cross section. (Adapted from Shi, X. et al., *J. Chem. Phys.*, 104, 1855, 1996.)

TABLE 52.8
Momentum Transfer Cross Sections

Energy (eV)	10	15	20	25	30
Q_M (10^{-20} m^2)	19.71	16.50	12.77	10.18	8.65

52.6 TOTAL IONIZATION CROSS SECTION

The ionization potential of CH_3Cl is 11.22 eV. Table 52.9 shows the total ionization cross section in CH_3Cl (see Vallance et al., 1997) and Figure 52.4 includes a graphical presentation.

CH$_3$Cl

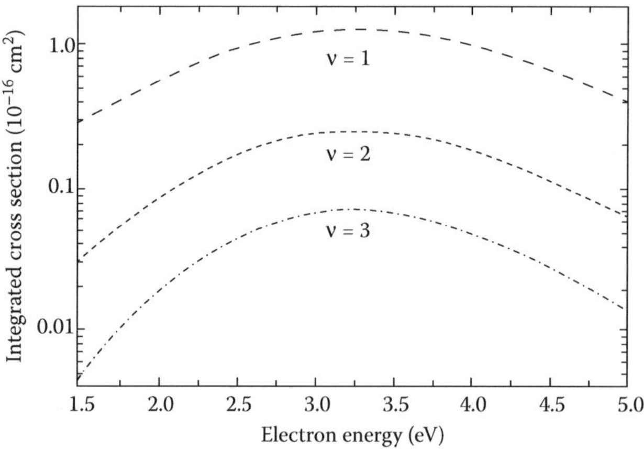

FIGURE 52.3 Total vibrational excitation cross sections for CH$_3$Cl. See Table 52.1 for the modes of vibration. (Adapted from Fabrikant, I. I., *J. Phys. B: At. Mol. Opt. Phys.*, 27, 4325, 1994.)

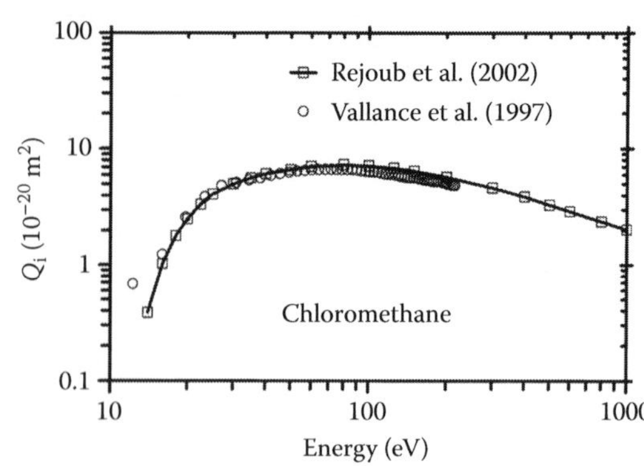

FIGURE 52.4 Total ionization cross sections for CH$_3$Cl. (—□—) Rejoub et al. (2002); (○) Vallance et al. (1997).

TABLE 52.9
Ionization Cross Section in CH$_3$Cl

Energy (eV)	Q_T (10^{-20} m^2)	Energy (eV)	Q_T (10^{-20} m^2)	Energy (eV)	Q_T (10^{-20} m^2)
12	0.68	82	6.68	153	5.66
16	1.23	86	6.65	156	5.59
20	2.61	90	6.59	160	5.51
23	3.91	94	6.50	164	5.46
27	4.85	97	6.42	167	5.41
31	5.17	101	6.39	171	5.36
34	5.43	105	6.36	175	5.30
38	5.68	108	6.31	179	5.26
42	5.93	112	6.22	182	5.26
45	6.16	116	6.14	186	5.26
49	6.33	119	6.09	190	5.23
53	6.48	123	6.06	193	5.17
57	6.57	127	6.01	197	5.10
60	6.65	130	5.92	201	5.03
64	6.68	134	5.83	204	4.98
68	6.70	138	5.75	208	4.94
71	6.70	142	5.71	212	4.90
75	6.68	145	5.69	215	4.89
79	6.66	149	5.69		

Source: Courtesy of Professor P. Harland, 2006.

52.7 POSITIVE ION APPEARANCE POTENTIALS

Positive ion appearance potentials are shown in Table 52.10.

52.8 PARTIAL IONIZATION CROSS SECTIONS

Absolute partial and total ionization cross sections measured by Rejoub et al. (2002) are shown in Table 52.11. Figure 52.5 also presents these data.

TABLE 52.10
Positive Ion Appearance Potentials (eV)

	Method		
Process	Photoionization[a]	Electron Impact[b]	Electron Impact[c]
$CH_3^+ + Cl^-$	<10.5		9.8
CH_3Cl^+	11.0	11.3	11.3
$CH_2Cl^+ + H$	13.0		
$CH_3^+ + Cl$	14.0	13.5	13.6
$CHCl^+ + H_2$	14.0		
$Cl^+ + C_3H$		16.6	16.6
$CH_2^+ + HCl$	15.5	15.3	
$CCl^+ + H_2 + H$	18.5		
$H^+ + HCl + CH$	21.5		
$CH^+ + Cl + 2H$	22.0	22.4	
$Cl^+ + CH_2 + H$	22		
$H_2^+ + Cl + C + H$	28		

[a] Olney et al. (1996).
[b] Branson and Smith (1953).
[c] Tsuda et al. (1964).

TABLE 52.11
Total and Partial Cross Sections for Selected Species (CH$_3$Cl)

Energy (eV)	Q_i (CH_nCl^+) (10^{-20} m^2)	Q_i (CH_n^+) (10^{-20} m^2)	Q_i (Total) (10^{-20} m^2)
14	0.385		0.385
16	1.03		1.03
18	1.45	0.344	1.794
20	1.81	0.698	2.508
22.5	2.24	1.12	3.36
25	2.51	1.58	4.0992
30	2.71	2.20	5.0177
35	2.90	2.49	5.6502
40	2.99	2.67	6.13
50	3.06	2.83	6.644
60	3.14	2.85	7.065
80	3.15	2.92	7.297

TABLE 52.11 (continued)
Total and Partial Cross Sections for Selected Species (CH_3Cl)

Energy (eV)	Q_i (CH_nCl^+) (10^{-20} m^2)	Q_i (CH_n^+) (10^{-20} m^2)	Q_i (Total) (10^{-20} m^2)
100	3.12	2.87	7.163
125	3.01	2.75	6.866
150	2.85	2.62	6.472
200	2.61	2.28	5.741
300	2.15	1.89	4.654
400	1.81	1.63	3.904
500	1.56	1.37	3.31
600	1.39	1.21	2.914
800	1.15	0.986	2.378
1000	0.983	0.844	2.0226

Note: $n = 0$–3.

52.9 ATTACHMENT RATE

Potential curves for a methyl chloride molecule and its negative ion are shown in Figure 52.6 (Barszczewska et al., 2003). The vertical attachment energy (VAE) for the molecule is 3.45 eV (see Aflatooni and Burrow, 2001).

Dissociative attachment occurs according to the reaction,

$$e + CH_3Cl \rightarrow CH_3 + Cl^- \quad (52.1)$$

The energy threshold for the process is −0.03 eV (see Petrovic et al., 1989).

Ion pair production occurs according to the reaction (Devins and Leblanc, 1960)

$$CH_3Cl + e \rightarrow CH_3^+ + Cl^- + e \quad (52.2)$$

CH$_3$Cl

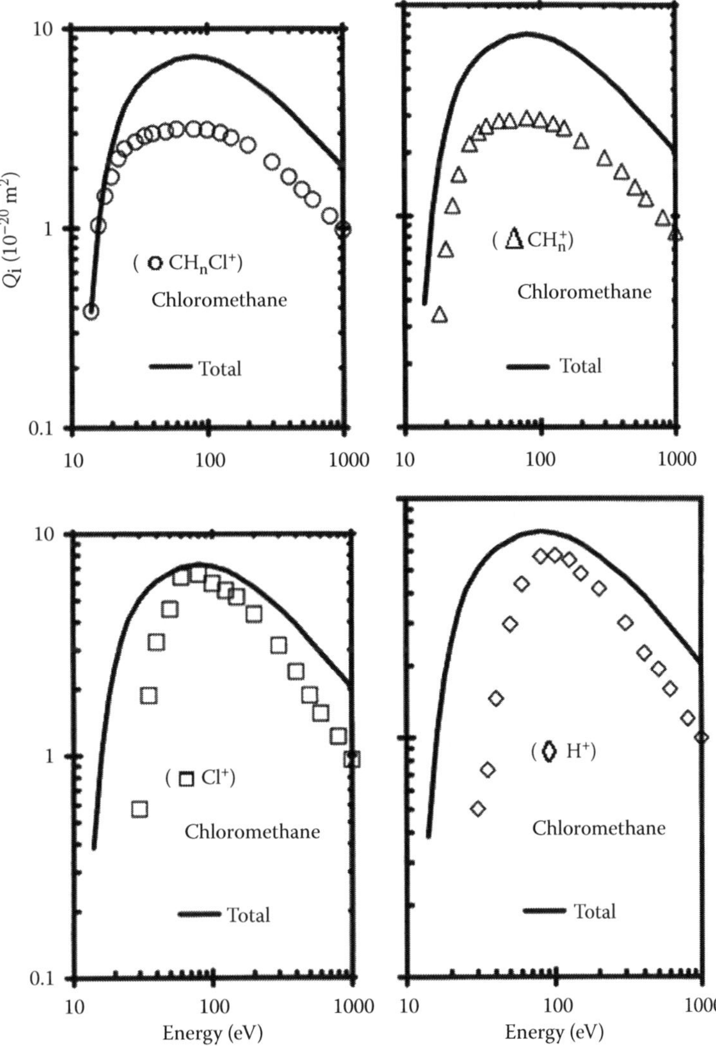

FIGURE 52.5 Partial ionization cross sections for CH_3Cl. Total (—) ionization cross sections have also been plotted to show the relative magnitude of the cross sections for dissociative fragments. (Adapted from Rejoub, R., B. G. Lindsay, and R. F. Stebbings, *J. Chem. Phys.*, 117, 6450, 2002.)

CH$_3$Cl

The peak cross section (5.5×10^{-24} m^2) for Reaction 52.1 occurs at an energy of 0.17 eV (see Petrovic et al., 1989).

Thermal attachment rates are shown in Table 52.12.

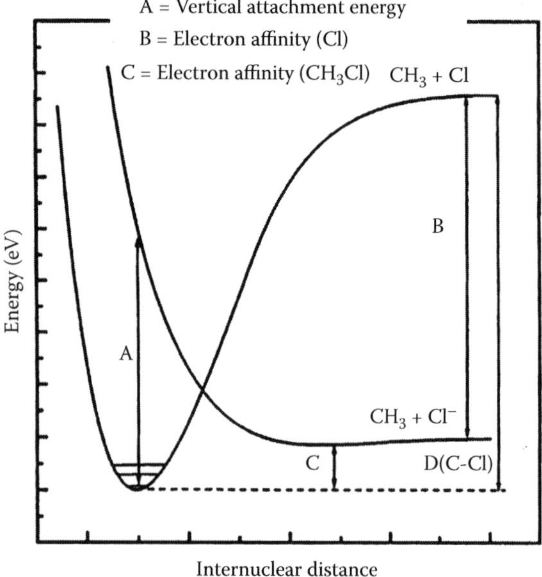

FIGURE 52.6 Potential energy for a molecule of CH$_3$Cl. VAE = vertical attachment energy; EA = electron affinity; D = dissociation energy. (Adapted from Barszczewska, W. et al., *J. Phys. Chem.*, 107, 11427, 2003.)

TABLE 52.12
Thermal Attachment Rate Constants (CH$_3$Cl)

Rate (m^3/s)	Method	Reference
1.0×10^{-19}	Swarm	Christophorou and Hadjiantoniou (2006)
2.0×10^{-19}	Swarm	Petrović et al. (1989)

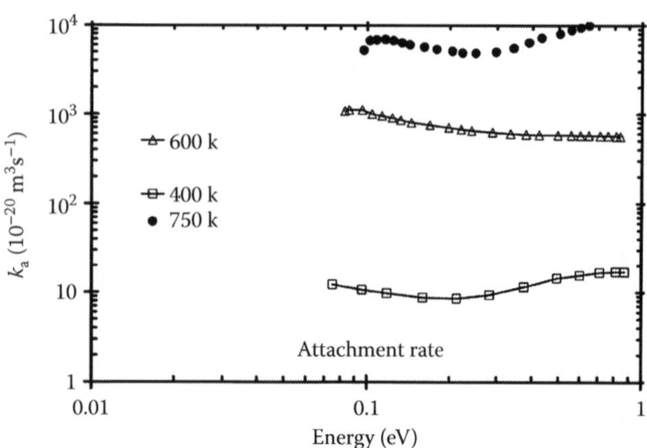

FIGURE 52.7 Attachment rate constants for CH$_3$Cl as a function of electron mean energy at various constant temperatures. Note the very large increase with temperature at the same mean energy. (Adapted from P. G. Datskos, L. G. Christophorou, and J. G. Carter, *Chem. Phys. Lett.*, 168, 324, 1990.)

Attachment rate constants as a function of mean energy increase very significantly with increasing temperature in the range from 300 to 750 K (see Datskos et al., 1990) as shown in Figure 52.7.

52.10 ATTACHMENT CROSS SECTIONS

Below 6 eV the negative ion generated is entirely Cl$^-$ (see Pearl and Burrow, 1994). The attachment cross section at 300 K is very small, ~4×10^{-27} m^2 for an energy of 0.6 eV (Pearl et al., 1995). The cross section increases with temperature as determined by Datskos et al. (1990) and as shown in Figure 52.8. The increase is attributed to thermal decomposition of the molecule (Pearl and Burrow, 1993).

52.11 GAS CONSTANTS

The gas constants in the expression

$$\frac{\alpha_{\text{eff}}}{N} = F \exp\left(-\frac{GN}{E}\right) \tag{52.3}$$

where α_{eff} is the effective ionization coefficient, E is the electric field, N is the gas number density, and F and G are the gas constants. Table 52.13 shows the constants in selected gases (Devins and Leblanc 1960).

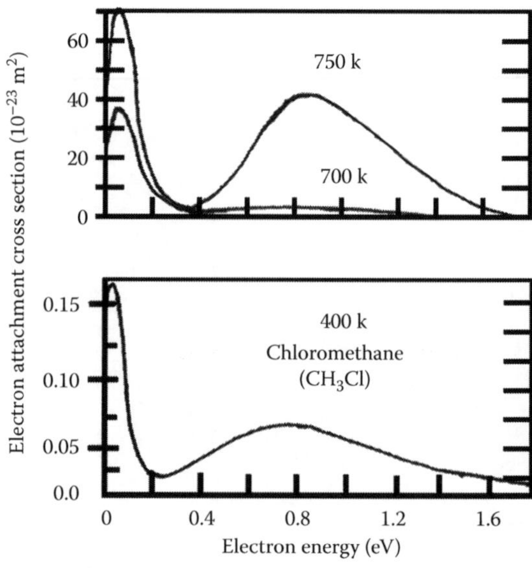

FIGURE 52.8 Swarm determined electron attachment cross section for CH$_3$Cl as a function of electron energy at various constant temperatures. Note the very large increase in cross section from 400 to 750 K. (Adapted from Datskos, P. G., L. G. Christophorou, and J. G. Carter, *Chem. Phys. Letters*, 168, 324, 1990.)

TABLE 52.13
Gas Constants in Selected Hydrocarbon Gases ($120 \leq E/N \leq 600$ Td)

Gas					
Name	**Formula**	**ε_i (eV)**	**F × 10^{-20} m^2**	**G(Td)$^{-1}$**	**Reference**
Methane	CH_4	13.04		353	Devins and Crowe (1956)
Ethane	C_2H_6	11.76		408	
Propane	C_3H_8	11.21		543	
Butane	C_4H_{10}	10.80		680	
Pentane	C_5H_{12}	10.55		768	
Hexane	C_6H_{14}	10.43		843	
2 Methyl propane		10.80		674	
2 Methyl butane		10.55		781	
2,2 Dimethyl propane		10.55		788	
Cyclopentane	C_5H_{10}	10.8		671	
Cyclohexane	C_6H_{12}	10.4		742	
Methyl chloride	CH_3Cl	11.28	4.16	756	Devins and Crowe (1956)
Chloroethane	C_2H_5Cl	10.97	5.00	1007	
Chloropropane	n-C_3H_7Cl	10.8	6.43	1203	
Chlorobutane	n-C_4H_9Cl	10.7	8.82	1460	
Chloropentane	n-$C_5H_{11}Cl$	10.6	12.55	1674	

REFERENCES

Aflatooni, K. and P. D. Burrow, *Int. J. Mass Spectrom.*, 205, 149, 2001.

Barszczewska, W., J. Kopyra, J. Wnorowska, and I. Szamrej, *J. Phys. Chem.*, 107, 11427, 2003.

Benitez, A., J. H. Moore, and J. A. Tassel, *J. Chem. Phys.* 88, 6691, 1988.

Blaunstein, R. P. and L. G. Christophorou, *J. Chem. Phys.*, 49, 1526, 1968.

Branson, H. and C. Smith, *J. Am. Chem. Soc.*, 75, 4133, 1953.

Christophorou, L. G. and D. Hadjiantoniou, *Chem. Phys. Lett.*, 419, 405, 2006.

Chu, S. C. and P. D. Burrow, *Chem. Phys. Lett.*, 172, 17, 1990.

Datskos, P. G., L. G. Christophorou, and J. G. Carter, *Chem. Phys. Lett.*, 168, 324, 1990.

Devins, J. C. and R. W. Crowe, *J. Chem. Phys.*, 25, 1053, 1956.

Devins, J. C. and O. H. Leblanc, Jr., *Nature*, 4735, 409, 1960.

Fabrikant, I. I., *J. Phys. B: At. Mol. Opt Phys.*, 24, 2213, 1991.

Fabrikant, I. I., *J. Phys. B: At. Mol. Opt. Phys.*, 27, 4325, 1994.

Karwasz, G. P., R. S. Brusa, A. Piazza, and A. Zecca, *Phys. Rev. A*, 59, 1341, 1999.

Kimura, M., O. Sueoka, C. Makochekanwa, H. Kawatw, and M. Kawada, *J. Chem. Phys.*, 115, 7442, 2001.

Krzysztofowicz, A. M. and C. Szmytkowski, *Chem. Phys. Lett.*, 219, 86, 1994.

Krzysztofowicz, A. M. and C. Szmytkowski, *J. Phys. B: At. Rcd. Phys.*, 28, 1593, 2005.

Olney, T. N., G. Cooper, W. F. Chan, G. R. Burton, C. E. Brion, and K. H. Tan, *Chem. Phys.*, 205, 421, 1996.

Pearl, D. M. and P. D. Burrow, *J. Chem. Phys.*, 101, 2940, 1994.

Pearl, D. M., P. D. Burrow, I. I. Fabrikant, and G. A. Gallup, *J. Chem. Phys.*, 102, 2737, 1995.

Petrović, Z. Lj., W. C. Wang, and L. C. Lee, *J. Chem. Phys.*, 90, 3145, 1989.

Pshenichnyuk, S. A., I. A. Pshenichnyuk, E. P. Nafikova, and N. L. Asfandiarov, *Rapid Commun. Mass Spectrom.*, 20, 1097, 2006.

Rejoub, R., B. G. Lindsay, and R. F. Stebbings, *J. Chem. Phys.*, 117, 6450, 2002.

Scheunemann, H.-U., E. Illenberger, and H. Baumgartel, *Ber Bunsenges, Phys. Chem.*, 84, 580, 1980.

Shi, X., T. M. Stephan, and P. D. Burrow, *J. Chem. Phys.*, 96, 4037, 1992.

Shi, X., V. K. Chan, G. A. Gallup, and P. D. Burrow, *J. Chem. Phys.*, 104, 1855, 1996.

Shimanouchi, T., *Tables of Vibrational Frequencies Consolidated Volume I*, NSRDS-NBS 39, 1972.

Tsuda, S., C. E. Melton, and W. H. Hamill, *J. Chem. Phys.*, 41, 689, 1964.

Vallance, C., S. A. Harris, J. E. Hudson, and P. W. Harland, *J. Phys. B: At. Mol. Opt. Phys.*, 30, 2465, 1997.

Wan, H., J. H. Moore, and J. A. Tossell, *J. Chem. Phys.*, 94, 1868, 1991.

Wataleuse, A. P. P., M. H. F. Bettega, L. G. Frriera, and M. A. P. Lima, *Phys. Rev. A.* 59, 879, 1999.

53

CHLOROTRIFLUOROMETHANE

CONTENTS

Chlorotrifluoromethane ($CClF_3$), also known as freon 13, is an electron-attaching gas that has 50 electrons and generally classified as chlorinated fluorocarbon (CFC). It has electronic polarizability of 6.364×10^{-40} F m^2, dipole moment of 0.5 D, and ionization potential of 13.1 eV (Irikura et al., 2003). The electronic excitation threshold is 7.1 eV (Doucet et al., 1973). The vibrational modes of the molecule are given in Table 53.1 (Mann and Linder, 1992).

TABLE 53.1
Vibrational Energies of $CClF_3$ Molecule

Mode	Nuclear Motion	Energy (meV)
ν_1	CF_3 symmetric stretch	137.4
ν_2	CF_3 symmetric deformation	97.2
ν_3	C–Cl symmetric stretch	59.1
ν_4	CF_3 asymmetric stretch	150.5
ν_5	CF_3 asymmetric deformation	69.7
ν_6	F_3–C–Cl bending	43.2

53.1 SELECTED REFERENCES

See Table 53.2.

TABLE 53.2
Selected References for $CClF_3$

Parameter	Range, eV (Td)	Reference
Q_i	20–85	**Sierra et al. (2004)**
Q_i	13–4000	**Irikura et al. (2003)**
Q_a	0–5	**Aflatooni and Burrow (2001)**
Q_i	12–216	**Bart et al. (2001)**
Q_T	0–2	**Field et al. (2001)**
Q_M	10–30	Natalense et al. (1999)
k_a	—	**Burns et al. (1996)**
Q_T	10–1000	**Jiang et al. (1995)**
Q_a	0–6	Underwood-Lemons et al. (1995)
Q_T	0–12	Underwood-Lemons et al. (1994)
Q_v	0–0.25	**Randell et al. (1993)**
Q_T	75–4000	**Zecca et al. (1992)**

continued

CClF$_3$

TABLE 53.2 (continued)
Selected References for CClF$_3$

Parameter	Range, eV (Td)	Reference
Q_T	0.6–50	**Jones (1986)**
k_a	0–5	**Spyrou and Christophorou (1985)**
Q_{el}	10–400	**Daimon et al. (1983)**
Q_a	0.01–0.8	**McCorkle et al. (1980, 1982)**
k_a	—	**Schumacher et al. (1978)**
Q_a	0–8	**Illenberger et al. (1979)**
k_a	Thermal	**Davis et al. (1973)**
k_a	Thermal	**Fessenden and Bansal (1970)**
Q_i	20, 35, 70	**Beran and Kevan (1969)**

Note: k_a = attachment rate; Q_a = attachment; Q_i = ionization; Q_M = momentum transfer; Q_T = total; Q_v = vibrational. Bold font indicates experimental study.

TABLE 53.3
Total Scattering Cross Sections for CClF$_3$

Energy (eV)	Q_T (10^{-20} m^2)	Energy (eV)	Q_T (10^{-20} m^2)	Energy (eV)	Q_T (10^{-20} m^2)
Jones (1986)				Zecca et al. (1992)	
0.6	17.67	3.9	24.26	75	24.0
0.7	17.83	4.0	24.82	80	23.5
0.8	17.81	4.2	25.94	90	22.5
0.9	17.72	4.4	27.07	100	22.7
1.0	18.03	4.6	28.51	110	22.0
1.1	18.93	4.8	30.05	125	20.8
1.2	19.74	5.0	31.24	150	19.3
1.3	21.09	5.2	32.21	175	18.1
1.4	22.93	5.4	32.98	200	17.0
1.5	25.04	5.6	33.35	225	16.0
1.6	27.20	5.8	33.68	250	15.0
1.7	29.42	6.0	33.66	275	14.4
1.8	30.67	6.2	33.42	300	13.9
1.9	31.62	6.4	33.26	350	12.4
2.0	31.77	6.6	33.28	400	11.5
2.1	31.49	6.8	32.98	450	10.7
2.2	30.80	7.0	32.88	500	9.92
2.3	29.62	7.5	32.98	600	8.90
2.4	28.49	8.0	33.62	700	7.98
2.5	27.18	8.5	34.13	800	7.27
2.6	26.27	9.0	34.39	900	6.79
2.7	25.22	9.5	34.58	1000	6.45
2.8	24.53	10.0	34.67	1100	6.05
2.9	23.93	10.5	34.86	1250	5.56
3.0	25.54	11.0	35.05	1500	4.86
3.1	23.39	11.5	35.46	1750	4.19
3.2	23.14	12.0	35.65	2000	3.72
3.3	23.15	12.5	35.51	2250	3.29
3.4	23.06	13.0	35.28	2500	3.05
3.5	23.27	13.5	34.76	2750	2.82
3.6	23.38	14.0	34.52	3000	2.63
3.7	23.75	14.5	33.84	3250	2.50
3.8	23.92	15.0	33.22	3500	2.35

TABLE 53.3 (continued)
Total Scattering Cross Sections for CClF$_3$

Energy (eV)	Q_T (10^{-20} m^2)	Energy (eV)	Q_T (10^{-20} m^2)	Energy (eV)	Q_T (10^{-20} m^2)
Jones (1986)				Zecca et al. (1992)	
16.0	32.27	25.0	29.06	4000	2.04
17.0	31.74	27.5	28.38		
18.0	30.82	30.0	27.65		
19.0	30.54	32.5	27.14		
20.0	29.97	35.0	26.45		
21.0	29.90	37.5	26.15		
22.0	29.57	40.0	25.70		
23.0	29.65	45.0	25.20		
24.0	29.31	50.0	24.34		

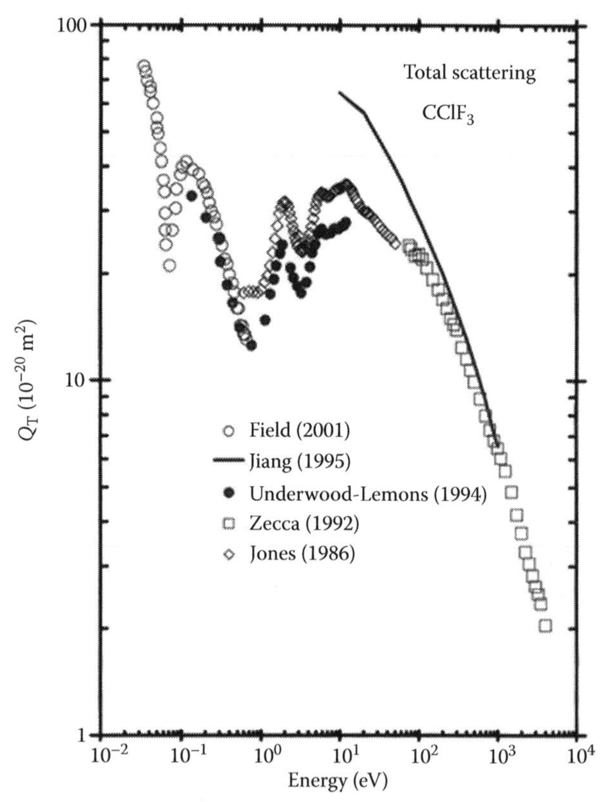

FIGURE 53.1 Total scattering cross sections for CClF$_3$. Note the Ramsauer–Townsend minimum at 0.1 eV. Other two minima at 1.4 and 4.4 eV are due to electron attachment. Cross sections except those from Jiang et al. (1995) are experimental. (○) Field et al. (2001); (——) Jiang et al. (1995); (•) Underwood-Lemons et al. (1994); (□) Zecca et al. (1992); (◊) Jones (1986).

53.2 TOTAL SCATTERING CROSS SECTIONS

Table 53.3 and Figure 53.1 show the total scattering cross sections in CClF$_3$. Figure 53.2 shows expanded views near the peaks in cross sections.

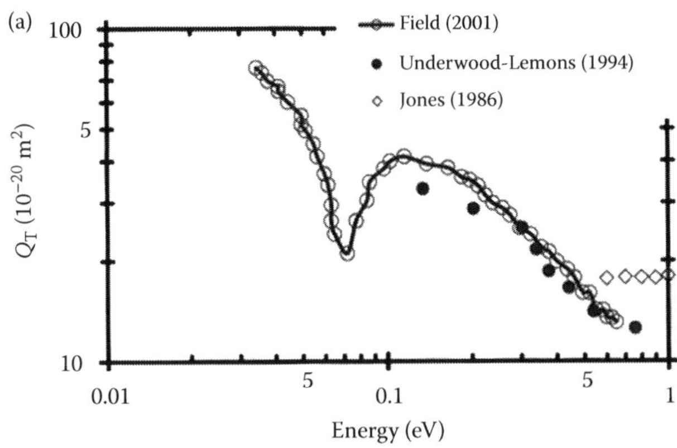

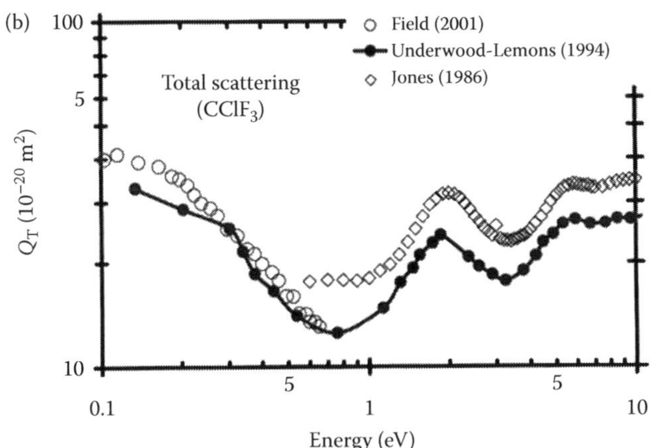

FIGURE 53.2 Total scattering cross sections for $CClF_3$. (a) details near Ramsauer–Townsend minimum. (b) details near 2.0 and 6 eV peaks due to electron attachment.

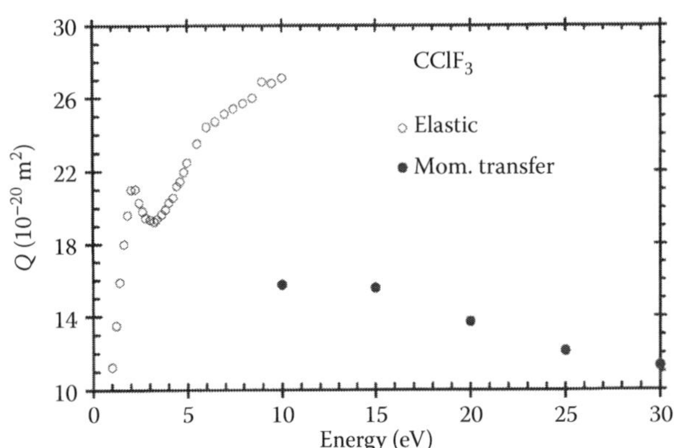

FIGURE 53.3 Elastic and momentum transfer cross sections for $CClF_3$. (○) Elastic, Mann and Linder (1992); (•) Natalense et al. (1999).

TABLE 53.4
Elastic Scattering Cross Sections for $CClF_3$

Energy (eV)	Q_{el} (10^{-20} m^2)	Energy (eV)	Q_{el} (10^{-20} m^2)	Energy (eV)	Q_{el} (10^{-20} m^2)
1.0	11.18	4.5	21.34	8.0	25.69
1.5	16.71	5.0	22.47	8.5	25.98
2.0	20.84	5.5	23.51	9.0	26.89
2.5	20.28	6.0	24.41	9.5	26.81
3.0	19.34	6.5	24.70	10.0	27.10
3.5	19.48	7.0	25.11		
4.0	20.29	7.5	25.40		

Source: Adapted from Mann, A. and F. Linder, *J. Phys. B: At. Mol. Opt. Phys.*, 25, 1621, 1992.

TABLE 53.5
Momentum Transfer Cross Sections ($CClF_3$)

Energy (eV)	10	15	20	25	30
Q_M (10^{-20} m^2)	15.72	15.55	13.70	12.12	11.31

Source: Adapted from Natalense, A. P. P. et al., *Phys. Rev.*, 59, 879, 1999.

The highlights of the cross sections are

1. There is a dominant peak at 2.0 eV that is attributed to electron attachment, possibly due to Cl^- ion (Illenberger et al., 1979).
2. A broad peak near 6 eV which is also attributed to dissociative attachment.
3. A Ramsauer–Townsend minimum at ~0.1 eV (Randell et al., 1993); the relatively less pronounced characteristic is attributed to the dipole moment of the molecule.
4. The cross section toward zero energy increases due to rotational excitation (Randell et al., 1993) reaching 80×10^{-20} m^2 at about 0.05 eV. Field et al. (2001) however suggest that rotational excitation is suppressed in the energy range, 50–100 meV. The dipole moment of the molecule is possibly effective in increasing the cross section toward zero energy.
5. At energies greater than 75 eV the cross section decreases monotonically, decreasing by a factor of 10. The monotonic decrease in the high-energy range is common to all gases, though the relative decrease is different.

53.3 ELASTIC SCATTERING CROSS SECTIONS

Elastic scattering cross sections measured by Mann and Linder (1992) are shown in Table 53.4 and Figure 53.3.

53.4 MOMENTUM TRANSFER CROSS SECTIONS

Momentum transfer cross sections calculated by Natalense et al. (1999) are shown in Table 53.5 and included in Figure 53.3.

53.5 ENERGY LOSS SPECTRUM

King and McConkey (1978) have observed peaks shown in Table 53.6 in electron energy loss spectroscopy.

TABLE 53.6
Energy Features in $CClF_3$ Spectrum

Energy (eV) (Gilbert et al. 1974)	Energy (eV) (King and McConkey, 1978)	Feature (King and McConkey, 1978)
9.68	9.68	Broad peak
10.59	10.65	Broad peak
11.58	11.61	Broad peak
	12.10	Shoulder
	12.55	Shoulder
	12.91	Shoulder
	13.30	Peak
13.48	13.42	Shoulder
	14.13	Shoulder
14.75	14.83	Shoulder
16.11	16.29	Broad peak
	17.1	Broad peak
	18.2	Broad peak
	18.7	Broad peak
	22.3	Broad peak

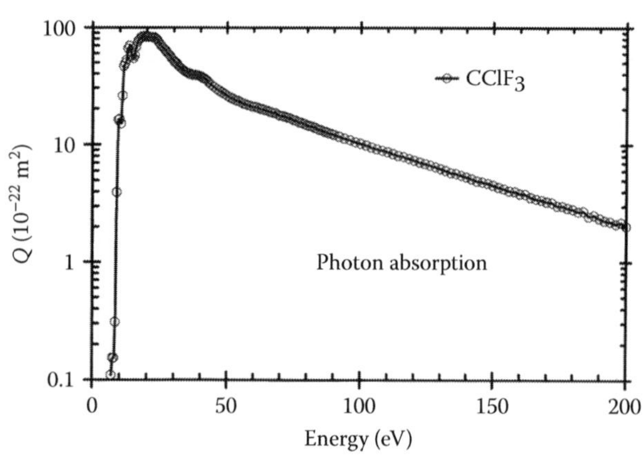

FIGURE 53.4 Photon absorption cross sections in $CClF_3$. The threshold is 7.0 eV. (Adapted from Au, J. W. *Chem. Phys.* 221, 181, 1997.)

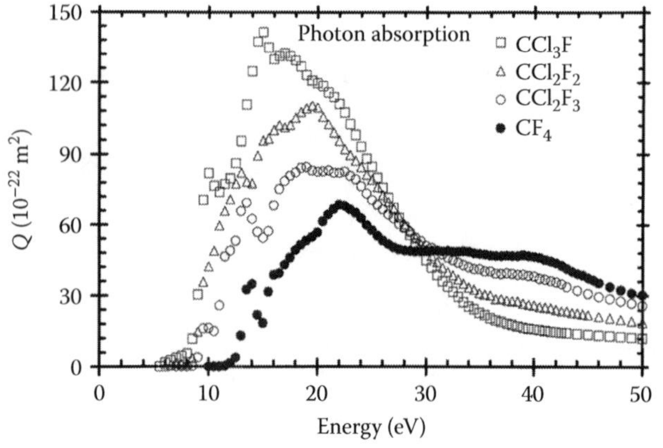

FIGURE 53.5 Photon absorption cross sections in fluorocarbons. Note the increase of absorption cross section peak as the number of chlorine atoms replacing fluorine increases. (Adapted from Au, J. W. *Chem. Phys.* 221, 181, 1997.)

53.6 ELECTRONIC EXCITATION CROSS SECTIONS

Au et al. (1997) have measured the photoabsorption cross sections in the photon energy range 5–200 eV as shown in Figures 53.4 and 53.5. Data for CCl_3F, CCl_2F_2, and CF_4 are also added for comparison.

53.7 POSITIVE ION APPEARANCE POTENTIALS

Ajello et al. (1976) have determined the positive ion appearance potentials by photoionization method and the appearance potentials are shown in Table 53.7.

53.8 TOTAL IONIZATION CROSS SECTIONS

Total ionization cross sections measured by Bart et al. (2001) and Sierra et al. (2004) are shown in Table 53.8 and Figure 53.6.

TABLE 53.7
Positive Ion Appearance Potentials

Ion	Process	Potential (eV)
CF_3Cl^+	$CF_3Cl + h\nu \rightarrow CF_3Cl^+ + e$	12.39
CF_3^+	$CF_3Cl + h\nu \rightarrow CF_3^+ + Cl + e$	12.65
CF_2^+	$CF_3Cl + h\nu \rightarrow CF_2^+ + F + Cl + e$	18.84

Source: Adapted from Ajello, J. M., W. T. Huntress, and P. Rayermann, *J. Chem. Phys.*, 64, 4746, 1976.

TABLE 53.8
Total Ionization Cross Sections for $CClF_3$

Bart et al. (2001)				Sierra et al. (2004)	
Energy (eV)	Q_i (10^{-20} m^2)	Energy (eV)	Q_i (10^{-20} m^2)	Energy (eV)	Q_i (10^{-20} m^2)
12	0.23	79	6.86	20	1.646
16	0.67	82	6.90	25	2.442
20	1.41	86	6.92	30	3.867
23	2.27	90	6.93	35	4.799
27	3.09	94	6.93	40	5.421
31	3.71	97	6.92	45	5.893
34	4.24	101	6.90	50	6.206
38	4.69	105	6.88	55	6.475
42	5.11	108	6.86	60	6.601
45	5.48	112	6.84	65	6.649
49	5.75	116	6.82	70	6.87
53	6.00	119	6.81	75	6.994
57	6.18	123	6.80	80	7.063
60	6.37	127	6.78	85	7.098
64	6.49	130	6.74		
68	6.60	134	6.68		
71	6.71	138	6.63		
75	6.80	142	6.59		

continued

TABLE 53.8 (continued)
Total Ionization Cross Sections for $CClF_3$

Bart et al. (2001)				Sierra et al. (2004)	
Energy (eV)	Q_i (10^{-20} m²)	Energy (eV)	Q_i (10^{-20} m²)	Energy (eV)	Q_i (10^{-20} m²)
145	6.57	182	6.16		
149	6.55	186	6.14		
153	6.52	190	6.11		
156	6.48	193	6.06		
160	6.42	197	6.00		
164	6.36	201	5.94		
167	6.29	204	5.88		
171	6.24	208	5.84		
175	6.20	212	5.82		
179	6.18	215	5.81		

Note: Tabulated values of Bart et al. (2001). Courtesy of Professor P. Harland.

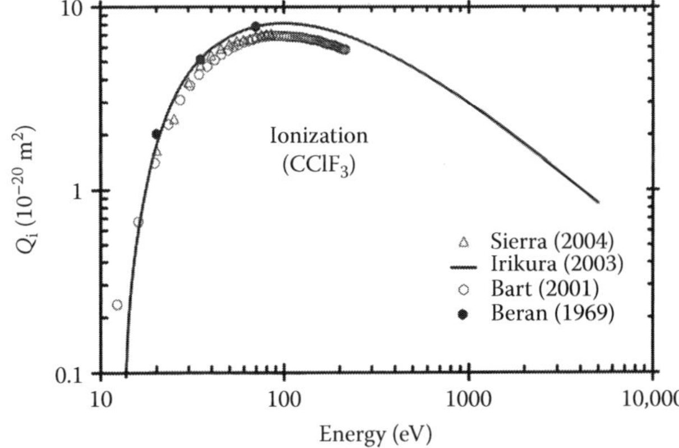

FIGURE 53.6 Total ionization cross sections for $CClF_3$. (Δ) Sierra et al. (2004); (——) Irikura et al. (2003). The tabulated values are taken from the web site http://Physics.nist.gov./ionxsec; (Adapted from Kim, Y.-K. et al., *Electron-Impact Ionization Cross Section for Ionization and Excitation database (version 3.0)*, http://Physics.nist.gov./ionxsec, National Institute of Standards and Technology, Gaithersberg, MD, USA, 2004.) (○) Bart et al. (2001); (•) Beran and Kevan (1969).

The earlier measurements of Beran and Kevan (1969) at three energies, 20, 35, and 70 eV are also included in the figure. The theoretical calculations are due to Irikura et al. (2003).

53.9 PARTIAL IONIZATION CROSS SECTIONS

Partial ionization cross sections measured by Sierra et al. (2004) are shown in Tables 53.9 and 53.10. Figure 53.7 provides a graphical presentation.

53.10 ATTACHMENT RATES

Electron attachment in $CClF_3$ occurs by the processes (Illenberger et al., 1979)

TABLE 53.9
Partial Ionization Cross Sections for $CClF_3$

Energy (eV)	Q_i (10^{-20} m²) CF_3^+	$CClF_2^+$	Cl^+	CF_2^+	CF^+	C^+
20	1.246	0.259	0.034	0.0529	0.011	
25	1.765	0.387	0.078	0.123	0.035	
30	2.591	0.652	0.182	0.262	0.104	0.003
35	2.944	0.859	0.324	0.364	0.208	0.011
40	3.146	1.000	0.407	0.412	0.271	0.025
45	3.252	1.109	0.501	0.46	0.301	0.047
50	3.346	1.18	0.563	0.493	0.307	0.062
55	3.39	1.257	0.624	0.523	0.321	0.076
60	3.426	1.277	0.63	0.545	0.336	0.081
65	3.38	1.317	0.653	0.558	0.331	0.084
70	3.433	1.365	0.709	0.577	0.347	0.094
75	3.452	1.396	0.729	0.605	0.35	0.107
80	3.446	1.412	0.759	0.611	0.352	0.108
85	3.434	1.439	0.742	0.623	0.352	0.121

Source: Adapted from Sierra, B. et al., *Int. J. Mass Spectrom.*, 235, 223, 2004.

TABLE 53.10
Partial Ionization Cross Sections for $CClF_3$ (contd.)

Energy (eV)	Q_i (10^{-20} m²) CCl^+	F^+	$CClF_3^+$	$CClF^+$	$CClF_2^{2+}$	CF_3^{2+}
20	0.003		0.04			
25	0.003		0.042	0.009		
30	0.013		0.047	0.013		
35	0.018	0.008	0.048	0.014	0.018	
40	0.053	0.011	0.054	0.015	0.026	0.001
45	0.074	0.025	0.06	0.018	0.039	0.007
50	0.077	0.037	0.062	0.022	0.048	0.009
55	0.079	0.051	0.06	0.018	0.062	0.014
60	0.077	0.059	0.06	0.02	0.069	0.021
65	0.083	0.066	0.057	0.024	0.075	0.021
70	0.085	0.068	0.066	0.024	0.079	0.023
75	0.089	0.073	0.06	0.026	0.081	0.026
80	0.104	0.078	0.057	0.025	0.084	0.026
85	0.091	0.088	0.066	0.026	0.09	0.026

Source: Adapted from Sierra, B. et al., *Int. J. Mass Spectrom.*, 235, 223, 2004.

$$CF_3Cl + e \rightarrow CF_2 + F + Cl^- \quad (53.1)$$

$$CF_3Cl + e \rightarrow CF_3 + Cl^- \quad (53.2)$$

$$CF_3Cl + e \rightarrow CF_2Cl + F^- \quad (53.3)$$

$$CF_3Cl + e \rightarrow CF_2 + FCl^- \quad (53.4)$$

$$CF_3Cl + e \rightarrow F + CF_2Cl^- \quad (53.5)$$

The negative ion appearance potentials and the peak position in cross section are shown in Table 53.11.

CClF$_3$

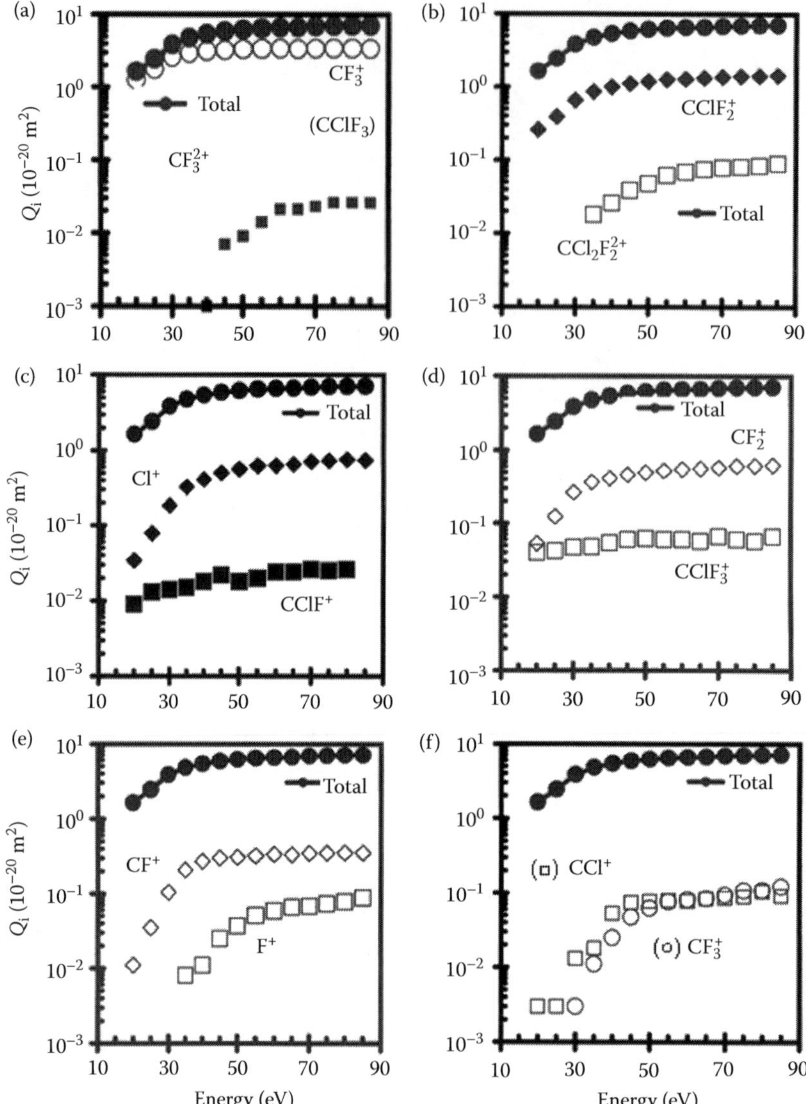

FIGURE 53.7 Partial ionization cross sections. Ion species as shown. Note that the cross section for the parent ion $CClF_3^+$ shown in frame (d) is about 1% of the total. The largest contribution is for the generation of CF_3^+ ion, shown in frame (a). (Adapted from Sierra, B. et al., *Int. J. Mass Spectrom.*, 205, 149, 2001.)

TABLE 53.11
Negative Ion Appearance Potentials

Ion	Appearance Potential (eV)	Peak Position (eV) Illenberger (1979)	Peak Position (eV) Spyrou (1985)
F^-	3.0	4.1	4.3
Cl^-	0.7	1.3	1.4
	3.4	4.8	5.0
FCl^-	3.0	3.9	4.7
CF_2Cl^-	3.5	4.2	4.4

Source: Adapted from Illenberger, E., H.-U. Scheunemann, and H. Baumgärtel, *Chem. Phys.*, 37, 21, 1979.

TABLE 53.12
Thermal Attachment Rate Constants (CClF$_3$)

Rate (k_a) (m^3/s)	Temperature (K)	Method	Reference
4.2×10^{-19}	293	ECR	Burns et al. (1996)
4.9×10^{-18}	300	Electron swarm	Spyrou and Christophorou (1985)
2.0×10^{-19}	300	Electron swarm	McCorkle et al. (1980)
7.0×10^{-20}	293	ECR	Schumacher et al. (1978)
$<3.9 \times 10^{-19}$	293	Electron swarm	Davis et al. (1973)
5.2×10^{-20}	300	Microwave conductivity	Fessenden and Bansal (1970)

Note: ECR = electron cyclotron resonance.

Table 53.12 shows the thermal attachment coefficients for $CClF_3$.

Table 53.13 and Figure 53.8 show the dependence of the attachment cross sections as a function of the mean energy of the electron swarm at various gas temperatures (Spyrou and Christophorou, 1985).

Table 53.14 shows the attachment cross sections at various temperatures are from Burns et al. (1996).

TABLE 53.13
Attachment Rates at Selected Gas Temperatures ($CClF_3$)

	k_a (10^{-17} m³/s)		
	Temperature (K)		
Mean Energy (eV)	**300**	**500**	**700**
0.590		2.80	9.10
1.068	3.88	7.50	13.30
1.50	4.20	7.08	10.90
2.14	3.97	6.10	9.20
3.00	5.30	6.92	9.36
4.03	5.67	6.68	
4.81	5.30	6.07	

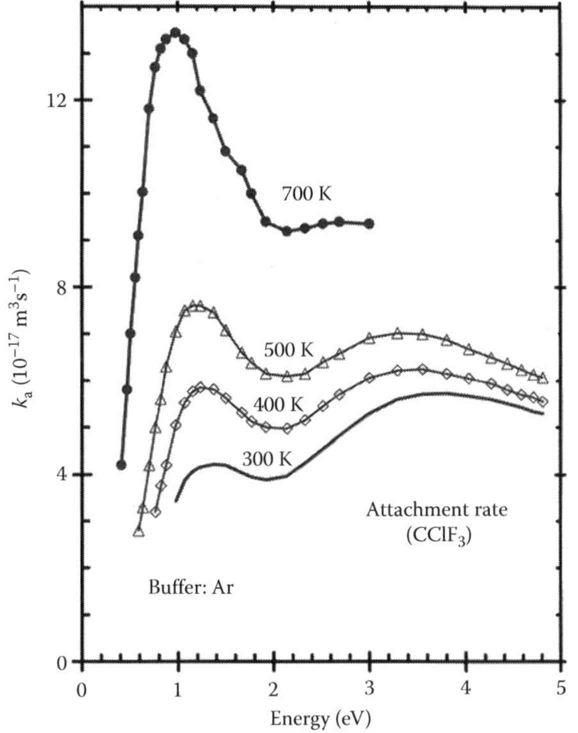

FIGURE 53.8 Attachment rates as a function of mean energy of electron swarm at various gas temperatures as shown Buffer gases are argon or N_2. (Adapted from Spyrou, S. M. and L. G. Christophorou, *J. Chem. Phys.*, 82, 2620, 1985).

53.11 ATTACHMENT CROSS SECTIONS

Figure 53.9 shows the total dissociative attachment cross sections at 300 K from swarm derived and beam methods. Figure 53.10 shows the temperature dependence of the attachment cross sections derived from swarm data (Spyrou and Christophorou, 1985).

TABLE 53.14
Attachment Rates at Various Gas Temperatures ($CClF_3$)

T (K)	293	467	579	777
k_a (m³/s)	4.2×10^{-19}	1.4×10^{-17}	2.4×10^{-16}	9.5×10^{-16}

Source: Adapted from Burns, S. J., J. M. Matthews, and D. L. McFadden, *J. Phys. Chem.*, 100, 19437, 1996.

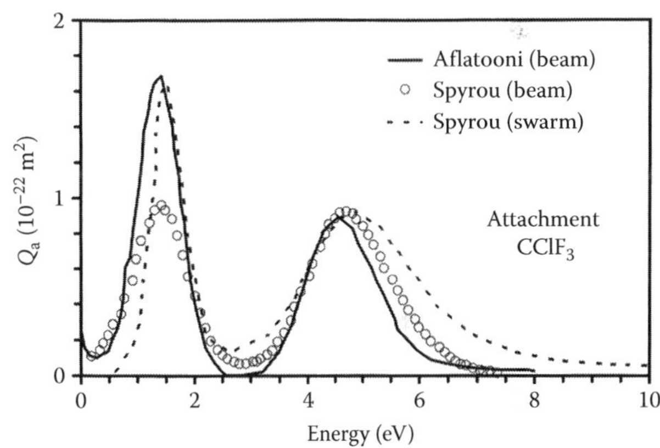

FIGURE 53.9 Total dissociative attachment cross sections for $CClF_3$. The low-energy peak is predominantly due to Cl^- and the second, predominantly due to F^- ion. The swarm-derived cross sections of Spyrou and Christophorou (1985) give better agreement with the more recent beam data of Aflatooni and Burrow (2001).

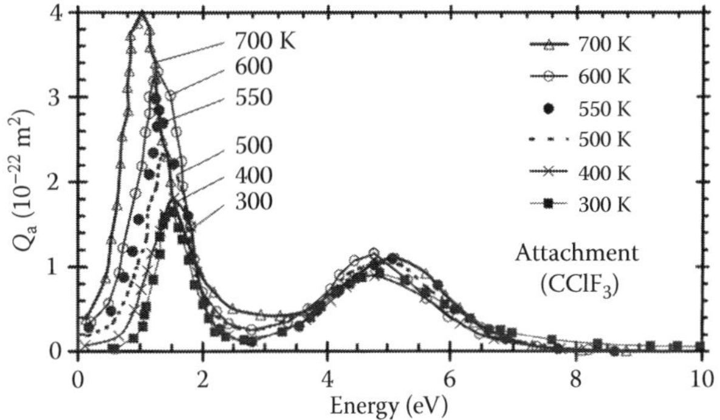

FIGURE 53.10 Attachment cross sections for $CClF_3$ as a function of energy at various temperatures determined by swarm method. (Adapted from Spyrou, S. M. and L. G. Christophorou, *J. Chem. Phys.*, 82, 2620, 1985.)

CClF$_3$

TABLE 53.15
Electron Drift Velocity for CClF$_3$ at Selected Intervals of *E/N*

E/N (Td)	Mean Energy (eV)	Drift Velocity (10^3 m/s)	*E/N* (Td)	Mean Energy (eV)	Drift Velocity (10^3 m/s)
0.0217	0.412	1.097	0.528	1.77	2.54
0.0528	0.590	1.38	1.097	2.52	3.04
0.109	0.822	1.68	2.17	3.55	3.63
0.217	1.15	2.03	3.42	4.43	4.18
0.466	1.67	2.46	4.35	4.81	4.90

Source: Adapted from Spyrou, S. M. and L. G. Christophorou, *J. Chem. Phys.*, 82, 2620, 1985.

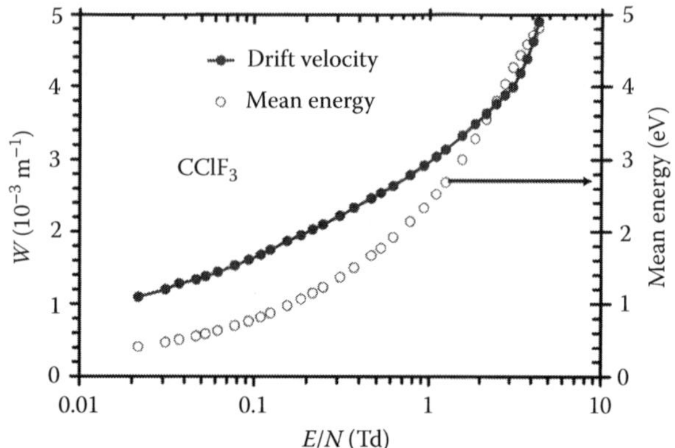

FIGURE 53.11 Drift velocity (W) and mean energy ($\bar{\varepsilon}$) as a function of *E/N* at 300 K for CClF$_3$ Buffer gas is argon or N$_2$. (Adapted from Spyrou, S. M. and L. G. Christophorou, *J. Chem. Phys.*, 82, 2620, 1985.)

53.12 ELECTRON DRIFT VELOCITY

Electron drift velocities have been determined by Spyrou and Christophorou (1985) in buffer gas argon as shown in Table 53.15 and Figure 53.11.

53.13 ELECTRON MEAN ENERGY

Mean energies of electron swarm at low values of the reduced electric field (*E/N*) have been determined by Spyrou and Christophorou (1985). These data are also included in Figure 53.11.

REFERENCES

Aflatooni, K. and P. D. Burrow, *Int. J. Mass Spectrom.*, 205, 149, 2001.

Ajello, J. M., W. T. Huntress, and P. Rayermann, *J. Chem. Phys.*, 64, 4746, 1976.

Au, J. W., G. R. Burton, and C. E. Brion, *Chem. Phys.*, 221, 151, 1997.

Bart, M., P. W. Harland, J. E. Hudson, and C. Vallance, *Phys. Chem. Chem. Phys.*, 3, 800, 2001.

Beran, J. A. and L. Kevan, *J. Phys. Chem.*, 73, 3866, 1969.

Burns, S. J., J. M. Matthews, and D. L. McFadden, *J. Phys. Chem.*, 100, 19437, 1996.

Daimon, H., T. Kondow, and K. Kuchitsu, *J. Phys. Soc. Jpn.*, 52, 84, 1983.

Davis, F. J., R. N. Compton, and D. R. Nelson, *J. Chem. Phys.*, 59, 2324, 1973.

Doucet, J., P. Sauvageau, and C. Sandorfy, *J. Chem. Phys.*, 58, 3708, 1973; the quoted value is from Jones (1986).

Fessenden, R. W. and K. M. Bansal, *J. Chem. Phys.*, 53, 3468, 1970.

Field, D., N. C. Jones, S. L. Lunt, J.-P. Ziesel, and R. J. Gulley, *J. Chem. Phys.*, 115, 3045, 2001.

Gilbert, R., P. Sauvageau, and C. Sandorfy, *J. Chem. Phys.*, 60, 4820, 1974.

Illenberger, E., H.-U. Scheunemann, and H. Baumgärtel, *Chem. Phys.*, 37, 21, 1979.

Irikura, K. K., M. A. Ali, and Y.-K. Kim, *Int. J. Mass Spectrom.*, 222, 189, 2003.

Jiang, Y., J. Sun, and L. Wan, *Phys. Rev.*, 52, 398, 1995. There is a misprint in the designation of the target molecule.

Jones, R. K., *J. Chem. Phys.*, 84, 813, 1986.

Kim, Y.-K., K. K. Irikura, M. E. Rudd, M. A. Ali, P. M. Stone, J. Chang, J. S. Coursey et al., *Electron-Impact Ionization Cross Section for Ionization and Excitation database (version 3.0)*, http://Physics.nist.gov./ionxsec, National Institute of Standards and Technology, Gaithersberg, MD, USA, 2004.

King, G. C. and J. W. McConkey, *J. Phys. B: At. Mol. Phys.*, 11, 1861, 1978.

Mann, A. and F. Linder, *J. Phys. B: At. Mol. Opt. Phys.*, 25, 1621, 1992.

McCorkle, D. L., A. A. Christodoulides, L. G. Christophorou, and I. Szamrej, *J. Chem. Phys.*, 72, 4049, 1980; erratum: *J. Chem. Phys.*, 76, 753, 1982.

Natalense, A. P. P., M. H. F. Bettega, L. G. Ferriera, and M. A. P. Lima, *Phys. Rev.*, 59, 879, 1999.

Randell, J., J. -P. Ziesel, S. L. Lunt, G. Mortzek, and D. Field, *J, Phys. B: At. Mol. Opt. Phys.*, 26, 3423, 1993.

Schumacher, R., H.-R. Sprünken, A. A. Christoudoulides, and R. N. Schlinder, *J. Phys. Chem.*, 82, 2248, 1978, cited by Burns et al. (1996).

Sierra, B., R. Martínez, C. Redondo, and F. Castaño, *Int. J. Mass Spectrom.*, 235, 223, 2004.

Spyrou, S. M. and L. G. Christophorou, *J. Chem. Phys.*, 82, 2620, 1985.

Underwood-Lemons, T., T. J. Gergel, and J. H. Moore, *J. Chem. Phys.*, 100, 9117, 1994.

Underwood-Lemons, T., T. J. Gergel, and J. H. Moore, *J. Chem. Phys.*, 102, 119, 1995.

Zecca, A., G. P. Karwasz, and R. S. Brusa, *Phys. Rev.*, 46, 3877, 1992.

54

DEUTERATED METHANE

CONTENTS

Deuterated methane (CD_4) (synonym perdeuteromethane) has deuterium substituted for hydrogen atoms in methane. It is a stable compound, highly flammable, and readily forms explosive mixture with air. CD_4 molecule has four fundamental vibrational modes as shown in Table 54.1.

54.1 SELECTED REFERENCES FOR DATA

Table 54.2 shows selected references for data on CD_4.

TABLE 54.1
Fundamental Vibrational Modes of CD_4 and Comparison with Those of Methane

		Energy (eV)	
Designation	**Mode**	**CD_4**	**CH_4**
v_1	Symmetric stretch	0.261	0.362
v_2	Degenerate deformation	0.135	0.190
v_3	Degenerate stretch	0.280	0.374
v_4	Degenerate deformation	0.123	0.162

TABLE 54.2
Selected References

Quantity	Energy Range (eV)	Reference
Q_i	0–200	Tarnovsky et al. (1996)
Q_M, Q_V	0–10	Pollock (1968a)
Q_a	0–114	Sharp and Dowell (1967)
	Swarm Properties	
D_r/μ	0.2–25	Millican and Walker (1987)
W_e, D_r/μ	0–10	Kleban and Davis (1978)
W_e, D_r/μ	0–10	Pollock (1968b)

Note: Q_M = momentum transfer; Q_i = ionization; Q_v = vibrational excitation; D_r = radial diffusion coefficient (m^2/s); μ = mobility ($m^2/V\ s$); W_e = drift velocity (m/s).

54.2 IONIZATION CROSS SECTION

See Table 54.3.

54.3 ATTACHMENT CROSS SECTION

The dissociative attachment cross sections in CD_4 as a function of electron energy are shown in Table 54.4. See Figure 54.2 for graphical presentation.

CD$_4$

TABLE 54.3
Ionization Cross Section of CD_4 and Dissociated Fragments

Energy	Q_i (10^{-20} m^2)				
(eV)	CD_4	CD_3	CD_2	CD	CH_4
13				0.17	
14	0.35	0.21	0.13	0.24	
15	0.48	0.31	0.28	0.33	0.235
20	1.17	0.94	0.91	0.80	1.421
25	1.86	1.68	1.63	1.26	2.281
30	2.41	2.25	2.02	1.57	2.788
35	2.74	2.57	2.12	1.73	3.095
40	2.96	2.74	2.28	1.82	3.338
50	3.33	2.87	2.39	1.89	3.717
60	3.40	2.90	2.45	1.93	3.902
70	3.43	2.99	2.50	1.94	4.010
80	3.40	3.02	2.48	1.93	4.011
90	3.32	2.94	2.44	1.91	3.993
100	3.32	2.92	2.46	1.88	3.944
120	3.08	2.86	2.40	1.83	
125					3.742
140	2.94	2.77	2.28	1.76	
150					3.516
160	2.74	2.62	2.15	1.64	
180	2.64	2.51	2.04	1.51	
200	2.49	2.44	1.95	1.42	3.153

Source: Adapted from Tarnovsky, V., A. Levin, H. Deutsch, and K. Becker, *J. Phys. B: At. Mol. Opt. Phys.*, 29, 139, 1996.

Note: See Figure 54.1 for graphical presentation. Cross section for methane added for comparison.

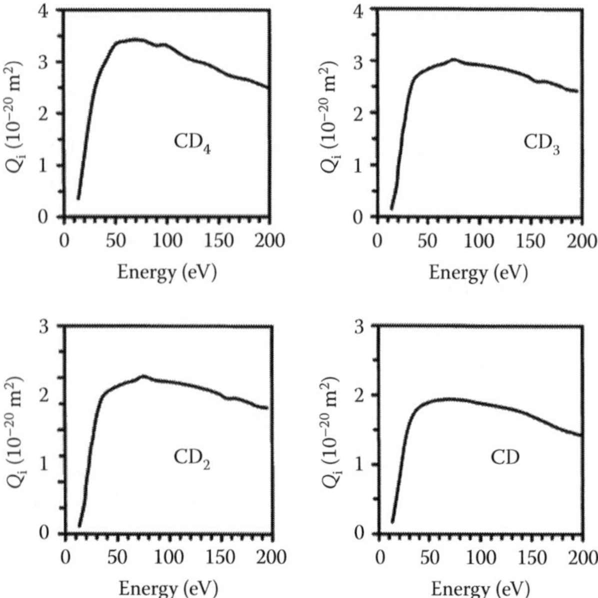

FIGURE 54.1 Total ionization cross sections of CD_4 and dissociated fragments as a function of electron energy.

TABLE 54.4
Dissociative Attachment Cross Section in CD_4

Energy	7.50	8.00	8.50	9.00	9.50	10.00	10.50
Q_a	0.005	0.030	0.325	0.897	0.940	0.879	0.668
Energy	11.00	11.50	12.00	12.50	13.00	13.50	
Q_a	0.372	0.211	0.107	0.045	0.0066	0.0065	

Source: Adapted from Sharp, T. E. and J. T. Dowell, *J. Chem. Phys.*, 46, 1530, 1967.

Note: Electron energy in eV and cross section in 10^{-23} m^2.

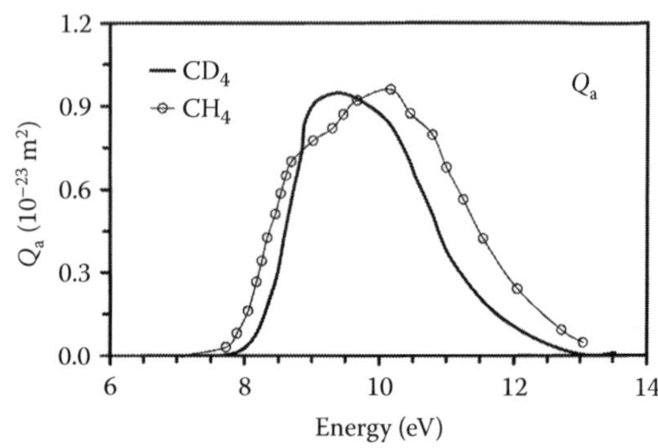

FIGURE 54.2 Dissociative attachment cross sections in CD_4. Methane cross sections are shown for the sake of comparison. (Adapted from Sharp, T. E. and J. T. Dowell, *J. Chem. Phys.*, 46, 1530, 1967.)

TABLE 54.5
Drift Velocity of Electrons in CD_4

	W_e (10^4 m/s)	
E/N (Td)	CD_4 [a,b]	CH_4 [a,c]
0.5	2.25	2.32
1	4.63	5.26
2	8.47	9.10
3	8.44	10.35
4	8.11	10.10
5	7.64	9.75
6	6.88	9.22
7	6.38	8.74
8	6.09	8.25
9	5.73	7.82
10	5.04	7.40
12		6.87
14		6.52

[a] Digitized and interpolated.
[b] Pollock (1968a).
[c] Hunter et al. (1986).

54.4 DRIFT VELOCITY

Table 54.5 presents the experimental drift velocity of electrons in CD_4 (see Pollock 1968a) and the corresponding values in methane are presented for the purpose of comparison. See Figure 54.3 for graphical presentation.

54.5 CHARACTERISTIC ENERGY

Table 54.6 presents the characteristic energy in CD_4.

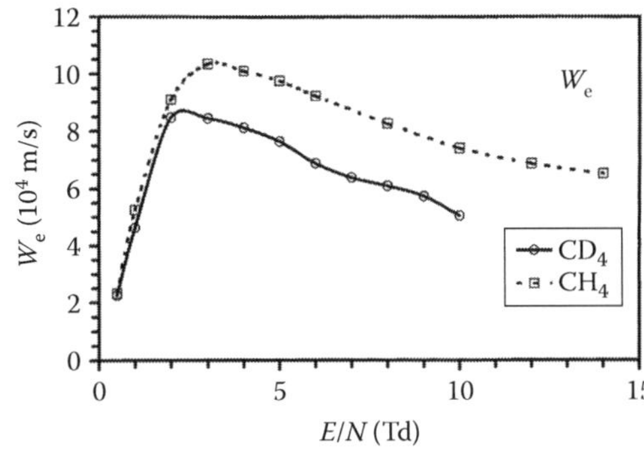

FIGURE 54.3 Experimental drift velocity of electrons in CD_4. Methane curve (Hunter et al. 1986) is added for the sake of comparison. (—○—) CD_4; (—□) CH_4. (Adapted from Pollock, W. J., *J. Chem. Soc. Faradat Trans.*, 64, 2919, 1968a.)

TABLE 54.6
Characteristic Energy in CD_4

E/N (Td)	D_r/μ (V)	*E/N* (Td)	D_r/μ (V)	*E/N* (Td)	D_r/μ (V)
0.20	0.025	1.2	0.074	7	0.854
0.25	0.025	1.4	0.089	8	1.02
0.30	0.025	1.7	0.111	10	1.33
0.35	0.028	2	0.135	12	1.68
0.4	0.032	2.5	0.193	14	1.98
0.5	0.035	3	0.254	17	2.45
0.6	0.039	4	0.394	20	2.84
0.7	0.042	5	0.534	25	3.42
1	0.061	6	0.693		

Source: Adapted from Millican, P. G. and J. C. Walker, *J. Phys. D: Appl. Phys.*, 20, 193, 1987.

REFERENCES

Hunter, S. R., J. G. Carter, and L. G. Christophorou, *J. Appl. Phys.*, 60, 24, 1986.
Kleban, P. and H. T. Davis, *J. Chem. Phys.*, 68, 2999, 1978.
Millican, P. G. and J. C. Walker, *J. Phys. D: Appl. Phys.*, 20, 193, 1987.
Pollock, W. J., *J. Chem. Soc. Faradat Trans.*, 64, 2919, 1968a.
Pollock, W. J., *J. Trans. Farad. Soc.*, 64, 2919, 1968b.
Sharp, T. E. and J. T. Dowell, *J. Chem. Phys.*, 46, 1530, 1967.
Tarnovsky, V., A. Levin, H. Deutsch, and K. Becker, *J. Phys. B: At. Mol. Opt. Phys.*, 29, 139, 1996.

55

DIBROMODIFLUORO-METHANE

CBr$_2$F$_2$

CONTENTS

Dibromodifluoromethane (CBr$_2$F$_2$) is a polar, electron-attaching gas that has 94 electrons. The electronic polarizability is 10.0×10^{-40} F m^2 (Barszczewska et al., 2004 quote a value of 8.0×10^{-40} F m^2), dipole moment 0.66 D, and ionization potential 11.03 eV.

55.1 ATTACHMENT RATES

Dissociative attachment occurs according to the reactions (Kalamarides et al., 1990)

$$CBr_2F_2 + e \rightarrow CBr_2F_2^* \rightarrow \begin{cases} CBrF_2 + Br^- \\ CF_2 + Br_2^- \end{cases} \quad (55.1)$$

the Br$^-$ ion being more abundant (~0.9:0.1). The appearance potential for both ions is 0 eV and both species show a peak at this energy (Alajajian et al., 1988).

Table 55.1 shows the attachment rate constants.

Figure 55.1 shows the attachment rate constants for CBr$_2$F$_2$ as a function of mean energy (Sunagawa and Shimamori, 1997).

TABLE 55.1
Thermal Attachment Rate Constants for CBr$_2$F$_2$

Method	Rate Constant (10^{-13} m^3/s)	Reference
Swarm	2.7	Barszczewska et al. (2004)
PR-MW	1.7	Sunagawa and Shimamori (1997)
RA	3.0	Kalamarides et al. (1990)
FA/LP	3.0 (298 K)	Smith et al. (1990)
	3.7 (380 K)	
	4.0 (475 K)	
Swarm	2.6	Blaunstein and Christophorou (1968)

Note: FA/LP = flowing afterglow/Langmuir probe; PR-MW = pulse-radiolysis/microwave cavity; RA = Rydberg atoms; thermal energy unless otherwise specified.

55.2 ATTACHMENT CROSS SECTIONS

Table 55.2 and Figure 55.2 show the attachment cross sections for CBr$_2$F$_2$.

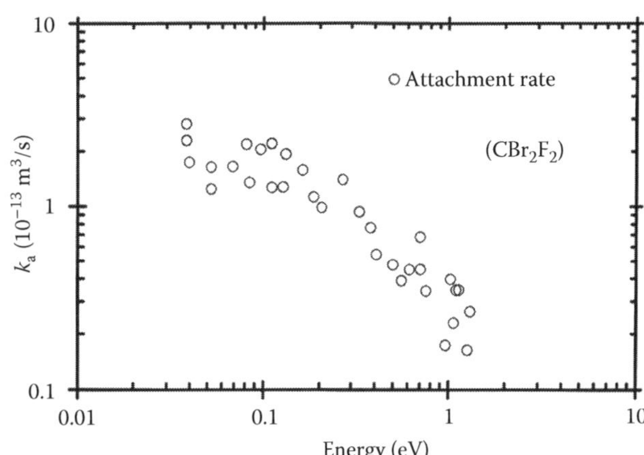

FIGURE 55.1 Attachment rate constant for CBr$_2$F$_2$. (Adapted from Sunagawa, T. and H. Shimamori, *J. Chem. Phys.*, 107, 7876, 1997.)

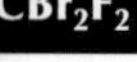

TABLE 55.2
Attachment Cross Sections for CBr_2F_2

Energy (meV)	Q_a (10^{-20} m^2)	Energy (meV)	Q_a (10^{-20} m^2)
1	1218	60	95.2
2	848	80	77.9
3	658	100	68.2
4	556	200	35.7
5	493	400	7.00
6	440	600	1.60
8	366	800	0.50
10	331	1000	0.20
20	207	1200	0.10
40	133	1500	0.05

Source: Digitized from Sunagawa, T. and H. Shimamori, *J. Chem. Phys.*, 107, 7876, 1997.

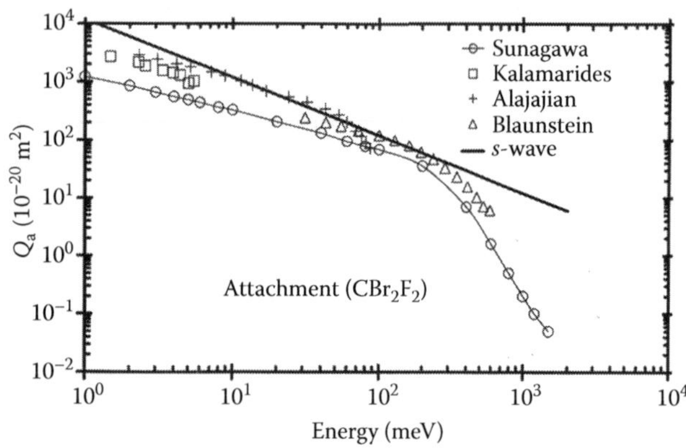

FIGURE 55.2 Attachment cross sections for CBr_2F_2. (—○—) Sunagawa and Shimamori (1997); (□) Kalamarides et al. (1990); (+) Alajajian et al. (1988); (Δ) Blaunstein and Christophorou (1968); (——) *s*-wave maximum cross section.

REFERENCES

Alajajian, S. H., M. T. Bernius, and A. Chutjian, *J. Phys. B: At. Mol. Opt. Phys.*, 21, 4021, 1988.

Barszczewska, W., J. Kopyra, J. Wnorowska, and I. Szamrej, *Int. J. Mass Spectrom.*, 233, 199, 2004.

Blaunstein, R. P. and L. G. Christophorou, *J. Chem. Phys.*, 49, 1526, 1968.

Kalamarides, A., R. W. Marawar, X. Ling, C. W. Walter, B. G. Lindsay, K. A. Smith, and F. B. Dunning, *J. Chem. Phys.*, 92, 1672, 1990.

Smith, D., C. R. Herd, N. G. Adams, and J. F. Paulson, *Int. J. Mass Spectrom. Ion. Proc.*, 96, 341, 1990.

Sunagawa, T. and H. Shimamori, *J. Chem. Phys.*, 107, 7876, 1997.

56

DIBROMOMETHANE

CONTENTS

Dibromomethane (CH_2Br_2) is a strongly polar molecule that has 78 electrons. Its electronic polarizability is 10.37×10^{-40} F m^2, dipole moment 1.42 D, and ionization potential 10.50 eV. The molecule has nine vibrational modes as shown in Table 56.1 (Shimanouchi, 1972).

56.1 IONIZATION CROSS SECTIONS

Bart et al. (2001) have measured the ionization cross sections as shown in Table 56.2 and Figure 56.1.

56.2 ATTACHMENT CROSS SECTIONS

Attachment cross sections unfolded from Swarm experiments by Sunagawa and Shimamori (1997) are shown in Table 56.3 and Figure 56.2.

56.3 ATTACHMENT RATE CONSTANTS

Electron attachment occurs according to the following process (Parthasarathy et al., 1998).

TABLE 56.1
Vibrational Modes of CH_2Br_2

Designation	Type of Motion	Energy (meV)
v_1	CH_2 symmetrical-stretch	373.1
v_2	CH_2 scissor	171.3
v_3	CBr_2 symmetrical-stretch	72.9
v_4	CBr_2 scissor	21.0
v_5	CH_2 twist	135.8
v_6	CH_2 asymmetrical-stretch	381.0
v_7	CH_2 rock	100.7
v_8	CH_2 wag	148.2
v_9	CBr asymmetrical-stretch	81.0

TABLE 56.2
Ionization Cross Sections for CH_2Br_2

Energy (eV)	Q_i (10^{-20} m^2)	Energy (eV)	Q_i (10^{-20} m^2)
12	1.32	116	10.80
16	2.30	119	10.73
20	4.38	123	10.66
23	6.16	127	10.57
27	7.49	130	10.45
31	8.26	134	10.30
34	8.98	138	10.17
38	9.55	142	10.09
42	9.91	145	10.04
45	10.35	149	9.99
49	10.68	153	9.92
53	11.01	156	9.82
57	11.26	160	9.69
60	11.47	164	9.55
64	11.57	167	9.43
68	11.62	171	9.33
71	11.67	175	9.27
75	11.67	179	9.24
79	11.62	182	9.23
82	11.59	186	9.21
86	11.53	190	9.16
90	11.45	193	9.08
94	11.36	197	8.97
97	11.27	201	8.85
101	11.17	204	8.74
105	11.06	208	8.66
108	10.96	212	8.63
112	10.87	215	8.61

Source: Adapted from Bart, M. et al., *Phys. Chem. Chem. Phys.*, 3, 800, 2001.

Note: Tabulated values courtesy of Professor Harland.

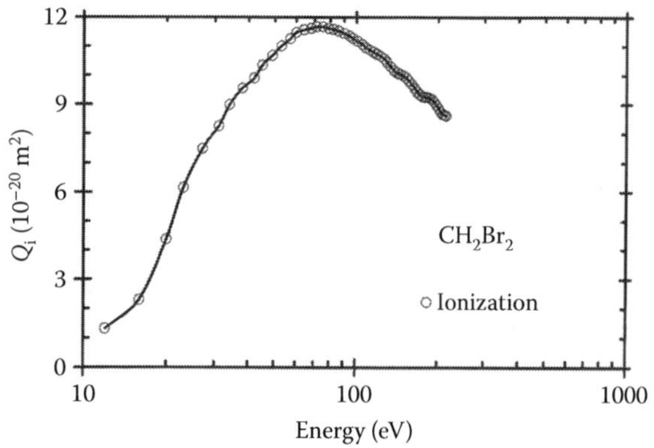

FIGURE 56.1 Ionization cross sections for CH_2Br_2. Tabulated values courtesy of Professor Harland (2006).

TABLE 56.3
Attachment Cross Sections for CH_2Br_2

Energy (eV)	Q_a (10^{-20} m²)	Energy (eV)	Q_a (10^{-20} m²)
0.001	500.0	0.08	42.83
0.002	340.7	0.1	34.29
0.003	273.8	0.2	11.05
0.004	247.3	0.4	2.45
0.005	219.8	0.6	1.02
0.006	191.4	0.8	0.33
0.008	157.4	1.0	0.29
0.01	145.9	1.2	0.33
0.02	110.8	1.5	0.17
0.04	69.72	2.0	0.12
0.06	55.67		

Source: Digitized from Sunagawa, T. and H. Shimamori, *J. Chem. Phys.*, 107, 7876, 1997.

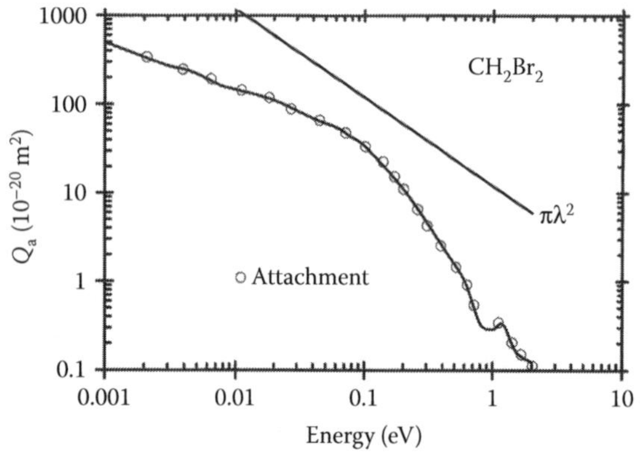

FIGURE 56.2 Attachment cross sections for CH_2Br_2. The maximum attachment cross section for *s*-wave is also shown. (Adapted from Sunagawa, T. and H. Shimamori, *J. Chem. Phys.*, 107, 7876, 1997.)

TABLE 56.4
Attachment Rate Constants for CH_2Br_2

Energy (Temperature) K (eV)	Rate Constant (10^{-14} m³/s)	Method	Reference
	9.60		Christophorou and Hadjiantoniou (2006)
	7.1	Swarm	Barszczewska et al. (2004)
	9.0	MW	Sunagawa and Shimamori (1997)
293	3.5	FA/ECR	Burns et al. (1996)
467	4.2		
777	4.5		
	9.0	MW	Shimamori et al. (1992)
	10.6	PI	Alajajian et al. (1988)
205	8.1	FA/LP	Alge et al. (1984)
300	9.3		
452	10.6		
585	22.0		
	3.2	Swarm	Blaunstein et al. (1968)

Note: FA/ECR = flowing afterglow/electron cyclotron resonance; FA/LP = flowing afterglow/Langmuir probe; MW = microwave cavity; P = photoionization. Thermal energy unless otherwise mentioned.

$$CH_2Br_2 + e \rightarrow CH_2Br_2^{-*} \rightarrow CH_2Br + Br^- \qquad (56.1)$$

The lifetime of the excited species is very short, dissociation occurring immediately after electron capture into the antibonding orbital. The electron affinity of the molecule is −0.05 eV (Christophorou and Hadjiantoniou, 2006).

Table 56.4 shows the attachment rate constants for CH_2Br_2.

REFERENCES

Alajajian, S. H., M. T. Bernius, and A. Chutjian, *J. Phys. B: At. Mol. Opt. Phys.*, 21, 4021, 1988.

Alge, E., N. G. Adams, and D. Smith, *J. Phys. B: At. Mol. Phys.*, 17, 3827, 1984.

Barszczewska, W., J. Kopyra, J. Wnorowska, and I. Szamrej, *Int. J. Mass. Spectrom.* 233, 199, 2004.

Bart, M., P. W. Harland, J. E. Hudson, and C. Vallance, *Phys. Chem. Chem. Phys.*, 3, 800, 2001.

Burns, S. J., J. M. Matthews, and D. L. McFadden, *J. Phys. Chem.*, 100, 19436, 1996.

Christophorou, L. G. and D. Hadjiantoniou, *Chem. Phys. Lett.*, 419, 405, 2006.

Parthasarathy, R., C. D. Finch, J. Wolfgang, P. Nordlander, and F. B. Dunning, *J. Chem. Phys.*, 109, 8829, 1998.

Shimamori, H., Y. Tatsumi, Y. Ogawa, and T. Sunagawa, *Chem. Phys. Lett.*, 194, 223, 1992.

Shimanouchi, T. *Tables of Molecular Vibrational Frequencies Consolidated Volume I*, NSRDS-NBS 39, Washington (DC), 1972.

Sunagawa, T. and H. Shimamori, *J. Chem. Phys.*, 107, 7876, 1997.

57

DICHLORODIFLUORO-METHANE

CCl_2F_2

CONTENTS

Dichlorodifluoromethane (CCl_2F_2, Freon-12) is an electron-attaching, polar gas with 58 electrons. The electronic polarizability is 8.82×10^{-40} F m^2, the dipole moment is 0.51 D, and the ionization potential is 12.05 eV. Extensive review of literature in the gas has been published by Christophorou and Olthoff (2004) which is referred to extensively. The electronic excitation potential is ~7.0 eV (King and McConkey, 1978). The molecule has nine vibrational modes (Mann and Linder, 1992) as shown in Table 57.1.

TABLE 57.1
Vibrational Modes of CCl_2F_2

Symbol	Nuclear Motion	Energy (meV)
ν_1	CF_2 symmetric stretch	136.1
ν_2	CF_2 bending	82.7
ν_3	CCl_2 symmetric stretch	56.6
ν_4	CCl_2 bending	32.5
ν_5	Torsion	39.9
ν_6	CF_2 asymmetric stretch	144.7
ν_7	CF_2 plane rocking	55.3
ν_8	CCl_2 asymmetric stretch	114.4
ν_9	CCl_2 plane rocking	53.9

Source: Adapted from Mann, A. and F. Linder, *J. Phys. B: At. Mol. Opt. Phys.*, 25, 1633, 1992.

57.1 SELECTED REFERENCES FOR DATA

See Table 57.2.

57.2 TOTAL SCATTERING CROSS SECTIONS

Total cross sections suggested by Christophorou and Olthoff (2004) are shown in Table 57.3 and Figure 57.1 which include

TABLE 57.2
Selected References

Quantity	Range (eV)	Reference
Q_i	15–1000	Lindsay et al. (2004)
Review	0–4000	Christophorou and Olthoff (2004)
Q_i	Onset–5000	Irikura et al. (2003)
Q_a	0–2	Matejčik et al. (2003)
Q_a	0.001–1	Skalny et al. (2003)
Q_a	0–8	Aflatooni and Burrow (2001)
Q_i	15–215	Bart et al. (2001)
Q_T	0.01–2.0	Field et al. (2001)
Q_i	15–215	Hudson et al. (2001)
Q_a	0–2	Skalny et al. (2001)
Q_a	0–15	Langer et al. (2000)
Q_M	10–30	Natalense et al. (1999)
Q_a	0–10	Wang et al. (1998)
k_a		Le Garrec (1997)
k_a		Burns et al. (1996)
Q_a	0.2	Kiendler et al. (1996)
Q_T	10–1000	Jiang et al. (1995)
Q_a	0.5	Underwood-Lemons (1995)
Q_T	0.4–10	Underwood-Lemons (1994)
Q_V	0–1	Randell et al. (1993)
Q_{el}	0.5–10	Mann and Linder (1992)
Q_T	75–4000	Zecca et al. (1992)
Q_i	15–180	Leiter et al. (1989)
Q_a	0–0.16	Chutjian and Alajajian (1987)
Q_T	0.7–50	Jones (1986)
Q_a	Thermal	Smith et al. (1984)
Q_a	0.04–1.2	McCorkle et al. (1980)
Q_i	15–250	Pejcev et al. (1979)
Q_i	20, 35, 70	Beran Kevan (1979)
	Swarm Properties	
η/N	240–600	Fréchette (1986)
η/N	210–300	Siddagangappa et al. (1983)
D/μ, α/N	330–640	Maller and Naidu (1975)
α/N, η/N	400–3000	Maller and Naidu (1974)
α/N, η/N	240–3000	Rao and Govinda Raju (1973)
α/N, η/N	400–700	Boyd et al. (1970)
W	350–640	Naidu and Prasad (1969)
α/N, η/N	400–800	Moruzzi (1963)
α/N, η/N	250–600	Harrison and Geballe (1953)

Note: k_a = attachment rate; Q_a = attachment; Q_{el} = elastic; Q_i = ionization; Q_T = total; D = diffusion coefficient; W = drift velocity; α = ionization coefficient; η = attachment coefficient; μ = mobility.

the data of Jones (1986) and Zecca et al. (1992). The highlights of the cross sections are

1. The cross section toward zero energy increases to ~370×10^{-20} m^2 due to dissociative attachment with the formation of Cl^-, combined with ro-vibrational excitation (Field et al., 2001). However, Randell et al. (1993) state in an earlier paper that attachment does not occur in the gas below 100 meV energy. See Table 57.16 for low-energy details.

TABLE 57.3
Total Scattering Cross Sections in CCl_2F_2

Christophorou and Olthoff (2004)		Zecca et al. (1992)		Jones (1986)	
Energy (eV)	Q_T (10^{-20} m^2)	Energy (eV)	Q_T (10^{-20} m^2)	Energy (eV)	Q_T (10^{-20} m^2)
0.10	98.1	75	30.9	0.7	33.58
0.20	57.6	80	29.9	0.8	36.91
0.30	41.5	90	28.6	0.9	39.78
0.35	37.2	100	27.4	1.0	41.54
0.40	34.1	110	26.2	1.1	40.80
0.45	32.3	125	24.6	1.2	38.82
0.50	31.3	150	22.9	1.3	35.98
0.60	30.9	175	21.7	1.4	33.06
0.70	32.6	200	20.7	1.5	30.47
0.80	35.4	225	19.1	1.6	29.39
0.90	37.7	250	18.4	1.7	29.18
1.0	39.0	275	17.7	1.8	29.41
1.5	29.4	300	17	1.9	30.46
2.0	31.0	350	15.2	2.0	31.82
2.5	38.1	400	14.1	2.1	33.48
3.0	38.0	450	13.1	2.2	35.19
3.5	38.2	500	12.2	2.3	36.46
4.0	39.9	600	10.8	2.4	37.76
4.5	40.6	700	9.77	2.5	38.50
5.0	41.6	800	8.98	2.6	38.68
6.0	43.4	900	8.21	2.7	38.65
7.0	44.1	1000	7.61	2.8	38.50
8.0	46.3	1100	7.01	2.9	38.28
9.0	48.5	1250	6.34	3.0	38.04
10.0	50.5	1500	5.45	3.1	37.67
12.5	49.0	1750	4.8	3.2	37.73
15.0	45.3	2000	4.32	3.3	37.74
20.0	40.5	2250	3.85	3.4	38.00
25.0	38.0	2500	3.5	3.5	38.20
30.0	36.2	2750	3.2	3.6	38.52
35.0	34.7	3000	2.96	3.7	38.83
40.0	33.5	3250	2.74	3.8	39.34
45.0	32.5	3500	2.59	3.9	39.68
50.0	31.6	4000	2.31	4.0	39.86
60.0	30.1			4.2	40.17
70.0	28.9			4.4	40.41
80.0	27.9			4.6	40.68
90.0	27.0			4.8	40.96
100	26.2			5.0	41.59
150	23.0			5.2	42.13
200	20.4			5.4	42.50
250	18.3			5.6	42.91
300	16.6			5.8	43.44
350	15.1			6.0	43.38
400	13.9			6.2	43.62
450	12.8			6.4	43.57
500	11.9			6.6	43.76
600	10.4			6.8	43.93
700	9.4			7.0	44.07
800	8.5			7.5	45.18
900	7.7			8.0	46.32
1000	7.3			8.5	47.44

TABLE 57.3 (continued)
Total Scattering Cross Sections in CCl_2F_2

Christophorou and Olthoff (2004)		Zecca et al. (1992)		Jones (1986)	
Energy (eV)	Q_T (10^{-20} m²)	Energy (eV)	Q_T (10^{-20} m²)	Energy (eV)	Q_T (10^{-20} m²)
2000	4.3			9.0	48.52
3000	3.0			9.5	49.73
4000	2.2			10.0	50.47
				10.5	50.67
				11.0	50.87
				11.5	50.52
				12.0	49.97
				12.5	49.05
				13.0	48.20
				13.5	47.67
				14.0	46.54
				14.5	45.80
				15.0	45.09
				16.0	43.94
				17.0	43.00
				18.0	42.30
				19.0	41.13
				20.0	40.31
				21.0	39.88
				22.0	39.15
				23.0	38.88
				24.0	38.33
				25.0	38.03
				27.5	36.89
				30.0	36.26
				32.5	34.99
				35.0	34.23
				37.5	33.42
				40.0	32.94
				45.0	31.78
				50.0	31.14

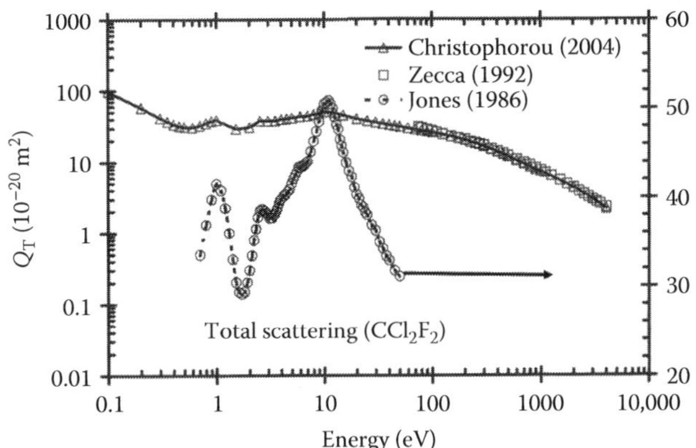

FIGURE 57.1 Total scattering cross sections in CCl_2F_2. (—Δ—) Christophorou and Olthoff (2004); (□) Zecca et al. (1992); (—○—) Jones (1986), note the secondary ordinate.

2. The cross sections pass through a minimum at ~0.6 eV attributed to Ramsauer–Townsend effect (Mann and Linder 1992). However, Randell et al. (1993) indicate an R–T minimum at 40–60 meV. The shallow minimum is attributed to the weak contribution of the *s*-wave to the total scattering cross section (Field et al., 2001).
3. The low-energy cross section shows an energy dependence according to $Q_T(\varepsilon) \propto \varepsilon^{-n}$: $n \cong 1$ at $10 \le \varepsilon \le 300$ meV, $n < 1$ at $\varepsilon < 10$ meV, and $n > 1$ at $\varepsilon > 500$ meV (Skalny et al., 2001).
4. A discernible shoulder at 3.5 eV attributed to the formation of FCl^-, F^-, and $CFCl_2^-$ ions (Christophorou and Olthoff, 2004).
5. A peak at 10 eV due to direct scattering and onset of a number of inelastic processes.
6. A monotonic decrease above 10 eV energy, as observed in most gases.

57.3 DIFFERENTIAL SCATTERING CROSS SECTIONS

Mann and Linder (1992) have measured the differential scattering cross sections as a function of electron energy in the

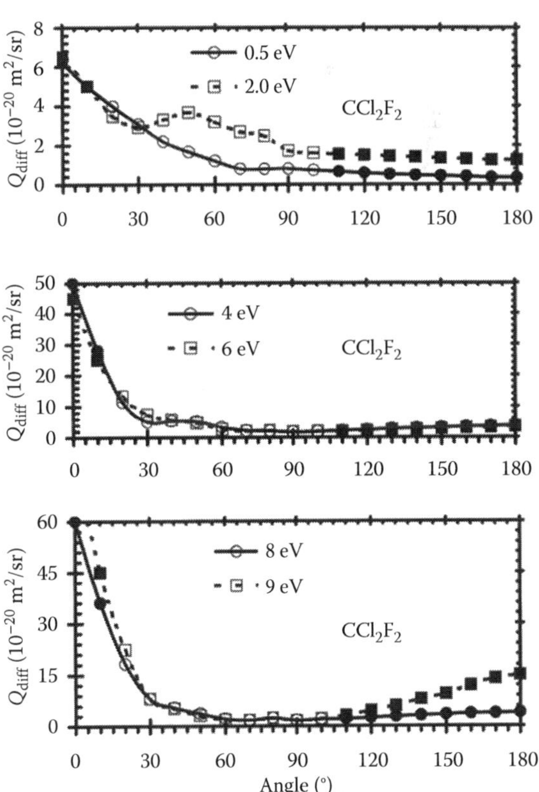

FIGURE 57.2 Differential cross sections in CCl_2F_2 as a function of scattering angle at selected energies. Open symbols are measured and closed symbols are extrapolated. (Adapted from Mann, A. and F. Linder *J. Phys. B: At. Mol. Opt. Phys.*, 25, 1633, 1992.)

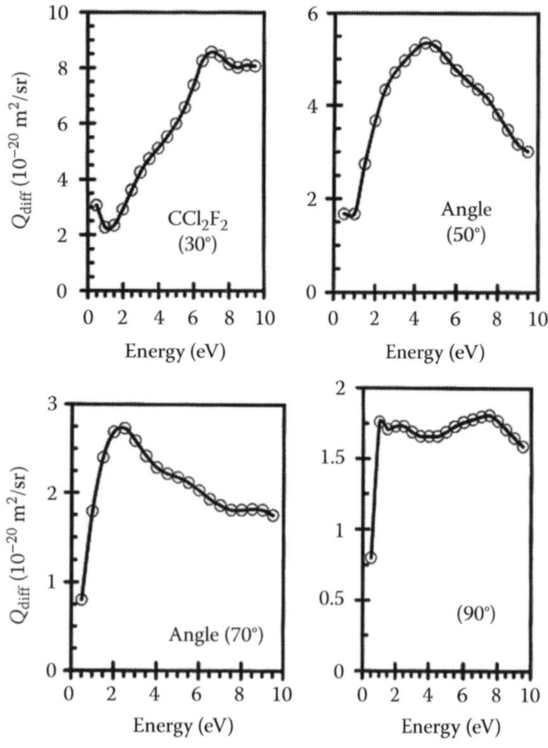

FIGURE 57.3 Differential scattering cross sections in CCl_2F_2 as a function of electron energy at selected angles. (Adapted from Mann, A. and F. Linder *J. Phys. B: At. Mol. Opt. Phys.*, 25, 1633, 1992.)

TABLE 57.4
Elastic Scattering Cross Sections in CCl_2F_2

	Q_{el} (10^{-20} m^2)			Q_{el} (10^{-20} m^2)	
Energy (eV)	**Mann and Linder (1992)**	**Field et al. (2001)**	**Energy (eV)**	**Mann and Linder (1992)**	**Field et al. (2001)**
0.015		380.0	0.65		29.38
0.02		350.0	0.7	16.9	30.80
0.025		290.0	0.8	18.0	33.91
0.03		260	0.9	19.0	35.85
0.04		220.0	1.0	20.0	36.27
0.05		180.0	1.5	24.1	27.34
0.06		150.0	2.0	27.2	30.34
0.07		120.0	2.5	30.2	
0.08		110.0	3.0	32.6	
0.09		100.0	3.5	34.4	
0.10		100.0	4.0	35.7	
0.15		70.0	4.5	36.7	
0.20		50.0	5.0	37.4	
0.30		40.0	6.0	38.2	
0.4		30.0	7.0	39.9	
0.5	14.7	29.06	8.0	40.8	
0.55		28.43	9.0	42.6	
0.6	15.8	28.55	9.5	44.2	

range from 0.5 to 9.5 eV. Figures 57.2 and 57.3 show these data. The highlights of the scattering cross sections are

1. The cross sections as function of scattering angle at a constant energy decrease monotonically except at 3.0 eV, showing significant forward scattering (Figure 57.2).
2. The cross sections as a function of electron energy at a constant scattering angle show a minimum at 1.5 and 7.0 eV (Figure 57.3).

57.4 ELASTIC SCATTERING CROSS SECTIONS

Table 57.4 shows the elastic scattering cross sections derived by Christophorou and Olthoff (2004) from the measurements of Mann and Linder (1992) and more recent measurements of Field et al. (2001). Figure 57.4 provides a graphical presentation.

57.5 MOMENTUM TRANSFER CROSS SECTIONS

Measured momentum transfer cross sections are not available in the literature. By extrapolating (with some "artistic license") and integrating, the derived cross sections are shown in Table 57.5 and Figure 57.5 which also include the data of Hayashi (1987) and Novak and Fréchette (1985).

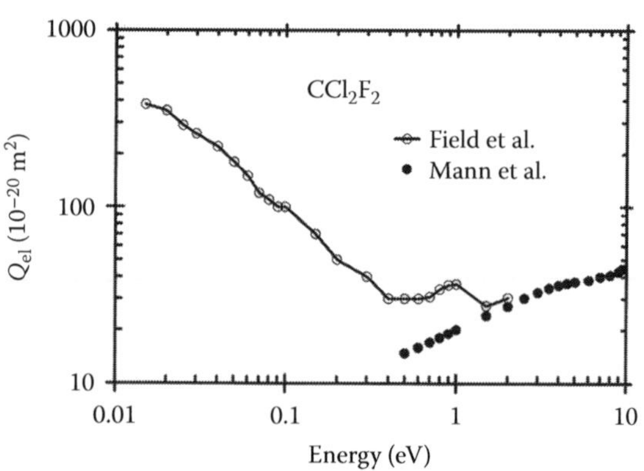

FIGURE 57.4 Elastic scattering cross sections in CCl_2F_2. (—○—) Field et al. (2001); (●) Mann and Linder (1992).

57.6 RO-VIBRATIONAL EXCITATION CROSS SECTIONS

CCl_2F_2 molecule possesses a dipole moment and data on quadrupole moment are not available. Threshold ro-vibrational excitation occurs below 100–200 meV (see Rendell et al., 1993) and below this energy the cross section increases rapidly due to pure rotational excitation. The total vibrational cross sections given by Mann and Linder (1992) are tabulated by Christophorou and Olthoff (2004) as shown in Table 57.6 and Figure 57.6.

TABLE 57.5
Momentum Transfer Cross Sections in CCl_2F_2

Energy (eV)	Q_M (10^{-20} m^2)	Energy (eV)	Q_M (10^{-20} m^2)
0.5	8.72	6.5	31.24
1	18.66	7.0	33.35
1.5	21.14	7.5	35.42
2	22.28	8	34.97
2.5	23.72	8.5	34.8
3	26.69	9	65.57
3.5	30.78	9.5	52.98
4	33.64	10	22.74
4.5	32.67	15	19.25
5	30.94	20	15.40
5.5	31.16	25	13.32
6	31.15	30	12.04

Note: Integrated from data shown in Figures 57.2 and 57.3 (±50%). Cross sections above 9.5 eV are theoretical from Natalense et al. (1999).

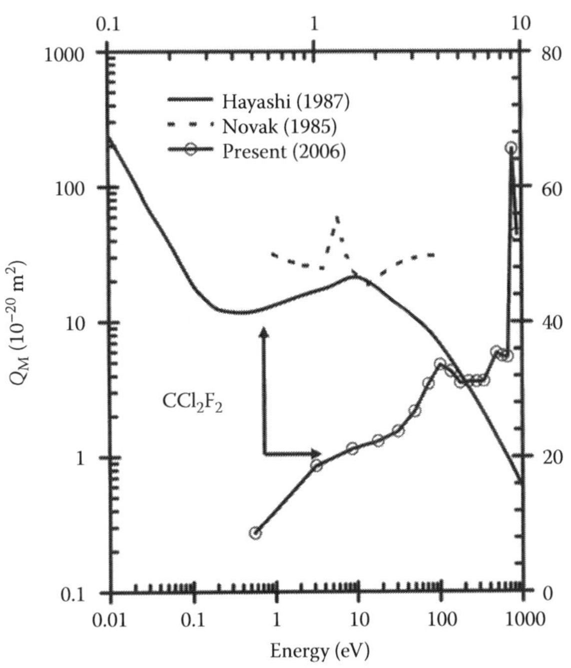

FIGURE 57.5 Momentum transfer cross sections in CCl_2F_2. (—○—) Present author, note the expanded scales for this curve; (——) Hayashi (1987); (– – –) Novak and Fréchette (1985). (Adapted from Mann, A. and F. Linder, *J. Phys. B: At. Mol. Opt. Phys.*, 25, 1633, 1992.)

57.7 ELECTRONIC EXCITATION CROSS SECTIONS

Experimental data on electronic excitation cross sections are not available. Novak and Fréchette have derived excitation cross sections with thresholds of 7.0 and 9.8 eV. Total excitation cross sections are obtained by summation according to

$$Q_{ex} + Q_T - Q_V - Q_i \quad (57.1)$$

TABLE 57.6
Total Direct Vibrational Cross Sections for CCl_2F_2

Energy (eV)	Q_V (10^{-20} m^2)	Energy (eV)	Q_V (10^{-20} m^2)	Energy (eV)	Q_V (10^{-20} m^2)
0.030	0.20	0.25	11.1	2.5	2.73
0.035	0.25	0.30	10.6	3.0	2.38
0.040	0.29	0.35	10.0	3.5	2.08
0.045	0.34	0.40	9.48	4.0	1.84
0.050	0.38	0.45	8.91	4.5	1.65
0.060	0.43	0.50	8.39	5.0	1.48
0.070	0.48	0.6	7.53	6.0	1.27
0.080	0.52	0.7	6.83	7.0	1.09
0.090	0.61	0.8	6.29	8.0	0.94
0.100	0.88	0.9	5.78	9.0	0.85
0.125	3.92	1.0	5.39	10.0	0.80
0.150	7.59	1.5	4.05		
0.20	10.9	2.0	3.23		

Source: Adapted from Mann, A. and F. Linder *J. Phys. B: At. Mol. Opt. Phys.*, 25, 1633, 1992.

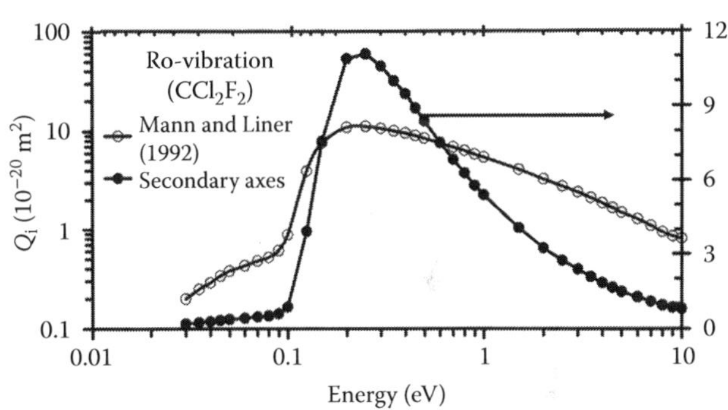

FIGURE 57.6 Total vibration cross sections for CCl_2F_2. Note the secondary axis for closed symbols. (Adapted from Mann, A. and F. Linder, *J. Phys. B: At. Mol. Opt. Phys.*, 25, 1633, 1992.)

where Q_{ex} is the total excitation cross section, Q_T the total, Q_V the ro-vibrational, and Q_i the total ionization cross sections, respectively. Figure 57.7 shows the electronic excitations referred to.

57.8 TOTAL IONIZATION CROSS SECTIONS

The measured total ionization cross sections are shown in Tables 57.7 and 57.8 and in Figure 57.8.

57.9 POSITIVE ION APPEARANCE POTENTIALS

Positive ion appearance potentials and peak cross sections are shown in Table 57.9.

Double ionization cross sections measured by Leiter et al. (1989) and quoted by Christophorou and Olthoff (2004) are shown in Table 57.10 and Figure 57.9.

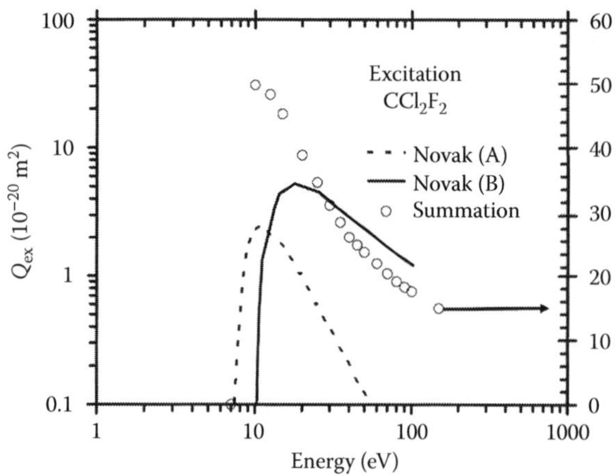

FIGURE 57.7 Excitation cross sections in CCl_2F_2. (– – –) 7.0 eV threshold, (——) 9.8 eV threshold, derived by Novak and Fréchette (1985); (○) calculated by summation method, present author (2006). Note the secondary ordinate for this curve.

TABLE 57.7
Total Ionization Cross Sections in CCl_2F_2

	Q_i (10^{-20} m²)		
Energy (eV)	Pejčev et al. (1979)	Leiter et al. (1989)	Lindsay et al. (2004)
12.24	0.00	0.00	0.00
15	1.02		0.72
16	1.76		
18	3.28		
20	4.65	1.67	2.63
22	5.76	2.56	
24	6.85	3.34	
25			4.27
26	7.95	4.03	
28	9.00	4.63	
30	9.82	5.14	5.45
32	10.33	5.60	
34	10.67	6.03	
35			6.24
36	10.98	6.51	
38	11.30	7.04	
40	11.63	7.44	6.81
44	12.31	7.80	
48	12.69	7.89	
50	12.82	7.94	7.72
55	13.15	8.06	
60	13.43	8.21	8.45
65	13.67	8.40	
70	13.92	8.59	
75	14.08	8.73	
80	14.12	8.81	9.04
85	14.10	8.82	
90	14.10	8.79	
95	14.12	8.73	
100	14.11	8.68	9.22
120	14.20	8.66	

TABLE 57.7 (continued)
Total Ionization Cross Sections in CCl_2F_2

	Q_i (10^{-20} m²)		
Energy (eV)	Pejčev et al. (1979)	Leiter et al. (1989)	Lindsay et al. (2004)
125			9.02
140	13.27	8.49	
150			8.70
160	12.98	8.23	
180	12.55	7.96	
200	12.05	8.06	8.28
225	11.42	8.21	
250	10.85	8.40	
300			6.75
400			5.78
500			4.98
600			4.53
800			3.78
1000			3.20

Note: Digitized and interpolated.

TABLE 57.8
Ionization Cross Sections for CCl_2F_2

Energy (eV)	Q_i (10^{-20} m²)	Energy (eV)	Q_i (10^{-20} m²)	Energy (eV)	Q_i (10^{-20} m²)
12	0.71	82	9.57	153	8.59
16	1.29	86	9.57	156	8.51
20	2.51	90	9.54	160	8.43
23	3.87	94	9.51	164	8.34
27	5.06	97	9.45	167	8.26
31	5.92	101	9.41	171	8.19
34	6.61	105	9.35	175	8.14
38	7.16	108	9.30	179	8.10
42	7.67	112	9.26	182	8.06
45	8.12	116	9.22	186	8.02
49	8.48	119	9.18	190	7.97
53	8.77	123	9.13	193	7.89
57	8.98	127	9.07	197	7.81
60	9.18	130	8.99	201	7.73
64	9.27	134	8.91	204	7.66
68	9.37	138	8.83	208	7.61
71	9.48	142	8.76	212	7.58
75	9.50	145	8.70	215	7.56
79	9.55	149	8.65		

Source: Adapted from Bart, M. et al., *Phys. Chem. Chem. Phys.*, 3, 800, 2001.
Note: Tabulated values are from courtesy of Professor P. Harland.

57.10 PARTIAL IONIZATION CROSS SECTIONS

The largest contribution to the total ionization cross section is from the ionization of dissociation fragment $CClF_2^+$. Partial ionization cross sections measured by Leiter et al. (1989) and quoted by Christophorou and Olthoff (2004) are shown in Figures 57.10 and 57.11. See Tables 57.11 and 57.12.

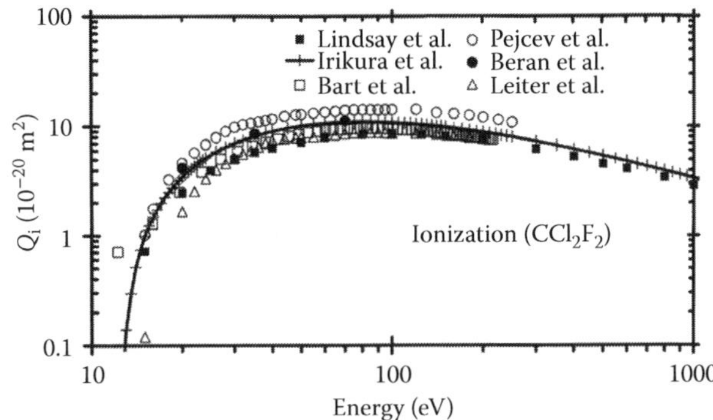

FIGURE 57.8 Ionization cross sections in CCl_2F_2. (—+—) Irikura et al. (2003); (Δ) Leiter et al. (1989); (○) Pejcev et al. (1979); (•) Beran and Kevan (1979); (■) Lindsay et al. (2004).

TABLE 57.9
Positive Ion Appearance Potentials and Peak Cross Section

Species	Energy Range (eV)	Reference	Peak Energy (eV)	Peak Cross Section (10^{-20} m^2)
			Lindsay et al. (2004)	
$CFCl_2^+$	11.5–12.10	Zhang et al. (1991)	80	4.21
CCl_2F^+	13.30–14.15	Schenk et al. (1979)	100	0.614
CF^+	17.3–19.84	Zhang et al. (1991)	60	0.635
Cl^+	18.5–29.3	Schenk et al. (1979) Zhang et al. (1991)	100	1.83
C^+	20.5–27.6	Zhang et al. (1991)	125	0.266
CCl^+	21.6–24.0	Schenk et al. (1979) Zhang et al. (1991)	100	1.17
F^+	25.7–33.8	Zhang et al. (1991) Christophorou and Olthoff (2004)	200	0.254

TABLE 57.10
Double Ionization Cross Sections in CCl_2F_2 for Selected Ion Species[a]

Energy (eV)	Q_i (10^{-22} m^2) $CClF_2^{2+}$	CCl_2^{2+}	CCl_2F^{2+}
40	0.09	0.06	0.34
45	0.39	0.45	0.99
50	0.71	1.07	1.50
55	1.00	1.52	1.88
60	1.23	1.88	2.10
65	1.39	2.08	2.29
70	1.51	2.25	2.44
75	1.60	2.37	2.58
80	1.67	2.46	2.66
90	1.76	2.61	2.82
100	1.80	2.70	2.90
110	1.82	2.78	2.93
120	1.80	2.81	2.92
130	1.78	2.83	2.90
140	1.76	2.79	2.88
150	1.72	2.77	2.85
160	1.69	2.74	2.78
170	1.65	2.68	2.71
180	1.60	2.62	2.64

[a] Measured by Leiter et al. (1989) and quoted by Christophorou and Olthoff (2004).

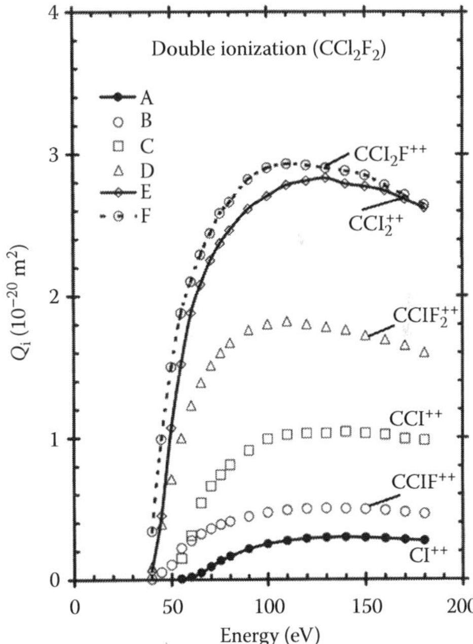

FIGURE 57.9 Double ionization cross section for CCl_2F_2. Measurements by Leiter et al. (1989), quoted by Christophorou and Olthoff (2004). Table 57.10 gives the tabulated values. Cross sections in increasing order are for: (—●—) Cl^{2+} (A); (○) $CClF^{2+}$ (B); (□) CCl^{2+} (C); (Δ) $CClF_2^{2+}$ (D); (—◊—) CCl_2^{2+} (E); (—○—) CCl_2F^{2+} (F).

57.11 ATTACHMENT CROSS SECTIONS

Thermal energy electrons captured by molecules in a two-body process form an excited negative ion according to the following reaction (Barszczewska et al., 2004):

$$AB + e \rightarrow (AB)^{*-} \quad (57.2)$$

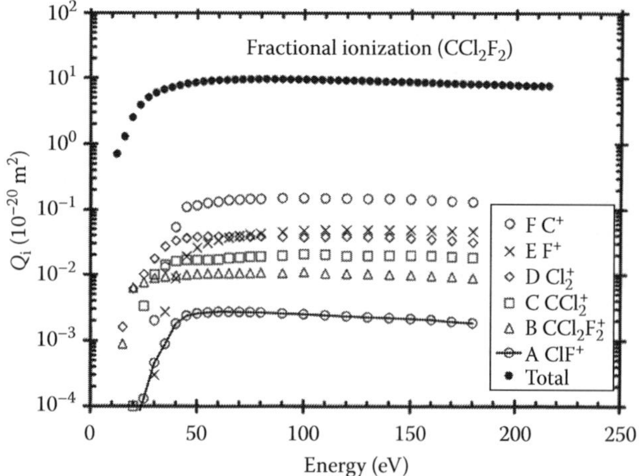

FIGURE 57.10 Partial ionization cross sections in CCl_2F_2. Measured by Leiter et al. (1989) and tabulated by Christophorou and Olthoff (2004). Tabulated values are given in Table 57.11. Ionized fragments are shown by the letters A through F.

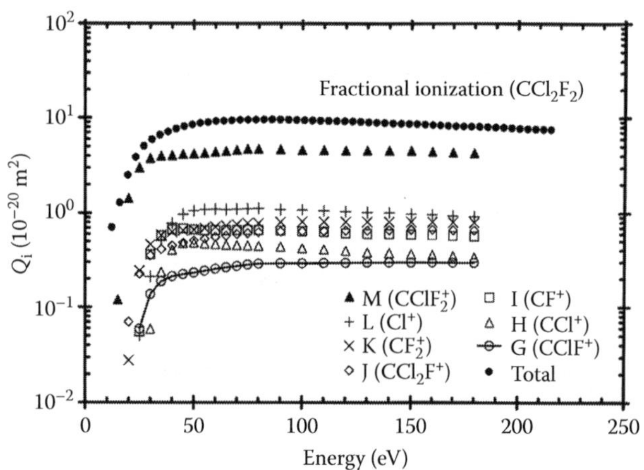

FIGURE 57.11 Partial ionization cross sections in CCl_2F_2. Measured by Leiter et al. (1989) and tabulated by Christophorou and Olthoff (2004). Tabulated values are given in Table 57.12. Ionized fragments are shown by the letters G through M.

TABLE 57.11
Partial Ionization Cross Sections in CCl_2F_2

	Q_i (10^{-20} m^2)			
Energy (eV)	F^+	C +	$CCl^+ + CF_2^+$	Cl^+
15				
20			0.049	
25			0.263	0.046
30			0.538	0.19
35		0.0238	0.742	0.443
40	0.026	0.0712	0.858	0.731
50	0.057	0.129	1.02	1.19
60	0.0866	0.169	1.1	1.48
80	0.16	0.225	1.15	1.75
100	0.219	0.257	1.17	1.83

TABLE 57.11 (continued)
Partial Ionization Cross Sections in CCl_2F_2

	Q_i (10^{-20} m^2)			
Energy (eV)	F^+	C +	$CCl^+ + CF_2^+$	Cl^+
125	0.252	0.266	1.11	1.76
150	0.251	0.264	1.06	1.72
200	0.254	0.237	1.01	1.59
300	0.195	0.186	0.826	1.18
400	0.165	0.151	0.706	0.968
500	0.131	0.123	0.591	0.793
600	0.12	0.106	0.542	0.709
800	0.0891	0.086	0.453	0.568
1000	0.0765	0.0723	0.377	0.442

Source: Adapted from Lindsay, B. G. et al., *J. Chem. Phys.*, 121, 1350, 2004.

Note: See Figure 57.12 for graphical presentation.

TABLE 57.12
Partial Ionization Cross Sections in CCl_2F_2

	Q_i (10^{-20} m^2)			
Energy (eV)	$CClF^+$	CCl_2F^+	CF^+	$CClF_2^+$
15				0.72
20		0.153		2.43
25	0.074	0.3	0.09	3.5
30	0.159	0.399	0.344	3.82
35	0.192	0.431	0.528	3.88
40	0.22	0.461	0.592	3.85
50	0.252	0.522	0.568	3.98
60	0.29	0.564	0.585	4.18
80	0.318	0.587	0.635	4.21
100	0.326	0.614	0.614	4.19
125	0.312	0.609	0.618	4.09
150	0.303	0.599	0.594	3.91
200	0.296	0.602	0.507	3.78
300	0.25	0.528	0.421	3.16
400	0.226	0.467	0.34	2.76
500	0.192	0.411	0.298	2.44
600	0.166	0.391	0.267	2.23
800	0.141	0.342	0.217	1.88
1000	0.123	0.303	0.182	1.62

Source: Adapted from Lindsay, B. G. et al., *J. Chem. Phys.*, 121, 1350, 2004.

Note: See Figure 57.13 for graphical presentation.

The modes of decay of the ion are

$$AB^{*-} \rightarrow AB + e(\text{autodetachment}) \quad (57.3)$$

$$AB^{*-} \rightarrow AB^{-} + h\nu \ (\text{radiative stabilization}) \quad (57.4)$$

$$AB^{*-} \rightarrow AB^{-} \ (\text{internal stabilization}) \quad (57.5)$$

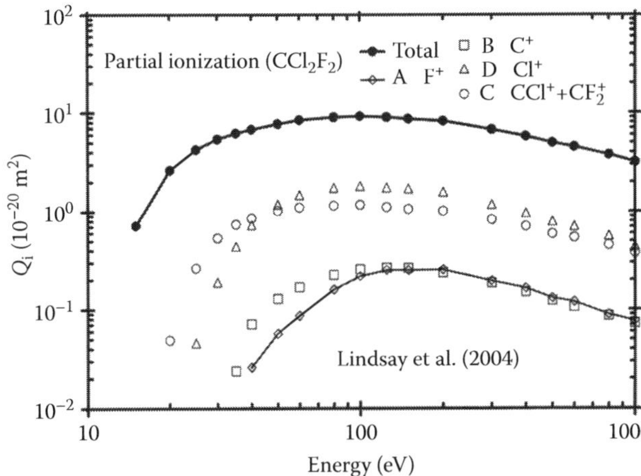

FIGURE 57.12 Partial ionization cross sections in CCl_2F_2. See Table 57.13 for tabulated values. The total cross sections are obtained by adding all partial cross sections. (—●—) Total; (—◊—) F^+; (□) C^+; (○) $CCl^+ + CF_2^+$; (Δ) Cl^+. (Adapted from Lindsay, B. G. et al., *J. Chem. Phys.*, 121, 1350, 2004.)

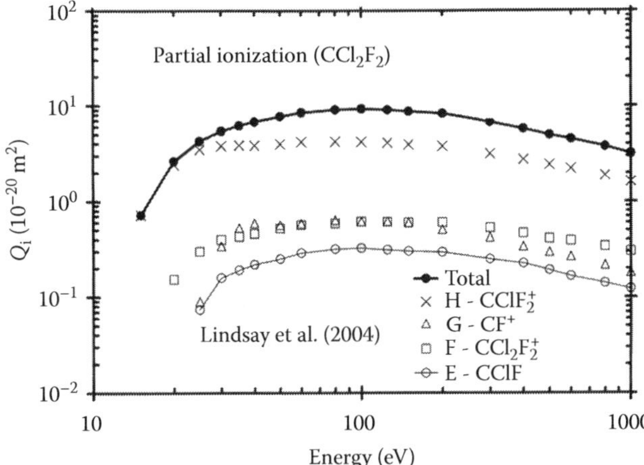

FIGURE 57.13 Partial ionization cross sections in CCl_2F_2. See Table 57.14 for tabulated values. The total cross sections are obtained by adding all partial cross sections. (—●—) Total; (—○—) $CClF^+$; (□) CCl_2F^+; (Δ) CF^+; (×) $CClF_2^+$. (Adapted from Lindsay, B. G. et al., *J. Chem. Phys.*, 121, 1350, 2004.)

$$AB^{*-} \rightarrow A + B^- \text{ or } A^- + B \text{ (dissociation)} \quad (57.6)$$

$$AB^* \rightarrow AB^* + e \text{ (excitation)} \quad (57.7)$$

The decay may also occur through a three-body process according to

$$AB^{*-} + M \rightarrow AB^- + M \text{ (three body process)} \quad (57.8)$$

Reaction 57.6 is usually associated with *s*-wave capture of the electron.

If the electron has greater energy, dissociative attachment in CCl_2F_2 occurs according to the following reactions (Mann and Linder 1992):

$$CCl_2F_2 + e \rightarrow \left.\begin{array}{l} CClF_2 + Cl^- + 0.3\,eV \\ CF_2 + Cl_2^- \\ CFCl + FCl^- \\ CF_2 + Cl + Cl^- \\ CF_2 + Cl_2^- \\ F + CFCl_2^- \\ CFCl + F + Cl^- \end{array}\right\} \quad (57.9)$$

Appearance potentials and peak for negative ion formation are shown in Table 57.13.

Features of the negative ion states and the methods of determination are shown in Table 57.14. Data are extracted from Christophorou and Olthoff (2004).

TABLE 57.13
Appearance Potentials (ε_A) of and Peak Energy (ε_P) due to Negative Ions

Ion	ε_A (eV)	Reference	ε_P (eV)	Peak Type	Reference
Cl^-			~0	M	Skalny et al. (2001)
Cl^-	0.5	Hickam and Berg (1959)	0.55	D	Illenberger et al. (1978a, 1978b, 1982) Randell et al. (1993)
Cl_2^-			0.65	D	Illenberger et al. (1978)
FCl^-			2.85		Illenberger et al. (1978)
F^-			3.1	D	Illenberger et al. (1982)
$CFCl_2^-$			3.55	D	Randell et al. (1993)
Cl^-			3.9	D	Mann and Linder (1992)
Total			0.9, 3.5	D	Christophorou and Olthoff (2004)

Note: D = distinct; M = monotonic rise or slow approach to zero. For a more detailed list, see Christophorou and Olthoff (2004).

TABLE 57.14
Negative Ion Features for CCl_2F_2

Energy	Method	Reference
0.97	Vertical attachment energy	Aflatooni and Burrow (2001)
~0 (Cl^-)	Molecular beam	Langer et al. (2000)
0.3	Electron transmission spectroscopy	Underwood-Lemons et al. (1995)
0.85		

TABLE 57.14 (continued)
Negative Ion Features for CCl_2F_2

Energy	Method	Reference
		continued
5.0–5.5	Differential scattering cross section	Mann and Linder (1992)
~0 (Cl^-)	Photoionization method	Chutjian and Alajajian (1987)
1.02	Total scattering cross section	Jones (1986)
2.64		
4.0		
0.55 (Cl^-)	Mass spectrometer method	Illenberger et al. (1979)
0.65 (Cl_2^-)		
2.85 (FCl^-)		
3.1 (F^-)		
3.55 ($CFCl_2^-$)		

CCl_2F_2

TABLE 57.15
Total Attachment Cross Sections in CCl_2F_2

ε (eV)	Q_a (10^{-20} m^2)	ε (eV)	Q_T (10^{-20} m^2)
Christophorou and Olthoff (2004)		Field et al. (2001), Digitized	
0.010	4.42	0.015	380.0
0.015	3.85	0.02	350.0
0.020	3.48	0.025	290.0
0.025	3.16	0.03	260.0
0.030	2.90	0.04	220.0
0.035	2.67	0.05	180.0
0.040	2.47	0.06	150.0
0.045	2.31	0.7	20.0
0.050	2.17	0.08	110.0
0.060	1.96	0.09	100.0
0.070	1.79	0.1	100.0
0.080	1.58	0.15	70.0
0.090	1.38	0.2	50.0
0.10	1.23	0.3	40.0
0.15	0.680	0.4	30.0
0.20	0.597	0.5	30.0
0.25	0.665	0.6	30.0
0.30	0.691	0.7	30.80
0.35	0.650	0.8	33.91
0.40	0.592	0.9	35.85
0.45	0.504	1	36.27
0.50	0.441	1.5	27.34
0.60	0.362	2	30.34
0.70	0.411	2.5	30.20
0.80	0.509	3	32.60
0.90	0.616	3.5	34.40
1.00	0.615	4	35.70
1.25	0.269	4.5	36.70
1.50	0.073	5	37.40
1.75	0.024	6	38.20
2.0	0.015	7	39.90
2.5	0.019	8	40.80
3.0	0.043	9.000	42.600
3.5	0.066	9.50	44.20
4.0	0.047		
4.5	0.023		
5.0	0.009		
6.0	0.001		

Source: Adapted from Christophorou, L. G. and J. K. Olthoff, *Fundamental Electrons Interactions with Plasma Processing Gases*, Kluwer Academic/Plenum Publishers, New York, NY, 2004.

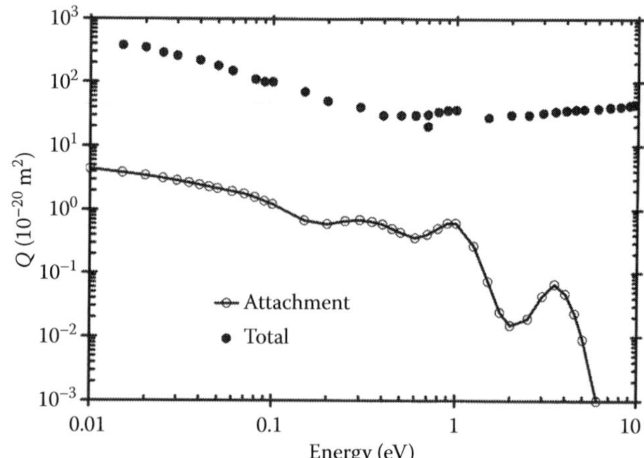

FIGURE 57.14 Total attachment cross section in CCl_2F_2. The total scattering cross sections are from Field et al. (2001). (Adapted from Christophorou, L. G. and J. K. Olthoff, *Fundamental Electrons Interactions with Plasma Processing Gases*, Kluwer Academic/Plenum Publishers, New York, NY, 2004.)

Total attachment cross sections suggested by Christophorou and Olthoff (2004) are shown in Table 57.15 and Figure 57.14 which also include the total cross section of Field et al. (2001) for comparison. Figure 57.15 shows the relative contribution of ion production processes.

57.12 ATTACHMENT RATES

Extensive discussion of attachment rates has been given by Christophorou and Olthoff (2004) and only representative data as a function of mean energy from electron studies, as a function of gas temperature and thermal energy, are given here. Tables 57.16 and 57.17 present these rates.

The influence of gas temperature on the attachment cross sections is shown in Figure 57.16 (Kiendler et al., 1996).

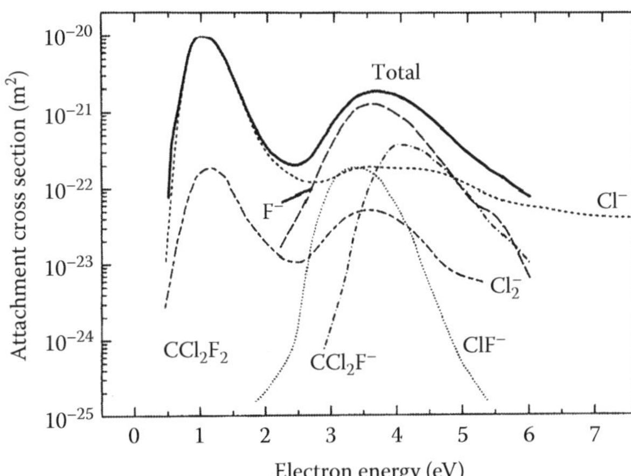

FIGURE 57.15 Attachment cross sections as a function of electron energy for ion production in CCl_2F_2. Relative values are normalized to 9×10^{-21} m² at 0.9 eV for total cross section. (Adapted from McCorkle, D. L. et al., *J. Chem. Phys.*, 72, 4049, 1980.)

TABLE 57.16
Attachment Rate Constants as a Function of Electron Mean Energy from Swarm Measurements

$\bar{\varepsilon}$ (eV)	k_a (10^{-15} m³/s)	$\bar{\varepsilon}$ (eV)	k_a (10^{-15} m³/s)	$\bar{\varepsilon}$ (eV)	k_a (10^{-15} m³/s)
0.05	1.77	0.80	1.85	2.6	0.86
0.06	1.75	0.90	1.75	2.8	0.85
0.07	1.73	1.0	1.59	3.0	0.83
0.08	1.71	1.1	1.42	3.2	0.80
0.09	1.68	1.2	1.26	3.4	0.77
0.10	1.66	1.3	1.10	3.6	0.75
0.20	1.44	1.4	0.97	3.8	0.74
0.30	1.39	1.5	0.86	4.0	0.77
0.40	1.53	1.6	0.79	4.2	0.76
0.50	1.72	1.7	0.77	4.4	0.68
0.60	1.86	1.8	0.76	4.6	0.67
0.65	1.90	2.0	0.79	4.8	0.62
0.70	1.90	2.2	0.85	5.0	0.50
0.75	1.89	2.4	0.87		

Source: Adapted from Christophorou, L. G. and J. K. Olthoff, *Fundamental Electrons Interactions with Plasma Processing Gases*, Kluwer Academic/Plenum Publishers, New York, NY, 2004.

Note: $\bar{\varepsilon}$ = mean energy; k_a = attachment rate.

Note the increase in the zero-energy cross section as the gas temperature increases (see Table 57.18).

57.13 DRIFT VELOCITY OF ELECTRONS

Drift velocity of electrons measured by Naidu and Prasad (1969) is shown in Table 57.19 and Figure 57.17.

57.14 DIFFUSION COEFFICIENT

Radial diffusion coefficients expressed as a ratio D_r/μ of electrons have been measured by Naidu and Prasad (1969) and Maller and Naidu (1975). Suggested values (Christophorou and Olthoff, 2004) are shown in Table 57.20 and Figure 57.18.

TABLE 57.17
Total Attachment Rate Constant in CCl_2F_2 at Thermal Energy

k_a (10^{-16} m³/s)	T (K)	Method	Reference
2.7×10^{-15}	~295	Analysis	Skalny et al. (2003)
16.6×10^{-16}	298	Electron swarm	Wang et al. (1998)
1.9×10^{-15}	293	ECR	Burns et al. (1996)
1.8×10^{-15}	~295	ECR	Marotta et al. (1989)
1.36×10^{-15}	298	Microwave conductivity	Bansal and Fessenden (1973)
18×10^{-16}	293	Electron cyclotron resonance	Marotta et al. (1989)
3.2×10^{-15}	300	FLAP	Smith et al. (1984)
12.3×10^{-16}	298	Electron swarm	McCorkle et al. (1980)
22×10^{-16}	298	Electron Swarm CRESU	Christophorou et al. (1974)

Source: Adapted from Skalny, J. D. et al., *Int. J. Mass Spectrom.*, 223–224, 217, 2003.

Note: ECR = electron cyclotron resonance; FALP = flowing after-glow/Langmuir probe; CRESU = supersonic jet crossed with electron beam/Langmuir probe. Average value recommended by Christophorou and Olthoff (2004) = $(15.7 \pm 7.1) \times 10^{-16}$ m³/s.

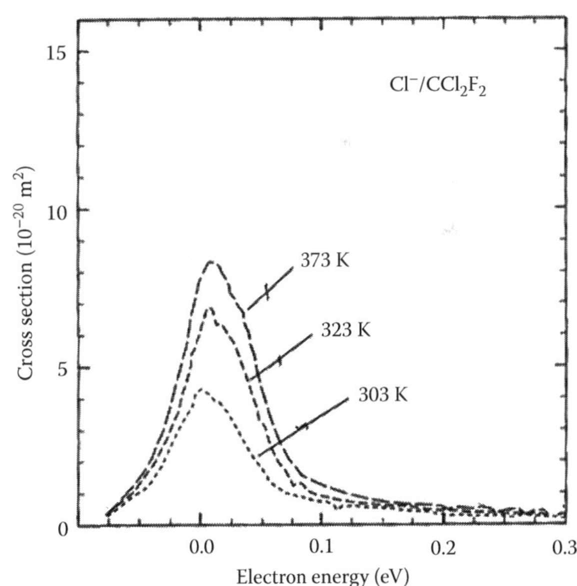

FIGURE 57.16 Attachment cross sections as a function of gas temperature. Note the increase in zero-energy cross section as the gas temperature increases. (Adapted from Kiendler, A. et al., *J. Phys. B: At. Mol. Opt. Phys.*, 29, 6217, 1996.)

57.15 REDUCED IONIZATION COEFFICIENTS

Reduced ionization coefficients recommended by Christophorou and Olthoff (2004) are shown in Table 57.21

CCl$_2$F$_2$

TABLE 57.18
Attachment Rates as a Function of Gas Temperature in CCl$_2$F$_2$

k_a (10^{-16} m^3/s)	T (K)	Reference
16.6	298	Wang et al. (1998)
60	400	
<140	500	
0.125	74	Le Garrec et al. (1997)
0.233	82	
0.234	123	
0.771	159	
2.44	168	
19	293	Burns et al. (1996)
140	467	
240	579	
420	777	
<10	205	Smith et al. (1984)
32	300	
160	455	
530	590	

Source: Adapted from Christophorou, L. G. and J. K. Olthoff, *Fundamental Electrons Interactions with Plasma Processing Gases*, Kluwer Academic/Plenum Publishers, New York, NY, 2004.

TABLE 57.19
Drift Velocity of Electrons

E/N (Td)	W (10^4 m/s)	E/N (Td)	W (10^4 m/s)
350	17.8	525	22.7
375	18.7	550	23.4
400	19.7	575	24.1
425	20.6	600	25.0
450	21.3	625	26.0
475	22.6	640	26.6
500	22.1		

[a] Measured by Naidu and Prasad (1969) and tabulated by Christophorou and Olthoff (2004)

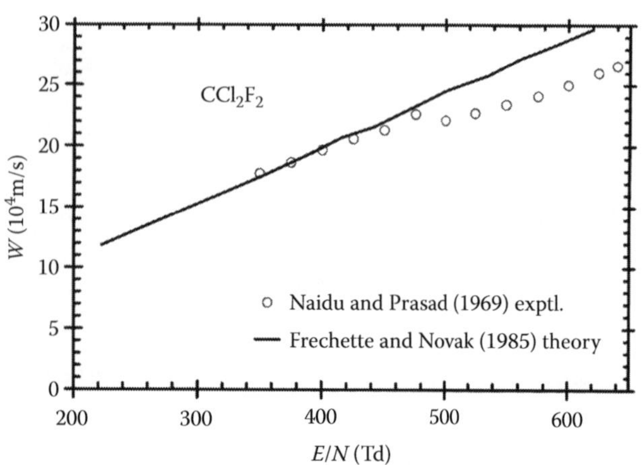

FIGURE 57.17 Drift velocity of electrons in CCl$_2$F$_2$. (○) Naidu and Prasad (1969), experimental; (——) Novak and Fréchette (1985), Boltzmann solution.

TABLE 57.20
Diffusion Coefficients Expressed as a Ratio D_r/μ of Electrons

E/N (Td)	D_r/μ (eV)	E/N (Td)	D_r/μ (eV)
335	3.82	550	3.88
350	3.81	575	3.90
400	3.76	600	3.93
425	3.77	625	3.96
450	3.79	650	4.01
475	3.82	675	4.07
500	3.84	700	4.13
525	3.86	725	4.19

Source: Adapted from Christophorou, L. G. and J. K. Olthoff, *Fundamental Electrons Interactions with Plasma Processing Gases*, Kluwer Academic/Plenum Publishers, New York, NY, 2004.

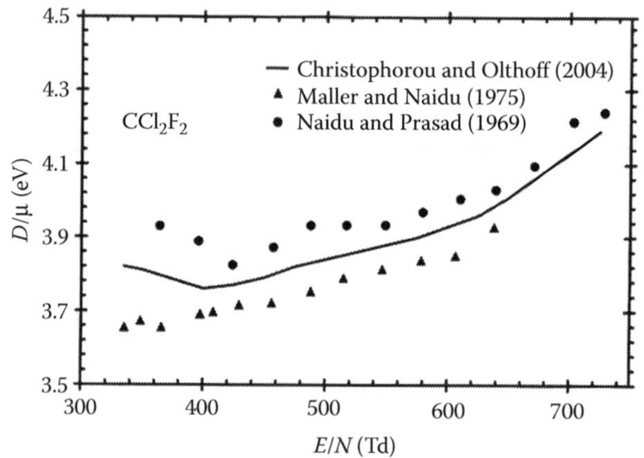

FIGURE 57.18 D/μ ratios in CCl$_2$F$_2$. (——) Suggested by Christophorou and Olthoff, (2004); (▲) Maller and Naidu (1975), experimental; (●) Naidu and Prasad (1969), experimental.

TABLE 57.21
Reduced Ionization Coefficients in CCl$_2$F$_2$

E/N (Td)	α/N (10^{-20} m^2)	E/N (Td)	α/N (10^{-20} m^2)
250	0.066	800	1.18
300	0.13	850	1.27
350	0.21	900	1.36
400	0.29	950	1.46
450	0.40	1000	1.56
500	0.50	1250	2.08
550	0.61	1500	2.65
600	0.73	2000	3.85
650	0.84	2500	5.15
700	0.96	3000	6.51
750	1.07		

Source: Adapted from Christophorou, L. G. and J. K. Olthoff, *Fundamental Electrons Interactions with Plasma Processing Gases*, Kluwer Academic/Plenum Publishers, New York, NY, 2004.

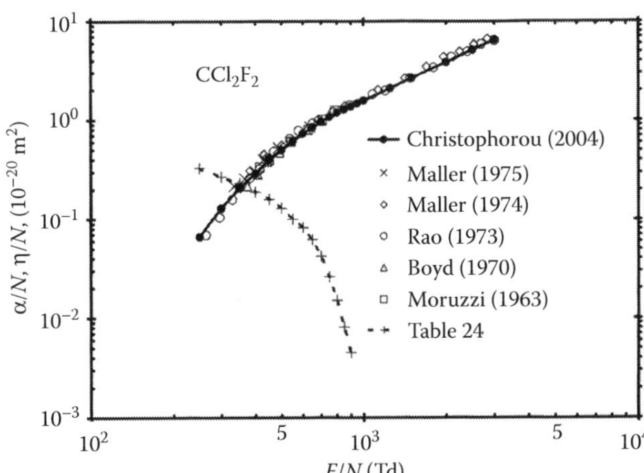

FIGURE 57.19 Reduced ionization coefficients in CCl_2F_2 (2004) recommended by Christophorou and Olthoff (2004). 10^{-17} V cm^2 = 1 Td. (—●—) Christophorou and Olthoff (2004); (×) Maller and Naidu (1975); (◊) Maller and Naidu (1974); (○) Rao and Govinda Raju (1973); (Δ) Boyd et al. (1970); (□) Moruzzi (1963); (— + —) Table 57.22, attachment coefficients are added for comparison sake.

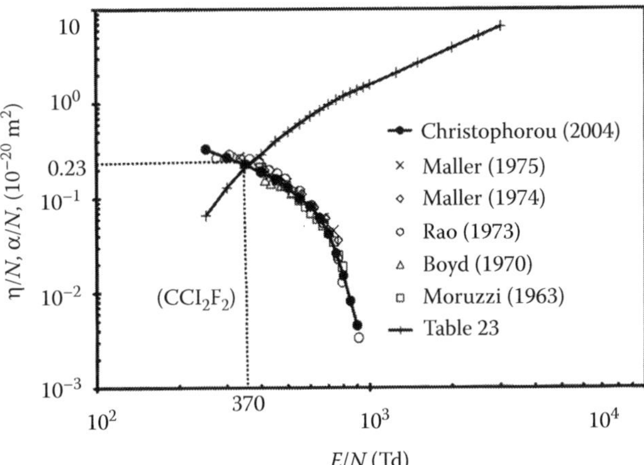

FIGURE 57.20 Reduced attachment coefficients in CCl_2F_2. (—●—) Christophorou and Olthoff (2004); (×) Maller and Naidu (1975); (◊) Maller and Naidu (1974); (○) Rao and Govinda Raju (1973); (Δ) Boyd et al. (1970). (□) Moruzzi (1963); (— × —) Table 57.21 ionization coefficients are added for comparison sake.

TABLE 57.22
Reduced Attachment Coefficients in CCl_2F_2

E/N (Td)	η/N (10^{-20} m²)	*E/N* (Td)	η/N (10^{-20} m²)
250	0.33	600	0.082
300	0.27	650	0.062
350	0.23	700	0.042
400	0.19	750	0.026
450	0.16	800	0.015
500	0.13	850	0.0081
550	0.1	900	0.0045

Source: Adapted from Christophorou, L. G. and J. K. Olthoff, *Fundamental Electrons Interactions with Plasma Processing Gases*, Kluwer Academic/Plenum Publishers, New York, NY, 2004.

and Figure 57.19 which also include selected previously published data.

57.16 REDUCED ATTACHMENT COEFFICIENTS

Reduced attachment coefficients recommended by Christophorou and Olthoff (2004) are shown in Table 57.22 and Figure 57.20 which also include selected previously published data.

57.17 LIMITING *E/N*

Limiting *E/N* values defined as the condition for α/N = η/N are shown in Table 57.23.

TABLE 57.23
$(E/N)_{Lim}$ of CCl_2F_2

$(E/N)_{Lim}$ (Td)	Method	Reference
360	Current growth	Dincer and Govinda Raju (1985)
372	—	Siddagangappa et al. (1983)
379	Sparking voltage	Nema et al. (1982)
390	—	Christophorou et al. (1981)
357	—	Wooton et al. (1980)
373	Current growth	Maller and Naidu (1975)
375	—	Rao and Govinda Raju (1973)
372	—	Boyd et al. (1970)
373	—	Moruzzi (1963)

REFERENCES

Aflatooni, K. and P. D. Burrow, *Int. J. Mass Spectrom.*, 205, 149, 2001.

Bansal, K. M. and R. W. Fessenden, *J. Chem. Phys.*, 59, 1760, 1973.

Barszczewska, W., J. Kopyra, J. Wnorowska, and I. Szamrej, *Int. J. Mass Spectrom.*, 233, 199, 2004.

Bart, M., W. Harland, J. E. Hudson, and C. Vallance, *Phys. Chem. Chem. Phys.*, 3, 800, 2001.

Beran, J. A. and L. Kevan, *J. Phys. Chem.*, 73, 3866, 1979.

Boyd, H. A., G. C. Chrichton, and T. M. Nielsen, 1970, cited by Christophorou and Olthoff 2004, p. 741.

Burns, S. J., J. M. Matthews, and D. L. McFadden, *J. Phys. Chem.*, 100, 19436, 1996.

Christophorou, L. G. and J. K. Olthoff, *Fundamental Electrons Interactions with Plasma Processing Gases*, Kluwer Academic/Plenum Publishers, New York, NY, 2004.

Christophorou, L. G., R. A. Mathis, D. R. James, and D. L. McCorkle, *J. Phys. D: Appl. Phys.*, 14, 1889, 1981.

Chutjian, A. and S. H. Alajajian, *J. Phys. B: At. Mol. Phys.*, 20, 839, 1987.

Dincer, M. S. and G. R. Govinda Raju, *IEEE Trans. Electr. Insul.*, EI-20, 595, 1985.

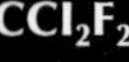

Field, D., N. C. Jones, S. L. Lunt, J.-P. Ziesel, and R. J. Gulley, *J. Chem. Phys.*, 115, 3045, 2001.

Fréchette, M. F., *J. Appl. Phys.*, 59, 3684, 1986.

Harrison, M. A. and R. Geballe, *Phys. Rev.*, 91, 1, 1953.

Hayashi, M. in L. C. Pitchford, B. V. McKoy, A. Chutjian, and S. Trajmar (Eds.), *Swarm Studies and Inelastic Electron-Molecule Collisions*, Springer, New York, NY, 1987, p. 167, cited by Christophorou and Olthoff, 2004.

Hickam W. M. and D. Berg, *Adv. Mass Spectr.*, 1, 458, 1959, cited by Christophorou and Olthoff, 2004.

Hudson, J. E., C. Vallance, M. Bart, and P. W. Harland, *J. Phys. B: At. Mol. Opt. Phys.*, 34, 3025, 2001.

Illenberger, E., H.-U. Scheunemann, and H. Baumgärtel, *Chem. Phys.*, 34, 161, 1978a.

Illenberger, E., H.-U. Scheunemann, and H. Baumgärtel, *Phys. Chem.*, 82, 1154, 1978b.

Illenberger, E., H.-U. Scheunemann, and H. Baumgärtel, *Chem. Phys.*, 37, 21, 1979.

Illenberger, E., H.-U. Scheunemann, and H. Baumgärtel, *Phys. Chem.*, 86, 252, 1982 (cited by Christophorou and Olthoff, 2004).

Irikura, K. K., M. A. Ali, and Y.-K. Kim, *Int. J. Mass Spectr.*, 222, 189, 2003.

Jiang, Y., J. Sun, and L. Wan, *Phys. Rev. A*, 52, 398, 1995.

Jones, R. K., *J. Chem. Phys.*, 84, 813, 1986.

Kiendler, A., S. Matejcik, J. D. Skalny, A. Stamatovic, and T. D. Märk, *J. Phys. B: At. Mol. Opt. Phys.*, 29, 6217, 1996.

King, G. C. and J. W. McConkey, *J. Phys. B: At. Mol. Opt. Phys.*, 11, 1861, 1978.

Langer, J., S. Mutt, M. Meinke, P. Tegeder, A. Stamatovic, and E. Illenberger, *J. Chem. Phys.*, 113, 11063, 2000.

Le Garrec, J. L., O. Sidko, J. L. Queffelec, S. Hamon, J. B. A. Mitchell, and B. R. Rowe, *J. Chem. Phys.*, 107, 54, 1997.

Leiter, K., P. Scheier, G. Walder, and T. D. Märk, *Int. J. Mass Spectrom. Ion Processes*, 87, 209, 1989.

Lindsay, B. G., K. F. McDonald, W. S. Yu, R. F. Stebbings, and F. B. Yousif, *J. Chem. Phys.*, 121, 1350, 2004.

Maller, V. N. and M. S. Naidu 1974, cited by Christophorou and Olthoff, 2004, p. 741.

Maller, V. N. and M. S. Naidu, *IEEE Trans. Plasma Sci.*, PS-3, 205, 1975.

Mann, A. and F. Linder, *J. Phys. B: At. Mol. Opt. Phys.*, 25, 1633, 1992.

Marotta, C. J., C.-P. Tsai, and D. L. McFadden, *J. Chem. Phys.*, 91, 2194, 1989.

Matejčik, Š., V. Foltin, M. Stano, and J. D. Skalný, *Int. J. Mass Spectrom.*, 223, 9, 2003.

McCorkle, D. L., A. A. Christodoulides, L. G. Christophorou, and I. Szamrej, *J. Chem. Phys.*, 72, 4049, 1980.

Moruzzi, J. L., *Br. J. Appl. Phys.*, 14, 938, 1963.

Naidu, M. S. and A. N. Prasad, *Br. J. Appl. Phys.*, 2, 1431, 1969.

Natalense, A. P. P., M. H. F. Bettega, L. G. Ferriera, and M. A. P. Lima, *Phys. Rev. A*, 59, 879, 1999.

Nema, R. S., S. V. Kulkarni, and E. Husain, *IEEE Trans. Electr. Insul.*, EI-17, 434, 1982.

Novak, J. P. and M. F. Fréchette, *J. Appl. Phys.*, 57, 4368, 1985.

Pejcev, V. M., M. V. Kurepa, and I, M, Cadez, *Chem. Phys. Lett.*, 63, 301, 1979.

Randell, J., J.-P. Ziesel, S. L. Lunt, G. Mrotzek, and D. Field, *J. Phys. B: At. Mol. Opt. Phys.*, 26, 3423, 1993.

Rao, C. R. and G. R. Govinda Raju, *Int. J. Electron.*, 35, 49, 1973.

Schenk, H., H. Oertel, and H. Bunsenges, *Phys. Chem*, 83, 683, 1979.

Siddagangappa, M. C., C. S. Lakshminarasimha, and M. S. Naidu, *J. Phys. D: Appl. Phys.*, 16, 763, 1983.

Skalny, J. D., S. Matejcik, T. Mikoviny, and T. D. Märk, *Int. J. Mass Spectrom.*, 223–224, 217, 2003.

Skalny, J. D., S. Matejcik, T. Mikoviny, J. Vencko, G. Senn, A. Stamatovic, and T. D. Märk, *Int. J. Mass Spectrom.*, 205, 77, 2001.

Smith, D., N. G. Adams, and E. Alge, *J. Phys. B: At. Mol. Phys.*, 17, 461, 1984.

Underwood-Lemons, T., D. C. Winkler, J. A. Tossell, and J. H. Moore, *J. Chem. Phys.*, 10, 9117, 1994.

Underwood-Lemons, T., T. J. Gergel, and J. H. Moore, *J. Chem. Phys.*, 102, 119, 1995.

Wang, Y., L. G. Christophorou, and J. K. Verbugge, *J. Chem. Phys.*, 1.9, 8304, 1998.

Wooton, R. E., S. J. Dale, and N. J. Zimmerman, in L. G. Christophorou (Ed.), *Gaseous Dielectrics II*, Pergamon, New York, NY, 1980, p. 137.

Zecca, A., G. P. Karwasz, and R. S. Brusa, *Phys. Rev. A*, 46, 3877, 1992.

Zhang, W., G. Cooper, T. Ibuki, and C. E. Brion, *Chem. Phys*, 151, 357, 1991.

58

DICHLOROMETHANE AND DIFLUOROMETHANE

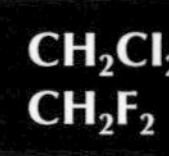

CONTENTS

Dichloromethane (CH_2Cl_2) and difluoromethane (CH_2F_2) are molecules in which two hydrogen atoms of methane are substituted by chlorine and fluorine atoms, respectively. The substitution renders both the molecules polar and electron-attaching. They have 42 and 26 electrons, respectively. Selected properties of the molecules are shown in Table 58.1.

Vibrational excitation modes and energies of the molecules are shown in Table 58.2. Theoretical integral elastic and momentum transfer cross sections in CH_2F_2 are given by Nishimura (1998). Differential cross sections measured by Tanaka et al. (1997) for selected energies are shown in Figure 58.1. Integrated values shown are given by the author by extrapolation at either end of the angular range.

TABLE 58. 1
Selected Molecular Properties

Property	CH_2Cl_2	CH_2F_2
Polarizability (α_e)	7.21×10^{-40} F m^2, 8.12×10^{-40} (alternate value)	—
Dipole moment (μ)	1.6 D	1.98 D
Ionization potential	11.32	12.71
Bond strength	4.22 eV	4.47 eV

Note: Vibrational excitation modes and energies of the molecules are shown in Table 58.2.

TABLE 58.2
Vibrational Excitation Modes and Energies

CH_2Cl_2[a]			CH_2F_2[b]	
Designation	Mode	Energy (meV)	Designation	Energy (meV)
ν_1	CH_2 s-stretch	371.8	ν_1	365.6
ν_2	CH_2 scissoring	181.9	ν_2	
ν_3	CCl_2 s-stretch	88.9	ν_3	132.7
ν_4	CCl_2 scissoring	35.0	ν_4	65.6
ν_5	CH_2 twist	143.0	ν_5	
ν_6	CH_2 a-stretch	376.9	ν_6	373.8
ν_7	CH_2 rock	111.3	ν_7	144.4
ν_8	CH_2 wag	157.2	ν_8	177.3
ν_9	CCl_2 a-stretch	94.0	ν_9	135.1

Note: Vibrational excitation modes and energies. a = Asymmetrical; s = symmetrical.

[a] Shimanouchi, 1972

[b] Sverdlov et al., 1974

58.1 TOTAL CROSS SECTIONS

Figure 58.1 shows the total scattering cross section in CH_2Cl_2 from Wan et al. (1991) in the energy range up to 12 eV and from Karwasz et al. (1999) for the range from 75 to 4000 eV. Table 58.3 shows the total scattering cross sections of CH_2Cl_2 and similar data for $CHCl_3$ (trichloromethane) are also included for comparison.

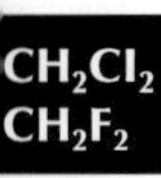

58.2 IONIZATION CROSS SECTIONS

Ionization cross sections in CH_2Cl_2 (see Hudson et al., 2005) are shown in Table 58.4. Cross sections for dibromomethane (CH_2Br_2) (Bart et al., 2001), tribromomethane ($CHBr_3$) (Bart

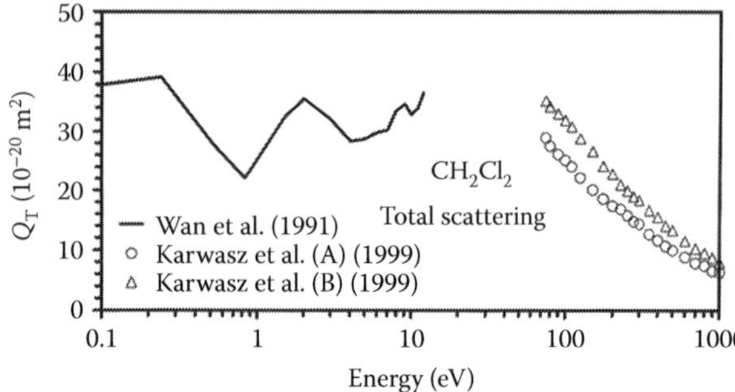

FIGURE 58.1 Total scattering cross sections in CH_2Cl_2. (——) Wan et al. (1991); (○) Karwasz et al. (1999). (Δ) Trichloromethane ($CHCl_3$), Karwasz et al. (1999) plotted for comparison. Substitution of a third chlorine atom in place of hydrogen increases the cross section in the high-energy range.

TABLE 58.3
Total Scattering Cross Sections in CH_2Cl_2 and $CHCl_3$

	Q_T (10^{-20} m²)			Q_T (10^{-20} m²)	
Energy (eV)	CH_2Cl_2	$CHCl_3$	Energy (eV)	CH_2Cl_2	$CHCl_3$
75	28.9	35.2	600	8.81	11.5
80	27.5	34.2	700	7.82	10.2
90	26.1	33.0	800	7.36	9.49
100	25.1	31.9	900	6.51	8.69
110	24.0	30.8	1000	6.24	7.91
125	22.1	28.8	1100	5.70	7.33
150	20.0	26.6	1250	5.15	6.86
175	18.5	24.1	1500	4.34	5.87
200	17.3	22.7	1750	3.80	5.14
225	16.8	20.9	2000	3.40	4.75
250	15.7	19.9	2250	3.10	4.36
275	14.7	18.9	2500	2.83	3.84
300	14.3	18.3	2750	2.64	3.47
350	12.6	16.6	3000	2.49	3.14
400	11.6	15.5	3250	2.37	2.91
450	10.6	14.0	3480	2.19	2.89
500	9.90	13.3	4000	1.82	–

Source: Adapted from Karwasz, G. P. et al., *Phys. Rev. A*, 59, 1341, 1999.

TABLE 58.4
Ionization Cross Sections in Substituted Methanes

	Ionization Cross Section (10^{-20} m²)				
Energy (eV)	CHF_3	CH_2Cl_2	CH_2Br_2	$CHBr_3$	CCl_4
12	0.07	0.93	1.32	1.45	1.76
16	0.21	1.66	2.30	2.53	2.79
20	0.52	3.16	4.38	4.73	4.97
23	0.91	4.65	6.16	6.71	7.31
27	1.36	5.84	7.49	8.22	9.30
31	1.76	6.58	8.26	9.13	10.55
34	2.14	7.19	8.98	10.10	11.56
38	2.47	7.67	9.55	10.84	12.36
42	2.75	8.07	9.91	11.48	13.16
45	3.01	8.43	10.35	12.03	13.77
49	3.23	8.69	10.68	12.45	14.25
53	3.42	8.95	11.01	12.82	14.69
57	3.58	9.10	11.26	13.12	14.92
60	3.72	9.20	11.47	13.36	15.16
64	3.82	9.29	11.57	13.57	15.27
68	3.92	9.34	11.62	13.67	15.45
71	4.01	9.35	11.67	13.75	15.52
75	4.09	9.35	11.67	13.75	15.45
79	4.15	9.33	11.62	13.75	15.44
82	4.19	9.30	11.59	13.74	15.40
86	4.23	9.25	11.53	13.71	15.32
90	4.26	9.18	11.45	13.64	15.23
94	4.28	9.10	11.36	13.56	15.11
97	4.30	9.02	11.27	13.48	14.98
101	4.31	8.94	11.17	13.40	14.84
105	4.31	8.86	11.06	13.31	14.71
108	4.31	8.78	10.96	13.21	14.59
112	4.32	8.71	10.87	13.12	14.47
116	4.32	8.64	10.80	13.02	14.36
119	4.32	8.58	10.73	12.93	14.24
123	4.32	8.51	10.66	12.85	14.12
127	4.32	8.43	10.57	12.78	13.98
130	4.30	8.32	10.45	12.67	13.80
134	4.27	8.21	10.30	12.52	13.61
138	4.24	8.11	10.17	12.38	13.46
142	4.23	8.04	10.09	12.27	13.36
145	4.22	8.00	10.04	12.20	13.29
149	4.22	7.96	9.99	12.15	13.22
153	4.21	7.90	9.92	12.10	13.12
156	4.19	7.81	9.82	12.03	12.98
160	4.16	7.71	9.69	11.91	12.81
164	4.12	7.60	9.55	11.77	12.62
167	4.09	7.50	9.43	11.62	12.45
171	4.06	7.43	9.33	11.49	12.32
175	4.03	7.38	9.27	11.39	12.22
179	4.02	7.34	9.24	11.32	12.15
182	4.01	7.32	9.23	11.28	12.11
186	4.00	7.29	9.21	11.26	12.06
190	3.99	7.25	9.16	11.20	11.98
193	3.96	7.17	9.08	11.11	11.87
197	3.92	7.08	8.97	10.99	11.73
201	3.88	6.99	8.85	10.85	11.59
204	3.85	6.91	8.74	10.73	11.46

TABLE 58.4 (continued)
Ionization Cross Sections in Substituted Methanes

	Ionization Cross Section (10^{-20} m^2)				
Energy (eV)	CHF_3	CH_2Cl_2	CH_2Br_2	$CHBr_3$	CCl_4
208	3.82	6.86	8.66	10.64	11.37
212	3.81	6.83	8.63	10.60	11.33

Sources: Courtesy of Professor Harland; adapted from Bart, M. et al., *Phys. Chem. Chem. Phys.*, 3, 800, 2001.

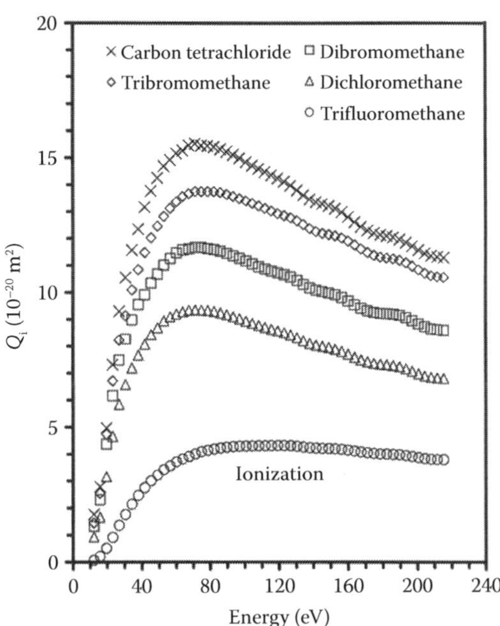

FIGURE 58.2 Ionization cross sections in selected substituted methanes. (○) Trifluoromethane (CHF_3, Bart et al., 2001); (Δ) dichloromethane (CH_2Cl_2, Hudson et al., 2001); (□) dibromomethane (CH_2Br_2, Bart et al., 2001); (◊) tribromomethane ($CHBr_3$, Bart et al., 2001); (×) carbon tetrachloride (CCl_4, Hudson et al., 2001). Tabulated values courtesy of Professor Harland.

et al., 2001), trifluoromethane (CHF_3) (Bart et al., 2001), and carbon tetrachloride (CCl_4) (Hudson et al., 2001) are included to demonstrate the influence of halogen substitution (Figure 58.2). Total and partial ionization cross sections in CH_2F_2 are given in the addendum.

58.3 ATTACHMENT CROSS SECTIONS

Dissociative attachment cross sections are measured by Wan et al. (1991) as shown in Figure 58.3. Table 58.5 provides a summary from Karwasz et al. (2001).

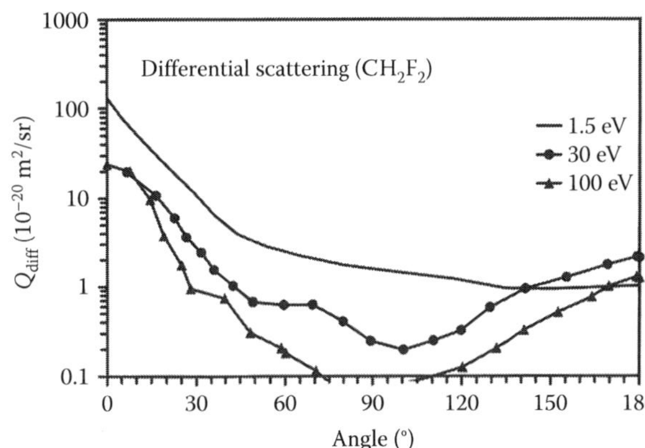

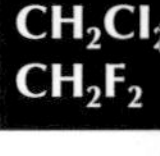

FIGURE 58.3 Dissociative attachment cross sections in chloromethanes. (Adapted from Wan, H., J. H. Moore, and J. A. Tossell, *J. Chem. Phys.*, 94, 1868, 1991.)

TABLE 58.5
Electron Attachment Cross Sections in Chloromethanes

	X-Section (m^2)		Peak Energy	
Molecule	Zero Energy	Peak	(eV)	Reference
CH_3Cl		2 (−25)	0.8	Pearl and Burrow (1993)
CH_2Cl_2	8.9 (−23)	2.4 (−22)	0.48	Chu and Burrow (1990)
$CHCl_3$	4.7 (−20)	3.7 (−20)	0.27	Chu and Burrow (1990)
CCl_4	1.3 (−18)	1.9 (−20)	0.80	Chu and Burrow (1990)

Source: Adapted from Karwasz, G. P., R. S. Brusa, and A. Zecca, *La Rivisita del Nuovo Cimento*, 24(3), 1, 2001.
Note: a (b) means a × 10^b.

TABLE 58.6
Elastic and Momentum Transfer Cross Sections in CH_2F_2

Energy (eV)	Nishimura (1998) Cross Section (10^{-20} m^2)		Energy (eV)	Tanaka et al. (1997) Cross Section (10^{-20} m^2)	
	Q_{el} (343 K)	Q_M (343 K)		Q_{el}	Q_M
3	60.51	25.59	1.5	47.19	18.97
5	48.29	22.09	30	14.83	8.10
7	40.16	18.74	100	7.98	3.11
10	36.65	16.87			
15	32.58	16.52			
20	28.15	14.74			
30	23.63	11.69			

Note: The differential scattering cross sections of Tanaka et al. (1997), Figure 58.4; integrated values are from the author.

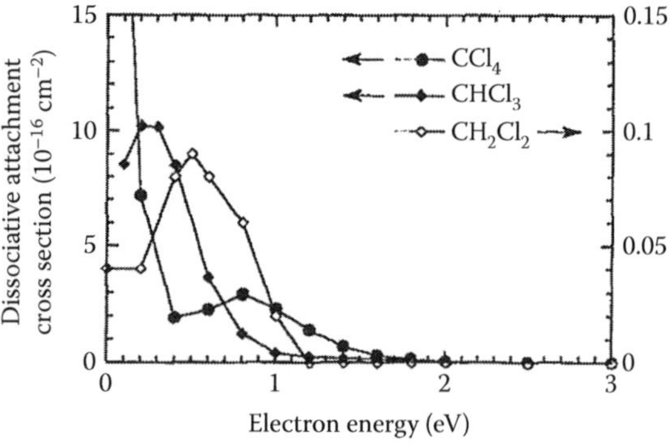

FIGURE 58.4 Differential scattering cross sections in CH_2F_2. The curves are extrapolated at either end of the angle scale by the author for purpose of integration. (Adapted from Tanaka, H. et al., *Phys. Rev. A*, 56, R3338, 1997.)

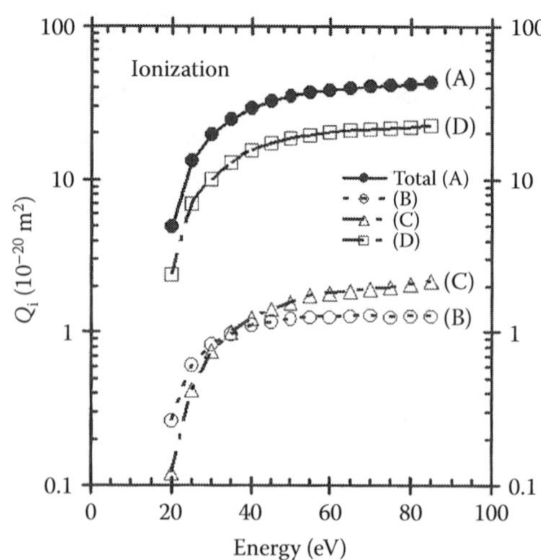

FIGURE 58.5 Total (A) and partial (B–D) ionization cross sections for CH_2F_2. (B) $CH_2F_2^+$; (C) CHF^+; (D) CH_2F^+. (Adapted from Torres, I. et al., *J. Phys. B: At. Mol. Opt. Phys.*, 33, 3615, 2000.)

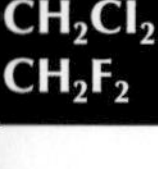

TABLE 58.7
Products of Ionization and Appearance Potentials for CH_2F_2

Products	Potential (eV)
$CH_2F_2^+$	12.6
$CHF_2^+ + H$	12.8
$CH_2F^+ + F$	14.9
$CHF^+ + HF$	13.3
$CF_2^+ + H_2$	13.5
$CF^+ + HF + H$	15.9
$CH^+ + HF + F$	19.2
$CH_2^+ + 2F$	23.9
$C^+ + 2H + 2F$	29.5
$F^+ + C + F + 2H$	34.3

58.4 ELASTIC AND MOMENTUM TRANSFER CROSS SECTIONS

See Table 58.6.

58.5 ADDENDUM

58.5.1 Ion Appearance Potentials

for selected ions are from Torres et al. (2000) as shown in Table 58.7.

58.5.2 Ionization Cross Section

Total ionization cross sections for CH_2F_2 from Torres et al. (2000) are shown in Figure 58.5. Partial ionization cross sections for selected ions are also included. Note that the dominant contribution is due to CH_2F^+ and the contribution from the parent ion $CH_2F_2^+$ is approximately 2.5–3% of the total.

REFERENCES

Bart, M., P. W. Harland, J. E. Hudson, and C. Vallance, *Phys. Chem. Chem. Phys.*, 3, 800, 2001. Also see the corrigendum, *J. Phys. B: At. Mol. Opt. Phys.*, 38, 1077, 2005.

Chu, S. C. and P. D. Burrow, *Chem. Phys. Lett.*, 171, 17, 1990.

Hudson, J. E., C. Vallance, M. Bart, and P. W. Harland, *J. Phys. B: At. Mol. Opt. Phys.*, 34, 3025, 2001. Also see the corrigendum, *J. Phys. B: At. Mol. Opt. Phys.*, 38, 1077, 2005.

Karwasz, G. P., R. S. Brusa, A. Piazza, and A. Zecca, *Phus. Rev. A*, 59, 1341, 1999.

Karwasz, G. P., R. S. Brusa, and A. Zecca, One century of experiments on electron–atom and molecule scattering: A critical review of integral cross-sections, *La Rivisita del Nuovo Cimento*, 24(3), 1, 2001.

Nishimura, T., *J. Phys. B: At. Mol. Opt. Phys.*, 31, 3471, 1998.

Pearl, D. M. and P. D. Burrow, *Chem. Phys. Lett.*, 206, 483, 1993.

Shimanouchi, T., *Tables of Molecular Vibrational Frequencies*, NSRDS-NBS 39, Washington (DC), 1972, p. 57.

Sverdlov, L. M., M. A. Kovner, and E. P. Krainov, *Vibrational Spectra of Polyatomic Molecules*, John Wiley & Sons, New York, 1974, p. 384.

Tanaka, H., T. Masai, M. Kimura, T. Nishimura, and Y. Itikawa, *Phys. Rev. A*, 56, R3338, 1997.

Torres, I., R. Martínez, M. N. Sánchez Rayo, and F. Castaño, *J. Phys. B: At. Mol. Opt. Phys.*, 33, 3615, 2000.

Wan, H., J. H. Moore, and J. A. Tossell, *J. Chem. Phys.*, 94, 1868, 1991.

59

FLUOROMETHANE

CONTENTS

Fluoromethane (CH_3F), synonym methyl fluoride, with 18 electrons, is a molecule in which one of the hydrogen atoms in methane is replaced with fluorine. The substitution renders the molecule both polar (polarizability = 3.30×10^{-40} F m^2; dipole moment = 1.858 D) and electron-attaching. The molecule has six vibrational modes in conformity with the other molecules of the same group. The energies and deformation details are shown in Table 63.1 in Chapter 63. The ionization potential is 12.47 eV.

59.1 SELECTED REFERENCES FOR DATA

See Table 59.1.

TABLE 59.1
Selected References for Data

Quantity	Energy Range (eV)	Reference
Q_i	14–1000	Rejoub et al. (2001)
Q_i	20.0–85.0	Torres et al. (2002)
Q_M	10–30	Natalense et al. (1999)
Q_{diss}	10–500	Motlagh and Moore (1998)
Q_{diff}	1.5–100	Tanaka et al. (1997)
Q_i	10.5–220	Vallance et. al. (1997)
Q_T	0.35–250	Krzysztofowicz et. al. (1995)
Q_T	0.5–8.0	Benitez et al. (1988)
k_a	Thermal	Fessenden and Bansal (1970)

Note: k_a = Attachment rate; Q_{diff} = differential; Q_{diss} = dissociation; Q_i = ionization; Q_M = momentum transfer; Q_T = total cross section.

59.2 TOTAL SCATTERING CROSS SECTION

Total scattering cross sections in CH_3F are shown in Table 59.2 and Figure 59.1. The highlights of Figure 59.1 are

TABLE 59.2
Total Scattering Cross Section in CH_3F

Energy (eV)	Q_T (10^{-20} m^2)	Energy (eV)	Q_T (10^{-20} m^2)	Energy (eV)	Q_T (10^{-20} m^2)
0.35	38.8	5.0	23.1	40	15.7
0.45	36.2	5.5	23.7	45	15.2
0.55	33.8	6.5	24.5	50	14.7
0.65	31.8	7.5	24.4	60	14.2
0.75	30.4	8.5	23.6	70	13.8
0.85	27.8	9.5	23.0	80	12.8
1.05	25.0	10.5	22.8	90	12.5
1.25	23.1	12	21.8	100	11.8
1.45	21.8	14	21.0	110	11.4
1.7	19.7	16	19.8	120	11.2
2.0	19.3	18	18.9	140	10.2
2.2	19.3	20	18.4	160	9.21
2.5	19.3	22	18.1	180	8.69
3.0	19.6	25	17.6	200	8.07
3.5	20.6	27	17.4	220	7.66
4.0	21.2	30	16.8	250	6.67
4.5	22.6	35	16.1		

Source: Adapted from Krzysztofowicz, A. M. and Cz. Szmytkowski, *J. Phys. D: At. Mol. Opt. Phys.*, 28, 1593, 1995.

1. The total cross section increases toward zero energy due to the dipole moment of the molecule.
2. There is a broad maximum at about 6.5 eV, possibly due to resonance or dipole potential.
3. Beyond this peak, the total scattering cross section decreases with increasing energy.
4. The total cross section increases in the order $CH_3I > CH_3Br > CH_3Cl > CH_3F$ (Figure 59.2).

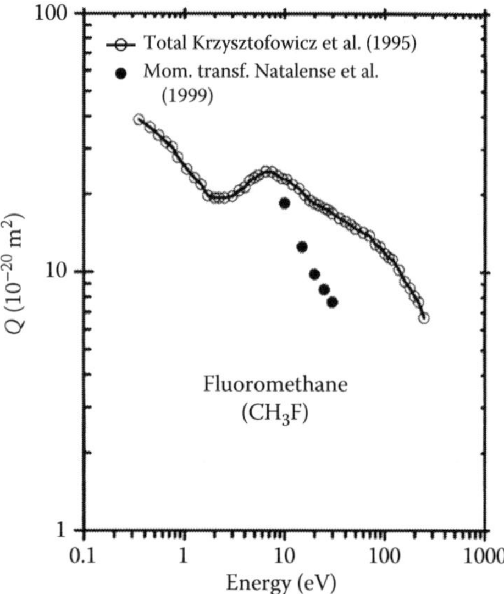

FIGURE 59.1 Total scattering and momentum transfer cross sections in CH_3F. Total cross section: (—○—) KS (Krzysztofowicz and Szmytkowski, 1995); momentum transfer: (•) Natalense et al. (1999).

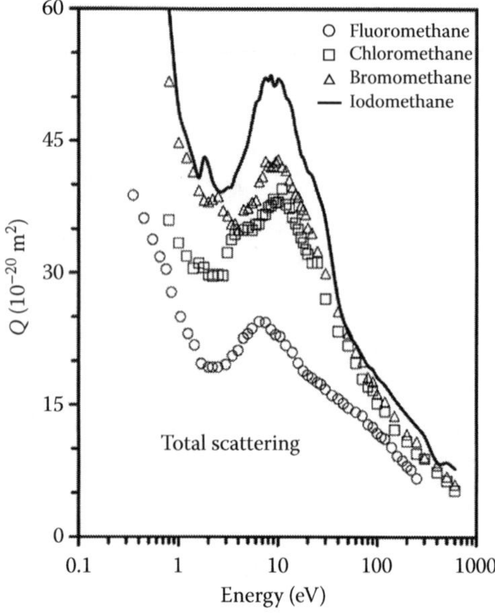

FIGURE 59.2 Total scattering cross sections in methyl halides. (○) fluoromethane (Krzysztofowicz and Szmytkowski, 1995); (Δ) bromomethane, (□) chloromethane, (——) iodomethane (Kimura et al. 2001).

5. Krzysztofowicz and Szmytkowski (1995) observe an energy dependence according to $\varepsilon^{-0.5}$ and polarizability dependence according to $\alpha^{0.5}$ in the energy range $50 \le \varepsilon \le 250$ eV for the molecules in the group.

59.3 DIFFERENTIAL SCATTERING CROSS SECTIONS

Figure 59.3 shows the differential scattering cross sections at selected energies.

59.4 MOMENTUM TRANSFER CROSS SECTION

Only theoretical cross sections are available in a limited range of electron energy (Natalense et al., 1999) as shown in Table 59.3. These data are also included in Figure 59.1.

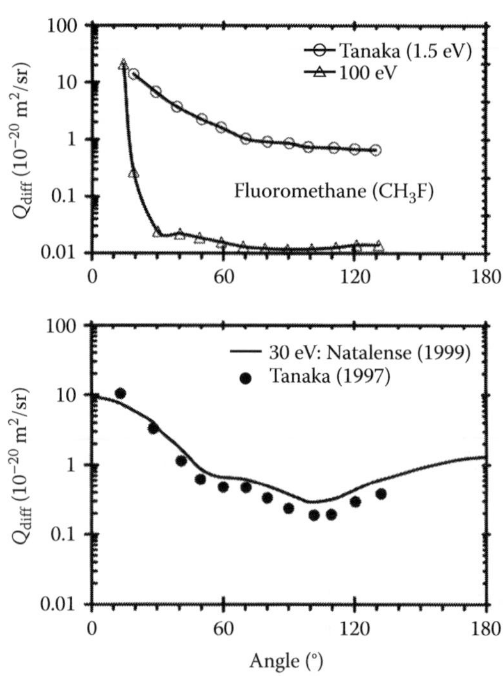

FIGURE 59.3 Differential cross sections in CH_3F. Top: (—○—) 1.5 eV, (—Δ—) 100 eV electron energy, Tanaka et al. (1997), experimental. Lower: 30 eV energy: (——) Natalense et al. (1999), theory; (•) Tanaka et al. (1997), experimental.

TABLE 59.3
Momentum Transfer Cross Sections for CH_3F

eV	10	15	20	25	30
Q_M (10^{-20} m^2)	18.41	12.51	9.80	8.54	7.64

59.5 DISSOCIATION CROSS SECTION

Dissociation cross sections of the molecule into radical CH_2F are shown in Table 59.4 and Figure 59.4. For the purpose of comparison, the dissociation cross sections of selected fluoromethyl radicals are also shown. Note that the number of pathways for generation of radicals is different. For example, CF_3 from CF_4 may be generated by four different pathways, any one of the four bonds may be broken; CF_3 is generated from CHF_3 by breaking a single CH bond.

59.6 POSITIVE ION APPEARANCE POTENTIALS

Positive ion appearance potentials measured by Torres et al. (2001) and previous references compiled by them are shown in Table 59.5.

TABLE 59.4
Neutral Dissociation into Radicals in Fluoromethyls

Parent	CH_3F	CH_2F_2	CHF_3		CF_4
Neutral	CH_2F	CHF_2	CHF_2	CF_3	CF_3
Energy (eV)	Q_{diss} (10^{-20} m^2)	Q_{diss} (10^{-20} m^2)	Q_{diss} (10^{-20} m^2)	Q_{diss} (10^{-20} m^2)	Q_{diss} (10^{-20} m^2)
10	0.07	0.03	0.05	0.01	0.03
12	0.13	0.10	0.08	0.04	0.04
15	0.21	0.16	0.09	0.07	0.06
18	0.26	0.21	0.14	0.09	0.09
20	0.27	0.30	0.22	0.11	0.09
25	0.32	0.34	0.35	0.14	0.33
30	0.34	0.40	0.49	0.21	0.67
35	0.36	0.46	0.58	0.25	0.83
40	0.40	0.53	0.64	0.30	1.08
45	0.44	0.56	0.71	0.32	1.23
50	0.45	0.60	0.76	0.32	1.24
55	0.45	0.63	0.79	0.34	1.31
60	0.45	0.65	0.80	0.36	1.39
65	0.45	0.66	0.83	0.36	1.44
70	0.46	0.66	0.86	0.36	1.46
75	0.47	0.67	0.89	0.37	1.48
80	0.48	0.68	0.91	0.38	1.49
90	0.48	0.66	0.90	0.39	1.49
100	0.48	0.68	0.91	0.39	1.50
125	0.46	0.67	0.91	0.39	1.43
150	0.44	0.63	0.87	0.39	1.39
200	0.40	0.62	0.83	0.37	1.35
250	0.37	0.60	0.77	0.35	1.27
300	0.35	0.55	0.72	0.33	1.17
350	0.34	0.50	0.70	0.30	1.10
400	0.32	0.46	0.68	0.27	1.06
450	0.30	0.43	0.67	0.25	1.04
500	0.30	0.44	0.65	0.26	1.01

Source: Adapted from Motlagh, S. and J. H. Moore, *J. Chem. Phys.*, 109, 432, 1998.
Note: Neutral dissociation cross sections. Digitized and interpolated from graphical data.

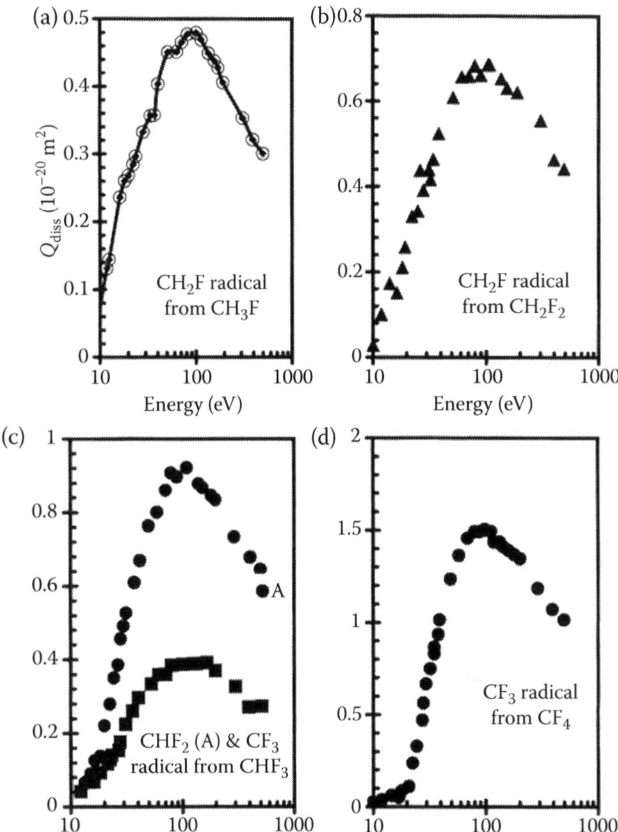

FIGURE 59.4 Neutral dissociation cross section of CH_3F into radical. Selected fluoromethyl radicals are also shown for comparison. (a) CH_2F from CH_3F; (b) CHF_2 from CH_2F_2; (c) CHF_2 and CF_3 from CHF_3; (d) CF_3 from CF_4. (Adapted from Motlagh, S. and J. H. Moore, *J. Chem. Phys.*, 109, 432, 1998.)

TABLE 59.5
Positive Ion Appearance Potentials

Reaction	Energy (eV)	Method	Reference
$CH_3F + e \rightarrow CH_3F^+$	13.1	TOF	Torres et al. (2001)
	12.53	PES	Weitzel et al. (1995)
	12.5	PES	Locht and Momigny (1986)
	13.05	PES	Brundle et al. (1970)
	13.07	PES	Pullen et al. (1970)
$CH_3F + e \rightarrow CH_2F^+ + H$	13.9	TOF	Torres et al. (2001)
	13.2	PES	Weitzel et al. (1995)
	13.35	PES	Locht et al. (1986)
	13.5	PES	Eland et al. (1976)
	13.37	MSPI	Krauss et al. (1968)
$CH_3F + e \rightarrow CH_3^+ + F$	16.7	TOF	Torres et al. (2001)
	14.51	PES	Weitzel et al. (1995)
	12.45	PES	Locht et al. (1986)
	14.50	EI	Tsuda et al. (1964)
	16.25	MSPI	Krauss et al. (1968)
$CH_3F + e \rightarrow CH_2^+ + H + F$	21.3	TOF	Torres et al. (2001)
	22.36	PES	Locht et al. (1986)
$CH_3F + e \rightarrow CH^+ + 2H + F$	26.8	TOF	Torres et al. (2001)

continued

TABLE 59.5 (continued)
Positive Ion Appearance Potentials

Reaction	Energy (eV)	Method	Reference
$CH_3F + e \rightarrow CF^+ + H_2 + H$	20.4	TOF	Torres et al. (2001)
$CH_3F + e \rightarrow CHF^+ + H_2$	13.8	TOF	Torres et al. (2001)
$CH_3F + e \rightarrow CHF^+ + 2H$	13.91	PES	Weitzel et al. (1995)
$CH_3F + e \rightarrow C^+ + 3H + F$	28.8	TOF	Torres et al. (2001)
$CH_3F + e \rightarrow F^+ + C + 3H$	>35	TOF	Torres et al. (2001)

Source: Adapted from Torres, I., R. Martínez, and F. Castaño, *J. Phys. B: At. Mol. Opt. Phys.*, 35, 4113, 2002.

Note: EI = electron impact; MSPI = mass spectrometric photo ionization; PES = photoelectron spectroscopy; TOF = time-of-flight mass spectrometer.

TABLE 59.6
Ionization Cross Section in CH_3F

	Q_i (10^{-20} m²)			Q_i (10^{-20} m²)
Energy (eV)	Vallance et al. (1997)	Torres et al. (2001)	Energy (eV)	Vallance et al. (1997)
12	0.15		97	3.74
16	0.42		101	3.72
20	0.86	0.488	105	3.70
23	1.35		108	3.68
25		1.223	112	3.67
27	1.80		116	3.65
30		1.766	119	3.63
31	2.14		123	3.62
34	2.45		127	3.60
35		2.176	130	3.57
38	2.68		134	3.53
40		2.537	138	3.50
42	2.89		142	3.48
45	3.08	2.789	145	3.46
49	3.24		149	3.45
50		3.002	153	3.43
53	3.38		156	3.41
55		3.148	160	3.38
57	3.48		164	3.34
60	3.57	3.243	167	3.30
64	3.62		171	3.27
65		3.314	175	3.24
68	3.67		179	3.23
70		3.408	182	3.22
71	3.71		186	3.21
75	3.74	3.448	190	3.19
79	3.75		193	3.16
80		3.543	197	3.13
82	3.76		201	3.09
85		3.592	204	3.05
86	3.78		208	3.03
90	3.79		212	3.02
94	3.75		215	3.01

Sources: Tabulated values courtesy of Professor P. W. Harland; adapted from Vallance, C. et al., *J. Phys. B: At. Mol. Opt. Phys.*, 30, 2465, 1997.

59.7 TOTAL IONIZATION CROSS SECTION

See Table 59.6 and Figure 59.5.

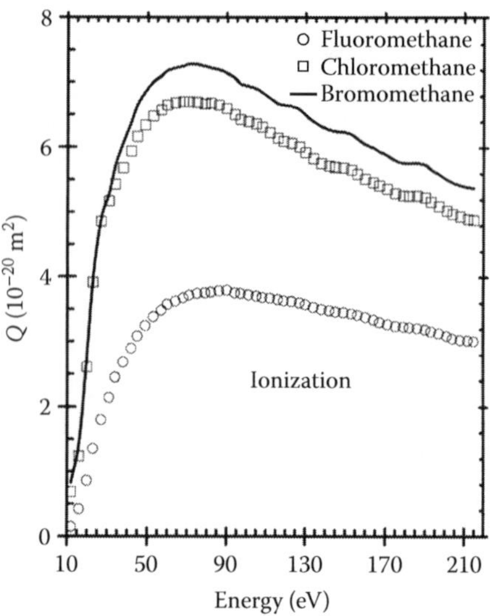

FIGURE 59.5 Ionization cross sections in selected methyl halides. Tabulated values courtesy of Professor P. Harland. (Adapted from Vallance, C. et al., *J. Phys. B: At. Mol. Opt. Phys.*, 30, 2465, 1997.)

TABLE 59.7
Partial Ionization Cross Sections in CH_3F for Selected Species

	Q_i (10^{-20} m²)					
	Torres et al. (2001)			Rejoub et al. (2002)		
Energy (eV)	CH_nF^+	CH_n^+	C^+ (×5)	Energy (eV)	CH_nF^+	CH_n^+
20.0	0.400	0.113	5×10^{-4}	14	0.070	
25.0	0.941	0.351	0.003	16	0.300	
30.0	1.302	0.563	0.017	18	0.513	
35.0	1.543	0.745	0.039	20	0.651	0.160
40.0	1.742	0.896	0.076	22.5	0.852	0.314
45.0	1.882	1.008	0.110	25	1.03	0.488
50.0	1.983	1.093	0.126	30	1.23	0.776
55.0	2.064	1.160	0.145	35	1.38	1.01
60.0	2.105	1.194	0.163	40	1.51	1.21
65.0	2.132	1.234	0.171	50	1.59	1.46
70.0	2.175	1.263	0.196	60	1.64	1.57
75.0	2.197	0.282	0.205	80	1.68	1.64
80.0	2.232	1.311	0.212	100	1.67	1.66
85.0	2.272	1.345	0.226	125	1.61	1.60
				150	1.56	1.57
				200	1.40	1.39
				300	1.191.19	1.21
				400	1.06	1.05
				500	0.911	0.897
				600	0.798	0.780
				800	0.645	0.638
				1000	0.546	0.558

Note: Partial ionization cross sections in CH_3F.

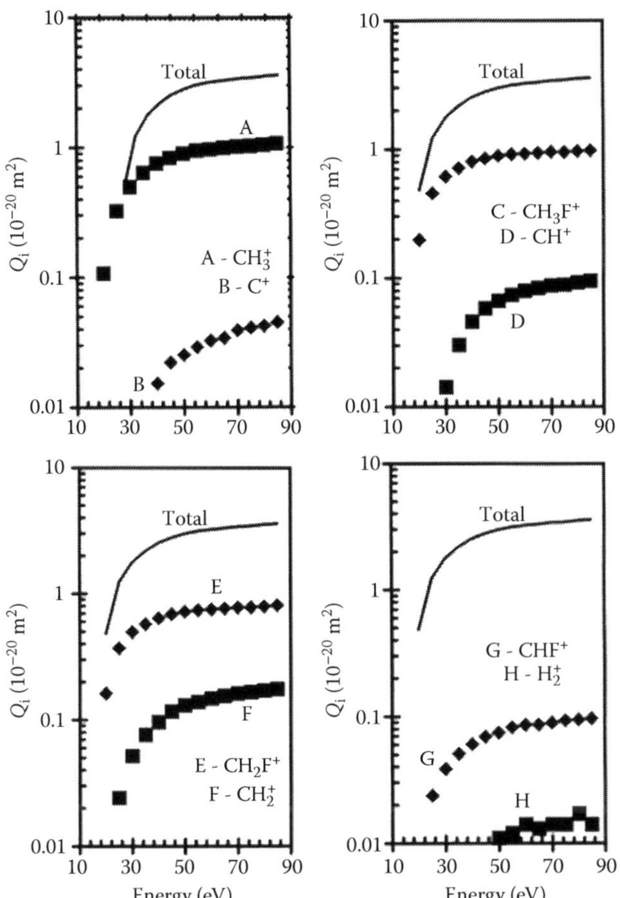

FIGURE 59.6 Partial ionization cross sections in CH_3F measured by Torres et al. (2001). The total cross section is plotted for comparison. The largest contributions are made by the parent ion CH_3F^+, CH_3^+, and CH_2F^+.

59.8 PARTIAL IONIZATION CROSS SECTION

Partial ionization cross sections for CH_3F have been measured by Torres et al. (2001) and Rejoub et al. (2002) as shown in Table 59.7 and Figure 59.6.

59.9 ATTACHMENT RATE

CH_3F is a very mildly attaching gas with a thermal attachment rate $<10^{-21}$ m^3/s (Fessenden and Bansal, 1970).

REFERENCES

Benitez, A., J. H. Moore, and J. A. Tossell, *J. Chem. Phys.*, 88, 6691, 1988.

Brundle, C. R., M. B. Robin, and H. Basch, *J. Chem. Phys.*, 53, 2196, 1970.

Eland, J. H. D., A. Kuestler, H. Schulte, and B. Brehm, *Int. J. Mass Spectrom. Ion Phys.*, 22, 155, 1976.

Fessenden, R. W. and K. M. Bansal, *J. Chem. Phys.*, 53, 3468, 1970.

Kimura, M., O. Sueoka, C. Makochekanwa, H. Kawate, and M. Kawada, *J. Chem. Phys.*, 115, 7442, 2001.

Krauss, M., J. A. Walker, and V. H. Diebler, *J. Res. Nat. Bur. Stand. A*, 72, 281, 1968.

Krzysztofowicz, A. M. and Cz. Szmytkowski, *J. Phys. D: At. Mol. Opt. Phys.*, 28, 1593, 1995.

Locht, R. and J. Momigny, *Int. J. Mass Spectrom. Ion Phys.*, 71, 141, 1986.

Motlagh, S. and J. H. Moore, *J. Chem. Phys.*, 109, 432, 1998.

Natalense, A. P. P., M. H. F. Bettega, L. G. Ferreira, and M. A. P. Lima, *Phys. Rev. A*, 59, 879, 1999.

Rejoub, R., B. G. Lindsay, and R. F. Stebbings, *J. Chem. Phys.*, 117, 6450, 2002.

Tanaka, H., T. Masai, M. Kimura, T. Nishimura, and Y. Itikawa, *Phys. Rev. A*, 56, R3338, 1997.

Torres, I., R. Martínez, and F. Castaño, *J. Phys. B: At. Mol. Opt. Phys.*, 35, 4113, 2001.

Tsuda, S., C. E. Melton, and W. H. Hamill, *J. Chem. Phys.*, 41, 689, 1964.

Vallance, C., S. A. Harris, J. E. Hudson, and P. W. Harland, *J. Phys. B: At. Mol. Opt. Phys.*, 30, 2465, 1997.

Weitzel, K. M., F. Güthe, J. Mähnert, R. Locht, and H. Baumgärtel, *Chem. Phys.*, 201, 287, 1995.

60

FORMIC ACID

CONTENTS

Formic acid (CH_2O_2) is a vapor with 24 electrons. Its electronic polarizability is 3.78×10^{-40} F m^2, dipole moment is 1.425 D, and ionization potential is 10.88 eV. The molecule has nine modes of vibration as shown in Table 60.1.

TABLE 60.1
Vibrational Modes and Energies of CH_2O_2 Molecule

Symbol	Deformation	Energy (meV)	Symbol	Deformation	Energy (meV)
v_1	OH stretch	442.6	v_6	CO stretch	137.0
v_2	CH stretch	364.9	v_7	OCO deform	77.5
v_3	C = O stretch	219.5	v_8	CH bend	128.1
v_4	CH bend	172.0	v_9	Torsion	79.1
v_5	OH bend	152.4			

Source: Adapted from Shimanouchi, T., *Tables of Molecular Vibrational Frequencies Consolidated Volume I.* NSRDS-NBS 39, Washington (DC), 1972.

60.1 TOTAL SCATTERING CROSS SECTION

Total scattering cross sections are given by Kimura et al. (2000) as shown in Table 60.1 and Figure 60.1. The highlights of the cross sections are

1. The cross section tends to increase toward zero energy as in several polar molecules. There is a minimum at 5.0 eV.
2. There is a moderately pronounced peak at 7.5 eV.
3. Beyond this energy, the cross section decreases monotonically, in common with many gases (see Table 60.2).

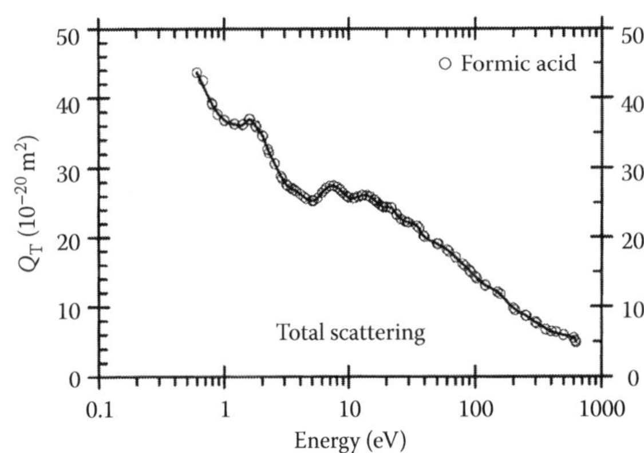

FIGURE 60.1 Total scattering cross section in CH_2O_2 digitized and interpolated from Kimura et al. (2000).

TABLE 60.2
Total Scattering Cross Section in CH_2O_2

Energy (eV)	Q_T (10^{-20} m^2)	Energy (eV)	Q_T (10^{-20} m^2)	Energy (eV)	Q_T (10^{-20} m^2)
0.6	43.64	3.7	26.74	10.0	25.78
0.8	39.05	4.0	26.29	11.0	25.68
1.0	36.84	4.5	25.61	12.0	25.89
1.2	36.29	5.0	25.31	13.0	26.09
1.4	36.17	5.5	25.67	14.0	26.05
1.6	36.97	6.0	26.44	15.0	25.82
1.8	35.84	6.5	27.10	16.0	25.45
2.0	34.67	7.0	27.43	17.0	25.01
2.2	32.70	7.5	27.46	18.0	24.61
2.5	30.63	8.0	27.24	19.0	24.41
2.8	28.83	8.5	26.88	20.0	24.40
3.1	27.66	9.0	26.45	22.0	24.24
3.4	27.13	9.5	26.05	24.0	23.36

continued

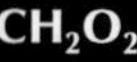

TABLE 60.2 (continued)
Total Scattering Cross Section in CH_2O_2

Energy (eV)	Q_T (10^{-20} m^2)	Energy (eV)	Q_T (10^{-20} m^2)	Energy (eV)	Q_T (10^{-20} m^2)
26.0	22.64	70.0	17.08	250.0	8.70
28.0	22.29	80.0	16.13	300.0	7.74
30.0	22.13	90.0	15.19	400.0	6.41
35.0	21.60	100.0	14.30	500.0	5.90
40.0	20.03	120.0	13.06	600.0	5.49
50.0	19.06	150.0	12.07	625.0	4.90
60.0	18.18	200.0	9.76		

Source: Digitized from M. Kimura, et al. *Advan. Chem. Phys.*, 111, 537, 2000.

REFERENCES

Kimura, M., O. Sueoka, A. Hamada, and Y. Itikawa, *Advan. Chem. Phys.*, 111, 537, 2000.

Shimanouchi, T., *Tables of Molecular Vibrational Frequencies Consolidated Volume I.* NSRDS-NBS 39, Washington (DC), 1972.

61

GERMANE

CONTENTS

Germane (GeH_4) is an inorganic molecule with 36 electrons. The average polarizability obtained from three different studies is 5.85×10^{-40} F m^2 (Karwasz, 1995). The ionization potentials obtained from two studies are 10.51 eV (photoionization mass spectrometry) and 11.31 eV (photoelectron mass spectrometry), respectively, quoted by Karwasz (1995).

61.1 SELECTED REFERENCES FOR DATA

See Table 61.1.

61.2 TOTAL SCATTERING CROSS SECTION

See Table 61.2.

TABLE 61.1
Selected References for Data

Quantity	Range: eV, (Td)	Reference
Ionization cross section	10.5–1000	Ali et al. (1997)
Total scattering cross section	0.75–250	**Możejko et al. (1996)**
Total scattering cross section	75–4000	**Karwasz (1995)**
Elastic cross section	1–100	Dillon et al. (1993)

Note: Bold font indicates experimental study.

TABLE 61.2
Total Scattering Cross Section

Energy (eV)	Q_T (10^{-20} m^2)	Energy (eV)	Q_T (10^{-20} m^2)	Energy (eV)	Q_T (10^{-20} m^2)
Karwasz et al. (2003)		Możeko et al. (1996)		Karwasz et al. (1995)	
0.8	12.7	0.75	12.5	75	19.4
0.9	14.2	0.85	12.8	80	19.3
1.0	16.7	1	17.3	90	18.2
1.2	23.6	1.2	23.3	100	17.3
1.5	33.7	1.5	34	110	16.6
1.7	37.4	1.7	37.2	125	14.6
2.0	41.9	2.0	42.3	150	13.1
2.5	49.2	2.4	47.5	175	12.1
3.0	55.0	3.1	55.4	200	11.1
3.5	58.0	3.5	57.9	225	10.3
4.0	57.9	3.8	58.8	250	9.36
4.5	55.4	4.3	56.2	300	8.59
5.0	53.5	4.8	54.4	350	7.63
6.0	51.1	5.8	51.1	400	7.02
7.0	49.6	7.3	49.4	500	6.43
8.0	48.4	8.5	48.2	600	6.07
9.0	47.3	9	47.2	700	4.73
10	46.1	10	46.5	800	4.34
12	43.7	12	43.4	900	3.97
15	39.9	15	39.9	1000	3.65
17	37.3	17	37.5	1250	3.05
20	34.0	21	33.2	1500	2.64
25	30.5	25	30.7	1750	2.29
30	28.1	30	27.9	2000	2.09
35	26.1	35	26.8	2250	1.88
40	24.5	40	24.5	2500	1.72
45	23.0	45	22.6	2750	1.57
50	21.7	50	21.7	3000	1.49
60	19.6	60	19.8	3250	1.38
70	18.0	70	18	3500	1.34
80	16.8	80	16.8	4000	1.16

continued

TABLE 61.2 (continued)
Total Scattering Cross Section

Energy (eV)	Q_T (10^{-20} m^2)	Energy (eV)	Q_T (10^{-20} m^2)	Energy (eV)	Q_T (10^{-20} m^2)
Karwasz et al. (2003)		Możeko et al. (1996)		Karwasz et al. (1995)	
90	15.9	90	15.8		
100	15.2	100	14.9		
120	14.1	110	14.1		
150	12.7	120	13.6		
170	11.9	140	12.7		
200	10.8	160	11.5		
250	9.46	180	10.5		
300	8.47	200	9.83		
350	7.66	220	9.41		
400	7.01	250	8.55		
450	6.48				
500	6.03				
600	5.32				
700	4.76				
800	4.32				
900	3.96				
1000	3.66				

Source: Adapted from Karwasz, G., R. Brusa, and A. Zecca, in *Interactions of Photons and Electrons with Molecules*, Vol. 17c, Springer-Verlag, New York, NY, 2003, Chapter 6.

Note: See Figure 61.1 for graphical presentation.

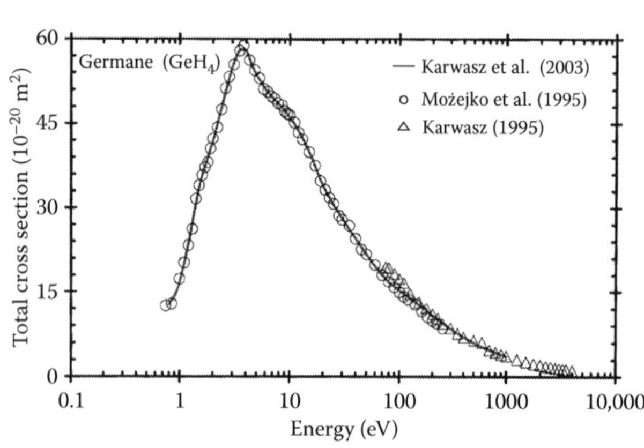

FIGURE 61.1 Total scattering cross section for GeH_4. (——) Recommended by Karwasz et al. (2003); (○) Możejko et al. (1996); (Δ) Karwasz (1995).

61.3 ELASTIC AND MOMENTUM TRANSFER CROSS SECTION

See Table 61.3.

61.4 IONIZATION CROSS SECTION

The only data available are theoretical values calculated by Ali et al. (1997) as shown in Figure 61.3.

TABLE 61.3
Elastic Scattering and Momentum Transfer Cross Section for GeH_4

Energy (eV)	Q_{el} (10^{-20} m^2)	Q_M (10^{-20} m^2)
1	8.4	7.11
2	26.45	26.03
2.5	28.76	27.31
3	34.07	27.67
5	45.48	26.87
7.5	43.4	21.72
10	39.42	18.54
15	30.14	11.48
20	23.63	6.52
60	7.47	1.44
100	6.36	1.6

Source: Adapted from Dillon, M. A. et al., *J. Phys. B: At. Mol. Opt. Phys.*, 26, 3147, 1993.

Note: See Figure 61.2 for graphical presentation.

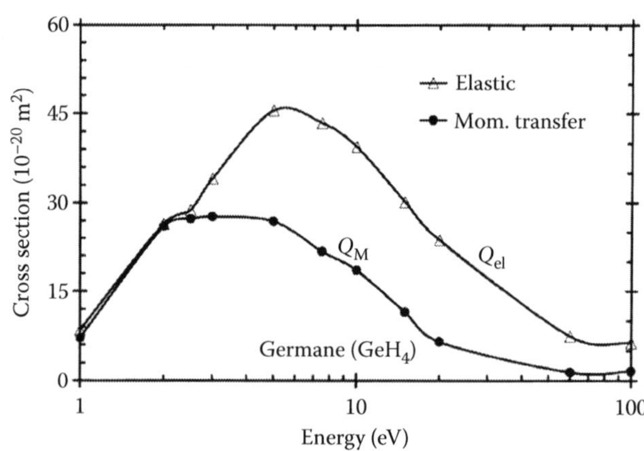

FIGURE 61.2 Elastic scattering and momentum transfer cross section for GeH_4. (Δ) Elastic scattering; (•) momentum transfer. (Adapted from Dillon, M. A. et al., *J. Phys. B: At. Mol. Opt. Phys.*, 26, 3147, 1993.)

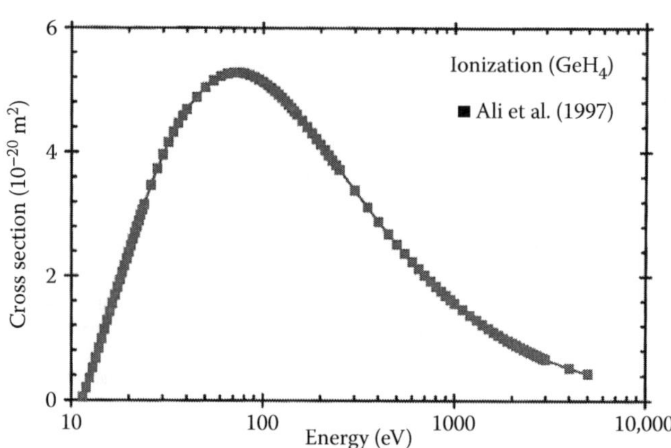

FIGURE 61.3 Ionization cross section for GeH_4. (Adapted from Ali, M. A. et al., *J. Chem. Phys.*, 106, 9602, 1997.)

REFERENCES

Ali, M. A., Y.-K. Kim, W. Hwang, N. M. Weinberger, and M. E. Rudd, *J. Chem. Phys.*, 106, 9602, 1997.

Dillon, M. A., L. Boesten, H. Tanaka, M. Kimura, and H. Sato, *J. Phys. B: At. Mol. Opt. Phys.*, 26, 3147, 1993.

Karwasz, G., R. Brusa, and A. Zecca, Cross sections for scattering- and excitation-processes in electron–molecule collisions, in *Interactions of Photons and Electrons with Molecules*, Vol. 17c, Springer-Verlag, New York, NY, 2003, Chapter 6.

Karwasz, G. P., *J. Phys. B: At. Mol. Opt. Phys.*, 28, 1593, 1995.

Możejko, P., G. Kasperski, and C. Szmytkowski et al., *J. Phys. B: At. Mol. Opt. Phys.,* 29, L571, 1996.

62

GERMANIUM TETRACHLORIDE

CONTENTS

Germanium tetrachloride ($GeCl_4$), also known as tetrachlorogermane, is a relatively large, nonpolar, electron-attaching molecule with 100 electrons; its electronic polarizability is 16.80×10^{-30} F m^2 and its ionization potential is 11.68 eV.

62.1 TOTAL SCATTERING CROSS SECTION

See Table 62.1 and Figure 62.1.

TABLE 62.1
Total Scattering Cross Section for $GeCl_4$

Energy (eV)	Q_T (10^{-20} m^2)	Energy (eV)	Q_T (10^{-20} m^2)	Energy (eV)	Q_T (10^{-20} m^2)
$GeCl_4$		GeH_4		CCl_4	
0.6	36.7	0.8	12.7	0.8	56.4
0.7	42.9	0.9	14.2	1.0	58.2
0.8	45.1	1.0	16.7	1.2	57.0
1.1	52.3	1.7	37.4	1.8	50.7
1.2	53.5	2.0	41.9	2.0	47.6
1.3	54.4	2.5	49.2	2.2	44.4
1.4	56.1	3.0	55.0	2.5	41.1
1.5	55.7	3.5	58.0	2.8	40.7
1.6	57.2	4.0	57.9	3.1	39.5
1.7	57.4	4.5	55.4	3.4	38.7
1.8	56.7	5.0	53.5	3.7	39.3
1.9	57.2	6.0	51.1	4.0	40.2
2.0	57.8	7.0	49.6	4.5	42.3
2.1	55.7	8.0	48.4	5.0	43.7
2.2	56.0	9.0	47.3	5.5	48.7
2.3	54.2	10	46.1	6.0	49.9
2.4	53.5	12	43.7	6.5	54.2
2.5	50.2	15	39.9	7.0	57.9
2.6	49.5	17	37.3	7.5	58.3
2.9	46.9	20	34.0	8.0	62.0

TABLE 62.1 (continued)
Total Scattering Cross Section for $GeCl_4$

Energy (eV)	Q_T (10^{-20} m^2)	Energy (eV)	Q_T (10^{-20} m^2)	Energy (eV)	Q_T (10^{-20} m^2)
$GeCl_4$		GeH_4		CCl_4	
3.3	49.5	25	30.5	8.5	60.0
3.6	47.9	30	28.1	9.0	59.5
4.1	50.3	35	26.1	9.5	58.7
4.6	53.0	40	24.5	10.0	59.7
5.1	60.4	45	23.0	11.0	57.5
5.6	67.2	50	21.7	12.0	55.0
6.1	71.4	60	19.6	13.0	55.0
6.6	73.4	70	18.0	14.0	54.4
7.1	73.9	80	16.8	15.0	54.0
7.6	75.3	90	15.9	16.0	53.0
8.1	75.4	100	15.2	17.0	50.7
8.6	76.3	120	14.1	18.0	50.3
9.1	78.0	150	12.7	19.0	50.5
9.6	79.7	170	11.9	20.0	50.3
10.1	79.6	200	10.8	22.0	47.8
10.6	79.6	250	9.46	25.0	45.6
11.6	78.4			30.0	41.9
12.6	77.8			35.0	40.5
15	73.8			40.0	38.4
17	71.0			50.0	36.5
19	69.1			60.0	35.2
21	66.2			70.0	33.1
23	63.8			80.0	32.5
26	62.6			90.0	31.5
28	61.3			100.0	31.1
30	60.3			120.0	29.5
35	58.2			150.0	28.6
40	56.0			200.0	27.5

continued

GeCl$_4$

TABLE 62.1 (continued)
Total Scattering Cross Section for GeCl$_4$

Energy (eV)	Q_T (10^{-20} m^2)	Energy (eV)	Q_T (10^{-20} m^2)	Energy (eV)	Q_T (10^{-20} m^2)
GeCl$_4$		GeH$_4$		CCl$_4$	
45	53.4			250.0	24.7
50	51.7				
60	49.4				
70	47.0				
80	44.3				
90	41.3				
100	39.6				
110	38.5				
120	36.8				
140	33.9				
160	31.9				
180	29.6				
200	28.6				
220	26.8				
250	25.2				

Source: Adapted from Szmytkowski, C., P. Możejko, and G. Kasperski, *J. Phys. B: At. Mol. Opt. Phys.*, 30, 4363, 1997.

Note: Cross sections for GeH$_4$ and CCl$_4$ are included to show the effects of substation of H in place of Cl and C in place of Ge, respectively.

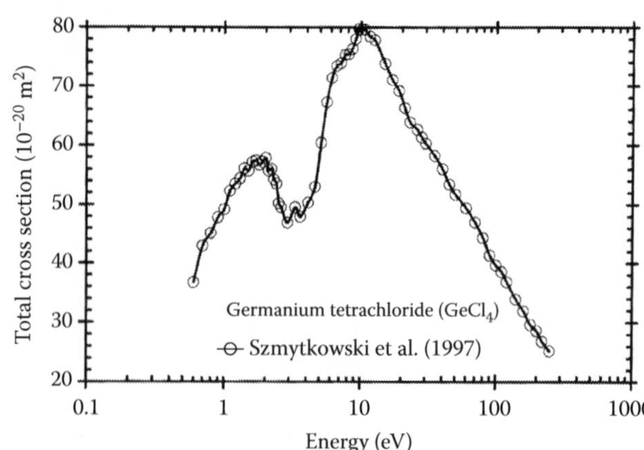

FIGURE 62.1 Total scattering cross section for GeCl$_4$. The resonance peak at 1.7 eV is attributed to electron attachment (see Appendix 8). (Adapted from Szmytkowski, C., P. Możejko, and G. Kasperski, *J. Phys. B: At. Mol. Opt. Phys.*, 30, 4363, 1997.).

REFERENCE

Szmytkowski, C., P. Możejko, and G. Kasperski, *J. Phys. B: At. Mol. Opt. Phys.*, 30, 4363, 1997.

63

IODOMETHANE

CONTENTS

Iodomethane (CH_3I), synonym methyl iodide, with 62 electrons, is a molecule in which one of the hydrogen atoms of methane is replaced with iodine. The substitution renders the molecule both polar (dipole moment = 1.641 D; electronic polarizability = 8.87×10^{-40} F m^2) and electron-attaching. The molecule has six vibrational modes in conformity with other halomethanes. The details of the modes and similar data for fluoromethane are given in Table 63.1 (Shimanouchi, 1972).

63.1 SELECTED REFERENCES FOR DATA

See Table 63.2.

63.2 TOTAL SCATTERING CROSS SECTION

The total scattering cross sections in CH_3I are measured by Kimura et al. (2001) as shown in Table 63.3 and Figure 63.1.

TABLE 63.1
Vibrational Modes and Energies in CH_3I and CH_3F

Iodomethane			Fluoromethane		
Designation	Deformation	Energy (meV)	Designation	Deformation	Energy (meV)
ν_1	CH_3 s-stretch	363.6	ν_1	CH_3 s-stretch	363.3
ν_2	CH_3 s-deform	155.2	ν_2	CH_3 s-deform	181.5
ν_3	CI stretch	66.1	ν_3	CF stretch	130.1
ν_4	CH_3 d-stretch	379.4	ν_4	CH_3 d-stretch	372.7
ν_5	CH_3 d-deform	178.0	ν_5	CH_3 d-deform	181.9
ν_6	CH_3 rock	109.4	ν_6	CH_3 rock	146.5

Source: Adapted from Shimanouchi, T. *Tables of Molecular Vibrational Frequencies Consolidated Volume I*, NSRDA-NBS, 39, 1972.

Note: d = Degenerate; s = symmetric.

TABLE 63.2
Selected References for Data

Quantity	Energy Range: eV, (K)	Reference
k_a	Thermal	Christophorou et al. (2006)
Q_i	13–1000	Rejoub et al. (2002)
Q_v	0–1	Allan et al. (2002)
Q_T	0.8–600	Kimura et al. (2001)
Q_a	0–180 m	Schramm et al. (1999)
Q_i	10.5–220	Vallance et al. (1997)
k_a	(293–777)	Burns et al. (1996)
Q_T	0.4–20	Szmytkowski et al. (1993)
Q_T	0.5–8	Benitez et al. (1988)

Note: k_a = Attachment rate; Q_a = attachment cross section; Q_T = total cross section; Q_i = ionization cross section; Q_v = vibrational excitation cross section.

63.3 IONIZATION CROSS SECTION

Total ionization cross sections have been measured by Vallance et al. (1997). Tabulated or graphical presentation of cross sections is not available. The peak of the ionization cross section occurs at 10.3 eV.

63.4 ATTACHMENT CROSS SECTION

At zero energy, the attachment cross section shows a peak (peak cross section, ~3.9×10^{-16} m^2; see Schramm et al., 1999) attributed to *s*-wave maximum and decreases toward higher energy, with a second peak at 57 meV (Alajajian et al., 1988).

CH$_3$I

TABLE 63.3
Total Scattering Cross Section in CH$_3$I

Energy (eV)	Q_T (10^{-20} m^2)	Energy (eV)	Q_T (10^{-20} m^2)	Energy (eV)	Q_T (10^{-20} m^2)
0.8	59.7	6.5	49.8	22	41.1
1.0	48.6	7.0	51.6	25	39.2
1.2	45.4	7.5	52.2	30	36.1
1.4	42.9	8.0	51.9	40	26.9
1.6	40.8	8.5	52.5	50	23.3
1.8	43.1	9.0	51.4	60	21.9
2.0	42.2	9.5	51.4	70	20.6
2.2	40.4	10	52.1	80	19.5
2.5	39.3	11	51.6	90	19.1
2.8	39.3	12	50.1	100	18.2
3.1	39.7	13	49.5	120	17.3
3.4	39.8	14	48.3	150	15.7
3.7	41.0	15	46.3	200	13.9
4.0	41.9	16	45.0	250	12.6
4.5	43.3	17	43.5	300	11.5
5.0	44.9	18	43.1	400	8.5
5.5	47.2	19	42.3	500	8.5
6.0	49.3	20	41.6	600	7.8

Source: Adapted from Kimura, M. et al., *J. Chem. Phys.*, 115, 7442, 2001.

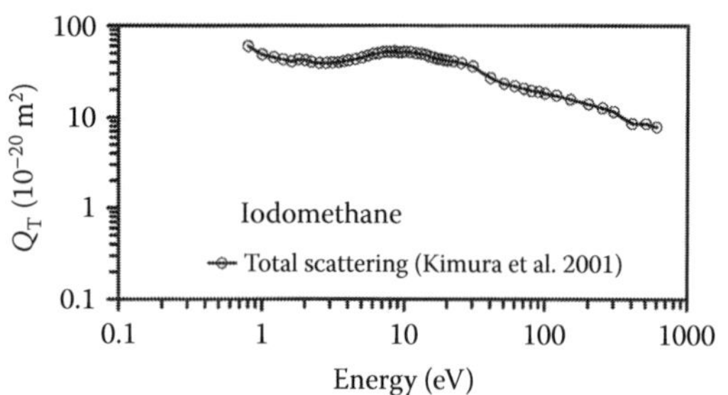

FIGURE 63.1 Total scattering cross section in CH$_3$I. (Adapted from Kimura, M. et al., *J. Chem. Phys.*, 115, 7442, 2001.)

Dissociative attachment occurs according to

$$e + CH_3I \rightarrow CH_3 + I^- \tag{63.1}$$

Dissociative attachment cross sections measured at low energy (Schramm et al., 1999) are shown in Figure 63.2. The figure also includes the cross section derived by Alajajian et al. (1988).

63.5 ATTACHMENT RATE CONSTANTS

Attachment rate constants for CH$_3$I are shown in Table 63.4.

63.6 ADDENDUM

Absolute total and partial cross sections for CH$_3$I have been measured by Rejoub et al. (2002) as shown in Table 63.5 and Figures 63.3 and 63.4.

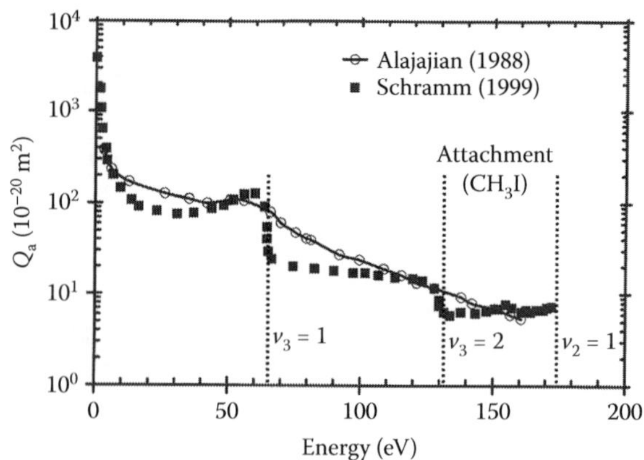

FIGURE 63.2 Dissociative attachment cross sections for I$^-$/CH$_3$I. (—○—) Alajajian et al. (1988); (■) Schramm et al. (1999). The vibrational modes are indicated. Note the energy scale and the increase toward zero energy. Permission from IOP is acknowledged.

TABLE 63.4
Attachment Rate Constants

Rate Constant (10^{-14} m^3/s)	Energy eV, (K)	Reference
9.3		Christophorou et al. (2006)
10.0	(293)	Burns et al. (1996)
11.0	(467)	
12.0	(579)	
12.0	(777)	
8.5	(205)	Alge et al. (1984)
12.0	(300)	
18.0	(452)	
11.0	(585)	
2.5		Blaunstein et al. (1968)

Note: Energy is thermal unless otherwise mentioned.

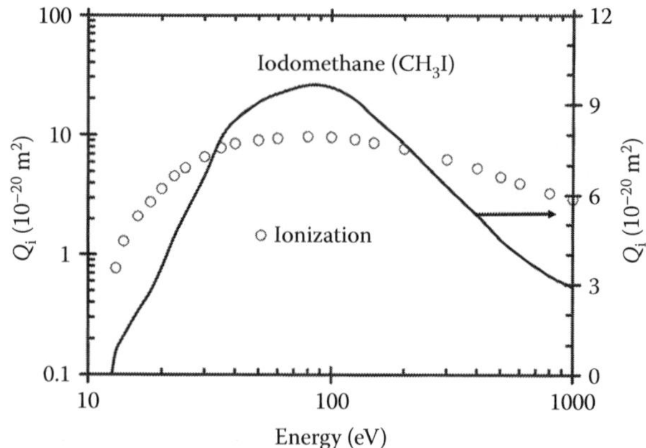

FIGURE 63.3 Absolute total ionization cross sections for CH$_3$I. Note the linear scale for the full curve. (Adapted from Rejoub, R., B. G. Lindsay, and R. F. Stebbings, *J. Chem. Phys.*, 117, 6450, 2002.)

TABLE 63.5
Total Ionization Cross Sections for CH_3I

Energy (eV)	Q_T (10^{-20} m^2)	Energy (eV)	Q_T (10^{-20} m^2)
13	0.77	80	9.64
14	1.29	100	9.57
16	2.10	125	9.16
18	2.77	150	8.59
20	3.57	200	7.71
22.5	4.59	300	6.30
25	5.37	400	5.33
30	6.64	500	4.51
35	7.88	600	3.98
40	8.49	800	3.30
50	9.07	1000	2.93
60	9.36		

Source: Adapted from Rejoub, R., B. G. Lindsay, and R. F. Stebbings, *J. Chem. Phys.*, 117, 6450, 2002.

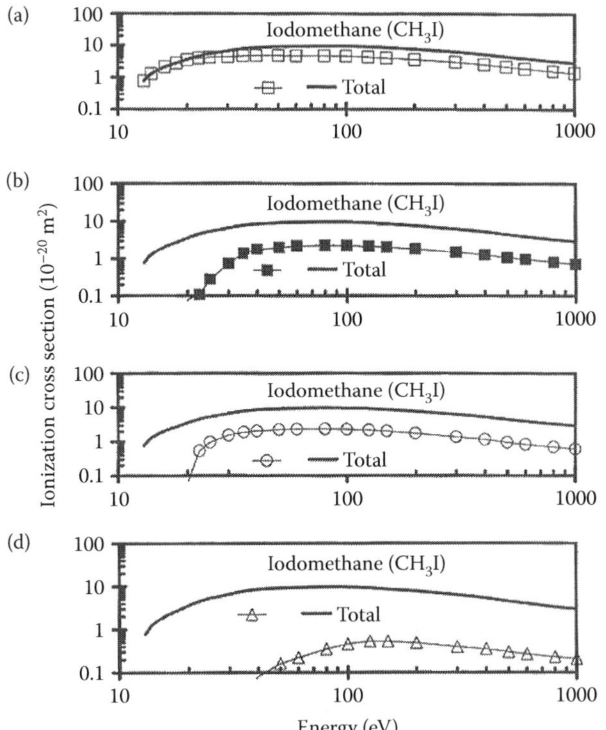

FIGURE 63.4 Partial ionization cross sections for CH_3I. Total cross sections are also plotted to show the relative contribution of the fragments. (a) (—□—) CH_nI^+; (b) (—■—) CH_n^+; (c) (—○—) I^+; (d) (—Δ—) H^+; $n = 0$ to 3. (Adapted from Rejoub, R., B. G. Lindsay, and R. F. Stebbings, *J. Chem. Phys.*, 117, 6450, 2002.)

REFERENCES

Alajajian, S. H., M. T. Bernius, and A. Chutjian, *J. Phys. B: At. Mol. Opt. Phys.*, 21, 4021, 1988.

Alge, E., N. G. Adams, and D. Smith, *J. Phys. B: At. Mol. Phys.*, 17, 3827, 1984.

Allan, M. and I. I. Fabrikant, *J. Phys. B: At. Mol. Opt. Phys.*, 35, 1025, 2002.

Benitez, A., J. H. Moore, and J. A. Tossell, *J. Chem. Phys.*, 88, 6691, 1988.

Blaunstein, R. P. and L. G. Christophorou, *J. Chem. Phys.*, 49, 1526, 1968.

Burns, S. J., J. M. Matthews, and D. L. McFadden, *J. Phys. Chem.*, 100, 19436, 1996.

Christophorou, L. G. and D. Hadjiantoniou, *Chem. Phys. Lett.*, 419, 405, 2006.

Kimura, M., O. Sueoka, C. Makochekanwa, H. Kawata, and M. Kawada, *J. Chem. Phys.*, 115, 7442, 2001.

Rejoub, R., B. G. Lindsay, and R. F. Stebbings, *J. Chem. Phys.*, 117, 6450, 2002.

Schramm, A., I. I. Fabrikant, J. M. Weber, E. Leber, M.-W. Ruf, and H. Hotop, *J. Phys. B: At. Mol. Opt. Phys.*, 32, 2153, 1999.

Shimanouchi, T., *Tables of Molecular Vibrational Frequencies Consolidated Volume I*, NSRDA-NBS, 39, Washington (DC), 1972.

Szmytkowski, Cz. and A. M. Krzysztofowicz, *Chem. Phys. Lett.*, 209, 474, 1993.

Vallance, C., S. A. Harris, J. E. Hudson, and P. W. Harland, *J. Phys. B: At. Mol. Opt. Phys.*, 30, 2465, 1997.

64

METHANE

CONTENTS

Methane (CH_4) is the simplest hydrocarbon molecule with a considerable number of papers on all aspects of electron interaction with gas molecules. It is nonpolar and possesses electronic polarizability of 2.885×10^{-40} F m^2.

The molecule is relatively stable and the strength of the H–CH_3 bond is 4.51 eV. The dissociation reaction is Savinov et al. (1999)

$$\left.\begin{array}{l} CH_4 + e^- \rightarrow CH_4^* + e^- \\ CH_4^* \rightarrow CH_3 + H \end{array}\right\} \quad (64.1)$$

The molecule has a highly symmetrical configuration and the ground state configuration is given by Mapstone and Newell (1992). It has four fundamental vibrational excitation modes (Cascella et al., 2001) as listed in Table 64.1. They cannot be individually resolved and measurements are for combined $\nu_1 + \nu_3$ (called hybrid stretching) and $\nu_2 + \nu_4$ (called hybrid bending) bands.

64.1 SELECTED REFERENCES FOR DATA

See Table 64.2.

TABLE 64.1
Vibrational Mode and Energy of CH_4 Molecule

Designation	Mode	Energy (eV)
ν_1	Symmetric stretch	0.362
ν_2	Twisting	0.190
ν_3	Asymmetric stretch	0.374
ν_4	Scissoring	0.162

TABLE 64.2
Selected References for Data

Quantity	Energy Range (eV)	Reference
	Cross Sections	
Q_i	0.01–0.175	Lunt (1998)
Q_{el}	300–1300	Maji et al. (1998)
Q_i	17.5–600	Tian and Vidal (1998)
Q_i	15–1000	Straub et al. (1997)
Q_{diff}	0–6	Bundschu et al. (1997)

continued

CH$_4$

TABLE 64.2 (continued)
Selected References for Data

Quantity	Energy Range (eV)	Reference
	Cross Sections	
Q_i	10–10^4	Hwang et al. (1996)
Q_{vib}	0–25	Lunt et al. (1994)
Q_i	10–3000	Nishimura and Tawara (1994)
Q_{diss}	10–700	Kanik et al. (1993)
Q_T	4–300	Kanik et al. (1992)
Q_{el}, Q_T	10–5000	Jain and Baluja (1992)
Q_M, Q_{el}	1.5–100	Boesten and Tanaka (1991)
Q_T	1–4000	Zecca et al. (1991)
Q_{diff}	75–700	Sakae et al. (1989)
Q_i	10–510	Orient and Srivastava (1987)
Q_T	0.1–20	Lohmann and Buckman (1986)
Q_{el}, Q_M	0.2–5	Sohn et al. (1986)
Q_T	1.0–400	Sueoka and Mori (1986)
Q_T	0.085–12.0	Ferch et al. (1985)
Q_T	5–400	Floeder et al. (1985)
Q_i	10–510	Orient and Srivastava (1987)
Q_T	1.3–50	Jones (1985)
Q_{rot}	0–10	Müller et al. (1985)
Q_{diff}	3–20	Tanaka et al. (1982)
Q_a	0–15	Rutkowsky et al. (1980)
Q_i	13–1000	Rapp and Englander–Golden (1965)
Q_{diff}	10–625	Hughes and McMillan (1933)
	Swarm Properties	
Quantity	**E/N Range (Td)**	**Reference**
α, W_+	100–3000	de Urquijo et al. (1999)
α, W_+	100–3000	de Urquijo et al. (1999)*
D, W_e, α	0–15	Alvarez-Pol et al. (1997)
W_e, α	0.1–2000	Yoshida et al. (1996)
W_e	0–15	Kunst et al. (1993)*
W_e	0–15	Zhao et al. (1994)*
W_e	0–15	Chang et al. (1992)*
W_e	0.02–14.0	Schmidt (1991)
D, W_e, α	0.01–15	Schmidt and Roncossek (1992)
D, W_e, α	10–1000	Davies et al. (1989)
W_e	0–15	Wong et al. (1988)*
D_r/μ	0.08–40	Millican and Walker (1987)
W_e	0.013–300	Hunter et al. (1986)
D, W_e, α	0.6–600	Ohmori et al. (1986)
W_e	0.28–848	Al-Amin et al.(1985)
W_e	0.02–1000	Haddad (1985)
W_e	0–15	Mathieson and Hakeem (1979)
W_e	0–15	Jean-Marie et al. (1979)*
W_e	0–6	Wagner et al. (1967)
W_e, α	360–3000	Schlumbohm (1965)
A	70–9000	Heylen (1963)

Note: D/μ = characteristic energy; D = diffusion coefficient; E/N = reduced electric field; W_e = electron drift velocity; W_+ = positive ion drift velocity; α = ionization coefficient. Asterisk (*) indicates gas mixtures of CH$_4$ with argon and CO$_2$. Q_a = attachment; Q_{diff} = differential; Q_{el} = elastic; Q_i = ionization; Q_M = momentum transfer; Q_{rot} = rotational excitation; Q_T = total; Q_{vib} = vibrational excitation (1994).

64.2 TOTAL SCATTERING CROSS SECTIONS

Tables 64.3 and 64.4 give the total scattering cross section. See Figure 64.1 for graphical presentation. The highlights of total cross section as a function of electron impact energy are

1. A relatively slow increase as the electron energy decreases toward thermal energy.
2. Ramsauer–Townsend minimum.
3. Shape resonance at about 8 eV.
4. Decrease of total cross section for increase of energy above 10 eV.

64.3 DIFFERENTIAL CROSS SECTIONS

The differential scattering cross sections for elastic scattering are shown in Figures 64.2 and 64.3. The highlights of the data are

1. At low energy (<10 eV) the differential scattering cross section shows two minima at 60° and 120° with a maximum at ~90°. This indicates a fairly large contribution from *d*-wave scattering (Mapstone and Newell, 1992).

TABLE 64.3
Total Scattering Cross Section at $\varepsilon \leq 20$ eV in Units of 10^{-20} m^2 in CH$_4$

Ferch et al. (1985)				Lohmann and Buckman (1986)			
Energy (eV)	Q_i (10^{-20} m^2)	Energy (eV)	Q_i (10^{-20} m^2)	Energy (eV)	Q_i (10^{-20} m^2)	Energy (eV)	Q_i (10^{-20} m^2)
0.085	5.08	0.8	1.86	0.10	4.179	0.80	1.794
0.09	4.79	0.85	1.97	0.11	3.474	0.85	1.859
0.095	4.52	0.9	2.09	0.12	3.312	0.90	1.942
0.1	4.27	0.95	2.21	0.13	3.109	0.95	2.006
0.11	3.82	1	2.34	0.18	2.246	1.00	2.134
0.12	3.53	1.1	2.59	0.20	2.051	1.20	2.544
0.13	3.30	1.2	2.83	0.25	1.751	1.50	3.536
0.14	3.07	1.3	3.12	0.28	1.527	2.00	5.242
0.15	2.90	1.4	3.39	0.30	1.462	3.00	9.057
0.16	2.76	1.5	3.67	0.32	1.405	4.00	14.07
0.17	2.61	1.6	4.00	0.34	1.358	5.00	18.81
0.18	2.48	1.7	4.33	0.36	1.314	6.00	22.40
0.19	2.37	1.8	4.65	0.38	1.327	7.00	24.56
0.2	2.27	1.9	4.97	0.40	1.343	8.00	25.31
0.25	1.87	2	5.31	0.42	1.371	9.00	25.17
0.3	1.58	3	9.25	0.44	1.387	10.00	24.68
0.35	1.42	4	14.00	0.46	1.396	12.00	23.27
0.4	1.36	5	19.00	0.48	1.398	14.00	21.89
0.45	1.39	6	22.40	0.50	1.403	16.00	20.79
0.5	1.43	7	24.20	0.52	1.422	18.00	19.86
0.55	1.49	8	24.70	0.55	1.451	20.00	18.97
0.6	1.54	9	24.40	0.60	1.489		
0.65	1.58	10	24.00	0.65	1.555		
0.7	1.66	12	22.70	0.70	1.618		
0.75	1.75			0.75	1.713		

TABLE 64.4
Total Scattering Cross Sections in CH_4 at $\varepsilon \geq 0.9$ eV

Sueoka and Mori (1986)				Zecca et al. (1991)			
Energy (eV)	Q_i (10^{-20} m^2)	Energy (eV)	Q_i (10^{-20} m^2)	Energy (eV)	Q_i (10^{-20} m^2)	Energy (eV)	Q_i (10^{-20} m^2)
1.0	2.4	12.0	21.0	0.90	2.0	45.00	12.8
1.2	3.2	13.0	20.6	1.00	2.3	50.00	12.3
1.4	4.1	14.0	19.7	1.20	2.9	60.00	11.7
1.6	4.6	15.0	19.4	1.40	4.0	70.00	11.0
1.8	5.7	16.0	18.5	1.70	4.9	75.00	10.8
2.0	6.2	17.0	17.9	2.00	6.3	80.00	10.30
2.2	6.9	18.0	17.9	2.50	8.2	90.00	9.60
2.5	8.3	19.0	17.5	3.00	10.3	100	9.00
2.8	9.8	20.0	17.0	3.50	12.6	150	7.35
3.1	11.2	25.0	15.3	4.00	15.1	200	6.31
3.4	12.5	30.0	14.1	4.50	17.9	250	5.36
3.7	13.8	35.0	13.2	5.00	19.9	300	4.76
4.0	15.5	40.0	12.6	5.50	21.8	350	4.28
4.5	17.6	50.0	10.8	6.00	23.9	400	3.90
5.0	20.1	60.0	10.3	6.50	25.0	450	3.53
5.5	21.3	70.0	9.4	7.00	26.5	500	3.18
6.0	22.9	80.0	8.9	7.50	26.5	600	2.71
6.5	22.8	90.0	8.5	8.00	27.4	700	2.49
7.0	23.9	100.0	8.0	8.50	26.8	800	2.21
7.5	23.9	120.0	7.3	9.00	26.5	900	1.98
8.0	23.4	150.0	6.6	9.50	26.1	1000	1.78
8.5	23.0	200.0	5.7	10.00	25.8	1250	1.45
9.0	23.3	250.0	5.0	10.50	25.2	1500	1.21
9.5	23.0	300.0	4.4	11.00	25.0	1750	1.03
10.0	22.5	350.0	4.1	13.00	23.9	2000	0.894
11.0	21.7	400.0	3.7	15.00	23.0	2250	0.803
				17.00	21.1	2500	0.717
				20.00	19.6	2750	0.647
				25.00	17.5	3000	0.588
				30.00	16.1	3250	0.554
				35.00	14.6	3500	0.517
				40.00	13.6	4000	0.441

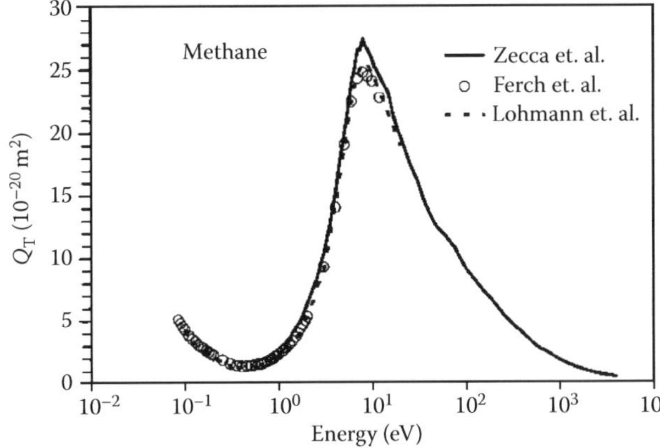

FIGURE 64.1 Total scattering cross section in CH_4. (——) Zecca et al. (1991); (○) Ferch et al. (1985); (— —) Lohmann and Buckman (1986). Note the deep Ramsauer–Townsend minimum at ~0.4 eV.

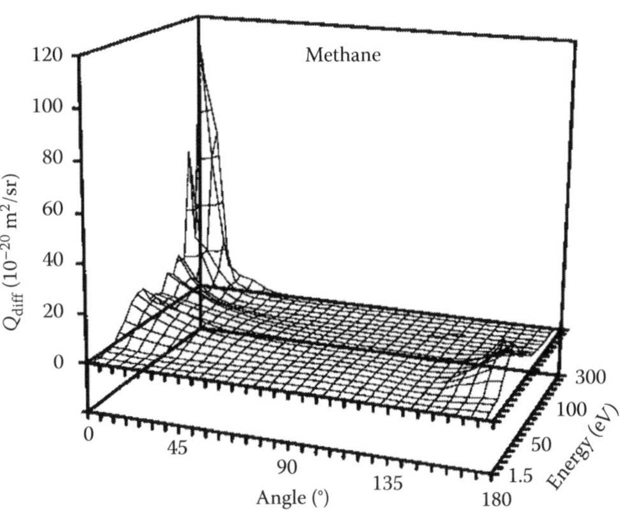

FIGURE 64.2 Differential scattering cross sections of electrons in CH_4 as a function of scattering angle at various electron energies. The cross sections at the extremities of the range are theoretical. $1.5 \leq \varepsilon \leq 100$ eV, 300 eV (Sakae et al., 1989). The integrated cross section for elastic scattering and momentum transfer are shown in the next section. (Adapted from Boesten, L. and H. Tanaka, *J. Phys. B: At. Mol. Opt. Phys.*, 24, 821, 1991.)

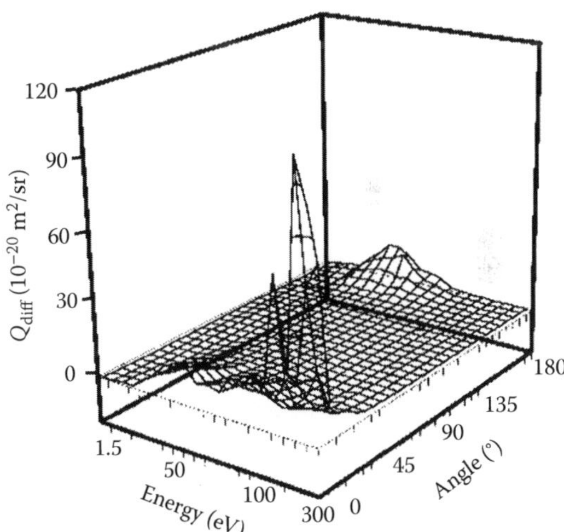

FIGURE 64.3 Integral elastic scattering cross sections of electrons in CH_4. (—○—) Boesten and Tanaka (1991); (— —) Sakae et al. (1989). For clarity sake only the more recent data are shown.

2. As the energy increases the minima coalesce into a single minimum centered around 90°. This indicates a decreasing influence of *d*-wave scattering at these energies.
3. As the energy increases the scattering cross section decreases.

64.4 ELASTIC AND MOMENTUM TRANSFER CROSS SECTION

The integral elastic scattering and momentum transfer cross sections of CH_4 are shown in Tables 64.5 through 64.9

CH_4

TABLE 64.5
Differential Cross Section in Units of 10^{-20} m² in CH_4 at Selected Electron Energy (Digitized and Interpolated)

Angle (°)	Energy (eV) 1.5	5	10	20	30	100	300
0	0.64	5.40	10.60	17.30	15.10	18.00	25.44
5	0.48	4.59	9.03	13.88	11.47	9.91	20.23
10	0.36	3.78	7.68	10.95	8.58	5.00	11.98
15	0.25	3.02	6.52	8.49	6.34	2.41	5.75
20	0.15	2.36	5.52	6.46	4.63	1.25	2.59
25	0.08	1.84	4.65	4.82	3.36	0.69	1.36
30	0.07	1.52	3.90	3.54	2.43	0.37	1.10
35	0.07	1.40	3.24	2.57	1.73	0.23	0.96
40	0.08	1.32	2.66	1.88	1.20	0.20	0.70
45	0.12	1.12	2.21	1.41	0.83	0.20	0.51
50	0.18	1.01	1.90	1.09	0.61	0.15	0.42
55	0.25	1.07	1.69	0.88	0.48	0.09	0.35
60	0.30	1.21	1.55	0.73	0.40	0.04	0.29
65	0.36	1.38	1.42	0.64	0.33	0.01	0.26
70	0.41	1.51	1.29	0.56	0.26	0.00	0.23
75	0.45	1.58	1.14	0.48	0.21	0.00	0.21
80	0.49	1.59	1.00	0.41	0.19	0.00	0.18
85	0.51	1.54	0.90	0.35	0.20	0.00	0.15
90	0.51	1.44	0.83	0.29	0.21	0.01	0.13
95	0.50	1.28	0.74	0.24	0.20	0.01	0.12
100	0.47	1.03	0.59	0.21	0.17	0.02	0.12
110	0.38	0.47	0.32	0.19	0.10	0.00	0.12
120	0.26	0.18	0.33	0.26	0.14	0.00	0.12
130	0.16	0.28	0.54	0.40	0.25	0.03	0.11
140	0.12	0.80	2.18	0.59	0.35	0.10	0.11
150	0.08	3.17	3.44	0.97	0.45	0.13	0.12
160	0.07	4.15	6.59	1.50	0.61	0.34	0.14
170	0.09	6.07	9.11	1.82	0.87	0.43	0.15
180	0.10	6.79	6.94	1.55	1.22	0.50	0.16

Note: $1.5 \leq \varepsilon \leq 100$ eV, Boesten and Tanaka (1991), 300 eV (Sakae et al. 1989). See Figures 64.2 and 64.3 for graphical presentation.

(Althorpe et al., 1995). See Figures 64.4 and 64.5 for graphical presentation.

64.5 MERT COEFFICIENTS

The four term MERT expression for phase shifts are (Raju, 2006)

$$s\text{-wave:}\quad \tan\eta_0 = -Ak - \left(\frac{\pi\alpha}{3a_0}\right)k^2 - Ak\left(\frac{4\alpha}{3a_0}\right)k^2\ln(ka_0) + Dk^3 + Fk^4 \quad (64.2)$$

$$p\text{-wave:}\quad \tan\eta_1 = \left(\frac{\pi\alpha}{15a_0}\right)k^2 - A_1k^3 \quad (64.3)$$

TABLE 64.6
Low-Energy Momentum Transfer Cross Section Derived from Transport Coefficients

Energy (eV)	Q_M (10^{-20} m²)	Energy (eV)	Q_M (10^{-20} m²)	Energy (eV)	Q_M (10^{-20} m²)
0.001	23.0	0.050	5.65	0.36	0.242
0.004	18.8	0.060	4.70	0.40	0.261
0.007	17.0	0.070	3.83	0.45	0.300
0.010	15.6	0.080	3.31	0.50	0.400
0.012	14.0	0.10	2.30	0.60	0.732
0.014	13.0	0.12	1.75	0.70	1.08
0.017	12.2	0.14	1.30	0.80	1.29
0.020	11.5	0.17	0.810	1.0	1.84
0.025	10.0	0.20	0.522	1.2	2.13
0.030	8.69	0.25	0.335	1.4	2.54
0.035	7.84	0.28	0.282	1.7	3.10
0.040	6.90	0.32	0.252	2.0	3.70

Source: Adapted from Schmidt, B., *J. Phys. B: At. Mol. Opt. Phys.*, 24, 4809, 1991. The low values of *E/N* and data on the longitudinal diffusion coefficient are of interest.
Note: See Figure 64.5 for graphical presentation.

TABLE 64.7
Momentum Transfer Cross Sections Derived from Transport Coefficients

Energy (eV)	Q_M (10^{-20} m²)	Energy (eV)	Q_M (10^{-20} m²)	Energy (eV)	Q_M (10^{-20} m²)
0.000	50.0	0.600	0.953	40.0	3.26
0.010	40.0	0.800	1.40	60.0	2.13
0.015	24.4	1.00	1.90	80.0	1.67
0.020	17.0	1.50	3.10	100.0	1.43
0.030	10.2	2.00	4.40	150	1.07
0.040	7.30	3.00	7.20	200	0.900
0.060	4.60	4.00	10.50	300	0.714
0.080	2.90	6.00	14.70	400	0.612
0.100	2.00	8.00	16.10	600	0.496
0.150	0.990	10.00	15.50	800	0.430
0.200	0.700	15.0	12.60	1000	0.384
0.300	0.560	20.0	8.80	3000	0.120
0.400	0.650	30.0	4.76	10000	0.016

Source: Adapted from Davies, D. K., L. E. Kline, and W. E. Bies, *J. Appl. Phys.*, 85, 3311, 1989.
Note: See Figure 64.5 for graphical presentation.

Higher waves:

$$\tan\eta_l = \frac{\pi\alpha k^2}{[(2l+3)(2l+1))(2l-1)]a_0} \quad \text{for } l \geq 2 \quad (64.4)$$

Further

$$Q_T(0) = Q_M(0) = 4\pi A^2; \quad A = \text{scattering length} \quad (64.5)$$

TABLE 64.8
Theoretical Low-Energy Elastic Scattering Cross Sections in CH_4

Energy (eV)	Q_{el} (10^{-20} m^2)	Q_T (10^{-20} m^2)	Energy (eV)	Q_{el} (10^{-20} m^2)	Q_T (10^{-20} m^2)
0.05	4.94	4.94	4.0	13.01	13.60
0.1	2.59	2.59	5.0	17.23	18.11
0.22	1.00	1.45	6.0	21.08	22.29
0.41	0.95	1.59	7.0	23.80	25.23
0.60	1.42	2.11	8.0	25.29	26.79
0.79	1.97	2.62	9.0	25.72	27.17
1.0	2.62	3.21	10.0	25.48	26.82
2.0	5.67	6.12	11.0	24.86	29.09
3.0	9.14	9.59	12.0	24.09	25.21

Source: Adapted from Althorpe, S. C., F. A. Gianturco, and N. Sanna, *J. Phys. B: At. Mol. Opt. Phys.*, 28, 4165, 1995.

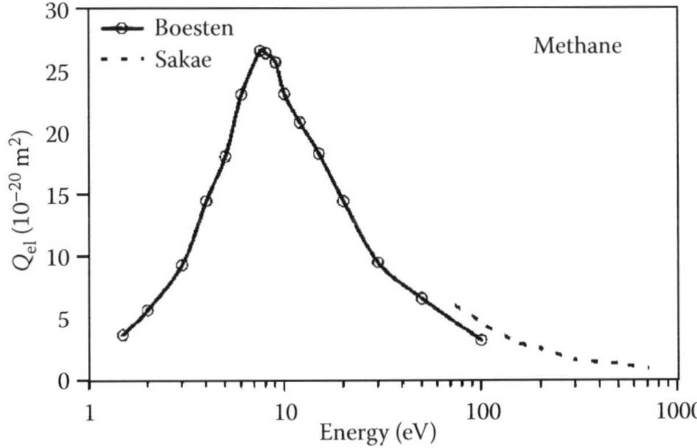

FIGURE 64.4 Momentum transfer cross section in CH_4. (——) Davies et al. (1989); (○) Schmidt; (Δ) Boesten and Tanaka (1991); (□) Sakae et al. (1989). Note that the Ramsauer–Townsend minimum is deeper than that in the total scattering cross section in agreement with theory.

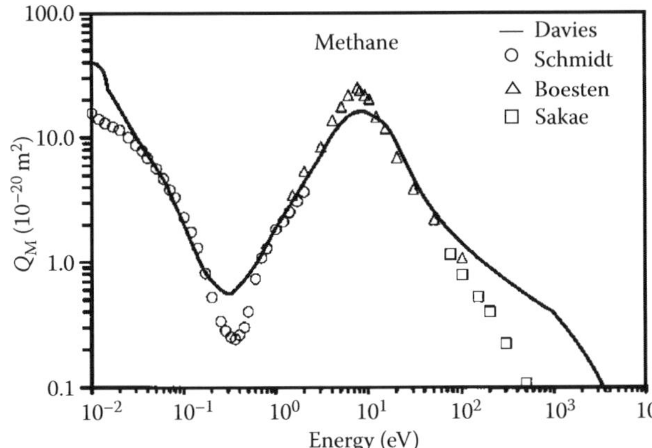

FIGURE 64.5 Vibrational excitation cross section of CH_4 derived from Swarm studies. (○) Vibrational band for v = 2 and v = 4 (Ohmori et al. 1986); (——) Davies et al. (1989); (□) Vibrational band for v = 1 and v = 3, (Ohmori et al. 1986); (— —) Davies et al. (1989); (—×—) Total Q_v (Davies et al. 1989).

TABLE 64.9
Integral Elastic and Momentum Transfer Cross Sections of CH_4

Energy (eV)	Q_{el} (10^{-20} m^2)	Q_M (10^{-20} m^2)	Energy (eV)	Q_{el} (10^{-20} m^2)	Q_M (10^{-20} m^2)
Boesten and Tanaka (1991)			Sakae et al. (1989)		
1.5	3.61	3.49	75	6.03	1.15
2	5.61	5.40	100	4.59	0.787
3	9.25	8.43	150	3.01	0.526
4	14.41	13.71	200	2.56	0.399
5	18.04	17.55	300	1.65	0.223
6	22.97	21.79	500	1.33	0.107
7.5	26.49	25.07	700	0.967	0.0663
8	26.30	23.76			
9	25.55	21.77			
10	23.02	20.21			
12	20.74	14.63			
15	18.27	11.78			
20	14.41	6.95			
30	9.49	3.87			
50	6.57	2.22			
100	3.20	1.08			

TABLE 64.10
MERT Parameters in CH_4

A/a_0	D/a_0^3	F/a_0^4	A_1/a_0^3	Reference
−2.048	253	−350	17.9	Lunt et al. (1998) from Q_T.
−2.58*	139	−55	11	Lunt et al. (1994) from Q_T
−2.54	153	−126	10	Lohmann and Buckman, (1986), from Q_T
−2.59	162	−129	13	Ferch et al. (1985), from Q_T
−2.755	171.2	−92.2	17.01	Schmidt (1991), from Q_M

Note: Negative sign for *A* denotes the existence of Ramsauer–Townsend minimum.

In the above equations η_0 is the wave shift (radians), α the electronic polarizability (F m^2), k the wave number related to the electron energy, $k = (2\varepsilon)^{1/2}$ and $a_0 = 5.2918 \times 10^{-11}$ m. The constants are shown in Table 64.10 (Lunt et al., 1994).

64.6 RAMSAUER–TOWNSEND MINIMUM

The Ramsauer–Townsend minimum cross section determined by various methods are shown in Table 64.11.

64.7 RO-VIBRATIONAL EXCITATION CROSS SECTION

The rotational constant of CH_4 molecule is 0.65 meV (Müller et al., 1985). The vibrational energies of the fundamental modes are shown in Table 64.1. Table 64.12 shows the vibrational cross sections. Figures 64.6 and 64.7 show graphical presentation.

64.8 EXCITATION CROSS SECTION

Excitation potentials for CH_4 are given in Table 111.2 of the Propylene (C_3H_6) chapter. All excited singlet states of CH_4 are unstable and the cross section for singlet state excitation is included in the dissociation cross section. If the triplet states are unstable as well, then the dissociation cross section will include the entire electronic excitation cross section (Floeder et al., 1985). However, dissociation cross sections corrected for dissociative ionization cross sections are given by Davies et al. (1989) and Table 64.13 gives these data.

64.9 TOTAL IONIZATION CROSS SECTION

Tables 64.14 and 64.15 present the total ionization cross section in CH_4. For graphical presentation see Figure 64.8.

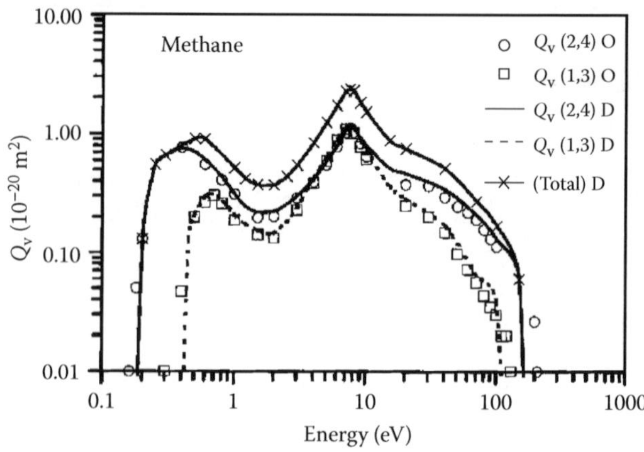

FIGURE 64.6 Theoretical vibrational excitation cross section of CH_4. Note the scale for Q_v. (Adapted from Althorpe, S. C., F. A. Gianturco, and N. Sanna, *J. Phys. B: At. Mol. Opt. Phys.*, 28, 4165, 1995.)

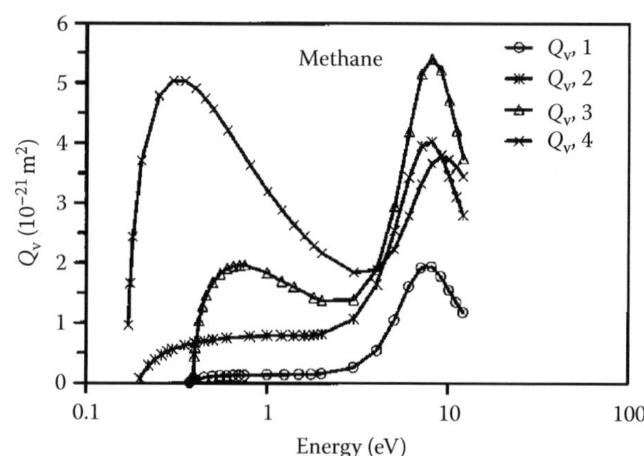

FIGURE 64.7 Total ionization cross section in CH_4. (——) Kim et al. (2004), (theory); (— —) Tian and Vidal (1998); (△) Straub et al. (1997); (□) Nishimura and Tawara (1994); (○) Orient and Srivastava (1987); (×) Rapp and Englander–Golden (1965). Theoretical results of Kim et al. (2004) are presented to show the degree of agreement that has been achieved.

TABLE 64.11
Ramsauer–Townsend Minimum Cross Section in CH_4

Energy (eV)	Q_{min} (10^{-20} m^2)	Method	Reference
0.22	1.45	Total cross section (theory)	Althorpe et al. (1995)
0.41	0.95	Elastic scattering (theory)	Althorpe et al. (1995)
0.36	0.242	Momentum transfer, Swarm method	Schmidt (1991)
0.3	0.56	Momentum transfer, Swarm method	Davies et al. (1989)
0.36	1.314	Total cross section measurement	Lohmann and Buckman (1986)
0.3	1.3	Swarm method	Ohmori et al. (1986)
0.6	0.92	Elastic scattering measurements	Sohn et al. (1986)
0.3	0.45	Momentum transfer measurements	Sohn et al. (1986)
0.4	1.36	Total cross section measurement	Ferch et al. (1985)
0.215	0.5	Total cross section measurement	Barbarito et al. (1979)
0.4	1.1	Total cross section measurement	Ramsauer et al. (1930)

TABLE 64.12
Vibrational Excitation Cross Section in CH_4

Ohmori et al. (Digitized)			Schmidt (1991)			
Energy (eV)	$Q_{v(2,4)}$ (10^{-20} m^2)	$Q_{v(1,3)}$ (10^{-20} m^2)	Energy (eV)	$Q_{v(2,4)}$ (10^{-20} m^2)	Energy (eV)	$Q_{v(1,3)}$ (10^{-20} m^2)
0.18	0.05		0.162	0.00	0.374	0.00
0.20	0.13		0.165	0.0427	0.380	0.143
0.40	0.76	0.05	0.17	0.241	0.400	0.330
0.5		0.20	0.18	0.384	0.450	0.440
0.60	0.55	0.27	0.200	0.527	0.500	0.495
0.7		0.30	0.230	0.459	0.550	0.442
0.80	0.41	0.26	0.300	0.300	0.600	0.360
1.0	0.31	0.19	0.400	0.181	0.700	0.308
1.5	0.20	0.14	0.500	0.159	0.800	0.273
2.0	0.20	0.13	0.600	0.165	1.00	0.215

Note: Data of Ohmori et al. (1986) and Schmidt (1991) are from swarm properties determination.

TABLE 64.13
Cross Section for Dissociation into Neutral Fragments of CH_4

Energy (eV)	Q_{diss} (10^{-20} m^2)	Energy (eV)	Q_{diss} (10^{-20} m^2)
Kanik et al. (1993)		Davies et al. (1989)	
10		10	0.3
15	1.1	11	0.3
20	1.75	12	0.3
30	1.75	13	1.2
50	1.2	15	1.6
100	0.67	20	2.16
150	0.47	25	2.16
200	0.37	30	2.16
300	0.26	100	2.16
500	0.17	150	1.94
700	0.13	200	1.80
		300	1.6

TABLE 64.13 (continued)
Cross Section for Dissociation into Neutral Fragments of CH_4

Energy (eV)	Q_{diss} (10^{-20} m^2)	Energy (eV)	Q_{diss} (10^{-20} m^2)
Kanik et al. (1993)		Davies et al. (1989)	
		400	1.5
		500	1.5
		1000	0.60
		2000	0.30
		5000	0.12
		10 000	0.06

Note: The cross sections of Davies et al. (1989) are derived from Monte Carlo simulation. For graphical presentation see Figure 64.11.

TABLE 64.14
Ionization Cross Sections of CH_4

	Q_i (10^{-20} m^2)				
Energy (eV)	Kim et al. (2004)	Straub et al. (1997)	Nishimura and Tawara (1994)	Orient et al. (1987)	Rapp and Englander-Golden (1965)
13.5					0.034
14					0.074
14.5	0.045				0.130
15	0.137	0.235	0.209	0.13	0.198
17.5	0.609	0.763	0.693	0.535 (*I*)	0.610
20	1.047	1.421	1.22	0.94	1.161 (*I*)
22.5	1.427	1.879	1.615 (*I*)	1.445 (*I*)	1.496
25	1.746	2.281	2.01	1.95	1.848 (*I*)
30	2.293	2.788	2.56	2.70	2.376
35	2.689	3.095	2.96	3.01 (*I*)	2.737
40	2.976	3.338	3.23	3.32	3.018
45	3.176	3.591	3.49	3.51	3.212
50	3.315	3.717	3.60	3.69	3.362
60	3.471	3.902	3.86	3.92	3.564
70	3.524	4.006	3.93	4.04	3.661
80	3.518	4.011	3.98	4.08	3.696
90	3.479	3.993	3.98	4.08	3.696
100	3.419	3.944	3.92	4.04	3.661
110	3.349	3.871			3.617
125	3.234	3.742	3.75		3.520
150	3.038	3.516	3.55	3.75 (*I*)	3.326
175	2.853	3.323	3.32		3.148
200	2.682	3.153	3.17	3.40	3.010
250	2.391	2.774	2.86	3.06 (*I*)	2.719
300	2.156	2.507	2.55	2.76	2.490
400	1.804	2.082	2.17	2.36	2.094
500	1.554	1.800	1.85	2.10	1.830
600	1.368	1.569	1.62		1.628
700	1.224	1.404	1.44		1.470
800	1.109	1.242	1.33		1.338
900	1.015	1.147	1.22		1.241
1000	0.937	1.050	1.13		1.179
1250			0.937		
1500			0.818		

TABLE 64.14 (continued)
Ionization Cross Sections of CH_4

	Q_i (10^{-20} m^2)				
Energy (eV)	Kim et al. (2004)	Straub et al. (1997)	Nishimura and Tawara (1994)	Orient et al. (1987)	Rapp and Englander-Golden (1965)
2000			0.660		
2500			0.552		
3000			0.435		

Note: (*I*) = Interpolated. For graphical presentation, see Figure 64.8.

TABLE 64.15
Ionization Cross Sections in CH_4 at Higher Energies

Energy (eV)	600	1000	2000	4000	7000	12,000
Q_i (10^{-20} m^2)	1.38	0.937	0.538	0.306	0.193	0.124

Source: Adapted from Schram, B. L. et al., *J. Chem. Phys.*, 44, 49, 1966.

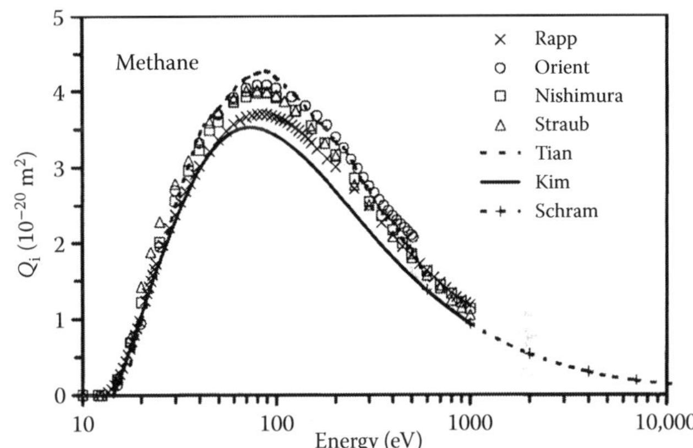

FIGURE 64.8 Partial ionization cross section in CH_4. (Adapted from Straub, H. C. et al., *J. Chem. Phys.*, 106, 4430, 1997.)

64.10 ION APPEARANCE POTENTIALS

Table 64.16 shows the ion appearance potentials (Chatham et al., 1984).

64.11 PARTIAL IONIZATION CROSS SECTIONS

Table 64.17 shows the partial ionization cross sections in CH_4. For graphical presentation, see Figure 64.9.

64.12 ATTACHMENT CROSS SECTION

The dissociative attachment processes in CH_4 are (Rutkowsky et al., 1980)

$$e(>6 \text{ eV}) + CH_4 \rightarrow CH_4^{-*} \rightarrow \begin{cases} CH_2^- + H_2 \\ H^- + CH_3 \\ CH^- + H_2 + H \end{cases} \quad (64.6)$$

TABLE 64.16
Ion Appearance Potentials

Ion Species	CH_4^+	CH_3^+	CH_2^+	CH^+	C^+	
Energy (eV)	12.6	14.3	15.1	22.2	25 ± 2	Chatham et al. (1984)
Energy (eV)	12.63	13.25[a]	15.06[b]	19.87[c]	19.56[d]	Plessis et al. (1983)

[a] $CH_4 \rightarrow CH_4^+$
[b] $CH_4^+ \rightarrow CH_3^+ + H^-$
[c] $CH_4^+ \rightarrow CH_2^+ + H_2$
[d] $CH_4^+ \rightarrow C^+ + 2H_2$

TABLE 64.17
Partial Ionization Cross Sections in CH_4 for Selected Ion Species

Energy (eV)	CH_4^+ (10^{-20} m²)	CH_3^+ (10^{-20} m²)	CH_2^+ (10^{-21} m²)
15	0.197	0.038	
17.5	0.519	0.238	0.055
20	0.892	0.512	0.169
22.5	1.11	0.74	0.285
25	1.29	0.923	0.559
30	1.46	1.12	1.41
35	1.51	1.19	2.18
40	1.55	1.22	2.65
45	1.61	1.27	2.97
50	1.63	1.29	3.07
60	1.65	1.33	3.18
70	1.66	1.35	3.21
80	1.64	1.34	3.23
90	1.62	1.33	3.14
100	1.60	1.31	3.12
110	1.56	1.29	3.05
125	1.52	1.25	2.90
150	1.44	1.18	2.70
175	1.37	1.13	2.48
200	1.31	1.08	2.34
250	1.17	0.970	2.01
300	1.07	0.886	1.78
400	0.918	0.742	1.40
500	0.799	0.655	1.19
600	0.703	0.574	1.02
700	0.636	0.518	0.892
800	0.562	0.463	0.796
900	0.524	0.427	0.725
1000	0.483	0.391	0.656

Source: Adapted from Straub, H. C. et al., *J. Chem. Phys.*, 106, 4430, 1997.

Dissociative attachment cross section in CH_4 is presented in Table 64.18.

64.13 CONSOLIDATED PRESENTATION

Figure 64.11 presents the consolidated presentation of all cross sections in CH_4.

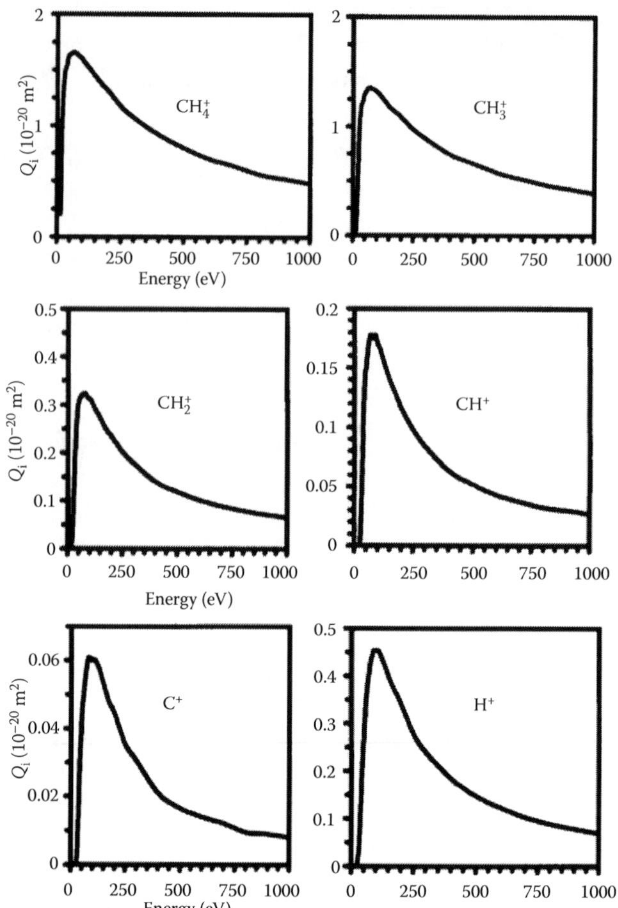

FIGURE 64.9 Dissociative attachment cross section in CH_4. (—○—) Rutkowsky et al. (1980); (●) Sharp and Dowell (1967).

64.14 ELECTRON MOBILITY AND DRIFT VELOCITY

Electron mobilities and drift velocities are shown in Table 64.19.

64.15 DIFFUSION COEFFICIENTS

Tables 64.19 and 64.20 show the ratio of diffusion coefficient to mobility as a function of E/N. See Table 64.21 also.

64.16 FIRST IONIZATION COEFFICIENTS

Townsend's first reduced ionization coefficients have been measured by the current growth method (Heylen, 1963), time-resolved current method (de Urquijo et al., 1999; Davies et al., 1989), Avalanche statistics method (Yoshida et al. 1996). These methods have been described by Raju (2006). See Table 64.22.

The gas constants according to expression

$$\frac{\alpha}{N} = F \exp\left(-\frac{GN}{E}\right) \tag{64.7}$$

are (Davies et al., 1989) $F = 2.51 \times 10^{-20}$ m² and $G = 652$ Td.

TABLE 64.18
Dissociative Attachment Cross Sections in CH_4

Energy (eV)	Q_a (10^{-23} m^2)
7.0	0.05
7.5	0.58
8.0	1.17
8.5	2.10
9.0	3.43
9.5	4.94
10	6.65
11	3.14
12	1.35
13	0.92
14	0.93

Source: Adapted from Rutkowsky, V. J., H. Drost, and H. J. Spangenberg, *Ann. Phys.* (Leipzig), 37, 259, 1980. (Dissociative and direct attachment cross sections in several hydrocarbon gases are measured.)

Note: For graphical presentation, see Figure 64.10.

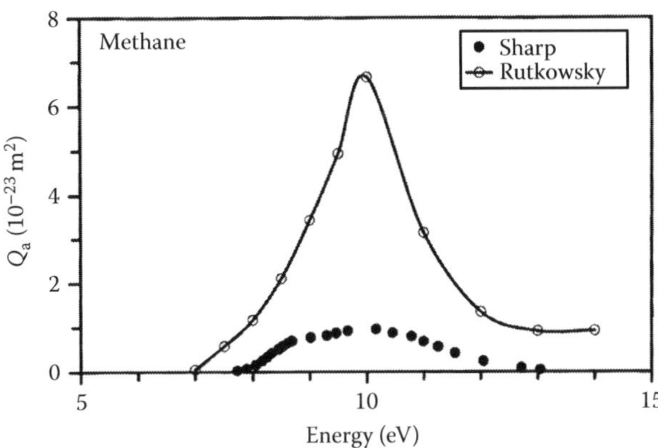

FIGURE 64.10 Consolidated cross sections in CH_4 showing the relative contributions to the total. (○) Total; (——) Momentum transfer; (—△—) Vibrational excitation (2,4); (—×—) Vibrational excitation (1,3); (—□—) Dissociation; (— —) Ionization; (—◊—) Dissociative attachment (multiply by 0.001).

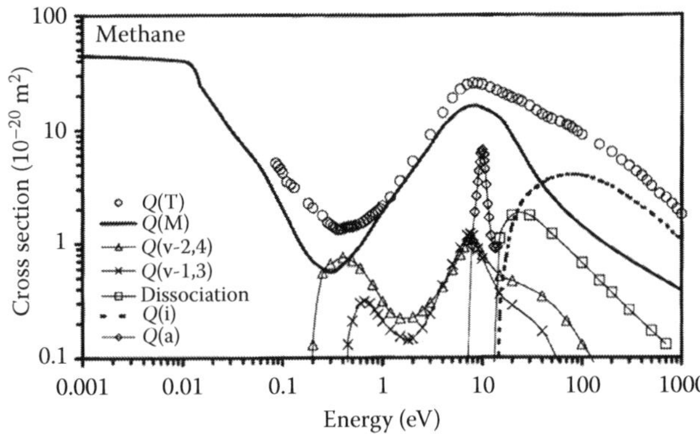

FIGURE 64.11 Electron mobility and drift velocity in CH_4. Note the negative slope in the latter. Drift velocity: (——) Davies et al. (1989); (○) Hunter et al. (1986). μN: (—△—) Hunter et al. (1986).

TABLE 64.19
Experimental Drift Velocity in CH_4

E/N (Td)	*W* (10^4 m/s)	*E/N* (Td)	*W* (10^4 m/s)	μN (10^{25} Vm s^{-1})	*E/N* (Td)	*W* (10^4 m/s)	μN (10^{25} (Vm s)$^{-1}$)
Davies et al. (1989)		Hunter et al. (1986)					
10.00	7.92	0.012	0.038	3.17	3.0	10.35	3.45
12.50	6.84	0.014	0.044	3.14	3.5	10.35	2.96
15.00	6.37	0.017	0.053	3.12	4.0	10.1	2.53
17.50	5.76	0.02	0.063	3.12	5.0	9.75	1.95
20.00	5.49	0.025	0.078	3.12	6.0	9.22	1.54
25.00	5.35	0.03	0.094	3.17	8.0	8.25	1.03
30.00	5.27	0.035	0.112	3.20	10.0	7.40	0.74
35.00	5.16	0.04	0.130	3.25	12.0	6.87	0.573
40.0	5.25	0.05	0.160	3.28	14.0	6.52	0.486
50.0	5.38	0.06	0.199	3.32	17.0	6.04	0.355
60.0	5.45	0.07	0.236	3.37	20.0	5.79	0.290
70.0	5.81	0.08	0.275	3.44	25.0	5.54	0.222
80.0	6.40	0.10	0.350	3.50	30.0	5.39	0.180
90.0	6.83	0.12	0.428	3.57	35.0	5.32	0.152
100.0	7.40	0.14	0.507	3.62	40.0	5.35	0.134
112.5	8.14	0.17	0.632	3.72	50.0	5.38	0.108
125.0	8.86	0.20	0.762	3.81	60.0	5.58	0.093
150.0	10.2	0.25	0.990	3.96	70.0	5.84	0.083
175.0	11.6	0.3	1.23	4.10	80.0	6.23	0.078
200.0	13.2	0.4	1.75	4.38	100.0	7.21	0.072
250.0	16.3	0.5	2.32	4.64	120.0	8.49	0.071
300.0	19.9	0.6	2.93	4.88	140.0	9.60	0.069
400	26.6	0.8	4.10	5.13	160.0	11.0	0.069
500	33.1	1.0	5.26	5.26	180.0	12.1	0.067
600	40.0	1.2	6.42	5.35	200.0	13.6	0.068
700	46.3	1.4	7.35	5.25	225.0	15.1	0.067
800	52.3	1.7	8.35	4.91	250.0	17.7	0.071
900	58.7	2.0	9.10	4.55	275	19.9	0.072
1000	66.2	2.5	10.05	4.02	300	21.8	0.073

Note: The reduced mobility (μN, $N = 2.688 \times 10^{25}$ m^3) is at 300 K. See Figure 64.12 for graphical presentation.

CH$_4$

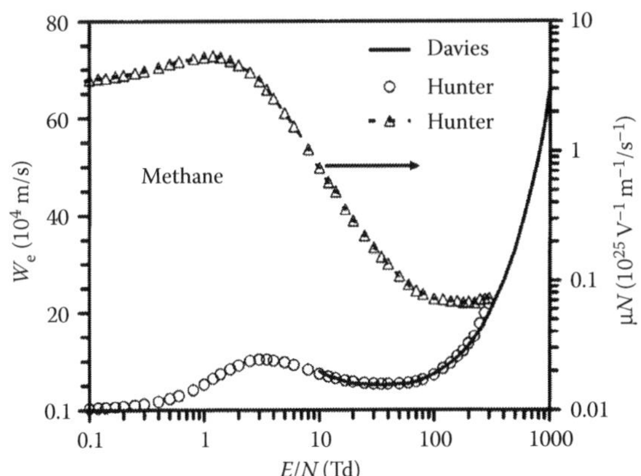

FIGURE 64.12 Radial and longitudinal diffusion coefficients expressed as ratios of mobility in CH_4. D_r/μ: (△) Lakshminarasimha and Lucas (1977); (□) Haddad (1985) *Austr. J. Phys.*, (1985); (○) Schmidt and Roncossek (1992); (——) Visual best fit, Table 64.19. The values have been multiplied by 10 for better presentation. D_L/μ: (×) Schlumbohm (1965); (△) Al-Amin et al. (1985); (□) Schmidt and Roncossek (1992); (——) Table 64.20.

CH$_4$

TABLE 64.20
Radial Diffusion Coefficients Expressed as Ratios of Mobility in CH$_4$

E/N (Td)	D_r/μ (V)	E/N (Td)	D_r/μ (V)	E/N (Td)	D_r/μ (V)
0.02	0.0270	0.50	0.0436 (0.049)	20	2.02 (2.15)
0.025	0.0281	0.60	0.0479 (0.051)	30	2.74 (3.14)
0.030	0.0283	0.70	0.0522 (0.056)	40	3.15 (3.77)
0.035	0.0285	0.80	0.0566	50	3.37
0.040	0.0289	1.0	0.0671 (0.069)	60	3.51
0.050	0.0289	1.2	0.0770 (0.085)	75	3.78
0.060	0.0289	1.4	0.0883 (0.095)	100	4.37
0.070	0.0292	1.7	0.112 (0.110)	200	4.65
0.080	0.0295 (0.0309)	2.0	0.125 (0.131)	300	4.77
0.10	0.0301	2.5	0.158 (0.165)	400	4.84
0.12	0.0307 (0.0308)	3.0	0.196 (0.202)	500	4.30
0.14	0.0313 (0.031)	3.5	0.240	600	4.73
0.17	0.0322 (0.033)	4.0	0.287 (0.290)	700	4.78
0.20	0.0331 (0.034)	5.0	0.334 (0.390)	800	4.83
0.25	0.0346 (0.036)	6.0	0.479 (0.490)	900	5.27
0.30	0.0363 (0.038)	7.0	0.585 (0.620)	1000	5.59
0.35	0.0381 (0.041)	8.0	0.69 (0.710)	1800	8.12
0.40	0.0399 (0.042)	10.0	0.94 (0.974)		

TABLE 64.21
Longitudinal Diffusion Coefficients Expressed as Ratios of Mobility in CH$_4$

E/N (Td)	D_L/μ (V)	E/N (Td)	D_L/μ (V)	E/N (Td)	D_L/μ (V)
0.0350	0.035	0.700	0.083	12.0	0.165
0.0400	0.035	0.800	0.088	14.0	0.168
0.0500	0.036	1.00	0.092	20.0	0.292
0.0600	0.036	1.20	0.095	30.0	0.395
0.0700	0.036	1.40	0.094	40.0	0.479
0.0800	0.037	1.60	0.091	50.0	0.598
0.100	0.037	1.80	0.088	60.0	0.806
0.120	0.038	2.00	0.086	75.0	1.30
0.140	0.040	2.50	0.082	100	2.32
0.160	0.042	3.00	0.083	200	3.15
0.180	0.043	3.50	0.087	300	3.42
0.200	0.045	4.00	0.093	400	4.46
0.250	0.050	4.50	0.098	500	5.26
0.300	0.054	5.00	0.103	600	5.92
0.350	0.058	6.00	0.109	700	6.63
0.400	0.062	7.00	0.115	800	7.77
0.500	0.071	8.00	0.126	900	9.73
0.600	0.078	10.0	0.156		

Note: See Figure 64.13 for graphical presentation.

64.17 ATTACHMENT COEFFICIENTS

Figure 64.15 shows the attachment coefficients as a function of E/N.

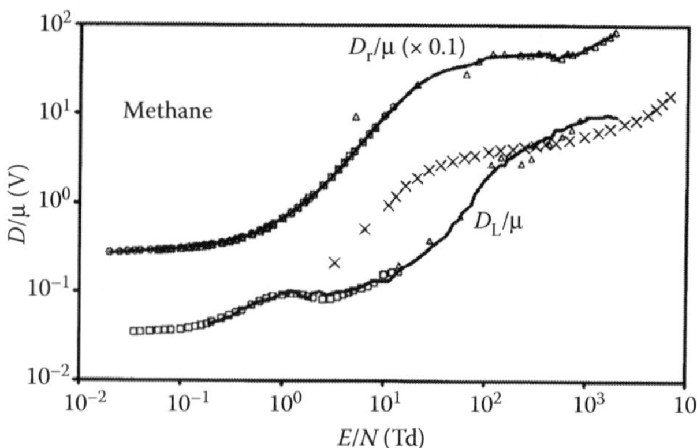

FIGURE 64.13 Reduced ionization coefficients as a function of reduced electric field in CH$_4$. (——) Heylen (1963); (○) Davies et al. (1989); (△) Yoshida et al. (1996); (□) de Urquijo et al. (1999).

TABLE 64.22
Reduced Ionization Coefficients as a Function of Reduced Electric Field in CH$_4$

	α/N (10^{-20} m^2)			
E/N (Td)	Heylen (1963)[a]	Davies et al. (1989)[b]	Yoshida (1992)[a]	de Urquijo (1999)[a]
70	8.00 (−4)			
80	1.80 (−3)	1.02 (−3)		
90	3.40 (−3)	2.41(−3)		
100	6.90 (−3)	5.00 (−3)		6.5 (−3)
125	2.32 (−2)	1.71 (−2)	2.26 (−2)	2.23 (−2)
150	4.91 (−2)	4.05 (−2)	5034	5.09 (−2)
175	8.31 (−2)	7.29 (−2)	9.05	8.07 (−2)
200	0.121	0.108	0.14	0.13
225	0.168		0.20	
250	0.223	0.198	0.24	0.23
300	0.336	0.310	0.34	0.36
350	0.352		0.48	0.45
400	0.576	0.52	0.59	0.61
450	0.693		0.72	0.83
500	0.813	0.72	0.85	0.93
600	1.066	0.92	1.13	1.33
700	1.315	1.11	1.40	1.74
800	1.559	1.25	1.64	1.92
900	1.787	1.34	1.73	2.08
1000	1.992	1.45	1.94	2.45
1250	2.441		2.34	
1500	2.802		2.51	3.48
2000	3.334		3.02	4.60
2500	3.746			5.10
3000	4.058			5.63
4000	4.498			
5000	4.800			
6000	5.009			
7000	5.162			

TABLE 64.22 (continued)
Reduced Ionization Coefficients as a Function of Reduced Electric Field in CH_4

	α/N (10^{-20} m^2)			
E/N (Td)	Heylen (1963)[a]	Davies et al. (1989)[b]	Yoshida (1992)[a]	de Urquiho (1999)[a]
8000	5.339			
9000	5.625			

Note: The values of de Urquijo et al. (1999) are effective ionization coefficients. a(b) means $a \times 10^b$. See Figure 64.14 for graphical presentation.

[a] Digitized from the original and interpolated.

[b] Tabulated in the original.

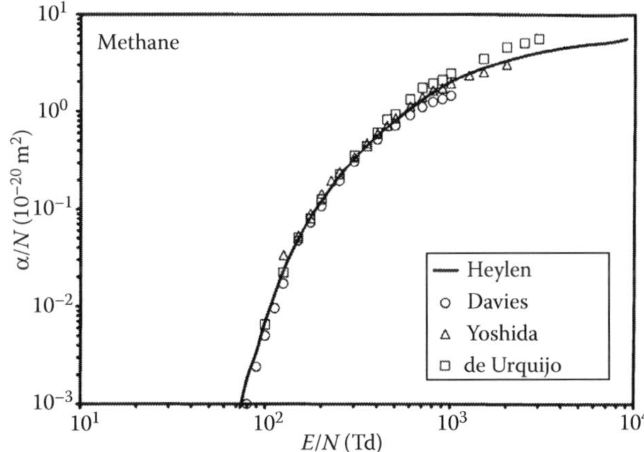

FIGURE 64.14 Attachment coefficient expressed as ratio of ionization coefficient in CH_4. (○) de Urquiho et al. (1999); (—□—) Hunter et al. (1986).

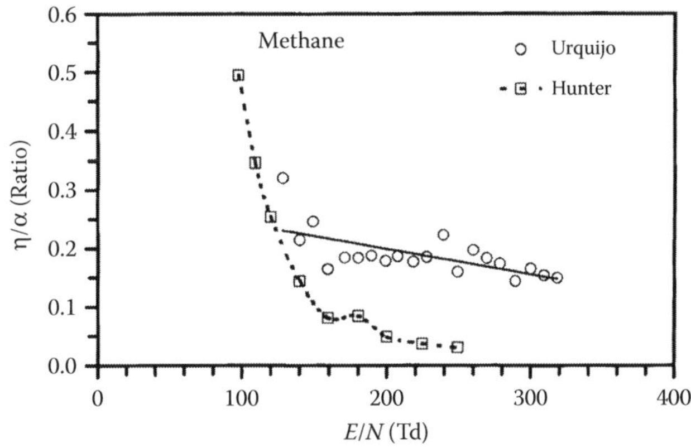

FIGURE 64.15 Reduced mobility of positive ions in CH_4. (——) de Urquiho et al. (1999); (○) Davies et al. (1989).

64.18 ION MOBILITY

Positive ion reduced mobilities at $N = 2.69 \times 10^{25}$ m^{-3} are shown in Table 64.23.

TABLE 64.23
Positive Ion Reduced Mobilities in CH_4

	μ_+ (10^{-4} m^2/V/s)	
E/N (Td)	Davies et al. (1989)[a]	De Urquijo et al. (1999)[b]
80.0	2.49	
90.0	2.64	
100	2.76	
112.5	2.85	2.78
125.0	2.97	2.90
150.0	3.16	3.17
175.0	3.31	3.43
200.0	3.40	3.38
250.0	3.42	3.50
300.0	3.35	3.39
400	3.18	3.25
450		3.15
500	3.05	3.01
600	2.93	2.84
700	2.82	2.71
800	2.71	2.60
900	2.63	2.54
1000	2.52	2.48
1500		2.18
2000		1.88
2500		1.51
3000		0.96

Note: See Figure 64.16 for graphical presentation.

[a] Tabulated in the original.

[b] Digitized from the original and interpolated.

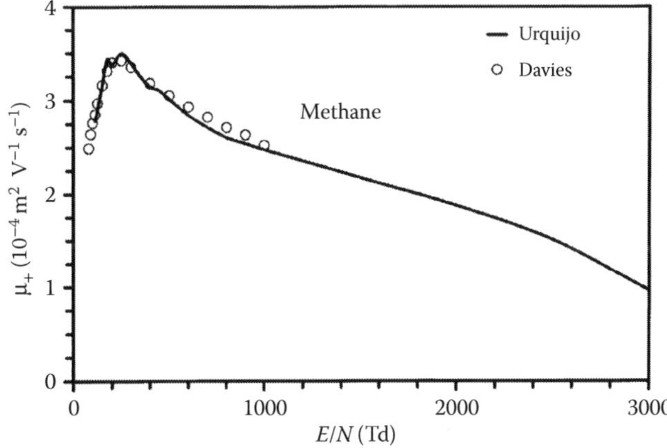

FIGURE 64.16 Reduced mobility of positive ions. (O) Davies et al. (1989); (——) de Urquiho et al. (1999).

REFERENCES

Al-Amin, S. A. J., H. N. Kücükarpaci, and J. Lucas, *J. Phys. D: Appl. Phys.*, 18, 1781, 1985.

Althorpe, S. C., F. A. Gianturco, and N. Sanna, *J. Phys. B: At. Mol. Opt. Phys.*, 28, 4165, 1995.

CH_4

Alvarez-Pol, H., J. Durant, and R. Lorenzo, *J. Phys. B: At. Mol. Opt. Phys.*, 30, 2455, 1997.

Barbarito, E., M. Basta, M. Calicchio, and G. Tessarf, *J. Chem. Phys.*, 71, 54, 1979.

Boesten, L. and H. Tanaka, *J. Phys. B: At. Mol. Opt. Phys.*, 24, 821, 1991.

Bundschu, C. T., J. C. Gibson, R. J. Gulley, M. J. Brunger, S. J. Buckman, N. Sanna, and F. A. Gianturco, *J. Phys. B: At. Mol. Opt. Phys.*, 30, 2239, 1997.

Cascella, M., R. Curik, and F. A. Gianturco, *J. Phys. B: At. Mol. Opt. Phys.*, 34, 705, 2001.

Chang, Y., M. E. Sarakinos, U. J. Becker, G. E. Sasser, and B. R. Smith, *Nuci. Instr. Methods A*, 311, 490, 1992.

Chatham, H., D. Hils, R. Robertson, and A. Gallagher, *J. Chem. Phys.*, 81, 1770, 1984.

Davies, D. K., L. E. Kline, and W. E. Bies, *J. Appl. Phys.*, 85, 3311, 1989.

de Urquijo, J., C. A. Arriaga, C. Cisneros, and I. Alvarez, *J. Phys. D: Appl. Phys.*, 32, 41, 1999.

de Urquijo, J., I. Alvarez, E. Basurto, and C. Cisneros, *J. Phys. D: Appl. Phys.*, 32, 1646, 1999.

Ferch, J., B. Granitza, and W. Raith, *J. Phys. B: At. Mol. Phys.*, 18, L445, 1985.

Floeder, K., D. Fromme, W. Raith, A. Schwab, and G. Sinapius, *J. Phys. B: At. Mol. Phys.*, 18, 3347, 1985.

Fraser, G. W. and E. Mathieson, *Nucl. Instr. Methods A*, 247, 544, 1986. Mixtures of argon and methane.

Haddad, G. N., *Austr. J. Phys.*, 38, 677, 1985.

Heylen, A. E. D. *J. Chem. Phys.*, 38, 765, 1963.

Hughes, A. L. and J. H. McMillan, *Phys. Rev.*, 44, 876, 1933.

Hunter, S. R., J. G. Carter, and L. G. Christophorou, *J. Appl. Phys.*, 60, 24, 1986.

Hwang, M., Y. K. Kim, and M. E. Rudd, *J. Chem. Phys.*, 104, 2956, 1996.

Jain, A. K. and K. L. Baluja, *Rev. A*, 45, 202, 1992. (Several fundamental parameters are given in the paper along with tabulated cross sections of 14 molecular gases.)

Jean-Marie, B., V. Lepeltier, and D. L'Hote, *Nucl. Instr. Methods*, 159, 213, 1979. (Drift velocity data in mixtures of methane with argon are given.)

Jones, R. K., *J. Chem. Phys.*, 82, 5424, 1985.

Kanik, I., B. Trajmar, and J. C. Nickel, *Chem. Phys. Lett.*, 193, 281–286, 1992.

Kanik, I., S. Trajmar, and J. C. Nickel, *J. Geophys. Res.*, 98, 7447, 1993.

Kim, Y.-K., K. K. Irikura, M. E. Rudd, M. A. Ali, P. M. Stone, J. chang, J. S. Coursey, Electron-Impact Ionization Cross Section for Ionization and Excitation Database (Version 3.0), http://Physics, nist.gov/ionxsec, NIST, Gaithesberg, MD, 2004.

Kunst, T., B. Götz, and B. Schmidt, *Nucl. Instrum. Methods A*, 324, 127, 1993. (Data for drift velocity in mixtures of argon and methane are provided.)

Lakshminarasimha, C. S. and J. Lucas, *J. Phys. D: Appl. Phys.*, 10, 313, 1977.

Lohmann, B. and S. J. Buckman, *J. Phys. B: At. Mol. Opt. Phys.*, 19, 2565, 1986.

Lunt, S. L., J. Randell, J. P. Ziesel, G. Mrotzek, and D. Field, *J. Phys. B: At. Mol. Opt. Phys.*, 27, 1407, 1994. (Low energy scattering from CH_4, C_2H_4 and C_2H_6 are studied.)

Lunt, S. L., J. Randell, J.-P. Ziesel, G. Mrotzek, and D. Field, *J. Phys. D: At. Mol. Opt. Phys.*, 31, 4225, 1998.

Maji, S., G. Basavaraju, S. M. Bharathi, K. G. Bhushan, and S. P. Khare, *J. Phys. B: At. Mol. Opt. Phys.*, 31, 4975, 1998. (The elastic scattering cross sections are measured in CO, CO_2, CH_4 and C_2H_6.)

Mapstone, B. and W. R. Newell, *J. Phys. B: At. Mol. Opt. Phys.*, 25, 491, 1992.

Mathieson, E. and El. Hakeem, *Nucl. Instr. Methods*, 159, 489, 1979.

Millican, P. G. and I. C. Walker, *J. Phys. D: Appl. Phys.*, 20, 193, 1987.

Müller, R., K. Jung, K.-H. Kochem, W. Sohn, and H. Ehrhardt, *J. Phys. B: At. Mol. Phys.*, 18, 3971, 1985.

Nishimura, H. and H. Tawara, *J. Phys. B: At. Mol. Opt. Phys.*, 27, 2063, 1994.

Ohmori, Y., K. Kitamori, M. Shimozuma, and H. Tagashira, *J. Phys. D: Appl. Phys.*, 19, 437, 1986.

Orient, O. J. and S. K. Srivastava, *J. Phys. B: At. Mol. Phys.*, 20, 3923, 1987.

Plessis, P., P. Marmet, and R. Dutil, *J. Phys. B: At. Mol. Phys.*, 16, 1283, 1983.

Raju, G. G., *Gaseous Electronics: Theory and Practice*, CRC Press, Boca Raton, FL, 2006, pp. 133–140.

Ramsauer, C. and R. Kollath, *Ann. Phys. Lpz.*, 4, 91, 1930.

Rapp, D. and P. Englander-Golden, *J. Chem. Phys.*, 43, 1464, 1965. (Total ionization cross section of 16 gases including C_2H_4 up to 1000 eV energy are measured.)

Rutkowsky, V. J., H. Drost, and H. J. Spangenberg, *Ann. Phys. (Leipzig)*, 37, 259, 1980. Dissociative and direct attachment cross sections in several hydrocarbon gases are measured.

Sakae, T., S. Sumiyoshi, E. Murakami, Y. Matsumoto, K. Ishibashi, and A. Katase, *J. Phys. B: At. Mol. Opt. Phys.*, 22, 1385, 1989.

Savinov, S. Y., H. Lee, H. K. Song, and B. K. Na, *Ind. Eng. Chem. Res.*, 38, 2540, 1999.

Schlumbohm, H. Z., *Phys.*, 182, 317, 1965.

Schmidt, B., *J. Phys. B: At. Mol. Opt. Phys.*, 24, 4809, 1991. (The low values of *E/N* and data on the longitudinal diffusion coefficient are of interest.)

Schmidt, B. and M. Roncossek, *Austr. J. Phys.*, 45, 351, 1992.

Schram, B. L., M. J. Van Der Weil, F. J. De Heer, and H. R. Moustafa, *J. Chem. Phys.*, 44, 49, 1966.

Sharp, T. E. and J. T. Dowell, *J. Chem. Phys.*, 46, 1536, 1967.

Sohn, W., K.-H. Kochem, K. -M. Scheuerlein, K. Jung, and H. Ehrhardt, *J. Phys. B: At. Mol. Phys.*,19, 3625, 1986.

Straub, H. C., D. Lin, B. G. Lindsay, K. A. Smith, and R. F. Stebbings, *J. Chem. Phys.*, 106, 4430, 1997.

Sueoka, O. and S. Mori, *At. Mol. Phys.*, 19, 4035, 1986.

Tanaka, H., T. Okada, L. Boesten, T. Suzuki, T. Yamamoto, and M. Kubo, *J. Phys. B: At. Mol. Opt. Phys.*, 18, 3305, 1982.

Tian, C. and C. R. Vidal, *J. Phys. B: At. Mol. Opt. Phys.*, 31, 895, 1998.

Wagner, E. B., F. J. Davis, and G. S. Hurst, *J. Chem. Phys.*, 47, 3138, 1967.

Wong, L., J. Armitage, and J. Waterhouse, *Nucl. Instr. Methods A*, 273, 476, 1988. (Drift velocity data in mixtures of methane with argon are given.)

Yoshida, K., T. Ohshima, H. Ohmori, and H. Tagashira, *J. Phys. D: Appl. Phys.*, 29, 1209, 1996.

Zecca, A., G. Karwasz, R. S. Brusa, and C. Szmytkowski, *J. Phys. B: At. Mol. Opt. Phys.*, 24, 2747, 1991.

Zhao, T., Y. Chen, S. Han, and J. Hersch, *Nucl. Instr. Methods A*, 340, 485, 1994.

65

SILANE

CONTENTS

Silane (SiH_4) is a weakly electron-attaching inorganic molecule with 18 electrons, electronic polarizability of 6.05×10^{-40} F m^2, and ionization potential 11.65 eV.

65.1 SELECTED REFERENCES FOR DATA

See Table 65.1

TABLE 65.1
Selected References for Data

Quantity	Range: eV, (Td)	Reference
Ionization cross section	10–2000	Joshipura et al. (2007)
Drift velocity	(435–3900)	**Lisovskiy et al. (2007)**
Transport coefficients	(3–3000)	Shimada et al. (2003)
Ionization cross section	16–900	**Basner et al. (2001)**
Ionization cross section	12–1000	Ali et al. (1997)
Ionization cross section	12–100	**Basner et al.(1997)**
Total scattering cross section	0.6–250	**Szmytkowski et al. (1997)**
Ionization cross section	12–1000	**Krishnakumar and Srivastava (1995)**
Total scattering cross section	0.7–400	**Sueoka and Morand Hamada (1994)**
Total scattering cross section	(75–4000)	**Zecca et al. (1992)**
Rotational excitation cross section	0.001–20	Jain and Thompson (1991)
Elastic scattering cross section	(1.8–100)	**Tanaka et al. (1990)**
Total scattering cross section	0.2–12.0	**Wan et al. (1989)**
Collision cross sections	0.01–100	Kurachi and Nakamura (1989)
Collision cross sections	0.01–100	Hayashi (1987)
Elastic scattering	0.1–10.0	Jain and Thompson (1987)
Total scattering cross section	0.1–30	Jain et al. (1987)
Diffusion coefficient	(1.2–170)	**Millican and Walker (1987)**
Swarm parameters	(0.6–900)	Ohmori et al. (1986)
Ionization coefficient	(180–810)	**Shimozuma and Tagashira (1986)**
Total scattering cross section	1–20	**Sueoka and Mori (1985)**
Ionization cross section	15–400	**Chatham et al. (1984)**
Drift velocity	(5–70)	**Pollock (1968)**
Drift velocity	(1–25)	**Cottrell and Walker (1965)**

Note: Bold font denotes experimental study.

65.2 TOTAL SCATTERING CROSS SECTION

Highlights of total scattering cross section are (Table 65.2)

1. As the energy decreases toward zero, there is a trend of total scattering cross section increasing.
2. Ramsauer–Townsend minimum at ~0.25 eV.

SiH_4

TABLE 65.2
Recommended Total Scattering Cross Section (SiH_4)

Energy (eV)	Q_T (10^{-20} m²)	Energy (eV)	Q_T (10^{-20} m²)	Energy (eV)	Q_T (10^{-20} m²)
Karwasz et al. (2003)		Szmytkowski et al. (1997)		Wan et al. (1989)	
0.15	3.86	0.6	4.65	0.2	1.9
0.17	2.84	0.7	4.54	0.4	2.7
0.20	2.00	0.8	7.99	0.6	4
0.25	1.53	0.9	8.75	0.8	6.2
0.30	1.55	1	10.6	1	9.5
0.35	1.77	1.1	15.8	1.5	25.7
0.40	2.08	1.2	18.9	2	44
0.45	2.45	1.3	27.7	2.5	53.4
0.50	2.87	1.4	30.2	3	54.7
0.60	4.00	1.5	34.9	3.5	53.1
0.70	5.05	1.6	40.2	4	51.6
0.80	6.13	1.7	43.2	4.5	50.1
0.90	7.81	1.8	44.8	5	48.9
1.0	11.0	2	49.5	5.5	48
1.2	17.9	2.2	53.1	6	47.1
1.5	29.3	2.4	55.1	6.5	45.7
1.7	36.4	2.6	56.2	7	45.2
2.0	46.1	2.9	56.6	7.5	44.1
2.5	53.1	3.3	56	8	43.3
3.0	54.6	3.6	54.6	8.5	42.4
3.5	54.1	4.1	53.4	9	41.5
4.0	52.9	4.6	51.7	9.5	40.5
4.5	51.4	5.1	50.5	10	40.2
5.0	49.9	5.6	49.5	10.5	39.6
6.0	47.3	6.1	48.7	11	39
7.0	45.2	6.6	47.9	11.5	38.3
8.0	43.5	7.6	46.5	12	38
9.0	42.0	8.6	45		
10.0	40.7	9.6	42.9		
12	38.0	10.6	40.9		
15	34.4	11.6	39.7		
17	32.6	12.6	38		
20	30.5	15	36.5		
25	27.8	17	35.2		
30	25.7	19	33.9		
35	24.1	21	31.8		
40	22.7	23	30.8		
45	21.6	25	29.6		
50	20.6	28	28.1		
60	18.8	30	25.9		
70	17.4	35	24.3		
80	16.4	40	23		
90	15.4	45	20.9		
100	14.6	50	20.1		
120	13.6	60	18.1		
150	12.1	70	16.9		
170	11.3	80	16.3		
200	10.20	90	14.9		
250	8.86	100	13.8		
300	7.82	110	12.8		
350	7.01	120	12.7		
400	6.35	140	11.3		
450	5.91	160	10.5		
500	5.44	180	9.99		
600	4.70	200	9.29		
700	4.14	220	8.61		
800	3.70	250	7.84		
900	3.34				
1000	3.05				

Source: Adapted from Karwasz, G., R. Brusa, and A. Zecca, in *Interactions of Photons and Electrons with Molecules*, Springer-Verlag, New York, NY, 2003.

Note: See Figure 65.1 for graphical presentation.

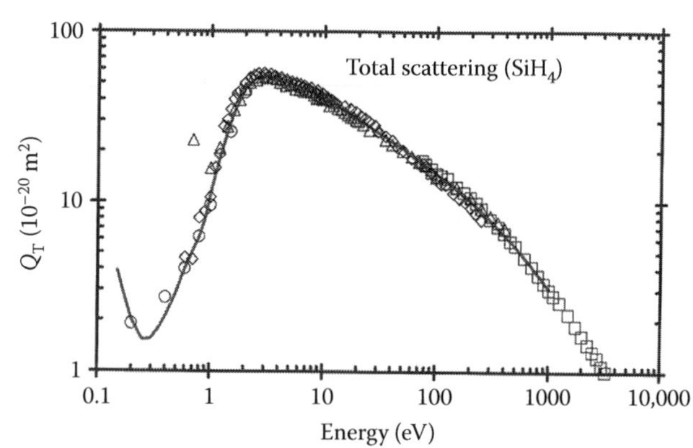

FIGURE 65.1 Total scattering cross section for SiH_4. (——) Karwasz et al. (2003), recommended; (◊) Szmytkowski et al. (1997); (Δ) Sueoka et al. (1994); (□) Zecca et al. (1992); (○) Wan et al. (1989).

3. A shape resonance at ~1.8–2.2 eV attributed to dissociative attachment and vibrational excitation (Karwasz, 2003).
4. Beyond resonance a monotonic decrease toward higher energy up to 3000 eV which is a common feature observed in many gases.

65.3 ELASTIC AND MOMENTUM TRANSFER CROSS SECTION

See Table 65.3.

65.4 VIBRATIONAL EXCITATION CROSS SECTION

The molecule does not have dipole or quadrupole moment, with negligible rotational excitation (Kurachi and Nakamura, 1989). The molecule has four modes of vibrational excitation as shown in Table 65.4 (Shimanouchi, 1972).

Only theoretically adjusted vibrational excitation cross sections for the stretching (v_1 and v_3) and bending modes (v_2 and v_4) are available as shown in Figure 65.3 (Ohmori et al., 1986).

65.5 IONIZATION CROSS SECTION

Positive ion appearance potentials are shown in Table 65.5 (Krishna kumar and Srivastav).

See Tables 65.6 and 65.7.

TABLE 65.3
Elastic and Momentum Transfer Cross Section

Energy (eV)	Q_{el} (10^{-20} m^2)	Q_M (10^{-20} m^2)
1.8	27.5	29.0
2.15	31.6	30.1
2.65	34.8	29.1
3	36.5	28.1
4	401	24.5
5	44.4	25.6
7.5	49.9	24.4
10	39.4	15.8
15	28.7	11.2
20	20.7	8.7
40	14.0	2.9
100	4.30	1.2

Source: Adapted from Tanaka, H. et al., *J. Phys. B: At. Mol. Opt. Phys.*, 23, 577, 1990.

Note: See Figure 65.2 for graphical presentation.

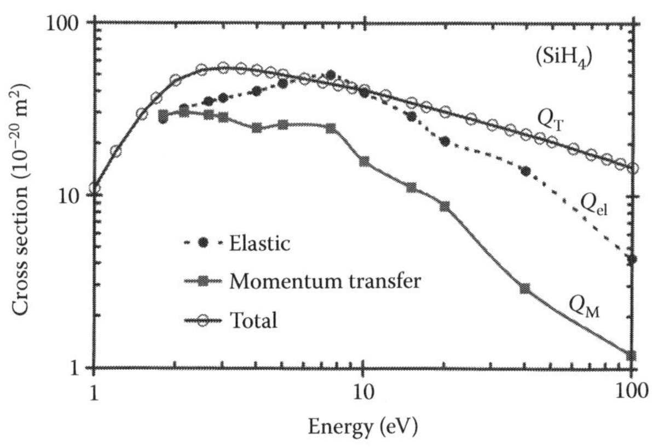

FIGURE 65.2 Elastic and momentum transfer cross section for SiH_4. Total scattering cross section is also shown for comparison. (Adapted from Tanaka, H. et al., *J. Phys. B: At. Mol. Opt. Phys.*, 23, 577, 1990.)

TABLE 65.4
Vibrational Excitation Modes for SiH_4

Designation	Type of Motion	Energy (meV)
v_1	Symmetrical stretch	271.2
v_2	Degenerate deformation	120.9
v_3	Degenerate stretch	271.7
v_4	Degenerate deformation	113.3

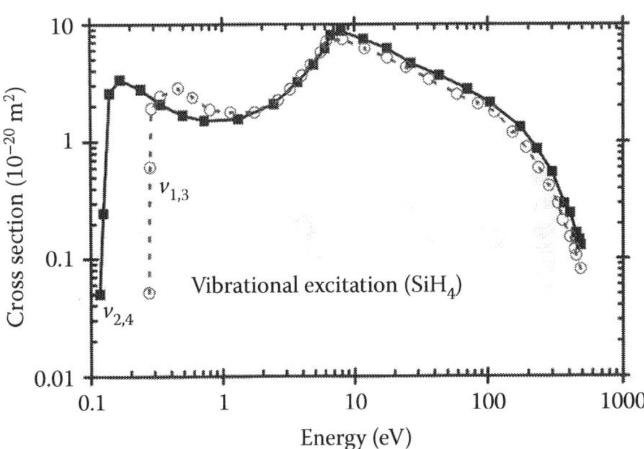

FIGURE 65.3 Vibrational excitation cross section for SiH_4. (Adapted from Ohmori, Y., M. Shimozuma, and H. Tagashira, *J. Phys. D: Appl. Phys.*, 19, 1029, 1986.)

TABLE 65.5
Ion Appearance Potentials

Ion	Appearance Potential (eV)	Ion	Appearance Potential (eV)
SiH_4^+	Not observed	Si^+	13.6
SiH_3^+	12.3	H_2^+	25.0
SiH_2^+	11.9	H^+	24.5
SiH^+	14.7		

TABLE 65.6
Recommended Ionization Cross Section for SiH_4

Energy (eV)	Q_i (10^{-20} m^2)	Energy (eV)	Q_i (10^{-20} m^2)	Energy (eV)	Q_i (10^{-20} m^2)
12	0.0085	50	4.288	190	2.698
13	0.0904	55	4.250	200	2.619
14	0.352	60	4.187	250	2.319
16	1.212	65	4.117	300	2.075
17	1.662	70	4.057	350	1.875
18	2.055	75	3.983	400	1.704
19	2.413	80	3.910	450	1.566
20	2.662	85	3.828	500	1.457
22	3.048	90	3.766	550	1.363
24	3.377	95	3.698	600	1.275
26	3.647	100	3.639	650	1.189
28	3.843	110	3.517	700	1.108
30	4.021	120	3.387	750	1.036
32	4.088	130	3.272	800	0.974
34	4.144	140	3.160	850	0.920
36	4.187	150	3.048	900	0.871
38	4.202	160	2.954	950	0.827
40	4.242	170	2.864	1000	0.787
45	4.296	180	2.777		

Source: Adapted from Krishnakumar, E. and S. K. Srivastava, *Contrib. Plasma Phys.*, 35, 395, 1995.

Note: See Figure 65.4 for graphical presentation.

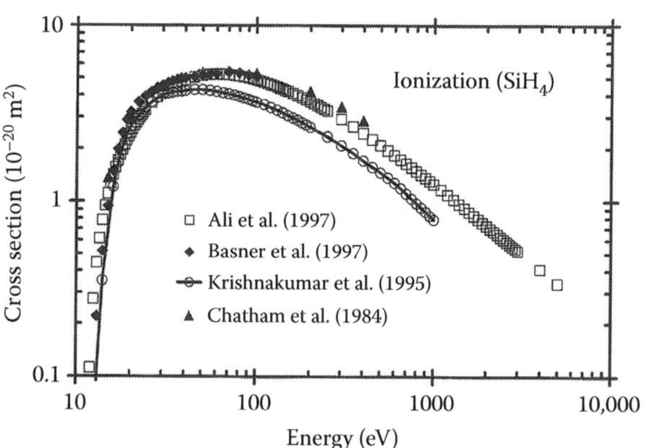

FIGURE 65.4 Ionization cross section for SiH_4.

TABLE 65.7
Partial Ionization Cross Sections at Selected Electron Energy for SiH_4

	Ionization Cross Section (10^{-22} m^2)					
Energy (eV)	H^+	H_2^+	Si^+	SiH^+	SiH_2^+	SiH_3^+
12					0.85	
15			0.69	44.4	26.2	71.8
20			3.73	9.5	151.3	101.7
30	5.80	1.57	29.0	56.0	191.8	117.9
50	25.9	3.21	45.4	57.3	176.9	120.1
100	25.4	1.85	31.4	35.4	153.8	116.0
200	15.5	1.18	15.5	18.9	120.8	90.0
300	11.2	0.88	10.4	13.3	98.2	73.5
500	7.27	0.63	6.46	8.92	69.2	53.2
1000	4.73	0.31	2.73	3.71	38.0	29.2

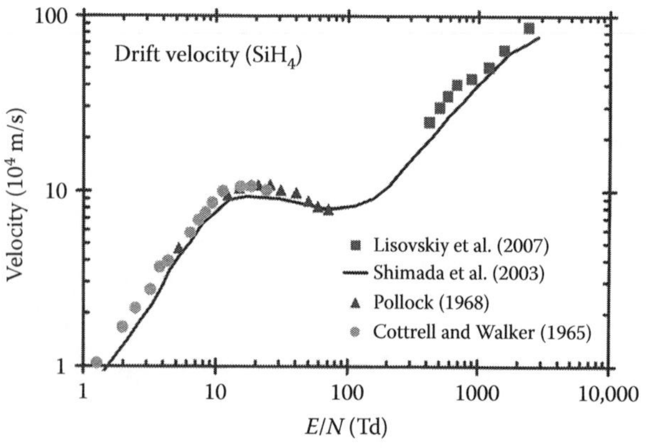

FIGURE 65.5 Drift velocity of electrons for SiH_4.

65.6 DRIFT VELOCITY OF ELECTRONS

Figure 65.5 shows the drift velocity of electrons.

65.7 DIFFUSION COEFFICIENTS

See Table 65.8.

TABLE 65.8
Characteristic Energy for SiH_4

	D_r/μ (V)			D_r/μ (V)	
***E/N* (Td)**	SiH_4	SiD_4	***E/N* (Td)**	SiH_4	SiD_4
1	0.0298		17	0.210	0.293
1.2	0.031	0.0265	20	0.266	0.388
1.4	0.031	0.0265	25	0.363	0.550
1.7	0.031	0.0278	30	0.482	0.723
2	0.031	0.0298	35	0.587	0.908
2.5	0.032	0.0298	40	0.700	1.05
3	0.034	0.030	50	0.853	1.30
4	0.037	0.034	60	1.01	1.59
5	0.043	0.039	70	1.16	1.75
6	0.048	0.046	80	1.25	1.85
7	0.056	0.056	100	1.49	1.87
8	0.066	0.069	120	1.53	1.88
10	0.089	0.105	140	1.62	
12	0.109	0.152	170	1.71	
14	0.145	0.206			

Source: Adapted from Millikan, P. C. and I. C. Walker, *J. Phys. D: Appl. Phys.*, 20, 193, 1987.

Note: Values for deuterated SiD_4 are also shown for comparison. See Figure 65.6 for graphical presentation. Products ND_L and ND_r are also shown in the figure.

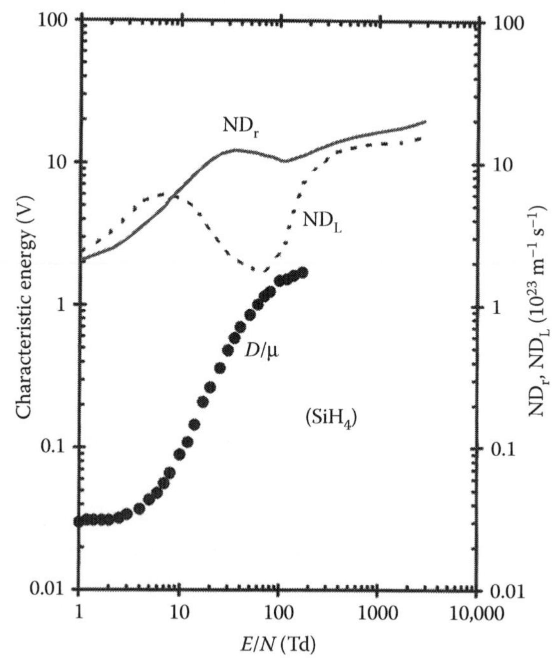

FIGURE 65.6 Diffusion coefficients for SiH_4. (●) Characteristic energy (D/μ), Millikan and Walker (1987), experimental; (——) ND_r, Shimada et al. (2003); (– – –) ND_L, Shimada et al. (2003). Note the appropriate ordinates.

65.8 IONIZATION COEFFICIENTS

See Table 65.9.

TABLE 65.9
Ionization and Effective Ionization Coefficients for SiH_4

E/N (Td)	α/N (10^{-20} m^2)	$(\alpha-\eta)/N$ (10^{-20} m^2)
165		−0.034
200	0.0057	
225	0.0099	
250	0.017	0.0152
275	0.0279	0.0347
300	0.0441	0.0472
325	0.0598	0.0594
350	0.0822	0.078
375	0.107	0.107
400	0.131	0.140
450	0.202	0.198
500	0.284	0.285
550	0.357	0.377
600	0.460	0.480
650	0.586	0.573
700	0.706	0.702
750	0.812	0.802
800	0.925	0.916
850	1.03	1.02
900	1.11	
950	1.20	
1000	1.34	

Source: Digitized from Shimozuma, M. and H. Tagashira, *J. Phys. D: Appl. Phys.*, 19, L179, 1986.
Note: See Figure 65.7 for graphical presentation.

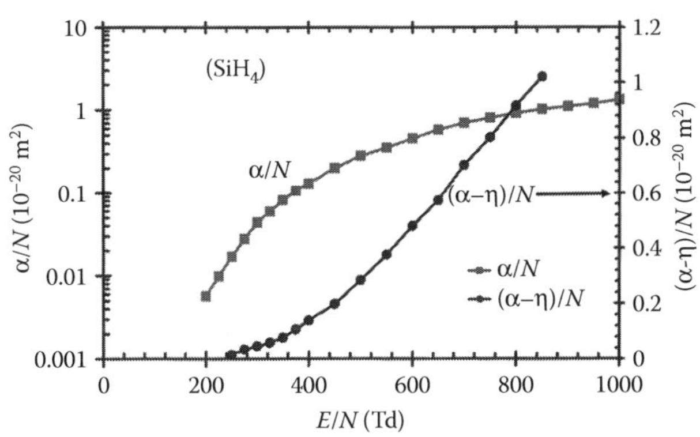

FIGURE 65.7 Density reduced ionization and effective ionization coefficients for SiH_4, Shimozuma and Tagashira (1986).

65.9 GAS CONSTANTS

Gas constants evaluated according to the expression

$$\frac{\alpha}{N} = F\exp\left(-\frac{GN}{E}\right) \tag{65.1}$$

are (Lisovskity et al., 2007) $F = 6.83 \times 10^{-20}$ m^2 and $G = 1640$ Td in the range $403 \leq E/N \leq 1860$ Td.

REFERENCES

Ali, M. A., Y.-K. Kim, M. Hwang, N. M. Weinberger, and M. E. Rudd, *J. Chem. Phys.*, 106, 9602, 1997.

Basner, R., M. Schmidt, E. Denisov, K. Becker, and H. Deutsch, *J. Chem. Phys.*, 114, 1170, 2001.

Basner, R., M. Schmidt, V. Tarnovsky, K. Becker, and H. Deutsch, *Int. J. Mass Spectrom. Ion Proc.*, 171, 83, 1997.

Chatham, H., D. Hils, R. Robertson, and A. Gallagher, *J. Chem. Phys.*, 81, 1770, 1984.

Cottrell, T. L. and I. C. Walker, *Trans. Farad. Soc.*, 61, 1585, 1965.

Hayashi, M. in L. C. Pitchford et al. (Eds.), *Swarm Studies and Inelastic Electron Molecule Collisions*, Springer-Verlag, New York, 1987, p. 167, Quoted by M. Kurachi and Y. Nakamura, *J. Phys. D: Appl. Phys.*, 22, 107, 1989.

Jain, A. and D. G. Thompson, *J. Phys. B; At. Mol. Phys.*, 20, 2861, 1987.

Jain, A. and D. G. Thompson, *J. Phys. B: At. Mol. Opt. Phys.*, 24, 1087, 1991.

Jain, A. K., A. N. Tripathi, and A. Jain, *J. Phys. B: At. Mol. Phys.*, 20, L389, 1987.

Joshipura, K. N., B. G. Vaishnav, and S. Gangopadhyay, *Int. J. Mass Spectrom.*, 261, 2007, 146.

Karwasz, G., R. Brusa, and A. Zecca, in *Interactions of Photons and Electrons with Molecules*, Springer-Verlag, New York, NY, 2003, Chapter 6.

Krishnakumar, E. and S. K. Srivastava, *Contrib. Plasma Phys.*, 35, 395, 1995.

Kurachi, M. and Y. Nakamura, *J. Phys. D: Appl. Phys.*, 22, 107, 1989.

Lisovskiy, V., J.-P. Booth, K. Landry, D. Douai, V. Cassagne, and V. Yegorenkov, *J. Phys. D: Appl. Phys.*, 40, 3408, 2007.

Millican, P. C. and I. C. Walker, *J. Phys. D: Appl. Phys.*, 20, 193, 1987.

Ohmori, Y., M. Shimozuma, and H. Tagashira, *J. Phys. D: Appl. Phys.*, 19, 1029, 1986.

Pollock, W. *Trans. Farad. Soc.*, 64, 2919, 1968.

Shimada, T., Y. Nakamura, Z. Lj. Petrović, and T. Makabe, *J. Phys. D: Appl. Phys.*, 36, 1936, 2003.

Shimanouchi, T. *Tables of Molecular Vibrational Frequencies Consolidated Volume I*, NSRDA-NBS 39, Washington (DC), 1972.

Shimozuma, M. and H. Tagashira, *J. Phys. D: Appl. Phys.*, 19, L179, 1986.

Sueoka, O. and S. Morand Hamada, *J. Phys. B: At. Mol. Opt. Phys.*, 27, 1453, 1994.

Sueoka, O. and S. Mori, *At. Coll. Res. Japan* 11, 19, 1985 cited in Jain et al. *J. Phys. B: At. Mol. Phys.*, 20, L389, 1987.

Szmytkowski, C., P. Możejko, and G. Kasperski, *J. Phys. B: At. Mol. Opt. Phys.*, 30, 1997, 4363.

Tanaka, H., L. Boesten, H. Sato, M. Kimura, M. A. Dillon, and D. Spence, *J. Phys. B: At. Mol. Opt. Phys.*, 23, 577, 1990.

Wan, H., J. H. Moore, and J. A. Tossell, *J. Chem. Phys.*, 91, 7340, 1989.

Zecca, A., G. P. Karwasz, and R. S. Brusa, *Phys. Rev.*, 45, 2777, 1992.

66

SILICON TETRAFLUORIDE

SiF

CONTENTS

Silicon tetrafluoride (SiF_4), synonym tetrafluorosilane, is an electron-attaching, nonattaching, inorganic molecule with 50 electrons. Its electronic polarizability is 6.06×10^{-40} F m^2 and ionization potential 15.24 eV.

66.1 SELECTED REFERENCES FOR DATA

See Table 66.1.

66.2 TOTAL SCATTERING CROSS SECTION

The highlights of total scattering cross section are

1. With decreasing energy toward zero, a trend for increasing cross section
2. Ramsauer–Townsend minimum at ~1.5 eV
3. A broad maximum in the range of 8–25 eV
4. Monotonic decrease >100 eV, a feature common to many gases (see Table 66.2)

TABLE 66.1
Selected References for Data

Quantity	Range: eV, (Td)	Reference
Ionization cross section	16.3–5000	Kim et al. (2005)
Total scattering cross section	0.6–3500	**Karwasz et al. (1998a,b)**
Dissociative attachment	0–50	**Iga et al. (1992)**
Swarm parameters	(0.5–300.0)	**Hunter et al. (1989)**
Total scattering cross section	2–50	**Wan et al. (1989)**
Ionization cross section	15–130	**Poll and Meichsner (1987)**

Note: Bold font indicates experimental study.

TABLE 66.2
Total Scattering Cross Section

Energy (eV)	Q_T (10^{-20} m^2)	Energy (eV)	Q_T (10^{-20} m^2)	Energy (eV)	Q_T (10^{-20} m^2)
Karwasz et al. (2003)		Karwasz et al. (1998)		Wan et al. (1989)	
0.5	9.26	0.6	9.3	0.2	19.9
0.6	8.97	0.7	8.6	0.4	13.7
0.7	8.65	0.8	8.3	0.6	9.9
0.8	8.30	0.9	7.9	0.8	7.7
0.9	7.93	1	7.5	1	6.5
		1.1	7.2	1.5	5.6
1.0	7.55	1.2	6.8	2	5.7
1.2	6.81	1.3	6.5	2.5	6.4
1.5	6.23	1.4	6.4	3	7.2
1.7	6.22	1.5	6.2	3.5	7.7
2.0	6.44	1.6	6.1	4	8.9
2.5	7.11	1.7	6.3	4.5	10.1
3.0	7.99	1.8	6.4	5	12.4
3.5	9.04	1.9	6.3	5.5	14.3
4.0	10.1	2	6.5	6	17.4
4.5	11.1	2.2	6.8	6.5	19.8

continued

SiF$_4$

TABLE 66.2 (continued)
Suggested Total Scattering Cross Section

Energy (eV)	Q_T (10^{-20} m^2)	Energy (eV)	Q_T (10^{-20} m^2)	Energy (eV)	Q_T (10^{-20} m^2)
Karwasz et al. (2003)		Karwasz et al. (1998)		Wan et al. (1989)	
5.0	12.8	2.4	6.8	7	21.6
6.0	17.7	2.6	7.4	7.5	22.9
7.0	22.7	2.8	7.2	8	23.7
8.0	25.2	3	8.2	8.5	24
9.0	25.7	3.4	8.7	9	24
10.0	25.9	3.6	9.2	9.5	23.9
12	27.3	4.1	10	10	23.8
15	28.5	4.6	12.1	10.5	24.2
17	28.7	5.1	12.8	11	24.6
20	29.3	5.6	15.5	11.5	24.9
25	30.2	6.1	18.1	12	25.6
30	30.1	6.6	21.3		
35	29.8	7.1	22.9		
40	29.4	8.1	25.3		
45	29.1	9.1	25.7		
50	28.8	10.1	25.9		
60	27.8	11.1	26.5		
70	26.6	12.1	27.5		
80	25.4	12.6	27.9		
90	23.9	13.6	28		
100	22.7	14.6	28.6		
120	20.7	15.6	28.5		
150	18.7	16.6	28.6		
170	17.7	17.6	28.7		
200	16.5	18.6	29		
250	14.7	19.6	29.2		
300	13.2	20.1	29.3		
350	12.0	21.6	29.7		
400	11.0	23.1	30		
450	10.2	24.6	30.3		
500	9.58	25.6	30.3		
600	8.51	28.1	30.2		
700	7.68	30	30.1		
800	7.01	40	29.6		
900	6.46	50	28.6		
1000	5.99	60	27.8		
		70	26.5		
		80	25.3		
		90	24.3		
		100	23.1		
		200	16.8		
		250	14.7		

Source: Adapted from Karwasz, G. P., R. Brusa, and A. Zecca, in *Interactions of Photons and Electrons with Molecules*, Vol. 17c, Springer-Verlag, New York, NY, 2003, Chapter 6.1.
Note: See Figure 66.1 for graphical presentation.

66.3 IONIZATION CROSS SECTION

See Table 66.3.

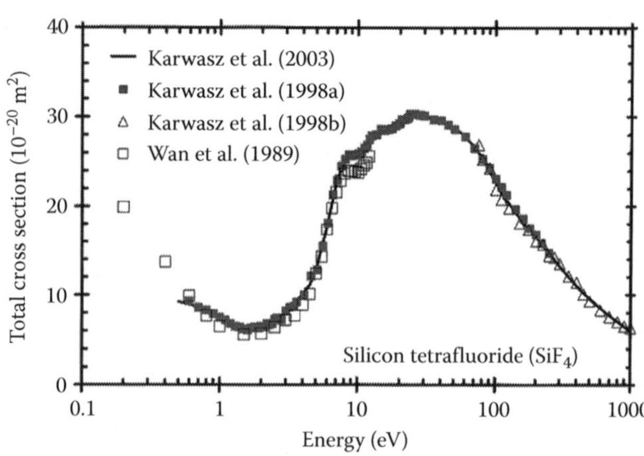

FIGURE 66.1 Total scattering cross section for SiF$_4$. (——) Recommended by Karwasz et al. (2003); (■) Karwasz et al. (1998a); (Δ) Karwasz et al. (1998b); (□) Wan et al. (1989).

TABLE 66.3
Ionization Cross Section

Energy (eV)	Q_i (10^{-20} m^2)	Energy (eV)	Q_i (10^{-20} m^2)	Energy (eV)	Q_i (10^{-20} m^2)
15	0.09	45	4.71	90	6.61
18	0.30	50	5.20	100	6.65
20	0.56	55	5.58	110	6.63
25	1.44	60	5.88	120	6.57
30	2.56	65	6.12	130	6.48
35	3.34	70	7.30		
40	4.10	80	6.51		

Source: Digitized from Poll, H. U. and J. Meichsner, *Contrib. Plas. Phys.*, 27, 359, 1987.
Note: See Figure 66.2 for graphical presentation.

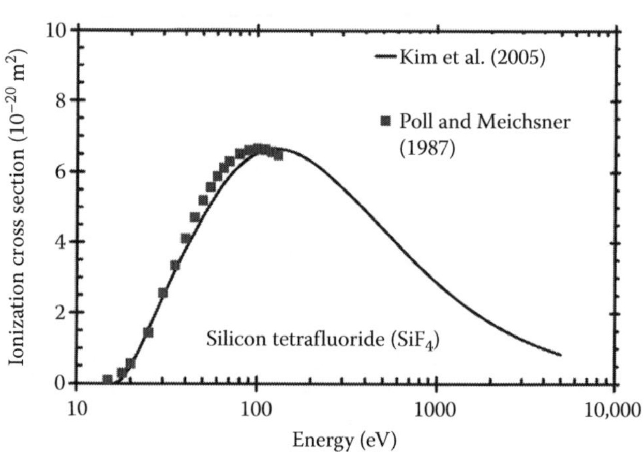

FIGURE 66.2 Ionization cross sections for SiF$_4$. (——) Kim et al. (2005); (■) Poll and Meichsner (1987).

66.4 ATTACHMENT CROSS SECTION

Dissociative attachment yields F^-, F_2^-, and SF_3^- ions with peak cross section close to threshold energy as shown in Table 66.4 (Iga et al., 1992).

66.5 ELECTRON DRIFT VELOCITY

See Table 66.5.

66.6 IONIZATION AND ATTACHMENT COEFFICIENTS

See Table 66.6.

TABLE 66.4
Attachment Cross Section

Species	Threshold (eV)	Peak (eV)	Q_a at Peak (10^{-22} m^2)
F^-	10.2	11.4	0.509
F_2^-	9.7	11.2	0.0178
SF_3^-	10.1	11.4	0.951

TABLE 66.5
Suggested Electron Drift Velocity

E/N (Td)	*W* (10^4 m/s)	*E/N* (Td)	*W* (10^4 m/s)	*E/N* (Td)	*W* (10^4 m/s)
0.5	0.062	10.0	1.28	80.0	6.99
0.6	0.075	12.0	1.55	90.0	7.29
0.8	0.099	14.0	1.95	100.0	7.65
1.0	0.124	17.0	2.27	120.0	8.55
1.2	0.148	20.0	2.85	140.0	9.28
1.5	0.185	25.0	4.30	160.0	10.05
2.0	0.248	30.0	6.15	180.0	10.7
2.5	0.314	35.0	7.58	200.0	11.7
3.0	0.377	40.0	7.98	225.0	12.7
4.0	0.506	45.0	7.69	250.0	13.9
5.0	0.632	50.0	7.56	275.0	15.0
6.0	0.76	60.0	7.14	300.0	16.0
8.0	1.02	70.0	7.00		

Source: Adapted from Hunter, S. R., J. G. Carter, and L. G. Christophorou, *J. Appl. Phys.*, 65, 1858, 1989.
Note: See Figure 66.3 for graphical presentation.

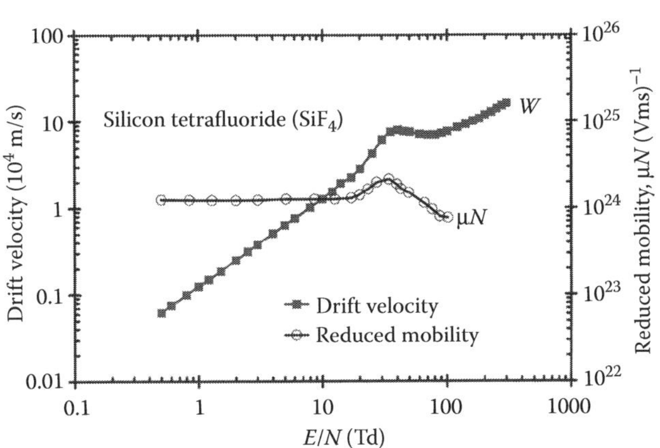

FIGURE 66.3 Drift velocity and reduced mobility for SiF_4. (Adapted from Hunter, S. R., J. G. Carter, and L. G. Christophorou, *J. Appl. Phys.*, 65, 1858, 1989.)

TABLE 66.6
Reduced Ionization and Attachment Coefficients

E/N (Td)	η/N (10^{-22} m^2)	α/N (10^{-22} m^2)	α_{eff}/N (10^{-22} m^2)
30	0.0006		−0.0006
35	0.0075		−0.0075
40	0.031		−0.031
45	0.091		−0.091
50	0.195		−0.195
60	0.525		−0.525
70	0.851	0.019	−0.832
80	1.16	0.075	−1.09
90	1.39	0.27	−1.12
100	1.59	0.58	−1.01
110	1.67	1.11	−0.56
120	1.81	1.71	−0.10
130	1.90	2.36	0.46
140	2.00	3.15	0.95
160	2.20	5.10	2.90
180	2.45	7.25	4.80
200	2.60	9.90	7.45
220	2.79	12.5	9.71
240	2.96	15.7	12.7
260	3.29	18.1	14.8
280	3.58	22.0	18.4
300	3.81	25.5	21.7

Source: Adapted from Hunter, S. R., J. G. Carter, and L. G. Christophorou, *J. Appl. Phys.*, 65, 1858, 1989.
Note: See Figure 66.4 for graphical presentation.

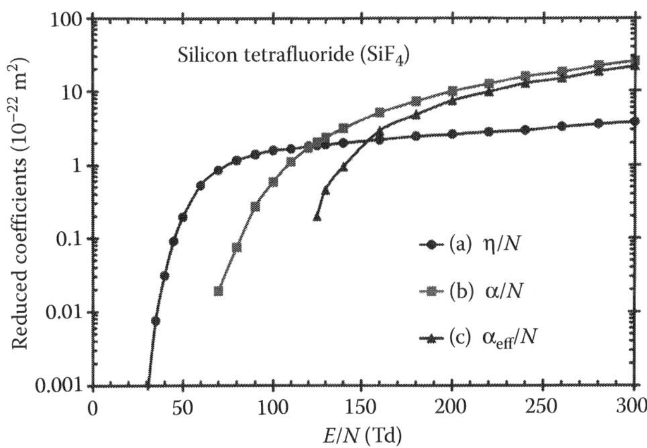

FIGURE 66.4 Ionization and attachment coefficients for SiF_4. (—●—) η/N, attachment; (—■—) α/N, ionization; (—▲—) effective ionization coefficients. (Adapted from Hunter, S. R., J. G. Carter, and L. G. Christophorou, *J. Appl. Phys.*, 65, 1858, 1989.)

66.7 ADDENDUM

Partial and total ionization cross sections have been measured by Basner et al. (2001) as shown in Table 66.7 and Figure 66.5.

SiF$_4$

TABLE 66.7
Absolute Partial (Counting) and Total Ionization Cross Section in Units of 10^{-20} m^2 for SiF$_4$ for Ion Species

	Ion			
Energy (eV)	SiF_n^+ ($n = 1$–4)	Si^+	F^+	Total
16	0.003			0.003
17	0.107			0.107
18	0.209			0.209
19	0.348			0.348
20	0.518			0.518
22	0.821			0.821
24	1.17			1.17
26	1.70			1.70
28	2.00			2.00
30	2.21			2.20
32	2.41			2.41
34	2.63			2.63
36	2.84	0.004	0.006	2.85
38	3.04	0.013	0.014	3.07
40	3.21	0.038	0.022	3.27
42	3.36	0.060	0.032	3.46
44	3.50	0.083	0.043	3.63
46	3.62	0.102	0.052	3.79
48	3.73	0.124	0.063	3.93
50	3.84	0.143	0.073	4.08
52	3.93	0.160	0.085	4.23
54	4.03	0.177	0.097	4.37
56	4.12	0.195	0.112	4.52
58	4.22	0.212	0.126	4.66
60	4.30	0.229	0.141	4.79
65	4.45	0.266	0.179	5.07
70	4.59	0.299	0.219	5.32
80	4.80	0.356	0.295	5.74
90	4.89	0.396	0.359	5.99
100	4.88	0.420	0.412	6.12
120	4.87	0.455	0.506	6.32
150	4.76	0.477	0.589	6.37
200	4.52	0.474	0.640	6.20
250	4.23	0.445	0.626	5.84
300	3.91	0.409	0.585	5.41
400	3.44	0.349	0.503	4.72
500	3.06	0.298	0.433	4.16
600	2.75	0.262	0.386	3.73
700	2.49	0.236	0.337	3.36
800	2.30	0.213	0.301	3.08
900	2.13	0.196	0.278	2.85

Source: Adapted from Basner, R. et al., *J. Chem. Phys.*, 114, 1170, 2001.

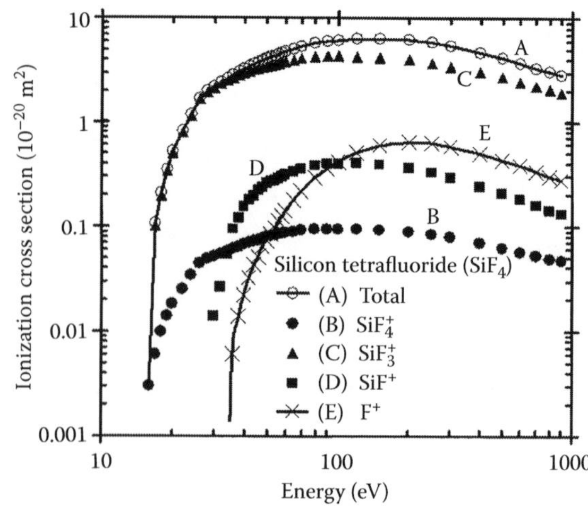

FIGURE 66.5 Total (A) and selected partial cross sections (B–E) for SiF$_4$. A—Total; B—SiF_4^+; C—SiF_3^+; D—SiF^+; E—F^+. Note the relatively low contribution of the parent ion SiF_4^+ and the largest contribution of SiF_3^+ to the total ionization cross section. (Adapted from Basner, R. et al., *J. Chem. Phys.*, 114, 1170, 2001.)

REFERENCES

Basner, R., M. Schmidt, E. Denisov, K. Becker, and H. Deutsch, *J. Chem. Phys.*, 114, 1170, 2001.

Hunter, S. R., J. G. Carter, and L. G. Christophorou, *J. Appl. Phys.*, 65, 1858, 1989.

Iga, I., M. V. V. S. Rao, S. K. Srivastava, and J. C. Nogueira, *Z. Phys. D*, 24, 111, 1992.

Karwasz, G. P., R. Brusa, and A. Zecca, Cross sections for scattering- and excitation-processes in electron–molecule collisions, in Y. Itikawa (Ed.), *Interactions of Photons and Electrons with Molecules*, Vol. 17c, Springer-Verlag, New York, NY, 2003, Chapter 6.1.

Karwasz, G. P., R. S. Brusa, A. Piazza, A. Zecca, P. Możeko, G. Kasperski, and C. Szmytkowski, *Chem. Phys. Lett.*, 284, 128, 1998. (a) and (b) denote measurements in two different laboratories.

Kim, Y.-K., K. K. Irikura, M. E. Rudd, M. A. Ali, P. M. Stone, J. Chang, J. S. Coursey et al., http://physics.nist.gov/PhysRefData/Ionization/index.html, 2005.

Poll, H. U. and J. Meichsner, *Contrib. Plas. Phys.*, 27, 359, 1987.

Wan, H.-X., J. H. Moore, and J. A. Tossell, *J. Chem. Phys.*, 91, 7349, 1989.

67

SULFURYL FLUORIDE

CONTENTS

Sulfuryl fluoride (SO_2F_2) is an abundant by-product of the decomposition of SF_6 gas (Datskos and Christophorou, 1989).

67.1 ION APPEARANCE POTENTIALS

Ion appearance potentials measured with mass spectrometer (Reese et al., 1958) are shown in Table 67.1.

67.2 ATTACHMENT CROSS SECTIONS

Electron attachment occurs at low energy (≤1.0 eV) both dissociatively and nondissociatively according to reactions

$$SO_2F_2 + e \rightarrow SO_2F_2^{-*} \rightarrow SO_2F_2^- \quad (67.1)$$

$$SO_2F_2 + e \rightarrow SO_2F_2^{-*} \rightarrow SO_2F^- + F \quad (67.2)$$

$$SO_2F_2 + e \rightarrow SO_2F_2^{-*} \rightarrow SO_2F + F^- \quad (67.3)$$

TABLE 67.1
Ion Appearance Potentials

Process	Potential (eV)
Positive ions	
$SO_2F_2 + e \rightarrow SO_2F_2^+ + 2e$	13.3
$SO_2F_2 + e \rightarrow SO_2F^+ + F + 2e$	15.1
$SO_2F_2 + e \rightarrow SOF^+ + F + O + 2e$	18.6
$SO_2F_2 + e \rightarrow SO_2^+ + 2F + 2e$	19.9
$SO_2F_2 + e \rightarrow SO^+ + O + 2e$	24.3
Negative ions	
$SO_2F_2 + e \rightarrow SO_2F_2^-$	~0.0
$SO_2F_2 + e \rightarrow SO_2F^- + F$	2.3
$SO_2F_2 + e \rightarrow SO_2 + F_2^-$	2.8
$SO_2F_2 + e \rightarrow SO_2F + F^-$	3.2

The energy for Reaction 67.1 is approximately 0 eV and the energy for Reactions 67.2 and 67.3 is ~0.5 eV. Electron attachment has been measured by Dastkos and Christophorou (1989).

The attachment cross sections are shown in Figure 67.1 at various gas temperatures. The highlights of the figure are

1. At constant energy the attachment cross section increases with temperature.
2. The attachment cross section as a function of energy shows a peak and the energy at which the peak is seen depends upon the temperature. The higher the temperature the lower the peak energy. At 300 K,

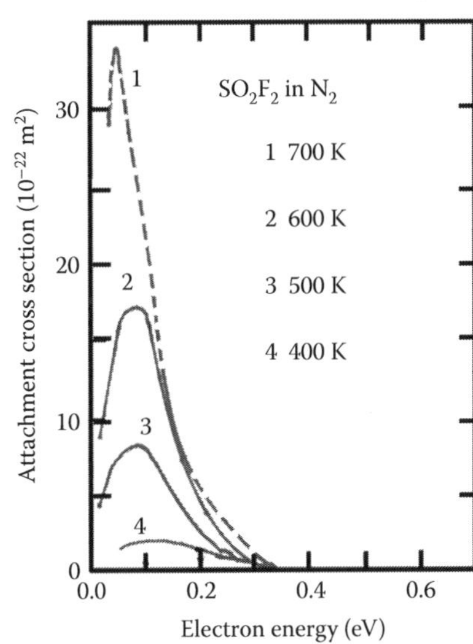

FIGURE 67.1 Attachment cross sections for SO_2F_2 at various constant gas temperatures. Note the shift of the peak energy to lower energies as the temperature increases. (Reprinted with permission from Datskos P. G. and L. G. Christophorou, *J. Chem. Phys.*, 90, 2626, 1989. Copyright (1989), American Institute of Physics.)

SO$_2$F$_2$

TABLE 67.2
Attachment Rates for SO$_2$F$_2$

Temperature (K)	Rate (10^{-15} m^3/s)
300	1.41
400	4.01
500	10.30
600	19.63
700	27.00

Source: Adapted from Datskos, P. G. and L. G. Christophorou, *J. Chem. Phys.*, 90, 2626, 1989.

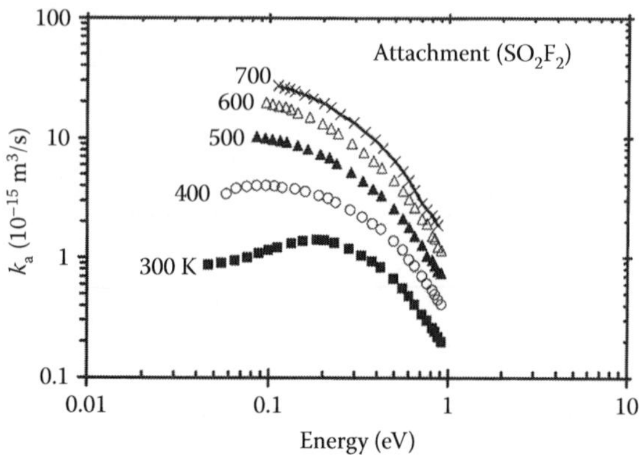

FIGURE 67.2 Attachment rates for SO$_2$F$_2$ at various constant temperatures as a function of electron mean energy. (Adapted from Datskos P. G. and L. G. Christophorou, *J. Chem. Phys.*, 90, 2626, 1989.)

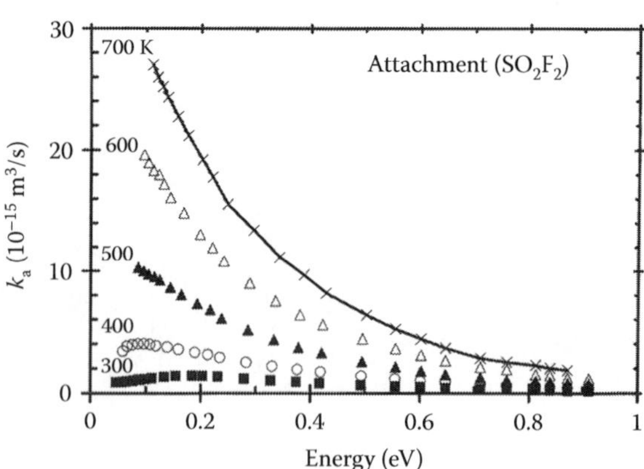

FIGURE 67.3 Attachment rates for SO$_2$F$_2$ at various constant temperatures as a function of electron mean energy with linear scales. (Adapted from Datskos, P. G. and L. G. Christophorou, *J. Chem. Phys.*, 90, 2626, 1989.)

the peak occurs at 0.22 eV, decreasing to 0.025 eV at 700 K.

67.3 ATTACHMENT RATES

The attachment rates are dependent on the gas temperature, increasing with increasing gas temperature as shown in Table 67.2 and Figure 67.2. Figure 67.3 also shows the same data with linear scales for the axes.

REFERENCES

Datskos, P. G. and L. G. Christophorou, *J. Chem. Phys.*, 90, 2626, 1989.

Reese, R. M., V. H. Diebler, and J. L. Franklin, *J. Chem. Phys.*, 29, 880, 1958.

68

TETRABROMO-METHANE

CONTENTS

Tetrabromomethane (CBr_4), synonym carbon tetrabromide, is a nonpolar, electron-attaching gas that has 146 electrons. The ionization potential is 10.31 eV. The molecule has four vibrational modes as shown in Table 68.1.

68.1 TOTAL SCATTERING CROSS SECTIONS

Calculated total scattering cross sections in the low-energy range (≤4.0 eV) are provided by Olthoff et al. (1986) as shown in Table 68.2 and Figure 68.1.

68.2 ATTACHMENT RATE CONSTANT

Dissociative attachment occurs due to reaction

$$CBr_4 + e \rightarrow CBr_2 + Br_2^- \quad (68.1)$$

TABLE 68.1
Vibrational Modes and Energies of CBr_4

Mode	Type of Motion	Energy (meV)
v_1	Symmetrical stretch	33.1
v_2	Degenerate deformation	15.1
v_3	Degenerate stretch	83.3
v_4	Degenerate deformation	22.6

Source: Adapted from Shimanouchi, T., *Tables of Molecular Virbrational Frequencies Consolidated Volume I*, NSRDS-NB539, U.S. Department of Commerce, Washington (DC), 1972.

Formation of Br^- ion has also been observed (Sunagawa and Shimamori, 1997). The appearance potential of Br^- is 0.0 eV (see Olthoff et al., 1986) and that of Br_2^- is also 0.0 eV (see DeCorpo and Franklin, 1971). The electron affinity of CBr_4 molecule is 2.06 eV (see Sunagawa and Shimamori, 1997).

The thermal attachment rate constant for CBr_4 is 2.5×10^{-14} m^3/s (see Sunagawa and Shimamori, 1997) and Figure 68.2 shows the attachment rate constants as a function of the mean energy of the electron swarm.

TABLE 68.2
Total Scattering Cross Sections for CBr_4

Energy (eV)	Q_T (10^{-20} m^2)	Energy (eV)	Q_T (10^{-20} m^2)
0.07	500	1.4	40.6
0.08	441	1.6	42.2
0.09	372	1.8	41.8
0.1	336	2	45.2
0.2	243	2.2	50.1
0.3	183	2.4	56.3
0.4	151	2.6	59.6
0.5	130	2.8	58.8
0.6	118	3	58.7
0.7	244	3.2	57.4
0.8	223	3.4	55.4
0.9	82.3	3.6	53.1
1	64.5	3.8	48.8
1.2	53.4	4	46.2

Source: Digitized and interpolated from Olthoff, J. K., J. H. Moore, and J. A. Tossell, *J. Chem. Phys.*, 85, 249, 1986.

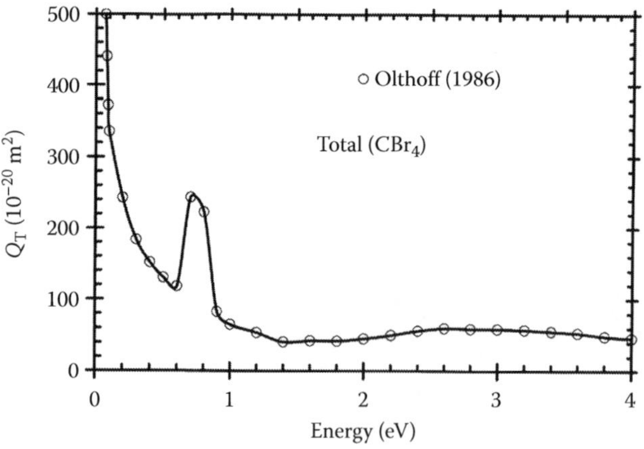

FIGURE 68.1 Total scattering cross sections for CBr_4 digitized from Olthoff et al. (1986).

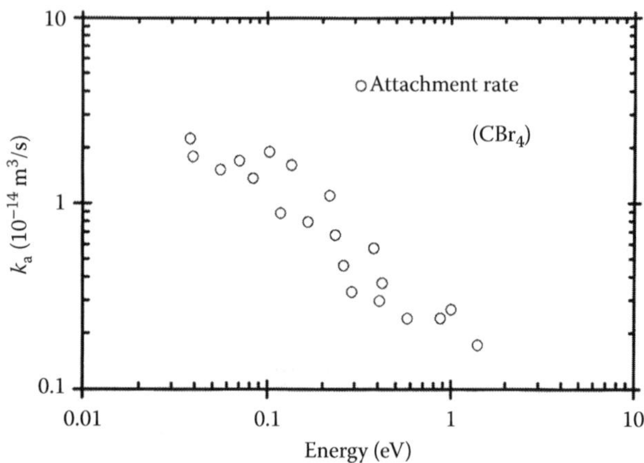

FIGURE 68.2 Attachment rate constants as function of mean energy for CBr_4. (Adapted from Sunagawa, T. and H. Shimamori, *J. Chem. Phys.*, 107, 7876, 1997.)

TABLE 68.3
Attachment Cross Sections for CBr_4

Energy (eV)	Q_a (10^{-20} m^2)	Energy (eV)	Q_a (10^{-20} m^2)
0.001	181.4	0.08	10.20
0.002	136.8	0.1	6.99
0.003	103.0	0.2	2.05
0.004	84.27	0.4	0.67
0.005	76.34	0.6	0.53
0.006	72.84	0.8	0.47
0.008	63.46	1.0	0.45
0.010	54.44	1.2	0.42
0.020	35.60	1.5	0.39
0.040	20.61	2.0	0.34
0.060	13.98	3.0	0.26

Source: Digitized and interpolated from Sunagawa, T. and H. Shimamori, *J. Chem. Phys.*, 107, 7876, 1997.

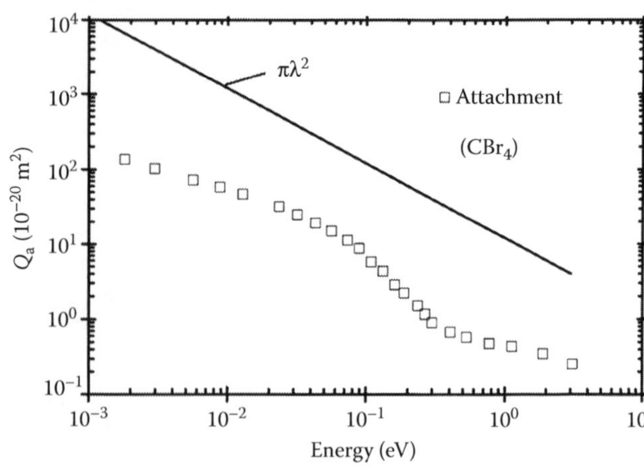

FIGURE 68.3 Attachment cross sections for CBr_4. The *s*-wave maximum cross section is also shown. (Adapted from Sunagawa, T. and H. Shimamori, *J. Chem. Phys.*, 107, 7876, 1997.)

68.3 ATTACHMENT CROSS SECTIONS

Attachment cross sections derived by Sunagawa and Shimamori (1997) are shown in Table 68.3 and Figure 68.3. The maximum *s*-wave cross section is also shown for comparison. The rise in the cross section toward zero energy is attributed to attachment.

REFERENCES

DeCorpo, J. J. and J. L. Franklin, *J. Chem. Phys.*, 54, 1885, 1971.

Olthoff, J. K., J. H. Moore, and J. A. Tossell, *J. Chem. Phys.*, 85, 249, 1986.

Shimanouchi, T., *Tables of Molecular Virbrational Frequencies Consolidated Volume I*, NSRDS-NB539, U.S. Department of Commerce, Washington (DC), 1972.

Sunagawa, T. and H. Shimamori, *J. Chem. Phys.*, 107, 7876, 1997.

69

TETRACHLOROSILANE

CONTENTS

Tetrachlorosilane ($SiCl_4$) is a nonpolar, electron-attaching molecule with 82 electrons and its ionization potential is 11.79 eV.

69.1 SELECTED REFERENCES FOR DATA

See Table 69.1.

69.2 TOTAL SCATTERING CROSS SECTION

Highlights of total scattering cross section are

1. A resonance at ~2.0 eV possibly due to dissociative attachment
2. A shoulder at 5.5 eV and resonance peak at ~10 eV due to electron attachment (see Appendix 8)
3. A monotonic decrease of cross section beyond 10 eV, which is a common feature in many gases (see Table 69.2)

69.3 IONIZATION CROSS SECTION

Appearance potentials for positive ions by ionization are shown in Tables 69.3 (Basner et al., 2005) and ionization cross sections in 69.4.

TABLE 69.1
Selected References for Data

Quantity	Range: eV, (Td)	Reference
Ionization cross section	12–900	**Basner et al. (2005)**
Total scattering cross section	0.3–1000	Karwasz et al. (2003)
Total scattering cross section	0.3–4000	**Możejko et al. (1999)**
Total scattering cross section	0.2–12.0	**Wan et al. (1989)**

Note: Bold font denotes experimental study.

TABLE 69.2
Total Scattering Cross Section for $SiCl_4$

Energy (eV)	Q_T (10^{-20} m^2)	Energy (eV)	Q_T (10^{-20} m^2)	Energy (eV)	Q_T (10^{-20} m^2)
(Karwasz et al. 2003)		Możejko et al. (1999)A		Możejko et al. (1999)B	
0.30	39.70	0.3	39.8	100	43.1
0.40	39.30	0.4	39.1	110	42.6
0.6	38.90	0.6	39.1	125	40.9
0.8	40.00	0.8	40.2	150	38.3
1.0	42.70	1	41.6	175	36.6
1.5	54.80	1.4	51.9	200	32.7
2.0	60.20	1.8	60.7	225	30.5
2.5	50.10	2	60	250	29.2
3.0	44.30	2.5	50.3	275	28.4
4.0	49.20	2.9	44.4	300	26.8
5	64.30	3.7	46.5	350	23.9
6	65.40	4.7	58.4	400	22.6
7	68.50	5.2	66.6	450	21.6
8	73.60	5.7	66.8	500	20.4
9	76.40	6.2	64.9	600	17.5
10	76.50	6.7	66.9	700	15.8
12	73.20	7.2	69.9	800	14.5
15	68.60	7.7	72.4	900	13
17	66.90	8.2	74	1000	11.8
20	64.30	8.7	75.6	1100	11.2
25	60.70	9.2	76.9	1200	10
30	57.80	9.7	76.8	1300	9.5
35	55.50	10.2	76.1	1500	8.6
40	53.60	11.2	74.3	1750	7.4
45	51.80	12.2	73.2	2000	6.6
50	50.00	14	69.7	2250	5.8
60	47.40	16	67.7	2500	4.91
70	45.20	18	66	3000	4.69
80	43.40	20	64.8	3250	4.23

continued

SiCl4

TABLE 69.2 (continued)
Total Scattering Cross Section for $SiCl_4$

Energy (eV)	Q_T (10^{-20} m²)	Energy (eV)	Q_T (10^{-20} m²)	Energy (eV)	Q_T (10^{-20} m²)
(Karwasz et al. 2003)		Możejko et al. (1999)A		Możejko et al. (1999)B	
90	41.80	25	60.7	3470	3.98
100	40.40	30	57.6	4000	3.67
120	37.90	40	53.4		
150	34.90	50	50.6		
170	33.30	60	48.9		
200	31.10	70	46.2		
250	28.10	80	43.5		
300	25.60	90	41.4		
350	23.60	100	39.5		
400	21.90	120	36.2		
450	20.40	140	33.8		
500	19.10	160	31.3		
600	16.90	180	29.1		
700	15.20	200	27.8		
800	13.80	250	24.9		
1000	12.00				

Note: See Figure 69.1 for graphical presentation.

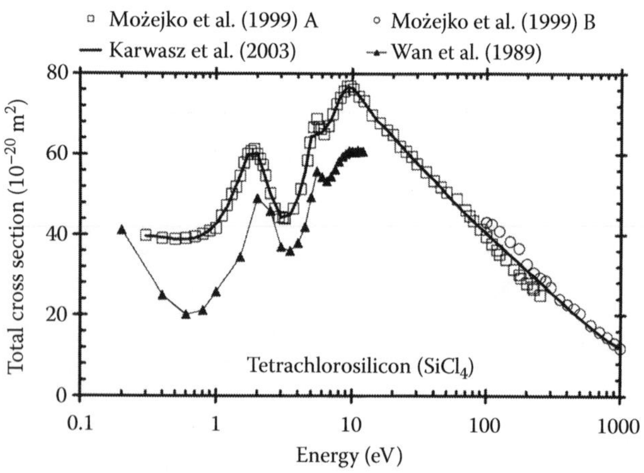

FIGURE 69.1 Total scattering cross sections for $SiCl_4$. (——) Karwasz et al. (2003); (□) Możejko et al. (1999)A; (○) Możejko et al. (1999)B; (▲) Wan et al. (1989).

TABLE 69.3
Ion Appearance Potentials

Ion	$SiCl_4^+$	$SiCl_3^+$	$SiCl_2^+$	$SiCl^+$	Si^+	Cl^+
Energy (eV)	11.7	12.7	17.5	19.4	22.3	21.8

TABLE 69.4
Total Ionization Cross Section

Energy (eV)	Q_i (10^{-20} m²)	Energy (eV)	Q_i (10^{-20} m²)	Energy (eV)	Q_i (10^{-20} m²)
12	0.193	36	16.36	120	18.89
14	1.78	38	16.77	140	18.48
16	3.64	40	17.18	160	18.10
18	5.69	42	17.51	180	17.46
20	8.03	44	17.77	200	16.86
22	9.43	46	18.00	300	12.65
24	10.85	48	18.39	400	10.69
26	12.19	50	18.55	500	9.51
28	13.60	60	18.93	600	8.48
30	14.47	70	19.59	700	7.64
32	15.50	80	19.72	800	6.90
34	15.92	100	19.46	900	6.21

Source: Adapted from Basner, R. et al., *J. Chem. Phys.*, 123, 054313, 2005.
Note: See Figure 69.2 for graphical presentation.

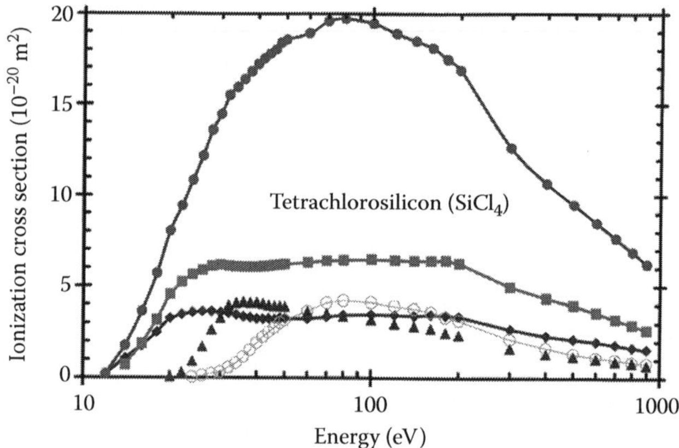

FIGURE 69.2 Partial and total ionization cross sections for $SiCl_4$. For clarity sake only selected fragments are shown. (——) Total; (◆) $SiCl_4^+$; (■) $SiCl_3^+$; (▲) $SiCl^+$; (—○—) Cl^+. (Adapted from Basner, R. et al., *J. Chem. Phys.*, 123, 054313, 2005.)

REFERENCES

Basner, R., M. Gutkin, J. Mahoney, V. Tarnovsky, H. Deutsch, and K. Becker, *J. Chem. Phys.*, 123, 054313, 2005.

Karwasz, G. P., R. Brusa, and A. Zecca, Cross sections for scattering- and excitation-processes in electron–molecule collisions, in Y. Itikawa (Ed.), *Interactions of Photons and Electrons with Molecules*, Vol. 17c, Springer Verlag, New York, NY, 2003, Chapter 6.

Możejko, P., G. Kasperski, Cz. Syzmytkowski, A. Zecca, G. P. Karwasz, L. Del Longo, and R. S. Brusa, *Eur. Phys. J. D*, 6, 481, 1999. A and B refer to measurements in two different laboratories.

Wan, H., J. H. Moore, and J. A. Tossell, *J. Chem. Phys.*, 91, 7340, 1989.

70

TETRAFLUORO-METHANE

CF

CONTENTS

Fluorocarbons, consisting of carbon and fluorine atoms, are industrial gases. We consider selected target particles with the general chemical formula C_nF_{2n+2}. They are fluorine-substituted derivatives of alkanes. The series for $n = 1–3$ has been extensively analyzed by Christophorou and Olthoff (2004) and we summarize their vast publications on tetrafluoromethane (CF_4), providing some additional data. Selected properties of target particles in the series are shown in Table 70.1.

The modes of vibrational excitation with energies are shown in Table 70.2 (Mann and Linder, 1992). Vibrations of similar type tend to have similar energies (Allan and Andric, 1996).

TABLE 70.1
Selected Properties of Fluorocarbons

Target					
Name	**Formula**	***z***	**α_e (10–40 Fm²)**	**μ (D)**	**ε_i (eV)**
CF_4	Carbon tetrafluoride	42	4.27	0	16.20
C_2F_6	Hexafluoroethane	66	7.588	0	14.6
C_3F_8	Perfluoropropane	90	7.20–10.46	0.07	13.3

Note: Z = Number of electrons; α_e = electronic polarizability; μ = dipole moment; ε_i = ionization potential.

70.1 SELECTED REFERENCES FOR DATA

See Table 70.3.

70.2 TOTAL SCATTERING CROSS SECTION

Table 70.4 and Figure 70.1 show the total scattering cross sections for CF_4 recommended by Christophorou and Olthoff (2004) and compared with later measurements by Ariyasinghe (2003). Figure 70.2 shows a comparison with selected fluorocarbons and Figure 70.3 shows the expanded view in the

CF$_4$

TABLE 70.2
Vibrational Energies for CF_4 and C_2F_6

CF_4 (Mann and Linder, 1992a)			C_2F_6 (Sverdlov et al. 1974)	
Mode	Deformation	Energy (meV)	Mode	Energy (meV)
ν_1	C–F symmetric stretch	112.6	ν_1	176
ν_2	Symmetric bend	53.9	ν_2	**100.1**
ν_3	C–F asymmetric stretch	**158.9**	ν_3	43.1
ν_4	Asymmetric bend	78.4		
			ν_5	138.4
			ν_6	88.5
			ν_7	155.0
			ν_8	76.7
			ν_9	46.1
			ν_{10}	155.0

Note: Bold font indicates strong intensity.

TABLE 70.3
Selected References for Data on CF_4

Parameter	Range: eV, (Td)	Reference
Q_i	16.2–5000	Kim et al. (2004)
Q_T	100–1500	**Ariyasinghe (2003)**
Q_i	20–85	**Torres et al. (2002)**
Q_i	18–1000	**Sieglaff et al. (2001)**
Q_i	12.2–215.5	**Bart et al. (2001)**
Q_i	10–40	**Fiegele et al. (2000)**
Q_i	16–3000	**Nishimura et al. (1999)**
Q_{diff}	1.5–100	**Tanaka et al. (1999)**
W	3–3000	**Lisovskiy and Yegorenkov (1999)**
Q_{diss}	10–1000	**Motlagh and Moore (1998)**
Q_T	1–400	**Sueoka et al. (1994)**
Q_T	75–4000	**Zecca et al. (1992)**
Q_{diff}	1.5–100	**Boesten et al. (1992)**
Q_{el}	0.3–20	**Mann and Linder (1992)**
Q_{el}, i_M	0.8	Pirgov et al. (1990)
D/μ	(0.14–23.0)	**Curtis et al. (1988)**
W	(0.03–500)	**Hunter et al. (1988)**
Q_{el}, Q_M	0–4	Stefanov et al. (1988)
α/N, η/N	(20–400)	**Hunter et al. (1987)**
k_a	(0.9–4.7)	**Hunter et al. (1984)**
α/N	(100–650)	**Shimozuma et al. (1983)**
k_a	0–12	**Spyrou et al. (1983)**
Q_{diss}	13–600	Winters and Inokuti (1982)

Note: Bold font denotes experimental study. k_a = Attachment rate; Q_{diff} = differential scattering, Q_{el} = elastic scattering; Q_T = total scattering; W = drift velocity; α/N = reduced ionization coefficient; η/N = reduced attachment coefficient.

TABLE 70.4
Total Scattering Cross Sections for CF_4

Energy (eV)	Q_T (10^{-20} m^2) Christophorou (2004)	Energy (eV)	Q_T (10^{-20} m^2) Christophorou (2004)	Q_T (10^{-20} m^2) Ariyasinghe (2003)
0.0030	12.7	4.5	13.8	
0.0035	12.2	5.0	14.0	
0.0040	11.9	6.0	15.3	
0.0045	11.5	7.0	18.6	
0.0050	11.2	8.0	21.1	
0.0060	10.6	9.0	21.8	
0.0070	10.1	10.0	20.8	
0.0080	9.69	12.0	18.5	
0.0090	9.26	15.0	17.9	
0.010	8.89	20.0	19.2	
0.015	7.40	25	20.4	
0.020	6.35	30	20.2	
0.025	5.41	35	19.9	
0.030	4.67	40	19.9	
0.035	4.12	45	19.9	
0.040	3.63	50	19.9	
0.045	3.21	60	19.9	
0.050	2.86	70	19.6	
0.060	2.30	80	19.2	
0.070	1.98	90	18.8	
0.080	1.76	100	18.3	18.37
0.090	1.62	150	16.2	
0.10	1.50	200	14.4	13.06
0.14	1.60	250	12.8	
0.15	2.17	300	11.6	10.27
0.16	3.05	350	10.8	
0.20	7.35	400	9.95	8.86
0.25	9.12	450	9.24	
0.30	9.26	500	8.60	7.77
0.35	9.28	600	7.57	6.81
0.40	9.25	700	6.80	6.14
0.45	9.23	800	6.21	5.71
0.50	9.27	900	5.70	5.17
0.60	9.45	1000	5.28	4.85
0.70	9.60	1100		4.55
0.80	9.75	1200		4.21
0.90	9.89	1300		3.94
1.0	10.0	1400		3.84
1.5	10.6	1500		3.73
2.0	11.3	2000	2.95	
2.5	12.0	3000	2.05	
3.5	13.0	4000	1.49	
4.0	13.5			

low-energy region up to 10 eV electron energy. The important features of the cross sections are

1. A deep Ramsauer–Townsend minimum at 0.15 eV for CF_4.
2. To the left of the minimum, the cross section increases, initially rapidly, but at lower energies < 0.01 eV, less rapidly.

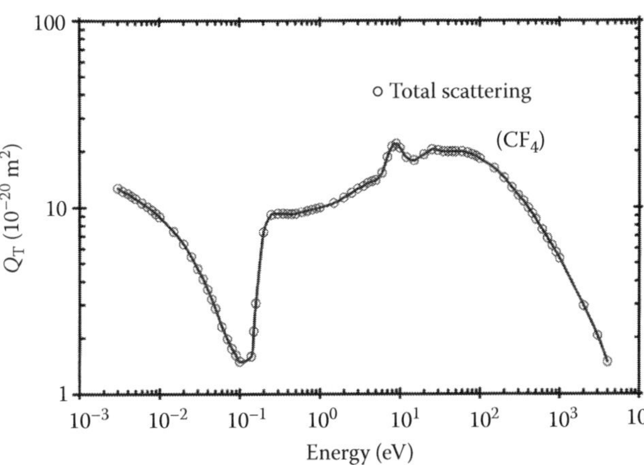

FIGURE 70.1 Total scattering cross sections for CF_4 recommended by Christophorou and Olthoff (2004).

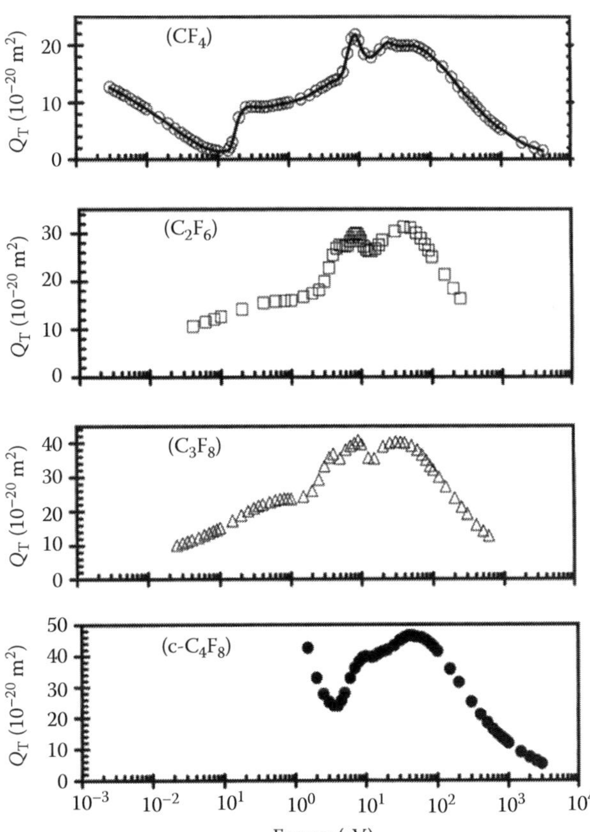

FIGURE 70.2 Comparison of total scattering cross section for selected fluorocarbon gases.

3. In the energy range of 6–50 eV, there is a peak at 8.5–9 eV attributed to shape resonance, with the formation of short-lived negative ions. This feature is common to CF_4, C_2F_6, and C_3F_8.
4. A smaller peak around 20 eV due to direct scattering and many inelastic processes including electronic excitation and ionization.

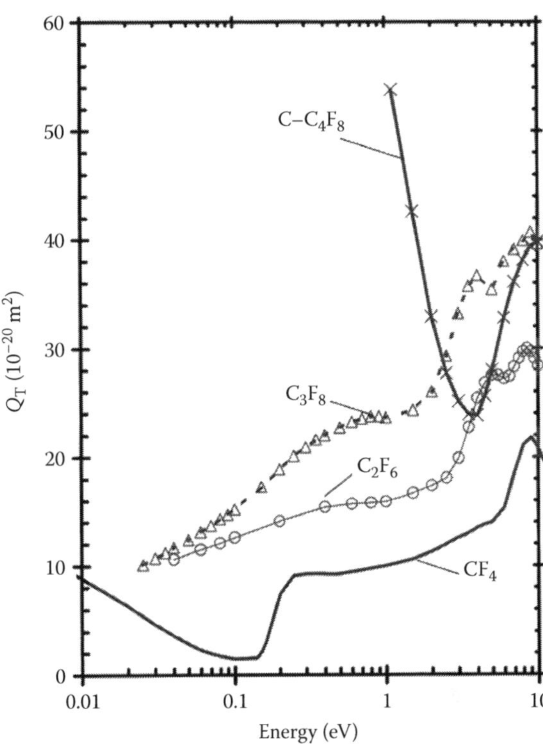

FIGURE 70.3 Total scattering cross sections for selected CF_4 on an expanded scale. Note the linear scale for the ordinate. Data for selected fluorocarbons added for comparison. (Adapted from Christophorou, L. G. and J. K. Olthoff, *Fundamental Electron Interactions with Plasma Processing Gases*, Kluwer Academic/Plenum Publishers, New York, NY, 2004.)

5. At energy >100 eV, there is a monotonic decrease of the cross section with the Born approximation method of calculating the cross section yielding satisfactory results. Features 3 and 5 are common to most target particles.
6. The total scattering cross section increases with increasing number of electrons of the target particle in the gases CF_4, C_2F_6, C_3F_8, and C–C_4F_8 (see Figures 70.2 and 70.3).

70.3 ELASTIC SCATTERING AND MOMENTUM TRANSFER CROSS SECTIONS

Table 70.5 gives the elastic scattering and momentum transfer cross sections recommended by Christophorou and Olthoff (2004). These are shown in Figures 70.4 and 70.5. Measured momentum transfer and elastic scattering cross sections are not available for C–C_4F_8 and theoretical values of Winstead and McKoy (2001) are included in Figure 70.5.

70.4 RAMSAUER–TOWNSEND MINIMUM

Ramsauer–Townsend minimum cross sections and the corresponding electron energy are shown in Table 70.6. Selected gases are included for comparison.

TABLE 70.5
Elastic Scattering and Momentum Transfer Cross Sections

	Q (10^{-20} m^2)			Q (10^{-20} m^2)	
Energy (eV)	Q_{el}	Q_M	**Energy (eV)**	Q_{el}	Q_M
0.0010		13.0	0.70	3.52	3.45
0.0015		12.3	0.80	4.02	4.01
0.0020		11.8	0.90	4.46	4.48
0.0025		11.3	1.0	4.86	4.92
0.0030	12.7	10.9	1.5	6.87	6.26
0.0035	12.2	10.6	2.0	8.48	6.92
0.0040	11.9	10.2	2.5	9.68	7.30
0.0045	11.5	9.93	3.0	10.5	7.53
0.0050	11.2	9.65	3.5	11.1	7.72
0.0060	10.6	9.14	4.0	11.6	7.89
0.0070	10.1	8.67	4.5	11.9	8.04
0.0080	9.68	8.25	5.0	12.1	8.21
0.0090	9.27	7.85	6.0	12.4	8.55
0.010	8.88	7.52	7.0	12.6	8.68
0.015	7.39	6.15	8.0	13.2	8.96
0.020	6.35	5.06	9.0	14.2	10.1
0.025	5.42	4.16	10.0	15.11	11.2
0.030	4.68	3.44	15.0	15.7	13.4
0.035	4.11	2.82	20.0	16.1	14.1
0.040	3.62	2.29	25.0	15.9	12.5
0.045	3.21	1.90	30.0	15.6	10.4
0.050	2.85	1.54	35.0	15.2	8.80
0.060	2.29	1.10	40.0	14.9	7.80
0.070	1.87	0.78	45.0	14.6	7.24
0.080	1.54	0.55	50.0	14.4	6.66
0.090	1.29	0.39	60.0	13.7	5.80
0.10	1.09	0.26	70.0	13.1	5.28
0.15	0.62	0.13	80.0	12.4	4.77
0.20	0.56	0.27	90.0	11.9	4.37
0.25	0.68	0.48	100.0	11.4	4.03
0.30	0.89	0.76	200.0	8.22	1.92
0.35	1.18	1.05	300.0	6.47	1.17
0.40	1.53	1.39	400.0	5.46	0.82
0.45	1.91	1.76	500.0	4.78	0.62
0.50	2.29	2.13	600.0	4.29	0.50
0.60	2.96	2.82	700.0	3.91	0.41

Source: Adapted from Christophorou, L. G. and J. K. Olthoff, *Fundamental Electron Interactions with Plasma Processing Gases*, Kluwer Academic/Plenum Publishers, New York, NY, 2004.

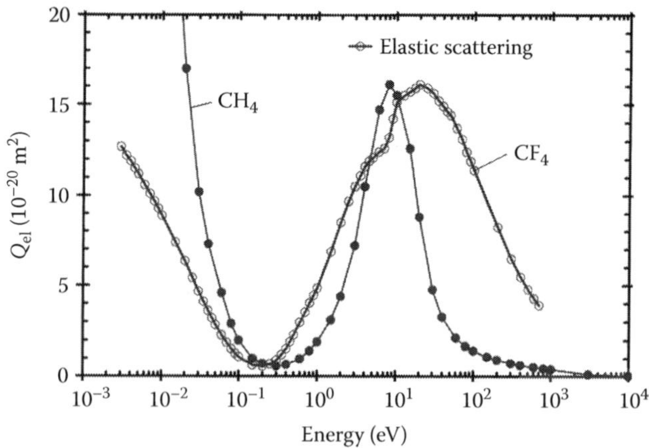

FIGURE 70.4 Elastic scattering and momentum transfer cross sections for CF_4. Data for CH_4 is given for comparison.

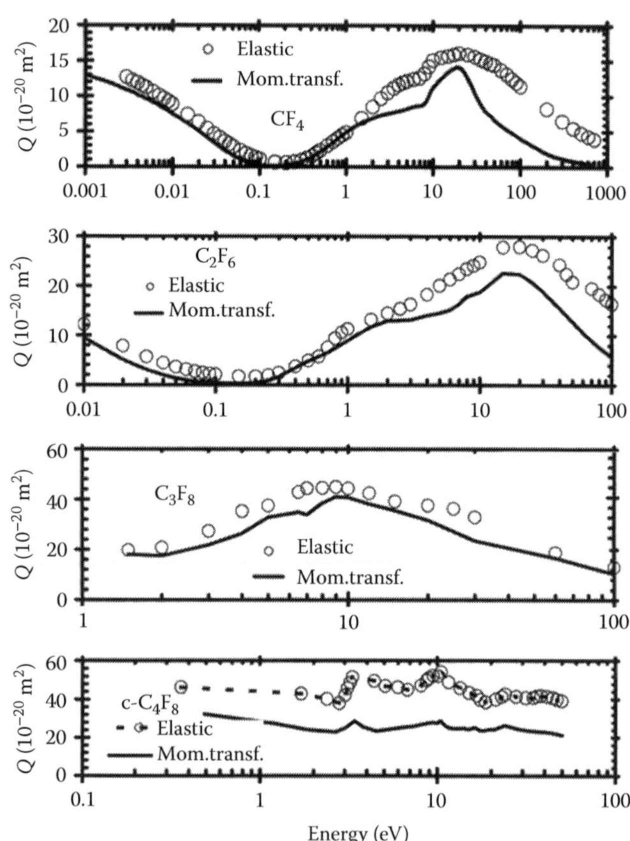

FIGURE 70.5 Comparison of elastic scattering and momentum transfer cross sections in selected fluorocarbons and selected fluorocarbons, top three frames, recommended by Christophorou and Olthoff (2004). The lowest frame is from Winstead and McKoy (2001) added for comparison.

70.5 VIBRATIONAL EXCITATION CROSS SECTIONS

Vibrational excitation cross sections given by Christophorou and Olthoff (2004) for CF_4 are shown in Table 70.7 (see Figure 70.6).

70.6 DISSOCIATION CROSS SECTIONS

The total excitation cross section for CF_4 is equal to the total dissociation cross section (Christophorou and Olthoff, 2004).

The dissociation processes, leaving apart the dissociative attachment, are (Moltagh and Moore, 1998)

$$e + CF_4 \rightarrow CF_3 + F + e \tag{70.1}$$

$$e + CF_4 \rightarrow CF_3 + F^+ + 2e \tag{70.2}$$

TABLE 70.6
Ramsauer–Townsend Minimum Cross Sections

Gas	Minimum Energy (eV)	Minimum Cross Section (10^{-20} m²)	Method	Reference
CF_4	0.1	1.50	Q_T	Christophorou et al. (2004)
	0.2	0.56	Q_{el}	Mann and Linder (1992)
	0.2	0.01	Q_M	Curtis et al. (1988)
	0.259	0.2	Q_M	Stefanov et al. (1988)
C_2F_6	0.2	1.66	Q_{el}	Christophorou et al. (2004)
	0.15	0.32	Q_M	Christophorou et al. (2004)
C_3F_8	0.07	0.8	Q_{el}	Pirgov et al. (1990)

TABLE 70.7
Vibrational Excitation Cross Sections for CF_4

Energy (eV)	Q_{vib} (10^{-20} m²)	Energy (eV)	Q_{vib} (10^{-20} m²)	Energy (eV)	Q_{vib} (10^{-20} m²)
0.08	0.14	0.35	8.10	7.0	1.09
0.10	0.39	0.40	7.74	8.0	0.98
0.12	0.43	0.50	7.01	9.0	0.90
0.14	0.43	1.0	4.67	10.0	0.81
0.16	1.86	1.5	3.54	15.0	0.59
0.20	7.43	2.0	2.88	20.0	0.46
0.25	8.40	3.0	2.12	35.0	0.29
0.275	8.46	5.0	1.43	50.0	0.22
0.30	8.39	6.0	1.24	60.0	0.19

Source: Adapted from Christophorou, L. G. and J. K. Olthoff, *Fundamental Electron Interactions with Plasma Processing Gases*, Kluwer Academic/Plenum Publishers, New York, NY, 2004.

Note: See Figure 70.6 for graphical presentation.

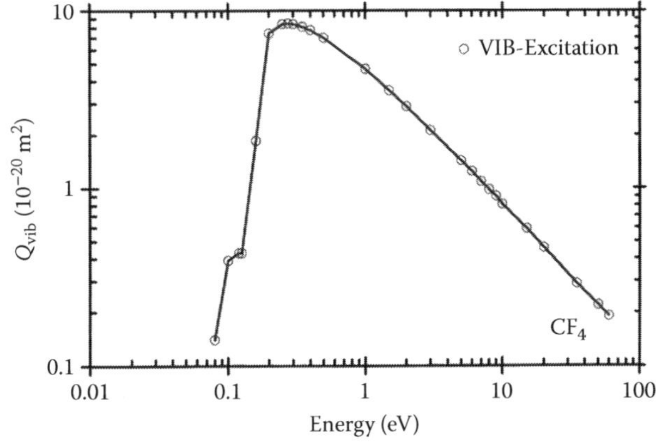

FIGURE 70.6 Vibrational excitation cross sections for CF_4.

Reaction 70.1 is neutral dissociation and Reaction 70.2 is dissociative ionization. The threshold for neutral dissociation is ~12.5 eV (Winters and Inokuti, 1982). The cross sections of Motlagh and Moore are given in Table 70.8 and Figure 70.7.

TABLE 70.8
Dissociation Cross Section for CF_4

Energy (eV)	Q_{diss} (10^{-20} m²) CF_3 (N.D.) + (D.I.)	F (N.D.) + (D.I.)	Total
10		0.51	0.51
14	0.06	0.45	0.51
16	0.04	0.79	0.83
18	0.08	1.04	1.12
20	0.10	1.13	1.24
22	0.19	1.59	1.79
25	0.39	2.49	2.89
30	0.65	3.24	3.89
35	0.84	3.90	4.75
40	1.08	4.41	5.49
50	1.22	5.06	6.28
60	1.38	6.11	7.49
70	1.44	6.09	7.53
80	1.47	6.19	7.66
90	1.47	6.60	8.07
100	1.50	6.84	8.34
125	1.42	6.70	8.12
150	1.39	6.67	8.05
175	1.36	6.66	8.02
200	1.34	6.71	8.05
250	1.23	6.43	7.66
300	1.20	5.92	7.11
350	1.15	5.49	6.63
400	1.08	5.17	6.25
450	1.02	4.96	5.98
500	1.00	4.81	5.81
550		4.65	4.65

Source: Digitized from Motlagh, S. and J. H. Moore, *J. Chem. Phys.*, 109, 432, 1998.

Note: N.D. = neutral dissociation; D. I. = dissociative ionization.

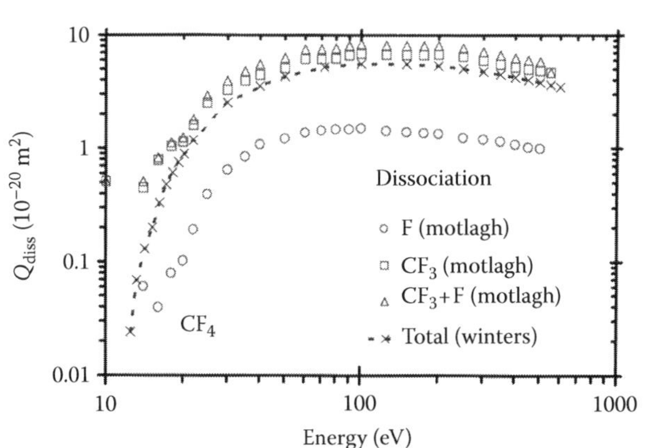

FIGURE 70.7 Total dissociation cross sections for CF_4.

CF

CF$_4$

70.7 POSITIVE ION APPEARANCE POTENTIALS

The parent ion CF_4^+ is unstable with a lifetime of 10 μs (see Fiegele et al., 2000) and the ion that appears at the lowest energy is CF_3^+. Ion appearance potentials are shown in Table 70.9 and a more detailed list for several paths for generation of C^+, F^+, and F_2^+ ions are given by Christophorou and Olthoff (2004).

70.8 IONIZATION CROSS SECTIONS

Ionization cross sections in fluorocarbons are shown in Table 70.10 and Figure 70.8. The data taken from Bart et al., (2001), which were not included in Christophorou and Olthoff (2004) are also shown.

TABLE 70.9
Positive Ion Appearance Potentials

Ion	Potential (eV)
CF_3^+	13.0
CF_2^+	21.47
CF^+	29.14
C^+	34.77

Source: Adapted from Jarvis, G. K. et al., *J. Phys. Chem. A*, 102, 3219, 1998.

TABLE 70.10
Ionization Cross Sections for CF$_4$

Energy (eV)	Q_i (10^{-20} m^2)	Energy (eV)	Q_i (10^{-20} m^2)	Energy (eV)	Q_i (10^{-20} m^2)
16	0.034	44	2.78	350	4.31
17	0.08	46	2.98	400	4.05
18	0.137	48	3.25	450	3.83
19	0.204	50	3.41	500	3.51
20	0.295	60	3.97	600	3.11
22	0.479	70	4.39	700	2.83
24	0.656	80	4.76	800	2.61
26	0.937	90	4.91	900	2.38
28	1.19	100	5.12	1000	2.23
30	1.41	125	5.31	1250	1.89
32	1.62	150	5.28	1500	1.64
34	1.83	175	5.31	1750	1.5
36	2.03	200	5.10	2000	1.34
38	2.18	250	4.78	2500	1.15
40	2.38	300	4.59	3000	0.99
42	2.6				

Source: Adapted from Nishimura, H. et al., *J. Chem. Phys.*, 110, 3811, 1999.

70.9 PARTIAL IONIZATION CROSS SECTIONS

The CF_4^+ ion is not observed and the sum of the partial ionization cross section is equal to the total ionization cross section (Christophorou and Olthoff, 2004). Partial ionization cross sections suggested by Christophorou and Olthoff (2004) are shown in Figure 70.9.

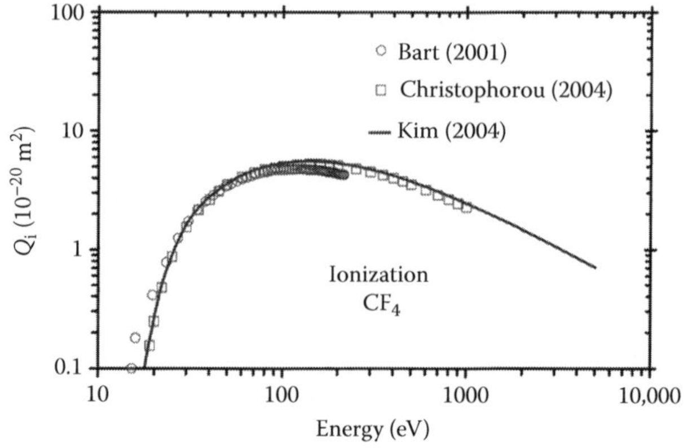

FIGURE 70.8 Ionization cross sections for fluorocarbons. Excellent agreement exists between the data shown.

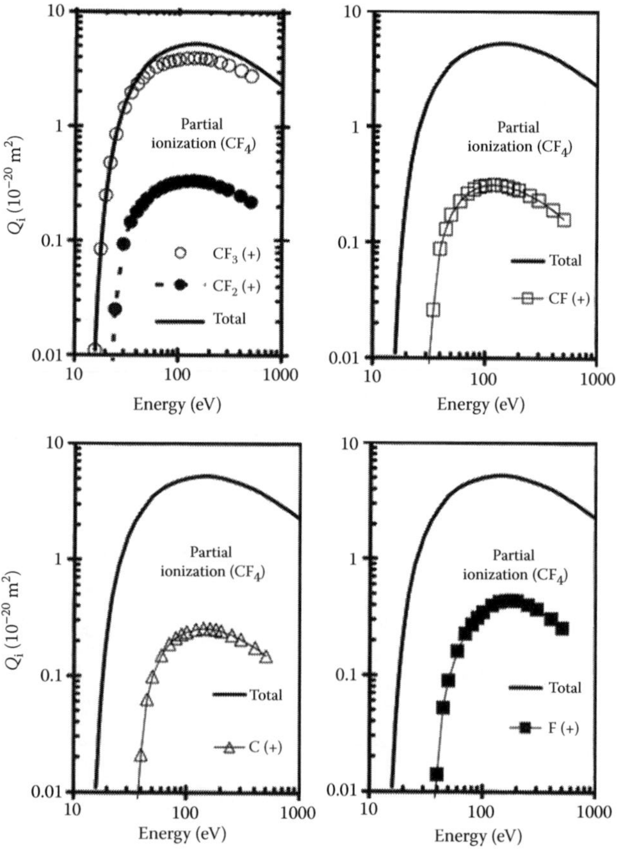

FIGURE 70.9 Partial ionization cross sections for CF$_4$. Note the absence of parent ion, CF_4^+. The largest contribution to the total is due to CF_3^+ shown in the first frame. (Adapted from Christophorou, L. G. and J. K. Olthoff, *Fundamental Electron Interactions with Plasma Processing Gases*, Kluwer Academic/Plenum Publishers, New York, NY, 2004.)

70.10 CROSS SECTIONS FOR ION PAIR PRODUCTION

Ion pairs (CF^+ and F^+) are generated during electron impact (Siegլaff et al., 2001) and the cross sections for ion pair production are shown in Table 70.11.

70.11 ATTACHMENT PROCESSES

Figure 70.10 shows the schematic potential energy levels of the CF_4 and CF_4^- species (Mann and Linder, 1992). Dissociative electron attachment processes are the only mechanisms of negative ion formation in CF_4, (Hunder and Christophorou, 1984) with F^- ion being the most abundant. The reactions are (Spyrou et al., 1983)

TABLE 70.11
Cross Sections for Ion Pair Production

Energy (eV)	Q_i (CF^+, F^+) (10^{-20} m²)	Q_i (CF_2^+, F^+) (10^{-20} m²)	Energy (eV)	Q_i (CF^+, F^+) (10^{-20} m²)	Q_i (CF_2^+, F^+) (10^{-20} m²)
60	0.016	0.0061	200	0.167	0.059
70	0.047	0.016	250	0.174	0.049
80	0.075	0.017	300	0.132	0.051
90	0.079	0.048	400	0.117	0.027
100	0.106	0.048	500	0.108	0.030
120	0.129	0.034	600	0.084	0.030
140	0.150	0.045	800	0.066	0.019
160	0.172	0.059	1000	0.046	0.013

Source: Adapted from Siegլaff, D. R. et al., *J. Phys. B; At. Mol. Opt. Phys.*, 34, 799, 2001.

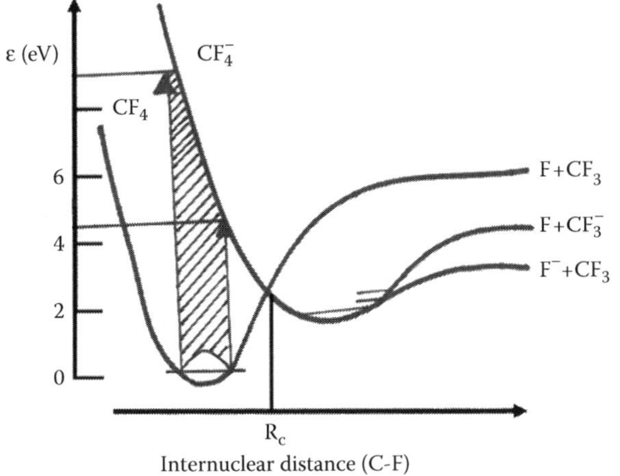

FIGURE 70.10 Schematic of the potential energy curves showing the dissociative attachment process and neutral fragmentation of the CF_4 molecule. The Gaussian curve is the distribution of vibrational energy of the ground state particles. The dashed line is the vertical attachment energy and the negative ions contributing to the ion current are also shown. R_c is the nuclear separation at which both curves have the same energy. (With permission from Institute of Physics, England.)

$$CF_4 + e \rightarrow CF_3 + F^- \quad (70.3)$$

$$CF_4 + e \rightarrow CF_3^- + F \quad (70.4)$$

$$CF_4 + e \rightarrow CF_2 + F + F^- \quad (70.5)$$

The appearance potentials and peak in the attachment cross sections are shown in Table 70.12.

70.12 ATTACHMENT RATES

Thermal electron attachment rates are shown in Table 70.13. Figure 70.11 shows the details for CF_4 as a function of *E/N*. Temperature dependence of the attachment rate in C_2F_6 is

TABLE 70.12
Negative Ions Due to Electron Impact on CF_4

Ion	Appearance Potential (eV)	Reaction Number	Peak (eV)	Reference
CF_3^-	5.1	4	7.1	Spyrou et al. (1983)
	5.4	4	—	Harland et al. (1974)
F^-	4.5	3	6.7	Spyrou et al. (1983)
	4.65	3	6.15	Harland et al. (1974)
	6.2	5	7.5	Harland et al. (1974)

TABLE 70.13
Thermal Attachment Rate Constants

Rate (m³/s)	Method	Reference
$<10^{-22}$	Swarm	Hunter et al. (1987)
$<3.1 \times 10^{-19}$	Swarm	Davis et al. (1973)
$<10^{-22}$	M.C.	Fessenden et al. (1970)
7.76×10^{-19}	Swarm	Blaunstein et al. (1968)

Note: M.C. = Microwave conductivity.

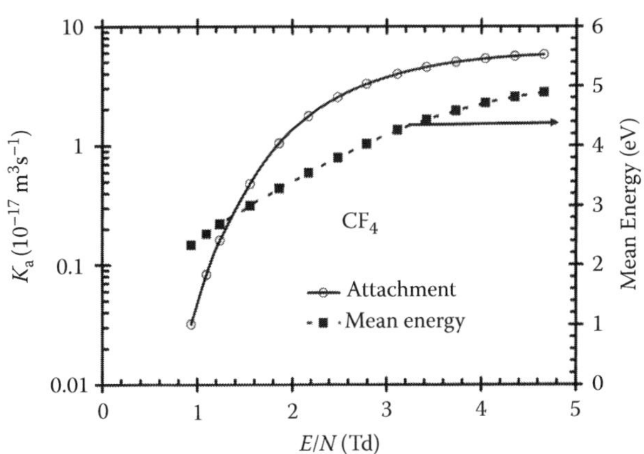

FIGURE 70.11 Attachment rate and mean energy for electrons impacting on CF_4. Note that the mean energy refers to argon buffer gas. (Adapted from Spyrou, S. M., I. Sauers, and L. G. Christophorou, *J. Chem. Phys.*, 78, 7200, 1983.)

CF$_4$

reviewed by Christophorou and Olthoff (2004). The total attachment rate moderately increases with temperature due to the fact that dissociative attachment onset, at ~2.1 eV is well above thermal energy.

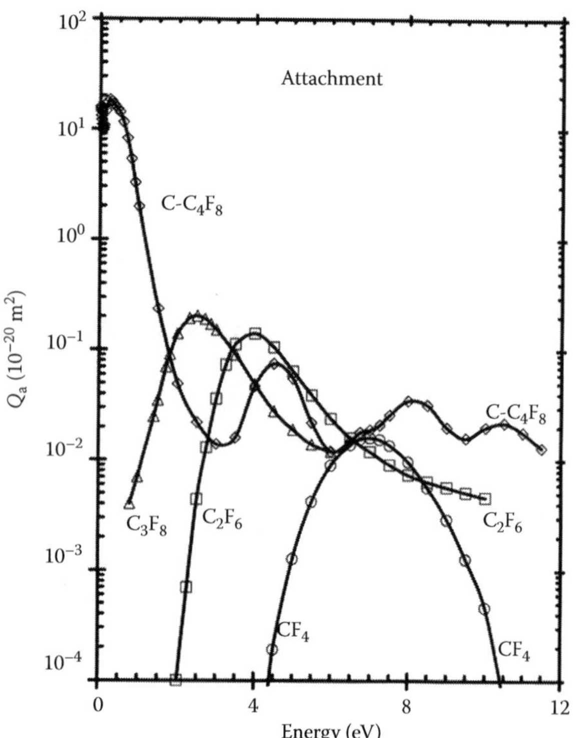

FIGURE 70.12 Total attachment cross sections for CF_4. Data for selected fluorocarbons added for comparison.

TABLE 70.14
Total Attachment Cross Sections in CF_4

Energy (eV)	Q_a (10^{-20} m^2)
4.3	4 (−5)
4.5	19 (−5)
5.0	127 (−5)
5.5	4.2 (−3)
6.0	8.98 (−3)
6.5	13.8 (−3)
6.75	15.7 (−3)
7.0	16.1 (−3)
7.25	15.5 (−3)
7.5	13.7 (−3)
8.0	9.69 (−3)
8.5	5.67 (−3)
9.0	2.91 (−3)
9.5	1.27 (−3)
10.0	4.6 (−4)
10.5	7 (−5)

Note: a (b) means a × 10^b.

70.13 ATTACHMENT CROSS SECTIONS

Total attachment cross sections are shown in Table 70.16 (Christophorou and Olthoff, 2004) and Figure 70.12 show the cross sections in selected fluorocarbons (see Table 70.14).

70.14 DRIFT VELOCITY OF ELECTRONS

The drift velocities of electrons in fluorocarbons are shown in Table 70.15 and Figure 70.13.

TABLE 70.15
Drift Velocity of Electrons

E/N (Td)	*W* (10^4 m/s)	*E/N* (Td)	*W* (10^4 m/s)	*E/N* (Td)	*W* (10^4 m/s)
0.03	0.275	2.0	8.05	50.0	9.6
0.04	0.36	2.5	8.72	60.0	9.5
0.05	0.46	3.0	9.10	70.0	9.6
0.06	0.55	4.0	9.88	80.0	9.8
0.08	0.74	5.0	10.5	90.0	10.0
0.10	0.93	6.0	10.8	100.0	10.4
0.12	1.11	8.0	11.6	120.0	11.3
0.15	1.40	10.0	12.0	140.0	11.9
0.20	1.83	12.0	12.6	160.0	12.8
0.30	2.61	15.0	13.0	180.0	13.9
0.40	3.28	17.0	13.2	200.0	14.9
0.50	3.85	20.0	13.1	220.0	15.9
0.60	4.38	25.0	12.5	240.0	17.0
0.80	5.22	30.0	11.3	260.0	18.1
1.0	5.95	35.0	10.7	280.0	19.2
1.2	6.53	40.0	10.2	300.0	20.3
1.5	7.20				

Source: Adapted from Hunter, S. R., J. G. Carter, and L. G. Christophorou, *Phys. Rev.*, 38, 58, 1988.

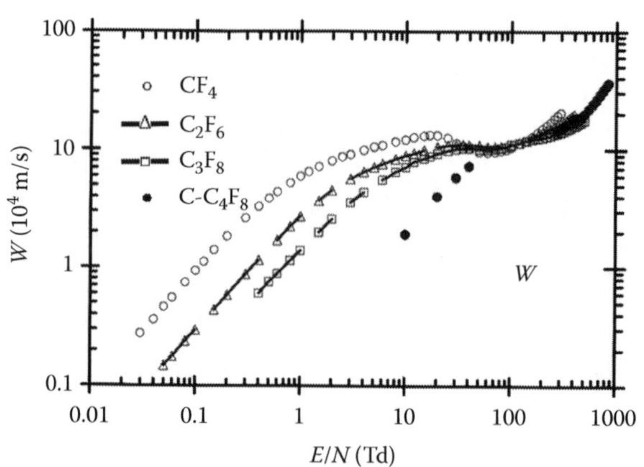

FIGURE 70.13 Drift velocity of electrons in CF_4. Data for selected fluorocarbons plotted for comparison. Note the prominent negative differential conductivity.

TABLE 70.16
D_r/μ Ratios for CF_4

E/N (Td)	D_r/μ (V)	*E/N* (Td)	D_r/μ (V)	*E/N* (Td)	D_r/μ (V)
0.14	0.025	3	0.036	40	1.01
0.17	0.026	4	0.037	45	1.29
0.20	0.027	5	0.039	50	1.58
0.25	0.028	6	0.041	60	2.16
0.30	0.029	8	0.046	70	2.68
0.35	0.030	10	0.052	80	3.12
0.40	0.031	12	0.062	90	3.49
0.50	0.032	15	0.084	100	3.81
0.60	0.033	20	0.155	150	4.78
0.80	0.034	25	0.293	200	5.16
1.00	0.035	30	0.492	250	5.29
2.00	0.035	35	0.736	300	5.39

Source: Adapted from Christophorou L. G. and J. K. Olthoff, *Fundamental Electron Interactions with Plasma Processing Gases*, Kluwer Academic/Plenum Publishers, New York, NY, 2004.

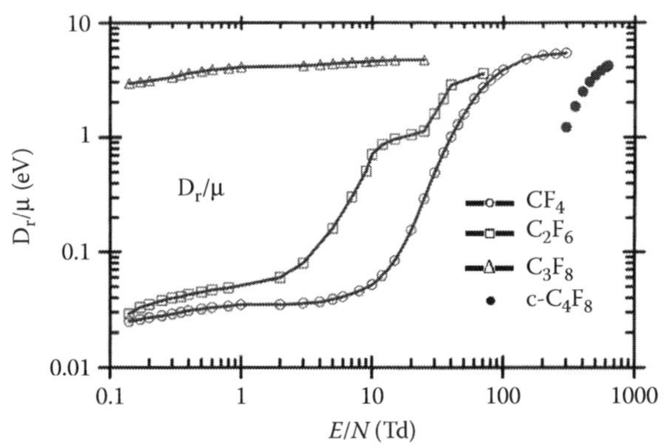

FIGURE 70.14 Ratios of D/μ for CF_4. Data for selected fluorocarbons plotted for comparison. (Adapted from Christophorou, L. G. and J. K. Olthoff, *Fundamental Electron Interactions with Plasma Processing Gases*, Kluwer Academic/Plenum Publishers, New York, NY, 2004.)

70.15 DIFFUSION COEFFICIENTS

Table 70.16 and Figure 70.14 show the radial diffusion coefficients expressed as ratios of mobilities in fluorocarbons.

70.16 IONIZATION AND ATTACHMENT COEFFICIENTS

The ionization and attachment coefficients for CF_4 are shown in Table 70.17 and Figure 70.15.

TABLE 70.17
Ionization and Attachment Coefficients for CF_4

	CF_4		C_2F_6	
	α/N	η/N	α/N	η/N
E/N (Td)	10^{-22} m²		10^{-22} m²	
20		0.0010		
22		0.0027		0.0009
25		0.0134		0.0082
27				0.0264
30		0.093		0.099
35		0.306		0.49
40		0.636		1.40
45		1.05		2.82
50		1.48		4.61
55		1.92		
60		2.38		8.7
70		3.10		12.7
80	0.11	3.64		16.2
90	0.32	4.08		18.8
100	0.44	4.36		20.9
110	1.10	4.50		
120	2.08	4.52		23.6
140	4.45	5.45	0.6	24.5
160	7.37	4.23	1.0	24.3
180	10.7	4.11	3.2	23.3
200	14.7	3.77	5.7	22.0
220	18.3			
240	22.8			
260	27.8			
280	32.9			
300	38.1			
320	43.4			
340	48.8			
360	54.2			
380	59.3			
400	64.2			
420	69.1			
440	74.0			
460	79.1			
480	84.3			
500	89.7			
550	101.4			
600	116.9			

Sources: Adapted from Hunter, S. R., J. G. Carter, and L. G. Christophorou, *J. Chem. Phys.*, 86, 693, 1987; Christophorou, L. G. and J. K. Olthoff, *Fundamental Electron Interactions with Plasma Processing Gases*, Kluwer Academic/Plenum Publishers, New York, NY, 2004.

Note: Data for C_2F_6 also given for comparison.

CF4

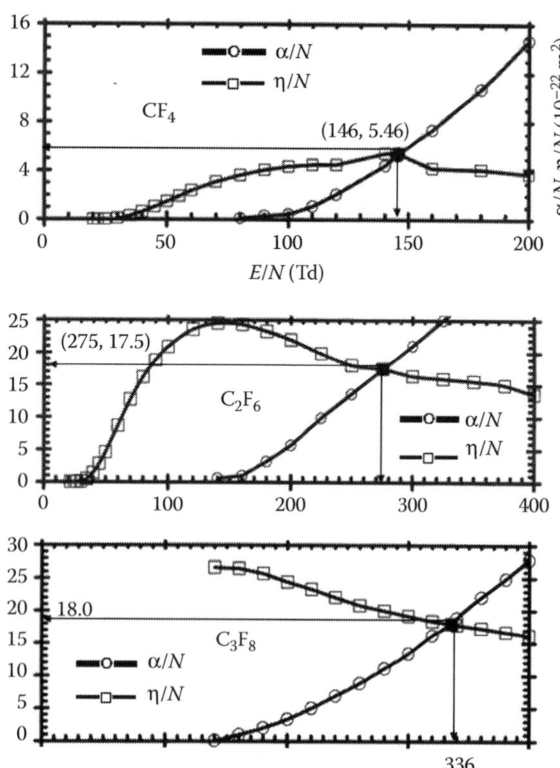

FIGURE 70.15 Ionization and attachment coefficients as a function of E/N in selected fluorocarbons. The limiting E/N at which $\alpha/N = \eta/N$ is also shown. The $(E/N)_{lim}$ determined from high N breakdown studies may deviate from these values by a few percent.

REFERENCES

Allan, M. and L. Andric, *J. Chem. Phys.*, 105, 3559, 1996.

Ariyasinghe, W. M., *Rad. Phys. Chem.* 68, 79, 2003.

Bart, M., P. W. Harland, J. E. Hudson, and C. Vallane, *Phys. Chem. Chem. Phys.*, 3, 800, 2001.

Blaunstein, R. P. and L. G. Christophorou, *J. Chem. Phys.*, 49, 1526, 1968.

Boesten, L., H. Tanaka, A. Kobayashi, M. A. Dillon, and M. Kimura, *J. Phys. B: At. Mol. Opt. Phys.*, 25, 1607, 1992.

Christophorou, L. G. and J. K. Olthoff, *Fundamental Electron Interactions with Plasma Processing Gases*, Kluwer Academic/Plenum Publishers, New York, NY, 2004.

Curtis, M. G., I. C. Walker, and K. J. Mathieson, *J. Phys. D: Appl. Phys.*, 21, 1271, 1988.

Davis, F. J., R. N. Compton, and D. R. Nelson, *J. Chem. Phys.*, 59, 2324, 1973.

Fessenden, R. W. and K. M. Bansal, *J. Chem. Phys.*, 53, 3468, 1970.

Fiegele, T., G. Hanel, I. Torres, M. Lezius, and T. D. Märk, *J. Phys. B: At. Mol. Opt. Phys.*, 33, 4263, 2000.

Harland, P. W. and J. L. Franklin, *J. Chem. Phys.*, 61, 1621, 1974.

Hunter, S. R., J. G. Carter, and L. G. Christophorou, *J. Chem. Phys.*, 86, 693, 1987.

Hunter, S. R., J. G. Carter, and L. G. Christophorou, *Phys. Rev.*, 38, 58, 1988.

Hunter, S. R. and L. G. Christophorou, *J. Chem. Phys.*, 80, 6150, 1984.

Jarvis, G. K., K. J. Boyle, C. A. Mayhew, and R. P. Tuckett, *J. Phys. Chem.* A, 102, 3219, 1998.

Kim, Y.-K., K. K. Irikura, M. E. Rudd, M. A. Ali, P. M. Stone, J. Chang, J. S. Coursey et al., *Electron-Impact Ionization Cross Section for Ionization and Excitation database* (Version 3.0), http://Physics.nist.gov./ionxsec, National Institute of Standards and Technology, Gaithesberg, MD, 2004.

Lisovskiy, V. A. and V. D. Yogorenkov, *J. Phys. D: Appl. Phys.*, 32, 2645, 1999.

Mann, A. and F. Linder, (a) *J. Phys. B: At. Mol. Opt. Phys.*, 25, 545, 1992; (b) *J. Phys. B: At. Mol. Opt. Phys.*, 25, 533, 1992

Mann, A. and F. Linder, (b) *J. Phys. B: At. Mol. Opt. Phys.*, 25, 533, 1992.

Motlagh, S. and J. H. Moore, *J. Chem. Phys.*, 109, 432, 1998.

Nishimura, H., W. M. Huo, M. A. Ali, and Y.-K. Kim, *J. Chem. Phys.*, 110, 3811, 1999.

Pirgov, P. and B. Stefanov, *J. Phys. B: At. Mol. Phys.*, 23, 2879, 1990.

Shimozuma, M., H. Tagashira, and H. Hasegawa, *J. Phys. D: Appl. Phys.*, 16, 971, 1983.

Sieglaff, D. R., R. Rejoub, B. G. Lindsay, and R. F. Stebbings, *J. Phys. B; At. Mol. Opt. Phys.*, 34, 799, 2001.

Spyrou, S. M., I. Sauers, and L. G. Christophorou, *J. Chem. Phys.*, 78, 7200, 1983.

Stefanov, B., N. Paprokova, and L. Zarkova, *J. Phys. B: At. Mol. Opt. Phys.*, 21, 3989, 1988.

Sueoka, O., S. Mori, and A. Hamada, *J. Phys. B: At. Mol. Opt. Phys.*, 27, 1453, 1994.

Sverdlov, L. M., M. A. Kovner, and E. P. Krainov, *Vibrational Spectra of Polyatomic Molecules*, John Wiley & Sons, New York, NY, 1974, p. 397.

Tanaka, H., T. Masai, M. Kimura, T. Nishimura, and Y. Itikawa, *Phys. Rev. A*, 56, R3338, 1997.

Torres, I., R. Martinez, and F. Castãno, *J. Phys. B: At. Mol. Opt. Phys.*, 35, 2423, 2002.

Winstead, C. and V. M. McKoy, *J. Chem. Phys.*, 114, 7407, 2001.

Winters, H. F. and M. Inokuti, *Phys. Rev. A*, 25, 1420, 1982.

Zecca, A., G. P. Kaewasz, and R. S. Brusa, *Phys. Rev.*, 46, 3877, 1992.

71

TRIBROMOFLUORO-METHANE

CBr3

CONTENTS

Tribromofluoromethane (CBr_3F) is a polar, electron-attaching gas that has 120 electrons. Its electronic polarizability is 10.80×10^{-40} F m^2 (Barszczewska et al., 2004).

71.1 ATTACHMENT RATE CONSTANTS

Dissociative attachment of electrons occurs with the formation of both Br^- and Br_2^- ions (Sunagawa and Shimamori, 1997) and the peak of the cross section occurs at zero energy. Table 71.1 gives the attachment rate constants. Figure 71.1 shows the attachment rate constants as a function of mean energy of the electron swarm.

71.2 ATTACHMENT CROSS SECTIONS

Table 71.2 and Figure 71.2 show the attachment cross sections for CBr_3F derived by Sunagawa and Shimamori (1997). *s*-Wave maximum cross section is also shown for comparison.

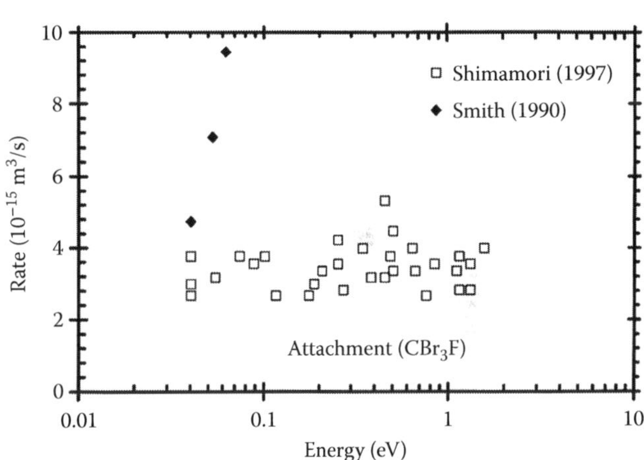

FIGURE 71.1 Attachment rate constants for CBr_3F. (□) Sunagawa and Shimamori (1997); (◆) Smith et al. (1990).

TABLE 71.1
Attachment Rate Constants for CBr_3F

Method	Rate Constant (10^{-15} m^3/s)	Reference
Swarm	4.4	Barszczewska et al. (2004)
PR-MW	3.0	Sunagawa and Shimamori (1997)
FA/LP	4.8 (298 K)	Smith et al. (1990)
	7.4 (380 K)	
	9.6 (475 K)	

Note: PR-MW = pulse radiolysis/microwave cavity; FA/LP = Flowing afterglow/Langmuir probe. Thermal energy unless otherwise specified.

TABLE 71.2
Attachment Cross Sections for CBr_3F

Energy (eV)	Q_a (10^{-20} m^2)	Energy (eV)	Q_a (10^{-20} m^2)
0.001	16.10	0.02	3.37
0.002	11.63	0.04	2.83
0.003	9.14	0.06	2.27
0.004	7.82	0.08	1.84
0.005	7.00	0.10	1.69
0.006	6.47	0.20	1.41
0.008	5.86	0.40	1.00
0.01	5.35	0.60	0.86

continued

TABLE 71.2 (continued)
Attachment Cross Sections for CBr_3F

Energy (eV)	Q_a (10^{-20} m^2)	Energy (eV)	Q_a (10^{-20} m^2)
0.80	0.74	1.50	0.56
1.00	0.65	2.00	0.49
1.20	0.60		

Source: Digitized from Sunagawa, T. and H. Shimamori, *J. Chem. Phys.*, 107, 7876, 1997.

CBr_3F

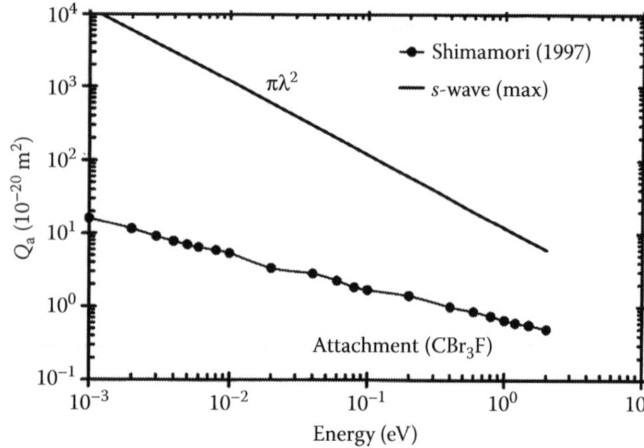

FIGURE 71.2 Attachment cross sections for CBr_3F. (—●—) Sunagawa and Shimamori (1997); (——) *s*-Wave maximum cross section.

REFERENCES

Barszczewska, W., J. Kopyra, J. Wnorowska, and I. Samrej, *Int. J. Mass Spectrom.*, 233, 199, 2004.

Smith, D., C. R. Herd, N. G. Adams, and J. F. Paulson, *Int. J. Mass. Spectrom. Ion Proc.*, 96, 341, 1990.

Sunagawa, T. and H. Shimamori, *J. Chem. Phys.*, 107, 7876, 1997.

72

TRIBROMOMETHANE

CHBr

CONTENTS

Tribromomethane ($CHBr_3$), synonym bromoform, is an electron-attaching, polar molecule having 112 electrons. It has electronic polarizability of 13.13×10^{-40} F m^2, dipole moment of 0.99 D, and ionization potential of 10.48 eV. It has six vibrational modes (Shimanouchi, 1972) as shown in Table 72.1.

72.1 IONIZATION CROSS SECTIONS

Ionization cross sections measured by Bart et al. (2001) are shown in Table 72.2 and Figure 72.1.

72.2 ATTACHMENT RATE COEFFICIENTS

Dissociative attachment produces Br^- ions at 0 eV and peaks at 0.5 eV and the measured attachment energy is 0.85 eV (Modelli et al. 1992).

Thermal attachment rate coefficients measured are 2.86×10^{-14} m^3/s (see Sunagawa and Shimamori, 1997). Table 72.3 and Figure 72.2 show the attachment rate as a function of mean electron energy.

TABLE 72.1
Vibrational Modes for $CHBr_3$

Designation	Type of Motion	Energy (meV)
v_1	CH stretch	377.2
v_2	CBr_3 symmetrical -stretch	67.1
v_3	CBr_3 symmetrical -deform	27.5
v_4	CH bend	142.5
v_5	CBr_3 degenerate -stretch	82.9
v_6	CBr degenerate -deform	19.2

TABLE 72.2
Ionization Cross Sections of $CHBr_3$

Energy (eV)	Q_{Ti} (10^{-20} m^2)	Energy (eV)	Q_{Ti} (10^{-20} m^2)	Energy (eV)	Q_{Ti} (10^{-20} m^2)
12	1.45	82	13.74	153	12.10
16	2.53	86	13.71	156	12.03
20	4.73	90	13.64	160	11.91
23	6.71	94	13.56	164	11.77
27	8.22	97	13.48	167	11.62
31	9.13	101	13.40	171	11.49
34	10.10	105	13.31	175	11.39
38	10.84	108	13.21	179	11.32
42	11.48	112	13.12	182	11.28
45	12.03	116	13.02	186	11.26
49	12.45	119	12.93	190	11.20
53	12.82	123	12.85	193	11.11
57	13.12	127	12.78	197	10.99
60	13.36	130	12.67	201	10.85
64	13.57	134	12.52	204	10.73
68	13.67	138	12.38	208	10.64
71	13.75	142	12.27	212	10.60
75	13.75	145	12.20	215	10.57
79	13.75	149	12.15		

Source: Tabulated values courtesy of Professor Harland.

72.3 ATTACHMENT CROSS SECTIONS

Attachment cross sections derived by Sunagawa and Shimamori (1997) are shown in Table 72.4 and Figure 72.2.

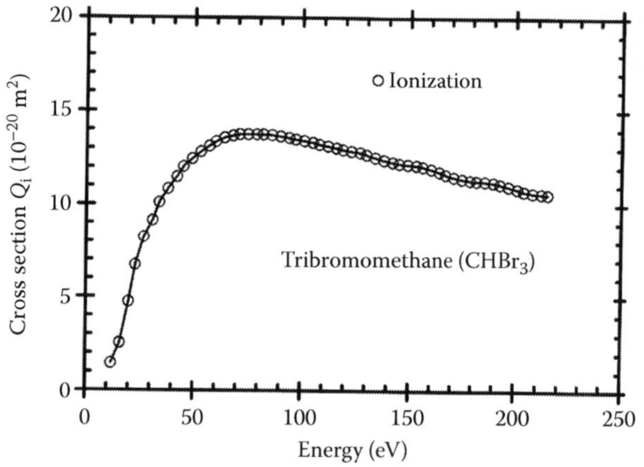

FIGURE 72.1 Ionization cross sections for $CHBr_3$. Tabulated values courtesy of Professor Harland (2006). (Adapted from Bart, M. et al., *Phys. Chem. Chem. Phys.*, 3, 800, 2001.)

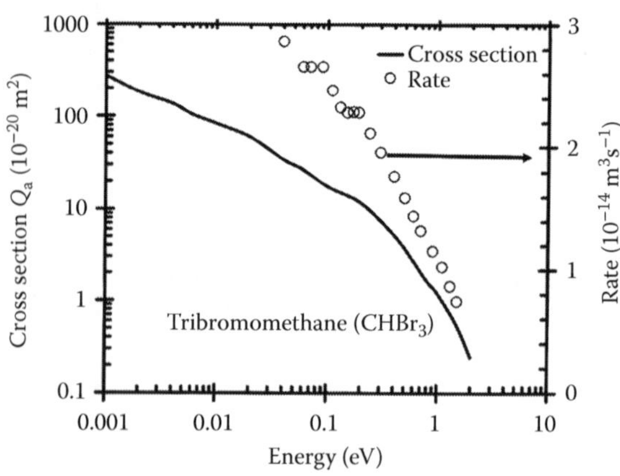

FIGURE 72.2 Attachment rates and cross sections for $CHBr_3$. Note the linear scale for the ordinate for the rate constant. (Adapted from Sunagawa, T. and H. Shimamori, *J. Chem. Phys.*, 107, 7876, 1997.)

TABLE 72.3
Attachment Rate Constants for $CHBr_3$

Mean Energy (eV)	Rate (10^{-14} m³/s)	Mean Energy (eV)	Rate (10^{-14} m³/s)
0.04	2.86	0.30	1.96
0.06	2.65	0.40	1.77
0.07	2.65	0.50	1.59
0.09	2.65	0.60	1.44
0.11	2.46	0.70	1.32
0.13	2.32	0.90	1.15
0.15	2.28	1.1	1.02
0.17	2.28	1.3	0.87
0.19	2.28	1.5	0.74
0.24	2.11		

Source: Digitized from Sunagawa, T. and H. Shimamori, *J. Chem. Phys.*, 107, 7876, 1997.

TABLE 72.4
Attachment Cross Sections for $CHBr_3$

Energy (eV)	Q_a (10^{-20} m²)	Energy (eV)	Q_a (10^{-20} m²)
0.001	274	0.1	17.72
0.002	182	0.2	11.73
0.004	137	0.4	5.24
0.006	105	0.6	2.68
0.01	83.3	0.8	1.62
0.02	60.1	1	1.19
0.04	34.3	1.5	0.54
0.06	26.8	2	0.24

Source: Digitized from Sunagawa, T. and H. Shimamori, *J. Chem. Phys.*, 107, 7876, 1997.

REFERENCES

Bart, M., P. W. Harland, J. E. Hudson, and C. Vallance, *Phys. Chem. Chem. Phys.*, 3, 800, 2001.

Modelli, A. and F. Scagnolari, *J. Chem. Phys.*, 96, 2061, 1992.

Shimanouchi, T. *Tables of Molecular Vibrational Frequencies*, NSRDS-NBS 39, Washington (DC), 1972, p. 55.

Sunagawa, T. and H. Shimamori, *J. Chem. Phys.*, 107, 7876, 1997.

73

TRICHLOROFLUORO-METHANE

CCl$_3$

CONTENTS

Trichlorofluoromethane (CCl_3F) (Freon-11) is a molecule with 66 electrons. It has an electronic polarizability of 10.54×10^{-40} F m^2, a relatively low dipole moment of 0.46 D, and an ionization potential of 11.77 eV. It is strongly electron-attaching. The vibrational energies of the molecule (Sverdlov et al., 1974) are shown in Table 73.1. Vibrations of similar type (similar local motion) tend to have similar energies associated with them (Allan and Andric, 1996).

73.1 SELECTED REFERENCES

See Table 73.2.

TABLE 73.1
Vibrational Energies

Mode	Energy (meV)
v_1	134.5
v_2	66.8
v_3	43.4
v_4	104.9
v_5	49.7
v_6	29.9

Source: Adapted from Sverdlov, L. M., M. A. Kovner, and E. P. Krainov, *Vibrational Spectra of Polyatomic Molecules*, John Wiley and Sons, New York, NY, 1974, p. 394.

TABLE 73.2
Selected References for Data on CCl_3F

Quantity	Energy Range (eV)	Reference
k_a	Thermal	Christophorou and Hadjiantoniou (2006)
Q_i	20–85	**Sierra et al. (2004)**
Q_i	11.8–5000	**Irikura et al. (2003)**
k_a	Thermal	**Lu et al. (2003)**
Q_a	0–5	**Aflatooni et al. (2001)**
Q_a	10^{-4}–4.0	**Klar et al. (2001)**
Q_{el}, Q_M	10–30	Natalense et al. (1999)
Q_a	0–0.2	**Finch et al. (1997)**
k_a	—	**Burns et al. (1996)**
Q_T	10–1000	Jiang et al. (1995)
Q_T	0.01–1.0	**Randell et al. (1993)**
Q_a	0–2	**Shimamori et al. (1992)**
Q_T	75–4000	**Zecca et al. (1992)**
Q_T	0.8–50.0	**Jones (1986)**
Q_a	0–1.0	**Chutjian et al. (1984, 1985)**
Q_a	0–1.0	**Zollars et al. (1984)**
Q_a	0.4–1.3	**McCorkle et al. (1982)**
Q_a	0–7	**Illenberger et al. (1979)**
Q_{ex}	5–20	**King et al. (1978)**
Q_i	11.27–19.07	**Ajello et al. (1976)**
Q_i	20, 35, 70	**Beran and Kevan (1969)**

Note: Bold font denotes experimental study. k_a = Attachment rate; Q_a = attachment; Q_{el} = elastic; Q_{ex} = excitation; Q_M = momentum transfer; Q_i = ionization.

73.2 TOTAL SCATTERING CROSS SECTIONS

Tables 73.3 and 73.4 present the total scattering cross sections in the energy range from 0.8 to 4000 eV. Figures 73.1 and 73.2 are graphical presentations. The highlights of the cross section are

1. The trend toward zero energy is to increase the cross section due to dissociative attachment and formation of Cl^- ions; see Figure 73.3 (Chutjian et al., 1985).
2. A resonance peak is observed at 1.7 eV, probably due to the formation of Cl_2^- ions.
3. A second resonance peak at 4.0 eV, also due to attachment and formation of F^-, CCl_3^-, and Cl_2^- ions.
4. A relatively broad peak in the 8.5–10.5 eV range, as observed in many gases.
5. Monotonic decrease of the cross section for energies >10 eV which is also common to many gases.
6. There is lack of data in the energy region 50–75 eV, and the theoretical curve of Jiang et al. (1995) does not display any sharp variations.

TABLE 73.3
Total Scattering Cross Sections in CCl_3F

Energy (eV)	Q_T (10^{-20} m²)	Energy (eV)	Q_T (10^{-20} m²)	Energy (eV)	Q_T (10^{-20} m²)
Jones (1986)				Zecca et al. (1992)	
0.8	39.75	6.2	50.89	110	31.0
0.9	36.02	6.4	52.17	125	29.9
1.0	34.80	6.6	53.54	150	27.3
1.1	36.33	6.8	54.93	175	26.1
1.2	39.46	7.0	56.18	200	23.9
1.3	43.92	7.5	58.56	225	22.5
1.4	47.67	8.0	59.60	250	21.2
1.5	51.27	8.5	59.97	275	20.7
1.6	53.34	9.0	59.71	300	20.0
1.7	53.99	9.5	59.92	350	17.7
1.8	53.97	10.0	60.09	400	16.7
1.9	53.40	10.5	60.10	450	16.0
2.0	52.57	11.0	59.95	500	14.6
2.1	51.36	11.5	59.60	600	13.0
2.2	49.96	12.0	58.78	700	11.8
2.3	48.66	12.5	57.97	800	10.6
2.4	47.43	13.0	57.00	900	9.65
2.5	45.95	13.5	56.10	1000	9.02
2.6	44.85	14.0	55.12	1100	8.58
2.7	43.85	14.5	54.41	1250	7.72
2.8	42.86	15.0	53.57	1500	6.71
2.9	42.56	16.0	52.18	1750	5.88
3.0	42.46	17.0	50.77	2000	5.17
3.1	42.39	18.0	49.66	2250	4.63
3.2	42.64	19.0	48.62	2500	4.21
3.3	43.24	20.0	47.86	2750	4.01
3.4	43.60	21.0	47.00	3000	3.59
3.5	44.02	22.0	45.90	3250	3.39
3.6	44.42	23.0	45.54	3500	3.19
3.7	44.73	24.0	44.74	4000	2.87
3.8	45.04	25.0	44.23		
3.9	44.98	27.5	42.81		
4.0	44.94	30.0	41.59		
4.2	44.79	32.5	40.51		
4.4	44.79	35.0	39.29		
4.6	44.69	37.5	38.46		
4.8	44.86	40.0	37.66		
5.0	44.97	45.0	36.28		
5.2	45.75	50.0	35.31		
5.4	46.47	75	35.0		
5.6	47.41	80	24.0		
5.8	48.50	90	33.0		
6.0	49.53	100	31.3		

Figure 73.2 shows the expanded total cross section in the low-energy region for the sake of clarity. The cross sections in methane (Ferch et al., 1985) are also added for comparison. It is recalled that CH_4 is nonpolar. The points with reference to this figure are

1. The dipole moment decreases the resonance energy and increases the total cross section.
2. As chlorine and fluorine atoms replace hydrogen atom, the molecule becomes larger, increasing the total cross section.

TABLE 73.4
A Comparison of Total Scattering Cross Sections and Peak Energy with Progressive Substitution of Fluorine Atoms for Chlorine Atoms

Formula	Name (Chapter)	Peak Energy (eV)	Peak Q_T (10^{-20} m²)
CCl_4	Carbon tetrachloride (50)	8.0	62.0
CCl_3F	Trichlorofluoromethane (73)	1.7	53.99
CCl_2F_2	Dichlorodifluoromethane (57)	1.0	39.0
		10.0	50.5
$CClF_3$	Trifluorochloromethane (53)	2.0	31.77
CF_4	Tetrafluoromethane (70)	9.0	21.8

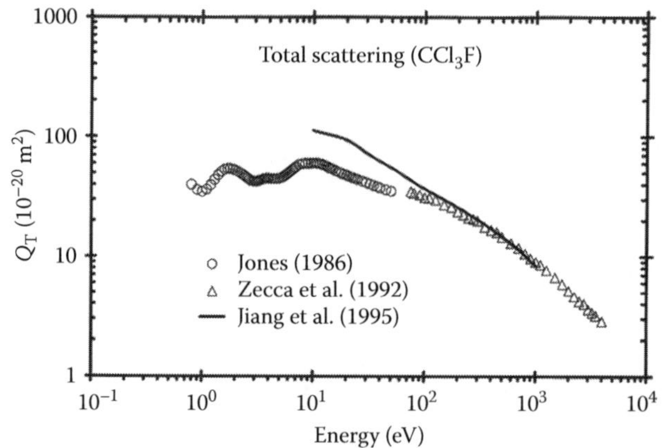

FIGURE 73.1 Total scattering cross sections in CCl_3F. (○) Jones (1986); (Δ) Zecca et al., (1992); (——) Jiang et al. (1995), theory.

73.3 MOMENTUM TRANSFER CROSS SECTIONS

The differential scattering and momentum transfer cross sections calculated by Natalense et al. (Watalense et al., 1999) are shown in Figure 73.3 and Table 73.5.

73.4 POSITIVE ION APPEARANCE POTENTIALS

Curran (1961) has determined the positive ion appearance potentials from electron beam studies as: CCl_2F^+—11.97 eV; CCl_3^+—12.77 eV; $CClF^+$—17.41 eV. Ajello et al. (1976) have determined the threshold energy by photoionization method as shown below

$$CFCl_3 + h\nu \rightarrow CFCl_2^+ + Cl + e; \quad \text{appearance potential} = 11.57\text{ eV} \quad (73.1)$$

$$CFCl_3 + h\nu \rightarrow CCl_3^+ + F + e; \quad 13.25\text{ eV} \quad (73.2)$$

$$CFCl_3 + h\nu \rightarrow CFCl^+ + Cl^- + Cl + e; \quad 16.02\text{ eV} \quad (73.3)$$

$$CFCl_3 + h\nu \rightarrow CFCl^+ + Cl_2 + e; \quad 17.61\text{ eV} \quad (73.4)$$

$$CFCl_3 + h\nu \rightarrow CCl_2^+ + Cl + F + e; \quad 17.12\text{ eV} \quad (73.5)$$

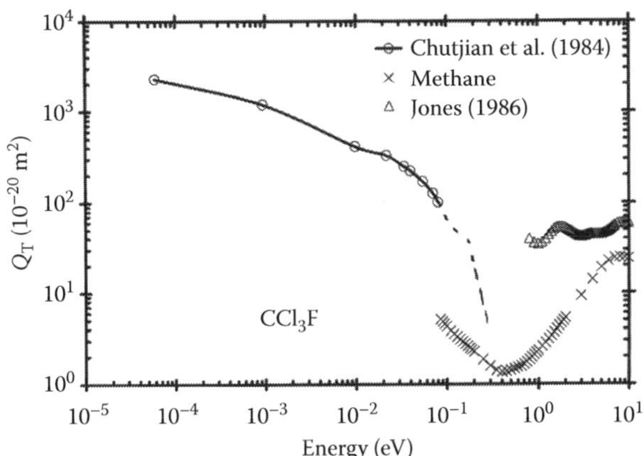

FIGURE 73.2 Total cross sections in CCl_3F. (—○—) Chutjian et al. (1984, 1985); (Δ) Jones (1986); (×) methane, Ferch et al. (1985). The broken line indicates data unavailable.

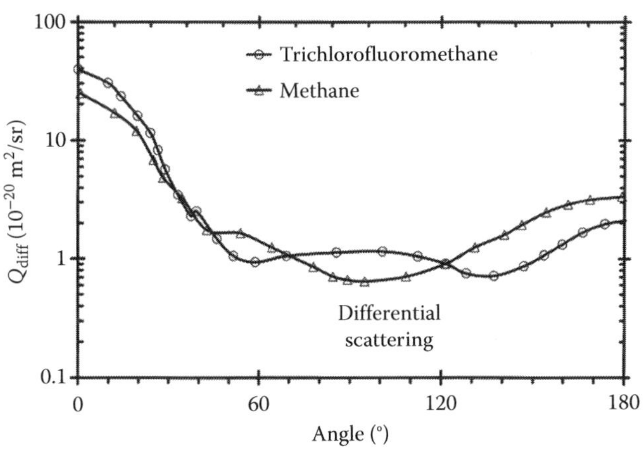

FIGURE 73.3 Differential scattering cross sections in CCl_3F. Methane data added for comparison. (Adapted from Natalense, A. P. P. et al., *Phys. Rev. A.*, 59, 879, 1999.)

TABLE 73.5
Momentum Transfer Cross Sections in CCl_3F

Energy (eV)	10	15	20	25	30
Q_M (10^{-20} m²)	22.80	21.96	18.56	18.33	17.13

Source: Adapted from Natalense, A. P. P. et al., *Phys. Rev. A.*, 59, 879, 1999.

73.5 IONIZATION CROSS SECTIONS

Limited experimental data are available as shown in Table 73.6. Figures 73.4 and 73.5 show the ionization cross sections shown in Table 73.6 and those calculated by Irikura et al. (2003)

TABLE 73.6
Ionization Cross Sections in CCl_3F

	Q_i (10^{-20} m²)	
Energy (eV)	Berran and Kevin (1969)	Sierra et al. (2004)
20	6.26	1.646
25		2.442
30		3.867
35	11.7	4.799
40		5.421
45		5.893
50		6.206
55		6.475
60		6.601
65		6.649
70	14.6	6.870
75		6.994
80		7.063
85		0.198

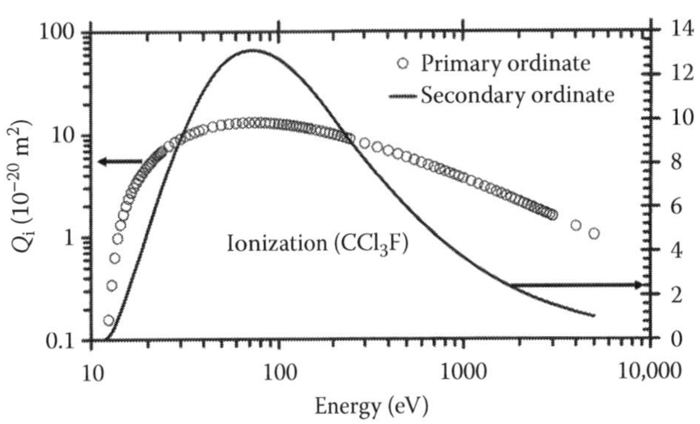

FIGURE 73.4 Ionization cross sections in CCl_3F. Tabulated values are taken from Y.-K. Kim et al. (2004). (Adapted from Irikura, K. K. M. Asgar Ali, and Y. Kim, *Int. J. Mass Spectrom.*, 222, 189, 2003.)

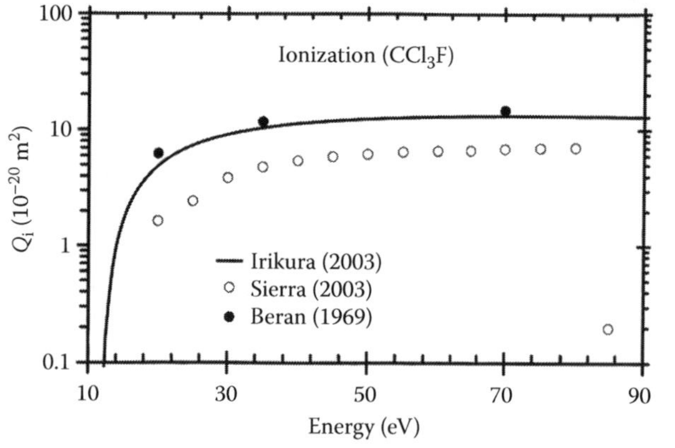

FIGURE 73.5 Ionization cross sections for CCl_3F on an expanded scale. Table 73.6 gives the measured values.

73.6 PARTIAL IONIZATION CROSS SECTIONS

Sierra et al. (2004) have measured the partial ionization cross sections in the energy range $20 \leq \varepsilon \leq 85$ eV as shown in Tables 73.7 for selected species. See Figure 73.6 for graphical presentation.

73.7 ATTACHMENT RATES AND CROSS SECTIONS

Very low-energy attachment including thermal energy dissociative electron attachment occurs with release of energy (Smith et al., 1984) according to the reaction

$$CCl_3F + e \rightarrow CCl_2F + Cl^- + 0.4eV \tag{73.6}$$

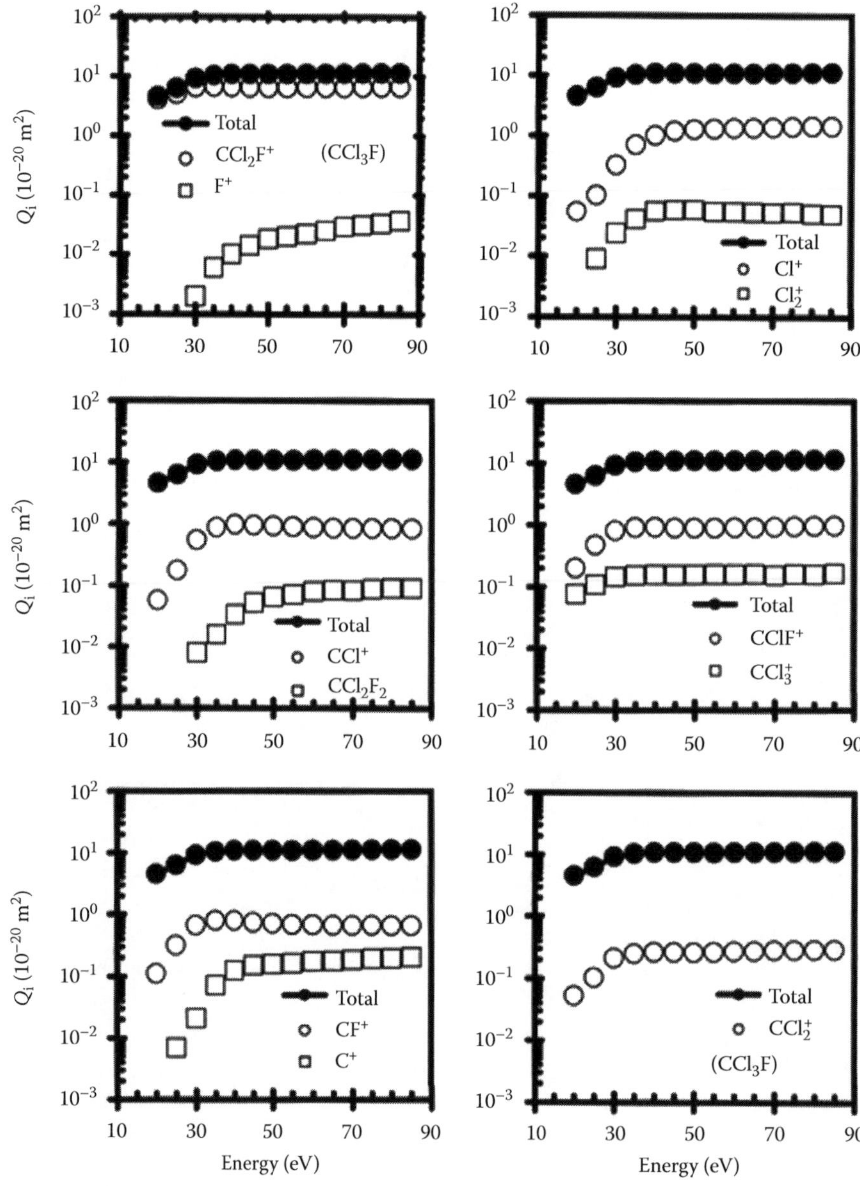

FIGURE 73.6 Partial ionization cross sections for CCl_3F. Note the absence of the parent ion CCl_3F^+. Values are given in Tables 73.7 and 73.8. Note the sequence of tabulation. (Adapted from Sierra, B. et al., *Int. J. Mass Spectrom.*, 235, 223, 2004.)

TABLE 73.7
Partial Ionization Cross Sections for CCl_3F

	Q_i (10^{-20} m^2)		
Energy (eV)	CCl_2F^+	Cl^+	CCl^+
20	4.037	0.055	0.058
25	5.057	0.103	0.176
30	6.524	0.331	0.549
35	6.642	0.718	0.879
40	6.596	1.027	0.988
45	6.461	1.207	0.96
50	6.445	1.291	0.934
55	6.388	1.319	0.903
60	6.449	1.357	0.885
65	6.449	1.371	0.867
70	6.503	1.399	0.858
75	6.549	1.434	0.856
80	6.605	1.458	0.861
85	6.597	1.469	0.849

Source: Adapted from Sierra, B. et al., *Int. J. Mass Spectrom.*, 235, 223, 2004.

TABLE 73.8
Negative Ion Appearance Potentials and Peak Position

	Appearance Potential (eV)		
Ion	Illenberger (1979)	Curran (1961)	Peak Position
Cl^-	0	0	0
F^-	1.6	1.8	3.0
Cl_2^-	0.6		1.6
	≤ 2.5		~ 3.0
CCl_3^-	2.7	2.75	3.3

Selected attachment reactions in chloromethanes in which the energy is released are (Smith et al., 1984)

$$CCl_4 + e \rightarrow CCl_3 + Cl^- + 0.6eV \quad (73.7)$$

$$CCl_2F_2 + e \rightarrow CClF_2 + Cl^- + 0.3eV \quad (73.8)$$

$$CHCl_3 + e \rightarrow CHCl_2 + Cl^- + 0.2eV \quad (73.9)$$

Attachment reactions with energy release in selected halomethanes are (Alge et al., 1984)

$$CH_3Br + e \rightarrow CH_3 + Br^- + 0.36eV \quad (73.10)$$

$$CF_3Br + e \rightarrow CF_3 + Br^- + 0.41eV \quad (73.11)$$

$$CH_2Br_2 + e \rightarrow CH_3Br + Br^- + 0.65eV \quad (73.12)$$

$$CH_3I + e \rightarrow CH_3 + I^- + 0.61eV \quad (73.13)$$

TABLE 73.9
Attachment Rates in CCl_3F as a Function of Electron Energy

McCorkle et al. (1982)		Shimamori et al. (1992)		Klar et al. (2001)	
Energy (eV)	k (m^3/s)	Energy (eV)	k (m^3/s)	Energy (eV)	k (m^3/s)
0.040	2.19×10^{-13}	0.036	19.05×10^{-14}	0.003	6.14×10^{-13}
0.054	1.97	0.038	14.40	0.005	5.23
0.064	1.69	0.069	14.29	0.007	4.78
0.074	1.48	0.095	10.07	0.01	4.07
0.093	1.24	0.11	9.23	0.024	2.90
0.112	1.03	0.12	7.52	0.026	2.82
0.130	090	0.18	4.99	0.03	2.65
0.164	0.74	0.20	5.53	0.04	2.22
0.195	0.62	0.25	3.79	0.06	1.72
0.226	0.54	0.26	2.95	0.08	1.36
0.319	0.38	0.39	2.41	0.1	1.11
0.408	0.29	0.45	2.15	0.2	0.61
0.547	0.20	0.55	1.91	0.3	0.40
0.638	0.15	0.69	1.34	0.4	0.28
0.703	0.12	0.87	1.23	0.5	0.22
0.744	0.11	0.97	0.98×10^{-14}	0.6	0.18
0.778	0.10×10^{-13}			0.7	0.14
				0.8	0.12
				0.9	0.13×10^{-13}

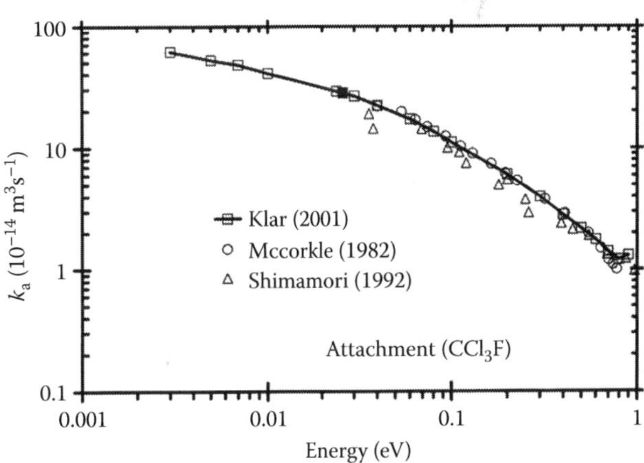

FIGURE 73.7 Attachment rates in CCl_3F as a function of mean energy of electron swarm. (—□—) Klar et al. (2001), symbol ■ shows thermal energy; (Δ) Shimamori et al. (1993); (○) McCorkle et al. (1982).

In addition to the process according to Reaction 1.6 above, dissociative attachments that occur in CCl_3F are

$$CCl_3F + e \rightarrow CCl_2F + Cl^- \quad (73.14)$$

$$CCl_3F + e \rightarrow CClF + Cl_2^- \quad (73.15)$$

TABLE 73.10
Attachment Reactions at Thermal Energy

Attachment Rate ($\times 10^{-13}$ m^3/s)	Method	Reference
2.82	LPA	Klar et al. (2001)
2.4	ECR	Burns et al. (1996)
1.8	PRMC	Shimamori et al. (1992)
3.1	ECR	Christopher et al. (1989)
2.38	CS	Orient et al. (1989)
2.6	FA/LP	Smith et al. (1984)
2.37	CS	Crompton et al. (1982)
2.2	Swarm	McCorkle et al. (1982)
1.0	Swarm	Christophorou and Stockdale (1968) and Blaunstein and Christophorou (1968)

Source: Adapted from Christophorou, L. G. and D. Hadjiantoniou, *Chem. Phys. Lett.*, 419, 405, 2006.

Note: CSM = Cavelleri sampling; ECR = electron cyclotron resonance; FA/LP = flowing afterglow/Langmuir probe; LPA = laser photoelectron attachment; PRMC = pulse radiolysis microwave cavity; recommended value = 2.2×10^{-13} m^{-3}/s.

CCl_3F

TABLE 73.11
Attachment Rate Coefficients at Various Gas Temperatures

Temperature (K)	Rate (10^{-13} m^3/s)	Reference
293	2.4	Burns et al. (1996)
467	1.8	
579	2.1	
777	1.9	
294	2.38	Orient et al. (1989)
404	2.16	
496	2.01	
205	2.2	Smith et al. (1984)
293	2.6	
455	3.6	
590	3.3	

$$CCl_3F + e \rightarrow CCl_3 + F^- \quad (73.16)$$

$$CCl_3F + e \rightarrow CCl_3^- + F \quad (73.17)$$

The appearance potential and peak position of the electron attachment cross section as a function of electron energy are shown in Table 73.8. The largest contribution is due to Cl^- followed by Cl_2^-.

The attachment rates in CCl_3F as a function of energy are shown in Table 73.9. See Figure 73.7 for graphical presentation. Table 73.10 presents thermal attachment cross sections determined by various methods.

Attachment rates at thermal energy measured by various methods are shown in Table 73.10.

Table 73.11 shows the attachment rate coefficients as a function of gas temperature.

TABLE 73.12
Attachment Cross Sections in CCl_3F

Energy (eV)	Q_a (10^{-20} m^2)	Energy (eV)	Q_a (10^{-20} m^2)
0.01	766	0.16	21.8
0.02	496	0.18	22.1
0.03	282	0.20	19.1
0.04	170	0.25	11.3
0.05	98.9	0.30	5.3
0.06	53.5	0.35	2.7
0.07	33.8	0.40	1.8
0.08	25.0	0.50	1.7
0.09	20.7	0.60	1.5
0.10	19.7	0.70	0.6
0.12	18.1	0.80	0.1
0.14	19.9		

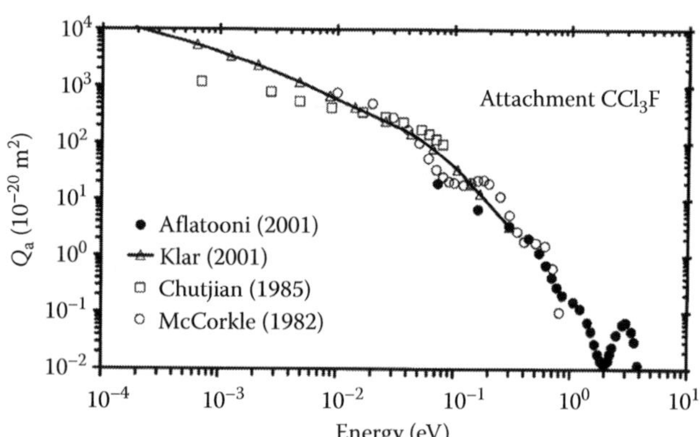

FIGURE 73.8 Attachment cross sections in CCl_3F as a function of electron energy.

TABLE 73.13
Mean Energy of Electrons

E/N (Td)	$\bar{\varepsilon}$ (eV)	*E/N* (Td)	$\bar{\varepsilon}$ (eV)
0.06	0.045	0.6	0.226
0.09	0.054	0.9	0.319
0.12	0.064	1.2	0.408
0.15	0.074	1.8	0.547
0.2	0.093	2.4	0.638
0.25	0.112	3	0.703
0.3	0.13	3.5	0.744
0.4	0.164	4.0	0.778
0.5	0.195		

Source: Adapted from McCorkle, D. L. et al., *J. Chem. Phys.*, 72, 4049, 1982.

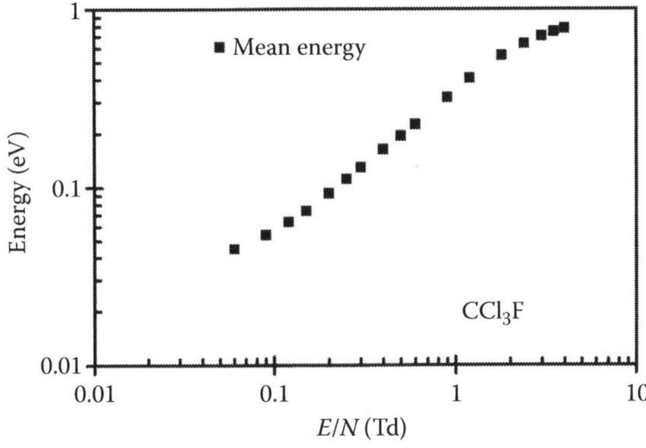

FIGURE 73.9 Mean energy of the electron swarm. Buffer gas is N_2. (Adapted from McCorkle, D. L. et al., *J. Chem. Phys.*, 72, 4049, 1980.)

Attachment cross sections as a function of electron energy (McCorkle et al., 1982) are shown in Table 73.12 and Figure 73.8 which also includes more recent data over the energy range $10^{-4} \leq \varepsilon \leq 10$ eV.

73.8 ELECTRON MEAN ENERGY

Table 73.13 and Figure 73.9 show the mean energy of electron swarm as a function of reduced electric field (*E*/*N*) with N_2 as buffer gas.

REFERENCES

Aflatooni, K. and P. D. Burrow, *Int. J. Mass Spectrom*, 205, 149, 2001.

Ajello, J. M., W. T. Huntress, Jr., and P. Rayermann, *J. Chem. Phys.*, 64, 4746, 1976.

Alge, E., N. G. Adams and D. Smith, *J. Phys. B: At. Mol. Opt. Phys.*, 17, 3827, 1984.

Allan, M. and L. Andric, *J. Chem. Phys.*, 105, 3559, 1996.

Beran, J. A. and L. Kevan, *J. Phys. Chem.*, 73, 3866, 1969.

Burns, S. J., J. M. Matthews, and D. L. McFadden, *J. Phys. Chem.*, 100, 19437, 1996.

Christopher, J., M. C. Tsai, and D. L. McFadden, *J. Chem. Phys.*, 91, 2194, 1989.

Christophorou, L. G. and D. Hadjiantoniou, *Chem. Phys. Lett.*, 419, 405, 2006.

(a) Christophorou, L. G. and J. A. D. Stockdale, *J. Chem. Phys.*, 48, 1956, 1968. The quoted value for attachment rate is cited by Burns et al. (1996). The original paper gives the attachment cross section at the peak (0 eV) as 9.5 × 10–19 m². (b) R. P. Blaunstein and L. G. Christophorou, *J. Chem. Phys.*, 49, 1526, 1968.

Chutjian, A., S. H. Aljajian, J. M. Ajello, and O. J. Orient, *J. Phys. B: At. Mol. Phys.*, 17, L745, 1984. Corrigendum: A. Chutjian, S. H. Aljajian, J. M. Ajello, and O. J. Orient, *J. Phys. B: At. Mol. Phys.*, 18, 3025, 1985.

Crompton, R. W., G. N. Haddad, and R. Hegerberg, *J. Phys. B: At. Mol. Phys.*, L483, 15, 1982.

Curran, R. K. *J. Chem. Phys.*, 34, 2007, 1961.

Ferch, J., B. Granitza, and W. Raith, *J. Phys. B: At. Mol. Phys.*, 18, L445, 1985.

Finch, C. D., R. Parthasarathy, H. C. Akpati, P. Norlander, and F. B. Dunning, *J. Chem. Phys.*, 1.6, 9594, 1997.

Illenberger, E., H.-U. Scheunemann, and H. Baumgärtel, *Chem. Phys.*, 37, 21, 1979.

Irikura, K. K., M. Asgar Ali, and Y. Kim, *Int. J. Mass Spectrom.*, 222, 189, 2003.

Jiang, Y., J. Sun, and L. Wan, *Phys. Rev.*, 52, 398, 1995.

Jones, R. K. *J. Chem. Phys.*, 84, 813, 1986.

King, G. C. and J. W. McConkey, *J. Phys. B: At. Mol. Phys.*, 11, 1861, 1978.

Klar, D., M.-W. Ruf, I. I. Fabrikant, and H. Hotop, *J. Phys. B: At. Mol. Opt. Phys.*, 34, 3855, 2001.

Lu, Q. B. and L. Sanche, *J. Chem. Phys.*, 119, 2658, 2003.

McCorkle, D. L., A. A. Christodoulides, L. G. Christophorou, and I. Szamrej, *J. Chem. Phys.*, 72, 4049, 1980; Erratum in *J. Chem. Phys.*, 76, 753, 1982.

Natalense, A. P. P., M. H. F. Bettega, L. G. Ferriera, and M. A. P. Lima, *Phys. Rev. A.*, 59, 879, 1999.

Orient, O. J., A. Chutjian, R. W. Crompton, and B. Cheung, *Phys. Rev. A*, 39, 4494, 1989.

Randell, J., J.-P. Ziesel, S. L. Lunt, G. Mrotzek, and D. Field, *J. Phys. B: At. Mol. Opt. Phys.*, 26, 3423, 1993.

Shimamori, H., Y. Tatsumi, Y. Ogawa, and T. Sunagawa, *J. Chem. Phys.*, 97, 6335, 1992.

Sierra, B., R. Martínez, C. Redondo, and F. Castaño, *Int. J. Mass Spectrom.*, 235, 223, 2004.

Smith, D., N. G. Adams, and E. Alge, *J. Phys. B: At. Mol. Opt. Phys.*, 17, 461, 1984.

Smith, D., N. G. Adams, and E. Alge, *J. Phys. B: At. Mol. Phys.*, 17, 461, 1984.

Sverdlov, L. M., M. A. Kovner, and E. P. Krainov, *Vibrational Spectra of Polyatomic Molecules*, John Wiley and Sons, New York, NY, 1974, p. 394.

Zecca, A., G. P. Karwasz, and R. S. Brusa, *Phys. Rev.*, 46, 3877, 1992.

Zollars, B. G., K. A. Smith, and F. B. Dunning, *J. Chem. Phys.*, 81, 3158, 1984.

74

TRICHLOROMETHANE

CHCl

CONTENTS

Trichloromethane ($CHCl_3$), synonym chloroform, is an electron-attaching, weakly polar gas with 58 electrons. Two values for electronic polarizability are available in the literature, 10.6×10^{-40} F m^2 and 9.16×10^{-40} F m^2. The dipole moment is 1.04 D and the ionization potential is 11.22 eV. The Cl–$CHCl_2$ bond dissociation energy is 3.32 eV and H–CCl_3 bond dissociation energy is 4.07 eV.

The molecule has six modes of vibrational frequencies as shown in Table 74.1 (Shimanouchi, 1972).

74.1 SELECTED REFERENCES FOR DATA

Selected references for data are shown in Table 74.2.

74.2 TOTAL SCATTERING CROSS SECTIONS

Total cross sections have been measured by Wan et al. (1991) in the range 0–12 eV and Karwasz et al. (1999) in the range 75–3480 eV as shown in Table 74.3 and Figure 74.1.

TABLE 74.1
Vibrational Modes and Energies

Designation	Type of Motion	Energy (meV)
ν_1	CH stretch	376.2
ν_2	CCl_3 s-stretch	84.3
ν_3	CCl_3 s-deform	45.0
ν_4	CH bend	151.3
ν_5	CCl_3 d-stretch	96.0
ν_6	CCl_3 d-deform	32.4

Note: d = degenerate; s = symmetrical.

TABLE 74.2
Selected References

Quantity	Energy (Temperature) eV (K)	Reference
Q_a	0–25	Denifl et al. (2007)
k_a	Thermal	Christophorou and Hadjiantoniou (2006)
k_a	Thermal	Onanang et al. (2006)
k_a	0.28–0.88 (300)	**Tabrizchi and Abedi (2004)**
Q_a	0–2	**Matejčík et al. (2003)**
k_a	Thermal	**Barszczewska et al. (2003)**
k_a	0–2	Skalny et al. (2001)
Q_i	12.5–215.0	**Hudson et al. (2001)**
k_a	(300–600)	**Sunagawa and Shimamori (2001)**
Q_T	75–3480	**Karwasz et al. (1999)**
Q_M	10–30	Natalense et al. (1999)
Q_a	0–2 (300–436)	Matejčík et al. (1997)
k_a	(293–777)	**Burns et al. (1996)**
k_a	(0.03–0.3)	**Spanel et al. (1995)**
k_a	Thermal	**Shimamori et al. (1992)**
Q_T, Q_a	0–12	**Wan et al. (1991)**
Q_a	0–10	**Chu and Burrow (1990)**
k_a	Thermal	**Christopher et al. (1989)**
k_a	Thermal	**Marotta et al. (1989)**
ε_a	0–12	**Benitez et al. (1988)**
k_a	(200–600)	**Smith et al. (1984)**
k_a	Thermal	Christophorou et al. (1981)
Q_a	(300–1200)	**Spence and Schulz (1973)**
k_a	Thermal	**Bansal and Fessenden (1973)**
k_a	Thermal	Blaunstein and Christophorou (1968)

Note: Bold font indicates experimental study. k_a = attachment rate; Q_a = attachment cross section; Q_i = ionization; Q_T = total scattering cross section; ε_a = attachment energy.

CHCl$_3$

TABLE 74.3
Total Scattering Cross Sections

Energy (eV)	Q_T (10^{-20} m^2)	Energy (eV)	Q_T (10^{-20} m^2)	Energy (eV)	Q_T (10^{-20} m^2)
Karwasz et al. (1999)				Wan et al. (1991)	
75	35.2	600	11.5	0.04	35.51
80	34.2	700	10.2	0.27	39.80
90	33.0	800	9.49	0.45	31.84
100	31.9	900	8.69	0.66	24.49
110	30.8	1000	7.91	0.83	22.04
125	28.8	1100	7.33	1.04	24.49
150	26.6	1250	6.86	1.56	32.45
175	24.1	1500	5.87	2.02	34.90
200	22.7	1750	5.14	2.49	33.06
225	20.9	2000	4.75	2.96	31.84
250	19.9	2250	4.36	3.48	28.78
275	18.9	2500	3.84	4.00	30.61
300	18.3	2750	3.47	4.53	27.55
350	16.6	3000	3.14	5.00	26.94
400	15.5	3250	2.91	5.49	28.16
450	14.0	3480	2.89	5.96	28.16
500	13.3			6.48	28.78
				6.98	28.16
				7.47	29.39
				8.00	30.00
				8.46	30.61
				9.04	31.84
				9.48	31.22
				10.00	30.61
				10.47	31.22
				10.99	31.84
				11.49	30.61
				12.04	30.00

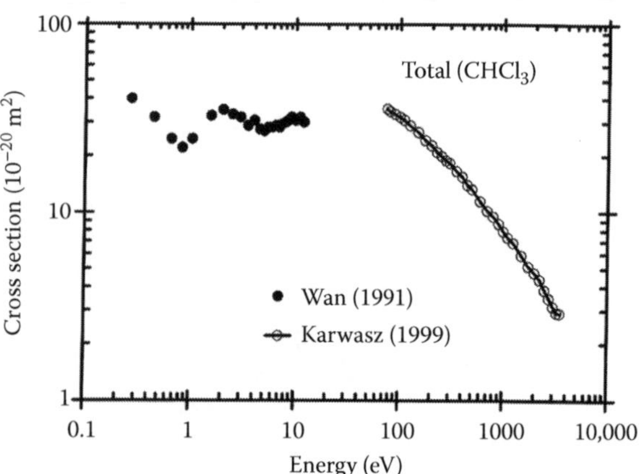

FIGURE 74.1 Total scattering cross sections of CHCl$_3$. (—○—) Karwasz et al. (1999); (●) Wan et al. (1991).

74.3 TOTAL IONIZATION CROSS SECTIONS

Total ionization cross sections have been measured by Hudson et al. (2001) as shown in Table 74.4 and Figure 74.2.

TABLE 74.4
Total Ionization Cross Sections

Energy (eV)	Q_i (10^{-20} m^2)	Energy (eV)	Q_i (10^{-20} m^2)
12	1.80	116	11.79
16	2.73	120	11.70
20	4.68	123	11.61
23	6.25	127	11.48
27	8.01	130	11.35
30	8.95	134	11.20
34	9.86	138	11.07
38	10.51	142	11.00
42	11.04	145	10.96
45	11.45	149	10.89
49	11.86	153	10.78
53	12.25	156	10.67
57	12.37	160	10.49
60	12.53	164	10.31
64	12.66	168	10.17
68	12.73	171	10.10
71	12.73	175	10.06
75	12.74	178	10.04
79	12.73	182	10.03
82	12.69	186	10.00
86	12.60	193	9.83
90	12.50	193	9.83
94	12.39	197	9.69
97	12.29	201	9.55
101	12.15	204	9.45
105	12.03	208	9.35
108	11.96	212	9.32
112	11.88	216	9.27

Note: Tabulated values courtesy of Professor Harland (2006).

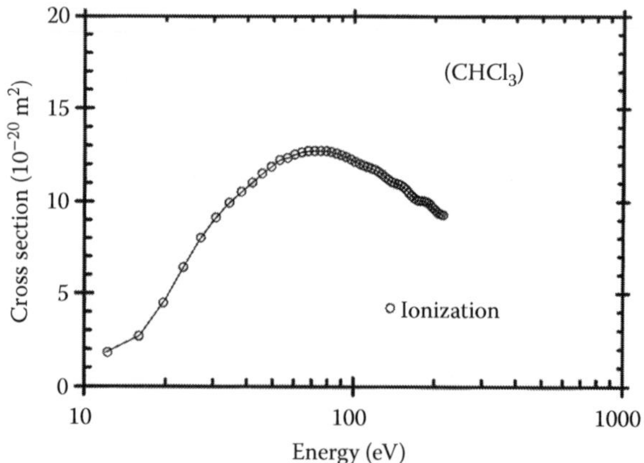

FIGURE 74.2 Ionization cross sections of CHCl$_3$, Hudson et al. (2001). Tabulated values courtesy of Professor Harland.

74.4 ATTACHMENT CROSS SECTIONS

Negative ions identified by beam studies are (Denifl et al., 2007)

$$CHCl_3 + e \rightarrow CHCl_3^{-*} \rightarrow Cl^- + \text{neutral fragments} \quad (74.1)$$

$$CHCl_3 + e \rightarrow CHCl_3^{-*} \rightarrow CCl^- + \text{neutral fragments} \quad (74.2)$$

$$CHCl_3 + e \rightarrow CHCl_3^{-*} \rightarrow CHCl^- + \text{neutral fragments} \quad (74.3)$$

$$CHCl_3 + e \rightarrow CHCl_3^{-*} \rightarrow Cl_2^- + \text{neutral fragments} \quad (74.4)$$

$$CHCl_3 + e \rightarrow CHCl_3^{-*} \rightarrow HCl_2^- + \text{neutral fragments} \quad (74.5)$$

$$CHCl_3 + e \rightarrow CHCl_3^{-*} \rightarrow CCl_2^- + \text{neutral fragments} \quad (74.6)$$

$$CHCl_3 + e \rightarrow CHCl_3^{-*} \rightarrow CHCl_2^- + \text{neutral fragments} \quad (74.7)$$

$$CHCl_3 + e \rightarrow CHCl_3^{-*} \rightarrow CH^- + \text{neutral fragments} \quad (74.8)$$

$$CHCl_3 + e \rightarrow CHCl_3^{-*} \rightarrow C^- + \text{neutral fragments} \quad (74.9)$$

$$CHCl_3 + e \rightarrow CHCl_3^{-*} \rightarrow H^- + \text{neutral fragments} \quad (74.10)$$

Attachment of electrons below 2.0 eV occurs according to the reaction (Matejčík et al., 1997)

$$CHCl_3 + e \rightarrow CHCl_3^{-*} \rightarrow CHCl_2 + Cl^- + 0.16\text{eV} \quad (74.11)$$

The peaks observed are 0 and 0.3 eV for Reaction 74.11, 0 and 1.71 eV for Reaction 74.5, and 1.56 eV for Reaction 74.6.

Figure 74.3 shows the attachment cross sections at low energies measured by the beam method (Matejčík et al., 1997) at various gas temperatures. Cross sections measured by Sunagawa and Shimamori using the microwave heating method are also shown. The essential features are

1. The cross section decreases with increasing electron energy at 300 K.
2. The attachment cross section varies according to ε^{-1} which is characteristic of a polar molecule, as compared with $\varepsilon^{-1/2}$ for a nonpolar molecule (Klar et al., 2001).
3. The cross section increases with increasing temperature, particularly at low energies, < 0.1 eV, attributed to vibrationally excited states. (Sunagawa and Shimamori, 2001) The earlier study of Spence and Schulz (1973) in the temperature range from 290 to 1400 K confirms the increase of attachment cross section with temperature.

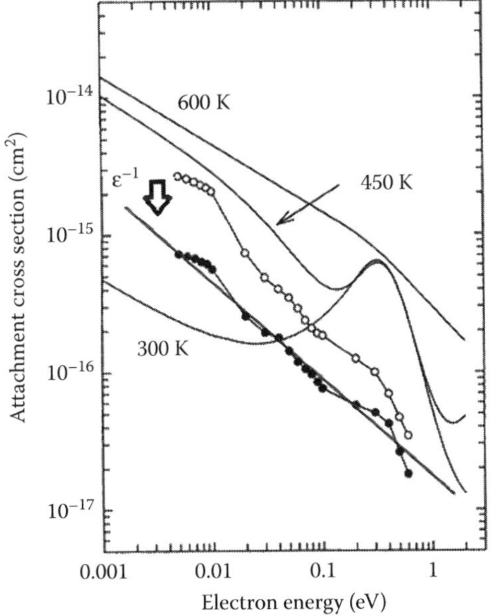

FIGURE 74.3 Attachment cross sections for electrons on $CHCl_3$. (●) Matejčík et al. (1997), 310 K; (○) Matejčík et al. (1997), 436 K; Full lines: Sunagawa and Shimamori (2001) at temperatures shown.

74.5 ATTACHMENT RATE CONSTANTS

See Table 74.5.

TABLE 74.5
Attachment Rate Constants at Various Temperatures (Electron Energy)

Method	Rate (m^3/s)	Temperature K (eV)	Reference
	3.1 (−15)		Christophorou and Hadjiantoniou (2006)
Swarm	**2.7 (−15)**		Barszczewska et al. (2004)
Drift tube	1.29 (−14)	(0.28)	Abrizchi and Abedi (2004)
	1.28 (−14)	(0.36)	
	1.19 (−14)	(0.44)	
	1.12 (−14)	(0.53)	
	9.4 (−15)	(0.61)	
	9.2 (−15)	(0.69)	
	9.4 (−15)	(0.78)	
	7.9 (−15)	(0.88)	
Beam	9.41 (−15)		Gallup et al. (2003)
Microwave heating	2.0 (−15)		Sunagawa and Shimamori (2001)
FA-ECR	4.7 (−15)	293	Burns et al. (1996)
	6.2 (−15)	467	
	1.7 (−14)	579	
	2.3 (−14)	777	
Microwave heating	2.0 (−15)		Shimamori et al. (1992)
ECR	3.8 (−15)		Marotta et al. (1989)
FA-LP	<1.0 (−15)	205	Smith et al. (1984)
	4.4 (−15)	300	
	1.7 (−14)	455	
	3.6 (−14)	590	
	3.76 (−15)		Christophorou et al. (1981)
Microwave conductance	2.0 (−15)		Bansal (1973)

Note: Unless otherwise mentioned the temperature is thermal (~0.026 eV). Bold value indicates the average value. FA-ECR = flowing afterglow-electron cyclotron resonance; ECR = electron cyclotron resonance; FA-LP = flowing afterglow-Langmuir probe. See Figures 74.4 and 74.5 for graphical presentation of rate constants as functions of temperature and mean energy, respectively.

$CHCl_3$

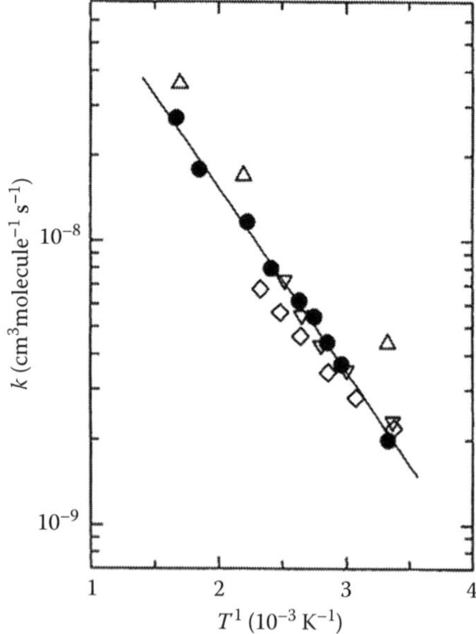

FIGURE 74.4 Attachment rate constants of $CHCl_3$ as a function of temperature. (●) Sunagawa and Shimamori (2001), microwave heating; (Δ) Smith et al. (1984) FA-LP method; (∇) Schultes et al. (1975) ECR method; (◊) Warman and Sauer (1971), microwave method. See Table 74.5 for abbreviations.

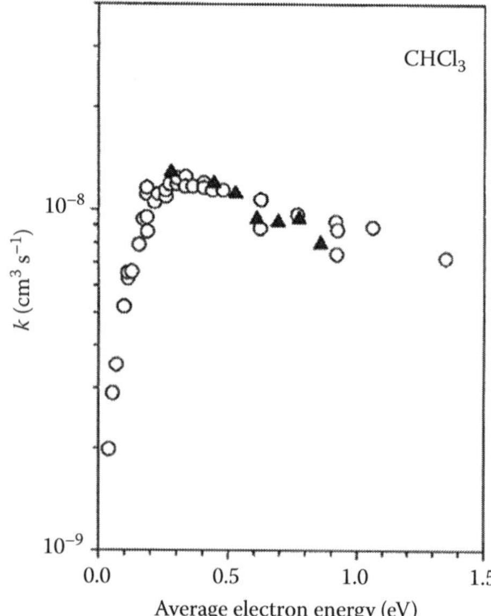

FIGURE 74.5 Attachment rate constant of $CHCl_3$ as a function of average energy of the swarm at 300 K. (▲) Tabrizchi and Abedi (2004); (○) Sunagawa and Shimamori (2001).

REFERENCES

Bansal, K. M. and R. W. Fessenden, *J. Chem. Phys.*, 59, 1760, 1973.

Barszczewska, W., J. Kopyra, J. Wnorowska, and I. Szamrej, *J. Phys. Chem. A*, 107, 11427, 2003.

Barszczewska, W., J. Kopyra, J. Wnorowska, and I. Szamrej, *Int. J. Mass Spectrom.*, 233, 199, 2004.

Benitez, A., J. H. Moore, and J. A. Tossell, *J. Chem. Phys.*, 88, 6691, 1988.

Blaunstein, R. P. and L. G. Christophorou, *J. Chem. Phys.*, 49, 1526, 1968.

Burns, S. J., J. M. Matthews, and D. L. McFadden, *J. Phys. Chem.*, 100, 19437, 1996.

Christopher, J. M., T. Cheng-Ping, and L. M. David, *J. Chem. Phys.*, 91, 2194, 1989.

Christophorou, L. G., R. A. Mathis, D. R. James, and D. L. McCorkle, *J. Phys. D: Appl. Phys.*, 14, 1889, 1981.

Christophorou, L. G. and D. Hadjiantoniou, *Chem. Phys. Lett.*, 419, 405, 2006.

Chu, S. C. and P. D. Burrow, *Chem. Phys. Lett.*, 172, 17, 1990.

Denifl, S., A. Mauracher, P. Sulzer, A. Bacher, T. D. Mark, and P. Scheier, *Int. J. Mass Spectrom.*, 265, 139, 2007.

Gallup, G. A., K. Aflatooni, and P. D. Burrow, *J. Chem. Phys.*, 118, 2562, 2003.

Hudson, J. E., C. Vallance, M. Bart, and P. W. Harland, *J. Phys. B: At. Mol. Opt. Phys.*, 34, 3025, 2001.

Karwasz, G. P., R. S. Brusa, A. Piazza, and A. Zecca, *Phys. Rev. A*, 59, 1341, 1999.

Klar, D., M.-W. Ruf, and H. Hotop, *Int. J. Mass Spectrom.*, 205, 93, 2001.

Marotta, C. J., C.-P. Tsai, and D. L. McFadden, *J. Chem. Phys.*, 91, 2194, 1989.

Matejčík, S., G. Senn, P. Scheire, A. Kindler, A. Stamatovic, and T. D. Mark, *J. Chem. Phys.*, 107, 8955, 1997.

Matejčík, S., V. Foltin, M. Stano, and J. D. Skalny, *Int. J. Mass Spectrometry*, 223, 9, 2003.

Natalense, A. P. P., M. H. F. Bettega, L. G. Ferriera, and M. A. P. Lima, *Phys. Rev. A*, 59, 879, 1999.

Onanong, S., P. D. Burrow, S. D. Comfort, and P. J. Shea, *J. Phys. Chem. A*, 110, 4363, 2006.

Schultes, E., A. A. Christodoulides, and R. N. Schindler, *Chem. Phys.*, 8, 354, 1975, cited by Sunagawa and Shimamori, 2001.

Shimamori, H., Y. Tatsumi, Y. Ogawa, and T. Sunagawa, *J. Chem. Phys.*, 97, 6335, 1992.

Shimanouchi, T. *Tables of Molecular Vibrational Frequencies Consolidated Volume I*, NSRDA-NBS 39, Washington (DC), 1972, p. 54.

Skalny, J. D., S. Matejčík, T. Mikoviny, J. Vencko, G. Senn, A. Stamatovic, and T. D. Märk, *Int. J. Mass Spectrom.*, 205, 77, 2001.

Smith, D., N. G. Adams, and E. Alge, *J. Phys. B: At. Mol. Phys.*, 17, 461, 1984.

Spanel, P., S. Matejčík, and D. Smith, *J. Phys. B: At. Mol. Opt. Phys.*, 28, 2941, 1995.

Spence, D. and G. J. Schulz, *J. Chem. Phys.*, 58, 1800, 1973.

Sunagawa, T. and H. Shimamori, *Int. J. Mass Spectrom.*, 205, 285, 2001.

Tabrizchi, M. and A. Abedi, *J. Phys. Chem. A*, 108, 6319, 2004.

Wan, H.-X., J. H. Moore, and J. A. Tossell, *J. Chem. Phys.*, 94, 1868, 1991.

Warman, J. M. and M. C. Sauer, Jr., *Int. J. Radiat. Phys. Chem.*, 3, 273, 1971, cited by Sunagawa and Shimamori, 2001.

75

TRIFLUOROMETHANE

CHF

CONTENTS

Trifluoromethane (CHF_3), synonym fluoroform, molecule is derived by substitution of three atoms of hydrogen in methane with fluorine atoms. The substitution renders the molecule both polar and electron attaching. It has 34 electrons, polarizability of 3.92×10^{-40} F m^2, dipole moment of 1.65 D, and ionization potential of 13.86 eV. Extensive review of the electron collision with the molecule has been published by Christophorou and Olthoff (2004) and the data provided here are extracted substantially from their analysis. The CF_3–H bond dissociation energy is 4.46 eV (Christophorou and Olthoff, 2004). The vibrational energy and deformations are shown in Table 75.1.

TABLE 75.1
Vibrational Modes of CHF_3

Designation	Energy (meV)	Deformation	Designation	Energy (meV)	Deformation
ν_1	376.4	CH stretch	ν_4	170.1	CH bend
ν_2	138.5	CF_3 s-stretch	ν_5	142.8	CH_3 d-stretch
ν_3	86.8	CF_3 s-deform	ν_6	62.9	CF_3 d-deform

Source: Adapted from Shimanouchi, T. *Tables of Molecular Vibrational Frequencies Consolidated Volume I* NSRDS-NBS, 399, Washington (DC), 1972.

75.1 SELECTED REFERENCES FOR CROSS SECTIONS

See Table 75.2.

75.2 TOTAL SCATTERING CROSS SECTION

Total scattering cross section for CHF_3 recommended by Christophorou and Olthoff (2004) are given in Table 75.3 and

TABLE 75.2
Selected References for Cross Sections

Quantity	Range: eV, (Td)	Reference
A. P.	—	Olivet et al. (2007)
Q_{iz}	20–85	**Torres et al. (2002)**
Q_{iz}	12–215	**Bart et al. (2001)**
Q_{el}	10–30	Natalense et al. (1999)
Q_T	0–20	**Sanabia et al. (1998)**
Q_T	0.8–600	**Sueoka et al. (1998)**
Q_{diff}	1.5–100	**Tanaka et al. (1997)**
Swarm coefficients	(0.1–1000)	Christophorou and Olthoff (2004)

Note: Bold font denotes experimental study. A. P. = appearance potentials; Q_{diff} = differential; Q_{el} = elastic; Q_{iz} = ionization; Q_T = total cross section.

TABLE 75.3
Total Scattering Cross Sections for CHF_3 and CF_3I

Energy (eV)	Q_T (CHF_3) (10^{-20} m²)	Q_T (CF_3I) (10^{-20} m²)	Energy (eV)	Q_T (CHF_3) (10^{-20} m²)	Q_T (CF_3I) (10^{-20} m²)
0.005	3321.0		8.5		52.0
0.01	1661.0		9.0	24.1	51.6
0.02	830.0		9.5		51.7
0.05	332.0		10.0	23.5	51.6
0.10	166.0		11		51.4
0.15	111.0		12		50.6
0.2	83.0		13		49.2
0.3	55.5		14		48.1
0.4	44.4		15.0	21.6	47.6
0.5	38.2		16		46.6
0.6	35.3		17		45.7
0.7	33.3		18		44.8
0.8	31.9	20.3	19		44.8
0.9	31.0		20.0	21.7	43.8
1.0	30.6	21.2	22		42.8
1.2		21.7	25.0	21.5	40.2
1.4		23.7	30.0	21.0	34.9
1.5	29.3		35		32.2
1.6		25.3	40	19.8	30.1
2.0	27.9	27.9	50	19.0	28.5
2.2		29.5	60	18.2	27.4
2.5	26.1	31.7	70	17.3	25.8
2.8		38.4	80	16.5	24.2
3.0	24.9		90	15.6	23.1
3.1		41.2	100	14.8	22.3
3.4		44.1	120		20.7
3.7		46.3	150	12.0	19.4
4.0	24.1	48.0	200	10.4	17.1
4.5		52.8	250		15.9
5.0	24.5	53.0	300	8.38	15.0
5.5		52.7	400	7.03	12.4
6.0	24.9	52.2	500	6.18	10.8
6.5		52.4	600		10.1
7.0	25.0	51.8	1000	3.65	
7.5		51.9	2000	1.84	
8.0	24.7	51.0	3000	1.19	

Figure 75.1. Table 75.3 also includes the total cross section for trifluoroiodomethane (CF_3I) to demonstrate the effect of substituting the hydrogen atom with iodine (Kawada et al., 2000). The highlights of the cross section are

1. A large increase toward zero energy due to the dipole moment of the molecule, possibly due to strong rotational excitation.
2. A minimum at ~3.0 eV and a peak at 7.0 eV.
3. Monotonic decrease toward higher energy up to 3000 eV.
4. Substitution of hydrogen atom with iodine lowers the total cross section <2 eV and for higher energy CF_3I shows higher cross section in addition to a broader peak in the 4–20 eV range.

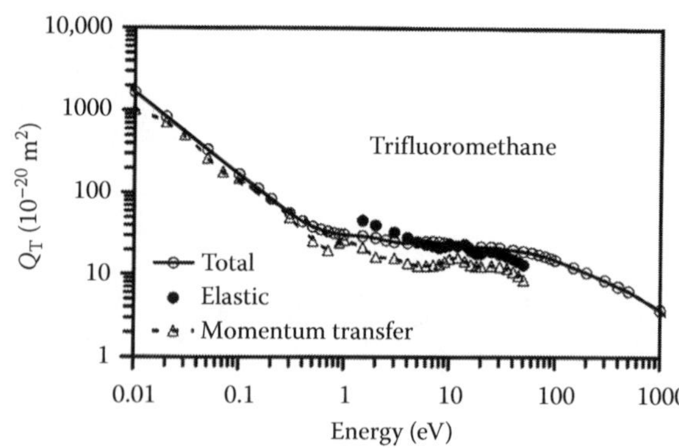

FIGURE 75.1 Total scattering cross section (—○—) in CHF_3. Elastic scattering (●) and momentum transfer (—△—) cross sections are plotted for comparison.

TABLE 75.4
Elastic Scattering and Momentum Transfer Cross Sections in CHF_3

Energy (eV)	Cross Section (10^{-20} m²) Q_{el}	Q_M	Energy (eV)	Cross Section (10^{-20} m²) Q_{el}	Q_M
0.01		1024	6.0	23.18	12.6
0.02		721	7.0	22.23	12.8
0.03		505	8.0	21.09	13.3
0.05		264	9.0	21.86	14.3
0.07		178	10.0	22.64	15.5
0.90		25	12.0	22.38	16.1
0.10		150	14.0	23.24	14.4
0.30		49.5	16.0	19.47	13.0
0.50		25.6	18.0	17.64	12.5
0.70		19.7	20.0	17.69	12.5
1.0		26.4	25.0	18.74	12.7
1.5	45.36	21.4	30.0	17.42	12.4
2.0	39.72	16.3	35.0	16.32	11.7
3.0	32.57	15.7	40.0	15.23	10.70
4.0	27.96	13.6	45.0	14.03	9.60
5.0	24.85	12.6	50.0	12.90	8.50

Source: Digitized from Christophorou, L. G. and J. K. Olthoff, *Fundamental Electron Interactions with Plasma Processing Gases*, Kluwer Academic/Plenum Publishers, New York, NY, 2004, p. 362.

75.3 ELASTIC SCATTERING AND MOMENTUM TRANSFER CROSS SECTIONS

Table 75.4 provides the elastic scattering and momentum transfer cross sections obtained by digitizing the cross sections, given by Christophorou and Olthoff (2004). Figure 75.1 also shows these cross sections.

75.4 DISSOCIATION CROSS SECTION

The molecule CHF_3 dissociates into fragments upon electronic excitation and the total dissociation cross section is equal to the total electronic excitation cross section (Christophorou and Olthoff, 2004). The dissociation cross section in several molecules determined by Winters and Inokuti (1982) are shown in Table 75.5 and Figure 75.2 (Margan et al., 2001).

75.5 POSITIVE ION APPEARANCE POTENTIALS

Positive ion appearance potentials due to electron impact are shown in Table 75.6.

TABLE 75.5
Dissociation Cross Section in CHF_3 and Selected Molecules

Energy	Cross Section (10^{-20} m^2)					
(eV)	CF_4	CF_3H	CH_4	C_2F_6	C_2D_6	C_3F_8
22	1.17	2.4	2.7	3.1	3.5	4.39
30	2.50					
40	3.50					
50	4.30		3.8		7.0	
60						
72	5.20	5.5	4.0	8.1	7.6	11.0
80					7.65	
90					7.6	
100	5.55	5.8	4.0	8.5	7.5	11.6
125		5.7	3.8	8.6	7.3	11.8
150	5.51		3.6		7.1	
175			3.5		6.9	
200	5.32	5.4	3.3	8.1	6.7	11.1
250	5.02		3.0		6.4	
300	4.72	4.9	2.8	7.3	6.0	10.0
350	4.45		2.6		5.6	
400	4.20		2.5		5.3	
450	3.98		2.4		4.9	
500	3.78		2.3		4.6	
550	3.60				4.3	
600	3.45		2.1		4.1	

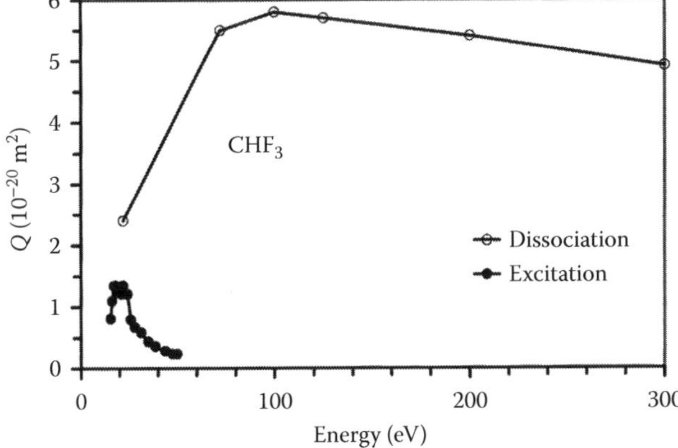

FIGURE 75.2 Excitation and dissociation cross sections in CHF_3. The former ones are derived by simulation. (Adapted from Morgan, W. L., C. Winstead, and V. McKoy, *J. Appl. Phys.*, 90, 2009, 2001; Winters, H. F. and M. Inokuti, *Phys. Rev. A*, 25, 1420, 1982.)

TABLE 75.6
Positive Ion Appearance Potentials due to Electron Impact

Reaction	Potential (eV)	Reference
$CHF_3 + e \rightarrow CHF_3^+$	13.41	Olivet et al. (2007)
$CHF_3 + e \rightarrow CF_3^+ + H$	13.9	Torres et al. (2002)
	14.43	Fiegele et al. (2000)
	15.2	Goto et al. (1994)
	14.42	Lifschitz and Long (1965)
	14.67	Hobrock and Kiser (1964)
	14.53	Farmer et al. (1956)
$CHF_3 + e \rightarrow CHF_2^+ + F$	15.7	Torres et al. (2002)
	15.23	Fiegele et al. (2000)
	16.8	Goto et al. (1994)
	16.4	Hobrock and Kiser (1964)
	15.75	Lifschitz and Long (1965)
$CHF_3 + e \rightarrow CF_2^+ + H + F$	19.5	Torres et al. (2002)
	15.04	Fiegele et al. (2000)
	17.6	Goto et al. (1994)
	20.2	Lifschitz and Long (1965)
	17.5	Hobrock and Kiser (1964)
$CHF_3 + e \rightarrow CHF^+ + 2F$	20.7	Torres et al. (2002)
	19.8	Fiegele et al. (2000)
$CHF_3 + e \rightarrow CF^+ + H + F_2$	20.9	Torres et al. (2002)
	17.89	Fiegele et al. (2000)
	20.9	Goto et al. (1994)
	20.75	Lifschitz and Long (1965)
	20.2	Hobrock and Kiser (1964)
$CHF_3 + e \rightarrow HF^+ + C + F_2$	28.8	Torres et al. (2002)
$CHF_3 + e \rightarrow CH^+ + 3F$	29.9	Torres et al. (2002)
	33.5	Goto et al. (1994)
$CHF_3 + e \rightarrow C^+ + H + 3F$	33.8	Torres et al. (2002)
$CHF_3 + e \rightarrow F^+ + C + H + F_2$	37.0	Torres et al. (2002)
	37.0	Goto et al. (1994)

Source: Adapted from Torres, I., R. Martinez, and F. Castaño, *J. Phys. B: At. Mol. Opt. Phys.*, 35, 2423, 2002.

Note: Stated accuracy for Torres et al (2002) entries is ±0.5 eV for all except the first, for which it is ±0.6 eV.

75.6 IONIZATION CROSS SECTIONS

Ionization cross sections recommended by Christophorou and Olthoff (2004) are shown in Table 75.7 and Figure 75.3. Additional tabulated values from Bart et al. (2001) are given in the subsection on ionization cross sections in dichlorofluoromethane (CH_2Cl_2). Partial ionization cross sections are given by Christophorou and Olthoff (2004) (Figure 75.4).

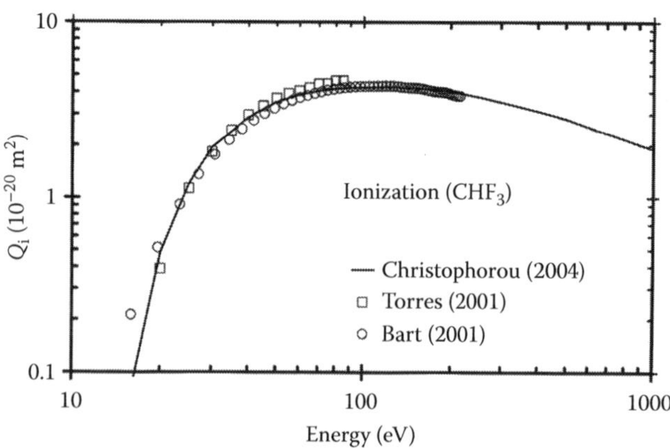

FIGURE 75.3 Ionization cross sections in CHF_3. Only more recent data are shown. (——) Christophorou and Olthoff (2004); (□) Torres et al. (2002); (○) Bart et al. (2001), courtesy of Professor P. Harland. Tabulated vaules are shown in Table 75.7.

75.7 ATTACHMENT CROSS SECTIONS

Parent negative ion formation according to the reaction

$$CHF_3 + e \rightarrow CHF_3^- \tag{75.1}$$

has been observed in Swarm studies only, with a threshold of 0.04 eV (Christophorou and Olthoff, 2004).

Dissociative attachment occurs mainly by the formation of F^- ion due to the following reactions (Christophorou and Olthoff, 2004)

$$CHF_3 + e \rightarrow \begin{cases} CHF_2 + F^- \\ CHF + F + F^- \\ CF_2 + H + F^- \\ CH + 2F + F^- \end{cases} \tag{75.2}$$

Absolute attachment cross sections are not available. Relative cross sections show that dissociative attachment occurs at 4.5, 10.1, and 12.3 eV. The attachment rate coefficient at $E/N < 50$ Td is 13×10^{-20} m^3/s.

75.8 ATTACHMENT RATES

Thermal energy attachment rates are shown in Table 75.8.

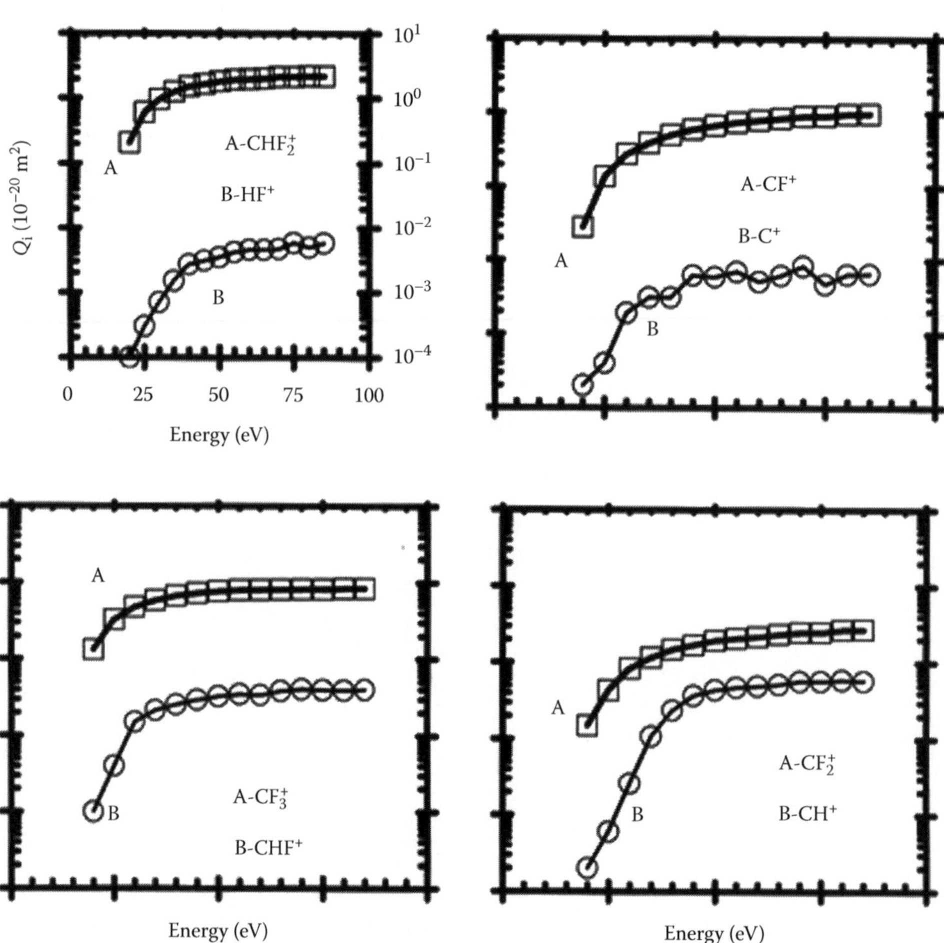

FIGURE 75.4 Partial ionization cross sections in CHF_3 measured by Torres et al. (2002). Ion appearance potentials are given in Table 75.6.

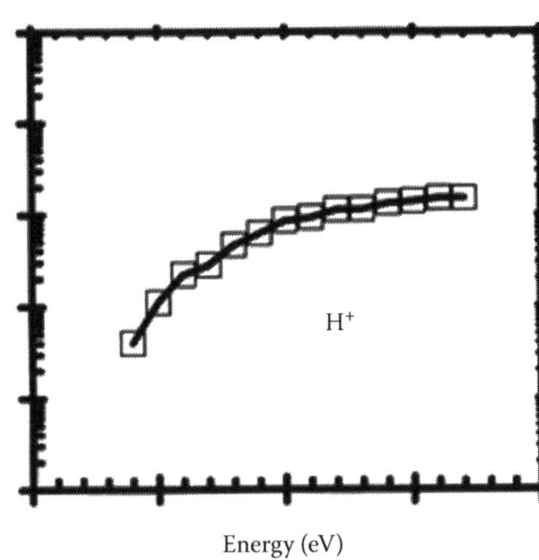

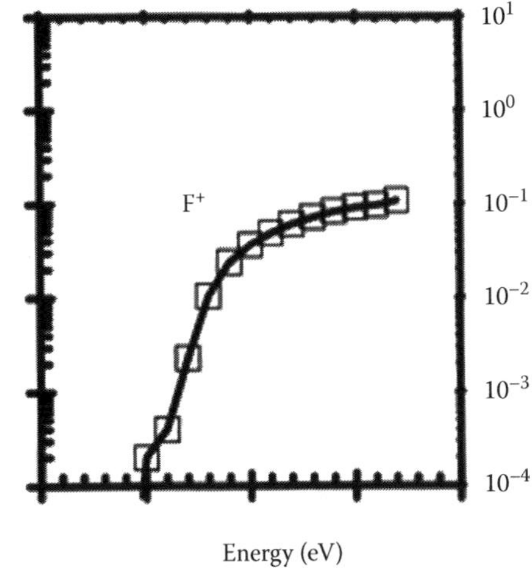

FIGURE 75.4 Continued.

TABLE 75.7
Ionization Cross Sections in CHF_3 Recommended by Christophorou and Olthoff

Energy (eV)	Q_i (10^{-20} m^2)	Energy (eV)	Q_i (10^{-20} m^2)	Energy (eV)	Q_i (10^{-20} m^2)
16	0.080	80	4.18	400	3.12
20	0.480	90	4.24	500	2.82
25	1.20	100	4.26	600	2.54
30	1.94	120	4.27	700	2.34
40	2.86	150	4.22	800	2.17
50	3.51	200	4.05	900	2.02
60	3.87	250	3.77	1000	1.91
70	4.06	300	3.51		

TABLE 75.8
Attachment Rates for CHF_3 at Thermal Energy

Rate (10^{-20} m^3/s)	Method	Reference
4.6	Microwave conductivity	Fessenden and Bansal (1970)
<6.2	Electron Swarm	Davis et al. (1973)
3.6		Christoulides et al. (1978)

Source: Adapted from Christophorou, L. G. and J. K. Olthoff, *Fundamental Electron Interactions with Plasma Processing Gases*, Kluwer Academic/Plenum Publishers, New York, NY, 2004, p. 362.

TABLE 75.9
Drift Velocity of Electrons in CHF_3

E/N (Td)	*W* (10^4 m/s)	*E/N* (Td)	*W* (10^4 m/s)	*E/N* (Td)	*W* (10^4 m/s)
0.40	0.022	5.0	0.208	70	6.92
0.45	0.024	6.0	0.252	80	7.49
0.50	0.026	7.0	0.296	90	8.12
0.60	0.030	8.0	0.342	100	8.66
0.70	0.034	9.0	0.390	150	11.6
0.80	0.038	10.0	0.439	200	14.3
0.90	0.042	15	0.720	250	16.9
1.0	0.046	20	1.09	300	19.6
1.5	0.065	25	1.52	400	24.5
2.0	0.085	30	2.02	500	28.9
2.5	0.105	40	3.38	600	33.0
3.0	0.125	50	4.98		
4.0	0.166	60	6.10		

Source: Adapted from Christophorou, L. G. and J. K. Olthoff, *Fundamental Electron Interactions with Plasma Processing Gases*, Kluwer Academic/Plenum Publishers, New York, NY, 2004, p. 362.

75.9 ELECTRON DRIFT VELOCITY

The drift velocity as a function of *E/N* is shown in Table 75.9 and Figure 75.5.

75.10 DIFFUSION COEFFICIENT

Density normalized longitudinal diffusion coefficients (ND_L) are shown in Table 75.10 and Figure 75.5.

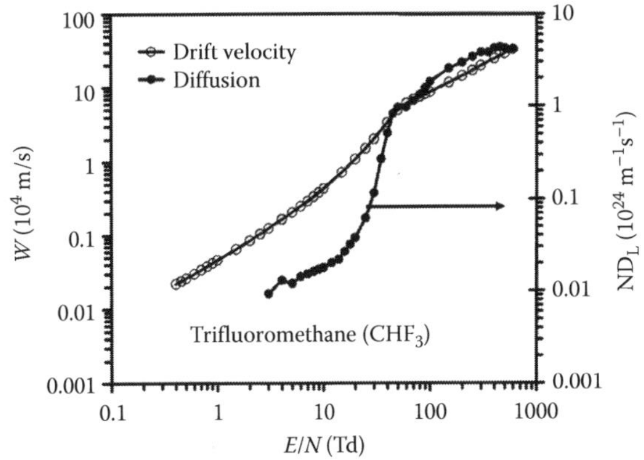

FIGURE 75.5 Drift velocity (Table 75.9) and ND_L (Table 75.10) in CHF_3. (Adapted from Christophorou, L. G. and J. K. Olthoff, *Fundamental Electron Interactions with Plasma Processing Gases*, Kluwer Academic/Plenum Publishers, New York, NY, 2004, p. 362.)

CHF$_3$

75.11 REDUCED IONIZATION COEFFICIENT

Density reduced effective ionization coefficients measured by de Urquijo et al. (1999) are shown in Table 75.11 and Figure 75.6.

75.12 GAS CONSTANTS

The gas constants in the expression is

$$\frac{\alpha_{eff}}{N} = F\exp\left(-\frac{GN}{E}\right) \quad (75.3)$$

determined by de Urquijo et al. (1999) are: $F = 2.6 \times 10^{-20}$ m^2 and $G = 566$ Td^{-1}.

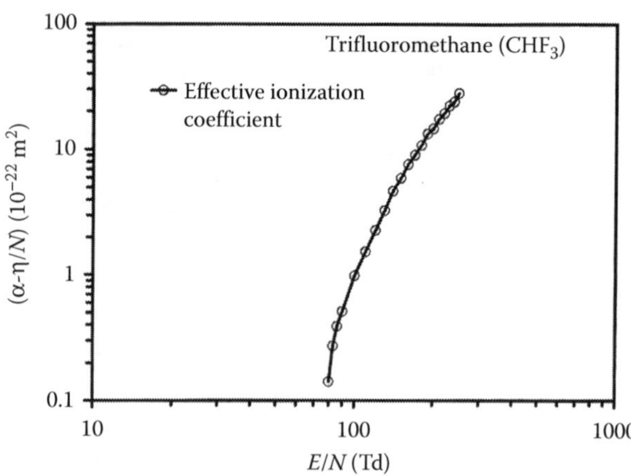

FIGURE 75.6 Effective ionization coefficients in CHF$_3$ measured by de Urquijo et al. (1999). Tabulated values are given in Table 75.11.

TABLE 75.10
Longitudinal Diffusion Coefficients

E/N (Td)	ND$_L$ (10^{24} m^{-1}s^{-1})	*E/N* (Td)	ND$_L$ (10^{24} m^{-1}s^{-1})	*E/N* (Td)	ND$_L$ (10^{24} m^{-1}s^{-1})
3	0.0093	20	0.0375	150	2.54
4	0.0130	25	0.0612	200	2.95
5	0.0120	30	0.116	250	3.40
6	0.0142	35	0.270	300	3.81
7	0.0152	40	0.511	350	3.82
8	0.0162	45	0.826	400	4.24
9	0.0170	50	0.966	450	4.37
10	0.0178	60	0.973	500	4.24
12	0.0201	70	1.132	550	4.09
14	0.0218	80	1.318	600	4.21
16	0.0265	90	1.557		
18	0.0318	100	1.819		

Source: Digitized from Christophorou, L. G. and J. K. Olthoff, *Fundamental Electron Interactions with Plasma Processing Gases*, Kluwer Academic/Plenum Publishers, New York, NY, 2004, p. 362.

TABLE 75.11
Reduced Effective Ionization Coefficients in CHF$_3$

E/N (Td)	(α − η)/*N* (10^{-22} m^2)	*E/N* (Td)	(α − η)/*N* (10^{-22} m^2)	*E/N* (Td)	(α − η)/*N* (10^{-22} m^2)
80	0.142	130	3.27	200	14.6
83	0.273	140	4.68	210	17.3
86	0.391	150	5.93	220	19.3
90	0.512	160	7.61	230	22.1
100	0.985	170	9.04	240	23.7
110	1.53	180	10.8	250	27.7
120	2.27	190	13.3		

TABLE 75.12
Summary of Selected Inelastic Processes with Corresponding Energies

Process	Reaction	Energy (eV)
Molecular ionization	$CHF_3 + e \to CHF_3^+ + 2e$	13.86
Dissociative ionization		
	$CHF_3 + e \to CF_3^+ H + 2e$	13.9
	$CHF_3 + e \to CHF_2^+ F + 2e$	15.7
	$CHF_3 + e \to CF_2^+ HF + 2e$	17.6
	$CHF_3 + e \to CHF^+ + F_2 + 2e$	19.8
	$CHF_3 + e \to CF^+ + HF + F + 2e$	20.9
	$CHF_3 + e \to CH^+ + F_2 + F + 2e$	33.5
Electron attachment		
	$CHF_3 + e \to CHF_3^-$	0.04
	$CHF_3 + e \to CHF_2 + F^-$	2.9
	$CHF_3 + e \to CF_2 + H + F^-$	6.3
	$CHF_3 + e \to CHF + F + F^-$	8.2
	$CHF_3 + e \to CH + 2F + F^-$	12.3
Neutral dissociation		
	$CHF_3 + e \to CF_2 + HF + e$	9.1
	$CHF_3 + e \to CF_3 + H + e$	11.0
	$CHF_3 + e \to CHF_2 + F + e$	13.0
	$CHF_3 + e \to CF + HF + F + e$	14.6
	$CHF_3 + e \to CF_2 + H + F + e$	15.0

75.13 SUMMARY OF SELECTED INELASTIC PROCESSES

See Table 75.12.

REFERENCES

Bart, M., P. W. Harland, J. E. Hudson, and C. Vallance, *Phys. Chem. Chem. Phys.*, 3, 800, 2001.

Christophorou, L. G. and J. K. Olthoff, *Fundamental Electron Interactions with Plasma Processing Gases*, Kluwer Academic/Plenum Publishers, New York, NY, 2004, p. 362.
Christoulides, A. A., R. Schumacher, and R. N. Schindler, *Int. J. Chem. Kinet.*, 10, 1215, 1978, cited by Christophorou and Olthoff, 2004.
Davis, F. J., R. N. Compton, and D. R. Nelson, *J. Chem. Phys.*, 59, 2324, 1973.
Farmer, J. B., I. H. S. Henderson, F. P. Lossing, and D. G. H. Marsden, *J. Chem. Phys.*, 24, 348, 1956.
Fessenden, R. W. and K. M. Bansal, *J. Chem. Phys.*, 53, 3468, 1970.
Fiegele, T., G. Hanel, I. Torres, M. Lezius, and T. D. Märk, *J. Phys. B: At. Mol. Opt. Phys.*, 33, 4263, 2000.
Goto, M., K. Nakamura, H. Toyoda, and H. Sugai, *Jpn. J. Appl. Phys.*, 33, 3602, 1994, cited by Torres et al. (2002).
Hobrock, D. L. and R. W. Kiser, *J. Phys. Chem.*, 68, 575, 1964, cited by Torres et al. (2002).
Kawada, M. K., O. Sueoka, and M. Kimura, *Chem. Phys. Lett.*, 330, 34, 2000.
Lifschitz, C. and F. A. Long, *J. Phys. Chem.*, 69, 3731, 1965, cited by Torres et al. (2002).
Morgan, W. L., C. Winstead, and V. McKoy, *J. Appl. Phys.*, 90, 2009, 2001.
Natalense, A. P. P., M. H. F. Bettega, L. G. Ferriera, and M. A. P. Lima, *Phys. Rev. A*, 59, 879, 1999.
Olivet, A., D. Duque, and L. F. Verga, *J. Appl. Phys.*, 101, 023308, 2007.
Sanabia, J. E., G. D. Cooper, J. A. Tossell, and J. H. Moore, *J. Chem. Phys.*, 108, 389, 1998.
Shimanouchi, T. *Tables of Molecular Vibrational Frequencies Consolidated Volume I* NSRDS-NBS, 399, washington (DC), 1972.
Sueoka, O., H. Takaki A. Hamada, H. Sato, and M. Kimura, *Chem. Phys. Lett.*, 288, 124, 1998.
Tanaka, H., T. Masai, M. Kimura, T. Nishimura, and Y. Itikawa, *Phys. Rev. A*, 56, R3338, 1997.
Torres, I., R. Martinez, and F. Castaíño, *J. Phys. B: At. Mol. Opt. Phys.*, 35, 2423, 2002.
de Urquijo, J., I. Alvarez, and G. Cisneros, *Phys. Rev. E*, 60, 4990, 1999.
Winters, H. F. and M. Inokuti, *Phys. Rev. A*, 25, 1420, 1982.

Section VI

6 ATOMS

76

DIBROMOETHENE

C_2H_2Br

CONTENTS

Dibromoethene ($C_2H_2Br_2$) is an electron-attaching gas that has 78 electrons. There are two isomers of the molecule as shown in Table 76.1.

76.1 ATTACHMENT RATE

Thermal attachment rate constant is 1.7×10^{-14} m^3/s.[1] Figure 76.1 shows the attachment rates as a function of mean energy obtained by the same authors.

TABLE 76.1
Isomers of $C_2H_2Br_2$

Name	Formula
cis-1,2-Dibromoethene	*cis*-1,2-$C_2H_2Br_2$
trans-1,2-Dibromoethene	*trans*-1,2-$C_2H_2Br_2$

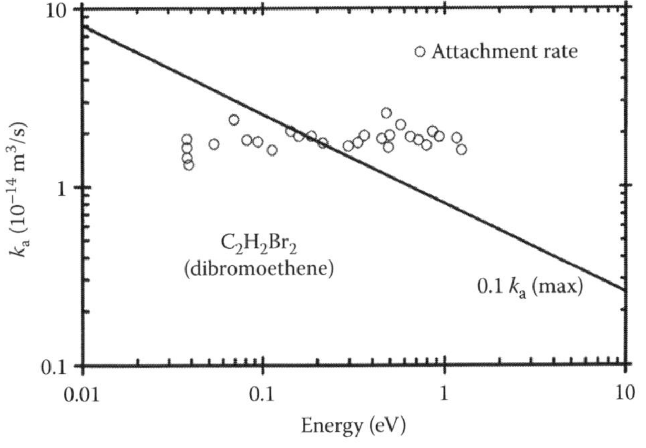

FIGURE 76.1 Attachment rate constant for dibromoethene. *s*-Wave maximum rate constant is also shown for comparison. (Adapted from Sunagawa, T. and H. Shimamori, *Int. J. Mass Spectrom. Ion Proc.*, 149/150, 123, 1995.)

76.2 ATTACHMENT CROSS SECTIONS

Attachment cross sections derived by Sunagawa and Shimamori (1995) from the measurement of attachment rates are shown in Table 76.2 and Figure 76.2.

TABLE 76.2
Attachment Cross Sections for $C_2H_2Br_2$

Energy (meV)	Q_a (10^{-20} m^2)	Energy (meV)	Q_a (10^{-20} m^2)
1.00	83.1	90.0	9.61
2.00	65.7	95.0	9.31
3.00	53.5	100	9.06
4.00	39.9	150	7.95
6.00	34.5	200	7.22
8.00	28.9	250	6.38
10.0	24.6	300	5.68
15.0	22.3	350	5.23
20.0	20.6	400	4.96
25.0	18.8	450	4.80
30.0	17.0	500	4.69
35.0	15.7	550	4.56
40.0	14.8	600	4.40
45.0	14.1	650	4.23
50.0	13.5	700	4.06
55.0	12.9	750	3.90
60.0	12.3	800	3.75
65.0	11.8	850	3.63
70.0	11.3	900	3.51
75.0	10.8	1000	3.33
80.0	10.3	1500	2.87
85.0	9.95	2000	2.24

Source: Digitized from Sunagawa, T. and H. Shimamori, *Int. J. Mass Spectrom. Ion Proc.*, 149/150, 123, 1995.

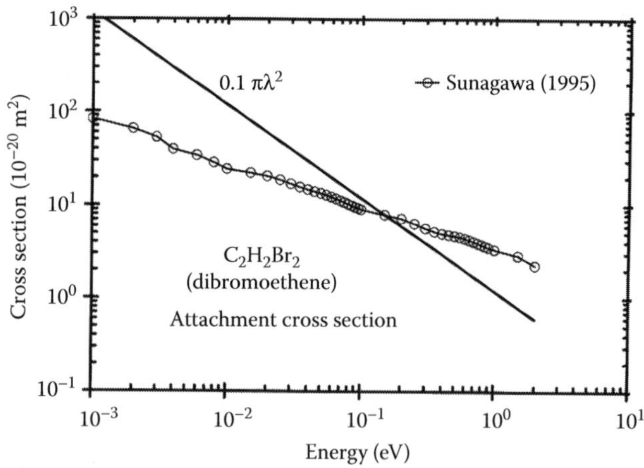

FIGURE 76.2 Attachment cross sections for $C_2H_2Br_2$. *s*-Wave maximum cross section shown for comparison. (Adapted from Sunagawa, T. and H. Shimamori, *Int. J. Mass Spectrom. Ion Proc.*, 149/150, 123, 1995.)

REFERENCE

Sunagawa, T. and H. Shimamori, *Int. J. Mass Spectrom. Ion Proc.*, 149/150, 123, 1995.

77

DICHLOROETHENE

CONTENTS

Dichloroethene ($C_2H_2Cl_2$), also called dichloroethylene, is a polar molecule with 48 electrons and a derivative of ethene (C_2H_4) with two hydrogen atoms replaced with two chlorine atoms. Three isomers are known and their selected properties are shown in Table 77.1.

The three isomers of $C_2H_2Cl_2$ have, each, 12 modes of vibrational excitation as shown in Tables 77.2 and 77.3 (Shimanouchi, 1972).

TABLE 77.1
Selected Properties of Isomers

Name	Formula	α_e	μ	ε_i
1,1-Dichloroethene	1,1-$C_2H_2Cl_2$	8.71	1.34	9.81
cis-1,2-Dichloroethene	*cis*-1,2-$C_2H_2Cl_2$	9.07	1.90	9.66
trans-1,2-Dichloroethene	*trans*-1,2-$C_2H_2Cl_2$	8.93		9.64

Note: α_e = electronic polarizability (10^{-40} F m^2); μ = dipole moment (*D*); ε_i = ionization potential (eV).

TABLE 77.2
Vibrational Modes and Energies of $C_2H_2Cl_2$

	1,1-Dichloroethene			*cis*-1,2-Dichloroethene	
Mode	Type of Motion	Energy (meV)	Mode	Type of Motion	Energy (meV)
ν_1	CH_2 s-stretch	376.3	ν_1	CH stretch	381.5
ν_2	CC stretch	201.7	ν_2	CC stretch	196.8
ν_3	CH_2 scissor	173.6	ν_3	CH bend	146.2
ν_4	CCl_2 s-stretch	74.8	ν_4	CCl stretch	88.2
ν_5	CCl_2 scissor	37.1	ν_5	CCCl deform	21.4
ν_6	Torsion	85.1	ν_6	CH bend	108.6
ν_7	CH_2 a-stretch	388.1	ν_7	Torsion	50.3
ν_8	CH_2 rock	135.8	ν_8	CH stretch	380.9
ν_9	CCl_2 a-stretch	99.2	ν_9	CH bend	161.6
ν_{10}	CCl_2 rock	46.1	ν_{10}	CCl stretch	106.3
ν_{11}	CH_2 Wag	108.5	ν_{11}	CCCl deform	70.8
ν_{12}	CCl_2 Wag	57.0	ν_{12}	CH bend	86.4

Source: Adapted from Shimanouchi, T. *Tables of Vibrational Frequencies Consolidated Volume I*, NSRDS-NBS 39, Washington (DC), 1972, pp. 78–82.

Note: a = asymmetric; s = symmetric.

TABLE 77.3
Vibrational Modes and Energies of *trans*-1,2-Dichloroethene (*trans*-1,2-$C_2H_2Cl_2$)

Mode	Type of Motion	Energy (meV)
ν_1	CH stretch	381.0
ν_2	CC stretch	195.6
ν_3	CH bend	158.0
ν_4	CCl stretch	104.9
ν_5	CCCl deform	43.4
ν_6	CH bend	111.6
ν_7	Torsion	28.1
ν_8	CH bend	94.6
ν_9	CH stretch	383.1
ν_{10}	CH bend	148.8
ν_{11}	CCl stretch	102.7
ν_{12}	CCl deform	31.0

Source: Adapted from Shimanouchi, T. *Tables of Vibrational Frequencies Consolidated Volume I*, NSRDS-NBS 39, Washington (DC), 1972, pp. 78–82.

77.1 IONIZATION CROSS SECTIONS

Ionization cross sections for CHCl=CHCl measured by Hudson et al. (2001) are shown in Table 77.4 and Figure 77.1.

TABLE 77.4
Ionization Cross Sections for CHCl=CHCl

Energy (eV)	Q_i (10^{-20} m^2)	Energy (eV)	Q_i (10^{-20} m^2)
16	2.26	119	13.75
20	5.08	123	13.65
23	7.58	127	13.36
27	9.19	130	13.35
31	10.37	134	13.30
34	11.05	138	13.19
38	11.61	142	13.11
42	12.18	145	12.97
45	12.76	149	12.80
49	13.17	153	12.55
53	13.57	156	12.42
57	13.79	160	12.39
60	14.12	164	12.38
64	14.24	167	12.38
68	14.32	171	12.36
71	14.26	175	12.22
75	14.43	179	12.08
79	14.47	182	11.90
82	14.38	186	11.65
86	14.42	190	11.49
90	14.27	193	11.37
94	14.22	197	11.37
97	14.25	201	11.36
101	14.21	204	11.42
105	14.02	208	11.42
108	13.84	212	11.44
112	13.87	215	11.37
116	13.79	219	11.13

Source: Adapted from Hudson, J. E. et al., *J. Phys. B: At. Mol. Opt. Phys.*, 34, 3025, 2001.

Note: Tabulated values are courtesy of Professor Harland (2006).

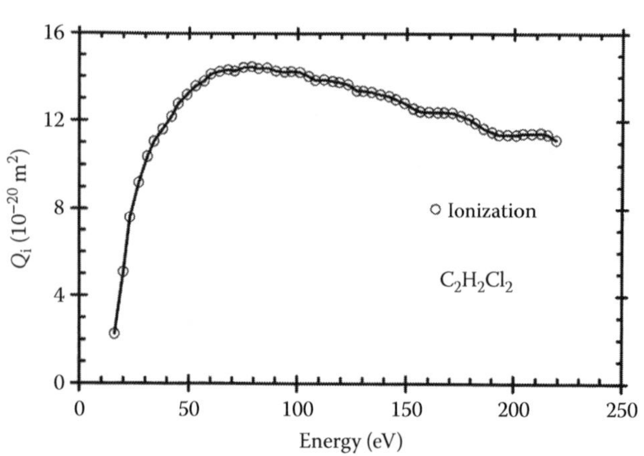

FIGURE 77.1 Ionization cross sections for $C_2H_2Cl_2$. Tabulated values are courtesy of Professor Harland (2006). (Adapted from Hudson, J. E. et al., *J. Phys. B: At. Mol. Opt. Phys.*, 34, 3025, 2001.)

77.2 ATTACHMENT CROSS SECTIONS

Attachment processes for the three isomers of $C_2H_2Cl_2$ have been studied by Johnson et al. (1977). The processes are

$$C_2H_2Cl_2 + e \rightarrow C_2H_2Cl_2^{-*} \rightarrow C_2H_2Cl_2{}^{*} + e \quad (77.1)$$

$$C_2H_2Cl_2 + e \rightarrow C_2H_2Cl_2^{-*} \rightarrow C_2H_2Cl + Cl^- \quad (77.2)$$

$$C_2H_2Cl_2 + e \rightarrow C_2H_2Cl_2^{-*} \rightarrow C_2H_2Cl^- + Cl \quad (77.3)$$

$$C_2H_2Cl_2 + e \rightarrow C_2H_2Cl_2^{-*} \rightarrow C_2H_2 + Cl_2^- \quad (77.4)$$

Reaction 77.1 is autodetachment, and the remaining reactions are dissociative attachment processes. The relative magnitudes of the three species of ions are shown in Figures 77.2 through 77.4 for the three isomers. The structures of the yield curves for the three isomers are shown in Table 77.5.

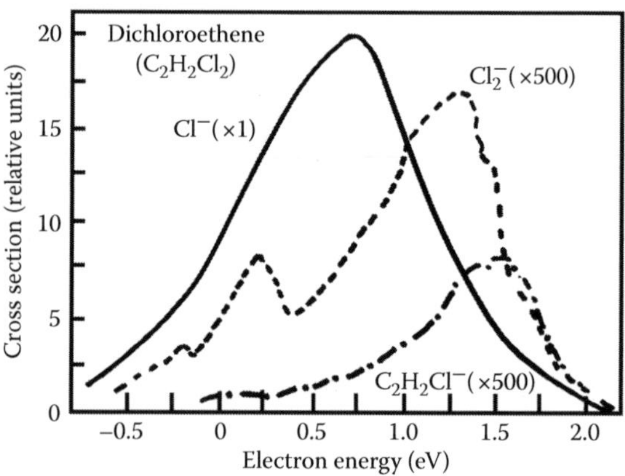

FIGURE 77.2 Relative attachment cross sections due to dissociative attachment for 1,1-dichloroethene ($C_2H_2Cl_2$). Note the multiplication factor for each species. (Adapted from Johnson, J. P., L. G. Christophorou, and J. G. Carter, *J. Chem. Phys.*, 67, 2196, 1977.)

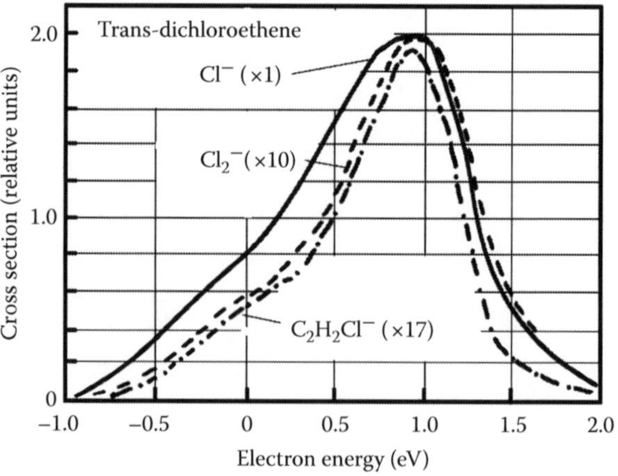

FIGURE 77.3 Relative abundance of ions due to dissociative attachment for *trans*-1,2-dichloroethene ($C_2H_2Cl_2$). Note the multiplication factor for each species. (Adapted from Johnson, J. P., L. G. Christophorou, and J. G. Carter, *J. Chem. Phys.*, 67, 2196, 1977.)

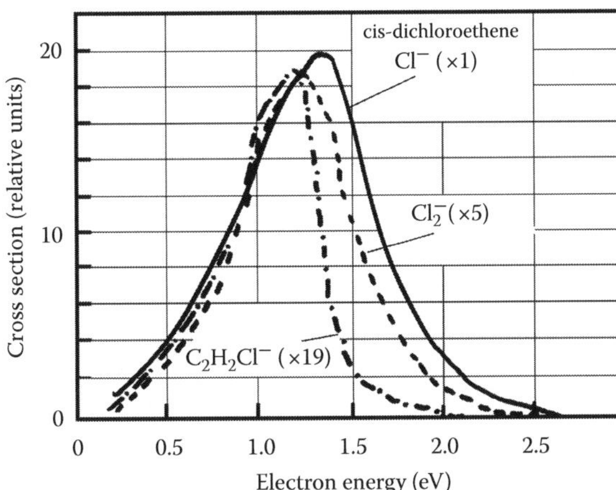

FIGURE 77.4 Relative abundance of ions due to dissociative attachment for *cis*-1,2-dichloroethene ($C_2H_2Cl_2$). Note the multiplication factor for each species. (Adapted from Johnson, J. P., L. G. Christophorou, and J. G. Carter, *J. Chem. Phys.*, 67, 2196, 1977.)

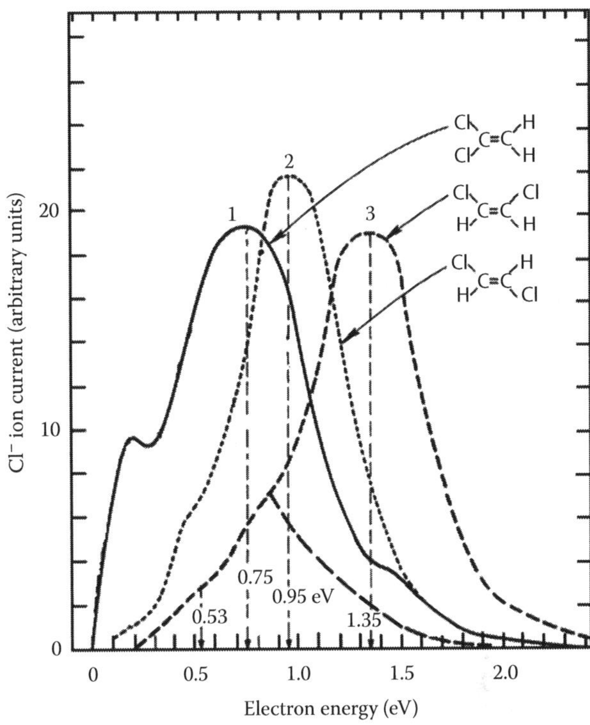

FIGURE 77.5 Comparison of unfolded Cl^- ion current. (1) 1,1-$C_2H_2Cl_2$; (2) *trans*-1,2-$C_2H_2Cl_2$; (3) *cis*-1,2-$C_2H_2Cl_2$. (Adapted from Johnson, J. P., L. G. Christophorou, and J. G. Carter, *J. Chem. Phys.*, 67, 2196, 1977.)

TABLE 77.5
Negative Ion Yield for $C_2H_2Cl_2$ Isomers

Molecule	Ion Species	Peak Position (eV)	Relative Yield	Comments
1,1-Dichloroethene	Cl^-	0.65	1000	Broad peak
	Cl_2^-	~1.2	2	Main peak
	C_2HCl^-	~1.5	1	Main peak
trans-1,2-Dichloroethene	Cl^-	0.85	1000	Main peak
	Cl_2^-	0.95	104	Main peak
	C_2HCl^-	0.95	58	Main peak
cis-1,2-Dichloroethene	Cl^-	1.25	1000	Main peak
	Cl_2^-	1.15	210	Main peak
	C_2HCl^-	1.15	54	Main peak

Source: Adapted from Johnson, J. P., L. G. Christophorou, and J. G. Carter, *J. Chem. Phys.*, 67, 2196, 1977.

Attachment cross sections derived swarm studies for the three isomers of CH_2Cl_2 are shown in Figure 77.5 (Johnson et al., 1977).

REFERENCES

Hudson, J. E., C. Vallance, M. Bart, and P. W. Harland, *J. Phys. B: At. Mol. Opt. Phys.*, 34, 3025, 2001.

Johnson, J. P., L. G. Christophorou, and J. G. Carter, *J. Chem. Phys.*, 67, 2196, 1977.

Shimanouchi, T. *Tables of Vibrational Frequencies Consolidated Volume I*, NSRDS-NBS 39, Washington (DC), 1972, pp. 78–82.

78

ETHYLENE

CONTENTS

Ethylene (C_2H_4), also known as ethene, belongs to the alkene group C_nH_{2n} with 16 electrons. It has an electronic polarizability of 4.73×10^{-40} F m^2, nonpolar dissociation energy of 4.82 eV, and ionization potential of 10.514 eV. The molecule has 12 modes of vibration with the energies shown in Table 78.1 (Herzberg, 1945).

The excitation potentials of the molecule are given in Table 111.2, propylene section. The electronic configuration

TABLE 78.1
Vibrational Modes and Energies of Ethane Molecule

Normal Mode	Vibrational Frequency (meV)	Activity
ν_4	102.3	Inactive
ν_8	116.9	Raman (weak)
ν_7	117.8	IR (very strong)
ν_{10}	123.4	IR (medium)
ν_6	130.2	Raman (not observed)
ν_3	166.4	Raman (very srtrong)
ν_{12}	179.0	IR (strong)
ν_2	201.2	Raman (very strong)
ν_{11}	370.7	IR (strong)
ν_1	374.3	Raman (very strong)
ν_9	385.1	IR (strong)
ν_5	405.7	Raman (very weak)

Note: IR = infrared.

TABLE 78.2
Selected References for Data on C_2H_4

Quantity	Energy Range (eV)	Reference
Q_i	10.51–5000	Y.-Kim et al. (2004)
Q_{el}, Q_M	1–100	**Panajotovic et al. (2003)**
Q_T	0.6–370	**Szmytkowski et al. (2003)**
Q_{vib}	3.2–15.4	**Mapstone et al. (2000)**
Q_{el}	300–1300	Maji et al. (1998)
Q_i	12–12,000	Hwang et al. (1996)
Q_T	0–6	**Lunt et al. (1994)**
Q_i	12–3000	**Nishimura and Tawara (1994)**
Q_{diff}	3–15	**Mapstone and Newell (1992)**
Q_T	4–500	**Nishimura and Tawara (1991)**
Q_T	1.0–400	**Sueoka and Mori (1986)**
Q_T	5–400	**Floeder et al. (1985)**
Q_a	0–15	**Rutkowsky et al. (1980)**
Q_{vib}	1–11	**Walker et al. (1978)**
Q_{diff}	100–1000	**Fink et al. (1975)**
Q_i	600–12,000	**Schram et al. (1966)**
Q_i	10.5–145.0	**Rapp and Englander-Golden (1965)**
Q_{diff}	10–225	**Hughes and Mcmillen (1933)**
Q_T	0.8–50	**Brüche (1929)**
	Swarm Parameters	
W, D_r/μ, D_L/μ	0.02–8.0	**Schmidt and Roncossek (1992)**
α/N	100–6000	**Heylen (1978)**

continued

TABLE 78.2 (continued)
Selected References for Data on C_2H_4

Quantity	Energy Range (eV)	Reference
W	0.06–30	**Bowman and Gorden (1967)**
W	0–1.5	Christophorou et al. (1966)
α/N	120–9000	**Heylen (1963)**

Note: Bold font denotes experimental study. Q_a = attachment; Q_{diff} = differential scattering; Q_{el} = elastic; Q_i = ionization cross section; Q_{vib} = vibrational; D_r/μ = ratio of radial diffusion coefficient to mobility; D_L/μ = ratio of longitudinal diffusion coefficient to mobility; W = drift velocity; α/N = reduced ionization coefficient.

C_2H_4

TABLE 78.3
Total Cross Section in C_2H_4

Energy (eV)	Q_T (10^{-20} m^2)	Energy (eV)	Q_T (10^{-20} m^2)	Energy (eV)	Q_T (10^{-20} m^2)
0.05	4.56	0.60	12.31	2.50	20.77
0.08	6.88	0.80	11.00	3.00	18.95
0.10	8.07	1.00	10.66	4.00	18.80
0.20	10.96	1.50	18.31	5.00	19.67
0.40	12.76	1.95	24.26	6.00	21.24
0.44	13.05	2.00	24.20		

Source: Digitized and interpolated from Lunt, S. L., J. Randell, J. P. Ziesel, G. Mrotzek, and D. Field, *J. Phys. B: At. Mol. Opt. Phys.*, 27, 1407, 1994.

of the ground state of the molecule is given by Dance and Walker. (1973) The electron affinity of the molecule is approximately 0.6 eV.

78.1 SELECTED REFERENCES FOR DATA

See Table 78.2.

78.2 TOTAL SCATTERING CROSS SECTION

The highlights of total scattering cross section are

1. A pronounced low energy peak at 400 meV due to vibrational modes (Lunt et al., 1994)
2. A peak at approximately 2 eV due to vibrational excitation which is an inelastic process
3. A broad shape resonance at approximately 8 eV. An additional peak at the intermediate energy of 6.5 eV observed by Dance and Walker (1973) is not evident in the total cross section.

Total cross sections in C_2H_4 are shown in Tables 78.3 and 78.4. See Figure 78.1 for graphical presentation.

TABLE 78.4
Total Cross Section in C_2H_4

Energy (eV)	Q_T (10^{-20} m^2)	Energy (eV)	Q_T (10^{-20} m^2)	Energy (eV)	Q_T (10^{-20} m^2)
0.6	14.6	4.5	24.3	40	21.7
0.8	14.6	5.0	26.2	45	21.1
1.0	16.8	5.5	27.6	50	20.5
1.2	20.6	6.0	29.0	60	19.2
1.4	23.7	6.5	30.0	70	17.9
1.5	24.9	7.0	31.0	80	17.0
1.6	26.4	7.5	31.6	90	16.0
1.7	27.9	8.0	31.9	100	15.0
1.8	28.8	8.5	31.9	110	14.2
1.9	28.9	9.0	31.7	120	13.4
2.0	28.6	9.5	31.6	140	12.4
2.1	28.1	10	31.6	160	11.5
2.2	27.1	11	31.0	180	10.8
2.4	25.4	12	30.7	200	10.0
2.6	23.5	14	30.1	220	9.40
2.8	21.7	16	29.4	250	8.73
3.0	20.9	18	28.3	275	8.14
3.2	20.3	20	27.4	300	7.77
3.5	20.2	25	25.4	350	6.90
3.7	20.8	30	24.2	370	6.65
4.0	22.2	35	22.8		

Source: Adapted from Szmytkowski, C., S. Kwitnewski, and E. Ptasińska, *Phys. Rev. A*, 68, 032715, 2003.

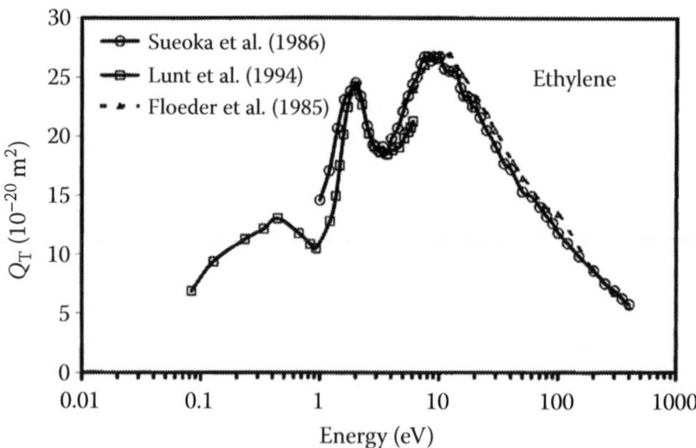

FIGURE 78.1 Total scattering cross section in C_2H_4. (—□—) Lunt et al. (1994); (—○—) Sueoka and Mori (1986); (– – –) Floeder et al. (1985).

78.3 ELASTIC SCATTERING CROSS SECTION

Measured elastic scattering cross sections are shown in Table 78.5 (Panajotovic et al., 2003). These are evaluated from measured differential cross sections such as those shown in Figures 78.2 and 78.3.

78.4 MOMENTUM TRANSFER CROSS SECTION

Limited data on low-energy momentum transfer cross section derived from drift velocity studies are shown in

TABLE 78.5
Elastic Scattering Cross Sections in C_2H_4

Energy (eV)	Q_{el} (10^{-20} m^2)	Energy (eV)	Q_{el} (10^{-20} m^2)	Energy (eV)	Q_{el} (10^{-20} m^2)
1.0	12.7	4.1	18.2	10.1	24.5
1.5	16.3	4.5	17.0	15	23.6
1.8	18.4	4.6	20.1	15	22.8
2.0	21.87	5.0	17.0	20	19.5
2.0	21.3	5.1	22.9	20	18.5
2.2	19.4	6.0	19.2	30	12.3
2.5	18.6	7.5	22.2	60	6.14
3.0	16.3	8.0	23.2	100	3.73
3.1	19.5	8.1	25.6		

Source: Adapted from Panajotovic, R. et al., *J. Phys. B: At. Mol. Opt. Phys.*, 36, 1615, 2003.
Note: See Figure 78.4 for graphical presentation.

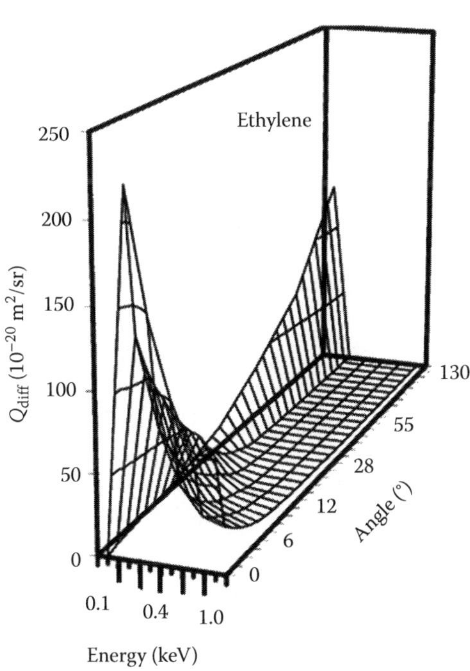

FIGURE 78.2 Differential cross sections in C_2H_4. (Plotted from tabulated data of Fink, M., K. Jost, and D. Hermann, *J. Chem. Phys.*, 63, 1985, 1975.)

Table 78.6. Measured momentum transfer cross sections are shown in Table 78.7 and Figure 78.4.

78.5 RAMSAUER–TOWNSEND MINIMUM

Ramsauer–Townsend minimum occurs at 0.1 eV electron energy. Also see Table 138.4 in the chapter on pentane.

78.6 MERT COEFFICIENTS

See Table 78.8.

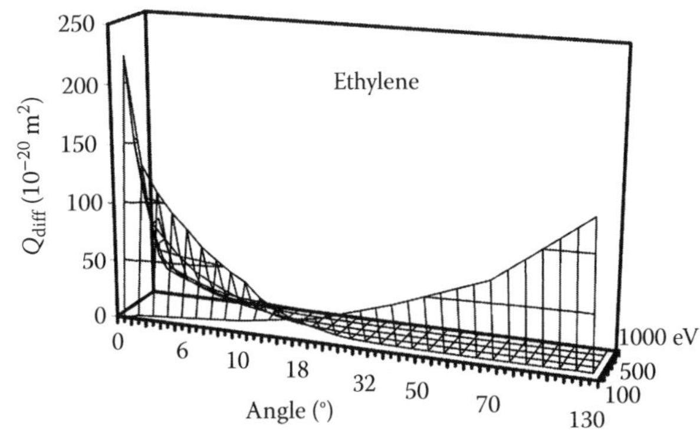

FIGURE 78.3 Differential cross sections in C_2H_4. (Plotted from tabulated data of Fink, M., K. Jost, and D. Hermann, *J. Chem. Phys.*, 63, 1985, 1975.)

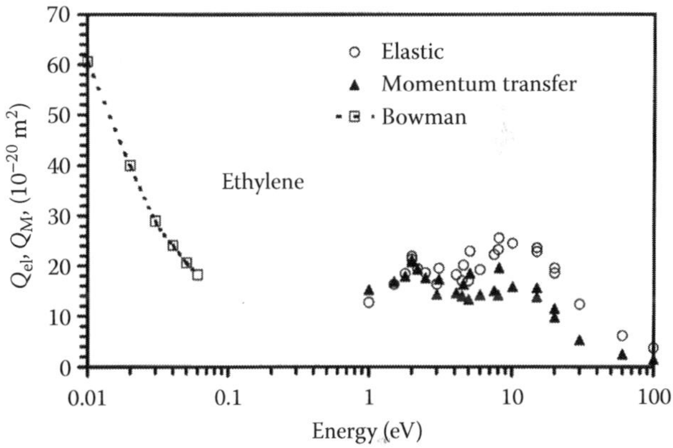

FIGURE 78.4 Integral elastic and momentum transfer cross sections in C_2H_4. (○) Integral elastic scattering; (▲) momentum transfer; (—□—) Bowman and Gordon (1967), momentum transfer cross section derived from drift velocity studies. (Adapted from Panajotovic, R. et al., *J. Phys. B: At. Mol. Opt. Phys.*, 36, 1615, 2003.)

TABLE 78.6
Derived Momentum Transfer Cross Sections in C_2H_4

Energy (eV)	0.01	0.02	0.03	0.04	0.05	0.06
Q_M (10^{-20} m^2)	60.54	39.98	28.94	24.11	20.61	18.23

Source: Adapted from Bowman, C. R. and D. E. Gordon, *J. Chem. Phys.*, 46, 1878, 1967.

78.7 IONIZATION CROSS SECTION

Tables 78.9 through 78.11 present the ionization cross sections in C_2H_4 for the energy range 10.5–12,000 eV. See Figure 78.5 for graphical presentation.

78.8 ATTACHMENT CROSS SECTION

Dissociative attachment process in alkenes occurs according to (Rutkowsky et al., 1980)

TABLE 78.7
Momentum Transfer Cross Sections in C_2H_4

Energy (eV)	Q_{el} (10^{-20} m²)	Energy (eV)	Q_{el} (10^{-20} m²)	Energy (eV)	Q_{el} (10^{-20} m²)
1.0	15.2	4.1	14.6	10.1	15.8
1.5	16.8	4.5	14.0	15	15.6
1.8	17.9	4.6	16.3	15	13.8
2.0	21.15	5.0	13.3	20	11.5
2.0	20.8	5.1	18.5	20	9.8
2.2	19.3	6.0	14.2	30	5.3
2.5	17.6	7.5	15.0	60	2.5
3.0	14.3	8.0	14.1	100	1.5
3.1	17.3	8.1	19.6		

Source: Adapted from Panajotovic, R. et al., *J. Phys. B: At. Mol. Opt. Phys.*, 36, 1615, 2003.
Note: See Figure 78.4 for graphical presentation.

TABLE 78.8

Scattering Length (A)	Method	Reference
6.19×10^{-10} (m)	Cross section measured	Lunt et al. (1994)
1.11×10^{-9} (m)	Drift velocity measured	Bowman and Gordon (1967)

TABLE 78.9
Ionization Cross Section in C_2H_4

Energy (keV)	0.6	1	2	4	7	12
Q_i (10^{-20} m²)	2.31	1.58	0.921	0.517	0.335	0.207

Source: Adapted from Schram, B. L. et al., *J. Chem. Phys.*, 14, 49, 1966.

TABLE 78.10
Total Ionization Cross Section in C_2H_4

Energy (eV)	Q_i (10^{-20} m²)	Energy (eV)	Q_i (10^{-20} m²)	Energy (eV)	Q_i (10^{-20} m²)
12	0.097	80	5.76	500	2.45
15	0.581	90	5.79	600	2.19
17.5	1.15	100	5.70	700	1.96
20	1.71	125	5.58	800	1.75
25	3.01	150	5.2	900	1.63
30	3.52	175	4.8	1000	1.52
35	4.17	200	4.58	1250	1.28
40	4.52	250	3.92	1500	1.11
45	4.82	300	3.56	1750	1.03
50	5.11	350	3.18	2000	0.908
60	5.48	400	2.87	2500	0.767
70	5.74	450	2.64	3000	0.678

Source: Adapted from Nishimura, H. and H. Tawara, *J. Phys. B: At. Mol. Opt. Phys.*, 27, 2063, 1994.

TABLE 78.11
Total Ionization Cross Section in C_2H_4

Energy (eV)	Q_i (10^{-20} m²)	Energy (eV)	Q_i (10^{-20} m²)	Energy (eV)	Q_i (10^{-20} m²)
10.5	0.01	21	1.94	65	5.73
11	0.05	21.5	2.04	70	5.81
11.5	0.09	22	2.14	75	5.88
12	0.14	22.5	2.23	80	5.92
12.5	0.20	23	2.32	85	5.94
13	0.27	23.5	2.40	90	5.95
13.5	0.35	24	2.49	95	5.94
14	0.44	26	3.01	100	5.91
14.5	0.54	28	3.31	105	5.89
15	0.65	30	3.59	110	5.85
15.5	0.77	32	3.83	115	5.82
16	0.88	34	4.06	120	5.77
16.5	0.98	36	4.26	125	5.72
17	1.08	38	4.43	130	5.67
17.5	1.18	40	4.59	135	5.62
18	1.29	45	4.90	140	5.53
18.5	1.40	50	5.17	145	5.50
19	1.52	55	5.39		
19.5	1.62	60	5.58		

Source: Adapted from Rapp, D. and P. Englander-Golden, *J. Chem. Phys.*, 43, 1464, 1965.

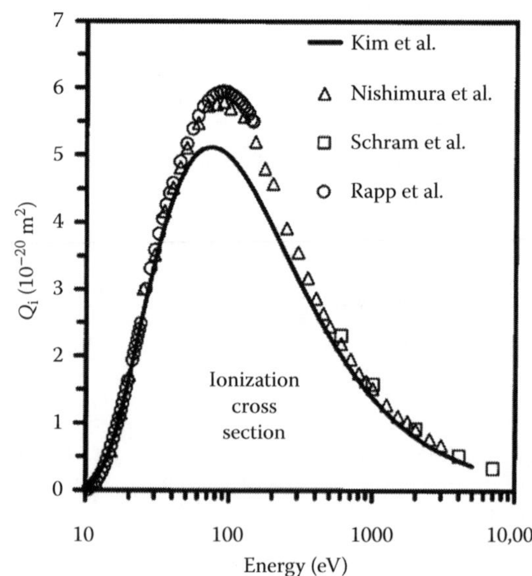

FIGURE 78.5 Total ionization cross section in C_2H_4. (——) Kim et al. (2004); (△) Nishimura and Tawara (1994); (□) Schram et al. (1966); (○) Rapp and Englander-Golden, 1965.

$$\text{e(between 5 and 7 eV)} + C_nH_{2n} \rightarrow C_nH_{2n}^{-*} \rightarrow \begin{cases} C_nH_{2n}^{-} \rightarrow H^- + C_nH_{2n-1} \\ CH^- + C_{n-1}H_{2n-1} \\ CH_2^- + C_{n-1}H_{2n-2} \end{cases} \quad (78.1)$$

$$e(>7eV) + C_nH_{2n} \rightarrow C_nH_{2n}^{-*} \rightarrow \begin{cases} C_2H^- + C_{n-2}H_{2n-1} \\ C_2H_3^- + C_{n-2}H_{n-3} \end{cases} \quad (78.2)$$

with additional process in C_2H_4 occurring according to

$$e\,(> 9eV) + C_2H_4 \rightarrow C_2H_4^{-*} \rightarrow \begin{cases} C_2H^- + H_2 + H \\ H^- + C_2H_3 \end{cases} \quad (78.3)$$

The dissociative attachment cross section in C_2H_4 is shown in Table 78.12 and the graphical presentation in Figure 78.6.

78.9 DRIFT VELOCITY OF ELECTRONS

The density-reduced mobility of electrons at low values of reduced electric field are shown in Table 78.13 (Bowman and Gordon, 1967).

TABLE 78.12
Dissociative Attachment Cross Section in C_2H_4

Energy (eV)	Q_a (10^{-24} m^2)	Energy (eV)	Q_a (10^{-24} m^2)
8.80	0.00	10.25	4.53
9.00	0.32	10.50	1.05
9.25	0.65	10.75	0.69
9.50	0.96	11.00	0.05
9.75	2.25	11.20	0.01
10.00	4.85		

Source: Adapted from Rutkowsky, V. J., H. Drost, and H.-J. Spangenberg, *Ann. Phys. Leipzig*, 37, 259, 1980. Digitized and interpolated by author.

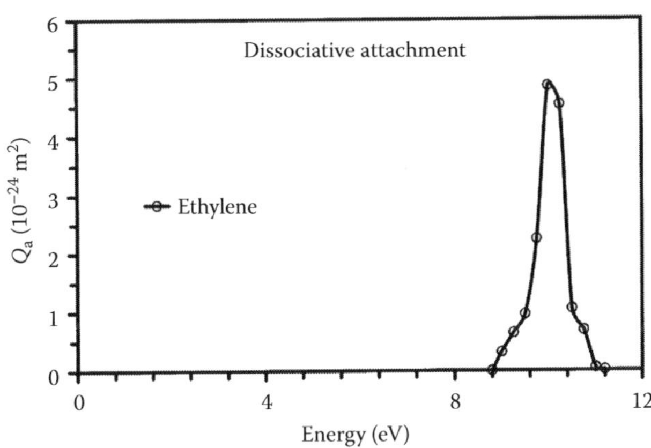

FIGURE 78.6 Dissociative attachment cross section in C_2H_4. Note the low value. (Adapted from Rutkowsky, V. J., H. Drost, and H.-J. Spangenberg, *Ann. Phys., Leipzig*, 37, 259, 1980.)

TABLE 78.13
Reduced Mobility in C_2H_4 at $N = 3.22 \times 10^{22}$ m^{-3}

Temperature (K)	μN ($\times 10^{25}$ m^{-1} V^{-1} s^{-1})
229	2.10
298	2.93
367	3.22

Drift velocity of electrons in C_2H_4 at low values of *E*/*N* is shown in Tables 78.13 and 78.14. Data for selected hydrocarbons are also included for comparison. Further data are also shown in Table 78.14 and the graphical presentation is shown in Figure 78.7. See Table 78.15 also.

78.10 DIFFUSION COEFFICIENTS

Longitudinal and radial diffusion coefficients (D_L and D_r) expressed as ratio of mobility (D/μ) are shown in Table 78.16 and Figure 78.7. Drift velocity is shown as additional data.

TABLE 78.14
Drift Velocity of Selected Hydrocarbons (Digitized) at Low Values of *E*/*N*

	Drift Velocity (10^4 m/s)				
E/*N* (Td)	C_2H_4	C_3H_8	C_4H_{10}	C_5H_{12}	C_6H_6
0.30	0.93	0.55	0.43	0.31	0.09
0.45		0.78		0.39	0.12
0.60	1.28	1.04	0.78	0.52	0.11
0.90	1.55	1.43	1.08	0.75	0.16
1.20	2.06	1.74	1.32	0.98	0.21
1.50	2.58	1.93	1.55	1.13	0.26

Source: Adapted from Christophorou, L. G., G. S. Hurst, and A. Hadjiantoniou, *J. Chem. Phys.*, 44, 3506, 1966.

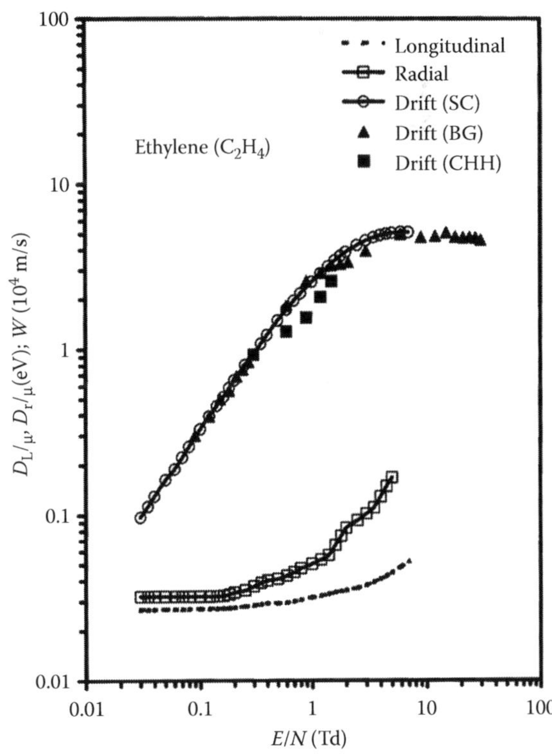

FIGURE 78.7 Swarm parameters in C_2H_4. (- - -) D_L/μ (Schmidt and Roncossek, 1992); (—□—) D_r/μ (Schmidt and Roncossek, 1992); Drift velocity: (—○—) (Schmidt and Roncossek, 1992); (▲) Bowman and Gordon (1967); (■) Christophorou et al. (1966).

C_2H_4

TABLE 78.15
Drift Velocity of Electrons in C_2H_4

E/N (Td)	W (10^3 m/s)	E/N (Td)	W (10^3 m/s)	E/N (Td)	W (10^3 m/s)
0.09	2.99	0.60	18.23	9.00	47.87
0.12	3.91	0.90	25.78	12.00	48.52
0.15	4.99	1.20	28.90	15.00	50.86
0.18	5.64	1.50	32.22	18.00	48.03
0.21	6.88	1.80	32.69	21.00	47.59
0.24	7.56	2.10	33.66	24.00	47.71
0.27	8.43	3.00	39.57	27.00	47.45
0.30	9.34	6.00	49.68	30.00	45.96

Source: Adapted from Bowman, C. R. and D. E. Gordon, *J. Chem. Phys.*, 46, 1878, 1967.

TABLE 78.16
Diffusion Coefficients and Drift Velocity as Functions of E/N in C_2H_4 at 297 K

E/N (Td)	D_L/μ (eV)	D_r/μ (eV)	W (10^4 m/s)
0.0300	0.0268		0.097
0.0350	0.0268		0.113
0.0400	0.0268		0.130
0.0500	0.0269	0.0321	0.163
0.0600	0.0269	0.0321	0.189
0.0700	0.0269	0.0321	0.222
0.0800	0.0270	0.0321	0.257
0.100	0.0271	0.0321	0.328
0.120	0.0272	0.0321	0.393
0.140	0.0272	0.0321	0.455
0.160	0.0273	0.0321	0.515
0.180	0.0274	0.0321	0.584
0.200	0.0275	0.0322	0.651
0.250	0.0280	0.0324	0.807
0.300	0.0284	0.0330	0.931
0.350	0.0288	0.0339	1.08
0.400	0.0292	0.0349	1.22
0.500	0.0293	0.0370	1.49
0.600	0.0295	0.0387	1.73
0.700	0.0300	0.0400	1.96
0.800	0.0306	0.0410	2.17
1.00	0.0317	0.0428	2.55
1.20	0.0325	0.0449	2.88
1.40	0.0332	0.0474	3.18
1.60	0.0340	0.0504	3.46
1.80	0.0345	0.0536	3.69
2.00	0.0349	0.0569	3.89
2.50	0.0361	0.0656	4.28
3.00	0.0376	0.0746	4.57
3.50	0.0393	0.0837	4.77
4.00	0.0410	0.0929	4.91
4.50	0.0428	0.102	5.01
5.00	0.0447	0.111	5.07
6.00	0.0485	0.129	5.13
7.00	0.0523	0.149	5.12
8.00		0.168	

Source: Adapted from Schmidt, B. and M. Roncossek, *Aust. J. Phys.*, 45, 351, 1992.

78.11 FIRST IONIZATION COEFFICIENT

Reduced first ionization coefficient as a function of E/N (Heylen, 1963) is shown in Table 78.17. For graphical presentation see Figure 78.8 in chapter on butane.

78.12 GAS CONSTANTS

Gas constants according to the expression

$$\frac{\alpha}{N} = F\exp\left(-\frac{GN}{E}\right) \tag{78.4}$$

are shown in Table 78.18. Graphical presentation of Equation 78.4 is shown in Figure 78.8 for selected hydrocarbons.

TABLE 78.17
Reduced Ionization Coefficients in C_2H_4

E/N (Td)	α/N (10^{-20} m²)	E/N (Td)	α/N (10^{-20} m²)
		900	1.610
125	0.0031	1000	1.883
150	0.0095	1250	2.539
175	0.0202	1500	3.037
200	0.0359	2000	3.652
225	0.0588	2500	4.115
250	0.0886	3000	4.516
300	0.158	4000	5.143
350	0.247	5000	5.547
400	0.351	6000	5.832
450	0.457	7000	6.089
500	0.576	8000	6.332
600	0.836	9000	6.567
700	1.103	900	1.610
800	1.353	1000	1.883

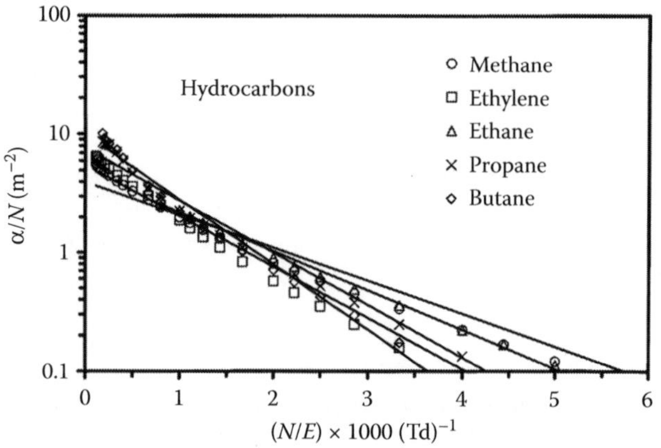

FIGURE 78.8 Reduced ionization coefficient as a function of reduced electric field plotted according to Equation 78.4. (○) Methane; (□) ethylene; (△) ethane; (×) propane; (◊) butane (Heylen, 1963, 1975, 1978).

TABLE 78.18
Gas Constants in Selected Hydrocarbons

Gas	F ($\times 10^{-20}$ m^2)	G (Td)	E/N range (Td)	Reference	Method
Ethylene	2.7	840	121–434	Heylen (1963)	A
Ethylene	3.1	897	212–406	Heylen (1963)	B
Ethylene	3.0	877	102–479	Heylen (1978)	A
Acetylene	4.2	939	132–804	Heylen (1963)	A
Acetylene	4.2	959	212–415	Heylen and Lewis (1956, 1958)	B
Butane	7.6	1136	166–1889	Heylen (1975)	A
Butene-1	5.4	1198	149–1120	Heylen (1978)	A
Ethane	3.50	677	141–536	Leblanc and Devins (1960)	A
Ethane	5.6	649	152–336	Heylen and Lewis (1956, 1958)	B
Methane	2.51	624	71–667	Davies et al. (1989)	A
Methane	2.15	564	93–434	Heylen (1963)	A
Methane	2.10	575	120–240	Schlumbohm (1959)	C
Methane	2.5	592	113–465	Leblanc et al. (1960)	A
Methane	2.28	599	82–405	Cookson et al. (1966)	C
Methane	2.0	544	132–279	Heylen et al. (1956, 1958)	B
Propane	6.5	973	100–1889	Heylen (1978)	A
Propylene	3.7	981	127–818	Heylen (1978).	A

Note: A = conductivity; B = breakdown; C = statistical.

REFERENCES

Bowman, C. R. and D. E. Gordon, *J. Chem. Phys.*, 46, 1878, 1967.

Brüche, E., *Ann. Phys. Leipzig*, 2, 909, 1929.

Christophorou, L. G., G. S. Hurst, and A. Hadjiantoniou, *J. Chem. Phys.*, 44, 3506, 1966.

Cookson, A. H., B. W. Ward, and T. J. Lewis, *Br. J. Appl. Phys.*, 17, 891, 1966.

Dance, D. F. and I. C. Walker, *Proc. Roy. Soc. Lond. A*, 334, 259, 1973.

Davies, D. K., L. E. Kline, and W. E. Bies, *J. Appl. Phys.*, 65, 3311, 1989.

Fink, M., K. Jost, and D. Hermann, *J. Chem. Phys.*, 63, 1985, 1975.

Floeder, K., D. Fromme, W. Raith, A. Schwab, and O. Sinapius, *J. Phys. B: At. Mol. Phys.*, 18, 3347, 1985.

Herzberg, G., *Molecular Spectra and Molecular Structure, Vol. II. Infrared and Raman Spectra of Polyatomic Molecules*, Van Nostrand Reinhold, Princeton (NJ), 1945.

Heylen, A. E. D. and T. J. Lewis, *Br. J. Appl. Phys.*, 7, 411, 1956; *Can. J. Phys.*, 36, 721, 1958.

Heylen, A. E. D., *J. Chem. Phys.*, 38, 765, 1963.

Heylen, A. E. D., *Int. J. Electronics*, 39, 653, 1975.

Heylen, A. E. D., *Int. J. Electronics*, 44, 367, 1978.

Hughes, A. L. and J. H. McMillen, *Phys. Rev.*, 44, 876, 1933.

Hwang, W., Y.-K. Kim, and M. E. Rudd, *J. Chem. Phys.*, 104, 2956, 1996.

Kim, Y.-K., K. K. Irikura, M. E. Rudd, M. A. Ali, P. M. Stone, J. Chang, J. S. Coursey et al., *Electron-Impact Ionization Cross Section for Ionization and Excitation Database (version 3.0)*, National Institute of Standards and Technology, Gaithesberg, MD, 2004.

LeBlanc, H., Jr. and J. C. Devins, *Nature*, 188, 219, 1960.

Lunt, S. L., J. Randell, J. P. Ziesel, G. Mrotzek, and D. Field, *J. Phys. B: At. Mol. Opt. Phys.*, 27, 1407, 1994.

Maji, S., G. Basavaraju, S. M. Bharathi, K. G. Bhushan, and S. P. Khare, *J. Phys. B: At. Mol. Opt. Phys.*, 31, 4975, 1998.

Mapstone, B. and N. R. Newell, *J. Phys. B: At. Mol. Opt. Phys.*, 25, 491, 1992.

Mapstone, B., M. J. Brunger, and W. R. Newell, *J. Phys. B: At. Mol. Opt. Phys.*, 33, 23, 2000.

Nishimura, H. and H. Tawara, *J. Phys. B: At. Mol. Opt. Phys.*, 24, L363, 1991.

Nishimura, H. and H. Tawara, *J. Phys. B: At. Mol. Opt. Phys.*, 27, 2063, 1994.

Panajotovic, R., M. Kitajima, H. Tanaka, M. Jelisavcic, J. Lower, L. Campbell, M. J. Brunger, and S. J. Buckman, *J. Phys. B: At. Mol. Opt. Phys.*, 36, 1615, 2003.

Rapp, D. and P. Englander-Golden, *J. Chem. Phys.*, 43, 1464, 1965.

Rutkowsky, V. J., H. Drost, and H.-J. Spangenberg, *Ann. Phys., Leipzig*, 37, 259, 1980.

Schlumbohm, H., *Z. Angew. Phys.*, 11, 156, 1959.

Schmidt, B. and M. Roncossek, *Aust. J. Phys.*, 45, 351, 1992.

Schram, B. L., M. J. van der Wiel, F. J. de Heer, and H. R. Moustafa, *J. Chem. Phys.*, 14, 49, 1966.

Sueoka, O. and S. Mori, *J. Phys. B: At. Mol. Phys.*, 19, 4015. 1986.

Szmytkowski, C., S. Kwitnewski, and E. Ptasińska, *Phys. Rev. A*, 68, 032715, 2003.

Walker, I. C., A. Stamatovic, and S. F. Wong, *J. Chem. Phys.*, 69, 5532, 1978.

79

METHANETHIOL

CONTENTS

Methanethiol (CH_3SH), synonym methyl mercaptan is a gaseous product of degradation of organic matter and is found in earth's atmosphere and interstellar space (Szmytkowski, 1995). It has 26 electrons, derived electronic polarizability of $5.6–6.0 \times 10^{-40}$ F m^2, and dipole moment of 1.52 D. The molecule is electron attaching due to the dissociative attachment occurring at 0.68 and 1.0 eV, resulting in the formation of CH_3S^-, $CH_2S,^-$ and S^- negative ions.

79.1 TOTAL SCATTERING CROSS SECTION

The total scattering cross sections measured by Szmytkowski et al. (1995) are shown in Table 79.1 and Figure 79.1. The highlights of the cross section are

1. A monotonic decrease of the cross section from 0.6 to 1.6 eV attributed to the direct scattering process

TABLE 79.1
Total Scattering Cross Sections in CH_3SH

Energy (eV)	Q_T (10^{-20} m^2)	Energy (eV)	Q_T (10^{-20} m^2)	Energy (eV)	Q_T (10^{-20} m^2)
	CH_3SH				CH_3OH
0.60	47.22	3.0	38.84	0.8	28.2
0.70	43.87	3.2	39.54	1.0	25.6
0.80	41.05	3.4	40.95	1.2	23.6
0.90	39.56	3.6	40.85	1.4	22.1
1.0	38.03	3.9	42.19	1.6	21.3
1.1	37.02	4.2	41.81	1.9	19.7
1.2	35.45	4.6	42.37	2.0	20.1
1.3	36.04	5.2	41.92	2.2	19.7
1.4	33.99	5.6	42.99	2.5	19.6
1.5	33.06	6.1	42.99	3.0	19.8
1.6	32.78	6.6	43.92	3.5	20.2
1.7	33.37	7.1	43.53	4	21.6
1.8	33.09	7.6	43.89	4.5	22.3
2.0	32.89	8.1	44.05	5	23.7
2.2	33.39	8.6	43.26	5.5	24.7
2.4	33.81	9.1	42.73	6.5	25.8
2.6	35.66	9.6	42.31	7.5	26.5
2.8	37.23	10.0	41.34	8.5	26.8

continued

CH$_3$SH

TABLE 79.1 (continued)
Total Scattering Cross Sections in CH$_3$SH

Energy (eV)	Q_T (10^{-20} m^2)	Energy (eV)	Q_T (10^{-20} m^2)	Energy (eV)	Q_T (10^{-20} m^2)
CH$_3$SH				CH$_3$OH	
10.5	40.53	50	25.00	9.5	26.2
11.5	39.10	60	23.45	10.5	25.9
12.5	37.55	70	21.92	11.5	24.9
14	36.40	80	19.26	13.5	24.1
16	35.30	90	17.96	15.5	23.1
18	34.51	100	17.46	17.5	22.4
20	33.33	110	16.14	20	21.8
23	32.67	120	15.67	22	21.1
25	31.14	140	14.13	25	20.0
28	31.63	160	13.45	27	19.2
30	30.61	180	12.29	30	18.4
35	29.27	200	11.64	35	17.5
40	27.49	220	11.18	40	16.7
45	27.29	250	10.00	45	16.1
				50	15.7
				60	14.9
				70	14.3
				80	13.5
				90	12.9
				100	12.4
				110	11.9
				120	11.4
				140	10.7
				160	10.1
				180	9.4
				200	8.81
				220	8.28
				250	7.55

Source: Adapted from Szmytkowski, C., G. Kasperski, and P. Możejko, *J. Phys. B: At. Mol. Opt. Phys.*, 28, L629, 1995.
Note: Data for CH$_3$OH included to show the effect of substitution of O atom (eight electrons) in place of the larger S atom (16 electrons).

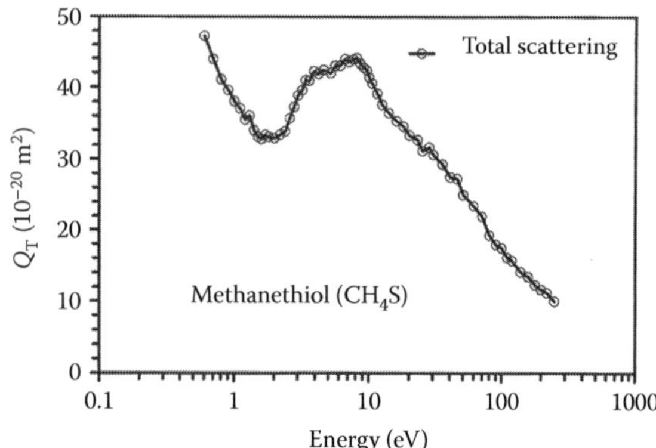

FIGURE 79.1 Total scattering cross sections in CH$_3$SH, (Adapted from Szmytkowski, C., G. Kasperski, and P. Moz'ejko, *J. Phys. B: At. Mol. Opt. Phys.*, 28, L629, 1995.)

rather than the attachment process or the dipole moment.

2. A minor peak at about 4 eV and a pronounced peak at 8 eV. The latter is attributed to a number of processes that occur at this energy level.
3. A monotonic decrease of the cross section beyond 10 eV, approximately according to $\varepsilon^{-\gamma}$, $\gamma = 0.5$ in the present case attributed to the dipole moment according to theory (Vogt and Wannier, 1954).

REFERENCES

Szmytkowski, C., G. Kasperski, and P. Możejko, *J. Phys. B: At. Mol. Opt. Phys.*, 28, L629, 1995.
Vogt, E. and H. Wannier, *Phys. Rev.*, 95, 1190, 1954.

80

METHANOL

CONTENTS

Methanol (CH_3OH), synonym methyl alcohol, is a molecule with 18 electrons. It has a polarizability of 3.66×10^{-40} F m^2, a dipole moment of 1.70 D, and an ionization potential of 10.85 eV. The molecule has 12 vibrational modes as shown in Table 80.1 (Shimanouchi, 1972).

80.1 TOTAL SCATTERING CROSS SECTIONS

Table 80.2 and Figure 80.1 show the total scattering cross sections measured by Szmytkowski and Krzysztofowicz (1995). The highlights of the cross sections are

1. In the range from 0.8 to 2 eV, the cross section decreases rapidly with increasing energy due to direct scattering on polar molecules.
2. A maximum centered around 9 eV, due to several possible mechanisms, is characteristic of many targets. Short-lived negative ion resonance is considered to be the most likely mechanism.
3. For energies >10 eV, the cross section decreases with increasing energy; the dipole moment of the target cannot be excluded for this behavior.

TABLE 80.1
Vibrational Modes and Energies of Methanol Molecule

Symbol	Deformation	Energy (meV)	Symbol	Deformation	Energy (meV)
ν_1	OH stretch	456.4	ν_7	CH_3 rock	131.4
ν_2	CH_3 d-stretch	372.0	ν_8	CO stretch	128.1
ν_3	CH_3 s-stretch	352.6	ν_9	CH_3 d-stretch	367.0
ν_4	CH_3 d-deform	183.1	ν_{10}	CH_3 d-deform	183.1
ν_5	CH_3 s-deform	180.4	ν_{11}	CH3 rock	144.4
ν_6	OH bend	166.8	ν_{12}	Torsion	36.6 (24.8)

Note: s = symmetrical; d = degenerate.

TABLE 80.2
Total Scattering Cross Sections in Methanol

Energy (eV)	Q_T (10^{-20} m^2)	Energy (eV)	Q_T (10^{-20} m^2)	Energy (eV)	Q_T (10^{-20} m^2)
0.8	28.2	7.5	26.5	45	16.1
1.0	25.6	8.5	26.8	50	15.7
1.2	23.6	9.0	26.2	60	14.9
1.4	22.1	10.5	25.9	70	14.3
1.6	21.2	11.5	24.9	80	13.5
1.9	19.7	13.5	24.1	90	12.9
2.2	19.7	15.5	23.1	100	12.4
2.5	19.6	17.5	22.4	110	11.9
3.0	19.8	20	21.8	120	11.4
3.5	20.2	22	21.1	140	10.7
4.0	21.6	25	20.0	160	10.1
4.5	22.3	27	19.2	180	9.4
5.0	23.7	30	18.4	200	8.81
5.5	24.7	35	17.5	220	8.28
6.5	25.8	40	16.7	250	7.55

Source: Adapted from Szmytkowski, Cz. and A. M. Krzysztofowicz, *J. Phys. B: At. Mol. Opt. Phys.*, 28, 4291, 1995.

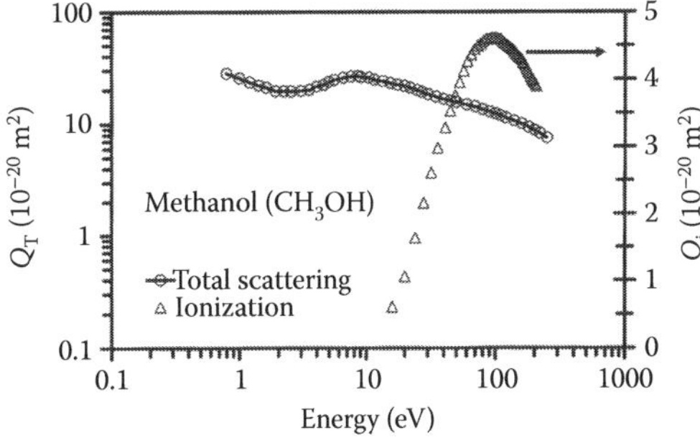

FIGURE 80.1 Total scattering and ionization cross sections in CH_3OH). (—○—) Total (Szmytkowski and Krzysztofowicz, 1995); (Δ) ionization. (Tabulated values courtesy of P. W. Harland, Professor of Chemistry, University of Canterbury, Christ Church, New Zealand. Private communication, 2006).

CH_3OH

TABLE 80.3
Ionization Cross Sections in Selected Alcohols

	Q_i (10^{-20} m^2)						
Energy (eV)	Methanol	Ethanol	*n*-Propanol	2-Propanol	1-Butanol	*i*-Butyl alcohol	*t*-Butyl alcohol
16	0.61	0.75	1.30	1.70	1.70	1.75	1.20
20	1.06	1.71	2.74	3.30	3.20	3.40	2.82
24	1.63	2.72	4.02	4.97	5.09	5.06	4.71
28	2.16	3.65	5.18	6.16	6.55	6.53	6.20
32	2.61	4.36	6.14	7.13	7.70	7.75	7.51
36	2.98	4.94	6.85	7.79	8.65	8.80	8.70
41	3.28	5.44	7.37	8.28	9.47	9.71	9.75
45	3.53	5.86	7.84	8.72	10.17	10.48	10.62
49	3.76	6.22	8.29	9.10	10.75	11.11	11.05
53	3.95	6.51	8.67	9.39	11.20	11.58	11.53
57	4.12	6.74	8.97	9.55	11.60	11.90	11.89
61	4.26	6.94	9.23	9.66	11.98	12.18	12.20
65	4.36	7.14	9.44	9.81	12.23	12.42	12.57
69	4.44	7.31	9.60	9.95	12.40	12.60	12.76
73	4.49	7.40	9.76	10.07	12.57	12.74	12.97
77	4.52	7.45	9.90	10.15	12.70	12.85	13.15
81	4.56	7.49	10.00	10.18	12.81	12.95	13.23
85	4.59	7.55	10.07	10.20	12.85	13.01	13.26
89	4.60	7.60	10.12	10.22	12.83	13.04	13.33
93	4.61	7.60	10.14	10.24	12.82	13.06	13.29
97	4.61	7.57	10.15	10.23	12.81	13.04	13.26
101	4.61	7.54	10.14	10.21	12.77	13.03	13.27
105	4.60	7.51	10.12	10.17	12.76	13.03	13.27
109	4.59	7.49	10.07	10.12	12.71	12.95	13.17
113	4.56	7.47	10.03	10.06	12.60	12.85	13.07
117	4.53	7.41	10.01	10.00	12.54	12.82	13.06
122	4.50	7.35	9.99	9.95	12.54	12.81	13.02
126	4.48	7.30	9.95	9.90	12.50	12.76	13.00
130	4.46	7.28	9.91	9.83	12.40	12.65	12.95
134	4.44	7.25	9.85	9.75	12.27	12.56	12.83
138	4.41	7.21	9.79	9.69	12.21	12.49	12.75
142	4.38	7.16	9.72	9.65	12.19	12.41	12.61
146	4.36	7.12	9.66	9.62	12.13	12.29	12.49
150	4.33	7.09	9.60	9.56	12.01	12.13	12.46
154	4.31	7.06	9.54	9.50	11.89	11.99	12.38
158	4.29	7.01	9.49	9.45	11.76	11.90	12.21
162	4.25	6.96	9.43	9.37	11.68	11.86	12.19
166	4.23	6.92	9.36	9.29	11.63	11.81	12.16
170	4.19	6.87	9.27	9.22	11.56	11.73	12.05
174	4.15	6.81	9.20	9.15	11.44	11.62	12.04
178	4.11	6.76	9.12	9.08	11.28	11.47	11.93
182	4.08	6.71	9.05	9.01	11.14	11.34	11.87
186	4.04	6.65	9.01	8.96	11.02	11.25	11.80
190	4.00	6.59	8.97	8.89	10.93	11.21	11.69
194	3.98	6.53	8.92	8.83	10.81	11.20	11.56
198	3.95	6.49	8.87	8.75	10.72	11.20	11.48
203	3.92	6.43	8.82	8.68	10.65	11.15	11.41
207	3.89	6.35	8.76	8.59	10.55	11.08	11.28

Source: Courtesy of P. W. Harland, Professor of Chemistry, University of Canterbury, Christ Church, New Zealand. Private communication (2006).

80.2 IONIZATION CROSS SECTIONS

Ionization cross sections are measured by Harland (P. W. Harland, Professor of Chemistry, University of Canterbury, Christ Church, New Zealand. Private communication, 2006) as shown in Table 80.3 and Figure 80.1. The cross sections of selected alcohols are also shown for comparison (Figure 80.2).

80.3 ADDENDUM

Ionization cross sections measured by Rejoub et al. (2003) up to 1000 eV for simple alcohols are shown in Table 80.4 and Figure 80.3.

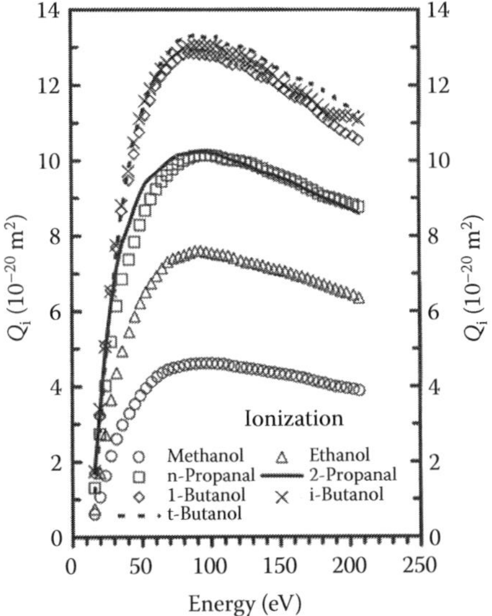

FIGURE 80.2 Ionization cross sections in selected alcohols. (o) Methanol; (Δ) ethanol; (□) propanol; (——) 2-propanal; (◊) 1-butanol; (×) i-butanol; (– –) t-butanol. (Tabulated values courtesy of P. W. Harland, Professor of Chemistry, University of Canterbury, Christ Church, New Zealand. Private communication, 2006.)

TABLE 80.4
Ionization Cross Sections for Simple Alcohols

Energy (eV)	Methanol (CH_3OH)	Ethanol (C_2H_5OH)	1-Propanol (C_3H_7OH)
	Cross Section (10^{-20} m^2)		
13	0.151		
14	0.335		
16	0.624	1.680	1.830
18	0.976	2.360	2.420
20	1.335	2.810	3.670
22.5	1.814	3.280	4.060
25	2.248	3.750	5.450
30	2.859	4.700	7.450
35	3.402	5.405	8.700
40	3.821	6.090	9.310
50	4.396	6.982	10.341
60	4.693	7.545	10.729
80	5.026	8.000	10.964
100	4.975	7.910	10.661
125	4.893	7.543	10.315
150	4.707	6.946	9.875
200	4.224	6.200	9.043
300	3.486	5.313	7.539
400	2.919	4.543	6.254
500	2.523	4.035	5.479
600	2.260	3.677	5.011
800	1.838	3.131	4.575
1000	1.563	2.782	4.217

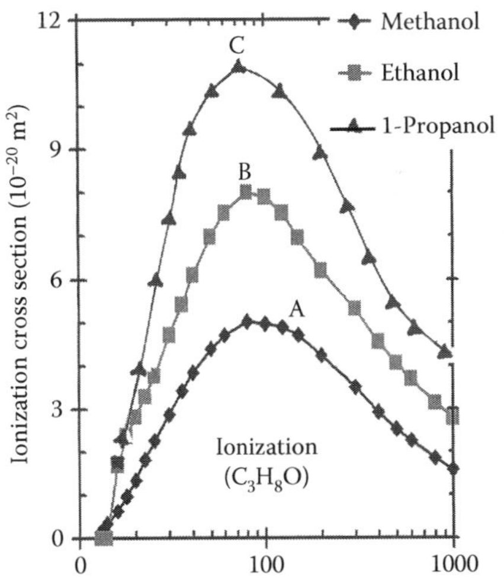

FIGURE 80.3 Ionization cross section for simple alcohols. A–Methanol; B—ethanol; C—1-propanol. (Adapted from Rejoub, R. et al., *J. Chem. Phys.*, 118, 1756, 2003.)

REFERENCES

Rejoub, R., C. D. Morton, B. G. Lindsay, and R. F. Stebbings, *J. Chem. Phys.*, 118, 1756, 2003.

Shimanouchi, T. *Tables of Molecular Vibrational Frequencies Consolidated Volume I*, NSRDS-NBS 39, Washington (DC), 1972.

Szmytkowski, Cz. and A. M. Krzysztofowicz, *J. Phys. B: At. Mol. Opt. Phys.*, 28, 4291, 1995.

81

TETRACHLOROETHENE

CONTENTS

Tetrachloroethene (C_2Cl_4), synonym perchloroethylene, is an electron-attaching molecule that has 80 electrons. It has 12 modes of vibrational excitation as shown in Table 81.1 (Shimanouchi, 1976).

81.1 SELECTED REFERENCES

See Table 81.2.

81.2 IONIZATION CROSS SECTIONS

Table 81.3 and Figure 81.1 show the ionization cross sections measured by Hudson et al. (2001).

TABLE 81.1
Vibrational Modes and Energies for C_2Cl_4

Designation	Type of Motion	Energy (meV)
ν_1	CC stretch	194.8
ν_2	CCl_2 symmetrical -stretch	55.4
ν_3	CCl_2 scissor	29.4
ν_4	CCl_2 twist	13.6
ν_5	CCl_2 asymmetrical -stretch	124.0
ν_6	CCl_2 rock	43.0
ν_7	CCl_2 wag	35.7
ν_8	CCl_2 wag	63.5
ν_9	CCl_2 asymmetrical -stretch	112.6
ν_{10}	CCl_2 rock	21.8
ν_{11}	CCl_2 symmetrical -stretch	96.3
ν_{12}	CCl_2 scissor	38.4

Source: Adapted from Shimanouchi, T. *Tables of Molecular Vibrational Frequencies*, NSRDS-NBS, 39, 1972, p. 76.

TABLE 81.2
Selected References (C_2Cl_4)

Quantity	Range: eV, (Td)	Reference
Attachment process	—	**Liu et al. (2005)**
Attachment process	—	**Suess et al. (2003)**
Attachment cross section	0–9.5	**Drexel et al. (2003)**
Ionization cross section	16–219	**Hudson et al. (2001)**
Attachment cross section	0.001–0.01	**Marawar et al. (1988)**
Attachment cross section	0–0.04	**Alajajian and Chutjian (1987)**
Attachment cross section	0–2	**Johnson et al. (1977)**

Note: A P = attachment process; Q_a = attachment cross section; Q_i = ionization. Bold font indicates experimental study.

81.3 ATTACHMENT CROSS SECTIONS

The potential diagram of the ground state molecule and negative ion are shown in Figure 81.2 (Drexel et al., 2003). The electron affinity of the molecule is observed to be positive.

Attachment occurs according to processes (Suresh et al., 2003)

$$C_2Cl_4 + e \rightarrow C_2Cl_4^{-*} \rightarrow C_2Cl_3 + Cl^- \quad (81.1)$$

$$C_2Cl_4 + e \rightarrow C_2Cl_4^{-*} \rightarrow C_2Cl_4 + e \quad (81.2)$$

$$C_2Cl_4 + e \rightarrow C_2Cl_4^{-*} \rightarrow C_2Cl_4^- \quad (81.3)$$

Reaction 81.1 is dissociative attachment, Reaction 81.2 is autodetachment, and Reaction 81.3 is direct attachment

C_2Cl_4

TABLE 81.3
Measured Ionization Cross Sections

Energy (eV)	Q_i (10^{-20} m²)	Energy (eV)	Q_i (10^{-20} m²)
16	3.61	119	20.56
20	7.94	123	20.42
23	11.55	127	20.00
27	13.67	130	19.93
31	15.53	134	19.81
34	16.99	138	19.70
38	17.59	142	19.56
42	18.41	145	19.37
45	19.14	149	19.16
49	19.68	153	18.81
53	20.42	156	18.55
57	20.66	160	18.46
60	21.18	164	18.45
64	21.24	167	18.46
68	21.47	171	18.41
71	21.40	175	18.28
75	21.64	179	18.01
79	21.74	182	17.73
82	21.51	186	17.40
86	21.59	190	17.10
90	21.49	193	16.94
94	21.34	197	16.88
97	21.29	201	16.82
101	21.37	204	16.87
105	21.01	208	16.92
108	20.78	212	17.00
112	20.79	215	16.80
116	20.62	219	16.58

Source: Adapted from Hudson, J. E. et al., *J. Phys. B: At. Mol. Opt. Phys.*, 34, 3025, 2001.
Note: Tabulated values are by courtesy of Professor Harland (2006).

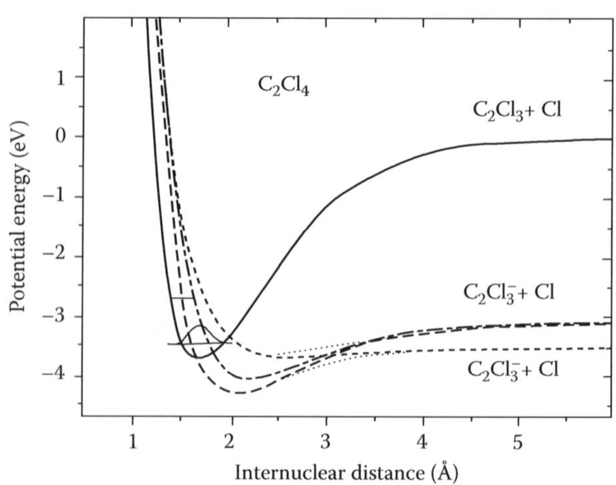

FIGURE 81.2 Potential energy diagram for the ground state and negative ion states of C_2Cl_4. (Adapetd from Drexel, H. et al., *J. Chem. Phys.*, 118, 7394, 2003.)

TABLE 81.4
Relative Negative Ion Yield (C_2Cl_4)

Species	Relative Yield
C^-	1
Cl_2^-	3×10^{-2}
C_2Cl^-	2.5×10^{-4}
CCl_2^-	1.0×10^{-4}
$C_2Cl_2^-$	1.5×10^{-4}
$C_2Cl_3^-$	5×10^{-4}
$C_2Cl_4^-$	2.5×10^{-2}

Source: Adapted from Drexel, H. et al., *J. Chem. Phys.*, 118, 7394, 2003.

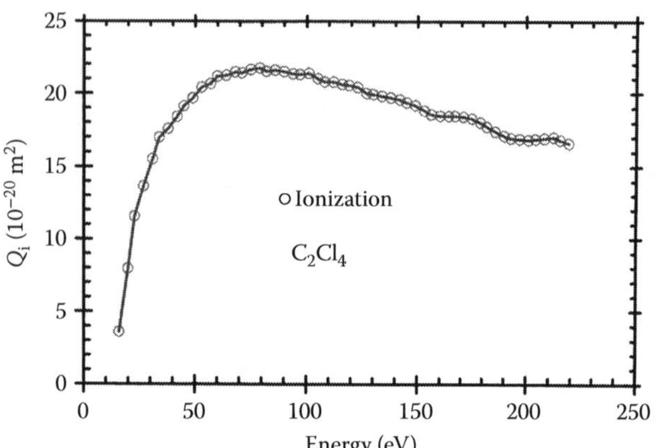

FIGURE 81.1 Ionization cross sections (C_2Cl_4). Tabulated values are by courtesy of Professor Harland (2006). (Adapted from Hudson, J. E. et al., *J. Phys. B: At. Mol. Opt. Phys.*, 34, 3025, 2001.)

leading to parent negative ion. The lifetime of $C_2Cl_4^-$ is in the range from 30 to 130 μs (Sues et al., 2003). The molecule is unusual in that, all three processes are known to occur. Below 10 eV, three species of ions, C^-, Cl_2^-, and $C_2Cl_4^-$ in the ratio 1000:4:31 are observed, respectively (Johnson et al., 1997). At higher electron energy (~30 eV) a variety of fragment-negative ions, C^-, C_2^-, Cl^-, CCl^-, C_2Cl^-, Cl_2^-, and $C_2Cl_4^-$ are observed by Macneil and Thynne (1968). The ratios of dissociative attachment yield for each species of negative ion are shown in Table 81.4.

Figure 81.3 shows the attachment cross sections for the formation of $C_2Cl_4^-$ and Cl^- ions (Marawar et al., 1988). Note that the cross section for the former ion is greater by a factor of about 4.

81.4 ATTACHMENT RATE CONSTANT

Table 81.5 shows the attachment rate constants for C_2Cl_4 at thermal energy.

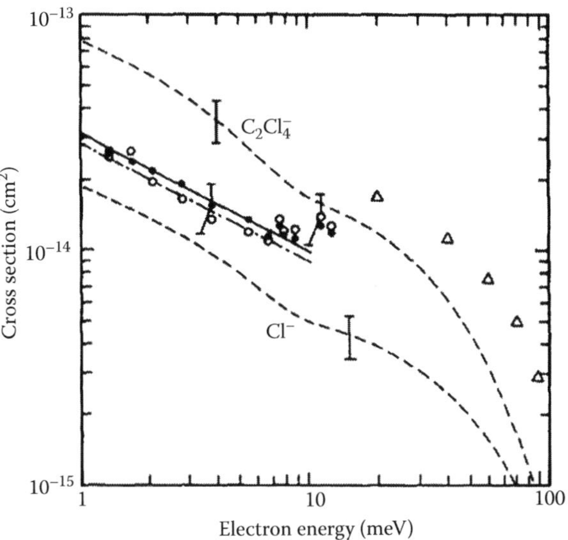

FIGURE 81.3 Low-energy electron attachment cross sections for C_2Cl_4. (– – –) Alajajian and Chutjian, 1987; (Δ) Swarm results, McCorkle[10] et al. (1984); (○) and (•) Marawar et al. (1988). (Reprinted with permission from Marawar, R. W. et al., *J. Chem. Phys.*, 88, 176, 1988. Copyright (1988), American Institute of Physics.)

TABLE 81.5
Attachment Rate Constant

Rate (m^3/s)	Method	Reference
9.0 (−14)	Rydberg atom	Suess et al. (2003)
1.04 (−13)	Swarm	McCorkle et al. (1984)

REFERENCES

Alajajian, S. H. and A. Chutjian, *J. Phys. B: At. Mol. Phys.*, 20, 2117, 1987.

Drexel, H., W. Sailer, V. Grill, P. Scheier, E. Illenberger, and T. D. Märk, *J. Chem. Phys.*, 118, 7394, 2003.

Hudson, J. E., C. Vallance, M. Bart, and P. W. Harland, *J. Phys. B: At. Mol. Opt. Phys.*, 34, 3025, 2001.

Johnson, J. P., L. G. Christophorou, and J. G. Carter, *J. Chem. Phys.*, 67, 2196, 1977.

Liu, Y., L. Suess, and F. B. Dunning, *J. Chem. Phys.*, 123, 054327, 2005.

Marawar, R. W., C. W. Walter, K. A. Smith, and F. B. Dunning, *J. Chem. Phys.*, 88, 176, 1988.

McCorkle, D. L., A. A. Christodoulides, and L. G. Christophorou, Basic physics of gaseous dielectrics, in L. G. Christophorou and M. O. Pace (Eds.), *Gaseous Dielectrics IV*, Pergamon Press, New York, NY, 1984, p. 12.

McNeil, K. A. G. and J. C. J. Thynne, *Trans. Faraday, Soc.*, 64, 2112, 1968.

Shimanouchi, T. *Tables of Molecular Vibrational Frequencies*, NSRDS-NBS, 39, Washington (DC), 1972, p. 76.

Suess, L., R. Parthasarathy, and F. B. Dunning, *J. Chem. Phys.*, 118, 6205, 2003.

82

TETRAFLUOROETHENE

CONTENTS

Tetrafluoroethene (C_2F_4), also known as tetrafluoroethylene, is a molecule that has 48 electrons and that is electron-attaching and non-dipolar. The electronic polarizability is 4.45×10^{-40} F m^2 (see Beran and Kevan, 1969) and the ionization potential is 10.52 eV (Coggiola et al., 1976). The vibrational modes of the molecule are given in Table 82.1 (Shimanouchi, 1972).

TABLE 82.1
Vibrational Modes and Energies for C_2F_4

Mode	Type of Motion	Energy (meV)
ν_1	CC stretch	232.1
ν_2	CF_2 symmetrical-stretch	96.5
ν_3	CF_2 scissor	48.8
ν_4	CF_2 twist	23.6
ν_5	CF_2 asymmetrical-stretch	166.1
ν_6	CF_2 rock	68.3
ν_7	CF_2 wag	50.3
ν_8	CF_2 wag	63.0
ν_9	CF_2 asymmetrical-stretch	165.8
ν_{10}	CF_2 rock	27.0
ν_{11}	CF_2 symmetrical-stretch	147.0
ν_{12}	CF_2 scissor	69.2

82.1 SELECTED REFERENCES FOR DATA

See Table 82.2.

TABLE 82.2
Selected References

Quantity	Range: eV, (Td)	Reference
Q_{el}, Q_M	1–100	Panajotovic et al. (2004)
Q_T	0.6–370	Szmytkowski et al. (2003)
Q_{el}, Q_v	0–50	Winstead and McKoy, (2002)
W, α/N, η/N	(15–2000)	Yoshida et al. (2002)
Q_i	12.2–215	Bart et al. (2001)
W, α/N, η/N	(7–1000)	Goyette et al. (2001)
Q_T	30–3000	Jiang et al. (2000)
Q_i	10–27	Jarvis et al. (1998)
Q_{diff}	25, 40	Coggiola et al. (1976)
Q_a	1–13	Lifshitz and Grajower (1972/73)
Q_i	70	Beran and Kevan (1969)
Q_i	<25	Lifshitz and Long (1964)

Note: Q_{diff} = differential; Q_{el} = elastic; Q_i = ionization cross section; Q_v = vibrational excitation; W = drift velocity; α/N = ionization coefficient; η/N = attachment coefficient.

82.2 TOTAL SCATTERING CROSS SECTIONS

Table 82.3 shows the total scattering cross sections measured by Szmytkowski et al. (2003). Figure 82.1 is a graphical presentation that also includes the theoretical values of Jiang et al. (2000). The highlights of the total scattering cross section are

1. The scattering cross section at zero energy shows a trend of increase as the energy is lowered. Since the molecule is nonpolar, and antibonding for the lowest unoccupied molecular orbit, the increase is attributed to the vibrational cross sections, particularly "rocking" and "wagging" (Panajotovic et al., 2004).
2. A minor peak at 2.8 eV attributed probably to the formation of short-lived negative ions.
3. A broad peak in the region of 22–30 eV.
4. At higher energies, the cross section decreases monotonically.

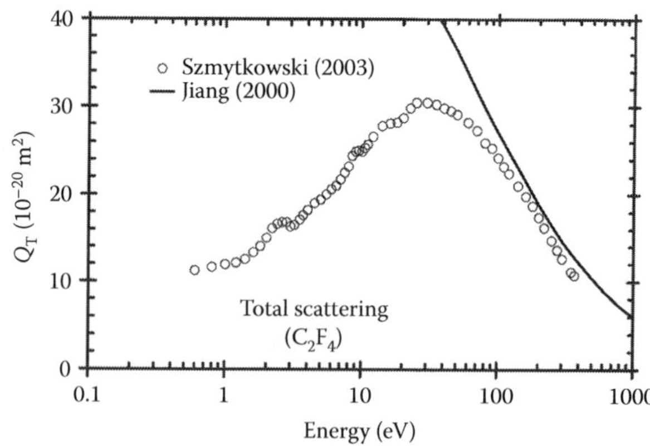

FIGURE 82.1 Total elastic scattering cross sections for C_2F_4. (○) Szmytkowski et al. (2003), measurement; (———) Jiang et al. (2000), theory.

TABLE 82.3
Total Scattering Cross Sections

Energy (eV)	Q_T (10^{-20} m^2) C_2F_4	Q_T (10^{-20} m^2) C_2H_4	Energy (eV)	Q_T (10^{-20} m^2) C_2F_4	Q_T (10^{-20} m^2) C_2H_4
0.6	11.2	14.6	11	25.7	31.0
0.8	11.6	14.6	12	26.6	30.7
1.0	11.9	16.8	14	27.8	30.1
1.2	12.1	20.6	16	28.1	29.4
1.4	12.5	23.7	18	28.2	28.3
1.6	13.3	26.4	20	28.7	27.4
1.8	14.0	28.8	22.5	29.8	
2.0	15.0	28.6	25	30.4	25.4
2.2	16.1	27.1	30	30.4	24.2
2.4	16.6	25.4	35	30.2	22.8
2.6	16.8	23.5	40	29.8	21.7
2.8	16.8	21.7	45	29.5	21.1
3.0	16.3	20.9	50	29.1	220.5
3.2	16.5	20.3	60	28.2	19.2
3.5	17.1	20.2	70	27.3	17.9
3.7	17.6	20.8	80	25.9	17.0
4.0	18.2	22.2	90	25.3	16.0
4.5	19.0	24.3	100	24.2	15.0
5.0	19.4	26.2	110	23.2	14.2
5.5	20.0	27.6	120	22.4	13.4
6.0	20.6	29.0	140	21.0	12.4
6.5	21.0	30.0	160	19.8	11.5
7.0	21.7	31.0	180	18.7	10.8
7.5	22.5	31.6	200	17.4	10.0
8.0	23.2	31.9	220	16.2	9.40
8.5	24.4	31.9	250	14.7	8.73
9.0	24.9	31.7	275	13.6	8.14
9.5	25.0	31.6	300	12.6	7.77
10	24.9	31.6	350	11.1	6.90
10.5	25.3		370	10.7	6.65

Note: Data for C_2H_4 are given for the sake of comparison.

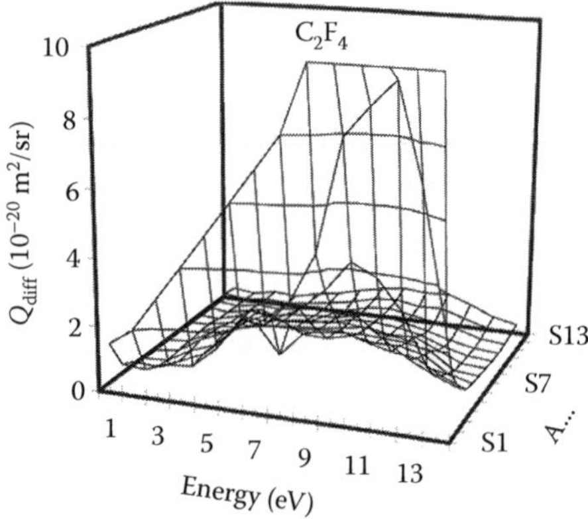

FIGURE 82.2 3D view of the differential scattering cross section as a function of energy and angle of scattering for C_2F_4. (Adapted from Panajotovic, R. et al., *J. Chem. Phys.*, 121, 4559, 2004.)

82.3 DIFFERENTIAL SCATTERING CROSS SECTIONS

Differential scattering cross sections as a function of electron energy and scattering angle have been measured by Panajotovic et al. (2004) and to a limited extent by Coggiola et al. (1976). The experimental differential cross sections of these authors, supplemented by theoretical values, are shown in Figures 82.2 and 82.3 in the 8–20 eV range. The significant points are

1. At the same energy, the scattering cross section decreases continually as the angle is increased.
2. At the same angle, the differential cross section as a function of energy shows peaks in the range 3–30 eV depending on the angle.

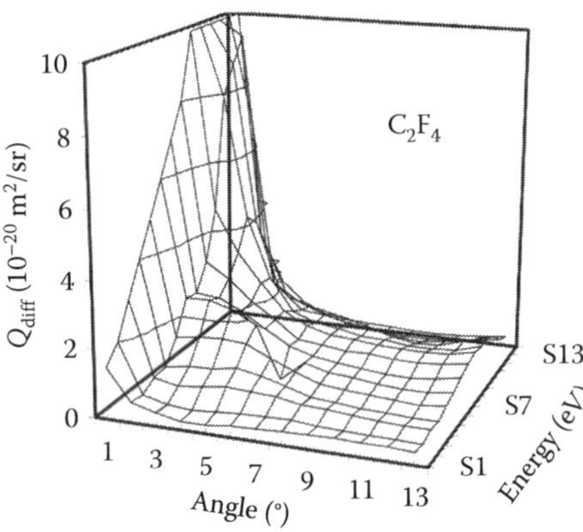

FIGURE 82.3 3D view of the differential scattering cross section as a function of angle of scattering and energy for C_2F_4. (Adapted from Panajotovic, R. et al., *J. Chem. Phys.*, 121, 4559, 2004.)

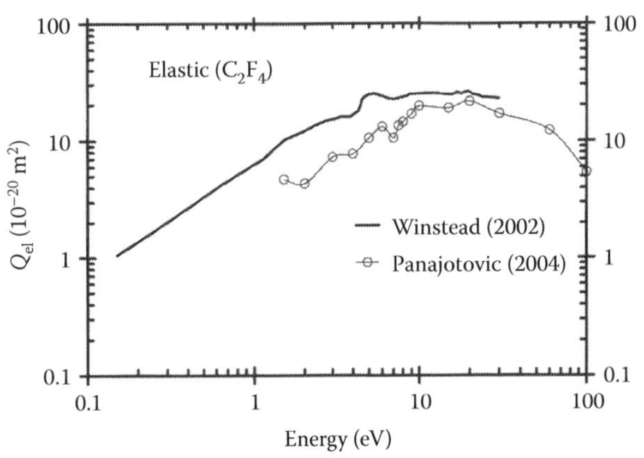

FIGURE 82.4 Elastic scattering cross section for C_2F_4. (-○-) Panajotovic (2004), average of measurements in two laboratories; (——) Winstead and McKoy (2002), theory.

TABLE 82.4
Elastic Scattering and Momentum Transfer Cross Sections for C_2F_4

Energy (eV)	Q_{el} (10^{-20} m^2)	Q_M (10^{-20} m^2)
1.5	4.705 (4.86, 4.55)	4.01 (3.89, 4.13)
2.0	4.34	3.70
3.0	7.25 (7.37, 7.13)	5.45 (5.06, 5.84)
4.0	7.73	5.90
5.0	10.5	7.08 (6.78, 7.38)
6.0	13.1	8.37
7.0	10.5	7.62
7.5	13.4	9.24
8.0	14.4	11.6
9.0	16.8	12.5
10.0	19.6 (19.4, 19.8)	15.45 (15.1, 15.8)
15	18.85 (19.0, 18.7)	11.85 (11.0, 12.7)
20	21.5 (22.2, 20.8)	15.95 (15.8, 16.1)
30	16.9	9.64
60	12.2	4.66
100	5.43	2.72

Source: Adapted from Panajotovic, R. et al., *J. Chem. Phys.*, 121, 4559, 2004.
Note: Average from measurements in two laboratories are shown.

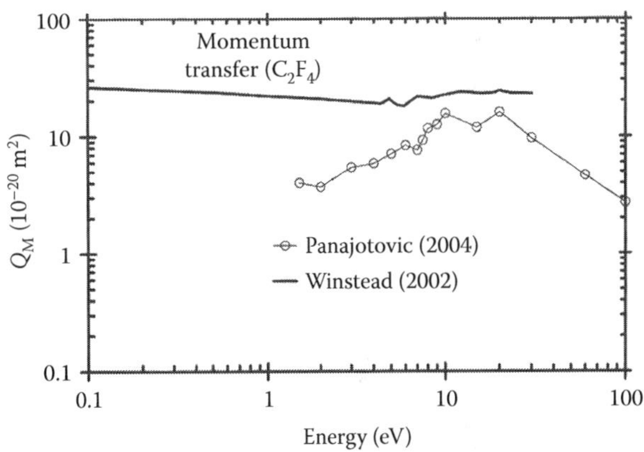

FIGURE 82.5 Momentum transfer cross section for C_2F_4. (-○-) Panajotovic (2004), average of measurements in two laboratories; (——) Winstead and McKoy (2002), theory.

82.4 ELASTIC SCATTERING AND MOMENTUM TRANSFER CROSS SECTIONS

Table 82.4 shows the measured elastic scattering and momentum transfer cross sections. Figures 82.4 and 82.5 present these data graphically, in addition to the computed values of Winstead and McKoy (2002).

82.5 VIBRATIONAL CROSS SECTIONS

Total vibrational excitation cross sections measured by Panajotovic et al. (2004) are shown in Figure 82.6. The significant points are

1. The vibrational cross sections are higher at lower energies than at higher ones.
2. The peaks seen at 3 and 7.5 eV are due to resonant mechanisms, the latter possibly due to C=C double bond (ν_1) mode.

82.6 TOTAL EXCITATION CROSS SECTIONS

Dissociation products and energies are shown in Table 82.5 (Yoshida et al. 2002).

The total excitation cross sections derived from swarm studies are shown in Figure 82.7.

C_2F_4

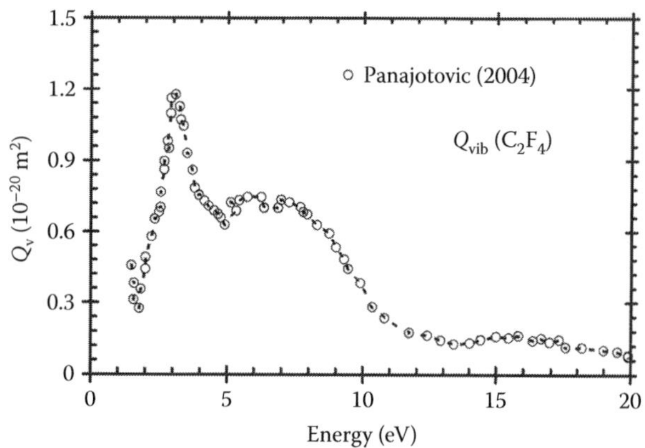

FIGURE 82.6 Measured vibrational cross sections for C_2F_4. Note the peak at 3.0 and 7.5 eV. (Adapted from Panajotovic, R. et al., *J. Chem. Phys.*, 121, 4559, 2004.)

TABLE 82.5
Dissociation Products

Dissociation Products	Energy (eV)
$CF_2 + CF_2$	3.06
$CF_3 + CF$	4.52
$C_2F_3 + F$	5.19
$C_2F_2 + F_2$	7.09
$C_2F_2 + F + F$	8.13
$CF_2 + CF + F$	8.13
$CF_3 + C + F$	9.79

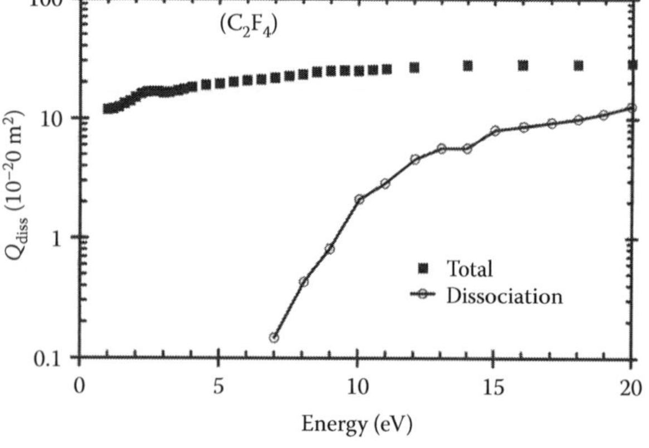

FIGURE 82.7 Dissociation cross sections for C_2F_4. Total scattering cross section has been shown for comparison sake. (—○—) Yoshida et al. (2002), dissociation; (■) Szmytkowski et al. (2003), total scattering. (Adapted from Yoshida, K. et al., *J. Appl. Phys.*, 91, 2637, 2002.)

82.7 POSITIVE ION APPEARANCE POTENTIALS

Positive ion appearance potentials due to parent and dissociative ionization are shown in Table 82.6 (Jarvis et al., 1998).

TABLE 82.6
Positive Ion Appearance Potentials

Dissociation Channel	Energy (eV)
$C_2F_4^+$	10.1
$C_2F_3^+ + F$	<15.85
$CF_3^+ + CF$	13.20
$CF_2^+ + CF_2$	13.20
$CF^+ + CF_3$	13.81
$CF^+ + CF_2 + F$	13.50

TABLE 82.7
Total Ionization Cross Sections for C_2F_4

Energy (eV)	Q_i 10^{-20} m²	Energy (eV)	Q_i 10^{-20} m²	Energy (eV)	Q_i 10^{-20} m²
12	0.24	82	5.71	153	5.79
16	0.44	86	5.81	156	5.77
20	0.91	90	5.86	160	5.74
24	1.42	94	5.87	164	5.69
27	1.83	97	5.88	168	5.63
31	2.35	101	5.89	171	5.59
35	2.90	105	5.90	175	5.55
38	3.26	108	5.90	179	5.52
42	3.68	112	5.90	182	5.50
46	4.04	116	5.90	186	5.48
49	4.29	119	5.91	190	5.46
52	4.52	123	5.92	193	5.42
57	4.86	127	5.92	197	5.38
60	5.03	130	5.91	201	5.33
64	5.20	134	5.88	204	5.29
68	5.32	138	5.84	208	5.27
72	5.47	142	5.81	212	5.25
75	5.56	145	5.80	216	5.25
79	5.65	149	5.79		

Source: Adapted from Bart, M. et al., *Phys. Chem. Chem. Phys.*, 3, 800, 2001.
Note: Tabulated values courtesy of Professor Harland.

82.8 TOTAL IONIZATION CROSS SECTIONS

Total ionization cross sections have been measured by Bart et al. (2001) as shown in Table 82.7 and Figure 82.8.

82.9 ATTACHMENT CROSS SECTIONS

Formation of $C_2F_4^-$ at zero electron energy is controversial (Goyette et al., 2001; McNeil et al., 1979). Dissociative attachment occurs via the following reactions (Lifshitz and Grajower, 1972/73)

$$C_2F_4 + e \rightarrow F + C_2F_3^- \quad (82.1)$$

$$C_2F_4 + e \rightarrow C_2F_3 + F^- \quad (82.2)$$

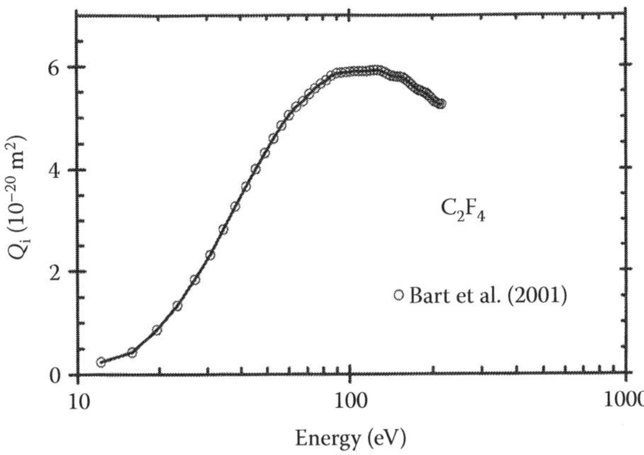

FIGURE 82.8 Total ionization cross sections for C_2F_4. Tabulated values are courtesy of Professor P. Harland. (Adapted from Bart, M. et al., *Phys. Chem. Chem. Phys.*, 3, 800, 2001.)

TABLE 82.8
Negative Ion Appearance Potentials

Ion	Appearance Potential (eV)	Reference
F^-	1.8	Lifshitz and Grajower (1972/73)
	1.8	Thynne and Macneil (1970)
	2.2	Bibby and Carter (1963)
$C_2F_3^-$	3.2	Lifshitz and Grajower (1972/73)
	3.2	Thynne and Macneil (1970)

TABLE 82.9
Resonance Peak Maxima for Negative Ions

Ion	F^-			$C_2F_3^-$
Energy (eV)	3.0	6.0	11	3.85

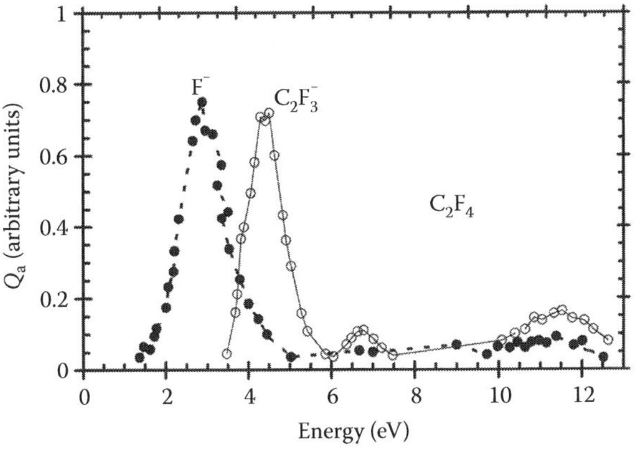

FIGURE 82.9 Attachment cross sections for C_2F_4 for the formation of F^- and $C_2F_3^-$ ions. Note the arbitrary scale for the ordinate. Data are not available for normalization. (Adapted from Lifshitz, C. and R. Grajower, *Int. J. Mass. Spectrom. Ion Phys.*, 10, 25, 1972/73.)

Additional dissociative mechanisms occur for the formation of F^- ions. Appearance potentials of negative ions are shown in Table 82.8. The resonance peak maxima for negative ions are shown in Table 82.9 (Lifshitz and Grajower, 1972/73).

The attachment rate at thermal energy is ~8.5×10^{-17} m^3/s (see Goyette et al., 2007). The relative attachment cross sections measured by Lifshitz and Grajower (1972/73) are shown in Figure 82.9. The author has not found data to normalize these cross sections. It is not clear whether part of the cross sections above 10 eV is due to ion pair production.

82.10 DRIFT VELOCITY OF ELECTRONS

Table 82.10 and Figure 82.10 show the drift velocity.

TABLE 82.10
Measured Drift Velocity of Electrons

Yoshida et al. (2001) (Digitized)		Goyette et al. (2001)	
E/N (Td)	*W* (10^4 m/s)	*E/N* (Td)	*W* (10^4 m/s)
60	12.63	7	2.31
70	13.09	8	2.65
80	13.49	9	3.06
90	13.88	10	3.51
100	14.29	12	4.37
120	15.22	14	5.41
140	16.36	16	6.56
160	17.68	18	7.54
180	19.06	20	8.27
200	20.18	23	9.12
230	21.01	26	9.71
260	21.53	30	10.3
300	23.18	35	10.7
350	27.49	40	10.9
400	30.29	45	11.1
450	33.09	50	11.3
500	36.10	60	11.7
600	40.45	70	12.1
700	44.68	80	12.5
800	49.59	90	12.9
900	55.01	100	13.3
1000	61.63	120	14.2
1200	69.27	140	15
1400	78.00	160	15.8
1600	87.60	180	16.7
1800	101.00	200	17.5
2000	106.00	230	18.8
		260	20.2
		300	21.3
		350	23.1
		400	25.4
		450	27.2
		500	28.9
		600	33.6
		700	35.9
		800	39.2
		900	43.2
		1000	47.6

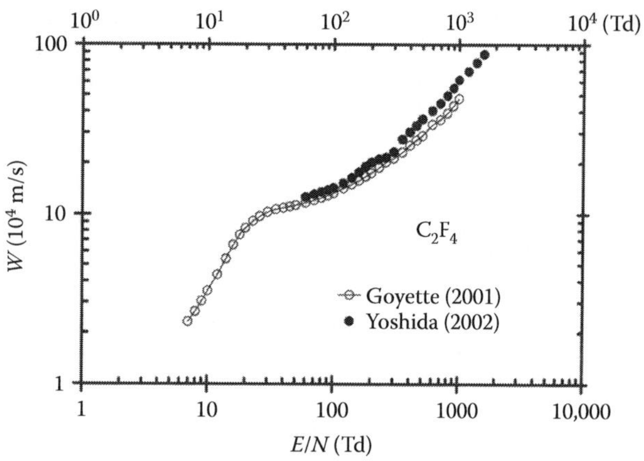

FIGURE 82.10 Drift velocity of electrons in C_2F_4 as a function of E/N. (•) Yoshida et al. (2002); (–○–) Goyette et al. (2001).

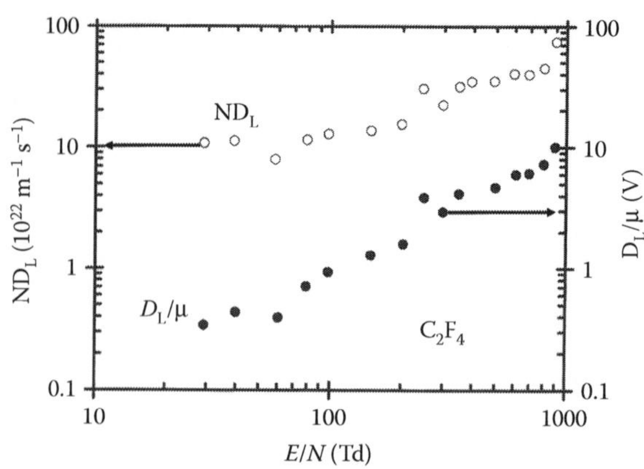

FIGURE 82.11 Density-normalized longitudinal diffusion coefficients and characteristic energy. (○) ND_L; (•) D_L/μ. (Adapted from Yoshida, K. et al., *J. Appl. Phys.*, 91, 2637, 2002.)

TABLE 82.11
Effective Ionization Coefficients

E/N (Td)	α_{eff}/N (10^{-20} m^2)	E/N (Td)	α_{eff}/N (10^{-20} m^2)
7.00	−0.298	80.0	−0.0101
8.00	−0. 244	90.0	−0.0085
9.00	−0.197	100	−0.00637
10.0	−0.157	120	−0.00156
12.0	−0.110	140	0.00563
14.0	−0.0832	160	0.0153
16.0	−0.0586	180	0.0279
18.0	−0.0415	200	0.0422
20.0	0.0322	230	0.0691
23.0	−0.0274	260	0.0998
26.0	−0.0229	300	0.149
30.0	−0.0207	350	0.226
35.0	−0.0213	400	0.306
40.0	−0.0169	450	0.396
45.0	−0.0138	500	0.498
50.0	−0.0132	600	0.678
60	−0. 0120	700	0.887
70.0	−0.0111	800	1.070

Source: Adapted from Goyette, A. N. et al., *J. Chem. Phys.*, 114, 8932, 2001.

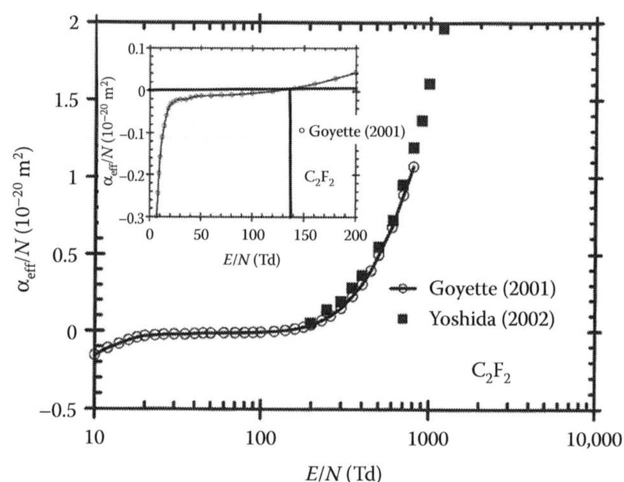

FIGURE 82.12 Effective reduced ionization coefficient in C_2F_4. (■) Yoshida et al. (2002); (–○–) Goyette et al. (2001). The inset shows the details at $(\alpha-\eta)/N = 0$.

82.11 DIFFUSION COEFFICIENTS

Yoshida et al. (2002) have measured the longitudinal diffusion coefficients as a function of E/N, as shown in Figure 82.11.

82.12 EFFECTIVE IONIZATION COEFFICIENTS

Effective ionization coefficients have been measured by Yoshida et al. (2002) and Goyette et al. (2001) as shown in Table 82.11 and Figure 82.12.

REFERENCES

Bart, M., P. W. Harland, J. E. Hudson, and C. Vallance, *Phys. Chem. Chem. Phys.*, 3, 800, 2001.

Beran, J. A. and L. Kevan, *J. Phys. Chem.*, 73, 3866, 1969.

Bibby, M. M. and G. Carter, *Trans. Farad. Soc.*, 59, 2455, 1963, cited by Lifshitz and Grajower, 1972/73.

Coggiola, M. J., W. M. Flicker, O. A. Mosher, and A. Kuppermann, *J. Chem. Phys.*, 65, 2655, 1976.

Goyette, A. N., J. de Urquijo, Y. Wang, L. G. Christophorou, and J. K. Olthoff, *J. Chem. Phys.*, 114, 8932, 2001.

Jarvis, G. K., K. J. Boyle, C. A. Mayhew, and R. P. Tuckett, *J. Phys. Chem. A*, 102, 3230, 1998.

Jiang, Y., J. Sun, and L. Wan, *Phys. Rev. A*, 62, 062712, 2000.

Lifshitz, C. and R. Grajower, *Int. J. Mass. Spectrom. Ion Phys.*, 10, 25, 1972/73.

Lifshitz, C. and F. A. Long, *J. Chem. Phys.*, 41, 2468, 1964.

McNeil, R. I., F. Williams, and M. B. Yim, *Chem. Phys. Lett.*, 61, 293, 1979.

Panajotovic, R., M. Jelisavcic, R. Kajita, T. Tanaka, M. Kitajima, H. Cho, H. Tanaka, and S. J. Buckman, *J. Chem. Phys.*, 121, 4559, 2004.

Shimanouchi, T. *Tables of Molecular Vibrational Frequencies Consolidated Volume*, NSRDS-NBS 39, Washington (DC), 1972, p. 75.

Szmytkowski, C., S. Kwitnewski, and E. Ptasińska-Denga, *Phys. Rev. A*, 68, 032715, 2003.

Thynne, J. C. J. and K. A. G. Macneil, *Int. J. Mass Spectrom. Ion Phys.*, 5, 455, 1970, cited by Lifshitz and Grajower, 1972/73.

Winstead, C. and V. Mckoy, *J. Chem. Phys.*, 116, 1380, 2002.

Yoshida, K., S. Goto, H. Tagashira, C. Winstead, B. V. McKoy, and W. L. Morgan, *J. Appl. Phys.*, 91, 2637, 2002.

83

TRIBROMOETHENE

CONTENTS

Tribromoethene (C_2HBr_3) is an electron-attaching gas that has 118 electrons.

83.1 ATTACHMENT RATE CONSTANT

The thermal attachment rate constant for C_2HBr_3 is 1.2×10^{-13} m^3/s (Sunagawa and Shimamori, 1995). Figure 83.1 shows the attachment rates as a function of mean energy of the electron swarm, measured by the same authors.

83.2 ATTACHMENT CROSS SECTION

Table 83.1 and Figure 83.2 show the dissociative attachment cross section as a function of mean energy derived by Sunagawa and Shimamori (1995). The attachment cross section, as in most bromine-containing compounds has a peak at zero energy.

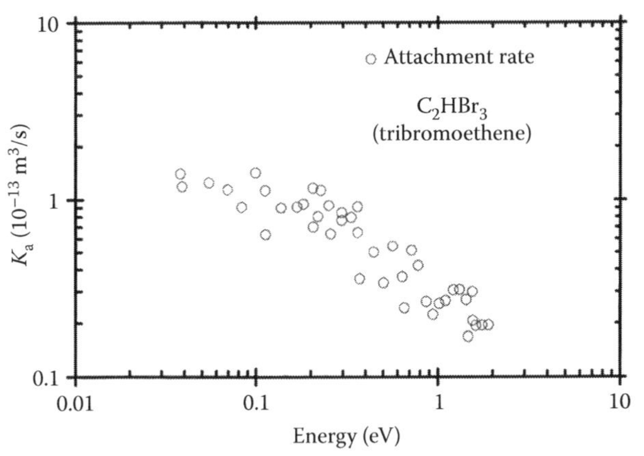

FIGURE 83.1 Dissociative attachment rate for C_2HBr_3. (Adapted from Sunagawa, T. and H. Shimamori, *Int. J. Mass Spectrom. Ion Proc.*, 149/150, 123, 1995.)

TABLE 83.1
Dissociative Attachment Cross Section for C_2HBr_3

Energy (meV)	Q_a (10^{-20} m^2)	Energy (meV)	Q_a (10^{-20} m^2)
1.00	788.46	90.0	52.71
2.00	587.38	95.0	50.75
3.00	468.77	100	48.92
4.00	400.42	150	34.05
6.00	322.22	200	24.20
8.00	278.85	250	18.70
10.0	248.28	300	15.19
15.0	194.02	350	12.34
20.0	164.88	400	10.03
25.0	152.83	450	8.25
30.0	142.72	500	6.98
35.0	123.81	550	6.16
40.0	103.13	600	5.57
45.0	89.54	650	5.05
50.0	81.43	700	4.58
55.0	75.26	750	4.14
60.0	70.17	800	3.75
65.0	65.99	850	3.39
70.0	62.54	900	3.07
75.0	59.64	1000	2.58
80.0	57.12	1500	1.36
85.0	54.83	2000	0.97

Source: Digitized from Sunagawa, T. and H. Shimamori, *Int. J. Mass Spectrom. Ion Proc.*, 149/150, 123, 1995.

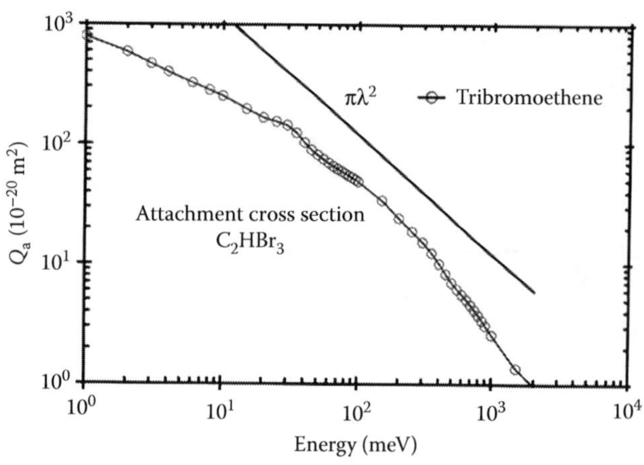

FIGURE 83.2 Dissociative attachment cross section for C_2HBr_3. *s*-Wave maximum cross section added for comparison. (Adapted from Sunagawa, T. and H. Shimamori, *Int. J. Mass Spectrom. Ion Proc.*, 149/150, 123, 1995.)

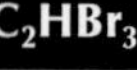

REFERENCE

Sunagawa, T. and H. Shimamori, *Int. J. Mass Spectrom. Ion Proc.*, 149/150, 123, 1995.

84

TRICHLOROETHENE

CONTENTS

Trichloroethene (C_2HCl_3) is an electron-attaching, weakly polar molecule that has 64 electrons. Its electronic polarizability is 11.16×10^{-40}F m^2, dipole moment 0.8 D, and ionization potential is 9.46 eV.

84.1 IONIZATION CROSS SECTIONS

Ionization cross sections have been measured by Hudson et al. (2001) as shown in Table 84.1 and Figure 84.1.

TABLE 84.1
Ionization Cross Sections for C_2HCl_3

Energy (eV)	Q_i (10^{-20} m^2)	Energy (eV)	Q_i (10^{-20} m^2)
16	2.80	97	17.66
20	6.31	101	17.65
23	9.39	105	17.37
27	11.31	108	17.17
31	12.82	112	17.15
34	14.00	116	17.09
38	14.49	119	17.04
42	15.15	123	16.99
45	15.78	127	16.62
49	16.22	130	16.53
53	16.86	134	16.44
57	17.08	138	16.37
60	17.51	142	16.22
64	17.58	145	16.10
68	17.75	149	15.92
71	17.72	153	15.60
75	17.84	156	15.41
79	17.99	160	15.29
82	17.80	164	15.28
86	17.88	167	15.27
90	17.75	171	15.27
94	17.67	175	15.22

TABLE 84.1 (continued)
Ionization Cross Sections for C_2HCl_3

Energy (eV)	Q_i (10^{-20} m^2)	Energy (eV)	Q_i (10^{-20} m^2)
179	14.98	201	14.01
182	14.78	204	13.99
186	14.50	208	14.05
190	14.26	212	14.07
193	14.09	215	13.97
197	14.03	219	13.79

Source: Adapted from Hudson, J. E. et al., *J. Phys. B: At. Mol. Opt. Phys.*, 34, 3025, 2001.
Note: Tabulated values by courtesy of Professor Harland (2006).

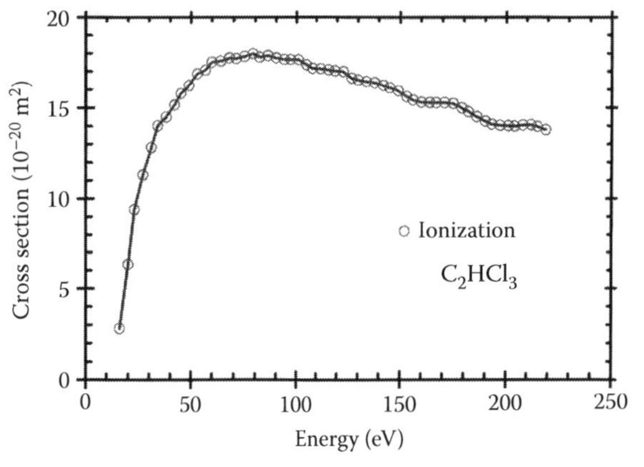

FIGURE 84.1 Ionization cross sections for C_2HCl_3. Tabulated values by courtesy of Professor Harland (2006). (Adapted from Hudson, J. E. et al., *J. Phys. B: At. Mol. Opt. Phys.*, 34, 3025, 2001.)

C_2HCl_3

84.2 ATTACHMENT CROSS SECTIONS

Attachment studies by Johnson et al. (1977) have shown three species of negative ions, Cl^-, Cl_2^-, and $C_2HCl_2^-$. The relative abundance of the ions are shown in Figure 84.2. The details are shown in Table 84.2.

The attachment cross sections derived from swarm measurements are shown in Table 84.3 and Figure 84.3 (Johnson et al., 1977).

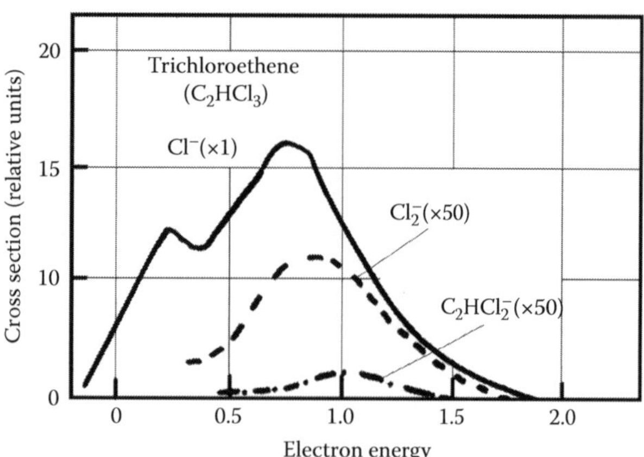

FIGURE 84.2 Relative abundance of negative ions for C_2HCl_3. Note the two peaks for the Cl^- ion and the multiplication factor for each curve. (Adapted from Johnson, J. P., L. G. Christophorou, and J. G. Carter, *J. Chem. Phys.*, 67, 2196, 1977.)

TABLE 84.2
Peak Position and Relative Abundance of Negative Ions

Species	Peak (eV)	Relative Yield
Cl^-	0.15	1000
	0.7	
Cl_2^-	0.85	11
$C_2HCl_2^-$	1.0	2

Source: Adapted from Johnson, J. P., L. G. Christophorou, and J. G. Carter, *J. Chem. Phys.*, 67, 2196, 1977.

TABLE 84.3
Attachment Cross Sections for C_2HCl_3

Energy (eV)	Q_a (10^{-20} m²)	Energy (eV)	Q_a (10^{-20} m²)
0.10	0.40	0.34	0.45
0.12	0.92	0.36	0.42
0.13	1.91	0.38	0.42
0.16	2.62	0.40	0.45
0.17	2.91	0.42	0.50
0.19	2.82	0.45	0.61
0.21	2.39	0.5	1.23
0.24	1.77	0.55	2.04
0.26	1.21	0.6	3.47
0.29	0.70	0.65	5.26
0.31	0.53	0.7	6.55

TABLE 84.3 (continued)
Attachment Cross Sections for C_2HCl_3

Energy (eV)	Q_a (10^{-20} m²)	Energy (eV)	Q_a (10^{-20} m²)
0.75	7.10	0.95	2.07
0.8	6.64	1.0	1.08
0.85	4.63	1.1	0.12
0.9	3.23	1.17	0.01

Source: Digitized from Johnson, J. P., L. G. Christophorou, and J. G. Carter, *J. Chem. Phys.*, 67, 2196, 1977.

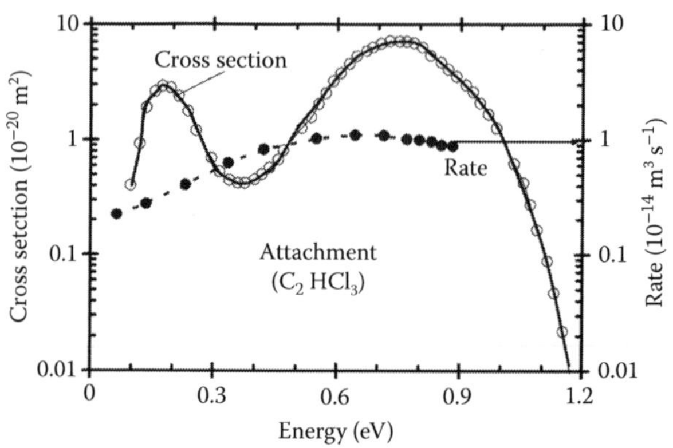

FIGURE 84.3 Attachment cross sections and rates as functions of mean energy. (Adapted from Johnson, J. P., L. G. Christophorou, and J. G. Carter, *J. Chem. Phys.*, 67, 2196, 1977.)

TABLE 84.4
Thermal Attachment Rate Constants for C_2HCl_3

Rate Constant (10^{-14} m³/s)	Method	Reference
0.22	Swarm	Johnson et al. (1977)
0.19	Swarm	Blaunstein and Christophorou (1968)

84.3 ATTACHMENT RATES

Thermal attachment rates are shown in Table 84.4 and included in Figure 84.3 as a function of mean energy of the electron swarm.

REFERENCES

Blaunstein, R. P. and L. G. Christophorou, *J. Chem. Phys.*, 49, 1526, 1968.

Hudson, J. E., C. Vallance, M. Bart, and P. W. Harland, *J. Phys. B: At. Mol. Opt. Phys.*, 34, 3025, 2001.

Johnson, J. P., L. G. Christophorou, and J. G. Carter, *J. Chem. Phys.*, 67, 2196, 1977.

Section VII

7 ATOMS

85

ALLENE

CONTENTS

Allene (C_3H_4), an isomer of cyclopropene with the structure $H_2C{=}C{=}CH_2$, is an open-chain molecule with two adjacent carbon–carbon double bonds and with two CH_2 groups twisted with respect to each other by 90° (Szmytkowski and Kwitnewski, 2002). Its ground-state configuration is given by Nakano et al. (2002). The geometrical length of the molecule is ~0.385 nm and a width of 0.179 nm (Nakano et al., 2002). The ground-state designation of the molecule is X 1A (Nakano et al., 2002). and the ionization potential is 9.69 eV (Lide, 2005–2006). C_3H_4 has 11 vibrational modes as shown in Table 85.1.

TABLE 85.1
Vibrational Modes and Energy of C_3H_4 Molecule

Mode	Energy (meV)	Type of Motion
ν_1	373.8	CH_2 symmetrical stretch
ν_2	178.9	CH_2 scissoring
ν_3	133.0	CC stretch
ν_4	107.2	CH_2 twisting
ν_5	372.8	CH_2 symmetrical stretch
ν_6	242.6	CC stretch
ν_7	173.3	CH_2 scissoring
ν_8	382.6	CH_2 antisymmetrical stretch
ν_9	123.9	CH_2 rocking
ν_{10}	104.3	CH_2 wagging
ν_{11}	44.0	CCC deformation

Source: Adapted from Shimanouchi, T. *Tables of Molecular Vibrational Frequencies, Consolidated Volume 1*, National Standards Reference Data Set, National Bureau of Standards (US), NSRDS-NBS, 39, 1972, p. 115.

85.1 SELECTED REFERENCES FOR DATA

See Table 85.2.

85.2 TOTAL SCATTERING CROSS SECTION

See Table 85.3.

85.3 ELASTIC SCATTERING AND MOMENTUM TRANSFER CROSS SECTION

See Tables 85.4 and 85.5.

TABLE 85.2
Selected References for Data

Type	Energy Range (eV)	Reference
Q_{el}, Q_M	0–40	Lopez and Bettega (2003)
Q_T	0.5–370	Szmytkowski and Kwitnewski (2002)
Q_{diff}, Q_v	0–100	Nakano et al. (2002)
Q_{ex}	20, 40, 60	Mosher et al. (1973)

Note: Q_{diff} = differential; Q_{el} = elastic; Q_M = momentum transfer; Q_T = total, Q_i = ionization cross section.

TABLE 85.3
Total Scattering Cross Sections of C_3H_4

	Q_T (10^{-20} m^2)			Q_T (10^{-20} m^2)	
Energy (eV)	Allene (C_3H_4)	Propyne (C_3H_4)	Energy (eV)	Allene (C_3H_4)	Propyne (C_3H_4)
0.5	16.3	18.9	9.5	38.4	38.0
0.6	16.4	18.7	10	38.5	37.5
0.8	16.8	19.1	11	37.9	36.4
1.0	17.3	19.6	12	37.5	35.4
1.2	18.7	20.2	13	36.8	35.2
1.4	21.7	20.8	14	36.8	35.0
1.6	24.3	22.0	15	36.3	34.6
1.8	27.8	22.8	16	35.3	34.0
2.0	31.5	24.2	18	34.5	32.8
2.1	36.8		20	33.7	31.5
2.2	39.2	25.9	25	31.9	30.4
2.3	39.9		30	30.7	29.1
2.4	39.7	28.6	35	29.7	28.3
2.5	38.4		40	28.1	27.3
2.6	35.4	32.0	45	27.0	26.4
2.8	31.2	35.4	50	26.6	25.8
3.0	29.3	39.3	60	25.3	24.1
3.2	27.7	41.2	70	23.9	23.1
3.5	27.1	41.6	80	22.8	21.9
3.7	27.5	40.0	90	21.3	20.7
4.0	28.3	38.0	100	20.4	19.7
4.5	29.4	35.8	110	19.2	18.8
5.0	30.6	35.1	120	18.3	17.8
5.5	31.9	35.2	140	16.7	16.4
.0	32.8	36.0	160	15.4	15.0
6.2	32.8		180	14.2	13.7
6.5	33.3	36.7	200	12.9	12.4
6.7	33.4		220	11.8	11.4
7.0	34.5	37.7	250	10.3	10.1
7.5	35.6	38.4	275	9.60	9.31
8.0	36.5	38.8	300	8.76	8.54
8.5	36.9	38.9	350	7.47	7.56
9.0	37.6	38.5	370	7.00	7.09

Source: Adapted from Szmytkowski, Cz. and S. Kwitnewski, *J. Phys. B: At. Mol. Opt. Phys.*, 35, 3781, 2002.

Note: Cross sections for propyne (isomer) are given for comparison. See Figure 85.1 for graphical presentation.

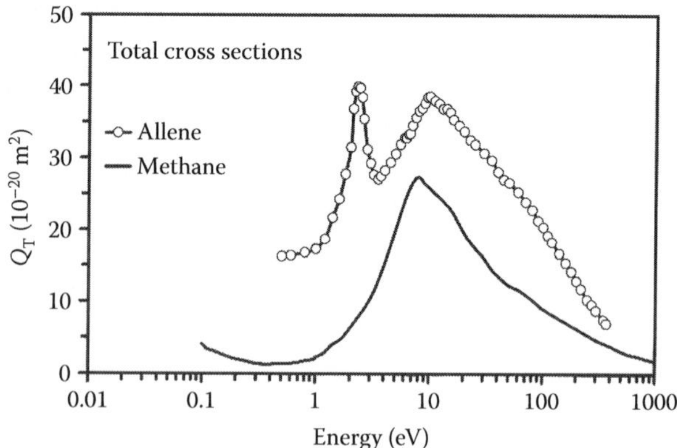

FIGURE 85.1 Total scattering cross section in C_3H_4. Cross sections in methane (CH_4) are plotted for comparison. (—○—) C_3H_4 (Szmytkowski and Kwitnewski, 2002); (——) methane, see section on methane for references.

TABLE 85.4
Integral Elastic Scattering and Momentum Transfer Cross Sections of C_3H_4

Energy (eV)	Q_{el} (10^{-20} m^2)	Q_M (10^{-20} m^2)	Energy (eV)	Q_{el} (10^{-20} m^2)	Q_M (10^{-20} m^2)
0.5	25.8	24.2	7.0		19.1
1.0	25.9	19.2	8.0	29.1	17.7
1.5		15.5	10	30.9	17.1
2.0	24.0	15.0	15	30.5	16.5
2.5		14.9	20	26.3	13.5
3.0		15.0	25	22.3	10.8
4.0	40.7	25.7	30	20.7	9.6
4.5	43.3		35	17.4	7.1
5.0		21.0	40	16.7	6.6
6.0	32.0	19.5			

Source: Digitized from Lopez, A. R. and M. H. F. Bettega, *Phys. Rev. A*, 67, 032711, 2003.

Note: See Figures 85.2 and 85.3 for graphical presentation.

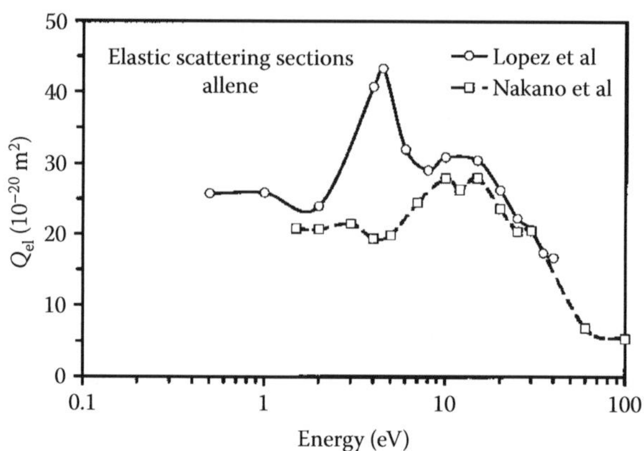

FIGURE 85.2 Elastic scattering cross sections of C_3H_4. (—○—) Lopez and Bettega (2003), theory; (—□—) Nakano et al. (2002), integral values are derived by the present author from differential cross section measurements.

85.4 DIFFERENTIAL SCATTERING CROSS SECTION

See Table 85.6.

85.5 IONIZATION CROSS SECTION

Only theoretical ionization cross sections of C_3H_4 calculated by the Binary Encounter Bethe (BEB) method (Kim et al.) are available. See Figure 85.6 for graphical presentation.

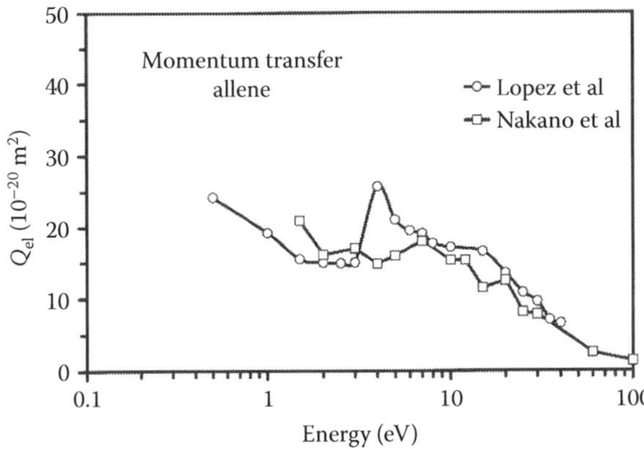

FIGURE 85.3 Momentum cross sections of C_3H_4. (—○—) Lopez and Bettega (2003), theory; (—□—) Nakano et al. (2002), integral values are derived by the present author from differential cross section measurements.

TABLE 85.5
Experimental Integral Elastic Scattering Cross Sections and Momentum Transfer

Energy (eV)	Q_{el} (10^{-20} m^2)	Q_M (10^{-20} m^2)	Energy (eV)	Q_{el} (10^{-20} m^2)	Q_M (10^{-20} m^2)
1.5	20.80	20.93	12.0	26.38	15.31
2.0	20.71	16.10	15.0	28.01	11.46
3.0	21.51	16.97	20.0	23.68	12.48
4.0	19.40	14.79	25.0	20.43	8.17
5.0	19.90	15.92	30.0	20.52	7.82
7.0	24.53	18.00	60.0	6.89	2.53
10.0	28.00	15.31	100.0	5.41	1.35

Note: Digitized and integrated by the author from differential scattering measurements of Nakano et al. (2002). See Figures 85.2 and 85.3 for graphical presentation.

TABLE 85.6
Differential Cross Section of C_3H_4 (10^{-20} m^2/sr) as Functions of Scattering Angle and Selected Electron Energy

Angle (deg)	Energy (eV) 1.5	5	20	100
0	10.00	5.50	40.00	16.00
10	4.92	4.18	24.71	9.80
20	2.78	3.22	11.56	0.10
30	2.21	2.48	5.03	1.04
40	1.32	2.34	2.57	0.32
50	1.12	2.21	1.59	0.21
60	1.06	2.03	1.03	0.20
70	1.40	1.73	0.53	0.15
80	1.64	1.60	0.55	0.07
90	1.43	1.23	0.60	0.08
100	1.37	1.13	0.57	0.08

TABLE 85.6 (continued)
Differential Cross Section of C_3H_4 (10^{-20} m^2/sr) as Functions of Scattering Angle and Selected Electron Energy

Angle (deg)	Energy (eV) 1.5	5	20	100
120	1.47	0.83	0.49	0.08
140	1.65	1.14	1.13	0.06
160	2.80	1.57	2.00	0.09
180	4.50	3.00	4.00	0.10

Source: Adapted from Nakano, Y. et al., *Phy. Rev. A*, 66, 032714, 2002.
Note: See Figures 85.4 and 85.5 for graphical presentation.

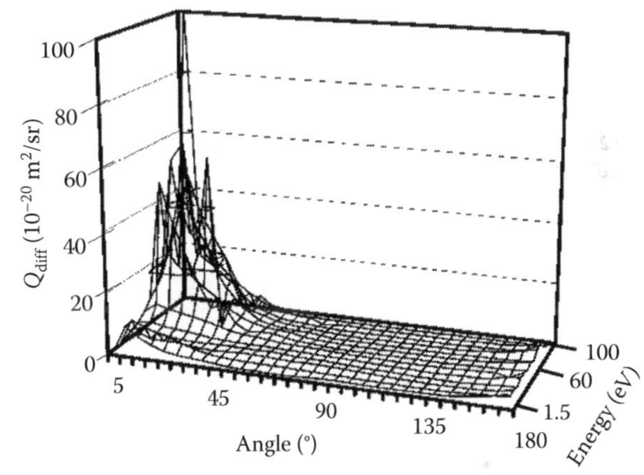

FIGURE 85.4 Three-dimensional view of differential scattering cross section as a function of scattering angle at various electron impact energies in the range from 0 to 100 eV. Measured data of Nakano et al. (2002) supplemented with their theoretical calculation at extremities of the angular range.

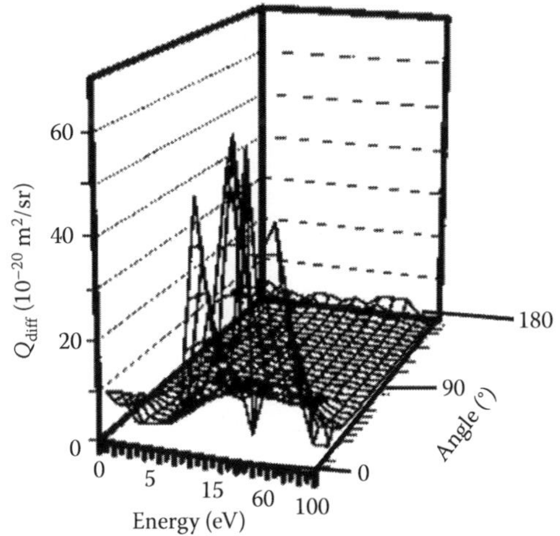

FIGURE 85.5 Three-dimensional view of differential cross section of C_3H_4 as a function of electron energy and scattering angle derived from measurements of Nakano et al. (2002) supplemented with their theoretical calculation at extremities of the angular range.

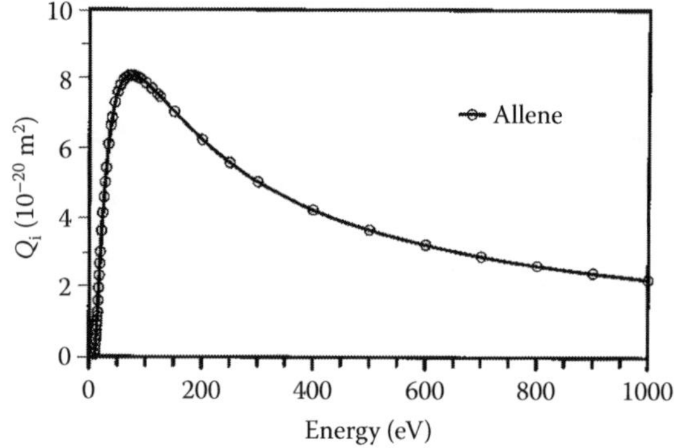

FIGURE 85.6 Theoretical ionization cross section of C_3H_4 by the BEB method. (Adapted from Kim, Y.-K. et al. http://physics.nist.gov/PhysRevData/contebts-miss.html.)

C_3H_4

85.6 NOTES ON CROSS SECTION DATA

1. C_3H_4 has shape resonance centered around 2.3 eV in comparison with 3.4 eV for propyne. There is also a broad structure around 10 eV for C_3H_4 and around 8.5 eV for propyne. Both isomers have similar total cross section–energy curves (Szmytkowski and Kwitnewski, 2002).
2. The differential scattering cross section shows resonance behavior below 4 eV (Nakano et al., 2002). The differential scattering at 100 eV shows undulations characteristic of interference of the scattering waves from the composite atoms in the molecule.
3. C_3H_4 has 11 fundamental vibrational modes. Three compound peaks are observed at 0.12 eV (C–H bending vibration), 0.24 eV (C–C stretching vibration), and 0.38 eV (C–H stretching vibration). The scattering is isotropic and dominated by the *s*-wave only near zero energies. As the energy increases above 5 meV, the backward scattering decreases to half the value of the forward scattering—a typical characteristic of *p*-wave scattering.
4. Experimental data on absolute excitation and ionization cross sections are required.

REFERENCES

Kim, Y.-K., K. K. Irikura, M. E. Rudd, M. A. Ali, and P. M. Stone, http://physics.nist.gov/PhysRevData/contents-misc.html.

Lide, D. R. (Ed.), *Handbook of Chemistry and Physics* (86th edn), CRC Press, Boca Raton, FL, 2005–2006.

Lopez, A. R. and M. H. F. Bettega, *Phys. Rev. A*, 67, 032711, 2003.

Mosher, O. A., W. M. Flicker, and A. K. Kupperman, *J. Chem Phys.*, 62, 2600, 1973.

Nakano, Y., M. Hoshino, M. Kitajima, H. Tanaka, and M. Kimura, *Phys. Rev. A*, 66, 032714, 2002.

Shimanouchi, T. *Tables of Molecular Vibrational Frequencies, Consolidated Volume 1*, National Standards Reference Data Set, National Bureau of Standards (US), NSRDS-NBS, 39, Washington (DC), 1972, p. 115.

Szmytkowski, Cz. and S. Kwitnewski, *J. Phys. B: At. Mol. Opt. Phys.*, 35, 3781, 2002.

86

CYCLOPROPENE

CONTENTS

Cyclopropene (C_3H_4) is a gas with 22 electrons and belongs to the group of cycloalkenes. The three compounds with the same chemical formula and selected properties are shown in Table 86.1.

86.1 SELECTED REFERENCES FOR CROSS SECTIONS

See Table 86.2.

TABLE 86.1
Selected Properties of C_3H_4 Molecules

Gas	α_e (10^{-40} F m^2)	μ (*D*)	ε_i (eV)
Cyclopropene		0.784	9.67
Allene		0.454	9.692
Propyne	6.88		10.37

Note: α_e = Electronic polarizability; μ = dipole moment; ε_i = ionization energy.

TABLE 86.2
Selected References for Cross Sections

Type	Energy (eV)	Reference
Q_{diff}, Q_{el}, Q_M	0–40	Lopez and Bettaga. (2003)
Q_{diff}, Q_v	0–100	Nakano et. al. (2002)

Note: Q_{diff} = Differential; Q_M = momentum transfer; Q_T = total; Q_v = vibrational excitation.

86.2 TOTAL CROSS SECTIONS

Data on total cross sections are not available.

86.3 INTEGRAL ELASTIC AND MOMENTUM TRANSFER CROSS SECTIONS

See Tables 86.3 and 86.4.

TABLE 86.3
Theoretical Integral Elastic and Momentum Cross Sections of C_3H_4

Energy (eV)	Q_{el} (10^{-20} m^2)	Q_M (10^{-20} m^2)	Energy (eV)	Q_{el} (10^{-20} m^2)	Q_M (10^{-20} m^2)
0.3		39.7	6.0	33.0	27.1
0.5	50.1	41.1	7.0		25.1
1.0	42.1	34.5	8.0	33.8	18.9
1.5		25.6	10	32.2	17.5
2.0	29.8	21.9	15	30.0	18.3
2.5		19.4	20	26.7	15.3
3.0		21.8	25	21.9	12.7
3.9		26.4	30	20.2	11.5
4.0	33.8	19.4	35	19.2	10.1
5.0		21.8	40	18.3	9.3

Source: Digitized from Lopez, A. R. and M. H. F. Bettega, *Phys. Rev. A*, 67, 032711, 2003.
Note: See Figure 86.1 for graphical presentation.

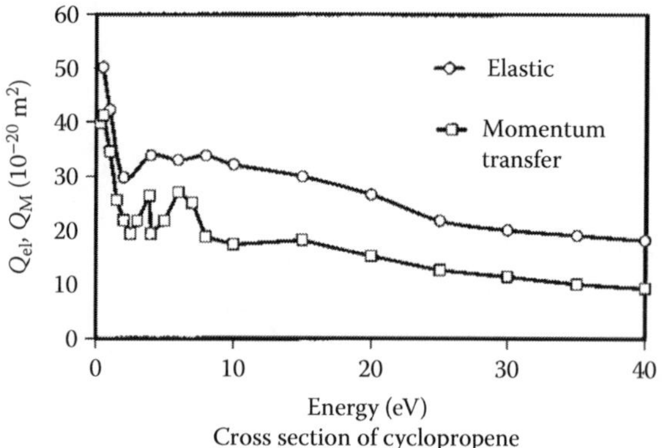

FIGURE 86.1 Theoretically calculated integral elastic scattering and momentum transfer cross sections of C_3H_4. (Adapted from Lopez, A. R. and M. H. F. Bettega, *Phys. Rev. A*, 67, 032711, 2003.)

TABLE 86.4
Differential Cross Sections of C_3H_4 in Units of 10^{-20} m²/sr

	Energy (eV)						
Angle°	**5**	**7**	**10**	**15**	**20**	**25**	**30**
0	11.55	14.88	21.00	25.00	26.00	25.00	22.00
5	11.03	15.23	17.96	21.43	22.83	21.59	20.39
10	10.35	14.44	15.93	18.89	21.70	19.78	18.88
15	9.52	12.85	14.20	16.19	16.59	16.77	16.36
20	8.60	10.85	12.11	12.47	11.89	12.35	11.83
25	7.61	8.78	9.30	8.69	8.26	7.72	7.06
30	6.58	7.01	6.55	5.83	5.20	4.79	4.56
35	5.57	5.58	5.05	3.77	3.77	3.44	3.05
40	4.59	4.31	3.77	2.47	2.44	1.63	1.93
45	3.70	3.21	2.62	1.83	1.63	1.24	1.25
50	2.93	2.40	1.93	1.52	1.28	1.14	1.03
55	2.31	1.95	1.60	1.35	1.14	0.93	0.92
60	1.83	1.68	1.43	1.27	1.13	0.90	0.83
65	1.48	1.38	1.30	1.22	1.14	0.96	0.76
70	1.20	1.20	1.25	1.17	1.08	0.96	0.70
75	0.97	1.12	1.25	1.10	0.97	0.90	0.65
80	0.89	1.06	1.27	1.04	0.87	0.82	0.62
85	0.91	1.09	1.34	1.05	0.81	0.78	0.59
90	0.98	1.18	1.46	1.12	0.79	0.76	0.56
95	1.08	1.32	1.55	1.21	0.80	0.75	0.53
100	1.21	1.54	1.61	1.28	0.82	0.75	0.51
110	1.58	2.12	1.89	1.36	0.88	0.78	0.56
120	1.86	2.15	2.06	1.36	0.99	0.90	0.67
130	2.02	2.11	1.92	1.33	1.11	1.04	0.78
140	2.25	1.95	1.77	1.42	1.16	1.18	0.90
150	2.58	2.09	1.68	1.79	1.28	1.32	1.10
160	2.70	2.35	1.96	2.41	1.71	1.50	1.29
170	2.79	1.57	2.34	2.99	1.83	1.77	1.28
180	3.30	3.00	2.69	3.50	2.30	2.43	1.70

Source: Adapted from Lopez, A. R. and M. H. F. Bettega, *Phys. Rev. A*, 67, 032711, 2003.

Note: Figures 86.2 and 86.3 are graphical presentations.

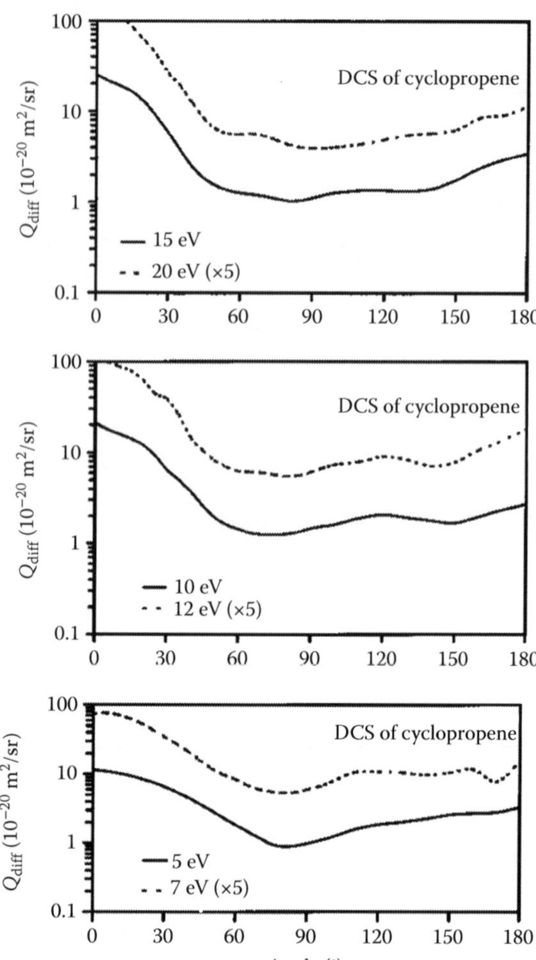

FIGURE 86.2 Differential cross section of C_3H_4 as a function of angle of scattering at various electron energies. Note the multiplying factor that is used for clarity of presentation. (The data are from Lopez and Bettega (2003) digitized and interpolated by the present author. Experimental values are not available for comparison. The shape of the curve is nearly independent of the electron energy.)

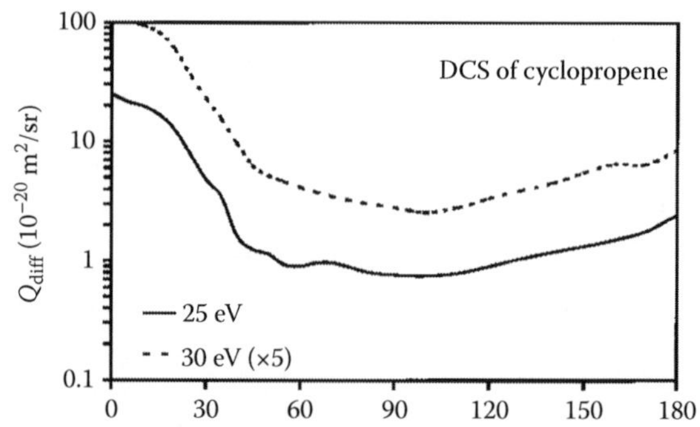

FIGURE 86.3 Same as Figure 86.2, except for electron energy.

REFERENCES

Lopez, A. R. and M. H. F. Bettega, *Phys. Rev. A*, 67, 032711, 2003.

Nakano, Y., M. Hoshino, M. Kitajima, H. Tanaka, and M. Kimura, *Phys. Rev. A*, 66, 032714, 2002.

87

ETHANAL

CONTENTS

Ethanal (C_2H_4O), synonym acetaldehyde, is a polar molecule that has 24 electrons. Its dipole moment is 2.75 D and ionization potential 10.229 eV. The molecule has 15 modes of vibrational excitation as shown in Table 87.1 (Shimanouchi, 1972).

87.1 IONIZATION CROSS SECTION

A single experimental study has been made by Vacher et al. (2006). These data are shown in Table 87.2 and Figure 87.1.

Mass number and ion species; 44-$C_2H_4O^+$; 43-$C_2H_3O^+$; 42-$C_2H_2O^+$; 41-C_2HO^+; 29-CHO^+; 26-$C_2H_2^+$.

TABLE 87.1
Vibrational Excitation Modes and Energies

Designation	Mode	Energy (meV)
ν_1	CH_3 degenerate-stretch	372.6
ν_2	CH_3 symmetrical-stretch	361.7
ν_3	CH stretch	349.9
ν_4	CO stretch	216.1
ν_5	CH_3 degenerate-deform	178.7
ν_6	CH bend	173.6
ν_7	CH_3 symmetrical-deform	167.6
ν_8	CC stretch	138.0
ν_9	CH_3 rock	113.9
ν_{10}	CCO deform	63.1
ν_{11}	CH_3 degenerate-stretch	367.9
ν_{12}	CH_3 degenerate-deform	176.1
ν_{13}	CH_3 rock	107.5
ν_{14}	CH bend	94.6
ν_{15}	Torsion	18.6

Source: Adapted from Shimanouchi, T., *Tables of Molecular Vibrational Frequencies Consolidated Volume I*, NSRDS-NBS 39, U. S. Department of Commerce, Washington (DC) 1972.

TABLE 87.2
Partial and Total Ionization Cross Sections

	Ionization Cross Section (10^{-20} m^2)						
Energy	Ion Mass Number						
(eV)	44	29	43	42	41	26	Total
14		1.10	0.164				1.264
15	0.028	1.10	0.173				1.301
16	0.133	1.24	0.370				1.743
18	0.366	1.15	0.528				2.044
20	0.526	1.29	0.628	0.0017			2.446
25	0.948	1.76	0.914	0.0934	0.005	0.009	3.730
30	1.130	1.71	0.981	0.144	0.030	0.022	4.018
35	0.836	1.40	0.712	0.181	0.077	0.028	3.234
40	0.828	1.23	0.720	0.158	0.120	0.060	3.116
50	1.100	0.96	0.698	0.162	0.105	0.070	3.093
60	1.340	1.27	0.796	0.267	0.159	0.089	3.921
70	1.020	1.28	0.800	0.257	0.112	0.077	3.546
80	0.845	1.33	0.805	0.206	0.109	0.062	3.357
86	1.030	1.29	0.801	0.243	0.126	0.069	3.559

Source: Tabulated values by courtesy of Dr. Vacher, J. R. et al., *Chem. Phys.*, 323, 587, 2006.

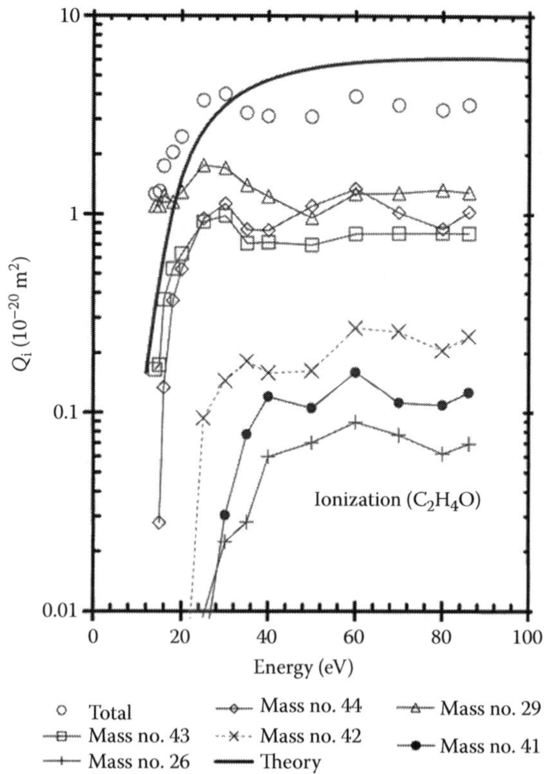

FIGURE 87.1 Selected partial and total ionization cross sections for C_2H_4O. (○) Total; (Δ) mass no. 29; (◊) mass no. 44; (□) mass no. 43; (×) mass no. 42; (&cirf;) mass no. 41; (+) mass no. 26; (——) theory. See Table 87.2 for ion species identification. (Adapted from Vacher, J. R. et al., *Chem.Phys.*, 323, 587, 2006.)

REFERENCES

Shimanouchi, T., *Tables of Molecular Vibrational Frequencies Consolidated Volume I*, NSRDS-NBS 39, U.S. Department of Commerce, 1972.

Vacher, J. R., F. Jorand, N. Blin-Simiand, and S. Pasquiers, *Chem. Phys.*, 323, 587, 2006.

88

METHYLAMINE

CONTENTS

Methylamine (CH_3NH_2), is a gas whose molecule has 18 electrons. It has a polarizability of 4.46×10^{-40} F m^2, dipole moment of 1.31 D, and ionization potential of 8.80 eV. The molecule has 15 vibrational modes (Shimanouchi, 1972) as shown in Table 88.1.

TABLE 88.1
Vibrational Modes in CH_3NH_2

Mode	Deformation	Energy (meV)	Mode	Deformation	Energy (meV)
ν_1	NH_2 symmetrical-stretch	416.7	ν_9	NH_2 wag	96.7
ν_2	CH_3 degenerate-stretch	367.1	ν_{10}	NH_2 antisymmetrical-stretch	424.9
ν_3	CH_3 symmetrical-stretch	349.6	ν_{11}	CH_3 degenerate-stretch	370.1
ν_4	NH_2 scissor	201.2	ν_{12}	CH_3 degenerate-deform	184.1
ν_5	CH_3 degenerate-deform	182.6	ν_{13}	NH_2 twist	175.9
ν_6	CH_3 symmetrical-deform	177.3	ν_{14}	CH_3 rock	148.2
ν_7	CH_3 rock	140.1	ν_{15}	Torsion	33.2
ν_8	CN stretch	129.4			

Source: Adapted from Shimanouchi, T. *Tables of Molecular Vibrational Frequencies Consolidated Volume I*, NSRDS-NBS, 39, Washington (DC), 1972.

88.1 TOTAL SCATTERING CROSS SECTION

The total scattering cross sections measured by Szmytkowski and Krzysztofowicz (1995) are shown in Table 88.2 and

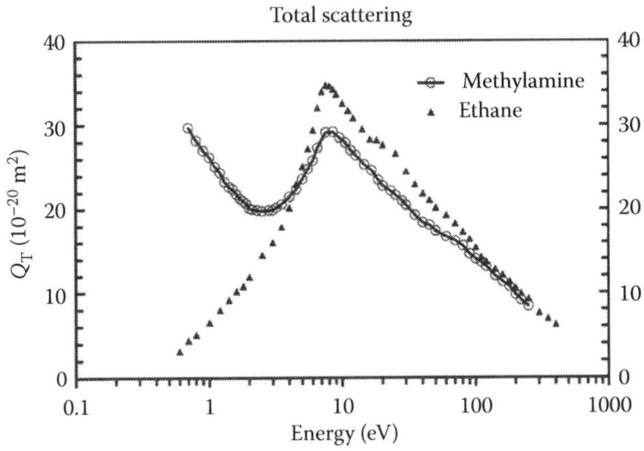

FIGURE 88.1 Total scattering cross section in CH_3NH_2. C_2H_6 data are added for comparison. (Adapted from Szmytkowski, C. and A. M. Krzysztofowicz, *J. Phys. B: At. Mol. Opt. Phys.*, 28, 4291, 1995.)

TABLE 88.2
Total Scattering Cross Section in CH_3NH_2

Energy (eV)	Q_T (10^{-20} m^2)	Energy (eV)	Q_T (10^{-20} m^2)
CH_3NH_2		C_2H_6	
0.7	29.7	0.6	3.11
0.8	28.2	0.7	4.37
0.9	27.0	0.8	5.05
1.0	26.2	1.0	6.50
1.1	25.1	1.2	7.97
1.2	24.4	1.4	9.13
1.3	23.3	1.6	10.2
1.4	22.7	1.8	10.8
1.5	22.4	2.0	11.9
1.6	21.8	2.5	14.5
1.7	21.3	3.0	16.0

continued

TABLE 88.2 (continued)
Total Scattering Cross Section in CH_3NH_2

Energy (eV)	Q_T (10^{-20} m²)	Energy (eV)	Q_T (10^{-20} m²)
CH_3NH_2		C_2H_6	
1.8	21.0	3.5	17.9
1.9	20.7	4	20.2
2.0	20.1	4.5	23.0
2.1	20.0	5	25.1
2.3	19.9	5.5	27.2
2.5	19.8	6	29.4
2.8	19.9	6.5	32.0
3.0	19.9	7	33.9
3.2	20.2	7.5	34.7
3.5	20.6	8	34.6
4.0	21.5	8.5	34.2
4.5	22.4	9	33.6
5.0	23.6	10	32.5
5.5	24.8	11	31.7
6.0	25.8	12	30.8
6.5	27.3	14	29.5
7.5	29.1	16	28.3
8.5	29.2	18	28.2
9.5	28.5	20	27.6
10.5	27.9	25	26.6
11.5	27.0	30	24.5
12.5	26.4	35	23.0
14.5	25.3	40	21.9
16.5	24.6	45	21.1
18.5	23.5	50	20.2
20	22.8	60	19.3
23	22.2	70	18.3
25	21.7	80	17.3
28	21.0	90	16.4
30	20.5	100	15.4
35	19.3	110	14.2
40	18.4	120	13.7
45	18.1	140	12.8
50	17.4	160	12.1
60	16.7	180	11.3
70	16.2	200	10.5
80	15.6	220	9.90
90	14.6	250	9.26
100	14.0		
110	13.6		
120	13.1		
140	12.0		
160	11.3		
180	10.7		
200	9.76		
220	9.09		
250	8.38		

Source: Adapted from Szmytkowski, C. and A. M. Krzysztofowicz, *J. Phys. B: At. Mol. Opt. Phys.*, 28, 4291, 1995.

Figure 88.1. Ethane (C_2H_6) data are given for comparison. The highlights of the cross section are

1. The cross section decreases rapidly from 0.7 eV reaching a minimum at 2.5 eV.
2. For higher energies it increases reaching a peak at 8.5 eV. This is attributed to negative ion formation, formed by both direct attachment and resonance process.
3. Beyond 8.5 eV the cross section decreases monotonically, as seen in many gases.

REFERENCES

Shimanouchi, T. *Tables of Molecular Vibrational Frequencies Consolidated Volume I*, NSRDS-NBS, 39, Washington (DC), 1972.

Szmytkowski, Cz. and A. M. Krzysztofowicz, *J. Phys. B: At. Mol. Opt. Phys.*, 28, 4291, 1995.

89

PROPYNE

CONTENTS

Propyne (C_3H_4) is an isomer of cyclopropene with the structure $H_3C–C≡CH$ with one triple bond and a single C–C bond (Szmytkowski and Kwitnewski, 2002). Its ground-state electronic configuration is given by Nakano et al. (2002) The geometric length of C_3H_4 molecule is approximately 0.484 nm and the width is ~0.18 nm (Nakano et al., 2002). It has a dipole moment of 0.78 D and a quadrupole moment of 6.1×10^{-10} D m. It has 10 fundamental vibrational modes classified as shown (Palmer et al., 1999) in Table 89.1

The ground-state designation is X $^1A_1$ (see Nakano et al., 2002) and the ionization potential is 10.37 eV (Lide, 2005–2006). Excitation potentials are given in Table 111.2.

89.1 SELECTED REFERENCES FOR DATA

See Table 89.2.

TABLE 89.1
Vibrational Excitation Energy of C_3H_4

Mode	Energy (eV)	Assignment
v_1	0.413	Acetylene C–H stretch
v_2	0.365	Symmetric methyl stretch
v_3	0.264	C≡C stretch
v_4	0.172	Methyl deformation
v_5	0.115	C–C stretch
v_6	0.369	Antisymmetric methyl C–H stretch
v_7	0.180	Methyl skeletal deformation
v_8	0.128	Methyl skeletal rock
v_9	0.078	C≡C–H bend
v_{10}	0.046	C–C≡C bend

TABLE 89.2
Selected References for Data

Type	Energy Range (eV)	Reference
Q_{diff}, Q_{el}, Q_M	0–40	Lopez et al. (2003)
Q_{diff}, Q_v	0–100	Nakano et al. (2002)
Q_T	0.5–370	Szmytkowski and Kwitnewski (2002)
Q_{ex}	20–100	Flicker et al. (1979)
Q_a	0–15	Rutkowsky et al. (1980)

Note: Q_{diff} = differential; Q_{el} = elastic scattering; Q_{ex} = excitation; Q_M = momentum transfer; Q_T = total; Q_v = vibrational excitation.

89.2 TOTAL SCATTERING CROSS SECTIONS

See Table 89.3.

89.3 ELASTIC AND MOMENTUM TRANSFER CROSS SECTIONS

See Tables 89.4 and 89.5.

89.4 DIFFERENTIAL CROSS SECTIONS

See Table 89.6.

89.5 ATTACHMENT CROSS SECTIONS

Electron attachment occurs in hydrocarbons due to both direct and dissociative attachment processes. The attachment cross section (see Table 89.7 and Figure 89.6) depends upon

C₃H₄

TABLE 89.3
Total Scattering Cross Sections of C_3H_4

Energy (eV)	Q_T (10^{-20} m^2) Allene (C_3H_4)	Q_T (10^{-20} m^2) Propyne (C_3H_4)	Energy (eV)	Q_T (10^{-20} m^2) Allene (C_3H_4)	Q_T (10^{-20} m^2) Propyne (C_3H_4)
0.5	16.3	18.9	9.5	38.4	38.0
0.6	16.4	18.7	10	38.5	37.5
0.8	16.8	19.1	11	37.9	36.4
1.0	17.3	19.6	12	37.5	35.4
1.2	18.7	20.2	13	36.8	35.2
1.4	21.7	20.8	14	36.8	35.0
1.6	24.3	22.0	15	36.3	34.6
1.8	27.8	22.8	16	35.3	34.0
2.0	31.5	24.2	18	34.5	32.8
2.1	36.8		20	33.7	31.5
2.2	39.2	25.9	25	31.9	30.4
2.3	39.9		30	30.7	29.1
2.4	39.7	28.6	35	29.7	28.3
2.5	38.4		40	28.1	27.3
2.6	35.4	32.0	45	27.0	26.4
2.8	31.2	35.4	50	26.6	25.8
3.0	29.3	39.3	60	25.3	24.1
3.2	27.7	41.2	70	23.9	23.1
3.5	27.1	41.6	80	22.8	21.9
3.7	27.5	40.0	90	21.3	20.7
4.0	28.3	38.0	100	20.4	19.7
4.5	29.4	35.8	110	19.2	18.8
5.0	30.6	35.1	120	18.3	17.8
5.5	31.9	35.2	140	16.7	16.4
6.0	32.8	36.0	160	15.4	15.0
6.2	32.8		180	14.2	13.7
6.5	33.3	36.7	200	12.9	12.4
6.7	33.4		220	11.8	11.4
7.0	34.5	37.7	250	10.3	10.1
7.5	35.6	38.4	275	9.60	9.31
8.0	36.5	38.8	300	8.76	8.54
8.5	36.9	38.9	350	7.47	7.56
9.0	37.6	38.5	370	7.00	7.09

Source: Adapted from Szmytkowski, Cz. and S. Kwitnewski, *J. Phys. B: At. Mol. Opt. Phys.*, 35, 3781, 2002.
Note: See Figure 89.1 for graphical presentation. Cross sections for allene (isomer) are given for comparison.

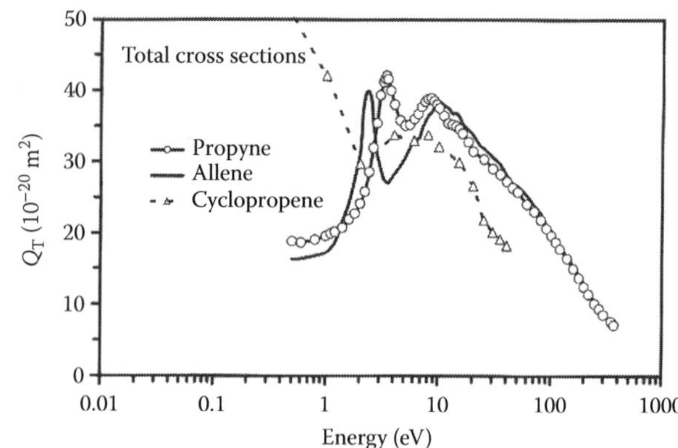

FIGURE 89.1 Total scattering cross sections in propene. The cross sections of allene and cyclopropene are also shown for comparison. Note that the cross sections of cyclopropene are integral elastic, theoretically obtained. (—○—) Propyne (Szmytkowski and Kwitnewski, 2002); (——) allene (Szmytkowski and Kwitnewski, 2002); (—△—) cyclopropene (Lopez and Bettega, 2003), theory.

TABLE 89.4
Theoretical Integral Elastic and Momentum Transfer Cross Sections of C_3H_4

Energy (eV)	Q_{el} (10^{-20} m^2)	Q_M (10^{-20} m^2)	Energy (eV)	Q_{el} (10^{-20} m^2)	Q_M (10^{-20} m^2)
0.5		24.7	6.0	38.5	23.4
1.0	25.7	19.2	8.0	33.3	21.3
1.5		17.2	10	33.5	21.8
2.0	25.8	16.8	15	33.1	19.4
2.5		17.2	20	27.1	14.2
3.0		17.8	25	25.2	13.3
4.0	32.4	20.5	30	22.5	12.2
5.0		29.8	35	21.6	10.7
5.1	43.6		40	20.4	9.9

Source: Digitized from Lopez, A. R. and M. H. F. Bettega, *Phys. Rev. A*, 67, 032711, 2003.
Note: See Figures 89.2 and 89.3 for graphical presentation.

the electron energy, the molecular size, and the number of unsaturated C–C bonds in the molecule. The dissociative attachment process is due to the following reactions (Rutkowsky et al., 1980)

$$e(>6\text{ eV}) + C_nH_{2n-2} \rightarrow C_nH_{2n-2}^{-*} \rightarrow \begin{cases} H^- + C_nH_{2n-2} \\ C_2^- + C_{n-2}H_{2n-2} \end{cases} \quad (89.1)$$

The dissociation energy of C_3H_4 (H–C_3H_3) is 4.551 eV (Lide, 2005–2006).

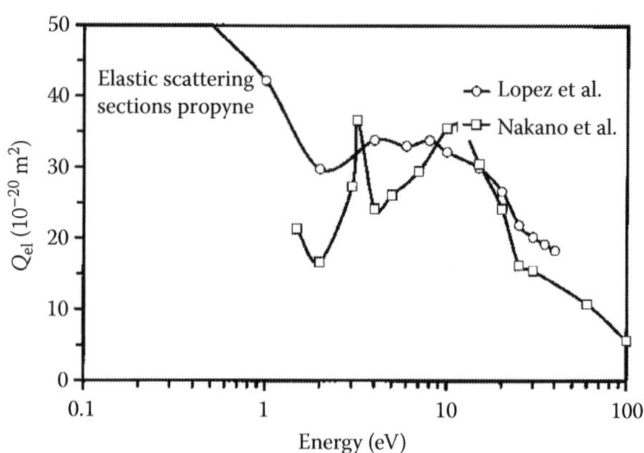

FIGURE 89.2 Elastic scattering cross sections of C_3H_4. (—○—) Lopez and Bettega (2003), theory; (—□—) integral values are evaluated by the present author from differential cross section measurements of Nakano et al. (2002).

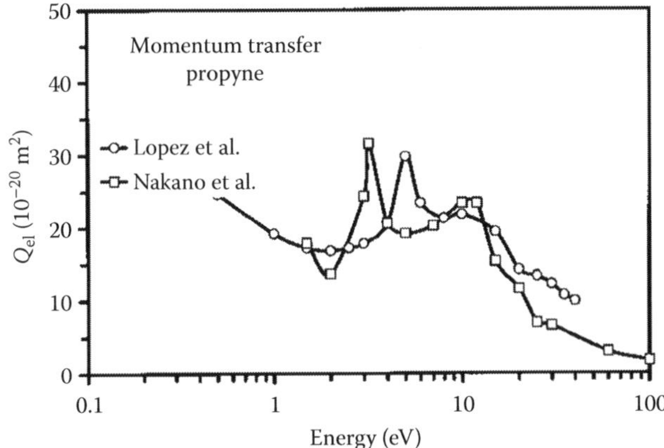

FIGURE 89.3 Momentum transfer cross sections of C_3H_4. (—○—) Lopez and Bettega (2003), theory; (—□—) integral values are evaluated by the present author from differential cross section measurements of Nakano et al. (2002).

TABLE 89.5
Integral Elastic and Momentum Transfer Cross Sections of C_3H_4[a]

Energy (eV)	Q_{el} (10^{-20} m^2)	Q_M (10^{-20} m^2)	Energy (eV)	Q_{el} (10^{-20} m^2)	Q_M (10^{-20} m^2)
1.5	21.32	17.88	12	35.79	23.38
2.0	16.66	13.66	15	30.53	15.30
3.0	27.33	24.37	20	24.22	11.59
3.2	36.58	31.60	25	16.19	6.94
4.0	24.20	20.68	30	15.49	6.57
5.0	26.17	19.17	60	10.81	3.01
7.0	29.50	20.25	100	5.66	1.75
10.0	35.51	23.39			

Note: See Figures 89.2 and 89.3 for graphical presentation.
[a] Derived by the Present Author from the Differential Cross Sections Measurements of Nakano et al. (2002).

TABLE 89.6
Differential Cross Sections of C_3H_4 (in Units of 10^{-20} m^2/sr) as a Function of Energy and Angle of Scattering

Angle (°)	1.5	3.2	5.0	10	20	30	60	100
0	7.74	22.50	15.00	40.00	57.80	28.04	40.00	23.82
5	6.52	18.03	9.82	28.45	32.50	26.14	34.30	14.31
10	5.31	12.98	6.93	21.22	28.26	22.81	23.50	9.46
15	4.15	9.09	5.59	16.21	16.21	16.92	9.74	5.76
20	3.13	6.56	5.03	11.98	12.10	7.60	5.97	3.20
25	2.31	5.29	4.71	8.81	8.75	3.91	1.94	1.51
30	1.76	4.68	4.35	6.82	5.52	3.74	1.62	0.71
35	1.50	4.13	3.80	5.65	3.56	1.75	1.67	0.68
40	1.47	3.58	3.27	4.86	2.58	0.72	1.00	0.64

TABLE 89.6 (continued)
Differential Cross Sections of C_3H_4 (in Units of 10^{-20} m^2/sr) as a Function of Energy and Angle of Scattering

Angle (°)	1.5	3.2	5.0	10	20	30	60	100
45	1.56	3.17	2.92	4.03	1.95	0.76	0.60	0.48
50	1.69	3.01	2.70	3.08	1.55	0.99	0.48	0.33
55	1.79	3.13	2.52	2.49	1.29	0.90	0.40	0.25
60	1.84	3.22	2.35	2.28	1.11	0.68	0.32	0.22
65	1.83	2.96	2.18	2.01	0.95	0.53	0.27	0.18
70	2.00	2.60	2.04	1.68	0.82	0.44	0.22	0.14
75	2.42	2.44	1.97	1.49	0.77	0.38	0.20	0.13
80	2.62	2.32	1.93	1.44	0.76	0.35	0.18	0.12
85	2.43	2.00	1.88	1.42	0.74	0.34	0.16	0.12
90	2.07	1.65	1.77	1.38	0.69	0.34	0.14	0.11
95	1.74	1.48	1.62	1.33	0.62	0.32	0.12	0.10
100	1.51	1.49	1.48	1.27	0.56	0.28	0.11	0.10
110	1.31	1.67	1.39	1.07	0.60	0.26	0.14	0.09
120	1.04	1.87	1.32	1.14	0.65	0.31	0.17	0.09
130	1.00	2.21	1.33	1.51	0.70	0.44	0.20	0.10
140	0.97	2.86	1.15	2.04	0.84	0.59	0.24	0.11
150	0.96	2.88	1.51	2.63	1.12	0.72	0.28	0.12
160	1.07	3.15	1.49	3.18	1.30	0.89	0.29	0.16
170	1.24	8.16	0.00	3.58	1.99	1.13	0.30	0.19
180	1.32	4.00	3.50	3.74	1.50	1.51	0.32	0.10

Energy (eV) columns: 1.5 to 100.

Note: See Figures 89.4 and 89.5 for graphical presentation. Digitized from measurements of Nakano et al. (2002) supplemented with their theoretical values at the extremities of the angular range.

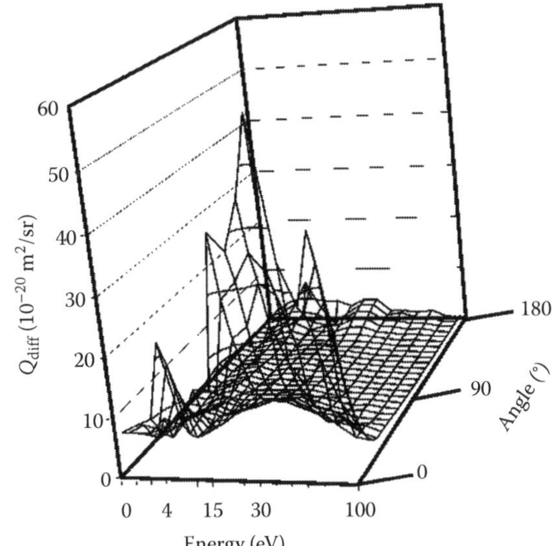

FIGURE 89.4 Three-dimensional view of differential scattering cross section as a function of energy and scattering angle. Measured data of Nakano et al. (2002) supplemented with their theoretical data at the extremities of the angular range.

C_3H_4

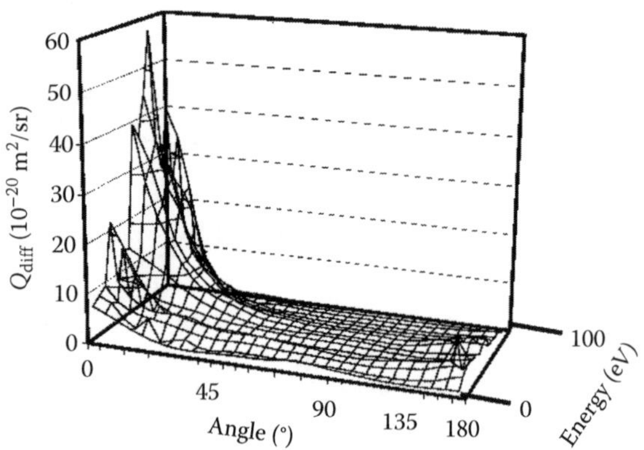

FIGURE 89.5 Three-dimensional view of differential scattering cross section as a function of the scattering angle at various impact energies in the range from 1.5 to 100 eV. Measured data of Nakano et al. (2002) supplemented with their theoretical data at the extremities of the angular range.

TABLE 89.7
Attachment Cross Sections in C_3H_4

Energy (eV)	Q_a (10^{-20} m²)	Energy (eV)	Q_a (10^{-20} m²)	Energy (eV)	Q_a (10^{-20} m²)
0.25	0.77	2.2	0.32	5.7	1.55
0.45	2.36	2.7	1.53	6	2.13
0.7	3.40	3	2.55	6.2	2.69
0.9	2.04	3.2	4.33	6.6	3.25
1.1	2.45	3.5	2.98	7.1	2.04
1.3	4.77	3.8	2.52	7.3	1.10
1.5	3.97	4.2	1.06	7.7	0.85
1.7	2.10	4.8	0.16	8.2	0.68
1.9	1.19	5.3	0.92	8.6	1.14

Source: Adapted from J. Rutkowsky, H. Drost, and H.-J Spangenberg, *Ann. Phys. Lpz.*, 37, 259, 1980.

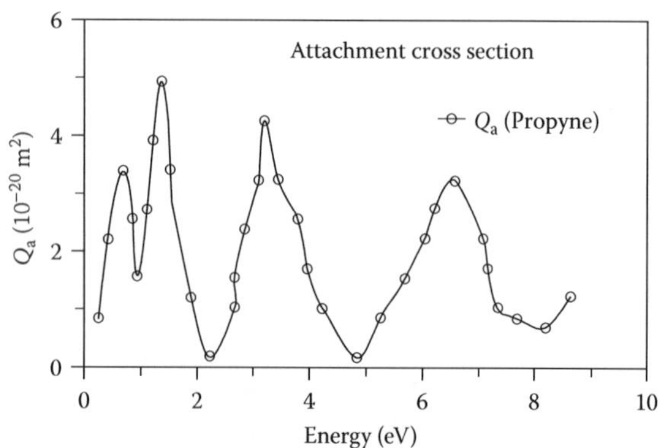

FIGURE 89.6 Attachment cross section in C_3H_4 as a function of electron energy. (Adapted from Rutkowsky, J., H. Drost, and H.-J Spangenberg, *Ann. Phys. Lpz.*, 37, 259, 1980.)

89.6 EXCITATION CROSS SECTION

The excitation potential and relative intensity are measured by Bowman and Miller (Table 89.8) (Bowman and Miller, 1965). Absolute excitation cross sections of C_3H_4 are not published.

TABLE 89.8
Excitation Potential and Relative Intensity of Propyne

Energy (eV)	Relative Intensity
2.8	<0.20
6.1	0.71
7.1	1.00
8.3	0.98

Source: Adapted from Bowman, C. R. and W. D. Miller, *J. Chem. Phys.*, 42, 681, 1965.

89.7 NOTES ON CROSS SECTION DATA

1. The total scattering cross sections of C_3H_4 and isomers (allene and cyclopropene) are similar.
2. C_3H_4 exhibits a shape resonance at 3.4 eV in contrast with the shape resonance at 2.3 eV for allene. The differential cross section shows a sharp peak at this energy for forward scattering ($\theta = 0$) and the highest cross section is obtained at 20 eV electron impact energy.
3. The differential cross sections of allene and C_3H_4 are much different signifying the isomeric effect more affirmatively (Nakano et al., 2002).
4. C_3H_4 has 10 fundamental modes of vibrational excitation. The vibrational excitation energy of C_3H_4 is 0.041 eV for CCC bending motion (ν_{10}). The low-energy differential scattering cross section shows structures at 1.5 eV and 80° due to *d*-wave scattering. This characteristic is also observed in acetylene (C_2H_2) and molecular nitrogen (N_2) which are isoelectronic to each other with 14 electrons (Nakano et al., 2002).
5. The attachment cross section shows several peaks including those below the dissociation energy and attributed to direct attachment.

REFERENCES

Bowman, C. R. and W. D. Miller, *J. Chem. Phys.*, 42, 681, 1965.

Flicker, W. M., O. A. Mosher, and A. Kupperman, *J. Chem. Phys.*, 69, 3311, 1979.

Lide, D. R. (Ed.), *Handbook of Chemistry and Physics*, 86th edn, CRC Press, Boca Raton, FL, 2005–2006.

Lopez, A. R. and M. H. F. Bettega, *Phys. Rev. A*, 67, 032711, 2003.

Nakano, Y., M. Hoshino, M. Kitajima, H. Tanaka, and M. Kimura, *Phys. Rev. A*, 66, 032714, 2002.

Palmer, M. H., C. C. Ballard, and I. C. Walker, *Chem. Phys.*, 249, 129, 1999.

Rutkowsky, J., H. Drost, and H.-J Spangenberg, *Ann. Phys. Lpz.*, 37, 259, 1980.

Szmytkowski, Cz. and S. Kwitnewski, *J. Phys. B: At. Mol. Opt. Phys.*, 35, 3781, 2002.

90

SULFUR HEXAFLUORIDE

SF$_6$

CONTENTS

Sulfur hexafluoride (SF_6) is a nonpolar, electron-attaching gas with 70 electrons. Its electronic polarizability is 7.277×10^{40} F m^2 and its ionization potential is 15.32 eV. Extensive data are provided by Christophorou and Olthoff (2000), Karwasz et al. (2003), Christophorou and Olthoff (2004), and Raju (2005).

90.1 SELECTED REFERENCES FOR DATA

See Table 90.1.

90.2 TOTAL SCATTERING CROSS SECTION

Highlights of the total cross section are (Figure 90.1)

1. Very high attachment cross section of 9500×10^{-20} m^2 at low energy of 0.0001 eV (Howe, 2001)
2. Several peaks due to attachment processes in the energy range of 1–20 eV
3. A monotonic decrease for energy >30 eV (see Table 90.2)

TABLE 90.1
Selected References for Data

Quantity	Range: eV, (Td)	Reference
Attachment cross section	0.007–0.042	**Ziesel et al. (2005)**
Ionization cross section	~45 eV	**Feil et al. (2004)**
Total scattering cross section	0.8–1000	**Makochekanwa et al. (2004)**
Attachment cross section	0–1.0	**Howe et al. (2001)**
Ionization cross section	18–1000	**Rejoub et al. (2001)**
Elastic scattering cross section	2.7–75	**Cho et al. (2000)**
Attachment coefficients	(250–1000)	Hayashi and Wang (2000)
Ionization cross section	(18–1000)	**Kim and Rudd (1999)**
Drift velocity	(150–3000)	**Lisovskiy et al. (1999)**
Swarm coefficients	(150–4000)	**Xiao et al. (1999)**
Ionization cross section	30–200	**Tarnovsky et al. (1998)**
Total scattering cross section	0.6–250	**Kasperski et al. (1997)**

continued

TABLE 90.1 (continued)
Selected References for Data

Quantity	Range: eV, (Td)	Reference
Total scattering cross section	100–700	Jiang et al. (1997)
Ionization cross section	16.5–100	**Rao and Srivastava (1997)**
Ionization cross section	18–6000	Hwang et al. (1996)
Attachment rate constant	0.2–1.0	**Christophorou and Dastkos (1995)**
Elastic scattering cross section	3–30	Gianturco et al. (1995)
Swarm coefficients	(45–285)	**Qiu and Xiao (1994)**
Total scattering cross section	0.002–1.0	**Wan et al. (1993)**
Attachment rate constant	0.001–0.2	**Klar et al. (1992)**
Elastic scattering cross section	0.05–1.0	**Randell et al. (1992)**
Attachment rate constant	0.005–1.0	**Shimamori et al. (1992)**
Total scattering cross section	75–4000	**Zecca et al. (1992)**
Detachment coefficient	(350–800)	**Hilmert and Schmidt (1991)**
Elastic scattering cross section	5–75	**Johnstone and Newell (1991)**
Ionization cross section	20–200	**Margreiter et al. (1990)**
Ion mobility	(20–510)	**de Urquijo-Carmona et al. (1990)**
Attachment rate constant	0.04–5 (0.003–5)	**Hunter et al. (1989)**
Elastic scattering cross section	75–700	**Sakae et al. (1989)**
Total scattering cross section	1–500	**Dababneh et al. (1988)**
Swarm coefficients	(450–6000)	**Hasegawa et al. (1988)**
Drift velocity	(20–700)	**Nakamura (1988)**
Momentum transfer cross section	(100–2000)	Phelps and van Brunt (1988)
Swarm coefficients	(300–600)	Fréchette (1986)
Swarm coefficients	(100–900)	**Ashwanden (1984)**
Total scattering cross section	0–24	**Romanyuk et al. (1984)**
Detachment coefficient	(350–800)	**Hansen et al. (1983)**
Ionization cross section	20–600	**Stanski and Adamczyk (1983)**
Elastic scattering cross section	5–75	**Trajmar et al. (1983)**
Total scattering cross section	0.036–1.0	**Ferch et al. (1982)**
Ionization coefficient	(250–600)	**Raju and Dincer (1982)**
Attachment coefficients	(250–550)	**Siddagangappa et al. (1982)**
Ionization coefficient	(250–500)	**Shimozuma et al. (1982)**
Attachment cross section	0–0.2	**Chutjian (1981)**
Drift velocity	(400–700)	**Urquiho-Carmona (1980)**
Ionization coefficient	(200–700)	**Itoh et al. (1979)**
Total scattering cross section	0.5–100.0	**Kennerly et al. (1979)**
Attachment coefficients	(300–500)	**Kline et al. (1979)**
Total scattering cross section	0.3–10.0	**Rohr (1979)**
Elastic scattering cross section	3.0–60.0	Benedict and Gemant (1978)
Elastic scattering cross section	0.2–20.0	Dehmer et al. (1978)
Attachment rate constant	(0.1–5.0)	**Gant (1976)**
Attachment coefficients	(300–700)	**Maller and Naidu (1976)**
Momentum transfer cross section	5–100	**Srivastava et al. (1976)**
Detachment coefficient	(400–500)	**O'Neil and Craggs (1973)**
Attachment rate constant	(0.1–4.0)	**Christophorou et al. (1972)**
Drift velocity	(300–700)	**Naidu and Prasad (1972)**
Ionization coefficient	(1000–6000)	**Teich and Sangi (1972)**
Drift velocity	(15–150)	**Harris and Jones (1971)**
Drift velocity	(400–700)	**Sangi (1971)**
Ionization cross section	16.5–600	**Rapp Englander-Golden (1965)**
Ionization cross section	(20–100)	**Asundi and Craggs (1964)**
Ionization coefficient	(250–450)	**Bhalla and Craggs (1962)**
Ionization coefficient	(250–450)	**Geballe and Harrison (1953)**

Note: Bold font denotes experimental study.

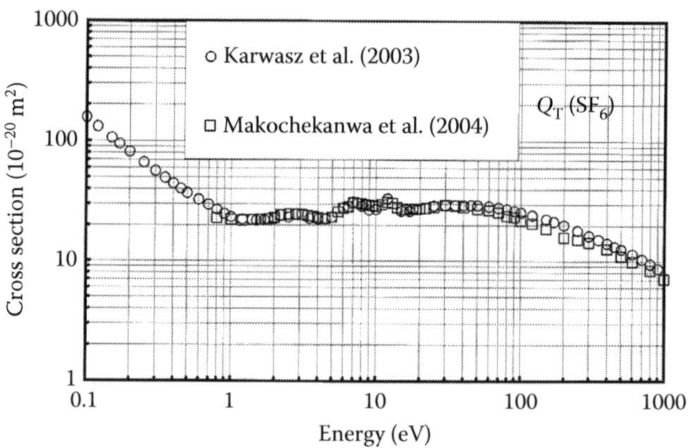

FIGURE 90.1 Total electron scattering, Q_T, of SF_6 as a function of electron energy. (a) (——) Recommended by Christophorou and Olthoff (2000); (×) Kasperski et al. (1997); (. . .) Jiang et al. (1995); (– – –) Wan et al. (1993); (○) Zecca et al. (1992); (●) Dababneh et al. (1988); (▼) Romanyuk (1984); (∇) Ferch et al. (1982); (- - -) Rohr (1979); (+) Kennerly et al. (1979); (b) expanded view.

TABLE 90.2
Total Scattering Cross Section

Energy (eV)	Q_T (10^{-20} m^2)	Energy (eV)	Q_T (10^{-20} m^2)	Energy (eV)	Q_T (10^{-20} m^2)
0.1	156	0.25	66.4	0.5	37.1
0.12	131	0.30	56.5	0.6	32.7
0.15	106	0.35	49.5	0.7	29.7
0.17	94.6	0.40	44.3	0.8	27.0 (22.9)
0.20	81.4	0.45	40.3	0.9	24.9

TABLE 90.2 (continued)
Total Scattering Cross Section

Energy (eV)	Q_T (10^{-20} m²)	Energy (eV)	Q_T (10^{-20} m²)	Energy (eV)	Q_T (10^{-20} m²)
1.0	23.5 (22.3)	7.0	31.3 (30.8)	45	29.4
1.2	22.4 (21.9)	7.5	(30.6)	50	29.3 (27.4)
1.4	(22.2)	8.0	28.9 (30.2)	60	28.8 (26.7)
1.5	21.9	8.5	(29.8)	70	28.0 (25.6)
1.6	(22.4)	9.0	27.0 (29.4)	80	27.3 (23.9)
1.7	22.1	9.5	(29.2)	90	26.4 (23.4)
1.8	(22.5)	10.0	27.1 (28.7)	100	25.6 (22.0)
2.0	22.6 (23.1)	11	(29.8)	120	24.3 (20.7)
2.2	(24.3)	12	33.3 (31.2)	150	22.5 (18.8)
2.5	23.3 (24.7)	13	(30.6)	170	21.5
2.8	(24.7)	14	(28.2)	200	20.1 (15.9)
3.0	23.1	15	26.3 (27.2)	250	18.1 (15.2)
3.1	(24.5)	16	(27.1)	300	16.4 (14.4)
3.4	(24.0)	17	26.2 (27.5)	350	15.2
3.5	22.8	18	(27.5)	400	14.2 (12.7)
3.7	(23.7)	19	(27.2)	450	13.3
4.0	22.4 (23.1)	20	27.3 (27.4)	500	12.6 (11.0)
4.5	22.5 (22.7)	22	(27.8)	600	11.5 (9.9)
5.0	23.5 (23.3)	25	29.1 (28.4)	700	10.4
5.5	(25.9)	30	29.3 (29.5)	800	9.47 (8.4)
6.0	27.9 (27.6)	35	29.3 (29.1)	900	8.69
6.5	(28.6)	40	29.3 (28.4)	1000	8.04 (7.1)

Source: Adapted from Karwasz, G. P., R. P. Brusa, and A. Zecca, in *Interactions of Photons and Electrons with Molecules*, Springer-Verlag, New York, NY, 2003, Chapter 6.

Note: See Figure 90.1 for graphical presentation. Numbers in brackets are from Makochekanwa et al. (2004).

TABLE 90.3
Suggested Elastic Scattering Cross Section

Energy (eV)	Q_{el} (10^{-20} m²)	Energy (eV)	Q_{el} (10^{-20} m²)	Energy (eV)	Q_{el} (10^{-20} m²)
Christophorou et al. (2003)		Sakae et al. (1989)		Johnson et al. (1991)	
0.30	45.6	75	20.0	5.0	15.08
0.35	33.0	100	18.7	7.5	22.72
0.40	26.2	150	15.6	10	19.41
0.45	21.8	200	13.1	12	28.26
0.50	18.7	300	10.6	15	17.07
0.60	14.8	500	7.59	20	18.10
0.70	12.5	700	6.02	30	16.41
0.80	11.1			40	12.38
0.90	10.2			50	11.23
1.0	9.72			75	9.37
1.2	9.73				
1.5	10.9				
2.0	14.8				
2.5	17.8				
3.0	19.3				
3.5	19.9				

TABLE 90.3 (continued)
Suggested Elastic Scattering Cross Section

Energy (eV)	Q_{el} (10^{-20} m²)	Energy (eV)	Q_{el} (10^{-20} m²)	Energy (eV)	Q_{el} (10^{-20} m²)
Christophorou et al. (2003)		Sakae et al. (1989)		Johnson et al. (1991)	
4.0	20.1				
4.5	20.6				
5.0	21.3				
6.0	23.6				
7.0	24.2				
8.0	24.6				
9.0	24.5				
10.0	24.8				
11.0	26.1				
12.0	26.6				
13.0	26.5				
14.0	26.1				
15.0	25.6				
16	25.2				
17	24.9				
18	24.8				
19	24.8				
20	24.7				
22	24.7				
25	24.7				
30	24.4				
35	24.0				
40	23.5				
45	22.8				
50	22.2				
60	21.3				
70	20.5				
75	20.2				
80	19.8				
90	19.1				
100	18.4				
125	16.9				
150	15.5				
200	13.3				
250	11.8				
300	10.6				
350	9.71				
400	8.95				
450	8.31				
500	7.74				
600	6.76				
700	5.94				

Note: See Figure 90.2 for graphical presentation.

90.3 ELASTIC SCATTERING CROSS SECTION

See Table 90.3.

SF$_6$

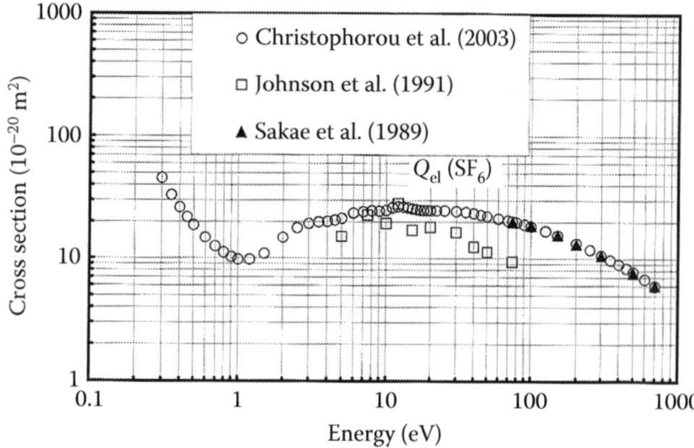

FIGURE 90.2 Elastic scattering cross section of SF_6 as a function of electron energy. Experimental: suggested: (——) Christophorou and Olthoff (2004); (▼) Cho et al. (2000); (– – –) Jiang et al. (1997); (• — • —) Gianturco et al. (1995); (+) Randell et al. (1992); (◆) Johnstone and Newell (1991); (■) Sakae et al. (1989); (●) Rohr (1979); (▲) Trajmar et al. (1983); (— —) Benedict and Gyemant (1978); (• • • — • • •) Dehmer et al. (1978).

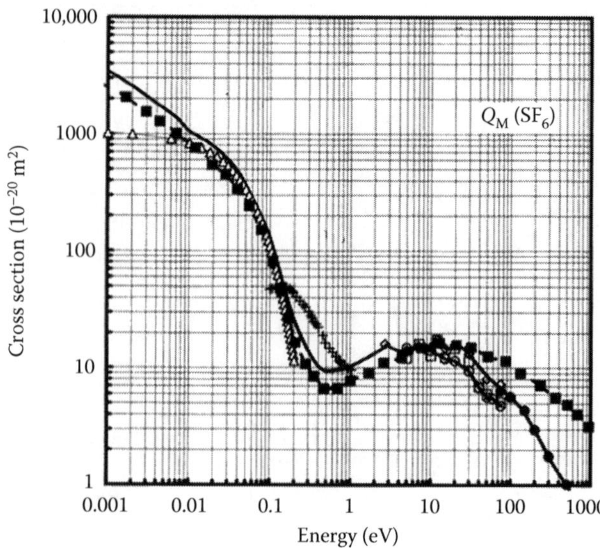

FIGURE 90.3 Momentum transfer cross sections of SF_6. (——) Christophorou and Olthoff (2004) supplemented by Raju (unpublished) in the range from 0 to 2.75 eV; (◊) Cho et al. (2000); (□) Johnstone and Newell (1991); (●) Sakae et al. (1989); (—■—) Phelps and van Brunt (1988); (—Δ—) derived from the attachment cross section of Chutjian (1981); (—○—) Srivastava et al. (1976) renormalized by Trajmar et al. (1983); (+) elastic collision cross section of Randell et al. (1992) plotted for comparison in the 0.1–1 eV range.

TABLE 90.4

Suggested Momentum Transfer Cross Section

Energy (eV)	Q_M (10^{-20} m^2)	Energy (eV)	Q_M (10^{-20} m^2)	Energy (eV)	Q_M (10^{-20} m^2)
0.001	3453.8	6.0	15.1	40	10.3
0.002	2542.9	7.0	15.5	45	9.37
0.004	1817.1	8.0	14.8	50	8.65
0.007	1371.7	9.0	14.4	60	7.69
0.01	1054.0	10	15.1	70	7.06
0.02	782.4	11	16.7	75	6.74
0.04	472.5	12	17.6	80	6.46
0.07	223.2	13	17.1	90	6.03
0.1	127.9	14	15.8	100	5.70
0.2	26.0	15	14.9	125	4.92
0.4	10.3	16	14.5	150	4.16
0.7	9.6	17	14.7	200	2.98
1	10.3	18	15.0	250	2.23
2	13.3	19	15.4	300	1.76
2.75	16.0 (15.7)	20	15.7	350	1.47
3.0	15.4	22	15.7	400	1.28
3.5	14.5	25	15.0	450	1.13
4.0	14.0	27	14.3	500	1.02
4.5	13.9	30	13.2	600	0.82
5.0	14.1	35	11.5	700	0.66

Source: Adapted from Raju, G. G. *Gaseous Electronics: Theory and Practice*, Taylor & Francis, London, 2005.

Note: See Figure 90.3 for graphical presentation.

90.4 MOMENTUM TRANSFER CROSS SECTION

See Table 90.4.

TABLE 90.5

Vibrational Excitation Modes

Designation	Type of Motion	Energy (meV)	Designation	Type of Motion	Energy (meV)
ν_1	Symmetrical stretch	96	ν_4	Degenerate Deform	76.4
ν_2	Degenerate stretch	79.6	ν_5	Degenerate Deform	65.1
ν_3	Degenerate stretch	117.5	ν_6	Deg enerate deform	43.0

Source: Adapted from Shimanouchi, T. *Tables of Molecular Vibrational Frequencies*, NSRDS-NBS 39, US Department of Commerce, Washington (DC), 1972.

90.5 VIBRATIONAL EXCITATION CROSS SECTION

See Tables 90.5 and 90.6.

90.6 ELECTRONIC EXCITATION

See Table 90.7.

90.7 IONIZATION CROSS SECTION

See Table 90.8.

TABLE 90.6
Vibrational Excitation Cross Section

Energy (eV)	Q_{vib}	Energy (eV)	Q_{vib}	Energy (eV)	Q_{vib} ($10^{-20}m^2$)
0.09	1.9	0.50	21.6	4.5	2.0
0.10	7.0	0.60	19.7	5.0	2.4
0.12	21.3	0.70	18.1	6.0	4.4
0.15	30.6	0.80	16.4	7.0	6.5
0.17	30.6	0.90	15.1	8.0	4.6
0.20	34.9	1.0	13.9	9.0	3.1
0.22	35.5	1.2	12.6	10.0	2.5
0.25	33.6	1.5	10.8	11.0	3.5
0.28	30.9	2.0	8.0	12.0	6.3
0.30	29.4	2.5	5.6	13.0	2.4
0.35	26.8	3.0	3.8	14.0	0.5
0.40	25.4	3.5	2.8		
0.45	23.5	4.0	2.3		

Source: Adapted from Christophorou, G. and J. K. Olthoff, *Fundamental Electron Interactions with Plasma Processing Gases*, Kluwer Academic/Plenum Publishers, New York, NY, 2004.
Note: Cross section in units of (10^{-20} m^2).

TABLE 90.7
Threshold Energy for Dissociation

Reaction	Threshold Energy (eV)
$SF_6 + e \rightarrow SF_5 + F + e$	9.6
$SF_6 + e \rightarrow SF_4 + F_2 + e$	11.3
$SF_6 + e \rightarrow SF_4 + 2F + e$	12.1
$SF_6 + e \rightarrow SF_3 + F + F_2 + e$	15.2
$SF_6 + e \rightarrow SF_3 + 3F + e$	16.0
$SF_6 + e \rightarrow SF_2 + 2F_2 + e$	17.0
$SF_6 + e \rightarrow SF_2 + 2F + F_2 + e$	17.8
$SF_6 + e \rightarrow SF_2 + 4F + e$	18.6
$SF_6 + e \rightarrow SF + F + 2F_2 + e$	21.1
$SF_6 + e \rightarrow SF + 3F + F_2 + e$	21.9
$SF_6 + e \rightarrow SF + 5F + e$	22.7

Source: Adapted from Ito, M. et al., *Contrib. Plas. Phys.*, 35, 405, 1995.
Note: Electronic excitation generally leads to dissociation.

90.8 ATTACHMENT PROCESSES

See Table 90.9. Selected attachment processes are

$$e + SF_6 \rightarrow SF_6^- \quad (90.1)$$

$$e + SF_6 \rightarrow SF_5^- + F \quad (90.2)$$

$$e + SF_6 \rightarrow SF_4^- + 2F \quad (90.3)$$

$$e + SF_6 \rightarrow SF_4^- + F_2 \quad (90.4)$$

TABLE 90.8
Ionization Cross Section

	Q_i (10^{-20} m^2)			Q_i (10^{-20} m^2)	
Energy (eV)	Rejoub et al. (2001)	Christophorou and Olthoff (2004)	Energy (eV)	Rejoub et al. (2001)	Christophorou and Olthoff (2004)
16.5		0.020	80	5.761	5.95
17.0		0.035	90	6.010	6.28
17.5		0.055	100	6.378	6.53
18	0.060	0.084	120	6.522	—
19.0		0.155	140	6.701	—
20	0.190	0.240	150		6.97
22.5	0.485	—	160	6.689	—
25	0.942	1.04	200	6.601	6.83
27.5	1.339	—	250	6.151	6.48
30	1.760	1.93	300	5.853	6.04
32.5	2.316	—	350		5.60
35	2.696	2.87	400	5.069	5.16
40	3.334	3.47	500	4.617	4.36
45	3.705	3.79	600	4.176	3.65
50	4.234	4.35	800	3.424	
60	5.012	5.09	1000	2.993	
70	5.482	5.65			

Note: See Figures 90.4 and 90.5 for graphical presentation.

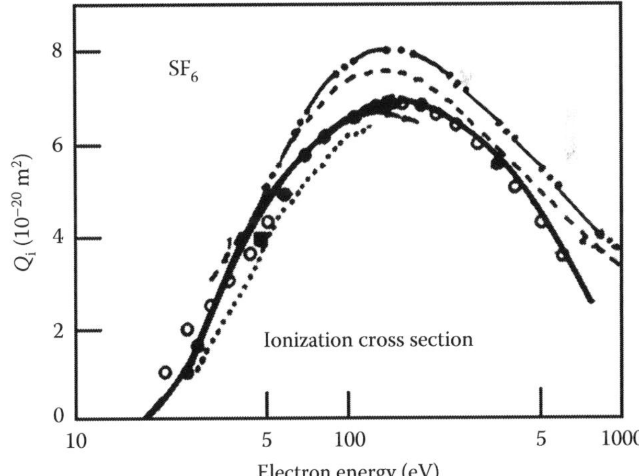

FIGURE 90.4 Ionization cross section of SF_6 as a function of electron energy. Measurements: (- - -) Asundi and Craggs (1964); (●) Rapp and Englander-Golden (1965); (○) Stanski et al. (1983); (– – –) Margreiter et al. (1990a); (. . . .) Margreiter et al. (1990b); (■) Rao and Srivastava (1997). Theory: (- — - —) Hwang et al. (1996); (— • — • —) Tarnovsky et al. (1998); (— ••• — ••• —) Kim and Rudd (1999). The measurements of Rejoub et al. (2001) agree very well with Rapp and Englander-Golden (1965). Recommended by Christophorou and Olthoff (2004).

$$e + SF_4 \rightarrow SF_3^- + F \quad (90.5)$$

$$e + SF_6 \rightarrow SF_6^- * (5.4\,eV) \rightarrow SF_4 + F_2^- \quad (90.6)$$

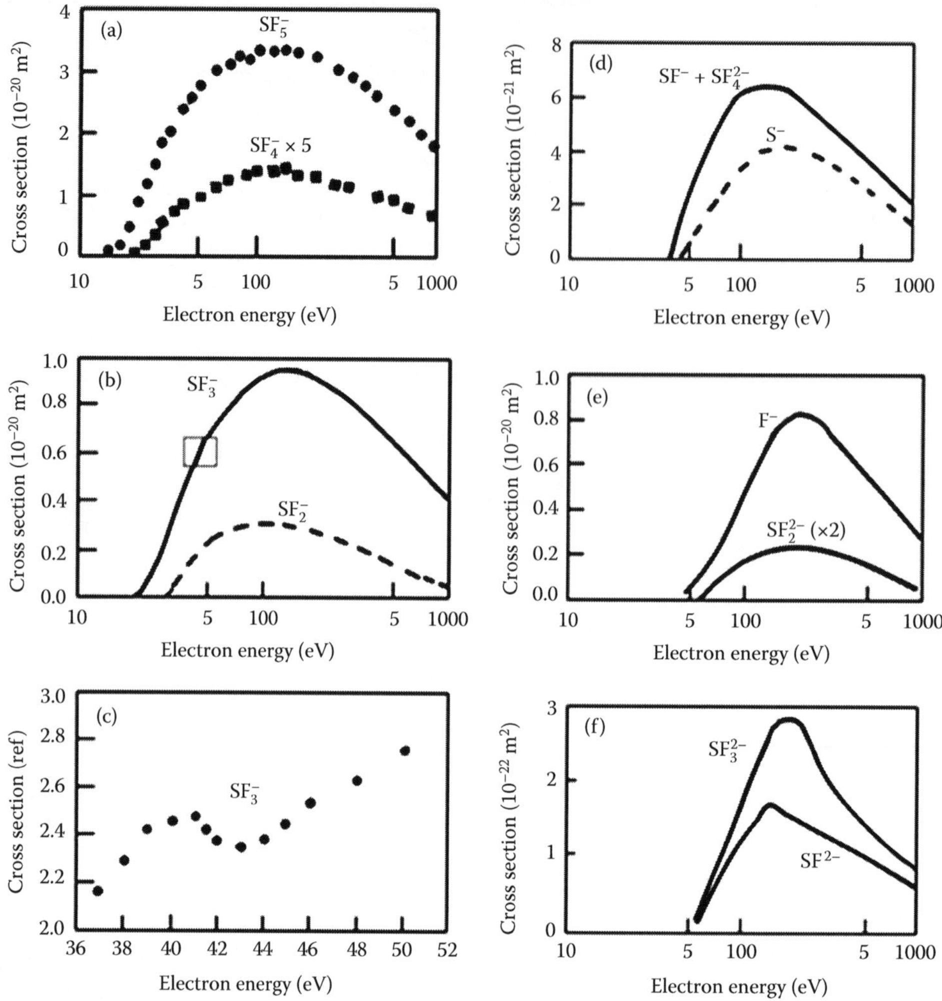

FIGURE 90.5 Partial impact ionization cross sections of SF_6. Structure is observed in the SF_3^+ cross section in the vicinity of 44 eV as shown in (c) with an enlarged abscissa (Rejoub et al. (2001). An explanation for this phenomenon has been experimentally provided by Feil et al. (2004). Appearance potentials of positive ions are (Christophorou and Olthoff, 2004): SF_5^+ – 11.2 eV; SF_4^+ – 14.5 eV; SF_3^+ – 17.0 eV; SF_2^+ – 21.8 eV; SF^+ – 27.0 eV.

TABLE 90.9
Attachment Processes

Ion	Threshold (eV)	Peak Energy (eV)	Q_a (Peak) (10^{-20} m²)	Reaction
SF_6^-	0.00	0.00	7617	90.1
SF_5^-	0.10	0.30	4.23	90.2
SF_4^-	2.8	5.0	4.57×10^{-3}	90.3, 90.4
SF_3^-	0.7	11.0	750×10^{-6}	90.5
F_2^-	2.00	8×10^{-5}		
	4.5	7.7×10^{-4}		
	11.5	9.54×10^{-4}	90.6	
F^-	0.2	5.0	0.0463	90.7
F^-	1.64	90.8		

Note: Peak energy and cross section are taken from Table 90.10. Neutral dissociation energies are given in Table 90.7.

TABLE 90.10
Attachment Cross Sections for SF_6 for Selected Ions

Energy (eV)	Total (10^{-20}m²)	SF_6^- (10^{-20} m²)	Energy (eV)	SF_5^- (10^{-20} m²)	Energy (ev)	F^- (10^{-20} m²)
0.0001	7617	7617	0.10	1.85	2.00	220 (−6)
0.0002	5283	5283	0.12	2.09	2.25	750 (−6)
0.0003	4284	4284	0.14	2.36	2.50	1490 (−6)
0.0004	3692	3692	0.15	2.48	3.0	980 (−6)
0.0005	3280	3280	0.16	2.61	3.5	1710 (−6)
0.0006	2968	2968	0.18	2.87	4.0	8560 (−6)
0.0007	2724	2724	0.20	3.15	4.5	26,900 (−6)
0.0008	2529	2529	0.22	3.45	5.0	46,300 (−6)
0.0009	2369	2369	0.25	3.86	5.5	43,900 (−6)
0.001	2237	2237	0.28	4.15	6.0	27,800 (−6)
0.002	1511	1511	0.30	4.24	6.5	13,700 (−6)
0.003	1202	1202	0.35	4.07	7.0	7490 (−6)

TABLE 90.10 (continued)
Attachment Cross Sections for SF_6 for Selected Ions

Energy (eV)	Total ($10^{-20}m^2$)	SF_6^- (10^{-20} m²)	Energy (eV)	SF_5^- (10^{-20} m²)	Energy (ev)	F^- (10^{-20} m²)
0.004	993	993	0.40	3.45	7.5	6150 (−6)
0.005	859	859	0.45	2.75	8.0	9770 (−6)
0.006	760	760	0.50	2.15	8.5	1420 (−6)
0.007	683	683	0.60	1.25	9.0	1570 (−6)
0.008	621	621	0.70	0.72	9.5	1390 (−6)
0.009	569	569	0.80	0.42	10.0	1130 (−6)
0.010	526	526	0.90	0.25	10.5	1310 (−6)
0.015	383	383	1.00	0.15	11.0	2100 (−6)
0.020	304	304	1.20	0.060	11.5	2350 (−6)
0.025	257	257	1.50	0.020	12.0	1950 (−6)
0.030	221	221			12.5	1220 (−6)
0.035	190	190			13.0	6290 (−6)
0.040	171	171			13.5	4000 (−6)
0.045	149	149			14.0	3400 (−6)
0.050	132	132			15.0	3000 (−6)
0.060	109	109				
0.070	92.7	92.7				
0.080	82.9	82.9				
0.090	74.3	74.3				
0.10	51.4	49.5				
0.12	32.9	30.8				
0.14	20.2	17.8				
0.15	16.7	14.2				
0.16	13.1	10.5				
0.18	8.72	5.85				
0.20	6.01	2.86				
0.22	4.69	1.24				
0.25	4.38	0.52				
0.28	4.40	0.25				
0.30	4.40	0.16				
0.35	4.12	0.05				
0.40	3.46	0.01				
0.45	2.75					
0.50	2.15					
0.60	1.25					
0.70	0.722					
0.80	0.416					
0.90	0.245					
1.00	0.147					
1.20	0.060					
1.50	0.020					
2.00	0.0043					
2.25	0.0020					
2.50	0.0019					
3.0	0.0010					
3.5	0.0018					
4.0	0.0092					
4.5	0.0290					
5.0	0.0514					
5.5	0.0493					
6.0	0.0317					
6.5	0.0162					
7.0	0.0088					

TABLE 90.10 (continued)
Attachment Cross Sections for SF_6 for Selected Ions

Energy (eV)	Total ($10^{-20}m^2$)	SF_6^- (10^{-20} m²)	Energy (eV)	SF_5^- (10^{-20} m²)	Energy (ev)	F^- (10^{-20} m²)
7.5	0.0066					
8.0	0.0099					
8.5	0.0143					
9.0	0.0159					
9.5	0.0144					
10.0	0.0120					
10.5	0.0142					
11.0	0.0227					
11.5	0.0252					
12.0	0.0206					
12.5	0.0128					
13.0	0.0066					
13.5	0.0041					
14.0	0.0035					
15.0	0.0030					

Source: Adapted from Christophorou, L. G. and J. K. Olthoff, *J. Phys. Chem. Ref. Data*, 29, 267, 2000.

Note: See Figure 90.6 for graphical presentation. More recent determination of Howe et al. (2001) gives the attachment cross section for SF_6^- at 0.0001 eV as 9500 × 10^{-20} m².

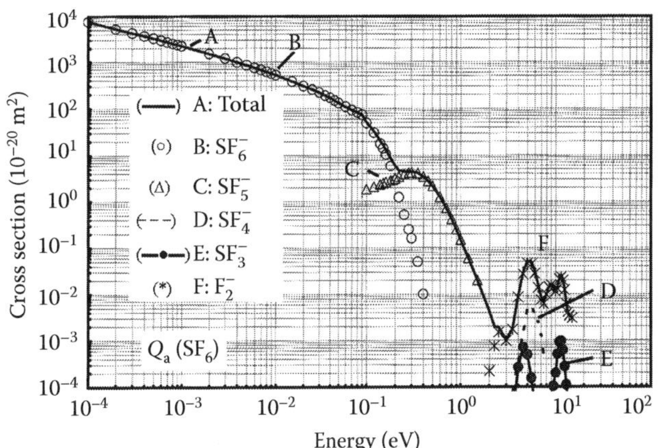

FIGURE 90.6 Direct and dissociative attachment cross sections of SF_6 as a function of energy. (Adapted from Christophorou, G. and J. K. Olthoff, *Fundamental Electron Interactions with Plasma Processing Gases*, Kluwer Academic/Plenum Publishers, New York, NY, 2004.)

$$e + SF_4 \rightarrow SF_3 + F^- \quad (90.7)$$

$$e + F_2 \rightarrow F + F^- \quad (90.8)$$

90.9 ATTACHMENT CROSS SECTIONS

See Table 90.10.

90.10 DRIFT VELOCITY OF ELECTRONS

See Table 90.11.

90.11 DIFFUSION COEFFICIENT

See Tables 90.12 and 90.13.

90.12 REDUCED IONIZATION COEFFICIENT

See Table 90.14.

TABLE 90.11
Recommended Drift Velocity of Electrons

E/N (Td)	W (10^4 m/s)	E/N (Td)	W (10^4 m/s)	E/N (Td)	W (10^4 m/s)
25	4.1	450	24.6	950	41.8
50	6.8	500	26.4	1000	43.4
100	10.2	550	28.2	1500	58.3
150	12.1	600	30.0	2000	71.7
200	13.5	650	31.7	2500	83.9
250	15.6	700	33.4	3000	95.4
275	17.0	750	35.1	3500	106.3
300	18.3	800	36.8	4000	116.8
350	20.5	850	38.5		
400	22.6	900	40.1		

Note: See Figure 90.7 for graphical presentation.

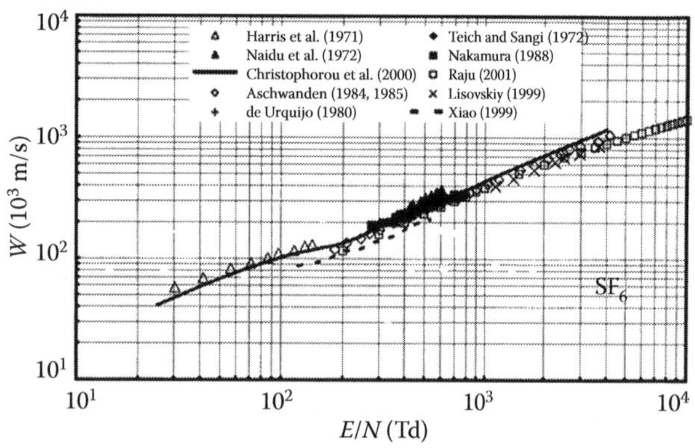

FIGURE 90.7 Drift velocity in SF_6 for the complete range. (——) Recommended by Christophorou and Olthoff (2004); (□) Raju (2001, unpublished); (×) Lisovskiy and Yegorenkov (1999); (– – –) Xiao et al. (1999); (■) Nakamura (1988); (◊) Aschwanden (1984); (+) de Urquijo (1980); (▲) Naidu and Prasad (1972); (Δ) Harris and Jones (1971); (◆) Sangi (1971).

TABLE 90.12
Characteristic Energy (D_r/μ) Suggested by Christophorou and Olthoff (2004)

E/N (Td)	D_r/μ (V)	E/N (Td)	D_r/μ (V)	E/N (Td)	D_r/μ (V)
365	4.86	500	5.21	650	5.82
400	4.99	550	5.32	700	6.22
450	5.13	600	5.52	725	6.44

TABLE 90.13
Product of Number Density and Longitudinal Diffusion Coefficient (ND_L)

E/N (Td)	ND_L (10^{24} m^{-1} s^{-1})	E/N (Td)	ND_L (10^{24} m^{-1} s^{-1})
85	2.05	350	2.90
100	2.10	400	3.01
150	2.28	450	3.10
200	2.46	500	3.19
250	2.65	550	3.27
300	2.79	600	3.36

Source: Adapted from Christophorou, G. and J. K. Olthoff, *Fundamental Electron Interactions with Plasma Processing Gases*, Kluwer Academic/Plenum Publishers, New York, NY, 2004 [3].

TABLE 90.14
Recommended Density-Reduced Effective Ionization Coefficient ($(\alpha-\eta)/N$)

E/N (Td)	$(\alpha-\eta)/N$ (10^{-22} m^2)	E/N (Td)	$(\alpha-\eta)/N$ (10^{-22} m^2)	E/N (Td)	$(\alpha-\eta)/N$ (10^{-22} m^2)
200	−55.3	600	63.8	1000	154
250	−32.8	650	75.2	1250	204
300	−16.1	700	87.0	1500	250
350	−2.43	750	98.8	2000	338
400	10.9	800	110	2500	413
450	25.8	850	122	3000	478
500	39.3	900	132	3500	531
550	51.9	950	143	4000	578

Source: Adapted from Christophorou, G. and J. K. Olthoff, *Fundamental Electron Interactions with Plasma Processing Gases*, Kluwer Academic/Plenum Publishers, New York, NY, 2004.
Note: See Figure 90.8 for graphical presentation.

9.13 ATTACHMENT COEFFICIENT

See Table 90.15.

90.14 ATTACHMENT RATE CONSTANT

See Tables 90.16 and 90.17.

90.15 DETACHMENT COEFFICIENT

Figure 90.11 shows the detachment coefficients as function of reduced electric field (Christophorou and Olthoff, 2004).

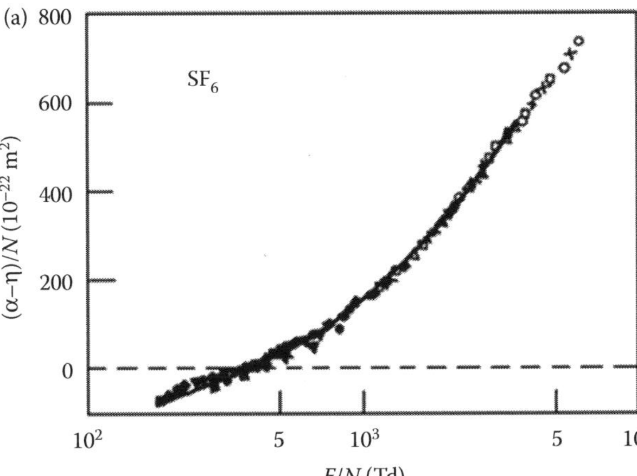

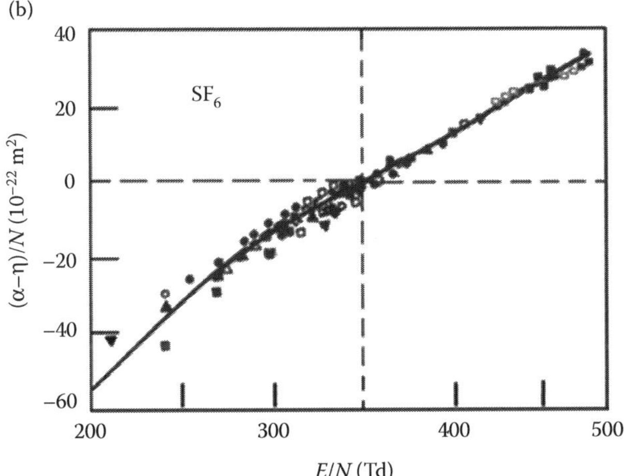

FIGURE 90.8 (a) Density-reduced effective ionization coefficients ($(\alpha-\eta)/N$) as a function of E/N for SF_6. (——) Recommended by Christophorou and Olthoff (2004); (*) Xiao et al. (1999); (×) Qiu and Xiao (1994); (+) Hasegawa et al. (1988); (∇) Fréchette (1986); (◊) Aschwanden (1984); (Δ) Raju and Dincer (1982); (□) Shimozuma et al. (1982); (○) Kline et al. (1979); (▼) Itoh (1979); (◆) Teich (1974); (■) Boyd and Crichton (1971); (▲) Bhalla and Craggs (1962); (●) Geballe and Harrison (1953); The point of intersection of the dotted line shows the $(E/N)_{Lim}$. (b) A close-up of this point.

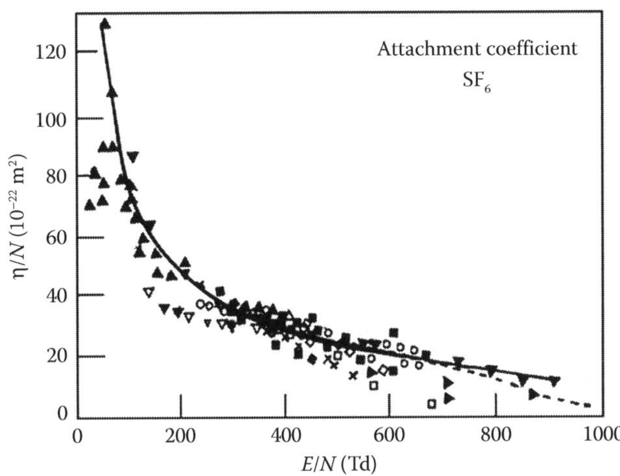

FIGURE 90.9 Density-reduced attachment coefficients as a function of E/N for SF_6. (——) Suggested (η/N) Christophorou and Olthoff (2000); (— — —) Christophorou and Olthoff (2000) $\alpha/N-(\alpha-\eta)/N$; (x) Qiu (1994); (◧) Hasegawa et al. (1988); (▶) Hayashi and Wang (2000); (0) de Urquiho-Carmona (1980); (×) Fréchette (1986); (∇) Siddagangappa et al. (1984); (▼) Aschwanden (1984); (+) Siddagangappa et al. (1982); (*) Shimozuma et al. (1982); (◊) Raju and Dincer (1982); (◆) Kline et al. (1979); (□) Maller and Naidu (1976); (■) Teich and Sangi (1972); (●) Bhalla and Craggs (1962); (○) Geballe and Harrison (1955).

TABLE 90.15

Recommended Reduced Attachment Coefficients

E/N (Td)	η/N (10^{-22} m²)	E/N (Td)	η/N (10^{-22} m²)	E/N (Td)	η/N (10^{-22} m²)
75	126	250	40.1	600	20.7
100	87	300	34.8	650	18.6
125	68	350	31.6	700	16.4
150	58.0	400	28.3	750	14.3
175	51.7	450	26.3	800	12.2
200	46.8	500	24.5	850	10.2
225	43.3	550	22.6	900	8.06

Source: Adapted from Christophorou, G. and J. K. Olthoff, *Fundamental Electron Interactions with Plasma Processing Gases*, Kluwer Academic/Plenum Publishers, New York, NY, 2004.

Note: See Figure 90.9 for graphical presentation.

TABLE 90.16

Attachment Rate Constant as Function of E/N

E/N (Td)	W (10^4 m/s)	η/N (10^{-22} m²)	k_a (10^{-18} m³/s)
100	10.2	87.0	887
150	12.1	58.0	702
200	13.5	46.8	632
250	15.6	40.1	626
300	18.3	34.8	637
350	20.5	31.6	648
400	22.6	28.3	640
450	24.6	26.3	647
500	26.4	24.5	647
550	28.2	22.6	637
600	30.0	20.7	621
650	31.7	18.6	590

Source: Adapted from Christophorou, G. and J. K. Olthoff, *Fundamental Electron Interactions with Plasma Processing Gases*, Kluwer Academic/Plenum Publishers, New York, NY, 2004.

Note: W = drift velocity; η/N = reduced attachment coefficient; k_a = attachment rate ($W \times \eta/N$).

SF$_6$

TABLE 90.17
Attachment Rate Constant as Function of Electron Mean Energy $\bar{\varepsilon}$ for SF_6

$\bar{\varepsilon}$ (eV)	k_a (10^{-13} m^3/s)	$\bar{\varepsilon}$ (eV)	k_a (10^{-13} m^3/s)
0.038	2.25	0.5	0.241
0.04	2.20	0.6	0.181
0.05	2.05	0.7	0.135
0.06	1.88	0.8	0.103
0.07	1.74	0.9	0.0833
0.08	1.62	1.0	0.0721
0.09	1.51	1.5	0.0406
0.10	1.41	2.0	0.0268
0.15	1.04	2.5	0.0197
0.20	0.801	3.0	0.0154
0.25	0.622	3.5	0.0125
0.30	0.487	4.0	0.0104
0.40	0.327		

Source: Adapted from Christophorou, G. and J. K. Olthoff, *Fundamental Electron Interactions with Plasma Processing Gases*, Kluwer Academic/Plenum Publishers, New York, NY, 2004.

Note: See Figure 90.10 for graphical presentation. Recommended thermal rate of attachment is 2.25×10^{-13} m^3/s.

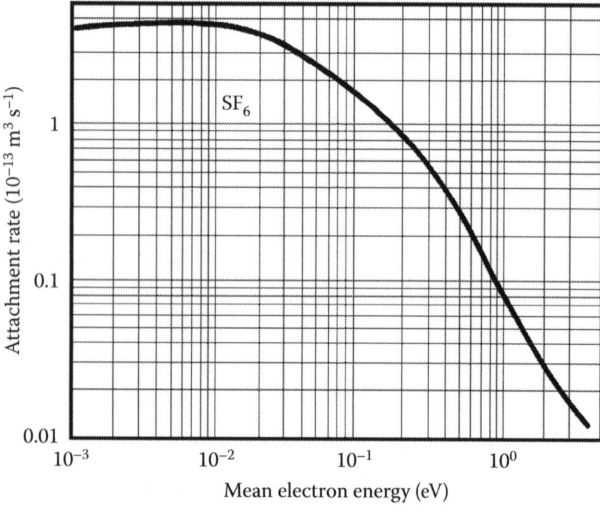

FIGURE 90.10 Total attachment rate constants as a function of mean energy for SF$_6$. Buffer gases are used for swarm measurements. Data used to draw the mean curve are: rates recommended by Christophorou and Olthoff (2000); Christophorou and Datskos (1995); Christophorou et al. (1971); (■, ◆) Gant (1976); Hunter et al. (1989); Shimamori et al. (1992); Klar et al. (1992); high-Rydberg-atom measurements.

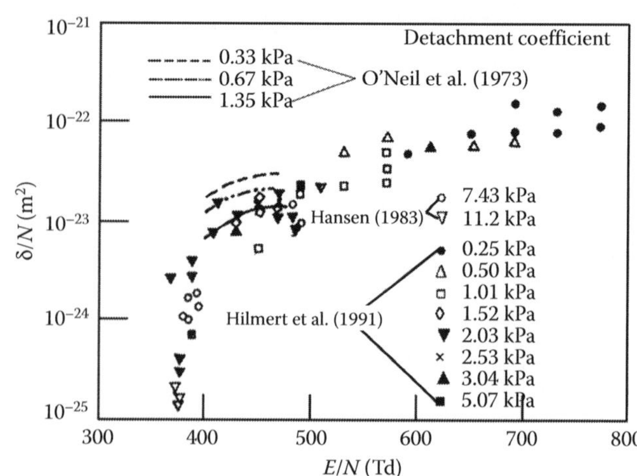

FIGURE 90.11 Detachment coefficients for SF$_6$ as a function of *E*/*N* at gas pressures shown in the legend. References are O'Neil and Craggs (195, 1973), Hansen et al. (196, 1983), and Hilmert and Schmidt (197, 199). (Adapted from L. G. Christophorou and J. K. Olthoff, *J. Phys. Chem. Ref. Data*, 29, 267, 2000.)

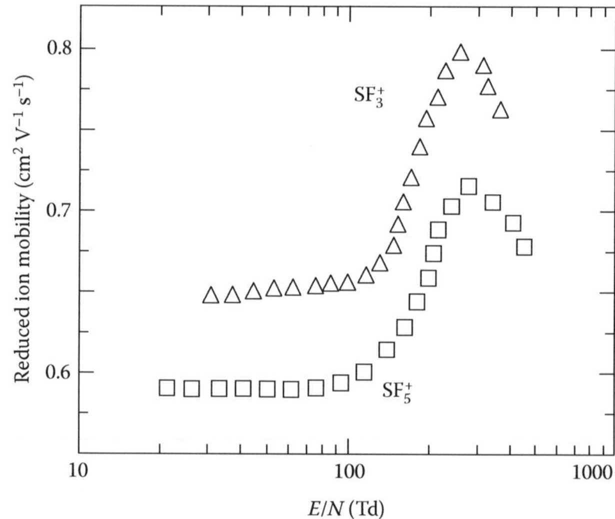

FIGURE 90.12 Mobility of SF_3^+ (•) and SF_5^+ ions (●) for SF$_6$. (From J. de Urquijo-Carmona, I. Alvarez, C. Cisneros, and H. Martínez, *J. Appl. Phys.*, 23, 778, 1990.)

90.16 POSITIVE ION MOBILITY

Figure 90.12 shows the reduced positive ion mobility ($N_o = 2.69 \times 10^{25}$ m^{-3}) for SF_3^+ and SF_5^+ ions (Urquijo-Carmona, 1990). The zero field mobilities for SF_3^+ and SF_5^+ ions are 0.651×10^{-4} and 0.591×10^{-4}, respectively.

REFERENCES

Aschwanden, Th., in L. G. Christophorou and M. O. Pace (Eds), *Gaseous Dieiectrics IV*, Pergamon, New York, NY, 1984, p. 24.

Asundi, R. K. and J. D. Craggs, *Proc. Phys. Soc.*, 83, 611, 1964.

Benedict, M. G. and I. Gyemant, *Int. J. Quantum Chem.* XIII, 597, 1978.

Bhalla, M. S. and J. D. Craggs, *Proc. Phys. Soc. London*, 80, 151, 1962.

Boyd, H.A. and G.C. Crichton, *Proc. IEEE*, 1118, 1872, 1971.

Cho, H., R. J. Gulley, K. W. Tranrham, L. J. Uhlmann, C. J. Dedman, and S. J. Buckman, *J. Phys. B: At. Mol. Opt. Phys.*, 33, 3531, 2000. Also see *Phys. B: At. Mol. Opt. Phys.*, 33, L309, 2000.

Christophorou, L. G. and P. G. Datskos, *Int. J. Mass. Spectrom. Ion Proc.*, 149/150, 59, 1995.

Christophorou, L. G. and J. K. Olthoff, *J. Phys. Chem. Ref. Data*, 29, 267, 2000.

Christophorou, G. and J. K. Olthoff, *Fundamental Electron Interactions with Plasma Processing Gases*, Kluwer Academic/Plenum Publishers, New York, NY, 2004.

Christophorou, L. G., D. L. McCorkle, and J. G. Carter, *J. Chem. Phys.*, 54 253, 1971; 57, 2228E, 1972.

Chutjian, A. *Phys. Rev. Lett.* 46, 1511, 1981.

Dababneh, M. S., Y.-F. Hseih, W. E. Kauppila, C. K. Kwan, S. J. Smith, T. S. Stein, and M. N. Uddin, *Phys. Rev. A*, 38, 1207, 1988.

de Urquijo-Carmona, J. PhD thesis, University of Manchester, 1980, as quoted in Christophoroi and Olthoff (2000) as their reference [302].

de Urquijo-Carmona, J., I. Alvarez, C. Cisneros, and H. Martínez, *J. Appl. Phys.*, 23, 778, 1990.

Dehmer, J. L., J. Siegel, and D. Dill, *J. Chem. Phys.* 69, 5205, 1978.

Feil, S., K. Gluch, P. Scheier, K. Becker, and T. D. Märk, *J. Chem. Phys.*, 120, 11465, 2004.

Ferch, J., W. Raith, and K. Schröder, *J. Phys. B: At. Mol. Phys.*, 15, L175, 1982.

Fréchette, M. F., *J. Appl. Phys.*, 59, 3684, 1986.

Gant, K. S., PhD dissertation, University of Tennessee, 1976.

Geballe, R. and M. A. Harrison, as reported in L. B. Loeb (Ed.), *Basic Processes of Gaseous Electronics*, University of California Press, Berkley, CA, 1965, Chapter 5, p. 415. Also see *Phys. Rev.* 91, 1, 1953.

Gianturco, F. A., R. R. Lucchese, and N. Sanna, *J. Chem. Phys.*, 102, 5743, 1995. Differential scattering cross sections in the range from 3.50 to 30 eV are given.

Govinda Raju, G. R. and M. S. Dincer, *J. Appl. Phys.*, 53, 8562, 1982.

Hansen, D., H. Jungblut, and W. F. Schmidt, *J. Phys. D*, 16, 1623, 1983.

Harris, F. M. and G. J. Jones, *J. Phys. B: At. Mol. Phys.*, 4, 1536, 1971.

Hasegawa, H., A. Taneda, K. Murai, M. Shimozuma, and H. Tagashira, *J. Phys. D: Appl. Phys.*, 21, 1745, 1988.

Hayashi and G. Wang, cited by Christophorou and Olthoff (2000) as their reference [182], private communication.

Hilmert, H. and W. F. Schmidt, *J. Phys. D: Appl. Phys.*, 24, 915, 1991.

Howe, P.-T., A. Kortyna, M. Darrach, and A. Chutjian, *Phys. Rev. A*, 64, 042706, 2001.

Hunter, S. R., J. G. Carter, and L. G. Christophorou, *J. Chem. Phys.*, 90, 4879, 1989.

Hwang, W., Y.-K. Kim, and M. E. Rudd, *J. Chem. Phys.*, 104, 2956, 1996. This paper gives important data such as molecular orbitals and binding energy for a number of molecules. Calculated ionization cross sections are given for 19 molecules.

Itoh, H., M. Shimozuma, H. Tagashira, and S. Sakamoto, *J. Phys. D: Appl. Phys.*, 12, 2167, 1979.

Ito, M., M. Goto, and H. Sugai, *Contrib. Plas. Phys.*, 35, 405, 1995.

Jiang, Y., J. Sun, and L. Wan, *Phys. Lett. A*, 231, 231, 1997.

Johnstone, W. M. and W. R. Newell, *J. Phys. B: At. Mol. Opt. Phys.*, 24, 473, 1991.

Karwasz, G. P., R. P. Brusa, and A. Zecca, Cross sections for scattering- and excitation-processes in electron–molecule collisions, in *Interactions of Photons and Electrons with Molecules*, Springer-Verlag, New York, NY, 2003, Chapter 6.

Kasperski, G., P. Możejko, and C. Szmytkowski, *Z. Phys. D*, 42, 187, 1997.

Kennerly, R. E., R. A. Bonham, and M. McMillan, *J. Chem. Phys.*, 70, 2039, 1979.

Kim, Y.-K. and M. E. Rudd, Comments, *At. Mol. Phys.*, 34, 309, 1999.

Klar, D., M.-W. Ruf, and H. Hotop, *Aust. J. Phys.*, 45, 263, 1992.

Kline, L. E., D. K. Davies, C. L. Chen, and P. J. Chantry, *J. Appl. Phys.*, 50, 6789, 1979.

Lisovskiy, V. A. and V. D. Yegorenkov, *J. Phys. D: Appl. Phys.*, 32, 2645, 1999.

Makochekanwa, C., M. Kimura, and O. Sueoka, *Phys. Rev. A*, 70, 022702, 2004.

Maller, V. N. and M. S. Naidu, *Proc. IEE*, 123, 107, 1976.

Margreiter, D., G. Walder, H. Deutsch, H. U. Poll, C. Winkler, K. Stephan, and T. D. Märk, *Int. J. Mass. Spectrom. Ion Proc.*, 100, 143, 1990 (a); D. Margreiter, H. Deutsch, M. Schmidt, and T. D. Märk, *Int. J. Mass. Spectrom. Ion Proc.*, 100, 157, 1990 (b).

Naidu, M. S. and A. N. Prasad, *J. Phys. D: Appl. Phys.*, 5, 1090, 1972.

Nakamura, Y., *J. Phys. D: Appl. Phys.*, 21, 67, 1988.

O'Neil, B. C. and J. D. Craggs, *J. Phys. B: At. Mol. Phys.*, 6, 2634, 1973.

Phelps, A. V. and R. J. Van Brunt, *J. Appl. Phys.*, 64, 4269, 1988.

Qiu, Y. and D. M. Xiao, *J. Phys. D: Appl. Phys.*, 27, 2663, 1994.

Raju, G. G., *Gaseous Electronics: Theory and Practice*, Taylor & Francis, London, 2005.

Randell, J., D. Field, S. L. Lunt, G. Mrotzek, and J. P. Ziesel, *J. Phys. B*, 25, 2899, 1992.

Rao, M. V. V. S. and S. K. Srivastava, *XX International Conference on the Physics of Electronic and Atomic Collisions*, Scientific Program and Abstracts of Contributed Papers, Vol. II, Vienna, Austria, July 23–29, 1997, Paper MO 151; cited by Christophorou and Olthoff (2000).

Rapp, D. and P. Englander-Golden, *J. Chem. Phys.*, 43, 4081, 1965.

Rejoub, R., D. R. Sieglaff, B. G. Lindsay, and R. F. Stebbings, *J. Phys. B: At. Mol. Opt. Phys.*, 34, 1289, 2001 .

Rohr, K., *J. Phys. B: At. Mol. Opt. Phys.*, 12, L185, 1979.

Romanyuk, N. I., I. V. Chernyshova, and O. B. Shpenik, *Sov. Phys. Tech. Phys.*, 29, 1204, 1984.

Sakae, T., S. Sumiyoshi, E. Murakami, Y. Matsumoto, Y. Ishibashi, and A. Katase, *J. Phys. B: At. Mol. Opt. Phys.*, 22, 1385, 1989. Differential cross sections in CH_4, CF_4, and SF_6 are studied and integrated elastic scattering and momentum transfer cross sections are given.

Sangi, B., PhD thesis, University of Manchester 1971, as quoted in Christophorou and Olthoff (2000) as their reference [299].

Shimamori, H., Y. Tatsumi, Y. Ogawa, and T. Sunagawa, *J. Chem. Phys.*, 97, 6335, 1992.

Shimanouchi, T., *Tables of Molecular Vibrational Frequencies*, NSRDS-NBS 39, US Department of Commerce, Washington (DC), 1972.

Shimozuma, M., H. Itoh, and H. Tagashira, *J. Phys. D: Appl. Phys.*, 15, 2443, 1982.

Siddagangappa, M. C., C. S. Lakshminarasimha, and M. S. Naidu, *J. Phys. D: Appl. Phys.*, 15, L83, 1982. Also see M. C. Siddagangappa, C. S. Lakshminarasimha, and M. S. Naidu, in L. G. Christophorou and M. O. Pace (Eds), *Gaseous Dielectrics IV*, Pergamon Press, New York, NY, 1984, p. 49.

Srivastava, S. K., S. Trajmar, A. Chutjian, and W. Williams, *J. Chem. Phys.*, 64, 2767, 1976.
Stanski, T. and B. Adamczyk, *Int. J. Mass Spectrom. Ion. Phys.*, 46, 31, 1983.
Tarnovsky, V., H. Deutsch, S. Matt, T. D. Märk, R. Basner, M. Schimdt, and K. Becker, in L. G. Christophorou and J. K. Olthoff (Eds), *Gaseous Electronics VIII*, Plenum, New York, NY, 1998, p. 3.
Teich, T. H. and R. Sangi, in F. Heidbromer (Ed.), *Proceedings of the First International Symposium on High Voltage Engineering, Munich*, Vol. 1, 1972, p. 391.
Trajmar, S., D. F. Register, and A. Chutjian, *Phys. Rep.*, 97, 219, 1983.
Wan, H.-X., J. H. Moore, J. K. Olthoff, and R. J. Brunt, *Plasma Chem. Plasma Process*. 13, 1, 1993.
Xiao, D. M., H. L. Liu, and Y. Z. Chen, *J. Appl. Phys.*, 86, 6611, 1999.
Zecca, A., G. Karwasz, and R. S. Brusa, *Chem. Phys. Lett.*, 199, 423, 1992.
Ziesel, J.-P., N. C. Jones, D. Field, and L. B. Madsen, *J. Chem. Phys.*, 122, 024309, 2005.

91

TUNGSTEN HEXAFLUORIDE

WF6

CONTENTS

Tungsten hexafluoride (WF_6) is an electron-attaching gas with 128 electrons. It has six modes of vibration as shown in Table 91.1 (Shimanouchi, 1972).

The parent negative ion WF_6^- forms at an energy of ~3 eV (see de Wall and Neuert, 1977). Electron impact excitation studies have been carried out by Rianda et al. (1979) who find excitation maxima at 7.5, 7.9, 8.5, and 9.85 eV.

TABLE 91.1
Vibrational Modes and Energies of WF_6 Molecule

Symbol	Mode	Energy (meV)	Symbol	Mode	Energy (meV)
ν_1	Symmetrical stretch	31.0	ν_4	Degenerate deformation	32.0
ν_2	Degenerate stretch	83.9	ν_5	Degenerate deformation	39.7
ν_3	Degenerate stretch	88.3	ν_6	Degenerate deformation	15.7

91.1 TOTAL SCATTERING CROSS SECTION

Total scattering cross sections measured by Szmytkowski et al. (2000) and Karwasz et al. (2000) are shown in Table 91.2.

TABLE 91.2
Total Scattering Cross Sections for WF_6

	Q_T (10^{-20} m^2)			Q_T (10^{-20} m^2)	
Energy (eV)	WF_6	SF_6	Energy (eV)	WF_6	SF_6
1.2	21.7	21.9	2.3	26.9	
1.3	22.0		2.7	30.1	
1.4	21.6	22.2	3.0	31.4	
1.5	21.8		3.5	27.4	
1.6	21.7	22.4	4.0	25.9	23.1
1.7	22.1		4.5	25.0	22.7
1.8	22.5	22.5	5.0	22.9	23.3
2.0	22.6	23.1	5.5	22.8	25.9

TABLE 91.2 (continued)
Total Scattering Cross Sections for WF_6

	Q_T (10^{-20} m^2)			Q_T (10^{-20} m^2)	
Energy (eV)	WF_6	SF_6	Energy (eV)	WF_6	SF_6
6.0	22.9	27.6	35	37.8	29.1
6.5	22.9		40	37.8	28.4
7.0	22.8	30.8	45	37.7	
7.5	23.3	30.6	50	36.8	27.4
8.0	24.2	30.2	60	36.3	26.7
8.5	24.0	29.8	70	35.0	25.6
9.0	24.4	29.4	80	34.4	23.9
9.5	25.6	29.2	90	33.7	23.4
10.0	26.8	28.7	100	32.5	22.0
11.0	27.2	29.8	110	31.8	
12.0	28.2	31.2	120	30.9	20.7
14.0	28.9	28.2	140	28.6	
16.0	32.8	27.1	160	27.6	
18.0	36.3	27.5	180	26.1	
20.0	36.5	27.4	200	25.3	15.9
22.5	36.7		220	24.4	
25.0	36.6	28.4	250	23.2	15.2
27.5	36.8				
30	37.1	29.5			

Source: Adapted from Szmytkowski, C. et al., *J. Phys. B: At. Mol. Opt. Phys.*, 33, 15, 2000.

Note: Data for SF_6 (Makochekanwa et al., 2004) shown for comparison.

Figure 91.1 is a graphical presentation. Cross sections for sulfur hexafluoride (SF_6) and hexafluoroethane (C_2F_6) are also plotted for comparison sake. The highlights of the cross section are

1. There is a very sharp peak at 3.0 eV, possibly due to resonance.
2. A structure in the vicinity of 8–10 eV, possibly due to excitation, and ionization onset and formation of ion fragments F^-, WF_4^-, and WF_5^-. This behavior is common to many gases. (SF_6, CO_2, O_2, etc.),
3. A broad peak in the range of 20–70 eV possibly due to high energy resonances and superposition of several direct processes allowed in this energy range (Szmytkowski et al., 2000). This behavior is also observed in several F_6 molecules (C_2F_6, C_6F_6, SF_6).
4. Szmytkowski et al. (2000) suggest that the energy dependence of the cross section in the 60–250 eV range is according to ε^{-n} with $n = 0.2$. This behavior is also observed in SF_6.

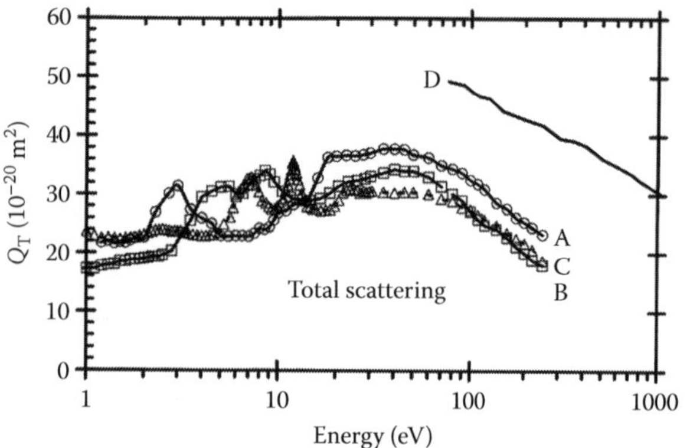

FIGURE 91.1 Total scattering cross sections in WF_6 and selected F_6 molecules. (—Δ—) A—WF_6 (Szmytkowski et al., 2000); (—□—) B—hexafluoroethane (Szmytkowski et al., 2000); (—○—) C—Kasperski et al. (1997); (——) D—Karwasz et al. (2000). (Adapted from Szmytkowski, C. et al., *J. Phys. B: At. Mol. Opt. Phys.*, 33, 15, 2000.)

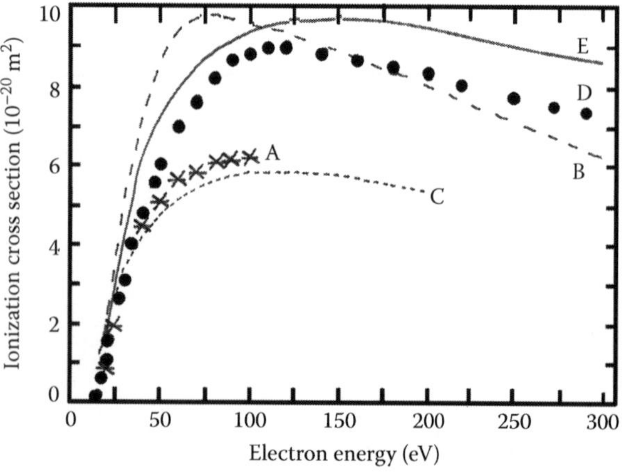

FIGURE 91.2 Ionization cross section for WF_6. A—Basner et al. (1996), experimental; B—Probst et al. (2001), theory; C, D—Kwitnewski et al. (2003), theory; E—Kim and Irikura (2000).

91.2 IONIZATION CROSS SECTION

Ionization cross sections have been derived or measured by the following: Basner et al., (1996) experimental; Kim and Irikura, (2003) theoretical; Probst et al., (2001) theoretical; and Kwitnewski et al., (2003) scaling law as shown in Figure 91.2.

91.3 ATTACHMENT RATE CONSTANT

Friedman et al. (2006) give the attachment rate constant at room temperature as $<10^{-18}$ m^3/s increasing to 2×10^{-16} m^3/s at 552 K.

REFERENCES

Basner, R., M. Schmidt, and H. Deutsch, *Bull. Am. Phys. Soc.*, 38, 2369, 1996, cited by Kwitnewski et al. (2003).

de Wall, R. and H. Neuert, *Z. Naturf.* a, 23, 968, 1977.

Friedman, J. F., A. E. Stevens, T. M. Miller, and A. A. Viggiano, *J. Chem. Phys.*, 124, 224306, 2006.

Karwasz, G. P., R. S. Brusa, L. D. Longo, and A. Zecca, *Phys. Rev. A*, 61, 024701, 2000.

Kasperski, G., P. Możejko, and Cz. Szmytkowski, *Z. Phys. D*, 42, 187, 1997.

Kim, Y.-K. and K. K. Irikura, in K. E. Berrington and K. L. Bell, (Eds.) *Atomic Molecular Data and Their Applications: AIP Conference Proceedings (USA)*, American Institute of Physics, New York, 543, 220, 2000, cited by Kwitnewski et al. (2003).

Kwitnewski, S., E. Ptasińska-Denga, and C. Szmytkowski, *Radiat. Phys. Chem.*, 68, 169, 2003.

Makochekanwa, C., M. Kimura, and O. Sueoka, *Phys. Rev. A*, 70, 022702, 2004.

Probst, M., H. Deutsch, H. Becker, K. Becker, and T. D. Märk, *Int. J. Mass Spectrom.*, 206, 13, 2001.

Rianda, R., R. P. Frueholz, and A. Kupperman, *J. Chem. Phys.*, 70, 1056, 1979.

Shimanouchi, T., *Tables of Molecular Vibrational Frequencies Consolidated Volume I*, NSRDA-NBS 39, Washington (DC), 1972.

Szmytkowski, Cz.. P. Możejko, G. Kasperski, and E. Ptasińska-Denga, *J. Phys. B: At. Mol. Opt. Phys.*, 33, 15, 2000.

92

URANIUM HEXAFLUORIDE

CONTENTS

Uranium hexafluoride (UF_6) is a large inorganic molecule with 146 electrons. It has an electronic polarizability of 13.91×10^{-40} F m^2, is nonpolar, and has an ionization potential of 14.00 eV. The dissociation energy (UF_5–F) is 3.0 eV (Compton, 1977). There are six vibrational modes as shown in Table 92.1.

92.1 DIFFERENTIAL SCATTERING CROSS SECTIONS

The measured differential scattering cross sections in the energy range 10–75 eV (Cartwright, 1983) are shown in Figures 92.1 and 92.2. These supersede the earlier measurements of Srivastava et al. (1976). The highlights of Figures 92.1 and 92.2 are

1. The differential scattering pattern remains essentially the same for all energies considered and over 0–180°.

TABLE 92.1
Vibrational Modes and Energies of UF_6

Symbol	Mode	Energy (meV)
v_1	Symmetric stretch	82.7
v_2	Degenerate stretch	66.1
v_3	Degenerate stretch	77.6
v_4	Degenerate deformation	23.1
v_5	Degenerate deformation	25.0
v_6	Degenerate deformation	17.6

Source: Adapted from Shimanouchi, T. *Tables of Molecular Vibrational Frequencies Consolidated Volume I*, NSRDS-NBS 39, (Washington, DC), 1972.

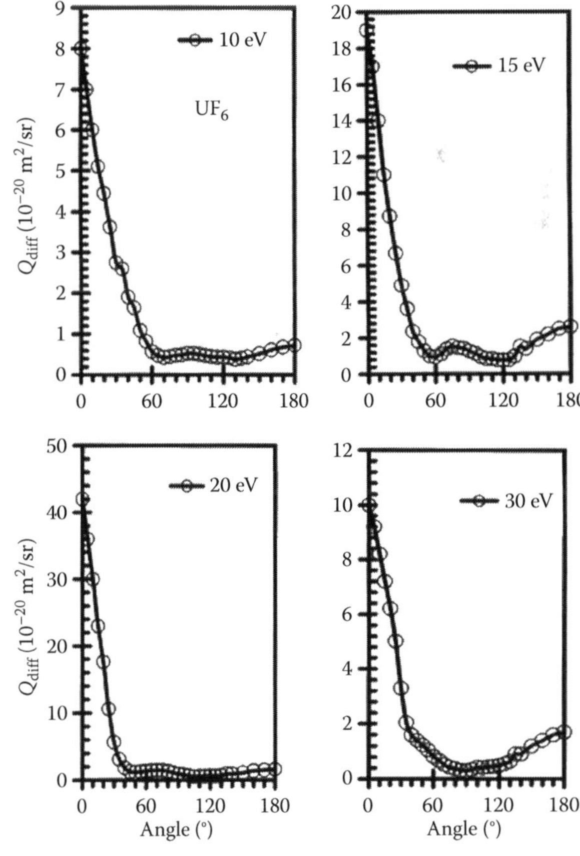

FIGURE 92.1 Differential cross sections for elastic scattering in UF_6 for electron energies. (Adapted from Cartwright, D. C. et al., *J. Chem. Phys.*, 79, 5483, 1983.)

2. There is significant forward scattering at all energies considered.
3. The differential cross section at the same angle for all energies shows a nonmonotonic variation.

92.2 ELASTIC AND MOMENTUM TRANSFER CROSS SECTIONS

Cartwright et al. (1983) have given integrated elastic scattering and momentum transfer cross sections as shown in Table 92.2.

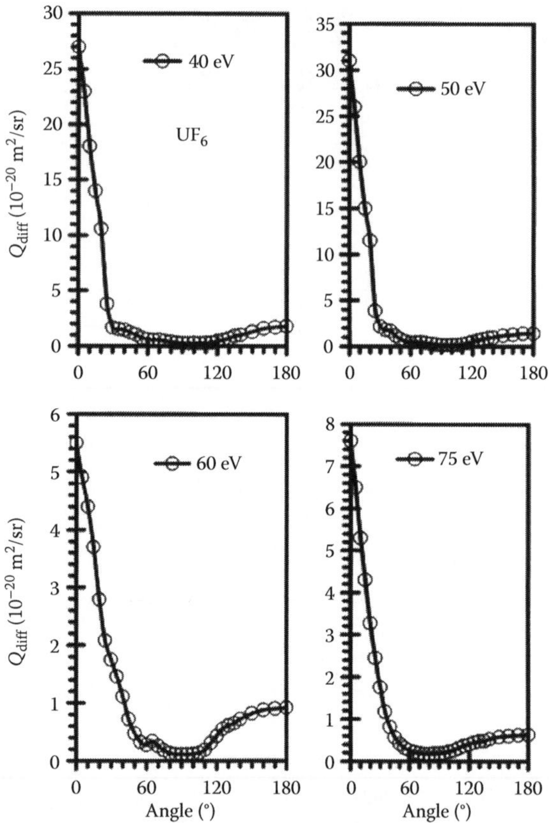

FIGURE 92.2 Differential cross sections for elastic scattering in UF_6 for electron energies. (Adapted from Cartwright, D. C. et al., *J. Chem. Phys.*, 79, 5483, 1983.)

TABLE 92.2
Elastic and Momentum Transfer Cross Sections

Energy (eV)	Q_{el} (10^{-20} m^2)	Q_M (10^{-20} m^2)
10	11.8	24.1
15	24.1	16.9
20	28.1	12.9
30	14.2	9.28
40	16.1	9.04
50	16.2	8.39
60	7.66	5.72
75	7.48	4.98

Source: Adapted from Cartwright, D. C. et al., *J. Chem. Phys.*, 79, 5483, 1983.

92.3 EXCITATION CROSS SECTIONS

Electron impact studies of Rianda et al. (1979) have shown that peaks exist at 11 energies below the ionization potential, viz., 3.26, 4.2, ~4.7, 5.8, 7.0, 7.86, 9.26, 11.01, 11.75, 12.5, and 13.2 eV. Since these energies are above the dissociation energy (UF_5–F) of 2.95 eV, excitation is purely dissociative. Cartwright et al. (1983) observed an additional feature of 5.41 eV and measured the differential scattering cross sections for the first nine levels as shown in Figures 92.3 through 93.8. Table 92.3 shows the integrated cross sections for these levels. The highlights of the cross sections are

1. There is monotonic decrease of the cross section at all energies studied.
2. The forward scattering is pronounced at all energies studied.
3. At 10 eV impact energy, the forward scattering cross section increases by a factor of ~8.
4. As the impact energy increases from 10 to 20 eV, the 5.41 eV excitation cross section increases dramatically by a factor of ~15. As the energy is further increased to 40 eV, the cross section reverts to its low value of ~5×10^{-23} m^2.

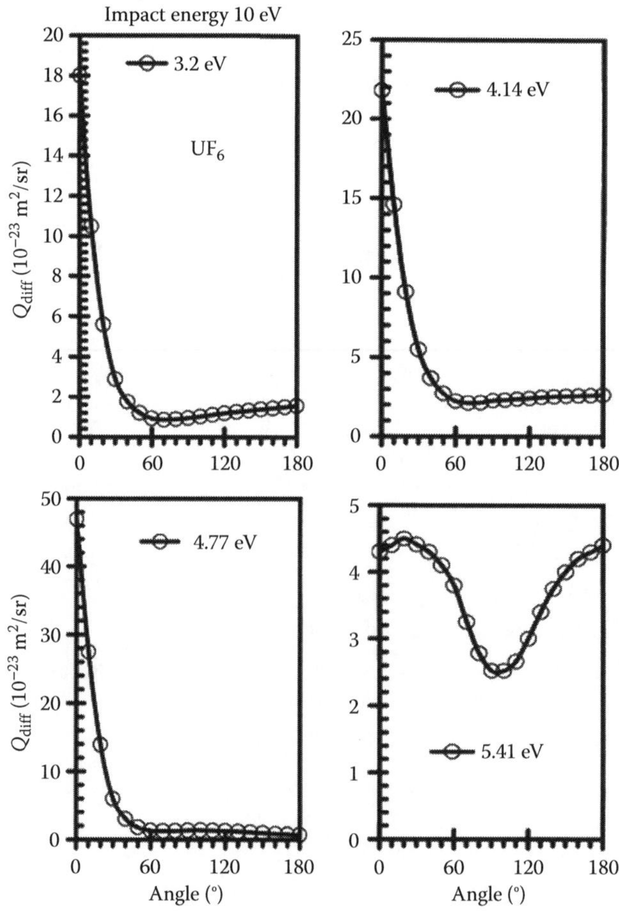

FIGURE 92.3 Differential excitation cross sections in UF_6 measured by Cartwright et al. (1983) at 10 eV electron impact energy. Note the energy displayed as label is excitation level.

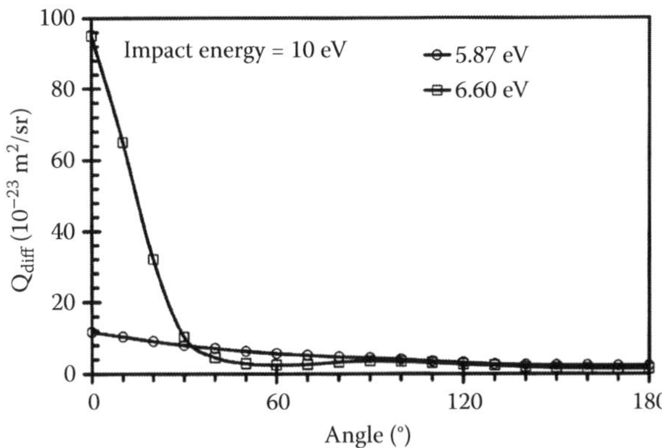

FIGURE 92.4 Differential excitation cross sections in UF_6 measured by Cartwright et al. (1983) at 10 eV electron impact energy.

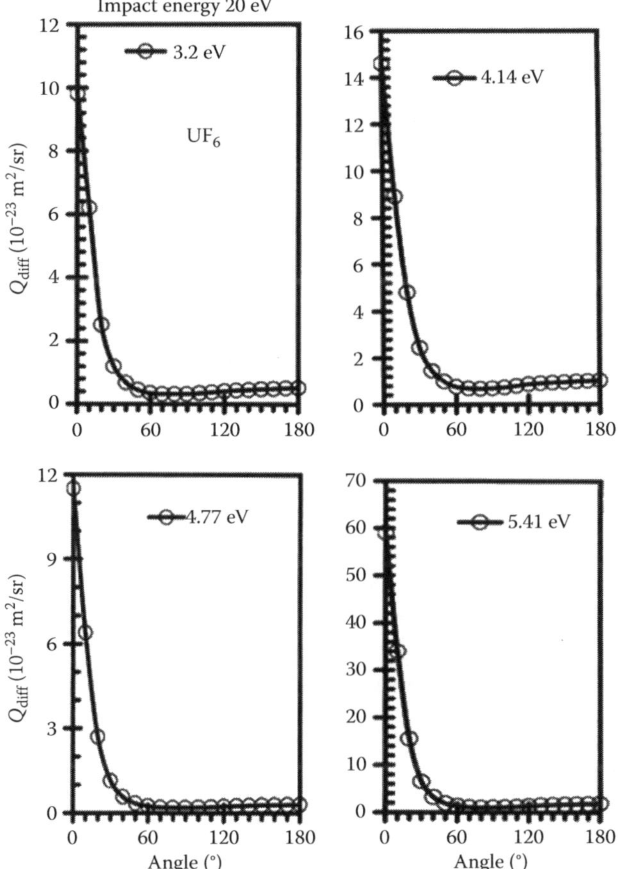

FIGURE 92.5 Differential excitation cross sections in UF_6 measured by Cartwright et al. (1983) at 20 eV electron impact energy. Note the energy displayed as label is excitation level.

5. The integrated excitation cross sections show a peak at 20 eV, being lower at 10 eV and 40 eV impact energy.

92.4 IONIZATION CROSS SECTIONS

The total ionization cross sections are measured by Compton (1977) up to 1000 eV impact energy. Though the ion fragments

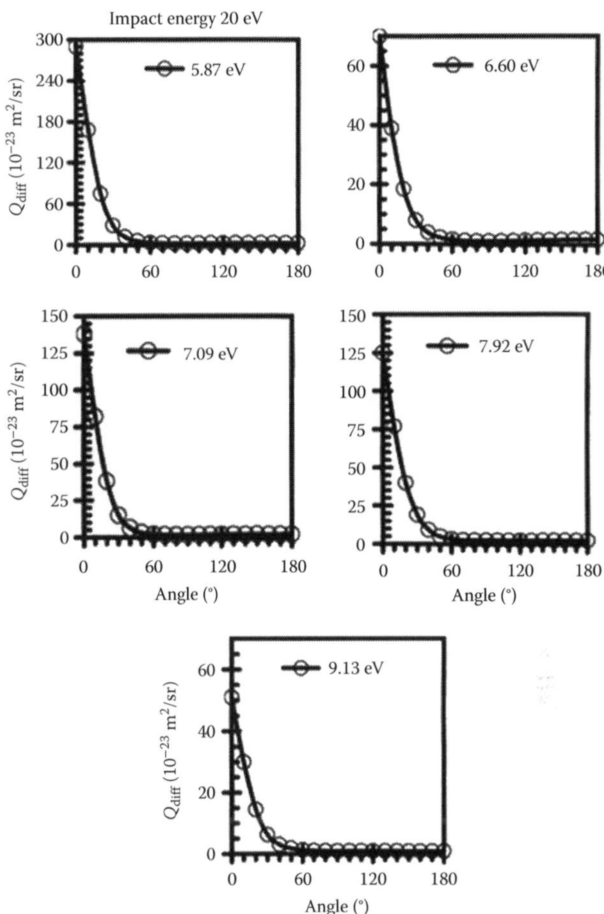

FIGURE 92.6 Differential excitation cross sections in UF_6 measured by Cartwright et al. (1983) at 20 eV electron impact energy. Note the energy displayed as label is excitation level.

are UF_n^+ ($n = 0$ to 6), UF_5^+ is observed to be the most dominant ion. Table 92.4 and Figure 92.9 show the total cross sections.

92.5 ATTACHMENT RATE COEFFICIENT

Electrons attach to uranium molecule at thermal energy, and the attachment rate for the formation of UF_6^- ion is given as ~4.6×10^{-16} m^3/s (Beauchamp, 1976).

Electron attachment occurs at thermal energy according to (Beauchamp, 1976)

$$UF_6 + e \rightarrow UF_6^- \tag{92.1}$$

and dissociative attachment occurs according to the reaction

$$UF_6 + e \rightarrow UF_5^- + F \tag{92.2}$$

with a threshold of 0.9 eV. The species UF_5^- reacts rapidly with UF_6 according to the charge transfer reaction

$$UF_5^- + UF_6 \rightarrow UF_6^- + UF_5 \tag{92.3}$$

UF$_6$

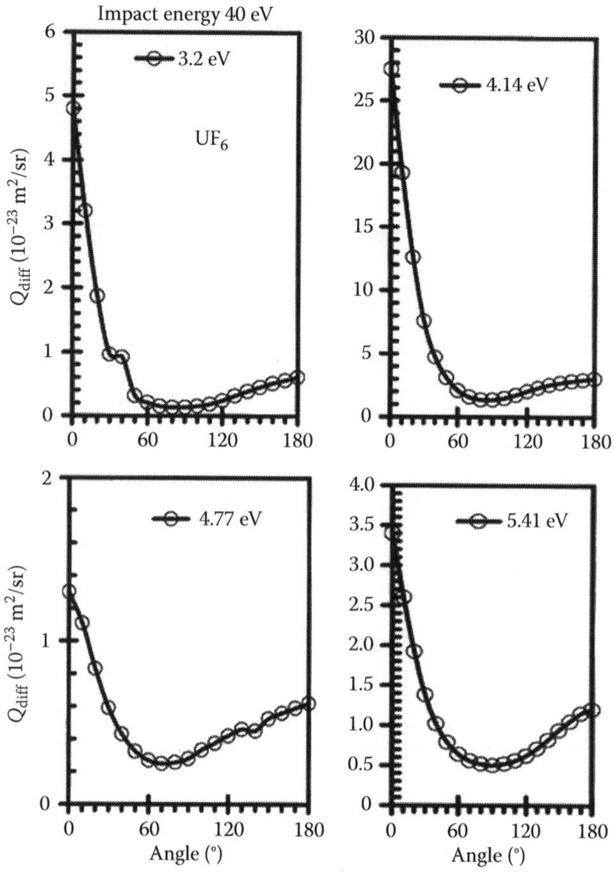

FIGURE 92.7 Differential excitation cross sections in UF$_6$ measured by Cartwright et al. (1983) at 40 eV electron impact energy. Note the energy displayed as label is excitation level.

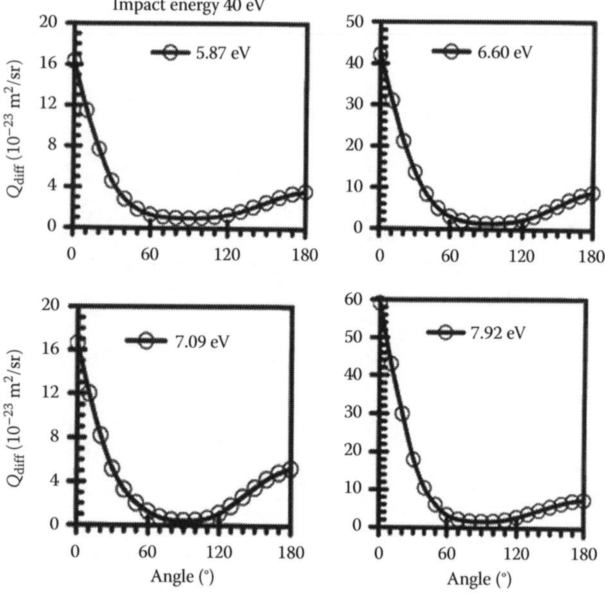

FIGURE 92.8 Differential excitation cross sections in UF$_6$ measured by Cartwright et al. (1983) at 40 eV electron impact energy. Note the energy displayed as label is excitation level.

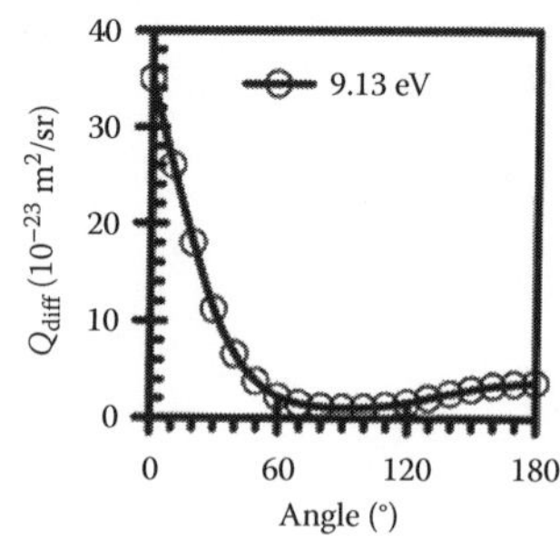

FIGURE 92.8 Continued.

TABLE 92.3
Excitation Cross Sections in UF$_6$

Excitation	Q_{ex} (10^{-23} m^2)		
Level (eV)	**10 eV**	**20 eV**	**40 eV**
3.22	1.915	0.764	0.4882
4.14	3.782	1.479	3.793
4.77	3.029	0.6014	0.5047
5.41	4.281	3.354	0.9777
5.87	5.934	12.64	2.595
6.60	6.479	3.935	5.979
7.09		6.760	2.561
7.92		7.819	7.254
9.13		2.937	4.364
	25.42 (Sum)	40.29 (Sum)	28.52 (Sum)

Source: Cartwright, D. C. et al., *J. Chem. Phys.*, 79, 5483, 1983. With permission.

TABLE 92.4
Total Ionization Cross Sections in UF$_6$

Energy (eV)	Q_i (10^{-20} m^2)	Energy (eV)	Q_i (10^{-20} m^2)	Energy (eV)	Q_i (10^{-20} m^2)
15	0.03	38	6.06	200	16.79
16	0.28	40	6.62	225	16.10
17	0.89	45	8.00	250	16.24
18	1.29	50	9.16	300	15.31
19	1.49	55	10.02	350	14.32
20	1.65	60	11.00	400	13.92
22	2.30	70	13.88	450	12.73
24	3.15	80	15.59	500	12.15
26	3.77	90	16.40	550	11.67
28	4.08	100	17.35	600	10.76
30	4.29	120	16.38	700	9.67
32	4.61	140	17.35	800	8.30
34	5.02	160	17.43	900	7.61
36	5.51	180	16.97	1000	6.05

Source: Digitized from Compton, R. N., *J. Chem. Phys.*, 66, 4478, 1977.

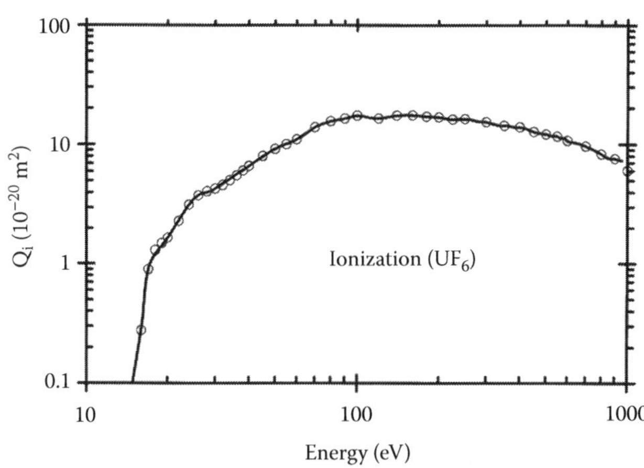

FIGURE 92.9 Total ionization cross sections in UF_6. (Adapted from Compton, R. N., *J. Chem. Phys.*, 66, 4478, 1977.)

TABLE 92.5
Electron Affinities of Hexafluoro Molecules

Molecule		
Name	**Formula**	**Affinity (eV)**
Platinum fluoride	PtF_6	6.5
Sulfur hexafluoride	SF_6	0.6
Selenium hexafluoride	SeF_6	3.0
Tellurium hexafluoride	TeF_6	3.2
Tungsten hexafluoride	WF_6	4.5
Uranium hexafluoride	UF_6	4.9

Source: Adapted from Beauchamp, J. L. *J. Chem. Phys.*, 64, 929, 1976.

Charge transfer may also take place from ions of other species such as

$$Cl^- + UF_6 \rightarrow UF_6^- + Cl \tag{92.4}$$

and

$$SF_6^- + UF_6 \rightarrow UF_6^- + SF_6 \tag{92.5}$$

The electron affinities of several hexafluoro molecules are shown in Table 92.5. UF_6^- ions may also form in collision with fast (0–50 eV) alkali atoms (Cs, K, or Na) according to

$$A\ (Cs, K, Na) + UF_6 \rightarrow A^+ + UF_6^- \tag{92.6}$$

REFERENCES

Beauchamp, J. L., *J. Chem. Phys.*, 64, 929, 1976.

Cartwright, D. C., S. Trajmar, A. Chutjian, and S. Srivastava, *J. Chem. Phys.*, 79, 5483, 1983.

Compton, R. N. *J. Chem. Phys.*, 66, 4478, 1977.

Rianda, R., R. P. Frueholz, and A. Kupperman, *J. Chem. Phys.*, 70, 1956, 1979.

Shimanouchi, T. *Tables of Molecular Vibrational Frequencies Consolidated Volume I*, NSRDS-NBS 39, Washington (DC), 1972.

Srivastava, S., A. Chutjian, and W. Williams, *J. Chem. Phys.*, 64, 2767, 1976.

Section VIII

8 ATOMS

93

BROMOFLUORO-ETHANE

C_2H_4Br

CONTENTS

Bromofluoroethane (C_2H_4BrF) is an electron-attaching molecule that has 60 electrons. Theoretically calculated electronic polarizability is 3.71×10^{-40} F m^2 (see Barszczewska et al., 2004).

93.1 ATTACHMENT RATE

Attachment rate coefficients are shown in Table 93.1 and Figure 93.1.

93.2 ATTACHMENT CROSS SECTIONS

Attachment cross sections are shown in Table 93.2 and Figure 93.2.

TABLE 93.1
Attachment Rate Coefficients for C_2H_4BrF

Method	Temperature: K, (eV)	Rate (10^{-15} m^3/s)	Reference
Swarm	293	0.05	Barszczewska et al. (2004)
PR/MW	293	1.3	Sunagawa and Shimamori (1995)
	(0.04)	1.35	
	(0.1)	1.5	
	(0.4)	1.9	
	(1.0)	1.0	

Note: PR/MW = pulse radiolysis/microwave heating.

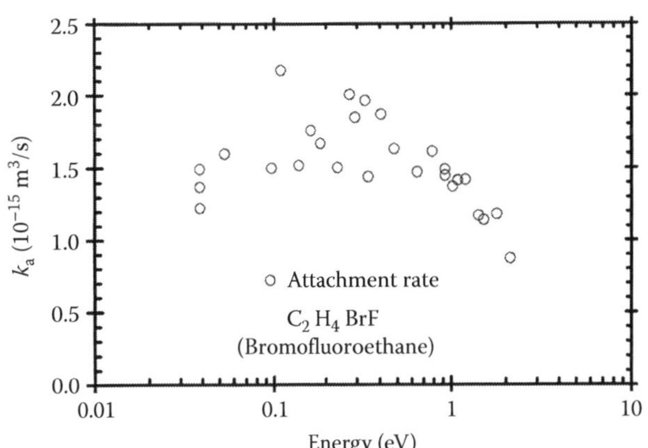

FIGURE 93.1 Attachment rate constants for C_2H_4BrF. (Adapted from Sunagawa, T. and H. Shimamori, *Int. J. Mass Spectrom. Ion Proc.*, 149, 123, 1995.)

TABLE 93.2
Attachment Cross Sections

Energy (meV)	Q_a (10^{-20} m^2)	Energy (meV)	Q_a (10^{-20} m^2)
1.00	6.16	35.0	1.13
2.00	4.39	40.0	1.07
3.00	3.61	45.0	1.04
4.00	3.09	50.0	1.01
6.00	2.52	55.0	0.99
8.00	2.23	60.0	0.97
10.0	2.01	65.0	0.95
15.0	1.66	70.0	0.92
20.0	1.46	75.0	0.90
25.0	1.31	80.0	0.87
30.0	1.20	85.0	0.84

continued

TABLE 93.2 (continued)
Attachment Cross Sections

Energy (meV)	Q_a (10^{-20} m^2)	Energy (meV)	Q_a (10^{-20} m^2)
90.0	0.82	550	0.35
95.0	0.80	600	0.31
100	0.78	650	0.28
150	0.71	700	0.25
200	0.70	750	0.23
250	0.65	800	0.22
300	0.59	850	0.21
350	0.53	900	0.20
400	0.48	1000	0.19
450	0.43	1500	0.12
500	0.39	2000	0.10

Source: Digitized from Sunagawa, T. and H. Shimamori, *Int. J. Mass Spectrom. Ion Proc.*, 149, 123, 1995.

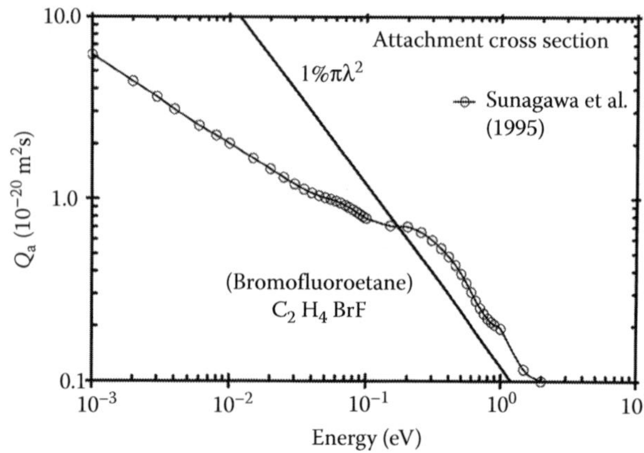

FIGURE 93.2 Attachment cross sections for C_2H_4BrF. *s*-Wave maximum cross section included for comparison. (Adapted from Sunagawa, T. and H. Shimamori, *Int. J. Mass Spectrom. Ion Proc.*, 149, 123, 1995.)

REFERENCES

Barszczewska, W., J. Kopyra, J. Wnorowska, and I. Szamrej, *Int. J. Mass Spectrom.*, 233, 199, 2004.

Sunagawa, T. and H. Shimamori, *Int. J. Mass Spectrom. Ion Proc.*, 149, 123, 1995.

94

BROMOTRIFLUOROETHANE

CONTENTS

Bromotrifluoroethane ($C_2H_2BrF_3$) is an electron-attaching gas that has 70 electrons. Theoretically calculated electronic polarizability is 4.33×10^{-40} F m^2 (Barszczewska, 2004).

94.1 ATTACHMENT RATES

Table 94.1 and Figure 94.1 show the attachment rate coefficients.

TABLE 94.1
Attachment Rate Coefficients for $C_2H_2BrF_3$

Method	Temperature: K, (eV)	Rate (10^{-14} m^3/s)	Reference
Swarm	293	0.14	Barszczewska et al. (2004)
PR/MW	293	15.0	Sunagawa and Shimamori and (1995)
	(0.04)	1.5	
	(0.1)	1.4	
	(0.4)	1.2	
	(1.0)	7.5	

Note: PR/MW = pulse radiolysis/microwave heating.

94.2 ATTACHMENT CROSS SECTIONS

Table 94.2 and Figure 94.2 show the attachment cross sections for $C_2H_2BrF_3$.

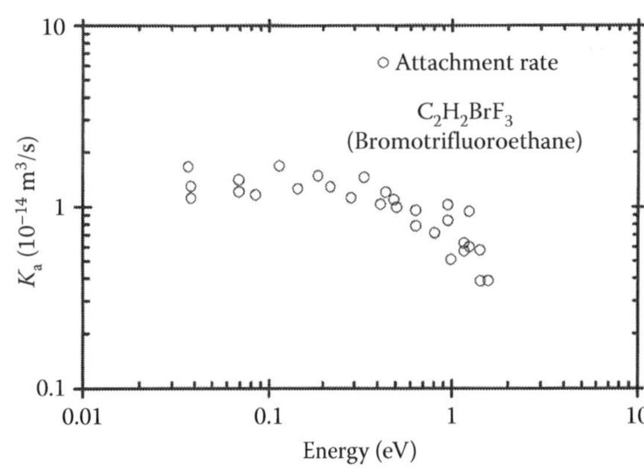

FIGURE 94.1 Attachment rate constants for $C_2H_2BrF_3$. (Adapted from Sunagawa, T. and H. Shimamori, *Int. J. Mass Spectrom. Ion Proc.*, 149/150, 123, 1995.)

$C_2H_2BrF_3$

TABLE 94.2
Attachment Cross Sections

Energy (meV)	Q_a (10^{-20} m^2)	Energy (meV)	Q_a (10^{-20} m^2)
1.00	87.0	90.0	8.40
2.00	57.1	95.0	8.09
3.00	44.6	100	7.77
4.00	40.3	150	6.18
6.00	35.9	200	5.64
8.00	29.9	250	4.91
10.0	26.2	300	4.23
15.0	21.0	350	3.69
20.0	18.0	400	3.22
25.0	16.4	450	2.81
30.0	15.1	500	2.46
35.0	13.9	550	2.16
40.0	12.8	600	1.90
45.0	11.8	650	1.70
50.0	11.0	700	1.54
55.0	10.4	750	1.41
60.0	9.91	800	1.30
65.0	9.60	850	1.20
70.0	9.38	900	1.10
75.0	9.19	1000	0.94
80.0	8.96	1500	0.46
85.0	8.70	2000	0.24

Source: Digitized from Sunagawa, T. and H. Shimamori, *Int. J. Mass Spectrom. Ion Proc.*, 149/150, 123, 1995.

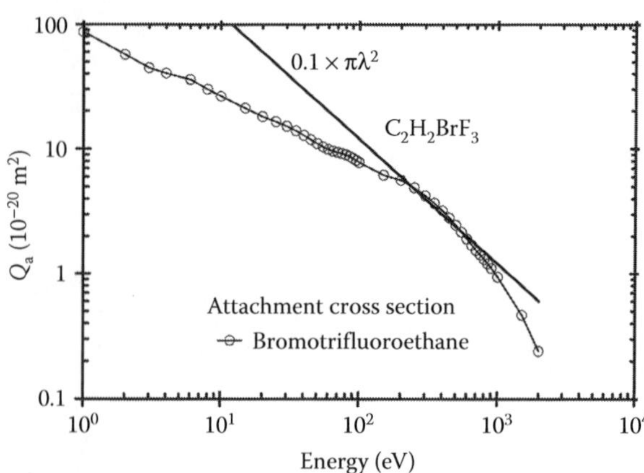

FIGURE 94.2 Attachment cross sections for $C_2H_2BrF_3$. *s*-Wave maximum cross section added for comparison. (Adapted from Sunagawa, T. and H. Shimamori, *Int. J. Mass Spectrom. Ion Proc.*, 149/150, 123, 1995.)

REFERENCES

Barszczewska, W., J. Kopyra, J. Wnorowska, and I. Szamrej, *Int. J. Mass Spectrom.*, 233, 199, 2004.

Sunagawa, T. and H. Shimamori, *Int. J. Mass Spectrom. Ion Proc.*, 149/150, 123, 1995.

95

CHLOROETHANE

C_2H_5Cl

CONTENTS

Chloroethane (C_2H_5Cl), synonym ethyl chloride, is a polar, electron-attaching gas that has 34 electrons. Three values for the electronic polarizability are available: 8.09×10^{-40}, 9.22×10^{-40}, and 7.12×10^{-40} F m^2. It has a dipole moment of 2.05 D and an ionization potential of 10.98 eV. The molecule has 18 vibrational modes as shown in Table 95.1 (Shimanouchi, 1972).

95.1 SELECTED REFERENCES FOR DATA

Table 95.2 gives the selected references for data for C_2H_5Cl.

TABLE 95.1
Vibrational Modes and Energies for C_2H_5Cl

Designation	Type of Motion	Energy (meV)
v_1	CH_2 s-stretch	367.9
v_2	CH_3 d-stretch	365.3
v_3	CH_3 s-stretch	357.2
v_4	CH_3 d-deform	181.4
v_5	CH_2 scissor	179.5
v_6	CH_3 s-deform	171.7
v_7	CH_2 wag	159.8
v_8	CH_3 rock	134.0
v_9	CC stretch	120.8
v_{10}	CCl stretch	83.9
v_{11}	CCCl deform	41.7
v_{12}	CH_2 a-stretch	373.7
v_{13}	CH_3 d-stretch	370.2
v_{14}	CH_3 d-deform	179.5
v_{15}	CH_2 twist	155.1
v_{16}	CH_3 rock	120.8
v_{17}	CH_2 rock	97.5
v_{18}	Torsion	31.1

Note: a = Asymmetrical; d = degenerate; s = symmetrical.

TABLE 95.2
Selected References for C_2H_5Cl

Quantity	Range: Energy (*E*/*N*) (eV (Td))	Reference
Q_i	16–219	**Hudson et al. (2001)**
k_a	Thermal	**Rosa et al. (2001)**
Q_a	0–4	**Pearl and Burrow (1994)**
k_a	0–4	**Pearl and Burrow (1993)**
k_a, Q_a	(0.1–2.0)	**Petrović et al. (1989)**
k_a, Q_a	(1–4)	**Christophorou et al. (1966)**

Note: Bold font indicates experimental study.

95.2 IONIZATION CROSS SECTIONS

Hudson et al. (2001) have measured the total ionization cross sections for C_2H_5Cl as shown in Table 95.3 and Figure 95.1.

95.3 ATTACHMENT CROSS SECTIONS

Dissociative attachment occurs according to the reaction

$$C_2H_5Cl + e \rightarrow C_2H_5 + Cl^- \tag{95.1}$$

The attachment cross section shown in Figure 95.2 attains a peak value at 0.8 eV. (Pearl and Burrow, 1994) This result is at variance with that of Petrović et al. (1989) who showed the peak at 0.12 eV. More recent measurements (Pshenichnyuk, 2006) show the dissociative attachment energy at 1.55 eV.

The temperature dependence of negative ion yield has been measured by Pearl and Burrow (1993).

C_2H_5Cl

TABLE 95.3
Total Ionization Cross Sections for C_2H_5Cl

Energy (eV)	Q_i (10^{-20} m^2)	Energy (eV)	Q_i (10^{-20} m^2)
16	1.32	119	9.74
20	3.00	123	9.71
23	4.86	127	9.68
27	6.18	130	9.60
31	7.05	134	9.49
34	7.74	138	9.39
38	8.28	142	9.30
42	8.74	145	9.22
45	9.19	149	9.14
49	9.52	153	9.11
53	9.82	156	9.07
57	10.00	160	9.02
60	10.19	164	8.93
64	10.33	167	8.83
68	10.39	171	8.71
71	10.48	175	8.59
75	10.54	179	8.48
79	10.49	182	8.39
82	10.53	186	8.34
86	10.54	190	8.32
90	10.44	193	8.30
94	10.35	197	8.28
97	10.21	201	8.25
101	10.10	204	8.23
105	10.07	208	8.19
108	10.03	212	8.15
112	9.94	215	8.09
116	9.83	219	7.97

Sources: Adapted from Hudson, J. E. et al., *J. Phys. B: At. Mol. Opt. Phys.*, 34, 3025, 2001; tabulated values courtesy of Professor Harland (2001).

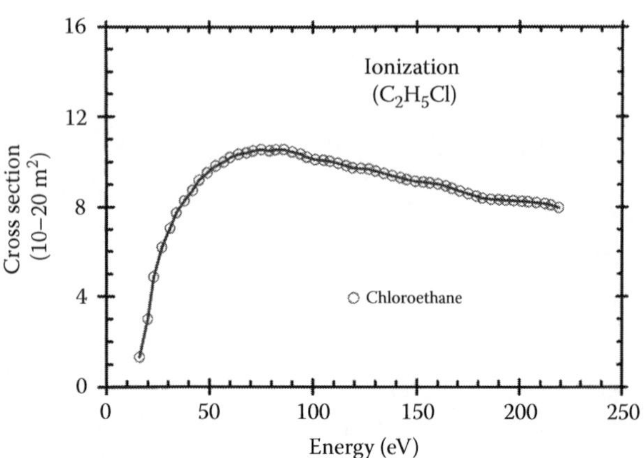

FIGURE 95.1 Total ionization cross sections for C_2H_5Cl. Tabulated values courtesy of Professor (Harland, 2006). (Adapted from Hudson, J. E. et al., *J. Phys. B: At. Mol. Opt. Phys.*, 34, 3025, 2001.)

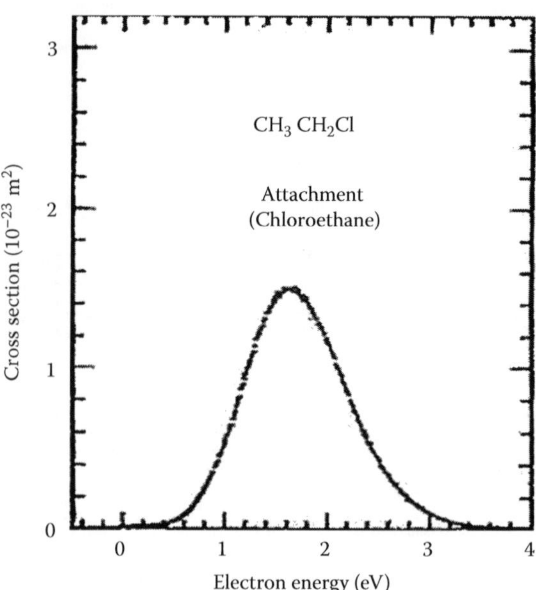

FIGURE 95.2 Attachment cross section for C_2H_5Cl. (Adapted from Pearl, D. M. and P. D. Burrow, *J. Chem. Phys.*, 101, 2940, 1994.)

TABLE 95.4
Thermal Attachment Rate Constants for C_2H_5Cl

Rate Constant (m^3/s)	Reference
3.4×10^{-20}	Rosa et al. (2001)
5×10^{-19}	Petrović et al. (1989)
1.6×10^{-21}	Bansal and Fessenden (1972)
3.48×10^{-19}	Christophorou et al. (1966)

TABLE 95.5
Attachment Coefficients and Rates for C_2H_5Cl

E/N (Td)	η/N (m^2)	k_a (m^3/s)
0.94	7.78×10^{-23}	3.48×10^{-19}
1.24	8.71×10^{-23}	4.37×10^{-19}
1.87	1.37×10^{-22}	7.67×10^{-19}
2.50	1.56×10^{-22}	1.04×10^{-18}
3.11	1.87×10^{-22}	1.42×10^{-18}
3.73	2.11×10^{-22}	1.81×10^{-19}

Source: Adapted from Christophorou, L. G. et al., *J. Chem. Phys.*, 45, 536, 1966.

95.4 ATTACHMENT RATE CONSTANTS

Thermal attachment rate constants are shown in Table 95.4. Table 95.5 shows the rate constants as a function of reduced electric field. (Christophorou et al., 1966) The attachment rate constant as a function of mean electron energy (Figure 95.3) shows a peak at 0.12 eV and falls off rapidly toward thermal energy.

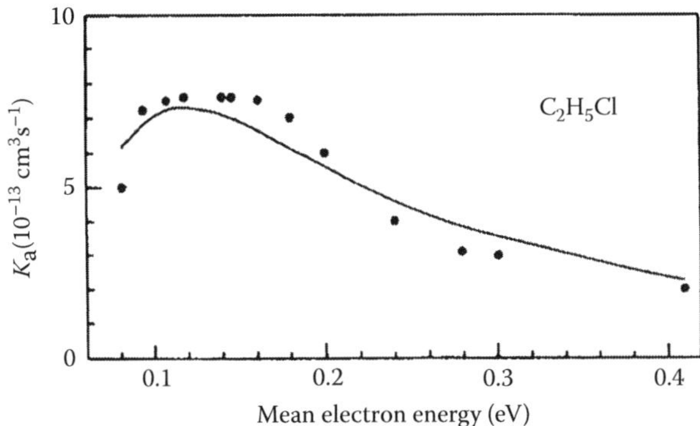

FIGURE 95.3 Attachment rate constants for C_2H_5Cl. Experimental data are shown as points and the solid line is calculated from derived attach cross sections. (Adapted from Petrović, Z. Lj., W. C. Wang, and L. C. Lee, *J. Chem. Phys.*, 90, 3145, 1989.)

REFERENCES

Bansal, K. M. and R. W. Fessenden, *Chem. Phys. Lett.*, 15, 21, 1972, cited by Petrović, et al., 1989.

Christophorou, L. G., R. N. Compton, G. S. Hurst, and P. W. Reinhardt, *J. Chem. Phys.*, 45, 536, 1966.

Hudson, J. E., C. Vallance, M. Bart, and P. W. Harland, *J. Phys. B: At. Mol. Opt. Phys.*, 34, 3025, 2001.

Pearl, D. M. and P. D. Burrow, *Chem. Phys. Lett.*, 206, 483, 1993.

Pearl, D. M. and P. D. Burrow, *J. Chem. Phys.*, 101, 2940, 1994.

Petrović, Z. Lj., W. C. Wang, and L. C. Lee, *J. Chem. Phys.*, 90, 3145, 1989.

Pshenichnyuk, S. A., I. A. Pshenichnyuk, E. P. Nafikova, and N. L. Asfandiarov, *Rapid Commum. Mass Spectrom.*, 20, 1097, 2006.

Rosa, A., W. Barszczewska, M. Foryć, and I. Szamrej, *Int. J. Mass Spectrom.*, 205, 85, 2001.

Shimanouchi, T. *Tables of Molecular Vibrational Frequencies Consolidated Volume 1*, NSRDS-NBS 39, Washington (DC), 1972, p. 104.

96

DIBROMODIFLUOROETHANE

CONTENTS

Dibromodifluoroethane ($C_2H_2Br_2F_2$) is an electron-attaching gas that has 102 electrons. The calculated electronic polarizability is 6.33×10^{-40} F m^2 (see Barszczewska et al., 2004).

96.1 ATTACHMENT RATES AND CROSS SECTIONS

Table 96.1 shows the attachment rate constants and Figure 96.1 shows the attachment rate constants as a function of mean energy (Sunagawa and Shimamori, 1995). The attachment cross sections are deconvoluted from the attachment rate constants.

The main ions observed are Br^- and Br_2^- with peak at zero energy (Sunagawa and Shimamori, 1995). Table 96.2 and Figure 96.2 show attachment cross sections.

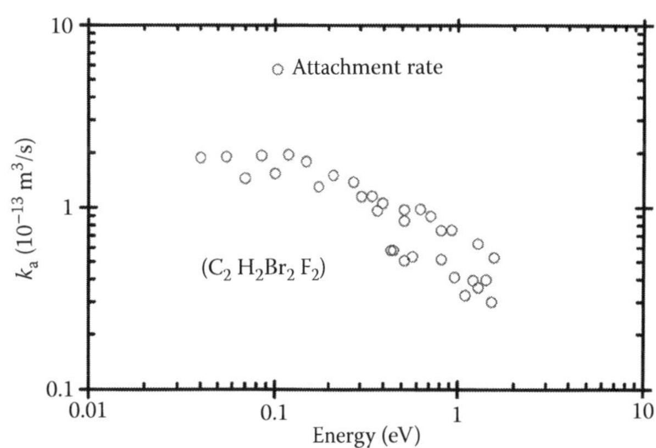

FIGURE 96.1 Attachment rate constants for $C_2H_2Br_2F_2$. (Adapted from Sunagawa, T. and H. Shimamori, *Int. J. Mass Spectrom. Ion Proc.*, 149/150, 123, 1995.)

TABLE 96.1
Attachment Rates

Method	Temperature, K (eV)	Rate (10^{-13} m^3/s)	Reference
Swarm	293	0.92	Barszczewska et al. (2004)
PR/MW	293	0.25[a]	Sunagawa and Shimamori (1995)
	293	1.7	
	(0.04)	1.7[a]	
	(0.10)	1.5	
	(0.40)	1.0	
	(1.00)	0.45	
FA/LP	300	0.15	Smith et al. (1990)

Note: FA/LP = flowing afterglow/Langmuir probe; PR/MW = pulse radiolysis/microwave heating.

[a] The authors do not discuss the inconsistency.

TABLE 96.2
Attachment Cross Sections for $C_2H_2Br_2F_2$

Energy (meV)	Q_a (10^{-20} m^2)	Energy (meV)	Q_a (10^{-20} m^2)
1.00	943	35.0	161
2.00	680	40.0	152
3.00	537	45.0	142
4.00	461	50.0	131
6.00	391	55.0	120
8.00	347	60.0	112
10.0	304	65.0	105
15.0	238	70.0	101
20.0	204	75.0	96.8
25.0	184	80.0	93.9
30.0	171	85.0	91.5

continued

TABLE 96.2 (continued)
Attachment Cross Sections for $C_2H_2Br_2F_2$

Energy (meV)	Q_a (10^{-20} m^2)	Energy (meV)	Q_a (10^{-20} m^2)
90.0	89.2	550	10.8
95.0	86.6	600	9.78
100.0	83.5	650	8.93
150	52.2	700	8.06
200	41.6	750	7.21
250	32.2	800	6.44
300	25.3	850	5.81
350	21.2	900	5.32
400	17.8	1000	4.62
450	14.7	1500	2.63
500	12.4	2000	1.71

Source: Digitized from Sunagawa, T. and H. Shimamori, *Int. J. Mass Spectrom. Ion Proc.*, 149/150, 123, 1995.

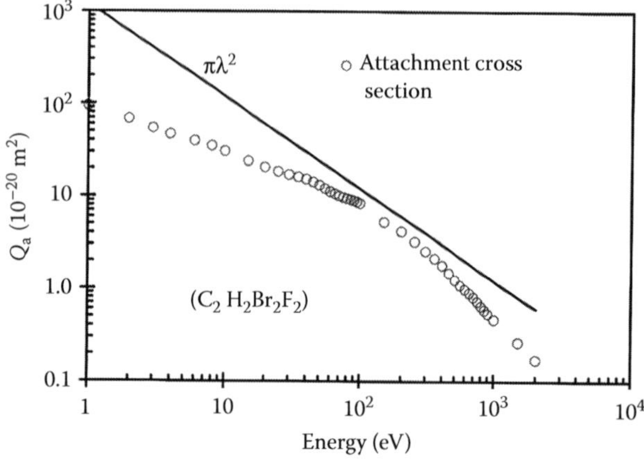

FIGURE 96.2 Attachment cross sections deconvoluted from Figure 96.1. *s*-Wave maximum cross sections are shown for comparison. (Adapted from Sunagawa, T. and H. Shimamori, *Int. J. Mass Spectrom. Ion Proc.*, 149/150, 123, 1995.) See Appendix A8.4 for method calculating $\pi\lambda^2$.

REFERENCES

Barszczewska, W., J. Kopyra, J. Wnorowska, and I. Szamrej, *Int. J. Mass Spectrom.*, 233, 199, 2004.

Smith, D., C. R. Herd, N. G. Adams, and J. F. Paulson, *Int. J. Mass Spectrom. Ion Proc.*, 96, 341, 1990.

Sunagawa, T. and H. Shimamori, *Int. J. Mass Spectrom. Ion Proc.*, 149/150, 123, 1995.

97

DIBROMOETHANE

$C_2H_4Br_2$

CONTENTS

Dibromoethane ($C_2H_4Br_2$) is an electron-attaching gas that has 76 electrons. There are two isomers: 1,1-dibromoethane (synonym ethylidene dibromide) and 1,2-dibromoethane (synonym ethylene dibromide). Selected properties are shown in Table 97.1.

97.1 ATTACHMENT RATES

The electron affinity of 1,1-$C_2H_4Br_2$ molecule is >–0.07 eV (see Christophorou, 1996) and the ion species observed are Br^- and Br_2^-. Table 97.2 and Figure 97.1 show the attachment rate constants for the isomers of $C_2H_4Br_2$.

TABLE 97.1
Selected Properties of $C_2H_4Br_2$ Isomers

Name	Formula	α_e (10^{-40} F m^2)	μ (D)	ε_i (eV)
1,1-Dibromoethane	$CHBr_2CH_3$	6.79[a]		
1,2-Dibromoethane	CH_2BrCH_2Br	11.9	1.19	10.35

Note: α_e = electronic polarizability; μ = dipole moment; ε_i = ionization potential.
[a] Barszczewska et al. (2004)

TABLE 97.2
Attachment Rate Constants for $C_2H_4Br_2$

Formula	Method	Temperature, K, (eV)	k_a (10^{-14} m^3/s)	Reference
1,1-$C_2H_4Br_2$	Swarm		3.5	Barszczewska et al. (2004)
	Swarm		1.5	Christophorou (1996)
	PR/MW		4.1	Sunagawa and Shimamori (1995)
		(0.04)	4.1	
		(0.1)	4.0	
		(0.4)	3.0	
		(1.0)	1.5	
	FA/LP	298	1.5	Smith et al. (1990)
		380	3.1	
		475	4.1	
1,2-$C_2H_4Br_2$	Swarm		1.8	Barszczewska et al. (2004)
		(0.04)	2.4	Sunagawa and Shimamori (1995)
		(0.1)	2.25	
		(0.4)	2.2	
		(1.0)	1.4	

Note: FA/LP = flowing afterglow/Langmuir probe; PR/MW = pulse radiolysis/microwave heating. Temperature is thermal unless otherwise mentioned.

97.2 ATTACHMENT CROSS SECTIONS

Table 97.3 and Figure 97.2 show the attachment cross sections deconvoluted by the present author from Sunagawa and Shimamori (1995). Like many bromine containing compounds, the attachment cross section shows a peak at zero energy. The isomer effect at lower energies is clearly evident.

$C_2H_4Br_2$

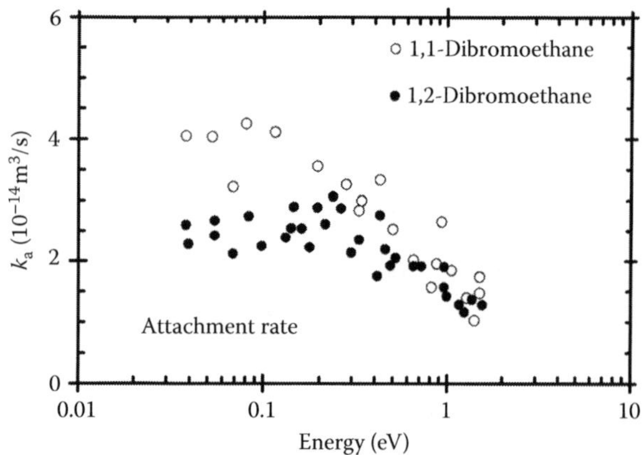

FIGURE 97.1 Attachment rate constants for $C_2H_4Br_2$. (○) 1,1-Dibromoethane; (●) 1,2-dibromoethane. Isomer effect is clearly seen. (Adapted from T. Sunagawa and H. Shimamori, *Int. J. Mass Spectrom. Ion Processes*, 149/150, 123, 1995.)

TABLE 97.3
Attachment Cross Sections for Isomers of $C_2H_4Br_2$

	Q_a (10^{-20} m^2)	
Energy (meV)	**1,1-Dibromoethane**	**1,2-Dibromoethane**
1.00	218.80	121.19
2.00	156.85	82.89
3.00	120.47	67.70
4.00	101.78	60.58
6.00	86.75	51.40
8.00	76.29	41.91
10.00	67.95	36.20
15.0	53.86	32.05
20.0	46.42	28.03
25.0	42.30	24.14
30.0	39.00	21.78
35.0	36.12	20.55
40.0	33.64	19.88
45.0	31.52	19.28
50.0	29.72	18.67
55.0	28.23	18.07
60.0	26.99	17.50
65.0	25.97	16.96
70.0	25.14	16.48
75.0	24.45	16.05
80.0	23.87	15.68
85.0	23.36	15.36
90.0	22.88	15.09
95.0	22.41	14.85
100.0	21.95	14.66
150	17.73	13.59
200	14.14	11.91
250	11.31	10.32
300	9.53	9.18
350	8.59	7.99
400	7.68	6.47
450	5.93	5.31
500	3.89	4.65
550	3.05	4.16
600	2.92	3.59
650	2.89	3.13
700	2.64	2.85
750	2.28	2.63
800	1.93	2.39
850	1.70	2.16
900	1.56	1.94
1000	1.46	1.63
1500	0.98	1.10
2000	0.77	0.74

Source: Digitized by the present author from Sunagawa, T. and H. Shimamori, *Int. J. Mass Spectrom. Ion Processes*, 149/150, 123, 1995.

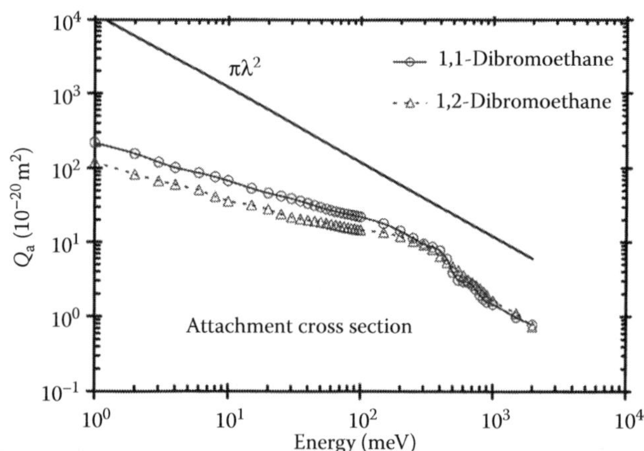

FIGURE 97.2 Attachment cross sections for $C_2H_4Br_2$. (—○—) 1,1-$C_2H_4Br_2$; 1,2-$C_2H_4Br_2$. The isomer effect at low energies is seen. *s*-Wave maximum cross section ($\pi\lambda^2$) is also shown for comparison. (Adapted from Sunagawa, T. and H. Shimamori, *Int. J. Mass Spectrom. Ion Processes*, 149/150, 123, 1995.)

REFERENCES

Barszczewska, W., J. Kopyra, J. Wnorowska, and I. Samrej, *Int. J. Mass Spectrom.*, 233, 199, 2004.

Christophorou, L. G. *Z. Phys. Chem.*, 195, 195, 1996.

Smith, D., C. R. Herd, N. G. Adams, and J. F. Paulson, *Int. J. Mass Spectrom. Ion Proc.* 96, 341, 1990.

Sunagawa, T. and H. Shimamori, *Int. J. Mass Spectrom. Ion Processes*, 149/150, 123, 1995.

98

DIBROMOTETRA-FLUOROETHANE

CONTENTS

Dibromotetrafluoroethane ($C_2Br_2F_4$), industrial name refrigerant 114B2, is an electron-attaching gas that has 118 electrons. Theoretically calculated electronic polarizability is 6.95×10^{-40} F m^2 (Barszczewska et al., 2004). The ionization potential is 11.1 eV.

98.1 ATTACHMENT RATES

Both Br^- and Br_2^- ions are observed with relative abundance of 80% and 20% at room temperature, respectively (Smith et al., 1990). Table 98.1 and Figure 98.1 show the attachment rate coefficients for $C_2Br_2F_4$.

TABLE 98.1
Attachment Rate Coefficients for $C_2Br_2F_4$

Method	Temperature: (K) (eV)	Rate (10^{-13} m^3/s)	Reference
PR/MW	293	1.3	Sunagawa and Shimamori (1995)
	(0.04)	1.3	
	(0.1)	1.2	
	(0.4)	0.62	
	(1.0)	0.3	
FA/LP	298	1.6	Smith et al. (1990)
	380	2.5	
	475	3.0	

Note: FA/LP = flowing afterglow/Langmuir probe; PR/MW = pulse radiolysis/microwave heating.

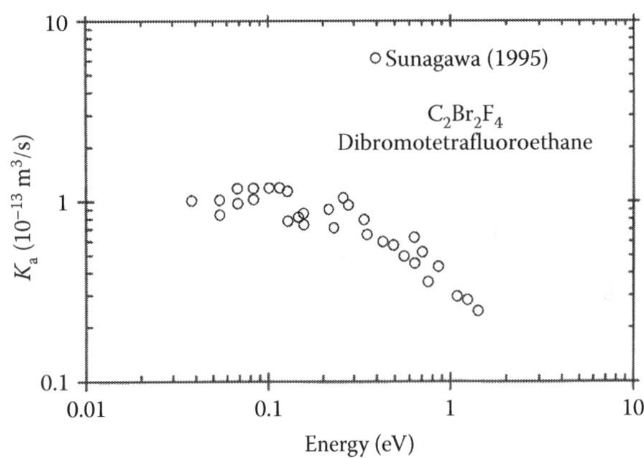

FIGURE 98.1 Attachment rate constants for $C_2Br_2F_4$. (Adapted from T. Sunagawa and H. Shimamori, *Int. J. Mass Spectrom. Ion Proc.*, 149/150, 123, 1995.)

TABLE 98.2
Attachment Cross Sections for $C_2Br_2F_4$

Energy (meV)	Q_a (10^{-20} m^2)	Energy (meV)	Q_a (10^{-20} m^2)
1.00	622	30.0	110
2.00	477	35.0	110
3.00	395	40.0	113
4.00	345	45.0	115
6.00	265	50.0	116
8.00	215	55.0	116
10.0	196	60.0	114
15.0	184	65.0	112
20.0	147	70.0	109
25.0	121	75.0	106

continued

TABLE 98.2 (continued)
Attachment Cross Sections for $C_2Br_2F_4$

Energy (meV)	Q_a (10^{-20} m^2)	Energy (meV)	Q_a (10^{-20} m^2)
80.0	102	500	14.7
85.0	98.3	550	12.4
90.0	94.7	600	10.9
95.0	91.4	650	9.77
100	88.4	700	8.79
150	69.6	750	7.93
200	55.8	800	7.18
250	44.5	850	6.56
300	35.7	900	6.06
350	28.6	1000	5.33
400	22.7	1500	2.93
450	18.0	2000	2.04

Source: Digitized from Sunagawa, T. and H. Shimamori, *Int. J. Mass Spectrom. Ion Proc.*, 149/150, 123, 1995.

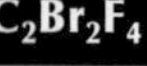

98.2 ATTACHMENT CROSS SECTIONS

Table 98.2 and Figure 98.2 show the attachment cross sections for $C_2Br_2F_4$.

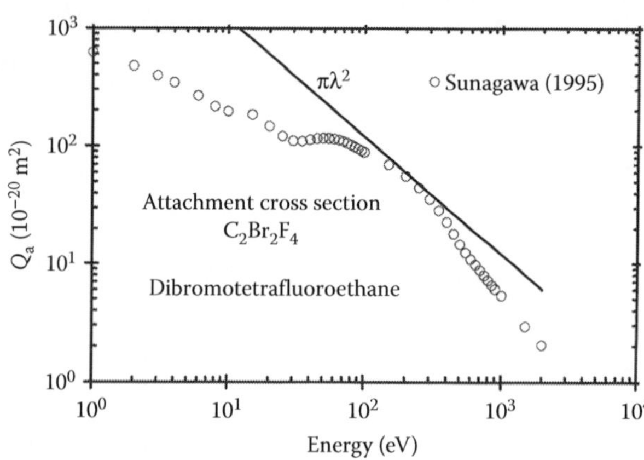

FIGURE 98.2 Attachment cross sections for $C_2Br_2F_4$. *s*-Wave maximum cross section is also shown for comparison. (Adapted from T. Sunagawa and H. Shimamori, *Int. J. Mass Spectrom. Ion Proc.*, 149/150, 123, 1995.)

REFERENCES

Barszczewska, W., J. Kopyra, J. Wnorowska, and I. Szamrej, *Int. J. Mass Spectrom.*, 233, 199, 2004.

Smith, D., C. R. Herd, N. G. Adams, and J. F. Paulson, *Int. J. Mass Spectrom. Ion Proc.*, 96, 341, 1990.

Sunagawa, T. and H. Shimamori, *Int. J. Mass Spectrom. Ion Proc.*, 149/150, 123, 1995.

99

DICHLOROETHANE

CONTENTS

Dichloroethane ($C_2H_4Cl_2$) is a polar, electron-attaching molecule that has 50 electrons. There are two isomers and selected properties are shown in Table 99.1.

99.1 SELECTED REFERENCES FOR DATA

See Table 99.2.

TABLE 99.1
Selected Properties of $C_2H_4Cl_2$

Name	Formula	α_e (F m^2)	μ (D)	ε_i (eV)
1,1-$C_2H_4Cl_2$	1,1-$C_2H_4Cl_2$	9.61×10^{-40}	2.06	11.04
1,2-$C_2H_4Cl_2$	1,2-$C_2H_4Cl_2$	8.90×10^{-40}	1.83	11.04

Note: α_e = electronic polarizability; μ = dipole moment; ε_i = ionization energy.

TABLE 99.2
Selected References

Parameter	Range, Energy, (E/N) eV, (Td)	Reference
k_a	Thermal	**Barszczewska et al. (2003)**
k_a	Thermal	**Gallup et al. (2003)**
Q_i	16–219	**Hudson et al. (2001)**
Q_a	0–3	**Aflatooni and Burrow (2000)**
Q_a	–	**Aflatooni et al. (1998)**
k_a, Q_a	0–2	**Johnson et al. (1977)**

Note: Bold font indicates experimental study.
k_a = attachment rate; Q_a = attachment cross section; Q_i = ionization cross section.

99.2 IONIZATION CROSS SECTIONS

Ionization cross sections for both isomers of $C_2H_4Cl_2$ measured by Hudson et al. (2001) are shown in Table 99.3 and Figure 99.1.

TABLE 99.3
Ionization Cross Sections for Isomers of $C_2H_4Cl_2$

	Cross Section (10^{-20} m^2)	
Energy (eV)	1,1-$C_2H_4Cl_2$	1,2-$C_2H_4Cl_2$
16	1.74	1.97
20	4.00	4.46
23	6.42	7.03
27	8.16	8.89
31	9.36	10.09
34	10.23	11.11
38	10.86	11.61
42	11.46	12.02
45	11.97	12.41
49	12.44	12.87
53	12.88	13.33
57	13.14	13.55
60	13.41	13.80
64	13.50	13.88
68	13.62	13.98
71	13.69	14.02
75	13.79	14.13
79	13.88	14.17
82	13.80	14.08
86	13.90	14.11
90	13.72	14.02
94	13.63	13.90
97	13.66	13.95
101	13.63	13.86

continued

TABLE 99.3 (continued)
Ionization Cross Sections for Isomers of $C_2H_4Cl_2$

	Cross Section (10^{-20} m²)	
Energy (eV)	**1,1-$C_2H_4Cl_2$**	**1,2-$C_2H_4Cl_2$**
105	13.40	13.67
108	13.24	13.51
112	13.25	13.52
116	13.25	13.44
119	13.16	13.33
123	13.07	13.23
127	12.83	13.02
130	12.77	12.98
134	12.73	12.91
138	12.68	12.86
142	12.57	12.72
145	12.40	12.54
149	12.28	12.40
153	12.10	12.21
156	12.02	12.17
160	11.94	12.10
164	11.96	12.11
167	11.89	12.06
171	11.82	12.00
175	11.73	11.89
179	11.54	11.78
182	11.38	11.60
186	11.22	11.38
190	11.11	11.25
193	10.98	11.15
197	10.93	11.10
201	10.89	11.03
204	10.89	11.02
208	10.85	11.01
212	10.85	11.02
215	10.78	10.93
219	10.60	10.77

Sources: Adapted from Hudson, J. E. et al., *J. Phys. B: At. Mol. Opt. Phys.*, 34, 3025, 2001; tabulated values due to courtesy of Professor Harland (2006).

99.3 ATTACHMENT CROSS SECTIONS

Johnson et al. (1977) have studied the attachment processes in both isomers of $C_2H_4Cl_2$ and the following processes occur.

A. 1,1-$C_2H_4Cl_2$

$$C_2H_4Cl_2 + e \rightarrow C_2H_4Cl_2^{-*} \rightarrow C_2H_4Cl_2{}^* + e \quad (99.1)$$

$$C_2H_4Cl_2^{-*} \rightarrow C_2H_4Cl + Cl^- \quad (99.2)$$

B. 1,2-$C_2H_4Cl_2$

$$C_2H_4Cl_2 + e \rightarrow C_2H_4Cl_2^{-*} \rightarrow C_2H_4Cl_2{}^* + e \quad (99.3)$$

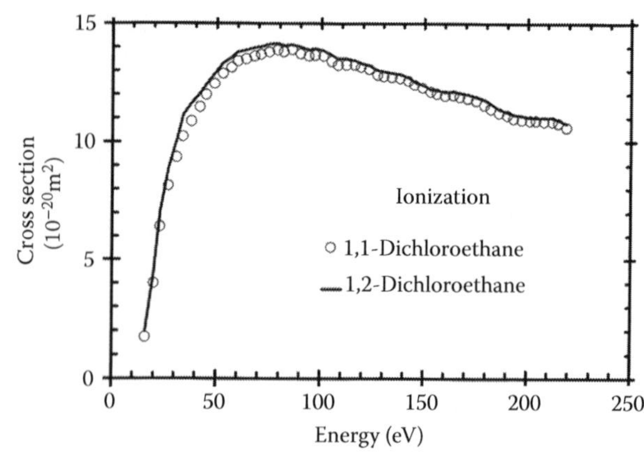

FIGURE 99.1 Total ionization cross sections for 1,1-$C_2H_4Cl_2$ (○) and 1,2-$C_2H_4Cl_2$ (——). Tabulated values courtesy of Professor Harland (2006). (Adapted from Hudson, J. E. et al., *J. Phys. B: At. Mol. Opt. Phys.*, 34, 3025, 2001.)

$$C_2H_4Cl_2^{-*} \rightarrow C_2H_4Cl + Cl^- \quad (99.4)$$

$$C_2H_4Cl_2^{-*} \rightarrow Cl + C_2H_4Cl^- \quad (99.5)$$

$$C_2H_4Cl_2^{-*} \rightarrow C_2H_4 + Cl_2^- \quad (99.6)$$

Reactions 99.1 and 99.3 are identical and categorized as autodetachment. No negative ion with sufficiently long time results. Reactions 99.2 and 99.4 are also identical and result in Cl^- ions. Further this is the only species observed below 10 eV for 1,1-$C_2H_4Cl_2$. The relative yield of ions for the isomers are shown in Figures 99.2 and 99.3. Figure 99.2 also include the relative ion yield (Cl^-) that is measured for trichloroethane demonstrating the effect of substituting a

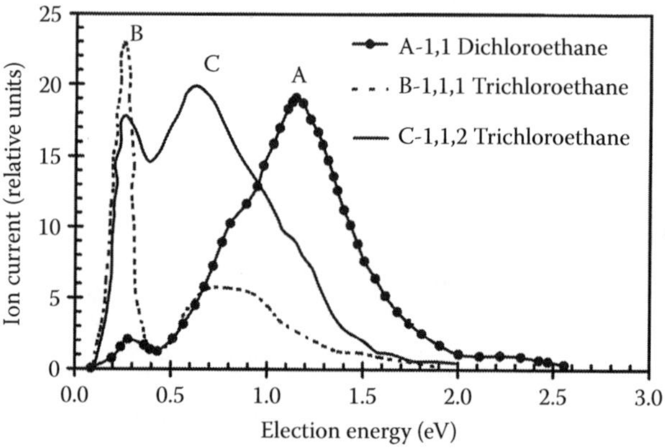

FIGURE 99.2 Negative ions in 1,1-$C_2H_4Cl_2$. Cl^- ion is the only species observed below 10 eV electron energy. Data are presented for 1,1,1-trichloroethane and 1,1,2-trichloroethane to demonstrate the effect of increasing the number of chlorine atoms from 2 to 3 and the influence of chemical structure. The curves are mean values derived from 5 to 6 measurements for each compound. (Adapted from Johnson, J. P., L. G. Christophorou, and J. G. Carter, *J. Chem. Phys.*, 67, 2196, 1977.)

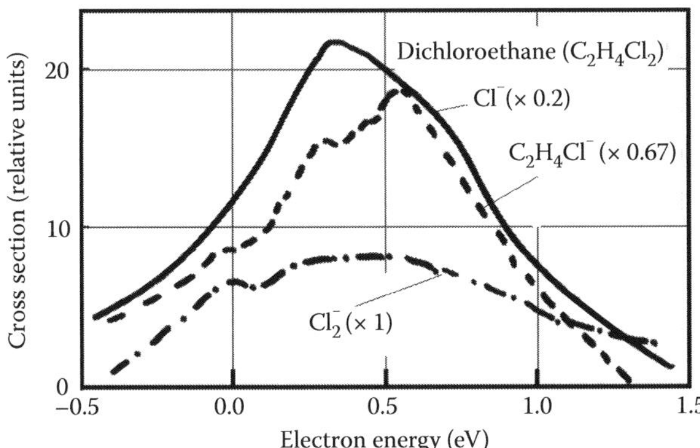

FIGURE 99.3 Negative ion yields in 1,2-$C_2H_4Cl_2$. (Adapted from Johnson, J. P., L. G. Christophorou, and J. G. Carter, *J. Chem. Phys.*, 67, 2196, 1977.)

TABLE 99.4
Total Attachment Cross Sections for Isomers of $C_2H_4Cl_2$

	Cross Section (10^{-21} m^2)	
Energy (eV)	**1,1-$C_2H_4Cl_2$**	**1,2-$C_2H_4Cl_2$**
0.00	1.54	0.42
0.01	0.97	0.39
0.03	0.52	0.37
0.05	0.54	0.39
0.07	0.56	0.44
0.10	0.47	0.51
0.30	0.32	0.88
0.50	1.17	0.82
0.70	2.71	0.50
1.00	3.80	0.22
1.20	2.87	0.13
1.40	1.65	0.08
1.60	0.73	0.06
1.80	0.32	0.02
2.00	0.18	0.02
2.20	0.08	0.09
2.40	0.04	0.29

Source: Digitized from K. Aflatooni and P. D. Burrow, *J. Chem. Phys.*, 113, 1455, 2000.

hydrogen atom with chlorine. The chemical structure for the same compound alters the ion yield profoundly.

Attachment cross sections for the isomers of $C_2H_4Cl_2$ are shown in Table 99.4 and Figure 99.4 (Aflatooni and Burrow, 2000).

Table 99.5 shows the selected features of the isomers of $C_2H_4Cl_2$.

99.4 ATTACHMENT RATE CONSTANT

Table 99.6 shows the thermal attachment rates for the isomers of $C_2H_4Cl_2$.

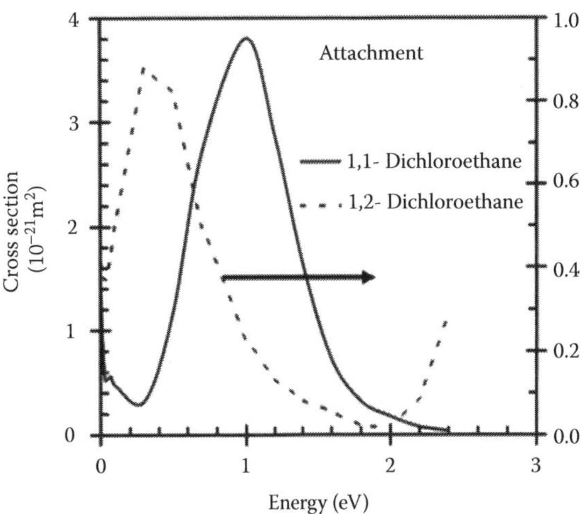

FIGURE 99.4 Attachment cross sections for 1,1-$C_2H_4Cl_2$ (——) and 1,2-$C_2H_4Cl_2$ (– – –). (Adapted from Aflatooni, K. and P. D. Burrow, *J. Chem. Phys.*, 113, 1455, 2000.)

TABLE 99.5
Attachment Energy Peak and Cross Section

Molecule	Peak Energy (eV)	Peak Magnitude ($\times 10^{-21}$ m^2)	Reference
1,1-$C_2H_4Cl_2$	0.96	3.94	Aflatooni and Burrow (2000)
	~1.05 (Cl^-)		Johnson et al. (1977)
1,2-$C_2H_4Cl_2$	0.37	0.93	Aflatooni and Burrow (2000)
	~0.3 (Cl^-)		Johnson et al. (1977)
	~0.4 (Cl_2^-)		
	~0.55 ($C_2H_4Cl^-$)		

TABLE 99.6
Thermal Attachment Rate Constants for $C_2H_4Cl_2$

Molecule	Rate (m^3/s)	Method	Reference
1,1-$C_2H_4Cl_2$	6.87×10^{-17}	Beam	Gallup et al. (2003)
	2.1×10^{-17}	Swarm	Christophorou et al. (1981)
1,2-$C_2H_4Cl_2$	2.9×10^{-17}	Swarm	Barszczewska et al. (2003)
	4.72×10^{-17}	Beam	Gallup et al. (2003)
	3.2×10^{-17}	Swarm	Christophorou et al. (1981)

REFERENCES

Aflatooni, K., and P. D. Burrow, *J. Chem. Phys.*, 113, 1455, 2000.

Aflatooni, K., G. A. Gallup, and P. D. Burrow, *Chem. Phys. Lett.*, 282, 398, 1998.

Barszczewska, W., J. Kopyra, J. Wnorowska, and I. Szamrej, *J. Phys. Chem. A*, 107, 11427, 2003.

Christophorou, L. G., R. A. Mathis, D. R. James, and D. L. McCorkle, *J. Phys. D: Appl. Phys.*, 14, 1889, 1981.

Gallup, G. A., K. Aflatooni, and P. D. Burrow, *J. Chem. Phys.*, 118, 2562, 2003.

Hudson, J. E., C. Vallance, M. Bart, and P. W. Harland, *J. Phys. B: At. Mol. Opt. Phys.*, 34, 3025, 2001.

Johnson, J. P., L. G. Christophorou, and J. G. Carter, *J. Chem. Phys.*, 67, 2196, 1977.

100

DISILANE

Si_2H_6

CONTENTS

Disilane (Si_2H_6) is a nonpolar molecule with 34 atoms, an electronic polarizability of 12.35×10^{-39} F m^2, and an ionization potential of 10.75 eV.

100.1 SELECTED REFERENCES FOR DATA

See Table 100.1.

100.2 TOTAL SCATTERING CROSS SECTION

See Table 100.2.

100.3 ELASTIC SCATTERING AND MOMENTUM TRANSFER CROSS SECTION

See Table 100.3.

TABLE 100.1
Selected References for Data for Si_2H_6

Quantity	Range: eV, (Td)	Reference
Transport coefficients	(1–2000)	Shimada et al. (2003)
Total scattering cross section	1–370	**Szmytkowski et al. (2001)**
Ionization cross section	15–5000	**Ali et al. (1997)**
Ionization cross section	15–1000	**Krishnakumar and Srivastava (1995)**
Elastic scattering cross section	2–100	**Dillon et al. (1994)**
Ionization coefficients	(180–810)	**Shimozuma and Tagashira (1986)**
Ionization cross section	15–400	**Chatham et al. (1984)**

Notes: Bold font denotes experimental study.

100.4 POSITIVE ION APPEARANCE POTENTIAL

See Table 100.4.

TABLE 100.2
Total Scattering Cross Section for Si_2H_6

Energy (eV)	Q_T (10^{-20} m^2)	Energy (eV)	Q_T (10^{-20} m^2)	Energy (eV)	Q_T (10^{-20} m^2)
1.0	49.6	5.3	70.6	35	44.3
1.1	51.8	5.8	69.0	40	41.5
1.2	56.4	6.3	69.7	45	38.4
1.3	60.2	6.8	68.1	50	36.6
1.4	61.8	7.3	68.4	60	34.0
1.5	66.8	7.8	69.0	70	32.3
1.6	70.4	8.3	67.8	80	30.4
1.7	71.8	8.8	67.0	90	27.5
1.8	74.4	9.3	66.6	100	26.2
1.9	75.0	9.8	64.7	110	25.8
2.0	76.1	10.3	64.5	120	24.7
2.1	77.1	11.3	63.7	140	22.8
2.2	75.3	12.3	62.5	160	21.4
2.3	76.6	14.3	60.5	180	19.4
2.6	76.4	16.3	58.5	200	18.0
2.8	77.3	18.3	56.6	220	15.8
3.0	75.3	20.3	55.6	250	14.9
3.3	73.3	22.8	53.4	275	14.2
3.8	71.0	25.3	50.4	300	13.1
4.3	69.3	27.8	48.0	350	11.5
4.8	69.0	30.0	46.2	370	11.4

Source: Adapted from Szmytkowski, C., P. Możejko, and G. Kasperski, *J. Phys. B: At. Mol. Opt. Phys.*, 34, 605, 2001.
Note: See Figure 100.1 for graphical presentation.

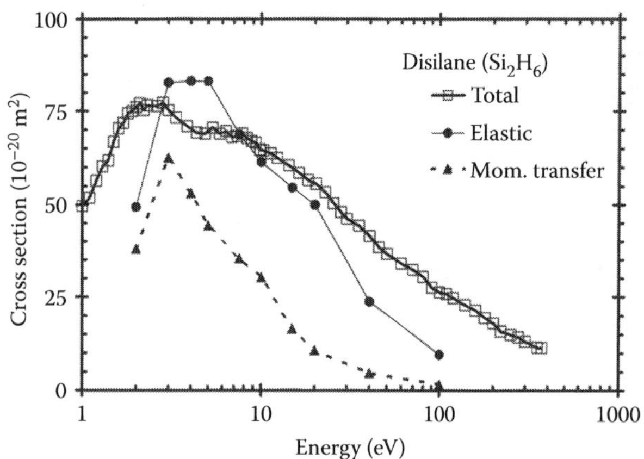

FIGURE 100.1 Scattering cross sections for Si_2H_6. (—□—) Total, Szmytkowski et al. (2001); (-●-) elastic, (– –Δ– –) momentum transfer (Dillon et al., 1994).

TABLE 100.3
Elastic Scattering and Momentum Transfer Cross Section

Energy (eV)	Q_{el} (10^{-20} m^2)	Q_M (10^{-20} m^2)	Energy (eV)	Q_{el} (10^{-20} m^2)	Q_M (10^{-20} m^2)
2	49.3	38	10	61.4	30.3
3	82.8	62.6	15	54.6	16.5
4	83.2	53.1	20	50	10.7
5	83.1	44.4	40	23.7	4.65
7.5	68.8	35.4	100	9.6	1.65

Source: Adapted from Dillon, M. A. et al. *J. Phys. B: At. Mol. Opt. Phys.*, 27, 1209, 1994.

Note: See Figure 100.1 for graphical presentation.

TABLE 100.4
Positive Ion Appearance Potentials

Ion	Appearance Potential (eV)	Ion	Appearance Potential (eV)
$Si_2H_6^+$	10.2	Si_2^+	14.5
$Si_2H_5^+$	11.7	SiH_3^+	12.0
$Si_2H_4^+$	11.0	SiH_2^+	12.0
$Si_2H_3^+$	12.8	SiH^+	14.7
$Si_2H_2^+$	12.2	Si^+	15.0
Si_2H^+	15.0		

Source: Adapted from Krishnakumar, E. and S. K. Srivastava, *Contrib. Plas. Phys.*, 35, 395, 1995.

100.5 IONIZATION CROSS SECTION

See Table 100.5.

TABLE 100.5
Ionization Cross Sections for Si_2H_6

Chatham et al. (1984) Energy (eV)	Chatham et al. (1984) Q_i (10^{-20} m^2)	Krishnakumar and Srivastava (1995) Energy (eV)	Q_i (10^{-22} m^2)	Energy (eV)	Q_i (10^{-22} m^2)
15	3.25	11	5.41	90	836.5
20	5.8	12	23.2	95	817.5
30	8.7	13	69.2	100	802.3
50	10.1	14	136.5	110	771.2
100	9.4	15	216.8	120	740.3
200	7.1	16	313.8	130	711.2
300	5.85	17	405.4	140	685.0
400	5.0	18	508.3	150	661.9
		19	618.5	160	638.0
		20	698.5	170	616.5
		22	866.5	180	598.3
		24	984.0	190	580.1
		26	1075.8	200	565.5
		28	1136.4	250	495.3
		30	1174.7	300	439.8
		32	1198.1	350	399.0
		34	1197.7	400	364.5
		36	1184.7	450	332.1
		38	1167.1	500	309.5
		40	1144.1	550	290.8
		45	1072.6	600	274.1
		50	1020.8	650	258.0
		55	979.5	700	245.3
		60	957.7	750	233.5
		65	921.5	800	222.7
		70	905.3	850	212.0
		75	885.4	900	202.7
		80	870.4	950	193.9
		85	852.4	1000	186.6

Note: See Figure 100.2 for graphical presentation.

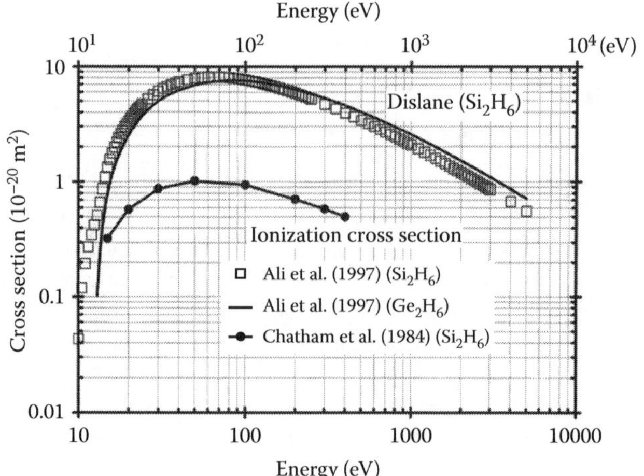

FIGURE 100.2 Ionization cross section for Si_2H_6. (□) Ali et al. (1997); (—●—) Chatham et al. (1984) ordinates shifted by factor of 10 for clarity; (——) Ali et al. (1997), cross sections for digermane (Ge_2H_6) are added for comparison.

REFERENCES

Ali, M. A., Y.-K. Kim, W. Hwang, N. M. Weinberger, and M. E. Rudd, *J. Chem. Phys.*, 106, 9602, 1997.

Chatham, H., D. Hils, R. Robertson, and A. Gallagher, *J. Chem. Phys.*, 81, 1770, 1984.

Dillon, M. A., L. Boesten, H. Tanaka, M. Kimura, and H. Sato, *J. Phys. B: At. Mol. Opt. Phys.*, 27, 1209, 1994.

Krishnakumar, E. and S. K. Srivastava, *Contrib. Plas. Phys.*, 35, 395, 1995.

Shimada, T., Y. Nakamura, Z. Lj. Petrović, and T. Makabe, *J. Phys. D: Appl. Phys.*, 36, 1936, 2003.

Shimozuma, M. and H. Tagashira, *J. Phys. D: Appl. Phys.*, 19, L179, 1986.

Szmytkowski, C., P. Możejko, and G. Kasperski, *J. Phys. B: At. Mol. Opt. Phys.*, 34, 605, 2001.

101

ETHANE

CONTENTS

Ethane (C_2H_6) is a saturated hydrocarbon that belongs to the series of C_nH_{2n+2}. It has 18 electrons, is nonpolar and possesses an electronic polarizability of 4.92×10^{-40} F m^2. The dissociation energy is 4.36 eV, and the ionization potential 11.56 eV. The ground state electronic configuration is given by Mapstone and Newell (1992). The lowest excitation energy to an optically allowed state is 9.4 eV. The molecule has 12 normal vibrational modes (ν_1–ν_{12}) some of which overlap considerably. Consequently their cross sections are given as bending composites ν_b ($\nu_b = \nu_2 + \nu_3 + \nu_6 + \nu_8 + \nu_9 + \nu_{11} + \nu_{12}$) centered around 160 meV, and stretching composites ν_s ($\nu_1 + \nu_5 + \nu_7 + \nu_{10}$) centered around 360 meV (Mapstone et al., 2000). The elastic peak energy coincides with that of ν_4. Energy assignments are shown in Table 101.1 (Shishikura et al., 1997).

Selected references for data are shown in Table 101.2.

TABLE 101.1
Vibrational Modes of C_2H_6 Molecule

Designation	Mode	Energy (meV)
ν_1	Stretching	360
ν_2	Bending	201.2
ν_3	Bending	166.4
ν_4	Torsional oscillation	102.3
ν_5	Stretching	366
ν_6	Bending	130.2
ν_7	Stretching	117.8
ν_8	Bending	116.9
ν_9	Bending	109
ν_{10}	Stretching	123.4
ν_{11}	Bending	370.7
ν_{12}	Bending	179.0

Source: Adapted from Lunt, S. L. et al., *J. Phys. B: At. Mol. Opt. Phys.*, 27, 1407, 1994.

101.1 SELECTED REFERENCES FOR DATA

101.2 TOTAL SCATTERING CROSS SECTION

Tables 101.3 and 101.4 show the total scattering cross section in C_2H_6. See Figure 101.1 for graphical presentation.

TABLE 101.2
Selected References for C_2H_6

Quantity	Energy Range (eV)	Reference
	Cross Sections	
Q_{diff}	3.2–15.4	Mapstone et al. (2000)
Q_{el}	300–1300	Maji et al. (1998)
Q_T	0.01–0.175	Lunt and Linder (1998)
Q_{el}	0.4–10	Merz and Lindner (1998)
Q_i	10–1000	Hwang et al. (1996)
Q_T	0.6–250	Szmytkowski and Krzysztofowicz (1995)
Q_v	0.05–10	Lunt et al. (1994)
Q_i	10–3000	Nishimura (1994)
Q_T	4–500	Nishimura (1991)
Q_v	0–20	Boesten (1990)
Q_{diff}	2–100	Tanaka (1988)
Q_T	0.7–400	Sueoka (1986)
Q_T	5–400	Floeder (1985)
Q_i	15–400	Chatham et al. (1984)
Q_a	0–14	Rutkowsky (1980)
Q_{el}	100–1000	Fink (1975)
Q_i	600–12000	Schram et al. (1966)
Quantity	***E/N* Range (Td)**	**Reference**
	Swarm Parameters	
W_e, D_L	0.03–300	Shishikura (1997)
W_e, D_r, D_L	0.02–14	Schmidt (1992)
μ_e	0.01–50.0	Floriano (1986)
W_e	0.005–10	Gee (1983)
Q_{diss}	15–100	Winters
W_e,	0.03–30	McCorkle et al. (1978)
α/N	10–2000	Watts and Heylen (1978)
α/N	10–4500	Heylen (1975)
W_e	0.03–30	Bowman (1967)

Note: Q_a = Attachment cross section, Q_M = momentum transfer, Q_i = ionization, Q_T = total, Q_v = vibrational excitation, D_L = Longitudinal diffusion coefficient, D_r = radial diffusion coefficient (m^2/s), μ_e = mobility ($m^2/V/s$), W_e = drift velocity (m/s), $\bar{\varepsilon}$ = mean energy.

TABLE 101.3
Total Scattering Cross Section in C_2H_6 ($\varepsilon \leq 1.00$ eV)[a]

Energy (eV)	Q_T (10^{-20} m^2)	Energy (eV)	Q_T (10^{-20} m^2)	Energy (eV)	Q_T (10^{-20} m^2)
0.02	21.56	0.20	8.14	0.5	0.72
0.04	17.95	0.25	8.705	0.6	2.03
0.06	14.2	0.30	6.20	0.7	3.40
0.08	12.80	0.35	4.41	0.8	4.26
0.1	11.4	0.40	2.52	0.9	6.54
0.15	9.15	0.45	1.31	1	7.75

[a] Digitized and interpolated from Lunt, S. L. et al., *J. Phys. B: At. Mol. Opt. Phys.*, 27, 1407, 1994.

TABLE 101.4
Total Scattering Cross Section in C_2H_6

Energy (eV)	Q_T (10^{-20} m^2)	Energy (eV)	Q_T (10^{-20} m^2)
Szmytkowski and Krzysztofowicz (1995)		Sueoka and Mori (1986)	
0.6	3.11	1.0	7.7
0.7	4.37	1.2	8.1
0.8	5.05	1.4	9.0
1.0	6.50	1.6	9.7
1.2	7.97	1.8	10.7
1.4	9.13	2.0	11.4
1.6	10.2	2.2	12.1
1.8	10.8	2.5	13.4
2.0	11.9	2.8	14.7
2.5	14.5	3.1	16.0
3.0	16.0	3.4	17.3
3.5	17.9	3.7	18.9
4	20.2	4.0	20.3
4.5	23.0	4.5	22.5
5	25.1	5.0	24.6
5.5	27.2	5.5	26.9
6	29.4	6.0	29.2
6.5	32.0	6.5	30.0
7	33.9	7.0	31.4
7.5	34.7	7.5	32.5
8	34.6	8.0	31.8
8.5	34.2	8.5	30.9
9	33.6	9.0	31.3
10	32.5	9.5	30.9
11	31.7	10.0	30.2
12	30.8	11.0	28.5
14	29.5	12.0	27.9
16	28.3	13.0	27.3
18	28.2	14.0	27.2
20	27.6	15.0	25.9
25	26.6	16.0	25.6
30	24.5	17.0	25.6
35	23.0	18.0	25.3
40	21.9	19.0	25.2
45	21.1	20.0	24.6
50	20.2	22.0	23.9
60	19.3	25.0	22.6
70	18.3	30.0	20.9
80	17.3	35.0	19.6
90	16.4	40.0	18.5
100	15.4	50.0	16.7
110	14.2	60.0	16.1
120	13.7	70.0	15.0
140	12.8	80.0	14.5
160	12.1	90.0	13.6
180	11.3	100	12.7
200	10.5	120	11.8
220	9.90	150	10.5
250	9.26	200	9.2
		250	8.2
		300	7.6
		350	6.9
		400	6.2

Note: See Figure 101.1 for graphical presentation.

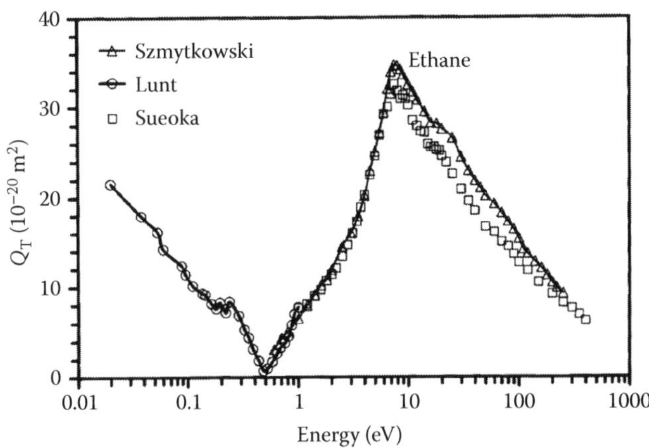

FIGURE 101.1 Total cross section in C_2H_6. (—○—) Lunt et al. (1994); (—△—) Szmytkowski and Krzysztofowicz (1995); Sueoka and Mori (1986).

The highlights of total scattering cross section are

1. Increase of cross section toward zero energy, due to vibrational excitation.
2. Ramsauer–Townsend minimum at approximately 0.4 eV.
3. Shape resonance at about 8.0 eV.
4. Decrease of cross section with energy beyond the resonance peak.

101.3 DIFFERENTIAL CROSS SECTION

The highlights of differential cross section as a function of electron energy and angle of scattering are (Figure 101.2)

1. At lower energies there are two minima, one at 45° and the other at about 100°.

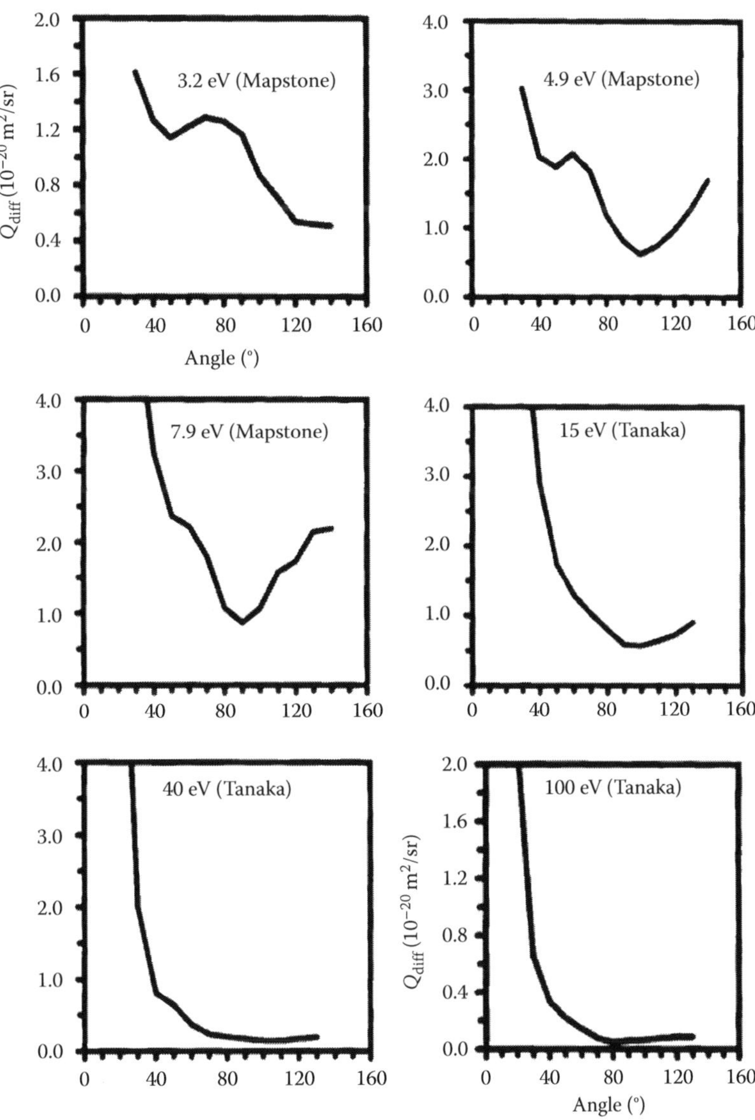

FIGURE 101.2 Differential scattering cross section in C_2H_6. $3.2 \le \varepsilon \le 7.9$ eV, Mapstone and Newell (1992); $15 \le \varepsilon \le 100$ eV, Tanaka et al. (1988).

C_2H_6

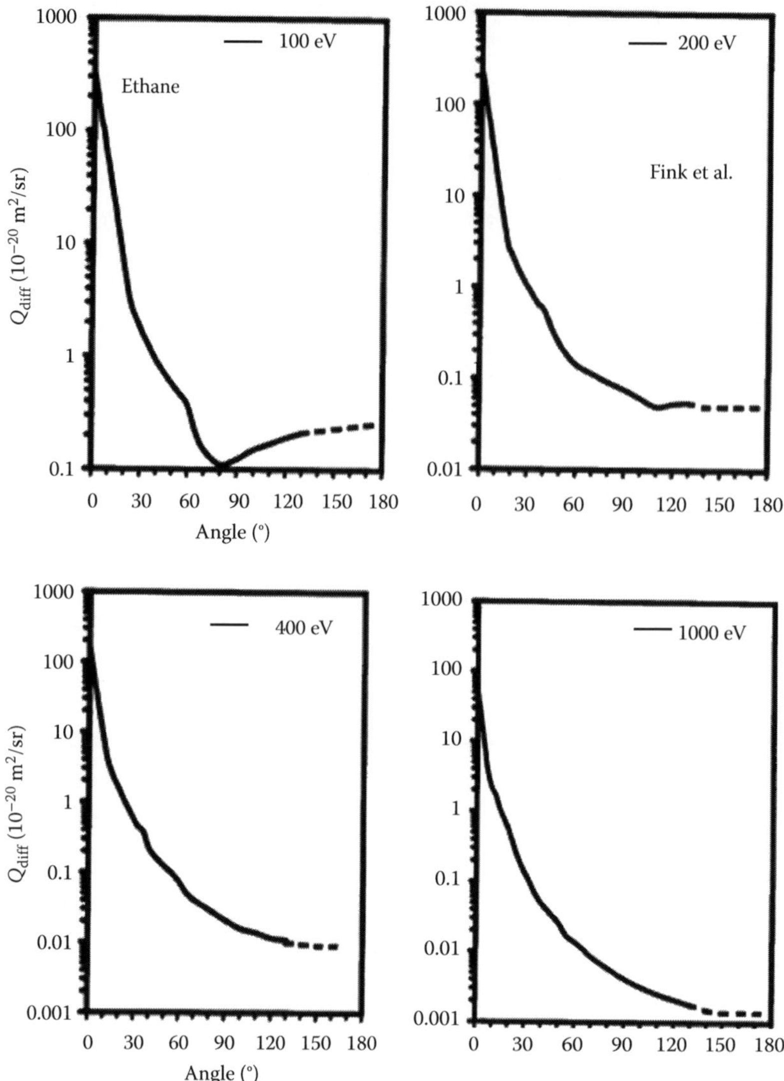

FIGURE 101.3 Differential cross sections (relative) in C_2H_6 at high energies. Dotted portion denotes extrapolation for integration purposes. (Adapted from Fink, M., K. Jost, and D. Hermann, *J. Chem. Phys.*, 63, 1985, 1975.)

2. There is a local maximum at approximately 65°. These are characteristic of f-wave scattering.
3. The minimum at higher angles becomes broader and less deep toward higher energies.
4. The low-energy minimum disappears at energies greater than 15 eV.
5. At higher energy ($40 \leq \varepsilon \leq 1000$) the differential scattering cross section is structureless, decreasing rapidly up to 60° and reaching a plateau for higher energies (Figure 101.3).

101.4 MOMENTUM TRANSFER AND INTEGRAL ELASTIC CROSS SECTIONS

See Tables 101.5 and 101.6.

101.5 MERT COEFFICIENTS

The four term MERT expression for phase shifts is (Raju, 2006)

s-wave:

$$\tan \eta_0 = -Ak - \left(\frac{\pi\alpha}{3a_0}\right)k^2 - Ak\left(\frac{4\alpha}{3a_0}\right)k^2 \ln(ka_0) + Dk^3 + Fk^4 \tag{101.1}$$

p-wave:

$$\tan \eta_1 = \left(\frac{\pi\alpha}{15a_0}\right)k^2 - A_1k^3 + Hk^5 \tag{101.2}$$

d-wave:

$$\frac{\pi\alpha k^2}{105a_0} - A_2k_5 \tag{101.3}$$

TABLE 101.5
Momentum Transfer Cross Section in C_2H_6

Energy (eV)	Q_M (10^{-20} m^2)	Energy (eV)	Q_M (10^{-20} m^2)	Energy (eV)	Q_M (10^{-20} m^2)	Energy (eV)	Q_M (10^{-20} m^2)
Merz and Lindner (1998)				Tanaka (1988)		Fink (1975)	
0.01	16.96	0.22	2.55	2	9.74	100	1.41
0.02	9.59	0.26	3.62	3	12.10	200	0.521
0.03	6.16	0.30	4.67	4	15.78	400	0.157
0.04	4.49	0.35	5.77	5	19.82	600	0.093
0.05	2.70	0.40	6.54	6	22.20	1000	0.039
0.06	2.24	0.45	6.85	7.5	23.48		
0.07	2.11	0.50	6.88	8.5	20.10		
0.08	1.85	0.60	7.20	10	16.25		
0.09	1.44	0.70	8.21	15	10.88		
0.10	1.15	0.80	8.87	20	6.99		
0.14	1.05	0.90	8.99	40	3.76		
0.18	1.86	1.00	9.75	100	1.41		

Note: $0.01 \le \varepsilon \le 1.0$, average of swarm derivation and beam experiments (Merz and Lindner, 1998). The relative differential cross sections of Fink et al. (1975), shown in Figure 101.4 is integrated by the author and normalized to the cross section of Tanaka et al. at 100 eV energy. For graphical presentation, see Figure 101.4.

TABLE 101.6
Integral Elastic Scattering Cross Section in C_2H_6

Energy (eV)	Q_{el} (10^{-20} m^2)	Energy (eV)	Q_{el} (10^{-20} m^2)	Energy (eV)	Q_{el} (10^{-20} m^2)	Energy (eV)	Q_{el} (10^{-20} m^2)
Merz and Lindner (1998)				Tanaka (1998)		Fink et al. (1975)	
0.05	5.42	1	10.63	2	10.82	100	6.61
0.07	4.26	2	12.57	3	14.7	200	3.154
0.1	3.19	4	22.25	4	19.13	400	1.524
0.2	3.56	6	27.68	5	24.34	600	0.98
0.3	5.05	8	31.49	6	26.76	1000	0.385
0.4	6.57	10	30.38	7.5	31.68		
0.5	7.80	15	25.85	8.5	30.11		
0.6	8.72	20	24.66	10	28.75		
0.7	9.36	25	23.83	15	25.26		
0.8	9.84	30	20.18	20	21.82		
0.9	10.24			40	14.1		
				100	6.6		

Note: $0.01 \le \varepsilon \le 30.0$, average of swarm derivation and beam experiments (Merz and Lindner, 1998). The relative differential cross sections of Fink et al. (1975), shown in Figure 101.3 is integrated by the author and normalized to the cross section of Tanaka et al. at 100 eV energy. For graphical presentation see Figure 101.5.

Higher waves:

$$\tan \eta_l = \frac{\pi \alpha k^2}{\left[(2l+3)(2l+1))(2l-1)\right] a_0} \quad \text{for } l > 2 \qquad (101.4)$$

Further,

$$Q_T(0) = Q_M(0) = 4\pi A^2 \qquad (101.5)$$

A = Scattering length.

In the above equations A, D, F, A_1, H, A_2 are free fitting parameters. η_0 is the wave shift (rad), α the electronic polarizability (F m^2), k the wave number related to the electron energy, $k = (2\varepsilon)^{1/2}$ and $a_0 = 5.2918 \times 10^{-11}$ m. The constants are shown in Table 101.7 (Merz and Lindner, 1998).

101.6 RAMSAUER–TOWNSEND MINIMUM

See Table 101.8.

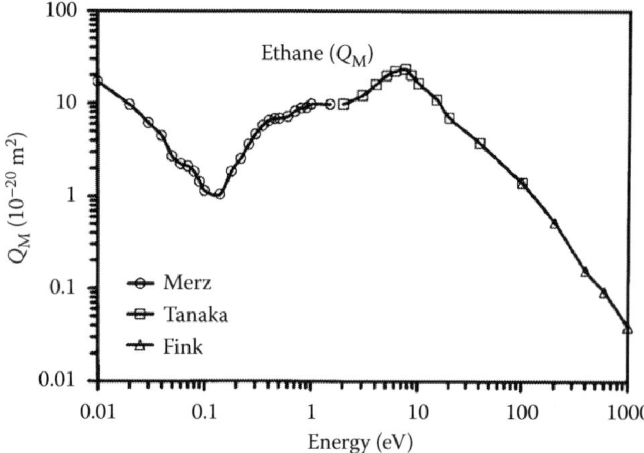

FIGURE 101.4 Momentum transfer cross section in C_2H_6. (—○—) Merz and Lindner (1998); (—□—) Tanaka et al. (1988); (—△—) Fink et al. (1975), integrated by the author.

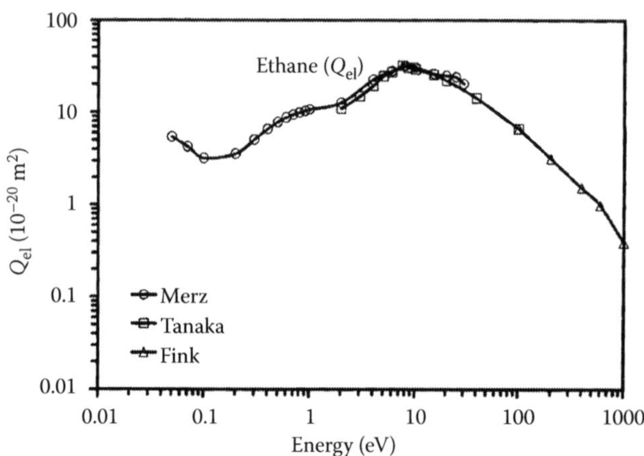

FIGURE 101.5 Elastic cross section in C_2H_6. (—○—) Merz and Lindner (1998); (—□—) Tanaka et al. (1988); (—△—) Fink et al. (1975), integrated by the author.

TABLE 101.7
MERT Parameters in C_2H_6

A (a_0)	A_1 (a_0^3)	D (a_0^3)	F (a_0^4)	H (a_0^5)	A_2 (a_0^5)	Reference
−3.01	12.69	384	−637	26	−30	Merz and Lindner (1998)[a]
−4.57	20.0	536	−984	—	—	Lunt et al. (1994)[a]
−4.09	31.1	426	−546			Lunt et al. (1994)[b]
−4.21	40	258	−53			McCorkle et al. (1978)[c]

Source: Adapted from Merz, R. and F. Lindner *J. Phys. B: At. Mol. Opt. Phys.*, 31, 4663, 1998.

Note: Negative value for the scattering length, *A*, denotes the existence of Ramsauer–Townsend minimum.

[a] From total cross section.
[b] From differential cross section.
[c] From momentum transfer cross section.

101.7 RO-VIBRATIONAL EXCITATION CROSS SECTION

See Table 101.9.

TABLE 101.8
Ramsauer–Townsend Minimum and Corresponding Cross Sections in C_2H_6

ε_{min} (eV)	Q_{min} (10^{-20}) m^2	Method	Reference
0.001	1.0	Total scattering	Merz and Lindner (1998)
0.430	0.7	Total scattering	Lunt et al. (1994)
0.270	0.2	Differential scattering	Lunt et al. (1994)
0.130	0.25	Theory	Sun et al. (1992)
0.120	3.4	Mobility measurement	Gee and Freeman (1983)
0.070	1.4	Swarm	Duncan and Walker (1974)
0.120	1.2	Momentum transfer	McCorkle et al. (1972)

TABLE 101.9
Vibrational Excitation Cross Sections in C_2H_6

	Q_v (10^{-20} m^2)				
	Boesten et al. (1990)		Shishikura et al. (1997)[a]		
Energy (eV)	$\nu_2 + \nu_3$	ν_1	ν_1	ν_2	ν_3
0.12					0.080
0.14					0.132
0.16				0.04	
0.20				0.42	0.041
0.30				0.64	0.033
0.36			0.070		
0.40			0.173	0.41	0.033
0.50			0.272	0.33	0.039
0.80			0.377	0.39	0.057
1.0			0.305	0.38	0.073
2.0			0.203	0.49	0.175
3	0.6038	0.3743	0.384	0.61	0.286
4			0.628	0.93	0.508
5	1.1845	1.0222	0.811	1.40	0.803
7.5	2.012	1.483	2.062	2.56	1.216
10	1.651	0.9518	1.561	2.62	1.146
15	1.158	0.5517	0.449	1.23	0.631
20	0.6482	0.2342	0.485	0.79	0.267
25			0.254	0.42	0.168
30			0.145	0.24	0.108
35			0.131	0.15	0.072
40			0.093	0.10	0.055
50			0.047	0.07	
55			0.009	0.04	

Note: For graphical presentation see Figure 101.6.

101.8 EXCITATION POTENTIALS AND CROSS SECTIONS

See Tables 101.10 and 101.11.

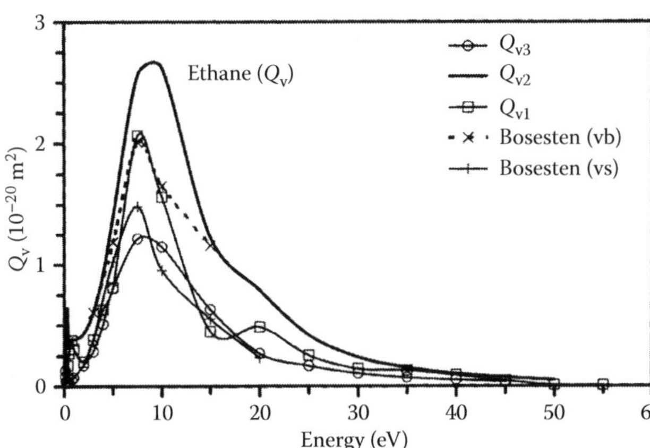

FIGURE 101.6 Vibrational excitation cross section in C_2H_6. Shishikura et al. (1997) from swarm data: (—□—) Q_{v1}; (——) Q_{v2}; (—○—) Q_{v3}. Boesten et al. (1990): (—×—) Q_{vb}; (—+— Q_{vs}). For definition of v_b and v_s, see the introductory paragraph.

TABLE 101.10
Excitation Potentials of C_2H_6

v′	C_2H_6 (eV)	C_2D_6 (eV)
0	8.73	8.89
1	8.86	9.01
2	9.00	9.12
3	9.13	9.23
4	9.26	9.34
5	9.40	9.45
6	9.54	9.56
7	9.69	9.67
8	9.84	9.78
9	9.98	9.90

Source: Adapted from Lassettre, E. N., A. Skerbele, and M. A. Dillon, *J. Chem. Phys.*, 49, 2382, 1968.
Note: Potentials of deuterated C_2H_6 are shown for comparison purposes.

TABLE 101.11
Dissociation Cross Section in C_2H_6, Including Partial Ionization

Energy (eV)	15	20	40	100
Q_{diss} (10^{-20} m²)	1.3	3.0	6.6	7.6

Source: Adapted from Winters, H. F. *Chem. Phys.*, 36, 353, 1979.

101.9 ION APPEARANCE POTENTIALS

See Table 101.12.

101.10 TOTAL IONIZATION CROSS SECTION

See Tables 101.13 and 101.14.

TABLE 101.12
Ion Appearance Potentials

Species	Chatham et al. (1984)	Suzuki and Maeda (1977)	Au et al. (1993)
$C_2H_6^+$	11.4	11.5	11.0
$C_2H_5^+$	12.1	12.0	12.0
$C_2H_4^+$	12.1	12.1	11.0
$C_2H_3^+$	14.5	14.6	13.5
$C_2H_2^+$	15.2	14.7	14.0
C_2H^+		25.6	27.0
C_2^+		31.5	40.0
CH_3^+	14.2	14.1	14.0
CH_2^+	17.0	17.3	25.0
CH^+		26.7	31.0
C^+		29.6	43.0
H_3^+			33.0
H_2^+			29.5
H^+			20.5

TABLE 101.13
Total Ionization Cross Section in C_2H_6 ($12 \leq \varepsilon \leq 3000$)

Chatham et al. (1984)		Nishimura and Tawara (1994)			
Energy (eV)	Q_i (10^{-20} m²)	Energy (eV)	Q_i (10^{-20} m²)	Energy (eV)	Q_i (10^{-20} m²)
		12	0.074	450	3.47
15	0.78	15	0.618	500	3.33
		17.5	1.39	600	3.03
20	2.25	20	2.24	700	2.71
		25	3.48	800	2.38
30	4.22	30	4.45	900	2.25
		35	4.94	1000	2.03
		40	5.41	1250	1.75
		45	5.84	1500	1.52
50	5.75	50	6.04	1750	1.37
		60	6.67	2000	1.22
		70	6.93	2500	1.08
		80	6.86	3000	0.899
		90	6.84		
100	6.04	100	6.89		
		125	6.53		
		150	6.32		
		175	5.98		
200	5.48	200	5.68		
		250	5.01		
300	4.50	300	4.60		
		350	4.18		
400	3.70	400	3.86		

Note: See Figure 101.7 for graphical presentation.

C_2H_6

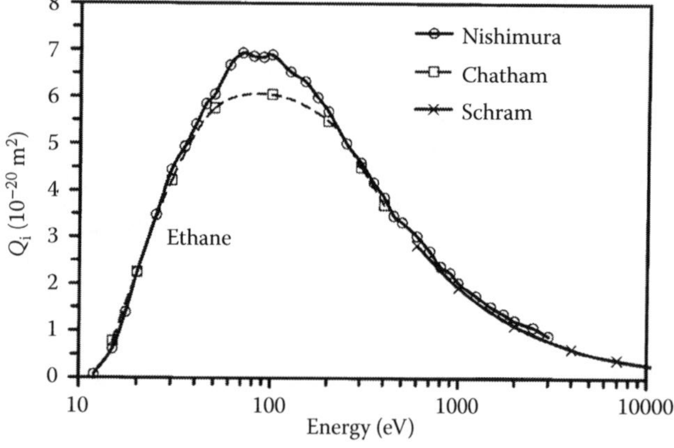

FIGURE 101.7 Total ionization cross section in C_2H_6. (—○—) Nishimura and Tawara (1994); (—□—) Chatham et al. (1984); (—×—) Schram et al. (1966).

101.11 PARTIAL IONIZATION CROSS SECTIONS

Figure 101.8 presents experimental partial ionization cross sections in C_2H_6 (Chatham et al., 1984).

101.12 ATTACHMENT CROSS SECTION

The dissociative attachment processes in C_2H_6 are (Rutkowsky et al., 1980)

$$e > (6\,\text{eV}) + C_2H_6 \rightarrow C_2H_6^{-*} \rightarrow \begin{cases} H^- + C_2H_5 \\ CH_2^- + CH_4 \end{cases} \quad (101.6)$$

TABLE 101.14
Ionization Cross Section at High Electron Energy

Energy (eV)	600	1000	2000	4000	7000	12,000
Q_i (10^{-20} m^2)	2.83	1.93	1.12	0.630	0.400	0.225

Source: Adapted from Schram, B. L. et al., *J. Chem. Phys.*, 44, 49, 1966.

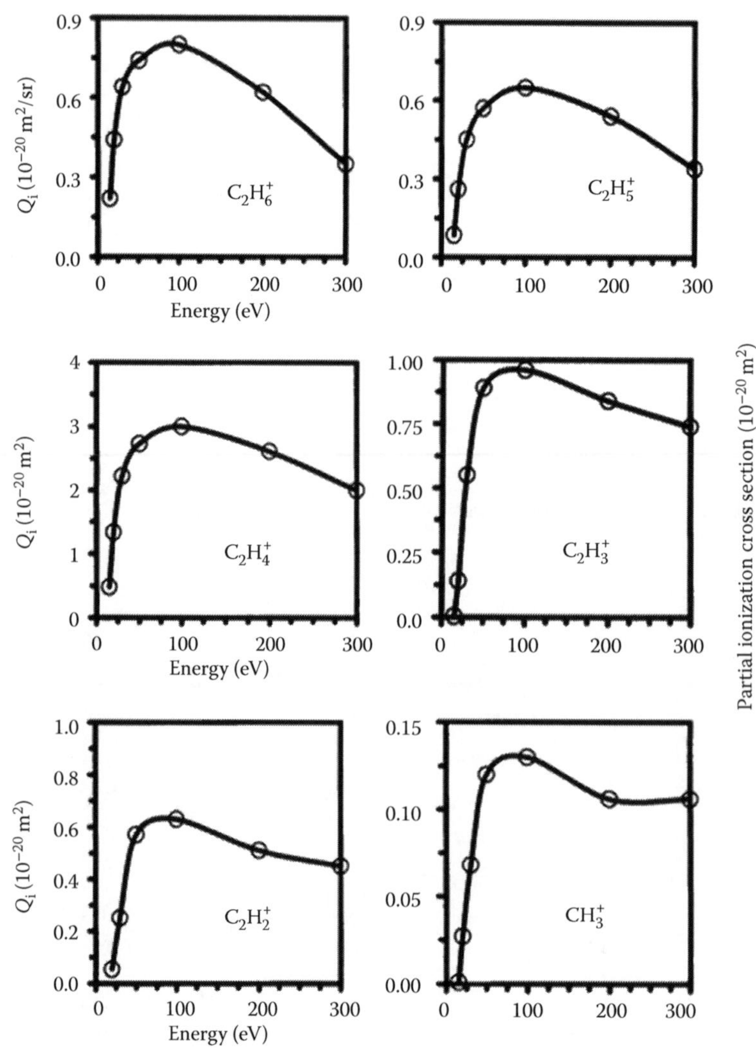

FIGURE 101.8 Partial ionization cross sections in C_2H_6. Note the relative magnitude for each ion species. Appearance potentials are given in Table 101.12. (Adapted from Chatham, H. et al., *J. Chem. Phys.*, 81, 1770, 1984.)

The dissociative cross section as a function of electron energy is shown in Table 101.15 and Figure 101.9.

101.13 CONSOLIDATED CROSS SECTIONS

Figure 101.10 shows the consolidated cross sections to demonstrate the relative magnitude of each cross section.

TABLE 101.15
Dissociative Attachment Cross Section in C_2H_6

Energy (eV)	Q_i (10^{-23} m^2)	Energy (eV)	Q_i (10^{-23} m^2)
6.8	0.08	9.5	1.75
7.0	0.20	10.0	0.72
7.5	0.56	11.0	0.39
8.0	1.04	12.0	0.32
8.5	1.66	12.5	0.43
9.0	2.17		

Source: Adapted from Rutkowsky, Von J., H. Drost, and H.-J. Spangenberg, *Ann. Phys. Leipzig*, 37, 259, 1980.

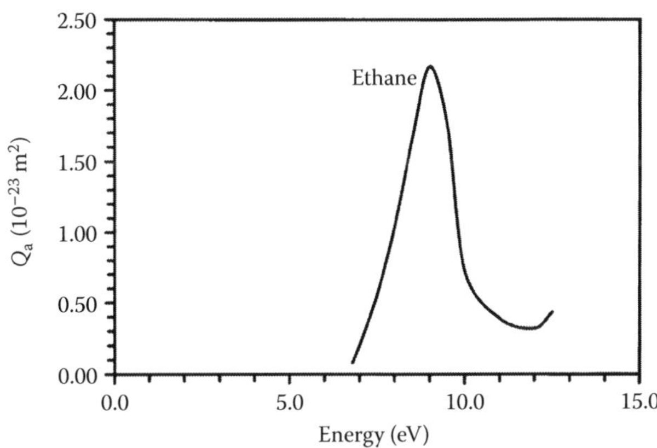

FIGURE 101.9 Dissociative attachment cross section in C_2H_6. (Adapted from Rutkowsky, Von J., H. Drost, and H.-J. Spangenberg, *Ann. Phys. Leipzig*, 37, 259, 1980.)

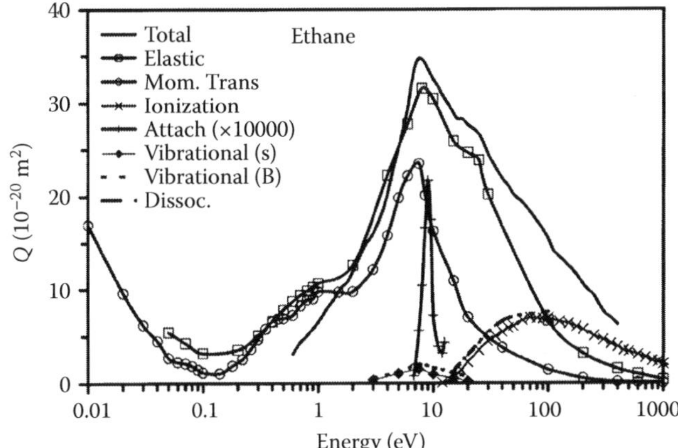

FIGURE 101.10 Consolidated presentation of all cross sections in C_2H_6. (——) Total; (—□—) elastic scattering; (—○—) momentum transfer; (—×—) ionization; (—+—) attachment, multiplied by 10^4; (—◊—) vibrational, stretching; (— —) vibrational, bending; (—•—) dissociation.

101.14 DRIFT VELOCITY OF ELECTRONS

See Tables 101.16 and 101.17.

TABLE 101.16
Drift Velocity of Electrons in C_2H_6 at 298 K

E/N (Td)	W_e (10^4 m/s)	*E/N* (Td)	W_e (10^4 m/s)	*E/N* (Td)	W_e (10^4 m/s)
McCorkle et al. (1978)		Schmidt and Roncossek (1992)		Shishikura et al. (1997)[a]	
0.03	0.11	0.08	0.289	0.03	0.110
0.06	0.24	0.10	0.366	0.05	0.191
0.09	0.36	0.12	0.445	0.07	0.270
0.1	0.49	0.14	0.526	0.1	0.390
0.2	0.75	0.16	0.609	0.2	0.80
0.2	1.02	0.18	0.693	0.4	1.54
0.3	1.27	0.20	0.777	0.7	3.03
0.5	1.97	0.25	0.988	1	3.49
0.6	2.56	0.30	1.19	2	4.95
0.8	3.00	0.35	1.41	4	5.52
0.9	3.36	0.40	1.62	7	5.83
1.1	3.88	0.50	2.00	10	5.93
1.2	4.14	0.60	2.36	15	5.79
1.6	4.49	0.70	2.68	20	5.66
1.9	4.57	0.80	2.98	40	5.27
2.2	5.03	1.00	3.47	70	6.18
3.1	5.06	1.20	3.83	100	6.68
4.7	5.13	1.40	4.11	150	9.08
7.8	5.28	1.60	4.33	200	11.54
12.4	5.10	1.80	4.50	300	17.33
21.8	4.93	2.00	4.64		
28.0	4.75	2.50	4.89		
		3.00	5.06		
		3.50	5.18		
		4.00	5.27		
		4.50	5.33		
		5.00	5.38		
		6.00	5.44		
		7.00	5.47		
		8.00	5.48		

Note: See Figure 101.11 for graphical presentation. Figure 101.12 shows reduced mobility.

[a] Digitized and interpolated.

TABLE 101.17
Mean Energy and Temperature Dependence of Drift Velocity of Electrons in C_2H_6

E/N	W_e (10^4 m/s)				Mean
(Td)	373 K	473 K	573 K	673 K	Energy (eV)
0.03	0.14	0.14	0.14	0.13	0.038
0.06	0.28	0.27	0.26	0.24	0.038
0.09	0.41	0.4	0.38	0.35	0.038
0.12	0.52	0.53	0.51	0.47	0.038

continued

C_2H_6

TABLE 101.17 (continued)
Mean Energy and Temperature Dependence of Drift Velocity of Electrons in C_2H_6

E/N	W_e (10^4 m/s)				Mean
(Td)	373 K	473 K	573 K	673 K	Energy (eV)
0.19	0.82	0.76	0.74	0.7	0.039
0.25	1.11	1.04	0.99	0.89	0.040
0.3	1.37	1.28	1.2	1.18	0.043
0.5	1.96	1.85	1.85	1.66	0.058
0.6	2.62	2.45	2.18	2.08	0.071
0.8	2.97	2.8	2.62	2.48	0.083
0.9	3.48	3.26	3.06	2.79	0.095
1.1	3.86	3.39	3.38	3.1	0.106
1.2	4.05	3.8	3.63	3.48	0.115
1.6	4.5	4.07	4.07	3.75	0.135
1.9	4.69	4.55	4.29	4	0.153
2.2	4.92	4.85	4.5	4.25	0.171
3.1	5.23	5.16	4.93	4.71	0.222
4.7	5.29	5.41	5.17	5.23	0.293
7.8	5.52	5.64	5.32	4.93	0.423
12	5.39	5.53	5.39	5.11	0.560
22	5.11	5.45	4.96	5.01	
28	4.95	5.23	4.81	5.07	

Source: Adapted from McCorkle, D. L. et al., *J. Phys. B: At. Mol. Phys.*, 11, 3067, 1978.
Note: For graphical presentation see Figure 101.13.

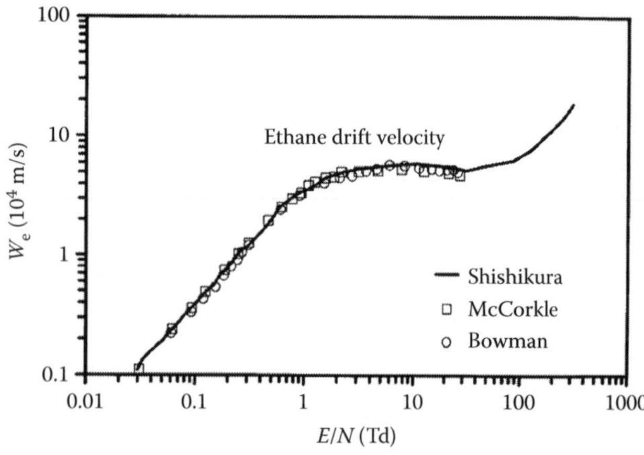

FIGURE 101.11 Drift velocity of electrons in C_2H_6. (——) Shishikura et al. (1997); (□) McCorkle et al. (1978); (○) Bowman and Gordon (1967).

101.15 DIFFUSION COEFFICIENT AND MEAN ENERGY

See Table 101.18.

101.16 REDUCED IONIZATION COEFFICIENT

See Table 101.19.

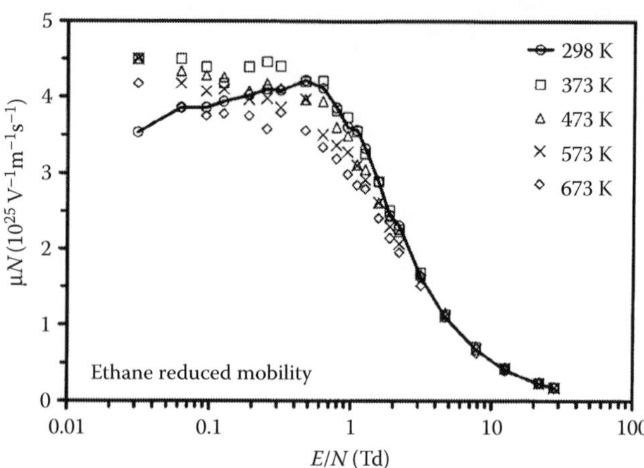

FIGURE 101.12 Normalized electron mobility in C_2H_6 at various temperatures McCorkle et al., 1978). (—○—) 293 K; (□) 373 K; (△) 473 K; (×) 573 K; (◊) 673 K.

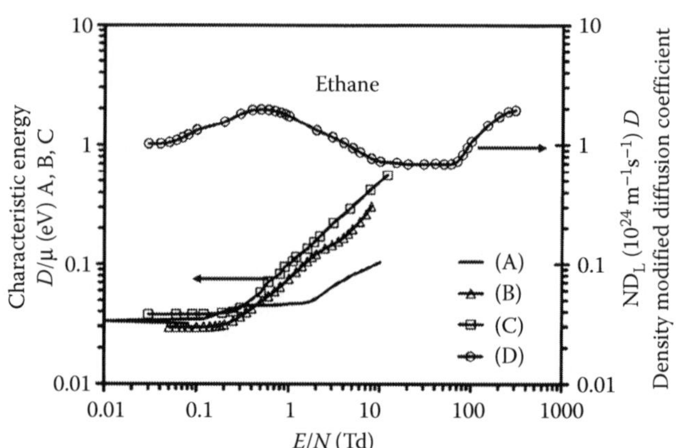

FIGURE 101.13 Diffusion coefficients and mean energy of electron swarm in C_2H_6. (A) D_L/μ (Schmidt and Rossenchek, 1992); (B) D_r/μ (Schmidt and Rossenchek, 1992); (C) $\bar{\varepsilon}$ (McCorkle et al., 1978); (D) ND_L (Shishikura et al., 1997). *Note:* The appropriate ordinate for the quantities shown.

TABLE 101.18
Longitudinal and Radial Diffusion Coefficients in C_2H_6

Schmidt and Roncossek (1992)				Shishikura et al. (1997)		McCorkle et al. (1978)	
E/N (Td)	D_L/μ (eV)	*E/N* (Td)	D_r/μ (eV)	*E/N* (Td)	ND_L^a (10^{24} (ms)$^{-1}$)	*E/N* (Td)	(eV)
0.080	0.0317	0.050	0.0295	0.03	1.02	0.03	0.038
0.010	0.0333	0.060	0.0295	0.04	1.02	0.06	0.038
0.120	0.0348	0.070	0.0295	0.05	1.05	0.09	0.038
0.140	0.0362	0.080	0.0295	0.06	1.09	0.12	0.038
0.160	0.0376	0.100	0.0295	0.08	1.21	0.19	0.039
0.180	0.0390	0.120	0.0296	0.1	1.33	0.25	0.040
0.200	0.0403	0.140	0.0298	0.2	1.55	0.3	0.043
0.250	0.0430	0.160	0.0301	0.3	1.80	0.5	0.058
0.300	0.0443	0.180	0.0305	0.4	1.94	0.6	0.071
0.350	0.0452	0.200	0.0311	0.5	1.97	0.8	0.083

TABLE 101.18 (continued)
Longitudinal and Radial Diffusion Coefficients in C_2H_6

Schmidt and Roncossek (1992)				Shishikura et al. (1997)		McCorkle et al. (1978)	
E/N (Td)	D_L/μ (eV)	*E/N* (Td)	D_r/μ (eV)	*E/N* (Td)	ND_L^a (10^{24} (ms) $^{-1}$)	*E/N* (Td)	(eV)
0.400	0.0454	0.250	0.0333	0.6	1.95	0.9	0.095
0.500	0.0454	0.300	0.0363	0.8	1.86	1.1	0.106
0.600	0.0454	0.350	0.0394	0.9	1.79	1.2	0.115
0.700	0.0456	0.400	0.0425	1.0	1.73	1.6	0.135
0.800	0.0460	0.500	0.0483	2.0	1.33	1.9	0.153
1.00	0.0467	0.600	0.0539	3.0	1.17	2.2	0.171
1.20	0.0469	0.700	0.0593	4.0	1.05	3.1	0.222
1.40	0.0472	0.800	0.0647	5.0	0.95	4.7	0.293
1.60	0.0481	1.00	0.0753	6.0	0.87	7.8	0.423
1.80	0.0496	1.20	0.0859	8.0	0.77	12	0.560
2.00	0.0516	1.40	0.0962	10	0.73		
2.50	0.0582	1.60	0.106	15	0.71		
3.00	0.0644	1.80	0.115	20	0.69		
3.50	0.0695	2.00	0.122	30	0.69		
4.00	0.0731	2.50	0.136	40	0.70		
4.50	0.0772	3.00	0.145	50	0.69		
5.00	0.0812	3.50	0.156	60	0.69		
6.00	0.0862	4.00	0.167	80	0.83		
7.00	0.0910	4.50	0.181	90	0.95		
8.00	0.0955	5.00	0.196	100	1.06		
10.0	0.104	6.00	0.227	150	1.45		
		7.00	0.263	200	1.71		
		8.00	0.305	300	1.94		

Note: See Figure 101.13 for graphical presentation.
[a] Digitized and interpolated.

TABLE 101.19
Reduced Ionization Coefficients in C_2H_6

E/N (Td)	α/N (10^{-20} m²)	*E/N* (Td)	α/N (10^{-20} m²)	*E/N* (Td)	α/N (10^{-20} m²)	*E/N* (Td)	α/N (10^{-20} m²)
80	0.0003	225	0.168	600	1.18	2000	4.40
90	0.0009	250	0.222	700	1.51	2500	5.09
100	0.0023	300	0.358	800	1.78	3000	5.48
125	0.012	350	0.490	900	2.03	4000	6.02
150	0.036	400	0.622	1000	2.28	4500	6.53
175	0.068	450	0.759	1250	2.88		
200	0.113	500	0.904	1500	3.41		

Source: Adapted from Heylen, A. E. D. *Int. J. Electron.*, 39, 653, 1975.
Note: See Figure 101.14 for graphical presentation.

101.17 ION DRIFT VELOCITY

The reduced positive ion mobilities are shown in Table 101.20 (see Watts and Heylen, 1979) and Figure 101.15.

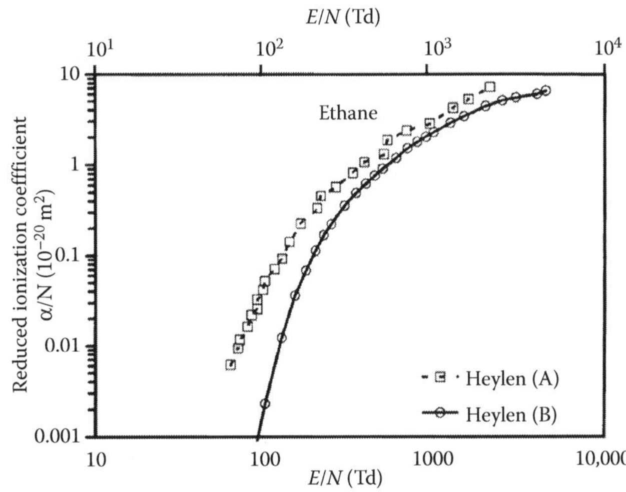

FIGURE 101.14. Reduced ionization coefficients in C_2H_6. (—□—) Watts and Heylen (1979), current pulse method; (—○—) Heylen (1975), current growth method.

TABLE 101.20
Positive Ion Drift Velocity

E/N (Td)	W^+ (10^3 m s^{-1})	*E/N* (Td)	W^+ (10^3 m s^{-1})	*E/N* (Td)	W^+ (10^3 m s^{-1})
110	0.221	800	3.306	1666	4.97
150	0.445	900	3.527	1736	5.08
200	0.726	1000	3.738	2134	5.69
250	1.006	1100	3.939	2486	6.19
300	1.286	1200	4.132	2792	6.60
350	1.566	1300	4.318	3052	6.93
400	1.846	1400	4.498	3161	7.07
450	2.074	1442	4.59.	4012	8.06
500	2.305	1481	4.66.		
600	2.822	1596	4.85.		
700	3.072	1616	4.89		

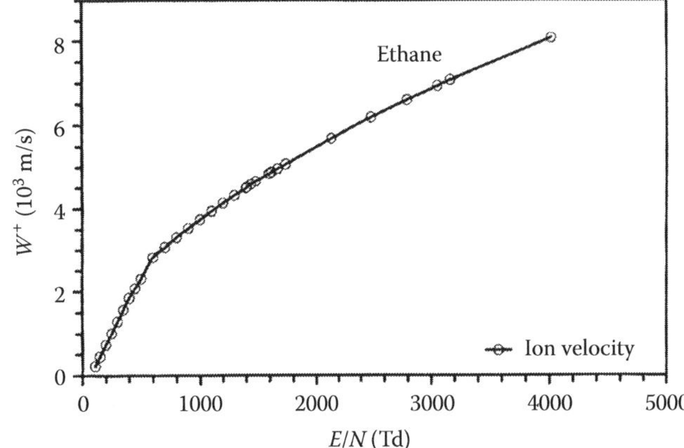

FIGURE 101.15 Positive ion drift velocity in C_2H_6. (Adapted from Watts, M. P. and A. E. D. Heylen, *J. Phys. D: Appl. Phys.*, 12, 695, 1979.)

REFERENCES

Au, J. W., G. Cooper, and C. E. Brion, *Chem. Phys.*, 173, 241, 1993.

Boesten, L., H. Tanaka, M. Kubo, H. Sato, M. Kimura, M. A. Dillon, and D. Spence, *J. Phys. B: At. Mol. Opt. Phys.*, 23, 1905, 1990.

Bowman, C. R. and D. E. Gordon, *J. Chem. Phys.*, 46, 1878, 1967.

Chatham, H., D. Hils, R. Robertson, and A. Gallagher, *J. Chem. Phys.*, 81, 1770, 1984.

Duncan, C. W. and I. C. Walker, *J. Chem. Soc., Farad. Trans., II*, 70, 1974.

Fink, M., K. Jost, and D. Hermann, *J. Chem. Phys.*, 63, 1985, 1975.

Floeder, K., D. Fromme, W. Raith, A. Schwab, and G. Sinapius, *J. Phys. B: At. Mol. Opt. Phys.*, 18, 3347, 1985.

Floriano, M. A., N. Gee, and G. R. Freeman, *J. Chem. Phys.*, 84, 6799, 1986.

Gee, N. and G. R. Freeman, *J. Chem. Phys.*, 78, 1951, 1983.

Heylen, A. E. D. *Int. J. Electron.*, 39, 653, 1975.

Hwang, W., Y.-K. Kim, and M. E. Rudd, *J. Chem. Phys.*, 104, 2956, 1996.

Lassettre, E. N., A. Skerbele, and M. A. Dillon, *J. Chem. Phys.*, 49, 2382, 1968.

Lunt, S. L., J. Randell, J. P. Ziese, O. Mroztek, and D. Field, *J. Phys. B: At. Mol. Opt. Phys.*, 27, 1407, 1994.

Lunt, S. L., J. Randell, J. P. Ziesel, G. Mrotzek, and D. Field, *J. Phys. B: At. Mol. Opt. Phys.*, 31, 4225, 1998.

Maji, S., G. Basavaraju, S. M. Bharathi, K. G. Bhushan, and S. P. Khare, *J. Phys. B: At. Mol. Opt. Phys.*, 31, 4975, 1998.

Mapstone, B., M. J. Brunger, and W. R. Newell, *J. Phys. B: At. Mol. Opt. Phys.*, 33, 23, 2000.

Mapstone, B. and W. R. Newell, *J. Phys. B: At. Mol. Opt. Phys.*, 25, 491, 1992.

McCorkle, D. L., L. G. Christophorou, D. V. Maxey, and J. G. Carter, *J. Phys. B: At. Mol. Phys.*, 11, 3067, 1978.

Merz, R. and F. Lindner, *J. Phys. B: At. Mol. Opt. Phys.*, 31, 4663, 1998.

Nishimura, H. and H. Tawara, *J. Phys. B: At. Mol. Opt. Phys.*, 24, L363, 1991.

Nishimura, H. and H. Tawara, *J. Phys. B: At. Mol. Opt. Phys.*, 27, 2063, 1994.

Raju, G. G., *Gaseous Electronics: Theory and Practice*, CRC Press, Boca Raton, FL, 2006, pp. 133–140.

Schmidt, B. and M. Roncossek, *Aust. J. Phys.*, 45, 351, 1992.

Schram, B. L., M. J. van der Wiel, F. J. de Heer, and H. R. Moustafa, *J. Chem. Phys.*, 44, 49, 1966.

Shishikura, Y., K. Asano, and Y. Nakamura, *J. Phys. D: Appl. Phys.*, 30, 1610, 1997.

Sueoka, O. and S. Mori, *J. Phys. B: At. Mol. Phys.*, 19, 4035, 1986.

Sun, W., C. W. McCurdy, and B. H. III Lengsfield, *J. Chem. Phys.*, 97, 5480, 1992.

Suzuki, I. H. and K. Maeda, *Int. J. Mass. Spectrom. Ion Phys.*, 24, 147, 1977.

Szmytkowski, Cz. and A. Krzysztofowicz, *J. Phys. B: At. Mol. Opt. Phys.*, 28, 4291, 1995.

Tanaka, H., L. Boesten, D. Matsunga, and T. Kudo, *J. Phys. B: At. Mol. Opt. Phys.*, 21, 1255, 1988.

Von J. Rutkowsky, H. Drost, and H.-J. Spangenberg, *Ann. Phys. Leipzig*, 37, 259, 1980.

Watts, M. P. and A. E. D. Heylen, *J. Phys. D: Appl. Phys.*, 12, 695, 1979.

Winters, H. F., *Chem. Phys.*, 36, 353, 1979.

102

HEXACHLOROETHANE

CONTENTS

Hexachloroethane (C_2Cl_6), synonym perchloroethane, is a polar, electron-attaching gas that has 114 electrons. The ionization potential is 11.1 eV.

102.1 TOTAL IONIZATION CROSS SECTIONS

Total ionization cross sections have been measured by Hudson et al. (2001) as shown in Table 102.1. Figure 102.1 provides a graphical presentation with total ionization cross sections for C_2H_6 and C_2F_6 added for comparison. Note the relative change in the cross section as Cl and F are substituted for H.

TABLE 102.1
Total Ionization Cross Sections for C_2Cl_6

Energy (eV)	Q_i (10^{-20} m^2)	Energy (eV)	Q_i (10^{-20} m^2)
16	3.29	105	26.21
20	7.59	108	25.86
23	11.69	112	25.88
27	14.59	116	25.73
31	17.15	119	25.63
34	19.37	123	25.48
38	20.81	127	25.02
42	21.99	130	24.89
45	22.99	134	24.74
49	23.72	138	24.67
53	25.24	142	24.44
57	25.10	145	24.17
60	25.93	149	23.97
64	25.67	153	23.62
68	26.07	156	23.41
71	26.04	160	23.29
75	26.34	164	23.22
79	26.57	167	23.17
82	26.36	171	23.08
86	26.48	175	23.04
90	26.54	179	22.74
94	26.42	182	22.43
97	26.39	186	22.09
101	26.42	190	21.82

TABLE 102.1 (continued)
Total Ionization Cross Sections for C_2Cl_6

Energy (eV)	Q_i (10^{-20} m^2)	Energy (eV)	Q_i (10^{-20} m^2)
193	21.58	208	21.27
197	21.40	212	21.38
201	21.32	215	21.31
204	21.24	219	21.10

Source: Adapted from Hudson, J. E. et al., *J. Phys. B: At. Mol. Opt. Phys.*, 34, 3025, 2001.
Note: Tabulated values by courtesy of Professor Harland (2006).

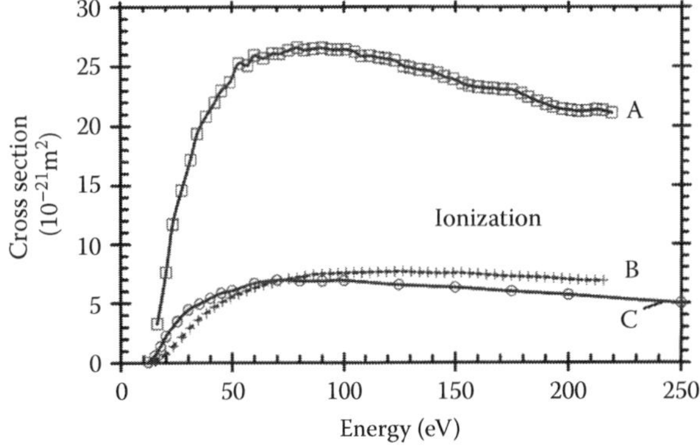

FIGURE 102.1 Ionization cross sections for C_2Cl_6. Selected gases added for comparison. (—□—, A) C_2Cl_6, Hudson et al. (2001); (— + —, B) C_2F_6, Bart et al. (2001); (—○—, C) Nishimura and Tawara (1994). Tabulated values for curves A and B by courtesy of Professor Harland (2001).

REFERENCES

Bart, M., P. W. Harland, J. E. Hudson, and C. Vallance, *Phys. Chem. Chem. Phys.*, 3, 800, 2001.

Hudson, J. E., C. Vallance, M. Bart, and P. W. Harland, *J. Phys. B: At. Mol. Opt. Phys.*, 34, 3025, 2001.

Nishimura, H. and H. Tawara, *J. Phys. B: At. Mol. Opt. Phys.*, 27, 2063, 1994.

103

HEXAFLUOROETHANE

C_2F

CONTENTS

Fluorocarbons, consisting of carbon and fluorine atoms are industrial gases. We consider selected target particles with the general chemical formula C_nF_{2n+2}. They are fluorine-substituted derivatives of alkanes. The series for $n = 1$–3 has been extensively analyzed by Christophorou and Olthoff (2004) and we summarize their vast publication for the sake of completeness, providing some additional data. Selected properties of target particles are shown in Table 103.1.

The modes of vibrational excitation with energies are shown in Table 103.2 (Mann and Linder, 1992). Vibrations of similar type tend to have similar energies (Allan and Andric, 1996).

103.1 SELECTED REFERENCES FOR DATA

See Table 103.3.

TABLE 103.1
Selected Properties of Fluorocarbons

Target					
Name	Formula	Z	α_e (10^{-40} Fm^2)	μ (D)	ε_i (eV)
CF_4	Carbon tetrafluoride	42	4.27	0	16.20
C_2F_6	Hexafluoroethane	66	7.588	0	14.6
C_3F_8	Perfluoropropane	90	7.20–10.46	0.07	13.3

TABLE 103.2
Vibrational Energies for (C_2F_6)

C_2F_6 (Sverdlov et al. 1974)	
Mode	Energy (meV)
ν_1	176
ν_2	**100.1**
ν_3	43.1
ν_5	138.4
ν_6	88.5
ν_7	155.0
ν_8	76.7
ν_9	46.1
ν_{10}	155.0

Note: Bold font indicates strong intensity.

103.2 TOTAL SCATTERING CROSS SECTION

Table 103.4 and Figures 103.1 through 103.4 show the total scattering cross sections for fluorocarbons recommended by Christophorou and Olthoff (2004). Figure 103.2 shows the expanded view in the low-energy region up to 10 eV electron energy. Figures 103.3 and 103.4 provide

TABLE 103.3
Selected References for Data on C_2F_6

Parameter	Range, eV (Td)	Reference
Q_T	30–5000	Shi et al. (2006)
Q_T	100–1500	**Aryasinghe (2003)**
Q_T	0.8–600	**Sueoka et al. (2002)**
Q_i	12.2–215.5	**Bart et al. (2001)**
Q_i	16–3000	**Nishimura et al. (1999)**
Q_{el}	2–100	**Takagi et al. (1994)**
W	(0.03–500)	**Hunter et al. (1988)**
α/N, η/N	(20–400)	**Hunter et al. (1987)**
k_a	0.412–4.81	**Spyrou et al. (1985)**
k_a	(0.9–4.7)	**Hunter et al. (1984)**
k_a	0–12	**Spyrou et al. (1983)**
Q_{diss}	13–600	Winters and Inokuti (1982)

Note: Bold font indicates experimental data.
k_a = attachment rate; Q_{diss} = dissociation; Q_T = total scattering; W = drift velocity; α/N = reduced ionization coefficient; η/N = reduced attachment coefficient.

TABLE 103.4
Total Scattering Cross Sections for C_2F_6

	Q_T (10^{-20} m²)		
Energy (eV)	Christophorou and Olthoff (2004)	Ariyasighe (2003)	Sueoka (2002)
0.040	10.6		
0.060	11.5		
0.080	12.1		
0.20	14.1		
0.80	15.8		15.2
1.0	15.9		15.7
1.2			16.0
1.4			16.2
1.5	16.7		
1.6			16.2
1.8			16.5
2.0	17.4		16.9
2.2			17.2
2.5	18.1		18.6
2.8			19.2
3.0	19.9		
3.1			19.9
3.4			21.2
3.5	22.7		
4.0	25.4		25.5
4.5	26.8		26.7
5.0	27.5		27.1
5.5	27.5		27.6
6.0	27.2		27.2
6.5	27.4		27.3
7.0	28.3		28.6
7.5	29.1		29.1
8.0	29.7		29.5
8.5	30.0		30.8
9.0	29.7		30.6
9.5	29.1		29.8

TABLE 103.4 (continued)
Total Scattering Cross Sections for C_2F_6

	Q_T (10^{-20} m²)		
Energy (eV)	Christophorou and Olthoff (2004)	Ariyasighe (2003)	Sueoka (2002)
10.0	28.4		29.4
11.0	27.2		28.0
12.0	26.5		27.1
13.0	26.2		27.0
14.0	26.2		27.0
15.0			27.8
16.0	26.7		27.3
17.0			27.8
18.0	27.5		28.2
19.0			28.8
20.0	28.5		29.6
22			29.9
25			30.5
30	30.4		30.2
35			30.4
40	31.3		30.3
50	31.1		29.5
60	30.0		28.6
70	28.9		27.4
80	27.6		25.9
90	26.4		24.6
100	25.0	25.65	24.0
150	21.3		20.2
200	18.4	17.99	17.7
250	16.3		15.8
300		15.17	14.5
350			13.3
400		12.83	12.5
500		11.57	10.7
600		10.40	10.0
700		9.20	
800		8.55	
900		7.82	
1000		7.36	
1100		6.90	
1200		6.59	
1300		6.15	
1400		5.92	
1500		5.72	

additional information. The important features of the cross sections are

1. The total scattering cross section increases with increasing number of electrons of the target particle.
2. Near 4.5 and 8 eV the increase in cross section is due to vibrational excitation (Takagi et al., 1994).
3. In the energy range of 6–50 eV there is a peak at 8.5–9 eV attributed to shape resonance, with the formation of short-lived negative ions. This feature is common to CF_4, C_2F_6, and C_3F_8.

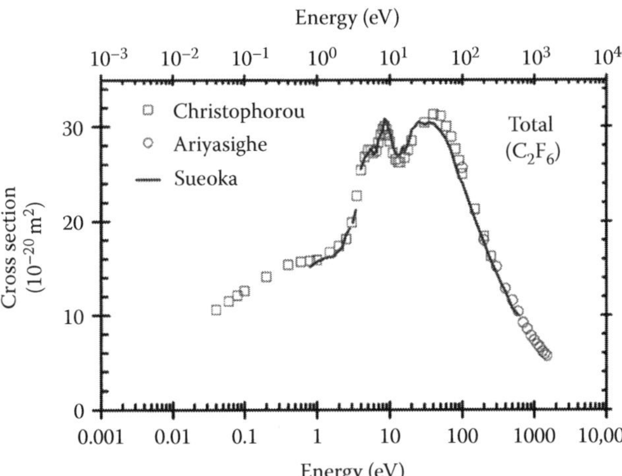

FIGURE 103.1 Total scattering cross sections for C_2F_6. (□) Recommended by Christophorou and Olthoff (2004); (○) Ariyasinghe (2003); (——) Sueoka et al. (2002).

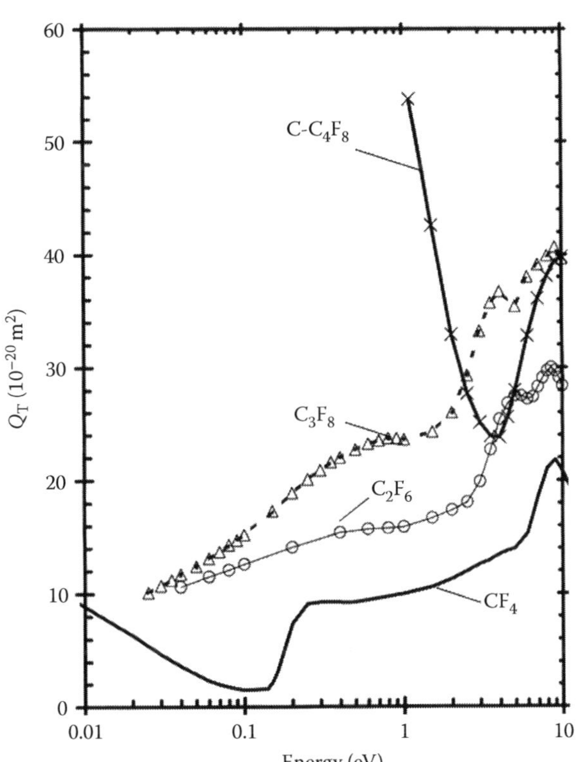

FIGURE 103.2 Total scattering cross sections for C_2F_6 and selected fluorocarbon gases in the low-energy region. With larger number of atoms the cross section increases in this energy range, but no definitive pattern is observed for Ramsauer–Townsend minimum.

4. At energy greater than 100 eV there is a monotonic decrease of the cross section with the Born approximation method of calculating the cross section yielding satisfactory results. This feature is common to most target particles.

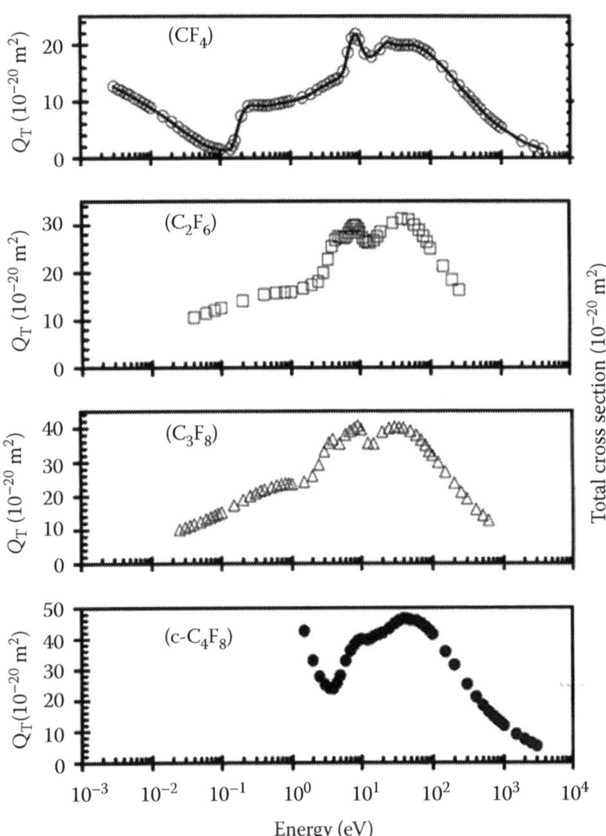

FIGURE 103.3 Total scattering cross sections for several fluorocarbon gases for the entire electron range.

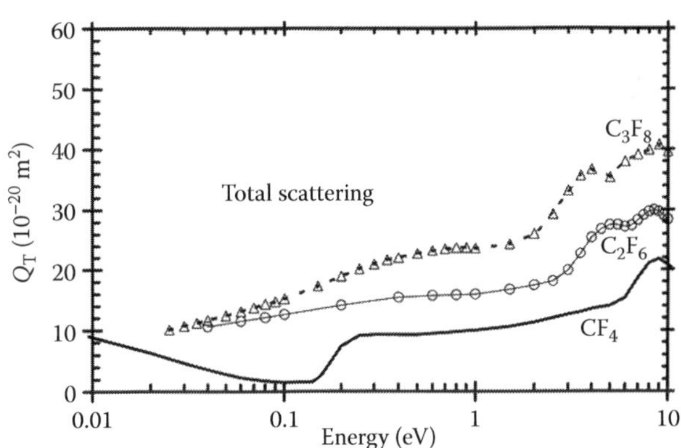

FIGURE 103.4 Total scattering cross sections for C_2F_6 on an expanded scale. Note the linear scale for the ordinate. Data for CF_4 and c-C_4F_8 added for comparison. (Adapted from Christophorou, L. G. and J. K. Olthoff, *Fundamental Electron Interactions with Plasma Processing Gases*, Kluwer Academic/Plenum Publishers, New York, NY, 2004.)

103.3 ELASTIC SCATTERING AND MOMENTUM TRANSFER CROSS SECTIONS

Table 103.5 shows the elastic scattering and momentum transfer cross sections recommended by Christophorou and Olthoff (2004). These are shown in Figure 103.5. Figure 103.6 compares these cross sections with selected fluorocarbon gases.

C₂F₆

TABLE 103.5
Elastic Scattering and Momentum Transfer Cross Sections (C_2F_6)

Energy (eV)	Q_{el} (10^{-20} m²)	Q_M (10^{-20} m²)	Energy (eV)	Q_{el} (10^{-20} m²)	Q_M (10^{-20} m²)
0.010	12.2	9.47	2.0	14.5	13.0
0.020	7.86	5.08	2.5	15.4	13.1
0.030	5.68	3.06	3.0	16.3	13.2
0.040	4.41	1.99	4.0	18.3	14.1
0.060	3.07	1.01	5.0	20.1	14.6
0.070	2.70	0.78	7.0	22.5	16.4
0.080	2.43	0.63	8.0	23.5	17.9
0.090	2.23	0.53	9.0	24.2	18.4
0.10	2.07	0.46	10.0	24.9	18.8
0.15	1.69	0.32	15.0	27.9	22.7
0.20	1.66	0.47	25.0	27.3	20.8
0.25	1.91	0.93	30.0	26.3	18.9
0.30	2.39	1.66	45.0	22.5	
0.40	3.70	3.45	50.0	20.9	12.8
0.50	4.91	4.82	60.0		10.6
0.60	5.71	5.71	70.0	19.6	8.96
0.80	9.39	7.39	80.0	18.4	7.67
0.90	10.6	8.20	90.0	17.3	6.66
1.0	11.3	8.96	100.0	16.4	5.86
1.5	13.2	11.8			

Source: Adapted from Christophorou, L. G. and J. K. Olthoff, *Fundamental Electron Interactions with Plasma Processing Gases*, Kluwer Academic/Plenum Publishers, New York, NY, 2004.

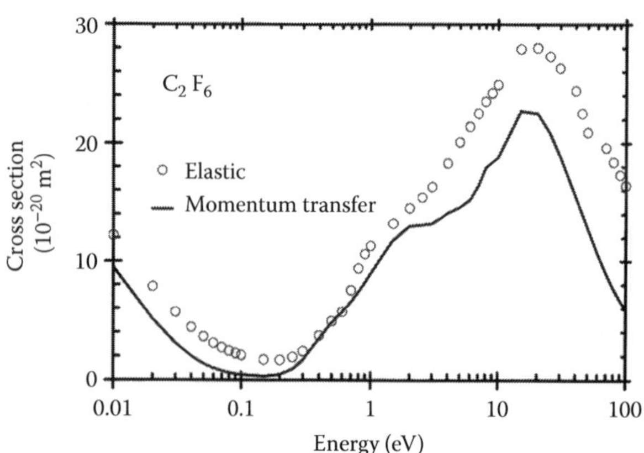

FIGURE 103.5 Elastic scattering and momentum transfer cross sections for C_2F_6. (Adapted from Christophorou, L. G. and J. K. Olthoff, *Fundamental Electron Interactions with Plasma Processing Gases*, Kluwer Academic/Plenum Publishers, New York, NY, 2004.)

103.4 RAMSAUER–TOWNSEND MINIMUM

Ramsauer–Townsend minimum cross sections and the corresponding mean energy are shown in Table 103.6.

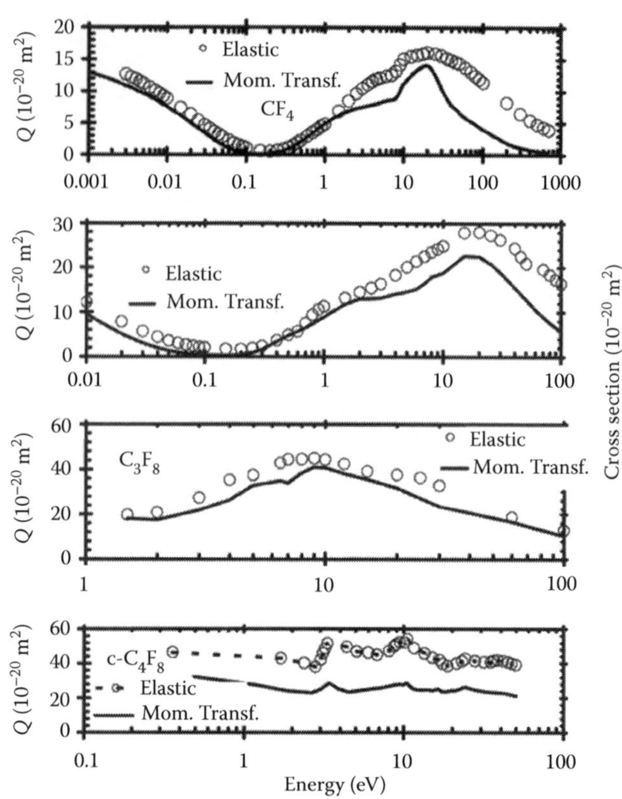

FIGURE 103.6 A comparison of elastic collision scattering and momentum transfer cross sections in selected fluorocarbons.

TABLE 103.6
Ramsauer–Townsend Minimum Cross Section (C_2F_6)

Minimum Energy (eV)	Cross Section (10^{-20} m²)	Method	Reference
0.2	1.66	Q_{el}	Christophorou and Olthoff (2004)
0.15	0.32	Q_M	Christophorou and Olthoff (2004)

103.5 VIBRATIONAL EXCITATION CROSS SECTIONS

Vibrational excitation cross sections given by Pirgov and Stefanov (1990) are shown in Table 103.7.

103.6 POSITIVE ION APPEARANCE POTENTIALS

Ion appearance potentials are shown in Table 103.8 and a more detailed list for several paths for generation of C^+, F^+, and F_2^+ ions are given by Christophorou and Olthoff (2004).

103.7 IONIZATION CROSS SECTIONS

Ionization cross sections are shown in Table 103.9 and Figure 103.7. The data of Bart et al. (2007) which was not included

TABLE 103.7
Vibrational Excitation Cross Sections for C_2F_6

Energy (eV)	Q_{vib} (10^{-20} m^2)
0.090	3.01
0.1056	5.57
0.20	10.67
0.35	15.47
0.70	16.45
1.00	12.63
1.39	11.33
1.96	13.89
2.56	17.47
2.90	18.09
3.24	17.29
4.00	12.11
5.00	7.35
6.00	5.22
8.00	4.18

TABLE 103.8
Positive Ion Appearance Potentials

Ion	Potential (eV)
CF_3^+	13.4
CF_2^+	14.0
$C_2F_5^+$	15.4
CF^+	16.6

TABLE 103.9
Ionization Cross Sections for C_2F_6

Energy (eV)	Q_i (10^{-20} m^2)	Energy (eV)	Q_i (10^{-20} m^2)	Energy (eV)	Q_i (10^{-20} m^2)
17	0.0889	45	5.14	351	7.13
18	0.211	47	5.52	401	6.55
19	0.375	49	5.77	451	6.21
21	0.782	51	6.19	501	5.89
23	1.18	61	6.82	601	5.17
25	1.59	71	7.57	701	4.72
27	2.11	81	7.84	801	4.4
29	2.49	91	8.17	901	3.96
31	2.81	101	8.39	1001	3.77
33	3.16	126	8.77	1251	3.19
35	3.49	151	8.75	1501	2.79
37	3.86	176	8.76	1751	2.44
39	4.17	201	8.57	2001	2.28
41	4.54	251	8.17	2501	1.88
43	4.85	301	7.41	3001	1.67

Source: Adapted from Nishimura, H. et al., *J. Chem. Phys.*, 110, 3811, 1999.

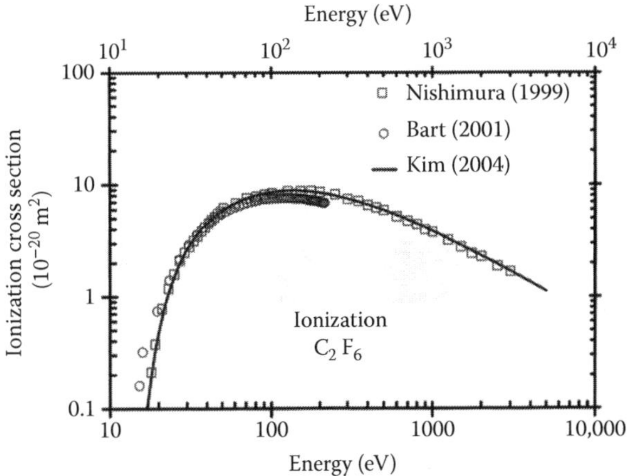

FIGURE 103.7 Total ionization cross sections for C_2F_6. Excellent agreement exists between the data shown. (Tabulated values for Bart et al. (2001) due to courtesy of Professor Harland.)

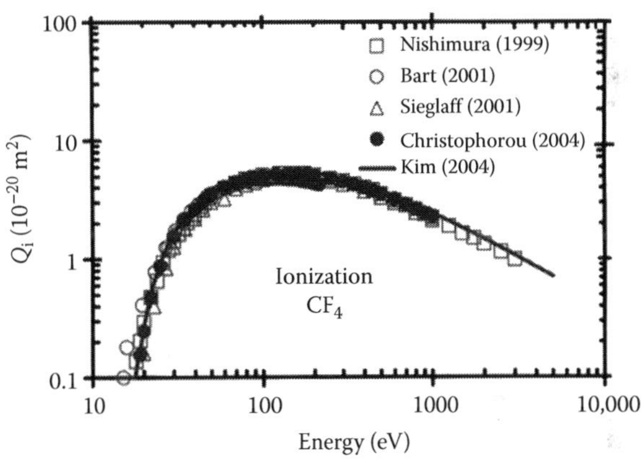

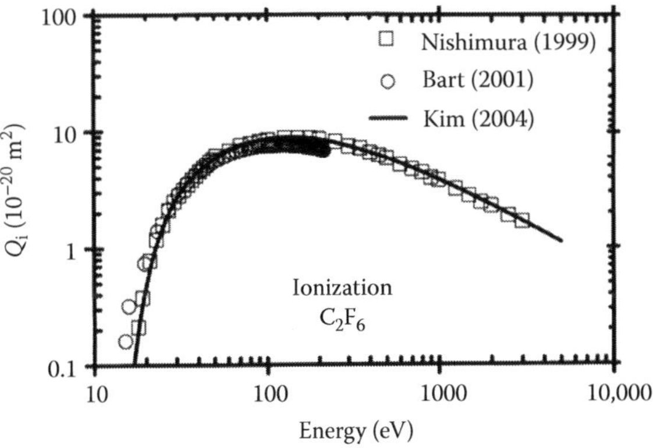

FIGURE 103.8 Total ionization cross section for C_2F_6 compared with those of CF_4.

in Christophorou and Olthoff (2004) are also shown in the figure. A comparison with ionization cross sections for CF_4 is provided in Figure 103.8.

103.8 ATTACHMENT PROCESSES

Dissociative attachment to the molecule occurs according to the reactions (Harland and Franklin, 1974)

$$C_2F_6 + e \rightarrow C_2F_5 + F^- \quad (103.1)$$

$$C_2F_6 + e \rightarrow C_2F_5^* + F^- \quad (103.2)$$

$$C_2F_6 + e \rightarrow CF_2 + CF_3 + F^- \quad (103.3)$$

$$C_2F_6 + e \rightarrow C_2F_4 + F + F^- \quad (103.4)$$

$$C_2F_6 + e \rightarrow CF_3 + CF_3^- \quad (103.5)$$

$$C_2F_6 + e \rightarrow C_2F_5^- + F \quad (103.6)$$

C_2F_6

The most abundant ion is due to Reaction 103.1 and the least, due to Reaction 103.6. The appearance potentials and peak in the attachment cross sections are shown in Table 103.10.

103.9 ATTACHMENT RATES

Thermal electron attachment rates are shown in Table 103.11. Temperature dependence of the attachment rate in C_2F_6 is reviewed by Christophorou and Olthoff (2004). The total attachment rate moderately increases with temperature due to the fact that dissociative attachment onset, at ~2.1 eV is well above thermal energy.

103.10 ATTACHMENT CROSS SECTIONS

Total attachment cross sections for C_2F_6 are shown in Table 103.12 (see Christophorou and Olthoff, 2004) and Figure 103.9 which also include data for CF_4 and C_3F_8. Notice the shift of the peak to higher energies for smaller molecules.

TABLE 103.10
Appearance Potentials and Peak Energy for Attachment for C_2F_6

Ion	Reaction Number	Appearance Potential (eV)	Peak (eV)	Reference
F^-	(103.1)	2 ± 0.1	3.9	Spyrou et al. (1983)
	(1031)	2.1 ± 0.2	4.3	Harland and Franklin (1974)
	(103.2)	4.9 ± 0.2		Harland and Franklin (1974)
CF_3^-	(103.5)	2.4 ± 0.1	4.0	Spyrou et al. (1983)
	(103.5)	2.2 ± 0.2	4.4	Harland and Franklin (1974)
$C_2F_5^-$	(103.6)	3.5 ± 0.1	4.8	Spyrou et al. (1983)

TABLE 103.11
Thermal Attachment Rate Constants (C_2F_6)

Rate (m³/s)	Method	Reference
$<10^{-22}$	Swarm	Spyrou and Christophorou (1985)
9.2×10^{-17}	Swarm	Spyrou and Christophorou (1985)
$<1.6 \times 10^{-19}$	Swarm	Davis et al. (1973)
$<10^{-22}$	MC	Fessenden and Bansall (1970)

Note: MC = Microwave conductivity.

TABLE 103.12
Total Attachment Cross Sections in C_2F_6

Energy (eV)	Q_a (10^{-20} m²)	Energy (eV)	Q_a (10^{-20} m²)
2.0	1 (−4)	5.5	0.039
2.25	7 (−4)	6.0	0.024
2.5	4.4 (−3)	6.5	0.016
2.75	0.013	7.0	0.012
3.0	0.036	7.5	9.1 (−3)
3.25	0.073	8.0	7.4 (−3)
3.5	0.112	8.5	6.4 (−3)
4.0	0.140	9.0	5.7 (−3)
4.5	0.105	9.5	5.1 (−3)
5.0	0.064	10.0	4.6 (−3)

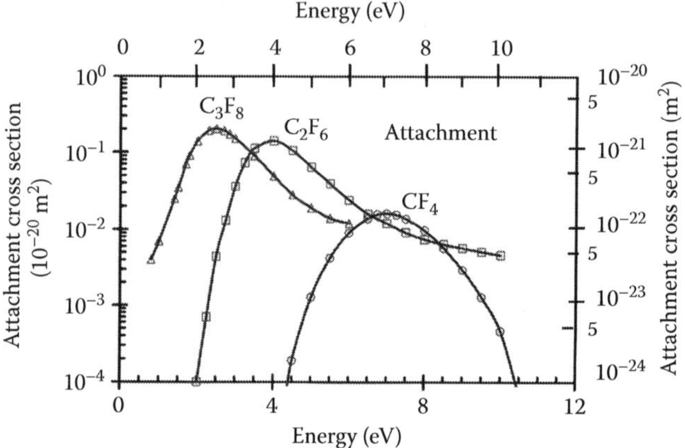

FIGURE 103.9 Attachment cross sections for C_2F_6 and selected fluorocarbons. Notice the shift of the peak toward higher energies for smaller molecules.

TABLE 103.13
Drift Velocity of Electrons (C_2F_6)

E/N (Td)	*W* (10^4 m/s)	*E/N* (Td)	*W* (10^4 m/s)	*E/N* (Td)	*W* (10^4 m/s)
0.05	0.147	6.0	7.62	120.0	11.4
0.06	0.176	8.0	8.36	140.0	11.9
0.08	0.237	10.0	8.92	160.0	12.6
0.10	0.295	12.0	9.3	180.0	13.3
0.15	0.435	15.0	9.9	200.0	14.0
0.20	0.58	20.0	10.5	220.0	14.5
0.30	0.87	25.0	10.8	240.0	15.0
0.40	1.15	30.0	10.9	260.0	15.4
0.60	1.69	35.0	11.0	280.0	16.0
0.80	2.23	40.0	10.9	300.0	16.6
1.0	2.71	50.0	10.8	320.0	17.1
1.5	3.69	60.0	10.5	340.0	17.6
2.0	4.54	70.0	10.5	360.0	18.4
3.0	5.68	80.0	10.6	380.0	19.1
4.0	6.51	100.0	10.9	400.0	19.8
5.0	7.13				

Source: Adapted from Hunter, S. R., J. G. Carter, and L. G. Christophorou, *Phys. Rev.*, 38, 58, 1988.

103.11 DRIFT VELOCITY OF ELECTRONS

The drift velocities of electrons are shown in Table 103.13 and Figure 103.10 which also include data-selected fluorocarbons.

103.12 DIFFUSION COEFFICIENTS

Table 103.14 and Figure 103.11 show the radial diffusion coefficients expressed as ratios of mobilities.

103.13 IONIZATION AND ATTACHMENT COEFFICIENTS

The ionization and attachment coefficients for C_2F_6 are shown in Table 103.15 and Figure 103.12.

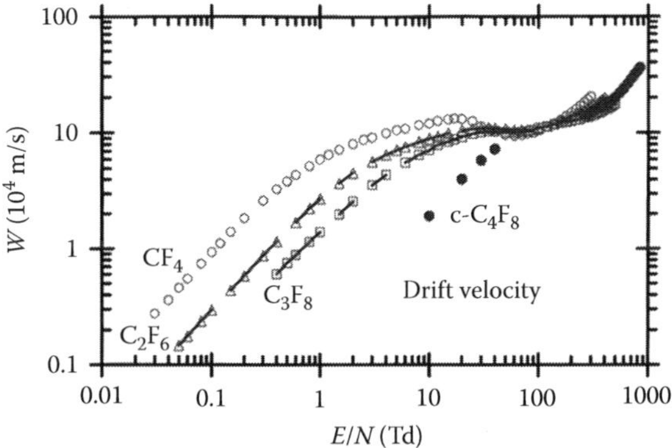

FIGURE 103.10 Electron drift velocity C_2F_6 and selected fluorocarbons. Notice the negative differential coefficient analogous to methane. (Adapted from Christophorou, L. G. and J. K. Olthoff, *Fundamental Electron Interactions with Plasma Processing Gases*, Kluwer Academic/Plenum Publishers, New York, NY, 2004.)

TABLE 103.14
D_r/μ Ratios for C_2F_6

E/N (Td)	D_r/μ (V)	E/N (Td)	D_r/μ (V)	E/N (Td)	D_r/μ (V)
1.00	0.029	15	0.060	150	1.59
2.00	0.033	20	0.080	200	2.15
3	0.035	30	0.161	250	2.83
4	0.038	40	0.305	300	3.56
5	0.040	50	0.511	350	4.00
6	0.041	60	0.712	400	4.43
7	0.043	70	0.868	450	4.77
8	0.045	80	0.967	500	4.93
9	0.047	90	1.05	550	5.07
10	0.049	100	1.13	600	5.20

Source: Adapted from Christophorou, L. G. and J. K. Olthoff, *Fundamental Electron Interactions with Plasma Processing Gases*, Kluwer Academic/Plenum Publishers, New York, NY, 2004.

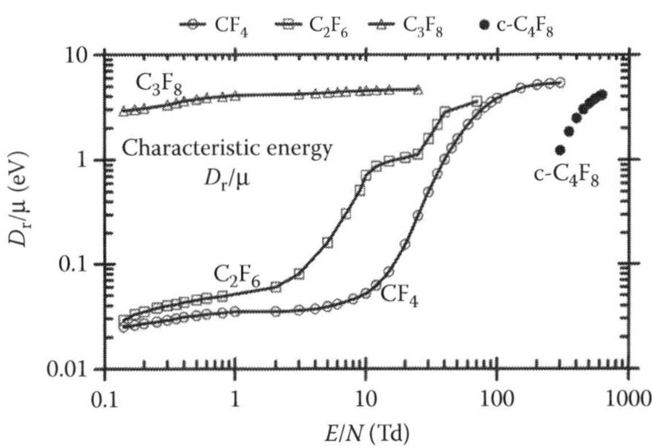

FIGURE 103.11 Ratios of D_r/μ for C_2F_6. Data for selected fluorocarbons added for comparison. (Adapted from Christophorou, L. G. and J. K. Olthoff, *Fundamental Electron Interactions with Plasma Processing Gases*, Kluwer Academic/Plenum Publishers, New York, NY, 2004.)

TABLE 103.15
Ionization and Attachment Coefficients

E/N (Td)	α/N (10^{-22} m²)	η/N (10^{-22} m²)
22		0.0009
25		0.0082
27		0.0264
30		0.099
35		0.49
40		1.40
45		2.82
50		4.61
70		12.7
80		16.2
90		18.8
100		20.9
120		23.6
140	0.6	24.5
160	1.0	24.3
180	3.2	23.3
200	5.7	22.0
225	9.9	19.9
250	13.6	18.1
275	17.5	17.5
300	21.1	16.4
325	25.0	16.0
350	29.4	15.6
375	35.0	15.0
400	40.4	13.6

Source: Adapted from Hunter, S. R., J. G. Carter, and L. G. Christophorou, *J. Chem. Phys.*, 86, 693, 1987.

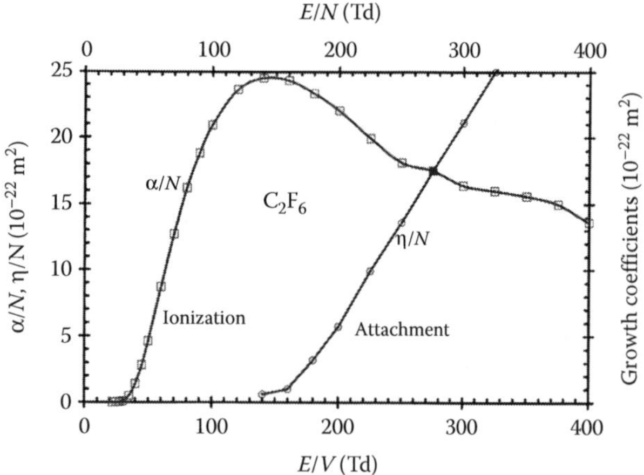

FIGURE 103.12 Density reduced ionization and attachment coefficients for C_2F_6. $\alpha/N = \eta/N = 17.5 \times 10^{-20}$ m^2 at $E/N = 275$ Td.

REFERENCES

Allan, M. and L. Andric, *J. Chem. Phys.*, 105, 3559, 1996.

Ariyasinghe, W. M., *Rad. Phys. Chem.*, 68, 79, 2003.

Bart, M., P. W. Harland, J. E. Hudson, and C. Vallane, *Phys. Chem. Chem. Phys.*, 3, 800, 2001.

Christophorou, L. G. and J. K. Olthoff, *Fundamental Electron Interactions with Plasma Processing Gases*, Kluwer Academic/Plenum Publishers, New York, NY, 2004.

Davis, F. J., R. N. Compton, and D. R. Nelson, *J. Chem. Phys.*, 55, 2324, 1973.

Fessenden, R. W. and D. K. Bansall, *J. Chem. Phys.*, 53, 3468, 1970.

Harland, P. W. and J. L. Franklin, *J. Chem. Phys.*, 61, 1621, 1974.

Hunter, S. R., J. G. Carter, and L. G. Christophorou, *J. Chem. Phys.*, 86, 693, 1987.

Hunter, S. R., J. G. Carter, and L. G. Christophorou, *Phys. Rev.*, 38, 58, 1988.

Hunter, S. R. and L. G. Christophorou, *J. Chem. Phys.*, 80, 6150, 1984.

Mann, A. and F. Linder, (a) *J. Phys. B: At. Mol. Opt. Phys.*, 25, 545, 1992; (b) *J. Phys. B: At. Mol. Opt. Phys.*, 25, 533, 1992.

Nishimura, H., W. M. Huo, M. A. Ali, and Y.-K. Kim, *J. Chem. Phys.*, 110, 3811, 1999.

Pirgov, P. and Stefanov, B., *J. Phys. B: At. Mol. Phys.* 23, 2879, 1990.

Shi, D. H., J. F. Sun, Y. F Liu, and Z. L. Zhu, *Chem. Phys. Lett.*, 429, 271, 2006.

Spyrou, S. M., I. Sauers, and L. G. Christophorou, *J. Chem. Phys.*, 78, 7200, 1983.

Spyrou, S. M. and L. G. Christophorou, *J. Chem. Phys.*, 82, 2620, 1985.

Sueoka, O., C. Makochekanwa, and H. Kawate, *Nucl. Instrum. Meth. Phys. Res. B*, 192, 206, 2002.

Sverdlov, L. M., M. A. Kovner, and E. P. Krainov, *Vibrational Spectra of Polyatomic Molecules*, John Wiley & Sons, New York, NY, 1974, p. 397.

Takagi, T., L. Boesten, H. Tanaka, and M. A. Dillon, *J. Phys. B: At. Mol. Phys.*, 27, 5389, 1994.

Winters, H. F. and M. Inokuti, *Phys. Rev. A*, 25, 1420, 1982.

104

PENTACHLORO-ETHANE

CONTENTS

Pentachloroethane (C_2HCl_5), synonym refrigerant 120, is a polar, electron-attaching gas that has 98 electrons. Its electronic polarizability is 15.60×10^{-40} F m^2, dipole moment is 0.92 D, and ionization potential is 11.0 eV.

104.1 IONIZATION CROSS SECTION

Table 104.1 shows the ionization cross sections for C_2HCl_5 (Hudson et al., 2001). Figure 104.1 is a graphical presentation

TABLE 104.1
Ionization Cross Sections for C_2HCl_5

Energy (eV)	Q_i (10^{-20} m^2)	Energy (eV)	Q_i (10^{-20} m^2)
16	299	94	23.31
20	6.90	97	23.28
23	10.78	101	23.30
27	13.55	105	22.98
31	15.81	108	22.73
34	17.74	112	22.75
38	18.77	116	22.64
42	19.69	119	22.51
45	20.41	123	22.36
49	21.03	127	21.92
53	21.97	130	21.88
57	22.18	134	21.76
60	22.79	138	21.61
64	22.91	142	21.50
68	23.10	145	21.20
71	23.16	149	21.00
75	23.48	153	20.68
79	23.61	156	20.48
82	23.42	160	20.43
86	23.51	164	20.41
90	23.42	167	20.37
171	20.36	197	18.78
175	20.18	201	18.73
179	19.97	204	18.71
182	19.67	208	18.76
186	19.32	212	18.83
190	19.04	215	18.72
193	18.89	219	18.47

Sources: Adapted from Hudson, J. E. et al., *J. Phys. B: At. Mol. Opt. Phys.*, 34, 3025, 2001; tabulated values courtesy of Professor Harland (2006).

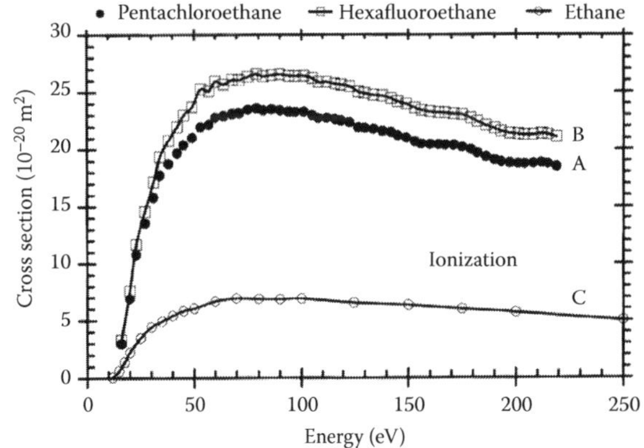

FIGURE 104.1 Ionization cross sections for C_2HCl_5. Selected gases added for comparison. (●, A) Pentachloroethane (C_2HCl_5); (—□—, B) hexachloroethane (C_2Cl_6); (—○—, C) ethane. Curves A and B, Harland et al. (2001), tabulated values courtesy of Professor Harland (2006); C, Nishimura and Tawara (1994).

with data for ethane (C_2H_6) and hexachloroethane added for comparison. The larger the molecule, the larger is the ionization cross section.

REFERENCES

Hudson, J. E., C. Vallance, M. Bart, and P. W. Harland, *J. Phys. B: At. Mol. Opt. Phys.*, 34, 3025, 2001.

Nishimura, H. and H. Tawara, *J. Phys. B: At. Mol. Opt. Phys.*, 27, 2063, 1994.

105

TETRABROMOETHANE

$C_2H_2Br_4$

CONTENTS

Tetrabromoethane ($C_2H_2Br_4$), synonym acetylene tetrabromide, is an electron-attaching gas that has 154 electrons. The theoretically calculated electronic polarizability is 10.18×10^{-40} F m^2 (Barszczewska et al., 2004) and the dipole moment is 1.38 D.

105.1 ATTACHMENT RATES

The ion species observed is Br^-. The attachment rate coefficient increases with increasing number of substituted bromine atoms, bromoethane (C_2H_5Br) < dibromoethane ($C_2H_4Br_2$) < tribromoethane ($C_2H_3Br_3$) < tetrabromoethane ($C_2H_2Br_4$) (Sunagawa and Shimamori, 1995). Measured values are shown in Table 105.1 and Figure 105.1.

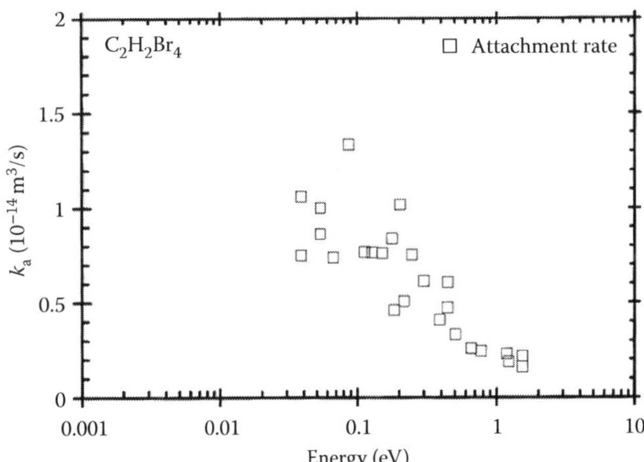

FIGURE 105.1 Attachment rates for $C_2H_2Br_4$. (Adapted from Sunagawa, T. and H. Shimamori, *Int. J. Mass Spectrom. Ion Proc.*, 149/150, 123, 1995.)

105.2 ATTACHMENT CROSS SECTIONS

Table 105.2 and Figure 105.2 show attachment cross sections for $C_2H_2Br_4$.

TABLE 105.1
Attachment Rate Coefficients for $C_2H_2Br_4$

Method	Temperature, K, (eV)	Rate (10^{-13} m^3/s)	Reference
Swarm	293	0.95	Barszczewska et al. (2004)
PR/MW	293	1.2	Sunagawa and Shimamori (1995)
	(0.04)	1.1	
	(0.1)	0.77	
	(0.4)	0.40	
	(1.0)	0.24	
TPI	293	0.69	Alajajian et al. (1988)
Swarm	293	0.18	Blaunstein and Christophorou (1968)

Note: PR/MW = pulse radiolysis/microwave heating; TPI = threshold photoionization.

TABLE 105.2
Attachment Cross Sections for $C_2H_2Br_4$

Energy (meV)	Q_a (10^{-20} m^2)	Energy (meV)	Q_a (10^{-20} m^2)
1.00	640	50.0	85.5
2.00	450	55.0	81.2
3.00	361	60.0	77.7
4.00	313	65.0	74.7
6.00	262	70.0	72.1
8.00	223	75.0	69.7
10.0	204	80.0	67.6
15.0	160	85.0	65.5
20.0	141	90.0	63.5
25.0	127	95.0	61.5
30.0	115	100	59.6
35.0	105	150	39.0
40.0	97.2	200	33.4
45.0	90.7	250	26.6

continued

TABLE 105.2 (continued)
Attachment Cross Sections for $C_2H_2Br_4$

Energy (meV)	Q_a (10^{-20} m^2)	Energy (meV)	Q_a (10^{-20} m^2)
300	19.6	700	7.14
350	16.0	750	6.45
400	14.4	800	5.77
450	12.4	850	5.25
500	10.2	900	4.91
550	8.84	1000	4.52
600	8.14	1500	3.16
650	7.70	2000	2.47

Source: Digitized from Sunagawa, T. and H. Shimamori, *Int. J. Mass Spectrom. Ion Proc.*, 149/150, 123, 1995.

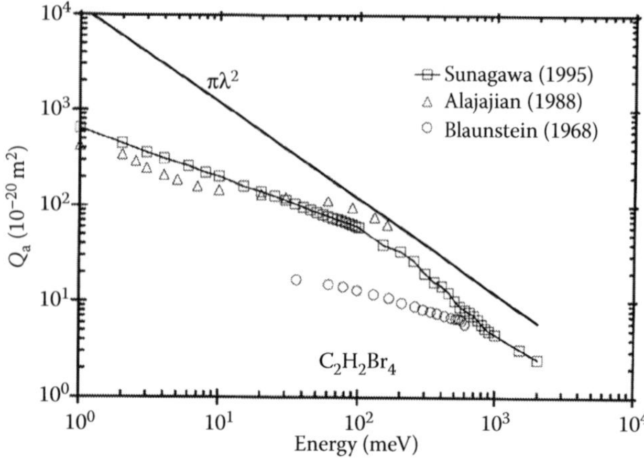

FIGURE 105.2 Attachment cross sections for $C_2H_2Br_2$ (—□—) Sunagawa and Shimamori (1995); (Δ) Alajajian et al. (1988); (○) Blaunstein and Christophorou (1968); (——) *s*-wave maximum cross section.

REFERENCES

Alajajian, S. H., M. T. Bernius, and A. Chutjian, *J. Phys. B: At. Mol. Opt. Phys.*, 21, 4021, 1988.

Barszczewska, W., J. Kopyra, J. Wnorowska, and I. Szamrej, *Int. J. Mass Spectrom.*, 233, 199, 2004.

Blaunstein, R. P. and L. G. Christophorou, *J. Chem. Phys.*, 49, 1526, 1968.

Sunagawa, T. and H. Shimamori, *Int. J. Mass Spectrom. Ion Proc.*, 149/150, 123, 1995.

106

TETRACHLOROETHANE

$C_2H_2Cl_4$

CONTENTS

Tetrachloroethane ($C_2H_2Cl_4$) is a polar, electron-attaching molecule that has 82 electrons. There are two isomers. Selected properties of the two isomers are shown in Table 106.1.

106.1 TOTAL IONIZATION CROSS SECTIONS

Total ionization cross sections for the isomers of $C_2H_2Cl_4$ are shown in Table 106.2 and Figure 106.1 (Hudson et al., 2001). Appreciable dependence on isomerism is observed.

TABLE 106.1
Selected Properties of $C_2H_2Cl_4$ Isomers

Name	Formula	α_e (10^{-40} F m^2)	μ(D)	ε_i(eV)
1,1,1,2-Tetrachloroethylene	1,1,1,2-$C_2H_2Cl_4$			11.1
1,1,2,2-Tetrachloroethylene (acetylene tetrachloride)	1,1,2,2-$C_2H_2Cl_4$	13.46	1.32	≤11.62

TABLE 106.2
Total Ionization Cross Sections for the Isomers of $C_2H_2Cl_4$

	Ionization Cross Section Q_i (10^{-20} m^2)	
Energy (eV)	1,1,1,2-$C_2H_2Cl_4$	1,1,2,2-$C_2H_2Cl_4$
16	3.04	2.72
20	7.07	6.10
23	10.65	9.53
27	12.94	11.99
31	15.07	13.76
34	16.69	15.10
38	17.65	16.03
42	18.17	16.67
45	18.60	17.21
49	19.19	17.74
53	19.90	18.29
57	20.19	18.61
60	20.71	19.03
64	20.75	19.10
68	20.93	19.32
71	20.89	19.39
75	21.11	19.48
79	21.25	19.65
82	20.97	19.50
86	21.13	19.49
90	20.99	19.46
94	20.84	19.30
97	20.88	19.35
101	20.84	19.35
105	20.56	19.04
108	20.22	18.76
112	20.32	18.81
116	20.17	18.78
119	20.07	18.57
123	19.91	18.40
127	19.52	18.12
130	19.42	18.02
134	19.34	17.94

continued

TABLE 106.2 (continued)
Total Ionization Cross Sections for the Isomers of $C_2H_2Cl_4$

	Ionization Cross Section Q_i (10^{-20} m^2)	
Energy (eV)	1,1,1,2-$C_2H_2Cl_4$	1,1,2,2-$C_2H_2Cl_4$
138	19.24	17.84
142	19.06	17.69
145	18.80	17.45
149	18.64	17.34
153	18.35	17.11
156	18.19	16.95
160	18.10	16.82
164	18.10	16.84
167	18.10	16.78
171	18.06	16.74
175	17.90	16.62
179	17.65	16.43
182	17.35	16.20
186	17.19	15.95
190	16.86	15.73
193	16.72	15.58
197	16.67	15.51
201	16.55	15.47
204	16.56	15.41
208	16.67	15.40
212	16.67	15.44
215	16.59	15.32
219	16.33	15.17

Sources: Adapted from Hudson, J. E. et al., *J. Phys. B: At. Mol. Opt. Phys.*, 34, 3025, 2001; tabulated values from courtesy of Professor Harland (2006).

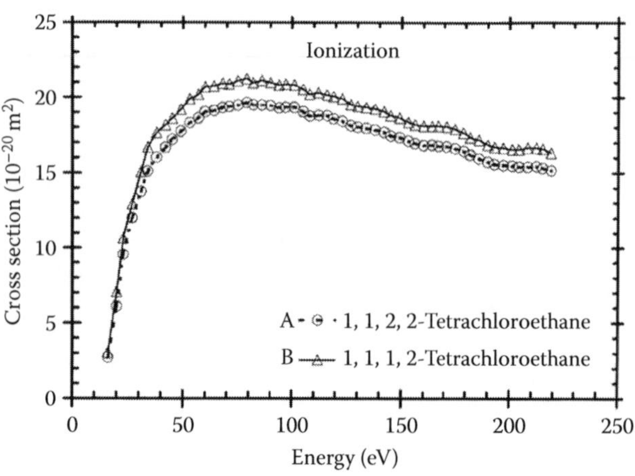

FIGURE 106.1 Total ionization cross sections for isomers of $C_2H_2Cl_4$. Tabulated values courtesy of Professor Harland (2006). Note the dependence on isomerism.

TABLE 106.3
Ion Species and Peak Energy for 1,1,2,2-$C_2H_2Cl_4$

Ion Species	Peak (eV)	Relative Intensity
Cl^-	0.2	425
Cl_2^-	0.11	42.5
$C_2H_2Cl_3^-$	0.06	50
HCl_2^-	0.02	11.5

Source: Adapted from Pshenichnyuk, S. A. et al., *Rapid Commn. Mass Spectrom.*, 20, 1097, 2006.

106.2 ATTACHMENT RATE

Attachment of electrons in 1,1,2,2-tetrachloroethane has been studied by Pshenichnyuk et al. (2006). The observed ions and peak of the intensity of ion yield are shown in Table 106.3.

There is a single study of attachment rate constant in 1,1,2,2-tetrachloroethane (Barszczewska et al., 2003) with a value of 3.5×10^{-14} m^3/s.

REFERENCES

Barszczewska, W., J. Kopyra, J. Wnorowska, and I. Szamrej, *J. Phys. Chem. A*, 107, 11427, 2003.

Hudson, J. E., C. Vallance, M. Bart, and P. W. Harland, *J. Phys. B: At. Mol. Opt. Phys.*, 34, 3025, 2001.

Pshenichnyuk, S. A., I. A. Pshenichnyuk, E. P. Nafikova, and N. L. Asfandiarov, *Rapid Commn. Mass Spectrom.*, 20, 1097, 2006.

107

TRIBROMOETHANE

$C_2H_3Br_3$

CONTENTS

Tribromoethane ($C_2H_3Br_3$) is an electron-attaching gas that has 114 electrons. The electronic polarizability is calculated as 8.49×10^{-40} F m^2 (see Barszczewska et al., 2004).

107.1 ATTACHMENT RATE CONSTANT

Thermal attachment rate constant is 9.2×10^{-14} m^3/s (Sunagawa and Shimamori 1995). Figure 107.1 shows the rate constants as a function of mean energy of the electron swarm, measured by the same authors.

107.2 ATTACHMENT CROSS SECTIONS

The ion species is Br^- formed by dissociative attachment. Table 107.1 and Figure 107.2 show the attachment cross sections deconvoluted from Figure 107.1 (Sunagawa and Shimamori, 1995). Significant points to note are

1. Zero energy attachment cross section has a peak at zero energy. This feature is common to most bromine-containing molecules.

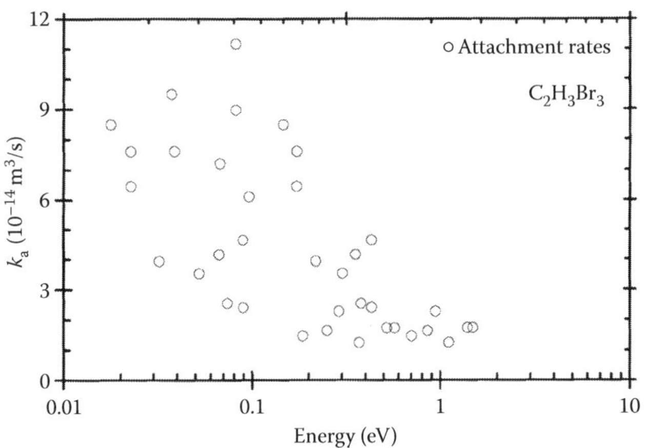

FIGURE 107.1 Attachment rate constants as a function of mean energy for $C_2H_3Br_3$.

TABLE 107.1
Attachment Cross Sections for $C_2H_3Br_3$

Energy (meV)	Q_a (10^{-20} m^2)	Energy (meV)	Q_a (10^{-20} m^2)
1.00	515	90.0	43.7
2.00	372	95.0	41.8
3.00	301	100.0	40.0
4.00	260	150	28.23
6.00	213	200	17.64
8.00	181	250	12.08
10.00	159	300	9.23
15.0	129	350	6.20
20.0	115	400	4.27
25.0	103	450	3.46
30.0	93.5	500	2.96
35.0	85.4	550	2.46
40.0	78.7	600	1.99
45.0	73.2	650	1.63
50.0	68.5	700	1.38
55.0	64.3	750	1.22
60.0	60.5	800	1.12
65.0	57.0	850	1.06
70.0	53.8	900	1.01
75.0	50.9	1000	0.90
80.0	48.2	1500	0.40
85.0	45.9	2000	0.23

Source: Digitized from Sunagawa, T. and H. Shimamori, *Int. J. Mass Spectrom.*, 149/150, 123, 1995.

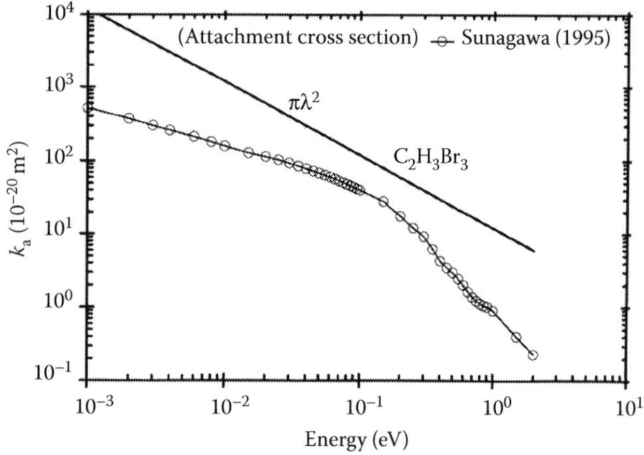

FIGURE 107.2 Attachment cross sections for $C_2H_3Br_3$. *s*-Wave maximum cross section ($\pi\lambda^2$) is also shown for comparison. (Adapted from Sunagawa and Shimamori, 1995.)

2. The low-energy cross section has the same slope as the *s*-wave maximum cross section suggesting that capture of the electron to the *s*-wave is the most probable process.
3. A comparison with chlorine-containing molecules shows that the cross sections are higher in the bromine-containing targets. A further difference is that the chlorine-containing compounds have a peak in the attachment cross sections in the 0.5–1.0 eV range (Shimamori and Sunagawa, 1995).

REFERENCES

Barszczewska, W., J. Kopyra, J. Wnorowska, and I. Szamrej, *Int. J. Mass Spectrom.*, 223, 199, 2004.

Sunagawa, T. and H. Shimamori, *Int. J. Mass Spectrom*, 149/150, 123, 1995.

108

TRICHLOROETHANE

$C_2H_3Cl_3$

CONTENTS

Trichloroethane ($C_2H_3Cl_3$) is an electron-attaching polar molecule with three chlorine atoms substituted in place of hydrogen atom in ethane (C_2H_6) molecule. Two isomers are known, 1,1,1-trichloroethane and 1,1,2-trichloroethane. Table 108.1 shows the selected properties of the molecule.

108.1 SELECTED REFERENCES FOR DATA

See Table 108.2.

108.2 IONIZATION CROSS SECTIONS

Ionization cross sections measured by Hudson et al. (2007) are shown in Table 108.3 and Figure 108.1.

108.3 ATTACHMENT CROSS SECTIONS

The attachment processes for electrons in $C_2H_3Cl_3$ are (Johnson, 1977):

$$C_2H_3Cl_3 + e \rightarrow C_2H_3Cl_3^{-*} \rightarrow C_2H_3Cl_3^{*} + e \quad (108.1)$$

$$\rightarrow C_2H_3Cl_3^{-*} \rightarrow C_2H_3Cl_2 + Cl^- \quad (108.2)$$

$$\rightarrow C_2H_3Cl_3^{-*} \rightarrow Cl + C_2H_3Cl_2^- \quad (108.3)$$

$$\rightarrow C_2H_3Cl_3^{-*} \rightarrow C_2H_3Cl + Cl_2^- \quad (108.4)$$

Reaction 108.1 is autodetachment and Reactions 108.2 through 108.4 are dissociative attachment processes.

1. 1,1,1-Trichloroethylene (1,1,1-$C_2H_3Cl_3$): Figure 108.2 shows the relative abundance of Cl^- and Cl_2^- ions for electron energy below 10 eV. Note the multiplication factor for the latter species. Cl^- shows a peak at 0.1 eV and a shoulder at 0.5 eV. Cl_2^- ion yield shows

TABLE 108.1
Selected Properties of $C_2H_3Cl_3$ Molecule

Name	Formula	α_e (10^{40} F m^2)	μ (*D*)	ε_i (eV)
1,1,1-Trichloroethane	1,1,1-$C_2H_3Cl_3$	11.9	1.755	11.0
1,1,2-Trichloroethane	1,1,2-$C_2H_3Cl_3$	5.95[a]	1.4	11.0

[a] Barszczewska et al. (2003).

TABLE 108.2
Selected References for Data

Parameter	Range, Energy (*E*/*N*)	Reference
k_a	Thermal	**Barszczewska et al. (2003)**
k_a	Thermal	**Gallup et al. (2003)**
Q_i	16.219	**Hudson et al. (2001)**
Q_a	0–3.5	**Aflatooni et al. (2000)**
Q_a	—	**Aflatooni and Burrow (1998)**
k_a	Thermal	**Shimamori et al. (1992)**
k_a	Thermal	**Christophorou et al. (1981)**
k_a, Q_a	0–1.5	**Johnson et al. (1977)**
k_a	0–1.0	**Blaunstein and Christophorou (1968)**

Note: k_a = Attachment rate; Q_a = attachment cross section. Bold font indicates experimental study.

$C_2H_3Cl_3$

TABLE 108.3
Ionization Cross Sections for CH_3CCl_3

Energy (eV)	Q_i (10^{-20} m^2)	Energy (eV)	Q_i (10^{-20} m^2)
16	5.74	119	16.24
20	10.50	123	16.03
23	14.53	127	15.81
27	16.93	130	15.71
31	16.44	134	15.72
34	14.37	138	15.77
38	14.09	142	15.47
42	14.53	145	15.18
45	15.07	149	15.01
49	15.53	153	14.99
53	15.93	156	14.98
57	16.18	160	14.74
60	16.50	164	14.63
64	16.69	167	14.56
68	16.82	171	14.40
71	16.89	175	14.11
75	17.03	179	13.99
79	17.00	182	13.88
82	17.02	186	13.85
86	17.06	190	13.67
90	16.87	193	13.47
94	16.86	197	13.38
97	16.78	201	13.33
101	16.75	204	13.33
105	16.66	208	13.21
108	16.33	212	13.20
112	16.26	215	12.97
116	16.27	219	12.75

Source: Adapted from Hudson, J. E. et al., *J. Phys. B: At. Mol. Opt. Phys.*, 34, 3025, 2001. Tabulated values are by courtesy of Professor Harland (2006).

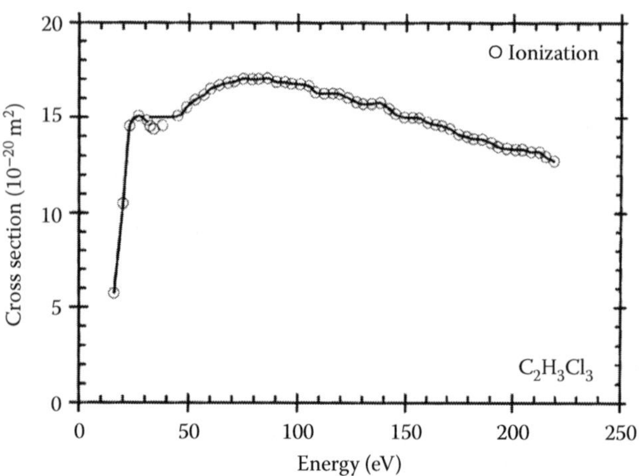

FIGURE 108.1 Ionization cross sections for $CH_2ClCHCl_2$. Tabulated values are by courtesy of Professor Harland 2006. (Adapted from Hudson, J. E. et al., *J. Phys. B: At. Mol. Opt. Phys.*, 34, 3025, 2001.)

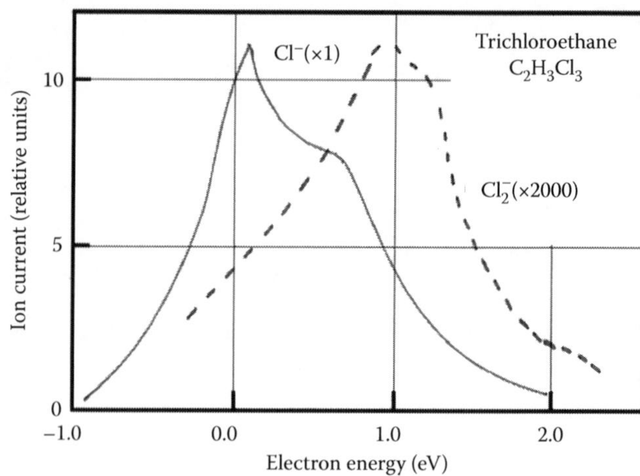

FIGURE 108.2 Relative abundance of negative ions in 1,1,1.trichloroethane ($C_2H_3Cl_3$). Note the multiplication factors. (Reprinted with permission from Johnson, J. P., L. G. Christophorou, and J. G. Carter, *J. Chem. Phys.*, 67, 2196, 1977. Copyright (1977) American Institute of Physics.)

TABLE 108.4
Attachment Cross Sections and Rates for 1,1,1.$C_2H_3Cl_3$

Energy (eV)	Q_a (10^{-20} m^2)	Energy (eV)	Q_a (10^{-20} m^2)
0.08	1.48	0.42	0.71
0.10	2.04	0.45	0.85
0.12	2.45	0.50	1.05
0.13	2.60	0.55	1.31
0.16	2.24	0.60	2.10
0.17	1.90	0.65	2.45
0.19	1.48	0.70	2.79
0.21	1.38	0.75	2.81
0.24	0.98	0.80	2.65
0.26	0.76	0.85	2.21
0.29	0.64	0.90	1.70
0.31	0.59	0.95	1.36
0.34	0.59	1.00	1.03
0.36	0.61	1.10	0.24
0.38	0.65	1.20	0.01
0.40	0.68		

Source: Digitized and interpolated from Johnson, J. P., L. G. Christophorou, and J. G. Carter, *J. Chem. Phys.*, 67, 2196, 1977.

a peak at ~1 eV, though the relative abundance is in the ratio 2000:1. The derived attachment cross sections are shown in Table 108.4 and Figure 108.3.

2. 1,1,2-Trichloroethylene (1,1,2-$C_2H_3Cl_3$): Figure 108.4 shows the relative abundance of Cl^-, Cl_2^-, and $C_2H_3Cl_2^-$ ions for electron energy below 10 eV. Note the multiplication factor for the species. The relative abundance increases to a peak at ~0.5 eV at the same

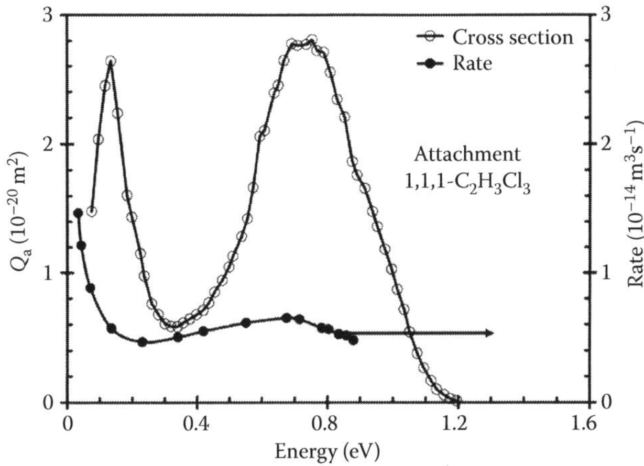

FIGURE 108.3 Attachment cross sections and rates for 1,1,1.trichloroethane ($C_2H_3Cl_3$). (Adapted from Johnson, J. P., L. G. Christophorou, and J. G. Carter, *J. Chem. Phys.*, 67, 2196, 1977.)

TABLE 108.5
Attachment Cross Sections for 1,1,2.$C_2H_3Cl_3$

Energy (eV)	Q_a (10^{-20} m^2)	Energy (eV)	Q_a (10^{-20} m^2)
0.12	0.03	0.45	0.54
0.14	0.04	0.50	0.57
0.16	0.06	0.55	0.66
0.17	0.10	0.60	0.93
0.19	0.20	0.65	1.27
0.21	0.30	0.70	1.68
0.24	0.87	0.75	2.03
0.26	1.13	0.80	2.04
0.29	1.32	0.85	1.94
0.31	1.25	0.90	1.65
0.34	1.04	0.95	1.27
0.36	0.91	1.00	0.87
0.38	0.79	1.10	0.14
0.40	0.67	1.20	0.09
0.42	0.59		

Source: Digitized and interpolated from Johnson, J. P., L. G. Christophorou, and J. G. Carter, *J. Chem. Phys.*, 67, 2196, 1977.

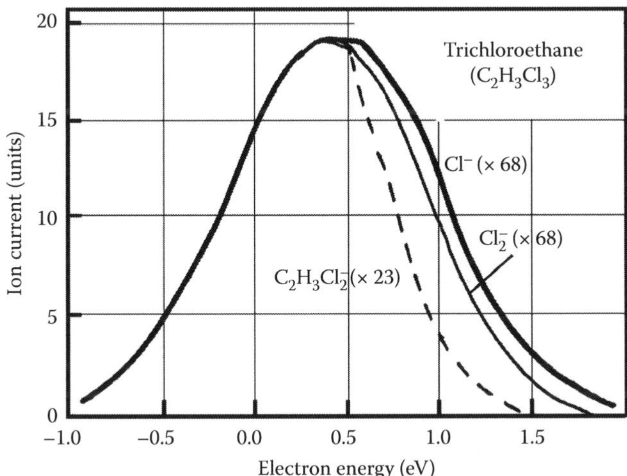

FIGURE 108.4 Relative abundance of negative ions in 1,1,2.trichloroethane ($C_2H_3Cl_3$). Note the multiplication factors. (Reprinted with permission from Johnson, J. P., L. G. Christophorou, and J. G. Carter, *J. Chem. Phys.*, 67, 2196, 1977. Copyright (1977), American Institute of Physics.)

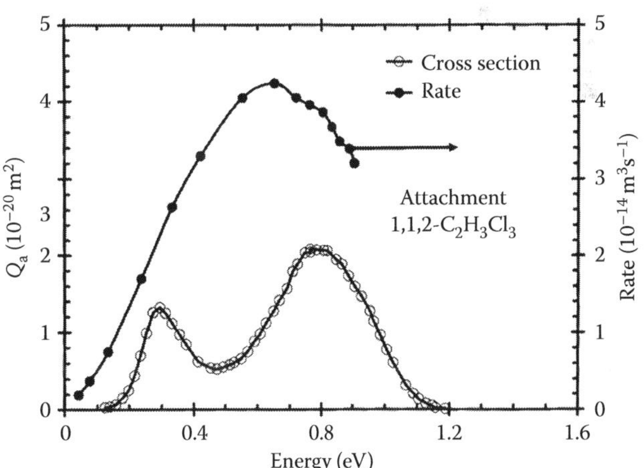

FIGURE 108.5 Attachment cross sections and rates for 1,1,2.trichloroethylene ($C_2H_3Cl_3$). (Adapted from Johnson, J. P., L. G. Christophorou, and J. G. Carter, *J. Chem. Phys.*, 67, 2196, 1977.)

rate for all the ions, though the rate of decline beyond the peak is different for each species. The derived attachment cross sections are shown in Table 108.5 and Figure 108.5.

Figure 108.6 shows the relative yield Cl^- ions in both isomers. The cross section details for both isomers are summarized in Table 108.6.

Details of negative ion formation are given in Table 108.6 and relative yields of the negative ion Cl^- in the two isomers are shown in Figure 108.6.

108.4 ATTACHMENT RATE CONSTANTS

Attachment rate constants in $C_2H_3Cl_3$ are shown in Table 108.7.

$C_2H_3Cl_3$

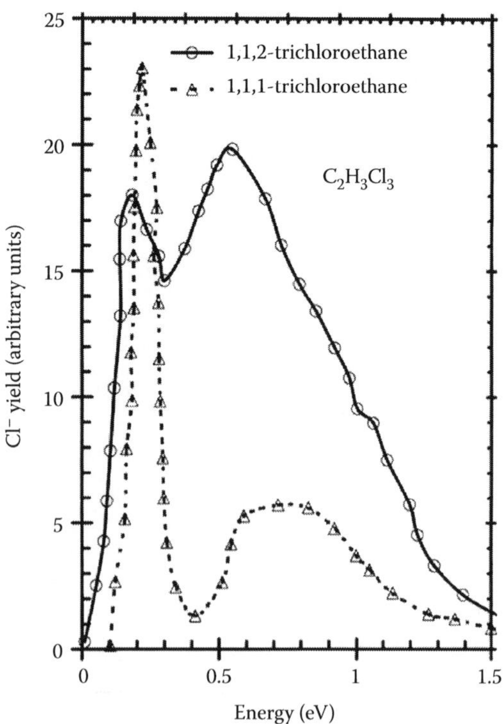

FIGURE 108.6 Relative yield of Cl^- ion in $C_2H_3Cl_3$. (Adapted from Johnson, J. P., L. G. Christophorou, and J. G. Carter, *J. Chem. Phys.*, 67, 2196, 1977.)

TABLE 108.6
Observed Negative Ions in $C_2H_3Cl_3$

Molecule	Ion	Peak Energy (eV)	Reference
1,1,1.$C_2H_3Cl_3$	Cl^-	0.61	Aflatooni and Burrow (2000)
	Cl^-	0.1	Johnson et al. (1977)
		0.5 (shoulder)	
	Cl_2^-	~1.0	
1,1,2-$C_2H_3Cl_3$		0.36	Aflatooni et al. (1998)
		0.88	Pearl and Burrow (1994)
	Cl^-	~0.4	Johnson et al. (1977)
	Cl_2^-	~0.4	
	$C_2H_3Cl_2^-$	~0.4	

TABLE 108.7
Attachment Rate Constants ($C_2H_3Cl_3$)

Molecule	Rate (m^3/s)	Temperature (K) Energy (eV)	Method	Reference
1,1,1-$C_2H_3Cl_3$	5.88×10^{-15}		Beam	Gallup et al. (2003)
	1.1×-10^{-14}		PR/MC	Shimamori et al. (1992)
	1.47×10^{-14}		Swarm	Johnson et al. (1977)
	1.5×10^{-14}		Swarm	Christodoulides and Christophorou (1971)
	1.6×10^{-14}		Swarm	Blaunstein and Christophorou (1968)
1,1,2-$C_2H_3Cl_3$	1.4×10^{-16}		Swarm	Barszczewska et al. (2003)
	1.8×10^{-16}		Swarm	Christophorou et al. (1981)
	1.9×10^{-16}		Swarm	Johnson et al. (1977)
	1.5×10^{-16}		Swarm	Blaunstein and Christophorou (1968)

Note: PR/MC = pulse-radiolysis/microwave cavity. Unless otherwise mentioned, energy is thermal.

REFERENCES

Aflatooni, K. and P. D. Burrow, *J. Chem. Phys.*, 113, 1455, 2000.

Aflatooni, K., G. A. Gallup, and P. D. Burrow, *Chem. Phys. Lett.*, 282, 398, 1998.

Barszczewska, W., J. Kopyra, J. Wnorowska, and I. Szamrej, *J. Phys. Chem.A*, 107, 11427, 2003.

Blaunstein, R. P. and L. G. Christophorou, *J. Chem. Phys.*, 49, 1526, 1968.

Christodoulides, A. A. and L. G. Christophorou, *J. Chem Phys.*, 54, 4691, 1971.

Christophorou, L. G., R. A. Mathis, D. R. James, and D. L. McCorkle, *J. Phys. D: Appl. Phys.*, 14, 1889, 1981.

Gallup, G. A., K. Aflatooni, and P. D. Burrow, *J. Chem. Phys.*, 118, 2562, 2003.

Hudson, J. E., C. Vallance, M. Bart, and P. W. Harland, *J. Phys. B: At. Mol. Opt. Phys.*, 34, 3025, 2001.

Johnson, J. P., L. G. Christophorou, and J. G. Carter, *J. Chem. Phys.*, 67, 2196, 1977.

Pearl, D. M. and P. D. Burrow, *J. Chem. Phys.*, 101, 2940, 1994.

Shimamori, H., Y. Tatsumi, Y. Ogawa, and T. Sunagawa, *J. Chem. Phys.*, 97, 6335, 1992.

109

1,1,1-TRIFLUORO-ETHANE

CONTENTS

1,1,1-Trifluoroethane ($C_2H_3F_3$) is a polar molecule with 72 electrons. It has an electronic polarizability of 4.9×10^{-40} F m^2, a dipole moment of 2.347 D, and an ionization potential of 13.3 eV.

109.1 TOTAL SCATTERING CROSS SECTIONS

Total electron scattering cross sections have been measured by Sueoka et al. (2002) as shown in Table 109.1 and Figure 109.1.

TABLE 109.1
Total Scattering Cross Sections for $C_2H_3F_3$

Energy	Q_T (10^{-20} m^2)			Energy	Q_T (10^{-20} m^2)			Energy	Q_T (10^{-20} m^2)			Energy	Q_T (10^{-20} m^2)		
(eV)	$C_2H_3F_3$	C_2H_6	C_2F_6	(eV)	$C_2H_3F_3$	C_2H_6	C_2F_6	(eV)	$C_2H_3F_3$	C_2H_6	C_2F_6	(eV)	$C_2H_3F_3$	C_2H_6	C_2F_6
0.8	30.9	5.05	15.2	5.0	33.5	25.1	27.1	15	25.7		27.8	90	18.2	16.4	24.6
1.0	31.5	6.50	15.7	5.5	35.5	27.2	27.6	16	25.2	28.3	27.3	100	17.7	15.4	24.0
1.2	31.0	7.97	16.0	6.0	34.5	29.4	27.2	17	25.5		27.8	125	15.8		
1.4	30.5	9.13	16.2	6.5	34.8	32.0	27.3	18	24.5	28.2	28.2	150	14.4		20.2
1.6	30.1	10.2	16.2	7.0	33.4	33.9	28.6	19	26.5		28.8	200	12.7	10.5	17.7
1.8	31.0	10.8	16.5	7.5	32.3	34.7	29.1	20	26.3	27.6	29.6	250	11.6	9.26	15.8
2.0	30.8	11.9	16.9	8.0	31.6	34.6	29.5	22	26.0		29.9	300	10.5		14.5
2.2	29.6		17.2	8.5	31.9	34.2	30.8	25	25.8	26.6	30.5	400	8.8		12.5
2.5	28.9	14.5	18.6	9.0	30.9	33.6	30.6	30	24.5	24.5	30.2	500	7.7		10.7
2.8	29.0		19.2	9.5	29.9		29.8	35	22.5	23.0	30.4	600	7.1		10.0
3.1	28.7		19.9	10.0	29.9	32.5	29.4	40	22.6	21.9	30.3				
3.4	28.9		21.2	11	28.9	31.7	28.0	50	20.5	20.2	29.5				
3.7	29.7			12	27.0	30.8	27.1	60	20.1	19.3	28.6				
4.0	30.4	20.2	25.5	13	26.9		27.0	70	20.3	18.3	27.4				
4.5	32.9	23.0	26.7	14	26.0	29.5	27.0	80	19.1	17.3	25.8				

Source: Adapted from Sueoka, O., C. Makochekanwa, and H. Kawate, *Nucl. Instrum. Methods Phys. Res. B*, 192, 206, 2002.
Note: Data for C_2H_6 and C_2F_6 given to show the effects of substitution of H and F atoms.

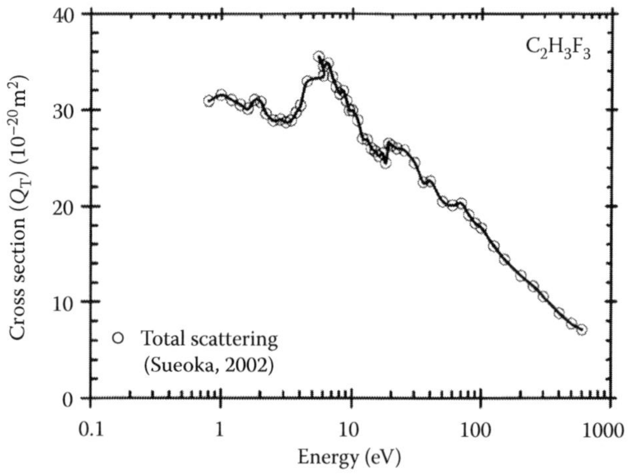

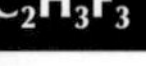

FIGURE 109.1 Total scattering cross sections for $C_2H_3F_3$. (Adapted from Sueoka, O., C. Makochekanwa, and H. Kawate, *Nucl. Instrum. Methods Phys. Res. B*, 192, 206, 2002.)

109.2 ATTACHMENT PROCESSES

Macneil and Thynne (1969) have studied the formation of negative ions in the gas.

Electron attachment to neutral particles occurs by three processes.

1. Resonance capture that occurs with low-energy electrons, 0–2 eV range, according to

$$XY + e \rightarrow XY^{-*} \rightarrow XY^- \quad (109.1)$$

2. Dissociative capture that occurs in the energy range 0–10 eV according to

$$XY + e \rightarrow X + Y^- \quad (109.2)$$

3. Ion pair formation that occurs usually at energies >10 eV according to

$$XY + e \rightarrow X^+ + Y^- + e \quad (109.3)$$

The following processes have been suggested to occur:

$$C_2H_3F_3 + e \rightarrow C_2H_3F_2 + F^- \text{ (1.5 eV)} \quad (109.4)$$

$$C_2H_3F_3 + e \rightarrow CF_2 + CH_3 + F^- \text{ (2.4 eV)} \quad (109.5)$$

$$C_2H_3F_3 + e \rightarrow C_2H_2F_2 + H + F^- \text{ (3.6 eV)} \quad (109.6)$$

$$C_2H_3F_3 + e \rightarrow CF_2 + CH_2 + H + F^- \text{ (5.2 eV)} \quad (109.7)$$

$$C_2H_3F_3 + e \rightarrow CF + F + CH_3 + F^- \quad (109.8)$$

Ion pair formation at 20.0 and 22.5 eV with low cross section are also suggested (Macneil and Thynne, 1969).

109.3 ATTACHMENT RATE CONSTANT

Thermal attachment rate constant is 4.34×10^{-20} m^3/s (Sueoka et al., 2002) with excellent agreement with Fessender and Bansal (1970).

REFERENCES

Fessenden, R. W. and K. M. Bansal, *J. Chem. Phys.*, 53, 3468, 1970.

Macneil, K. A. G. and J. C. J. Thynne, *Int. J. Mass Spectrom. Ion Phys.*, 2, 1, 1969.

Sueoka, O., C. Makochekanwa, and H. Kawate, *Nucl. Instrum. Methods Phys. Res. B*, 192, 206, 2002.

Section IX

9 ATOMS

110

HEXAFLUOROPROPENE

CONTENTS

Hexafluoropropene (1-C_3F_6), also known as perfluoropropene, is a molecule with all the hydrogen atoms of propene replaced with fluorine atoms. The substitution renders the molecule electron attaching. It has nine atoms, 132 electrons, electronic polarizability (α_e) of 6.68×10^{-40} F m^2 (see Beran and Kevan, 1969), and ionization potential of 10.60 eV.

110.1 TOTAL SCATTERING CROSS SECTIONS

Table 110.1 and Figure 110.1 show the total cross section.

TABLE 110.1
Total Scattering Cross Section

Energy (eV)	Q_T (10^{-20} m^2)	Energy (eV)	Q_T (10^{-20} m^2)
30	36.9	120	29.8
35	36.9	140	28.3
40	37.3	160	26.9
45	37.7	180	25.0
50	37.7	200	24.1
60	37.1	220	23.1
70	36.1	250	21.4
80	35.0	275	20.1
90	33.4	300	19.2
100	32.4	350	17.3
110	31.3	370	16.5

Source: Adapted from Szmytkowski, Cz., P. Możejko, and S. Kwitnewski, *J. Phys. B: At. Mol. Opt. Phys.*, 35, 1267, 2002.

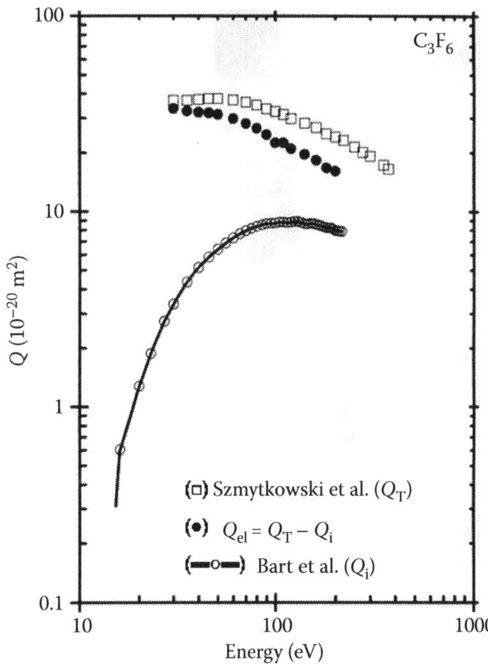

FIGURE 110.1 Scattering cross sections for 1-C_3F_6. (□) Szmytkowski et al. (2002), total; (•) elastic, ($Q_T - Q_i$); (—○—) Bart et al. (2001). Elastic scattering cross sections are approximate values.

110.2 ELASTIC SCATTERING CROSS SECTIONS

Elastic scattering cross sections calculated as the difference between the total cross sections (Table 110.1) and ionization

1-C_3F_6

TABLE 110.2
Elastic Scattering Cross Sections

Energy (eV)	Q_i (10^{-20} m²)	Energy (eV)	Q_i (10^{-20} m²)
30	33.54	90	24.74
35	32.55	100	22.48
40	32.13	110	22.48
45	31.86	120	20.99
50	31.31	140	19.64
60	29.81	160	18.32
70	28.15	180	16.73
80	26.62	200	16.12

cross sections (Bart et al., 2001) are shown in Table 110.2. These are only approximate values since all other cross sections are neglected due to unavailability. Figure 110.1 also includes the elastic scattering cross sections.

110.3 IONIZATION CROSS SECTIONS

See Table 110.3.

TABLE 110.3
Ionization Cross Sections

Energy (eV)	Q_i (10^{-20} m²)	Energy (eV)	Q_i (10^{-20} m²)
16	0.60	115	8.76
20	1.27	120	8.81
23	1.88	125	8.88
27	2.74	130	8.87
30	3.36	135	8.76
35	4.35	140	8.66
40	5.17	145	8.63
45	5.84	150	8.69
50	6.39	155	8.67
55	6.88	160	8.58
60	7.29	165	8.50
65	7.66	170	8.42
70	7.95	175	8.33
75	8.19	180	8.27
80	8.38	185	8.24
85	8.53	190	8.22
90	8.66	195	8.11
95	8.68	200	7.98
100	8.72	205	7.93
105	8.82	210	7.91
110	8.82	215	7.89

Source: Adapted from Bart, M. et al., *Phys. Chem. Chem. Phys.*, 3, 800, 2001.

Note: Tabulated values courtesy of Professor Harland. Beran and Kevan (1969) give a single value of 9.16×10^{-20} m² at 70 eV electron energy.

110.4 ATTACHMENT PROCESSES

Parent negative ions $C_3F_6^-$ with a comparatively long lifetime of ~10^{-6} s are formed due to electron attachment (Hunter et al., 1983). The attachment processes proposed by them are

$$C_3F_6 + e \rightarrow C_3F_6^{-*} \rightarrow \left.\begin{array}{ll} C_3F_6^- + \text{energy (Parent negative ion formation)} & \text{A} \\ C_3F_6 + e^* \text{ (Indirect elastic and inelastic scattering)} & \text{B} \\ C_xF_y^- + C_{3-x} + F_{6-y} \text{ (Dissociative attachment, } \geq 2\text{ eV)} & \text{C} \\ C_3F_6 + e \text{ (Collisional detachment)} & \text{D} \end{array}\right\} \quad (110.1)$$

The excited negative ion may collide with a neutral third body forming a short-lived (~10^{-12} s) dimer, with the possible processes

$$C_3F_6^{-*} + C_3F_6 \rightarrow (C_3F_6)_2^{-*} \begin{cases} \rightarrow C_3F_6 + C_3F_6 + e & \text{E} \\ \rightarrow (C_3F_6)_2^- + \text{energy} & \text{F} \end{cases} \quad (110.2)$$

Process F is further explained in more detail as

$$e + 1\text{-}C_3F_6 \rightarrow (C_3F_6)^{-*} \quad (110.3)$$

$$(C_3F_6)^{-*} + 1\text{-}C_3F_6 \rightarrow (C_3F_6)_2^{-*} \quad (110.4)$$

$$(C_3F_6)_2^{-*} + (\text{Ar or N}_2) \rightarrow (C_3F_6)_2^- + (\text{Ar or N}_2) + \text{energy} \quad (110.5)$$

Further, process A is negligible in comparison with E and F combined (Hunter et al., 1983). Jarvis et al. (1996) suggest an alternative process according to

$$(C_3F_6)^{-*} + 1\text{-}C_3F_6 \rightarrow (C_6F_{11})^- + F \quad (110.6)$$

110.5 ATTACHMENT RATES

Hunter et al. (1983) have measured the total attachment cross section by the swarm technique using buffer gas and found that the total attachment rates are dependent on both the partial number density of 1-C_3F_6 and the buffer gas. Figures 110.2 and 110.3 show the attachment rates both as a function of buffer gas number density and 1-C_3F_6. Approximate attachment rate constants (mean energy ~0.04 eV) at different temperatures are shown in Table 110.4. The features to note in Figures 110.2 and 110.3 are the large increase in k_a with increase in gas pressure, not withstanding the fact that the increase in buffer gas pressure decreases the percentage of 1-C_3F_6 in the mixture since the pressure (number density) of the attaching component has remained constant. Further, the attachment rate decreases with temperature (Table 110.4) which is in contrast with that of several other attaching gases. This is

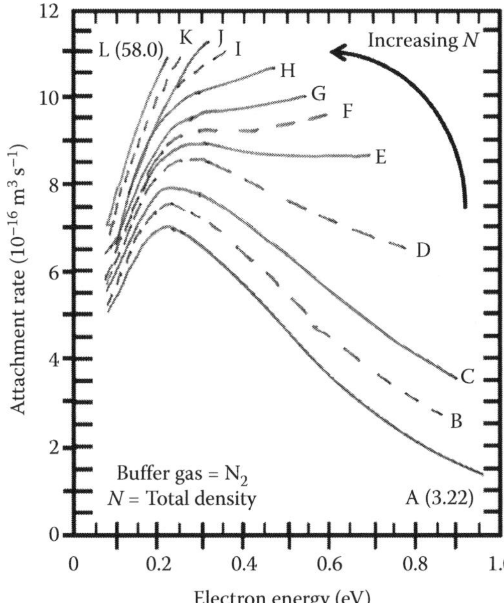

FIGURE 110.2 Attachment rates for (1-C_3F_6) as a function of mean energy with N_2 as buffer gas. Partial number density of 1-C_3F_6 is 2.5×10^{21} m^{-3}. Labels on curves show the total gas number density in 10^{25} m^{-3}. A—3.22; B—4.83; C—6.44; D—9.66; E—12.9; F—16.1; G—19.3; H—25.8; I—32.2; J—38.7; K—48.3; L—58.0. (Adapted from Hunter, S. R. et al., *J. Phys. D: Appl. Phys.*, 16, 573, 1983.)

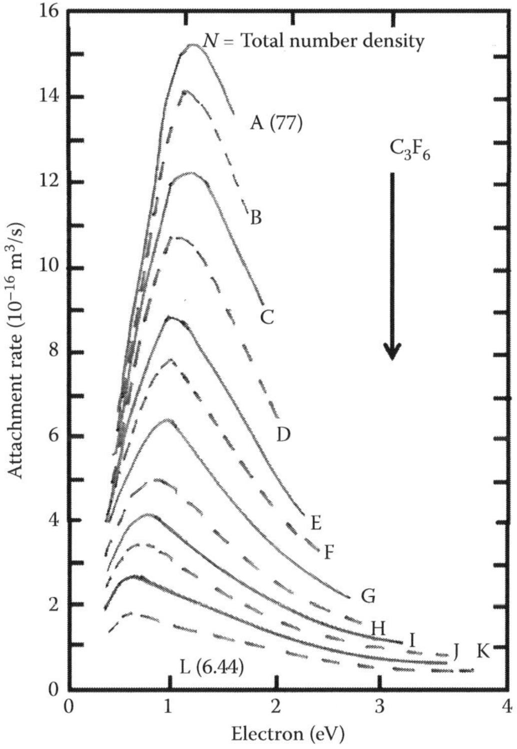

FIGURE 110.3 Attachment rates in 1-C_3F_6 with argon gas as buffer. Partial number density of 1-C_3F_6 4.94×10^{20} m^{-3}. Total gas number density in 10^{25} m^{-3}. A— 77.3; B— 67.6; C—57.9; D—48.3; E—38.6; F—32.2; G—25.8; H—19.3; I—16.1; J—12.9; K—9.66; L—6.44. (Adapted from Hunter, S. R. et al., *J. Phys. D: Appl. Phys.*, 16, 573, 1983.)

TABLE 110.4
Attachment Rates for 1-C_3F_6

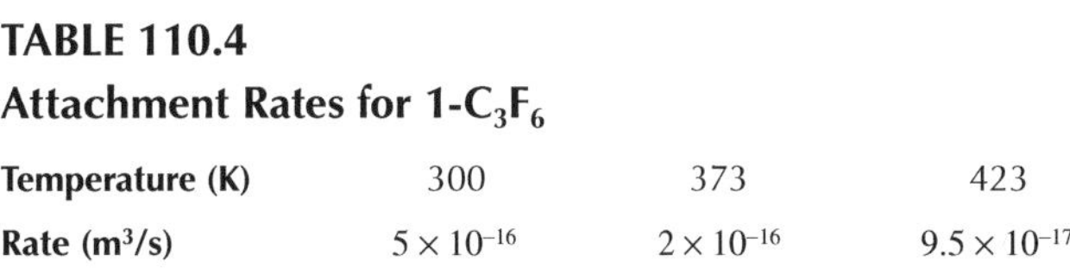

Temperature (K)	300	373	423
Rate (m^3/s)	5×10^{-16}	2×10^{-16}	9.5×10^{-17}

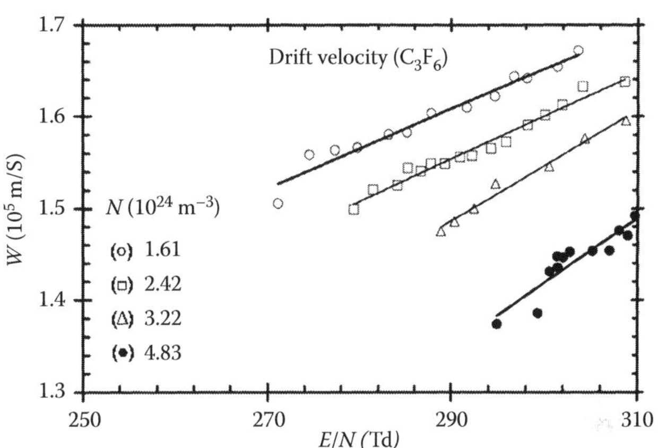

FIGURE 110.4 Drift velocity of electrons in C_3F_6 as a function of *E*/*N*. Note the decrease in drift velocity with increasing *N* at constant *E*/*N*. The values of *N* (10^{24} m^{-3}) are: (○) 1.61; (□) 2.42; (Δ) 3.22; (●) 4.83. (Adapted from Aschwanden, Th. et al., in L. G. Christophorou (Ed.), *Gaseous Dielectrics III*, Pergamon Press, New York, NY, 1982, p. 23, 32.)

attributed to the decrease of attachment cross sections with temperature. In SF_6, CCL_4, and CH_2Br_2 the attachment cross section remains constant as the temperature is increased up to 1200 K, whereas in several other attaching gases ($CFCl_3$, CH_3I, $CHCl_3$, CF_3Br, and CH_3Br), the attachment cross section decreases with temperature (Spence and Schulz, 1973).

110.6 DRIFT VELOCITY

Aschwanden and Biasiutti (1981) and Aschwanden et al. (1982) have measured the drift velocities of electrons in C_3F_6 and observed that they are dependent on both *E*/*N* and *N*. With increasing *E*/*N*, the drift velocity increases; with increasing *N*, it decreases at the same *E*/*N*, due to the increasing attachment rate at high gas number densities. Data are shown in Figure 110.4.

110.7 IONIZATION COEFFICIENTS

Aschwanden et al. (1982) have measured the net ionization coefficients as a function of *E*/*N* as shown in Figure 110.5.

110.8 LIMITING *E*/*N*

Sparking potentials have been measured by Hunter et al. (1982) for Bruce profile electrodes (Figure 110.6) and Paschen

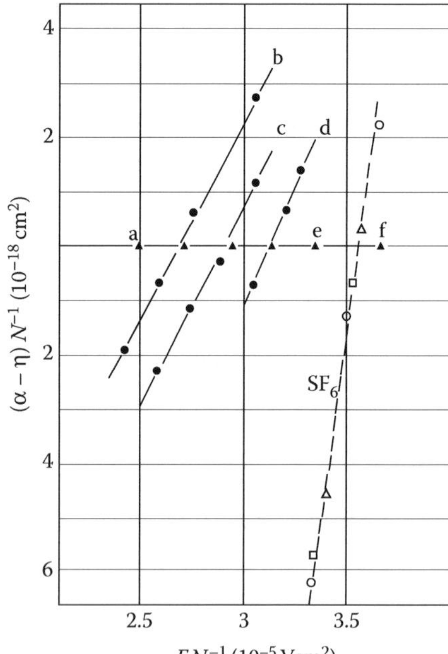

FIGURE 110.5 Density-reduced effective ionization coefficients for 1-C_3F_6 as a function of *E*/*N*. Number densities (10^{24} m^{-3}) are (a) 0.6; (b) 1.6; (c) 4.9; (d) 9.9; (e) 12; (f) 23. Effective ionization coefficients for SF_6 are shown for comparison. The symbols are (○) 1.6; (Δ) 3.3; (□) 16. (Adapted from Aschwanden, Th. et al., in L. G. Christophorou (Ed.), *Gaseous Dielectrics III*, Pergamon Press, New York, NY, 1982, p. 23, 32.)

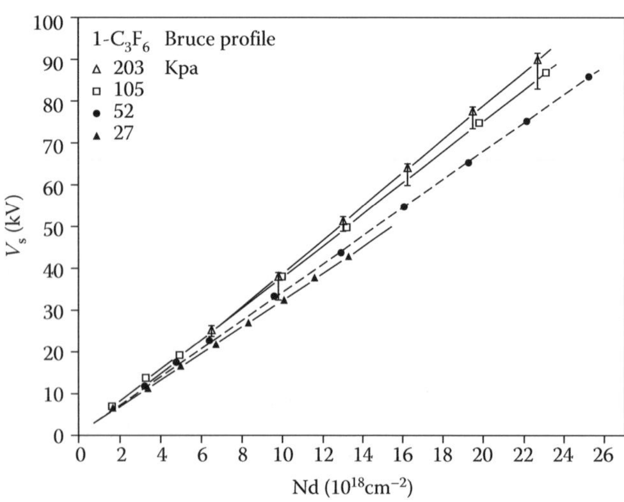

FIGURE 110.6 Sparking potentials in 1-C_3F_6 as a function of Nd at various values of *N* and electrode separation for several gas pressures. (Adapted from Hunter, S. R. et al., *J. Phys. D: Appl. Phys.*, 16, 573, 1983.)

Law is not observed in 1-C_3F_6. The limiting *E*/*N* is found to be a function of pressure, increasing with gas number density as shown in Table 110.5 and Figure 110.7, attributed to increasing attachment at high number densities.

TABLE 110.5
Limiting *E*/*N* and Gas Number Density

N (10^{24} m^{-3})	*E*/N_{Lim} (Td)	Method
1.71	275	Current growth
2.45	283	Current growth
3.18	289	Current growth
4.89	297	Current growth
6.12	303	Breakdown
7.10	310	Current growth
10.28	326	Current growth
12.24	328	Current growth
12.24	338	Breakdown
22.02	368	Current growth
24.23	366	Current growth
24.23	373	breakdown
36.95	394	Current growth
36.95	400	Breakdown
48.70	414	Breakdown

Source: Adapted from Aschwanden, Th. and G. Biasiutti, *J. Phys. D: Appl. Phys.*, 14, L-189, 1981.

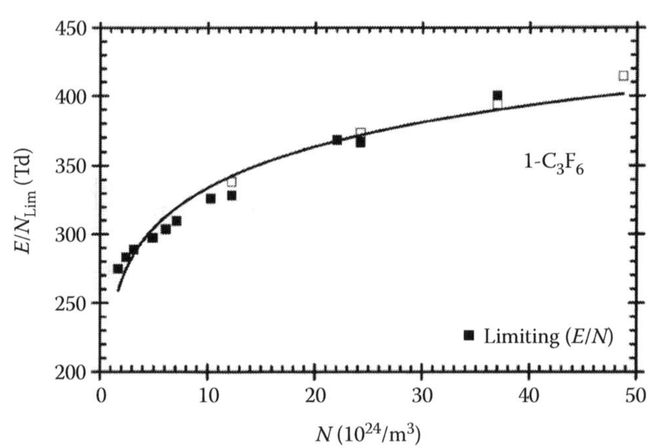

FIGURE 110.7 Limiting *E*/*N* for 1-C_3F_6 as a function of gas number density. Open symbols—breakdown experiments; closed symbols—current growth experiments. The values shown correspond to 1 bar gas pressure. (Adapted from Aschwanden, Th. and G. Biasiutti, *J. Phys. D: Appl. Phys.*, 14, L-189, 1981.)

REFERENCES

Aschwanden, Th. and G. Biasiutti, *J. Phys. D: Appl. Phys.*, 14, L-189, 1981.

Aschwanden, Th., D. Böttcher, M. Jungblut, and W. F. Schmidt, in L. G. Christophorou (Ed.), *Gaseous Dielectrics III*, Pergamon Press, New York, NY, 1982, p. 23, 32.

Bart, M., P. W. Harland, J. E. Hudson, and C. Vallance, *Phys. Chem. Chem. Phys.*, 3, 800, 2001.

Beran, J. A. and L. Kevan, *J. Phys. Chem.*, 73, 3866, 1969.

Hunter, S. R., L. G. Christophorou, D. L. McCorkle, I. Sauers, H. W. Ellis, and D. R. James, *J. Phys. D: Appl. Phys.*, 16, 573, 1983.

Jarvis, G. K., R. Peverall, and C. A. Mayhew, *J. Phys. B: At. Mol. Opt. Phys.*, 29, L713–L718, 1996.

Spence, D. and G. J. Schulz, *J. Chem. Phys.*, 58, 1800, 1973.

Szmytkowski, Cz., P. Możejko, and S. Kwitnewski, *J. Phys. B: At. Mol. Opt. Phys.*, 35, 1267, 2002.

111

PROPYLENE AND CYCLOPROPANE

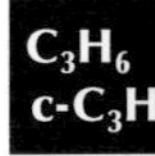

CONTENTS

Propylene (C_3H_6), also called propene, is a hydrocarbon belonging to the category of alkenes (C_3H_6) with 24 electrons. It has an isomer called cyclopropane (c-C_3H_6). Selected properties of the two molecules are shown in Table 111.1. Excitation potentials of C_3H_6 and selected hydrocarbons are shown in Table 111.2.

The vibrational modes and corresponding energies of c-C_3H_6 are shown in Table 111.3. The type of motion may generally be discerned because vibrations of similar type (similar local motion) tend to have similar frequencies (energies) (Allan and Andric, 1996). Refer to Table 117.1 in the chapter on propane.

TABLE 111.1
Selected Properties of Propylene and Cyclopropane

Property	Symbol	Propylene	Cyclopropane	Units
Electronic polarizability	α_e	6.97×10^{-40}	6.30×10^{-40}	F m^2
Dipole moment	μ	0.366, 0.07[a]	—	D
Bond dissociation energy	ε_d	3.844	4.608	eV
Ionization potential	ε_i	9.73	9.86	eV

[a] C_3H_6 has two dipole moments; the second value is from Gordy and Cook (1984).

111.1 SELECTED REFERENCES FOR DATA

Selected references for data are provided in Table 111.4.

TABLE 111.2
Excitation Potentials of C_3H_6 and Selected Hydrocarbons

	Excitation Potentials (eV)	
Gas	Bowman and Miller (1965)	Dance and Walker (1973)
Propylene	1.8, 4.4, 7.8, 8.8	1.8, 2.2, 4.35, 6.7, 7.1, 7.7, 8.9
Methane	10.2, 11.8	
Ethylene	1.7, 4.4, 7.7, 9.2	1.7, 1.87, 4.2, 6.5, 7.2, 7.7, 8.1, 9.2
Acetylene	2.0, 6.2, 7.7, 7.9, 8.2, 9.2, 10.0	
Propyne	2.8, 6.1, 7.1, 8.3	
1-Butene		2.0
cis-2-Butene		2.3, 4.3, 6.1, 7.0, 8.1
trans-2-Butene		2.4, 4.4, 6.0, 7.1, 8.7
1-Butyne	2.4, 6.3, 7.3, 8.2	4.25, 5.8–6.2

Source: Adapted from Bowman C. R. and W. D. Miller, *J. Chem. Phys.*, 42, 681, 1965.

TABLE 111.3
Vibrational Modes in Gas Phase of c-C_3H_6

Mode	Energy (meV)	Type of Motion
ν_{14}	91.6	CH_2 twist; CH_2 rock
ν_7	105.9	CH2 rock
ν_{11}	107.6	C–C skeletal stretch
ν_{10}	127.6	CH2 wag; C–C skeletal stretch
ν_5	132.7	CH2 wag
ν_4	139.6	CH2 twist
ν_3	147.3	C–C ring breathing
ν_{13}	147.3	CH2 rock; CH2 twist
ν_9	178.3	CH2 scissoring
ν_2	183.3	CH2 scissoring; C–C ring breathing
ν_8	375.1	CH2 symmetric stretch (out-of-phase)
ν_1	376.7	CH2 symmetric stretch (in-phase)
ν_{12}	382.1	CH2 asymmetric stretch (out-of-plane)
ν_6	384.6	CH2 asymmetric stretch (in-phase)

Source: Adapted from Göötz, B., H. Winterling, and P. Swiderek, *J. Electron Spectrom.*, 105, 1, 1999.

Note: The last column is from Levin and Pearce (1978).

C_3H_6 c-C_3H_6

TABLE 111.4
Selected References for Data

Quantity	Range (eV)	Gas	Reference
		Cross Sections	
Q_i	10–1000	Propene	Feil et al. (2006)
Q_{el}, Q_M	1.5–100	Propylene, cyclopropane	Makochekanwa et al. (2006)
Q_i	9.95–5000	Propylene	Y. -K. Kim et al. (2004)
Q_T	0.5–370	Propylene, cyclopropane	Szmytkowski and Kwitnewski (2002)
Q_{el}	1–15	Cyclopropane	Curic and Gianturco (2002a)
Q_v	1–15	Cyclopropane	Curic and Gianturco (2002b)
Q_v	0–5	Cyclopropane	Göötz et al. (1999)
Q_T	0.01–4.0	Propylene	Lunt et al. (1998)
Q_i	10–1000	Propylene	Jiang et al. (1997)
Q_i	10–3000	Propylene, cyclopropane	Nishimura et al. (1994)
Q_v	0–20	Cyclopropane	Allan and Andric (1993)
Q_T	4–500	Propylene, cyclopropane	Nishimura et al. (1991)
Q_T	5–400	Propylene, cyclopropane	Floeder et al. (1985)
Q_a	0–15	Propylene	Rutkowsky et al. (1980)
Q_{ex}	0–16	Propylene	Dance and Walker (1973)
Q_{ex}	0–15	Propylene	Bowman and Miller (1965)
Q_i	600–12,000	Propylene	Schram et al. (1966)
Quantity	***E/N* Range (Td)**	**Gas**	**Reference**
		Swarm Properties	
W, D_L, D_r	0.02–8.0	Propylene, cyclopropane	Schmidt and Roncossek (1992)

TABLE 111.4 (continued)
Selected References for Data

Quantity	*E/N* Range (Td)	Gas	Reference
W	0.01–10	Cyclopropane	Gee and Freeman (1983)
α	120–6000	Propylene	Heylen (1978)
W	0.3–30	Propylene	Bowman and Gordon (1967)

Note: Q_a = attachment; Q_{el} = elastic; Q_{ex} = excitation; Q_i = ionization; Q_M = momentum transfer; Q_T = total; Q_v = vibrational; D_L = longitudinal diffusion; D_r = radial diffusion; W = drift velocity; α = ionization coefficient.

TABLE 111.5
Total Scattering Cross Sections

Energy (eV)	Propylene Q_T (10^{-20} m^2)	Cyclopropane Q_T (10^{-20} m^2)	Energy (eV)	Propylene Q_T (10^{-20} m^2)	Cyclopropane Q_T (10^{-20} m^2)
0.5	17.5	18.9	9.0	43.7	36.1
0.6	18	18.8	9.5	44	36.3
0.8	19	19	10.0	43.7	36.1
1	19.2	19.4	11	43.3	35.9
1.2	20.1	19.5	12	42.4	35.5
1.4	22.2	19.8	14	41.6	34.4
1.6	25.9	20.2	16	40.6	33.9
1.8	30.1	20.3	18	38.8	33.3
2.0	32.6	20.6	20	37.6	32.3
2.2	33.4	20.7	22.5	36.4	
2.4	32.3	20.8	25	35.9	31.3
2.6	31.3	21	30	34.1	30.6
2.8	29.4	20.9	35	32.9	29.6
3.0	27.6	20.8	40	31.3	28.6
3.2	27.6	20.7	45	30.4	28.1
3.5	27.4	20.7	50	29.1	27.4
3.7	27.9	21.2	60	27.2	25.8
4.0	28.2	22.3	70	25.9	24.8
4.5	30.2	24.8	80	24.4	23.1
5	32.1	28.9	90	23.2	22.3
5.5	35.1	31.3	100	22.3	21.0
5.7		32.1	110	21.4	20.3
6.0	36.8	32.5	120	20.5	19.6
6.2		32.7	140	18.7	18.0
6.5	38.7	33.0	160	17.2	17.0
6.7		33.6	180	16.3	15.8
7.0	40.4	33.9	200	15.2	14.9
7.2		34.1	220	14.6	14.2
7.5	41.4	34.8	250	13.5	13.3
7.7		35.2	275	12.5	12.4
8.0	42.7	35.4	300	11.9	11.8
8.2		35.5	350	10.6	10.6
8.5	43.6	35.7	370	10.1	10.0

Source: Adapted from Szmytkowski C. and S. Kwitnewski, *J. Phys. B: At. Mol. Opt. Phys.*, 35, 2613, 2002.

111.2 TOTAL SCATTERING CROSS SECTION

The highlights of total scattering cross sections in C_3H_6 and c-C_3H_6 are

1. A deep Ramsauer–Townsend minimum at 0.1 eV (see Lunt et al., 1998).
2. Two distinct maxima, at 2.2 and 9.5 eV for C_3H_6 and 2.6 and 9.5 eV for c-C_3H_6. The low-energy maximum is possibly due to capture of an electron and a short-lived negative ion (shape resonance). The process leads to vibrational excitation of the molecule. The second broad maximum occurs just below the ionization potential for both molecules, and it is associated mainly with elastic scattering.
3. The total cross section curve of C_3H_6 lies above that of c-C_3H_6 due to isomeric effect.

Table 111.5 gives the total cross sections in both gases. See Figures 111.1 and 111.2 for graphical presentation.

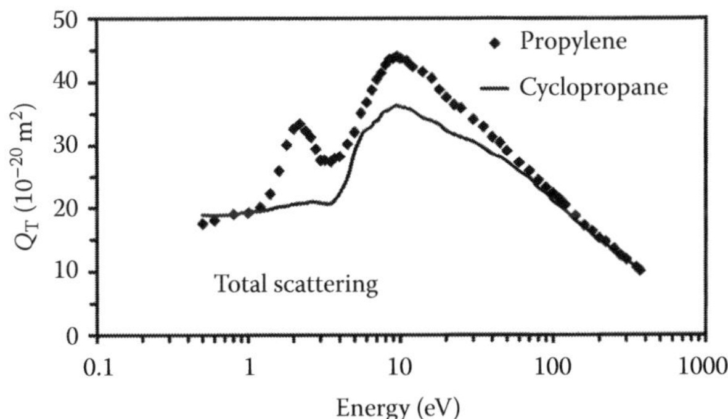

FIGURE 111.1 Total scattering cross sections in C_3H_6 and c-C_3H_6 demonstrating isomer effect. (◊) C_3H_6; (—) c-C_3H_6. (Adapted from Szmytkowski, C. and S. Kwitnewski, *J. Phys. B: At. Mol. Opt. Phys.*, 35, 2613, 2002.)

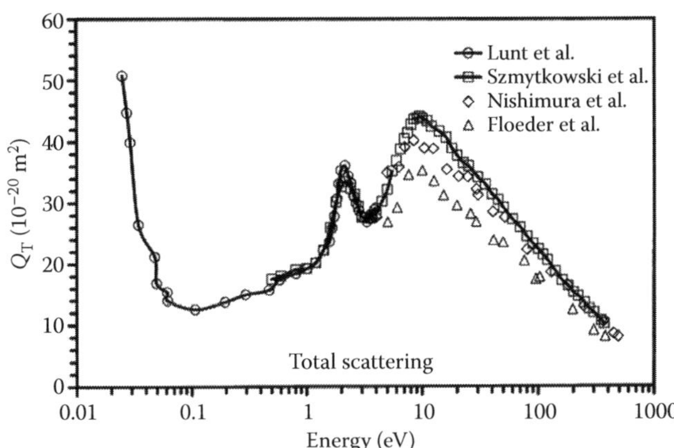

FIGURE 111.2 Total scattering cross section in C_3H_6. (○—○) Lunt et al. (1998), arbitrary cross sections are normalized to Szmytkowski and Kwitnewski (2002) at 4 eV; (—□—) Szmytkowski and Kwitnewski (2002); (◊) Nishimura and Tawara (1991); (D) Floeder et al. (1985). The large cross section in the 30–40 meV range is attributed to rotational inelastic scattering.

111.3 ELASTIC SCATTERING CROSS SECTION

Recent experimental results of Makochekanwa et al. (2006) are the only data available (Table 111.6).

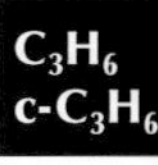

TABLE 111.6
Integral Elastic Scattering and Momentum Transfer Cross Sections in C_3H_6 and c-C_3H_6

Energy (eV)	Q_{el} (10^{-20} m^2)	Q_M (10^{-20} m^2)	Energy (eV)	Q_{el} (10^{-20} m^2)	Q_M (10^{-20} m^2)
	Propylene			Cyclopropane	
1.5	20.9	21.5	1.5	17.7	19.1
1.8	22.8	21.8	2.0	17.2	17.6
2.0	28.8	25.8	3.0	17.8	17.1
2.3	25.7	22.4	4.0	19.5	18.5
2.6	26.8	24.4	5.0	28.9	26.4
3.0	20.9	18.9	7.0	30.8	24.7
5.0	25.6	20.4	10.0	30.0	21.3
8	28.6	19.9	12	28.0	17.7
10	33.5	19.1	15	27.2	15.4
20	25.1	13.7	20	25.5	11.5
30	19.8	8.9	25	23.8	10.6
60	11.6	8.1	30	22.6	9.8
100	6.1	4.3	60	11.8	8.6
			100	6.4	4.7

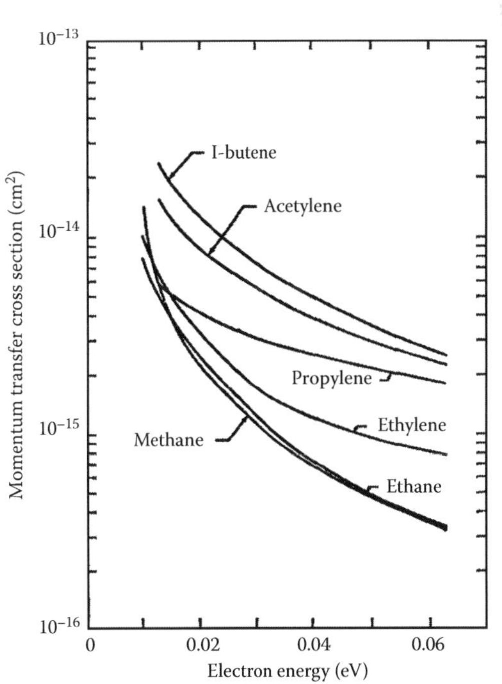

FIGURE 111.3 Momentum transfer cross sections in C_3H_6 and selected hydrocarbons derived from drift velocity by Bowman and Gordon (1967).

C_3H_6
$c\text{-}C_3H_6$

111.4 MOMENTUM TRANSFER CROSS SECTION

Low-energy momentum transfer cross sections derived from drift velocity studies are given by Bowman and Gordon (1967). These are shown in Figure 111.3 and Table 111.7. The highlight is the increasing momentum transfer cross section with decreasing electron energy due to low-lying rotational excitation levels.

Gee and Freeman (1983) have derived, from drift velocity studies, the relationship for momentum transfer cross section

$$Q_M \propto \varepsilon^{-0.2} \quad (0.03 \leq \varepsilon \leq 0.15 \text{ eV}) \quad (111.1)$$

TABLE 111.7
Momentum Transfer Cross Section in C_3H_6[a]

Energy (eV)	0.01	0.02	0.03	0.04	0.05	0.06
Q_M (10^{-20} m²)	60.54	39.98	28.74	24.11	20.61	18.23

[a] Derived by Bowman and Gordon (1967).

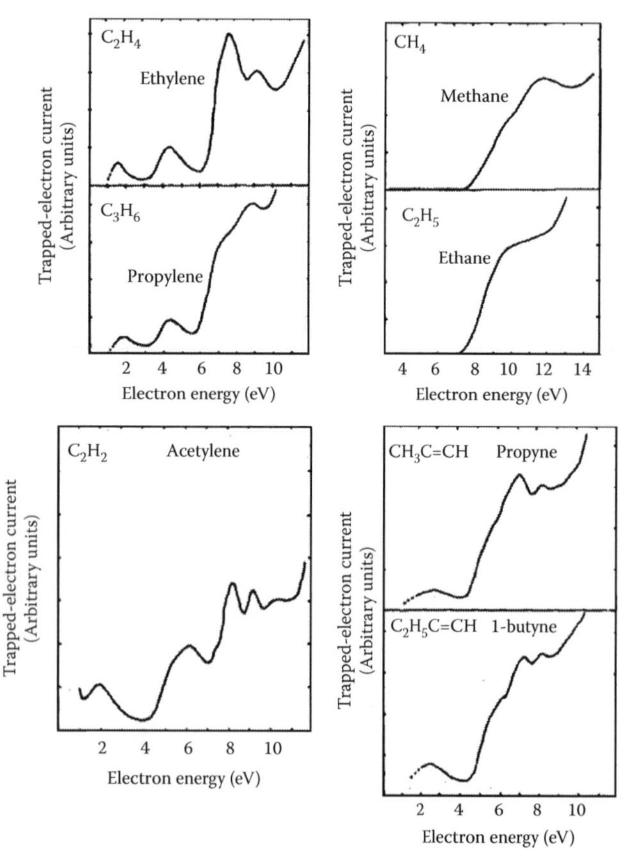

FIGURE 111.4 Electron impact excitation cross section (arbitrary units) in polypropylene and selected hydrocarbons. Top left: ethylene and propylene; top right: methane and ethane; lower left: acetylene; lower right: propyne and 1-butyne. (Adapted from Bowman, C. R. and W. D. Miller, *J. Chem. Phys.*, 42, 681, 1965.)

The derived values of the exponent for selected hydrocarbons, by these authors, are shown in Table 138.4 in the chapter on pentane.

111.5 EXCITATION CROSS SECTION

Only relative cross sections in several hydrocarbons measured by trapped electron technique at low energies (Bowman and Miller, 1965) are available as shown in Figure 111.4.

111.6 IONIZATION CROSS SECTION

Ionization cross sections in C_3H_6 and c-C_3H_6 are shown in Table 111.8. Graphical presentation is given in Figure 111.5. A comparison with other selected hydrocarbons is given in Figure 111.6.

111.7 ATTACHMENT CROSS SECTION

The dissociative attachment processes in C_3H_6 are (Rutkowsky et al., 1980):

$$\text{e (between 5 and 7 eV)} + C_nH_{2n}$$
$$\rightarrow C_nH_{2n}^{-*} \rightarrow \begin{cases} C_nH_{2n}^{-} \rightarrow H^- + C_nH_{2n-1} \\ CH^- + C_{n-1}H_{2n-1} \\ CH_2^- + C_{n-1}H_{2n-2} \end{cases} \quad (111.2)$$

TABLE 111.8
Ionization Cross Sections in C_3H_6 and c-C_3H_6

Energy (eV)	Q_i (10^{-20} m²) Propylene	Q_i (10^{-20} m²) Cyclopropane	Energy (eV)	Q_i (10^{-20} m²) Propylene	Q_i (10^{-20} m²) Cyclopropane
10	0.093	0.034	175	7.83	7.29
12	0.417	0.256	200	7.34	7.25
12.5	0.498	0.312	250	6.78	6.59
15	1.26	1.04	300	5.99	5.88
17.5	2.15	1.97	350	5.48	5.34
20	3.07	2.61	400	4.95	4.88
25	4.54	4.27	450	4.66	4.56
30	5.54	5.36	500	4.50	4.33
35	6.42	6.13	600	3.96	3.77
40	7.18	6.71	700	3.61	3.51
45	7.54	7.42	800	3.18	3.12
50	8.00	7.84	900	2.98	2.77
60	8.42	8.27	1000	2.79	2.58
70	8.82	8.48	1250	2.34	2.17
80	9.04	8.83	1500	2.06	1.97
90	9.17	8.87	1750	1.87	1.76
100	9.02	8.27	2000	1.67	1.56
125	8.62	8.30	2500	1.42	1.36
150	8.14	8.10	3000	1.24	1.16

Source: Adapted from Nishimura, H. and H. Tawara, *J. Phys. D: Appl. Phys.*, 27, 2063, 1994.

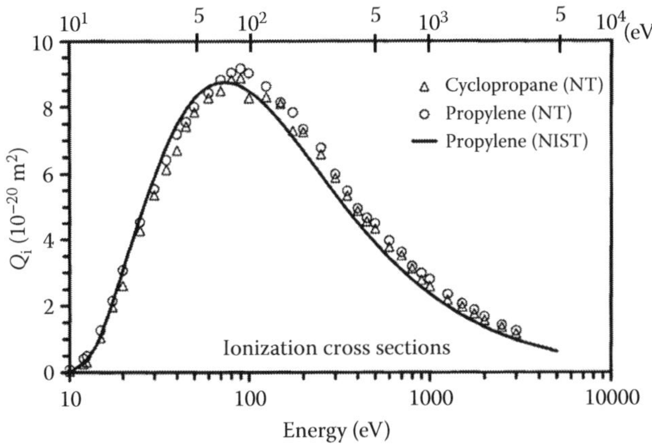

FIGURE 111.5 Ionization cross sections in C_3H_6 and c-C_3H_6. (○) C_3H_6 (NT) Nishimura and Tawara (1994); (——) C_3H_6 (NIST) Kim et al. (2004); (Δ) c-C_3H_6 (NT) Nishimura and Tawara (1994).

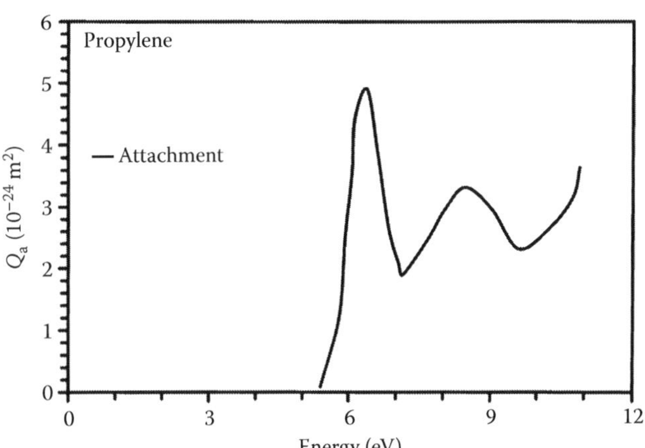

FIGURE 111.7 Dissociative attachment cross section in C_3H_6. Note the low magnitude of the cross section. (Adapted from Rutkowsky, J., H. Drost, and H.-J. Spangenberg, *Ann. Phys. Leipzig*, 37, 259, 1980.)

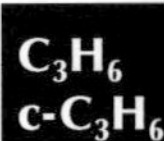

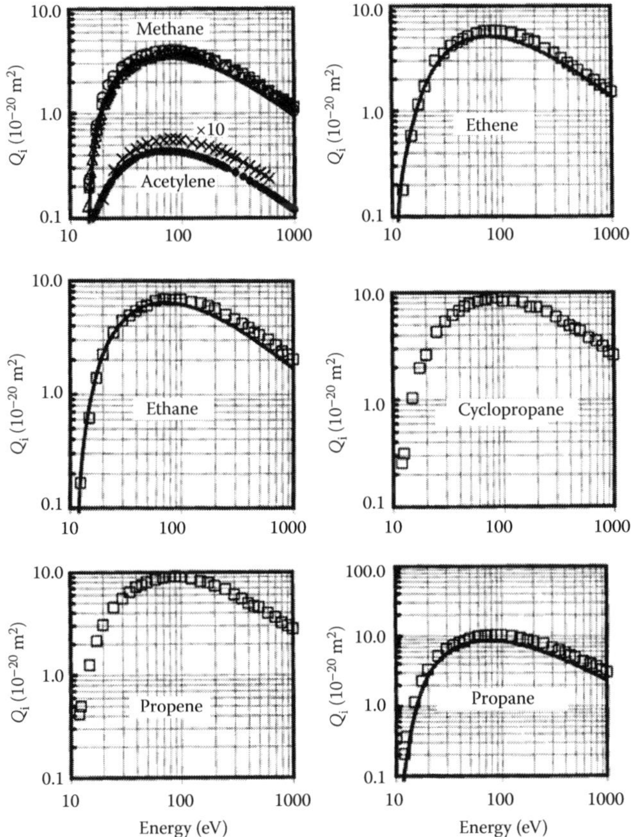

FIGURE 111.6 Ionization cross section in selected hydrocarbon gases drawn for comparison purpose. Reference for each gas is given in the relevant section.

$$e(>7\ \text{eV}) + C_nH_{2n} \rightarrow C_nH_{2n}^{-*} \rightarrow \begin{cases} C_2H^- + C_{n-2}H_{2n-1} \\ C_2H_3^- + C_{n-2}H_{n-3} \end{cases} \tag{111.3}$$

Figure 111.7 shows the attachment cross sections as a function of electron energy in C_3H_6.

TABLE 111.9
Drift Velocity of Electrons

Gee and Freeman (1983) Cyclopropane		Bowman and Gordon (1967) Propylene			
		298 K		367 K	
E/N (Td)	*W* (10^3 m/s)	*E/N* (Td)	*W* (10^3 m/s)	*E/N* (Td)	*W* (10^3 m/s)
0.01	0.26	0.9	5.00	0.9	6.42
0.02	0.52	1.2	6.26	1.2	8.79
0.04	1.05	1.5	8.43	1.6	10.43
0.06	1.57	1.8	9.85	1.8	12.86
0.09	2.33	2.1	11.51	2.4	16.39
0.18	4.48	2.4	14.21	3.1	18.44
0.36	8.28	2.8	17.23	3.8	21.54
0.5	10.7	3.4	18.09	4.9	24.26
0.9	17.5	3.7	19.73		
1.0	18.2	4.6	25.17		
1.2	20.3	6.1	33.24		
1.7	25.2	7.8	36.12		
2.6	30.7	9.3	37.27		
3.6	32.4	12.4	40.46		
5.3	31.8	15.7	39.54		
		19.0	40.07		
		21.6	40.67		
		24.9	42.75		
		30.3	42.56		

111.8 DRIFT VELOCITY

Measured drift velocities are shown in Table 111.9 and Figure 111.8. Notice the large difference between the two isomers. Additional data are given in Table 111.10.

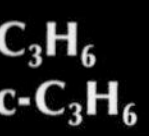

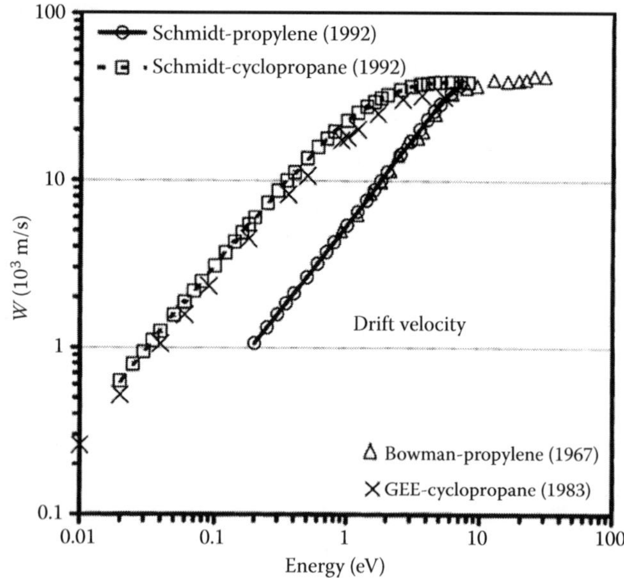

FIGURE 111.8 Drift velocity in C_3H_6 and c-C_3H_6. (D) C_3H_6, Bowman and Gordon (1967); (—○—) Schmidt and Roncossek (1992); (—□—) c-C_3H_6, Schmidt and Roncossek (1992); (×) c-C_3H_6, Gee and Freeman (1983).

TABLE 111.10
Swarm Parameters in Propene and Cyclopropane at 297 K

	W (10^4 m/s)		D_L/μ (eV)		D_r/μ (eV)	
E/N (Td)	Cyclo-propane	Propene	Cyclo-propane	Propene	Cyclopro-pane	Propene
0.020	0.063		0.0267			
0.025	0.079		0.0267			
0.030	0.094		0.0266			
0.035	0.110		0.0266			
0.040	0.125		0.0266			
0.050	0.156		0.0266		0.0259	
0.060	0.187		0.0266		0.0260	
0.070	0.218		0.0265		0.0260	
0.080	0.248		0.0265		0.0261	
0.10	0.309		0.0267		0.0263	
0.12	0.369		0.0269		0.0265	
0.14	0.429		0.0272		0.0268	
0.16	0.487		0.0274		0.0272	
0.18	0.544		0.0276		0.0276	
0.20	0.601	0.106	0.0277	0.0244	0.0280	0.0269
0.25	0.738	0.132	0.0276	0.0242	0.0290	0.0269
0.30	0.871	0.158	0.0279	0.0241	0.0301	0.0269
0.35	1.00	0.184	0.0286	0.0241	0.0313	0.0270
0.40	1.12	0.211	0.0293	0.0242	0.0326	0.0270
0.50	1.36	0.264	0.0295	0.0248	0.0353	0.0271
0.60	1.59	0.319	0.0297	0.0256	0.0381	0.0273
0.70	1.79	0.374	0.0297	0.0264	0.0408	0.0278
0.80	1.97	0.429	0.0296	0.0272	0.0433	0.0285
1.00	2.29	0.540	0.0295	0.0287	0.0472	0.0300

TABLE 111.10 (continued)
Swarm Parameters in Propene and Cyclopropane 2t 297 K

	W (10^4 m/s)		D_L/μ (eV)		D_r/μ (eV)	
E/N (Td)	Cyclo-propane	Propene	Cyclo-propane	Propene	Cyclo-propane	Propene
1.20	2.56	0.652	0.0300	0.0301	0.0501	0.0308
1.40	2.78	0.765	0.0304	0.0315	0.0529	0.0311
1.60	2.97	0.883	0.0311	0.0330	0.0556	0.0316
1.80	3.13	1.00	0.0319	0.0346	0.0585	0.0323
2.00	3.27	1.13	0.0327	0.0363	0.0614	0.0332
2.50	3.52	1.44	0.0346	0.0409	0.0687	0.0376
3.00	3.67	1.73	0.0362	0.0459	0.0762	0.0417
3.50	3.78	2.03	0.0374	0.0508	0.0840	0.0468
4.00	3.84	2.33	0.0387	0.0545		0.0515
4.50	3.89	2.62	0.0403	0.0559		0.0564
5.00	3.91	2.88	0.0422	0.0564		0.0612
6.00	3.92	3.36	0.0472	0.0585		0.0707
7.00	3.91	3.76	0.0535	0.0677		0.0803
8.00	3.89		0.0598			

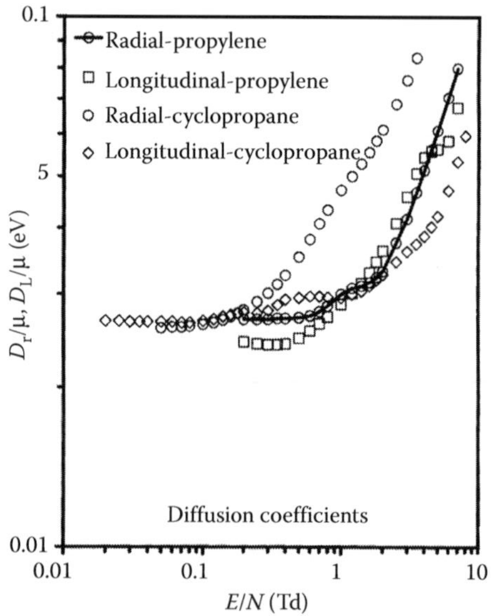

FIGURE 111.9 Radial and longitudinal diffusion coefficients, as ratios of mobility, in C_3H_6 and c-C_3H_6 as function of E/N. (Adapted from Schmidt, B. and M. Roncossek, *Aust. J. Phys.*, 45, 351, 1992.)

111.9 DIFFUSION COEFFICIENTS

Radial and longitudinal diffusion coefficients expressed as ratio of mobility (D_r/μ, D_L/μ) are shown in Table 111.10 and Figure 111.9 (see Schmidt and Roncossek, 1992). Additional data on drift velocity measured in the same apparatus are also included.

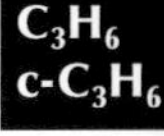

TABLE 111.11
Reduced Ionization Coefficients as a Function of *E/N* in C_3H_6

E/N (Td)	α/*N* (10^{-20} m²)	*E/N* (Td)	α/*N* (10^{-20} m²)
125	0.0012	800	1.33
150	0.0054	900	1.63
175	0.0130	1000	1.91
200	0.0262	1250	2.61
225	0.0447	1500	3.26
250	0.0692	2000	4.14
300	0.132	2500	4.93
350	0.216	3000	5.68
400	0.318	3500	6.33
450	0.424	4000	6.86
500	0.531	4500	7.26
600	0.766	5000	7.52
700	1.03	5500	7.63

Source: Adapted from Heylen, A. E. D. *Int. J. Electron.*, 44, 367, 1978.

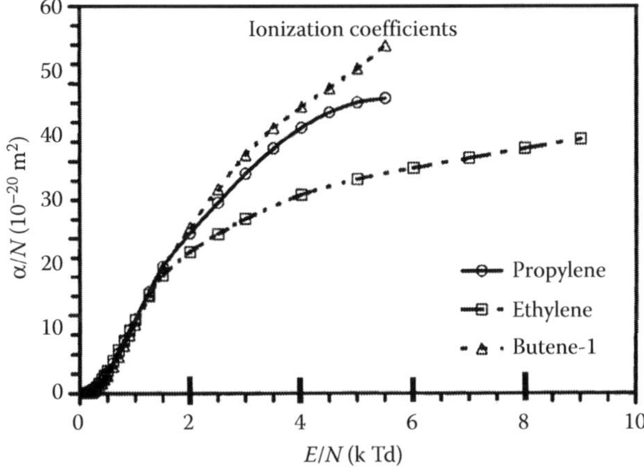

FIGURE 111.10 Reduced ionization coefficients as a function of reduced electric field in C_3H_6 and selected hydrocarbons. (—○—) C_3H_6; (—Δ—) butene-1; (— —□— —) ethylene. (Adapted from Heylen, A. E. D. *Int. J. Electron.*, 44, 367, 1978.)

111.10 FIRST IONIZATION COEFFICIENT

Reduced ionization coefficients as function of *E/N* are given in Table 111.11 and Figure 111.10.

REFERENCES

Allan, M. and L. Andric, *J. Chem. Phys.*, 105, 3559, 1996.

Bowman, C. R. and D. E. Gordon, *J. Chem. Phys.*, 46, 1878, 1967.

Bowman, C. R. and W. D. Miller, *J. Chem. Phys.*, 42, 681, 1965.

Curik, R. and F. A. Gianturco, *J. Phys. B: At. Mol. Opt. Phys.*, 35, 717, 2002a.

Curik, R. and F. A. Gianturco, *J. Phys. B: At. Mol. Opt. Phys.*, 35, 1235, 2002b.

Dance, D. F. and I. C. Walker, *Proc. Roy. Soc. Lond. A*, 334, 259, 1973.

Feil, S., A. Bacher, K. Gluch, S. Matt-Leubner, P. Scheier, and T. D. Märk, *Int. J. Mass Spectrom.*, 253, 122, 2006.

Floeder, K., D. Fromme, W. Raith, A. Schwab, and G. Sinapius, *J. Phys. B: At. Mol. Phys.*, 18, 3347, 1985.

Gee, N. and G. R. Freeman, *J. Chem. Phys.*, 78, 1951, 1983.

Gordy, W. and R. I. Cook, *Microwave Molecular Spectra*, Wiley Inter-Science, London, 1984.

Göötz, B., H. Winterling, and P. Swiderek, *J. Electron Spectrom.*, 105, 1, 1999.

Heylen, A. E. D., *Int. J. Electron.*, 44, 367, 1978.

Jiang, Y., J. Sun, and L. Wan, *J. Phys. B: At. Mol. Opt. Phys.*, 30, 5025, 1997.

Kim, Y.-K., K. K. Irikura, M. E. Rudd, M. A. Ali, P. M. Stone, J. Chang, J. S. Coursey et al., *Electron-Impact Ionization Cross Section for Ionization and Excitation Database (version 3.0)*, National Institute of Standards and Technology, Gaithesberg, MD, USA, 2004. Available at http://Physics.nist.gov./ionxsec.

Levin, I. W. and R. A. R. Pearce, *J. Chem. Phys.*, 69, 2196, 1978.

Lunt, S. L., J. Randell, J.-P Ziesel, G. Mrotzek, and D. Field, *J. Phys. B: At. Mol. Phys.*, 31, 4225, 1998.

Makochekanwa, C., H. Kato, M. Hoshino, H. Tanaka, H. Kubo, M. H. F. Bettega, A. R. Lopes, M. A. P. Lima, and L. G. Ferreira, *J. Chem. Phys.*, 124, 024323, 2006.

Nishimura, H. and H. Tawara, *J. Phys. D: Appl. Phys.*, 24, L363, 1991.

Nishimura, H. and H. Tawara, *J. Phys. D: Appl. Phys.*, 27, 2063, 1994.

Rutkowsky, J., H. Drost, and H.-J. Spangenberg, *Ann. Phys. Leipzig*, 37, 259, 1980.

Schmidt, B. and M. Roncossek, *Aust. J. Phys.*, 45, 351, 1992.

Schram, B. L., M. J. Van der Wiel, F. J. de Heer, and H. R. Moustafa, *J. Chem. Phys.*, 44, 49, 1966.

Szmytkowski, C. and S. Kwitnewski, *J. Phys. B: At. Mol. Opt. Phys.*, 35, 2613, 2002.

Section X

10 ATOMS

112

ACETONE

CONTENTS

Acetone (C_3H_6O), synonym 2-propanone, is a powerful solvent the molecule of which has 32 electrons. The molecule has a polarizability of 7.04×10^{-40} F m^2, a dipole moment of 2.88 D, and an ionization potential of 9.70 eV. The molecule has 24 vibrational modes (Shimanouchi, 1972) as shown in Table 112.1.

112.1 TOTAL SCATTERING CROSS SECTION

Measured total scattering cross sections are given by Kimura et al. (2000) and shown in Table 112.2. Figure 112.1 provides a graphical presentation. Highlights of the cross section are

1. There are two peaks, at 1.5 and 8.0 eV.
2. The 8.0 eV peak is attributed to shape resonance, common to all gases.
3. The 1.5 eV peak tends to saturation on the lower energy side (<1.5 eV) over a short energy interval. In view of lack of data, it is not possible to comment on the zero energy cross section.
4. At energies >8.0 eV, the cross section decreases monotonically, as found in majority of gases.

TABLE 112.1
Vibrational Modes and Energies of C_3H_6O

Symbol	Deformation	Energy (meV)	Symbol	Deformation	Energy (meV)
v_1	CH_3 d-stretch	374.3	v_{13}	CH_3 d-stretch	374.3
v_2	CH_3 s-stretch	364.1	v_{14}	CH_3 s-stretch	364.1
v_3	CO stretch	214.6	v_{15}	CH_3 d-deform	174.8
v_4	CH_3 d-deform	177.9	v_{16}	CH_3 s-deform	169.1
v_5	CH_3 s-deform	169.1	v_{17}	CC stretch	150.8
v_6	CH_3 rock	132.2	v_{18}	CH_3 rock	10.0
v_7	CC stretch	96.3	v_{19}	CO-ip bend	65.7
v_8	CCC deform	47.7	v_{20}	CH_3 d-stretch	368.5
v_9	CH_3 d-stretch	367.4	v_{21}	CH_3 d-deform	180.3
v_{10}	CH_3 d-deform	176.8	v_{22}	CH_3 rock	135.3
v_{11}	CH_3 rock	108.7	v_{23}	CO-op bend	60.0
v_{12}	CH_3 torsion	13.0	v_{24}	Torsion	13.5

Note: a = asymmetrical; d = degenerate; ip = in plane; op = out of plane; s = symmetrical.

TABLE 112.2
Total Scattering Cross Sections in C_3H_6O

Energy (eV)	Q_T (10^{-20} m^2)	Energy (eV)	Q_T (10^{-20} m^2)	Energy (eV)	Q_T (10^{-20} m^2)
0.8	52.24	6.5	54.07	22	43.26
1.0	52.03	7.0	55.31	24	42.42
1.2	52.02	7.5	56.14	26	41.59
1.4	52.41	8.0	56.26	28	40.60
1.6	52.84	8.5	55.85	30	39.50
1.8	49.94	9.0	55.11	35	37.26
2.0	47.55	9.5	54.25	40	36.02
2.2	46.76	10.0	53.47	50	31.25
2.5	45.38	11	52.22	60	31.23
2.8	43.17	12	51.25	70	29.42
3.1	48.15	13	49.72	80	27.43
3.4	46.98	14	48.69	90	25.63
3.7	47.64	15	47.50	100	24.50
4.0	48.69	16	46.42	120	22.48
4.5	49.71	17	46.16	150	20.34
5.0	51.51	18	45.88	200	17.37
5.5	53.23	19	45.32	250	15.22
6.0	53.75	20	44.60	300	13.82

Source: Digitized from Kimura, M. et al., *Adv. Chem. Phys.*, 111, 537, 2000.

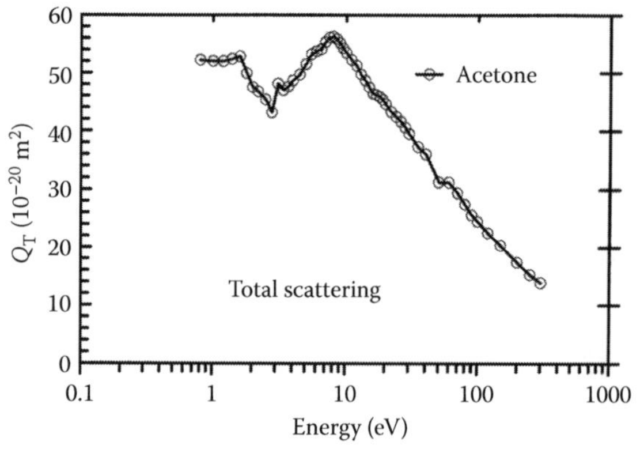

FIGURE 112.1 Total scattering cross section in C_3H_6O. (Adapted from Kimura, M. et al., *Adv. Chem. Phys.*, 111, 537, 2000.)

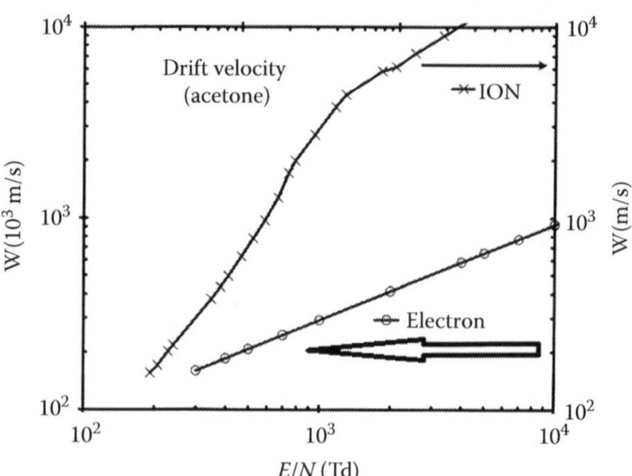

FIGURE 112.2 Experimental drift velocity of electrons and ions in C_3H_6O. Note the ordinate for each species. The electron velocity is two orders of magnitude higher. (Adapted from Schlumbohm, H. Z., *Phys.*, 182, 317, 1965.)

TABLE 112.3
Measured Electron Drift Velocity in Selected Gases

Gas	Formula	E/N Range (Td)	C_1 (ms^{-1} Td$^{-\nu}$)	ν-Equation 112.1
Acetone	C_3H_6O	600–12,000	9.24×10^3	0.50
Benzene	C_6H_6	600–9000	9.58×10^3	0.50
Carbon dioxide	CO_2	450–6000	8.08×10^3	0.591
Diethyl ether	$C_4H_{10}O$	600–15,000	5.04×10^3	0.555
Methane	CH_4	360–3000	2.45×10^4	0.758
Methylal	$C_3H_8O_2$	300–9000	3.06×10^3	0.635
Nitrogen	N_2	360–9000	1.87×10^4	0.50
Oxygen	O_2	300–24,000	2.13×10^4	0.50

Source: Adapted from Schlumbohm, H. Z., *Phys.*, 182, 317, 1965.

112.2 ELECTRON DRIFT VELOCITY

Schlumbohm (1965) has measured the drift velocity of electrons in several gases and obtained the relationship

$$W_e = C_1\left(\frac{E}{N}\right)^{\nu} \quad (112.1)$$

in which C_1 and ν are constants and *E*/*N* is the reduced electric field expressed in Td.

Table 112.3 provides electron drift velocity in selected gases according to Equation 112.1. Figure 112.2 provides a graphical presentation.

112.3 POSITIVE ION DRIFT VELOCITY

Schlumbohm (1965) has measured the drift velocity of positive ions in C_3H_6O and obtained the relationship

$$W_+ = C_2\left(\frac{E}{N}\right)^{\gamma} \quad (112.2)$$

in which C_2 and γ are constants. Table 112.4 lists the constants in selected gases. Figure 112.2 also includes a graphical presentation for C_3H_6O.

TABLE 112.4
Measured Positive Ion Drift Velocities in Selected Gases

Gas	Formula	E/N Range (Td)	C_2 (ms^{-1} Td$^{-\gamma}$)	γ-Equation 112.2
Acetone	C_3H_6O	270–900	0.563	1.0
Acetone	C_3H_6O	4500–30,000	62.48	0.5
Benzene	C_6H_6	450–1800	0.868	1.0
Benzene	C_6H_6	3000–60,000	43.66	0.5
Carbon dioxide	CO_2	135–660 (S1962)	2.78	1.0
Carbon dioxide	CO_2	1500–15,000	96.40	0.5
Cyclohexane	C_6H_{12}	300–900	1.12	1.0
Cyclohexane	C_6H_{12}	1500–39,000	49.33	0.5
Diethyl ether	$C_4H_{10}O$	300–4200	0.916	1.0
Diethyl ether	$C_4H_{10}O$	5400–48,000	62.38	0.5
Freon-12	CCl_2F_2	360–420 (S1962)	1.65	1.0
Hydrogen	H_2	600–1200	0.322	1.0
Methane	CH_4	150–600	6.75	1.0
Methane	CH_4	900–15,000	181.5	0.5
Methylal	$C_3H_8O_2$	270–870	1.08	1.0
Methylal	$C_3H_8O_2$	1800–30,000	59.54	0.5
Nitrogen	N_2	900–12,000	62.38	0.5
Oxygen	O_2	600–60,000	82.79	0.5
Sulfur dioxide	SO_2	330–360 (S1962)	0.97	1.0

Source: Adapted from Schlumbohm, H. Z., *Phys.*, 182, 317, 1965.

REFERENCES

Kimura, M., O. Sueoka, A. Hamada, and Y. Itikawa, *Adv. Chem. Phys.*, 111, 537, 2000.

Schlumbohm, H. Z. *Phys.*, 166, 192, 1962.

Schlumbohm, H. Z. *Phys.*, 182, 317, 1965.

Shimanouchi, T., *Tables of Molecular Vibrational Frequencies Consolidated Volume I*, NSRDS-NBS, 39, Washington (DC), 1972.

113

CYCLOBUTENE, 1,3-BUTADIENE, AND 2-BUTYNE

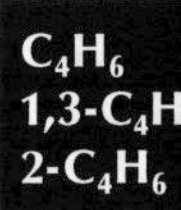

CONTENTS

Cyclobutene (C_4H_6) is a molecule with 30 electrons and there are five compounds with the same chemical formula. Selected properties of the molecules are shown in Table 113.1.

The molecules each have large number of vibrational modes. 1,3-Butadiene (1,3-C_4H_6) has 24 modes and 2-butyne (2-C_4H_6) has 16 modes (Shimanouchi, 1972). According to Allan and Andric (1996) vibrations of similar type (local motion) tend to have similar frequencies (energies). See Table 113.3 under propylene for general designation of modes of vibration. Excitation potentials of 1-butyne (1-C_4H_6) are 2.4, 6.3, 7.3, and 8.2 eV (Bowman and Miller, 1965). Table 113.2

TABLE 113.1
Selected Properties of Isomers of C_4H_6

Molecule	α_e (10^{-40} F m²)	μ (D)	ε_i (eV)	ε_d (eV)
Cyclobutene		0.132		3.79
1-Butyne		0.782	10.19	
2-Butyne	8.24		9.59	
1,2-Butadiene			9.03	
1,3-Butadiene	9.62	0.403	9.082	

Note: α_e = electronic polarizability; μ = dipole moment; ε_i = ionization energy; ε_d = dissociation energy.

TABLE 113.2
Vibrational Modes and Energies of C_4H_6

	1,3-C_4H_6			2-C_4H_6	
Mode	Type of Motion	Energy (meV)	Mode	Type of Motion	Energy (meV)
ν_1	CH_2 a-stretch	382.7	ν_1	CH_3 s-stretch	361.5
ν_2	CH stretch	372.3	ν_2	C≡C stretch	277.7
ν_3	CH_2 s-stretch	371.0	ν_3	CH_3 s-deform.	171.1
ν_4	C=C stretch	202.1	ν_4	C-C stretch	89.9
ν_5	CH_2 scissoring	178.3	ν_5	CH_3 torsion	*
ν_6	CH bending	158.7	ν_6	CH_3 s-stretch	364.3
ν_7	C–C stretch	148.3	ν_7	CH_3 s-deform.	171.3
ν_8	CH_2 rock	110.8	ν_8	C–C stretch	142.8
ν_9	CCC deform.	63.5	ν_9	CH_3 d-stretch	368.6
ν_{10}	CH bend	125.6	ν_{10}	CH_3 d-deform.	180.5
ν_{11}	CH_2 wag	112.6	ν_{11}	CH_3 rock	130.7
ν_{12}	CH_2 twist	64.7	ν_{12}	CCC deform.	26.4
ν_{13}	C–C torsion	20.1	ν_{13}	CH_3 d-stretch	367.7
ν_{14}	CH bend	121.0	ν_{14}	CH_3 d-deform	179.5
ν_{15}	CH_2 wag	113.1	ν_{15}	CH_3 rock	127.6
ν_{16}	CH_2 twist	95.5	ν_{16}	CCC deform	46.0
ν_{17}	CH_2 a-stretch	384.5			

continued

TABLE 113.2 (continued)
Vibrational Modes and Energies of C_4H_6

	1,3-C_4H_6			2-C_4H_6	
Mode	Type of Motion	Energy (meV)	Mode	Type of Motion	Energy (meV)
ν_{18}	CH stretch	378.8			
ν_{19}	CH_2 s-stretch	370.0			
ν_{20}	C=C stretch	197.7			
ν_{21}	CH_2 scissoring	171.2			
ν_{22}	CH bend	160.4			
ν_{23}	CH_2 rock	122.7			
ν_{24}	CCC deform.	373.0			

Source: Adapted from Shimanouchi, T. *Tables of Molecular Vibrational Frequencies Consolidated Volume 1*, NSRDS-NBS, 39, 1972.
Note: a = asymmetrical; d = degenerate; s = symmetrical; * = free rotation.

TABLE 113.3
Total Scattering Cross Sections for 1,3-C_4H_6 and 2-C_4H_6

Energy	Q_T (10^{-20} m²)		Energy	Q_T (10^{-20} m²)	
(eV)	1,3-C_4H_6	2-C_4H_6	(eV)	1,3-C_4H_6	2-C_4H_6
0.5	37.1		8.5	50.4	51.2
0.6	37.5	18.2	9.0	51.2	50.3
0.7	37.3		9.5	51.3	49.5
0.8	37.6	18.2	10	51.0	49.0
0.9	37.7		11	49.9	48.7
1.0	37.1	18.8	12	48.8	48.0
1.1	36.5		14	47.4	46.5
1.2	35.7	19.1	16	45.5	45.0
1.4	34.9	19.6	18	44.0	43.5
1.8	35.5	21.6	20	43.0	43.0
2.0	36.5	23.1	25	41.0	40.4
2.3	38.3		30	39.1	39.2
2.5	40.4		35	37.7	37.5
2.7	41.8		40	36.8	35.9
2.9	43.0		50	33.8	33.2
3.0	43.6	39.1	60	31.9	31.7
3.1	43.9		70	30.3	30.3
3.2	44.0	41.3	80	28.6	28.8
3.3	43.8		90	27.4	27.1
3.4	43.6	41.8	100	25.9	25.6
3.5	43.1		110	25.1	24.5
3.7	42.5		120	23.2	23.2
4.0	41.7	39.8	140	22.0	21.5
4.2	41.2	39.0	160	19.8	20.2
4.5	41.3	38.5	180	18.3	18.1
4.7	41.7	38.9	200	17.0	16.8
5.0	43.1	41.0	220	16.1	15.9
5.5	45.0	44.7	250	13.9	13.5
6.0	46.6	47.3	275	12.5	12.5
6.5	47.9	49.3	300	11.3	11.2
7.0	48.8	50.8	350	9.80	9.88
7.5	49.4	51.2	370	9.25	9.22
8.0	49.5	51.4			

Source: Adapted from Szmytkowski, Cz. and S. Kwitnewski, *J. Phys. B: At. Mol. Opt. Phys.*, 36, 2129, 2003.

presents the vibrational modes and energies. Note that all CH stretch motion has energy of approximately 360 eV, and the lowest energy is C–C torsion at 20.1 meV.

113.1 TOTAL SCATTERING CROSS SECTIONS OF 1,3-C_4H_6

Total scattering cross sections for 1,3-C_4H_6 and 2-C_4H_6 have been measured by Szmytkowski and Kwitnewski (2003) as shown in Table 113.3 and Figure 113.1. The essential features of the cross sections are

1. A small increase in the total cross section at 1.0 eV, attributed to the formation of temporary negative ion.
2. A peak in the cross section in the vicinity of 3.2 eV, also attributed to the formation of temporary negative ion.
3. A relatively broad peak at 9.5 eV attributed to elastic scattering with contributions from weak inelastic components (Szmytkowski and Kwitnewski, 2003). This feature is common to several gases.
4. A monotonic decrease for energies >10 eV. This feature is also common to several gases.
5. Evidence that the cross section decreases for energies <0.5 eV (Szmytkowski and Kwitnewski, 2003).

113.2 TOTAL SCATTERING CROSS SECTIONS OF 2-C_4H_6

Total scattering cross sections for 2-C_4H_6 have been measured by Szmytkowski and Kwitnewski (2003) as shown in Table 113.3 and Figure 113.2. The essential features are

1. A shape resonance at approximately 3.2 eV.
2. A distinct peak at 8.0 eV which is common to many gases.
3. A monotonic decrease for energies >10 eV, which is also common to many gases.

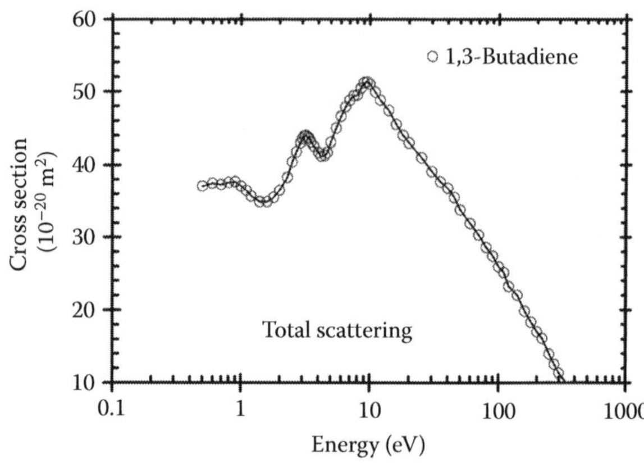

FIGURE 113.1 Total scattering cross sections for 1,3-C_4H_6. (Adapted from Szmytkowski, Cz. and S. Kwitnewski, *J. Phys. B: At. Mol. Opt. Phys.*, 36, 2129, 2003.)

113.3 POSITIVE ION APPEARANCE POTENTIALS

Franklin and Mogenis (1967) have reported relative abundance and appearance potentials of principle positive ions for 1,3-C_4H_6.

113.4 IONIZATION CROSS SECTIONS OF 1,3-C_4H_6

Ionization cross section of 1,3-C_4H_6 in the high-energy range are given by Schram et al. (1966) as shown in Table 113.4.

113.5 ATTACHMENT CROSS SECTIONS OF 1,3-C_4H_6

Rutkowsky et al. (1980) have measured the dissociative attachment cross sections for 1,3-C_4H_6. Attachment occurs according to the reactions:

$$e + C_4H_6 \rightarrow \begin{cases} C_4H_6^{-}* \rightarrow H - + C_4H_5 \\ C_2^{-} + C_2H_6 \end{cases} \quad (113.1)$$

The measured dissociative attachment cross sections are shown in Figure 113.3. Note the low values of the ordinate.

113.6 ADDENDUM

The study by Lopes et al. (2004) has been found after completion of this section. The integral elastic, momentum transfer and ionization cross sections for isomers of C_4H_6 have been calculated in the energy range, 10–60 eV that partially covers the gap in the cross sections. Figures 113.4, 113.5, and 113.6 show these data.

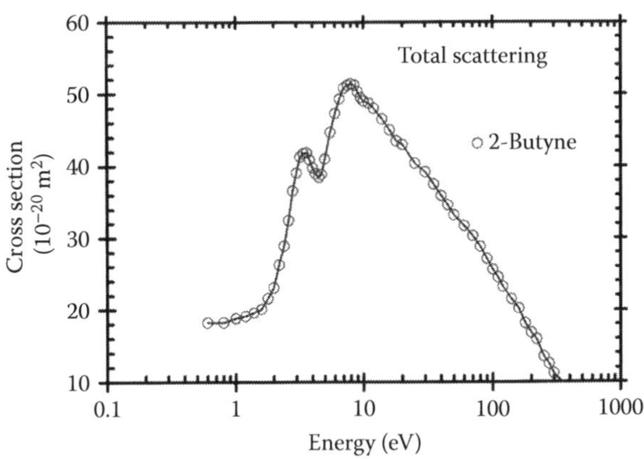

FIGURE 113.2 Total scattering cross sections for 2-C_4H_6. (Adapted from Szmytkowski, Cz. and S. Kwitnewski, *J. Phys. B: At. Mol. Opt. Phys.*, 36, 2129, 2003.)

TABLE 113.4
Ionization Cross Sections in 1,3-C_4H_6

Energy (keV)	0.6	1	2	4	7	12
Q_i (10^{-20} m^2)	4.49	3.06	1.81	1.06	0.665	0.421

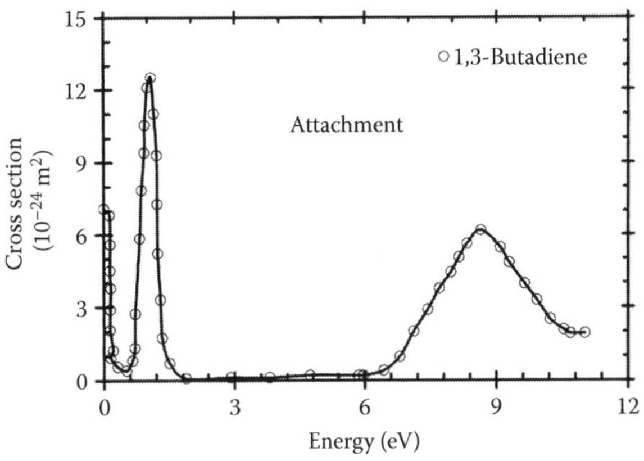

FIGURE 113.3 Dissociative attachment cross sections for 1,3-C_4H_6. (Adapted from Rutkowsky, J., H. Drost, and H.-J. Spangenberg, *Ann. Phys. (Leipzig)*, 37, 259, 1980.)

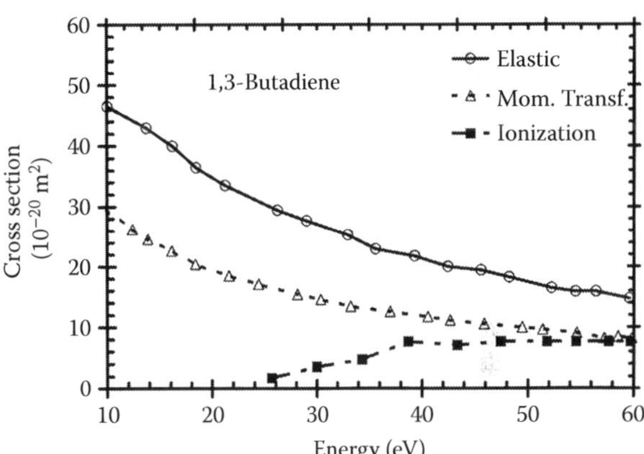

FIGURE 113.4 Cross sections for 1,3-C_4H_6. (—○—) Integral elastic; (- - Δ- -) momentum transfer; (—■—) ionization cross sections. (Adapted from Lopes, A. R. et al., *Phys. Rev.*, 69, 01472, 2004.)

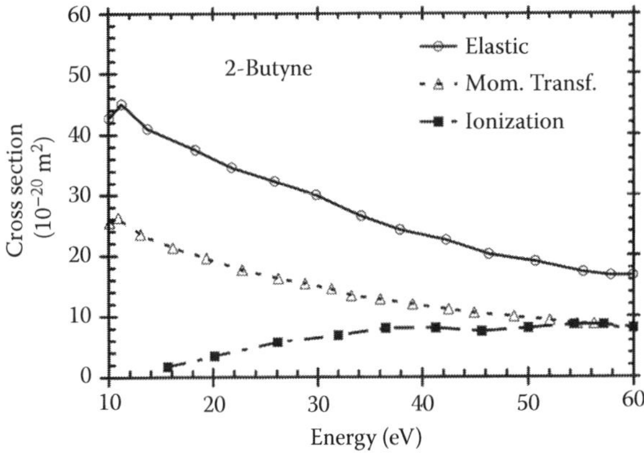

FIGURE 113.5 Cross sections for 2-C_4H_6. (—○—) Integral elastic; (- - Δ- -) momentum transfer; (—■—) ionization cross sections. (Adapted from Lopes, A. R. et al., *Phys. Rev.*, 69, 01472, 2004.)

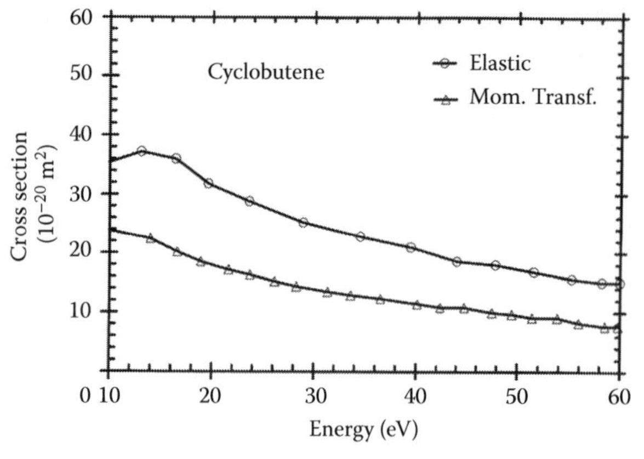

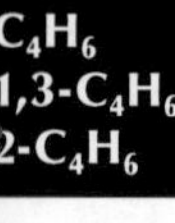

FIGURE 113.6 Cross sections for C_4H_6. (—○—) Integral elastic; (- Δ- -) momentum transfer cross sections. (Adapted from Lopes, A. R. et al., *Phys. Rev.*, 69, 01472, 2004.)

REFERENCES

Allan, M. and L. Andric, *J. Chem. Phys.*, 105, 3559, 1996.

Bowman, C. R. and W. D. Miller, *J. Chem. Phys.*, 42, 681, 1965.

Franklin, J. L. and A. Mogenis, *J. Phys. Chem.*, 71, 2820, 1967, cited by Cz. Szmytkowski and S. Kwitnewski (2003).

Lopes, A. R., M. A. P. Lima, L. G. Ferriera, and M. H. F. Bettega, *Phys. Rev.*, 69, 01472, 2004.

Rutkowsky, J., H. Drost, and H.-J. Spangenberg, *Ann. Phys. (Leipzig)*, 37, 259, 1980.

Schram, B. L., M. J. van der Wiel, F. J. de Heer, and H. R. Moustafa, *J. Chem. Phys.*, 44, 49, 1966.

Shimanouchi, T. *Tables of Molecular Vibrational Frequencies Consolidated Volume 1*, NSRDS-NBS, 39, Washington (DC), 1972.

Szmytkowski, Cz. and S. Kwitnewski, *J. Phys. B: At. Mol. Opt. Phys.*, 36, 2129, 2003.

114

HEXAFLUORO-CYCLOBUTENE, HEXAFLUORO-1, 3-BUTADIENE, AND HEXAFLUORO-2-BUTYNE

CONTENTS

Hexafluorocyclobutene (C_4F_6) is a chemical analog of cyclobutene (C_6H_6) with the hydrogen replaced with fluorine atoms. The molecule has 138 electrons and it is electron attaching. There are three compounds with the same chemical formula:

Hexafluorocyclobutene	C_4F_6
Hexafluoro-1,3-butadiene	1,3-C_4F_6
Hexafluoro-2-butyne	2-C_4F_6

114.1 SELECTED REFERENCES

See Table 114.1.

114.2 TOTAL SCATTERING CROSS SECTIONS

Total scattering cross sections measured by Szmytkowski and Kwitnewski (2003a,b) are shown in Tables 114.2, 114.3, and Figure 114.1. Essential features of the cross section variation are 2-C_4F_6 (see Szmytkowski and Kwitnewski, 2003a):

1. A peak at 8 eV for which the type of collision is not specified.
2. A second peak, broader than the first, in the energy range 25–40 eV.
3. A monotonic decrease of the cross section for energies >50 eV according to a power law, $\varepsilon^{-0.46}$.

TABLE 114.1
Selected References for C_4F_6 and Isomers

Quantity	Isomer	Range: ε (*EN*) eV, (Td)	Reference
Q_T	$2.C_4F_6$	0.6–375	**Szmytkowski and Kwitnewski (2003a)**
Q_T	$1,3\text{-}C_4F_6$	0.6–375	**Szmytkowski and Kwitnewski (2003b)**
Q_i	$2\text{-}C_4F_6$, $1,3\text{-}C_4F_6$	12.5–215	**Bart et al. (2001)**
k_a	$c\text{-}C_4F_6$	(0.2–1.0)	**Christophorou and Datskos (1995)**
k_a, Q_a	$c\text{-}C_4F_6$	0.2–1.0	**Datskos et al. (1993)**
Q_a	$2\text{-}C_4F_6$	0–0.07	**Chutjian et al. (1984)**
k_a, Q_a	$c\text{-}C_4F_6$, $1,3\text{-}C_4F_6$, $2\text{-}C_4F_6$	Thermal (0–1.5)	**Christodoulides et al. (1979)**
k_a	$c\text{-}C_4F_6$, $1,3\text{-}C_4F_6$, $2\text{-}C_4F_6$	0–12	**Sauers et al. (1979)**
k_a	$c\text{-}C_4F_6$	Thermal	**Bansal and Fessenden (1973)**

Note: k_a = Attachment rate; Q_T = total scattering; Q_i = ionization. Bold font indicates experimental data.

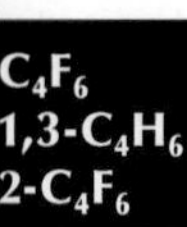

TABLE 114.2
Total Scattering Cross Sections for $2\text{-}C_4F_6$

Energy (eV)	Q_T (10^{-20} m²)	Energy (eV)	Q_T (10^{-20} m²)	Energy (eV)	Q_T (10^{-20} m²)
0.6	27.0	6.0	44.1	35	48.3
0.8	27.2	6.2	44.7	40	48.1
1.0	27.3	6.5	44.7	45	47.6
1.2	27.4	6.7	45.2	50	46.8
1.4	27.5	7.0	46.0	60	45.3
1.6	27.7	7.5	46.5	70	43.5
1.8	27.7	8.0	46.6	80	41.5
2.0	28.3	8.5	46.4	90	39.6
2.2	28.8	9.0	45.6	100	38.1
2.4	29.8	9.5	45.1	110	37.0
2.6	30.3	10	44.4	120	35.6
2.8	30.8	11	44.3	140	33.2
3.0	30.5	12	44.2	160	31.2
3.2	31.0	14	44.8	180	28.8
3.5	31.5	16	45.7	200	26.9
3.7	32.2	18	46.1	220	25.0
4.0	34.6	20	47.4	250	23.2
4.5	38.2	22.5	48.2	275	21.3
5.0	41.0	25	48.6	300	20.3
5.2	41.6	27.5	48.6	350	18.4
5.5	43.1	30	48.4	375	17.8
5.7	43.6				

Source: Szmytkowski, Cz. and S. Kwitnewski, *J. Phys B. At. Mol. Opt. Phys.*, 36, 2129, 2003a. With kind permission from Institute of Physics, England.

$1,3\text{-}C_4F_6$ (see Szmytkowski and Kwitnewski, 2003b):

1. A maximum around 1 eV attributed to the formation of temporary negative ion $C_4F_6^{-*}$ (see Section 4).

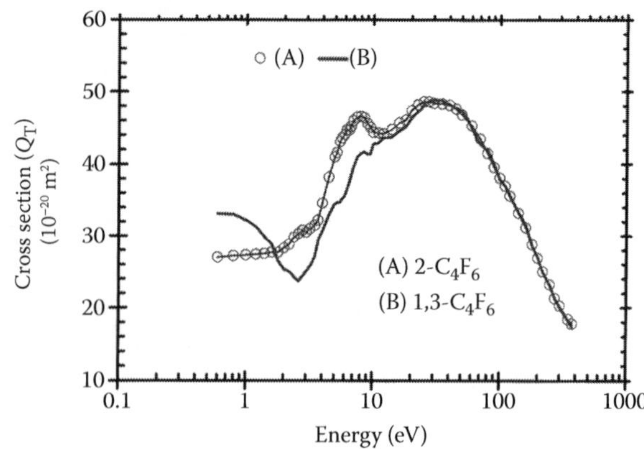

FIGURE 114.1 Total scattering cross sections for $2\text{-}C_4F_6$. (Adapted from Szmytkowski, Cz. and S. Kwitnewski, *J. Phys B. At. Mol. Opt. Phys.*, 36, 2129, 2003a.)

TABLE 114.3
Total Scattering Cross Sections for $1,3\text{-}C_4F_6$

Energy (eV)	Q_T (10^{-20} m²)	Energy (eV)	Q_T (10^{-20} m²)	Energy (eV)	Q_T (10^{-20} m²)
0.6	33.13	6.5	37.43	50	47.34
0.8	33.05	7.0	39.61	55	46.30
1.0	32.22	7.5	40.94	60	44.93
1.2	31.30	8.0	41.53	65	43.63
1.4	29.87	8.5	41.70	70	42.79
1.6	28.78	9.0	41.43	80	41.96
1.8	26.40	9.5	41.42	90	40.01
2.0	25.32	10	42.83	100	37.87
2.2	25.00	11	42.94	110	37.00
2.4	24.25	12	43.63	120	35.61
2.6	23.65	14	43.76	140	33.30
2.8	24.33	16	44.54	160	31.58
3.0	25.04	18	45.16	180	28.60
3.3	25.86	20	46.31	200	26.94
3.5	27.12	22	47.07	220	24.94
3.8	29.62	24	47.70	250	23.12
4.0	30.66	26	48.54	275	21.29
4.5	32.92	30	48.81	300	20.02
5.0	34.52	34	48.81	350	18.24
5.5	34.77	40	48.19	375	17.37
6.0	35.62	45	47.74		

Source: Digitized from Szmytkowski Cz. and S. Kwitnewski, *J. Phys. B: At. Mol. Opt. Phys.*, 36, 4865, 2003b.

2. A broad peak centered near 30 eV.
3. A deep minimum at 2.6 eV separating the two peaks.
4. For energies >30 eV, both isomers show substantially the same cross section.

114.3 IONIZATION CROSS SECTIONS

Ionization cross sections measured by Bart et al. (2001) for $2\text{-}C_4F_6$ are shown in Table 114.3. Ionization cross sections measured by Bart et al. (2001) for $1,3\text{-}C_4F_6$ are shown in Tables 114.4 and 114.5. Figure 114.2 shows these cross sections on the same plot to demonstrate, the isomer effect.

TABLE 114.4
Ionization Cross Sections for 2-C_4F_6

Energy (eV)	Q_i (10^{-20} m^2)	Energy (eV)	Q_i (10^{-20} m^2)	Energy (eV)	Q_i (10^{-20} m^2)
12.5	0.35	82	9.98	153	10.02
16	0.70	86	10.11	156	9.94
20	1.64	90	10.17	160	9.80
23	2.45	94	10.19	164	9.71
27	3.61	97	10.19	167	9.66
31	4.54	101	10.24	171	9.61
34	5.24	105	10.28	175	9.54
38	6.04	108	10.27	179	9.49
42	6.76	112	10.23	182	9.47
46	7.36	116	10.20	186	9.47
49	7.76	119	10.24	190	9.42
53	8.19	123	10.30	193	9.34
57	8.59	127	10.30	197	9.23
60	8.84	130	10.23	201	9.13
64	9.14	134	10.12	204	9.09
68	9.37	138	10.03	208	9.05
71	9.55	142	10.00	212	9.00
75	9.72	145	10.00	215	8.97
79	9.87	149	10.04		

Source: Adapted from Bart, M. et al., *Phys. Chem. Chem. Phys.*, 3, 800, 2001.
Note: Tabulated values are by courtesy of Professor Harland.

TABLE 114.5
Ionization Cross Sections for 1,3-C_4F_6

Energy (eV)	Q_i (10^{-20} m^2)	Energy (eV)	Q_i (10^{-20} m^2)	Energy (eV)	Q_i (10^{-20} m^2)
12.5	0.59	82	10.13	153	10.22
16	1.01	86	10.25	156	10.16
20	2.04	90	10.33	160	10.04
23	2.83	94	10.37	164	9.96
27	3.85	97	10.36	167	9.92
31	4.68	101	10.41	171	9.85
34	5.37	105	10.47	175	9.80
38	6.19	108	10.50	179	9.73
42	6.87	112	10.47	182	9.71
46	7.43	116	10.42	186	9.70
49	7.84	119	10.43	190	9.65
53	8.29	123	10.48	193	9.58
57	8.74	127	10.49	197	9.48
60	8.95	130	10.45	201	9.37
64	9.27	134	10.34	204	9.33
68	9.51	138	10.24	208	9.28
71	9.70	142	10.19	212	9.25
75	9.87	145	10.20	215	9.24
79	10.02	149	10.24		

Source: Adapted from Bart, M. et al., *Phys. Chem. Chem. Phys.*, 3, 800, 2001.
Note: Tabulated values are by the courtesy of Professor Harland.

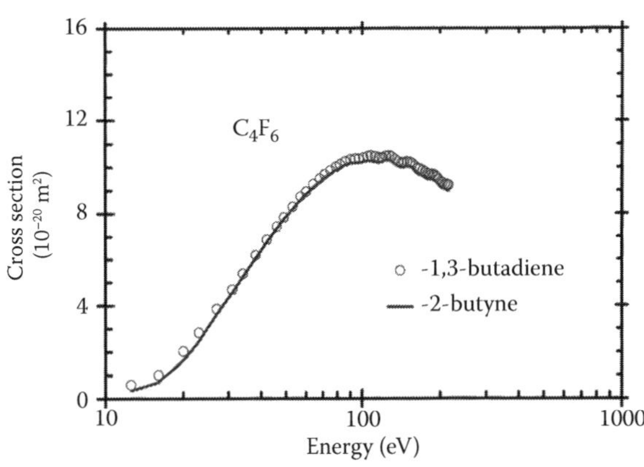

FIGURE 114.2 Ionization cross sections for 2-butyne and 1,3-butadiene. Tabulated values are by courtesy of Professor Harland. (Adapted from Bart, M. et al., *Phys. Chem. Chem. Phys.*, 3, 800, 2001.)

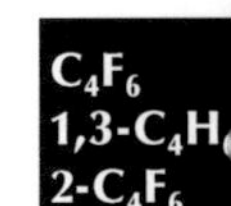

114.4 ATTACHMENT PROCESSES

The processes that yield negative ions are (Christodoulides, et al., 1979)

$$e + C_4F_6 \rightarrow C_4F_6^{-*} \quad (114.1)$$

In this reaction, the energy of the incoming electron tends to vibrationally excite the molecule. If the electron energy is zero, then the internal energy of the molecule does not change. If the negative ion formed survives for more than 1 μs, it is generally considered to be stabilized. The contribution, to the total attachment cross section, of this process are C_4F_6—97%; 2-C_4F_6—96%; and 1,3-C_4F_6—2% (Sauers et al., 1979).

$$e + C_4F_6 \rightarrow C_4F_6^{-*} \rightarrow C_4F_6^{*} + e \quad (114.2)$$

This reaction is usually referred to as autoionization, though auto detachment is a more accurate description. The electron after collision has less energy than the incoming electron.

$$e + C_4F_6 \rightarrow C_4F_6^{-*} \rightarrow C_4F_5 + F^{-} \quad (114.3)$$

This reaction is known as dissociative attachment, in which the electron attaches to the dissociation fragment, F atom.

$$e + C_4F_6 \rightarrow C_4F_6^{-*} \quad (114.4)$$

$$C_4F_6^{-*} + A \rightarrow C_4F_6^{-} + A + \text{energy} \quad (114.5)$$

where A is a neutral particle of another gas, usually nitrogen or argon, less often CO_2; it can also be a parent molecule. This reaction is known as collisional stabilization.

Negative ions of fragments, $C_4F_5^-$, $C_4F_4^-$, $C_3F_3^-$, CF_3^- with relatively smaller abundance are also formed due to dissociative attachment (Sauers et al., 1979).

114.5 ATTACHMENT CROSS SECTIONS

Low-energy attachment cross sections from Chutjian et al. (1984) are shown in Table 114.6.

Swarm unfolded attachment cross sections for C_4F_6 and isomers (Christodoulides et al., 1979) are shown in Table 114.7 and Figures 114.3 and 114.4.

TABLE 114.6
Low-Energy Attachment Cross Sections for 2-C_4F_6

Energy (meV)	5	20	45	70
Q_a (10^{-19} m^2)	77.0	3.0	1.50	0.8

Source: Adapted from Chutjian, A. et al., *J. Phys. B: At. Mol. Phys.*, 17, L745, 1984.

TABLE 114.7
Attachment Cross Sections for C_4F_6 and Isomers

	Q_a(10^{-19} m^2)		
Energy (eV)	**C_4F_6**	**1,3-C_4F_6**	**2-C_4F_6**
0.04	9.57	10.00	5.09
0.06	4.23	5.08	2.94
0.08	2.60	3.38	1.99
0.10	2.16	2.94	1.64
0.12	2.06	2.95	1.55
0.14	2.00	3.08	1.56
0.16	1.85	3.19	1.60
0.18	1.55	3.15	1.63
0.20	1.22	2.97	1.63
0.22	0.89	2.66	1.59
0.24	0.63	2.32	1.51
0.26	0.44	1.97	1.40
0.28	0.30	1.64	1.29
0.30	0.21	1.34	1.17
0.34	0.11	0.89	0.97
0.38	0.07	0.60	0.84
0.42	0.05	0.44	0.77
0.46	0.047	0.33	0.75
0.50	0.046	0.26	0.77
0.55	0.053	0.22	0.84
0.60	0.062	0.19	0.96
0.65	0.068	0.17	1.09
0.70	0.073	0.17	1.21
0.75	0.070	0.17	1.30
0.80	0.058	0.16	1.35
0.85	0.041	0.17	1.32
0.90	0.028	0.18	1.24
0.95	0.017	0.19	1.11
1.00	0.008	0.20	0.93
1.05	0.003	0.20	0.69
1.10		0.19	0.46
1.15		0.17	0.25
1.20		0.14	0.12
1.25		0.10	0.04
1.30		0.06	0.011
1.35		0.028	0.002
1.40		0.011	
1.45		0.004	

Source: Adapted from Christodoulides, A. et al., *J. Chem.. Phys.*, 70, 1156, 1979.

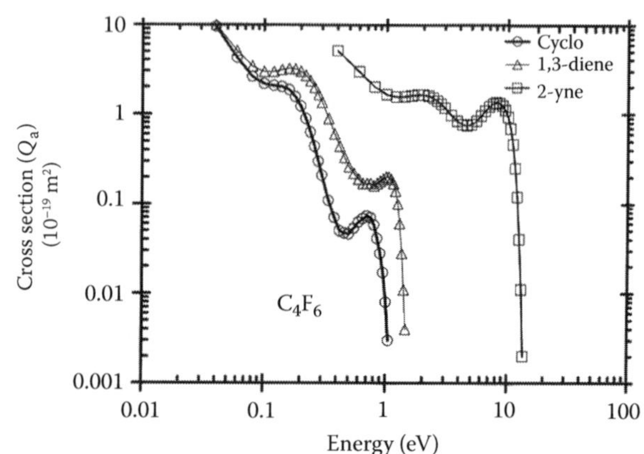

FIGURE 114.3 Attachment cross sections for hexafluorocyclobutene (c-C_4F_6), 1,3-C_4F_6, and 2-C_4F_6. The curve for the last mentioned isomer has been displaced to the right by one decade for clarity. (Adapted from Christodoulides, A. A. et al., *J. Chem. Phys.*, 70, 1156, 1979.)

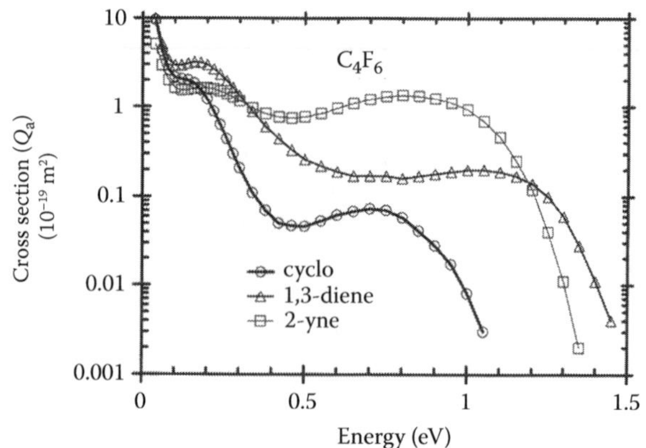

FIGURE 114.4 Attachment cross sections for hexafluorocyclobutene (c-C_4F_6), 1,3-C_4F_6, and 2-C_4F_6. Note the linear scale for the abscissa. (Adapted from Christodoulides, A. A. et al., *J. Chem. Phys.*, 70, 1156, 1979.)

The essential features of Figures 114.3 and 114.4 are

1. A very large cross section toward zero energy for all the isomers considered.

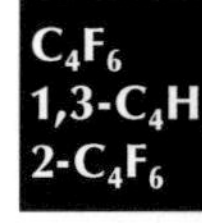

TABLE 114.8
Attachment Cross Section Details

Molecule	Ion Species							
	($C_4F_6^{-*}$)			F^-			$C_3F_3^-$	
	Threshold (eV)	Peak (eV)	Abundance (%)	Threshold (eV)	Peak (eV)	Abundance (%)	Peak (eV)	Abundance (%)
2-C_4F_6		0.0	96	3.75	5.3	0.1	1.5	2.5
c-C_4F_6		0.0	97	2.7	4.1	0.1	1.85	0.9
1,3-C_4F_6		0.0	2	0.5	0.5	32	1.45	64.5

Source: Adapted from Sauers, I., L. G. Christophorou, and J. G. Carter, *J. Chem. Phys.*, 71, 3016, 1979.

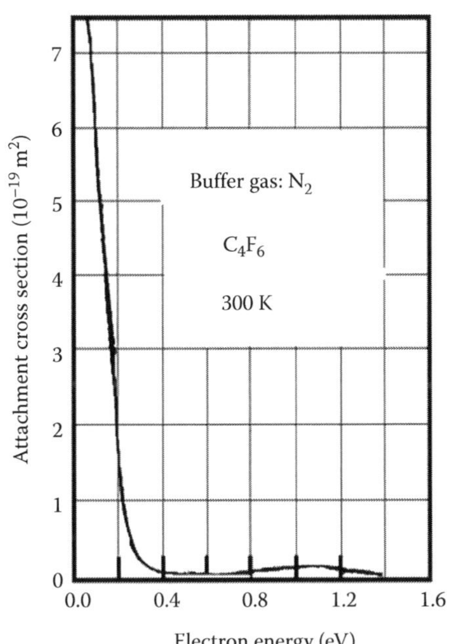

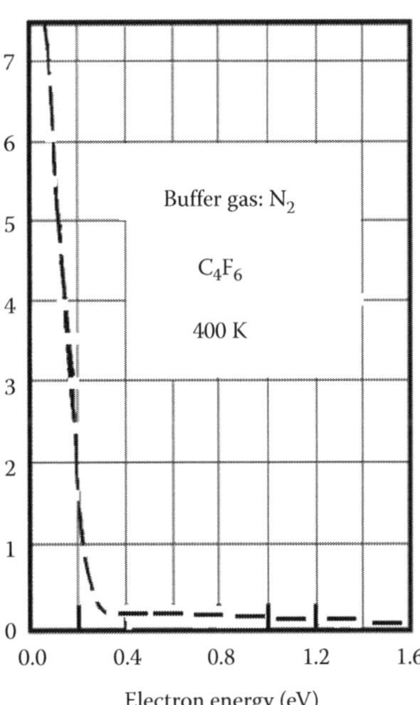

FIGURE 114.5 Electron attachment cross sections as functions of electron energy at various constant temperatures. Note the sharp peak toward zero electron energy. The temperature increase makes marginal difference to the rate constant. (Adapted from Datskos, P. G., L. G. Christophorou, and J. G. Carter, *J. Chem. Phys.*, 99, 8607, 1993.)

2. Three peaks (including the first at zero energy), as identified in Table 114.8. See Table 114.6 for energy in the meV range.
3. The cross sections are relatively small energy >2 eV (see Table 114.8).

Figure 114.5 shows the influence of temperature on the low-energy attachment cross sections for c-C_4F_6 (Datskos et al., 1993).

114.6 ATTACHMENT RATES

Thermal attachment rate constants for the three isomers of C_4F_6 are shown in Table 114.9.

Figure 114.6 shows the attachment rates in the three isomers as a function of electron mean energy with nitrogen as buffer gas. Figure 114.7 shows the influence of temperature

TABLE 114.9
Thermal Attachment Rate Constants for Isomers of C_4F_6

Name	Formula	Rate (m^3/s)	Reference
Hexafluoro-1, 3-butadiene	1,3-C_4F_6	1.26×10^{-13}	Christodoulides et al. (1979)
Hexafluoro-2-butyne	2-C_4F_6	5.4×10^{-14}	Christodoulides et al. (1979)
Perfluorocyclobutene	c-C_4F_6	1.43×10^{-13}	Christophorou and Datskos (1995)
	c-C_4F_6	1.4×10^{-13}	Bansal and Fessenden (1973)

Source: Adapted from Christodoulides, A. A. et al., *J. Chem. Phys.*, 70, 1156, 1979.

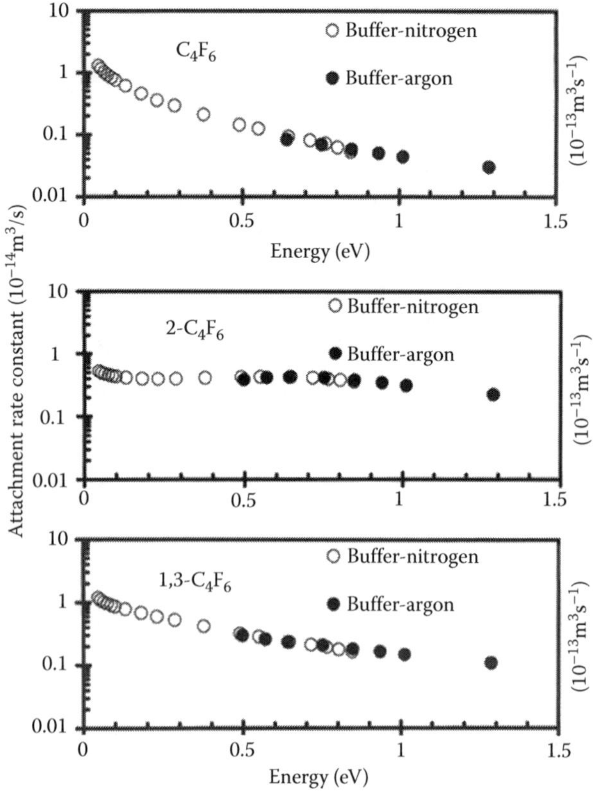

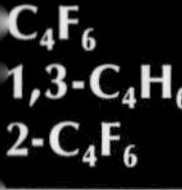

FIGURE 114.6 Attachment rate constants for the three isomers as a function of mean energy. (Adapted from Christodoulides, A. A. et al., *J. Chem. Phys.*, 70, 1156, 1979.)

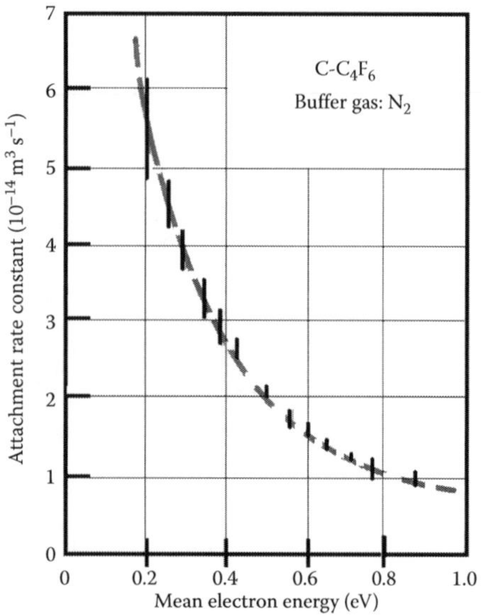

FIGURE 114.7 Electron attachment rate constant as a function of electron mean energy at various constant temperatures for c.C_4F_6. The vertical lines show the spread in the temperature range 300–600 K. At energies >0.4 eV the temperature makes even smaller difference. (Adapted from Christophorou L. G. and P. G. Datskos, *Int. J. Mass Spectrom. Ion Proc.*, 149/150, 59, 1995.)

on the attachment rates for c-C_4F_6 (Datskos et al., 1993). The temperature makes marginal difference to the rate constant.

REFERENCES

Bansal, K. M. and R. W. Fessenden, *J. Chem. Phys.*, 59, 1760, 1973.

Bart, M., P. W. Harland, J. E. Hudson, and C. Vallance, *Phys. Chem. Chem. Phys.*, 3, 800, 2001.

Christodoulides, A. A., L. G. Christophorou, R. Y. Pai, and C. M. Tung, *J. Chem. Phys.*, 70, 1156, 1979.

Christophorou, L. G. and P. G. Datskos, *Int. J. Mass Spectrom. Ion Proc.*, 149/150, 59, 1995.

Chutjian, A., S. H. Alajajian, J. M. Ajello, and O. J. Orient, *J. Phys. B: At. Mol. Phys.*, 17, L745, 1984.

Datskos, P. G., L. G. Christophorou, and J. G. Carter, *J. Chem. Phys.*, 99, 8607, 1993.

Sauers, I., L. G. Christophorou, and J. G. Carter, *J. Chem. Phys.*, 71, 3016, 1979.

Szmytkowski, Cz. and S. Kwitnewski, *J. Phys. B: At. Mol. Opt. Phys.*, 36, 2129, 2003a.

Szmytkowski, Cz. and S. Kwitnewski, *J. Phys. B: At. Mol. Opt. Phys.*, 36, 4865, 2003b.

Section XI

11 ATOMS

115

CHLOROPROPANE

C_3H_7Cl

CONTENTS

Chloropropane is a polar, electron-attaching gas that has 42 electrons. There are two isomers, 1-chloropropane (synonym propyl chloride) and 2-chloropropane (synonym isopropyl chloride). Selected properties are shown in Table 115.1.

115.1 IONIZATION CROSS SECTIONS

Total ionization cross sections have been measured by Hudson et al. for both isomers as shown in Table 115.2 and Figure 115.1. No evidence of the effects of isomerism on the cross section is observed.

TABLE 115.1
Selected Properties of Chloropropane

Molecule	Formula	α_e (10^{-40} F m^2)	μ (D)	ε_i (eV)
1-Chloropropane	1-C_3H_7Cl	11.13	2.05	10.81
2-Chloropropane	2-C_3H_7Cl	—	2.17	10.79

TABLE 115.2
Total Ionization Cross Sections (C_3H_7Cl)

	Ionization Cross Section (10^{-20} m^2)	
Energy (eV)	1-C_3H_7Cl	2-C_3H_7Cl
16	1.68	2.05
20	3.98	4.71
23	6.62	7.38
27	8.15	8.95
31	9.37	10.13
34	10.23	10.56
38	10.97	11.19
42	11.53	11.81
45	12.10	12.35
49	12.59	12.76
53	12.99	13.26
57	13.27	13.50

TABLE 115.2 (continued)
Total Ionization Cross Sections (C_3H_7Cl)

	Ionization Cross Section (10^{-20} m^2)	
Energy (eV)	1-C_3H_7Cl	2-C_3H_7Cl
60	13.56	13.79
64	13.70	13.93
68	13.88	14.03
71	13.92	14.01
75	14.08	14.20
79	14.15	14.28
82	14.10	14.17
86	14.21	14.22
90	14.02	14.08
94	13.95	14.03
97	14.01	14.09
101	14.00	14.08
105	13.83	13.87
108	13.73	13.68
112	13.71	13.77
116	13.63	13.68
119	13.61	13.61
123	13.50	13.48
127	13.28	13.29
130	13.23	13.21
134	13.15	13.15
138	13.06	13.08
142	12.99	12.93
145	12.92	12.81
149	12.67	12.68
153	12.48	12.47
156	12.39	12.34
160	12.33	12.30
164	12.29	12.27
167	12.26	12.25
171	12.21	12.19
175	12.07	12.10

continued

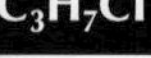

TABLE 115.2 (continued)
Total Ionization Cross Sections (C_3H_7Cl)

	Ionization Cross Section (10^{-20} m^2)	
Energy (eV)	1-C_3H_7Cl	2-C_3H_7Cl
179	11.92	11.94
182	11.76	11.79
186	11.62	11.57
190	11.46	11.42
193	11.37	11.30
197	11.37	11.27
201	11.34	11.23
204	11.29	11.22
208	11.29	11.22
212	11.26	11.24
215	11.10	11.13
219	10.99	11.03

Source: Adapted from Hudson, J. E. et al., *J. Phys. B: At. Mol. Opt. Phys.*, 34, 3025, 2001.
Note: Tabulated values courtesy of Professor Harland (2006).

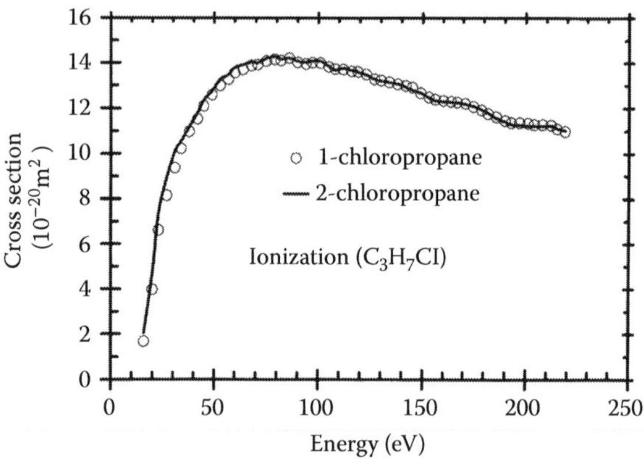

FIGURE 115.1 Ionization cross sections for 1—chloropropane and 2—chloropropane. (Adapted from Hudson, J. E. et al., *J. Phys. B: At. Mol. Opt. Phys.*, 34, 3025, 2001.) Tabulated values courtesy of Professor Harland (2006).

TABLE 115.3
Attachment Rates for Chloropropane

Molecule	Rate Constant (m^3/s)
1-Cyclopropane	3.6×10^{-19}
2-Cyclopropane	3.8×10^{-18}

115.2 ATTACHMENT RATES

Attachment rate constants measured by Barszczewska et al. (2003) for the isomers of chloropropane by the swarm method are shown in Table 115.3.

REFERENCES

Barszczewska, W., J. Kopyra, J. Wnorowska, and I. Szamrej, *J. Phys. Chem. A*, 107, 11427, 2003.

Hudson, J. E., C. Vallance, M. Bart, and P. W. Harland, *J. Phys. B: At. Mol. Opt. Phys.*, 34, 3025, 2001.

116

PERFLUOROPROPANE

C_3F_8

CONTENTS

Fluorocarbons, consisting of carbon and fluorine atoms are industrial gases. We consider perfluoropropane (C_3F_8) which is a member of the family with the general chemical formula C_nF_{2n+2}. They are fluorine-substituted derivatives of alkanes. The series for $n = 1–3$ has been extensively analyzed by Christophorou and Olthoff (2004) and we summarize their vast publication for the sake of completeness, providing some additional data. Selected properties of target particles for $n = 1–3$ in the series are shown in Table 116.1.

116.1 SELECTED REFERENCES FOR DATA

See Table 116.2.

TABLE 116.1
Selected Properties of Fluorocarbons

Target					
Name	**Formula**	Z	α_e **(10^{-40} Fm²)**	D	ε_i **(eV)**
CF_4	Carbon tetrafluoride	42	4.27	0	16.20
C_2F_6	Hexafluoroethane	66	7.588	0	14.6
C_3F_8	Perfluoropropane	90	7.20.10.46	0.07	13.3

116.2 TOTAL SCATTERING CROSS SECTION

Table 116.3 and Figure 116.1 show the total scattering cross sections for C_3F_8 recommended by Christophorou and

TABLE 116.2
Selected References for Data on C_3F_8

Parameter	Range: eV, (Td)	Reference
Q_i	12.2.215.5	Bart et al. (2001)
Q_i	16–3000	**Nishimura et al. (1999)**
Q_T	0.7–600	**Tanaka et al. (1999)**
Q_{el}, Q_M	0.8	Pirgov and Stefanov (1990)
W	(0.03–500)	**Hunter et al. (1988)**
k_a	(0.9–4.7)	**Hunter and Christophorou (1984)**
k_a	0–12	**Spyrou et al. (1983)**
Q_{diss}	13–600	**Winters and Inokuti (1982)**
D_r/μ	270–650	**Naidu and Prasad (1972)**
α/N, η/N	(300–900)	**Moruzzi and Craggs (1963)**

Note: Bold font indicates experimental study.
k_a = attachment rate; Q_{diff} = differential scattering, Q_{el} = elastic scattering; Q_T = total scattering; W = drift velocity; α/N = reduced ionization coefficient; η/N = reduced attachment coefficient.

TABLE 116.3
Total Scattering Cross Sections for C_nF_{2n+2}, $n = 3$, $n = 1$, see CF_4

	C_3F_8	CF_4
Energy (eV)	Q_T (10^{-20} m^2)	Q_T (10^{-20} m^2)
0.025	10.1	5.41
0.030	10.7	4.67
0.035	11.2	4.12
0.040	11.7	3.63
0.060	13.1	2.30
0.070	13.7	1.98
0.080	14.3	1.76
0.090	14.7	1.62
0.10	15.2	1.50
0.15	17.3	2.17
0.20	18.9	7.35
0.25	20.1	9.12
0.30	20.9	9.26
0.35	21.6	9.28
0.40	22.0	9.25
0.50	22.7	9.27
0.60	23.2	9.45
0.70	23.5	9.60
0.80	23.7	9.75
0.90	23.7	9.89
1.5	24.3	10.6
2.0	26.0	11.3
2.5	29.3	12.0
3.0	33.2	
3.5		13.0
4.0	36.7	13.5
4.5		13.8
5.0	35.4	14.0
7.0	39.1	18.6
8.0	39.9	21.1
9.0	40.7	21.8
10.0	39.6	20.8
12.5	35.6	18.5
15.0	35.4	17.9
20.0	38.9	19.2
25.0	39.9	20.4
30	40.3	20.2
35	40.2	19.9
40	40.0	19.9
50	39.1	19.9
60	37.9	19.9
70	36.5	19.6
80	35.0	19.2
90	33.2	18.8
100	32.1	18.3
120	30.0	
150	27.2	16.2
200	23.9	14.4
250	21.2	12.8
300	19.2	11.6
400	16.0	9.95
500	14.2	8.60
600	12.8	7.57

Source: Adapted from Christophorou, L. G. and J. K. Olthoff, *Fundamental Electron Interactions with Plasma Processing Gases*, Kluwer Academic, New York, NY, 2004.

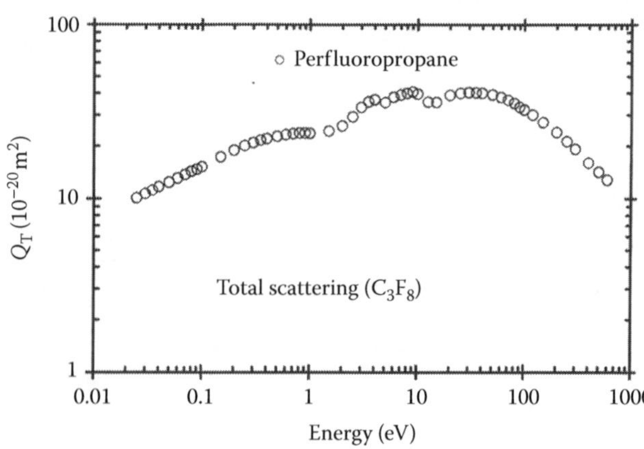

FIGURE 116.1 Total scattering cross sections for C_3F_8 recommended by Christophorou and Olthoff (2004).

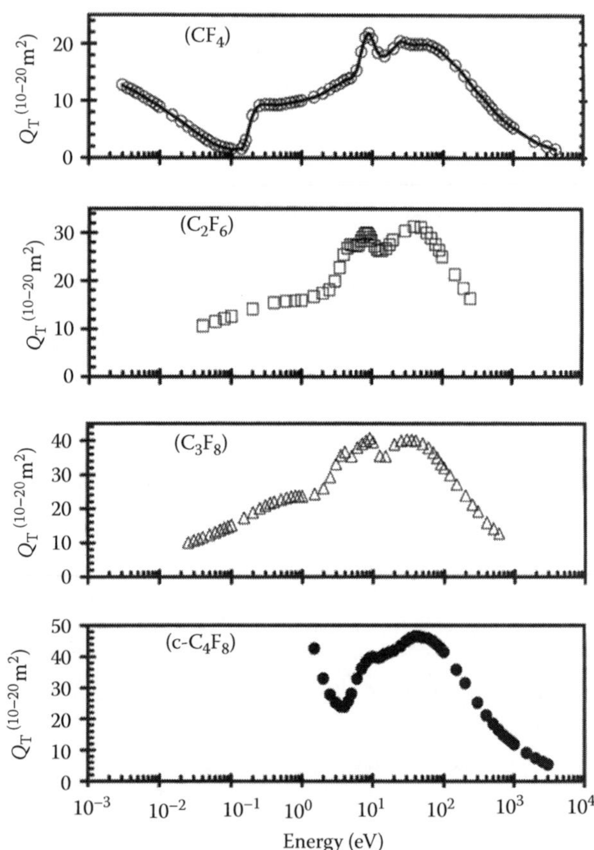

FIGURE 116.2 Total scattering cross sections for selected fluorocarbons on an expanded scale. Note the linear scale for the ordinate. (Adapted from Christophorou, L. G. and J. K. Olthoff, *Fundamental Electron Interactions with Plasma Processing Gases*, Kluwer Academic, New York, NY, 2004.)

Olthoff (2004). Figure 116.2 shows the expanded view in the low-energy region up to 10 eV electron energy. The important features of the cross sections are

1. In the energy range of 6–50 eV there is a peak at 8.5–9 eV attributed to shape resonance, with the

formation of short-lived negative ions. This feature is common to CF_4, C_2F_6, and C_3F_8.

2. At energy >100 eV there is a monotonic decrease of the cross section with the Born approximation method of calculating the cross section yielding satisfactory results. This feature is common to most target particles.

116.3 ELASTIC SCATTERING AND MOMENTUM TRANSFER CROSS SECTIONS

Table 116.4 gives the elastic scattering and momentum transfer cross sections for C_3F_8. These are shown in Figure 116.3. Figure 116.4 provides a comparison with selected gases demonstrating the effect of the size of the molecule.

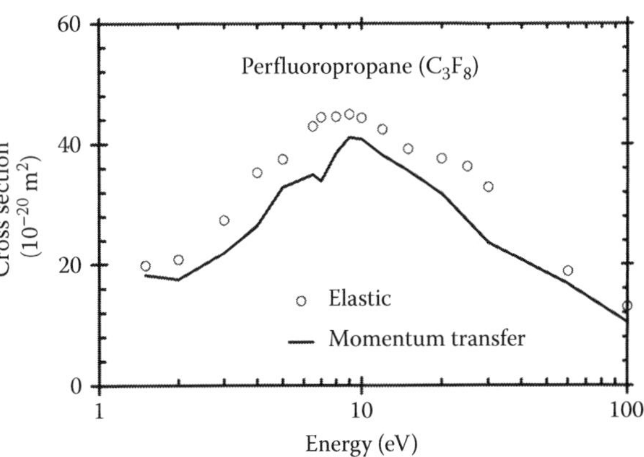

FIGURE 116.3 Elastic scattering and momentum transfer cross sections in C_3F_8 recommended by Christophorou and Olthoff (2004).

TABLE 116.4
Elastic Scattering and Momentum Transfer Cross Sections

Energy (eV)	Pirgov and Stefanov (1990) (Q_{el} (10^{-20} m^2))	Christophorou and Olthoff (2004)	
		Q_{el} (10^{-20} m^2)	Q_M (10^{-20} m^2)
0.0020	96.00		
0.010	63.23		
0.020	16.79		
0.040	1.89		
0.12	2.04		
0.20	6.57		
0.36	10.11		
0.64	10.91		
0.80	11.41		
1.0	13.21	13	
1.5	21.70	19.8	18.2
2.0		20.8	17.5
2.2	34.67		
3.0	39.63	27.4	21.9
4.0	30.91	35.3	26.5
5.0		37.5	32.9
6.0	10.05		
6.5		42.9	35.0
7.0		44.4	33.9
8.0	4.98	44.5	38.5
9.0		44.9	41.1
10.0		44.3	40.8
12.0		42.4	38.2
15.0		39.2	35.6
20.0		37.6	31.8
25.0		36.3	
30.0		32.9	23.6
60.0		18.8	16.7
100.0		13.0	10.4

Source: Adapted from Christophorou, L. G. and J. K. Olthoff, *Fundamental Electron Interactions with Plasma Processing Gases*, Kluwer Academic, New York, NY, 2004.

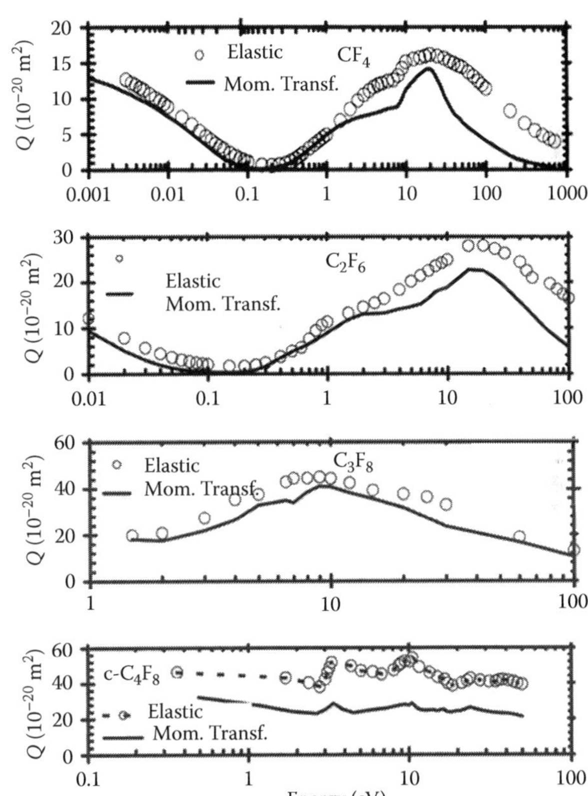

FIGURE 116.4 Elastic and momentum transfer cross sections in selected fluorocarbons. See relevant sections for references.

116.4 RAMSAUER–TOWNSEND MINIMUM

Ramsauer–Townsend minimum cross sections and the corresponding mean energy for selected fluorocarbons are shown in Table 116.5.

116.5 VIBRATIONAL EXCITATION CROSS SECTIONS

Vibrational excitation cross sections given by Pirgov and Stefanov (1990), are shown in Table 116.6.

C3F8

TABLE 116.5
Ramsauer–Townsend Minimum Cross Sections

	Minimum			
Gas	Energy (eV)	Cross Section (10^{-20} m²)	Method	Reference
CF_4	0.1	1.50	Q_T	Christophorou et al. (2004)
	0.2	0.56	Q_{el}	Mann and Linder (1992)
	0.2	.01	Q_M	Curtis et al. (1988)
	0.259	0.2	Q_M	Stefanov et al. (1988)
C_2F_6	0.2	1.66	Q_{el}	Christophorou et al. (2004)
	0.15	0.32	Q_M	Christophorou et al. (2004)
C_3F_8	0.07	0.8	Q_{el}	Pirgov and Stefanov (1990)

TABLE 116.6
Vibrational Excitation Cross Sections for CF_4

Energy (eV)	Q_{vib}(10^{-20} m²)	Energy (eV)	Q_{vib} (10^{-20} m²)
0.28	1.38	2.00	4.84
0.36	4.13	2.60	8.02
0.49	8.43	3.24	12.25
0.64	9.17	4.00	15.76
0.80	7.60	4.41	16.17
1.00	5.56	5.50	13.37
1.54	3.78	8.00	6.02

Source: Adapted from Christophorou, L. G. and J. K. Olthoff, *Fundamental Electron Interactions with Plasma Processing Gases*, Kluwer Academic, New York, NY, 2004.

TABLE 116.7
Positive Ion Appearance Potentials

Ion	Potential (eV)
CF_3^+	13.0
$C_2F_4^+$	13.0
$C_2F_5^+$	13.6
$C_3F_7^+$	15.4

Source: Adapted from Jarvis, G. K. et al., *J. Phys. Chem. A*, 102, 3219, 1998.

116.6 POSITIVE ION APPEARANCE POTENTIALS

The ion that appears at the lowest energy is CF_3^+. Ion appearance potentials are shown in Table 116.7 and a more detailed list for several paths for generation of C^+, F^+, and F_2^+ ions are given by Christophorou and Olthoff (2004).

116.7 IONIZATION CROSS SECTIONS

Ionization cross sections in fluorocarbons are shown in Table 116.8 and Figure 116.5. The data of Bart et al. (2001) which was not included in Christophorou and Olthoff (2004) are also shown.

116.8 ATTACHMENT PROCESSES

Dissociative attachment to the molecule occurs according to the reactions (Harland and Franklin, 1974).

TABLE 116.8
Ionization Cross Sections for C_3F_8

Energy (eV)	Q_T (10^{-20} m²)	Energy (eV)	Q_T (10^{-20} m²)	Energy (eV)	Q_T (10^{-20} m²)
16	0.129	44	7.43	350	10.5
17	0.316	46	7.85	400	9.8
18	0.562	48	8.31	450	9.31
19	0.815	50	8.99	500	8.61
20	1.13	60	10.4	600	7.8
22	1.81	70	11.3	700	6.99
24	2.31	80	11.9	800	6.41
26	3.08	90	12.5	900	5.82
28	3.58	100	12.8	1000	5.48
30	4.1	125	13.3	1250	4.64
32	4.57	150	13.4	1500	4.05
34	5.12	175	13.2	1750	3.62
36	5.58	200	12.8	2000	3.31
38	5.88	250	12.1	2500	2.78
40	6.54	300	11.2	3000	2.44
42	7.01				

Source: Adapted from Nishimura, H. et al., *J. Chem. Phys.*, 110, 3811, 1999.

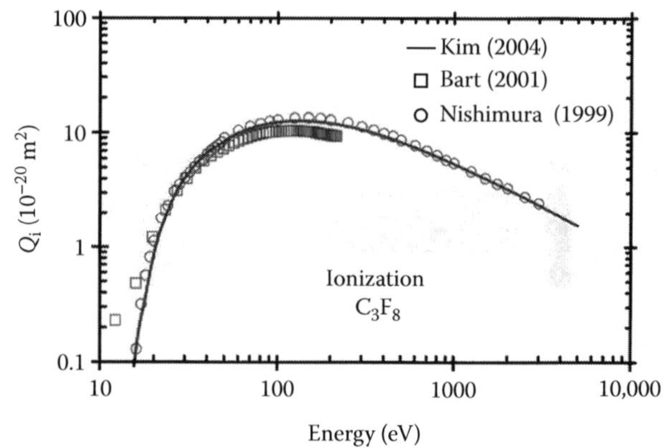

FIGURE 116.5 Ionization cross sections for C_3F_8. Tabulated values of Bart et al. (2001) by courtesy of Professor Harland (2006). (Adapted from Bart, M. et al., *Phys. Chem. Chem. Phys.*, 3, 800, 2001.)

$$C_3F_8 + e \rightarrow C_3F_7 + F^- \quad (116.1)$$

$$C_3F_8 + e \rightarrow n - C_3F_7 + F^- \quad (116.2)$$

$$C_3F_8 + e \rightarrow C_3F_7{}^* + F^- \quad (116.3)$$

$$C_3F_8 + e \rightarrow C_2F_5 + CF_3^- \quad (116.4)$$

Additional fragmentary negative ions are produced by the reaction, (Spyroll et al., 1983),

$$C_3F_8 + e \rightarrow CF_3 + C_2F_5^- \quad (116.5)$$

These authors also observe generation of $C_2F_3^-$ and $C_3F_7^-$ ions by unspecified reactions.

Table 116.9 summarizes the appearance potentials and peak energies, as summarized by Spyrou et al. (1983).

116.9 ATTACHMENT RATES

Thermal electron attachment rates are shown in Table 116.10.

116.10 ATTACHMENT CROSS SECTIONS

Total attachment cross sections for C_3F_8 are shown in Table 116.11 and Figure 116.6. Data for CF_4 and C_2F_6 added for comparison.

TABLE 116.9
Appearance Potentials and Peak Energy for Attachment for C_3F_8

Ion	Reaction Number	Appearance Potential (eV)	Peak (eV)	Reference
F^-	(116.1)	1.7	2.9	Spyrou et al. (1983)
F^-	(116.2)	2.0	3.15	Harland and Franklin (1974)
F^-	(116.3)	4.0		—
CF_3^-	(116.4)	2.4	3.4	Spyrou et al. (1983)
CF_3^-	(116.4)	2.55	3.65	Harland and Franklin (1974)
$C_2F_3^-$		1.1	3.3	Spyrou et al. (1983)
$C_2F_5^-$	(116.5)	2.1	3.2	—
$C_3F_7^-$		2.5	3.75	—

TABLE 116.10
Thermal Attachment Rate Constants

Gas	Rate (m^3/s)	Method	Reference
	$<1.24 \times 10^{-18}$	Swarm	Davis et al. (1973)
	$<10^{-21}$	MC	Fessenden and Bansal (1970)

Note: MC = microwave conductivity.
Note that the rate in C_3F_8 depends on gas number density.

TABLE 116.11
Total Attachment Cross Sections for C_3F_8

CF_4		C_2F_6		C_3F_8	
Energy (eV)	Q_a (10^{-20} m^2)	Energy (eV)	Q_a (10^{-20} m^2)	Energy (eV)	Q_a (10^{-20} m^2)
4.3	4 (−5)	2.0	1 (−4)	0.80	0.004
4.5	19 (−5)	2.25	7 (−4)	1.00	0.007
5.0	127 (−5)	2.5	4.4 (−3)	1.40	0.025
5.5	4.2 (−3)	2.75	0.013	1.50	0.035
6.0	8.98 (−3)	3.0	0.036	1.70	0.070
6.5	13.8 (−3)	3.25	0.073	1.80	0.091
6.75	15.7 (−3)	3.5	0.112	2.00	0.139
7.0	16.1 (−3)	4.0	0.140	2.30	0.192
7.25	15.5 (−3)	4.5	0.105	2.50	0.202
7.5	13.7 (−3)	5.0	0.064	2.70	0.19
8.0	9.69 (−3)	5.5	0.039	2.85	0.171
8.5	5.67 (−3)	6.0	0.024	3.00	0.151
9.0	2.91 (−3)	6.5	0.016	3.50	0.090
9.5	1.27 (−3)	7.0	0.012	4.00	0.049
10.0	4.6 (−4)	7.5	9.1 (−3)	4.50	0.028
10.5	7 (−5)	8.0	7.4 (−3)	5.00	0.019
		8.5	6.4 (−3)	5.50	0.014
		9.0	5.7 (−3)	6.00	0.012
		9.5	5.1 (−3)	6.50	0.010
		10.0	4.6 (−3)		

Note: a (b) means $a \times 10^b$.

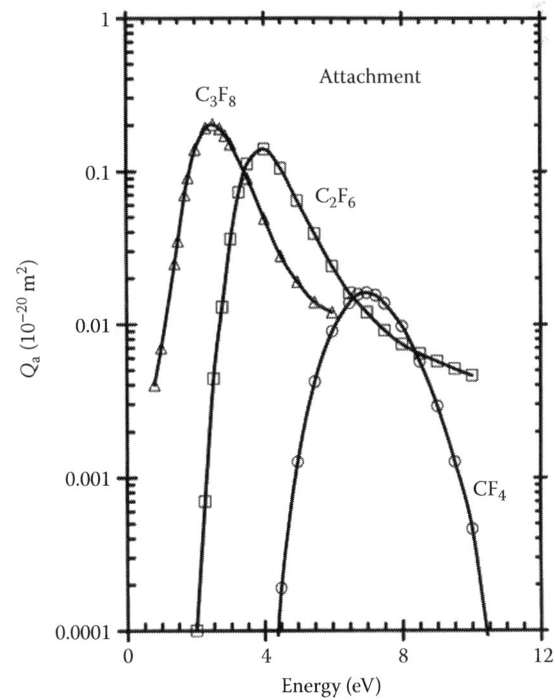

FIGURE 116.6 Total attachment cross sections in fluorocarbons. Note the shift of the peak to lower energy with increasing number of fluorine molecule.

C_3F_8

116.11 DRIFT VELOCITY OF ELECTRONS

The drift velocities of electrons for C_3F_8 are shown in Table 116.12 and Figure 116.7. Data for selected fluorocarbons included for comparison.

116.12 DIFFUSION COEFFICIENTS

Table 116.13 and Figure 116.8 show the radial diffusion coefficients expressed as ratios of mobilities for electrons in C_3F_8 fluorocarbon. Figure 116.9 provides a comparison with selected fluorocarbons demonstrating that larger molecules have lower characteristic energy at the same value of *E*/*N*.

TABLE 116.12
Drift Velocity of Electrons

E/*N* (Td)	*W* (10^4 m/s)	*E*/*N* (Td)	*W* (10^4 m/s)	*E*/*N* (Td)	*W* (10^4 m/s)
0.40	0.60	25.0	9.8	220.0	12.8
0.50	0.75	30.0	10.1	240.0	13.0
0.60	0.88	35.0	10.3	260.0	13.4
0.80	1.14	40.0	10.3	280.0	13.6
1.0	1.39	50.0	10.1	300.0	13.9
1.5	1.98	60.0	10.0	320.0	14.2
2.0	2.57	70.0	10.1	340.0	14.6
3.0	3.57	80.0	10.3	360.0	14.9
4.0	4.37	90.0	10.5	380.0	15.3
6.0	5.57	100.0	11.0	400.0	15.7
8.0	6.49	120.0	11.3	420.0	15.9
10.0	7.14	140.0	11.8	440.0	16.4
12.0	7.92	160.0	12.1	460.0	16.7
15.0	8.45	180.0	12.3	480.0	17.2
17.0	8.80	200.0	12.5	500.0	17.5
20.0	9.25				

Source: Adapted from Hunter, S. R., J. G. Carter, and L. G. Christophorou, *Phys. Rev.*, 38, 58, 1988.

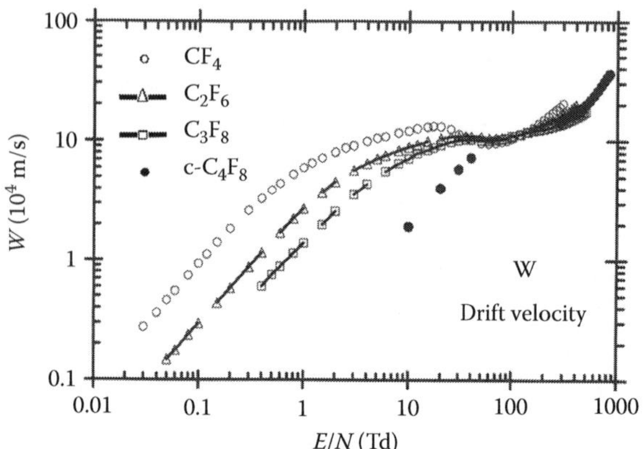

FIGURE 116.7 Drift velocity of electrons for C_3F_8 and selected fluorocarbons. Notice the prominent negative differential conductivity in CF_4, C_2F_6, and C_3F_8.

TABLE 116.13
D_r/μ Ratios for C_3F_8 Fluorocarbon

E/*N* (Td)	D_r/μ (V)	*E*/*N* (Td)	D_r/μ (V)
270	2.93	460	4.19
280	3.01	480	4.28
290	3.09	500	4.35
320	3.33	520	4.41
340	3.48	540	4.46
350	3.62	560	4.51
380	3.76	580	4.56
400	3.89	600	4.60
420	3.99	620	4.65
440	4.10	640	4.68

Source: Adapted from Naidu, M. S. and A. N. Prasad, *J. Phys. D: Appl Phys.*, 5, 983, 1972.

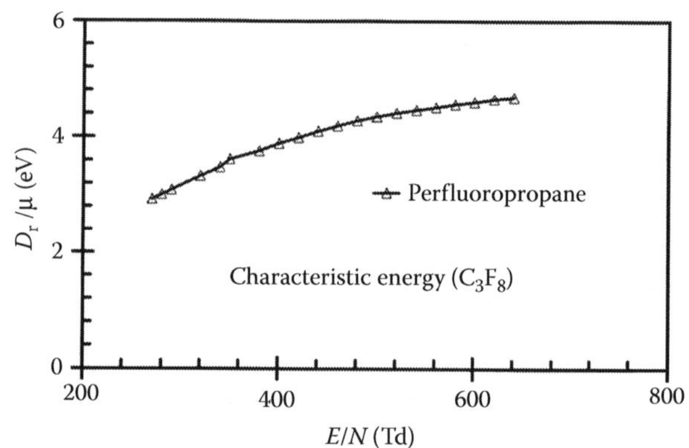

FIGURE 116.8 Ratio of D_r/μ for C_3F_8. (Adapted from Naidu, M. S. and A. N. Prasad, *J. Phys. D: Appl Phys.*, 5, 983, 1972.)

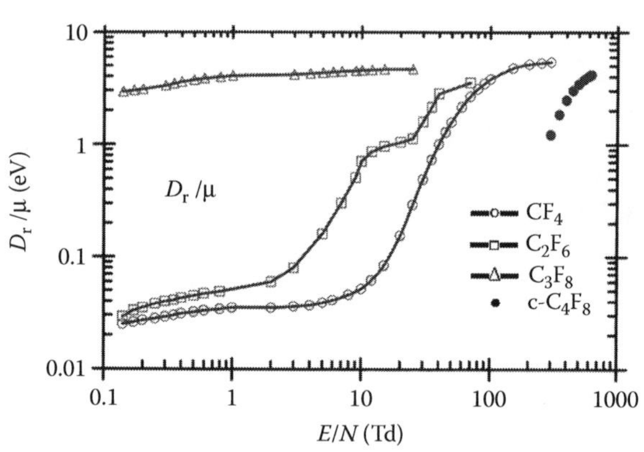

FIGURE 116.9 Ratio of D_r/μ for selected fluorocarbons. See relevant sections for references.

TABLE 116.14
Ionization and Attachment Coefficients for C_3F_8

E/N (Td)	α/N (10^{-22} m²)	η/N	E/N (Td)	α/N (10^{-22} m²)	η/N
140	0.12	26.62	400	27.8	16.20
160	0.99	26.49	420	30.7	
180	2.02	25.72	440	33.9	
200	3.34	24.44	460	37.7	
220	5.02	23.32	480	41.7	
240	6.88	22.08	500	45.4	
260	8.84	20.84	520	49.0	
280	11.1	20.00	540	52.9	
300	13.4	19.20	560	57.3	
320	16.2	18.4	580	62.0	
340	18.9	17.8	600	66.6	
360	22.1	17.30	620	71.3	
380	24.9	16.70			

Source: Adapted from Christophorou, L.G. and J.K. Olthoff, *Fundamental Electron Interactions with Plasma Processing Gases*, Kluwer Academic, New York, NY, 2004.

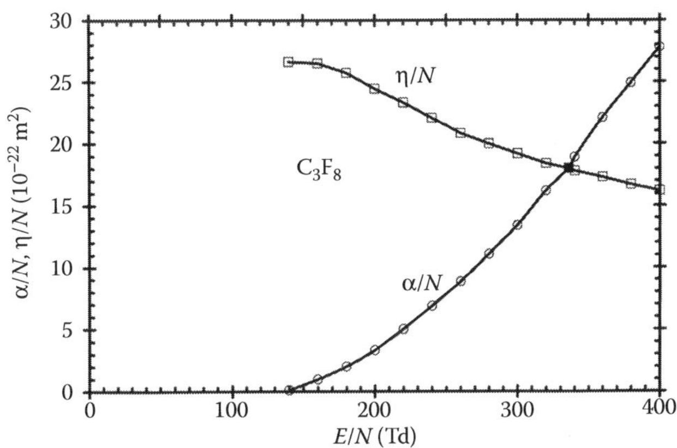

FIGURE 116.10 Density reduced ionization and attachment coefficients for C_3F_8. (Adapted from Naidu, M. S. and A. N. Prasad, *J. Phys. D: Appl. Phys.*, 5, 983, 1972.)

116.13 IONIZATION AND ATTACHMENT COEFFICIENTS

The ionization and attachment coefficients in fluorocarbons are shown in Table 116.14 and Figure 116.10. Figure 116.11 shows the limiting *E/N* in selected fluorocarbons.

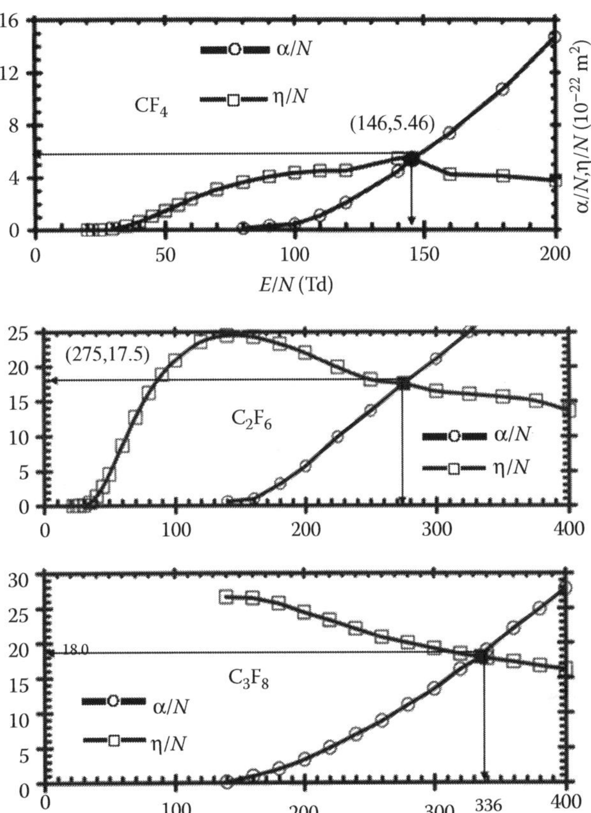

FIGURE 116.11 Ionization and attachment coefficients as a function of *E/N* in selected fluorocarbons. The limiting *E/N* at which $\alpha/N = \eta/N$ is also shown. The $(E/N)_{lim}$ determined from high *N* breakdown studies may deviate from these values by a few percent.

REFERENCES

Bart M., P. W. Harland, J. E. Hudson, and C. Vallane, *Phys. Chem. Chem. Phys.*, 3, 800, 2001.

Christophorou L. G. and J. K. Olthoff, *Fundamental Electron Interactions with Plasma Processing Gases*, Kluwer Academic Publishers, New York, NY, 2004.

Curtiss M. G., I. C. Walker, and K. J. Mathieson, *J. Phys. D: Appl. Phys.*, 21, 1271, 1988.

Davis, F. J., R. N. Compton, and D. R. Nelson, *J. Chem. Phys.*, 59, 2324, 1973.

Fessenden, R. W. and K. M. Bansal, *J. Chem. Phys.* 53, 3468, 1970.

Harland, P. W. and J. L. Franklin, *J. Chem. Phys.*, 61, 1621, 1974.

Hunter S. R. and L. G. Christophorou, *J. Chem. Phys.*, 80, 6150, 1984.

Hunter S. R., J. G. Carter, and L. G. Christophorou, *Phys. Rev.*, 38, 58, 1988.

Jarvis G. K., K. J. Boyle, C. A. Mayhew, and R. P. Tuckett, *J. Phys. Chem. A*, 102, 3219, 1998.

Mann A. and F. Linder, (a) *J. Phys. B: At. Mol. Opt. Phys.*, 25, 545; 1992; (b) *J. Phys. B: At. Mol. Opt. Phys.*, 25, 533; 1992;

Moruzzi J. L. and J. D. Craggs, *Proc. Phys. Soc.*, 82, 979, 1963.

Naidu M. S. and A. N. Prasad, *J. Phys. D: Appl. Phys.*, 5, 983, 1972.

Nishimura H., W. M. Huo, M. A. Ali, and Y-.K. Kim, *J. Chem. Phys.*, 110, 3811, 1999.

Pirgov P. and B. Stefanov, *J. Phys. B: At. Mol. Phys.*, 23, 2879, 1990.

Spyrou S. M., I. Sauers, and L. G. Christophorou, *J. Chem. Phys.*, 78, 7200, 1983.

Stefanov B., N. Paprokova, and L. Zarkova, *J. Phys. B: At. Mol. Opt. Phys.*, 21, 3989, 1988.

Tanaka H., Y. Tachibana, M. Kitajima, O. Sueoka, H. Takaki, A. Hamada, and M. Kimura, *Phys. Rev. A*, 59, 2006, 1999.

Winters H. F. and M. Inokuti, *Phys. Rev. A*, 25, 1420, 1982.

117

PROPANE

CONTENTS

Propane (C_3H_8) is a saturated hydrocarbon with 26 electrons and the ground state is given by Boesten et al. (1994). The bond strength (C_3H_7–H) is 4.37 eV. It is weakly polar with a dipole moment of 0.084 D and an electronic polarizability of 6.998×10^{-40} F m^2. The first ionization potential is 10.95 eV. The molecule has 27 vibrational modes out of which nine (ν_1 through ν_9) are totally symmetrical (Allan and Andric 1996). Vibrations of similar type tend to have similar frequencies. Selected vibrational energies of the molecule are shown in Table 117.1.

117.1 SELECTED REFERENCES FOR DATA ON C_3H_8

See Table 117.2.

117.2 TOTAL SCATTERING CROSS SECTION

Low-energy scattering cross section in C_3H_8 is shown in Table 117.3. The total cross section in the range from 1 to 600 eV is shown in Table 117.4. See Figure 117.1 for graphical presentation of the entire range. The highlights of the total scattering cross section are

1. A significant Ramsauer–Townsend minimum at 0.1–0.2 eV electron energy.
2. A dominant peak at 8–9 eV electron energy due to shape resonance and formation of short-lived negative ion. This aspect is common to many gases.
3. The cross section increases for energies lower than that at Ramsauer–Townsend minimum. This is attributed to the dipole moment of the target molecule, with the total cross section varying as $1/\varepsilon$,

TABLE 117.1
Vibrational Modes, Their Energies, and Relative Intensity

Mode	Energy (meV)	Relative Intensity	Type of Motion
ν_9	46	1.457	Torsion vibrations
ν_8	108	1.774	C–C stretch
ν_7	144	2.312	CH_3 rock
ν_6	173	0.276	H–C–H deformation
$\nu_{4,5}$	182	3.828	H–C–H deformation
$\nu_{1,2,3}$	369, 368, 358	4.588	C–H stretch

Source: Adapted from Boesten, L. et al., *J. Phys. B: At. Mol. Opt. Phys.*, 27, 1845, 1994.
Note: The last column is from Allan and Andric (1996).

TABLE 117.2
Selected References

Quantity	Energy Range (eV)	Reference
	Cross Sections	
Q_i	10.95–5000	**Kim et al. (2004)**
Q_T	0.5–400	**Szmytkowski and Kwitnewski (2002)**
Q_{diff}, Q_{el}, Q_T	0.8–600	**Tanaka et al. (1999)**
Q_T	0.01–2	**Lunt et al. (1998)**
Q_i	10–1000	**Hwang et al. (1996)**
Q_{diff}, Q_{el}, Q_v	2–100	**Boesten et al. (1994)**
Q_i	10–3000	**Nishimura and Tawara (1994)**
Q_i	Onset–950	**Grill et al. (1993)**
Q_i	Onset–240	**Duric et al. (1991)**
Q_T	4–500	**Nishimura and Tawara (1991)**
Q_T	5–400 eV	**Floeder et al. (1985)**
Q_a	0–15	**Rutkowsky et al. (1980)**
Q_i	600–12000	**Schram et al. (1966)**
Q_T	0.8–50	**Brüche (1930)**
	Swarm Parameters	
W, D_L, D_r	0.14–7.0	**Schmidt and Roncossek (1992)**
W	0.03–30	**Floriano et al. (1986)**
W	0.03–30	**Gee and Freeman (1983)**
W	0.03–27.0	**McCorkle et al. (1978)**
α/N	110–6000	**Heylen (1975)**

Note: Q_{diff} = differential; Q_{el} = elastic; Q_i = ionization; Q_T = total; D_L = longitudinal diffusion coefficient; D_r = radial diffusion coefficient; W = drift velocity. Bold font indicates experimental study.

TABLE 117.3
Total Scattering Cross Section for Electrons at Low Energy in C_3H_8

Energy (eV)	Q_T (10^{-20} m²)	Energy (eV)	Q_T (10^{-20} m²)
0.006	23.24	0.09	4.40
0.008	20.73	0.1	4.36
0.01	18.83	0.2	5.61
0.02	12.49	0.3	8.06
0.03	9.55	0.4	9.72
0.04	5.69	0.6	12.86
0.06	5.64	0.8	14.80
0.08	4.60	1	16.23

Source: Adapted from Lunt, S. L. et al., *J. Phys. B: At. Mol. Opt. Phys.*, 31, 4225, 1998.

Note: The relative values are converted to absolute values with the measurements of Tanaka et al. (1999) at 1.0 eV.

where ε, the electron energy, is below thermal energy (Christophorou et al., 1997).

4. For energies greater than about 9 eV, the cross section decreases monotonically up to the highest energy.

TABLE 117.4
Total Scattering Cross Section in C_3H_8

Energy (eV)	Q_T (10^{-20} m²) Tanaka et al. (1999)	Q_T (10^{-20} m²) Szmytkowski and Kwitnewski (2002)
0.5		13.2
0.6		13.8
0.8	15.33	14.1
1	16.18	14.9
1.2		15.7
1.5		16.6
2	17.88	18.1
2.5		19.8
3	22.21	21.8
3.5		23.7
4.0	27.84	26.6
4.5		31.2
5.0	33.54	33.5
5.5		37.8
6.0	41.42	41.3
6.5		43.6
7.0	44.49	45.7
7.5		47.6
8.0	47.74	48.8
8.5		49.6
9.0		48.9
9.5		48.1
10	44.74	47.2
11		45.6
12		44.0
14		42.4
16		40.8
18		40.2
20	37.81	
22.5		38.0
25		37.0
30	34.34	35.2
35		33.8
40	31.73	32.6
45		31.7
50	29.02	30.8
60	26.59	29.0
70	25.27	27.4
80	23.80	26.1
90	22.26	24.9
100	21.08	24.0
110		23.1
120		21.7
140		20.2
150	16.45	
160		18.8
180		18.0
200	14.46	16.7
220		16.0
250		14.2
275		13.6
300	11.55	12.6
350		11.3
370		10.7
400	9.78	
500	8.40	
600	6.40	

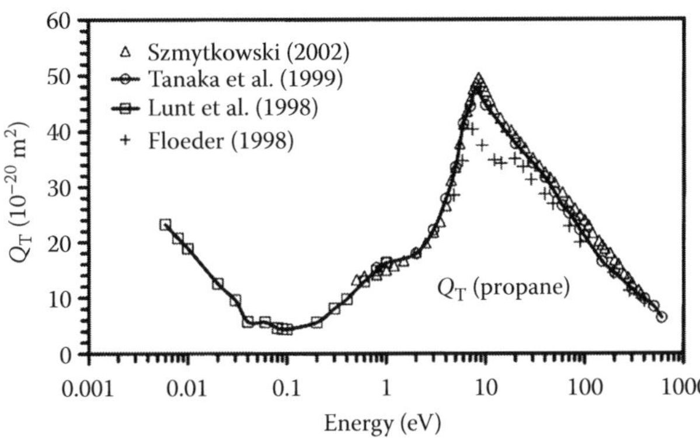

FIGURE 117.1 Total scattering cross section in C_3H_8. (Δ) Szmytkowski et al. (2002); (—○—) Tanaka et al. (1999); (—□—) Lunt et al. (1998); (+) Floeder et al. (1985).

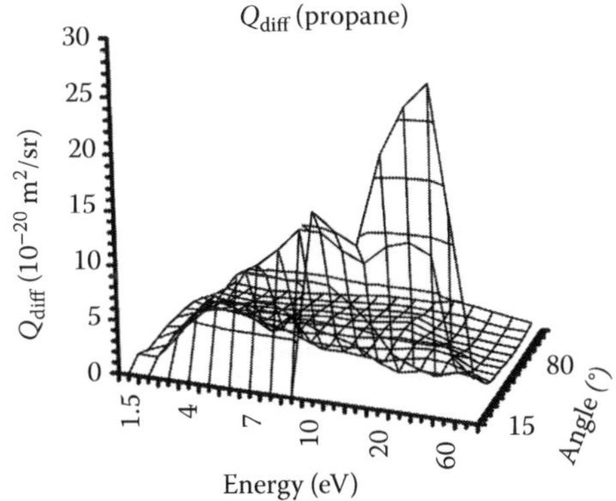

FIGURE 117.2 Differential scattering cross section in C_3H_8 as a function of electron energy and angle of scattering. (Adapted from Tanaka, et al. H., *Phys. Rev. A*, 59, 2006, 1999.)

117.3 DIFFERENTIAL SCATTERING CROSS SECTION

Figures 117.2 and 117.3 show the differential cross section as a function of electron energy and scattering angle. (Tanaka et al., 1999) Highlights of differential scattering cross section are

1. At electron energies up to 100 eV, the cross section decreases with increasing angle, suggesting considerable forward scattering.
2. At 5 eV and approximately 80°, a clear hump is seen which disappears for energies above 7 eV.
3. At higher energies up to 100 eV, the cross section reaches a plateau with a trace indication of backward scattering.

117.4 ELASTIC SCATTERING CROSS SECTION

See Table 117.5.

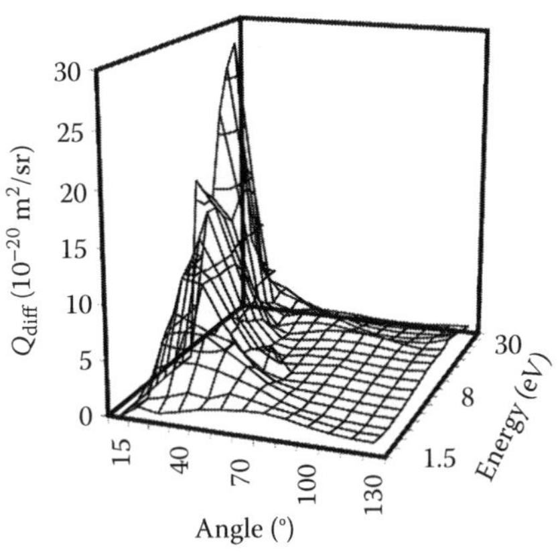

FIGURE 117.3 Differential scattering cross section in C_3H_8 as a function of scattering angle and electron energy. (Adapted from Tanaka, H. et al., *Phys. Rev. A*, 59, 2006, 1999.)

TABLE 117.5
Elastic Scattering Cross Section for Electrons in C_3H_8

Energy (eV)	Q_{el} (10^{-20} m^2) Tanaka et al. (1999)	Q_{el} (10^{-20} m^2) Boesten et al. (1994)
1.5	19.800	
2	20.817	17.04
3	27.401	19.82
4	35.317	26.97
5	37.503	34.09
6		39.95
6.5	42.877	
7	44.365	
7.5		44.46
8	44.513	
8.5		41.54
9	44.942	
10	44.335	38.36
12	42.379	
15	39.150	30.86
20	37.631	24.32
25	36.324	
30	32.869	
40		15.92
60	18.784	
100	13.001	8.21

Source: Adapted from Tanaka, H. et al., *Phys. Rev. A*, 59, 2006, 1999.
Note: See Figure 117.4 for graphical presentation.

117.5 MOMENTUM TRANSFER CROSS SECTION

Momentum transfer cross sections in C_3H_8 are shown in Tables 117.6 and 117.7. See Figure 117.4 for graphical presentation.

117.6 RAMSAUER–TOWNSEND MINIMUM

See Table 117.8.

C_3H_8

TABLE 117.6
Momentum Transfer Cross Section in C_3H_8 at Low Electron Energy

Energy (eV)	Q_M (10^{-20} m²)	Energy (eV)	Q_M (10^{-20} m²)
0.015	17.63	0.08	5.01
0.02	16.52	0.09	4.29
0.03	13.64	0.1	3.79
0.04	10.75	0.14	3.00
0.06	7.08	0.2	4.47
0.07	5.95	0.3	20.20

Source: Adapted from McCorkle, D. L. et al., *J. Phys. B: At. Mol. Phys.*, 11, 3067, 1978.

C_3H_8

TABLE 117.7
Momentum Transfer Cross Section in C_3H_8 at Intermediate Electron Energy

Energy (eV)	Q_M (10^{-20} m²) Tanaka et al. (1999)	Q_M (10^{-20} m²) Boesten et al. (1994)
1.5	18.244	
2	17.524	14.41
3	21.909	17.65
4	26.542	21.97
5	32.918	25.40
6		27.68
6.5	35.031	
7	33.888	
7.5		29.98
8	38.513	
8.5		25.98
9	41.088	
10	40.784	23.13
12	38.193	
15	35.61	16.73
20	31.745	13.73
25	26.921	
30	23.625	
40		6.34
60	16.713	
100	10.376	2.88

Source: Adapted from Tanaka, H. et al., *Phys. Rev. A*, 59, 2006, 1999.

117.7 ELECTRONIC EXCITATION CROSS SECTION

The lower electronic excitation potentials are 9.13, 9.84, 11.41, and 13.54 eV (Lassettre, 1964). Excitation cross sections in C_3H_8 are not available. For approximate values, photon absorption cross sections measured by Au et al. (1993) are shown in Figures 117.5 and 117.6. In the latter, photon absorption cross sections in selected hydrocarbons are shown. The larger molecule has higher cross section as expected.

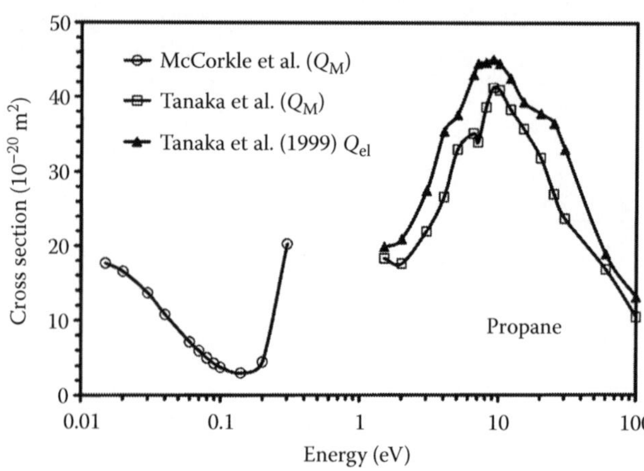

FIGURE 117.4 Elastic and momentum transfer cross section in C_3H_8. (—▲—) Tanaka et al. (1999), elastic; (—□—) Tanaka et al. (1999), momentum transfer; (—○—) McCorkle et al. (1978), momentum transfer.

TABLE 117.8
Ramsauer–Townsend Minimum

Energy (eV)	Cross Section (10^{-20} m²)	Method	Authors
0.14	3.0	Swarm	McCorkle et al. (1978)
0.13	8.2	Swarm	Gee and Freeman (1983)
0.1	4.36	Total	Lunt et al. (1998)

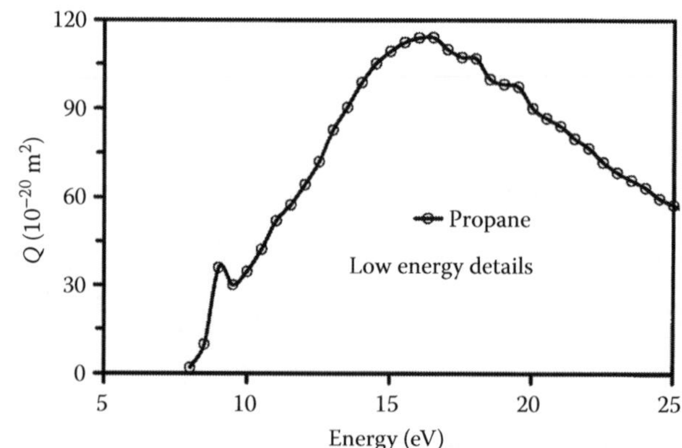

FIGURE 117.5 Low-energy details of photon absorption cross section in C_3H_8 as function of energy. (Adapted from Au, W. et al., *Chem. Phys.*, 173, 209, 1993.)

117.8 TOTAL IONIZATION CROSS SECTION

Tables 117.9 and 117.10 show the total ionization cross sections in C_3H_8. See Figure 117.7 for graphical presentation.

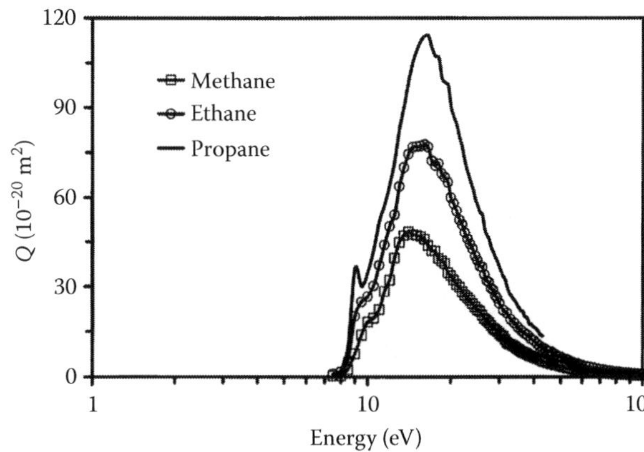

FIGURE 117.6 Photon absorption cross section in C_3H_8 as a function of energy. Methane and ethane data are included for the sake of comparison. (—□—) Methane; (—○—) ethane; (——) propane. (Adapted from Au, W. et al., *Chem. Phys.*, 173, 209, 1993.)

TABLE 117.9
Total Ionization Cross Sections in C_3H_8

Energy (eV)	Q_i (10^{-20} m²)	Energy (eV)	Q_i (10^{-20} m²)	Energy (eV)	Q_i (10^{-20} m²)
12.00	0.206	90.00	10.24	600.00	4.33
17.50	2.30	100.00	10.23	700.00	3.99
20.00	3.31	125.00	9.90	800.00	3.67
25.00	5.21	150.00	9.36	900.00	3.27
30.00	6.47	175.00	8.84	1000.00	3.05
35.00	7.37	200.00	8.35	1250.00	2.64
40.00	8.00	250.00	7.80	1500.00	2.27
45.00	8.54	300.00	6.84	1750.00	2.06
50.00	9.22	350.00	6.25	2000.00	1.88
60.00	9.79	400.00	5.78	2500.00	1.62
70.00	10.09	450.00	5.26	3000.00	1.39
80.00	10.20	500.00	4.93		

Source: Adapted from Nishimura, H. and H. Tawara, *J. Phys, B: At. Mol. Opt. Phys.*, 27, 2063, 1994.

TABLE 117.10
Total Ionization Cross Sections in C_3H_8 at High Electron Energies

Energy (eV)	0.6	1	2	4	7	12
Q_i (10^{-20} m²)	4.13	2.83	1.65	0.934	0.594	0.383

Source: Adapted from Schram, B. L. et al., *J. Chem. Phys.*, 44, 49, 1966.

117.9 PARTIAL IONIZATION CROSS SECTION

The fragments of ionization in C_3H_8 and relative intensity at 100 eV electron energy are shown in Table 117.11. Cross sections for selected fragments are shown in Figures 117.8 and 117.9 (Grill et al. 1993).

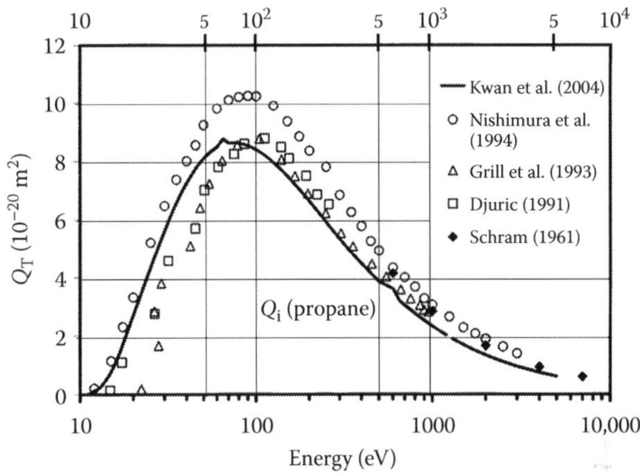

FIGURE 117.7 Total ionization cross section in C_3H_8. (——) Kim et al. (2004); (○) Nishimura and Tawara (1994); (Δ) Grill et al. (1991); (◆) Schram et al. (1966); (□) Duric et al. (1991).

TABLE 117.11
Partial Contribution for Ionization Cross Section at 100 eV Impact Energy

Species	Appearance Potential (eV) (Fiegele, 2000)	Q_i (10^{-20} m²)	Species	Q_i (10^{-20} m²)
$C_3H_8^+$	11.22	0.544	C^+	0.0256
$C_3H_7^+$	11.51	0.423	$C_3H_5^{2+}$	0.00069
$C_3H_6^+$	11.75	0.125	$C_3H_4^{2+}$	0.0168
$C_3H_5^+$	13.48	0.304	$C_3H_3^{2+}$	0.0151
$C_3H_4^+$	13.79	0.0756	$C_2H_2^{2+}$	0.0254
$C_3H_3^+$	16.50	0.752		
$C_3H_2^+$		0.349		
C_3H^+		0.215		
C_3^+		0.0391		
$C_2H_5^+$	11.91	2.060		
$C_2H_4^+$	11.94	1.025		
$C_2H_3^+$		1.346		
$C_2H_2^+$		0.537		
C_2H^+		0.0746		
C_2^+		0.0136		
CH_3^+		0.443		
CH_2^+		0.160		
CH^+		0.0554		

Source: Adapted from Grill,V. et al., *Z. Phys. D*, 25, 217, 1993.

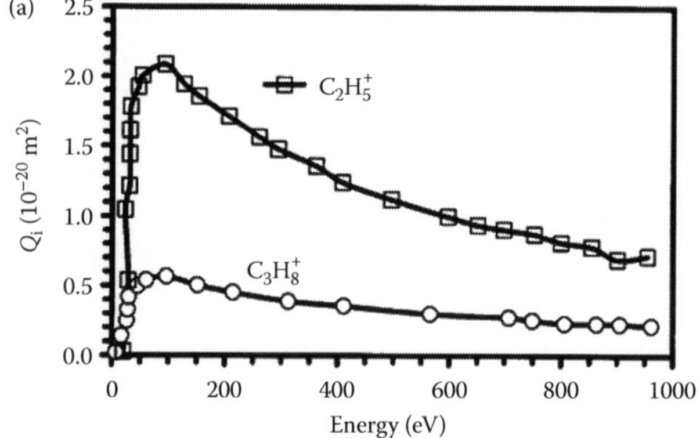

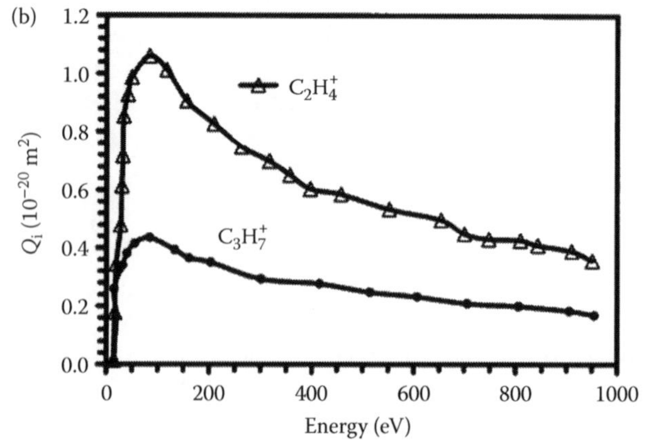

FIGURE 117.8 (a,b) Partial ionization cross section in C_3H_8 for selected fragments. (Adapted from Grill, V. et al., *Phys. D*, 25, 217, 1993.)

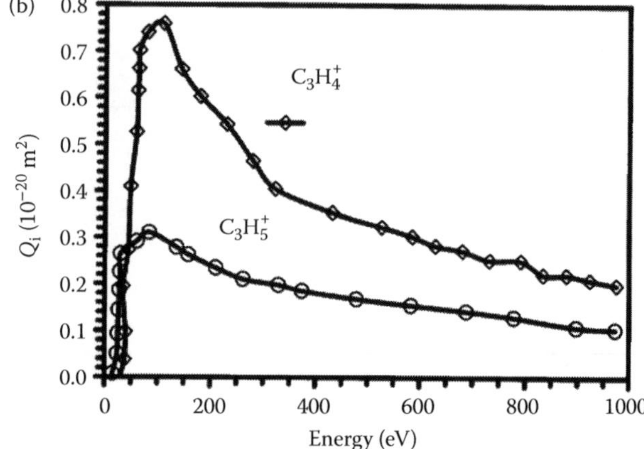

FIGURE 117.9 (a,b) Partial ionization cross section in C_3H_8 for selected fragments. (Adapted from Grill, V. et al., *Z. Phys. D*, 25, 217, 1993.)

117.10 ATTACHMENT CROSS SECTION

The dissociative attachment reaction in C_3H_8 occurs through the processes (Rutkowsky et al. 1980)

$$e(>6\,\text{eV}) + C_3H_8 \rightarrow C_3H_8^{-*} \rightarrow \begin{cases} H^- + C_3H_7 \\ CH_3^- + C_2H_5 \end{cases} \tag{117.1}$$

The dissociative attachment cross section is shown in Table 117.12. See Figure 117.10 for graphical presentation.

TABLE 117.12
Dissociative Attachment Cross Section in C_3H_8

Energy (eV)	Q_a (10^{-24} m^2)	Energy (eV)	Q_a (10^{-24} m^2)	Energy (eV)	Q_a (10^{-24} m^2)
6.0	0	8.0	3.60	10	1.54
6.5	0.25	8.5	4.61	11	0.20
7.0	0.99	9.0	4.01	12	0.53
7.5	2.37	9.5	3.12		

Source: Adapted from Rutkowsky, V. J., H. Drost, and H.-J. Spangenberg, *Ann. Der. Phys. Leipzig*, 37, 259, 1980.

Note: Note the relatively small cross section.

117.11 ELECTRON DRIFT VELOCITY

Tables 117.13 and 117.14 present drift velocity of electrons in C_3H_8.

117.12 DIFFUSION COEFFICIENTS

The transverse and longitudinal diffusion coefficients of electrons in C_3H_8 at 293 K expressed as ratios of mobility are shown in Table 117.15 (Schmidt and Roncossek, 1992). See Figure 117.12 for graphical presentation.

117.13 REDUCED IONIZATION COEFFICIENT

See Table 117.16.

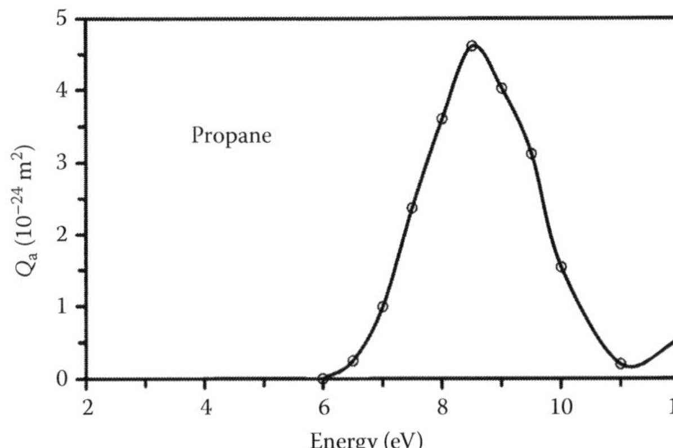

FIGURE 117.10 Dissociative attachment cross section in C_3H_8. (Adapted from Rutkowsky, V. J., H. Drost, and H.-J. Spangenberg, *Ann. Der. Phys. Leipzig*, 37, 259, 1980.)

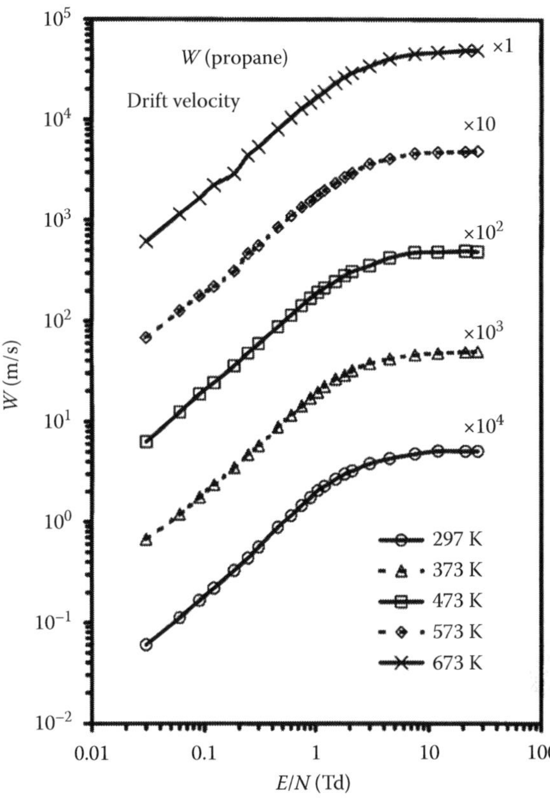

FIGURE 117.11 Drift velocity of electrons in C_3H_8 as a function of reduced electric field and temperature. (Adapted from McCorkle, D. L. et al., *J. Phys. B: At. Mol. Phys.*, 11, 3067, 1978.)

TABLE 117.13
Drift Velocity of Electrons in Pentane as a Function of Reduced Electric Field and Temperature

	Drift Velocity (10⁴ m/s)				
***E/N* (Td)**	**297 K**	**373 K**	**473 K**	**573 K**	**673 K**
0.03	0.060	0.067	0.063	0.068	0.061
0.06	0.112	0.119	0.124	0.126	0.114
0.09	0.167	0.176	0.186	0.178	0.165
0.12	0.221	0.235	0.241	0.220	0.223
0.18	0.330	0.349	0.353	0.313	0.291
0.24	0.438	0.470	0.474	0.462	0.437
0.30	0.561	0.577	0.592	0.556	0.535
0.45	0.878	0.885	0.874	0.840	0.802
0.60	1.15	1.16	1.14	1.10	1.04
0.75	1.45	1.43	1.42	1.34	1.28
0.90	1.74	1.71	1.69	1.54	1.48
1.05	2.06	1.95	1.93	1.77	1.70
1.2	2.26	2.22	2.13	1.98	1.88
1.5	2.67	2.63	2.46	2.33	2.30
1.8	3.02	2.86	2.83	2.68	2.62
2.1	3.24	3.18	3.10	2.94	2.90
3.0	3.83	3.74	3.55	3.63	3.40
4.5	4.30	4.16	4.22	4.09	4.04
7.5	4.78	4.57	4.76	4.63	4.52
12.0	5.11	4.74	4.82	4.74	4.68
21.0	5.09	4.89	4.93	4.82	4.91
27.0	5.10	4.95	4.87	4.88	4.93
(Td)	297 K	373 K	473 K	573 K	673 K

Source: Adapted from McCorkle, D. L. et al., *J. Phys. B: At. Mol. Phys.*, 11, 3067, 1978.

Note: See Figure 117.11 for graphical presentation.

TABLE 117.14
Drift Velocity of Electrons in C_3H_8 at 293 K

***EN* (Td)**	***W* (10⁴ m/s)**	***E/N* (Td)**	***W* (10⁴ m/s)**	***E/N* (Td)**	***W* (10⁴ m/s)**
0.14	0.228	0.60	1.08	2.50	3.40
0.16	0.262	0.70	1.26	3.00	3.67
0.18	0.297	0.80	1.44	3.50	3.88
0.20	0.332	1.00	1.78	4.00	4.05
0.25	0.421	1.20	2.09	4.50	4.17
0.30	0.514	1.40	2.36	5.00	4.25
0.35	0.610	1.60	2.61	6.00	4.45
0.40	0.705	1.80	2.82	7.00	4.59
0.50	0.894	2.00	3.00		

Source: Adapted from Schmidt B. and M. Roncossek, *Aust. J. Phys.*, 45, 351, 1992.

117.14 GAS CONSTANTS

Gas constants according to the expression

$$\frac{\alpha}{N} = F \exp -\left(\frac{GN}{E}\right) \tag{117.2}$$

are (Raju, 2005) $F = 6.5 \times 10^{-20}$ m^2 and $G = 973$ Td in the range $100 \le E/N \le 1900$ Td.

C_3H_8

TABLE 117.15
Diffusion Coefficients

E/N (Td)	D_r/μ (eV)	D_L/μ (eV)	*E/N* (Td)	D_r/μ (eV)	D_L/μ (eV)
0.120	0.0316		1.00	0.0547	0.0476
0.140	0.0319		1.20	0.0629	0.0481
0.160	0.0321	0.0276	1.40	0.0709	0.0484
0.180	0.0325	0.0282	1.60	0.0758	0.0487
0.200	0.0328	0.0288	1.80	0.0785	0.0491
0.250	0.0336	0.0306	2.00	0.0818	0.0494
0.300	0.0346	0.0327	2.50	0.102	0.0491
0.350	0.0356	0.0349	3.00	0.133	0.0503
0.400	0.0369	0.0371	3.50	0.179	0.0507
0.500	0.0400	0.0410	4.00	0.228	0.0500
0.600	0.0434	0.0441	4.50		0.0499
0.700	0.0464	0.0460	5.00		0.0508
0.800	0.0489	0.0468	6.00		0.0553

Note: D_r = radial diffusion coefficient; D_L = longitudinal diffusion coefficient; μ = mobility.

TABLE 117.16
Reduced Ionization Coefficients in C_3H_8

E/N (Td)	α/N (10^{-20} m^2)	*E/N* (Td)	α/N (10^{-20} m^2)
100	0.0003	700	1.398
125	0.0024	800	1.679
150	0.009	900	1.964
175	0.026	1000	2.259
200	0.051	1250	2.945
225	0.089	1500	3.550
250	0.134	2000	4.740
300	0.248	2500	6.002
350	0.378	3000	7.134
400	0.522	4000	8.116
450	0.658	4500	8.254
500	0.792	5000	8.373
600	1.100	5500	8.635

Source: Adapted from Heylen, A. E. D., *Int. J. Electr.*, 39, 653, 1975.
Note: See Figure 117.13 for graphical presentation and comparison with methane and ethane.

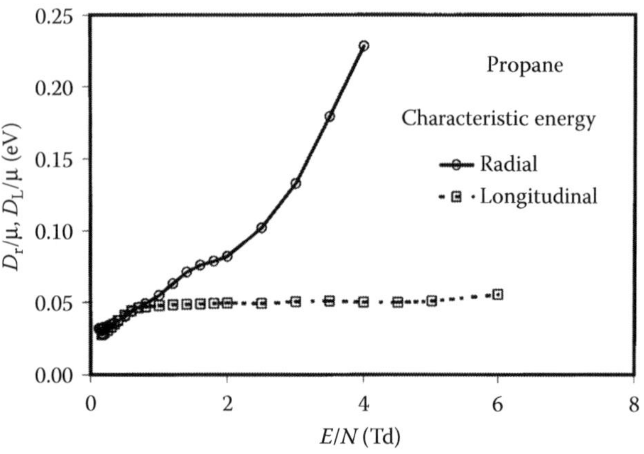

FIGURE 117.12 Radial and longitudinal diffusion energy in C_3H_8. (—) Radial diffusion; (—□—) longitudinal diffusion. (Adapted from Schmidt B. and M. Roncossek, *Aust. J. Phys.*, 45, 351, 1992.)

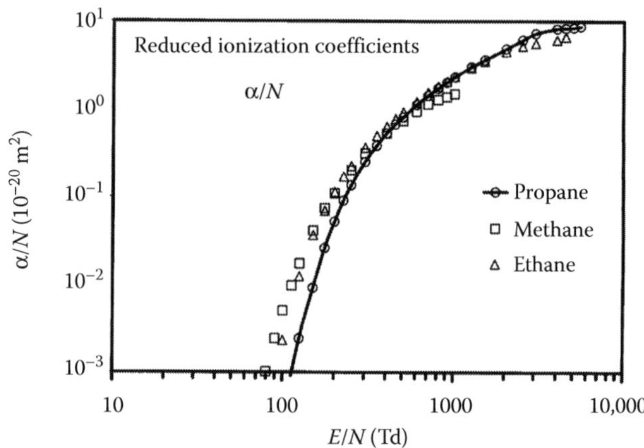

FIGURE 117.13 Reduced ionization coefficients in C_3H_8 (Heylen, 1975) and comparison with methane (Davies et al., 1989) and ethane (Heylen, 1975).

REFERENCES

Allan, M. and L. Andric, *J. Chem. Phys.*, 105, 3559, 1996.

Au, J. W., G. Cooper, G. R. Burton, T. N. Olney, and C. E. Brion, *Chem. Phys.*, 173, 209, 1993.

Boesten, L., M. A. Dillon, H. Tanaka, M. Kimura, and H. Sato, *J. Phys. B: At. Mol. Opt. Phys.*, 27, 1845, 1994.

Brüche, E., *Ann. Phys. Liepzig*, 4, 387, 1930.

Christophorou, L. G., J. K. Olthoff, and M. V. V. S. Rao, *J. Phys. Chem. Ref. Data*, 26, 1, 1997.

Davies, D. K., L. E. Kline, and W. E. Bies, *J. Appl. Phys.*, 85, 3311, 1989.

Duric, N., I. Cadež, and M. Kurepa, *Int. J. Mass Spectrom. Ion. Proc.*, 108, R1, 1991.

Fiegele, T., G. Hanel, I. Torres, M. Lezius, and T. D. Märk, *J. Phys. B: At. Mol. Phys.*, 33, 4263, 2000.

Floeder, K., D. Frommer, W. Raith, A. Schwab, and G. Sinapius, *J. Phys. B: At. Mol. Phys.*, 18, 3347, 1985.

Floriano, M. A., N. Gee, and G. R. Freeman, *J. Chem. Phys.*, 84, 6799, 1986.

Gee, N. and G. R. Freeman, *J. Chem. Phys.*, 78, 1951, 1983.

Grill, V., G. Walder, D. Margreiter, T. Rauth, H. U. Poll, P. Scheier, and T. D. Mark, *Z. Phys. D*, 25, 217, 1993.

Heylen, A. E. D., *Int. J. Electr.*, 39, 653, 1975.

Hwang, M., Y.-K. Kim, and M. E. Rudd, *J. Chem. Phys.*, 104, 2956, 1996.

Kim, Y.-K., K. K. Irikura, M. E. Rudd, M. A. Ali, P. M. Stone, J. Chang, J. S. Coursey, R. A. Dragoset, A. R. Kishore, K. J. Olsen, A. M. Sansoretti, G. G. Wiersma, D. S. Zucker, and M. A. Zucker, *Electron-Impact Ionization Cross Section for Ionization and Excitation Database (version 3.0)*, http://Physics.nist.gov./ionxsec, National Institute of Standards and Technology, Gaithesberg, MD, USA, 2004.

Lassettre, E. N. and S. A. Francis, *J. Chem. Phys.*, 40, 1208, 1964.

Lunt, S. L., J. Randell, J.-P. Ziesel, G. Mrotzek, and D. Field, *J. Phys. B: At. Mol. Opt. Phys.*, 31, 4225, 1998.

McCorkle, D. L., L. G. Christophorou, D. V. Maxey, and J. G. Carter, *J. Phys. B: At. Mol. Phys.*, 11, 3067, 1978.

Nishimura, H. and H. Tawara, *J. Phys. B: At. Hol. Opt. Phys.*, 24, L363, 1991.

Nishimura, H. and H. Tawara, *J. Phys. B: At. Mol. Opt. Phys.*, 27, 2063, 1994.

Raju, G. G. *Gaseous Electronics: Theory and Practice*, Taylor & Francis LLC, Boca Raton, FL, 2005.

Rutkowsky, V. J., H. Drost, and H.-J. Spangenberg, *Ann. Der. Phys. Leipzig*, 37, 259, 1980.

Schmidt, B. and M. Roncossek, *Aust. J. Phys.*, 45, 351, 1992.

Schram, B. L., M. J. Van der Wiel, F. J. de Heer, and H. R. Moustafa, *J. Chem. Phys.*, 44, 49, 1966.

Szmytkowski, C. and S. Kwitnewski, *J. Phys. B: At. Mol. Opt. Phys.*, 35, 3781, 2002.

Tanaka, H., Y. Tachibana, M. Kitajima, O. Sueoka, H. Takaki, A. Hamada, and M. Kimura, *Phys. Rev. A*, 59, 2006, 1999.

Section XII

12 ATOMS

118

BENZENE AND DEUTERATED BENZENE

C_6H_6
C_6D_6

CONTENTS

Benzene (C_6H_6) is an aromatic molecule with a ring structure, the six carbon atoms arranged in the form of a regular hexagon. The C–C bond length is 0.139 nm. The molecule has 42 electrons, a large electronic polarizability (Cho et al., 2001) of 11.48×10^{-40} F m^2 (9.84×10^{-40} F m^2 for C_6F_6), and an ionization potential of 9.244 eV. It does not have a dipole moment but possesses a quadrupole moment (Ritchie and Watson, 2000) of -30.4×10^{-40} C m^2 (in comparison, C_6F_6 has a quadrupole moment of 28.3×10^{-40} C m^2). Deuterated C_6H_6 (also designated as C_6H_6-d_6) has the hydrogen atoms replaced by its isotope deuterium, and data for C_6D_6 are provided as well in this chapter. The molecules each have 20 vibrational modes (Shimanouchi, 1972) as shown in Tables 118.1 and 118.2. CH and CD stretching modes are centered around 370 meV, whereas at the lower end, ring deformation is centered around 50 meV. The molecular orbital sequence of the 42 electrons is given by Boechat-Roberty et al. (2004).

Excitation level of 6.8 eV makes a strong contribution to the excitation cross section.

TABLE 118.1
Vibrational Modes and Energies of C_6H_6

	Type of Motion	Energy (meV)	Type of Motion	Energy (meV)
Mode	Shimanouchi (1972)		Gallup (1993)	
v_1	CH stretch	379.6	C–H stretch	380
v_2	Ring stretch	123.0	C–C stretch	123
v_3	CH bend	164.4	C–H ∥ bend	148
v_4	CH bend	83.4	C–H ⊥ bend	83
v_5	CH stretch	380.4	C–H stretch	380
v_6	Ring deformation	125.2	C–C–C ∥ bend	125
v_7	CH bend	123.4	C–H ⊥ bend	188
v_8	Ring deformation	87.2	C–C–C ⊥ bend	67
v_9	Ring stretch	162.4	C–C stretch	230
v_{10}	CH bend	142.6	C–H ∥ bend	142
v_{11}	CH bend	105.3	CH ⊥ bend	105
v_{12}	CH stretch	379.8	C–H stretch	384
v_{13}	Ring stret + deform	184.2	C–C stretch	184

continued

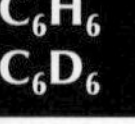

TABLE 118.1 (continued)
Vibrational Modes and Energies of C_6H_6

Mode	Type of Motion	Energy (meV)	Type of Motion	Energy (meV)
	Shimanouchi (1972)		Gallup (1993)	
ν_{14}	CH bend	128.7	C–H ‖ bend	129
ν_{15}	CH stretch	377.8	C–H stretch	376
ν_{16}	Ring stretch	197.9	C–C stretch	196
ν_{17}	CH bend	146.1	C–H ‖ bend	146
ν_{18}	Ring deformation	75.1	C–C–C ‖ bend	75
ν_{19}	CH bend	120.9	C–H ⊥ bend	144
ν_{20}	Ring deformation	50.8	C–C–C ⊥ bend	50

Source: Adapted from Shimanouchi, T., *Tables of Molecular Vibrational Frequencies Consolidated Volume 1*, NSRDS-NBS, 39, Washington (DC), 1972.

TABLE 118.2
Vibrational Modes and Energies of C_6D_6

Mode	Type of Motion	Energy (meV)	Mode	Type of Motion	Energy (meV)
ν_1	CD stretch	284.3	ν_{11}	CD bend	82.1
ν_2	Ring stretch	116.9	ν_{12}	CD stretch	283.6
ν_3	CD bend	128.6	ν_{13}	Ring stret + deform	165.5
ν_4	CD bend	61.6	ν_{14}	CD bend	100.9
ν_5	CD stretch	284.2	ν_{15}	CD stretch	280.8
ν_6	Ring deformation	120.1	ν_{16}	Ring stretch	192.4
ν_7	CD bend	102.5	ν_{17}	CD bend	142.5
ν_8	Ring deformation	74.5	ν_{18}	Ring deformation	71.5
ν_9	Ring stretch	159.4	ν_{19}	CD bend	98.6
ν_{10}	CD bend	102.2	ν_{20}	Ring deformation	43.6

Source: Adapted from Shimanouchi, T., *Tables of Molecular Vibrational Frequencies Consolidated Volume 1*, NSRDS-NBS, 39, Washington (DC), 1972.

118.1 SELECTED REFERENCES FOR DATA

See Table 118.3.

118.2 TOTAL SCATTERING CROSS SECTION

Table 118.4 shows the total scattering cross sections in C_6H_6 and deuterated C_6H_6. The highlights of the cross section curve (Figure 118.1) are

1. A slow increase toward lower energy, below 200 meV, attributed partially to rotationally inelastic excitations and virtual state scattering (Field et al., 2001)
2. A succession of peaks in the region 1.1–1.4 eV attributed to vibrational excitations (Gulley et al., 1998)
3. A resonance at 4.8 eV due to vibrational excitation (Wong and Schulz, 1975)
4. A broad maximum centered near 8.5 eV (see Możejko et al., 1996) attributed to short lifetime resonance, also observed in many gases

TABLE 118.3
Selected References

Cross Section	Energy (eV)	Reference
Q_i	Onset–5000	Kim et al. (2004)
Q_{diff}	1000	**Boechat-Roberty et al. (2004)**
Q_T	0.2–1000	**Makochekanwa et al. (2004)**
Q_T	0.2–1000	**Makochekanwa et al. (2003)**
Q_{el}	1.1–40	**Cho et al. (2001)**
Q_v	<200 meV	**Field et al. (2001)**
Q_{el}	8.5, 20	**Gulley and Buckman (1999)**
Q_T	0.01–2.0	**Gulley et al. (1998)**
Q_T	0.6–3500	**Możejko et al. (1996)**
Q_T	1–400	**Sueoka (1988)**
Q_{ex}	—	**Frueholz et al. (1979)**
Q_v	0–5	Wong and Schulz (1975)
Q_i	600–12000	**Schram et al. (1966)**
Q_i	—	**Diebler et al. (1957)**
	Swarm Parameters	
Parameter	***E/N* Range (Td)**	**Reference**
W	0–1.5	**Christophorou et al. (1966)**
W	600–9000	**Schlumbohm (1965)**

Note: Q_{diff} = differential; Q_{el} = elastic; Q_i = ionization cross section; Q_M = momentum transfer; Q_T = total; Q_v = vibrational excitation; W = drift velocity; α/N = reduced ionization coefficient. Bold font indicates experimental study.

TABLE 118.4
Total Scattering Cross Sections in C_6H_6

Energy (eV)	Q_i (10^{-20} m²) Makochekanwa et al. (2003)	Możejko et al. (1996)	Sueoka (1988)
0.4	23.5		
0.6	23.5	32.7	
0.7		32.2	
0.8	24.9	32.9	
0.9		31.9	
1.0	28.1	32.5	27.4
1.2	31.6	32.9	28.4
1.4	33.4	34.2	29.9
1.6	34.2	33.1	30.0
1.8	33.0	32.7	29.9
2.0	32.6	34.2	28.9
2.2	32.9		29.2
2.5	32.6	35.7	29.7
2.8	34.1		32.0
3.0		38.4	
3.1	35.7		33.0
3.4	37.0		34.0
3.5	40.2		
3.7	39.1		35.7
4.0	40.7	43.8	38.1
4.5	43.1	48.6	39.8
5.0	45.4	51.0	40.6
5.5	46.6	50.9	40.7

TABLE 118.4 (continued)
Total Scattering Cross Sections in C_6H_6

	Q_i (10^{-20} m^2)		
Energy (eV)	**Makochekanwa et al. (2003)**	**Możejko et al. (1996)**	**Sueoka (1988)**
6.0	47.6	52.3	43.5
6.5	48.6	54.1	43.7
7.0	49.4	55.6	46.9
7.5	51.7	56.6	49.2
8.0	54.8	57.6	48.5
8.5	57.8	58.1	48.2
9.0	60.0	58.0	48.6
9.5	59.8	57.8	47.6
10	59.3	56.7	46.6
11	59.7	55.0	44.4
12	58.4	54.1	42.5
13	55.3		43.0
14	63.5	51.9	42.3
15	52.7		41.2
16	51.0	49.4	40.2
17	49.8		39.3
18	48.8	46.6	39.5
19	48.8		38.8
20	47.7	45.9	39.3
22	45.7	44.4	38.3
25	45.5	44.0	35.7
27		43.2	
30	44.8	41.1	35.1
35	44.2	40.3	33.5
40	45.8	39.7	31.7
45		38.3	
50	41.9	37.6	30.1
60	39.3	35.8	29.4
70	36.7	34.1	26.9
80	33.4	31.8	27.1
90	33.3	29.4	25.5
100	30.3	29.2	24.3
110		28.2	
120	29.3	27.3	22.7
140		24.6	
150	27.1		20.6
160		23.0	
180		22.5	
200	24.5	21.6	18.9
220		20.4	
250	20.5	18.1	16.5
275		17.8	
300	19.1	16.8	15.1
350		14.5	14.0
400	14.3	13.4	13.0
450		12.3	
500	12.7	11.3	
600	10.4	9.71	
700		8.52	
800	7.68	7.62	
900		7.05	
1000	6.15	9.47	
1100		5.96	
1250		5.47	

TABLE 118.4 (continued)
Total Scattering Cross Sections in C_6H_6

	Q_i (10^{-20} m^2)		
Energy (eV)	**Makochekanwa et al. (2003)**	**Możejko et al. (1996)**	**Sueoka (1988)**
1500		4.67	
1750		4.01	
2000		3.50	
2250		3.07	
2500		2.83	
3000		2.38	
3250		2.18	
3500		2.04	

C_6H_6
C_6D_6

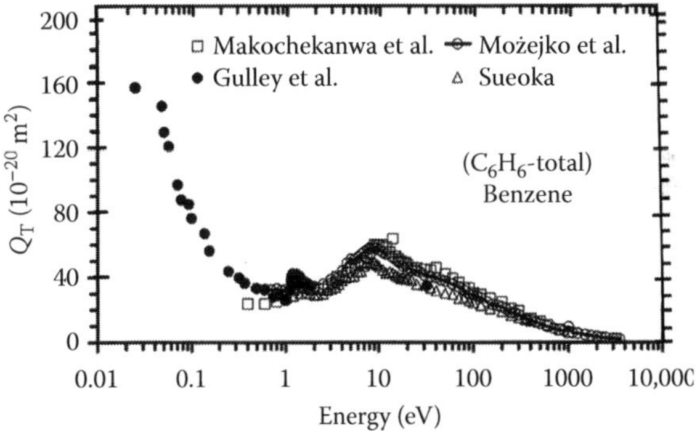

FIGURE 118.1 Total scattering cross section in C_6H_6. (□) Makochekanwa et al. (2003); (●) Gulley et al. (1998); Możejko et al. (1996); (Δ) Sueoka (1988). For details in the low-energy range, see Figure 118.2.

5. A monotonic decline with increase of energy beyond 10 eV

Figure 118.1 shows the total scattering cross section in C_6H_6. Figure 118.2 shows the expanded view of cross sections in the range from 0 to 2 eV (see Gulley et al., 1998). The effect of substituting the deuterium atoms is also shown. Table 118.5 shows the effect of deuteration on the resonance peaks.

118.3 DIFFERENTIAL SCATTERING CROSS SECTION

Differential cross sections in C_6H_6 in the energy range 1.1–40 eV are given by Cho et al. Figures 118.3 and 118.4 show the dependence of the differential scattering on the angle of scattering and electron energy. The highlights of the figures are

1. In the energy range 1.1–4.5 eV, oscillations are observed due to resonance phenomena (Figure 118.3).
2. For energy higher than about 8 eV, the cross section falls rapidly with increasing angle, suggesting forward scattering. The shape of the angle–cross section curves remains essentially the same for energy range 8.5–40 eV.

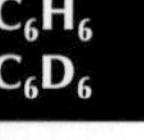

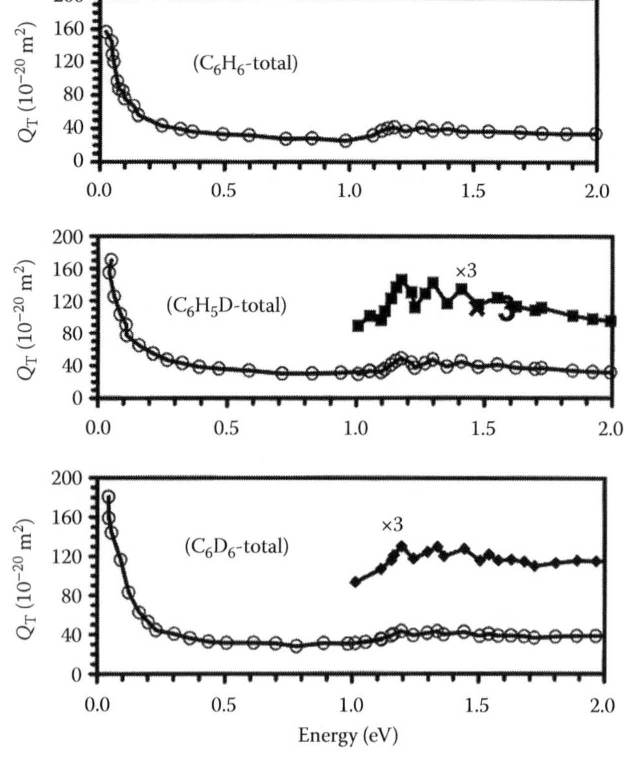

FIGURE 118.2 Total cross sections in the low-energy range for C_6H_6, mono-deuterated C_6H_6, and deuterated C_6H_6. (Adapted from Gulley, R. J. et al., *J. Phys. B: At. Mol. Opt. Phys.*, 31, 2735, 1998.)

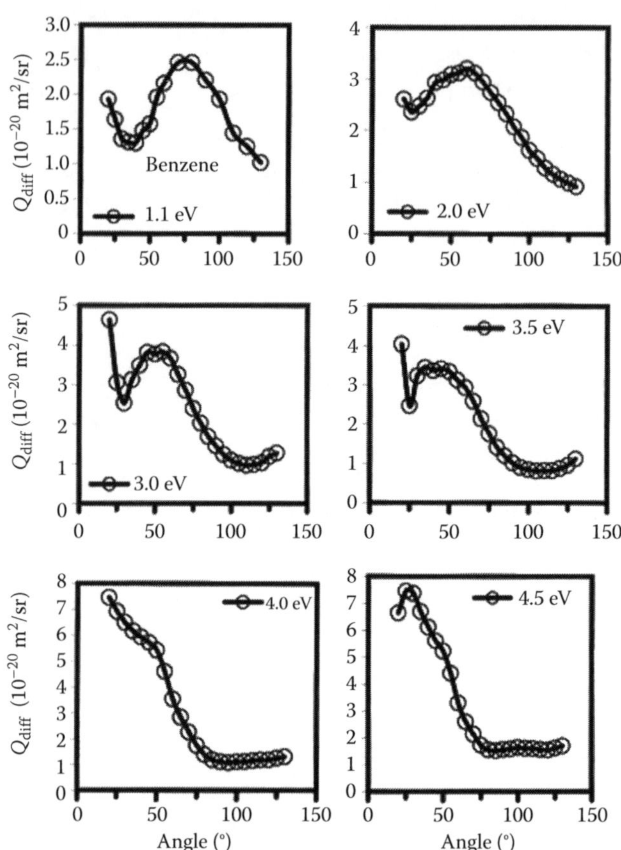

FIGURE 118.3 Differential cross sections in C_6H_6 at low electron energy. (Adapted from Cho, H. et al., *J. Phys. B: At. Mol. Opt. Phys.*, 34, 1019, 2001.)

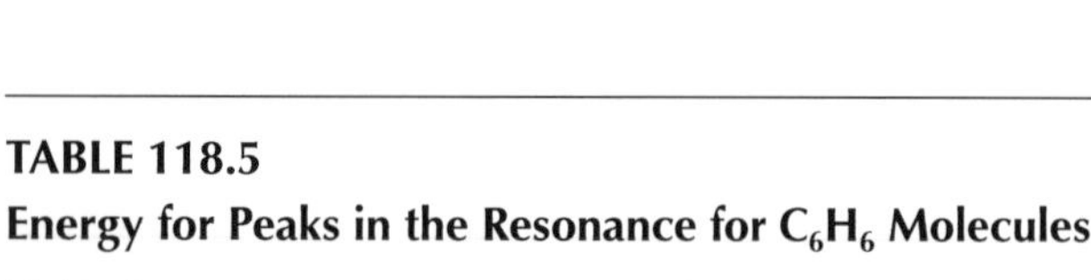

TABLE 118.5
Energy for Peaks in the Resonance for C_6H_6 Molecules

Molecule	Peak Energy (eV)		
C_6H_6	1.17	1.29	1.40
C_6D_6	1.196	1.31	1.42
C_6H_5D	1.175	1.295	1.40

118.4 ELASTIC AND MOMENTUM TRANSFER CROSS SECTIONS

Elastic and momentum transfer cross sections obtained from the measurements of Cho et al. (2001) are shown in Table 118.6. For graphical presentation, see Figure 118.5 in which the total scattering cross sections are plotted for comparison. The highlights of Figure 118.5 are

1. The cross sections decrease in the order: total, elastic, and momentum transfer.
2. The presence of resonance at 1.15, 4.5, and 8–9 eV is discernible.

118.5 EXCITATION CROSS SECTION

Frueholz et al. (1979) have measured the excitation potentials and relative cross sections by electron impact spectroscopy. The measured excitation potentials are 3.90, 5.59, 4.80, 6.25, 6.95, 8.14, 8.41, 8.72, 8.88, 9.50, 9.76, 9.90, 10.01, 10.51, 11.08, 11.61, 12.15, 13.08, 13.96, 14.26, 15.21, 15.81, and 17.45 eV.

118.6 IONIZATION CROSS SECTION

Measured ionization cross sections in C_6H_6 are available only in the high-energy range (Schram et al., 1966) as shown in Table 118.7. Theoretically calculated cross sections from onset energy to 5000 eV are reported by Kim et al. (2004) These data are shown in Figure 118.6. Positive ion appearance potentials and relative abundance are shown in Table 118.8.

118.7 ATTACHMENT PROCESSES

Mass spectrometer study has not revealed negative ion formation (Diebler et al., 1957).

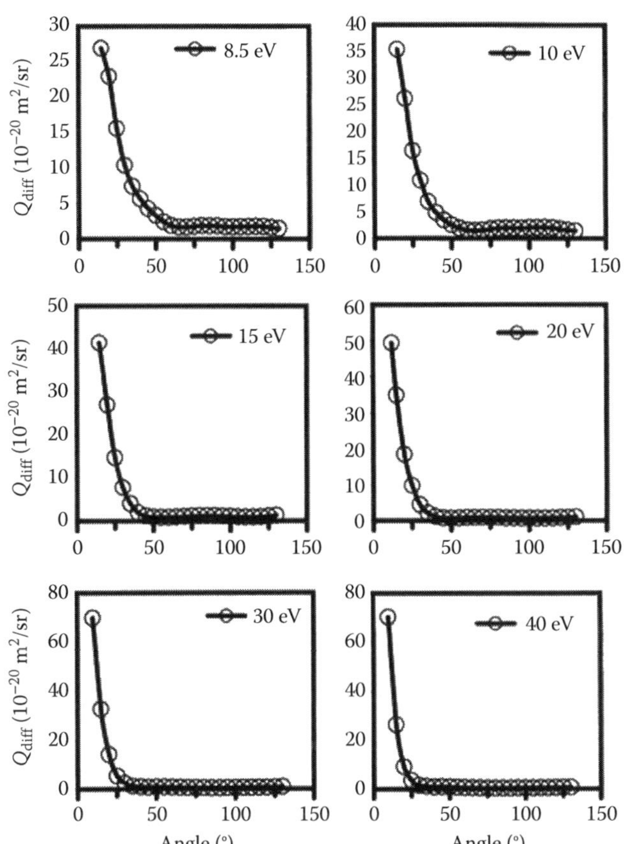

FIGURE 118.4 Differential scattering cross sections in C_6H_6 at intermediate electron energy. (Adapted from Cho, H. et al., *J. Phys. B: At. Mol. Opt. Phys.*, 34, 1019, 2001.)

TABLE 118.6
Integral Elastic and Momentum Transfer Cross Sections in C_6H_6

Energy (eV)	Q_{el} (10^{-20} m^2)	Q_M (10^{-20} m^2)
1.1	21.21	19.12
2.0	25.68	19.96
3.0	28.96	21.49
3.5	24.74	18.95
4.0	33.41	22.10
4.9	35.64	25.37
6.0	36.09	24.05
8.5	44.94	23.88
10.0	47.75	24.29
15.0	33.79	14.95
20.0	33.33	15.06
30.0	31.08	11.66
40.0	26.90	9.19

Source: Adapted from Cho, H. et al., *J. Phys. B: At. Mol. Opt. Phys.*, 34, 1019, 2001.

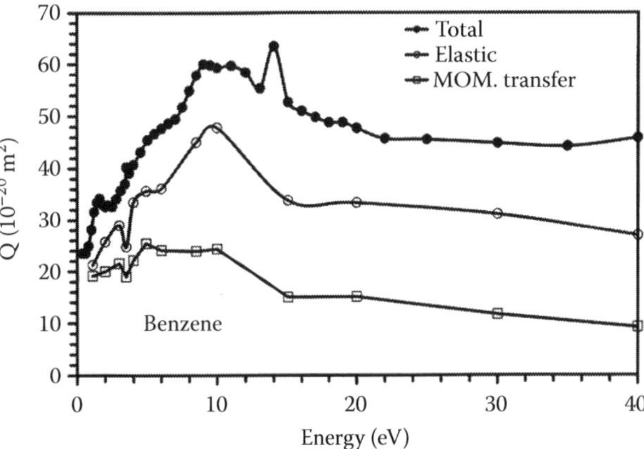

FIGURE 118.5 Integral elastic and momentum transfer cross sections in C_6H_6. Total scattering cross sections from Makochekanwa et al. (2003) are also shown for comparison. (Adapted from Cho, H. et al., *J. Phys. B: At. Mol. Opt. Phys.*, 34, 1019, 2001.)

TABLE 118.7
Ionization Cross Section in C_6H_6

Energy (keV)	0.6	1	2	4	7	12
Q_i (10^{-20} m^2)	7.83	5.37	3.12	1.77	1.11	0.698

Source: Adapted from Schram B. L. et al., *J. Chem. Phys.*, 44, 49, 1966

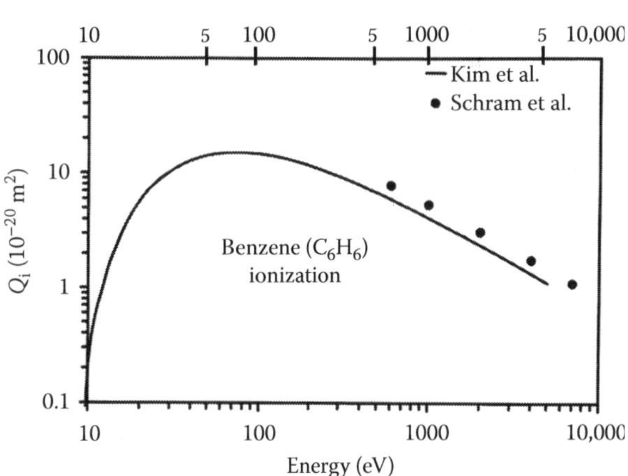

FIGURE 118.6 Ionization cross sections in C_6H_6. (—) Kim et al. (2004); (•) Schram et al. (1966).

118.8 DRIFT VELOCITY OF ELECTRONS

Schlumbhom (1965) has measured the drift velocity of electrons as a function of *E*/*N* and observed the relationship (see Figure 118.7)

$$W_e = 9.76 \times 10^3 \sqrt{\frac{E}{N}}(\text{Td}) \quad \text{m/s} \qquad (118.1)$$

in the range $600 \leq E/N \leq 9000$ Td.

C$_6$H$_6$ C$_6$D$_6$

TABLE 118.8
Positive Ion Appearance Potentials

Ion	Appearance Potential (eV)	Relative Abundance
$C_6H_6^+$	9.24	100
$C_6H_5^+$	14.5	13.7
$C_5H_3^+$	16.8	2.95
$C_5H_2^+$	19.1	0.64
C_5H^+	27.4	0.57
$C_4H_2^+$	18.3	15.6
$C_3H_3^+$	16.1	13.2
$C_3H_2^+$	23.0	5.40
C_3H^+	27.4	4.03
CH_3^+	—	0.93
CH^+	—	0.24
H^+	—	1.25

Source: Adapted from Diebler, V. H., R. M. Reese, and F. L. Mohler, *J. Chem. Phys.*, 26, 304, 1957.

TABLE 118.9
Drift Velocity of Electrons in C_6H_6

E/N (Td)	0.3	0.45	0.6	0.9	1.2	1.5
W (10^4 m/s)	0.09	0.12	0.11	0.16	0.21	0.26

Source: Adapted from Christophorou, L. G., G. S. Hurst, and A. Hadjiantoniou, *J. Chem. Phys.*, 44, 3506, 1966.

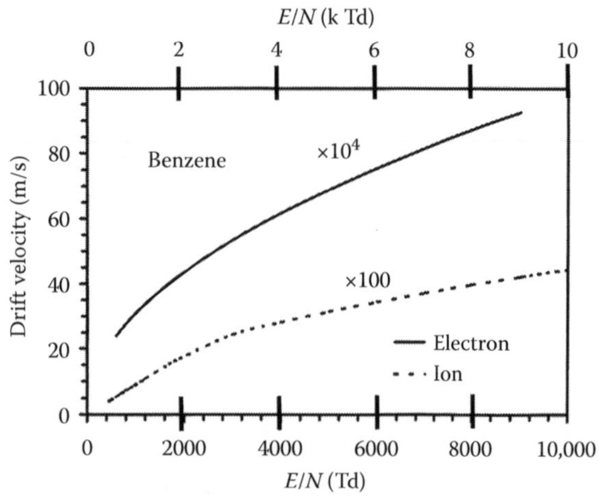

FIGURE 118.7 Measured drift velocity of electrons and ions in C_6H_6. (Adapted from Schlumbohm, H., *Z. Phys.*, 182, 317, 1965.)

118.9 DRIFT VELOCITY OF POSITIVE IONS

Schlumbhom (1965) has measured drift velocity of positive ions as a function of *E/N* and observed the relationship

$$W_+ = 0.90 \times \frac{E}{N}(\mathrm{Td})\ \mathrm{m/s} \quad 450 \le E/N \le 1800\,\mathrm{Td} \qquad (118.2)$$

$$W_+ = 44.5\sqrt{\frac{E}{N}(\mathrm{Td})}\ \mathrm{m/s} \quad 3000 \le E/N \le 60{,}000\,\mathrm{Td} \qquad (118.3)$$

These values are shown in Figure 118.7.

REFERENCES

Boechat-Roberty H. M., M. L. M. Rocco, C. A. Lucas, and G. G. B. de Souza, *J. Phys. B: At. Mol. Opt. Phys.*, 37, 1467, 2004.

Cho H., R. J. Gulley, K. Sunohara, M. Kitajima, L. J. Uhlmann, H. Tanaka, and S. J. Buckman, *J. Phys. B: At. Mol. Opt. Phys.*, 34, 1019, 2001.

Christophorou L. G., G. S. Hurst, and A. Hadjiantoniou, *J. Chem. Phys.*, 44, 3506, 1966.

Diebler V. H., R. M. Reese, and F. L. Mohler, *J. Chem. Phys.*, 26, 304, 1957.

Field D., J.-P. Ziesel, S. L. Lunt, R. Parthasarathy, L. Suess, S. B. Hill, F. B. Dunning, R. R. Lucchese, and F. A. Gianturco, *J. Phys. B: At. Mol. Opt. Phys.*, 34, 4371, 2001.

Frueholz R. P., W. M. Flicker, O. A. Mosher, and A. Kupperman, *J. Chem. Phys.*, 70, 3057, 1979.

Gallup G. A., *J. Chem. Phys.*, 99, 827, 1993.

Gulley R. J. and S. J. Buckman, *J. Phys. B: At. Mol. Opt. Phys.*, 32, L405, 1999.

Gulley R. J., S. L. Lunt, J.-P. Ziesel, and D. Field, *J. Phys. B: At. Mol. Opt. Phys.*, 31, 2735, 1998.

Kim Y.-K., K. K. Irikura, M. E. Rudd, M. A. Ali, P. M. Stone, J. Chang, J. S. Coursey et al., *Electron-Impact Ionization Cross Section for Ionization and Excitation Database (Version 3.0)*. National Institute of Standards and Technology, Gaithesberg, MD, USA, 2004. Available at http://Physics.nist.gov/ionxsec.

Makochekanwa C., O. Sueoka, and M. Kimura, *J. Phys. B: At. Mol. Opt. Phys.*, 37, 1841, 2004.

Makochekanwa C., O. Sueoka, and M. Kimura, *Phys. Rev. A*, 68, 032707, 2003.

Możejko P., G. Kasperski, C. Szmytkowski, G. P. Karwasz, R. S. Brusa, and A. Zecca, *Chem. Phys. Lett.*, 257, 309, 1996.

Ritchie G. L. D. and J. N. Watson, *Chem. Phy. Lett.*, 322, 143, 2000.

Schlumbohm H., *Z. Phys.*, 182, 317, 1965.

Schram B. L., M. J. Van der Wiel, F. J. De Heer, and H. R. Moustafa, *J. Chem. Phys.*, 44, 49, 1966.

Shimanouchi T., *Tables of Molecular Vibrational Frequencies Consolidated Volume 1*, NSRDS-NBS, 39, Washington (DC), 1972.

Sueoka O., *J. Phys. B: At. Mol. Opt. Phys.*, 21, L631, 1988.

Wong S. F. and G. J. Schulz, *Phys. Rev. Lett.*, 35, 1429, 1975.

119

BROMOBENZENE

CONTENTS

Bromobenzene (C_6H_5Br) is a polar, electron-attaching gas that has one hydrogen atom from benzene replaced with bromine atom and it has 76 electrons. The electronic polarizability is 16.36×10^{-40} F m^2, dipole moment 1.70 D, and ionization potential 9.00 eV. Resonances due to electron attachment are measured at 0.67 and 4.40 eV (Olthoff et al., 1985).

119.1 TOTAL SCATTERING CROSS SECTION

Integral electron scattering cross sections at low electron energies have been measured by Lunt et al. (1999). and shown in Figure 119.1. Low-energy scattering behavior is observed to be governed by the dipole moment of the target, the cross section increases as the energy decreases toward zero. Resonance is observed at 0.69 eV.

Electron attachment cross section at low energy (<0.1 eV) is not significant (Lunt et al., 1999).

119.2 ATTACHMENT RATE CONSTANT

The product ion is Br^-. Thermal attachment rate constant for C_6H_5Br is found to be 6.5×10^{-18} m^3/s (Shimamori et al., 1993), which is lower than that determined by Christophorou et al. (1966) 2.3×10^{-16} m^3/s. Figure 119.2 shows the attachment rate constant and Figure 119.3 shows the attachment cross section determined by unfolding procedure (Shimamori et al., 1995). For comparison, the *s*-wave maximum cross section has also been shown. The very small magnitude of the attachment cross section shows that attachment is not a significant process at zero energy.

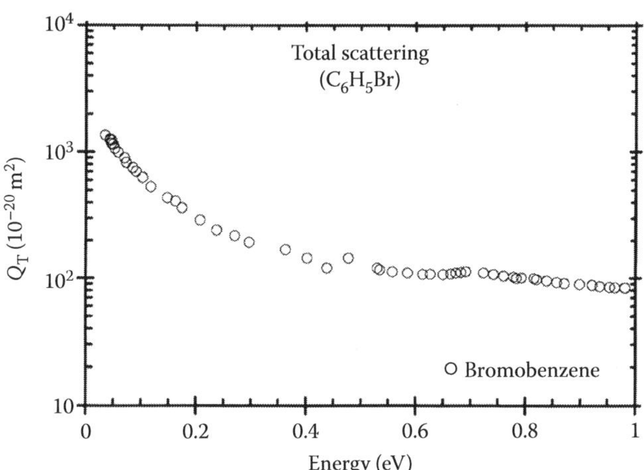

FIGURE 119.1 Total scattering cross sections for C_6H_5Br at low electron energies. (Adapted from Lunt S. L. et al., *J. Phys. B: At. Mol. Phys.*, 32, 2707, 1999.)

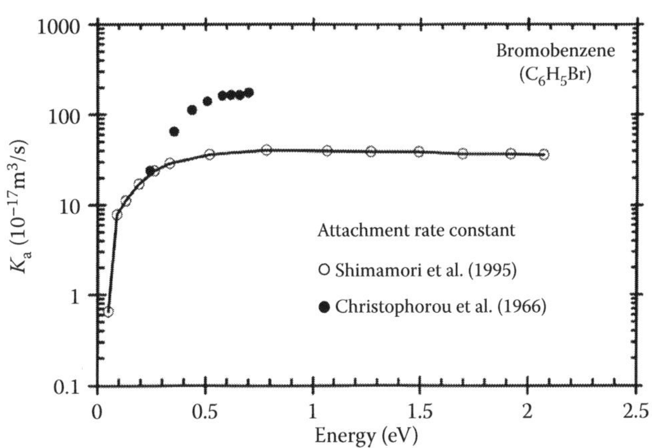

FIGURE 119.2 Attachment rate constant for C_6H_5Br as a function of mean electron energy. (—○—) Shimamori et al. (1995); (•) Christophorou et al. (1966).

C_6H_5Br

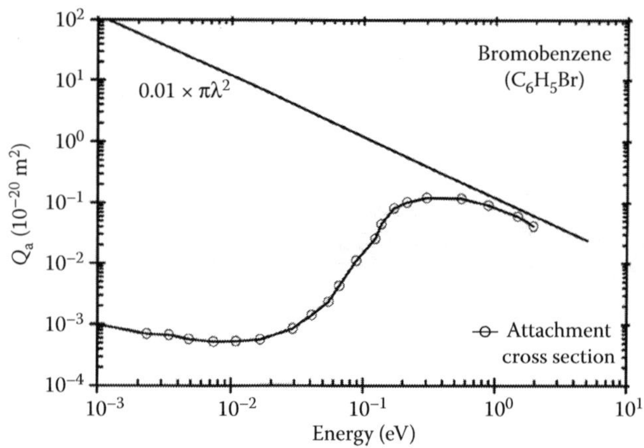

FIGURE 119.3 Attachment cross sections for C_6H_5Br. The *s*-wave maximum cross section is several orders of magnitude larger. (Adapted from Shimamori, H. et al., *Chem. Phys. Lett.*, 232, 115, 1995.)

TABLE 119.1
Reduced Attachment Coefficients as Function of Reduced Electric Field for C_6H_5Br

E/N (Td)	η/N (10^{-20} m²)
0.66	5.94
1.0	15.05
1.35	23.14
1.70	26.56
2.05	28.24
2.40	27.46
2.75	25.63
3.05	24.72

Source: Adapted from Christophorou, L. G. et al., *J. Chem. Phys.*, 45, 536, 1966.

119.3 ATTACHMENT COEFFICIENTS

Table 119.1 shows the attachment cross sections at low values of the reduced electric field (Christophorou et al., 1966).

REFERENCES

Christophorou, L. G., R. N. Compton, G. S. Hurst, and P. W. Reinhardt, *J. Chem. Phys.*, 45, 536, 1966.

Lunt, S. L., D. Field, S. V. Hoffmann, R. J. Gulley, and J.-P. Ziesel, *J. Phys. B: At. Mol. Opt. Phys.*, 32, 2707, 1999.

Olthoff, J. K., J. A. Tossell, and J. H. Moore, *J. Chem. Phys.*, 83, 5627, 1985.

Shimamori, H., Y. Tatsimu, and T. Sunagawa, *J. Chem. Phys.*, 99, 7787, 1993.

Shimamori, H., T. Sunagawa, Y. Ogawa, and Y. Tatsumi, *Chem. Phys. Lett.*, 232, 115, 1995.

120

BUTENE

CONTENTS

Butene (C_4H_8) is a hydrocarbon belonging to the class of alkenes (C_nH_{2n}) and there are three isomers: 1-butene (also called α-butylene), 2-butene (also called β-butylene), and isobutene (also called isobutylene) (Sharp, 2003). There are six gases having the same chemical formula each with 32 electrons. The electronic polarizability (α_e), dipole moment (D), and ionization energy (ε_i) are shown Table 120.1.

The vibrational modes and energies of selected butenes are shown in Table 120.2.

Excitation levels have been determined by Dance and Walker (1973) by the electron trap method, as shown in Table 120.3.

120.1 SELECTED REFERENCES FOR DATA

See Table 120.4.

TABLE 120.1
Selected Properties of C_4H_8 Molecules

Gas	α_e (10^{-40} F m^2)	μ (*D*)	ε_i (eV)
1-Butene	8.868	0.438	9.55
cis-2-Butene		0.253	9.11
trans-2-Butene	9.446		9.10
Isobutene		0.503	9.239
Cyclobutane			9.82
Methylcyclopropane	9.224	0.139	9.46

Source: Adapted from Lide, D. R., *Handbook of Chemistry and Physics* (86th edn), CRC Press, Taylor & Francis LLC, Boca Raton, FL, 2005–2006.

Note: α_e = Electronic polarizability, μ = dipole moment, ε_i = ionization energy.

TABLE 120.2
Vibrational Modes and Energies of Selected Butanes

Cyclobutane			2-Methylpropene		
Mode	Type of Motion	Energy (meV)	Mode	Type of Motion	Energy (meV)
ν_1	CH_2 s-stretch	358.9	ν_1	CH_2 s-stretch	370.6
ν_2	CH_2 scissoring	178.9	ν_2	CH_2 d-stretch	364.6
ν_3	Ring stretch	124.1	ν_3	CH_3 s-stretch	360.9
ν_4	CH_2 a-stretch	368.9	ν_4	C=C stretch	205.9
ν_5	CH_2 rock	91.9	ν_5	CH_3 d-deform	182.3
ν_6	Ring puckering	24.4	ν_6	CH_2 scissoring	175.6
ν_7	CH_2 wag	156.2	ν_7	CH_3 s-deform	169.4
ν_8	CH_2 twist	155.8	ν_8	CH_3 rock	131.9
ν_9	CH_2 wag	151.1	ν_9	C–C stretch	99.3
ν_{10}	Ring deformation	114.8	ν_{10}	C=CC_2 ip-deform	47.5
ν_{11}	CH_2 twist	151.5	ν_{11}	CH_3 d-stretch	368.2
ν_{12}	CH_2 s-stretch	358.7	ν_{12}	CH_3 d-deform	180.9
ν_{13}	CH_2 scissoring	178.9	ν_{13}	CH_3 rock	133.4
ν_{14}	Ring deformation	124.1	ν_{14}	CH_2 twist	121.6
ν_{15}	CH_2 a-stretch	370.3	ν_{15}	CH_3 torsion	23.9
ν_{16}	CH_2 rock	81.5	ν_{16}	CH_2 a-stretch	382.6
ν_{17}	CH_2 a-stretch	366.0	ν_{17}	CH_3 d-stretch	369.5
ν_{18}	CH_2 twist	151.6	ν_{18}	CH_3 s-stretch	358.7
ν_{19}	CH_2 rock	92.9	ν_{19}	CH_3 d-deform	180.8
ν_{20}	CH_2 s-stretch	357.9	ν_{20}	CH_3 s-deform	171.2
ν_{21}	CH_2 scissoring	179.4	ν_{21}	C–C stretch	158.9
ν_{22}	CH_2 wag	155.8	ν_{22}	CH_3 rock	129.3
ν_{23}	Ring deformation	111.3	ν_{23}	CH_2 rock	120.8
			ν_{24}	C=CC_2 ip-deform	53.3
			ν_{25}	CH_3 d-stretch	365.1
			ν_{26}	CH_3 d-deform	179.0

continued

TABLE 120.2 (continued)
Vibrational Modes and Energies of Selected Butanes

Cyclobutane			2-Methylpropene		
Mode	Type of Motion	Energy (meV)	Mode	Type of Motion	Energy (meV)
			ν_{27}	CH_3 rock	133.8
			ν_{28}	CH_2 wag	110.3
			ν_{29}	$C{=}CC_2$ op-deform	53.2
			ν_{30}	CH_3 torsion	24.3

Source: Adapted from Shimanouchi, T. *Tables of Molecular Vibrational Frequencies*, Consolidated Volume I, NSRDS-NBS, 39, Washington (DC), 1972.

Note: a = Asymmetrical, d = degenerate, ip = inplane, op = out of plane, s = symmetrical.

C_4H_8

TABLE 120.3
Excitation Potentials in Butenes

Gas	Excitation Potential (eV)
1-Butene	2.0 (vibrational, resonance), 4.25, 5.8–6.2
cis-2-Butene	2.3 (vibrational, resonance), 4.3, 6.1, 7.0, 8.1
trans-2-Butene	2.4 (vibrational, resonance), 4.4, 6.0, 7.1

Source: Adapted from Dance, D. F. and I. C. Walker, *Proc. Roy. Soc. Lond., A*, 334, 259, 1973.

TABLE 120.4
Cross Sections

Quantity	Range (eV)	Gas	Reference
		Cross Sections	
Q_i	Onset-5000	1-, *trans*-2-, isobutene	Kim et al. (2004)
Q_T	5–400	1-Butene	Floeder et al. (1985)
Q_a	0–15	Iso, *cis*-, *trans*-butene	Rutkowsky (1980)
Q_i	600–12,000	1-, *trans*-2-, *cis*-2-C_4H_8	Schram et al. (1966)

Quantity	*E/N* Range (Td)	Gas	Reference
		Swarm Properties	
W	0.01–10	*trans*-2-, *cis*-2-, i-C_4H_8	Gee and Freeman (1983)
α/N	150–6000	1-Butene	Heylen (1978)
W	0.6–6.0	1-Butene	Bowman and Gordon (1967)

Note: D_L = Longitudinal diffusion, D_r = radial diffusion, W = drift velocity, α/N = reduced ionization coefficient, Q_a = Attachment, Q_{el} = elastic, Q_{ex} = excitation, Q_i = ionization, Q_M = momentum transfer, Q_T = total, and Q_v = vibrational.

120.2 TOTAL SCATTERING CROSS SECTIONS

Table 120.5 shows the total scattering cross sections in 1-butene and Figure 120.1 shows the graphical presentation. The peak in total cross section at 9 eV is attributed to the onset of excitation processes.

TABLE 120.5
Total Scattering Cross Sections in 1-Butene

Energy (eV)	Q_T (10^{-20} m^2)	Energy (eV)	Q_T (10^{-20} m^2)	Energy (eV)	Q_T (10^{-20} m^2)
5	39.44	18	49.90	80	30.80
6	45.17	20	48.80	90	29.29
7	51.24	25	47.77	100	28.21
8	54.41	30	44.66	150	23.22
9	55.02	35	43.49	200	19.21
10	54.27	40	43.00	250	16.80
12	52.58	50	40.10	300	15.36
14	52.58	60	36.51	350	13.92
16	51.60	70	33.19	400	11.52

Source: Adapted from Floeder, K., et al., *J. Phys. B: At. Mol. Opt. Phys.*, 18, 3347, 1985.

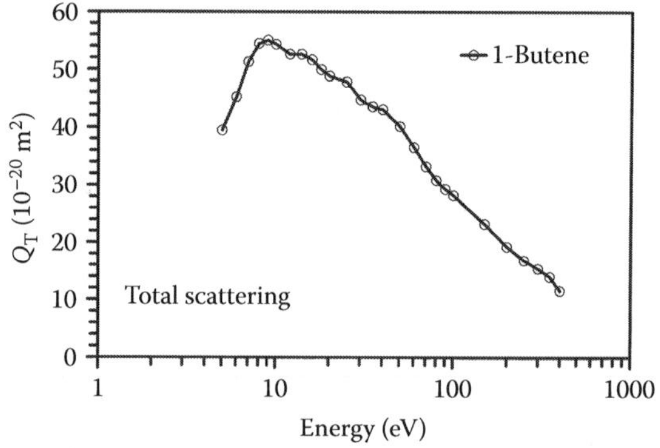

FIGURE 120.1 Total scattering cross section in 1-butene. (Adapted from Floeder, K. et al., *J. Phys. B: At. Mol. Opt. Phys.*, 18, 3347, 1985.)

TABLE 120.6
Ionization Cross Section in Butanes

	Q_i (10^{-20})		
Energy (keV)	1-Butene	*trans*-2-Butene	*cis*-2-Butene
0.6	5.65	5.25	5.65
1	3.61	3.56	3.78
2	2.11	2.06	2.20
4	1.20	1.06	1.28
7	0.749	0.731	0.781
12	0.471	0.459	0.490

Source: Adapted from Schram, B. L. et al., *J. Chem. Phys.*, 44, 49, 1966.

120.3 IONIZATION CROSS SECTION

Ionization cross sections in butenes at high energy are shown in Table 120.6 and Figure 120.2. The isomer effect is not significant in butenes.

120.4 ATTACHMENT CROSS SECTIONS

Table 120.7 and Figure 120.3 show attachment cross section in butanes (Von J. Rutkowsky, 1980). Both direct and dissociative attachments are observed. The attachment processes are

$$\text{e (between 5 and 7 eV)} + C_nH_{2n} \rightarrow C_nH_{2n}^{-*} \rightarrow \begin{cases} C_nH_{2n}^{-} \rightarrow H^- + C_nH_{2n-1} \\ CH^- + C_{n-1}H_{2n-1} \\ CH_2^- + C_{n-1}H_{2n-2} \end{cases} \quad (120.1a)$$

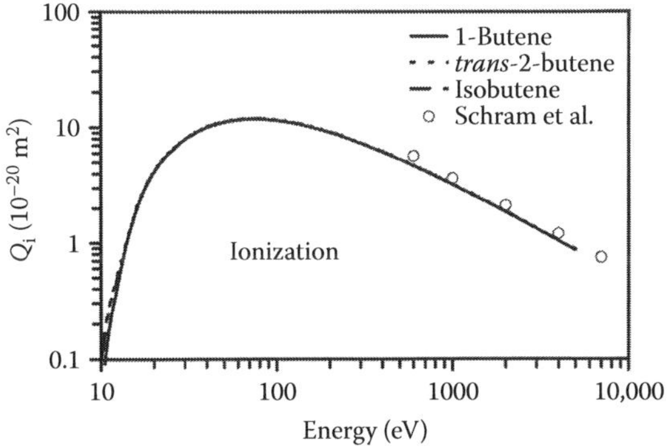

FIGURE 120.2 Ionization cross section in butanes. Symbols are experimental in 1-butene (Schram et al., 1966). Isomer effect is not observed. (Adapted from Kim, Y.-K. et al., *Electron-Impact Ionization Cross Section for Ionization and Excitation Database (version 3.0)*, http://Physics.nist.gov./ionxsec, National Institute of Standards and Technology, Gaithesberg, MD, USA, 2004.)

TABLE 120.7
Direct and Dissociative Attachment Cross Sections in Butanes

1-Butene Energy (eV)	1-Butene Q_a (10^{-20} m²)	*trans*-Butene Energy (eV)	*trans*-Butene Q_a (10^{-20} m²)	*cis*-Butene Energy (eV)	*cis*-Butene Q_a (10^{-20} m²)
0.00	8.44	0.00	1.86	0.00	0.62
0.20	13.08	0.10	2.5	0.40	0.70
0.40	0.07	0.20	1.13	0.60	3.15
0.60	1.76	0.30	9.02	0.80	5.07
0.80	4.30	0.40	12.04	1.00	4.62
1.00	3.01	0.60	8.68	1.20	2.09
1.20	1.07	0.80	3.69	1.40	0.67
1.40	0.17	1.00	1.20	1.70	0.10
1.70	0.01	1.20	0.47	2.00	0.05
2.00	0.16	1.40	0.53	3.50	0.15
2.50	0.46	1.70	0.71	4.00	0.12

TABLE 120.7 (continued)
Direct and Dissociative Attachment Cross Sections in Butanes

1-Butene Energy (eV)	1-Butene Q_a (10^{-20} m²)	*trans*-Butene Energy (eV)	*trans*-Butene Q_a (10^{-20} m²)	*cis*-Butene Energy (eV)	*cis*-Butene Q_a (10^{-20} m²)
3.00	0.23	2.00	0.79	4.50	0.04
3.50	0.07	2.50	0.75	5.00	0.12
4.00	0.07	3.00	0.50	5.50	2.97
4.50	0.08	3.50	0.07	6.00	7.31
5.00	0.31	4.00	0.01	6.50	9.21
5.50	1.96	4.50	0.04	7.00	5.15
6.00	7.87	5.00	1.56	7.50	1.17
6.50	14.11	5.50	3.28	8.00	0.65
7.00	9.29	6.00	5.14	8.50	0.75
7.50	5.07	6.50	5.15	9.00	2.13
8.00	3.70	7.00	3.58		
8.50	3.60	7.50	2.16		
9.00	4.08	8.00	0.91		
		8.50	1.30		
		9.00	2.90		

Source: Adapted from Von J. Rutkowsky, H. Drost, and H.-J. Spangenberg, *Ann. Phys.*, Leipzig, 37, 259, 1980. Digitized by the present author.

C_4H_8

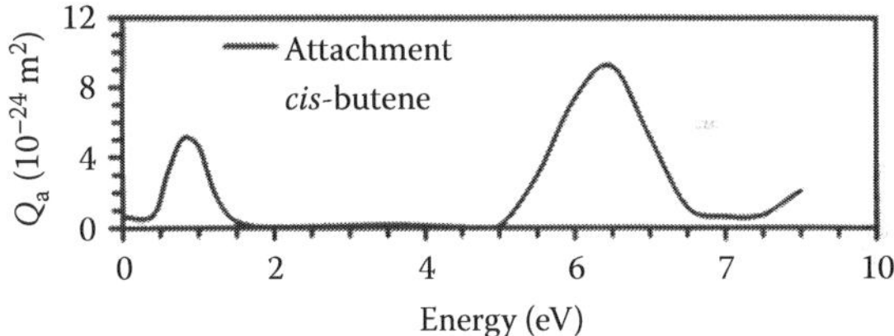

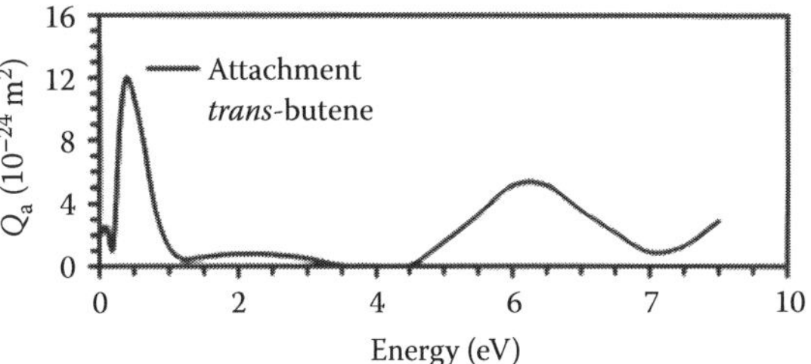

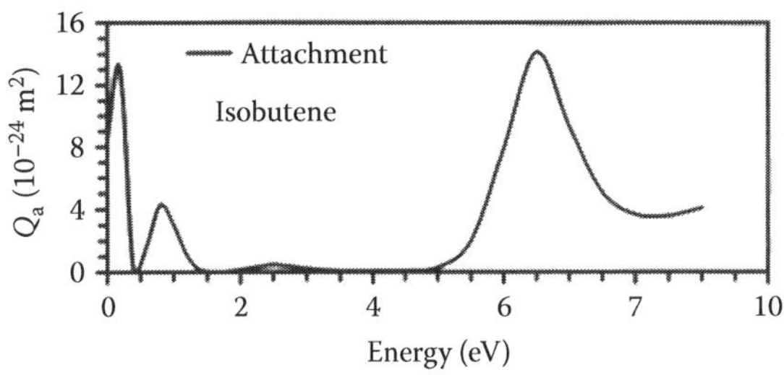

FIGURE 120.3 Attachment cross sections in butanes. Both direct and dissociative processes are observed. Note the low magnitude of the cross section. Marginal isomer effect is observed. (From Von J. Rutkowsky, H. Drost, and H.-J. Spangenberg, *Ann. Phys., Leipzig*, 37, 259, 1980.)

$$e(>7\,eV) + C_nH_{2n} \rightarrow C_nH_{2n}^{-*}$$
$$\rightarrow \begin{cases} C_2H^- + C_{n-2}H_{2n-1} \\ C_2H_3^- + C_{n-2}H_{n-3} \end{cases} \quad (120.1b)$$

120.5 REDUCED IONIZATION COEFFICIENTS

Table 120.8 shows the reduced ionization coefficients in 1-butene (Heylen, 1978).

C_4H_8

120.6 GAS CONSTANTS

Heylen (1978) gives the gas constants in the expression

$$\frac{\alpha}{N} = F\exp\left(-\frac{GN}{E}\right) \quad (120.2)$$

as $F = 5.37 \times 10^{-20}\,m^2$ and $G = 1198$ Td in the range $150 \le E/N \le 1120$ Td.

TABLE 120.8
Reduced Ionization Coefficients in 1-Butene

E/N (Td)	α/N (m²)	E/N (Td)	α/N (m²)	E/N (Td)	α/N (m²)
150	0.0014	500	0.4607	2500	5.2807
175	0.0051	600	0.6841	3000	6.1719
200	0.0125	700	0.9466	3500	6.857
225	0.0245	800	1.2279	4000	7.4018
250	0.042	900	1.5013	4500	7.8854
300	0.0952	1000	1.7644	5000	8.3868
350	0.1677	1250	2.4973	5500	8.9851
400	0.2564	1500	3.2405		
450	0.3574	2000	4.2876		

Source: Adapted from Heylen, A. E. D. *Int. J. Electron.*, 44, 367, 1978.
Note: See Figures 120.4 and 120.5.

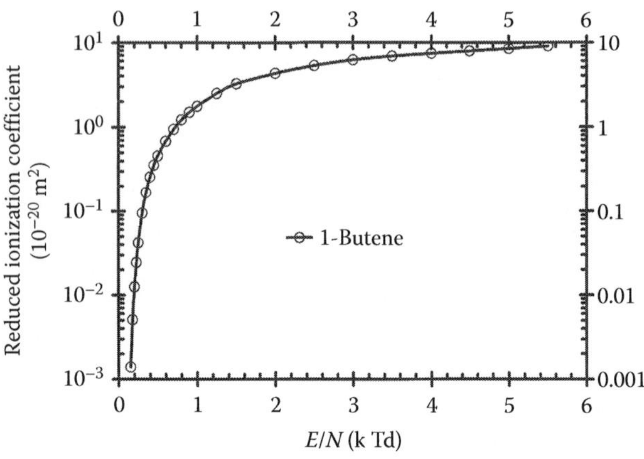

FIGURE 120.4 Reduced ionization coefficients for 1-butene as a function of *E/N*. (Adapted from Heylen, A. E. D. *Int. J. Electron.*, 44, 367, 1978.)

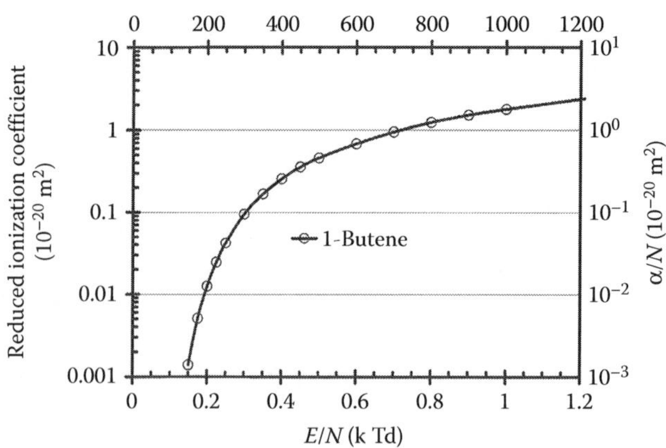

FIGURE 120.5 Same as Figure 120.4 except for the expanded x-axis values.

REFERENCES

Bowman, C. R. and D. E. Gordon, *J. Chem. Phys.*, 46, 1878, 1967.

Dance, D. F. and I. C. Walker, *Proc. Roy. Soc. Lond., A*, 334, 259, 1973.

Floeder, K., D. Fromme, W. Raith, A. Schwab, and G. Sinapius, *J. Phys. B: At. Mol. Opt. Phys.*, 18, 3347, 1985.

Gee, N. and G. R. Freeman, *J. Chem. Phys.*, 78, 1951, 1983.

Heylen, A. E. D., *Int. J. Electron.*, 44, 367, 1978.

Kim, Y.-K., K. K. Irikura, M. E. Rudd, M. A. Ali, P. M. Stone, J. Chang, J. S. Coursey, et al. *Electron-Impact Ionization Cross Section for Ionization and Excitation Database (version 3.0)*, http://Physics.nist.gov./ionxsec, National Institute of Standards and Technology, Gaithesberg, MD, USA, 2004.

Lide, D. R., *Handbook of Chemistry and Physics* (86th edn), CRC Press, Taylor & Francis LLC, Boca Raton, FL, 2005–2006.

Schram, B. L., M. J. Van der Wiel, F. J. De Heer, and H. R. Moustafa, *J. Chem. Phys.*, 44, 49, 1966.

Sharp, D. W. A., *The Penguin Dictionary of Chemistry* (IIIrd edn), Penguin Books, London, 2003.

Shimanouchi, T., *Tables of Molecular Vibrational Frequencies*, Consolidated Volume I, NSRDS-NBS, 39, Washington (DC), 1972.

Von J. Rutkowsky, H. Drost, and H.-J. Spangenberg, *Ann. Phys., Leipzig*, 37, 259, 1980.

121

CHLOROBENZENE

CONTENTS

Chlorobenzene (C_6H_5Cl), synonym phenyl chloride, is a 12-atom aromatic molecule that has one hydrogen atom replaced with chlorine in the benzene molecule. It is polar with a dipole moment of 1.69 D and a polarizability of 15.70×10^{-40} F m^2 (alternate value 13.70×10^{-40} F m^2). The ionization potential is 9.07 eV. Resonances determined by electron transmission technique are at 0.9, 1.74, 3.1, 4.68, and 8.22 eV (Mathur and Hasted, 1976).

121.1 SELECTED REFERENCES FOR DATA

Table 121.1 presents selected references for data.

121.2 TOTAL SCATTERING CROSS SECTION

Table 121.2 and Figure 121.1 show the total scattering cross section in C_6H_5Cl. Figure 121.2 shows the details

TABLE 121.1
Selected References

Quantity	Range, eV, (Td)	Reference
Q_T	0.4–1000	Makochekanwa (2003a)
Q_T	0–1	Lunt et al. (1999)
k_a, Q_a	0.04–2.0	Shimamori et al. (1995)
A	—	Olthoff et al. (1985)
Q_a	0–10	Mathur and Hasted (1976)
Q_a, η/N	0–2 (0.6–2.1)	Christophorou et al. (1966)
η/N	(0.6–2.1)	Stockdale et al. (1964)

Note: A = attachment process; k_a = attachment rate; Q_a = attachment; Q_T = total; η/N = density (N)-reduced attachment coefficient.

TABLE 121.2
Total Scattering Cross Section in C_6H_5Cl

Makochekanwa et al. (2003a)		Lunt et al. (1999)		Makochekanwa et al. (2003b)	
Chlorobenzene				Benzene	
Energy (eV)	Q_T (10^{-20} m^2)	Energy (eV)	Q_T (10^{-20} m^2)	Energy (eV)	Q_T (10^{-20} m^2)
0.4	44.0	0.03	1443.15	0.4	23.5
0.6	48.1	0.04	1124.26	0.6	23.5
0.8	49.1	0.06	962.57	0.8	24.9
1.0	48.7	0.07	732.79	1.0	28.1
1.2	46.5	0.08	656.84	1.2	31.6
1.4	45.3	0.10	581.46	1.4	33.4
1.6	44.0	0.11	483.95	1.6	34.2
1.8	45.1	0.13	408.57	1.8	33
2.0	46.3	0.14	377.46	2.0	32.6
2.2	46.3	0.16	324.22	2.2	32.9
2.5	47.6	0.17	272.10	2.5	32.6
2.8	46.8	0.18	264.81	2.8	34.1
3.1	47.5	0.20	233.15	3.1	35.7
3.4	47.7			3.4	37
3.7	46.6			3.5	40.2
4.0	47.7			3.7	39.1
4.5	50.3			4.0	40.7
5.0	51.3			4.5	43.1
5.5	50.9			5.0	45.4
6.0	54.7			5.5	46.6
6.5	58.2			6.0	47.6
7.0	61.3			6.5	48.6
7.5	65.1			7.0	49.4
8.0	65.8			7.5	51.7

continued

TABLE 121.2 (continued)
Total Scattering Cross Section in C_6H_5Cl

Makochekanwa et al. (2003a)		Lunt et al. (1999)		Makochekanwa et al. (2003a)	
Chlorobenzene				Benzene	
Energy (eV)	Q_T (10^{-20} m²)	Energy (eV)	Q_T (10^{-20} m²)	Energy (eV)	Q_T (10^{-20} m²)
8.5	67.7			8.0	54.8
9.0	66.2			8.5	57.8
9.5	65.3			9.0	60
10.0	64.3			9.5	59.8
11.0	64.6			10.0	59.3
12.0	61.9			11.0	59.7
13.0	59.8			12.0	58.4
14.0	59.3			13.0	55.3
15.0	57.2			14.0	63.5
16.0	55.9			15.0	52.7
17.0	55.9			16.0	51
18.0	55.3			17.0	49.8
19.0	53.5			18.0	48.8
20.0	52.5			19.0	48.8
22.0	50.6			20.0	47.7
25.0	49.6			22.0	45.7
30.0	45.5			25.0	45.5
35.0	44.0			30.0	44.8
40.0	44.9			35.0	44.2
50.0	42.8			40.0	45.8
60.0	38.5			50.0	41.9
70.0	35.7			60.0	39.3
80.0	35.1			70.0	36.7
90.0	32.9			80.0	33.4
100.0	30.8			90.0	33.3
120.0	29.0			100.0	30.3
150.0	27.1			120.0	29.3
200.0	22.6			150.0	27.1
250.0	21.3			200.0	24.5
300.0	18.6			250.0	20.5
400.0	15.9			300.0	19.1
500.0	13.3			400.0	14.3
600.0	11.6			500.0	12.7
800.0	9.5			600.0	10.4
1000.0	7.7			800.0	7.68
				1000.0	6.15

Source: Adapted from Makochekanwa, C., O. Sueoka, and M. Kimura, *J. Chem. Phys.*, 119, 12257, 2003a.

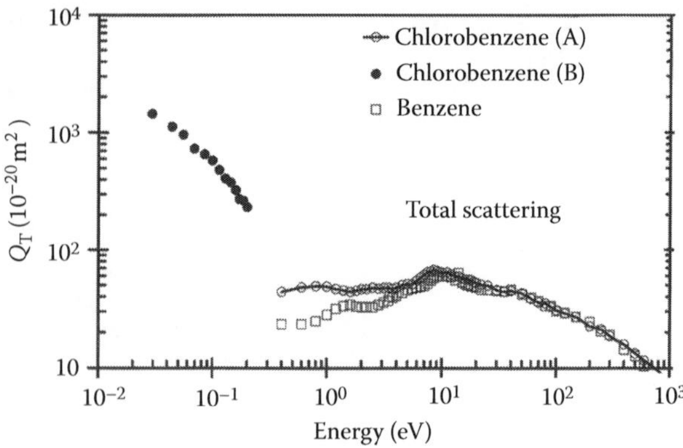

FIGURE 121.1 Total scattering cross section in C_6H_5Cl (A—Makochekanwa et al., 2003a; B—Lunt et al., 1999). Benzene cross sections from Makochekanwa et al. (2003b) are plotted for comparison.

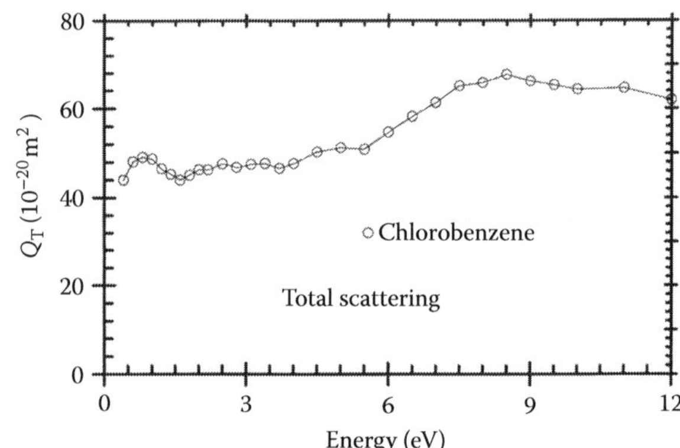

FIGURE 121.2 Details of the total scattering cross sections in the low-energy region for C_6H_5Cl from Makochekanwa et al. (2003a).

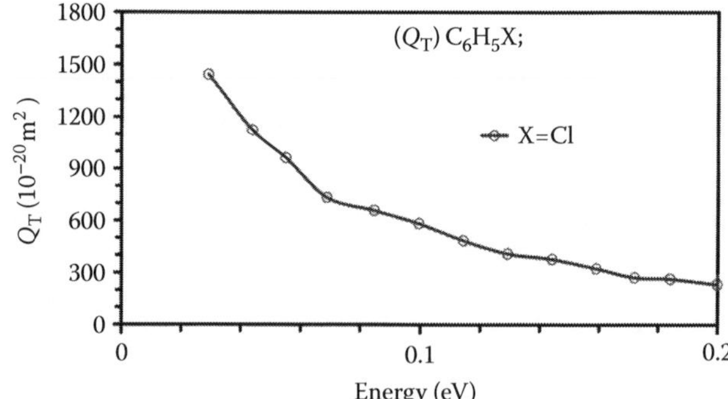

FIGURE 121.3 Low-energy scattering cross section in C_6H_5Cl. The rise toward zero energy is attributed to rotational inelastic scattering. (Adapted from Lunt, S. L. et al., *J. Phys. B: At. Mol. Opt. Phys.*, 32, 2707, 1999.)

in the low-energy region. The highlights of the cross section are

1. Peaks are observed at 0.8, 2.5, and 8.5 eV. A plateau is also observed at 2–2.2 eV. These energies agree reasonably well with those determined by Mathur and Hasted(1976).
2. A monotonic decrease beyond the peak energy.
3. The effect of substitution of a chlorine atom in place of a hydrogen atom increases the cross section up to about 50 eV, thereafter remaining the same as that of benzene. This effect is attributed to the enlargement of the target particle from 0.691 to 0.769 nm, (Makochekanwa et al., 2003b) resulting out of the substitution.
4. The cross section toward zero energy rises rapidly to a value approaching 1800×10^{-20} m² (see Figure 121.3), attributed to rotational inelastic collisions, dictated

by the dipole moment of the target molecule (Lunt et al., 1999).

5. The zero-energy cross section due to electron attachment tends toward a very low value of 10^{-24} m^2 (Shimamori et al., 1995).

121.3 RESONANCE ENERGY

Resonance peaks observed are shown in Table 121.3.

121.4 ATTACHMENT PROCESS

Low-energy attachment process is predominantly dissociative (Christophorou et al., 1966) according to

$$C_6H_5Cl + e \rightarrow C_6H_5 + Cl^- \quad (121.1)$$

121.5 ATTACHMENT RATES

The rate constants for benzene derivatives are shown in Table 121.4.

Attachment rates as a function of mean electron energy (Shimamori et al., 1995) are shown in Figure 121.4. Attachment rates determined by Christophorou et al. (1966) as a function of the reduced electric field (*E*/*N*, Td) are shown in Table 121.5 and Figure121.5.

121.6 ATTACHMENT CROSS SECTIONS

Attachment cross sections determined by unfolding method are shown in Figure 121.6.

TABLE 121.3
Resonance Energies for C_6H_5Cl

Energy	Reference
0.75	Lunt et al. (1999)
0.70	Shimamori et al. (1995)
0.73	Olthoff et al. (1985)
2.50	
4.50	
0.90	Mathur and Hasted (1976)
1.74	
3.1	
4.68	
8.22	
0.86	Compton et al. (1968)
0.86	Christophorou et al. (1966)

TABLE 121.4
Thermal Electron Attachment Rate Constants in Benzene Derivatives

Gas	C_6H_5I	C_6H_5Br	C_6H_5Cl	C_6H_5F
Rate (m³/s)	1.0×10^{-14}	6.5×10^{-18}	3.0×10^{-20}	$<<1.0 \times 10^{-19}$

Source: Adapted from Shimamori, H., Y. Tatsumi, and T. Sunagawa, *J. Chem. Phys.*, 99, 7787, 1993.

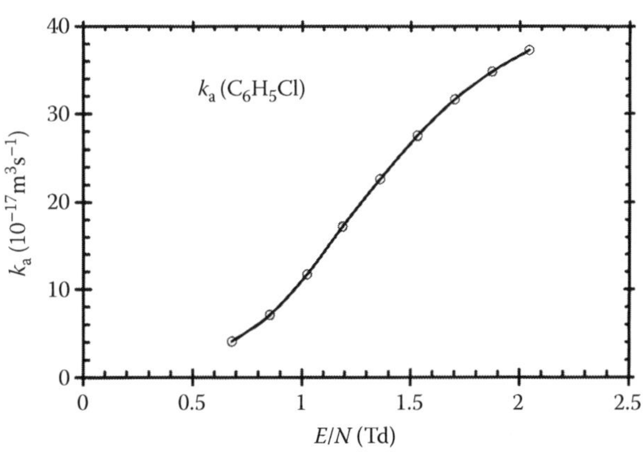

FIGURE 121.4 Attachment rate constants as a function of reduced electric field for C_6H_5Cl. (Adapted from Christophorou, L. G., et al., *J. Chem. Phys.*, 45, 536, 1966.)

TABLE 121.5
Attachment Rates as a Function of Reduced Electric Field for C_6H_5Cl

E/*N* (Td)	k_a (10^{-17} m³/s)	*E*/*N* (Td)	k_a (10^{-17} m³/s)
0.682	4.07	1.531	27.52
0.853	7.08	1.702	31.62
1.023	11.68	1.872	34.81
1.190	17.21	2.043	37.27
1.361	22.61		

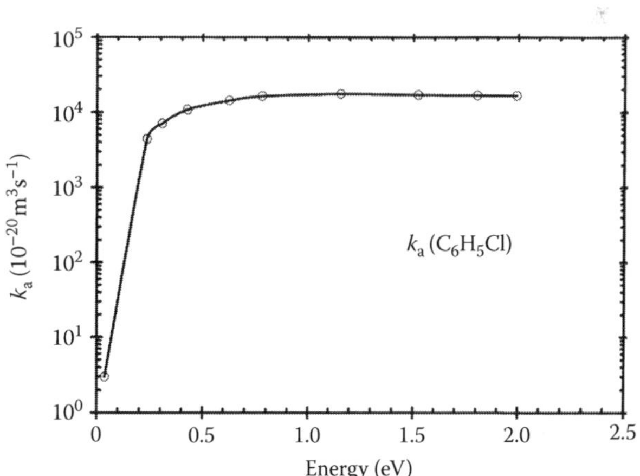

FIGURE 121.5 Attachment rate constants as a function of mean electron energy for C_6H_5Cl. The lowest energy point is from Shimamori et al. (1993). (Adapted from Shimamori, H. et al., *Chem. Phys. Lett.*, 232, 115, 1995.)

121.7 ATTACHMENT COEFFICIENTS

Density-reduced attachment coefficients as a function of reduced electric field for C_6H_5Cl are measured by Christophorou et al. (1966) (see Table 121.6).

C_6H_5Cl

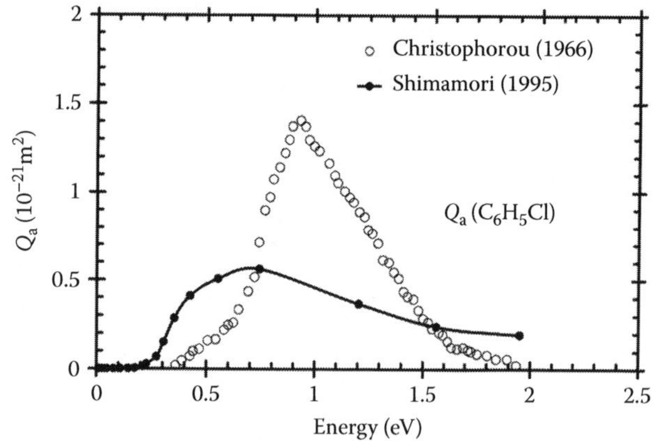

FIGURE 121.6 Attachment cross sections for C_6H_5Cl. (—●—) Shimamori et al. (1995); (○) Christophorou et al. (1966).

TABLE 121.6
Density-Reduced Attachment Coefficients

E/N (Td)	η/N (10^{-20} m^2)	*E/N* (Td)	η/N (10^{-20} m^2)
0.682	1.07	1.531	5.44
0.853	1.71	1.702	5.97
1.023	2.66	1.872	6.28
1.190	3.73	2.043	6.44
1.361	4.66		

REFERENCES

Christophorou, L. G., R. N. Compton, G. S. Hurst, and P. W. Reinhardt, *J. Chem. Phys.*, 45, 536, 1966.

Compton, R. N., R. H. Huebner, P. W. Reinhardt, and L. G. Christophorou, *J. Chem. Phys.*, 48, 901, 1968.

Lunt, S. L., D. Field, S. V. Hoffmann, R. J. Gulley, and J. P. Ziesel, *J. Phys. B: At. Mol. Opt. Phys.*, 32, 2707, 1999.

Makochekanwa, C., O. Sueoka, and M. Kimura, *Phys. Rev. A*, 68, 032707, 2003b.

Makochekanwa, C., O. Sueoka, and M. Kimura, *J. Chem. Phys.*, 119, 12257, 2003a.

Mathur, D. and J. B. Hasted, *J. Phys. B: At. Mol. Phys.*, 9, L31, 1976.

Olthoff, J. K., J. A. Tossell, and J. H. Moore, *J. Chem. Phys.*, 83, 5627, 1985.

Shimamori, H., T. Sunagawa, Y. Ogawa, and Y. Tatsumi, *Chem. Phys. Lett.*, 232, 115, 1995.

Shimamori, H., Y. Tatsumi, and T. Sunagawa, *J. Chem. Phys.*, 99, 7787, 1993.

Stockdale, J. A. and G. S. Hurst, *J. Chem. Phys.*, 41, 255, 1964.

122

CHLOROPENTA-FLUOROBENZENE

C_6F_5Cl

CONTENTS

Chloropentafluorobenzene (C_6F_5Cl) is 12 atom, 98 electron molecule, intermediate between C_6H_6 and C_6F_6 in the sense that chlorination may be seen as replacing a fluorine atom or halogens replacing hydrogen atoms. Chlorobenzene (C_6H_5Cl) is another gas with which comparison may be made to get insight into the influence of substituted atoms. C_6F_5Cl has ionization potential of 9.72 eV (Hikida et al., 1998). Total scattering cross sections in the range 0.8–1000 eV were measured by Makochekanwa et al. (2003a) and are shown in Table 122.1. Figure 122.1 graphically presents data and also the total cross sections of gases with substituted atoms. The characteristic features of the cross section are as follows

1. There is a minimum at 3 eV which is also observed in C_6F_6. In chlorobenzene, however, the cross section has a nearly constant value in the 1–3 eV region. To the left of the peak the cross section increases toward lower electron energy possibly due to negative ion formation.
2. The main resonance peak covers the range 9–15 eV possibly due to negative ion formation. In C_6F_6 the broad peak covers a much larger energy range 10–45 eV.
3. The peak cross section at 8.5 resonance decreases in the order $C_6H_5Cl < C_6H_6 < C_6F_5Cl < C_6F_6$ (Makochekanwa et al., 2003b). The energy at, and the magnitude of the resonance peak in the gases are summarized in Table 122.2.

122.1 TOTAL SCATTERING CROSS SECTION

TABLE 122.1
Total Scattering Cross Section in C_6F_5Cl

C_6F_5Cl		C_6H_5Cl		C_6F_6	
Energy (eV)	Q_T (10^{-20} m^2)	Energy (eV)	Q_T (10^{-20} m^2)	Energy (eV)	Q_T (10^{-20} m^2)
0.6	48.3	0.4	44	0.4	30.7
0.8	46.6	0.6	48.1	0.6	33.4
1.0	41.3	0.8	49.1	0.8	35.2
1.2	39.5	1.0	48.7	1.0	32.9
1.4	37.1	1.2	46.5	1.2	31.1
1.6	34.9	1.4	45.3	1.4	28.7
1.8	33.5	1.6	44	1.6	27.3
2.0	31.9	1.8	45.1	1.8	26.2
2.2	29.7	2.0	46.3	2.0	24.5
2.5	28.3	2.2	46.3	2.2	25.5
2.8	26.3	2.5	47.6	2.5	23.6
3.1	25.8	2.8	46.8	2.8	21.8
3.4	26.6	3.1	47.5	3.1	22.3
3.7	27.4	3.4	47.7	3.4	21.3
4.0	28.0	3.7	46.6	3.7	21.5
4.5	30.5	4.0	47.7	4.0	22.3
5.0	31.4	4.5	50.3	4.5	25.4
5.5	35.4	5.0	51.3	5.0	26.1
6.0	36.1	5.5	50.9	5.5	29.8
6.5	39.7	6.0	54.7	6.0	31.3

continued

TABLE 122.1 (continued)
Total Scattering Cross Section in C_6F_5Cl

C_6F_5Cl		C_6H_5Cl		C_6F_6	
Energy (eV)	Q_T (10^{-20} m^2)	Energy (eV)	Q_T (10^{-20} m^2)	Energy (eV)	Q_T (10^{-20} m^2)
7.0	41.0	6.5	58.2	6.5	33.1
7.5	44.7	7.0	61.3	7.0	33.3
8.0	45.8	7.5	65.1	7.5	34.2
8.5	48.8	8.0	65.8	8.0	36.5
9.0	48.3	8.5	67.7	8.5	38.4
9.5	49.8	9.0	66.2	9.0	39.2
10	49.2	9.5	65.3	6.5	40.6
11	51.2	10.0	64.3	10.0	41.3
12	50.0	11.0	64.0	11.0	43.3
13	51.1	12.0	61.9	12.0	44.7
14	50.6	13.0	59.8	13.0	45.6
15	49.8	14.0	59.3	14.0	46.2
16	49.7	15.0	57.2	15.0	46.7
17	48.8	16.0	55.9	16.0	46.5
18	49.7	17.0	55.9	17.0	46.6
19	48.5	18.0	55.3	18.0	46.7
20	48	19.0	53.5	19.0	46.7
22	48.8	20.0	52.5	20.0	46.7
25	47.9	22.0	50.6	22.0	47.5
30	47.4	25.0	49.6	25.0	47.4
35	46.0	30.0	45.5	30.0	46.2
40	44.1	35.0	44	35.0	47.4
50	39.5	40.0	44.9	40.0	46
60	38.3	50.0	42.8	50.0	43.1
70	37.6	60.0	38.5	60.0	41.3
80	38.2	70.0	35.7	70.0	38.9
90	34.9	80.0	35.1	80.0	37.6
100	34.3	90.0	32.9	90.0	35.3
120	30.7	100.0	30.8	100.0	33.3
150	28.5	120.0	29	120.0	30.7
200	25.4	150.0	27.1	150.0	28.7
250	23.5	200.0	22.6	200.0	25.2
300	21.1	250.0	21.3	250.0	23
400	18.4	300.0	18.6	300.0	20.3
500	16.3	400.0	15.9	400.0	18.1
600	13.9	500.0	13.3	500.0	15.6
		600.0	11.6	600.0	13.9
		800.0	9.5	800.0	11
		1000.0	7.7	1000.0	9.5

Source: Adapted from Makochekanwa, C., O. Sueoka, and M. Kimura, *J. Chem. Phys.*, 119, 12257, 2003a (C_6F_6); Makochekanwa, C., O. Sueoka, and M. Kimura, *Phys. Rev,*. A 68, 032707, 2003b.

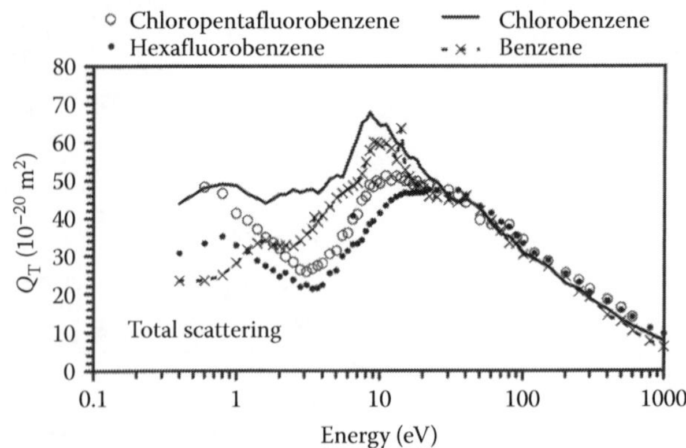

FIGURE 122.1 Total scattering cross sections in C_6F_5Cl and selected benzene-substituted gases. See Table 122.2 for references.

TABLE 122.2
Resonance Peak in Selected Benzene-Substituted Gases

Gas	Energy (eV)	Q_T (10^{-20} m^2)	Reference
Chlorobenzene (C_6H_5Cl)	8.5	67.7	Makochekanwa et al. (2003a)
Benzene (C_6H_6)	9.0	60.0	Makochekanwa et al. (2003b)
1,3-Difluorobenzene ($C_6H_4F_2$)	9.5	54.4	Makochekanwa et al. (2004)
Fluorobenzene (C_6H_5F)	8.5	51.2	Makochekanwa et al. (2004)
1,3-Difluorobenzene ($C_6H_4F_2$)	9.0	49.6	Makochekanwa et al. (2004)
Chloropentafluorobenzene (C_6F_5Cl)	8.5	48.8	Makochekanwa et al. (2003a)
Hexafluorobenzene (C_6F_6)	22.0	47.5	Makochekanwa et al. (2003b)

REFERENCES

Hikida, T., T. Ibuki, and K. Okada, *Chem. Phys. Lett.*, 292, 638, 1998.

Makochekanwa, C., O. Sueoka, and M. Kimura, *J. Chem. Phys.*, 119, 12257, 2003a.

Makochekanwa, C., O. Sueoka, and M. Kimura, *Phys. Rev,*. A 68, 032707, 2003b.

Makochekanwa, C., O. Sueoka, and M. Kimura, *J. Phys. B: At. Mol. Opt. Phys.*, 37, 1841, 2004.

123

1,3-DIFLUORO-BENZENE

CONTENTS

1,3-Difluorobenzene is a substituted benzene with two atoms of hydrogen replaced by fluorine atoms in the 1,3 position (Figure 123.1). The molecule is alternately designated as *m(eta)*-difluorobenzene (Lide, 2005–2006). It has polarizability of 11.5×10^{-40} F m^2, dipole moment of 1.51 D, and ionization potential of 9.33 eV. The total scattering cross sections are measured by Makochekanwa et al. (2004). and shown in Table 123.1. Figure 123.2 is a graphical presentation including data of selected gases for comparison sake.

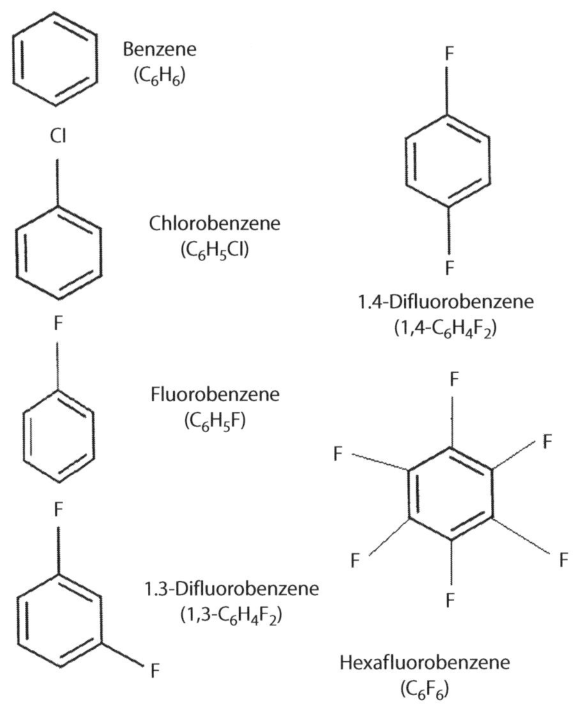

FIGURE 123.1 Selected substituted benzene molecules.

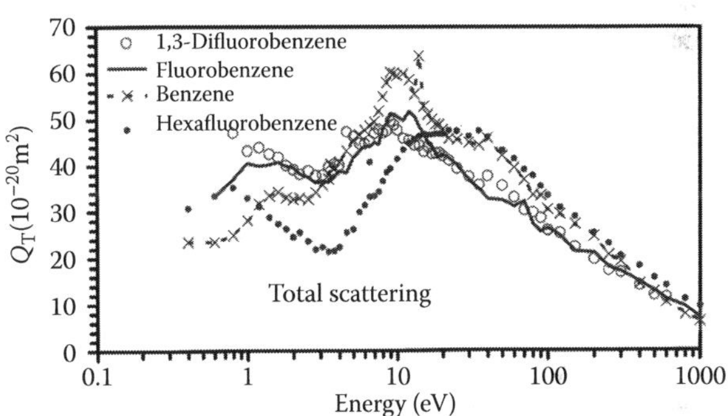

FIGURE 123.2 Total scattering cross sections in 1,3-dichlorobenzene and selected group molecules. (Adapted from Makochekanwa, C., O. Sueoka, and M. Kimura, *J. Phys. B: At. Mol. Opt. Phys.*, 37, 1841, 2004.)

123.1 TOTAL SCATTERING CROSS SECTION

The characteristic features of Figure 123.2 are

1. The general shape of the Q_T-energy curve remains the same as that of benzene and hexafluorobenzene, though the latter has a much broader resonance peak in the 8–10 eV range.
2. The peak of the major resonance at 49.6×10^{-20} m^2 is lower than that of benzene (see Table 122.2 of chloropentafluorobenzene (C_6F_5Cl)).
3. A minimum at 3 eV is common to all the molecules in this group and it is possibly due to increasing negative ion formation to the left of the minimum.
4. The small increasing trend seen toward zero energy is possibly due to electron attachment.

TABLE 123.1
Total Scattering Cross Sections in 1,3-Difluorobenzene

$1,3\text{-}C_6H_4F_2$

Energy (eV)	Q_T	Energy (eV)	Q_T	Energy (eV)	Q_T	Energy (eV)	Q_T
0.8	47.1	4.0	40.0	11	45.8	40	37.5
1.0	43.2	4.5	47.3	12	45.4	50	35.4
1.2	43.9	5.0	46.3	13	44.3	60	32.9
1.4	42.5	5.5	44.7	14	44.2	70	30.1
1.6	41.8	6.0	44.7	15	42.9	80	29.6
1.8	40.0	6.5	45.3	16	43.9	90	28.4
2.0	39.1	7	46.9	17	42.4	100	25.8
2.2	38.1	7.5	47.8	18	42.5	120	25.2
2.5	38.8	8	47.0	19	42.6	150	22.2
2.8	37.8	8.5	47.4	20	41.9	200	19.6
3.1	37.9	9	49.6	25	39.2	300	16.8
3.4	39.6	9.5	48.8	30	37.4	500	11.9
3.7	40.3	10	47.7	35	35.6	600	11.4

Source: Adapted from Makochekanwa, C., O. Sueoka, and M. Kimura, *J. Phys. B: At. Mol. Opt. Phys.*, 37, 1841, 2004.

Note: Energy in units of eV and Q_T in units of (10^{-20} m^2).

REFERENCES

Lide D. R., Editor in Chief, *CRC Handbook of Chemistry and Physics* (86th edition). Taylor & Francis LLC, Boca Raton, FL, 2005–2006.

Makochekanwa C., O. Sueoka, and M. Kimura, *J. Phys. B: At. Mol. Opt. Phys.*, 37, 1841, 2004.

124

1,4-DIFLUOROBENZENE

$C_6H_4F_2$

CONTENTS

1,4-Difluorobenzene is a substituted benzene with two atoms of hydrogen replaced by fluorine atoms in the 1,4 position (see Figure 123.1 of the 1,3-difluorobenzene chapter). The molecule is alternatively designated as *p(ara)*-difluorobenzene (Lide, 2005–2006). It is a polar molecule with polarizability of 10.90×10^{-40} m^2 and ionization potential of 9.14 eV.

TABLE 124.1
Total Scattering Cross Section in 1,4-Difluorobenzene

Energy (eV)	Q_T (10^{-20} m^2)	Energy (eV)	Q_T (10^{-20} m^2)	Energy (eV)	Q_T (10^{-20} m^2)
0.8	39.6	7.0	49.9	30	42.7
1	40.4	7.5	50.1	35	42.2
1.2	41.4	8.0	52.0	40	40.1
1.4	42.0	8.5	51.6	50	38.5
1.6	42.6	9.0	53.4	60	36.4
1.8	41.7	9.5	54.4	70	35
2.0	40.1	10	53.7	80	33.6
2.2	38.3	11	52.6	90	31.6
2.5	39.2	12	52.1	100	30.0
2.8	38.7	13	50.8	120	27.6
3.1	38.5	14	48.5	150	24.6
3.4	39.7	15	49.2	200	21.9
3.7	40.6	16	48.0	250	18.9
4.0	41.9	17	48.4	300	18.0
4.5	45.0	18	47.5	400	15.1
5.0	45.8	19	47.5	500	12.9
5.5	45.9	20	46.8	600	12
6.0	46.9	22	44.9		
6.5	48.4	25	43.7		

Source: Adapted from Makochekanwa, C., O. Sueoka, and M. Kimura, *J. Phys. B: At. Mol. Opt. Phys.*, 37, 1841, 2004.

124.1 TOTAL SCATTERING CROSS SECTION

The total scattering cross sections are measured by Makochekanwa et al. (2004) and Table 124.1 shows the data. Figure 124.1 is a graphical presentation, including data for selected gases in the group for comparison.

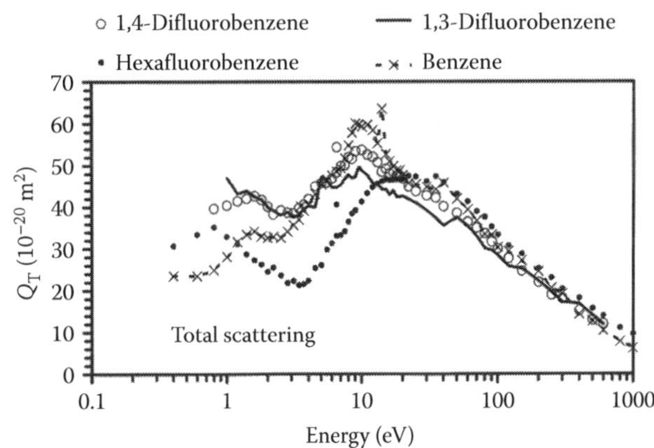

FIGURE 124.1 Total scattering cross sections in 1,4-difluorobenzene and selected molecules of the group. (Δ) 1,4-Difluorobenzene (Makochekanwa et al., 2004); (——) 1,3-difluorobenzene (Makochekanwa et al., 2004); (—□—) benzene (Makochekanwa et al., 2003); (○) hexafluorobenzene (Makochekanwa et al., 2003).

REFERENCES

Lide, D. R., editor in chief, *Handbook of Chemistry and Physics*, 86th edition, Taylor & Francis LLC, Boca Raton, FL, 2005–2006.

Makochekanwa, C., O. Sueoka, and M. Kimura, *J. Phys. B: At. Mol. Opt. Phys.*, 37, 1841, 2004.

Makochekanwa, C., O. Sueoka, and M. Kimura, *Phys. Rev. A*, 68, 032707, 2003.

125

FLUOROBENZENE

CONTENTS

Fluorobenzene (C_6H_5F) is a 12-atom, 20-electron, aromatic molecule, with one hydrogen replaced with a fluorine atom. It is polar with a dipole moment of 1.60 D; it has an electronic polarizability of 11.46×10^{-40} F m^2 and an ionization potential of 9.2 eV. The vibrational energy determined by electron transmission technique is 113–133 meV (Frazier et al., 1978). A comparison with a benzene molecule suggests that the motion is possibly C–H bend. Electronic excitation energies are 0.91, 1.27, 1.74, 3.65, 3.97, 4.34, 5.02, 5.32, 5.85, 6.5, 6.98, 7.27, 7.82, and 8.83 eV (Christophorou et al., 1974).

125.1 SELECTED REFERENCES FOR DATA

Table 125.1 presents selected references for data.

125.2 TOTAL SCATTERING CROSS SECTION

Table 125.2 presents the total cross sections and Figure 125.1 shows a comparison with benzene to decipher the influence of the substitution. The highlights of the total cross section variation are

1. Increasing cross section as the electron energy tends to zero due to rotational inelastic excitation (see Section 125.3).

TABLE 125.1
Selected References for Data

Cross Section	Energy (eV)	Reference
Q_T	0.4–1000	Makochekanwa et al. (2004)
Q_{ex}	25–75	Frueholz et al. (1979)
Q_a	1–10	Frazier et al. (1978)

Note: Q_a = attachment; Q_T = total.

TABLE 125.2
Total Scattering Cross Section in Fluorobenzene

Energy (eV)	Q_T (10^{-20} m^2)	Energy (eV)	Q_T (10^{-20} m^2)	Energy (eV)	Q_T (10^{-20} m^2)
0.4	33.3	6.5	44.9	30.0	34.8
0.6	37	7.0	44.5	35.0	32.6
0.8	40.5	7.5	47.9	40.0	32.3
1.0	39.9	8.0	49.1	50.0	31
1.2	40.2	8.5	51.2	60.0	32
1.4	40.7	9.0	51.1	70.0	27.1
1.6	39.8	9.5	50.5	80.0	25.4
1.8	39.6	10.0	49.7	90.0	26.3
2.0	38.2	11.0	51.6	100.0	24.7
2.2	37.4	12.0	50.6	120.0	21.3
2.5	36.2	13.0	48.3	150.0	21
2.8	36.5	14.0	46.7	200.0	17.9
3.1	36.4	15.0	44.7	250.0	17
3.4	37.5	16.0	44	300.0	14.8
3.7	38.7	17.0	42.5	400.0	12.9
4.0	38.4	18.0	41.7	500.0	11.1
4.5	41.7	19.0	42.3	600.0	9.46
5.0	43	20.0	41.4	800.0	6.99
5.5	44.1	22.0	39.9	1000.0	6.02
6.0	44	25.0	36.7		

Source: Adapted from Makochekanwa, C., O. Sueoka, and M. Kimura, *J. Phys. B: At. Mol. Opt. Phys.*, 37, 1841, 2004.

2. Resonances at 0.82, 1.40, and 4.66 eV due to negative ion resonance (Frazier et al., 1978).
3. A peak at 8.5 eV that is common to many gases.
4. Monotonic decrease of cross section for energies >8.5 eV.

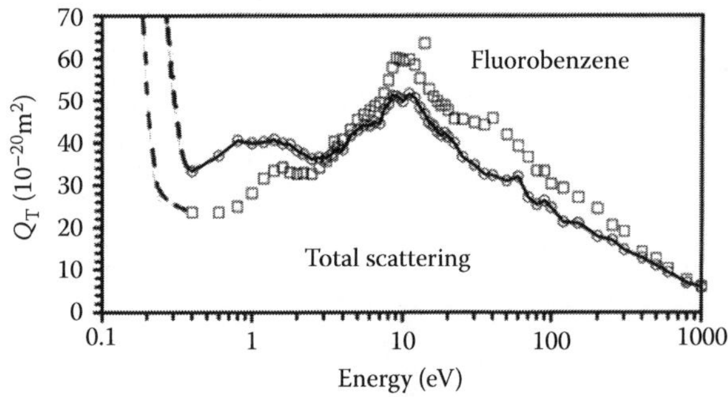

FIGURE 125.1 Total scattering cross sections in C_6H_5F. Benzene cross sections from Makochekanwa et al, (2003). are plotted for comparison. The dashed lines are schematic only; for experimental values at low energy, see Figure 125.2. (Adapted from Makochekanwa, C., O. Sueoka, and M. Kimura, *J. Phys. B: At. Mol. Opt. Phys.*, 37, 1841, 2004.)

5. A comparison with the cross section of benzene shows that the two curves intersect at about 5 eV. At lower energies, benzene has lower cross section; at higher energies, C_6H_5F has higher cross section.

Figure 125.2 gives a comparison of low-energy cross section for mono-substituted benzene molecules.

125.3 ZERO-ENERGY CROSS SECTION

One of the characteristic features of the total scattering cross section is the zero-energy cross section and the trend toward zero energy. The following principal mechanisms are related to the magnitude or trend. The cross section increases toward or from zero energy due to

A. Rotational excitation, generally observed in weakly polar molecules.

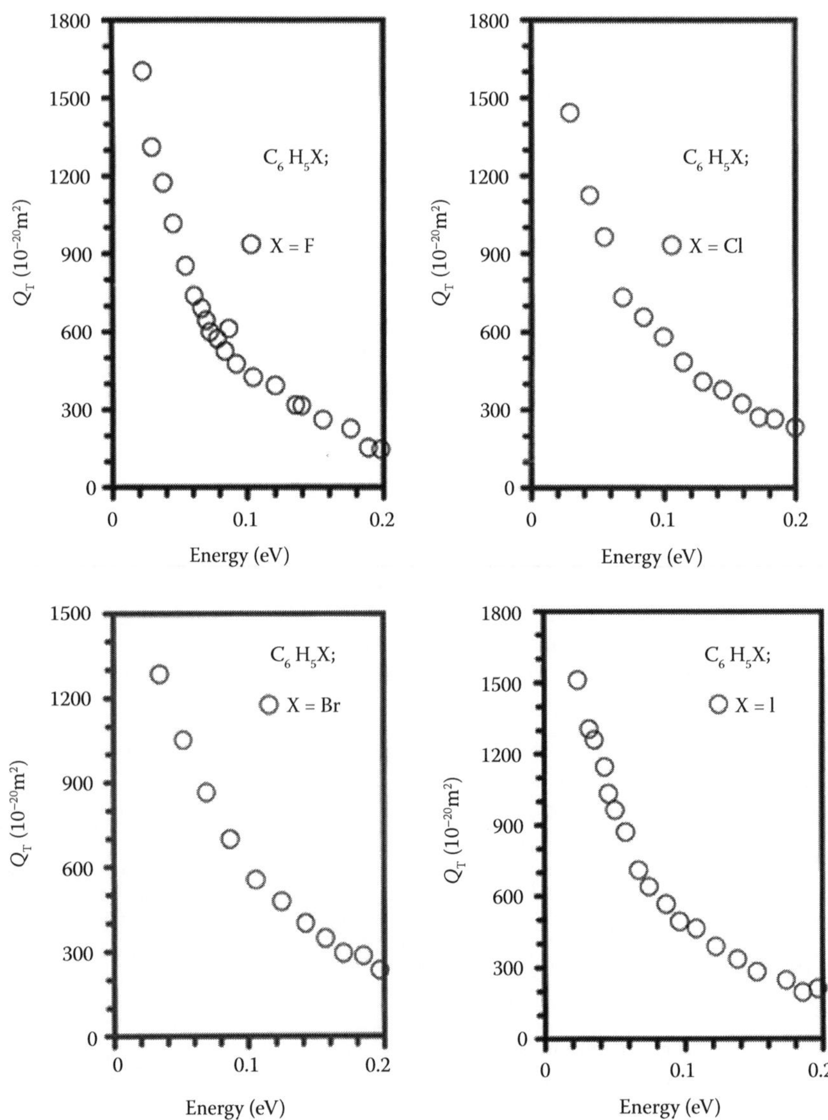

FIGURE 125.2 Total scattering cross sections in benzene derivatives at low electron energy. The authors attribute the large increase toward zero energy to rotational inelastic vibrations and not to electron attachment. (Adapted from Lunt, S. L. et al., *J. Phys. B: At. Mol. Opt. Phys.*, 32, 2707, 1999.)

B. Rotational and vibrational excitation, generally observed in nonpolar molecules.
C. Dipole moment of the target particle.
D. Electron attachment with zero- or low-energy threshold (thermal resonance). A stable negative ion may (SF_6^-) or may not form ($C_6H_6^-$). Dissociative attachment may occur.
E. The virtual state of an *s*-electron or low-energy Ramsauer–Townsend minimum.
F. The cross section changes slowly or remains essentially constant according to scattering theory of diffraction of partial waves.

Table 125.3 provides the details for selected gases.

TABLE 125.3
Zero-Energy Cross Section for Selected Gases

Target				
Name	**Formula**	**Q_T (10^{-20} m^2)**	**Mechanism**	**Reference**
Argon	Ar	10.3	F	Raju (2005)
Benzene	C_6H_6	170–180	E	Field et al. (2001)
Benzene-d_1	C_6H_5D	175	D	Gulley et al. (1998)
Benzene-d_6	C_6D_6	175	D	Gulley et al. (1998)
Boron trichloride	BCl_3	45	D	Christophorou and Olthoff (2004)
Bromobenzene	C_6H_5Br	>1500	A	Lunt et al. (1999)
Carbon dioxide	CO_2	600	E	Raju (2005)
Carbon monoxide	CO	60	A, C	Raju (2005)
Carbon tetrachloride	CCl_4	18,695	D	Randell et al. (1993)
Carbon tetrafluoride	CF_4	18.0	B	Lunt et al. (1998)
Chlorine	Cl_2	40	D	Gulley et al. (1998)
Chlorine dioxide	ClO_2	300	A	Gulley et al. (1998)
Chlorobenzene	C_6H_5Cl	>1500	A	Lunt et al. (1999)
Chlorotrifluoro methane	CF_3Cl	>30	A	Randell et al. (1993)
Ethane	C_2H_6	56	E	Lunt et al. (1998)
Fluorobenzene	C_6H_5F	>1500	A	Lunt et al. (1999)
Helium	He	4.5	F	Raju (2005)
Hydrogen	H_2	6.4	B	Randell et al. (1994)
Freon-12	CCl_2F_2	>100	A, D	Christophorou (2004)
Hexafluorobenzene	C_6F_6	1303	D	Chutjian and Alajajian (1985)
Hexafluoroethane	C_2F_6	90.0	E	Lunt et al. (1998)
Iodobenzene	C_6H_5I	>2000	A	Lunt et al. (1999)
Krypton	Kr	46.4	F	Raju (2005)
Methane	CH_4	21.0	B	Lunt et al. (1994)
Nitrous oxide	NO	20	C, D	Raju (2005)
Perfluoro-1,3-butadiene	1,3-C_4F_6	>50	D	Christodoulides et al. (1979)
Perfluoro-2-butene	2-C_4F_8	>50	D	Christodoulides et al. (1979)
Perfluoro-2-butyne	2-C_4F_6	>50	D	Christodoulides et al. (1979)
Perfluorocyclobutane	c-C_4F_8	~20	D	Christophorou (2004)
Perfluorocyclobutene	c-C_4F_6	>50	D	Christodoulides et al. (1979)
Perfluoropropane	C_3F_8	>46.0	A	Lunt et al. (1998)
Nitrogen	N_2	3.98	B	Randell et al (1994)
Oxygen	O_2	0.72	B	Randell et al (1994)
Ozone	O_3	130	D	Gulley et al (1998)
Propane	C_3H_8	>46.0	A	Lunt et al. (1998)
Propylene	C_3H_6	62.0	A	Lunt et al. (1998)
Sulfur hexafluoride	SF_6	520	D	Chutjian (1981)
Trichlorofluoro methane	CCl_3F	860	A	Randell et al. (1993)
		766	D	McCorkle et al. (1980)
		1.2×10^4	D	Klar et al. (2001)
Trifluoroiodomethane	CF_3I	>1000	D	Christophorou (2004)
Trifluoromethane	CHF_3	>3321	C, D	Christophorou (2004)
Xenon	Xe	35.2	F	Raju (2005)

Note: The scattering length A is given by the relation $Q_T = 4\pi A^2$. See Figure 125.3 for quick reference of schematic presentation.

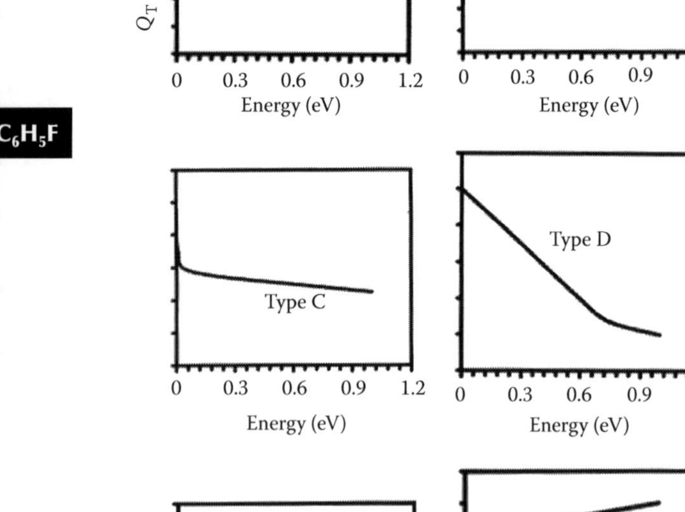

FIGURE 125.3 Schematic representation of the trend of total scattering cross section toward zero energy. For selected gases, see Table 125.3.

125.4 EXCITATION POTENTIALS

Frueholz et al. (1979). have measured the electronic excitation potentials and relative cross sections for C_6H_5F by electron impact spectroscopy. The excitation potentials are 3.90, 5.72, 4.78, 6.23, 6.99, 8.87, 9.73, 10.75, 11.38, 11.72, 12.33, 12.73, 13.93, 14.63, 15.27, and 16.97 eV.

REFERENCES

Christodoulides, A. A., L. G. Christophorou, R. Y. Pai, and C. M. Tung, *J. Chem. Phys.*, 70, 1156, 1979.

Christophorou L. G. and J. K. Olthoff, *Fundamental Electron Interactions with Plasma Processing Gases*, Kluwer Academic/Plenum Publishers, New York, NY, 2004.

Christophorou L. G., D. L. McCorkle, and J. G. Carter, *J. Chem. Phys.*, 60, 3779, 1974.

Chutjian A., *Phys. Rev. Lett.*, 46, 1511, 1981.

Chutjian A. and S. H. Alajajian, *J. Phys. B: At. Mol. Opt. Phys.*, 18, 4159, 1985.

Field D., J. P. Ziesel, S. L. Lunt, R. Parthasarathy, I. Suess, S. B. Hill, F. B. Dunning, R. R. Lucchese, and F. A. Gianturco, *J. Phys. B: At. Mol. Opt. Phys.*, 34, 4371, 2001.

Frazier J. R., L. G. Christophorou, J. G. Carter, and H. C. Schweinler, *J. Chem. Phys.*, 69, 3807, 1978.

Frueholz R. P., W. M. Flicker, O. A. Mosher, and A. Kupperman, *J. Chem. Phys.*, 70, 3057, 1979.

Gulley R. J., S. L. Lunt, J.-P. Ziesel, and D. Field, *J. Phys. B: At. Mol. Opt. Phys.*, 31, 2735, 1998.

Gulley R. J., T. A. Field, W. A. Steer, N. J. Mason, S. L. Lunt, J.-P. Ziesel, and D. Field, *J. Phys. B: At. Mol. Opt. Phys.*, 31, 2971, 1998.

Gulley R. J., T. A. Field, W. A. Steer, N. J. Mason, S. L. Lunt, J.-P. Ziesel, and D. Field, *J. Phys. B: At. Mol. Opt. Phys.*, 31, 5197, 1998.

Klar D., M.-W. Ruf, and H. Hotop, *Int. J. Mass Spectrom.* 205, 1993, 2001.

Lunt S. L., D. Field, S. V. Hoffmann, R. J. Gulley, and J. P. Ziesel, *J. Phys. B: At. Mol. Opt. Phys.*, 32, 2707, 1999.

Lunt S. L., J. Randell, J. P. Ziesel, G. Mrotzek, and D. Field, *J. Phys. B: At. Mol. Opt. Phys.*, 27, 1407, 1994.

Lunt S. L., J. Randell, J.-P. Ziesel, G. Mrotzek, and D. Field, *J. Phys. B: At. Mol. Opt. Phys.*, 31, 4525, 1998.

Makochekanwa C., O. Sueoka, and M. Kimura, *Phys. Rev. A*, 68, 032707, 2003

Makochekanwa C., O. Sueoka, and M. Kimura, *J. Phys. B: At. Mol. Opt. Phys.*, 37, 1841, 2004.

McCorkle D. L., A. A. Christodoulides, L. G. Christophorou, and I. Szamrej, *J. Chem. Phys.*, 72, 4049, 1980.

Raju G. G., *Gaseous Electronics: Theory and Practice*, Taylor & Francis LLC, Boca Raton, FL, 2005.

Randell J., J.-P. Ziesel, S. L. Luntl, G. Mmtzekl, and D. Field, *J. Phys B: At. Mol. Opt. Phys.*, 26, 3423, 1993.

Randell J., S. L. Lunt, G Mrotzek, J.-P. Ziesel, and D. Field, *Phys. B: At. Mol. Opt. Phys.* 27, 2369, 1994.

126

HEXAFLUOROBENZENE

C_6F_6

CONTENTS

Hexafluorobenzene (C_6F_6), synonym perfluorobenzene, is a nonpolar molecule with 90 electrons. It has an electronic polarizability of 10.66×10^{-40} F m^2 and an ionization potential of 9.89 eV. Limited data are available for the vibrational modes of the molecule (Delbouille, 1956), as shown in Table 126.1. A comparison with vibrational modes of benzene (C_6H_6) is interesting. Allan and Andric (1996) suggest that vibrations of similar local motion (mode) in hydrocarbons tend to have similar frequencies (energy). The electron affinity of the molecule is 0.83 eV (Weik and Illenberger, 1995).

Frazier et al. (1978) observe a dominant vibrational excitation at 50 meV in electron transmission studies.

TABLE 126.1
Vibrational Modes for C_6F_6

Mode	Energy (meV)
ν_1	184.7 (L)
ν_2	**69.4 (L)**
ν_4	26.7
ν_{11}	53.9 (L)
ν_{12}	**189.8**
ν_{13}	**123.5,** 126.0
ν_{14}	39.1
ν_{15}	205.8 (L)
ν_{16}	143.8 (L)
ν_{17}	45.6 (L)
ν_{18}	26.7 (L)

Source: Adapted from Delbouille, L., *J. Chem. Phys.*, 25, 182, 1956.
Note: (L) means liquid phase. Bold font indicates strong intensity.

126.1 SELECTED REFERENCES FOR DATA

See Table 126.2.

126.2 TOTAL SCATTERING CROSS SECTIONS

Total scattering cross sections in C_6H_6 have been measured by Makochekanwa et al. (2003) and Kasperski et al. (1997) as shown in Tables 126.3 and 126.4 and Figure 126.1. The highlights of the cross sections are

1. A peak in the cross section in the vicinity of ~1 eV attributed to the formation of the parent negative ion, $C_6F_6^-$, and autodetachment (Kasperski et al., 1997).
2. A relatively broad peak in the range from 10 to 40 eV with spikes at 14 and 30 eV attributed to dissociative attachment, with the formation of fragments, F^-, $C_5F_3^-$, and $C_6F_5^-$ ions (Makochekanwa et al., 2003).
3. A monotonic decrease above 40 eV.
4. Zero-energy cross section approaches a value $>1.3 \times 10^{-17}$ m^2 due to attachment.

C_6F_6

TABLE 126.2
Selected References for Data

Quantity	Range: eV, (Td)	Reference
Q_{el}, Q_v	0.015–0.5	Field et al. (2004)
Q_T	0.4–1000	Makochekanwa et al. (2003)
A	—	Suess et al. (2002)
Q_{el}, Q_M	1.1–100	Cho et al. (2001)
Q_T	30–3000	Jiang et al. (2000)
A	—	Finch et al. (1999)
Q_T	0.6–250	Kasperski et al. (1997)
Q_a	0–8	Ingólfsson and Illenberger (1995)
k_a, Q_a	0.2–1.2	Christophorou and Datskos (1995)
A	0–14	Weik and Illenberger (1995)
k_a	0–1.0	Dastkos et al. (1993)
Q_a	0.001–0.10	Marawar et al. (1988)
Q_a	0–0.150	Chutjian and Alajajian (1985)
k_a	0–1.0	Spyrou and Christophorou (1985)
Q_{ex}	25, 30, 50, 70	Frueholz et al. (1979)
A	0–5	Frazier et al. (1978)
k_a	≤3	Gant and Christophorou (1976)
A	—	Lifshitz et al. (1973)

Note: A = attachment processes; k_a = attachment rate; Q_a = attachment; Q_{el} = elastic; Q_v = vibrational excitation; Q_M = momentum transfer; Q_i = ionization; Q_T = Total scattering; T = theoretical. Data are experimental unless otherwise mentioned.

TABLE 126.3
Total Scattering Cross Sections for C_6F_6 in the Low-Energy Region

Energy (eV)	Q_T (10^{-20} m^2)	Energy (eV)	Q_T (10^{-20} m^2)
0.02	797	0.15	268
0.03	636	0.20	218
0.04	523	0.25	185
0.05	440	0.30	166
0.06	414	0.35	153
0.07	385	0.40	139
0.08	363	0.45	132
0.09	349	0.50	120
0.10	337		

Source: Adapted from Field, D., N. C. Jones, and J.-P. Ziesel, *Phys. Rev.*, A69, 052716, 2004.

126.3 DIFFERENTIAL SCATTERING CROSS SECTIONS

Differential scattering cross sections measured by Cho et al. (2001) are shown in Figure 126.2. The highlights are

1. Significant forward scattering at all energies measured.
2. The differential cross section decreases monotonically with angle except a marginal increase at the largest angles.
3. A peak in the vicinity of 10 eV for all angles.

TABLE 126.4
Total Scattering Cross Sections

Makochekanwa et al. (2003)		Kasperski et al. (1997)	
Energy (eV)	Q_T (10^{-20} m^2)	Energy (eV)	Q_T (10^{-20} m^2)
0.4	30.7	0.6	31.6
0.6	33.4	0.7	33.0
0.8	35.2	0.8	32.7
1.0	32.9	0.9	33.1
1.2	31.1	1.0	33.4
1.4	28.7	1.1	33.3
1.6	27.3	1.2	32.5
1.8	26.2	1.3	32.6
2.0	24.5	1.5	31.8
2.2	25.5	1.6	31.7
2.5	23.6	1.7	31.3
2.8	21.8	1.8	31.2
3.1	22.3	1.9	31.1
3.4	21.3	2.0	31.0
3.7	21.5	2.1	30.5
4.0	22.3	2.2	29.9
4.5	25.4	2.3	30.4
5.0	26.1	2.4	30.4
5.5	29.8	2.5	30.0
6	31.3	2.6	30.0
6.5	33.1	2.9	29.2
7	33.3	3.3	28.8
7.5	34.2	3.6	29.4
8	36.5	4.1	29.9
8.5	38.4	4.6	31.3
9	39.2	5.1	32.3
9.5	40.6	5.6	35.2
10	41.3	6.1	36.3
11	43.3	6.6	38.0
12	44.7	7.1	40.7
13	45.6	7.6	42.0
14	46.2	8.1	44.6
15	46.7	8.6	46.7
16	46.5	9.1	48.5
17	46.6	9.6	48.6
18	46.7	10.1	47.6
19	46.7	10.6	50.7
20	46.7	11.6	52.7
22	47.5	12.6	54.3
25	47.4	13.6	55.2
30	46.2	14.6	56.2
35	47.4	15.6	55.9
40	46.0	16.6	56.0
50	43.1	18.6	56.9
60	41.3	20.6	58.0
70	38.9	23.1	59.2
80	37.6	25.6	60.1
90	35.3	28	60.1
100	33.3	30	60.4
120	30.7	35	60.2
150	28.7	40	59.7
200	25.2	45	58.4

continued

TABLE 126.4 (continued)
Total Scattering Cross Sections

Makochekanwa et al. (2003)		Kasperski et al. (1997)	
Energy (eV)	Q_T (10^{-20} m^2)	Energy (eV)	Q_T (10^{-20} m^2)
250	23.0	50	56.6
300	20.3	60	53.8
400	18.1	70	51.0
500	15.6	80	49.3
600	13.9	90	47.0
800	11.0	100	46.0
1000	9.5	110	43.3
		120	42.0
		140	39.3
		160	36.8
		180	34.6
		200	33.6
		220	31.9
		250	29.9

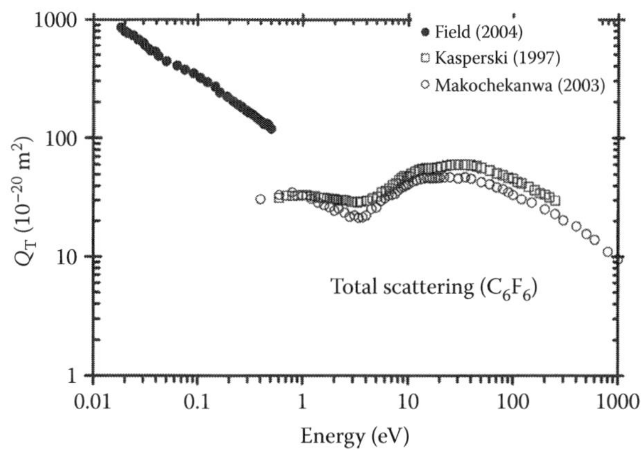

FIGURE 126.1 Total scattering cross sections for C_6F_6. (○) Makochekanwa et al. (2003); (□) Kasperski et al. (1977); (●) Field et al. (2004). Below 0.5 eV, the attachment cross sections from Field et al. (2004) are shown. Note the very large cross section toward zero-energy electrons.

126.4 ELASTIC SCATTERING CROSS SECTIONS

Elastic scattering cross sections measured by Cho et al. (2001) are shown in Table 126.5 and Figure 126.3.

The scattering length A_0 related to the cross section at zero energy is 1.3×10^{-9} m yielding a cross section of 475×10^{-20} m^2 (Field et al. 2004).

126.5 MOMENTUM TRANSFER CROSS SECTIONS

Momentum transfer cross sections measured by Cho et al. (2001) are shown in Table 126.6 and Figure 126.4.

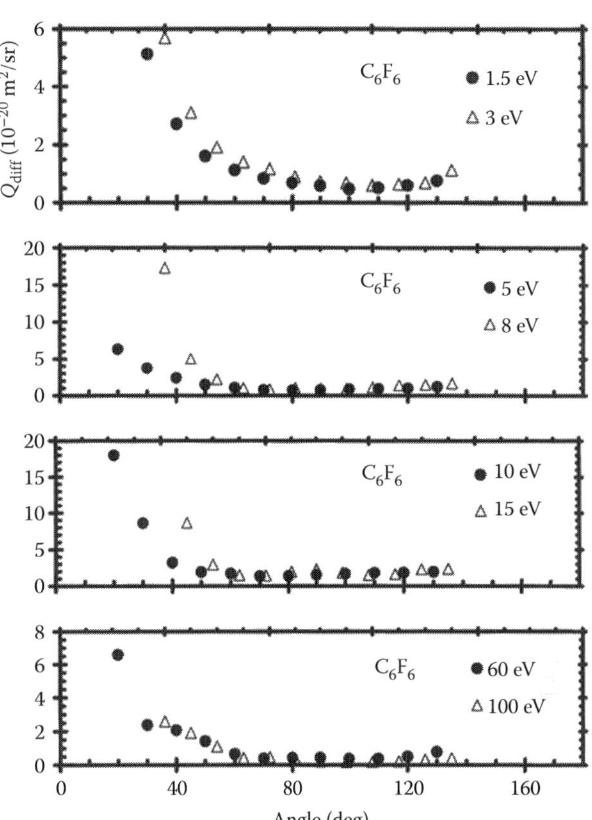

FIGURE 126.2 Differential scattering cross sections for C_6F_6 as a function of electron energy. Note the ordinate for each energy. (Adapted from Cho, H. et al., *J. Phys. B: At. Mol. Opt. Phys.*, 34, 1019, 2001.)

TABLE 126.5
Elastic Scattering Cross Sections

Energy (eV)	Q_{el} (10^{-20} m^2)	Energy (eV)	Q_{el} (10^{-20} m^2)
1.5	21.75	15	51.62
3.0	18.60	20	48.01
5.0	21.51	30	32.65
8.0	30.98	60	24.26
10	41.09	100	9.04

Source: Adapted from Cho, H., *J. Phys. B: At. Mol. Opt. Phys.*, 34, 1019, 2001.

126.6 ELECTRONIC EXCITATION CROSS SECTIONS

Frueholz et al. (1979) have measured the excitation potentials and relative cross sections by electron impact spectroscopy. The excitation potentials are 3.86, 4.80, 5.32, 6.36, 7.10, 9.82, 10.26, 10.99, 11.14, 11.39, 11.72, 12.49, 12.79, 13.46, 15.81, and 18.32 eV.

C_6F_6

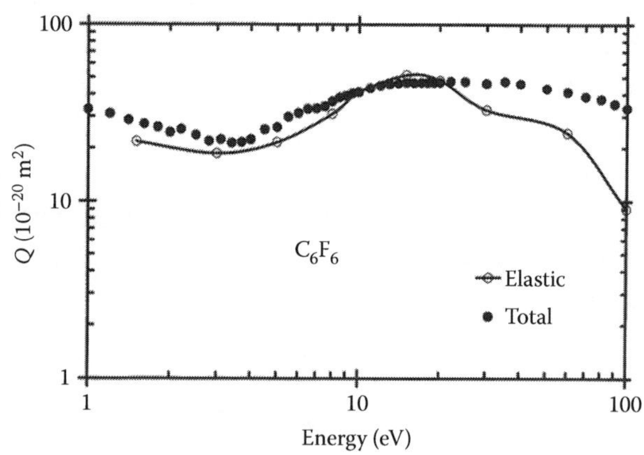

FIGURE 126.3 Elastic scattering cross sections for C_6F_6. Total scattering cross sections are shown for comparison. (—○—) Elastic scattering (Cho et al., 2001); (●) total (Makochekanwa et al., 2003).

TABLE 126.6
Momentum Transfer Cross Sections

Energy (eV)	Q_{el} (10^{-20} m^2)	Energy (eV)	Q_{el} (10^{-20} m^2)
1.5	11.49	15	29.93
3.0	14.25	20	26.35
5.0	16.54	30	16.86
8.0	18.50	60	11.58
10	24.40	100	5.63

Source: Adapted from Cho, H. et al., *J. Phys. B: At. Mol. Opt. Phys.*, 34, 1019, 2001.

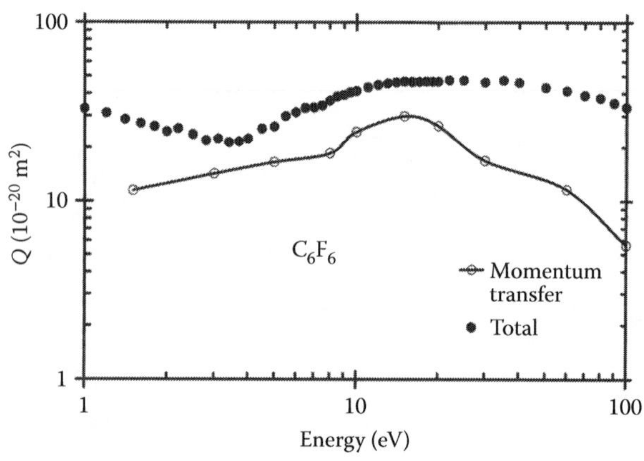

FIGURE 126.4 Momentum transfer cross sections for C_6F_6. Total scattering cross sections are shown for comparison. (—○—) Momentum transfer (Cho et al., 2001); (•) total (Makochekanwa et al., 2003).

126.7 APPEARANCE POTENTIALS OF POSITIVE IONS

Table 126.7 shows the appearance potentials and relative abundance of positive ions (Diebler et al. 1957).

TABLE 126.7
Appearance Potentials and Relative Abundance of Positive Ions

Ion	Potential (eV)	Abundance (rel) %	Ion	Potential (eV)	Abundance (rel) %
$C_6F_6^+$	9.9	100	$C_4F_2^+$	20.2	3.8
$C_6F_5^+$	16.9	10.4	$C_3F_3^+$	16.8	19.9
$C_5F_5^+$	17.2	14.1	$C_3F_2^+$	18.9	4.1
$C_5F_4^+$	16.1	8.3	CF_2^+	25.9	3.0
$C_5F_3^+$	15.8	45.6	CF_3^+	19.8	3.9
$C_5F_2^+$	24.3	5.9	CF^+	17.3	40.0
C_5F^+	29.0	5.1	F^+	29.2	3.2

TABLE 126.8
Appearance Potentials of Negative Ions

Ion	Appearance Potential (eV)
$C_6F_6^-$	~0
$C_6F_5^-$	4.5, 8.2
F^-	4.0, 8.2

126.8 ATTACHMENT PROCESSES

Attachment occurs through both nondissociative (low-energy) and dissociative (higher-energy) processes (Gant and Christophorou, 1976) according to

$$C_6F_6 + e \rightarrow C_6F_6^{-}* \rightarrow C_6F_6^- \quad (126.1)$$

Ingólfsson and Illenberger (1995) suggest the following attachment process:

$$C_6F_6 + e \rightarrow C_6F_6^{-}* \rightarrow \begin{cases} C_6F_5 + F^- \\ C_6F_5^- + F \\ CF_3 + C_5F_3^- \end{cases} \quad (126.2)$$

The lifetimes of $C_6F_6^{-*}$ has a distribution extending to values >50 μs (Finch et al., 1999). The appearance potentials are shown in Table 126.8.

Free electron attachment is characterized by three resonance regions near 0, 4.5, and 8.13 eV (Weik and Illenberger 1995).

Formation of negative ions by attachment to C_6F_6 clusters has been studied by Ingólfsson and Illenberger (Weik and Illenberger, 1995).

126.9 ATTACHMENT RATE COEFFICIENTS

The thermal attachment rate coefficient is 1.49×10^{-13} m^3/s (Christophorou and Datskos, 1995; Spyrou and Christophorou, 1985). With increasing electron energy the attachment rate decreases to 2.0×10^{-14} m^3/s at ~1.0 eV. The attachment rate constants as a function of mean energy determined by the swarm method by Spyrou and Christophorou (1985) are shown in Figure 126.5. The significant point to note is that the attachment rate coefficient in the low-energy range, contrary to the observation in many attaching gases, decreases with temperature. The results are attributed to the increased internal energy of the molecule reducing the attachment rates. Marawar et al. (1988) quote the rate constant as 4×10^{-14} m³/s.

126.10 ATTACHMENT CROSS SECTIONS

Attachment cross sections in the low-energy range are shown in Table 126.9. Figure 126.6 shows the cross sections up to 15 eV; the cross sections are due to

$0.001 \leq \varepsilon \leq 0.10$ eV (Chutjian and Alajajian, 1985)
$0.01 \leq \varepsilon \leq 1.3$ eV (Gant and Christophorou, 1976)
$3.0 \leq \varepsilon \leq 15.0$ eV (Weik and Illenberger, 1995)

The measurements of Weik and Illenberger are relative cross sections for F^- ions and they have been converted to absolute values so that the peak cross section at 4 eV is 2×10^{-20} m² (Figure 126.7) on the rationale that the total scattering cross sections at 4.0 eV is ~22×10^{-20} m² (Makochekanwa et al., 2003) and the elastic scattering cross section at the same energy is ~20×10^{-20} m² (Cho et al. 2001).

126.11 SUMMARY OF INELASTIC COLLISIONS

1. Parent negative ion $C_6F_6^-$ is formed at zero energy with a very large cross section.
2. Dissociative attachment is observed at 4.0 and 8.2 eV.
3. Parent molecule is ionized with the formation of $C_6F_6^+$ ion.

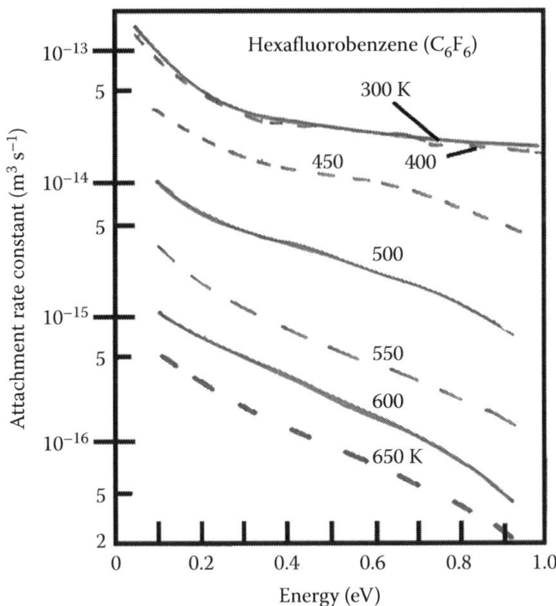

FIGURE 126.5 Attachment rate constants for C_6F_6 as a function of mean energy at various constant gas temperatures. (Adapted from Spyrou S. M. and L. G. Christophorou, *J. Chem. Phys.*, 82, 1048, 1985.)

TABLE 126.9
Attachment Cross Sections for C_6F_6

Energy (eV)	Q_{el} (10^{-20} m²)	Energy (eV)	Q_{el} (10^{-20} m²)
1.00E-03	1303	0.025	176
2.00E-03	894	0.030	145
3.00E-03	666	0.035	125
4.00E-03	514	0.040	107
5.00E-03	414	0.045	90.9
6.00E-03	362	0.050	78.6
7.00E-03	335	0.055	67.4
8.00E-03	313	0.060	57
9.00E-03	295	0.065	48
0.010	282	0.070	40.4
0.012	265	0.075	34.4
0.014	245	0.080	29.9
0.016	228	0.090	22.8
0.018	216	0.100	15.3
0.020	206		

Source: Adapted from Chutjian A. and S. H. Alajajian, *J. Phys. B: At. Mol. Opt. Phys.*, 18, 4159, 1985.

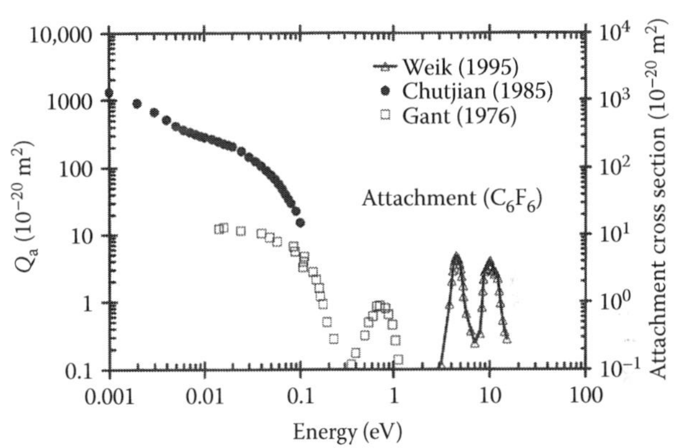

FIGURE 126.6 Attachment cross sections for (C_6F_6). (—Δ—) Weik and Illenberger (1995); (●) Chutjian and Alajajian (1985); (□) Gant and Christophorou (1976).

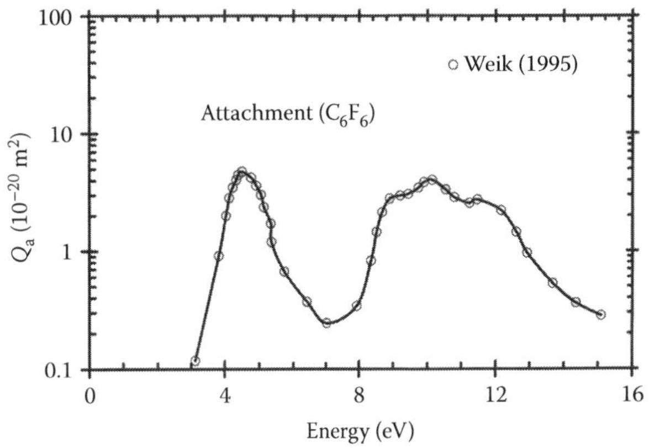

FIGURE 126.7 Details of dissociative attachment cross sections for F^- ions. Relative cross sections are converted to absolute values using cross section of 2×10^{-20} m². (Adapted from Weik F. and E. Illenberger, *J. Chem. Phys.*, 103, 1406, 1995.)

C_6F_6

REFERENCES

Allan, M. and L. Andric, *J. Chem. Phys.*, 105, 3559, 1996.

Cho, H. R., J. Gulley, K. Sunohara, M. Kitajima, L. J. Uhlmann, H. Tanaka, and S. J. Buckman, *J. Phys. B: At. Mol. Opt. Phys.*, 34, 1019, 2001.

Christophorou, L. G. and P. G. Datskos, *Int. J. Mass. Spectrom. Ion Proc.*, 149/150, 59, 1995.

Chutjian, A. and S. H. Alajajian, *J. Phys. B: At. Mol. Opt. Phys.*, 18, 4159, 1985.

Dastkos, P. G., L. G. Christophorou, and J. G. Carter, *J. Chem. Phys.*, 98, 7875, 1993.

Delbouille, L., *J. Chem. Phys.*, 25, 182, 1956.

Diebler, V. H., R. M. Reese, and F. L. Mohler, *J. Chem. Phys.*, 26, 304, 1957.

Field, D., N. C. Jones, and J.-P. Ziesel, *Phys. Rev.*, A69, 052716, 2004.

Finch, C. D., R. Parthasarathy, S. B. Hill, and F. B. Dunning, *J. Chem. Phys.*, 111, 7316, 1999.

Frazier, J. R., L. G. Christophorou, J. G. Carter, and H. C. Schweinler, *J. Chem. Phys.*, 69, 3807, 1978.

Frueholz, R. P., W. M. Flicker, O. A. Mosher, and A. Kupperman, *J. Chem. Phys.*, 70, 3057, 1979.

Gant, K. S. and L. G. Christophorou, *J. Chem. Phys.*, 65, 2977, 1976.

Ingólfsson, O. and E. Illenberger, *Int. J. Mass Spectrom. Ion Proc.*, 149/150, 79, 1995.

Jiang, Y., J. Sun, and L. Wan, *Phys. Rev.*, 62, 062712, 2000.

Kasperski, G., P. Możeko, and C. Szmytkowski, *Z. Phys. D*, 42, 187, 1997.

Lifshitz, C., T. O. Tiernan, and B. M. Hughes, *J. Chem. Phys.*, 59, 3182, 1973.

Makochekanwa, C., O. Sueoka, and M. Kimura, *Phys. Rev. A*, 68, 032707, 2003.

Marawar, R. W., C. W. Walter, K. A. Smith, and F. B. Dunning, *J. Chem. Phys.*, 88, 2853, 1988.

Spyrou, S. M. and L. G. Christophorou, *J. Chem. Phys.*, 82, 1048, 1985.

Suess, L., R. Parthasarathy, and F. B. Dunnung, *J. Chem. Phys.*, 117, 11222, 2002.

Weik, F. and E. Illenberger, *J. Chem. Phys.*, 103, 1406, 1995.

127

IODOBENZENE

CONTENTS

Iodobenzene (C_6H_5I) is a polar electron-attaching gas that has one hydrogen atom from benzene replaced with iodine atom. It has 94 electrons. The electronic polarizability is 17.25×10^{-40} F m^2, dipole moment 1.70 D and ionization potential 8.685 eV. Resonances due to electron attachment are measured at 0.59 and 4.3 eV (Olthoff et al., 1985).

127.1 TOTAL SCATTERING CROSS SECTION

Integral electron scattering cross sections have been measured by Lunt et al. (1999) and shown in Figure 127.1. The highlights are

1. Low-energy scattering is observed to be governed by the dipole moment of the target.
2. The cross section increases toward zero energy, $>2000 \times 10^{-20}$ m^2 due to rotationally inelastic scattering.
3. Resonance is not observed upto 1.0 eV energy.

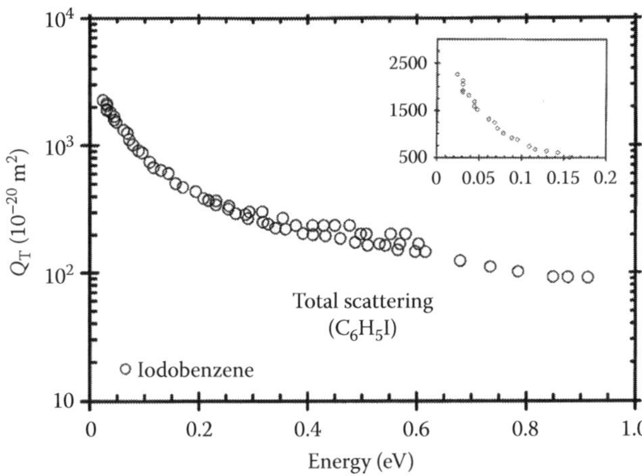

FIGURE 127.1 Total scattering cross sections for C_6H_5I. Inset shows details near zero energy on linear scale. (From Lunt, S. L. et al., *J. Phys. B. Appl. Phys.*, 32, 2707, 1999.)

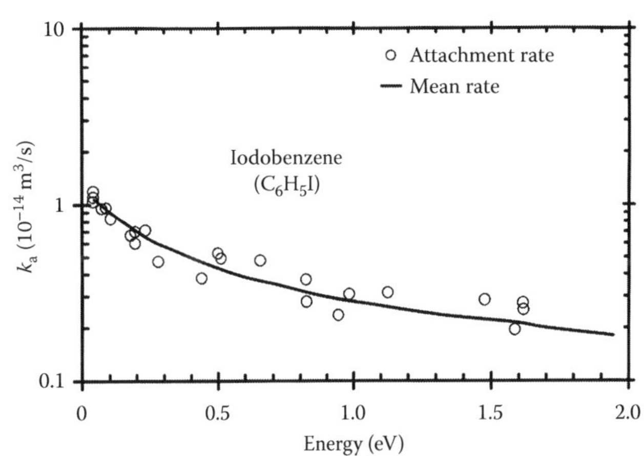

FIGURE 127.2 Attachment rate constants as a function of electron mean energy. (Adapted from Shimamori, H. et al., *Chem. Phys. Lett.*, 232, 115, 1995.)

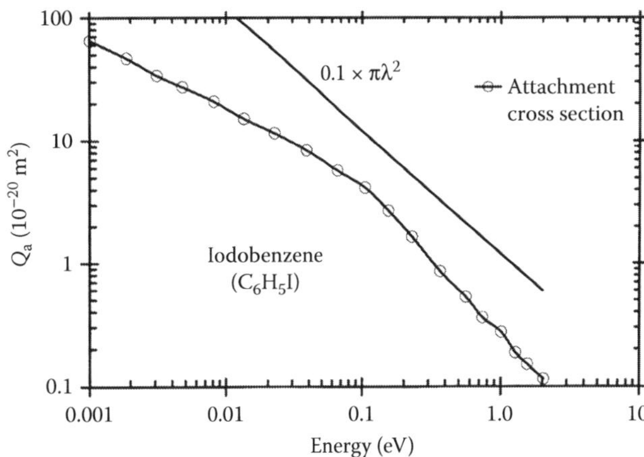

FIGURE 127.3 Attachment cross sections as a function of electron energy derived by Shimamori et al. s-Wave maximum cross section is also shown for comparison. (Adapted from Shimamori, H. *Chem. Phys. Lett.*, 232, 115, 1995).

127.2 ATTACHMENT RATE CONSTANT

The product ion is I^-. Thermal attachment rate constant measured by Shimamori et al. (1993) is 1.0×10^{-14} m^3/s. Figure 127.2 shows the attachment rate constants as a function of mean energy (Shimamori et al., 1995). Figure 127.3 shows the derived attachment cross section by these authors as a function of electron energy.

REFERENCES

Lunt, S. L., D. Field, S. V. Hoffmann, R. J. Gulley, and J-P. Ziesel, *J. Phys. B. Appl. Phys.*, 32, 2707, 1999.

Olthoff, J. K., J. A. Tossell, and J. H. Moore, *J. Chem. Phys.*, 83, 5627, 1985.

Shimamori, H., Y. Tatsumi, and T. Sunagawa, *J. Chem. Phys.*, 99, 7787, 1993.

Shimamori, H., T. Sunagawa, Y. Ogawa, and Y. Tatsumi, *Chem. Phys. Lett.*, 232, 115, 1995.

128

PERFLUOROCYCLOBUTANE, PERFLUORO-2-BUTENE, AND PERFLUOROISOBUTENE

CONTENTS

Fluorocarbons, consisting of carbon and fluorine atoms, are industrial gases. We consider target particles with the general chemical formula C_4F_8. There are three molecules with the same number of carbon and fluorine atoms, with 96 electrons, and their properties are shown in Table 128.1. Perfluorocyclobutane (c-C_4F_8) has already been dealt with in great detail by Christophorou and Olthoff (2004). The properties for the gas (c-C_4F_8) are summarized from their publication, with additional data provided as available. Selected properties of target particles are shown in Table 128.1.

128.1 SELECTED REFERENCES

See Table 128.2.

TABLE 128.1
Selected Properties of Fluorocarbons (C_nF_{2n+2})

Target		α_e	Electron	
Name	Formula	(10^{-40} Fm^2)	Affinity (eV)	ε_i (eV)
c-C_4F_8	Perfluorocyclobutane	8.2–13.87	≥0.4	12.1
2-C_4F_8	Perfluoro-2-butene	8.9[a]	≥0.7	
i-C_4F_8	Perfluoroisobutene		—	

[a] Beran and Kevan (1969).

TABLE 128.2
Selected References

Gas	Quantity	Range: eV, (Td)	Reference
c-C_4F_8	α/N	160–480	**Wu et al. (2006)**
c-C_4F_8	Q_{el}, Q_M	1.5–100	**Jelisavcic et al. (2004)**
c-C_4F_8	k_a	—	**Miller et al. (2004)**
c-C_4F_8	Q, Swarm properties	0.1–100 (1–1000)	**Yamaji and Nakamura (2004)**
2-C_4F_8	Q_i	12.5–250	**Bart et al. (2001)**
c-C_4F_8	Q_{el}, Q_M	1.5–50	**Winstead and McKoy (2001)**
c-C_4F_8	Swarm parameters	(12–600)	**de Urquiho et al. (2001)**
c-C_4F_8, 2-C_4F_8	—	10–27	**Jarvis et al. (1998a, b)**
c-C_4F_8	Q_{diss}	10–250	**Toyoda et al. (1997)**
c-C_4F_8	Q_T	1–20	**Sanabia et al. (1998)**
c-C_4F_8	Q, Swarm properties	0.1–3.0	Novak and Frechette (1988)
c-C_4F_8	k_a, Q_a	0.1–10	**Spyrou et al. (1985)**
2-C_4F_8	k_a	—	**Christophorou et al. (1981)**
c-C_4F_8	k_a	Thermal	**Woodin et al. (1980)**
c-C_4F_8, 2-C_4F_8	k_a, Q_a	0.04–1.2	**Christodoulides et al. (1979)**
c-C_4F_8, 2-C_4F_8	Q_a	0–10	**Sauers et al. (1979)**
c-C_4F_8, 2-C_4F_8	k_a	<2.0	**Christophorou et al. (1974)**
c-C_4F_8	—	—	**Harland and Franklin (1974)**
c-C_4F_8, 2-C_4F_8	k_a	—	**Bansal and Fessenden (1973)**
c-C_4F_8	k_a	Thermal	**Davis et al. (1973)**
c-C_4F_8, 2-C_4F_8	Q_a	0–12	**Lifshitz and Grajower (1972/73)**
c-C_4F_8, *i*-C_4F_8	Swarm parameters	(270–810)	**Naidu et al. (1972)**
2-C_4F_8	Q_i	70	**Beran and Kevan (1969)**

Note: k_a = attachment rate; Q = all cross sections; Q_a = attachment cross section; Q_{diss} = dissociation; Q_{el} = elastic; Q_M = momentum transfer; α/N = reduced ionization coefficient. Bold font indicates experimental study.

128.2 TOTAL SCATTERING CROSS SECTIONS

Table 128.3 and Figure 128.1 present the total scattering cross sections for c-C_4F_8 recommended by Christophorou and Olthoff (2004). The highlights of the total scattering cross sections are

1. There is a trend of increasing total cross section toward zero energy suggesting the possibility of ground-state negative ion, c-$C_4F_8^{-2}$ (Sanabia et al., 1998).
2. A minimum is observed at 4.0 eV, attributed to the decreasing contribution from electron attachment processes. Alternatively, Sanabia et al. (1998) suggest that the minimum is due to Ramsauer–Townsend minimum.
3. To the left of the minimum the cross section increases up to 1.1 eV due to increasing attachment.
4. The shape resonance occurs at 40 eV which is a much higher energy level compared to other gases.
5. At energy greater than 50 eV, the cross section decreases monotonically and the Born approximation method of calculating the cross section yields satisfactory results.

128.3 ELASTIC AND MOMENTUM TRANSFER CROSS SECTIONS

Elastic and momentum transfer cross sections have been measured by Jelisavcic et al. (2004) in two different

TABLE 128.3
Total Scattering Cross Sections for c-C_4F_8

Christophorou and Olthoff (2004)				Sanabia et al. (1998)	
Energy (eV)	Q_T (10^{-20} m^2)	Energy (eV)	Q_T (10^{-20} m^2)	Energy (eV)	Q_T (10^{-20} m^2)
1.1	53.8	40	46.5	0.84	49.85
1.5	42.6	45	46.4	1.32	40.02
2.0	32.9	50	46.1	1.59	35.03
2.5	27.7	60	45.8	1.80	30.84
3.0	25.1	70	44.9	2.07	27.45
3.5	23.9	80	43.8	2.46	24.23
4.0	23.8	90	42.7	3.12	21.66
4.5	25.6	100	41.4	3.96	21.03
5.0	28.0	150	35.7	4.92	23.31
6.0	32.8	200	31.4	5.24	25.57
7.0	36.1	300	25.1	5.75	29.13
8.0	38.1	400	21.0	6.33	31.88
9.0	39.4	500	18.3	6.84	33.34
10.0	39.8	600	16.3	7.42	35.12
12.0	39.6	700	14.8	8.58	37.88
13	39.8	800	13.6	9.80	38.38
15	40.7	900	12.7	11.23	38.25
17	41.3	1000	11.8	12.52	37.78
20	41.9	1500	9.00	14.21	37.81
25	43.3	2000	7.35	15.82	38.16
30	44.9	2500	6.20	17.18	38.99
35	45.9	3000	5.29	18.34	39.81
				19.90	40.00

Note: Data of Sanabia et al. (1998) are digitized.

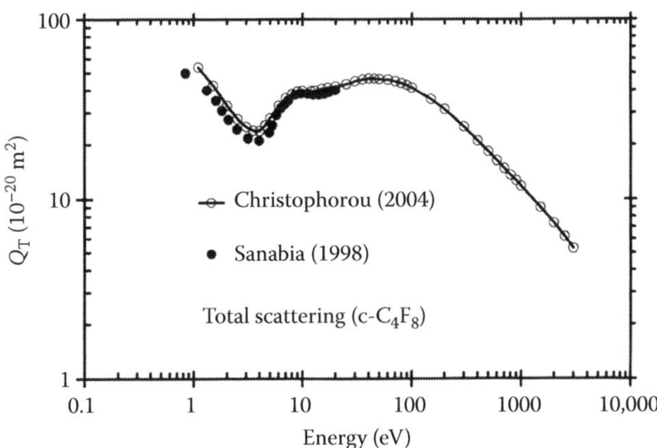

FIGURE 128.1 Total scattering cross sections for c-C_4F_8. (●) Sanabia et al. (1998), measurement; (—○—) Christophorou and Olthoff (2004), recommended.

laboratories and the average of the two values is shown in Table 128.4 Yamaji and Nakamura (2004) have also derived the momentum transfer cross sections. These data are shown in Tables 128.4 and 128.5 and Figure 128.2.

TABLE 128.4
Elastic and Momentum Transfer Cross Sections

	Q_T (10^{-20} m^2)	
Energy (eV)	Elastic	Momentum Transfer
1.5	17.85 (16.9, 18.8)	13.25 (12.5, 14.0)
2.0	17.65 (16.8, 18.5)	11.45 (10.8, 12.1)
2.6	18.1	11.0
3.0	18.7	12.9
4.0	21.4	16.6
5.0	21.90 (22.5, 21.3)	15.2 (14.2, 16.2)
6.0	22.8	11.9
7.0	24.9	13.3
8.0	30.5	18.5
10	35.24 (34.9, 35.6)	21.90 (21.5, 22.3)
15	34.7 (34.2, 35.2)	20.3 (19.0, 21.6)
20	35.35 (32.9, 37.8)	18.80 (17.2, 20.4)
30	31.3	15.6
60	16.1	6.21
100	11.0	3.68

Source: Adapted from Jelisavcic, M. et al., *J. Chem. Phys.*, 121, 5272, 2004.

TABLE 128.5
Momentum Transfer Cross Sections for c-C_4F_8

Energy (eV)	Q_M (10^{-20} m^2)	Energy (eV)	Q_M (10^{-20} m^2)
0.1	64.67	14	36.47
0.2	57.28	15	36.33
0.4	47.87	20	35.72
0.6	43.02	25	35.02
0.8	40.56	30	34.14
1	38.96	35	33.45
2	16.60	40	33.14
3	2.44	50	32.97
4	5.22	60	32.41
5	10.69	70	31.36
6	19.12	80	30.09
8	29.57	90	28.87
10	34.86	100	27.99
12	36.53		

Source: Digitized and interpolated from Yamaji, M. and Y. Nakamura, *J. Phys. D: Appl. Phys.*, 37, 1525, 2004.

128.4 VIBRATIONAL EXCITATION CROSS SECTIONS

Vibrational excitation cross sections evaluated by Novak and Frechette (1988), (Yamaji and Nakamura, 2004) are shown in Figure 128.3. Assuming isotropic scattering, Jelisavcic et al. (2004) have obtained, from measurements of differential scattering cross sections, a vibrational cross section of 2.5×10^{-20} m^2 at 1.5 and 7 eV.

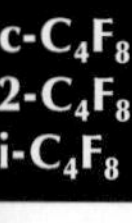

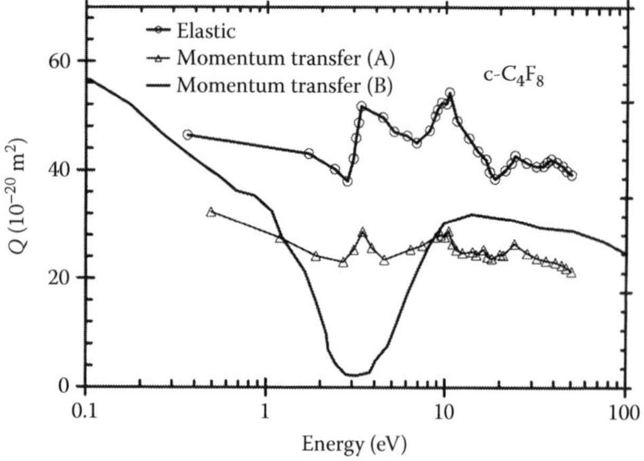

FIGURE 128.2 Elastic and momentum transfer cross sections for c-C_4F_8. (—○—) Elastic scattering (Winstead and McKoy (2001); (—△—) momentum transfer A (Winstead and McKoy, 2001); (———) momentum transfer B (Yamaji and Nakamura, 2004).

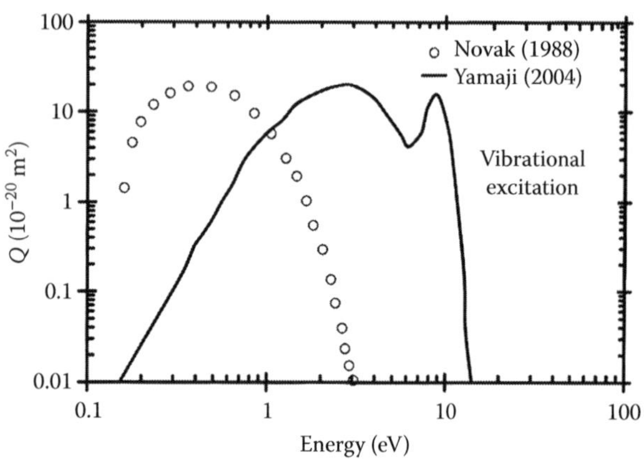

FIGURE 128.3 Vibrational cross sections derived from swarm coefficients for c-C_4F_8.

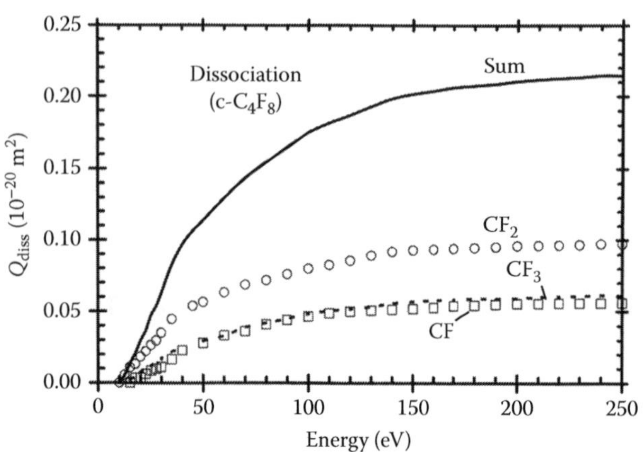

FIGURE 128.4 Neutral dissociation cross sections for c-C_4F_8. (□) CF; (– – –) CF_3; (○) CF_2; (———) sum. (Adapted from Toyoda, H., M. Ito, and H. Sugai, *Jpn. J. Appl. Phys.*, 36, 3730, 1997.)

128.5 NEUTRAL DISSOCIATION CROSS SECTIONS

A number of dissociation fragments are obtained depending upon the electron impact energy and the cross sections for the production of neutral fragments CF, CF_2, and CF_3 have been measured by Toyoda et al. (1997) Figure 128.4 shows the cross sections. As noted by Christophorou and Olthoff, (2004) the sum of the dissociation cross sections shown is the lower limit as there are several more fragments than shown.

128.6 POSITIVE ION APPEARANCE POTENTIALS

Positive ion appearance potentials are shown in Table 128.6.

128.7 IONIZATION CROSS SECTIONS

Ionization cross sections for (c-C_4F_8) recommended by Christophorou and Olthoff (2004) are shown in Table 128.7 and Figure 128.5. The data for perfluoro-2-butene (2-C_4F_8) are from Bart et al. (2001)

128.8 ATTACHMENT PROCESSES

128.8.1 Perfluorocyclobutane

There are two primary attachment mechanisms (Jelisavcic et al., 2004): (1) At energies below 1 eV, parent negative ions c-$C_4F_8^-$ are formed with a long half-life time of 6–500 μs (Sauers et al., 1979). (2) Above 1 eV, dissociative attachment is the dominant process resulting in the formation of F^-. The reactions are (Harland and Franklin, 1974)

$$\text{c-}C_4F_8 + e \rightarrow \text{c-}C_4F_7 + F^- \qquad (128.1)$$

$$\text{c-}C_4F_8 + e \rightarrow F + \text{c-}C_4F_6 + F^- \qquad (128.2)$$

TABLE 128.6
Positive Ion Appearance Potentials

Molecule	Ion Species	Potential (eV)	Reference
c-C_4F_8	$C_3F_5^+$	11.6	Jarvis et al. (1998b)
	$C_2F_4^+$	11.8	Jarvis et al. (1998b)
	$C_3F_5^+$	12.1	Lifshitz and Grajower (1972/73)
	$C_2F_4^+$	12.35	Lifshitz and Grajower (1972/73)
	CF_3^+	14.4	Lifshitz and Grajower (1972/73)
	CF_2^+	20.3	Toyoda et al. (1997)
	CF^+	18.4	Lifshitz and Grajower (1972/73)
2-C_4F_8	2-$C_4F_8^+$	11.1	Jarvis et al. (1998b)
	CF_3^+	12.2	Jarvis et al. (1998b)
	$C_3F_5^+$	12.4	Jarvis et al. (1998b)
	$C_4F_7^+$	14.2	Jarvis et al. (1998b)
	$C_3F_6^+$	14.8	Jarvis et al. (1998b)
	$C_2F_4^+$	14.8	Jarvis et al. (1998b)

TABLE 128.7
Total Ionization Cross Sections

c-C_4F_8 (Christophorou and Olthoff, 2004)		2-C_4F_8 (Bart et al. 2001)	
Energy (eV)	Q_i (10^{-20} m^2)	Energy (eV)	Q_i (10^{-20} m^2)
16	0.185	13	0.32
20	1.22	16	0.74
25	3.34	20	1.82
30	6.14	24	3.04
40	10.0	27	3.97
50	12.1	31	4.95
60	13.2	35	5.98
70	13.8	38	6.65
80	14.2	40	7.05
90	14.4	45	7.91
100	14.4	50	8.63
		55	9.27
		60	9.79
		65	10.19
		70	10.48
		75	10.74
		80	10.97
		85	11.16
		90	11.31
		95	11.36
		100	11.41
		105	11.55
		110	11.56
		115	11.49
		120	11.52
		125	11.62
		130	11.63
		135	11.48
		140	11.35
		145	11.32
		150	11.38
		155	11.36
		160	11.22
		165	11.12
		170	11.03
		175	10.91
		180	10.85
		185	10.81
		190	10.74
		195	10.61
		200	10.49
		205	10.40
		210	10.36
		215	10.33

$$c\text{-}C_4F_8 + e \rightarrow c\text{-}C_4F_7^* + F^- \quad (128.3)$$

$$c\text{-}C_4F_8 + e \rightarrow CF_2CF + C_2F_4 + F^- \quad (128.4)$$

$$c\text{-}C_4F_8 + e \rightarrow CF_2 + C_3F_5 + F^- \quad (128.5)$$

$$c\text{-}C_4F_8 + e \rightarrow C_3F_5 + CF_3^- \quad (128.6)$$

Sauers et al. (1979) list an additional reaction

$$c\text{-}C_4F_8 + e \rightarrow C_2F_5 + C_2F_3^- \quad (128.7)$$

Table 128.8 summarizes the appearance potentials and peak energies for c-C_4F_8.

128.8.2 Perfluoro-2-butene

There are two primary attachment mechanisms: (1) At energies below 1 eV, parent negative ions 2-$C_4F_8^-$ are formed with a peak at near zero energy. This is the most abundant ion and the cross section determined by Chutjian and Aljajian (1985)

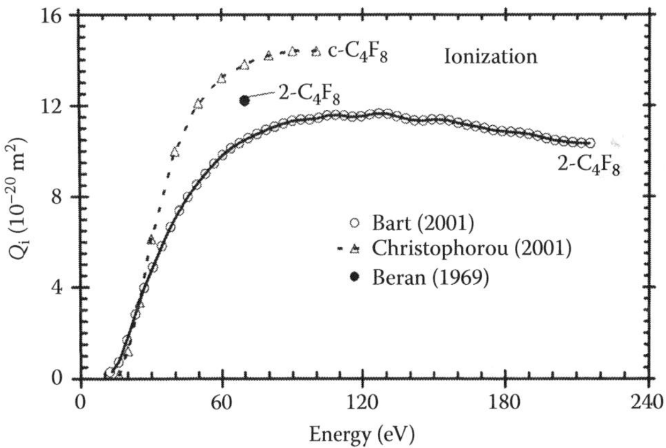

FIGURE 128.5 Total ionization cross sections for c-C_4F_8 and 2-C_4F_8. (—Δ—) c-C_4F_8, Christophorou and Olthoff; (○) 2-C_4F_8, Bart et al. (2001); 2-C_4F_8, Beran and Kevan (1969). Tabulated values of Bart et al. (2001) are courtesy of Professor Harland.

TABLE 128.8
Appearance Potentials and Peak Energy for Attachment for c-C_4F_8

Ion	Reaction Number	Appearance Potential (eV)	Peak (eV)	Reference
F^-	128.1	3.45	4.8	Sauers et al. (1979)
			6.5	Sauers et al. (1979)
			7.9	Sauers et al. (1979)
			10.2	Sauers et al. (1979)
CF_3^-	128.6	3.4	4.8	Sauers et al. (1979)
$C_2F_3^-$	128.7		4.9	Sauers et al. (1979)
			7.9	Sauers et al. (1979)
$C_3F_5^-$		3.0	4.1	Sauers et al. (1979)
F^-	128.2 or 128.3	3.7	4.95	Harland and Franklin (1974)
F^-	128.3 or 128.4	6.6	7.4	Harland and Franklin (1974)
F^-		7.5	8.6	Harland and Franklin (1974)
F^-	128.3 or 128.5	10.0	10.8	Harland and Franklin (1974)
CF_3^-	128.6	3.85	4.95	Harland and Franklin (1974)

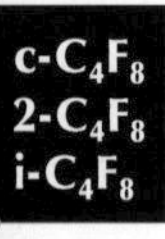

for 2-$C_4F_8^-$ at zero energy is 5×10^{-18} m^2. $C_4F_7^-$ and $C_4F_6^-$ ions are also observed at near zero energy (Sauers et al., 1979). (2) Dissociative attachment occurs in the energy range above 1 eV with several fragments including F^-.

Selected reactions are (Sauers et al., 1979):

$$2\text{-}C_4F_8 + e \rightarrow F^- + C_4F_7 \quad (128.8)$$

$$2\text{-}C_4F_8 + e \rightarrow CF_3^- + C_3F_5 \quad (128.9)$$

$$2\text{-}C_4F_8 + e \rightarrow C_2F_3^- + C_2F_5 \quad (128.10)$$

$$2\text{-}C_4F_8 + e \rightarrow C_3F_5^- + CF_3 \quad (128.11)$$

$$2\text{-}C_4F_8 + e \rightarrow C_3F_3^- + CF_4 + F \quad (128.12)$$

$$2\text{-}C_4F_8 + e \rightarrow C_3F_3^- + CF_3 + F_2 \quad (128.13)$$

$$2\text{-}C_4F_8 + e \rightarrow C_3F_3^- + CF_2 + F_2 + F \quad (128.14)$$

TABLE 128.9
Peak Energy for Negative Ions and Relative Yield for 2-C_4F_8

Ion Species	Onset Energy (eV)	Peak Energy (eV)	Relative Yield (%)
F^-	3.65	5.2	12.7
CF_3^-		5.3	0.6
$C_3F_3^-$	3.8	5.1	0.4
$C_3F_5^-$		2.3	0.9
$C_4F_6^-$		~0	0.4
$C_4F_7^-$		~0	1.5
$C_4F_8^-$		~0	83.4

TABLE 128.10
Thermal Attachment Rate Constants

Gas	Rate (m³/s)	Method	Reference
c-C_4F_8	9.1×10^{-15}	FA/LP	Miller et al. (2004)
	1.6×10^{-14}	Swarm	Spyrou et al. (1985)
	0.04×10^{-14}	ICRS	Woodin et al. (1980)
	1.4×10^{-14}	Swarm	Christodoulides et al. (1979)
	2.18×10^{-14}	Swarm	Christophorou et al. (1974)
	$1.05\text{–}1.17 \times 10^{-14}$	Swarm	Davis et al. (1973)
	1.21×10^{-14}	MC	Bansal and Fessenden (1973)
2-C_4F_8	4.87×10^{-14}	MC	Bansal and Fessenden (1973)
	4.60×10^{-14}	Swarm	Christodoulides et al. (1979)

Note: FA/LP = flowing afterglow/Langmuir probe; ICRS = ion cyclotron resonance spectroscopy; MC = microwave conductivity.

Table 128.9 shows the peak energies for various species of ions (Sauers et al., 1979).

128.9 ATTACHMENT RATE CONSTANTS

Thermal attachment rate constants are shown in Table 128.10. Attachment rates as a function of mean energy of the swarm are shown in Figure 128.6 (Christodoulides et al., 1979).

Table 128.11 shows the rate constants for attachment to c-C_4F_8 and detachment from c-$C_4F_8^-$ at various gas temperatures (Miller et al., 2004).

128.10 ATTACHMENT CROSS SECTIONS

Attachment cross sections for electrons impacting on c-C_4F_8 and 2-C_4F_8 arte shown in Table 128.12 and Figure 128.7 (Christodoulides et al., 1979). Revised values for c-C_4F_8 are given by Christophorou and Olthoff (2004) as shown in Figure 128.7. The cross section determined by Chutjian and Aljajian (1985) for 2-$C_4F_8^-$ at zero energy is 5×10^{-18} m^2.

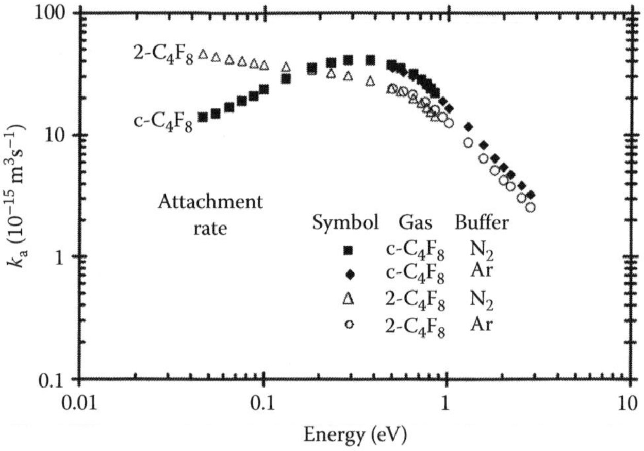

FIGURE 128.6 Attachment rates as a function of electron mean energy. Closed symbols: c-C_4F_8; open symbols: 2-C_4F_8. (■, Δ) N_2 buffer; (◆, ○) Ar buffer gas. (Adapted from Christodoulides, A. A. et al., *J. Chem. Phys.*, 70, 1156, 1979.)

TABLE 128.11
Attachment and Detachment Coefficients

T (K)	k_a (10–15 m³/s)	k_d (s^{-1})	Electron Affinity (meV)
298	9.3		
313	8.5		
349	8.3		
361	8.55	490	614
373	8.2	750	620
385	8.85	1126	630
400	9.0	1945	637

TABLE 128.12
Attachment Cross Sections of c-C_4F_8 and 2-C_4F_8

	Q_a (10^{-19} m^2)			Q_a (10^{-19} m^2)	
Energy (eV)	**c-C_4F_8**	**2-C_4F_8**	**Energy (eV)**	**c-C_4F_8**	**2-C_4F_8**
0.04	1.13	4.40	0.46	1.18	0.50
0.06	0.99	2.52	0.50	1.16	0.48
0.08	0.97	1.71	0.55	1.13	0.49
0.10	1.07	1.42	0.60	1.00	0.49
0.12	1.23	1.37	0.65	0.92	0.48
0.14	1.43	1.40	0.70	0.78	0.47
0.16	1.65	1.44	0.75	0.62	0.43
0.18	1.86	1.47	0.80	0.46	0.37
0.20	1.98	1.45	0.85	0.32	0.31
0.22	2.03	1.38	0.90	0.23	0.25
0.24	1.99	1.28	0.95	0.15	0.19
0.26	1.89	1.16	1.00	0.09	0.13
0.28	1.78	1.04	1.05	0.05	0.08
0.30	1.64	0.92	1.10	0.024	0.045
0.34	1.42	0.72	1.15	0.009	0.020
0.38	1.29	0.60	1.20	0.003	0.007
0.42	1.21	0.53	1.25		0.002

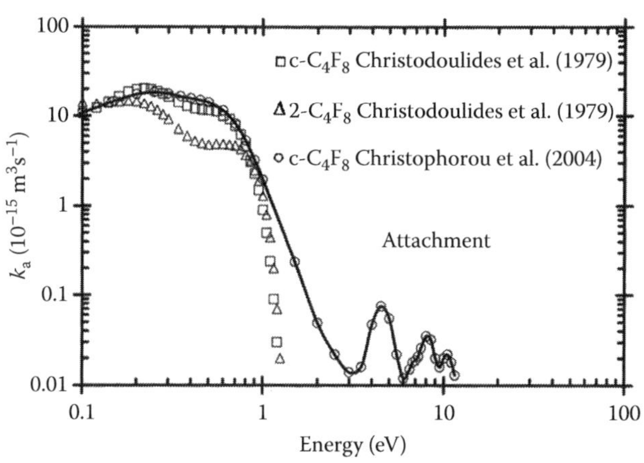

FIGURE 128.7 Attachment cross sections determined by swarm method (Christodoulides et al. 1979) and recommended by Christophorou and Olthoff (2004). Data of Spyrou et al. (1985) are not shown as they agree very well with Christophorou et al. (2004).

128.11 DRIFT VELOCITIES OF ELECTRONS

The drift velocity of electrons for c-C_4F_8 has been measured by Yamaji and Nakamura (2004) in the range from 110 to 1000 Td. de Urquijo and Basurto (2001) provide data in the low *E/N* range (12–43 Td) and in the high *E/N* range (275–600 Td). There is a slight pressure dependence of the drift velocity in the lower range. The velocities recommended by Christophorou and Olthoff (2004) are shown in Table 128.13 and Figure 128.8. The drift velocities of CF_4, C_2F_6, and C_3F_8 are also shown for comparison. The prominent negative differential conductivity observed in these gases is not observed in c-C_4F_8.

TABLE 128.13
Drift Velocities of Electrons for c-C_4F_8

E/N (Td)	*W* (10^4 m/s)	*E/N* (Td)	*W* (10^4 m/s)
10.0	1.91	500.0	19.9
20.0	4.02	550	21.8
30.0	5.83	600	23.9
40.0	7.28	650	26.1
275.0	14.2	700	28.7
300.0	14.9	750	31.3
350.0	16.2	800	33.8
400.0	17.5	840	35.9
450.0	18.6	1000	31.0[a]

Source: Adapted from Christophorou, L. G. and J. K. Olthoff. *Foundamental Electron Interaction with Plasma Gases,* Kluwer Academic, New York, NY, 2004.

[a] From Yamaji and Nakamura (2004).

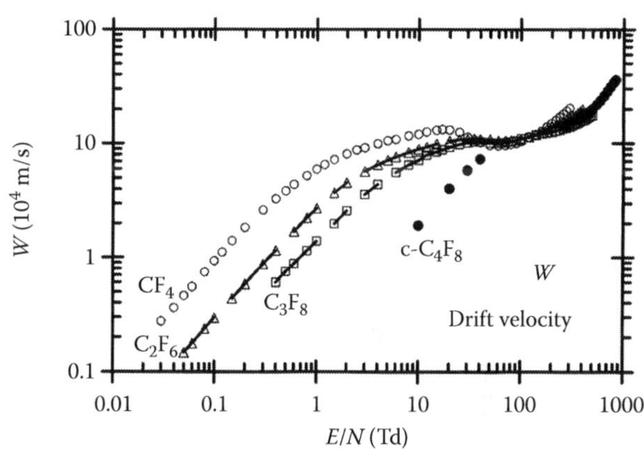

FIGURE 128.8 Drift velocities of electrons in c-C_4F_8 recommended by Christophorou and Olthoff (2004). Selected perfluorocarbons are also included for comparison. Note the negative differential conductivity for the gases.

TABLE 128.14
Drift Velocities of Electrons for *i*-C_4F_8

E/N (Td)	*W* (10^4 m/s)	*E/N* (Td)	*W* (10^4 m/s)
450	20.20	650	27.60
475	21.00	675	28.60
500	22.00	700	29.50
525	23.10	725	30.60
550	24.10	750	31.80
575	25.00	775	32.60
600	26.00	800	33.60
625	26.80	825	34.80

Source: Adapted from Naidu, M. S., A. N. Prasad, and J. D. Craggs, *J. Phys. D: Appl. Phys.*, 5, 741, 1972.

Table 128.14 and Figure 128.9 show the drift velocities in *i*-C_4F_8 (see Naidu et al., 1972) and the influence of isomerism.

128.12 CHARACTERISTIC ENERGY (D/μ)

Characteristic energies for c-C_4F_8 and *i*-C_4F_8 as a function of reduced electric field are shown in Table 128.15 and Figure 128.10 (Naidu et al., 1972). Slight dependence on gas number density (D/μ) increases with N is observed (not shown).

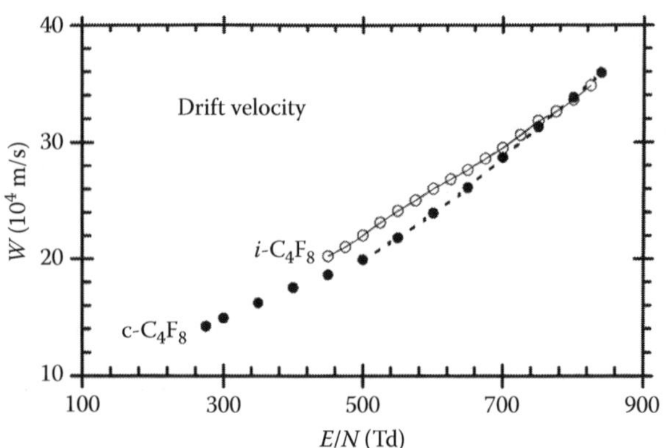

FIGURE 128.9 Drift velocity of *i*-C_4F_8 measured by Naidu et al. (1972). Data for c-C_4F_8 are also shown to demonstrate the effect of isomerism.

TABLE 128.15
Characteristic Energy for c-C_4F_8 and i-C_4F_8

E/N (Td)	c-C_4F_8 *D*/μ (V)	*i*-C_4F_8 *D*/μ (V)
300	1.22	
350	1.84	
400	2.47	
450	3.00	1.68
460		
475		1.88
500	3.42	2.09
520		
525		2.25
550		2.49
560	3.76	
575		2.25
600	4.03	2.82
625	4.14	2.97
650		3.11
675		3.22
700		3.33
725		3.45
750		3.53
800		3.67
850		3.80

Source: Adapted from Naidu, M. S., A. N. Prasad, and J. D. Craggs, *J. Phys. D: Appl. Phys.*, 5, 741, 1972.

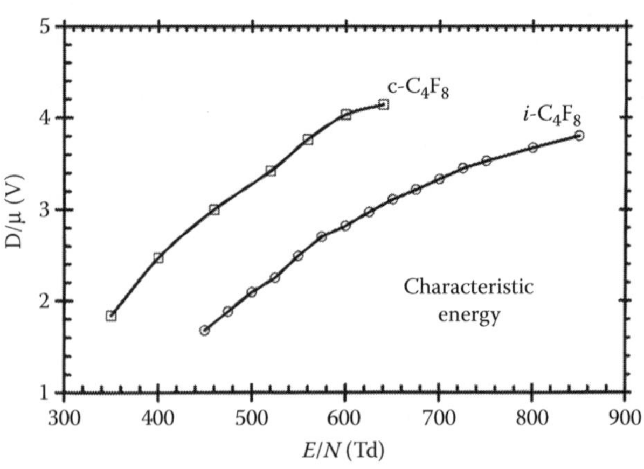

FIGURE 128.10 Characteristic energy for c-C_4F_8 and *i*-C_4F_8. (Adapted from Naidu, M. S., A. N. Prasad, and J. D. Craggs, *J. Phys. D: Appl. Phys.*, 5, 741, 1972.)

TABLE 128.16
Longitudinal Diffusion Coefficients for c-C_4F_8

E/N (Td)	ND_L (×10^{23} m^{-1} s^{-1})	*E/N* (Td)	ND_L (×10^{23} m^{-1} s^{-1})
170	4.76	500	12.83
200	7.92	600	13.46
250	10.85	700	12.81
300	12.14	800	13.46
350	12.46	900	14.38
400	12.83	1000	13.33

Source: Adapted from Yamaji, M. and Y. Nakamura, *J. Phys. D: Appl. Phys.*, 37, 1525, 2004.

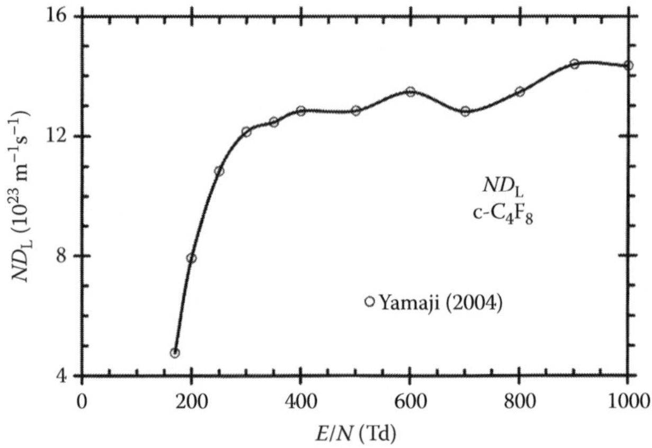

FIGURE 128.11 ND_L as a function of *E/N* for c-C_4F_8. (Adapted from Yamaji, M. and Y. Nakamura, *J. Phys. D: Appl. Phys.*, 37, 1525, 2004.)

128.13 LONGITUDINAL DIFFUSION COEFFICIENTS

Products of gas number density and longitudinal diffusion coefficients (ND_L) are shown in Table 128.16 and Figure 128.11 for c-C_4F_8 (Yamaji and Nakamura, 2004).

128.14 IONIZATION AND ATTACHMENT COEFFICIENTS

Ionization and attachment coefficients for c-C_4F_8 are shown in Table 128.16 and Figure 128.11. The effective ionization coefficient up to 10,000 Td is measured by Yamaji and Nakamura (2004) in c-C_4F_8. Data for *i*-C_4F_8 are shown in Table 128.17 and Figure 128.12.

Density-reduced ionization and attachment coefficients for *i*-C_4F_8 are shown in Table 128.18 and Figure 128.13 (Naidu et al., 1972).

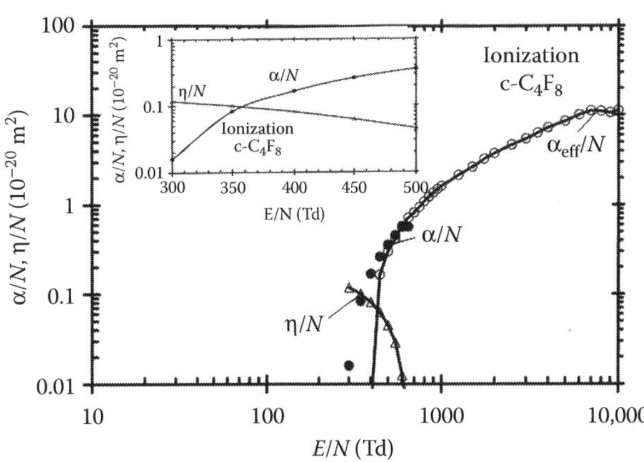

FIGURE 128.12 Ionization and attachment coefficients for c-C_4F_8. (●) α/N Naidu et al. (1972); (Δ) η/N Naidu et al. (1972); (—○—) α_{eff}/N Yamaji and Nakamura (2004). Inset shows the details at $\alpha_{eff}/N = (\alpha - \eta)/N$ Naidu et al. (1972).

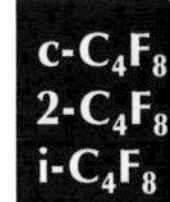

TABLE 128.17
Ionization and Attachment Coefficients for c-C_4F_8

E/N (Td)	Naidu et al. (1972)		Yamaji and Nakamura (2004)	
	α/N (10^{-20} m^2)	η/N (10^{-20} m^2)	*E/N* (10^{-20} m^2)	$(\alpha-\eta)/N$ (10^{-20} m^2)
300	1.60×10^{-2}	11.8×10^{-2}	400	−0.060
350	8.27×10^{-2}	9.98×10^{-2}	450	0.163
400	0.167	8.10×10^{-2}	500	0.299
450	0.261	6.22×10^{-2}	550	0.420
500	0.355	4.49×10^{-2}	600	0.578
550	0.455	2.87×10^{-2}	650	0.717
600	0.562	1.25×10^{-2}	700	0.821
650	0.666	4.23×10^{-3}	750	0.947
			800	1.095
			850	1.238
			900	1.373
			950	1.502
			1000	1.625
			1250	2.150
			1500	2.653
			1750	3.265
			2000	3.755
			2500	4.668
			3000	5.452
			3500	6.335
			4000	7.261
			5000	8.559
			6000	10.062
			7000	11.154
			8000	11.004
			9000	10.597
			10000	11.234

TABLE 128.18
Density-Reduced Ionization and Attachment Coefficients for i-C_4F_8

E/N (Td)	α/N (10^{-20} m^2)	η/N (10^{-20} m^2)
425		0.13
450		0.12
475		0.11
500	0.08	0.09
525	0.12	0.08
550	0.17	0.07
575	0.22	0.06
600	0.26	0.05
625	0.30	0.04
650	0.34	0.02
675	0.37	0.02
700	0.40	0.01
725	0.44	
750	0.48	
775	0.50	
800	0.53	
825	0.56	
850	0.58	

Source: Adapted from Naidu, M. S., A. N. Prasad, and J. D. Craggs, *J. Phys. D: Appl. Phys.*, 5, 741, 1972.
Note: α/N and η/N (i-C_4F_8).

128.15 LIMITING *E/N*

The limiting value of *E/N* defined by the condition $\alpha/N = \eta/N$ is shown in Table 128.19.

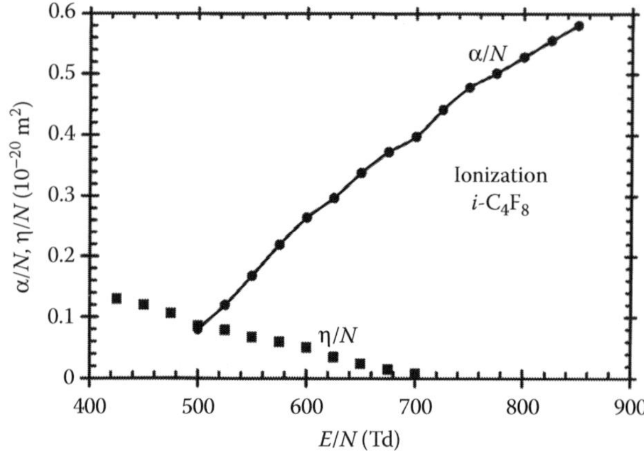

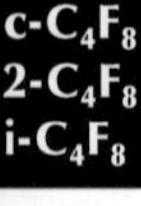

FIGURE 128.13 Ionization and attachment coefficients for i-C_4F_8. (—•—) α/N; (■) η/N. (Adapted from Naidu, M. S., A. N. Prasad, and J. D. Craggs, *J. Phys. D: Appl. Phys.*, 5, 741, 1972.)

TABLE 128.19
Limiting Value of *E/N*

Gas	$(E/N)_{lim}$ (Td)	Reference
c-C_4F_8	360	Wu et al. (2006)
	440	de Urquijo and Basurto (2001)
	358	Naidu et al. (1972)
i-C_4F_8	500	Naidu et al. (1972)

REFERENCES

Bansal, K. M. and R. W. Fessenden, *J. Chem. Phys.*, 59, 1760, 1973.

Bart, M., P. W. Harland, J. E. Hudson, and C. Vallance, *Phys. Chem. Chem. Phys.*, 3, 800, 2001.

Beran, J. A. and L. Kevan, *J. Phys. Chem.*, 73, 3866, 1969.

Christodoulides, A. A., L. G. Christophorou, R. Y. Pai, and C. M. Tung, *J. Chem. Phys.*, 70, 1156, 1979.

Christophorou, L. G. and J. K. Olthoff, *Fundamental Electron Interactions with Plasma Processing Gases*, Kluwer Academic, New York, NY, 2004.

Christophorou, L. G., D. L. McCorkle, and D. Pittmann, *J. Chem. Phys.*, 60, 1183, 1974.

Christophorou, L. G., R. A. Mathis, D. R. James, and D. L. McCorkle, *J. Phys. D: Appl. Phys.*, 14, 1889, 1981.

Chutjian, A. and S. H. Alajajian, *J. Phys. B: At. Mol. Phys.*, B20, 839, 1987.

Davis, F. J., R. N. Compton, and D. R. Nelson, *J. Chem. Phys.*, 59, 2324, 1973.

de Urquijo, J. and E. Basurto, *J. Phys. D: Appl. Phys.*, 34, 1952, 2001.

Harland, P. W. and J. L. Franklin, *J. Chem. Phys.*, 61, 1621, 1974.

Jarvis, G. K., K. J. Boyle, C. A. Mayhew, and R. P. Tuckett, *J. Phys. Chem. A*, 102, 3219, 1998a; *J. Phys. Chem. A*, 102, 3230, 1998b.

Jelisavcic, M., R. Panajotovic, M. Kitajima, M. Hoshino, H. Tanaka, and S. J. Buckman, *J. Chem. Phys.*, 121, 5272, 2004.

Lifshitz, C. and R. Grajower, *Int. J. Mass Spectrom. Ion Phys.*, 10, 25, 1972/73.

Miller, T. M., J. F. Friedman, and A. A. Vijjiano, *J. Chem. Phys.*, 120, 7024, 2004.

Naidu, M. S., A. N. Prasad, and J. D. Craggs, *J. Phys. D: Appl. Phys.*, 5, 741, 1972.

Novak, J. P. and M. F. Frechette, *J. Appl. Phys.*, 63, 2570, 1988.

Sanabia, J. E., G. D. Cooper, J. A. Tossell, and J. H. Moore, *J. Chem. Phys.*, 108, 389, 1998.

Sauers, I., L. G. Christophorou, and J. G. Carter, *J. Chem. Phys.*, 71, 3016, 1979.

Spyrou, S. M., S. R. Hunter, and L. G. Christophorou, *J. Chem. Phys*, 83, 641, 1985.

Toyoda, H., M. Ito, and H. Sugai, *Jpn. J. Appl. Phys.*, 36, 3730, 1997.

Winstead, C. and V. McKoy, *J. Chem. Phys.*, 114, 7407, 2001.

Woodin, R. L., M. S. Foster, and J. L. Beauchamp, *J. Chem. Phys.*, 72, 4223, 1980.

Wu, B., D. Xiao, Z. Liu, L. Zhang, and X. Liu, *J. Phys. D: Appl. Phys.*, 39, 4204, 2006.

Yamaji, M. and Y. Nakamura, *J. Phys. D: Appl. Phys.*, 37, 1525, 2004.

129

1-PROPANOL 2-PROPANOL

CONTENTS

2-Propanol (2-C_3H_8O) is a polar, electron-attaching molecule that has 34 electrons. Its dipole moment is 1.58 D and ionization potential 10.17 eV. In comparison, the dipole moment of 1-propanal is 1.55 D and ionization potential 10.18 eV.

129.1 IONIZATION CROSS SECTION

Partial and total ionization cross sections in 1-propanol (1-C_3H_8O) have been measured by Rejoub et al. (2003) as shown in Table 129.1 and Figure 129.1. A single study in 2-C_3H_8O is due to Vacher et al., (2006) as shown in Tables 129.2 and 129.3 and Figures 129.2 and 129.3.

TABLE 129.1
Ionization Cross Section in Simple Alcohols

	Cross Section (10^{-20} m^2)		
Energy (eV)	Methanol (CH_3OH)	Ethanol (C_2H_5OH)	1-Propanol (C_3H_7OH)
13	0.151		
14	0.335		
16	0.624	1.68	1.83
18	0.976	2.36	2.42
20	1.335	2.81	3.67
22.5	1.814	3.28	4.06
25	2.248	3.75	5.45
30	2.859	4.7	7.45
35	3.402	5.405	8.7
40	3.821	6.09	9.31
50	4.396	6.982	10.341
60	4.693	7.545	10.729
80	5.026	8	10.964
100	4.975	7.91	10.661
125	4.893	7.543	10.315
150	4.707	6.946	9.875
200	4.224	6.2	9.043
300	3.486	5.313	7.539
400	2.919	4.543	6.254
500	2.523	4.035	5.479
600	2.26	3.677	5.011
800	1.838	3.131	4.575
1000	1.563	2.782	4.217

Source: Adapted from Rejoub, R. et al., *J. Chem. Phys.*, 118, 1756, 2003.

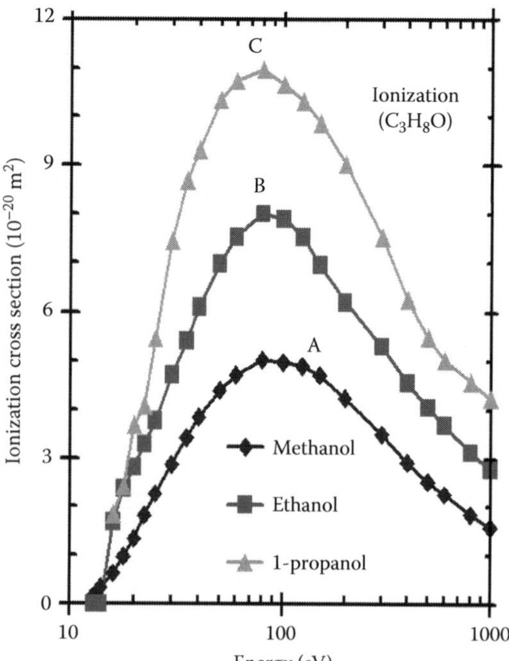

FIGURE 129.1 Ionization cross section for simple alcohols. A—Methanol; B—ethanol; C-1-C_3H_8O. (Adapted from Rejoub, R. et al., *J. Chem. Phys.*, 118, 1756, 2003.)

TABLE 129.2
Selected Partial Ionization Cross Sections for $2\text{-}C_3H_8O$

Energy (eV)	Ionization Cross Section (10^{-20} m²) Ion Mass Number 27	29	39	41	42	43	44
12							
12.5						2.44 (−2)	
13						2.72 (−2)	
16	2.64 (−3)	5.13 (−3)		1.76 (−2)		5.74 (−2)	
17	3.04 (−3)	7.55 (−3)		1.95 (−2)	7.95 (−2)	7.92 (−2)	0.145
18	7.74 (−3)	1.26 (−2)	1.99 (−3)	2.82 (−2)	7.47 (−2)	0.107	0.142
20	2.56 (−2)	2.63(−2)	4.76 (−3)	5.13 (−2)	7.43 (−2)	0.171	0.132
22	5.70 (−2)	4.48(−2)	8.39 (−3)	8.40 (−2)	7.82 (−2)	0.247	0.138
25	8.83 (−2)	6.33(−2)	1.99 (−2)	0.122	8.02 (−2)	0.387	0.145
30	0.142	0.101	5.02 (−2)	0.167	8.38 (−2)	0.700	0.156
35	0.188	0.137	9.96 (−2)	0.193	9.12 (−2)	0.872	0.150
40	0.223	0.170	0.160	0.263	0.120	1.070	0.168
50	0.263	0.207	0.201	0.279	0.161	1.100	0.197
60	0.275	0.093	0.124	0.239	0.106	0.937	0.167
70	0.183	0.126	0.173	0.246	0.143	0.761	0.211
80	0.261	0.222	0.284	0.353	0.213	0.856	0.234
86	0.319	0.286	0.333	0.390	0.240	0.993	0.245

Note: Tabulated values by courtesy of Dr. Vacher (2007). a (b) means $a \times 10^{-b}$.

TABLE 129.3
Selected Partial and Total Ionization Cross Sections for $2\text{-}C_3H_8O$

Energy (eV)	Ionization Cross Section (10^{-20} m²) Ion Mass Number 45	59	60	Total
12	5.36			5.360
12.5	5.02			5.044
13	4.47	3.74 (−2)		4.535
16	4.95	6.29 (−2)	3.50 (−2)	5.131
17	6.04	6.98 (−2)	3.26 (−2)	6.476
18	5.68	7.53 (−2)	3.18 (−2)	6.161
20	5.18	9.48 (−2)	3.16 (−2)	5.792
22	4.96	0.107	3.34 (−2)	5.758
25	3.78	0.124	3.46 (−2)	4.844
30	3.02	0.135	3.59 (−2)	4.591
35	2.31	0.136	3.70 (−2)	4.214
40	2.04	0.148	4.91 (−2)	4.411
50	1.97	0.178	4.96 (−2)	4.606
60	1.86	0.186	5.54 (−2)	4.043
70	1.56	0.214	6.23 (−2)	3.679
80	1.36	0.208	5.84 (−2)	4.049
86	1.10	0.201	5.71 (−2)	4.164

Note: Tabulated values by courtesy of Dr. Vacher (2007). a (b) means $a \times 10^{-b}$.

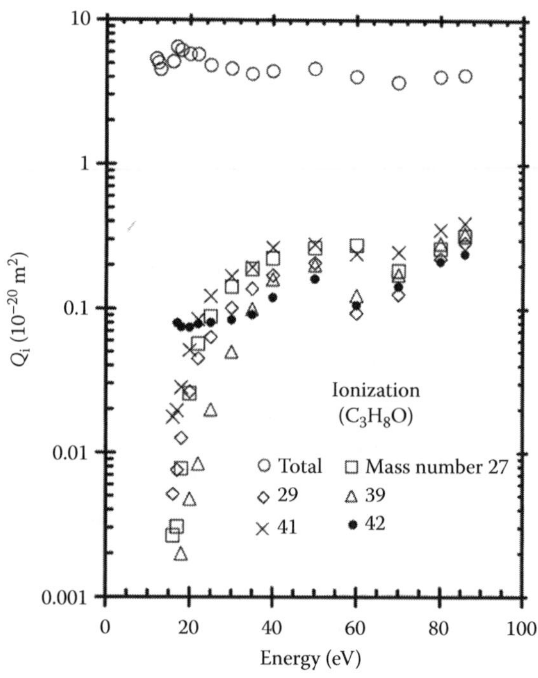

FIGURE 129.2 Partial ionization cross sections for selected fragments for $2\text{-}C_3H_8O$. Mass number and ion species: (□) 27-$C_2H_3^+$; (◊) 29-$C_2H_5^+$; (Δ) 39-$C_3H_3^+$; (×) 41-$C_3H_5^+$; (•) 42-$C_3H_6^+$; (o-top most) total, included for comparison. (Adapted from Vacher, J. R. et al., *Chem. Phys. Lett.*, 323, 587, 2006.)

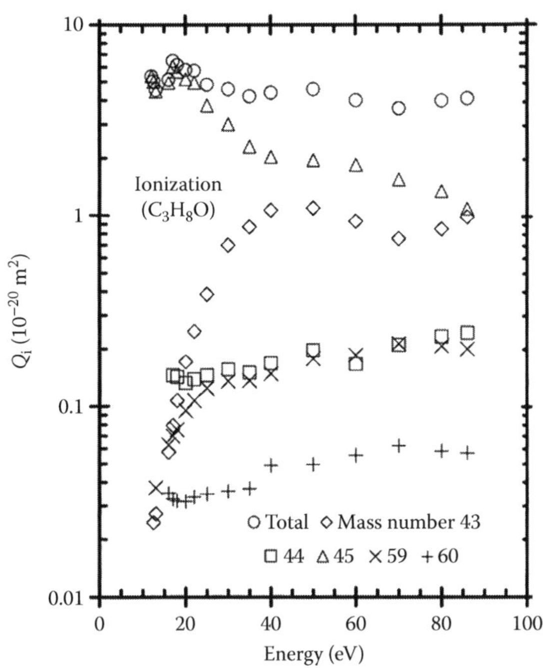

FIGURE 129.3 Partial ionization cross sections for selected fragments for 2-C_3H_8O. Mass number and ion species: (◊) 43-$C_2H_3O^+$; (□) 44-$C_3H_8^+$; (Δ) 45-$C_2H_5^+O$; (×) 59-$C_3H_7^+O$; (+) 60-$C_3H_8O^+$; (○) total, included for comparison. (Adapted from Vacher, J. R. et al., *Chem. Phys. Lett.*, 323, 587, 2006.)

REFERENCES

Rejoub, R., C. D. Morton, B. G. Lindsay, and R. F. Stebbings, *J. Chem. Phys.*, 118, 1756, 2003.

Vacher, J. R., F. Jorand, N. Blin-Simiand, and S. Pasquiers, *Chem. Phys. Lett.*, 323, 587, 2006.

Section XIII

MORE THAN 12 ATOMS

130

BUTANE

C_4H_{10}

CONTENTS

Butane (C_4H_{10}) is a saturated hydrocarbon (C_nH_{2n+2}) with 34 electrons and is the lowest member of the series to exhibit isomerism. The ground-state configuration is given by Au et al. (1993).

The molecule is nonpolar and has an electronic polarizability of 9×10^{-40} F m^2 and an ionization potential of 10.53 eV. The dissociation potential is 3.854 eV.

130.1 SELECTED REFERENCES FOR DATA

See Table 130.1.

TABLE 130.1
Selected References for Data

Type	Energy Range (eV)	Reference
	Cross Sections	
Q_i	200	Wang and Vidal (2002)
Q_T	5–400	Floeder (1985)
Q_a	0–15	Rutkowsky et al. (1980)
Q_M	0.02–0.25	McCorkle et al. (1978)
Q_i	600–12,000	Schram et al. (1966)
Q_T	0.8–50	Brüche (1930a)
Q_T	1–50	Brüche (1930b)
Type	***E/N* Range (Td)**	**Reference**
	Swarm Parameters	
α/N	125–5500	Heylen (1975)
W	0.6–21	McCorkle et al. (1978)
W	0.01–0.3	Gee et al. (1983)
W	0.01–0.5	Floriano et al. (1986)

Note: Q_a = attachment; Q_i = ionization; Q_M = momentum transfer; Q_T = total; W = drift velocity; α/N = reduced ionization coefficient.

130.2 TOTAL SCATTERING CROSS SECTION

See Tables 130.2 and 130.3.

130.3 MOMENTUM TRANSFER CROSS SECTION

See Table 130.4.

130.4 RAMSAUER–TOWNSEND MINIMUM

See Table 130.5.

TABLE 130.2
Total Scattering Cross Section (Digitized and Interpolated) in C_4H_{10}

Energy (eV)	Q_T (10^{-20} m^2)	Energy	Q_T (10^{-20} m^2)
1.0	12.40	14.0	42.99
2.0	25.94	16.0	40.65
3.0	35.94	18.0	38.67
4.0	42.89	20.0	37.02
5.0	47.29	25.0	33.90
6.0	49.62	30.0	31.40
7.0	50.36	35.0	29.33
8.0	50.02	40.0	27.69
9.0	49.09	45.0	26.49
10.0	47.96	50.0	25.72
12.0	45.51		

Sources: Adapted from Brüche, E. *Ann. Phys. Leipzig*, 4, 387, 1930a; Brüche, E. *Ann. Phys. Leipzig*, 5, 281, 1930b.
Note: See Figure 130.1 for graphical presentation.

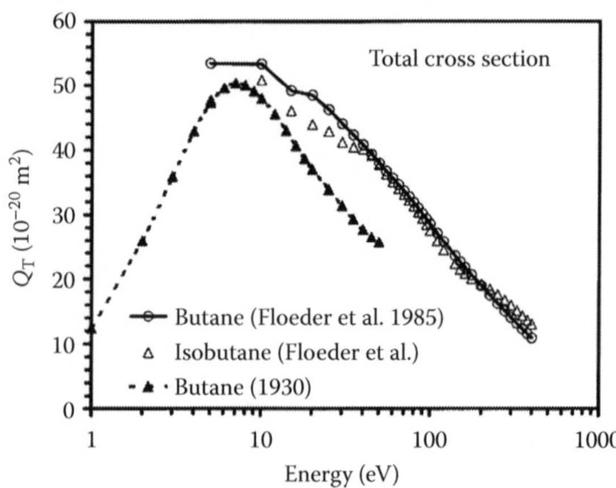

FIGURE 130.1 Total scattering cross section in butane and isobutane. (—○—) Butane, Floeder et al. (1985); (—△—) isobutane, Floeder et al. (1985); (—▲—) Brüche (1930b).

TABLE 130.3
Total Scattering Cross Section (Digitized and Interpolated) in Butane and Isobutane

Energy (eV)	Q_T (10^{-20} m^2) Butane	Q_T (10^{-20} m^2) Isobutane	Energy (eV)	Q_T (10^{-20} m^2) Butane	Q_T (10^{-20} m^2) Isobutane
5.0	53.46	47.69	95.0	29.36	28.47
10.0	53.35	50.82	100.0	28.57	27.54
15.0	49.18	46.04	110.0	27.11	25.89
20.0	48.46	43.93	120.0	25.80	24.49
25.0	46.21	42.89	140.0	23.57	22.33
30.0	44.02	41.19	150.0	22.62	21.52
35.0	42.32	40.47	160.0	21.76	20.86
40.0	40.80	40.16	175.0	20.61	20.09
45.0	39.33	39.20	200.0	18.96	19.18
50.0	37.96	37.76	225.0	17.50	18.40
55.0	36.74	36.31	250.0	16.21	17.59
60.0	35.65	35.10	275.0	15.06	16.76
65.0	34.65	34.08	300.0	14.04	15.92
70.0	33.71	33.17	325.0	13.13	15.11
75.0	32.80	32.29	350.0	12.31	14.34
80.0	31.90	31.37	375.0	11.57	13.64
85.0	31.03	30.40	400.0	10.88	13.01
90.0	30.18	29.43			

Source: Adapted from Floeder, K. et al., *J. Phys. B: At. Mol. Phys.*, 18, 3347, 1985.
Note: See Figure 130.1 for graphical presentation.

TABLE 130.4
Momentum Transfer Cross Section

Energy (eV)	Q_M (10^{-20} m^2)	Energy (eV)	Q_M (10^{-20} m^2)
0.015	26.29	0.080	6.92
0.020	26.05	0.090	6.10
0.030	21.84	0.100	5.5438
0.040	16.45	0.015	26.288
0.085	6.47	0.200	6.8368
0.060	10.10	0.270	18.45
0.070	8.19		

Source: Adapted from McCorkle, D. L. et al., *J. Phys. B: At Mol. Phys.*, 11, 3067, 1978.
Note: See Figure 130.2 for graphical presentation.

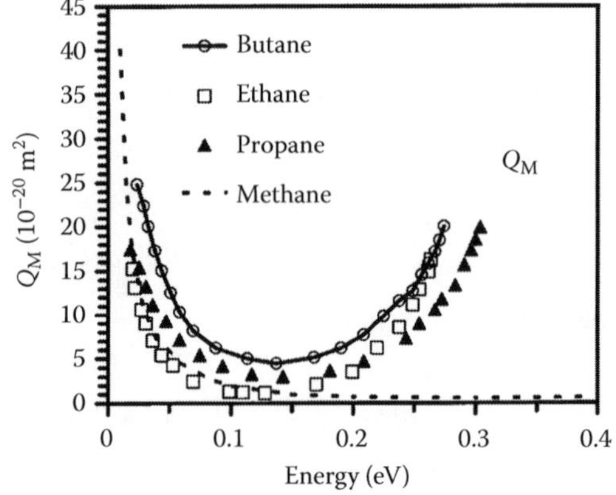

FIGURE 130.2 Low-energy momentum transfer cross section in C_4H_{10}. Selected hydrocarbons are shown for the purpose of comparison. Data for methane are from Davies et al. (1988) and for other gases the data are from McCorkle et al. (1978). (—○—) Butane; (□) ethane; (▲) propane; (- - -) methane.

TABLE 130.5
Ramsauer–Townsend Minimum

Gas	Minimum Energy	Cross Section	Method	Reference
Butane	0.14	4.50	Swarm	McCorkle (1978)
n-Butane	0.14	5.0	Swarm	Gee et al. (1983)
i-Butane	0.19	5.0	Swarm	Gee et al. (1983)

Note: Energy in Units of eV and Cross Section in 10^{-20} m².

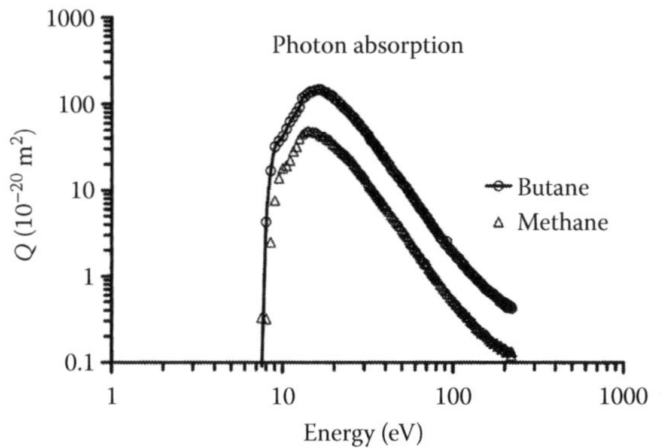

FIGURE 130.3 Photon absorption cross section in C_4H_{10} as a function of energy. (—○—) Butane; (△) methane. Curve for methane is included for comparison. (Adapted from Au, J. W., G. Cooper, and C. E. Brion, *Chem. Phys.*, 173, 241, 1993.)

TABLE 130.6
Ion Appearance Potentials

Ion Species	Potential (eV)	Species	Potential (eV)	Species	Potential (eV)
$C_4H_{10}^+$	10.0	$C_3H_7^+$	10.0	$C_2H_4^+$	10.0
$C_4H_9^+$	10.5	$C_3H_6^+$	10.0	$C_2H_3^+$	13.5
$C_4H_8^+$	10.5	$C_3H_5^+$	12.5	$C_2H_2^+$	20.5
$C_4H_7^+$	14.5	$C_3H_4^+$	14.0	CH_3^+	23.5
$C_4H_5^+$	26.0	$C_3H_3^+$	17.0	CH_2^+	29.5
$C_4H_4^+$	28.0	$C_3H_2^+$	34.0	H^+	22.0
$C_4H_2^+$	33.0	C_3H^+	39.0		
C_4H^+	30.0	$C_2H_5^+$	11.5		

130.5 ELECTRONIC EXCITATION CROSS SECTION

Electronic excitation cross sections are not available. Photon absorption cross sections (Au et al., 1993) shown in Figure 130.3 may be used as an approximation.

130.6 ION APPEARANCE POTENTIALS

Positive ion appearance potentials determined by the photoionization method (Au et al., 1993) are shown in Table 130.6.

130.7 IONIZATION CROSS SECTION

Limited data are available on electron impact ionization cross section (Table 130.7). Wang and Vidal have measured a total cross section of 9.52×10^{-20} m² at a single electron energy of 200 eV. Total photoionization cross section (Table 130.8) may be used as a guide.

TABLE 130.7
Ionization Cross Sections in Butane and Isobutane

	Q_i (10^{-20} m²)			Q_i (10^{-20} m²)	
Energy (keV)	Butane	Isobutane	Energy (keV)	Butane	Isobutane
0.6	5.45	5.53	4.0	1.24	1.23
1.0	3.74	3.72	7.0	0.775	0.779
2.0	2.19	2.17	12.0	0.502	0.497

Source: Adapted from Schram, B. L. et al., *J. Chem. Phys.*, 44, 49, 1966.

TABLE 130.8
Photoionization Cross Section in Butane and Selected Hydrocarbons

	Q_i (10^{-22} m²)				Q_i (10^{-22} m²)		
Energy (eV)	Butane	Ethane	Propane	Energy (eV)	Butane	Ethane	Propane
10.0	1.70			27.0	62.29	31.85	38.51
10.5	7.70		1.31	27.5	61.20	31.34	37.34
11.0	20.01	1.49	5.53	28.0	58.20	29.30	35.10
11.5	35.52	2.96	13.65	28.5	54.97	28.64	33.80
12.0	58.21	17.31	22.27	29.0	53.04	26.87	32.54
12.5	77.11	31.95	34.65	29.5	51.41	25.98	31.61
13.0	102.80	49.77	43.15	30.0	48.83	24.30	29.94
13.5	116.46	57.96	53.56	31.0	45.15	21.99	27.55
14.0	131.85	61.98	65.87	32.0	40.46	21.07	24.95
14.5	138.58	64.92	75.09	33.0	36.95	18.96	22.76
15.0	138.59	70.50	80.66	34.0	35.13	17.77	21.19
15.5	143.45	76.24	86.74	35.0	32.37	16.41	19.81
16.0	146.88	77.64	89.74	36.0	29.59	15.30	18.41
16.5	147.24	76.84	90.50	37.0	27.54	14.37	17.07
17.0	142.43	72.25	87.35	38.0	25.64	13.61	16.19
17.5	137.42	70.67	85.28	39.0	24.32	12.53	14.93
18.0	135.65	71.25	85.26	40.0	22.19	11.87	14.07
18.5	127.97	67.98	79.67	41.0	20.92	10.96	13.31
19.0	126.59	66.67	78.59	42.0	19.61	10.20	12.24
19.5	123.87	65.00	77.78	43.0	17.79	9.59	11.64
20.0	114.30	59.87	72.33	44.0	16.84	9.12	10.85
20.5	111.25	57.83	69.76	45.0	15.95	8.55	10.13
21.0	106.80	55.46	67.42	46.0	15.16	8.33	9.75
21.5	99.98	52.33	63.94	47.0	14.44	7.78	9.33
22.0	98.09	50.86	61.72	48.0	13.75	7.35	8.88
22.5	91.58	48.00	57.66	49.0	13.06	6.85	8.23

continued

C$_4$H$_{10}$

TABLE 130.8 (continued)
Photoionization Cross Section in Butane and Selected Hydrocarbons

Energy (eV)	Q_i (10^{-22} m^2) Butane	Ethane	Propane	Energy (eV)	Q_i (10^{-22} m^2) Butane	Ethane	Propane
23.0	88.50	45.93	55.04	50.0	12.17	6.60	7.98
23.5	85.62	44.22	53.16	55.0		4.94	6.01
24.0	82.43	42.64	50.97	60.0		3.84	4.89
24.5	78.26	40.14	48.08	65.0		3.16	3.86
25.0	74.75	38.66	46.52	70.0		2.58	3.20
25.5	71.97	36.78	44.55	75.0		2.11	2.61
26.0	70.73	36.45	44.02	80.0		1.79	2.16
26.5	65.98	33.82	40.23				

Source: Adapted from Au, J. W., G. Cooper, and C. E. Brion, *Chem. Phys.*, 173, 241, 1993.
Note: See Figure 130.4 for graphical presentation.

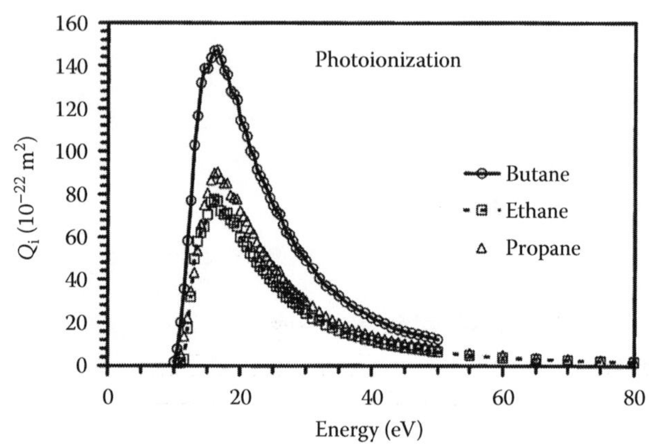

FIGURE 130.4 Photoionization cross section in C$_4$H$_{10}$ and selected hydrocarbons. (—○—) Butane; (□) ethane; (△) propane. (Adapted from Au, J. W., G. Cooper, and C. E. Brion, *Chem. Phys.*, 173, 241, 1993.)

130.8 ATTACHMENT CROSS SECTION

The dissociative attachment process in C$_4$H$_{10}$ occurs according to (Rutkowsky et al., 1980)

$$e\,(>eV) + C_4H_{10} \rightarrow C_4H_{10}^{*} \rightarrow \begin{cases} H^- + C_4H_9 \\ CH_3^- + C_3H_7 \end{cases} \quad (130.1)$$

The attachment cross section is shown in Table 130.9 and Figure 130.5 (Rutkowsky et al., 1980).

130.9 DRIFT VELOCITY OF ELECTRONS

Table 130.10 shows the drift velocity of electrons in C$_4$H$_{10}$ as a function of *E*/*N* and temperature (McCorkle et al., 1978). See Figure 130.6 for graphical presentation. Table 130.11 shows the drift velocity of electrons in isobutane.

TABLE 130.9
Attachment Cross Section in C$_4$H$_{10}$

Energy (eV)	Q_a (10^{-24} m^2)	Energy (eV)	Q_a (10^{-24} m^2)
5.7	0.05	9.1	4.14
6.2	0.58	9.4	3.33
6.6	1.20	9.7	2.52
6.8	2.23	10.2	1.50
7.2	3.17	10.4	0.89
7.4	3.68	11.4	0.61
7.8	4.31	12.3	0.63
8.4	4.53		

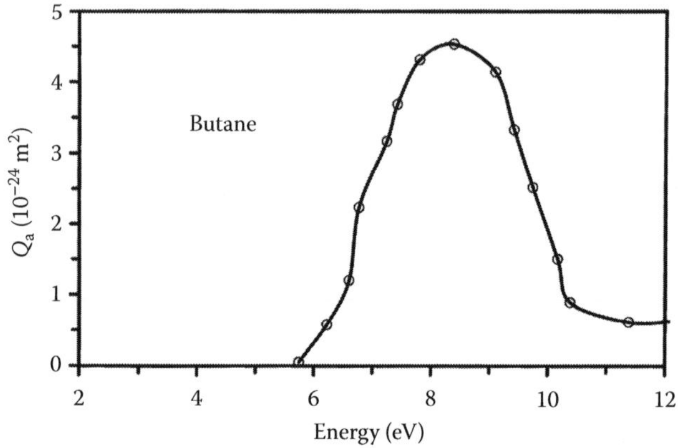

FIGURE 130.5 Attachment cross section in C$_4$H$_{10}$.

TABLE 130.10
Drift Velocity

E/*N* (Td)	*W* (10^4 m/s) 299 K	373 K	473 K	573 K	673 K
0.06	0.073		0.079	0.078	0.076
0.09	0.111	0.122	0.121	0.115	0.11
0.12	0.152	0.155	0.162	0.154	0.147
0.18	0.23	0.236	0.241	0.227	0.217
0.24	0.321	0.327	0.316	0.301	0.286
0.3	0.396	0.397	0.391	0.375	0.356
0.45	0.594	0.6	0.592	0.546	0.523
0.6	0.835	0.795	0.753	0.72	0.702
0.75	1.00	1.00	0.938	0.907	0.856
0.9	1.23	1.16	1.13	1.05	0.999
1.05	1.37	1.4	1.32	1.21	1.17
1.2	1.54	1.51	1.49	1.36	1.32
1.5	1.82	1.79	1.76	1.63	1.55
1.8	2.1	2.06	1.99	1.89	1.86
2.1	2.4	2.3	2.23	2.08	2.03
3	2.89	2.74	2.68	2.78	2.69
4.5	3.32	3.2	3.25	3.25	3.21
7.5	3.64	3.7	3.75	3.77	4.13
12	3.83	3.96	4.02	4.15	4.28
21			4.5	4.51	4.72

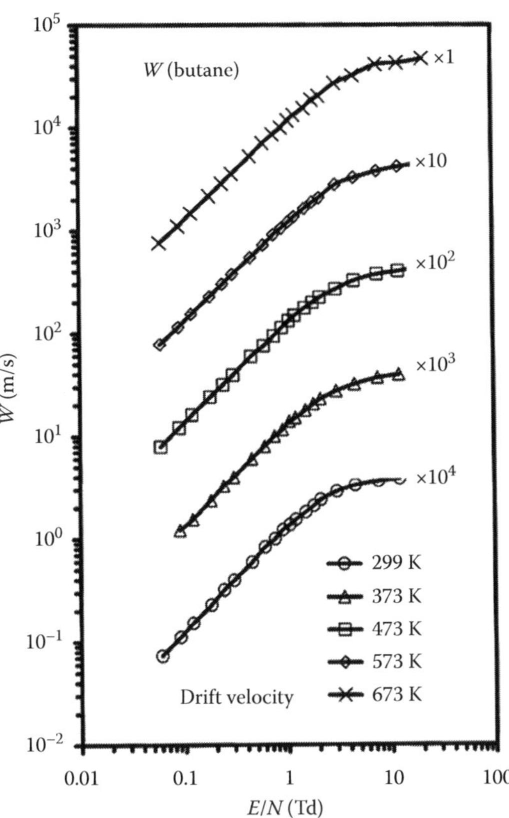

FIGURE 130.6 Drift velocity of electrons as a function of *E/N* and temperature in C_4H_{10}. (Adapted from McCorkle, D. L. et al., *J. Phys. B: At Mol. Phys.*, 11, 3067, 1978.)

TABLE 130.11
Drift Velocity of Electrons in Isobutane

E/N (Td)	*W* (103 m/s)	*E/N* (Td)	*W* (103 m/s)	*E/N* (Td)	*W* (103 m/s)
0.03	0.195	0.6	3.9	12.0	55
0.06	0.390	0.8	5.2	15	55
0.09	0.59	1.0	6.5	20	55
0.12	0.78	2.0	13.0	25	55
0.24	1.56	4.0	26.0	27	55
0.36	2.35	6.0	39.0		
0.5	3.12	8.0	42.0		

Source: Adapted from Floriano, M. A., N. Gee, and G. R. Freeman, *J. Chem. Phys.*, 84, 6799, 1986.
Note: Note the saturation above 12 Td.

130.10 REDUCED IONIZATION COEFFICIENT

Reduced ionization coefficients as a function of reduced electric field are shown in Table 130.12. For graphical presentation and comparison with selected hydrocarbons, see Figure 130.7.

TABLE 130.12
Reduced Ionization Coefficients as a Function of Reduced Electric Field in C_4H_{10}

E/N (Td)	α/*N* (10^{-20} m^2)	*E/N* (Td)	α/*N* (10^{-20} m^2)
125	0.0002	800	1.644
150	0.0025	900	1.939
175	0.0092	1000	2.238
200	0.0254	1250	3.004
225	0.0488	1500	3.712
250	0.079	2000	4.965
300	0.1743	2500	6.303
350	0.2943	3000	7.397
400	0.4231	4000	8.347
450	0.5631	4500	8.645
500	0.7107	5000	9.143
600	1.0178	5500	10.079
700	1.3402		

Source: Adapted from Heylen, A. E. D. *Int. J. Electron.*, 39, 653, 1975.

C_4H_{10}

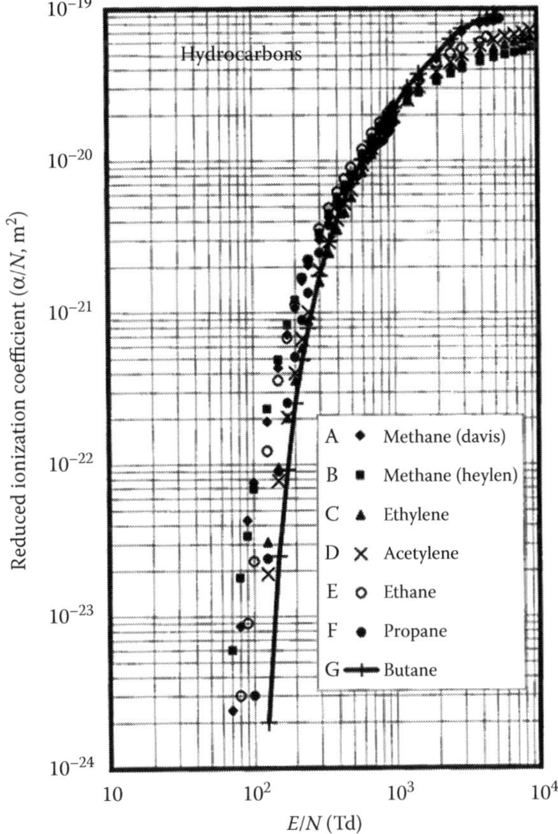

FIGURE 130.7 Reduced ionization coefficients as a function of reduced electric field in C_4H_{10} and comparison with selected hydrocarbons. (A, ◆) Methane, Davies et al. (1989); (B, ■) methane, Heylen (1975); (C, ▲) ethylene, Heylen (1978); Heylen (1975); (D, ×) acetylene, Heylen (1963); (E, ○) ethane, Heylen (1975); (F, ●) propane, Heylen (1975); (G, —+—) butane, Heylen (1975).

130.11 GAS CONSTANTS

Gas constants according to the expression

$$\frac{\alpha}{N} = F\exp-\left(\frac{GN}{E}\right) \quad (130.2)$$

are (Raju, 2005) $F = 7.6 \times 10^{-20}$ m^2 and $G = 1136$ Td applicable for the range $165 \leq E/N \leq 1900$ Td.

C_4H_{10}

REFERENCES

Au, J. W., G. Cooper, and C. E. Brion, *Chem. Phys.*, 173, 241, 1993.
Brüche, E., *Ann. Phys. Leipzig*, 4, 387, 1930a.
Brüche, E., *Ann. Phys. Leipzig*, 5, 281, 1930b.
Davies, D. K., L. E. Kline, and W. E. Bies, *J. Appl. Phys.*, 65, 3311, 1989.
Floeder, K., D. Fromme, W. Raith, A. Schwab, and G. Sinapius, *J. Phys. B: At. Mol. Phys.*, 18, 3347, 1985.
Floriano, M. A., N. Gee, and G. R. Freeman, *J. Chem. Phys.*, 84, 6799, 1986.
Gee, N. and G. R. Freeman, *J. Chem. Phys.*, 78, 1951, 1983.
Heylen, A. E. D., *Int. J. Electron.*, 39, 653, 1975.
Heylen, A. E. D., *Int. J. Electron.*, 44, 367, 1978.
Heylen, A. E. D., *J. Chem. Phys.*, 38, 765, 1963.
McCorkle, D. L., L. G. Christophorou, D. V. Maxey, and J. G. Carter, *J. Phys. B: At. Mol. Phys.*, 11, 3067, 1978.
Raju, G. G. *Gaseous Electronics: Theory and Practice* Taylor & Francis LLC, Boca Raton, FL, 2005.
Rutkowsky, J., H. Drost, and H. J. Spengenberg, *Ann. Phys. Liepzig*, 37, 259, 1980.
Schram, B. L., M. J. Van der Wiel, F. J. de Herr, and H. R. Moustafa, *J. Chem. Phys.*, 44, 49, 1966.
Wang, P. and C. R. Vidal, *Chem. Phys.*, 280, 309, 2002.

131

CHLOROBUTANE, 1-CHLOROBUTANE, 2-CHLOROBUTANE, AND *t*-CHLOROBUTANE

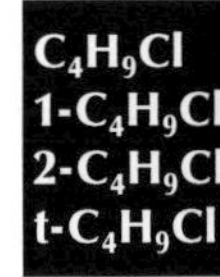

CONTENTS

Chlorobutane (C_4H_9Cl) is a derivative of butane (C_4H_{10}) molecule, with one of the hydrogen atoms replaced with a chlorine atom. It has 50 electrons. Of several isomers, we consider three, and their selected properties are shown in Table 131.1.

131.1 IONIZATION CROSS SECTIONS

Ionization cross sections measured by Hudson et al. (2001) are shown in Table 131.1 and Figure 131.1. There is no

TABLE 131.1
Selected Properties of Isomers of C_4H_9Cl

Molecule	α_e (10^{-40} m^2)	μ (D)	ε_i (eV)
1-Chlorobutane (butyl chloride)	12.57	2.05	10.67
2-Chlorobutane	13.80	2.04	10.53
t-Chlorobutane	13.91	2.13	10.61

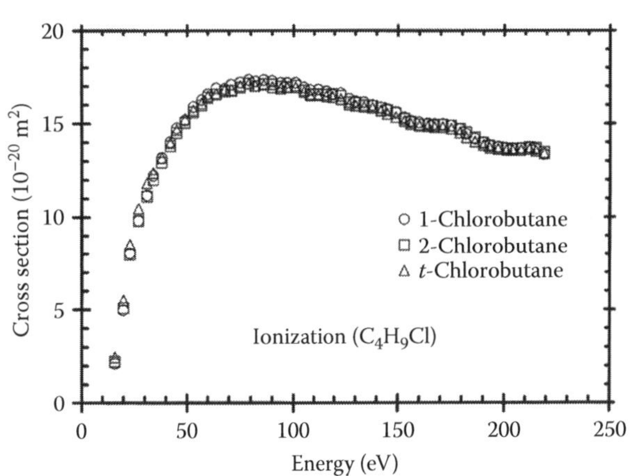

FIGURE 131.1 Ionization cross sections for isomers of C_4H_9Cl. (○) 1-Chlorobutane; (□) 2-chlorobutane; (Δ) *t*-chlorobutane. Tabulated values are courtesy of Professor Harland (2006).

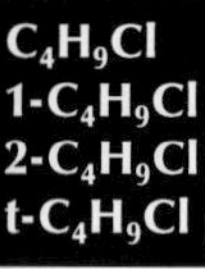

TABLE 131.2
Ionization Cross Sections for Selected Isomers of C_4H_9Cl

	Ionization Cross Section (10^{-20} m^2)		
Energy (eV)	$1\text{-}C_4H_9Cl$	$2\text{-}C_4H_9Cl$	$t\text{-}C_4H_9Cl$
16	2.12	2.22	2.46
20	4.97	5.08	5.51
23	8.03	7.97	8.52
27	9.81	9.80	10.47
31	11.17	11.11	11.84
34	12.23	12.01	12.38
38	13.18	12.93	13.18
42	14.02	13.80	13.98
45	14.75	14.50	14.67
49	15.17	15.04	15.24
53	15.89	15.61	15.71
57	16.26	15.96	16.08
60	16.57	16.38	16.46
64	16.89	16.56	16.60
68	16.86	16.72	16.80
71	17.08	16.75	16.81
75	17.20	16.93	17.02
79	17.34	17.15	17.14
82	17.26	17.01	17.04
86	17.33	17.12	17.08
90	17.28	17.10	16.94
94	17.18	16.98	16.88
97	17.17	16.95	16.95
101	17.18	17.03	16.94
105	16.93	16.77	16.70
108	16.79	16.56	16.50
112	16.82	16.56	16.52
116	16.72	16.55	16.46
119	16.55	16.58	16.42
123	16.62	16.48	16.30
127	16.31	16.13	16.02
130	16.18	16.11	15.96
134	16.13	15.96	15.89
138	15.98	15.92	15.84
142	15.84	15.83	15.69
145	15.75	15.70	15.51
149	15.62	15.54	15.34
153	15.32	15.24	15.13
156	15.13	15.07	14.97
160	15.05	14.93	14.87
164	14.99	14.93	14.85
167	14.95	14.88	14.81
171	14.95	14.92	14.80
175	14.87	14.87	14.72
179	14.69	14.69	14.50
182	14.43	14.51	14.25
186	14.18	14.23	14.03
190	13.98	13.96	13.84
193	13.83	13.80	13.70
197	13.74	13.72	13.64
201	13.69	13.65	13.59
204	13.66	13.63	13.58
208	13.67	13.66	13.58
212	13.72	13.73	13.64
215	13.63	13.68	13.54
219	13.37	13.46	13.40

Sources: Adapted from Hudson, J. E. et al., *J. Phys. B: At. Mol. Opt. Phys.*, 34, 3025, 2001; tabulated values courtesy of Professor P. Harland (2006).

evidence for dependence of cross section on isomerism (see Table 131.2).

REFERENCE

Hudson, J. E., C. Vallance, M. Bart, and P. W. Harland, *J. Phys. B: At. Mol. Opt. Phys.*, 34, 3025, 2001.

132

CHLOROPENTANE

CONTENTS

Chloropentane ($C_5H_{11}Cl$) is a polar, electron-attaching molecule with 58 electrons and several isomers. The electronic polarizability of 1-chloropentane is 13.35×10^{-40} F m^2 and the dipole moment is 2.16 D.

TABLE 132.1
Ionization Cross Sections for $C_5H_{11}Cl$

Energy (eV)	Q_i (10^{-20} m^2)	Energy (eV)	Q_i (10^{-20} m^2)
16	2.89	119	18.69
20	6.37	123	18.58
23	9.83	127	18.22
27	11.99	130	18.18
31	13.44	134	18.04
34	14.34	138	17.92
38	15.01	142	17.82
42	15.95	145	17.74
45	16.66	149	17.67
49	17.26	153	17.41
53	17.89	156	17.21
57	18.29	160	17.08
60	18.69	164	17.06
64	18.90	167	16.97
68	19.16	171	16.97
71	19.16	175	16.90
75	19.40	179	13.72
79	19.49	182	16.40
82	19.40	186	16.13
86	19.48	190	15.91
90	19.30	193	15.74
94	19.27	197	15.65
97	19.29	201	15.58
101	19.29	204	15.54
105	19.05	208	15.55
108	18.81	212	15.54
112	18.84	215	15.43
116	18.74	219	15.27

Source: Adapted from Hudson, J. E. et al., *J. Phys. B: At. Mol. Opt. Phys.*, 34, 3025, 2001.
Note: Tabulated values courtesy of Professor Harland (2006).

132.1 IONIZATION CROSS SECTIONS

The ionization cross section for t-$C_5H_{11}Cl$ has been measured by Hudson et al. (2001) as shown in Table 132.1 and Figure 132.1.

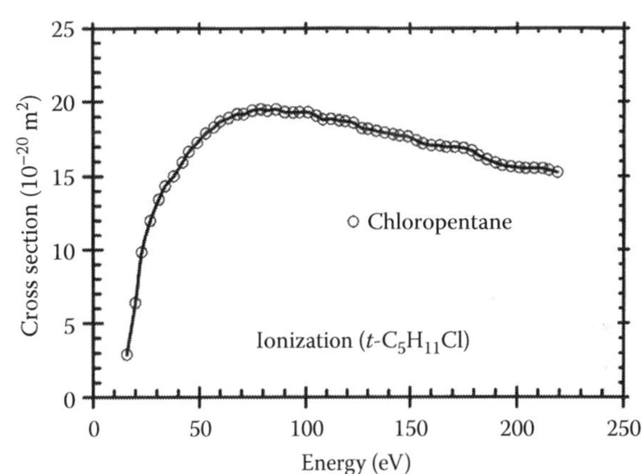

FIGURE 132.1 Ionization cross sections for $C_5H_{11}Cl$. (Adapted from Hudson, J. E. et al., *J. Phys. B: At. Mol. Opt. Phys.*, 34, 3025, 2001.)

REFERENCE

Hudson, J. E., C. Vallance, M. Bart, and P. W. Harland, *J. Phys. B: At. Mol. Opt. Phys.*, 34, 3025, 2001.

133

CYCLOPENTANE

C_5H_{10}

CONTENTS

133.1 SELECTED PROPERTIES OF ISOMERS OF CYCLOPENTANE

Cyclopentane (C_5H_{10}) is a cycloalkane gas belonging to the C_nH_{2n} group with 40 electrons. There are 12 compounds with the same chemical formula and we refer to only those for which data are available. Table 133.1 shows selected properties of the molecule. Twenty-nine vibrational modes have been measured by Kruse and Scott (1966) and the energies are (Lifson and Warshell, 1966) 357, 184, 110, 147, 163, 368, 95, 357, 181, 179, 163, 118, 110, 365, 150, 106, 103, 357, 183, 159, 129, 128, 76, 68, 368, 156, 127, 122, and 35 meV. Assignment of vibration modes is tentative, but according to Allan and Andric, (1996) vibrations of similar type (local motion) tend to have similar frequencies (energies). A prominent group of three vibrations in the energy range 100–200 meV comprises of C–C stretch, CH_2 rocking, and CH_2 scissoring vibrations. A weaker group of unresolved motions in the 200–350 meV range comprises of overtones and combinations of the 100–200 meV range. The C–H stretch occurs at 365 meV. See Table 111.3 under propylene for general designation of modes of vibration.

Mobility data for thermal electrons are given by Christophorou et al. (1973).

TABLE 133.1
Selected Properties of Isomers of C_5H_{10}

Molecule	α_e (10^{-40} F m^2)	μ (D)	ε_i (eV)	ε_d (eV)
Cyclopentane	10.18		10.33	4.14
1-Pentene	10.74	~0.5	9.51	
2-Pentene	10.95		9.01	

Note: α_e = Electronic polarizability; μ = dipole moment; ε_i = ionization energy; ε_d = dissociation energy.

REFERENCES

Allan, M. and L. Andric, *J. Chem. Phys.*, 105, 3559, 1996.

Christophorou, L. G., R. P. Blaunstein, and D. Pitman, *Chem. Phys. Lett.*, 18, 509–514, 1973.

Kruse, F. H. and D. W. Scott, *J. Mol. Spectrom.*, 20, 276, 1966.

Lifson, S. and A. Warshell, *J. Chem. Phys.*, 49, 5116, 1966.

134

HEXANE AND CYCLOHEXANE

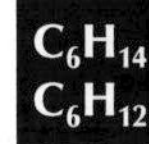

CONTENTS

Hexane (C_6H_{14}) is a linear hydrocarbon molecule with 50 electrons and cyclohexane (C_6H_{12}) has a ring structure like benzene without double bonds and has 48 electrons. Limited data are available for the two gases and selected properties of the two molecules are shown in Table 134.1.

C_6H_{12} has 32 modes of vibration owing to its large size, as shown in Table 134.2 (Shimanouchi, 1972). Similar modes of vibration have approximately the same energy as noted by Allan and Andric (1996).

134.1 TOTAL SCATTERING CROSS SECTION

Total scattering cross sections are measured and reported by Sueoka et al. (2005). Tables 134.3 and 134.4 show the values for both the gases. See Figure 134.1 for graphical presentation. The highlights of the total cross section are

1. At low energy, <0.4 eV, the behavior cannot be deciphered due to lack of data.
2. For both gases there is a large peak at 8 eV attributed to shape resonance, that is, the process of short-lived negative ion formation followed by autodetachment of the electron.

TABLE 134.1
Selected Properties of Normal and C_6H_{12}

Property	*n*-Hexane	Cyclohexane
Electronic polarizability	13.24×10^{-40} F m^2	12.23×10^{-40} F m^2
Ionization potential	8.86 eV	9.86 eV

TABLE 134.2
Vibrational Modes and Energies in C_6H_{12}

Symbol	Deformation	Energy (meV)	Symbol	Deformation	Energy (meV)
v_1	CH_2 a-stretch	363.3	v_{17}	CH_2 a-stretch	363.3
v_2	CH_2 s-stretch	353.6	v_{18}	CH_2 s-stretch	359.2
v_3	CH_2 scissor	181.6	v_{19}	CH_2 scissor	178.9
v_4	CH_2 rock	143.4	v_{20}	CH_2 wag	167.0
v_5	CC stretch	99.4	v_{21}	CH_2 twist	157.0
v_6	CCC deform +CC torsion	47.5	v_{22}	CC stretch	127.3
v_7	CH_2 twist	171.5	v_{23}	CH_2 rock	97.3
v_8	CH_2 wag	143.4	v_{24}	CCC deform + CC torsion	52.8
v_9	CC stretch + CC torsion	131.1	v_{25}	CH_2 a-stretch	363.6
v_{10}	CH_2 wag	178.2	v_{26}	CH_2 s-stretch	355.0
v_{11}	CH_2 twist	135.1	v_{27}	CH_2 scissor	180.6
v_{12}	CC stretch	361.4	v_{28}	CH_2 wag	168.0
v_{13}	CH_2 s-stretch	354.6	v_{29}	CH_2 twist	156.3
v_{14}	CH_2 scissor	178.2	v_{30}	CH_2 rock	112.5
v_{15}	CH_2 rock	127.7	v_{31}	CC stretch	107.0
v_{16}	CCC deform	64.8	v_{32}	CCC deform + CC torsion	30.7

Source: Adapted from Shimanouchi, T., *Tables of Molecular Vibrational Frequencies Consolidated Volume I*, NSRDS-NBS 39, Washington (DC), 1972.

Note: a = Asymmetrical; d = degenerate; s = symmetrical.

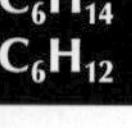

TABLE 134.3

Total Scattering Cross Sections in *n*-Hexane

Energy (eV)	Q_T (10^{-20} m²)	Energy (eV)	Q_T (10^{-20} m²)	Energy (eV)	Q_T (10^{-20} m²)
0.4	29.2	6.5	66.7	30	53.3
0.6	29.2	7.0	67.0	35	51.5
0.8	31.2	7.5	69.2	40	47.6
1.0	31.5	8.0	68.5	50	45.0
1.2	31.8	8.5	68.4	60	42.8
1.4	31.9	9.0	68.5	70	39.8
1.6	32.3	9.5	67.0	80	37.3
1.8	33.7	10	65.5	90	35.2
2.0	34.4	11	68.8	100	34.4
2.2	35.0	12	62.4	120	31.4
2.5	36.1	13	61.9	150	27.6
2.8	38.6	14	61.4	200	24.1
3.1	40.2	15	61.1	250	21.8
3.4	41.7	16	60.6	300	19.3
3.7	44.0	17	59.9	400	16.3
4.0	45.6	18	60.4	500	14.1
4.5	52.8	19	59.2	600	12.0
5.0	56.1	20	58.9	800	9.0
5.5	60.0	22	58.5	1000	7.1
6.0	64.2	25	57.7		

Source: Adapted from Sueoka, O. et al., *Phys. Rev. A*, 72, 042705, 2005.

TABLE 134.4

Total Scattering Cross Sections in C_6H_{12}

Energy (eV)	Q_T (10^{-20} m²)	Energy (eV)	Q_T (10^{-20} m²)	Energy (eV)	Q_T (10^{-20} m²)
0.4	25.0	6.5	55.2	30	47.7
0.6	24.6	7.0	58.9	35	46.0
0.8	25.9	7.5	59.4	40	44.7
1.0	27.3	8.0	60.3	50	41.1
1.2	27.8	8.5	59.1	60	39.4
1.4	28.2	9.0	58.5	70	37.0
1.6	28.8	9.5	57.1	80	34.7
1.8	29.4	10	55.6	90	32.5
2.0	31.8	11	53.1	100	31.2
2.2	32.4	12	51.1	120	28.5
2.5	33.8	13	50.0	150	25.5
2.8	35.1	14	49.4	200	22.2
3.1	36.3	15	49.5	250	20.1
3.4	37.9	16	50.8	300	18.1
3.7	39.1	17	50.1	400	14.9
4.0	43.4	18	50.2	500	13.1
4.5	45.2	19	50.1	600	11.7
5.0	47.0	20	50.2	800	9.6
5.5	51.5	22	50.2	1000	7.0
6.0	53.7	25	49.4		

Source: Adapted from Sueoka, O. et al., *Phys. Rev. A*, 72, 042705, 2005.

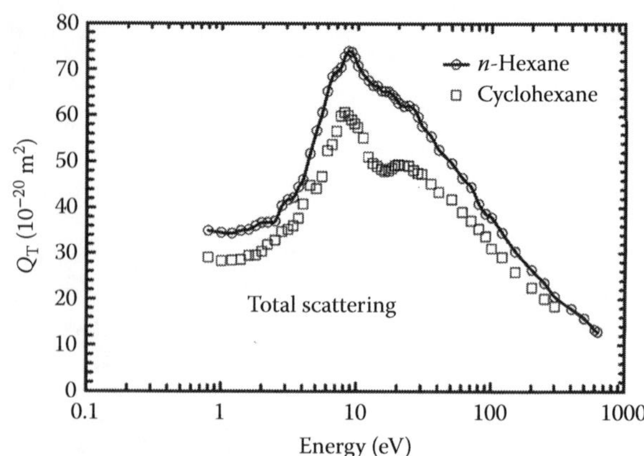

FIGURE 134.1 Total scattering cross sections in *n*-hexane and C_6H_{12}. (—○—) *n*-Hexane; (□) cyclohexane (Kimura et al., 2000). (From Kimura et al., 2000.)

TABLE 134.5

Ion Drift Velocity in C_6H_{12}

E/N Range (Td)	*C* (ms^{-1} Td$^{-\gamma}$)	γ (Equation 134.1)
300–900	1.12	1.0
1500–39,000	49.33	0.5

Source: Adapted from Schlumbohm, H., *Z. Phys.*, 182, 317, 1965.

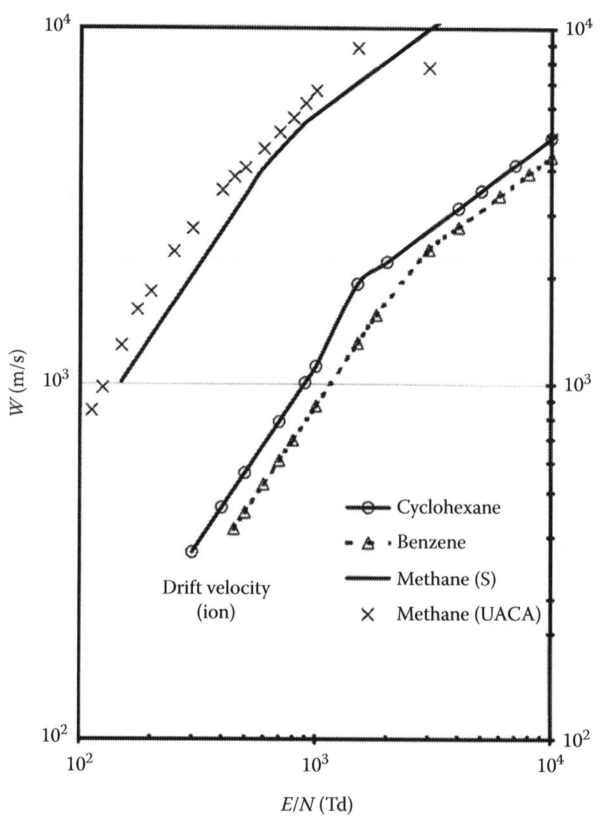

FIGURE 134.2 Ion drift velocity in C_6H_{12} and selected hydrocarbons. Methane results are from de Urquiho et al. (1999). Other gases are from Schlumbohm (1965).

3. A smaller broad structure at ~15–20 eV attributed various resonances due to processes that are possible at these energies (excitation, dissociation, ionization, etc.).

134.2 ION DRIFT VELOCITY

Ion drift velocities measured by Schlumbohm (1965) may be represented by an equation of the type

$$W_+ = C\left(\frac{E}{N}\right)^{\gamma} \qquad (134.1)$$

where W_+ is the drift velocity (m/s), E/N is the reduced electric field (Td), and C and γ are constants. The values obtained by Schlumbohm (1965) are shown in Table 134.5. See the chapter on acetone for additional data. Figure 134.2 compares the ion drift velocity with selected gases.

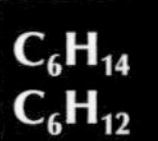

REFERENCES

Allan, M. and L. Andric, *J. Chem. Phys.*, 105, 3559, 1996.

de Urquiho, J., C. A. Arriaga, C. Cisneros, and I. Alvarez, *J. Phys. D: Appl. Phys.*, 32, 41, 1999.

Kimura, M., O. Sueoka, A. Hamada, and Y. Itikawa, *Adv. Chem. Phys.*, 111, 537, 2000.

Schlumbohm, H., *Z. Phys.*, 182, 317, 1965.

Shimanouchi, T. *Tables of Molecular Vibrational Frequencies Consolidated Volume I*, NSRDS-NBS 39, Washington (DC), 1972.

Sueoka, O., C. Makochekanwa, H. Tanino, and M. Kimura, *Phys. Rev. A*, 72, 042705, 2005.

135

ISOBUTANE

i-C_4H_{10}

CONTENTS

Isobutane (i-C_4H_{10}) is an isomer of butane and Figure 135.1 shows the structural formula of selected hydrocarbons. Table 135.1 shows selected molecular parameters in comparison with those of butane. Total cross sections are shown in Table 135.3 and Figure 135.2 (also see the chapter on butane, Table 130.7).

Target gas	Molecular formula	Structural formula
Methane	CH_4	–C–
Ethane	C_2H_6	–C–C–
Ethene	C_2H_4	>C=C<
Propane	C_3H_8	–C–C–C–
Propene	C_3H_6	>C=C–C–
Cyclopropane	C_3H_6	–C—C– (ring with C)
n-Butane	C_4H_{10}	–C–C–C–C–
Isobutane	C_4H_{10}	–C–C–C– (with –C– branch)
1-Butene	C_4H_8	>C=C–C–C–
Acetylene	C_2H_2	H – C ≡ C – H

FIGURE 135.1 Structural formula of selected hydrocarbons.

TABLE 135.1
Selected Molecular Parameters

Parameter	*i*-Butane	Butane	Units
Polarizability	9.06	9.12	F m[2]
Dipole moment	0.132	≤0.05[a]	D
Ionization potential	10.57	10.57	eV
Dissociation potential	3.823	3.854	eV

[a] Gee and Freeman (1983).

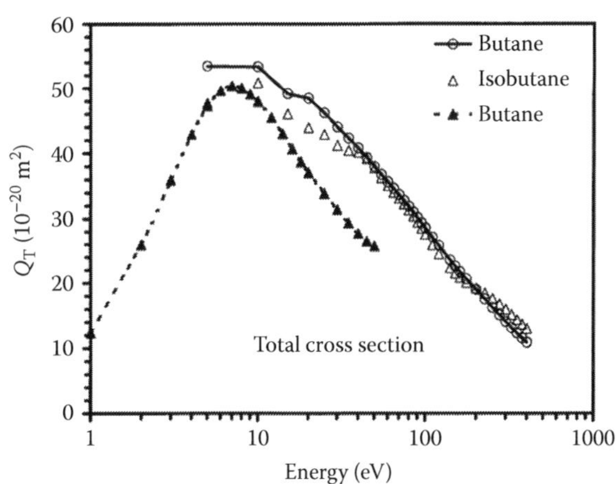

FIGURE 135.2 Total scattering cross section in butane and i-C_4H_{10}. (—○—) Butane, Floeder et al. (1985); (—△—) isobutane, Floeder et al. (1985); (—▲—) butane, Brüche (1930).

135.1 SELECTED REFERENCES FOR DATA

Selected references are shown in Table 135.2.

135.2 TOTAL SCATTERING CROSS SECTION

See Table 135.3.

TABLE 135.2
Selected References for Data

Quantity	Range	Reference
W	0.2–30 Td	Floriano et al. (1986)
Q_T	5–400 eV	Floeder et al. (1985)
Q_T	1–50 eV	Brüche (1930)

Note: Q_T = total; W = drift velocity.

i-C_4H_{10}

TABLE 135.3
Total Cross Section for i-C_4H_{10}

Energy (eV)	Q_T (10^{-20} m^2)	Energy (eV)	Q_T (10^{-20} m^2)
5.0	47.69	95.0	28.47
10.0	50.82	100.0	27.54
15.0	46.04	110.0	25.89
20.0	43.93	120.0	24.49
25.0	42.89	140.0	22.33
30.0	41.19	150.0	21.52
35.0	40.47	160.0	20.86
40.0	40.16	175.0	20.09
45.0	39.20	200.0	19.18
50.0	37.76	225.0	18.40
55.0	36.31	250.0	17.59
60.0	35.10	275.0	16.76
65.0	34.08	300.0	15.92
70.0	33.17	325.0	15.11
75.0	32.29	350.0	14.34
80.0	31.37	375.0	13.64
85.0	30.40	400.0	13.01
90.0	29.43		

Source: Adapted from Floeder, K. et al., *J. Phys. B: At. Mol. Phys.*, 18, 3347, 1985.

REFERENCES

Brüche, E. *Ann. Phys. Leipzig*, 5, 281, 1930.

Gee, N. and G. R. Freeman, *J. Chem. Phys.*, 78, 1951, 1983.

Floeder, K., D. Fromme, W. Raith, A. Schwab, and G. Sinapius, *J. Phys. B: At. Mol. Phys.*, 18, 3347, 1985.

Floriano, M., N. Gee, and G. R. Freeman, *J. Chem. Phys.*, 84, 6799, 1986.

136

ISOOCTANE

CONTENTS

Isooctane (i-C_8H_{18}), synonym 2,2,4-trimethylpentane, is a nonpolar, saturated hydrocarbon molecule with 66 electrons. Its ionization potential is 9.86 eV. A single study of partial and total ionization cross sections has been carried out by Bouamra et al., (2003) as shown in Table 136.1 and Figures 136.1 and 136.2.

TABLE 136.1
Total Ionization Cross Sections for i-C_8H_{18}

Energy (eV)	Q_i (10^{-20} m^2)	Energy (eV)	Q_i (10^{-20} m^2)
17	9.10	45	15.03
18	9.98	50	14.98
20	11.10	55	15.12
22	11.87	60	15.61
25	12.99	65	16.14
30	13.94	70	18.20
35	15.83	75	18.95
40	16.09	80	20.96

Source: Digitized and interpolated from Bouamra, K. et al., *Chem. Phys. Lett.*, 373, 237, 2003.

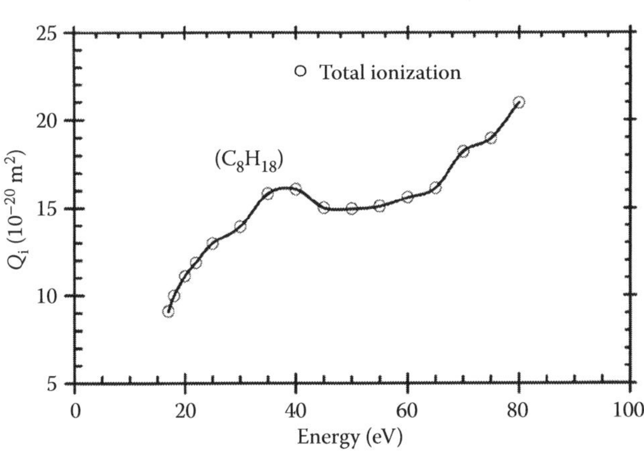

FIGURE 136.1 Total ionization cross sections for i-C_8H_{18}. (Adapted from Bouamra, K. et al., *Chem. Phys. Lett.*, 373, 237, 2003.)

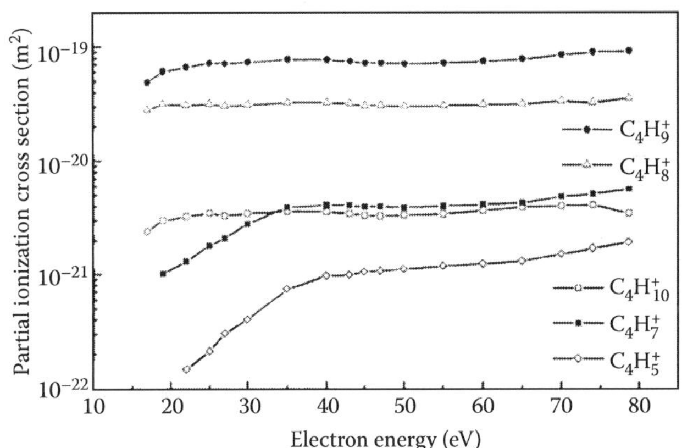

FIGURE 136.2 Partial ionization cross sections for selected fragments for i-C_8H_{18}. (Adapted from Bouamra, K. et al., *Chem. Phys. Lett.*, 373, 237, 2003.)

136.1 IONIZATION CROSS SECTION

See Table 136.1.

REFERENCE

Bouamra, K., J. R. Vacher, F. Jorand, N. Simiand, and S. Pasquiers, *Chem. Phys. Lett.*, 373, 237, 2003.

137

OCTANE

CONTENTS

Octane (C_8H_{18}) is a saturated hydrocarbon with 66 electrons. There are a large number of molecules having the same chemical formula and only limited data are available. It has an electronic polarizability of 17.69 × 10^{-40} F m^2 and an ionization potential of 9.80 eV, and it is nonpolar. There is only one study of total scattering cross sections in the 0.4–1000 eV energy range in the gas (Sueoka et al., 2005). Table 137.1 and Figure 137.1 show these data and also provide a comparison of total cross section in several saturated hydrocarbons from methane to C_8H_{18}.

TABLE 137.1
Total Scattering Cross Section in C_8H_{18}

Energy (eV)	Q_T (10^{-20}m²)	Energy (eV)	Q_T (10^{-20} m²)	Energy (eV)	Q_T (10^{-20} m²)
0.4	37.44	7.0	79.10	30	61.76
0.6	39.43	7.5	76.62	35	59.58
0.8	40.38	8.0	79.67	40	55.91
1.0	41.35	8.5	89.72	50	52.55
1.2	41.91	9.0	89.71	60	50.83
1.4	41.98	9.5	66.94	70	45.99
1.6	41.78	10	64.65	80	43.31
1.8	42.54	11	78.53	90	40.89
2.0	43.55	12	73.18	100	38.51
2.2	44.39	13	69.35	120	36.68
2.5	45.15	14	70.40	150	32.56
2.8	46.41	15	70.91	200	28.69
3.1	48.92	16	69.81	250	25.60
3.4	51.83	17	69.41	300	23.14
3.7	54.59	18	69.42	400	19.50
4.0	55.79	19	69.42	500	17.21
4.5	69.43	20	69.32	600	15.24
5.0	72.54	22	68.75	800	11.09
5.5	75.02	24	67.59	1000	8.28
6.0	79.51	26	65.48		
6.5	79.82	28	63.06		

Source: Adapted from Sueoka, O. et al., *Phys. Rev. A*, 72, 042705, 2005.

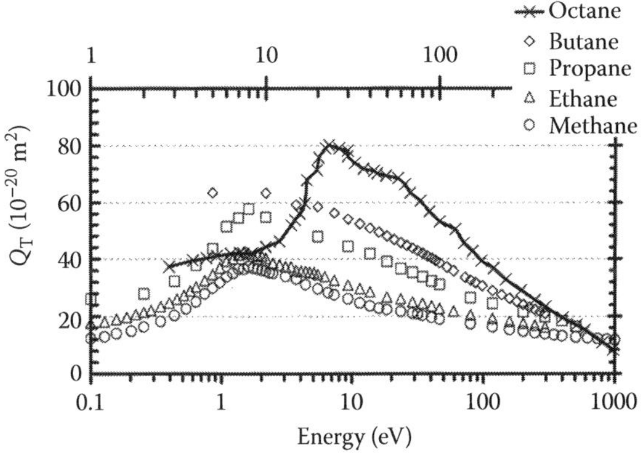

FIGURE 137.1 Total scattering cross sections in saturated hydrocarbon gases. (—×—) Octane, Sueoka et al. (2005); (◊) butane, Floeder et al. (1985); (□) propane, Tanaka et al. (1999); (Δ) ethane, Sueoka and Mori (1986); (○) methane, Zecca et al. (1991).

137.1 TOTAL SCATTERING CROSS SECTIONS

See Table 137.1.

137.2 IONIZATION CROSS SECTIONS

Ionization cross sections and dissociation fragments produced by electron impact in the energy range from 10 to 70 eV have been measured by Jiao et al. (2001). The major neutral fragments produced from electron impact ionization at low energies (≤16 eV) are C_2H_5 and C_2H_6. Figure 137.2 shows the total ionization cross sections and cross sections for selected fragment ions. Note the relative contribution of the parent ion $C_8H_{18}^+$ to the total.

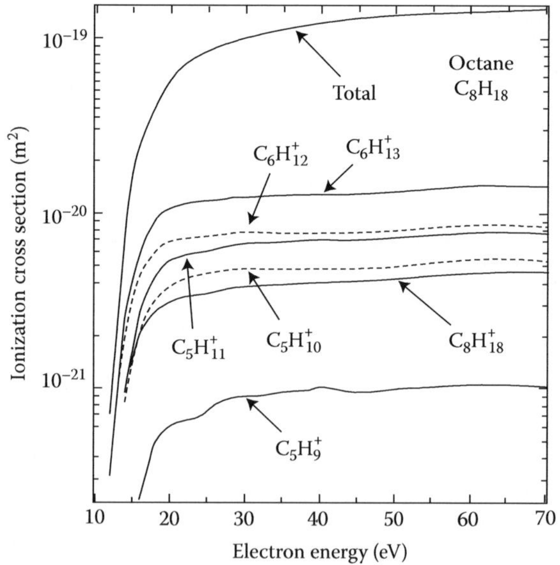

FIGURE 137.2 Total ionization cross section and formation of fragment ions in *n*-octane (C_8H_{18}) by electron impact. Note the relatively large total cross section and the contribution of the parent ion $C_8H_{18}^+$. (Adapted from Jiao, C. Q., C. A. DeJoseph, Jr., and A. Garscadden, *J. Chem. Phys.*, 114, 2166, 2001.)

REFERENCES

Floder, K., D. Fromme, W. Raith, A. Schwab, and G. Sinapius, *J. Phys. B: At. Mol. Phys.*, 18, 3347, 1985.

Jiao, C. Q., C. A. DeJoseph, Jr., A. Garscadden, *J. Chem. Phys.*, 114, 2166, 2001.

Sueoka, O. and S. Mori, *J. Phys. B: At. Mol. Phys.*, 19, 4035, 1986.

Sueoka, O., C. Makochekanwa, H. Tanino, and M. Kimura, *Phys. Rev. A*, 72, 042705, 2005.

Tanaka, H., Y. Tachibana, M. Kitajima, O. Sueoka, H. Takaki, A. Hamada, and M. Kimura, *Phys. Rev. A.*, 59, 2006, 1999.

Zecca, A., G. Karwasz, R. S. Brusa, and C. Szmytkowski, *J. Phys. B: At. Mol. Opt. Phys.*, 24, 2747, 1991.

138

PENTANE

C_5H_{12}

CONTENTS

Pentane (C_5H_{12}) is a saturated hydrocarbon (C_nH_{2n+2}) with 42 electrons and it has two isomers, isopentane and neopentane. We consider all the three gases in this chapter. Selected properties of the three molecules are shown in Table 138.1.

138.1 SELECTED REFERENCES FOR DATA

Selected references for data on C_5H_{12} and its isomers are shown in Table 138.2.

138.2 MOMENTUM TRANSFER CROSS SECTIONS

See Table 138.3. Gee and Freeman (1983) suggest a relationship of the form

$$Q_M = Q_0 \varepsilon^{-n} \quad (138.1)$$

where Q_0 is a constant and n the index, both depending upon the gas. The range of electron energy is between 0.03 and 0.15 eV. Table 138.4 shows the value of the derived index for C_5H_{12}, and values for other hydrocarbons are also included for comparison. Data on the Ramsauer–Townsend minimum are also given (see Section 138.3).

TABLE 138.1
Selected Properties of C_5H_{12} and Its Isomers

Molecule	α_e	M	ε_{dis}	ε_i
Pentane	11.12		3.85	10.28
Neopentane	11.35			10.32
Isopentane		0.13	3.84	≤10.20

Note: α_e = Electronic polarizability (10^{-40} F m^2); μ = dipole moment (D); ε_{dis} = dissociation energy (eV); ε_i = ionization energy (eV).

TABLE 138.2
Selected References for Data

Cross Sections

Quantity	Energy Range (eV)	Gas	Reference
Q_i	600–12000	Pentane, isopentane, neopentane	Schram et al. (1966)

Swarm Coefficients

Quantity	E/N Range (Td)	Gas	Reference
W	0.3–30	Pentane, isopentane, neopentane	Floriano et al. (1986)
W	0.01–10	Pentane, isopentane, neopentane	Gee and Freeman (1983)
W	0.06	Neopentane	McCorkle et al. (1978)

Note: Q_i = ionization; W = drift velocity.

TABLE 138.3
Momentum Transfer Cross Sections

Energy (eV)	Q_M (10^{-20} m^2)	Energy (eV)	Q_M (10^{-20} m^2)
0.10	20.56	0.35	8.00
0.15	9.70	0.40	12.00
0.20	6.80	0.45	18.20
0.25	5.80	0.50	24.44
0.30	7.50		

Source: Adapted from McCorkle, D. L. et al., *J. Phys./B; At. Mol. Phys.*, 11, 3067, 1978.
Note: See Figure 138.1 for graphical presentation.

C_5H_{12}

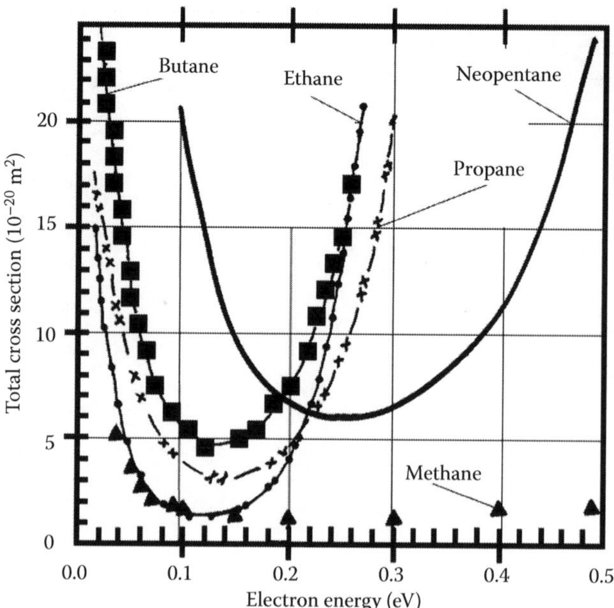

FIGURE 138.1 Momentum transfer cross section in neopentane and selected hydrocarbons, derived from drift velocity data by McCorkle et al. (1978). The data for methane are from Duncan and Walker (1972). (With permission from Institute of Physics, England.)

TABLE 138.4
Index *n* in Equation 138.1 and Ramsauer–Townsend Minimum for Selected Hydrocarbons According to Gee and Freeman (1983)

Formula	Name	*n*	R–T Minimum (eV)
	Alkanes		
CH_4	Methane	0.9	0.23
C_2H_6	Ethane	1.1	0.12
C_3H_8	Propane	1.2	0.13
n-C_4H_{10}	Butane	1.0	0.14
i-C_4H_{10}	Isobutane	1.3	0.19
n-C_5H_{12}	Pentane	0.9	0.14
i-C_5H_{12}	Isopentane	1.2	0.14
Neo-C_5H_{12}	Neopentane	1.7	0.21
n-C_6H_{14}	Hexane	1.0	0.13
	Cycloalkanes		
c-C_3H_6	Cyclopropane	0.2	0.10
c-C_5H_{10}	Cyclopentane	1.3	0.17
c-C_6H_{12}	Cyclohexane	1.0	0.13
	Alkenes		
C_2H_4	Ethene (ethylene)	0.5	0.1
C_4H_8-1	1-Butene	1.2	0.12
trans-C_4H_8-2	*trans*-2-Butene	0.5	0.09
cis-C_4H_8-2	*cis*-2-Butene	1.6	0.13
i-C_4H_8	Isobutene	1.7	0.16

138.3 RAMSAUER–TOWNSEND MINIMUM

Ramsauer–Townsend minimum energy and cross section in C_5H_{12} are shown in Table 138.5.

138.4 IONIZATION CROSS SECTIONS

See Table 138.6.

138.5 DRIFT VELOCITY

Drift velocity of electrons as a function of *E*/*N* and temperature in neopentane is shown in Table 138.7 (McCorkle et al., 1978). See Figure 138.2 for graphical presentation. Table 138.8 shows the dependence of drift velocity on gas number density in the same gas (Figure 138.3).

TABLE 138.5
Ramsauer–Townsend Minimum Energy and Cross Section

Gas	Minimum Energy (eV)	Cross Section (10^{-20} m²)	Method	Reference
Neopentane	0.25	5.80	Momentum transfer	McCorkle et al. (1978)

TABLE 138.6
Ionization Cross Section

Energy (keV)	Q_i (10^{-20} m²) Pentane	Isopentane	Neopentane
0.6	7.65	7.37	6.68
1.0	5.23	5.02	4.70
2.0	3.08	2.82	2.75
4.0	1.75	1.69	1.57
7.0	1.08	1.06	0.991
12.0	0.697	0.684	0.627

Source: Adapted from Schram, B. L. et al., *J. Chem. Phys.*, 44, 49, 1966.

TABLE 138.7
Drift Velocity

E/*N* (Td)	*W* (10^4 m/s) 298 K	373 K	473 K	573 K	673 K
0.06		0.026	0.030	0.035	0.037
0.09	0.030	0.036	0.047	0.053	0.058
0.12	0.041	0.050	0.062	0.070	0.076
0.18	0.057	0.073	0.091	0.107	0.117
0.24	0.080	0.095	0.119	0.140	0.157
0.30	0.099	0.118	0.151	0.175	0.193
0.45	0.151	0.182	0.229	0.268	0.286
0.6	0.207	0.250	0.303	0.345	0.377
0.75	0.272	0.319	0.381	0.426	0.476
0.90	0.342	0.389	0.466	0.513	0.568
1.05	0.398	0.455	0.536	0.596	0.664
1.20	0.471	0.539	0.630	0.680	0.749
1.50	0.662	0.692	0.796	0.865	0.910
1.80	0.833	0.897	0.984	1.071	1.145

TABLE 138.7 (continued)
Drift Velocity

	W (10^4 m/s)				
E/N (Td)	**298 K**	**373 K**	**473 K**	**573 K**	**673 K**
2.10	1.034	1.071	1.18	1.24	1.34
3.0	1.70	1.74	1.85	2.14	2.03
4.5	2.80	2.83	2.89	2.89	3.05
7.5	4.66	4.83	4.88	4.74	4.75
12.0	6.09	6.08	6.52	6.08	5.90
16.5	6.64	6.34	6.60	6.21	6.32
21.0	6.59	6.36	6.57	6.52	6.12
27.0	6.00	6.10	6.06	5.98	5.88

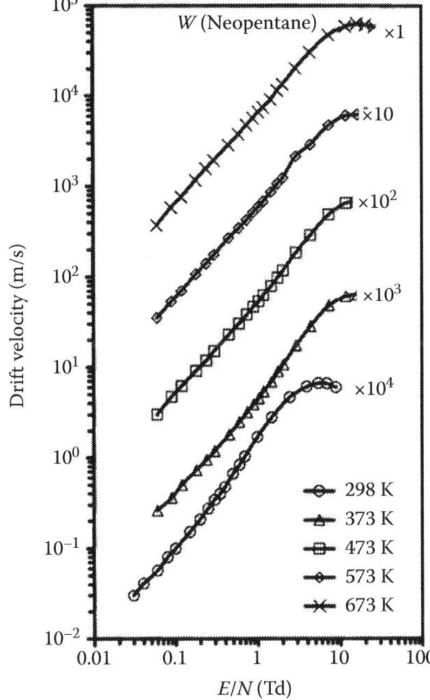

FIGURE 138.2 Drift Velocity of electrons in neo pentane as a function of reduced Electric field at various temperatures. (—○—) 298 K; (—△—) 373 K; (—□—) 473 K; (—◊—) 573 K; (—×—) 673 K. Ordinates should be multiplied by the factors shown. (Adapted from McCorkle, D. L. et al., *J. Phys. B: At. Mol. Phys.*, 11, 3067, 1978).

TABLE 138.8
Drift Velocity of Electrons as a Function of E/N and Gas Number Density

	W (10^4 m/s)			
E/N (Td)	**N = 2.59 × 10^{23} m^{-3}**	**2.59 × 10^{24}**	**2.70 × 10^{25}**	**2.13 × 10^{26}**
0.15	0.062	0.061	0.058	0.051
0.18	0.074	0.073	0.069	0.061
0.21	0.083	0.084	0.080	0.073
0.24	0.095	0.095	0.091	0.081
0.27	0.106	0.107	0.103	0.091
0.30	0.123	0.118	0.115	0.104
0.45	0.180	0.182	0.176	0.157

TABLE 138.8 (continued)
Drift Velocity of Electrons as a Function of E/N and Gas Number Density

	W (10^4 m/s)			
E/N (Td)	**N = 2.59 × 10^{23} m^{-3}**	**2.59 × 10^{24}**	**2.70 × 10^{25}**	**2.13 × 10^{26}**
0.60	0.249	0.250	0.240	0.211
0.75	0.314	0.317	0.305	0.268
0.90	0.390	0.389	0.372	0.327
1.05	0.451	0.451	0.440	0.388
1.20	0.541	0.542	0.515	0.45
1.50	0.692	0.690	0.677	0.577
1.80	0.853	0.897	0.85	0.707
2.10	1.01	1.05	1.03	0.842
2.40	1.24	1.25	1.23	0.995
2.70	1.44	1.46	1.43	1.15
3.00	1.64	1.72	1.65	1.31
3.60	2.11	2.09	2.12	
4.20	2.52	2.59	2.58	
6.00	3.66	3.75	3.99	
7.50	4.83	4.8	5.05	
9.00	5.45	5.49	5.60	

Source: Adapted from McCorkle, D. L. et al., *J. Phys. B: At. Mol. Phys.*, 11, 3067, 1978.

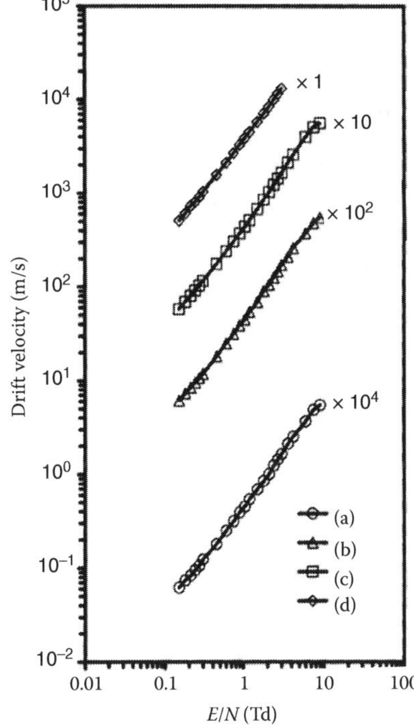

FIGURE 138.3 Drift velocity of electrons in neopentane as a function of reduced electric field at various gas number densities. (a) (—○—) $N = 2.59 \times 10^{23}$ m^{-3}; (b) (—△—) $N = 2.59 \times 10^{24}$ m^{-3}; (c) (—□—) $N = 2.70 \times 10^{25}$ m^{-3}; (d) (—◊—) $N = 2.13 \times 10^{26}$ m^{-3}. Ordinates should be multiplied by the factors shown. (Adapted from McCorkle, D. L. et al., *J. Phys. B: At. Mol. Phys.*, 11, 3067, 1978.)

C$_5$H$_{12}$

TABLE 138.9
Mean Energy in Isopentane and Selected Hydrocarbons

	Mean Energy (eV)			
E/N (Td)	Ethane	Propane	Butane	Isopentane
0.03	0.038	0.039	0.038	0.039
0.06	0.038	0.039	0.039	0.04
0.09	0.038	0.039	0.04	0.041
0.12	0.038	0.04	0.042	0.041
0.18	0.039	0.04	0.043	0.041
0.24	0.040	0.042	0.044	0.041
0.30	0.043	0.042	0.047	0.042
0.45	0.058	0.046	0.052	0.043
0.60	0.071	0.051	0.056	0.044
0.75	0.083	0.057	0.062	0.046
0.90	0.095	0.063	0.066	0.047
1.05	0.105	0.071	0.074	0.051
1.20	0.115	0.077	0.086	0.056
1.50	0.135	0.092	0.096	0.061
1.80	0.153	0.105	0.107	0.078
2.10	0.171	0.119	0.137	0.111
3.00	0.222	0.151	0.177	
4.50	0.293	0.202	0.237	
7.50	0.423	0.294		
12.00	0.560	0.422		

Source: Adapted from McCorkle, D. L. et al., *J. Phys. B: At. Mol. Phys.*, 11, 3067, 1978.
Note: See Figure 138.4 for graphical presentation.

138.6 MEAN ENERGY

Mean energies of electron swarm as a function of reduced electric field at 298 K are shown in Table 138.9. Data for selected hydrocarbons are included for comparison.

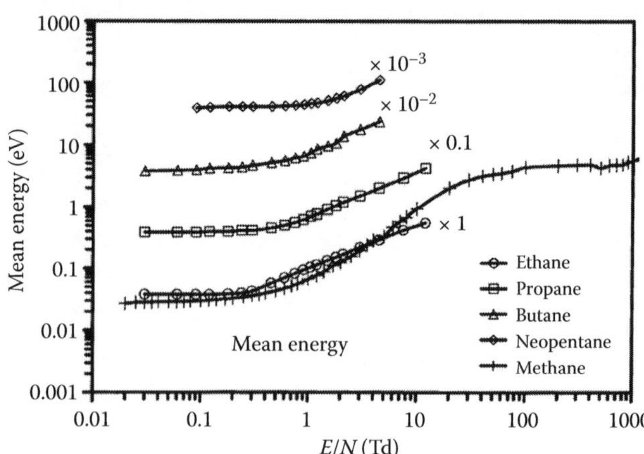

FIGURE 138.4 Mean energy in isopentane and selected hydrocarbon gases. (—○—) Ethane; (—□—) propane; (—△—) butane; (—◊—) isopentane. The ordinates should be multiplied by the factors shown. Data for methane (— + —) show the characteristic energy tabulated in Table 64.20 of the chapter on methane. (Adapted from McCorkle, D. L. et al., *J. Phys. B: At. Mol. Phys.*, 11, 3067, 1978.)

REFERENCES

Duncan, C. W. and I. C. Walker, *J. Chem. Soc. Farad. II*, 68, 1514, 1972.
Floriano, M. A., N. Gee, and G. R. Freeman, *J. Chem. Phys.*, 84, 6799, 1986.
Gee, N. and G. R. Freeman, *J. Chem. Phys.*, 78, 1951, 1983.
McCorkle, D. L., L. G. Christophorou, D. V. Mazey, and J. G. Carter, *J. Phys. B: At. Mol. Phys.*, 11, 3067, 1978.
Schram, B. L., M. J. Van der Wiel, F. J. de Heer, and H. R. Moustafa, *J. Chem. Phys.*, 44, 49, 1966.

139

PERFLUOROBUTANE PERFLUOROISO-BUTANE

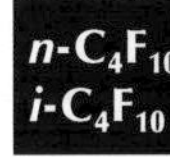

CONTENTS

Perfluorobutane (n-C_4F_{10}) (synonym decaflurobutane) is a colorless electron-attaching gas with 96 electrons and belongs to the category of perfluoroalkanes. Perfluoroisobutane (i-C_4F_{10}) also has the same chemical formula with a different chemical structure (isomer). The electronic polarizability of n-C_4F_{10} is derived by Beran and Kevan (1969) as 9.46×10^{-40} F m^2 and the ionization potential of n-C_4F_{10} for the largest ion yield ($C_3F_6^+$, CF_3^+) is determined by threshold photoelectron-photon coincidence spectroscopy as 12.6 eV (Jarvis et al., 1998).

139.1 SELECTED REFERENCES FOR DATA

Table 139.1 shows selected references for data on n-C_4F_{10} and i-C_4F_{10}.

TABLE 139.1
Selected References for Data

Parameter	Range: eV, (Td)	Reference
ε_i	12–25	Jarvis et al. (1998)
W	(5–500)	Hunter et al. (1988)
α, η	(5–400)	Hunter et al. (1987)
Q_a	0.2–10	Spyrou and Christophorou (1985)
k_a	<10 eV	Spyrou et al. (1985)
k_a	0.04–4.9	Hunter and Christophorou (1984)
$(E/N)_{Lim}$	(~400)	Nakanishi et al. (1984)
k_a	0–10	Spyrou et al. (1983)
Swarm parameters	(300–810)	Naidu and Prasad (1972)
k_a	Thermal	Fessenden and Bansal (1970)
Q_i	70, 35, 20	Beran and Kevan (1969)
α, η	(240–450)	Razzak and Goodyear (1968)

Note: W = drift velocity; k_a = attachment rate; α = ionization coefficient; η = attachment coefficient; ε_i = ionization potential.

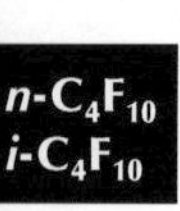

139.2 POSITIVE ION APPEARANCE POTENTIALS

Table 139.2 shows the dissociative ionization potentials and positive ions produced as determined by Jarvis et al. (1998). Parent positive ion is not observed in their study at any energy.

139.3 IONIZATION CROSS SECTIONS

Total ionization cross sections measured by Beran and Kevan (1969) are shown in Table 139.3.

139.4 ATTACHMENT PROCESSES

139.4.1 n-C_4F_{10}

Parent negative ions $C_4F_{10}^{-*}$ with low abundance are observed due to the attachment in n-C_4F_{10} according to Spyrou et al. (1983) due to the reaction

$$e + n\text{-}C_4F_{10} \rightarrow C_4F_{10}^{-}* \quad (139.1)$$

Possible dissociative attachment processes are

$$e + n\text{-}C_4F_{10} \rightarrow C_4F_9 + F^- \quad (139.2)$$

$$e + n\text{-}C_4F_{10} \rightarrow 2\text{-}C_4F_8 + F + F^- \quad (139.3)$$

TABLE 139.2
Positive Ion Appearance Potentials for n-C_4F_{10}

Dissociation Products	Appearance Energy (eV)
$C_3F_6^+ + CF_4$	12.6
$C_2F_4^+ + C_2F_6$	12.6
$CF_3^+ + C_3F_7$	12.6
$C_2F_5^+ + C_2F_5$	13.0
$C_4F_9^+ + F$	15.7

Source: Adapted from Jarvis, G. K. et al., *J. Phys. Chem. A*, 102, 3219, 1998.

TABLE 139.3
Ionization Cross Sections for n-C_4F_{10}

Energy (eV)	20	35	70
Q_i (10^{-20} m²)	1.62	7.25	13.5

Source: Adapted from Beran, J. A. and L. Kevan, *J. Phys. Chem.*, 73, 3866, 1969.

$$e + n\text{-}C_4F_{10} \rightarrow C_3F_6 + CF_3 + F^- \quad (139.4)$$

$$e + n\text{-}C_4F_{10} \rightarrow C_2F_5 + C_2F_4 + F^- \quad (139.5)$$

$$e + n\text{-}C_4F_{10} \rightarrow C_3F_7 + CF_3^- \quad (139.6)$$

$$e + n\text{-}C_4F_{10} \rightarrow C_2F_4 + CF_3 + CF_3^- \quad (139.7)$$

$$e + n\text{-}C_4F_{10} \rightarrow C_2F_5 + C_2F_5^- \quad (139.8)$$

$$e + n\text{-}C_4F_{10} \rightarrow CF_3 + CF_2 + C_2F_5^- \quad (139.9)$$

$$e + n\text{-}C_4F_{10} \rightarrow CF_3 + C_3F_7^- \quad (139.10)$$

$$e + n\text{-}C_4F_{10} \rightarrow F + C_4F_9^- \quad (139.11)$$

139.4.2 i-C_4F_{10}

Parent negative ions $C_4F_{10}^{-*}$ with high abundance are observed due to attachment in i-C_4F_{10} according to Spyrou et al. (1983) due to the reaction

$$e + i\text{-}C_4F_{10} \rightarrow C_4F_{10}^{-}* \quad (139.12)$$

Possible dissociative attachment processes are (Spyrou et al., 1983)

$$e + i\text{-}C_4F_{10} \rightarrow C(CF_3)_3 + F^- \quad (139.13)$$

$$e + i\text{-}C_4F_{10} \rightarrow CF_2CF(CF_3)_2 + F^- \quad (139.14)$$

$$e + i\text{-}C_4F_{10} \rightarrow i\text{-}C_3F_7 + CF_3^- \quad (139.15)$$

$$e + i\text{-}C_4F_{10} \rightarrow CF_3 + C_3F_7^- \quad (139.16)$$

$$e + i\text{-}C_4F_{10} \rightarrow F + C_4F_9^- \quad (139.17)$$

139.5 NEGATIVE ION APPEARANCE POTENTIALS

Negative ion appearance potentials and peak of the ion yields for n-C_4F_{10} are shown in Tables 139.4 and 139.5 (Spyrou et al., 1983). Bold letters indicate the most abundant ion.

139.6 ATTACHMENT RATES

Thermal attachment rate coefficients are shown in Table 139.6. Table 139.7 and Figure 139.1 show the electron attachment rate for n-C_4F_{10} as a function of mean energy in a buffer gas of nitrogen or argon.

TABLE 139.4
Negative Ion Appearance Potentials for n-C_4F_{10}

Ion	Energy (eV) Appearance	Energy (eV) Peak of Cross Section	Reaction
F^-	1.6	2.65	139.2
CF_3^-	2.2	3.2	139.6
	>3.2	4.6	139.7
$C_2F_5^-$	1.8	2.8	139.8
$C_3F_7^-$	1.2	2.85	139.10
$C_4F_9^-$	1.0	1.85	139.11
$C_4F_{10}^{-*}$	~0	0.6	139.11

Source: Adapted from Spyrou, S. M., I. Sauers, and L. G. Christophorou, *J. Chem. Phys.*,78, 7200, 1983.

TABLE 139.5
Negative Ion Appearance Potentials for i-C_4F_{10}

Ion	Energy (eV) Appearance	Energy (eV) Peak of Cross Section	Reaction
F^-	1.2	1.85	139.13
	2.1	3.7	139.14
CF_3^-	2.5	4.0	139.15
$C_3F_7^-$	0.9	1.7	139.16
$C_4F_9^-$	0.9	1.85	139.17
$C_4F_{10}^{-*}$	~0	0.6	139.12

Source: Adapted from Spyrou, S. M., I. Sauers, and L. G. Christophorou, *J. Chem. Phys.*,78, 7200, 1983.

TABLE 139.6
Thermal Attachment Rate Coefficients

Gas	Rate (m^3/s)	Reference
i-C_4F_{10}	~1.0×10^{-16}	Spyrou et al. (1985)
n-C_4F_{10}	4.1×10^{-17}	Hunter and Christophorou (1984)
n-C_4F_{10}	9.6×10^{-18}	Fessenden and Bansal (1970)

TABLE 139.7
Electron Attachment Rate for n-C_4F_{10}

TABLE 139.7 (continued)
Electron Attachment Rate for n-C_4F_{10}

Buffer Gas: Nitrogen		Buffer Gas: Argon	
Mean Energy ($\bar{\varepsilon}$) (eV)	Attachment Rate (10^{-16} m^3/s)	Mean Energy ($\bar{\varepsilon}$) (eV)	Attachment Rate (10^{-15} m^3/s)
0.0310	0.41	0.335	0.102
0.0466	0.41	0.364	0.118
0.0621	0.41	0.412	0.148
0.0931	0.41	0.473	0.188
0.124	0.42	0.509	0.216
0.155	0.43	0.559	0.262
0.186	0.44	0.590	0.293
0.217	0.46	0.634	0.352
0.248	0.47	0.702	0.46
0.310	0.50	0.764	0.59
0.373	0.53	0.822	0.74
0.457	0.58	0.876	0.89
0.528	0.61	0.976	1.20
0.621	0.66	1.068	1.47
0.776	0.76	1.15	1.70
0.931	0.90	1.23	1.90
1.087	1.09	1.37	2.17
1.24	1.32	1.50	2.34
1.55	1.93	1.67	2.43
1.86	2.7	1.77	2.43
2.17	3.4	1.92	2.40
2.48	4.1	2.14	2.28
3.10	5.3	2.33	2.15
3.73	6.3	2.52	2.03
4.66	7.4	2.69	1.92
5.28	7.9	3.00	1.71
6.21	8.7	3.29	1.56
7.76	9.4	3.55	1.43
9.31	10.0	3.80	1.32
10.87	10.1	4.03	1.22
		4.26	2.16
		4.43	1.10
		4.58	1.05

Source: Adapted from Hunter, S. R. and L. G. Christophorou, *J. Chem. Phys.*, 80, 6150, 1984.

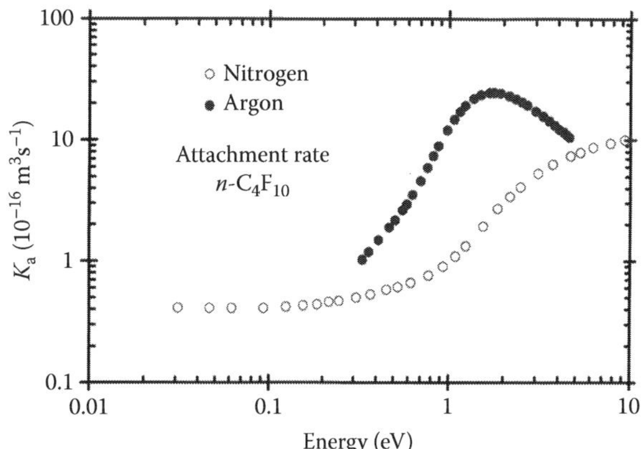

FIGURE 139.1 Attachment rates for n-C_4F_{10} as a function of mean energy of electrons in a buffer gas as shown. (Adapted from Hunter, S. R. and L. G. Christophorou, *J. Chem. Phys.*, 80, 6150, 1984.)

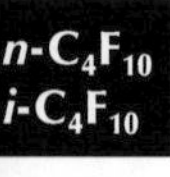

TABLE 139.8
Electron Attachment Rate for i-C_4F_{10}

Buffer Gas: Nitrogen		Buffer Gas: Argon	
Mean Energy (ε̅) (eV)	Attachment Rate (10^{-15} m³/s)	Mean Energy (ε̅) (eV)	Attachment Rate (10^{-15} m³/s)
0.0873	0.113	0.300	0.42
0.099	0.113	0.335	0.50
0.111	0.114	0.364	0.58
0.133	0.121	0.412	0.75
0.154	0.136	0.473	0.99
0.184	0.176	0.509	1.16
0.203	0.205	0.559	1.37
0.231	0.266	0.590	1.51
0.279	0.41	0.634	1.69
0.327	0.57	0.702	1.95
0.374	0.77	0.764	2.17
0.417	0.97	0.822	2.35
0.493	1.36	0.876	2.51
0.555	1.69	0.976	2.69
0.601	1.96	1.068	2.79
0.647	2.16	1.15	2.83
0.711	2.46	1.23	2.84
0.759	2.65	1.37	2.84
0.812	2.78	1.50	2.82
0.839	2.83	1.67	2.82
0.872	2.91	1.77	2.81
0.911	2.97	1.92	2.79
0.938	3.02	2.14	2.78
		2.33	2.75
		2.52	2.72
		2.69	2.66
		3.00	2.53
		3.29	2.37
		3.55	2.26
		3.80	2.16
		4.03	2.06
		4.26	1.99
		4.43	1.93
		4.58	1.86
		4.71	1.79
		4.81	1.75

Source: Adapted from Spyrou, S. M., S. R. Hunter, and L. G. Christophorou, *J. Chem. Phys.*, 83, 641, 1985.

Table 139.8 and Figure 139.2 show the attachment rates for i-C_4F_{10} as a function of mean energy of the swarm in a buffer gas of nitrogen or argon.

139.7 ATTACHMENT CROSS SECTIONS

Table 139.9 and Figure 139.3 show attachment cross sections as a function of electron energy for n-C_4F_{10}. Table 139.10 shows attachment cross sections as a function of electron energy for i-C_4F_{10}.

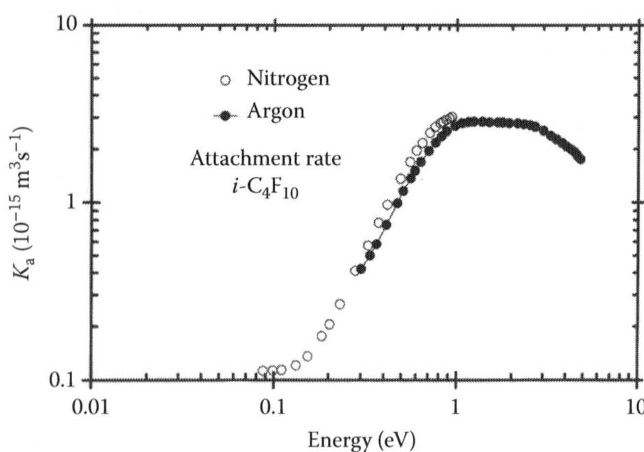

FIGURE 139.2 Attachment rates for i-C_4F_{10} as a function of mean energy of electrons in a buffer gas as shown. (Adapted from Spyrou, S. M., S. R. Hunter, and L. G. Christophorou, *J. Chem. Phys.*, 83, 641, 1985.)

TABLE 139.9
Attachment Cross Sections for n-C_4F_{10}

Energy (eV)	Q_a (10^{-21} m²)	Energy (eV)	Q_a (10^{-21} m²)
0.30	0.209	2.3	5.0
0.35	0.235	2.5	4.0
0.40	0.285	2.7	3.05
0.5	0.40	3.0	1.85
0.6	0.55	3.3	1.13
0.7	0.67	3.6	0.72
0.8	0.74	4.0	0.45
0.9	0.84	4.5	0.287
1.0	1.02	5.0	0.215
1.1	1.29	5.5	0.170
1.2	1.69	6.0	0.139
1.3	2.02	6.5	0.116
1.4	2.88	7.0	0.099
1.5	3.66	7.5	0.086
1.6	4.4	8.0	0.076
1.7	5.1	8.5	0.068
1.8	5.7	9.0	0.061
1.9	6.0	9.5	0.055
2.0	6.0	10.0	0.049
2.1	5.8		

Source: Adapted from Hunter, S. R. and L. G. Christophorou, *J. Chem. Phys.*, 80, 6150, 1984.

139.8 DRIFT VELOCITY

Drift velocities of electrons have been measured by Naidu and Prasad (1972) and Hunter et al. (1988) as shown in Tables 139.11 and 139.12 and Figure 139.4. Hunter et al. (1988) report a pressure dependence of the drift velocity as shown in Figure 139.5. The decrease in drift velocity with increasing pressure is attributed to the pressure dependence of the

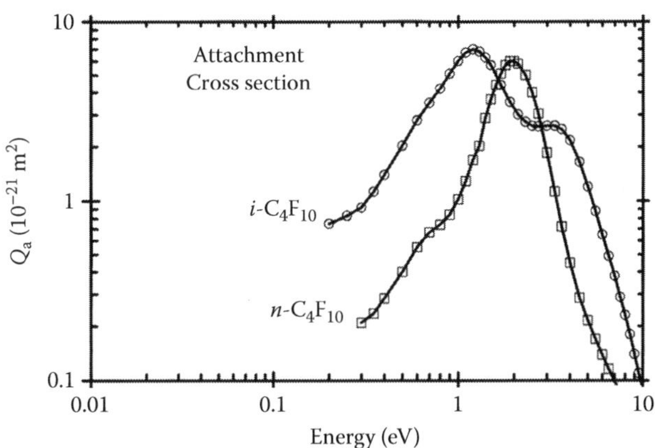

FIGURE 139.3 Attachment cross sections as a function of electron energy. (—□—) n-C_4F_{10} (Hunter and Christophorou, 1984); (—○—) i-C_4F_{10} (Spyrou et al. 1985).

TABLE 139.10
Attachment Cross Sections for i-C_4F_{10}

Energy (eV)	Q_a (10^{-21} m^2)	Energy (eV)	Q_a(10^{-21} m^2)
0.20	0.75	2.3	2.75
0.25	0.83	2.5	2.61
0.30	0.92	2.7	2.59
0.35	1.13	3.0	2.61
0.40	1.40	3.3	2.61
0.5	2.03	3.6	2.50
0.6	2.8	4.0	2.17
0.7	3.5	4.5	1.65
0.8	4.2	5.0	1.20
0.9	5.1	5.5	0.88
1.0	6.0	6.0	0.65
1.1	6.7	6.5	0.49
1.2	7.0	7.0	0.38
1.3	6.8	7.5	0.29
1.4	6.3	8.0	0.23
1.5	5.7	8.5	0.18
1.7	4.4	9.0	0.14
1.9	3.54	9.5	0.11
2.1	3.02	10.0	0.09

Source: Adapted from Spyrou, S. M., S. R. Hunter, and L. G. Christophorou, *J. Chem. Phys.*, 83, 641, 1985.

attachment coefficient in the gas which increases with increasing gas number density.

139.9 CHARACTERISTIC ENERGY

The characteristic energies (D/μ) as a function of E/N for n-C_4F_{10} are shown in Table 139.13 and Figure 139.6. The characteristic energy is observed to be larger at higher gas number densities (Naidu and Prasad, 1972).

TABLE 139.11
Drift Velocities of Electrons for n-C_4F_{10}

E/N (Td)	W (10^5 m/s)	E/N (Td)	W (10^5 m/s)
230	1.28	375	1.54
240	1.30	400	1.59
250	1.32	450	1.66
275	1.37	500	1.77
300	1.42	550	1.85
325	1.47	575	1.93
350	1.51		

Source: Adapted from Naidu, M. S. and A. N. Prasad, *J. Phys. D: Appl. Phys.*, 5, 741, 1972.

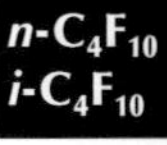

TABLE 139.12
Drift Velocities of Electrons for n-C_4F_{10}

E/N (Td)	W (10^4 m/s)	E/N (Td)	W (10^4 m/s)
5.0	4.10	160.0	10.9
6.0	4.45	180.0	11.2
8.0	5.08	200.0	11.4
10.0	5.46	220.0	11.6
12.0	5.85	240.0	11.8
15.0	6.38	260.0	12.0
20.0	7.11	280.0	12.3
25.0	7.54	300.0	12.6
30.0	7.94	320.0	12.8
35.0	8.18	340.0	13.2
40.0	8.45	360.0	13.4
50.0	8.75	380.0	13.7
60.0	9.0	400.0	14.0
70.0	9.2	420.0	14.4
80.0	9.4	440.0	14.9
90.0	9.6	460.0	15.2
100.0	9.9	480.0	15.6
120.0	10.3	500.0	15.8
140.0	10.6		

Source: Adapted from Hunter, S. R., J. G. Carter, and L. G. Christophorou, *Phys. Rev.*, 38, 58, 1988.

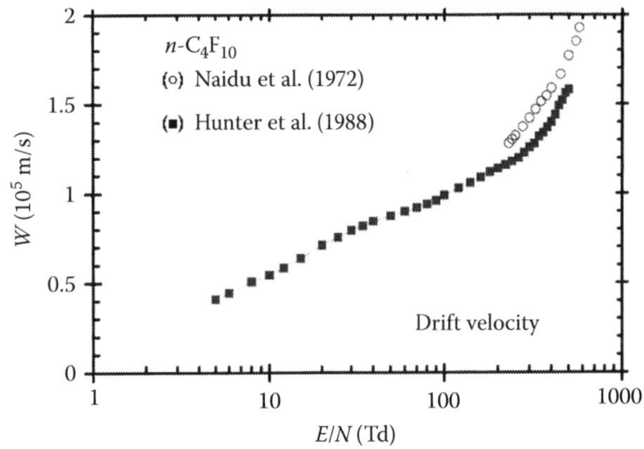

FIGURE 139.4 Drift velocities in n-C_4F_{10}. (■) Hunter et al. (1988); (○) Naidu and Prasad (1972).

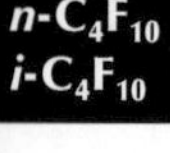

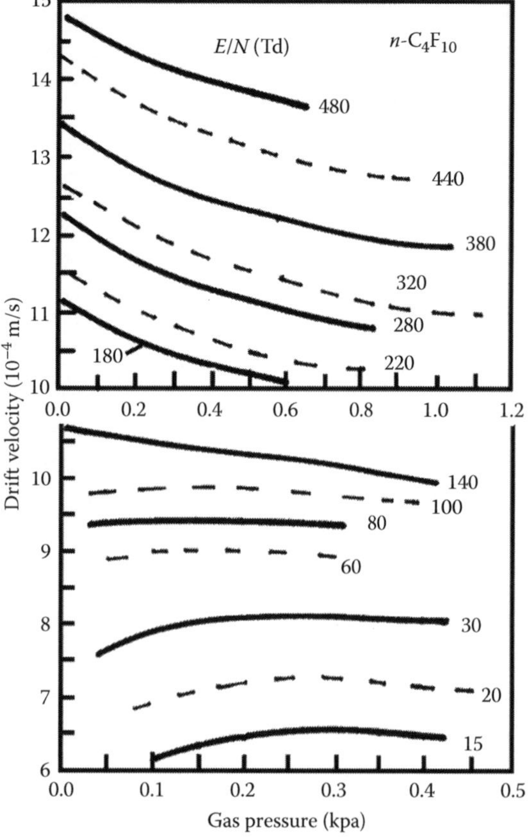

FIGURE 139.5 Pressure dependence of the drift velocity in n-C_4F_{10}. The decrease in W with increase of gas number density is more intense at higher E/N. (Adapted from Hunter, S. R., J. G. Carter, and L. G. Christophorou, *Phys. Rev.*, 38, 58, 1988.)

TABLE 139.13
Characteristic Energy for n-C_4F_{10}

E/N (Td)	Energy (V)	E/N (Td)	Energy (V)
Low Number Density ($N \leq 3.22 \times 10^{22}$ m^{-3})		High Number Density ($N \geq 5.2 \times 10^{22}$ m^{-3})	
250	2.65	280	2.88
280	2.83	300	3.06
300	2.97	320	3.22
320	3.12	340	3.36
340	3.26	360	3.47
360	3.37	380	3.58
380	3.48	400	3.70
400	3.59		
420	3.69		
440	3.78		
460	3.87		
480	3.99		
500	4.08		
520	4.10		

Source: Adapted from Naidu, M. S. and A. N. Prasad, *J. Phys. D: Appl. Phys.*, 5, 741, 1972.

139.10 IONIZATION AND ATTACHMENT COEFFICIENTS

Table 139.14 and Figure 139.7 show the ionization and attachment coefficients for n-C_4F_{10}. The attachment coefficients are dependent on gas number density, higher density resulting in higher coefficients (Hunter et al., 1988; Razzak and Goodyear, 1968).

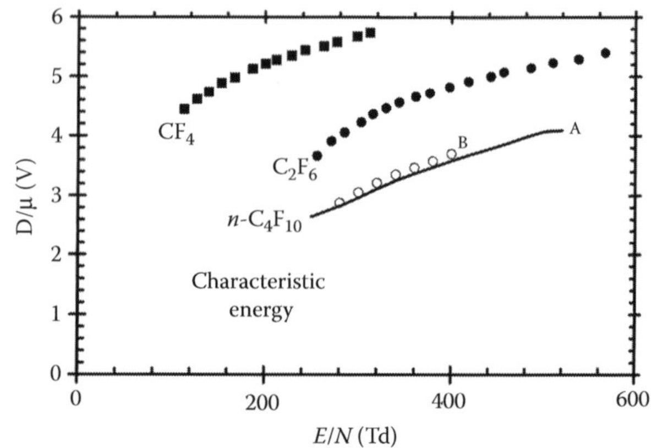

FIGURE 139.6 Characteristic energy for n-C_4F_{10}. A, low density; B, high density; see Table 139.13. Data for CF_4 and C_2F_6 are added for comparison. (Adapted from Naidu, M. S. and A. N. Prasad, *J. Phys. D: Appl. Phys.*, 5, 741, 1972.)

TABLE 139.14
Ionization and Attachment Coefficients

	Hunter et al. (1988)			Naidu and Prasad (1972)
	High N	Low N		
	η/N	η/N	α/N	$(\alpha - \eta/N)$
E/N (Td)		(10^{-21} m^2)		
22	0.51			
23		0.29		
30	0.51	0.29		
40	0.68	0.35		
50	1.06	0.41		
60		0.58		
70	2.51	0.73		
80	3.12	0.81		
90	3.68	0.95		
100	4.09	1.04		
120	4.38	1.10		
140	4.41	1.12	0.03	
160	4.28	1.10	0.06	
180	4.06	1.04	0.12	
200	3.80	0.97	0.20	
220	3.50	0.88	0.29	
240	3.22	0.83	0.41	
260	2.89	0.71	0.56	
280	2.58	0.64	0.73	−1.14

TABLE 139.14 (continued)
Ionization and Attachment Coefficients

	Hunter et al. (1988)			Naidu and Prasad (1972)
	High N	Low N		
E/N (Td)	η/N	η/N (10^{-21} m^2)	α/N	$(\alpha - \eta/N)$
300	2.29	0.57	0.92	
310				−0.56
320	2.02	0.54	1.10	
340	1.82	0.47	1.32	0.066
360	1.61	0.49	1.63	
370				0.656
380	1.43	0.46	1.88	
400	1.25	0.40	2.21	1.408
420			2.58	
430				1.945
450			3.13	
460				2.705
490				3.373
520				3.956
550				4.53

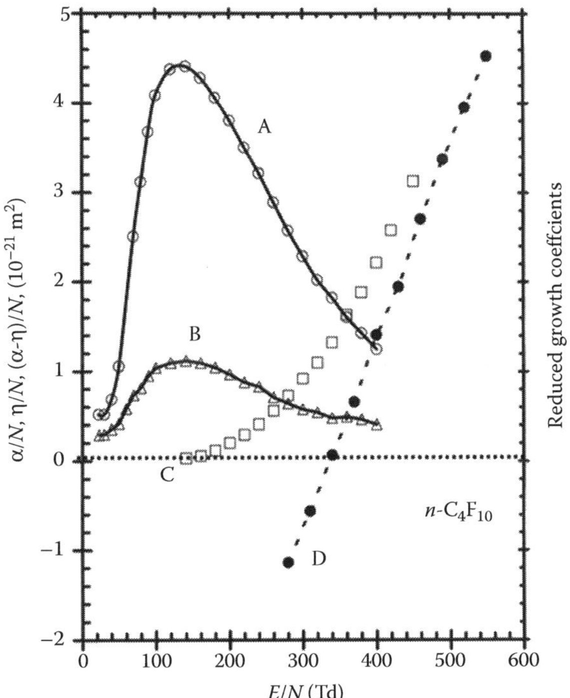

FIGURE 139.7 Ionization and attachment coefficients for n-C_4F_{10}. (A) Attachment coefficients (η/N) at high N; (B) attachment coefficients at low N (η/N); (C) ionization coefficients (α/N); (D) effective ionization coefficients $(\alpha-\eta)/N$. A, B, and C are from Hunter et al. (1988); D is from Naidu and Prasad (1972).

REFERENCES

Beran, J. A. and L. Kevan, *J. Phys. Chem.*, 73, 3866, 1969.

Fessenden, R. W. and M. Bansal, *J. Chem. Phys.*, 53, 3468, 1970.

Hunter, S. R. and L. G. Christophorou, *J. Chem. Phys.*, 80, 6150, 1984.

Hunter, S. R., J. G. Carter, and L. G. Christophorou, *Phys. Rev.*, 38, 58, 1988.

Hunter, S. R., J. G. Carter, and L. G. Christophorou, *J. Chem. Phys.*, 86, 293, 1987.

Jarvis, G. K., K. J. Boyle, C. A. Mayhew, and R. P. Tuckett, *J. Phys. Chem. A*, 102, 3219, 1998.

Naidu, M. S. and A. N. Prasad, *J. Phys. D: Appl. Phys.* 5, 741, 1972.

Nakanishi, K., D. R. James, H. Rodrogo, and L. G. Christophorou, *J. Phys. D: Appl. Phys.*, 17, L73, 1984.

Razzak, S. A. A. and C. C. Goodyear, *J. Phys. D: Appl. Phys.*, 1, 1215, 1968.

Spyrou, S. M. and L. G. Christophorou, *J. Chem. Phys.*, 83, 2829, 1985.

Spyrou, S. M., I. Sauers, and L. G. Christophorou, *J. Chem. Phys.*, 78, 7200, 1983.

Spyrou, S. M., S. R. Hunter, and L. G. Christophorou, *J. Chem. Phys.*, 83, 641, 1985.

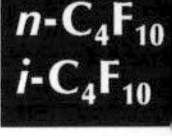

140

TETRAHYDROFURAN AND α-TETRAHYDRO-FURFURYL ALCOHOL

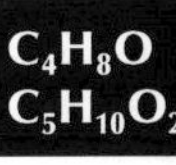

CONTENTS

Tetrahydrofuran (THF: C_4H_8O), synonym tetramethylene oxide, is a common organic solvent used in industry. The interest in the large molecules, tetrahydrofurfuryl alcohol (THFA: $C_5H_{10}O_2$) and THF arises out of their structural similarity to DNA and RNA (Zecca et al., 2005). The molecular structure is closely related to cyclopentane (C_5H_{10}) as shown in Figure 140.1, with one of the functional groups $-CH_2$ being replaced with O-atom. It has melting point of −108.44°C and boiling point of 64°C. The electronic polarizability is ~9.1×10^{-40} F m^2 and dipole moment of 1.75 D. Zecca et al. have measured the total scattering cross section in the vapor phase as shown in Figure 140.2 and Table 140.1.

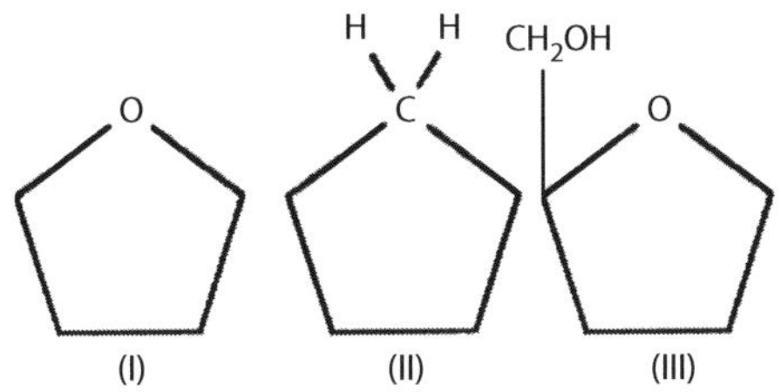

FIGURE 140.1 C_4H_8O (I), C_5H_{10} (II), and α-$C_5H_{10}O_2$ (III) molecules.

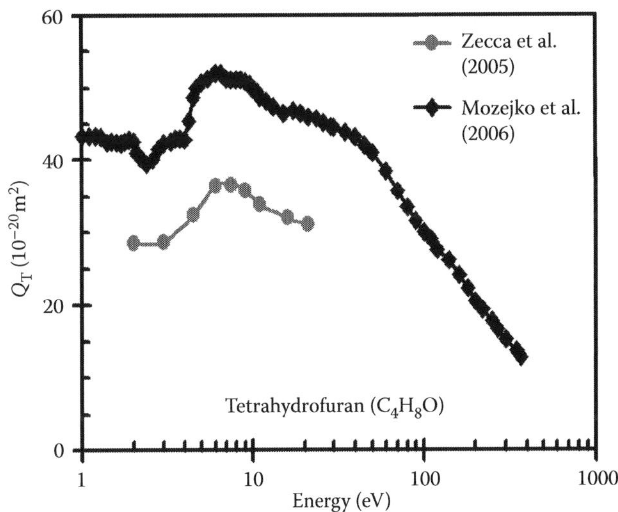

FIGURE 140.2 Total scattering cross section in C_4H_8O, P.- Możejko et al. (2006b). The characteristic feature, common to most gases, of peak in the 8–10 eV region is seen. (Adapted from Zecca, A., C. Perazzolli, and M. J. Brunger, *J. Phys. B: At. Mol. Opt. Phys.*, 38, 2079, 2005.)

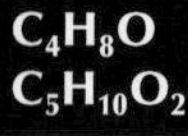

TABLE 140.1
Total Scattering Cross Section in C_4H_8O and $C_5H_{10}O_2$

Energy (eV)	Q_T (10^{-20} m^2) C_4H_8O	Q_T (10^{-20} m^2) $C_5H_{10}O_2$	Energy (eV)	Q_T (10^{-20} m^2) C_4H_8O	Q_T (10^{-20} m^2) $C_5H_{10}O_2$
1.0	43.4	58.6	10.5	49.3	66.9
1.1	43.3	58.4	11.0	48.6	64.8
1.2	43.3	58.2	12.0	48.2	63.5
1.3	43.1	58.1	13.0	47.3	61.3
1.4	42.5	58.1	15.0	46.5	58.6
1.5	42.6	58.6	17.0	46.9	57.0
1.6	42.5	58.6	19.0	46.4	56.9
1.7	42.3	58.7	21.0	45.9	55.8
1.8	42.7	59.5	23.5	45.7	54.6
1.9	42.8	60.1	26.0	45.2	53.3
2.0	42.7	60.6	28.5	44.7	52.4
2.1	41.0	60.7	30	44.4	52.0
2.2	40.5	60.5	35	43.9	50.4
2.4	39.5	60.0	40	43.2	49.7
2.6	40.1	60.0	45	42.1	48.6
2.8	41.6	60.4	50	41.0	48.0
3.0	42.2	60.2	60	38.4	46.8
3.3	42.6	60.3	70	35.6	45.7
3.5	43.0	60.6	80	33.4	44.0
3.7	42.9	61.5	90	31.6	42.5
4.0	42.8	62.8	100	30.0	40.2
4.2	45.3		110	29.1	38.4
4.5	48.6	64.9	120	27.6	37.2
4.7	50.0		140	26.1	34.1
5.0	50.7	67.3	160	24.2	32.1
5.5	51.2	69.2	180	22.4	30.0
6.0	52.0	69.1	200	20.5	28.1
6.5	51.9	67.7	220	19.3	25.8
7.0	51.3	67.0	250	17.7	23.5
7.5	51.1	66.8	270	16.7	21.9
8.0	51.1	67.0	300	15.1	20.1
8.5	51.1	67.9	350	13.7	18.2
9.0	50.8	68.3	370	12.8	16.7
9.5	50.5	68.2			
10.0	49.8	67.3			

Sources: Adapted from Možejko et al., *Phys. Rev. A*, 74, 012708, 2006b; Možejko et al., *Chem. Phys. Lett.*, 429, 378, 2006a.
Note: See Figure 140.2 for graphical presentation.

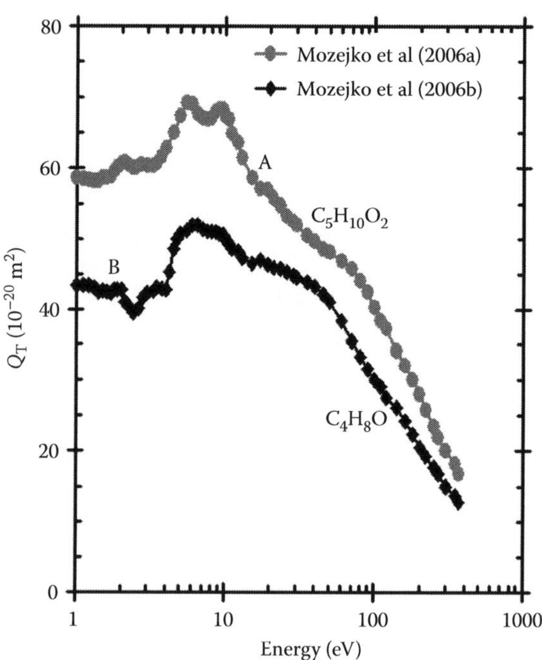

FIGURE 140.3 Total cross section for $C_5H_{10}O_2$ measured by Možejko et al. (2006a). Cross section for C_4H_8O measured by Možejko et al. (2006b) are shown for comparison purpose. Note the larger cross section for the larger molecule.

140.1 TOTAL CROSS SECTION

140.1.1 Tetrahydrofuran

See Table 140.1 and Figure 140.2.

140.1.2 α-Tetrahydrofurfuryl Alcohol

Total cross section for $C_5H_{10}O_2$ (Možejko et al., 2006b) are also shown in Table 140.1 and Figure 140.3.

REFERENCES

Možejko, P., A. Domaracka, E. Ptasińska-Denga, and C. Szmytkowski, *Chem. Phys. Lett.*, 429, 378, 2006a

Možejko, P., E. Ptasińska-Denga, A. Domaracka, and C. Szmytkowski, *Phys. Rev. A*, 74, 012708, 2006b.

Zecca, A., C. Perazzolli, and M. J. Brunger, *J. Phys. B: At. Mol. Opt. Phys.*, 38, 2079, 2005.

141

TOLUENE

CONTENTS

Toluene (C_7H_8) is a hydrocarbon molecule with 50 electrons. Its dipole moment is 0.375 D and ionization potential 8.828 eV. The only experimental data available for partial and total ionization cross sections are from Vacher et al. (2007) as shown in Table 141.1 and Figure 141.1.

141.1 IONIZATION CROSS SECTIONS

See Table 141.1 and Figure 141.1.

TABLE 141.1
Ionization Cross Sections for C_7H_8

	Ionization Cross Section (10^{-20} m^2)						
	Ion Species						
Energy (eV)	$C_3H_3^+$	$C_4H_3^+$	$C_5H_3^+$	$C_5H_5^+$	$C_7H_7^+$	$C_7H_8^+$	Total
12					0.0057	1.17	1.176
13					0.019	1.50	1.519
14					0.035	1.97	2.005
15					0.127	2.39	2.517
16					0.295	2.81	3.105
17					0.513	3.22	3.733
18					0.839	3.55	4.389
19					1.350	3.68	5.030
20				0.0075	1.890	3.76	5.657
21				0.026	2.310	3.92	6.256
23	0.008	0.030		0.165	3.210	3.88	7.292
25	0.030	0.091	0.0129	0.415	3.680	4.15	8.379
30	0.239	0.463	0.130	0.822	4.810	4.13	10.594
35	0.478	0.772	0.294	0.998	5.240	4.21	11.992
40	0.683	1.050	0.561	1.280	5.790	3.84	13.204
50	1.000	1.330	1.140	1.590	5.680	3.87	14.610
60	1.300	1.430	1.510	1.440	5.810	3.70	15.190
70	0.837	1.040	0.856	1.300	7.060	4.50	15.593
78	0.678	0.872	0.822	1.130	7.240	4.66	15.402

Note: Tabulated values by courtesy of Dr. J. R. Vacher, 2007.

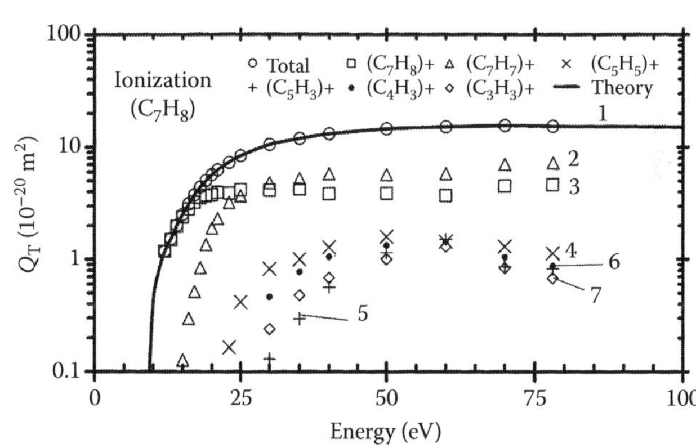

FIGURE 141.1 Total and selected partial ionization cross sections for C_7H_8. (1) Total (○); (2) $C_7H_7^+$ (Δ); (3) $C_7H_8^+$ (□); (4) $C_5H_5^+$ (×); (5) $C_4H_3^+$ (+); (6) $C_5H_3^+$ (●); (7) $C_3H_3^+$ (◊); (——) theory. (Adapted from Vacher, J. R. et al., *Chem. Phys. Lett.*, 434, 188, 2007.)

REFERENCE

Vacher, J. R., F. Jorand, N. Blin-Simiand, and S. Pasquiers, *Chem. Phys. Lett.*, 434, 188, 2007.

Section XIV

GAS MIXTURES

142

AIR

CONTENTS

142.1 SELECTED REFERENCES FOR DATA

See Table 142.1.

TABLE 142.1
Selected References

Quantity	Range: eV, (Td)	Reference
Negative ion formation	—	**Nagato et al. (2006)**
Swarm and transport properties	(0.3–2000)	**Raju (2005)**
Swarm and transport properties	(0.3–2000)	Liu and Raju (1993)
Drift velocity	(50–400)	**Roznerski and Leja (1984)**
Ionization coefficients	(70–170)	**Risbud and Naidu (1979)**
Swarm parameters	(20–1420)	**Lakshminarasimha and Lucas (1977)**
Drift velocity	(0.7–2.0)	**Milloy et al. (1975)**
Swarm parameters	(0.7–120)	Huxley and Crompton (1974)
Drift velocity	(0.3–12.0)	**Rees (1973)**
Ionization coefficients	(80–720)	**Rajapandiyan and Raju (1972)**
Attachment coefficients	(100–275)	**Parr and Moruzzi (1972)**
Ionization coefficients	(150–3000)	**Rao and Raju (1971)**
Ionization coefficients	(600–1800)	**Bhiday et al. (1970)**
Drift velocity	(0.5–110)	**Hessenauer (1967)**
Attachment coefficients	(50–90)	**Ryzko (1966)**
Attachment coefficients	(70–170)	**Crompton et al. (1965)**
Attachment processes	(0.15–0.70)	Dutton et al. (1963)
Ionization coefficients	75–200)	**Prasad (1959)**
Attachment coefficients	(2.5–75)	**Kuffel (1959)**

Note: Bold font indicates experimental study.

142.2 DRIFT VELOCITY

See Table 142.2.

142.3 CHARACTERISTIC ENERGY (D_r/μ)

See Table 142.3.

142.4 ELECTRON MOBILITY IN HUMID AIR

See Table 142.4.

142.5 REDUCED IONIZATION COEFFICIENTS

Table 142.5 and Figure 142.2 show the ionization coefficients for dry air (Raju, 2005). See Table 142.5.

TABLE 142.2
Recommended Drift Velocity for Dry Air

E/N (Td)	*W* (10^3 m/s)	*E/N* (Td)	*W* (10^3 m/s)	*E/N* (Td)	*W* (10^3 m/s)
0.3	3.60	8	18.26	400	383.6
0.4	4.00	10	21.00	500	496.2
0.6	4.91	20	40.00	600	556.7
0.8	5.78	40	62.50	700	650.5
1	6.58	60	77.60	800	726.0
2	9.52	80	97.10	900	788.3
3	11.36	100	112.80	1000	860
4	12.76	200	210.0	1300	1103
6	15.50	300	309.6	2000	1567

Source: Adapted from Raju, G. G., *Gaseous Electronics: Theory and Practice*, Taylor & Francis, Boca Raton, FL, 2005, p. 340.
Note: See Figure 142.1 for graphical presentation.

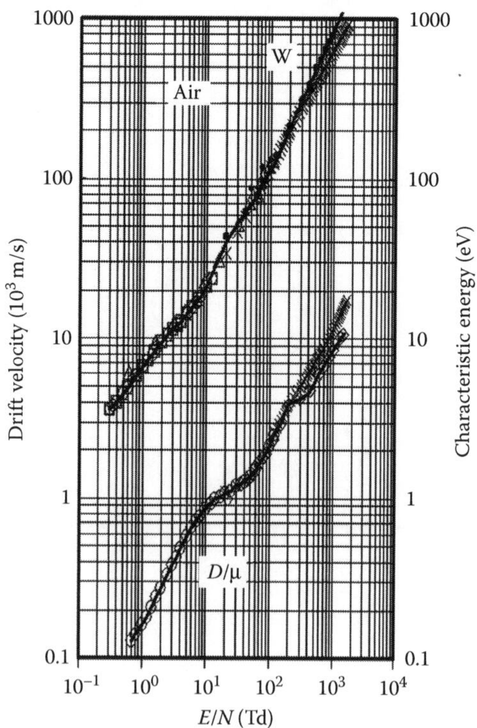

FIGURE 142.1 Drift velocity and characteristic energy for dry air. Drift velocity: (Δ) Hessenauer (1967); (□) Rees (1973); (○) Huxley and Crompton (1974); (●) Roznerski and Leja (1984); (◆) Liu and Raju (1993); (×) author's calculation from Bolsig (Morgan et al. 1996); (——) recommended. Characteristic energy: (○) Huxley and Crompton (1974); (◊) Lakshminarasimha and Lucas (1977); (×) author's calculation from Bolsig; (——) recommended.

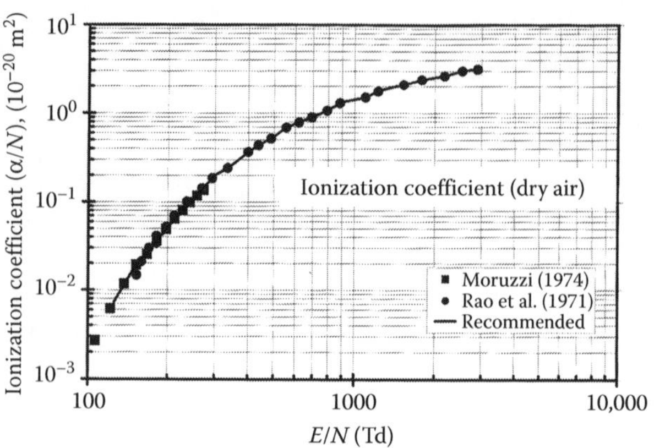

FIGURE 142.2 Density-reduced ionization coefficient for dry air. (■) Parr and Moruzzi (1972); (●) Rao and Raju (1971); (——) recommended.

TABLE 142.3
Recommended Characteristic Energy for Dry Air

E/N (Td)	D_r/μ (eV)	E/N (Td)	D_r/μ (eV)	E/N (Td)	D_r/μ (eV)
0.70	0.13	30.0	1.19	400	4.50
0.80	0.14	40.0	1.28	500	5.29
1.00	0.16	50.0	1.37	600	6.13
1.40	0.21	60.0	1.50	700	6.68
2.00	0.28	70.0	1.64	800	7.35
3.00	0.38	80.0	1.79	900	8.0
5.00	0.57	100	2.13	1000	8.60
10.0	0.86	150	2.97	1500	10.9
14.0	0.99	200	3.88		
20.0	1.07	300	4.25		

Source: Adapted from Raju, G. G., *Gaseous Electronics: Theory and Practice*, Taylor & Francis, Boca Raton, FL, 2005, p. 340.
Note: See Figure 142.1 for graphical presentation.

142.6 NEGATIVE ION FORMATION

Nagato et al (2006). have studied the negative ion species formed in a corona discharge by the mass spectrometer method. Table 142.6 shows the peak assignment. Selected reaction rates are shown in Table 142.7 (Nagato et al., 2006).

TABLE 142.4
Electron Mobility in Humid Air

H_2O (%)	$N\mu = W/(E/N)$ E/N = 0.7 Td	0.8	1.0	1.5	2.0
0.30	14.3	13.4	11.7	8.60	6.75
0.45	13.7	13.2	11.9	9.01	7.13
0.60	13.1	12.7	11.7	9.32	7.50
0.75	12.0	11.8	11.2	9.41	7.80
0.90	11.1	10.9	10.5	9.29	8.00
1.0	10.5	10.4	10.1	9.00	8.00
1.2	9.56	9.45	9.25	8.66	7.85
1.5	8.40	8.43	8.27	7.83	7.45

Source: Adapted from Milloy, H. B., I. D. Reid, and R. W. Crompton, *Aust. J. Phys.*, 28, 231, 1975.
Note: W in units of 10^3 m/s.

TABLE 142.5
Reduced Ionization Coefficients in Dry Air

E/N (Td)	α/N (10^{-20} m^2)	E/N (Td)	α/N (10^{-20} m^2)	E/N (Td)	α/N (10^{-20} m^2)
95	0.0001	200	0.0521	1100	1.51
100	0.0002	250	0.109	1200	1.71
105	0.0009	300	0.199	1300	1.88
110	0.0020	350	0.265	1400	1.98
120	0.0047	400	0.368	1500	2.06
125	0.0066	450	0.458	1600	2.17
130	0.0086	500	0.549	1700	2.29
135	0.0105	550	0.678	1800	2.40
140	0.0126	600	0.760	1900	2.48
150	0.0174	700	0.908	2000	2.54
160	0.0224	800	1.09	2200	2.68
170	0.0278	900	1.33	2500	2.99
175	0.0306	1000	1.43		

TABLE 142.6
Ion Species in Corona in Air

Mass Number	Assignment
32	O_2^-
46	NO_2^-
48	O_3^-
60	CO_3^-
61	HCO_3^-
62	NO_3^-
76	CO_4^-
77	HCO_4^-
108	$NO_3^- NO_2$
109	$NO_2^- HNO_3$
123	$CO_3^- HNO_3$
124	$HCO_3^- HNO_3$
125	$O_2^- HNO_3$

TABLE 142.7
Selected Reaction Rates for Formation of Ions in Air

Reaction	Rate (m³/s)
$O^- + NO_2 \rightarrow NO_2^- + O$	1.0×10^{-15}
$O_2^- + NO_2 \rightarrow NO_2^- + O_2$	7.0×10^{-16}
$NO_2^- + O_3 \rightarrow NO_3^- + O_2$	1.2×10^{-16}
$O^- + O_3 \rightarrow O_3^- + O$	8.0×10^{-16}
$O_2^- + O_3 \rightarrow O_3^- + O_2$	6.0×10^{-16}
$O_3^- + NO_2 \rightarrow NO_3^- + O_2$	2.8×10^{-16}
$O^- + CO_2 + M \rightarrow CO_3^- + M$	3.1×10^{-34}
$O_3^- + CO_2 \rightarrow CO_3^- + O_2$	5.5×10^{-16}
$CO_3^- + NO_2 \rightarrow NO_3^- + CO_2$	2.0×10^{-16}
$NO_2 + OH + M \rightarrow HNO_3 + M$	3.3×10^{-36}
$O^- + H_2O \rightarrow OH^- + OH$	6.0×10^{-19}
$OH^- + O_3 \rightarrow O_3^- + OH$	9.0×10^{-16}
$OH^- + NO_2 \rightarrow NO_2^- + OH$	1.1×10^{-15}
$OH^- + CO_2 + M \rightarrow HCO_3^- + M$	7.6×10^{-34}
$NO_3^- + H_2SO_4 \rightarrow HSO_4^- + HNO_3$	2.6×10^{-15}

TABLE 142.8
Reduced Attachment Coefficients

E/N (Td)	η/N (10^{-20} m²)	*E/N* (Td)	η/N (10^{-20} m²)
80	1.45×10^{-3}	125	1.98×10^{-3}
95	1.60×10^{-3}	140	2.10×10^{-3}
110	1.80×10^{-3}		

Source: Adapted from Raju, G. G., *Gaseous Electronics: Theory and Practice*, Taylor & Francis, Boca Raton, FL, 2005, p. 340.

142.7 REDUCED ATTACHMENT COEFFICIENTS

Table 142.8 and Figure 142.3 show the reduced attachment coefficients in dry and humid air. Figure 142.4 gives attachment and ionization coefficients for H_2O vapor. Figure 37.15 gives additional data for attachment coefficient for H_2O vapor.

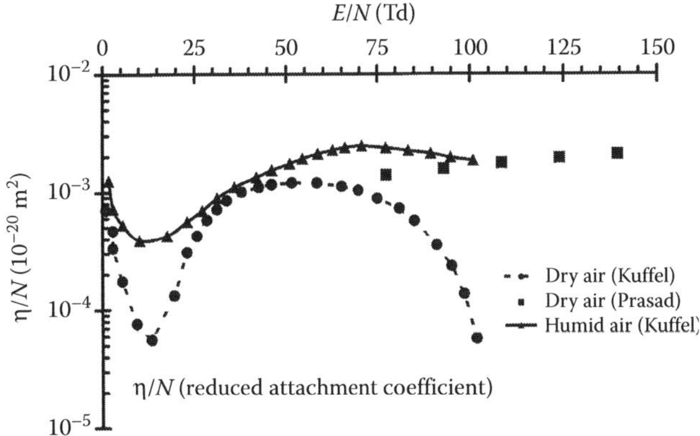

FIGURE 142.3 Attachment coefficients for dry and humid air.

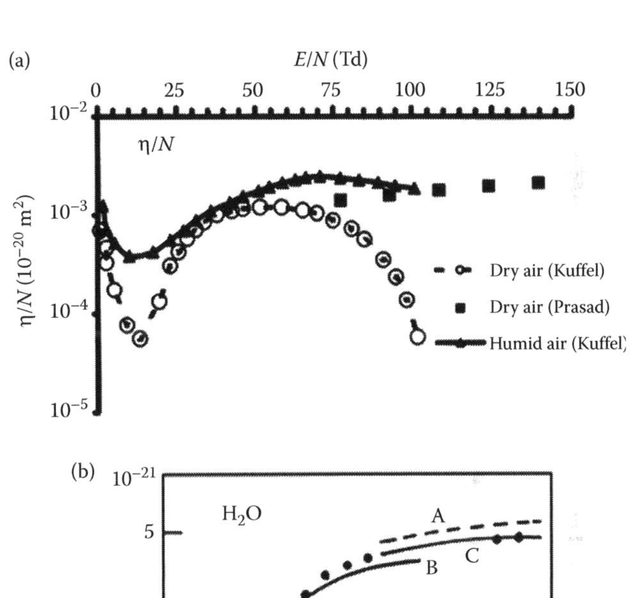

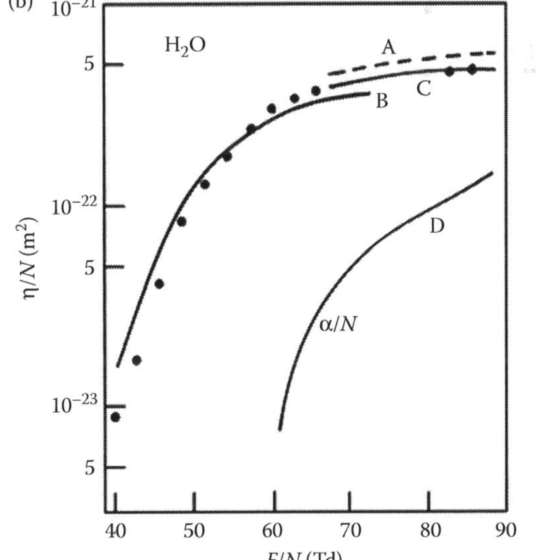

FIGURE 142.4 Ionization and attachment coefficients for water vapor. (A) Crompton et al. (1965); (B) Kuffel (1959); (C) and (D) Ryzko (1966); (•) Parr and Moruzzi (1972). Also see Figure 37.14 for additional data. (a) Dry and humid air (b) water vapor.

TABLE 142.9
Gas Constants for Dry Air

E/N Range (Td)	167–372	372–608	608–1686	1686–3035
F ($\times 10^{-20}$ m^2)	2.59	3.30	5.11	5.53
G (Td)	848	956	1228	1288
% error	±2.1	±2.0	±2.25	±0.7

Note: Percent error refers to measured values.

142.8 GAS CONSTANTS

Gas constants according to Townsend's semiempirical equation

$$\frac{\alpha}{N} = F\exp-\left(\frac{GN}{E}\right) \tag{142.1}$$

in which F and G are gas constants are shown in Table 142.9 (Raju, 2005).

REFERENCES

Bhiday, M. R., A. S. Paithankar, and B. L. Sharda, *J. Phys. Soc. Appl. Phys.*, 3, 943, 1970.

Crompton, R. W., J. A. Rees, and R. L. Jory, *Aust. J. Phys.*, 18, 541, 1965.

Dutton, J., F. M. Harris, and F. Llewelyn-Jones, *Proc. Phys. Soc. London*, 82, 51, 1963.

Hessenauer, H., *Z. Physik*, 204, 142, 1967.

Huxley, L. G. H. and R. W. Crompton, *The Diffusion and Drift of Electrons in Gases*, John Wiley and Sons, New York, NY, 1974.

Kuffel, E. *Proc. Phys. Soc. London*, 74, 297, 1959.

Lakshminarasimha, C. S. and J. Lucas, *J. Phys. D: Appl. Phys.*, 10, 313, 1977.

Liu, J. and G. R. Govinda Raju, *IEEE Trans. Elec. Insul.*, 28, 154, 1993.

Milloy, H. B., I. D. Reid, and R. W. Crompton, *Aust. J. Phys.*, 28, 231, 1975.

Morgan, W. L., J. P. Boeuf, L. C. Pitchford, www.siglo-kinema.com, 1996.

Nagato, K., Y. Matsui, T. Miyata, and T. Yamauchi, *Int. J. Mass Spectrom.*, 248, 142, 2006.

Parr, J. E. and J. L. Moruzzi, *J. Phys. D: Appl. Phys.*, 5, 514, 1972.

Prasad, A. N., *Proc. Phys. Soc. London*, 74, 33, 1959.

Rajapandiyan, S. and G. R. Govinda Raju, *J. Phys. D: Appl. Phys.*, 5, 16, 1972.

Raja Rao, C. and G. R. Govinda Raju, *J. Phys. D: Appl. Phys.*, 4, 494, 1971.

Raju, G. G., *Gaseous Electronics: Theory and Practice*, Taylor & Francis, Boca Raton, FL, 2005, p. 340.

Rees, J. A., *Aust. J. Phys.*, 26, 403, 1973.

Risbud, A. V. and M. S. Naidu, *J. Phys. (Paris) Colloq. C7*, 40, 77, 1979.

Roznerski, W. and K. Leja, J. *Phys. D: Appl. Phys.*, 17, 279, 1984.

Ryzko, H., *Arkiv Physik*, 32, 1, 1966.

Section XV

Appendices

Appendix 1: Fundamental Constants

CONTENTS

Quantity	Symbol	Value	Units
Avogadro constant	N_A	6.022×10^{23}	mol^{-1}
Boltzmann constant	k	1.381×10^{-23}	J/K
Boltzmann constant in eV	k/e	8.617×10^{-5}	eV/K
Electron mass	m	9.109×10^{-31}	kg
Elementary charge	e	1.602×10^{-19}	C
Electron charge-to-mass ratio	e/m	1.759×10^{11}	C/kg
Planck's constant	h	6.626×10^{-34}	J s
Permittivity of free space	ε_0	8.854×10^{-12}	F/m
Permeability of free space	μ_0	$4\pi \times 10^{-7}$	A^{-2}
Planck's constant in eV	h	4.136×10^{-15}	eV s
Bohr radius	a_0	0.529×10^{-10}	m
Proton mass	m_p	1.672×10^{-27}	kg
Proton-to-electron mass ratio	m_p/m_e	1836.152	
Rydberg constant in eV	R	13.606	eV
Speed of light	c	2.9979×10^8	m/s

A1.1 SELECTED CONVERSION FACTORS

A1.1.1 Energy

1 eV = 11,604 K = 8065.54 cm^{-1} = 8.066×10^5 m^{-1}
1 m^{-1} = 1.2398×10^{-6} eV
1 eV = 23.06 k Cal/mol
kT (77 K) = 0.0066 eV
kT (300 K) = 0.026 eV
1 Rydberg (R) = 13.595 eV = 1.097×10^7 m^{-1}

$$\text{Energy (eV)} = 2.843 \times 10^{-12}\ [\text{Drift velocity (m/s)}]^2$$

A1.1.2 Cross Section

$$a_0^2 = 0.28 \times 10^{-20} m^2;\ \pi\, a_0^2 = 8.791 \times 10^{-21} m^2$$

A1.1.3 Pressure

1 atmosphere = 760 torr = 101.31 k Pa = 1 bar

A1.1.4 Gas Number Density (*N*), (m^{-3})

$$N_{p(\text{torr}),T(\text{K})} = 3.54 \times 10^{22} \times 273 \frac{p}{T}$$

A1.1.5 Reduced Electric Field

1 Td = 10^{-17} V cm^2 = 10^{-21} V m^2

$$\frac{E_{\text{Volts/m}}}{N_{1/m^3}} = 2.82 \frac{E_{\text{Volts/cm}}}{p_{\text{Torr}}}\ \text{Td} \quad (273\,\text{K})$$

$$\frac{E_{\text{Volts/m}}}{N_{1/m^3}} = 3.1 \frac{E_{\text{Volts/cm}}}{p_{\text{Torr}}}\ \text{Td} \quad (293\,\text{K})$$

A1.1.6 Reduced Ionization Coefficient

$$\frac{\alpha_{1/\text{m}}}{N_{1/m^3}} = 2.82 \times 10^{-21} \frac{\alpha_{1/\text{cm}}}{p_{\text{Torr}}} \quad (273\,\text{K})$$

$$\frac{\alpha_{1/\text{m}}}{N_{1/m^3}} = 3.11 \times 10^{-21} \frac{\alpha_{1/\text{cm}}}{p_{\text{Torr}}} \quad (293\,\text{K})$$

Appendix 2: Target Particles (Namewise)

	Name	Formula	Section-Chapter	Appendix	Page
A					
1.	**Acetaldehyde**	C_2H_4O	**VII-87**		489
2.	**Acetone**	C_3H_6O	**X-112**	A5	593, 743
3.	Acetonitrile	C_2H_3N		A5, A7	593, 743
4.	Acetophenone	C_8H_8O		A4, A7	759, 773
5.	**Acetylene**	C_2H_2	**IV-38**		259
6.	**Air**	-	**XIV-142**		719
7.	**Allene**	C_3H_4	**VII-85**		483
8.	**Ammonia**	NH_3	**IV-39**	A5	269, 754
9.	*i*-Amyl acetate	$i\text{-}C_7H_{14}O_2$		A7	777
10.	*n*-Amyl acetate	$n\text{-}C_7H_{14}O_2$		A7	777
11.	*n*-Amyl formate	$n\text{-}C_6H_{12}O_2$		A7	777
12.	Anthracene	$C_{14}H_{10}$		A5	764
13.	Aniline	C_6H_7N		A5	764
14.	**Argon**	**Ar**	**I-1**		3
B					
15.	Benzaldehyde	C_7H_6O		A4	743
16.	Benz[a]anthracene [S1]	$C_{18}H_{12}$		A5	743
17.	1,2-Benzanthracene [S1]	$C_{18}H_{12}$		A4, A5	743
18.	**Benzene**	C_6H_6	**XII-118**	A4	635
19.	Benzoic acid	$C_7H_6O_2$		A5	764
20.	Benzophenone	$C_3H_{10}O$		A4	743
21.	Benzo[a]pyrene	$C_{20}H_{12}$		A5	764
22.	Bis(trifluoromethyl)sulfide	C_2F_6S		A5	762
23.	**Boron trichloride**	BCl_3	**IV-40**	A4, A5	277, 743
24.	**Boron trifluoride**	BF_3	**IV-41**	A5	279, 277
25.	**Bromine**	$\mathbf{Br_2}$	**II-10**	A5	759
26.	Bromoacetonitrile	C_2H_2BrN		A5	641, 743
27.	**Bromobenzene**	C_6H_5Br	**XII-119**	A4, A5	635, 743, 764
28.	Bromobutane	$n\text{-}C_4H_9Br$		A5	753
29.	1-Bromobutane	$1\text{-}C_4H_9Br$		A4, A5	743, 753
30.	2-Bromobutane	$2\text{-}C_4H_9Br$		A4	743
31.	4-Bromo-1-butene	C_4H_7Br		A4	743
32.	1-Bromo-4-chlorobenzene	$1,4\text{-}C_6H_4BrCl$		A4	743
33.	1-Bromo-2-chloroethane	$1,2\text{-}C_2H_4BrCl$		A4, A5	743, 760
34.	**Bromochloromethane**	CH_2BrCl	**V-46**	A5	297, 754
35.	1,2-Bromochloropropane	$1,2\text{-}C_3H_6BrCl$		A4	743
36.	1,3-Bromochloropropane	$1,3\text{-}C_3H_6BrCl$		A4	744
37.	Bromocyclohexene	C_6H_9Br		A4	744
38.	*n*-Bromodecane	$n\text{-}C_{10}H_{21}Br$		A4, A5	744, 764
39.	Bromodichloromethane	$CHCl_2Br$		A5	754
40.	Bromodimethylpropane	$n\text{-}C_5H_{11}Br$		A5	764
41.	**Bromoethane**	$\mathbf{C_2H_5Br}$		A4, A5	744, 760
42.	Bromoethene	C_2H_3Br		A4	744
43.	1-Bromo-2-fluoroethane	CH_2FCH_2Br	VIII-93	A5	519, 760
44.	Bromoethylbenzene	C_8H_9Br		A4	744
45.	2-Bromoethylbenzene	C_8H_9Br		A4	744

continued

continued

	Name	Formula	Section-Chapter	Appendix	Page
46.	Bromohexane	n-$C_6H_{13}Br$		A5	764
47.	**Bromomethane**	CH_3Br	**V-47**		299
48.	Bromomethylbenzene	C_7H_7Br		A5	764
49.	2-Bromo-2-methylpropane	C_4H_9Br		A4	744
50.	Bromooctane	n-$C_8H_{17}Br$		A5	764
51.	Bromopentafluorobenzene	C_6BrF_5		A5	763
52.	5-Bromo-1-pentene	C_5H_9Br		A4	744
53.	Bromopropane	n-C_3H_7Br		A4, A5	744, 755
54.	1-Bromopropane	1-C_3H_7Br		A4, A5	744
55.	2-Bromopropane	2-C_3H_7Br		A4, A5	744
56.	2-Bromopropene	2-C_3H_5Br		A4	744
57.	3-Bromopropene	3-C_3H_5Br		A4	744
58.	*cis*-1-Bromopropene	*cis*-C_3H_5Br		A4	744
59.	*trans*-1-Bromopropene	*trans*-C_3H_5Br		A4	744
60.	3-Bromopropylbenzene	$C_9H_{11}Br$		A4	744
61.	*o*-Bromotoluene	*o*-C_7H_7Br		A4	744
62.	**Bromotrichloromethane**	**$CBrCl_3$**	**V-48**	A5	305, 755
63.	**2-Bromo-1,1,1-trifluoroethane**	$C_2H_2BrF_3$	**VIII-94**	A5	521, 760
64.	**Bromotrifluoromethane [S21]**	$CBrF_3$	**V-49**		307
65.	**1,3-Butadiene**	1,3-C_4H_6	**X-113**		595
66.	**Butane**	C_4H_{10}	**XIII-130**	A5	679
67.	2,3-Butanedione	2,3-$C_4H_6O_2$		A5	763
68.	**Butene**	C_4H_8	**XII-120**		637
69	*i*-Butyl acetate	*i*-$C_4H_9OC(O)CH_3$		A7	778
70	*n*-Butyl acetate	*n*-$C_4H_9OC(O)CH_3$		A7	778
71.	*s*-Butyl acetate	*s*-$C_4H_9OC(O)CH_3$		A7	778
72.	*t*-Butyl acetate	*t*-$C_4H_9OC(O)CH_3$		A7	778
73.	Butyl alcohol (1-Butanol)	$C_4H_{10}O$	**VI-80**	A7	461, 779
74.	*i*-Butyl alcohol (2-Methyl-1-propanol)	$C_4H_{10}O$		A7	779
75.	*t*-Butyl alcohol (2-Methyl-2-propanol)	$C_4H_{10}O$		A7	779
76.	*t*-Butyl chloride [S3]	*t*-C_4H_9Cl		A4	744
77.	*n*-Butyl formate	*n*-$C_5H_{10}O_2$		A7	779
78.	**2-Butyne**	2-C_4H_6	**X-113**		595
79.	γ-Butyrolactone	$C_4H_6O_2$		A8	763
C					
80.	Carbon	C		A6	769
81.	**Carbon dioxide**	CO_2	**III-27**		181
82.	**Carbon disulfide**	CS_2	**III-28**		195
83.	**Carbon monoxide**	**CO**	**II-11**		83
84.	**Carbon oxysulfide**	COS	**III-29**		201
85.	Carbonyl chloride or Phosgene	CCl_2O		A5	754
86.	**Carbon tetrachloride [S4]**	CCl_4	**V-50**	A4	311, 744
87.	**Carbon tetrafluoride [S5]**	see Tetrafluoromethane			
88.	**Carbonyl sulfide**	COS	**III-29**		201
89.	**Cesium**	**Cs**	**I-2**		15
90.	**Chlorine**	**Cl_2**	**II-12**	A5	95
91.	**Chlorine dioxide**	ClO_2	**III-30**		205
92.	**Chlorine nitrate**	$ClNO_3$		A5	755
93.	Chloroacetone	C_3H_5ClO		A4	744
94.	Chloroacetonitrile	C_2H_2ClN		A5	759
95.	**Chlorobenzene**	C_6H_5Cl	**XII-121**	A4, A5	641, 744, 753
96.	1-Chloro-1-bromo-2,2,2-trifluoroethane	$C_2HBrClF_3$		A4	744
97.	**1-Chlorobutane**	**1-C_4H_9Cl**	**XIII-131**	A4	685, 745
98.	**2-Chlorobutane**	**2-C_4H_9Cl**	**XIII-131**	A4	685, 745
99.	***t*-chlorobutane [S3]**	***t*-C_4H_9Cl**	**XIII-131**		685
100.	Chlorocyanogen	CClN		A4	745
101.	**Chlorodibromomethane**	$CHBr_2Cl$	**V-51**	A5	319, 755

continued

	Name	Formula	Section-Chapter	Appendix	Page
102.	1-Chloro-1,1-difluoroethane	$C_2H_3ClF_2$		A5	760
103.	Chlorodifluoromethane	$CHClF_2$		A4, A5	745
104.	Chloroethylene [S6]	C_2H_3Cl		A5	755
105.	**Chloroethane (S7)**	C_2H_5Cl	**VIII-95**	A4, A5	523, 744, 755
106.	Chloroethene [S6]	C_2H_3Cl		A5	759
107.	Chlorofluoromethane	CH_2ClF		A4	745
108.	**Chloroform [S7]**	$CHCl_3$	**V-74**	A4, A5	431
109.	1-Chloroheptane	$C_7H_{15}Cl$		A4	745
110.	1-Chlorohexane	$C_6H_{13}Cl$		A4	745
111.	**Chloromethane**	CH_3Cl	**V-52**	A4, A5	321, 755
112.	Chloromethylbenzene	C_7H_7Cl		A5	764
113.	**Chloropentafluorobenzene**	C_6ClF_5	**XII-122**	A5	645, 753
114.	Chloropentafluoroethane	C_2ClF_5		A5	755
115.	**1-Chloropropane**	**1-C_3H_7Cl**	**XI-115**		607
116.	**2-Chloropropane**	**2-C_3H_7Cl**	**XI-117**	A4	607, 743
117.	1-Chlorononane	$C_9H_{19}Cl$		A4	745
118.	1-Chlorooctane	$C_8H_{17}Cl$		A4	745
119.	**1-Chloropentane**	$C_5H_{11}Cl$	**XIII-132**	A4	687, 743
120.	*o*-Chlorotoluene	*o*-C_7H_7Cl		A4, A5	745, 764
121.	**Chloropentafluoroethane**	C_2ClF_5	VIII	A5	760, 753
122.	Chlorotrifluoroethene	C_2ClF_3	A5		755
123.	**Chlorotrifluoromethane**	$CClF_3$	**V-59**	A4, A5	329, 745, 753
124.	Chrysene	$C_{18}H_{12}$		A5	764
125.	Cyanogen	C_2N_2		A4	745
126.	Cyanogen bromide	BrCN		A4, A5	745, 754
127.	Cyanogen chloride	ClCN		A4	745
128.	**Cyclobutane**	C_4H_8	**XII-120**		637
129.	**Cyclobutene**	C_4H_6	**X-113**		595
130.	**Cyclohexane**	C_6H_{12}	**XIII-134, 112**		593, 691
131.	Cyclohexyl chloride	$C_6H_{11}Cl$		A4	745
132.	Cyclooctatetraene	C_8H_8		A5	745, 764
133.	**Cyclopentane**	C_5H_{10}	**XIII-133**		689
134.	**Cyclopentanone**	C_5H_8O		A5	753
135.	Cyclopentyl bromine	C_5H_9Br		A4	745
136.	Cyclopentyl chloride	C_5H_9Cl		A4	745
137.	**Cyclopropane**	c-C_3H_6	**IX-111**		583
138.	**Cyclopropene**	C_3H_4	**VII-86**		487
D					
139.	Deuterated ammonia	ND_3	**IV-42**		283
140.	**Deuterated benzene**	C_6D_6	**IV-118**	A6	629, 763
141.	Deuterated bromobenzene	C_6D_5Br		A4, A5	745, 763
142.	Deuterated ethane	C_2D_6	**VIII-101**		547
143.	**Deuterated methane**	CD_4	**V-54**		337
144.	**Deuterated water [S9]**	D_2O	See Heavy water		
145.	**Deuterium**	$\mathbf{D_2}$	**II-13**		105
146.	**Deuterium bromide**	DBr	**II-14**	A5	109, 753
147.	**Deuterium chloride**	DCl	**II-15**	A5	111, 753
148.	**Deuterium iodide**	**DI**	**II-16**	A5	113, 753
149.	Deuterium sulfide	D_2S		A4	745
150.	Dibenz[a,h]anthracene	$C_{22}H_{14}$		A5	764
151.	Diborane	B_2H_6		A7	773
152.	Dibromobenzene	$C_6H_4Br_2$		A4	745
153.	Dibromodichloromethane	CBr_2Cl_2		A5	755
154.	**1,2-Dibromo-1,1-difluoroethane**	$C_2H_2Br_2F_2$	VIII-96	A5	527, 761
155.	**Dibromodifluoromethane**	CBr_2F_2	V-55	A5	341
156.	**1,1-Dibromoethane**	1,1-$C_2H_4Br_2$	VIII-97	A5	529, 761
157.	**1,2-Dibromoethane**	**1,2-$C_2H_4Br_2$**	**VIII-97**	A4	529, 745

continued

continued

	Name	Formula	Section-Chapter	Appendix	Page
158.	**Dibromoethene**	$C_2H_2Br_2$	**VI-76**	A5	445, 759
159.	**Dibromomethane**	CH_2Br_2	V-56	A5	343, 756
160.	**Dibromotetrafluoroethane**	$C_2Br_2F_4$	**VIII-98**	A5	531, 761
161.	1,1-Dibromotetrafluoroethane	$1,1\text{-}C_2Br_2F_4$		A5	761
162.	1,2-Dibromotetrafluoroethane	$1,2\text{-}C_2Br_2F_4$		A5	761
163.	2,2-Dibromotetrafluoroethane	$2,2\text{-}C_2Br_2F4$		A5	761
164.	Dichlorobenzene	$C_6H_4Cl_2$		A4	745
165.	*m*-Dichlorobenzene	$m\text{-}C_6H_4Cl_2$		A5	763
166.	*o*-Dichlorobenzene	$o\text{-}C_6H_4Cl_2$		A4, A5	745, 763
167.	*p*-Dichlorobenzene	$p\text{-}C_6H_4Cl_2$		A5	763
168.	1,3-Dichlorobutane	$1,3\text{-}C_4H_8Cl_2$		A4	745, 764
169.	1,4-Dichlorobutane	$1,4\text{-}C_4H_8Cl_2$		A5, A4	745, 764
170.	2,3-Dichlorobutane	$2,3\text{-}C_4H_8Cl_2$		A5, A4	745, 764
171.	*trans*-1-2-Dichlorocyclohexane	$C_6H_{10}Cl_2$		A4	746
172.	**Dichlorodifluoromethane**	CCl_2F_2	**V-57, 112**	A4, A5	345, 746, 753
173.	**1,1-Dichloroethane**	$1,1\text{-}C_2H_4Cl_2$	**VIII-99**	A4, A5	533, 761
174.	**1,2-Dichloroethane**	$1,2\text{-}C_2H_4Cl_2$	**VIII-99**	A4, A5	533, 761
175.	**1,1-Dichloroethene**	$1,1\text{-}C_2H_2Cl_2$	**VI-77**	A5	447, 761
176.	*cis*-1,2-Dichloroethene	*cis*-1,2- $C_2H_2Cl_2$	**VI-77**	A5	447, 761
177.	***trans*-1,2-Dichloroethene**	*trans*-1,2- $C_2H_2Cl_2$	**VI-77**	**A5**	447, 761
178.	Dichlorofluoromethane	$CHCl_2F$		A4, A5	746, 756
179.	1,6-Dichlorohexane	$1,6\text{-}C_6H_{12}Cl_2$		A4, A5	746, 764
180.	**Dichloromethane**	CH_2Cl_2	**V-58**	A4, A5	359, 746
181.	1,2-Dichloro-2-methylpropane	$1,2\text{-}Cl_2\text{-}2\text{-}C_4H_8$		A4, A5	746, 764
182.	1,8-Dichlorooctane	$1,8\text{-}C_8H_{16}Cl_2$		A5	764
183.	1,5-Dichloropentane	$1,5\text{-}C_5H_{10}Cl_2$		A4, A5	746, 764
184.	1,1-Dichloropropane	$1,1\text{-}C_3H_6Cl_2$		A4, A5	746, 763
185.	1,2-Dichloropropane	$1,2\text{-}C_3H_6Cl_2$		A4, A5	746, 763
186.	1,3-Dichloropropane	$1,3\text{-}C_3H_6Cl_2$		A4, A5	746, 763
187.	2,2-Dichloropropane	$C_3Cl_2H_6$		A4, A5	746, 763
188.	1,1-Dichloro-1,2,2,2-tetrafluoroethane	$1,1\text{-}C_2Cl_2F_4$		**A5**	761
189.	1,2-Dichloro-1,1,2,2-tetrafluoroethane	$1,2\text{-}C_2Cl_2F_4$		**A5**	761
190.	Diethyl ether	$C_4H_{10}O$		A5	763
191.	**Difluorobenzene**	$C_6H_4F_2$	**XII-123**	A4	647, 747
192.	**1,3-Difluorobenzene**	$1,3\text{-}C_6H_4F_2$	**XII-123**	A4	647
193.	**1,4-Difluorobenzene**	$1,4\text{-}C_6H_4F_2$	**XII-124**		649
194.	Difluorocyanophosphine	PF_3CN		A4	747
195.	Difluoroethane	$C_2H_4F_2$		A5	761
196.	Difluoroethene	$C_2H_2F_2$		A5	759
197.	**Difluoromethane**	CH_2F_2	**V-58**		359
198.	**Digermane**	Ge_2H_6	**VIII-100**		537
199.	Diiodobenzene	$C_6H_4I_2$		A4	747
200.	9,10-Dimethylanthracene	C_6H_{14}		A5	764
201.	Diiodomethane	CH_2I_2		A5	757
202.	Dinitrogen pentoxide	N_2O_5		A4, A5	747
203.	**Disilane**	Si_2H6	**VIII-100**		537
E					
204.	**Ethanal**	C_2H_4O	**VII-87**		489
205.	**Ethane**	C_2H_6	**VIII-101**		541
206.	**Ethene [S10]**	C_2H_4	See ethylene		
207.	Ethanol	C_2H_6O	**VI-80, XII-129**		461, 673
208.	Ethyl acetate	$C_4H_8O_2$		A7	773
209.	**Ethyl chloride [S7]**	C_2H_5Cl	**VIII-95**	**A5**	523, 760
210.	**Ethylene [S10]**	C_2H_4	**VI-78**		451
211.	Ethylene carbonate	$C_3H_4O_3$		A4	744
212.	Ethylene carbonate-d4	$C_3D_4O_3$		A4	744
213.	Ethyl formate	C_3H_6O2		A7	775

continued

	Name	Formula	Section-Chapter	Appendix	Page
F					
214.	**Fluorine**	F_2	**II-17**	**A5**	115, 753
215.	**Fluorobenzene**	C_6H_5F	**XII-125**	A4, A5	651, 747, 753
216.	Fluoroethane	C_2H_5F		A5	761
217.	**Fluoromethane**	CH_3F	**V-59**	**A5**	363, 753
218.	Fluorosulfonic acid	FHO_3S		A5	759
219.	Formaldehyde [S17]	CH_2O		A7	780
220.	**Formic acid**	CH_2O_2	**V-60**	A4	369, 747
221.	Fructose	$C_6H_{12}O_6$		A4	747
222.	Fullerene	C_{60}–C_{84}		A4	747
223.	Furan	C_4H_4O		A4	743
G					
224.	**Germane**	$\mathbf{GeH_4}$	**V-61**		371
225.	**Germanium tetrachloride [S18]**	$\mathbf{GeCl_4}$	**V-62**	A4	375
H					
226.	**Heavy water [S9]**	D_2O	**III-31**	**A5**	207, 754
227.	*n*-Heptane	C_7H_{16}		**A7**	780
228.	**Helium**	He	**I-3**		19
229.	**Hexachloroethane**	C_2Cl_6	**VIII-102**		553, 761
230.	**Hexafluorobenzene [S12]**	C_6F_6	**XII-126**	A5	655, 763
231.	**Hexafluoro-1,3-butadiene**	**1,3-**$\mathbf{C_4F_6}$	**X-114**	A5	599, 762
232.	**Hexafluoro-2-butyne**	**2-**$\mathbf{C_4F_6}$	**X-114**	A5	599, 762
233.	**Hexafluorocyclobutene**	c-C_4F_6	**X-114**	A5	599, 762
234.	**Hexafluoroethane**	C_2F_6	**VIII-103**	A5	555, 761
235.	**Hexafluoropropane [S13]**	C_3F_8	**XI-116**		609
236.	**Hexafluoropropene [S14]**	1-C_3F_6	**IX-110**	A5	579, 762
237.	**Hexafluoropropylene [S14]**	C_3F_6	**IX-110**	A5	579, 762
238.	**Hexane**	C_6H_{14}	**XIII-134, 112**		691
239.	Hydrogen	H		A6	769
240.	**Hydrogen**	$\mathbf{H_2}$	**II-18, 112**		121
241.	**Hydrogen bromide**	**HBr**	**II-19**	A4, A5	131, 747
242.	**Hydrogen chloride**	**HCl**	**II-20**	A5	135, 753
243.	**Hydrogen fluoride**	**HF**	**II-21**	A5	139, 753
244.	**Hydrogen iodide**	HI	**II-22**	A5	141, 753
245.	Hydrogen peroxide	H_2O_2		A4	747
246.	**Hydrogen sulfide**	H_2S	**III-32**	A5	211, 754
247.	Hydroxybenzene	C_6H_6O		A5	765
I					
248.	**Iodine**	I_2	**II-23**	A5	145, 753
249.	**Iodobenzene**	C_6H_5I	**XII-129**	A4, A5	661, 747, 763
250.	Iodoethane	C_2H_5I		A5	753
251.	**Iodomethane**	CH_3I	**V-63**	A5	377, 761
252.	Iodopentafluorobenzene	C_6F_5I		A5	763
253.	1-Iodopropane	1-C_3H_7I		A5	763
254.	2-Iodopropane	2-C_3H_7I		A5	763
255.	**Isobutane**	i-C_4H_{10}	**XIII-135**		695
256.	Isobutyl formate	i-$C_5H_{10}O_2$		A7	776
257.	**Isooctane**	C_8H_{18}	**XIII-136**		697
258.	Isopropyl acetate	i-$C_5H_{10}O_2$		A7	776
259.	Isopropyl formate	i-$C_4H_8O_2$		A7	776
K					
260.	**Krypton**	**Kr**	**I-4**		27
M					
261.	$MeCCl_3$ (Trichloroethane)	$(CH_3)CCl_3$		A4	747
262.	$MeCl_3Sn$ (Trichlorostannane)	$(CH_3)Cl_3Sn$		A4	747
263.	Me_2CCl_2 (Dichloropropane)	$(CH_3)_2CCl_2$		A4	747

continued

continued

	Name	Formula	Section-Chapter	Appendix	Page
264.	Me_2Cl_2Sn	$(CH_3)_2Cl_2Sn$		A4	747
265.	Me_3CCl (Chlorobutane)	$(CH_3)_3CCl$		A4	747
266.	Me_3ClSi (Trimethyl chlorosilane)	$(CH_3)_3ClSi$		A4	747
267.	Me_3ClSn (Chlorotrimethyl stannane)	$(CH_3)_3ClSn$		A4	747
268.	**Mercury**	**Hg**	**I-5**		37
269.	Methanal [S17]	CH_2O		A7	773
270.	**Methane**	CH_4	**V-64, 112**		381, 593
271.	**Methanol [S15]**	CH_3OH	**VI-80, XII-129**	**A5**	461, 673
272.	**Methanethiol**	CH_4S	**VI-79**	**79**	459
273.	1-Methoxy-2-nitrobenzene [S16]	$C_7H_7NO_3$		A4	743
274.	Methyl acetate	$C_3H_6O_2$		A7	773
275.	**Methylal**	C_3H_8O	**X-112**	112	593
276.	**Methyl alcohol [S15]**	CH_3OH	See methanol		
277.	**Methyl amine**	CH_5N	**VI-88**		491
278.	Methyl formate	$C_2H_4O_2$		A7	776
279.	Methyl iodide	CF_3I		A4	747
280.	Methyl vinyl ether	C_3H_6O		A4	747
281.	Molybdenum hexafluoride	MoF_6		A5	760
N					
282.	Naphthalene	$C_{10}H_8$		A5	764
283.	1,4-Naphthoquinone	$C_{10}H_6O_2$		A4, A5	748, 764
284.	**Neon**	**Ne**	**I-6**	**6**	43
285.	**Nitric oxide**	NO	II-24	**A5**	149
286.	2-Nitroanisole [S16]	$C_7H_7NO_3$		A4	748
287.	Nitrobenzene	$C_6H_5NO_2$		A4	748
288.	Nitroethane	$C_2H_5NO_2$		A4	748
289.	Nitrogen	N		A6	770
290.	**Nitrogen**	N_2	**II-25, 112**		159, 593
291.	**Nitrogen dioxide**	NO_2	**III-33**		215
292.	Nitrogen pentoxide	see dinitrogen pentoxide			
293.	Nitromethane	CH_3NO_2		A4, A5	748, 760
294.	2-Nitrophenol	$2\text{-}C_6H_5NO_3$		A4	748
295.	*m*-Nitrotoluene	$m\text{-}C_7H_7NO_2$		A4	748
296.	*o*-Nitrotoluene	$o\text{-}C_7H_7NO_2$		A4	748
297.	**Nitrogen trifluoride**	NF_3	**IV-43**	**A5**	285, 754
298.	**Nitrous oxide**	N_2O	**III**	**34**	219
O					
299.	**Octafluorocyclobutane [S18]**	$c\text{-}C_4F_8$	**XII-118**		663
300.	**Octane**	C_8H_{18}	**XIII-137**		699
301.	Oxygen	O		A6	770
302.	**Oxygen**	O_2	**II-26, 112**		169, 593
303.	**Ozone**	O_3	**III-35**		231
P					
304.	**Pentachloroethane**	C_2HCl_5	**VIII-104**	**A5**	563, 761
305.	Pentafluorobenzene	C_6HF_5		A5	763
306.	Pentafluorobutadiene	C_5F_8		A5	764
307.	Pentafluorodimethylether	C_2HF_5O		A5	762
308.	Pentafluoroiodobenzene	C_6F_5I		A5	763
309.	**Pentane**	C_5H_{12}	**XIII-138**		701
310.	2,3-pentanedione	$2,3\text{-}C_5H_8O_2$		A5	764
311.	**Perfluorobenzene [S12]**	C_6F_6	See Hexafluoriobenzene		
312.	Perfluoro-1,3-butadiene [S11]	$1,3\text{-}C_4F_6$		A4, A5	748, 753
313.	**Perfluorobutane.**	$n\text{-}C_4F_{10}$	**XIII-139**	A5	705
314.	**Perfluoro-2-butene**	$2\text{-}C_4F_8$	**XII-128**	A4, A5	663, 748, 753
315.	Perfluoro-2-butyne	$2\text{-}C_4F_6$		A4, A5	762, 753
316.	**Perfluorocyclobutane [S18]**	$c\text{-}C_4F_8$	**XII-128**	**A4**	663, 748

continued

	Name	Formula	Section-Chapter	Appendix	Page
317.	**Perfluorocyclobutene**	**c-C_4F_6**	X, A4, A5	**114**	599, 748, 753
318.	Perfluorocyclohexene	c-C_6F_{10}		A4, A5	749, 764
319.	Perfluorocyclopentane	c-C_5F_{10}		A5	749
320.	Perfluorocyclopentene	c-C_5F_8		A4, A5	764
321.	Perfluoro-1,2-dimethyl-cyclobutane	c-C_6F_{12}		A5, A8	764
322.	Perfluorodimethyl ether	C_2F_6O		A5	762
323.	Perfluoro-1,3-dimethylcyclohexane	C_8F_{16}		A4, A5	748
324.	**Perfluoroethane**	C_2F_6	**VIII-103**		555
325.	Perfluorohexane	n-C_6F_{14}		A5	764
326.	Perfluoro-1-heptene	1-C_7F_{14}		A5	764
327.	Perfluoroisobutane	i-C_4F_{10}		A5	764
328.	**Perfluoroisobutene**	i-C_4F_8	**XII-128**		663
329.	Perfluoromethylcyclohexane	c-C_7F_{14}		A5	764
330.	Perfluoronaphthelene	$C_{10}F_8$		A5	765
331.	Perfluoroneopentane	neo-C_5F_{12}			765
332.	Perfluoropentane	n-C_5F_{12}		A5	765
333.	**Perfluoropropane [S13]**	**C_3F_8**	See Hexafluoropropane		
334.	**Perfluoropropylene [S14]**	1-C_3F_6	**IX-110**	A5	579, 762
335.	Perfluoropyridine	C_5F_5N		A5,	763
336.	Perfluorotoluene	C_7F_8		A5	765
337.	Phenanthrene	$C_{14}H_{10}$		A5	765
338.	Phenol (Hydroxybenzene)	C_6H_6O		A5	765
339.	**Phosphine**	PH_3	**IV-44**		291
340.	Phosphorous (III) chloride	PCl_3		A5	759
341.	**Phosphorous (III) fluoride**	PF_3	**IV-45, A5**	A5	293
342.	Phosphorous (V) fluoride	PF_5		A5	759
343.	Phosphoric trichloride	$POCl_3$		A5	757
344.	Platinum hexafluoride	PtF_6	**VII-92**		511
345.	**Potassium**	**K**	**I-7**		53
346.	**Propane**	C_3H_8	**XI-117**		617
347.	**1-Propanol (Propyl alcohol)**	1-C_3H_8O	**VI-80, XII-129**	A5	461, 673, 753
348.	**2-Propanol (1sopropyl alcohol)**	2-C_3H_8O	**XII-129**		673
349.	n-Propyl acetate	n-$C_5H_{10}O_2$		A7	776
350.	**Propylene (Propene)**	C_3H_6	**IX-111**		583
351.	n-Propyl formate	n-$C_4H_8O_2$		A7	776
352.	**Propyne**	C_3H_4	**VII-89**		493
353.	Pyrene	$C_{16}H_{10}$		A5	765
354.	Pyridine	C_5H_5N		A5	763
Q					
355.	Quinoline	C_9H_7N		A5	765
R					
356.	Rhenium hexaflouride	ReF_6		A5	760
S				92	
357.	Selenium hexafluoride	SeF_6	**VII-92**	A5	511
358.	**Silane**	SiH_4	**V-65**		393
359.	Silane-d_4	SiD_4	**V-65**		393
360.	Silicon	Si		A6	771
361.	**Silicon tetrachloride [S19]**	$SiCl_4$	**V-69**	A4	407, 749
362.	**Silicon tetrafluoride**	SiF_4	**V-66**		399
363.	**Sodium**	**Na**	**I-8**	A4	55, 749
364.	Sodium bromide	NaBr		A4	749
365.	Sodium chloride	Nacl		A4	749
366.	Sodium fluoride	NaF		A4	749
367.	Sodium iodide	NaI		A4	749
368.	**Sulfurhexafluoride**	**SF_6**	**VII-90**	A5	497, 760
369.	Sulfur pentafluorochloride	SF_5Cl		A4, A5	749, 760

continued

continued

	Name	Formula	Section-Chapter	Appendix	Page
370.	Succinic anhydride	$C_4H_4O_3$		A5	763
371.	**Sulfur dioxide**	SO_2	**III-36**		239
372.	Sulfuric acid	H_2SO_4		A5	760
373.	Sulfur trioxide	SO_3		A5	754
374.	**Sulfuryl fluoride**	SO_2F_2	**V-67**	A4, A5	403, 749, 753
T					
375.	Tellurium hexafluoride	TeF_6	VII-92	A5, A6	511, 753, 769
376.	**1,1,2,2-Tetrabromoethane**	**1,1,2,2-$C_2H_2Br_4$**	**VIII-105**	A5	565
377.	**Tetrabromomethane**	CBr_4	**V-68**	A5	405
378.	**1,1,1,2-Tetrachloroethane**	1,1,1,2-$C_2H_2Cl_4$	**VIII-106**	A5	567
379.	**1,1,2,2-Tetrachloroethane**	1,1,2,2-$C_2H_2Cl_4$	**VIII-106**	A4, A5	567, 749
380.	**Tetrachloroethene**	C_2Cl_4	**VI-81**	A5	465
381.	**Tetrachlorogermane [S11]**	$GeCl_4$	**V-62**	A4	375
382.	**Tetrachloromethane [S4]**	CCl_4	See Carbon tetrachloride		
383.	**Tetrachlorosilane [S19]**	**$SiCl_4$**	**V-69**		407
384.	1,1,2,2- Tetrafluorodimethylether	$C_2H_2F_4O$		A5	762
385.	**Tetrafluoroethene [S20]**	C_2F_4	**VI-82**	A5	469, 759
386.	**Tetrafluoroethylene [S20]**	C_2F_4	See Tetrafluoroethene		
387.	**Tetrafluoromethane [S5]**	CF_4	**V-70**	A5	409
388.	Tetrafluoropyridine	C_5HF_4N		A5	763
389.	**Tetrahydrofuran**	C_4H_8O	**XIII-140**	A4	713, 750
390.	**Tetrahydrofuryl alcohol**	$C_5H_{10}O_2$	**XIII-140**		713
391.	Thionyl chloride	Cl_2OS	A5		754, 757
392.	Thymine	$C_5H_6N_2O_2$	A4		750
393.	Thymine (4D)	$C_5H_2D_4N_2O_2$	A4		750
394.	Tin (IV) chloride	$SnCl_4$	A4		750
395.	**Toluene**	C_7H_8	**XIII-141**	**A5**	715, 765
396.	**Tribromoethane**	**$C_2H_3Br_3$**	**XIII-107**	A5	569, 759
397.	**Tribromoethene**	**C_2HBr_3**	**VI-83**	A5	477, 759
398.	**Tribromofluoromethane**	CBr_3F	**VI-71**	A5	419
399.	**Tribromomethane**	**$CHBr_3$**	**V-72**	A5	421
400.	Trichloroacetonitrile	CCl_3CN		A7	773
401.	**1,1,1-Trichloroethane**	**1,1,1-$C_2H_3Cl_3$**	**VIII-108**	A4, A5	571, 750, 753
402.	**1,1,2-Trichloroethane**	**1,1,2-$C_2H_3Cl_3$**	**VIII-108**	A4, A8	571, 750,781
403.	**Trichloroethene**	**C_2HCl_3**	**VI-84**	A5	479
404.	**Trichlorofluoromethane**	CCl_3F	**V-73**	A4, A5	423, 750, 753
405.	**Trichloromethane [S8]**	**$CHCl_3$**	**V-74**	A4, A5	431, 750
406.	1,1,2-Trichloro-2-methylpropane	1,1,2-Cl_3-2-C_4H_8		A4, A5	750, 765
407.	1,2,3-Trichloropropane	1,2,3-$C_3H_5Cl_3$		A4, A5	750, 765
408.	1,1,1-Trichlorotrifluoroethane	1,1,1- $C_2Cl_3F_3$		A5	765
409.	1,1,2-Trichlorotrifluoroethane	1,1,2-$C_2Cl_3F_3$		A5	762
410.	Triflic acid	CHF_3SO_3		A4, A5	750, 762
411.	Triflic anhydride	$(CF_3SO_2)_2O$		A4	750
412.	Trifluoroacetonitrile	CF_3CN		A7	773
413.	Trifluorodimethylether	$C_2H_3F_3O$		A5	762
414.	2,2,2-Trifluorodimethylsulfide	$C_2H_3F_3S$		A5	762
415.	Trifluoroethane	$C_2H_3F_3$		A5	762
416.	Trifluoroiodomethane	CF_3I		A4	750
417.	*o*-Trifluoromethylbenzonitrile	*o*-$CF_3CNC_6H_4$		A5	765
418.	*m*-Trifluoromethylbenzonitrile	*m*-$CF_3CNC_6H_4$		A5	765
419.	*p*-Trifluoromethylbenzonitrile	*p*-$CF_3CNC_6H_4$		A5	765
420.	Trifluoromethyl sulfurpentafluoride	CF_3SF_5		A4, A5	750
421.	**Trifluorobromomethane [S21]**	$CBrF_3$	**V-49**		307
422.	**1,1,1-Trifluoroethane**	1,1,1-$C_2H_3F_3$	**VIII-109**	A5	575, 753
423.	**Trifluoromethane**	CHF_3	**V-75**	A5, A8	435, 753, 781
424.	**Tungsten hexafluoride**	**WF_6**	**VII-91**	A5	759, 753

continued

	Name	Formula	Section-Chapter	Appendix	Page
U					
425.	**Uranium hexafluoride**	$\mathbf{UF_6}$	**VII-92**		509, 760
426.	Urea	CH_4N_2O		A4	750
V					
427.	Vinyl chloride [S6]	C_2H_3Cl		A5	759
W					
428.	Water vapor	H_2O	**III-37**	A5	247, 753
X					
429.	**Xenon**	**Xe**	**I-9**		59
430.	Xenon difluoride	XeF_2		A5	754
		Addendum			
431.	Deuterated chloromethane	CD_3Cl	**V-52**		321
432.	Germanium	Ge		A6	776
433.	Sulfur	S		A6	771
434.	*s*-Amyl acetate	$s\text{-}C_7H_{14}O_2$		A7	777
435.	Sulfur chloride	S_2Cl_2		A10	787
436.	Dimethylamine	C_2H_7N		A10	787

Note: Bold font denotes major chapter. A = Appendix; S = Synonym.

Appendix 3: Target Particles (Formulawise)

	Formula	Name	Cross Reference to A2 (Numberwise)
1.	**Ar**	**Argon**	**14**
2.	**BCl_3**	**Boron trichloride**	23
3.	**BF_3**	**Boron trifluoride**	24
4.	BrCN	Cyanogen bromide	126
5.	**Br_2**	**Bromine**	25
6.	B_2H_6	Diborane	151
7.	C	Carbon	80
8.	C_{60}-C_{84}	Fullerene	222
9.	**$CBrCl_3$**	**Bromotrichloromethane**	62
10.	**$CBrF_3$**	**Bromotrifluoromethane [S2]**	64
11.	**$CBrF_3$**	**Trifluorobromomethane [S2]**	421
12.	CBr_2Cl_2	Dibromodichloromethane	153
13.	**CBr_2F_2**	**Dibromodifluoromethane**	155
14.	**CBr_3F**	**Tribromofluoromethane**	398
15.	**CBr_4**	**Tetrabromomethane**	377
16.	**$CClF_3$**	**Chlorotrifluoromethane**	**123**
17.	CClN	Chlorocyanogen	100
18.	**CCl_2F_2**	**Dichlorodifluoromethane**	**172**
19.	CCl_2O	Carbonyl chloride or Phosgene	**85**
20.	**CCl_3F**	**Trichlorofluoromethane**	404
21.	**CCl_4**	**Carbon tetrachloride [S4]**	86
22.	**CCl_4**	**Tetrachloro methane [S4]**	382
23.	**CD_4**	**Deuterated methane**	143
24.	CF_3I	Trifluoroiodomethane	416
25.	CF_3I	Methyl iodide	279
26.	CF_3NP	Difluorocyanophosphine	194
27.	CF_3SF_5	Trifluoromethyl sulfurpentafluoride	420
28.	**CF_4**	**Carbon tetrafluoride [S5]**	87
29.	**CF_4**	**Tetrafluoromethane [S5]**	387
30.	**$CHBr_2Cl$**	**Dibromochloromethane**	101
31.	**$CHBr_3$**	**Tribromomethane**	399
32.	$CHClF_2$	Chlorodifluoromethane	103
33.	$CHCl_2Br$	Bromodichloromethane	39
34.	$CHCl_2F$	Dichlorofluoromethane	178
35.	**$CHCl_3$**	**Trichloromethane [S8]**	405
36.	**$CHCl_3$**	**Chloroform [S8]**	108
37.	**CHF_3**	**Trifluoromethane**	423
38.	CHF_3SO_3	Triflic acid	410
39.	**CH_2BrCl**	**Bromochloromethane**	34
40.	**CH_2Br_2**	**Dibromomethane**	**159**
41.	CH_2ClF	Chlorofluoromethane	107
42.	**CH_2Cl_2**	**Dichloromethane**	180
43.	**CH_2F_2**	**Difluoromethane**	197
44.	CH_2I_2	Diiodomethane	201
45.	CH_2O	Methanal [S17]	269
46.	CH_2O	Formaldehyde [S17]	219
47.	**CH_2O_2**	**Formic acid**	220
48.	**CH_3Br**	**Bromomethane**	47

continued

continued

	Formula	Name	Cross Reference to A2 (Numberwise)
49.	**CH_3Cl**	**Chloromethane**	111
50.	CH_3Cl_3Sn	$MeCl_3Sn$	262
51.	**CH_3F**	**Fluoromethane**	217
52.	**CH_3I**	**Iodomethane**	251
53.	CH_3NO_2	Nitromethane	293
54.	**CH_3OH**	**Methanol [S15]**	271
55.	**CH_3OH**	**Methyl alcohol [S15]**	276
56.	**CH_4**	**Methane**	270
57.	CH_4N_2O	Urea	426
58.	**CH_4S**	**Methanethiol**	272
59.	**CH_5N**	**Methyl amine**	277
60.	**CO**	**Carbon monoxide**	83
61.	**COS**	**Carbon oxysulfide [S6]**	84
62.	**COS**	**Carbonyl sulfide [S6]**	88
63.	**CO_2**	**Carbon dioxide**	81
64.	**Cs**	**Cesium**	89
65.	**CS_2**	**Carbon disulfide**	82
66.	**$C_2Br_2F_4$**	**Dibromotetrafluoroethane**	160
67.	1,1-$C_2Br_2F_4$	Dibromotetrafluoroethane	161
68.	1,2-$C_2Br_2F_4$	Dibromotetrafluoroethane	162
69.	2,2-$C_2Br_2F_4$	2,2-Dibromotetrafluoroethane	163
70.	C_2ClF_3	Chlorotrifluoroethene	122
71.	C_2ClF_5	Chloropentafluoroethane	114
72.	1,1-$C_2Cl_2F_4$	Dichlorotetrafluoroethane	188
73.	1,2-$C_2Cl_2F_4$	Dichlorotetrafluoroethane	189
74.	1,1,1-$C_2Cl_3F_3$	1,1,1-Trichloro-2,2,2-trifluoroethane	408
75.	1,1,2-$C_2Cl_3F_3$	1,1,2-Trichloro-1,2,2-trifluoroethane	409
76.	C_2Cl_3N	Trichloroacetonitrile	400
77.	**C_2Cl_4**	**Tetrachloroethene**	380
78.	**C_2Cl_6**	**Hexachloroethane**	229
79.	**C_2D_6**	Deuterated ethane	142
80.	C_2F_3N	Trifluoroacetonitrile	412
81.	**C_2F_4**	**Tetrafluoroethene [S20]**	385
82.	**C_2F_4**	**Tetrafluoroethylene [S20]**	386
83.	C_2F_5Cl	Chloropentafluoroethane	114
84.	**C_2F_6**	**Hexafluoroethane**	234
85.	**C_2F_6**	**Perfluoroethane**	324
86.	C_2F_6O	Perfluorodimethyl ether	322
87.	C_2F_6S	Bis(trifluoromethyl)sulfide	22
88.	$(CF_3SO_2)_2O$	Triflic anhydride	411
89.	$C_2HBrClF_3$	1-Chloro-1-bromo-2,2,2-trifluoroethane	96
90.	**C_2HBr_3**	**Tribromoethene**	396
91.	**C_2HCl_3**	**Trichloroethene**	403
92.	**C_2HCl_5**	**Pentachloroethane**	304
93.	C_2HF_5O	Pentafluorodimethylether	307
94.	**C_2H_2**	**Acetylene**	5
95.	**$C_2H_2BrF_3$**	**Bromotrifluoroethane**	63
96.	C_2H_2BrN	Bromoacetonitrile	26
97.	**$C_2H_2Br_2$**	**Dibromoethene**	158
98.	**$C_2H_2Br_2F_2$**	**1,2-Dibromo-1,1-difluoroethane**	98
99.	**1,1,2,2-$C_2H_2Br_4$**	**1,1,2,2-Tetrabromoethane**	99
100.	C_2H_2ClN	Chloroacetonitrile	94
101.	**1,1-$C_2H_2Cl_2$**	**1,1-Dichloroethene**	175
102.	*cis*-$C_2H_2Cl_2$	*cis*-Dichloroethene	176
103.	***trans*-$C_2H_2Cl_2$**	***trans*-Dichloroethene**	177

continued

	Formula	Name	Cross Reference to A2 (Numberwise)
104.	**1,1,1,2-$C_2H_2Cl_4$**	**1,1,1,2-Tetrachloroethane**	378
105.	**1,1,2,2-$C_2H_2Cl_4$**	**1,1,2,2-Tetrachloroethane**	379
106.	$C_2H_2F_2$	Difluoroethene	196
107.	$C_2H_2F_4O$	1,1,2,2-Tetrafluorodimethylether	384
108.	C_2H_3Br	Bromoethene	42
109.	C_2H_3N	Acetonitrile	3
110.	C_2H_3Cl	Chloroethylene [S6]	104
111.	C_2H_3Cl	Chloroethene [S6]	106
112.	C_2H_3Cl	Vinyl chloride [S20]	427
113.	$C_2H_3ClF_2$	1-Chloro-1,1-difluoroethane	102
114.	$C_2H_3Cl_3$	MeCCl3 orTrichloroethane	261
115.	**1,1,1-$C_2H_3Cl_3$**	**1,1,1-Trichloroethane**	401
116.	**1,1,2-$C_2H_3Cl_3$**	**1,1,2-Trichloroethane**	402
117.	$C_2H_3F_3$	Trifluoroethane	415
118.	**$C_2H_3F_3$**	**1,1,1-Trifluoroethane**	**422**
119.	$C_2H_3F_3O$	Trifluorodimethylether	413
120.	$C_2H_3F_3S$	2,2,2-Trifluorodimethylsulfide	414
121.	C_2H_3N	Acetonitrile	3
122.	**C_2H_4**	**Ethylene [S10]**	214
123.	**C_2H_4**	**Ethene [S10]**	206
124.	1,2-C_2H_4BrCl	1,2-Bromochloroethane	33
125.	**1,2-C_2H_4BrF**	**1,2-Bromofluoroethane**	43
126.	**1,1-$C_2H_4Br_2$**	**1,1-Dibromoethane**	156
127.	**1,2-$C_2H_4Br_2$**	**1,2-Dibromoethane**	157
128.	**$C_2H_4Cl_2$**	**1,1-Dichloroethane**	173
129.	1,2-$C_2H_4Cl_2$	1,2-Dichloroethane	174
130.	$C_2H_4F_2$	Difluoroethane	195
131.	**C_2H_4O**	**Ethanal**	204
132.	**C_2H_4O**	**Acetaldehyde**	1
133.	$C_2H_4O_2$	Methyl formate	278
134.	C_2H_5Br	Bromoethane	41
135.	**C_2H_5Cl**	**Chloroethane [S7]**	105
136.	**C_2H_5Cl**	**Ethyl chloride [S7]**	209
137.	C_2H_5F	Fluoroethane	216
138.	C_2H_5I	Iodoethane	250
139.	$C_2H_5NO_2$	Nitroethane	288
140.	C_2H_5OH	Ethanol	207
141.	**C_2H_6**	**Ethane**	205
142.	$C_2H_6Cl_2Sn$	Me_2Cl_2Sn	264
143.	C_2H_6O	Ethanol	207
144.	C_2N_2	Cyanogen	125
145.	$C_3D_4O_3$	Ethylene carbonate-d4	212
146.	**C_3F_6**	**Hexafluoropropene [S14]**	236
147.	**C_3F_6**	**Perfluoropropylene [S14]**	334
148.	**C_3F_8**	**Perfluoropropane [S13]**	333
149.	**C_3F_8**	**Hexafluoropropane [S13]**	235
150.	**C_3H_4**	**Allene**	7
151.	**C_3H_4**	**Cyclopropene**	138
152.	**C_3H_4**	**Propyne**	352
153.	$C_3H_4O_3$	Ethylene carbonate	211
154.	2-C_3H_5Br	2-Bromopropene	55
155.	3-C_3H_5Br	3-Bromopropene	56
156.	*cis*-C_3H_5Br	*cis*-1-Bromopropene	58
157.	*trans*-C_3H_5Br	*trans*-1-Bromopropene	59
158.	C_3H_5ClO	Chloroacetone	93

continued

continued

	Formula	Name	Cross Reference to A2 (Numberwise)
159.	1,2,3-Trichloro propane	1,2,3-$C_3H_5Cl_3$	407
160.	**C_3H_6**	**Propene or propylene**	350
161.	**c-C_3H_6**	**cyclopropane**	137
162.	1,2-C_3H_6BrCl	1,2-Bromochloropropane	35
163.	1,3-C_3H_6BrCl	1,3-Bromochloropropane	36
164.	$C_3H_6Cl_2$	$MeCCl_2$ (Dichloropropane)	263
165.	1,1-$C_3H_6Cl_2$	1,1-Dichloropropane	184
166.	1,2-$C_3H_6Cl_2$	1,2-Dichloropropane	185
167.	1,3-$C_3H_6Cl_2$	1,3-Dichloropropane	186
168.	2,2-$C_3H_6Cl_2$	2,2-Dichloropropane	187
169.	**C_3H_6O**	**Acetone**	2
170.	C_3H_6O	Methyl vinyl ether	280
171.	$C_3H_6O_2$	Ethyl formate	213
172.	$C_3H_6O_2$	Methyl acetate	208
173.	*n*-C_3H_7Br	Bromopropane	53
174.	1-C_3H_7Br	1-Bromopropane	54
175.	2-C_3H_7Br	2-Bromopropane	55
176.	**1-C_3H_7Cl**	**1-Chloropropane**	115
177.	**2-C_3H_7Cl**	**2-Chloropropane**	116
178.	1-C_3H_7I	1-Iodopropane	253
179.	2- C_3H_7I	2-Iodopropane	254
180.	1-C_3H_7OH	**1-Propanol (Propyl alcohol)**	347
181.	**2-C_3H_7OH**	**2-Propanol (isopropyl alcohol)**	348
182.	**C_3H_8**	**Propane**	346
183.	**$C_3H_8O_2$**	**Methylal**	275
184.	C_3H_9ClSi	Me_3ClSi (Trimethylchlorosilane)	266
185.	C_3H_9ClSn	Me_3ClSn (Chlorotrimethylstannane)	267
186.	$C_3H_{10}O$	Benzophenone	20
187.	**C_4F_6**	**Hexafluorocyclobutene**	233
188.	**c-C_4F_6**	**Perfluorocyclobutene**	317
189.	**1,3-C_4F_6**	**Hexafluoro-1,3-butadiene [S11]**	230
190.	**1,3-C_4F_6**	**Perfluoro-1,3-butadiene [S11]**	312
191.	**2-C_4F_6**	**Perfluoro-2-butyne**	315
192.	**C_4F_8**	**Perfluoro-2-butene**	314
193.	**c-C_4F_8**	**Perfluorocyclobutane [S18]**	316
194.	**c-C_4F_8**	**Octafluorocyclobutane [S18]**	299
195.	***i*-C_4F_8**	**Perfluoroisobutene**	328
196.	***n*-C_4F_{10}**	**Perfluorobutane**	313
197.	*i*-C_4F_{10}	Perfluoroisobutane	327
198.	C_4H_4O	Furan	223
199.	$C_4H_4O_3$	Succinic anhydride	370
200.	**C_4H_6**	**Cyclobutene**	129
201.	**2-C_4H_6**	**2-Butyne**	78
202.	**1,3-C_4H_6**	**1,3-Butadiene**	65
203.	$C_4H_6O_2$	γ-butyrolactone	79
204.	2,3-$C_4H_6O_2$	2,3-Butanedione	67
205.	C_4H_7Br	4-Bromo-1-butene	31
206.	**C_4H_8**	**Cyclobutane**	128
207.	**C_4H_8**	**Butene**	68
208.	$C_4H_8Cl_2$	1,2-Dichloro-2-methylpropane	181
209.	1,3-$C_4H_8Cl_2$	1,3-Dichlorobutane	168
210.	1,4-$C_4H_8Cl_2$	1,4-Dichlorobutane	169
211.	2,3-$C_4H_8Cl_2$	2,3-Dichlorobutane	170
212.	$C_4H_8Cl_3$	1,1,2-Trichloro-2-methylpropane	406

continued

	Formula	Name	Cross Reference to A2 (Numberwise)
213.	**C_4H_8O**	**Tetrahydrofuran**	389
214.	$C_4H_8O_2$	Ethyl acetate	208
215.	*i*-$C_4H_8O_2$	Isopropyl formate	259
216.	*n*-$C_4H_8O_2$	*n*-propyl formate	351
217.	C_4H_9Br	2-Bromo-2-methylpropane	49
218.	1-C_4H_9Br	1-Bromobutane	29
219.	2-C_4H_9Br	2-Bromobutane	30
220.	*n*-C_4H_9Br	Bromobutane	28
221.	C_4H_9Cl	Me_3CCl (Chlorobutane)	265
222.	**1-C_4H_9Cl**	**1-Chlorobutane**	97
223.	**2-C_4H_9Cl**	**2-Chlorobutane**	98
224.	***t*-C_4H_9Cl**	***t*-Butyl chloride [S3]**	76
225.	***t*-C_4H_9Cl**	***t*-Chlorobutane [S3]**	99
226.	**C_4H_{10}**	**Butane**	66
227.	***i*-C_4H_{10}**	**Isobutane**	255
228.	*i*-$C_4H_{10}O$	*i*-Butyl alcohol (2-Methyl-1-Propanol)	74
229.	*i*-$C_4H_{10}O$	2-Methyl-2-Propanol (*t*-Butyl alcohol)	75
230.	*n*-$C_4H_{10}O$	Butyl alcohol (1-butanol)	73
231.	***t*-$C_4H_{10}O$**	*t*-Butyl alcohol	75
232.	**$C_4H_{10}O$**	**Diethyl ether**	190
233.	C_5F_8	Pentafluorobutadiene	306
234.	C_5F_5N	Perfluoropyridine	335
235.	c-C_5F_8	Perfluorocyclopentene	317
236.	c-C_5F_{10}	Perfluorocyclopentane	319
237.	*n*-C_5F_{12}	Perfluoropentane	332
238.	neo-C_5F_{12}	Perfluoroneopentane	331
239.	C_5HF_4N	Tetrafluoropyridine	388
240.	C_5H_5N	Pyridine	354
241.	$C_5H_2D_4N_2O_2$	Thymine-d4	393
242.	$C_5H_6N_2O_2$	Thymine	392
243.	C_5H_8O	Cyclopentanone	134
244.	2,3-C_5H_8O	2,3-Pentanedione	310
245.	C_5H_9Br	5-Bromo-1-pentene	52
246.	C_5H_9Br	Cyclopentyl bromine	135
247.	C_5H_9Cl	Cyclopentylchloride	136
248.	**C_5H_{10}**	**Cyclopentane**	133
249.	1,5-$C_5H_{10}Cl_2$	1.5-Dichloropentane	183
250.	$C_5H_{10}O_2$	**Tetrahydrofuryl alcohol**	390
251.	*i*-$C_5H_{10}O_2$	Isobutyl formate	256
252.	*i*-$C_5H_{10}O2$	Isopropyl acetate	258
253.	*n*-$C_5H_{10}O_2$	*n*-Butyl formate	77
254.	*n*-$C_5H_{10}O_2$	*n*-Propyl acetate	349
255.	*n*-$C_5H_{11}Br$	Bromodimethylpropane	40
256.	**1-$C_5H_{11}Cl$**	**1-Chloropentane**	119
257.	**C_5H_{12}**	**Pentane**	309
258.	C_6BrD_5	Deuterated bromobenzene	141
259.	C_6BrF_5	Bromopentafluorobenzene	51
260.	**C_6ClF_5**	**Chloropentafluorobenzene**	113
261.	C_6F_5I	Pentafluoroiodobenzene	308
262.	**C_6D_6**	**Deuterated benzene**	140
263.	C_6F_5I	Iodopentafluorobenzene	252
264.	**C_6F_6**	**Hexafluorobenzene [S12]**	232
265.	**C_6F_6**	**Perfluorobenzene [S12]**	311
266.	c-C_6F_{10}	Perfluorocyclohexene	318
267.	c-C_6F_{12}	Perfluoro-1,2-dimethylcyclobutane	321

continued

continued

	Formula	Name	Cross Reference to A2 (Numberwise)
268.	C_6HF_5	Pentafluorobenzene	305
269.	$C_6H_4Br_2$	Dibromo benzene	152
270.	1,4-C_6H_4BrCl	1-Bromo-4-chlorobenzene	32
271.	$C_6H_4Cl_2$	Dichlorobenzene	164
272.	*m*-$C_6H_4Cl_2$	*m*-Dichlorobenzene	165
273.	*o*-$C_6H_4Cl_2$	*o*-Dichlorobenzene	167
274.	*p*-$C_6H_4Cl_2$	*p*-Dichlorobenzene	166
275.	**$C_6H_4F_2$**	**Difluorobenzene**	191
276.	**1,3-$C_6H_4F_2$**	**1,3-Difluorobenzene**	192
277.	**1,4-$C_6H_4F_2$**	**1,4-Difluorobenzene**	193
278.	$C_6H_4I_2$	Diiodobenzene	199
279.	**C_6H_5Br**	**Bromobenzene**	27
280.	**C_6H_5Cl**	**Chlorobenzene**	95
281.	**C_6H_5F**	**Fluorobenzene**	215
282.	**C_6H_5I**	**Iodobenzene**	249
283.	$C_6H_5NO_2$	Nitrobenzene	287
284.	2-$C_6H_5NO_3$	2-Nitrophenol	284
285.	**C_6H_6**	**Benzene**	18
286.	**C_6H_6O**	Phenol	338
287.	**C_6H_6O**	Hydroxy benzene	247
288.	C_6H_7N	Aniline	13
289.	$C_6H_{10}Cl_2$	*Trans*-1-2-Dichlorocyclohexane	171
290.	$C_6H_{11}Cl$	Cyclohexyl chloride	131
291.	**C_6H_{12}**	**Cyclohexane**	130
292.	1,6-$C_6H_{12}Cl_2$	1,6-Dichlorohexane	179
293.	*i*-$C_6H_{12}O_2$	*i*-Butyl acetate	69
294.	*n*-$C_6H_{12}O_2$	*n*-Amyl formate	11
295.	*n*-$C_6H_{12}O_2$	*n*-Butyl acetate	70
296.	*s*-$C_6H_{12}O_2$	*s*-Butyl acetate	71
297.	*t*-$C_6H_{12}O_2$	*t*-Butyl acetate	72
298.	$C_6H_{12}O_6$	Fructose	221
299.	$C_6H_{13}Br$	Bromohexane	46
300.	1-$C_6H_{13}Cl$	1-Chlorohexane	110
301.	C_6H_{14}	9,10-Dimethylanthracene	200
302.	**C_6H_{14}**	**Hexane**	238
303.	*n*-C_6F_{14}	Perfluorohexane	325
304.	C_7F_8	Perfluorotoluene	336
305.	1-C_7F_{14}	Perfluoro-1-heptane	326
306.	c-C_7F_{14}	Perfluoromethyl-cyclohexane	329
307.	C_7H_7Cl	Chloromethylbenzene	112
308.	$C_7H_7NO_3$	2-Nitroanisole [S16]	286
309.	$C_7H_7NO_3$	1-Methoxy-2-nitrobenzene [S16]	273
310.	C_8F_{16}	Perfluoro-1,3-dimethylcyclohexane	323
311.	C_7H_6O	Benzaldehyde	15
312.	C_7H_7Br	Bromomethylbenzene	48
313.	$C_7H_6O_2$	Benzoic acid	19
314.	*o*-C_7H_7Br	*o*-Bromotoluene	61
315.	*o*-C_7H_7Cl	*o*-Chlorotoluene	120
316.	*m*-$C_7H_7NO_2$	*m*-Nitrotoluene	295
317.	*o*-$C_7H_7NO_2$	*o*-Nitrotoluene	296
318.	**C_7H_8**	**Toluene**	395
319.	*i*-$C_7H_{14}O_2$	Isoamyl acetate	9
320.	*n*-$C_7H_{14}O_2$	*n*-Amyl acetate	10
321.	1-$C_7H_{15}Cl$	1-Chloroheptane	109
322.	*n*-C_7H_{16}	*n*-Heptane	227
323.	*m*-$C_8H_4F_3N$	*m*-Trifluoromethylbenzonitrile	417
324.	*o*-$C_8H_4F_3N$	*o*-Trifluoromethylbenzonitrile	418

continued

	Formula	Name	Cross Reference to A2 (Numberwise)
325.	p-$C_8H_4F_3N$	*p*-Trifluoromethylbenzonitrile	419
326.	1,8-$C_8H_6Cl_2$	1,8-Dichlorooctane	182
327.	C_8H_8	Cyclooctatetraene	132
328.	C_8H_8O	Acetophenone	4
329.	C_8H_9Br	Bromoethylbenzene	44
330.	2-C_8H_9Br	2-Bromoethylbenzene	45
331.	n-$C_8H_{17}Br$	Bromooctane	50
332.	1-$C_8H_{17}Cl$	1-Chlorooctane	118
333.	$\mathbf{C_8H_{18}}$	**Isooctane**	257
334.	$\mathbf{C_8H_{18}}$	**Octane**	300
335.	C_9H_7N	Quinoline	355
336.	C_9H_9Br	Bromocyclohexene	37
337.	$C_9H_{11}Br$	3-Bromopropylbenzene	60
338.	$C_9H_{19}Cl$	1-Chlorononane	117
339.	$C_{10}F_8$	Perfluoronaphthelene	330
340.	$C_{10}H_6O_2$	1,4-Naphthoquinone	283
341.	$C_{10}H_8$	Naphthalene	282
342.	n-$C_{10}H_{21}Br$	Bromodecane	38
343.	$C_{14}H_{10}$	Anthracene	12
344.	$C_{14}H_{10}$	Phenanthrene	337
345.	$C_{16}H_{10}$	Pyrene	353
346.	$C_{18}H_{12}$	Benz[a]anthracene [S1]	16
347.	$C_{18}H_{12}$	1,2-Benzanthracene [S1]	17
348.	$C_{18}H_{12}$	Chrysene	124
349.	$C_{20}H_{12}$	Benzo[a]pyrene	21
350.	$C_{22}H_{14}$	Dibenz[a,h]anthracene	150
351.	$ClCN$	Cyanogenchloride	127
352.	$ClNO_3$	Chlorine nitrate	92
353.	$\mathbf{ClO_2}$	**Chlorine dioxide**	91
354.	$\mathbf{Cl_2}$	**Chlorine**	90
355.	Cl_2OS	Thionyl chloride	391
356.	$\mathbf{D_2}$	**Deuterium**	145
357.	**DBr**	**Deuterium bromide**	146
358.	**DCl**	**Deuterium chloride**	147
359.	**DI**	**Deuterium iodide**	148
360.	$\mathbf{D_2O}$	**Deuterated water [S 17]**	144
361.	$\mathbf{D_2O}$	**Heavy water [S 17]**	226
362.	D_2S	Deuterium sulfide	149
363.	FHO_3S	Fluorosulfonic acid	218
364.	$\mathbf{F_2}$	**Fluorine**	214
365.	$\mathbf{GeCl_4}$	**Tetrachlorogermane [S 11]**	381
366.	$\mathbf{GeCl_4}$	**Germanium tetrachloride [S11]**	225
367.	$\mathbf{GeH_4}$	**Germane**	224
368.	Ge_2H_6	Digermane	198
369.	H	Hydrogen	239
370.	**HBr**	**Hydrogen bromide**	241
371.	**HCl**	**Hydrogen chloride**	242
372.	**HF**	**Hydrogen fluoride**	243
373.	**Hg**	**Mercury**	268
374.	**HI**	**Hydrogen iodide**	244
375.	$\mathbf{H_2}$	**Hydrogen**	240
376.	$\mathbf{H_2O}$	**Water**	428
377.	H_2O_2	Hydrogen peroxide	245
378.	$\mathbf{H_2S}$	**Hydrogen sulphide**	246
379.	H_2SO_4	Sulfuric acid	372
380.	**He**	**Helium**	228
381.	$\mathbf{I_2}$	**Iodine**	248

continued

continued

	Formula	Name	Cross Reference to A2 (Numberwise)
382.	**K**	**Potassium**	345
383.	**Kr**	**Krypton**	260
384.	MoF_6	Molybdynum hexafluoride	281
385.	Nitrogen	N	289
386.	$\mathbf{N_2}$	**Nitrogen**	290
387.	**Na**	**Sodium**	363
388.	NaBr	Sodium bromide	364
389.	NaCl	Sodium chloride	365
390.	NaF	Sodium fluoride	366
391.	NaI	Sodium iodide	367
392.	$\mathbf{ND_3}$	**Deurerated ammonia**	139
393.	**Ne**	**Neon**	284
394.	$\mathbf{NF_3}$	**Nitrogen trifluoride**	297
395.	$\mathbf{NH_3}$	**Ammonia**	8
396.	**NO**	**Nitric oxide**	285
397.	$\mathbf{NO_2}$	**Nitrogen dioxide**	291
398.	$\mathbf{N_2O}$	**Nitrous oxide**	298
399.	$\mathbf{N_2O_5}$	Dinitrogen pentoxide	202
400.	$\mathbf{N_2O_5}$	Nitrogen pentoxide	See dinitrogen pentoxide
401.	**O**	Oxygen	301
402.	$\mathbf{O_2}$	**Oxygen**	302
403.	$\mathbf{O_3}$	**Ozone**	303
404.	PCl_3	Phosphorous (III) chloride	340
405.	$\mathbf{PF_3}$	**Phosphorous trifluoride**	341
406.	PF_5	Phosphorous (V)fluoride	342
407.	$\mathbf{PH_3}$	**Phosphine**	339
408.	$POCl_3$	Phosphoric trichloride	343
409.	PtF_6	Platinum hexafluoride	344
410.	ReF_6	Rhenium hexafluoride	356
411.	SF_5Cl	Sulfur pentafluorochloride	369
412.	$\mathbf{SF_6}$	**Sulfur hexafluoride**	368
413.	SeF_6	Selenium hexafluoride	357
414.	Silicon	Si	360
415.	$\mathbf{SiCl_4}$	**Tetrachlorosilane [S19]**	383
416.	$\mathbf{SiCl_4}$	**Silicon tetrachloride [S19]**	361
417.	$\mathbf{Si\text{-}d_4}$	**Silane-d4**	359
418.	$\mathbf{SiF_4}$	**Silicon tetrafluoride**	362
419.	$\mathbf{SiH_4}$	**Silane**	358
420.	$\mathbf{Si_2H_6}$	**Disilane**	203
421.	$SnCl_4$	Tin (IV) chloride	394
422.	$\mathbf{SO_2}$	**Sulphur dioxide**	371
423.	$\mathbf{SO_2F_2}$	**Sulfuryl fluoride**	374
424.	SO_3	Sulfur trioxide	373
425.	TeF_6	Tellurium hexafluoride	375
426.	$\mathbf{UF_6}$	**Uranium hexafluoride**	425
427.	$\mathbf{WF_6}$	**Tungsten hexafluoride**	424
428.	**Xe**	**Xenon**	429
429.	XeF_2	Xenondifluoride	430
430.		**Air**	6
		Addendum	
431	$s\text{-}C_7H_{14}O_2$	*s*-Amyl acetate	434
432	CD_3Cl	Deuterated chloromethane	321
433	Ge	Germanium	432
434	S	Sulfur	433
435	S_2Cl_2	Sulfur chloride	435
436	C_2H_7N	Dimethylamine	436

Note: Bold font indicates major section. S = Synonym.

Appendix 4: Attachment Peaks and Cross Sections

CONTENTS

Attachment cross sections and peak energy of occurrence have been measured in a number of compounds that are considered before. The peak cross sections shown in the table are in units of 1.0×10^{-21} m^2. Cross sections due to Modelli and Jones (2004) are normalized to 1.5×10^{-21} m^2.

TABLE A4.1
Peak Attachment Energy and Cross Sections

Name	Formula	Peak Energy (eV)	Cross Section 10^{-21} m^2	Reference
Acetaldehyde	C_2H_4O	1.2		Naff (1972)
		3.8		
		6.35		
		6.62		
Acetone	C_3H_6O	1.6		Naff (1972)
		4.15		
		6.13		
		7.5		
		8.1		
Acetophenone	C_8H_8O	0.95		Naff (1972)
Benzaldehyde	C_7H_6O	0.72		Naff (1972)
1,2-Benzanthracene	$C_{18}H_{12}$	0.05	0.52	Reese et al. (1958)
Benzene	C_6H_6	1.09		Olthoff et al. (1985)
		4.85		Olthoff et al. (1985)
Benzophenone	$C_3H_{10}O$	0.75		Naff (1972)
Boron trichloride	BCl_3	0.4	2.9	Buchel'nikova (1959)
Bromobenzene	C_6H_5Br	0.66	4.13	Modelli (2005)
		0.67		Olthoff et al. (1985)
		4.4		Olthoff et al. (1985)
		0.84	9.6	Christophorou et al. (1966)
		0.3	1.2	Shimamori et al. (1995)
1-Bromobutane	$CH_3(CH_2)_3Br$	0.63	0.24	Modelli and Jones (2004)
2-Bromobutane	$CH_3CH_2CH(Br)CH_3$	0.85	1.245	Modelli and Jones (2004)
4-Bromo-1-butene	$H_2C{=}CH(CH_2)_2Br$	0.70	1.5	Modelli and Jones (2004)
1-Bromo-4-chlorobenzene	1,4-C_6H_4BrCl	0.3 (Br^-)		Rosa et al. (2001)
		0.45 (Cl^-)		
1,2-Bromochloroethane	1,2-C_2H_4BrCl	0.2 $(CHClBr)^-$		Pshenichnyuk et al. (2006)
		0.05 $(ClBr)^-$		Pshenichnyuk et al. (2006)
		0.2 $(M\text{-}Cl)^-$		Pshenichnyuk et al. (2006)
		0.2 (Br^-)		Pshenichnyuk et al. (2006)
		0.23 (Cl^-)		Pshenichnyuk et al. (2006)
1,2-Bromochloropropane	1,2-C_3H_6BrCl	0.0 (M^-)		Pshenichnyuk et al. (2006)
		0.27 $(M\text{-}Cl)^-$		

continued

TABLE A4.1 (continued)
Peak Attachment Energy and Cross Sections

Name	Formula	Peak Energy (eV)	Cross Section 10^{-21} m^2	Reference
		0.18 (M-HCl)$^-$		
		0.12 (BrHCl$^-$)		
		0.18 (BrCl)$^-$		
		0.31 (Br)$^-$		
		0.22 (M-Br)$^-$		
		0.31 Cl$^-$		
1,3-Bromochloropropane	1,3-C_3H_6BrCl	0.75 (Cl$^-$)		Pshenichnyuk et al. (2006)
		6.8 (Cl$^-$)		
		~8.8 (Cl$^-$)		
		0.35 (Br$^-$)		
		4.8 (Br$^-$)		
		7.8 (Br$^-$)		
		9.3 (Br$^-$)		
Bromocyclohexene	C_6H_9Br	0.41	20.24	Modelli and Jones (2004)
n-Bromodecane	*n*-$C_6H_{21}Br$	0.05	0.23	Reese et al. (1958)
Bromoethane	CH_3CH_2Br	0.0 (Br$^-$)		Pshenichnyuk et al. (2006)
		0.6	0.225	Modelli and Jones (2004)
		6.8		Pshenichnyuk et al. (2006)
		9.0		Pshenichnyuk et al. (2006)
Bromoethene	$H_2C{=}CHBr$	1.16	4.68	Modelli and Jones (2004)
Bromomethylbenzene	$C_6H_5CH_2Br$	0.9	12.71	Modelli (2005)
		0.20	42.39	Modelli (2005)
2-Bromoethylbenzene	$C_6H_5CH_2CH_2Br$	0.60	6.61	Modelli (2005)
2-Bromo-2-methylpropane	$(CH_3)_3CBr$	0.86	1.80	Modelli and Jones (2004)
5-Bromo-1-pentene	$H_2C{=}CH(CH_2)_3Br$	0.5	0.55	Modelli and Jones (2004)
1-Bromopropane	$CH_3(CH_2)_2Br$	0.0 (Br$^-$)		Pshenichnyuk et al. (2006)
		0.65	0.293	Modelli and Jones (2004)
		6.8		Pshenichnyuk et al. (2006)
		8.5		Pshenichnyuk et al. (2006)
2-Bromopropane	$CH_3CH(Br)CH_3$	0.0 (Br$^-$)		Pshenichnyuk et al. (2006)
		0.9	1.12	Modelli and Jones (2004)
		6.38		Pshenichnyuk et al. (2006)
		8.4		Pshenichnyuk et al. (2006)
2-Bromopropene	$H_2C{=}C(Br)CH_3$	1.1	0.33	Modelli and Jones (2004)
3-Bromopropene	$H_2C{=}CHCH_2Br$	0.39	28.9	Modelli and Jones (2004)
cis-1-Bromopropene	*cis*-$CH_3CH{=}CHBr$	1.51	3.87	Modelli and Jones (2004)
trans-1-Bromopropene	*trans*-$CH_3CH{=}CHBr$	1.31	1.35	Modelli and Jones (2004)
o-Bromotoluene	*o*-C_7H_7Br	0.63		Christophorou et al. (1966)
3-Bromopropylbenzene	$C_6H_5CH_2CH_2CH_2Br$	0.7	1.59	Modelli (2005)
t-Butyl chloride	*t*-C_4H_9Cl	1.55	31.8×10^{-2}	Pearl and Burrow (1994)
Carbon tetrachloride	CCl_4	0.00		Modelli et al. (1998)
		0.77		Modelli et al. (1998)
Chloroacetone	C_3H_5ClO	0.5		Naff (1972)
Chlorobenzene	C_6H_5Cl	0.7	0.55	
		0.73		Olthoff et al. (1985)
		2.5		Olthoff et al. (1985)
		4.50		Olthoff et al. (1985)
		0.86	1.4	Christophorou et al. (1966)
1-Chloro-1-bromo-2,2,2-trifluoroethane	$CHBrClF_3$	4.7 (M-HFBr)$^-$		Pshenichnyuk et al. (2006)
		0.06 Br$^-$		
		6.33		

TABLE A4.1 (continued)
Peak Attachment Energy and Cross Sections

Name	Formula	Peak Energy (eV)	Cross Section 10^{-21} m^2	Reference
		0.21 Cl^-		
		5.34		
		7.05 F^-		
1-Chlorobutane	1-C_4H_9Cl	1.54	1.65×10^{-2}	Pearl and Burrow (1994)
2-Chlorobutane	2-C_4H_9Cl	1.51	17.8×10^{-2}	Pearl and Burrow (1994)
Chlorodifluoromethane	$CHClF_2$	2.76		Aflatooni and Burrow (2001)
		6.82		Aflatooni and Burrow (2001)
Chloroethane	C_2H_5Cl	1.55	1.49×10^{-2}	Pearl and Burrow (1994)
		0.80	0.007	Christophorou et al. (1966)
Chlorofluoromethane	$CClFH_2$	4.17		Aflatooni and Burrow (2001)
1-Chloroheptane	$C_7H_{15}Cl$	1.47	2.13×10^{-2}	Pearl and Burrow (1994)
1-Chlorohexane	$C_6H_{13}Cl$	1.43	1.99×10^{-2}	Pearl and Burrow (1994)
Chloromethane	$CClH_3$	5.43		Aflatooni and Burrow (2001)
		0.8	$< 2 \times 10^{-4}$	Pearl and Burrow (1994)
1-Chlorononane	$C_9H_{19}Cl$	1.34	3.01×10^{-2}	Pearl and Burrow (1994)
1-Chlorooctane	$C_8H_{17}Cl$	1.45	2.30×10^{-2}	Pearl and Burrow (1994)
1-Chloropentane	$C_5H_{11}Cl$	1.47	2.08×10^{-2}	Pearl and Burrow (1994)
1-Chloropropane	1-C_3H_7Cl	1.50	1.17×10^{-2}	Pearl and Burrow (1994)
2-Chloropropane	2-C_3H_7Cl	1.58	27.6×10^{-2}	Pearl and Burrow (1994)
o-Chlorotoluene	*o*-C_7H_7Cl	1.10	2.2	Christophorou et al. (1966)
Chlorotrifluoromethane	$CClF_3$	2.31		Aflatooni and Burrow (2001)
		6.26		Aflatooni and Burrow (2001)
Cyanogen	C_2N_2	0.8–0.9 (CN^-)		Deng and Souda (2002)
		3.50		
		7.5		
Cyanogen bromide	BrCN	0.0 (CN^-)		Bruning et al. (1996)
		1.8		
		5.5		
Cyanogen chloride	ClCN	0.5 (CN^-)		Bruning et al. (1996)
		1.6		
		6.0		
Cyclohexyl bromine		0.83	0.45	Modelli and Jones (2004)
Cyclohexyl chloride	$C_6H_{11}Cl$	1.43	4.13×10^{-2}	Pearl and Burrow (1994)
Cyclopentyl bromine		0.85	1.755	Modelli and Jones (2004)
Cyclopentyl chloride	C_5H_9Cl	1.52	28.3×10^{-2}	Pearl and Burrow (1994)
Cyclopropyl bromine		0.86	1.755	Modelli and Jones (2004)
	$CH_3C{=}CHCH_2Br$	0.40	2.20	Modelli and Jones (2004)
	$H_2C{=}CH(CH_2)_4Br$	0.5	0.15	Modelli and Jones (2004)
Deuterated bromobenzene	C_6D_5Br	0.80	10.4	Christophorou et al. (1966)
Deuterium sulfide	D_2S	2.2	0.0068	Fiquet-Fayard et al. (1972)
	HDS		0.012	Fiquet-Fayard et al. (1972)
Dibromobenzene	$C_6H_4Br_2$	0.2		Olthoff et al. (1985)
		1.50		Olthoff et al. (1985)
		4.0		Olthoff et al. (1985)
1,2-Dibromoethane	1,2-$C_2H_4Br_2$	0.1 (Br^-)		Pshenichnyuk et al. (2006)
Dichlorobenzene	$C_6H_4Cl_2$	0.3		Olthoff et al. (1985)
		2.25		Olthoff et al. (1985)
		4.1		Olthoff et al. (1985)
o-Dichlorobenzene	*o*-$C_6H_4Cl_2$	0.36	43	Christophorou et al. (1966)
1,3-Dichlorobutane	1,3-$C_4H_8Cl_2$	1.07	1.20	Aflatooni et al. (1998), Aflatooni and Burrow (2000)
1,4-Dichlorobutane	$C_4H_8Cl_2$	~0	0.03	Aflatooni et al. (1998), Aflatooni and Burrow (2000)

continued

TABLE A4.1 (continued)
Peak Attachment Energy and Cross Sections

Name	Formula	Peak Energy (eV)	Cross Section 10^{-21} m^2	Reference
		1.09	0.149	Aflatooni et al. (1998), Aflatooni and Burrow (2000)
2,3-Dichlorobutane	2,3-$C_4H_8Cl_2$	~0	0.5	Aflatooni et al. (1998), Aflatooni and Burrow (2000)
		0.89	3.34	Aflatooni et al. (1998), Aflatooni and Burrow (2000)
trans-1,2-Dichlorocyclohexane	$C_6H_{10}Cl_2$	0.94	3.58	Aflatooni et al. (1998), Aflatooni and Burrow (2000)
Dichlorodifluoromethane	CCl_2F_2	1.26		Aflatooni and Burrow (2001)
		2.75		Aflatooni and Burrow (2001)
		4.5		Aflatooni and Burrow (2001)
1,1-Dichloroethane	$C_2H_4Cl_2$	~0	1.6	Aflatooni et al. (1998), Aflatooni and Burrow (2000)
		0.96	3.94	Aflatooni et al. (1998), Aflatooni and Burrow (2000)
1,2-Dichloroethane	$C_2H_4Cl_2$	~0	0.4	Aflatooni et al. (1998), Aflatooni and Burrow (2000)
		0.37	0.93	Aflatooni et al. (1998), Aflatooni and Burrow (2000)
Dichlorofluoromethane	$CHCl_2F$	1.38		Aflatooni and Burrow (2001)
		4.38		Aflatooni and Burrow (2001)
1,6-Dichlorohexane	$C_6H_{12}Cl_2$	~0	0.05	Aflatooni et al. (1998), Aflatooni and Burrow (2000)
		1.23	0.114	Aflatooni et al. (1998), Aflatooni and Burrow (2000)
Dichloromethane	CCl_2H_2	1.53		Aflatooni and Burrow (2001)
		3.90		Aflatooni and Burrow (2001)
		~0	0.22	Aflatooni et al. (1998), Aflatooni and Burrow (2000)
		0.43	0.517	Aflatooni et al. (1998), Aflatooni and Burrow (2000)
		0.8 (300 K)	0.31	Pinnaduwage et al. (1999)
		1.7 (400 K)	0.80	
		0.55 (500 K)	1.7	
1,2-Dichloro-2-methylpropane	1,2-$C_4H_8Cl_2$	~0	0.9	Aflatooni et al. (1998), Aflatooni and Burrow (2000)
		0.87	5.98	Aflatooni et al. (1998), Aflatooni and Burrow (2000)
1,8-Dichlorooctane	$C_8H_{16}Cl_2$	1.25	0.057	Aflatooni et al. (1998), Aflatooni and Burrow (2000)
1,5-Dichloropentane	$C_5H_{10}Cl_2$	~0	0.01	Aflatooni et al. (1998), Aflatooni and Burrow (2000)
		1.17	0.075	Aflatooni et al. (1998), Aflatooni and Burrow (2000)
1,1-Dichloropropane	1,1-$C_3H_6Cl_2$	~0	3.05	Aflatooni et al. (1998), Aflatooni and Burrow (2000)
		0.90	2.11	Aflatooni et al. (1998), Aflatooni and Burrow (2000)
2,2-Dichloropropane	2,2-$C_3H_6Cl_2$	~0	1.2	Aflatooni et al. (1998), Aflatooni and Burrow (2000)
		1.16	6.68	Aflatooni et al. (1998), Aflatooni and Burrow (2000)
1,2-Dichloropropane	1,2-$C_3H_6Cl_2$	~0	0.4	Aflatooni et al. (1998), Aflatooni and Burrow (2000)
		0.76	1.52	Aflatooni et al. (1998), Aflatooni and Burrow (2000)

TABLE A4.1 (continued)
Peak Attachment Energy and Cross Sections

Name	Formula	Peak Energy (eV)	Cross Section 10^{-21} m²	Reference
1,3-Dichloropropane	$C_3H_6Cl_2$	~0	0.07	Aflatooni et al. (1998), Aflatooni and Burrow (2000)
		1.14	0.179	Aflatooni et al. (1998), Aflatooni and Burrow (2000)
Difluorobenzene	$C_6H_4F_2$	0.65		Olthoff et al. (1985)
		1.38		Olthoff et al. (1985)
		4.54		Olthoff et al. (1985)
Difluorocyanophosphine	PF_3CN	6.4 (PF_2^-)		Harland et al. (1974)
		6.4 (CN^-)		
		10.4 (F^-)		
Diiodobenzene	$C_6H_4I_2$	~0.0		Olthoff et al. (1985)
		3.9		Olthoff et al. (1985)
Dinitrogen pentoxide	N_2O_5	~0.0	0.20	Cicman et al. (2004)
Fluorobenzene	C_6H_5F	0.87		Olthoff et al. (1985)
		1.48		Olthoff et al. (1985)
		4.80		Olthoff et al. (1985)
Formic acid	CH_2O_2	1.4	0.14 ($HCOO^-$)	Prabhudesai et al. (2005)
		7.3	0.012 (H^-)	
		7.6	0.0061 ($O^- + OH^-$)	
Fructose	$C_6H_{12}O_6$	0.0		Aflatooni and Burrow (2000), Sulzer et al. (2006)
Fullerene	C_{60}	0.0		Ptasińska et al. (2006)
		1.0–8.4		Ptasińska et al. (2006)
	C_{70}	0.0		Ptasińska et al. (2006)
		5.0		Ptasińska et al. (2006)
	C_{76}	0.0		Ptasińska et al. (2006)
		5.0		Ptasińska et al. (2006)
	C_{84}	0.0		Ptasińska et al. (2006)
		4.7		Ptasińska et al. (2006)
Germanium tetrachloride	$GeCl_4$	0.00		Modelli et al. (1998)
		5.6		Modelli et al. (1998)
	$(CH_3)CCl_3$	0.00		Modelli et al. (1998)
		0.63		Modelli et al. (1998)
Hydrogen bromide	HBr	0.4	45.0	Abouaf and Teillet-Billy (1977)
Hydrogen peroxide	H_2O_2	0.4	6.8 (OH^-)	Nandi et al. (2003)
			1.7 (O^-)	
Iodobenzene	C_6H_5I	~0.0	640	
		0.59		Olthoff et al. (1985)
		4.3		Olthoff et al. (1985)
$MeCCl_3$	$(CH_3)CCl_3$	0.00		Modelli et al. (1998)
		0.63		Modelli et al. (1998)
$MeCl_3Sn$	$(CH_3)Cl_3Sn$	0.00		Modelli et al. (1998)
		5.2		Modelli et al. (1998)
Me_2CCl_2	$(CH_3)_2CCl_2$	0.00		Modelli et al. (1998)
		1.20		Modelli et al. (1998)
Me_2Cl_2Sn		0.00		Modelli et al. (1998)
		0.13		Modelli et al. (1998)
Me_3CCl	$(CH_3)_3CCl$	1.54		Modelli et al. (1998)
Me_3ClSi	$(CH_3)_3ClSi$	6.2		Modelli et al. (1998)
Me_3ClSn		5.6		Modelli et al. (1998)
Methyl vinyl ether	C_3H_6O	2.97 (H_2CC^-)		Bulliard el al. (2001)

continued

TABLE A4.1 (continued)
Peak Attachment Energy and Cross Sections

Name	Formula	Peak Energy (eV)	Cross Section 10^{-21} m^2	Reference
		3.05 (CH_3O^-)		
		6.60 (CH_3O^-)		
1-Methoxy-2-nitrobenzene	See 2-Nitroanisole			
Methyl nitrite	CH_3NO_2			
1,4-Naphthoquinone		0.05	0.022	Christophorou et al. (1971)
2-Nitroanisole	$C_7H_7NO_3$	0.00 (M^-)		Modelli and Marco Venuti (2001)
		0.75 (NO_2^-)		
		0.8 $(M–CH_3)^-$		
		3.8 (NO_2^-)		
Nitrobenzene	$C_6H_5NO_2$	0.03 ($C_6H_5NO_2^-$)		Christophorou et al. (1966)
		1.06 (NO_2^-)		
		3.53 (NO_2^-)		
Nitroethane	$C_2H_5NO_2$	0.0 ($C_2H_5NO_2^-$)	0.004	Pelc et al. (2003)
		0.75 (NO_2^-)	1.1	
		1.7 (CN^-)	0.5	
		4.5 (CNO^-)	0.0003	
		4.5 (HCN^-)	0.0002	
		5.7 (O^-)	0.005	
		5.7 (OH^-)	0.002	
		8.0 (NO^-)	0.0003	
Nitromethane	CH_3NO_2	0.62 (NO_2^-)	1.0	Sailer et al. (2002)
		1.7 (CN^-)	0.001	
		4.0 (OH^-)	0.001	
		4.0 (CNO^-)	0.0001	
		5.4 (O^-)	0.01	
2-Nitrophenol		0.00 (M^-)		Modelli and Marco Venuti (2001)
		0.00 ($M–OH^-$)		
		0.7 (NO_2^-)		
		0.7 $(M–OH)^-$		
		1.1 $(M–OH)^-$		
		2.8 (NO_2^-)		
		3.1 $(M–OH)^-$		
		3.6 (NO_2^-)		
m-Nitrotoluene	*m*-$C_7H_7NO_2$	0.03 ($C_7H_7NO_2^-$)		Christophorou et al. (1966)
		1.06 (NO_2^-)		
		3.50 (NO_2^-)		
o-Nitrotoluene	*o*-$C_7H_7NO_2$	0.03 ($C_7H_7NO_2^-$)		Christophorou et al. (1966)
		0.62 (NO_2^-)		
		3.22 (NO_2^-)		
Perfluoro-1,3-butadiene	1,3-C_4F_6	0.0	>600	Christodoulides et al. (1979)
		0.17	319	
		1.04	20.0	
Perfluorocyclobutane	c-C_4F_8	0.0	>600	Christodoulides et al. (1979)
		0.22	203	
		0.48	121	
Perfluoro-2-butene	2-C_4F_8	0.0	>600	Christodoulides et al. (1979)
		0.18	147	
		0.59	50	
Perfluorocyclobutene	c-C_4F_6	0.0	>600	Christodoulides et al. (1979)
		0.14	196	

TABLE A4.1 (continued)
Peak Attachment Energy and Cross Sections

Name	Formula	Peak Energy (eV)	Cross Section 10^{-21} m^2	Reference
		0.71	7	
Perfluoro-2-butyne	2-C_4F_6	0.0	>600	Christodoulides et al. (1979)
		0.19	164	
		0.80	135	
Perfluorocyclohexene	c-C_6F_{10}	0.0	>700	Pai et al. (1979)
		0.22	369	
		0.71	33	
Perfluorocyclopentene	c-C_5F_8	0.0	>700	Pai et al. (1979)
		0.24	335	
		0.99	43	
Perfluoro-1, 3-dimethylcyclohexane	C_8F_{16}	0.0	>700	Pai et al. (1979)
		0.18	560	
Perfluorotoluene	C_7F_8	0.0	>700	Pai et al. (1979)
		~1.1	0.24	
Silicon tetrachloride	$SiCl_4$	0.00		Modelli et al. (1998)
		1.8 (?)		
		7.8		
Sodium	Na	3.1 (Na^-)		Ziesel et al. (2001)
		4.2		
		2.55		
Sodium bromide	NaBr	3.36		
		3.75		
		4.38		
		4.08		
Sodium chloride	NaCl	4.22 (Na^-)		Ziesel et al. (2001)
		5.58 (Na^-)		
		5.20 (Cl^-)		
		5.67 ($NaCl^-$)		
		5.38 (Na_2Cl^-)		
Sodium fluoride	NaF	5.1 (Na^-)		Ziesel et al. (2001)
		6.75 (Na^-)		
		5.25 (F^-)		
		6.25 (F^-)		
		7.0 (NaF^-)		
		6.65 (Na_2F^-)		
Sodium iodide	NaI	2.55 (I^-)		Ziesel et al. (2001)
		3.8		
		4.6		
		4.0 (Na^-)		
		3.7 (Na_2I^-)		
Sulfur dioxide	SO_2	~5.0	0.41 (300 K)	Spyrou et al. (1986)
			0.43 (400 K)	
			0.55 (500 K)	
			0.68 (600 K)	
			0.96 (700 K)	
Sulfur pentafluorochloride	SF_5Cl	0.00		
Sulfuryl fluoride	SF_2O_2	0 ($SF_2O_2^-$)		Reese et al. (1958)
		2.3 (SO_2F^- + F)[a]		
		2.8 (F_2^- + SO_2)[a]		
		3.2 (F^- + SO_2F)[a]		
1,1,2,2-Tetrachloroethane	1,1,2,2-$C_2H_2Cl_4$	0.2 Cl^-		
		0.11 Cl_2^-		

continued

TABLE A4.1 (continued)
Peak Attachment Energy and Cross Sections

Name	Formula	Peak Energy (eV)	Cross Section 10^{-21} m^2	Reference
		0.02 $ClHCl^-$		
		0.06 $(M\text{-}Cl)^-$		
Tetrachloromethane	CCl_4	0.80	~45.0	Aflatooni et al. (1998), Aflatooni and Burrow (2000)
Tetrahydrofuran	C_4H_8O	1.25		Sulzer et al. (2006)
		7.5		Sulzer et al. (2006)
Thymine	$C_5H_6N_2O_2$	0.01		Ptasińska et al. (2005)
		0.72		
		0.84		
		1.04		
		1.26		
		1.31		
		1.44		
		1.64		
Thymine (4D)	$C_5H_2D_4N_2O_2$	0.01		Ptasińska et al. (2005)
		0.70		
		0.82		
		1.03		
		1.24		
		1.30		
		1.46		
		1.63		
Tin tetrachloride	$SnCl_4$	0.00		Modelli et al. (1998)
		0.7		Modelli et al. (1998)
		4.2		Modelli et al. (1998)
		5.4		Modelli et al. (1998)
1,1,2-Trichloroethane	$C_2H_3Cl_3$	0.36	1.90	Aflatooni et al. (1998), Aflatooni and Burrow (2000)
Trichlorofluoromethane	CCl_3F	0.57		Aflatooni and Burrow (2001)
		2.26		Aflatooni and Burrow (2001)
		4.20		Aflatooni and Burrow (2001)
Trichloromethane	CCl_3H	0.65		Aflatooni and Burrow (2001)
		2.7		Aflatooni and Burrow (2001)
		~0	120	Aflatooni et al. (1998), 18
		0.27	96.3	Aflatooni et al. (1998), 18
1,1,2-Trichloro-2 methylpropane		0.30	8.90	Aflatooni et al. (1998), 18
1,2,3-Trichloropropane	$C_3H_5Cl_3$	0.30	8.90	Aflatooni et al. (1998), 18
Triflic acid	CHF_3SO_3	0.0	3500	Aljajian et al. (1991)
Triflic anhydride	$(CF_3SO_2)_2O$	~0.0	4000	Aljajian et al. (1991)
Trifluoroiodo methane	CF_3I	~0.0	1.1×10^4	Aljajian et al. (1991)
Trifluoromethylsulfur pentafluoride	SF_5CF_3	~0.0	1000.0	Balog et al. (2003)
Urea	CH_4N_2O	2.1 (CN^-)		Naff (1972)
		2.3 (CNO^-)		
		5.8 (NH_2^-)		
		5.8 $(N_2H_3CO^-)$		
		~6.0 (OH^-)		
		6.4 (NH^-)		

[a] These are appearance potentials for the ion shown.

REFERENCES

Abouaf, R. and D. Teillet-Billy, *J. Phys. B*, 10, 2261, 1977.

Aflatooni, K. and P. D. Burrow, *J. Chem. Phys.*, 113, 1455, 2000.

Aflatooni, K. and P. D. Burrow, *Int. J. Mass Spectrom.*, 205, 149, 2001.

Aflatooni, K., G. A. Gallup, and P. D. Burrow, *Chem. Phys. Lett.*, 282, 398, 1998.

Aljajian, S. H., K.-F. Man, and A. Chutjian, *J. Chem. Phys.*, 94 (5), 3629, 1991.

Balog, R., M. Stano, P. Limão-Vieira, C. König, I. Bald, N. J. Mason, and E. Illenberger, *J. Chem. Phys.*, 119, 10396, 2003.

Bruning, F., I. Hahndorf, A. Atamatovic, and E. Illenberger, *J. Phys. Chem.*, 100, 19740, 1996.

Buchel'nikova, I. S., *Sov. Phys. JETP*, 35, 783, 1959.

Bulliard, C., M. Allan, and S. Grimme, *Int. J. Mass Spectrom.*, 205, 43, 2001.

Christodoulides, A. A., L. G. Christophorou, R. Y. Pai, and C. M. Tung, *J. Chem. Phys.*, 70, 1156, 1979.

Christophorou, L. G., D. L. McCorkle, and V. E. Anderson, *J. Phys. B: Atom. Molec. Phys.*, 4, 1163, 1971.

Christophorou, L. G., R. N. Compton, G. S. Hurst, and P. W. Reinhardt, *J. Chem. Phys.*, 45, 536, 1966.

Cicman, P., G. A. Buchanan, G. Marston, B. Gulejovà. J. D. Skalný, N. J. Mason, P. Scheier, and T. D. Märk, *J. Chem. Phys.*, 121, 9891, 2004.

Deng, Z. W. and R. Souda, *J. Chem. Phys.*, 116, 1725, 2002.

Fiquet-Fayard, F., J. P. Ziesel, R. Azria, M. Tronc, and J. Chiari, *J. Chem. Phys.*, 56, 2540, 1972.

Harland, P. W., D. W. H. Rankin, and J. C. J. Thynne, *Int. J. Mass Spectrom. Ion Phys.*, 13, 395, 1974.

Modelli, A., *J. Phys. Chem. A.*, 109, 6193, 2005.

Modelli, A. and D. Jones, *J. Phys. Chem. A*, 108, 417, 2004.

Modelli, A., and Marco Venuti, *Int. Jour. Mass Spectrom.*, 205, 7, 2001.

Modelli, A., M. Guerra, D. Jones, G. Distefano, and M. Tronc, *J. Chem. Phys.*, 108, 9004, 1998.

Naff, W. T., R. N. Compton, and C. D. Cooper, *J. Chem. Phys.*, 57, 1303, 1972.

Nandi, D., E. Krishnakumar, A. Rosa, W.-F. Schmidt, and E. Illenberger, *Chem. Phys. Lett.*, 373, 454, 2003.

Olthoff, J. K., J. A. Tossell, and J. H. Moore, *J. Chem. Phys.*, 83, 5627, 1985.

Pai, R. Y., L. G. Christophorou, and A. A. Christodoulides, *J. Chem. Phys.*, 70, 1169, 1979.

Pearl, D. M. and P. D. Burrow, *J. Chem. Phys.*, 101, 2940, 1994.

Pelc, A., W. Sailer, S. Matejcik, P. Scheier, and T. D. Märk, *J. Chem. Phys.*, 119, 7887, 2003.

Pinnaduwage, L. A., C. Tav, D. L. McCorkle, and W. X. Ding, *J. Chem. Phys.*, 110, 9011, 1999.

Prabhudesai, V. S., D. Nandi, A. H. Kelkar, R. Parajuli, and E. Krishnakumar, *Chem. Phys., Lett.*, 405, 172, 2005.

Pshenichnyuk, S. A., I. A. Pshenichnyuk. E. P. Nafikova, and N. L. Asfandiarov, *Rapid Comm. Mass Spectrom.*, 20, 1097, 2006.

Ptasińska, S., O. Echt, S. Denifl, M. Stano, P. Sulzer, F. Zappa, A. Stamatovic, P. Scheier, and T. D. Märk, *J. Phys. Chem. A*, 110, 8451, 2006.

Ptasińska, S., S. Denifl, B. MrÓz, M. Probst, V. Grill, E. Illenberger, P. Scheier, and T. D. Märk, *J. Phys. Chem. A*, 123, 124302, 2005.

Reese, R. M., V. H. Diebler, and J. L. Franklin, *J. Chem. Phys.*, 29, 880, 1958.

Rosa, A., W. Barszczewska, D. Nandi, V. Ashok, S. V. K. Kumar, E. Krishnakumar, F. Bruning, and E. Illenberger, *Chem. Phys. Lett.*, 342, 2001, 536.

Sailer, W., A. Pelc, S. Matejcik, E. Illenberger, P. Scheier, and T. D. Märk, *J. Chem. Phy.*, 117, 7989, 2002.

Shimamori, H., T. Sunagawa, Y. Ogawa, and Y. Tatsumi, *Chem. Phys. Lett.*, 232, 115, 1995.

Spyrou, S. M., I. Sauers, and L. G. Christophorou, *J. Chem. Phys.*, 84, 239, 1986.

Sulzer, P., S. Ptasinska, F. Zappa, B. Mielewska, I. Bald, S. Gohlke, M. A. Huels, and E. Illenberger, *J. Chem. Phys.*, 125, 044304, 2006.

Ziesel, J.-P., R. Azria, D. Teillet-Billy, *Int. J. Mass Spectrom.*, 205, 137, 2001.

ADDENDUM

1. Hasegawa et al. (Hasegawa, H., H. Date, M. Shimozuma, and H. Itoh, *Applied Physics Letter*, 95, 101504, 2009) have published attachment coefficients in CF_3I for E/N = 200–5000 Tol.
2. Han, H. et al. (Han, H., H. Feng, H. Li, H. Wang, H. Jiang, and Y. Chu, *Chinese J. Chem. Phys.*, 24, 218, 2011) have provided the most recent data on attachment in $CHCl_3$.
3. Electron attachment in $POCl_3$ (Phosphoryl Chloride) has been studied recently by Shuman et al. (Shuman N.S., T.M. Miller, A.A.Viggiano, and J.Troe, *J. Chem. Phys.*, 134, 094310, 2011).
4. Attachment Rate constants upto 1100 K in long Carkan molecules (CH_2Cl_2, CF_2Cl_2, CH_3Cl, and CF_3Cl) have been updated by Miller et al. (Miller, T.M., J.F. Friedman, L.C. Schaffer, and A.A. Viggiano, *J. Chem. Phys.*, 131, 084302, 2009).
5. Electron attachment to complex molecules, SF_3CN, $SF_3C_6F_5$, and SF_3 has been reported by Shuman, N.S., T.M. Miller, A.A. Viggiano, E.D. Luzik(Jr.), and N. Hazari, *J. Chem. Phys.*, 134, 044323, 2011.
6. Electron attachment to SOF_2, $SOCl_2$, SO_2F_2, SO_2Cl_2, and SO_2FCl have been measured at 300–900 K (Miller, T.M., J.F. Friedman, C.M. Caples, N.S. Shuman, J.M. Van Doren, M.F. Bardaro, Jr., P. Nguyen, C. Swieben, M.J. Campbell, and A.A. Viggiano, *J. Chem. Phys.*, 132, 241302, 2011.)
7. Total absolute cross sections for dissociative electron attachment for $(CH_2)_nCl$ groups, n = 1 to 4, have been measured by Atlatooni, K., G.A. Gallup, and P.D. Burrow, *J. Chem. Phys.*, 132, 094306, 2010.

Appendix 5: Attachment Rates

CONTENTS

TABLE A5.1
Attachment Rates

Target: Name	Target: Formula	k_a (m^3/s)	Temperature K (eV)	Reference
		Diatomic Molecules		
Bromine	Br_2	8.2 (−19)	296	Sides and Tiernan (1976)
		1.0 (−17)	350	
Chlorine	Cl_2	<1.0 (−15)	205	Smith et al. (1984)
		2.0 (−15)	300	
		3.3 (−15)	455	
		4.8 (−15)	590	
		12.2 (−16)	213	Christophorou and Olthoff (2004)
		13.5 (−16)	233	
		15.1 (−16)	253	
		16.7 (−16)	273	
		18.6 (−16)	298	
		21.4	323	
Deuterium bromide	DBr	3.1 (−16)		Christophorou et al. (1968)
Deuterium chloride	DCl	1.2 (−16)		Christophorou et al. (1968)
Deuterium iodide	DI	7.87 (−14)		Christophorou et al. (1968)
Fluorine	F_2	12 (−15)	233	McCorkle et al. (1986)
		18 (−15)	298	
		19 (−15)	373	
		3.1 (−15)	350	Sides and Tierman (1976)
		7.5 (−15)	500	
		4.6 (−15)	600	
Hydrogen bromide	HBr	1.0 (−16)		Wang and Lee (1988)
		≤3.0 (−18)	300	Smith and Adams (1987)
		3 (−16)	515	
		1.0 (−16)		Christophorou et al. (1968)
Hydrogen chloride	HCl	3.1 (−18)		Davis et al. (1973)
		4.35 (−16)		Christophorou et al. (1968)
Hydrogen fluoride	HF	<1.0 (−17)	300	Adams et al. (1986)
		<1.0 (−17)	510	
Hydrogen iodide	HI	~1.6 (−13)		Klar et al. (2001)
		3 (−13)		Smith and Adams (1987)
		1.86 (−13)		Christophorou et al. (1968)
Iodine	I_2	1.8 (−16)	(0.04)	Truby (1969)
		2.8 (−16)	(0.27)	

continued

TABLE A5.1 (continued)
Attachment Rates

Target				
Name	Formula	k_a (m^3/s)	Temperature K (eV)	Reference
		Triatomic Molecules		
Ammonia	NH_3	2.0 (−14)		Dunning (1995)
Cyanogen bromide	BrCN	4.2 (−15)		Alajajian et al. (1988)
		1.80 (−13)		Mothes et al. (1972)
Deuterated water	D_2O	4.3 (−16)	(6.3)	Stockdale et al. (1964)
		5.6 (−16)	(6.4)	
		7.4 (−16)	(6.5)	
		9.8 (−16)	(6.6)	
Hydrogen sulfide	H_2S	4.5 (−14)		Dunning (1995)
Water vapor	H_2O	2.3 (−18)		Compton and Christophorou (1967)
Xenon difluoride	XeF_2	2.4 (−15)		Sides and Tierman (1976)
		Four Atoms Molecules		
Boron trichloride	BCl_3	1.8 (−14)		Tav et al. (1998)
		2.8 (−15)		Stockdale et al. (1972)
Boron trifluoride	BF_3	<1.5 (−17)		Stockdale et al. (1972)
Carbonyl chloride or phosgene	CCl_2O	5 (−14)		Schultes et al. (1975)
Nitrogen trifluoride	NF_3	7 (−18)	300	Miller et al. (1995)
		12 (−18)	343	
		24 (−18)	363	
		39 (−18)	402	
		68 (−18)	434	
		170 (−18)	462	
		504 (−18)	504	
		420 (−18)	551	
		2.1 (−17)	300	Sides and Tierman (1977)
Phosphorous (III) chloride	PCl_3	6.4 (−14)	296	Miller et al. (1998)
Phosphorous (III) fluoride	PF_3	3.0 (−18)	300	Miller et al. (1995)
		7 (−18)	(0.025)	
		9 (−18)	(0.029)	
		24 (−18)	(0.031)	
		40 (−18)	(0.034)	
		68 (−18)	(0.037)	
		165 (−18)	(0.040)	
		380 (−18)	(0.043)	
		409 (−18)	(0.047)	
Sulfur trioxide	SO_3	3 (−15)	300	Miller et al. (1995)
Thionyl chloride	Cl_2OS	13 (−16)		Petrović et al. (1989)
		Molecules with Five Atoms		
Bromochloromethane	CH_2BrCl	7.1 (−15)		Sunagawa and Shimamori (1997)
Bromodichloromethane	$CHCl_2Br$	4.0 (−14)	300	Španěl and Smith (2001)
		1.2 (−13)	540	
		4.91 (−14)		Alajajian et al. (1988)
		6.6 (−14)		Barszczewska et al. (2004)
		6.0 (−18)		Burns et al. (1996)
		1.08 (−17)		Datskos et al. (1992)
		5.8 (−18)		Wang and Lee (1988)
		6.7 (−18)		Petrović and Crompton (1987)

continued

TABLE A5.1 (continued)
Attachment Rates

Target				
Name	**Formula**	**k_a (m^3/s)**	**Temperature K (eV)**	**Reference**
		6.0 (−18)	300	Alge et al. (1984)
		2.3 (−16)	452	
		25 (−16)	585	
		3.6 (−18)		Mothes et al. (1972)
Bromotrichloromethane	$CBrCl_3$	5.7 (−14)		Sunagawa and Shimamori (1997)
		4.91 (−14)		Alajajian et al. (1988)
		8.2 (−14)		Adams et al. (1988)
Bromotrifluoromethane	$CBrF_3$	1.4 (−14)		Christophorou and Hadjiantoniou (2006)
		1.3 (− 14)		Barszczewska et al. (2004)
		8.6 (−15)		Sunagawa and Shimamori (1997)
		1.2 (−14)	293	Burns et al. (1996)
		3.9 (−14)	615	
		1.2 (−13)	777	
		8.6 (−15)		Shimamori et al. (1992)
		1.5 (−14)		Christopher et al. (1989)
		1.5 (−14)		Marotta et al. (1989)
		5.3 (−15)	205	Alge et al. (1984)
		1.6 (−14)	300	
		4.9 (−14)	452	
		7.7 (−14)	585	
		1.36 (−14)		Blaunstein and Christophorou (1968)
Chlorine nitrate	$ClONO_2$	1.1 (−13)	300	Van Doren et al. (1996)
Chlorodibromomethane	$CHBr_2Cl$	2.7 (−14)	321	Ipolyi et al. (2004)
		3 (−14)	300	Španěl and Smith (2001)
		4.9 (−14)	540	
		1.2 (−13)		Sunagawa and Shimamori (1997)
Chlorodifluoromethane	$CHClF_2$	<3.3 (−19)		Christophorou and Hadjiantoniou (2006)
		1.1 (−19)		Szamrej et al. (1996)
Chlorotrifluoroethene	C_2ClF_3	6.6 (−18)		Christophorou (1996)
Chlorotrifluoromethane	$CClF_3$	1.5 (−19)		Christophorou and Hadjiantoniou (2006)
		2.1 (−19)		Barszczewska et al. (2003)
		1.3 (−19)		Rosa and Szamrej (2000)
		4.2 (−19)	293	Burns et al. (1996)
		1.4 (−17)	467	
		2.4 (−16)	579	
		9.5 (−16)	777	
		<3.43 (−17)	300	Spyrou and Christophorou (1985a)
		<3.2 (−17)	400	
		<2.80 (−17)	500	
		<3.60 (−17)	600	
		<4.20 (−17)	700	
		<1.6 (−19)		McCorkle et al. (1982)
		5.2 (−20)		Fessenden and Bansal (1970)
Chloromethane	CH_3Cl	1.0 (−19)		Christophorou and Hadjiantoniou (2006)
		1.0 (−20)		Datskos et al. (1990)
		2.0 (−19)		Petrović et al. (1989)

continued

TABLE A5.1 (continued)
Attachment Rates

Target			Temperature	
Name	Formula	k_a (m^3/s)	K (eV)	Reference
Dibromodichloromethane	CBr_2Cl_2	9.0 (−15)	300	Španěl and Smith (2001)
		4.5 (−14)	540	
		1.7 (−13)		Sunagawa and Shimamori (1997)
		1.5 (−13)		Adams et al. (1988)
		2.6 (−13)		Blaunstein and Christophorou (1968)
Dibromodifluoromethane	CBr_2F_2	**2.7 (−13)**		Barszczewska et al. (2004)
		1.7 (−13)		Sunagawa and Shimamori (1997)
		2.6 (−13)		Christophorou (1996)
		2.6 (−13)		Blaunstein and Christophorou (1968)
		2.8 (−13)		Alajajian et al. (1988)
Dibromomethane	CH_2Br_2	9.6 (−14)		Christophorou and Hadjiantoniou (2006)
		7.1 (−14)		Barszczewska et al. (2004)
		9.0 (−14)		Sunagawa and Shimamori (1997)
		3.5 (−14)	293	Burns et al. (1996)
		4.2 (−14)	467	
		4.5 (−14)	777	
		9.0 (−14)		Shimamori et al. (1992)
		1.06 (−13)		Alajajian et al. (1988)
		8.1 (−14)	205	Alge et al. (1984)
		9.3 (−14)	300	
		1.6 (−13)	452	
		2.2 (−13)	585	
		3.2 (−14)		Blaunstein and Christophorou (1968)
Dichlorodifluoromethane	CCl_2F_2	1.6 (−15)		Christophorou and Hadjiantoniou (2006)
		1.6 (−15)		Barszczewska et al. (2004)
		9.6 (−16)		Aflatooni and Burrow (2001)
		9.6 (−16)		Sunagawa and Shimamori (1997)
		1.9 (−15)	293	Burns et al. (1996)
		1.4 (−14)	467	
		2.4 (−14)	579	
		4.2 (−14)	777	
		1.8 (−15)		Marotta et al. (1989)
		1.7 (−15)		Wang and Lee (1987)
		<1.0 (−15)	205	Smith et al. (1984)
		3.2 (−15)	300	
		1.6 (−14)	455	
		5.3 (−14)	590	
		1.2 (−15)		McCorkle et al. (1982)
		1.23 (−15)		Christophorou et al. (1981)
		2.2 (−15)		Christophorou et al. (1974)
		1.3 (−15)		Bansal and Fessenden (1973)
		5.9 (−15)*		Chen and Chantry (1972)

continued

TABLE A5.1 (continued)
Attachment Rates

Target			Temperature	
Name	Formula	k_a (m^3/s)	K (eV)	Reference
Dichlorofluoromethane	$CHCl_2F$	6.1 (−18)		Christophorou and Hadjiantoniou (2006)
		1.5 (−18)		Christophorou and Hadjiantoniou (2006)
		5.5 (−18)		Barszczewska et al. (2004)
		5.0 (−18)		Aflatooni and Burrow (2001)
		6.1 (−18)	293	Burns et al. (1996)
		1.2 (−16)	467	
		6.6 (−16)	579	
		2.0 (−15)	777	
		7.4 (−19)		Sunagawa and Shimamori (1997)
Dichloromethane	CH_2Cl_2	1.60 (−17)*		Gallup et al. (2003)
		4.7 (−18)		Christophorou and Hadjiantoniou (2006)
		4.7 (−18)		Barszczewska et al. (2004)
		2.6 (−19)		Aflatooni and Burrow (2000)
		1.8 (−16)	467	Burns et al. (1996)
		6.4 (−16)	579	
		2.1 (−15)	777	
		6.5 (−19)		Ayala et al. (1981)
		4.7 (−18)		Fessenden and Bansal (1970)
Diiodomethane	CH_2I_2	5.0 (−15)		Christophorou and Hadjiantoniou (2006)
Fluoromethane	CH_3F	<1.0 (−21)		Fessenden and Bansal (1970)
Iodomethane	CH_3I	9.3 (−14)		Christophorou and Hadjiantoniou (2006)
		1.0 (−13)	293	Burns et al. (1996)
		1.1 (−13)	467	
		1.2 (−13)	579	
		1.2 (−13)	777	
		8.5 (−14)	205	Alge et al. (1984)
		1.2 (−13)	300	
		1.8 (−13)	452	
		1.1 (−13)	585	
Nitric acid	HNO_3	5 (−14)	300	Adams et al. (1986)
Phosphoric trichloride	$POCl_3$	2.5 (−13)	297	van Doren et al. (2006)
		2.0 (−13)		Williamson et al. (2000)
		1.8 (−13)		Miller et al. (1998)
Sulfuryl fluoride	SO_2F_2	8.4 (−16)	300	Datskos and Christophorou (1989)
		3.11 (−15)	400	
		1.09 (−14)		
		2.08 (−14)		
		3.07 (−16)		
		5.8 (−16)		Davis et al. (1973)
Tetrabromomethane	CBr_4	2.5 (−14)		Sunagawa and Shimamori (1997)
Tetrachloromethane	CCl_4	3.4 (−13)		Christophorou and Hadjiantoniou (2006)
		2.8 (−13)*		Onanong et al. (2006)

continued

TABLE A5.1 (continued)
Attachment Rates

Target Name	Formula	k_a (m^3/s)	Temperature K (eV)	Reference
		4.0 (−13)		Barszczewska et al. (2004)
		2.8 (−13)*		Gallup et al. (2003)
		3.6 (−13)	293	Burns et al. (1996)
		2.1 (−13)	467	
		1.4 (−13)	579	
		1.2 (−13)	777	
		4.0 (−13)		Shimamori et al. (1992)
		3.8 (−13)		Christopher et al. (1989)
		3.7 (−13)		Marotta et al. (1989)
		3.79 (−13)	294	Orient et al. (1989)
		2.96 (−13)	400	
		2.33 (−13)	500	
		4.1 (−13)	205	Smith et al. (1984)
		3.9 (−13)	300	
		3.7 (−13)	455	
		3.5 (−13)	590	
		4.2 (−13)		Ayala et al. (1981)
Tetrafluoromethane	CF_4	<1.0 (−22)		Christophorou and Hadjiantoniou (2006)
Tribromofluoromethane	CBr_3F	4.8 (−15)		Christophorou and Hadjiantoniou (2006)
		4.4 (−15)		Barszczewska et al. (2004)
		3.0 (−15)		Sunagawa and Shimamori (1997)
Tribromomethane	$CHBr_3$	4.3 (−14)		Sunagawa and Shimamori (1997)
Trichlorofluoromethane	CCl_3F	2.1 (−13)		Christophorou and Hadjiantoniou (2006)
		2.1 (−13)		Barszczewska et al. (2004)
		2.5 (−13)		Klar et al. (2001)
		2.4 (−13)	293	Burns et al. (1996)
		1.8 (−13)	467	
		2.1 (−13)	579	
		1.9 (−13)	777	
		1.8 (−13)		Shimamori et al. (1992)
		2.4 (−13)		Christopher et al. (1989)
		3.1 (−13)		Christopher et al. (1989)
		3.1 (−13)		Marotta et al. (1989)
		2.38 (−13)	294	Orient et al. (1989)
		2.16 (−13)	404	
		2.01 (−13)	496	
		3.0 (−15)		Sunagawa and Shimamori (1997)
		2.4 (−13)		Blaunstein and Christophorou (1968)
		2.2 (−13)	205	Smith et al. (1984)
		2.6 (−13)	300	
		3.6 (−13)	455	
		3.3 (−13)	590	

continued

TABLE A5.1 (continued)
Attachment Rates

Target				
Name	Formula	k_a (m^3/s)	Temperature K (eV)	Reference
		2.37 (−13)		Crompton et al. (1982)
		2.19 (−13)		McCorkle et al. (1982)
Trichloromethane	$CHCl_3$	9.41 (−15)*		Gallup et al. (2003)
		3.1 (−15)		Christophorou and Hadjiantoniou (2006)
		2.7 (−15)		Barszczewska et al. (2004)
		9.41 (−15)		Gallup et al. (2003)
		2.0 (−15)		Sunagawa and Shimamori (2001)
		4.7 (−15)	293	Burns et al. (1996)
		6.2 (−15)	467	
		1.7 (−14)	579	
		2.3 (−14)	777	
		3.8 (−15)		Marotta et al. (1989)
		2.0 (−15)		Shimamori et al. (1992)
		<1.0 (−15)	205	Smith et al. (1984)
		4.4 (−15)	300	
		1.7 (−14)	455	
		3.6 (−14)	590	
		3.1 (−15)		Johnson et al. (1977)
		2.7 (−15)		Ayala et al. (1981)
		2.0 (−15)		Bansal and Fessenden (1973)
		3.76 (−15)		Christophorou et al. (1981)
Trifluoroiodomethane	CF_3I	1.9 (−13)		Christophorou and Hadjiantoniou (2006)
		1.0 (−13)	293	Burns et al. (1996)
		1.1 (−13)	467	
		1.2 (−13)	579	
		1.2 (−13)	777	
		2.0 (−13)		Shimamori et al. (1992)
Trifluoromethane	CHF_3	4.8 (−20)		Christophorou and Hadjiantoniou (2006)
		Molecules with Six Atoms		
Acetonitrile	C_2H_3N	≤1.24 (−20)		Stockdale et al. (1974)
Bromoacetonitrile	C_2H_2BrN	1.9 (−13)	295	van Doren et al. (1995)
		2.3 (−13)	386	
		2.4 (−13)	476	
		2.4 (−13)	556	
Chloroacetonitrile	C_2H_2ClN	3.9 (−14)	295	van Doren et al. (1995)
		4.8 (−14)	368	
		6.5 (−14)	468	
		9.7 (−14)	556	
Chloroethene	C_2H_3Cl	See vinyl chloride		
Dibromoethene	$C_2H_2Br_2$	1.7 (−14)	(0.04)	Sunagawa and Shimamori (1995)
		1.8 (−14)	(0.1)	
		1.9 (−14)	(0.4)	
		2.0 (−14)	(1.0)	
cis-Dichloroethene	*cis*-1,2-$C_2H_2Cl_2$	<1.0 (−16)		Christophorou (1996)

continued

TABLE A5.1 (continued)
Attachment Rates

Target				
Name	**Formula**	k_a **(m³/s)**	**Temperature K (eV)**	**Reference**
Difluoroethene	$C_2H_2F_2$	<1.6 (−18)		Davis et al. (1973)
Fluorosulfonic acid	FHO_3S	6 (−14)	300	Adams et al. (1986)
		1.2 (−13)	510	
Methanol	CH_3OH	Very small		Christophorou and Hadjiantoniou (2006)
Phosphorous (V) fluoride	PF_5	10 (−17)	293	Miller et al. (1995)
		9 (−17)	300	
		11 (−17)	302	
		15 (−17)	331	
		27 (−17)	336	
		24 (−17)	364	
		32 (−17)	367	
		39 (−17)	408	
		45 (−17)	415	
		47 (−17)	442	
		35 (−17)	466	
		46 (−17)	475	
		34 (−17)	513	
		29 (−17)	545	
		19 (−17)	552	
Tetrachloroethene	C_2Cl_4	9.0 (−14)		Suess et al. (2003)
		1.04 (−13)		McCorkle et al. (1984)
Tetrafluoroethene	C_2F_4	8.5 (−17)		Goyette et al. (2001)
Tribromoethene	C_2HBr_3	1.2 (−13)	(0.04)	Sunagawa and Shimamori (1995)
		1.0 (−13)	(0.1)	
		6.5 (−14)	(0.4)	
		3.0 (−14)	(1.0)	
		1.8 (−14)	(2.0)	
Trichloroethylene	C_2HCl_3	6.64 (−15)		Christophorou (1996)
		2.2 (−15)		Johnson et al. (1977)
		1.97 (−15)		Blaunstein and Christophorou (1968)
Vinyl chloride	C_2H_3Cl	5.3 (−16)		Petrović et al. (1989)
		Molecules with Seven Atoms		
Dinitrogen pentoxide	N_2O_5	9.5 (−26)		Cicman et al. (2004)
Methyl nitrite	CH_3NO_2	Very small		Christophorou and Hadjiantoniou (2006)
Molybdenum hexafluoride	MoF_6	2.3 (−15)	297	Friedman et al. (2006)
		2.9(−15)	340	
		3.5 (−15)	385	
Rhenium hexafluoride	ReF_6	2.0 (−15)		Friedman et al. (2006)
Selenium hexafluoride	SeF_6	8.0 (−16)		Jarvis et al. (2001)
		1.27 (−15)		Davis et al. (1973)
Sulfur pentafluorochloride	SF_5Cl	1.30 (−14)		Mayhew et al. (2004)
Sulfurhexafluoride	SF_6	3.1 (−13)	205	Smith et al. (1984)
		3.1 (−13)	300	
		4.5 (−13)	455	
		4.0 (−13)	590	
		2.28 (−13)		Crompton et al. (1982)

continued

TABLE A5.1 (continued)
Attachment Rates

Target			Temperature	
Name	Formula	k_a (m^3/s)	K (eV)	Reference
		2.49 (−13)		Christophorou et al. (1981)
		2.21 (−13)		Fessenden and Bansal (1970)
Sulfuric acid	H_2SO_4	> 5 (−14)	510	Adams et al. (1986)
Tellurium hexafluoride	TeF_6	8.2 (−17)		Jarvis et al. (2001)
Tungsten hexafluoride	WF_6	~ 1.0 (−18)	297	Friedman et al. (2006)
		< 1.0 (−18)	302	
		1.5 (−17)	450	
		2.1 (−17)	522	
		Molecules with Eight Atoms		
Bromoethane	C_2H_5Br	**5.0 (−18)**		Barszczewska et al. (2004)
		5.3 (−18)		Rosa et al. (2001)
		1.4 (−16)		Christophorou (1996)
		9.0 (−17)		Christodoulides and Christophorou (1971)
1-Bromo-2-chloroethane	1,2-C_2H_4BrCl	4.5 (−16)		Rosa et al. (2001)
		1.0 (−15)	298	Smith et al. (1990)
		3.6 (−15)	380	
		9.7 (−16)	475	
1-Bromo-2-fluoroethane	C_2H_4BrF	5.0 (−17)		Barszczewska et al. (2004)
		1.3 (−15)	(0.04)	Sunagawa and Shimamori (1995)
		1.6 (−15)	(0.1)	
		1.8 (−15)	(0.4)	
		1.2 (−15)	(1.0)	
2-Bromo-1,1,1-trifluoroethane	$C_2H_2BrF_3$	1.4 (−15)		Barszczewska et al. (2004)
		1.5 (−14)	(0.04)	Sunagawa and Shimamori (1995)
		1.4 (−14)	(0.1)	
		1.2 (−14)	(0.4)	
		7.5 (−15)	(1.0)	
		5.0 (−15)	(2.0)	
1-Chloro-1,1-difluoroethane	1,1,1$C_2H_3ClF_2$	6.0 (−19)		Rosa et al. (2001)
		6.0 (−19)		Szamrej et al. (1996)
Chloroethane	C_2H_5Cl	3.4 (−20)		Rosa et al. (2001)
		5.0 (−19)		Petrović et al. (1989)
		3.5 (−19)		Christophorou et al. (1966)
Chloropentafluoroethane	C_2ClF_5	3.3 (−18)		Rosa et al. (2001)
		3.3 (−18)		Szamrej et al. (1996)
		< 6.2 (−18)		Davis et al. (1973)
1,2-Dibromo-1,1-difluoroethane	$C_2H_2Br_2F_2$	9.2 (−14)		Barszczewska et al. (2004)
		1.7 (−13)	(0.04)	Sunagawa and Shimamori (1995)
		1.5 (−13)	(0.1)	
		1.0 (−13)	(0.4)	
		4.5 (−14)	(1.0)	Sunagawa and Shimamori (1995)
1,1-Dibromoethane	1,1-$C_2H_4Br_2$	3.5 (−14)		Barszczewska et al. (2004)
		4.1 (−14)		Christophorou (1996)
		4.1 (−14)	(0.04)	Sunagawa and Shimamori (1995)
		4.0 (−14)	(0.1)	

continued

TABLE A5.1 (continued)
Attachment Rates

Target			Temperature	
Name	Formula	k_a (m^3/s)	K (eV)	Reference
		3.0 (−14)	(0.4)	
		1.5 (−14)	(1.0)	
1,2-Dibromoethane	1,2-$C_2H_4Br_2$	1.8 (−14)		Barszczewska et al. (2004)
1,1-Dibromo tetrafluoroethane	1,2-$C_2Br_2F_4$	<1.0 (−13)		Christophorou (1996)
1,2-Dibromo tetrafluoroethane	1,2-$C_2Br_2F_4$	1.4 (−13)		Barszczewska et al. (2004)
		1.3 (−13)	(0.04)	Sunagawa and Shimamori (1995)
		1.4 (−13)	(0.1)	
		8.5 (−14)	(0.4)	
		4.0(−14)	(1.0)	
		3.0 (−14)	(2.0)	
2,2-Dibromo tetrafluoroethane	2,2-$C_2Br_2F_4$	1.6 (−13)		Barszczewska et al. (2004)
1,1-Dichloroethane	1,1-$C_2H_4Cl_2$	2.1 (−17)		Christophorou et al. (1981)
1,2-Dichloroethane	1,2-$C_2H_4Cl_2$	**2.9 (−17)**		Barszczewska et al. (2004)
		4.72 (−17)*		Gallup et al. (2003)
		2.6 (−17)		Rosa et al. (2001)
		3.2 (−17)		Christophorou et al. (1981)
1,1-Dichloro tetrafluoroethane	1,1-$C_2Cl_2F_4$	4.8 (−15)		McCorkle et al. (1982)
		4.8 (−15)		Christophorou et al. (1981)
1,2-Dichloro tetrafluoroethane	1,2-$C_2Cl_2F_4$	7.0 (−16)		McCorkle et al. (1982)
Difluoroethane	$C_2H_4F_2$	<1.6 (−18)		Davis et al. (1973)
Fluoroethane	C_2H_5F	<5 (−21)		Fessenden and Bansal (1970)
Hexachloroethane	C_2Cl_6	7.3 (−15)		Christophorou (1996)
Hexafluoroethane	C_2F_6	< 1.0 (−22)		Spyrou and Christophorou (1985a)
Iodoethane	C_2H_5I	5.8 (−15)		Alajajian et al. (1988)
		4.8 (−15)		Blaunstein and Christophorou (1968)
Methyl nitrite	CH_3NO_2	Very small		Christophorou and Hadjiantoniou (2006)
Nitromethane	CH_3NO_2	9.9 (−13)		Stockdale et al. (1974)
Pentachloroethane	C_2HCl_5	5.64 (−15)		Christophorou (1996)
Tetrabromoethane	1,1,2,2-$C_2H_2Br_4$	**9.5 (−14)**		Barszczewska et al. (2004)
		9.2 (−14)		Sunagawa and Shimamori (1995)
		6.9 (−14)		Alajajian et al. (1988)
Tetrachloroethane	$C_2H_2Cl_4$	3.5 (−14)		Barszczewska et al. (2003)
1,1,2,2-Tetrachloroethane	1,1,2,2-$C_2H_2Cl_4$	6.97 (−15)		Christophorou (1996)
1,1,1,2-Tetrachloroethane	1,1,1,2-$C_2H_2Cl_4$	7.47 (−15)		Christophorou (1996)
Tribromoethane	$C_2H_3Br_3$	9.2 (−14)	(0.04)	Sunagawa and Shimamori (1995)
		8.0 (−14)	(0.1)	
		3.0 (−14)	(0.4)	
		1.5 (−14)	(1.0)	
1,1,1-Trichloroethane	1,1,1-$C_2H_3Cl_3$	**1.4 (−14)**		Barszczewska et al. (2004)
		5.88 (−15)*		Gallup et al. (2003)
		1.1 (−14)		Shimamori et al. (1992)
		1.5 (−14)		Christophorou et al. (1981)
		1.5 (−14)		Johnson et al. (1977)
		1.6 (−14)		Blaunstein and Christophorou (1968)
1,1,2-Trichloroethane	1,1,2-$C_2H_3Cl_3$	1.4 (−16)		Barszczewska et al. (2004)

continued

TABLE A5.1 (continued)
Attachment Rates

Target Name	Formula	k_a (m^3/s)	Temperature K (eV)	Reference
		~1.2 (−15)*		Gallup et al. (2003)
		2.0 (−16)		Barszczewska et al. (2004)
		1.8 (−16)		Christophorou et al. (1981)
		3.11 (−16)		Johnson et al. (1977)
		1.5 (−16)		Blaunstein and Christophorou (1968)
1,1,1-Trichloro trifluoroethane	1,1,1- $C_2Cl_3F_3$	**2.2 (−13)**		Barszczewska et al. (2004)
		2.0 (−13)		Klar et al. (2001)
		2.5(−13)		Rosa et al. (2001)
		2.5 (−13)		Szamrej et al. (1996)
		1.4 (−13)		Shimamori et al. (1992)
		2.25 (−13)		Alajajian and Chutjian (1987)
		2.25 (−13)		McCorkle et al. (1982)
1,1,2-Trichloro trifluoroethane	1,1,2-$C_2Cl_3F_3$	**1.1 (−14)**		Barszczewska et al. (2004)
		5.4 (−15)		Rosa et al. (2001)
		1.4 (−13)		Shimamori et al. (1992)
		1.2 (−14)		Marotta et al. (1989)
		1.1 (−14)		McCorkle et al. (1982)
		1.1 (−14)		Christophorou et al. (1981)
1,1,1-Trifluoroethane	$C_2H_3F_3$	4.3 (−20)		Fessenden and Bansal (1970)
		Molecules with Nine Atoms		
Bis(trifluoromethyl)sulfide	C_2F_6S	1.14 (−16)		Spyrou et al. (1984)
1,1-Dichloroethane	1,1-$C_2H_5Cl_2$	6.87 (−17)*		Gallup et al. (2003)
		2.1 (−17)		Christophorou et al. (1981)
Perfluoropropylene	1-C_3F_6	~4.5 (−16)		Hunter et al. (1983)
Pentafluoro dimethyl ether	C_2HF_5O	<7.0 (−18)		Spyrou et al. (1984)
Perfluorodimethyl ether	C_2F_6O	<1.0 (−18)		Hunter et al. (1985)
1,1,2,2- Tetrafluoro dimethylether	$C_2H_2F_4O$	<2.4 (−18)		Spyrou et al. (1984)
2,2,2-Trifluoro dimethylether	$C_2H_3F_3O$	<1.0 (−17)		Spyrou et al. (1984)
Trifluoromethane sulfonic acid	CHF_3O_3S	1.0 (−13)		Adams et al. (1986)
2,2,2-Trifluoro dimethylsulfide	$C_2H_3F_3S$	4.2 (−17)		Spyrou et al. (1984)
		Molecules with 10 Atoms		
Acetone	C_3H_6O	<1.66 (−19)		Christophorou (1996)
Hexafluoro-1,3-butadiene	1,3-C_4F_6	1.26 (−13)		Christophorou et al. (1981)
Hexafluoro-2-butyne	2-C_4F_6	5.4 (−14)		Christophorou et al. (1981)
Hexafluorocyclobutene	c-C_4F_6	1.43 (−13)		Christophorou et al. (1981)
		1.30 (−13)		Christodoulides et al. (1979a)
		1.4 (−13)		Bansal and Fessenden (1973)
Trifluoromethyl sulfur pentafluoride	SF_5CF_3	8.6 (−14)	296	Miller et al. (2004b)
		7.7 (−14)		Kennedy and Mayhew (2001)
		Molecules with 11 Atoms		
Bromopropane	*n*-C_3H_7Br	1.0 (−16)		Christodoulides and Christophorou (1971)
1-Bromopropane	1-C_3H_7Br	1.4 (−18)		Barszczewska et al. (2004)
2-Bromopropane	2-C_3H_7Br	1.1 (−17)		Barszczewska et al. (2004)
1-Chloropropane	C_3H_7Cl	3.6 (−19)		Barszczewska et al. (2003)
2-Chloropropane	C_3H_7Cl	3.8 (−18)		Barszczewska et al. (2003)
1,1-Dichloropropane	$C_3H_6Cl_2$	5.7 (−17)		Barszczewska et al. (2003)
	$C_3H_6Cl_2$	2.0 (−18)		Barszczewska et al. (2003)

continued

TABLE A5.1 (continued)
Attachment Rates

Target: Name	Target: Formula	k_a (m^3/s)	Temperature K (eV)	Reference
	1,1-$C_3H_6Cl_2$	3.15 (−16)*		Gallup et al. (2003)
1,2-Dichloropropane	$C_3H_6Cl_2$	6.3 (−18)		Barszczewska et al. (2003)
	1,2-$C_3H_6Cl_2$	2.70 (−17)*		Gallup et al. (2003)
1,3-Dichloropropane	$C_3Cl_2H_6$	8.1 (−18)		Barszczewska et al. (2003)
	1,3-$C_3H_6Cl_2$	4.94 (−18)*		Gallup et al. (2003)
2,2-Dichloropropane	$C_3Cl_2H_6$	2.0 (−16)		Barszczewska et al. (2003)
	2,2-$C_3H_6Cl_2$	5.73 (−17)*		Gallup et al. (2003)
1- Iodopropane	1-C_3H_7I	1.2 (−14)		Christophorou (1996)
2- Iodopropane	2-C_3H_7I	1.1 (−14)		Christophorou (1996)
Perfluoropropane	C_3F_8	1.6 (−18)		Hunter and Christophorou (1984)
Perfluoropyridine	C_5F_5N	1.8 (−13)	297	van Doren et al. (2005)
		1.27 (−13)	383	
		1.02 (−13)	393	
		9.82 (−14)	403	
		9.76 (−14)	413	
		9.98 (−14)	423	
		9.04 (−14)	433	
Pyridine	C_5H_5N	0.0 (?)		Christophorou (1996)
Succinic anhydride	$C_4H_4O_3$	4.98 (−16)		Christophorou (1996)
Tetrafluoropyridine	C_5HF_4N	7 (−17)	303	van Doren et al. (2005)
1,2,3-Trichloropropane	1,2,3-$C_3H_5Cl_3$	6.31 (−16)*		Gallup et al. (2003)
		Molecules with 12 Atoms		
Bromobenzene	C_6H_5Br	6.5 (−18)		Shimamori et al. (1993)
		2.3 (−16)		Christophorou et al. (1966)
Bromopentafluorobenzene	C_6BrF_5	1.9 (−13)		Shimamori et al. (1993)
2,3-Butanedione	2,3-$C_4H_6O_2$	3.4 (−14)		Bouby et al. (1965)
Chlorobenzene	C_6H_5Cl	3 (−20)		Shimamori et al. (1993)
		1.1 (−16)		Christophorou et al. (1966)
Chloropentafluorobenzene	C_6ClF_5	2.0 (−13)		Shimamori et al. (1993)
Deuterated bromobenzene	C_6D_5Br	2.1 (−16)		Christophorou et al. (1966)
m-Dichlorobenzene	*m*-$C_6H_4Cl_2$	7.8 (−16)		Christophorou et al. (1966)
p-Dichlorobenzene	*p*-$C_6H_4Cl_2$	1.3 (−16)		Christophorou et al. (1966)
o-Dichlorobenzene	*o*-$C_6H_4Cl_2$	6.5 (−16)		Christophorou et al. (1966)
Fluorobenzene	C_6H_5F	<1.0 (−19)		Shimamori et al. (1993)
Hexafluorobenzene	C_6F_6	2.1 (−13)		Shimamori et al. (1993)
		1.49 (−13)		Spyrou et al. (1985a)
		1.02 (−13)		Woodin et al. (1980)
		1.02 (−13)		Gant and Christophorou (1976)
Iodobenzene	C_6H_5I	1.0 (−14)		Shimamori et al. (1993)
Iodopentafluorobenzene	C_6F_5I	1.15 (−13)		Christophorou (1996)
Pentafluorobenzene	C_6HF_5	7 (−18)		Shimamori et al. (1993)
		100 (−18)		Wentworth et al. (1987)
Pentafluoroiodobenzene	C_6F_5I	2.0 (−13)		Shimamori et al. (1993)
Perfluoro-2-butene	2-C_4F_8	4.81 (−14)		Christoudoulides and Christophorou (1979b)
		4.9 (−14)		Bansal and Fessenden (1973)
Perfluorocyclobutane	c-C_4F_8	9.5 (−15)	298	Miller et al. (2004a)
		8.5 (−15)	313	

continued

TABLE A5.1 (continued)
Attachment Rates

Target			Temperature	
Name	Formula	k_a (m^3/s)	K (eV)	Reference
		8.7 (−15)	322	
		8.4; 8.2 (−15)	349	
		8.1; 9.0 (−15)	361	
		8.4 (−15)	373	
		9.3 (−15)	385	
		9.0 (−15)	400	
		1.1 (−14)		Woodin et al. (1980)
		1.25 (−14)		Christodoulides and Christophorou (1979b)
Molecules with More Than 12 Atoms				
Aniline	C_6H_7N	2.3 (−18)		Christophorou (1996)
Anthracene	$C_{14}H_{10}$	1.9 (−14)		Christophorou et al. (1971)
1,2-Benzanthracene	$C_{18}H_{12}$	8.7 (−15)		Christophorou et al. (1971)
Benzoic acid	$C_7H_6O_2$	3.15 (−15)		Christophorou (1996)
Benzo[a]pyrene	$C_{20}H_{12}$	6.64 (−15)		Christophorou (1996)
Bromobutane	n-C_4H_9Br	1.27 (−16)		Christodoulides and Christophorou (1971)
Bromodecane	n-$C_{10}H_{21}Br$	7.3 (−15)		Christodoulides and Christophorou (1971)
Bromodimethylpropane	n-$C_5H_{11}Br$	1.4 (−16)		Christodoulides and Christophorou (1971)
Bromohexane	n-$C_6H_{13}Br$	1.6 (−16)		Christodoulides and Christophorou (1971)
Bromomethyl benzene	o-C_7H_7Br	6.9 (−16)		Christophorou et al. (1966)
Bromooctane	n-$C_8H_{17}Br$	6.2 (−16)		Christodoulides and Christophorou (1971)
Chloromethylbenzene	C_7H_7Cl	7.30 (−15)		Christophorou (1996)
o-Chlorotoluene	o-C_7H_7Cl	6.4 (−17)		Christophorou and Christophorou (1971)
Chrysene	$C_{18}H_{12}$	5.15 (−15)		Christophorou (1996)
Cyclooctatetraene	C_8H_8	7.70 (−16)		Davis et al. (1973)
Dibenz[a,h]anthracene	$C_{22}H_{14}$	5.98 (−15)		Christophorou (1996)
1,4-Dichlorobutane	1,4-$C_4H_8Cl_2$	2.80 (−18)*		Gallup et al. (2003)
2,3-Dichlorobutane	2,3-$C_4H_8Cl_2$	3.11 (−17)*		Gallup et al. (2003)
1,2-Dichloro-2-methylpropane	1,2-Cl_2-2-C_4H_8	4.80 (−17)*		Gallup et al. (2003)
1,6-Dichlorohexane	1,6-$C_6H_{12}Cl_2$	4.00 (−18)*		Gallup et al. (2003)
1,5-Dichloropentane	1,5-$C_5H_{10}Cl_2$	9.00 (−19)*		Gallup et al. (2003)
9,10-Dimethylanthracene	C_6H_{14}	6.31 (−15)		Christophorou (1996)
2,3-Pentanedione	2,3-$C_5H_8O_2$	9.3 (−14)		Bouby et al. (1965)
Naphthalene	$C_{10}H_8$	3.32 (−15)		Christophorou (1996)
1,4-Naphthoquinone	$C_{10}H_6O_2$	5.0 (−14)		Davis et al. (1971)
Pentafluorobutadiene	C_5F_8	1.18 (−13)		Davies et al. (1973)
Perfluorobutane	n-C_4F_{10}	4.1 (−17)		Hunter and Christophorou (1984)
Perfluorocyclohexene	c-C_6F_{10}	3.69 (−13)		Pai et al. (1979)
		3.13 (−13)		Davis et al. (1973)
Perfluorocyclopentane	c-C_5F_{10}	4.2 (−14)		Spyrou et al. (1985)
Perfluorocyclopentene	c-C_5F_8	3.61 (−13)	(0.046)	Pai et al. (1979)
		1.18 (−13)		Davis et al. (1973)
Perfluoro-1,2-dimethyl-cyclobutane	c-C_6F_{12}	1.47 (−13)		Pai et al. (1979)

continued

TABLE A5.1 (continued)
Attachment Rates

Target				
Name	**Formula**	**k_a (m³/s)**	**Temperature K (eV)**	**Reference**
Perfluoro-1,3-dimethylcyclohexane	C_8F_{16}	7.46 (−14)		Pai et al. (1979)
Perfluorohexane	n-C_6F_{14}	2.3 (−16)		Hunter and Christophorou (1984)
Perfluoro-1-heptene	1-C_7F_{14}	3.8 (−15)		Christophorou et al. (1981)
		4.0 (−14)		Davis et al. (1973)
Perfluoroisobutane	i-C_4F_{10}	1.13 (−16)		Spyrou et al. (1985b)
Perfluoromethylcyclohexane	c-C_7F_{14}	2.0 (−14)		Walter et al. (1987)
		4.5 (−14)	205	Alge et al. (1984)
		6.8 (−14)	300	
		1.3 (−13)	452	
		1.6 (−13)	585	
		5.2 (−14)		Woodin et al. (1980)
		5.69 (−14)		Christodoulides and Christophorou (1979b)
		4.36 (−14)	(0.038)	Davis et al. (1973)
		4.87 (−14)	(0.051)	
		5.05(−14)	(0.063)	
		5.62 (−14)	(0.076)	
		6.19 (−14)	(0.090)	
		6.48 (−14)	(0.105)	
		7.39 (−14)	(0.246)	
Perfluoropentane	n-C_5F_{12}	<1.9 (−16)		Hunter and Christophorou (1984)
Perfluorotoluene	C_7F_8	2.44 (−13)		Woodin et al. (1980)
		2.61 (−13)		Pai et al. (1979)
		2.42 (−13)		Davis et al. (1973)
Perfluoronaphthelene	$C_{10}F_8$	2.0 (−13)		Suess et al. (2002)
Phenanthrene	$C_{14}H_{10}$	3.82 (−14)		Christophorou (1996)
Phenol	C_6H_6O	1.66 (−17)		Christophorou (1996)
Pyrene	$C_{16}H_{10}$	4.81 (−15)		Christophorou (1996)
Quinoline	C_9H_7N	5.81 (−15)		Christophorou (1996)
Toluene	C_7H_8	<1.0 (−19)		Christophorou (1996)
1,1,2-Trichloro-2-methylpropane	1,1,2-Cl_3-2-C_4H_8	2.9 (−16)		Onanong et al. (2006)
o-Trifluoromethyl-benzonitrile	o-$CF_3CNC_6H_4$	9.0 (−14)		Miller et al. (2004b)
m-Trifluoromethyl-benzonitrile	m-$CF_3CNC_6H_4$	5.5 (−14)		Miller et al. (2004b)
p-Trifluoromethyl-benzonitrile	p-$CF_3CNC_6H_4$	8.9 (−14)		Miller et al. (2004b)

Note: Attachment rate (k_a) for selected molecules (alphabetical format). a (b) means a × 10^b. Unless otherwise mentioned the energy (temperature) is thermal. Values in bold are best values quoted by Barszczewska et al (2004). Both electron beam and swarm values are shown. (*) indicates electron beam values. The table supplements the data in major sections.

REFERENCES

Adams, N. G., D. Smith, and C. R. Herd, *Int. J. Mass Spectrom. Ion Proc.*, 84, 243, 1988.

Adams, N. G., D. Smith, A. A. Viggiano, J. F. Paulson, and M. J. Henchman, *J. Chem. Phys.*, 84, 6728, 1986.

Aflatooni, K. and P. D. Burrow, *J. Chem. Phys.*, 113, 1455, 2000.

Aflatooni, K. and P. D. Burrow, *Int. J. Mass Spectrum*, 205, 149, 2001.

Alajajian, S. H. and A. Chutjian, *J. Phys. B: At. Mol. Phys.*, 20, 2117, 1987.

Alajajian, S. H., M. T. Bernius, and A. Chutjian, *J. Phys. B: At. Mol. Phys.*, 21, 4021, 1988.

Alge, E., N. G. Adams, and D. Smith, *J. Phys. B: At. Mol. Phys.*, 17, 3827, 1984.

Ayala, J. A., W. E. Wentworth, and E. C. M. Chen, *J. Phys. Chem.*, 85, 3989, 1981.

Bansal, K. M. and R. W. Fessenden, *J. Chem. Phys.*, 59, 1760, 1973.

Barszczewska, W., J. Kopyra, J. Wnorowska, and I. Szamrej, *J. Phys. Chem. A*, 107, 11427, 2003.

Barszczewska, W., J. Kopyra, J. Wnorowska, and I. Szamrej, *Int. J. Mass Spectrom.*, 233, 199, 2004.

Blaunstein, R. P. and L. G. Christophorou, *J. Phys. Chem.*, 49, 1526, 1968.

Bouby, L., F. Fiquet-Fayard, and H. Abgrall, *C. R. Acad. Sci. Paris*, 261, 4059, 1965, cited by Alajajian et al. *J. Phys. B: At. Mol. Phys.*, 20, 5567, 1987.
Burns, S. J., J. M. Mathews, and D. L. Mcfadden, *J. Phys. Chem.*, 100, 19436, 1996.
Chen, C. L. and P. J. Chantry, *Bull. Ann. Phys. Society.* 17, 2133, 1972 cited by Smith et al., 1984.
Christodoulides, A. A. and L. G. Christophorou, *J. Chem. Phys.*, 54, 4691, 1971.
Christodoulides, A. A., L. G. Christophorou, R. Y. Pai, and C. M. Tung, *J. Chem. Phys.*, 70, 1156, 1979a.
Christodoulides, and L. G. Christophorou, *Chem. Phys. Lett.*, 61, 553, 1979b.
Christopher, J. M., T. Cheng-ping, and L. M. David, *J. Chem. Phys.*, 91, 2194, 1989.
Christophorou, L. G., *Z. Phys. Chem.*, 195, 195, 1996.
Christophorou, L. G. and D. Hadjiantoniou, *Chem. Phys. Letters*, 419, 405, 2006.
Christophorou, L. G. and J. K. Olthoff, *Fundamental Electron Interactions with Plasma Processing Gases*, Kluwer Academic/Plenum Publishers, New York, 2004, p. 490.
Christophorou, L. G., R. N. Compton, and H. W. Dickson, *J. Chem. Phys.*, 48, 1949, 1968.
Christophorou, L. G., D. L. McCorkle, and D. Pittman, *J. Chem. Phys.*, 60, 1183, 1974.
Christophorou, L. G., D. L. McCorkle, and J. G. Carter, *J. Chem. Phys.*, 54, 253, 1971.
Christophorou, L. G., R. A. Mathis, D. R. James, and D. L. McCorkle, *J. Phys. D: Appl. Phys.*, 14, 1889, 1981.
Christophorou, L. G., R. N. Crompton, G. S. Hurst, and P. W. Reinhardt, *J. Chem. Phys.*, 45, 536, 1966.
Cicman, P., G. A. Buchanan, G. Marston, B. Gulejovà. J. D. Skalný, N. J. Mason, P. Scheier, and T. D. Märk, *J. Chem. Phys.*, 121, 9891, 2004.
Compton, R. N. and L. G. Christophorou, *Phys. Rev.*, 154, 110, 1967.
Crompton, R. W., G. N. Haddad, R. Hegerberg, and A. G. Robertson, *J. Phys. B: At. Mol. Phys.*, 15, L483, 1982.
Datskos, P. G. and L. G. Christophorou, *J. Chem. Phys.*, 90, 2626, 1989.
Datskos, P. G., L. G. Christophorou, and J. G. Carter, *Chem. Phys. Lett.*, 168, 324, 1990.
Datskos, P. G., L. G. Christophorou, and J. G. Carter, *J. Chem. Phys.*, 97, 9031, 1992.
Davis, F. J., R. N. Compton, and D. R. Nelson, *J. Chem. Phys.*, 59, 2324, 1973.
Dunning, F. B., *J. Phys. B: At. Mol. Opt. Phys.*, 28, 1645, 1995.
Fessenden, R. W. and K. M. Bansal, *J. Chem. Phys.*, 53, 3468, 1970.
Friedman, J. F., A. E. Stevens, T. M. Miller, and A. A. Viggiano, *J. Chem. Phys.*, 124, 224306, 2006.
Gallup, G. A., K. Aflatooni, and P. D. Burrow, *J. Chem. Phys.*, 118, 2562, 2003.
Gant, K. S. and L. G. Christophorou, *J. Chem. Phys.*, 65, 2977, 1976.
Goyette, A. N., J. de Urquiho, Y. Wang, L. G. Christophorou, and J. K. Olthoff, *J. Chem. Phys.*, 114, 8932, 2001.
Hunter, S. R. and L. G. Christophorou, *J. Chem. Phys.*, 80, 6150, 1984.
Hunter, S. R., J. G. Carter, and L. G. Christophorou, *J. Appl. Phys.*, 58, 3001, 1985.
Hunter, S. R., L. G. Christophorou, D. L. McCorkle, I. Sauers, H. W. Ellis, and D. R. James, *J. Phys. D: Appl. Phys.*, 16, 573, 1983.
Ipolyi, I., S. Matejcik, P. Lukac, J. D. Skalny, P. Mach, and J. Urban, *Int. J. Mass Spectr.*, 233, 193, 2004.
Jarvis, G. K., R. A. Kennedy, and C. A. Mayhew, *Int. J. Mass Spectrom.*, 205, 253, 2001.
Johnson, J. P., L. G. Christophorou, and J. G. Carter, *J. Chem. Phys.*, 67, 2196, 1977.
Kennedy, R. A. and C. A. Mayhew, *Int. J. Mass Spectrom.*, 206, I, 2001.
Klar, D., M.-W. Ruf, I. I. Fabrikant, and H. Hotop, *J. Phys. B: At. Mol. Opt. Phys.*, 34, 3855, 2001.
Lj, Z. Petrović, and R. W. Crompton, *J. Phys. B: At. Mol. Phys.*, 20, 5557. 1987.
Lj, Z., Petrović, W. C. Wang, and L. C. Lee, *J. Chem. Phys.*, 90, 3145, 1989.
Marotta, C. J., C. Tsai, and D. L. McFadden, *J. Chem. Phys.*, 91, 2194, 1989.
Mayhew, C. A., A. Critchley, and G. K. Jarvis, *Int. J. Mass Spectrom.*, 233, 259, 2004.
McCorkle, D. L., A. A. Christodoulides, and L. G. Christophorou, in L. G. Christophorou and M. O. Pace (Eds.), *Gaseous Dielectrics IV*, Pergamon, New York, 1984, p. 12.
McCorkle, D. L., A. A. Christoudoulides, L. G. Christophorou, and I. Szamrej, *J. Chem. Phys.*, 76, 753, 1982.
McCorkle, D. L., I. Szamrej, and L. G. Christophorou, *J. Chem. Phys.*, 77, 5542, 1982.
McCorkle, D. L., L. G. Christophorou, A. A. Christodoulides, and L. Pichiarella, *J. Chem. Phys.*, 85, 1966, 1986.
Miller, T. M., S. A. Arnold, and A. A. Viggiano, *J. Chem. Phys.*, 116, 6021, 2002.
Miller, T. M., J. F. Friedman, and A. A. Viggiano, *J. Chem. Phys.*, 120, 7024, 2004a.
Miller, T. M., A. A. Viggiano, S. A. Arnold, and J. T. Jayne, *J. Chem. Phys.*, 102, 6021, 1995.
Miller, T. M., A. A. Viggiano, J. F. Friedman, and J. M. Van Doren, *J. Chem. Phys.*, 121, 9993, 2004b.
Miller, T. M., J. F. Friedman, A. E. S. Miller, and J. F. Paulson, *Int. J. Mass Spectrom. Ion Proc.*, 149/150, 111, 1995.
Miller, T. M., J. V. Seely, W. B. Knighton, R. F. Meads, A. A. Viggiano, R. A. Morris, J. M. van Doren, J. Gu, and H. F. Schafer III, *J. Chem. Phys.*, 109, 578, 1998.
Mothes, K. G., E. Schultes, and R. N. Schindler, *J. Phys. Chem.*, 76, 3758, 1972.
Onanong, S., P. D. Burrow, S. D. Comfort, and P. J. Shea, *J. Phys. Chem. A*, 110, 4363, 2006.
Orient, O. J., A. Chutjian, R. W. Crompton, and B. Cheung, *Phys. Rev. A*, 39, 4494, 1989.
Pai, R. Y., L. G. Christophorou, and A. A. Christodoulides, *J. Chem. Phys.*, 70, 1169, 1979.
Petrović, Z. L. and R. W. Crompton, *J. Phys. B.*, 20, 5552, 1987.
Petrović, Z. L. W. C. Wang, and L. C. Lee, *J. Chem. Phys.*, 90, 3145, 1989.
Rosa, A. and I. Szamrej, *J. Phys. Chem.*, 104, 67, 2000.
Rosa, A., W. Barszczewska, M. Foryś, and I. Szamrej, *Int. J. Mass Spectrom.* 205, 85, 2001.
Schultes, E., A. A. Christodoulides, and R. N. Schindler, Studies by the electron cyclotron resonance technique, *Chem. Phys.*, 8, 354–365, 1975.
Shimamori, H., T. Sunagawa, and Y. Tatsumi, *J. Chem. Phys.*, 99, 7787, 1993.
Shimamori, H., Y. Tatsumi, Y. Ogawa, and T. Sunagawa, *J. Chem. Phys.*, 97, 6335, 1992.
Sides, G. D. and T. O. Tierman, *J. Chem. Phys.*, 65, 3392, 1976.
Sides, G. D. and T. O. Tierman, *J. Chem. Phys.*, 67, 2382, 1977.
Sides, G. D., T. O. Tierman, and R. J. Harnaharan, *J. Chem. Phys.*, 65, 1966, 1976.
Smith, D. and N. G. Adams, *J. Phys. B: At. Mol. Phys.*, 20, 4903, 1987.
Smith, D., N. G. Adams, and E. Alge, *J. Phys. B: At. Mol. Phys.*, 17, 461, 1984.

Smith, D., C. R. Herd, N. G. Adams, and J. F. Paulson, *Int. J. Mass Spectrom. Ion Proc.*, 96, 341, 1990.

Španěl, P. and D. Smith, *Int. J. Mass Spectrom.*, 205, 243, 2001.

Spyrou, S. M., and L. G. Christophorou, *J. Chem. Phys.*, 82, 2620, 1985a.

Spyrou, S. M., S. R. Hunter, and L. G. Christophorou, *J. Chem. Phys.* 81, 4481, 1984.

Spyrou, S. M., S. R. Hunter, and L. G. Christophorou, *J. Chem. Phys.*, 83, 641, 1985b.

Stockdale, J. A. and G. S. Hurst, *J. Chem. Phys.*, 41, 255, 1964.

Stockdale, J. A., F. J. Davis, R. N. Compton, and C. E. Klots, *J. Chem. Phys.*, 60, 4279, 1974.

Stockdale, J. A., D. R. Nelson, F. J. Davis, and R. N. Compton, *J. Chem. Phys.*, 56, 3336, 1972.

Suess, L., R. Parthasarathy, and F. B. Dunning, *J. Chem. Phys.*, 117, 11222, 2002.

Suess, L., R. Parthasarathy, and F. B. Dunning, *J. Chem. Phys.*, 118, 6205, 2003.

Sunagawa, T. and H. Shimamori, *Int. J. Mass Spectrom., Ion Proc.*, 149, 123, 1995.

Sunagawa, T. and H. Shimamori, *J. Chem. Phys.*, 107, 7876, 1997.

Sunagawa, T. and H. Shimamori, *Int. J. Mass Spectrom.*, 205, 285, 2001.

Szamrej, I., J. Jówko, and M. Foryś, *Radiat. Phys. Chem.*, 48, 65, 1996.

Szamrej, I., W. Tchórzewska, H. Kość, and M. Foryś, *Radiation Phy. Chem.*, 48, 65, 1996.

Tav, C., P. G. Datskos, and L. A. Pinnaduwage, *J. Appl. Phys.*, 84, 5805, 1998.

Truby, F. K., *Phys. Rev.*, 188, 508, 1969.

van Doren, J. M., D. M. Kerr, T. M. Miller, and A. A. Viggiano, *J. Chem. Phys.*, 123, 114303, 2005.

van Doren, J. M., J. F. Friedman, T. M. Miller, A. A. Viggiano, S. Denifl, P. Scheier, and T. D. Mark, *J. Chem. Phys.*, 124, 124322, 2006.

van Doren, J. M., J. McClellan, T. M. Miller, J. F. Paulson, and A. A. Viggiano, *J. Chem. Phys.*, 105, 104, 1996.

Van Doren, J. M., W. M. Foley, J. E. McClellan, T. M. Miller, A. D. Kowalak, and A. A. Viggiano, *Int. J. Mass Spectrom. Ion Proc.*, 149/150, 423, 428, 1995.

Walter, C. W., C. B. Johnson, A. Kalamarides, D. F. Gray, K. A. Smith, and F. B. Dunning, *J. Phys. Chem.*, 91, 4284, 1987.

Wang, W. C. and L. C. Lee, *IEEE Trans. Plasma Sci.* PS-15, 460, 1987.

Wang, W. C. and L. C. Lee, *J. Appl. Phys.*, 63, 4905, 1988.

Wentworth, W. E., T. Limero, and E. C. M. Chen, *J. Phys. Chem.*, 91, 241, 1987.

Williamson, D. H., C. A. Mayhew, W. B. Knighton, and E. P. Grimsrud, *J. Chem. Phys.*, 113, 11035, 2000.

Woodin, R. L., M. S. Foster, and J. L. Beauchamp, *J. Chem. Phys.*, 72, 4223, 1980.

Appendix 6: Atomic Ionization Cross Sections

CONTENTS

Atomic ionization cross sections are often required for theoretical calculation of ionization cross sections by the additivity rule. Tables A6.1 through A6.4 provide cross sections for selected atoms. The data supplement those given in Chapter 1. First entry in each table is the ionization potential.

TABLE A6.1
Ionization Cross Sections for Hydrogen Atoms in Units of 10^{-21} m²

Energy (eV)	Q_i	Energy (eV)	Q_i	Energy (eV)	Q_i
13.598	0.000	22.6	3.76	66.0	6.11
14.6	0.544	23.3	4.01	69.0	6.11
14.8	0.661	24.0	4.15	72.1	6.01
15.0	0.762	24.8	4.30	75.5	5.96
15.1	0.820	25.6	4.44	79.5	5.91
15.2	0.870	26.6	4.57	84.0	5.84
15.4	0.990	27.3	4.75	89.0	5.78
15.6	1.08	28.3	4.95	94.0	5.59
15.9	1.25	29.3	5.01	102.0	5.40
16.1	1.37	30.5	5.10	103.0	5.42
16.4	1.45	31.6	5.27	113.0	5.23
16.6	1.63	32.8	5.39	121.0	5.07
16.9	1.68	34.1	5.53	130.2	5.05
17.1	1.73	35.4	5.59	138.2	4.83
17.4	1.96	36.7	5.74	148.2	4.62
17.6	2.07	38.1	5.83	158.2	4.55
17.9	2.15	39.6	5.78	168.2	4.43
18.1	2.22	41.2	5.89	178.2	4.28
18.4	2.35	42.9	6.02	188.2	4.10
18.7	2.50	44.7	6.07	198.2	3.98
19.0	2.61	46.6	6.08	213.2	3.79
19.3	2.75	48.6	6.23	228.2	3.61
19.6	2.81	50.7	6.27	248.2	3.43
20.0	2.93	52.9	6.19	268.2	3.31
20.4	3.11	55.2	6.23	288.0	3.03
20.9	3.34	57.6	6.21	317.9	2.84
21.4	3.39	60.1	6.13	347.9	2.66
22.0	3.61	63.0	6.14	387.9	2.50
427.9	2.31	898.2	1.26	1998.1	0.631

TABLE A6.1 (continued)
Ionization Cross Sections for Hydrogen Atoms in Units of 10^{-21} m²

Energy (eV)	Q_i	Energy (eV)	Q_i	Energy (eV)	Q_i
467.9	2.15	998.2	1.13	2198.1	0.577
508.2	2.00	1100.0	1.05	2448.1	0.525
548.2	1.86	1200.0	0.982	2698.1	0.472
598.2	1.77	1300.0	0.914	2998.1	0.437
668.2	1.59	1506.7	0.807	3298.1	0.403
748.2	1.47	1662.7	0.721	3648.1	0.370
818.2	1.38	1848.1	0.673	3998.1	0.339

Source: Adapted from Shah, M. B., D. S. Elliot, and H. B. Gilbody, *J. Phys. B: At. Mol. Opt. Phys.*, 20, 3501, 1987.

Note: See Figure A6.1 for graphical presentation.

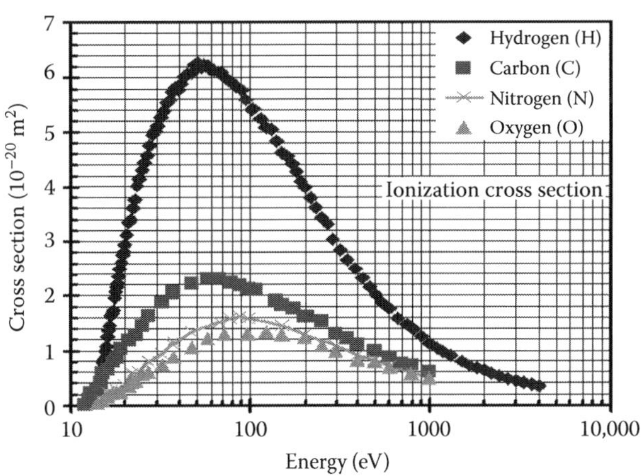

FIGURE A6.1 Ionization cross sections for selected atomic gases. Tabulated values are given in Tables A6.1 for H (Shaw et al., 1987) and A6.3 for the remaining gases (Brook et al., 1978).

TABLE A6.2
Ionization Cross Sections for Selected Atoms

	Ionization Cross Section (10^{-20} m^2)			
Energy (eV)	Bromine Hayes et al. (1978)	Chlorine Hayes et al. (1987)	Fluorine Hayes et al. (1987)	Iodine Hayes et al. (1987)
10.45				0.00
11				0.25
11.814	0.00			—
12	0.04			0.62
12.968	—	0.00		–
13	0.22	0.02		1.08
14	0.54	0.24		1.49
15	0.81	0.52		1.92
16	1.04	0.74		2.38
17	1.27	1.01		2.75
17.423	—	—	0.00	—
18	1.52	1.27	0.03	3.08
19	1.79	1.50	0.06	3.50
20	2.00	1.65	0.07	3.82
21	2.23	1.81	0.09	4.15
22	2.44	1.99	0.12	4.43
23	2.67	2.16	0.14	4.59
24	2.83	2.34	0.16	4.74
25	2.98	2.50	0.19	4.92
26	3.16	2.59	0.22	5.05
27	3.26	2.71	0.26	5.17
28	3.36	2.80	0.29	5.34
29	3.51	2.89	0.30	5.43
30	3.62	2.96	0.34	5.53
32	3.78	3.16	0.39	5.63
34	3.96	3.20	0.41	5.78
36	4.05	3.27	0.46	5.93
38	4.14	3.35	0.52	5.94
40	4.34	3.35	0.48	6.03
45	4.37	3.43	0.61	6.01
50	4.43	3.44	0.68	5.96
55	4.40	3.47	0.72	5.97
60	4.43	3.49	0.78	5.93
65	4.42	3.49	0.83	5.91
70	4.36	3.47	0.87	5.91
75	4.37	3.44	0.88	5.90
80	4.31	3.43	0.90	5.85
85	4.23	3.43	0.92	5.80
90	4.22	3.37	0.95	5.74
95	4.46	3.34	0.96	5.63
100	4.10	3.31	0.98	5.55
105	4.06	3.23	0.97	5.47
110	3.99	3.20	0.98	5.41
115	3.97	3.21	0.96	5.34
120	3.89	3.15	0.98	5.26
125	3.85	3.13	0.98	5.16
130	3.83	3.07	0.98	5.08
135	3.79	3.05	0.98	4.98
140	3.71	3.01	0.96	4.89
145	3.65	2.97	0.97	4.81
150	3.61	2.96	0.98	4.74
155	3.55	2.91	0.95	4.69
160	3.55	2.85	0.96	4.64
165	3.49	2.84	0.96	4.59

TABLE A6.2 (continued)
Ionization Cross Sections for Selected Atoms

	Ionization Cross Section (10^{-20} m^2)			
Energy (eV)	Bromine Hayes et al. (1987)	Chlorine Hayes et al. (1987)	Fluorine Hayes et al. (1987)	Iodine Hayes et al. (1987)
170	3.44	2.81	0.94	4.52
175	3.40	2.78	0.96	4.42
180	3.32	2.72	0.92	4.35
185	3.29	2.71	0.93	4.28
190	3.21	2.68	0.92	4.21
195	3.18	2.61	0.91	4.14
200	3.14	2.63	0.90	4.07

Note: See Figure A6.2 for graphical presentation. First entry for each gas is the ionization potential.

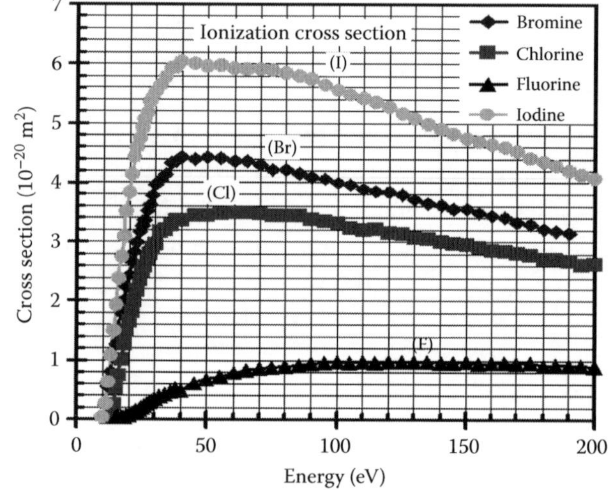

FIGURE A6.2 Ionization cross sections for selected atoms. (Adapted from Hayes, T. R., R. C. Wetzel, and R. S. Freund, *Phys. Rev.* A 35, 578, 1987.)

TABLE A6.3
Ionization Cross Sections of Selected Atoms in Units of (10^{-20} m^2),

Carbon (C)		Oxygen (O)		Nitrogen (N)	
Energy (eV)	Q_i (10^{-20} m^2)	Energy (eV)	Q_i (10^{-20} m^2)	Energy (eV)	Q_i (10^{-20} m^2)
11.26	0.00	13.61	0.00	14.53	0.00
11.9	0.010	13.9	0.057	14.9	0.122
12.6	0.116	14.9	0.093	15.9	0.248
12.9	0.225	15.9	0.170	16.9	0.232
13.9	0.283	16.9	0.231	17.9	0.280
14.4	0.412	17.9	0.246	19.9	0.372
14.9	0.581	18.9	0.241	21.9	0.568
15.9	0.653	20.9	0.353	23.9	0.608
16.9	0.800	21.9	0.427	26.9	0.799
17.9	0.872	22.9	0.439	31.9	0.945
18.9	1.037	24.9	0.576	36.9	1.157
20.9	1.184	26.9	0.596	37.0	1.117

TABLE A6.3 (continued)
Ionization Cross Sections of Selected Atoms in Units of (10^{-20} m²),

Carbon (C)		Oxygen (O)		Nitrogen (N)	
Energy (eV)	Q_i (10^{-20} m²)	Energy (eV)	Q_i (10^{-20} m²)	Energy (eV)	Q_i (10^{-20} m²)
22.9	1.282	31.9	0.735	47	1.341
24.9	1.451	36.9	0.879	67	1.513
26.9	1.630	47	1.067	77	1.590
31.9	1.882	57	1.223	87	1.608
36.9	2.073	72	1.303	97	1.586
37.0	2.042	97	1.301	147	1.445
47	2.215	122	1.334	197	1.289
57	2.309	147	1.281	297	1.097
67	2.315	197	1.239	397	0.914
77	2.234	247	1.093	497	0.816
87	2.193	297	0.994	597	0.697
97	2.118	397	0.816	797	0.587
107	2.103	497	0.784	997	0.490
137	1.909	597	0.686		
147	1.834	797	0.571		
172	1.765	997	0.482		
197	1.608				
207	1.614				
247	1.531				
297	1.310				
347	1.220				
397	1.117				
497	0.994				
597	0.874				
697	0.744				
797	0.727				
997	0.596				

Source: Adapted from E. Brook, M. F. A. Harrison, and A. C. H. Smith, *J. Phys. B: At. Mol. Phys.*, 11, 3115, 1978 [4].
Note: The first entry for each gas is the ionization potential.

TABLE A6.4
Ionization Cross Sections for Selected Atomic Gases

Germanim (Ge)		Silicon (Si)		Sulfur (S)	
Energy (eV)	Q_i (10^{-20} m²)	Energy (eV)	Q_i (10^{-20} m²)	Energy (eV)	Q_i (10^{-20} m²)
7.899	0.00				
8	0.40	8.152	0.00		
9	0.95	9	0.86		
10	1.62	10	1.63	10.36	0.00
11	2.34	11	2.56	11	0.47
12	3.10	12	3.34	12	0.65
13	3.85	13	4.12	13	1.06
14	4.46	14	4.65	14	1.45
15	4.90	15	5.12	15	1.72
16	5.34	16	5.52	16	2.09
17	5.80	17	5.85	17	2.39

TABLE A6.4 (continued)
Ionization Cross Sections for Selected Atomic Gases

Germanim (Ge)		Silicon (Si)		Sulfur (S)	
Energy (eV)	Q_i (10^{-20} m²)	Energy (eV)	Q_i (10^{-20} m²)	Energy (eV)	Q_i (10^{-20} m²)
18	6.08	18	6.05	18	2.59
19	6.27	19	6.20	19	2.84
20	6.40	20	6.34	20	3.21
21	6.59	21	6.48	21	3.43
22	6.75	22	6.56	22	3.56
23	6.96	23	6.53	23	3.66
24	7.07	24	6.63	24	3.77
25	7.11	25	6.64	25	3.95
26	7.19	26	6.64	26	4.00
27	7.28	27	6.69	27	4.08
28	7.32	28	6.77	28	4.20
29	7.41	29	6.78	29	4.26
30	7.52	30	6.89	30	4.29
32	7.68	32	7.00	32	4.39
34	7.72	34	7.13	34	4.43
36	7.78	36	7.15	36	4.52
38	7.81	38	7.12	38	4.48
40	7.81	40	7.18	40	4.52
45	7.86	45	7.10	45	4.66
50	7.80	50	7.02	50	4.78
55	7.77	55	6.89	55	4.89
60	7.64	60	6.82	60	4.91
65	7.52	65	6.66	65	4.93
70	7.36	70	6.55	70	4.87
75	7.27	75	6.38	75	4.88
80	7.15	80	6.23	80	4.92
85	7.07	85	6.08	85	4.88
90	6.92	90	6.01	90	4.84
95	6.83	95	5.87	95	4.79
100	6.73	100	5.78	100	4.67
105	6.64	105	5.69	105	4.71
110	6.55	110	5.63	110	4.63
115	6.50	115	5.52	115	4.55
120	6.39	120	5.40	120	4.55
125	6.29	125	5.36	125	4.41
130	6.19	130	5.27	130	4.36
135	6.09	135	5.17	135	4.32
140	6.02	140	5.12	140	4.27
145	5.94	145	5.05	145	4.22
150	5.85	150	4.94	150	4.23
155	5.77	155	4.92	155	4.13
160	5.70	160	4.82	160	4.10
165	5.62	165	4.76	165	4.04
170	5.55	170	4.71	170	4.00
175	5.46	175	4.66	175	3.98
180	5.36	180	4.58	180	3.86
185	5.26	185	4.47	185	3.71
190	5.13	190	4.42	190	3.73
195	5.04	195	4.33	195	3.65
200	4.95	200	4.30	200	3.58

Source: Adapted from Freund, R. S. et al., *Phys. Rev.* A 41, 3575, 1990.
Note: First entry for each gas is the ionization potential. See Figure A6.3 for graphical presentation.

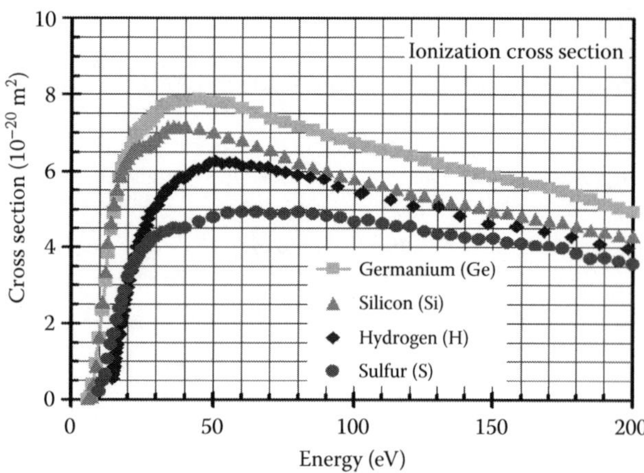

FIGURE A6.3 Ionization cross sections for selected atomic gases. The cross section for H is added for comparison (Shaw et al., 1987). (Adapted from Freund, R. S. et al., *Phys. Rev.* A41, 3575, 1990.)

REFERENCES

Brook, E., M. F. A. Harrison, and A. C. H. Smith, *J. Phys. B: At. Mol. Phys.*, 11, 3115, 1978.

Freund, R. S., R. C. Wetzel, R. J. Shul, and T. R. Hayes, *Phys. Rev.* A 41, 3575, 1990.

Hayes, T. R., R. C. Wetzel, and R. S. Freund, *Phys. Rev.* A 35, 578, 1987.

Shah, M. B., D. S. Elliot, and H. B. Gilbody, *J. Phys. B: At. Mol. Opt. Phys.*, 20, 3501, 1987.

Appendix 7: Ionization Cross Sections—Molecules

CONTENTS

TABLE A7.1
Tabulated Values in Nitriles Due to Courtesy of Professor Harland

Basner et al. (2003)		Nitriles (Aceto, Trifluoroaceto, Trichloroaceto) Bart et al. (2001)			
Diborane (B_2H_6)			CH_3CN	CF_3CN	CCl_3CN
Energy (eV)	Q_i (10^{-20} m^2)	Energy (eV)	Q_i (10^{-20} m^2)	Q_i (10^{-20} m^2)	Q_i (10^{-20} m^2)
18.5	3.583	12	0.45	0.18	1.02
19	3.852	16	0.83	0.38	1.92
19.5	4.183	20	1.89	0.99	3.93
20	4.531	23	2.92	1.68	6.02
21	4.983	27	3.76	2.36	7.79
22	5.46	31	4.10	2.85	8.94
23	5.857	34	4.47	3.35	9.94
24	6.237	38	4.82	3.82	10.71
25	6.569	42	5.16	4.23	11.39
26	6.864	45	5.46	4.56	11.96
28	7.356	49	5.66	4.84	12.46
30	7.889	53	5.84	5.08	12.87
32	8.307	57	5.97	5.30	13.22
36	8.984	60	6.10	5.49	13.51
38	9.295	64	6.16	5.66	13.75
42	9.75	68	6.23	5.78	13.90
46	10.065	71	6.28	5.87	14.04
50	10.318	75	6.30	5.95	14.08
55	10.51	79	6.30	6.02	14.09
60	10.639	82	6.33	6.12	14.12
65	10.704	86	6.34	6.18	14.15
70	10.74	90	6.32	6.21	14.10
75	10.732	94	6.27	6.22	13.95
80	10.678	97	6.23	6.24	13.80
90	10.416	101	6.18	6.27	13.74
100	10.104	105	6.12	6.31	13.67
110	9.844	108	6.07	6.32	13.54
120	9.601	112	6.04	6.31	13.39
140	9.16	116	5.99	6.29	13.27

TABLE A7.1 (continued)
Tabulated Values in Nitriles Due to Courtesy of Professor Harland

Basner et al. (2003)		Nitriles (Aceto, Trifluoroaceto, Trichloroaceto) Bart et al. (2001)			
Diborane (B_2H_6)			CH_3CN	CF_3CN	CCl_3CN
Energy (eV)	Q_i (10^{-20} m^2)	Energy (eV)	Q_i (10^{-20} m^2)	Q_i (10^{-20} m^2)	Q_i (10^{-20} m^2)
160	8.729	119	5.97	6.29	13.31
180	8.333	123	5.93	6.31	13.33
200	7.965	127	5.89	6.31	13.24
		130	5.81	6.26	13.00
		134	5.75	6.19	12.73
		138	5.69	6.13	12.53
		142	5.65	6.10	12.41
		145	5.63	6.10	12.41
		149	5.60	6.12	12.47
		153	5.57	6.11	12.42
		156	5.52	6.06	12.26
		160	5.49	6.00	12.01
		164	5.47	5.94	11.83
		167	5.43	5.89	11.69
		171	5.37	5.86	11.59
		175	5.29	5.82	11.50
		179	5.26	5.80	11.46
		182	5.27	5.79	11.46
		186	5.26	5.78	11.48
		190	5.21	5.75	11.44
		193	5.14	5.70	11.29
		197	5.07	5.63	11.06
		201	5.01	5.57	10.85
		204	4.96	5.53	10.73
		208	4.91	5.51	10.64
		212	4.87	5.49	10.55
		215	4.85		10.52

Note: See Figure A7.1 for graphical presentation.

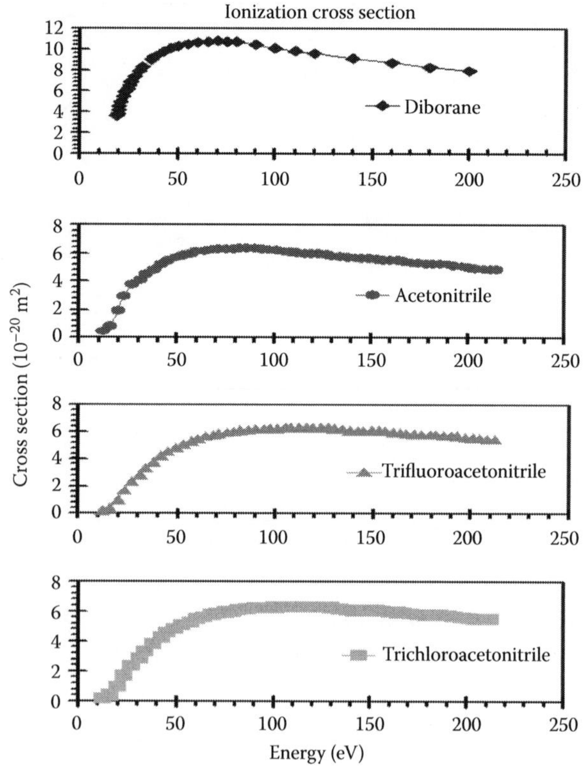

FIGURE A7.1 Ionization cross section for selected molecules (not covered in major sections).

TABLE A7.2
Ion Appearance Potentials in Diborane (B_2H_6)

Ion	Potential (eV)	Ion	Potential (eV)	Ion	Potential (eV)
$B_2H_6^+$	11.4	BH_3^+	15.1	B_2H^+	21.0
$B_2H_5^+$	11.5	H_2^+	16.0	B_2^+	23.5
$B_2H_4^+$	12.0	BH_2^+	16.6	H^+	30.0
$B_2H_2^+$	14.0	BH^+	19.0	H_2^+	36.0
$B_2H_3^+$	14.3	B^+	19.4		

Source: Adapted from Basner, R., M. Schmidt, and K. Becker, *J. Chem. Phys.*, 118, 2153, 2003.

TABLE A7.3
Ionization Cross Sections in Formates in Units of 10^{-20} m^2

	Formates				
Energy (eV)	Methyl $CH_3OC(O)H$	Ethyl $C_2H_5OC(O)H$	*n*-Propyl *n*-$C_3H_7OC(O)H$	Isopropyl *i*-$C_3H_7OC(O)H$	*n*-Butyl *n*-$C_4H_9OC(O)H$
15	0.52	1.94	2.48	2.49	3.23
20	1.30	3.27	4.27	4.20	5.10
25	2.22	4.50	5.93	5.77	7.00
30	3.13	5.50	7.27	7.06	8.65
35	3.90	6.39	8.42	8.13	10.03
40	4.59	7.10	9.43	9.02	11.11
45	5.15	7.76	10.25	9.82	11.91
50	5.62	8.24	10.75	10.44	12.56
55	5.97	8.70	11.30	10.90	13.14
60	6.32	9.12	11.78	11.38	13.68
65	6.56	9.37	12.17	11.72	14.15
70	6.82	9.60	12.47	12.06	14.51
75	7.01	9.83	12.78	12.32	14.81
80	7.17	10.05	13.03	12.53	15.07
85	7.30	10.18	13.24	12.70	15.28
90	7.40	10.27	13.42	12.80	15.44
95	7.43	10.33	13.51	12.88	15.54
100	7.47	10.37	13.56	12.97	15.56
105	7.52	10.41	13.60	12.99	15.55
110	7.54	10.42	13.56	13.00	15.53
115	7.56	10.44	13.54	13.00	15.54
120	7.56	10.44	13.49	12.99	15.54
125	7.55	10.40	13.53	12.98	15.51
130	7.54	10.35	13.47	12.96	15.45
135	7.54	10.33	13.43	12.88	15.40
140	7.52	10.24	13.34	12.79	15.32
145	7.51	10.19	13.24	12.71	15.23
150	7.47	10.17	13.13	12.66	15.16
155	7.45	10.12	13.10	12.61	15.07
160	7.40	10.11	13.02	12.56	14.98
165	7.36	10.06	12.93	12.51	14.91
170	7.32	9.96	12.88	12.45	14.84
175	7.30	9.87	12.84	12.35	14.75
180	7.26	9.82	12.76	12.31	14.68
185	7.19	9.76	12.69	12.19	14.58
190	7.13	9.69	12.63	12.12	14.45
195	7.07	9.62	12.49	12.06	14.31
200	7.03	9.57	12.39	11.96	14.20
205	6.96	9.50	12.31	11.83	14.08
210	6.91	9.38	12.22	11.74	13.96
215	6.84	9.29	12.09	11.59	13.84
220	6.80	9.22	11.91	11.47	13.71
225	6.75	9.13	11.88	11.41	13.56
230	6.68	9.04	11.76	11.29	13.41
235	6.64	8.95	11.66	11.20	13.29
240	6.60	8.93	11.56	11.14	13.18
245	6.53	8.87	11.42	11.04	13.08
250	6.48	8.81	11.34	10.92	12.99
255	6.46	8.77	11.30	10.84	12.89
260	6.40	8.70	11.20	10.75	12.78
265	6.35	8.61	11.15	10.68	12.68
270	6.32	8.54	11.07	10.61	12.57
275	6.25	8.48	11.02	10.52	12.47
280	6.21	8.41	10.88	10.44	12.36
285	6.16	8.34	10.78	10.41	12.27

Source: Adapted from Hudson, J. E. et al., *Int. J. Mass Spectrom.*, 248, 42, 2006.
Note: Tabulated values by courtesy of Professor Harland. See Figure A7.2 for graphical presentation.

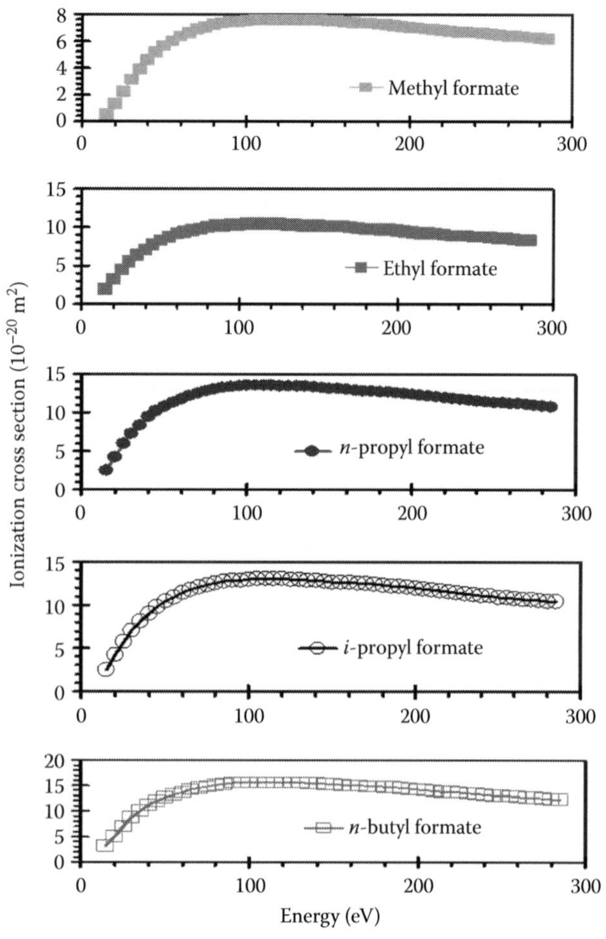

FIGURE A7.2 Ionization cross sections for selected formates (not covered in major sections). (Adapted from Bart, M. et al., *Phys. Chem. Chem. Phys.*, 3, 800, 2001.) Tabulated values are due to courtesy of Professor Harland.

TABLE A7.4
Ionization Cross Sections in Esters in Units of 10^{-20} m^2

Energy (eV)	Isobutyl formate $i\text{-}C_4H_9OC(O)H$	Acetate: Methyl $C_3H_6O_2$	Ethyl $C_4H_8O_2$	n-propyl $n\text{-}C_5H_{10}O_2$	Isopropyl $i\text{-}C_5H_{10}O_2$
15	2.98	0.75	1.35	3.00	2.92
20	5.12	1.97	2.65	5.14	4.94
25	7.05	3.27	4.47	7.06	6.76
30	8.65	4.56	6.15	8.60	8.21
35	10.03	5.68	7.57	9.94	9.41
40	11.10	6.64	8.79	11.10	10.51
45	11.89	7.48	9.70	11.96	11.33
50	12.57	8.08	10.53	12.68	12.04
55	13.13	8.42	11.14	13.33	12.74
60	13.68	8.90	11.67	13.79	13.29
65	14.12	9.21	12.09	14.26	13.65
70	14.45	9.49	12.55	14.71	14.11

TABLE A7.4 (continued)
Ionization Cross Sections in Esters in Units of 10^{-20} m^2

Energy (eV)	Isobutyl formate $i\text{-}C_4H_9OC(O)H$	Acetate: Methyl $C_3H_6O_2$	Ethyl $C_4H_8O_2$	n-propyl $n\text{-}C_5H_{10}O_2$	Isopropyl $i\text{-}C_5H_{10}O_2$
75	14.79	9.74	12.85	14.94	14.37
80	15.08	9.94	13.11	15.17	14.60
85	15.22	10.10	13.38	15.32	14.78
90	15.43	10.23	13.44	15.45	14.98
95	15.54	10.30	13.57	15.58	15.08
100	15.58	10.32	13.61	15.63	15.13
105	15.53	10.42	13.73	15.69	15.17
110	15.53	10.48	13.76	15.70	15.20
115	15.57	10.48	13.76	15.66	15.23
120	15.56	10.47	13.80	15.63	15.21
125	15.50	10.42	13.79	15.58	15.19
130	15.51	10.41	13.73	15.55	15.14
135	15.47	10.37	13.74	15.47	15.09
140	15.37	10.35	13.66	15.42	15.02
145	15.22	10.31	13.67	15.38	14.89
150	15.17	10.28	13.56	15.35	14.83
155	15.06	10.21	13.54	15.28	14.80
160	15.00	10.17	13.45	15.19	14.75
165	14.93	10.12	13.38	15.10	14.68
170	14.88	0.02	13.31	15.01	14.59
175	14.80	9.97	13.23	14.90	14.44
180	14.70	9.88	13.11	14.81	14.39
185	14.62	9.84	13.00	14.66	14.31
190	14.50	9.76	12.90	14.54	14.21
195	14.40	9.69	12.82	14.48	14.08
200	14.23	9.59	12.71	14.33	13.92
205	14.14	9.52	12.58	14.21	13.80
210	14.00	9.44	12.48	14.06	13.74
215	13.87	9.36	12.33	13.97	13.61
220	13.76	9.32	12.25	13.84	13.47
225	13.61	9.23	12.13	13.68	13.35
230	13.49	9.13	12.05	13.55	13.21
235	13.40	9.08	11.97	13.47	13.15
240	13.27	9.00	11.86	13.42	13.08
245	13.16	8.93	11.81	13.30	12.98
250	13.07	8.87	11.68	13.24	12.93
255	12.95	8.80	11.62	13.17	12.78
260	12.88	8.73	11.58	13.07	12.67
265	12.78	8.66	11.42	12.95	12.62
270	12.67	8.59	11.35	12.87	12.52
275	12.56	8.53	11.28	12.79	12.45
280	12.50	8.45	11.19	12.66	12.38
285	12.38	8.39	11.09	12.58	12.25

Source: Adapted from Hudson, J. E. et al., *Int. J. Mass Spectrom.*, 248, 42, 2006.

Note: Tabulated values courtesy of Prof. Harland. See Figure A7.3 for graphical presentation.

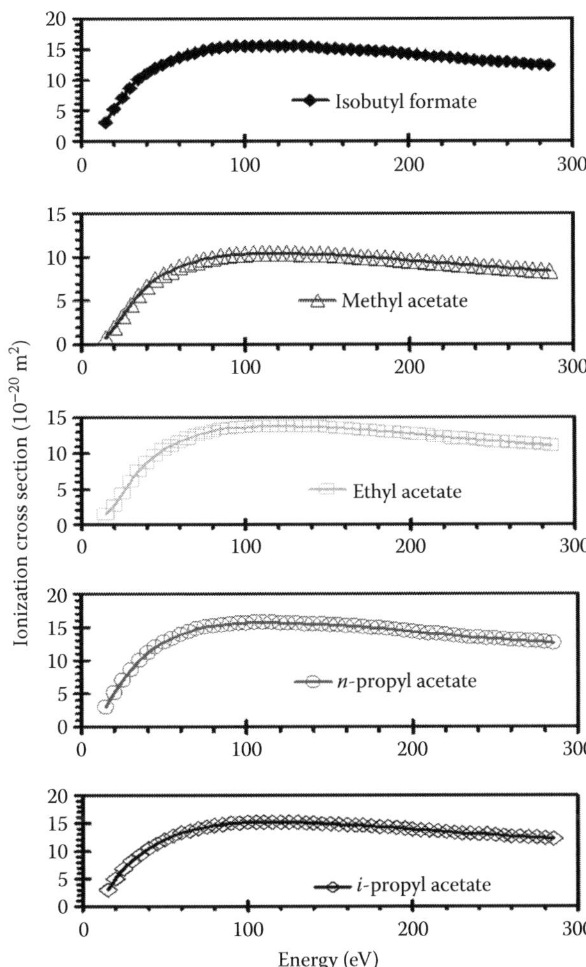

FIGURE A7.3 Ionization cross sections in esters (not included in main sections). (Adapted from Bart, M. et al., *Phys. Chem. Chem. Phys.*, 3, 800, 2001.) Tabulated values are due to courtesy of Professor Harland.

TABLE A7.5
Ionization Cross Sections in Complex Molecules

Energy (eV)	*n*-Butyl acetate	*i*-Butyl acetate	*s*-Butyl acetate	*t*-Butyl acetate	*n*-Amyl formate	*n*-Amyl acetate	*i*-Amyl acetate
15	3.61	3.55	3.58	3.74	3.68	4.12	3.87
20	5.98	6.02	6.08	6.26	6.22	6.94	6.55
25	8.10	8.18	8.31	8.46	8.46	9.44	8.95
30	9.90	9.89	10.05	10.29	10.24	11.49	10.89
35	11.48	11.35	11.54	11.75	11.76	13.16	12.58
40	12.98	12.62	12.79	13.00	13.01	14.55	14.02
45	14.06	13.66	13.87	14.00	14.13	15.73	15.19
50	15.00	14.50	14.69	14.85	15.02	16.74	16.13
55	15.60	15.32	15.40	15.57	15.80	17.61	17.03
60	16.13	15.94	16.05	16.26	16.50	18.36	17.83
65	16.64	16.46	16.51	16.76	17.03	18.95	18.40
70	17.02	16.94	16.97	17.19	17.43	19.44	18.97
75	17.41	17.30	17.34	17.58	17.89	19.87	19.27
80	17.73	17.53	17.61	17.86	18.16	20.20	19.52
85	17.92	17.80	17.85	18.13	18.45	20.45	19.72
90	18.06	18.05	18.01	18.24	18.59	20.65	19.87
95	18.17	18.13	18.06	18.42	18.66	20.76	20.04
100	18.23	18.18	18.11	18.46	18.67	20.84	20.06
105	18.27	18.23	18.12	18.49	18.73	20.91	20.22
110	18.29	18.29	18.16	18.54	18.75	20.93	20.42
115	18.29	18.30	18.19	18.53	18.74	20.94	20.52
120	18.25	18.30	18.19	18.52	18.71	20.92	20.45
125	18.27	18.29	18.13	18.50	18.64	20.78	20.34
130	18.19	18.26	18.05	18.44	18.54	20.62	20.24
135	18.14	18.21	18.00	18.45	18.45	20.52	20.24
140	18.01	18.10	17.91	18.29	18.37	20.43	20.09
145	17.95	18.02	17.76	18.17	18.29	20.33	19.92
150	17.79	17.91	17.61	18.05	18.15	20.24	19.72
155	17.69	17.89	17.57	17.92	18.11	20.16	19.70
160	17.62	17.86	17.47	17.85	18.05	20.09	19.76
165	17.55	17.80	17.40	17.84	17.94	19.97	19.67
170	17.45	17.75	17.35	17.78	17.78	19.79	19.62
175	17.32	17.63	17.23	17.68	17.64	19.66	19.50
180	17.19	17.59	17.12	17.58	17.56	19.58	19.34
185	17.06	17.53	16.99	17.44	17.46	19.46	19.18
190	16.93	17.38	16.90	17.32	17.43	19.31	19.09
195	16.78	17.27	16.68	17.15	17.18	19.15	19.00
200	16.62	17.11	16.49	17.01	17.03	18.95	18.93
205	16.45	16.92	16.35	16.81	16.84	18.76	18.80
210	16.27	16.76	16.16	16.73	16.69	18.60	18.67
215	16.13	16.57	16.09	16.59	16.49	18.43	18.59
220	15.97	16.40	15.93	16.44	16.35	18.27	18.38
225	15.81	16.18	15.78	16.23	16.15	18.13	18.20
230	15.68	16.02	15.66	16.09	16.03	17.97	18.01
235	15.54	15.89	15.57	15.89	15.39	17.20	17.56
240	15.39	15.77	15.38	15.77	15.86	17.79	17.89
245	15.33	15.68	15.22	15.63	15.79	17.62	17.77
250	15.26	15.59	15.11	15.50	15.66	17.49	17.66
255	15.12	15.48	14.99	15.44	15.52	17.37	17.59
					15.39	17.20	17.56
260	15.04	15.33	14.89	15.32	15.26	17.05	17.55
265	14.92	15.30	14.73	15.21			
270	14.82	15.19	14.61	15.13	15.11	16.91	17.51
275	14.71	15.10	14.50	15.00	15.00	16.77	17.39
280	14.64	15.06	14.41	14.94	14.90	16.64	17.23
					14.73	16.52	17.12
285	14.53	14.93	14.34	14.84	14.61	16.37	16.97

Source: Adapted from Hudson, J. E. et al., *Int. J. Mass Spectrom.*, 248, 42, 2006.
Note: Tabulated values courtesy of Professor Harland. *n* = normal; *i* = iso; *s* = secondary; *t* = tertiary. See Figures A7.4 and A7.5 for graphical presentation.

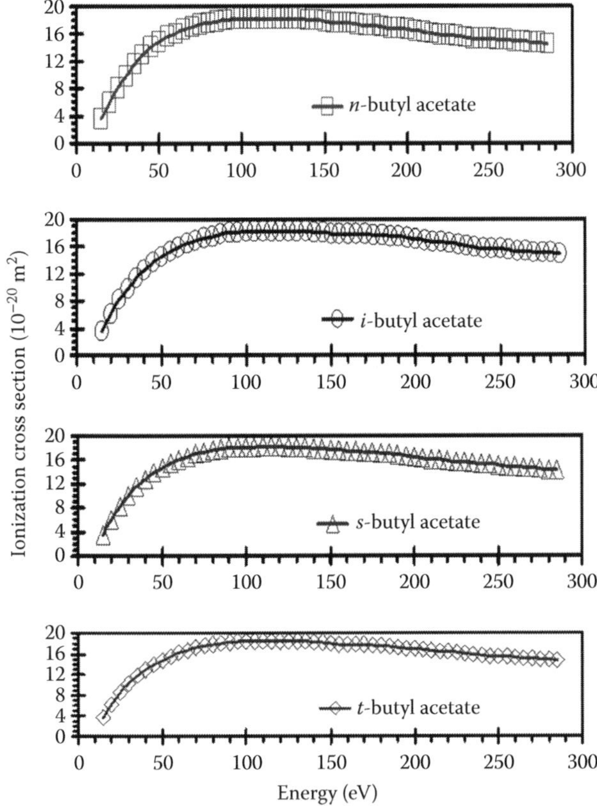

FIGURE A7.4 Ionization cross sections in complex molecules. (Adapted from Hudson, J. E. et al., *Int. J. Mass Spectrom.*, 248, 42, 2006.)

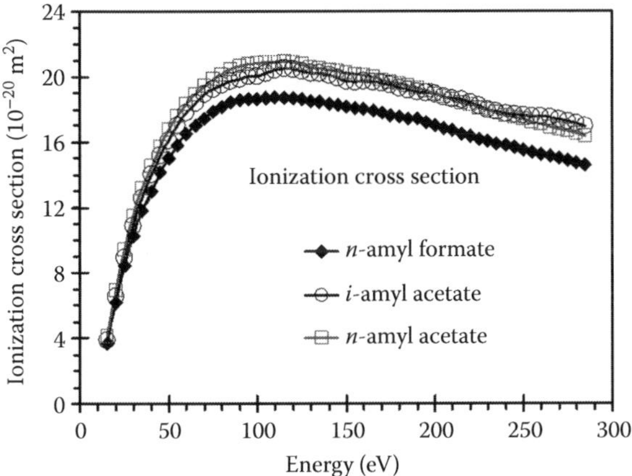

FIGURE A7.5 Ionization cross sections in complex molecules. (Adapted from Hudson, J. E. et al., *Int. J. Mass Spectrom.*, 248, 42, 2006.)

TABLE A7.6
Ionization Cross Sections for Selected Alcohols

Energy (eV)	Butyl alcohol (1-Butanol) n-$C_4H_{10}O$	2-Methyl-1-propanol i-$C_4H_{10}O$	2-Methyl-2-propanol t-$C_4H_{10}O$
16	1.70	1.75	1.20
20	3.20	3.40	2.82
24	5.09	5.06	4.71
28	6.55	6.53	6.20
32	7.70	7.75	7.51
36	8.65	8.80	8.70
41	9.47	9.71	9.75
45	10.17	10.48	10.62
49	10.75	11.11	11.05
53	11.20	11.58	11.53
57	11.60	11.90	11.89
61	11.98	12.18	12.20
65	12.23	12.42	12.57
69	12.40	12.60	12.76
73	12.57	12.74	12.97
77	12.70	12.85	13.15
81	12.81	12.95	13.23
85	12.85	13.01	13.26
89	12.83	13.04	13.33
93	12.82	13.06	13.29
97	12.81	13.04	13.26
101	12.77	13.03	13.27
105	12.76	13.03	13.27
109	12.71	12.95	13.17
113	12.60	12.85	13.07
117	12.54	12.82	13.06
122	12.54	12.81	13.02
126	12.50	12.76	13.00
130	12.40	12.65	12.95
134	12.27	12.56	12.83
138	12.21	12.49	12.75
142	12.19	12.41	12.61
146	12.13	12.29	12.49
150	12.01	12.13	12.46
154	11.89	11.99	12.38
158	11.76	11.90	12.21
162	11.68	11.86	12.19
166	11.63	11.81	12.16
170	11.56	11.73	12.05
174	11.44	11.62	12.04
178	11.28	11.47	00.93
182	11.14	11.34	11.87
186	11.02	11.25	11.80
190	10.93	11.21	11.69
194	10.81	11.20	11.56
198	10.72	11.20	11.48
203	10.65	11.15	11.41
207	10.55	11.08	11.28

Note: Tabulated values by courtesy of Professor Harland (private communication, 2006). For graphical presentation see Figure A7.6.

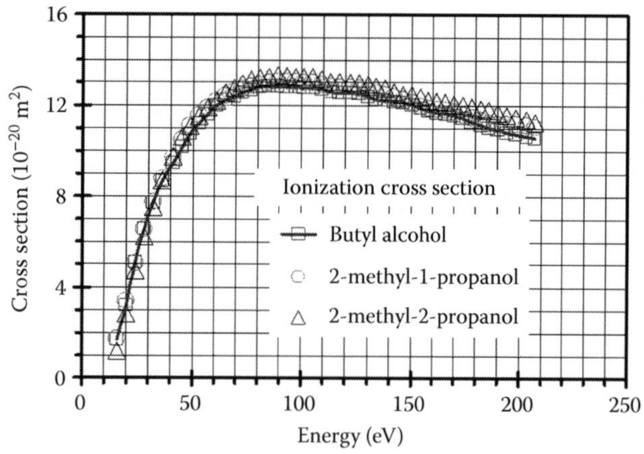

FIGURE A7.6 Ionization cross sections for selected alcohols. (P. W. Harland, private communication, 2006.)

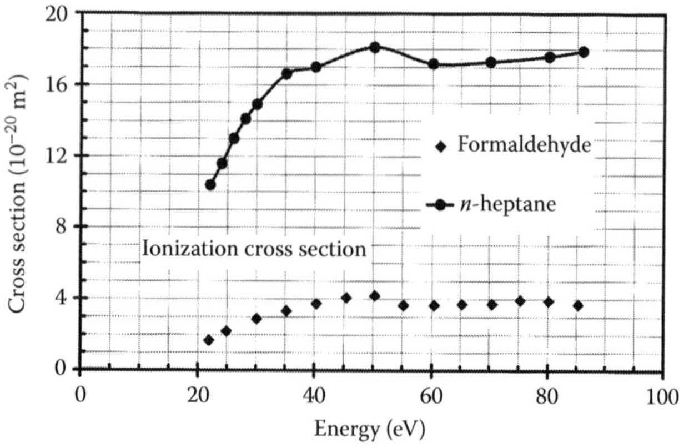

FIGURE A7.7 Ionization cross section for formaldehyde (Adapted from Vacher, J. R. et al., *Chem. Phys. Lett.*, 476, 178, 2009.) and n-heptane (Adapted from Vacher, J. R. et al., *Int. J. Mass Spectrom.*, 295, 78, 2010.)

REFERENCES

Bart, M., P. W. Harland, J. E. Hudson, and C. Vallance, *Phys. Chem. Chem. Phys.*, 3, 800, 2001. (Tabulated values are due to courtesy of Professor Harland).

Basner, R., M. Schmidt, and K. Becker, *J. Chem. Phys.*, 118, 2153, 2003.

Hudson, J. E., Z. F. Weng, C. Vallance, and P. W. Harland, *Int. J. Mass Spectrom.*, 248, 42, 2006.

Vacher, J. R., F. Jorland, N. Blin-Simiand, and S. Pasquiers, *Chem. Phys. Lett.*, 476, 178, 2009.

Vacher, J. R., F. Jorland, N. Blin-Simiand, and S. Pasquiers, *Int. J. Mass Spectrom.*, 295, 78, 2010.

Appendix 8: Important Relationships

CONTENTS

A8.1 DEFINITION OF TERMS FOR ATTACHMENT

As an example we consider the molecule chloromethane (CH_3Cl). The potential curve of the molecule and its negative ion are drawn to scale as shown in Figure A8.1 (Barszczewska et al., 2003). The following quantities are easy to identify (see Table A8.1).

A negative sign is usually assigned to the electron affinity (EA) of the molecule, if the asymptote of the negative ion curve falls below zero energy ($D <$ EA). See hydrogen iodide (HI) for this example. Care should be exercised in using the term "electron affinity" since both the atom and the molecule have different electron affinities.

Appearance potential refers to a particular species of ion and it is related to dissociation energy (according to Sauers et al., 1979)

$$\begin{aligned}\text{Appearance potential (AP)} = {} & \text{Dissociation energy (DE)} \\ & - \text{electron affinity (EA)} \\ & + \text{kinetic energy.}\end{aligned} \quad \text{(A8.1)}$$

In many cases the kinetic energy is negligible and one can write

$$\text{DE} \le \text{AP} + \text{EA} \quad \text{(A8.2)}$$

In the above example, the appearance potential of the Cl^- ion calculated according to Equation A8.1 is 0.53 eV (3.63–3.1 eV). The peak in the attachment cross section is higher, of course, than the appearance potential. In the case of CH_3Cl, the observed peak occurs at 0.8 eV (Chu and Burrow, 1990).

Three situations arise. (1) The dissociation energy of the molecule is higher than the electron affinity of the atom; see hydrogen fluoride (HF). (2) The dissociation energy of the molecule is approximately equal to the electron affinity of the atom; see hydrogen bromide (HBr). (3) The dissociation energy of the molecule is less than the electron affinity of the atom; see hydrogen iodide (HI). In this case, the appearance potential becomes negative according to Equation A8.2. A negative appearance potential implies that the attachment cross section peak occurs at approximately zero electron energy.

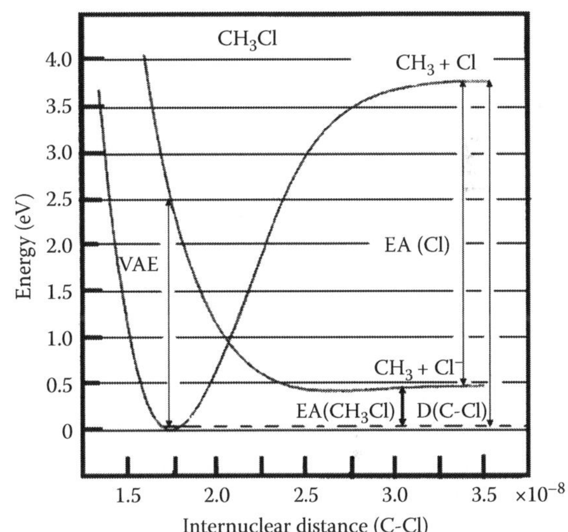

FIGURE A8.1 Potential energy curves for CH_3Cl and the negative ion (schematic and not for scale). (Adapted from Barszczewska, W. et al., *J. Phys. Chem. A*, 107, 11427, 2003.)

A8.2 ATTACHMENT RATE

Attachment of electrons to the neutral atom or molecule is a fundamental process and the attachment rate is the number

TABLE A8.1
Definition of EA

Quantity	Energy (eV)
Vertical attachment energy (VAE)	2.5
Adiabatic electron affinity (AEA or EA)	0.43
Electron affinity (chlorine atom)	3.1
Bond dissociation energy, D (Cl-CH_3)	3.63

Note: Other fragments of dissociation such as HCl are not shown.

of attaching collisions that occur per second in a unit volume of the gas. It is normally reduced by the number density of the gas, and expressed in units of m^3/s.

Attachment is quantified by two parameters. (1) The attachment rate coefficient (m^3/s) or (m^6/s) for two-body and three-body processes, respectively. Traditionally, the attachment rate coefficient is k_a is expressed as a function of mean energy of the swarm or energy of the electron in beam experiments. (2) The attachment coefficient (η/N, m^2) is defined as the number of attachment collisions per unit length along the electric field, reduced by the gas number density. It is expressed as a function of the reduced electric field, E/N (Td = 10^{-21} V m^2). Since the basic process is of the same kind, the relationship between the attachment rate and attachment coefficient is given by the relationship (Skalny et al., 2001; Raju, 2006)

$$k_a(<\varepsilon>) = \left(\frac{2e}{m}\right)^{1/2} \int_0^\infty Q_a(\varepsilon)\varepsilon^{1/2} f(\varepsilon, E/N)\mathrm{d}\varepsilon \quad \text{(A8.3)}$$

where

k_a = attachment rate coefficient
$<\varepsilon>$ = mean energy of the swarm
e = electronic charge
m = electron mass
Q_a = attachment coefficient
ε = electron energy
f = energy distribution function
E = electric field
N = gas number density

If one uses the Maxwellian distribution,

$$f_M(\varepsilon) = 2\pi^{-1/2}(kT)^{-3/2}\varepsilon^{1/2}\exp\left(-\frac{\varepsilon}{kT}\right) \quad \text{(A8.4)}$$

where T is the temperature, and substitution of Equation A8.4 into A8.3 results in an analytical equation for k_a, provided that Q_a can be expressed as an analytical function of ε. At low electron energy, a representation of the type

$$Q_a = \varepsilon^{-n} \quad \text{(A8.5)}$$

where the index n has values between –1.0 and 1.0 applies.

A8.3 ATTACHMENT CROSS SECTIONS

Attachment cross sections (Q_a) are a function of electron energy and the measurements under swarm conditions do not yield the cross section–energy curve in a straightforward manner. The quantity measured is ηW, where η is the attachment coefficient and W is the drift velocity at a given E/N, where E is the electric field and N is the gas number density. ηW is expressed in units of s^{-1}/torr. To convert the measured value of αW to attachment rate, the following relationships (Christodoulides et al., 1979) are required:

$$Q_a(\varepsilon) = \frac{\eta W(<\varepsilon>)}{N_o(2/m)^{1/2}\varepsilon^{1/2}} \quad \text{(A8.6)}$$

where

Q_a = attachment cross section (m^2)
$N_o = 3.22 \times 10^{22}$ m^{-3} at 300 K.

Substituting this value in Equation A8.6 one obtains

$$Q_a(\varepsilon) = 5.246 \times 10^{-29}\frac{\eta W}{(\varepsilon)^{1/2}} \quad \text{(A8.7)}$$

Example: For c-C_4F_6 the measured value of $\eta W = 3.7 \times 10^9$/s and the computed value of $<\varepsilon> = 0.054$ eV, what is the attachment cross section?

From Equation A8.7,

$$Q_a = 5.246 \times 10^{-29} \times \frac{3.7 \times 10^9}{(0.054)^{1/2}} = 8.35 \times 10^{-19}\ \mathrm{m}^2$$

A8.4 MAXIMUM SCATTERING CROSS SECTION

For an electron of energy ε (eV), the de Broglie wavelength is given by λ(m). The maximum scattering cross section is given by

$$Q_{max} = \pi\lambda^2 \quad \text{(A8.8)}$$

where λ is the de Broglie wavelength.

Example: What is the de Broglie wavelength of an electron that has 1 eV energy? What is the maximum scattering cross section at this energy?

The de Broglie wave length is given by Eisberg and Resnick (1985)

$$\lambda = \frac{h}{p} = \frac{h}{m\mathrm{v}} \quad \text{(A8.9)}$$

where h = Planck's constant, m = mass, and v = velocity. The energy of the electron is converted into velocity according to

$$\varepsilon = \frac{1}{2e}mv^2 \quad \text{(A8.10)}$$

where e is the electronic charge. Substituting Equation A8.10 into A8.9 one gets

$$\lambda = \frac{h}{\sqrt{2m\varepsilon}} = \frac{6.626 \times 10^{-34}}{\sqrt{2 \times 9.109 \times 10^{-31} \times 1.0 \times 1.602 \times 10^{-19}}} = 1.227 \times 10^{-9}\ \mathrm{m} \quad \text{(A8.11)}$$

The maximum cross section is given by

$$Q_{max} = \pi\lambda^2 = \pi \times (1.227 \times 10^{-9})^2 = 4.726 \times 10^{-18}\ \mathrm{m}^2 \quad \text{(A8.12)}$$

A short form for Equations A8.11 and A8.12 combined is

$$Q_{\max} = \frac{4.726\times10^{-18}}{\varepsilon}\,\mathrm{m}^2 \quad (A8.13)$$

For a slow electron, the effective range of electron molecule interaction is much shorter than the de Broglie wavelength meaning that the maximum cross section will be lower than that given by Equation A8.13. The maximum cross section is given by (Christodoulides et al., 1979)

$$Q_{\max} = \frac{1.197\times10^{-19}}{\varepsilon\ (\text{in eV})}\,\mathrm{m}^2 \quad (A8.14)$$

Equation A8.14 is used for calculating the maximum cross section shown in selected plots.

Often one desires to calculate the maximum *s*-wave attachment rate coefficient as a function of electron temperature. The equation applicable is (Spanel et al., 1977)

$$k_a(\max) = 5\times10^{-13}\left(\frac{T_e}{300}\right)^{-1/2}\ \mathrm{m^3/s} \quad (A8.15)$$

where T_e is the electron temperature in *K*. For electron temperature expressed in eV the *s*-wave maximum attachment rate is

$$K_a(\max) = 5\times10^{-13}\left(\frac{\varepsilon}{0.0259}\right)^{-1/2} \quad (A8.16)$$

A8.5 REDUCED ELECTRIC FIELD

The reduced electric field is expressed by *E/N* in units of V m², where *E* = electric field in V/m and *N* = gas number density in 1 m³. The relationships between measured gas pressure (*p*) in units of torr and temperature *T* (*K*) and *N* are given by Raju (2006):

$$N = 3.54\times10^{22}\times273\times\frac{p}{T}\,\mathrm{m}^{-3} \quad (A8.17)$$

$$\frac{E}{N} = 2.82\times10^{-21}\times\frac{E(\mathrm{V/cm})}{p(\mathrm{Torr})}\times\frac{T(K)}{273}\,\mathrm{V\,m^2} \quad (A8.18)$$

Substituting the conversion factor:

$$1\,\mathrm{Td}\ (\mathrm{Townsend}) = 1\times10^{-21}\ \mathrm{V\,m^2} \quad (A8.19)$$

We have the following relationships:

$$\left.\begin{aligned}\frac{E}{N}(\mathrm{Td}) &= 2.82\times\frac{E(\mathrm{V/cm})}{p(\mathrm{torr})}\quad \text{at } 273\,\mathrm{K}\\ \frac{E}{N}(\mathrm{Td}) &= 3.10\times\frac{E(\mathrm{V/cm})}{p(\mathrm{torr})}\quad \text{at } 300\,\mathrm{K}\end{aligned}\right\} \quad (A8.20)$$

A8.6 REDUCED TRANSPORT COEFFICIENTS

Mobility is defined by

$$\mu(\mathrm{V^{-1}cm^2/s}) = \frac{W(\mathrm{cm/s})}{E(\mathrm{V/cm})} \quad (A8.21)$$

Product of number density and mobility is defined by

$$\left.\begin{aligned}&\mu(\mathrm{V^{-1}\,cm^2/s})\times N(\mathrm{cm^{-3}}) = W(\mathrm{cm/s})\times\frac{N}{E}(\mathrm{V^{-1}/cm^2})\\ &\mu N(\mathrm{V^{-1}/m/s}) = 10^2\times\mu N(\mathrm{V^{-1}/cm/s})\end{aligned}\right\} \quad (A8.22)$$

Reduced mobility is defined by

$$\mu_0(\mathrm{V^{-1}cm^2/s}) = \frac{W(\mathrm{cm/s})}{E(\mathrm{V/cm})}\times\frac{N}{N_0}$$

where μ_0 = reduced mobility; N_0 = number density at 760 torr; $N_0 = 2.69\times10^{25}\ \mathrm{m^{-3}}$ at 1 atmospheric pressure and 273 K.

The most convenient form is

$$\mu N(\mathrm{m/s/Td}) = \mu_0 N_0 = W(\mathrm{m/s})\times\frac{N}{E}(\mathrm{Td^{-1}}) \quad (A8.23)$$

The density-normalized diffusion coefficient is defined as the product of diffusion coefficient (*D*, cm²/s) and gas pressure (*p*, torr). Conversion to SI units is carried out according to

$$ND(\mathrm{m/s}) = 3.22\times10^{22}\times10^{-4}\times D(\mathrm{cm^2/s/torr})\times p(\mathrm{torr}) \quad (A8.24)$$

As an example, the density-normalized diffusion coefficient in water (H_2O) is given as $Dp = 0.1$ cm² μ/s /torr at $E/p = 15$ V/cm/torr (Wilson et al., 1975). Substituting these values into Equation A8.24 we get the density-normalized diffusion coefficient as $ND = 3.22\times10^{23}\ \mathrm{m^{-1}/s}$ at 45 Td.

A8.7 DENSITY-REDUCED GROWTH COEFFICIENTS

The reduced ionization coefficient is defined in older units as α/p, where α is in units of cm⁻¹ and *p* is in units of Torr. The conversion factor is

$$\frac{\alpha}{N}(\mathrm{m^2}) = 3.11\times10^{-21}\times\frac{\alpha}{P}(\mathrm{cm^{-1}/torr}) \quad (A8.25)$$

A8.8 GAS CONSTANTS

The gas constants are defined by

$$\frac{\alpha}{p} = A\exp\left(-\frac{Bp}{E}\right)\text{cm}^{-1}/\text{torr} \qquad (A8.26)$$

where A and B are gas constants in units of (cm^{-1}/torr) and (V/cm/torr), respectively. The equation in SI units is

$$\frac{\alpha}{N} = F\exp\left(-\frac{GN}{E}\right)\text{m}^2 \qquad (A8.27)$$

where

$$\left.\begin{aligned} F &= \frac{A\times 100}{N_0} = 3.11\times 10^{-21}\times A\,\text{m}^2 \\ G &= 3.11\times B\,\text{Td} \end{aligned}\right\} \text{at } 300\,\text{K} \qquad (A8.28)$$

$$\left.\begin{aligned} F &= \frac{A\times 100}{N_0} = 2.82\times 10^{-21}\times A\,\text{m}^2 \\ G &= 2.82\times B\,\text{Td} \end{aligned}\right\} \text{at } 273\,\text{K} \qquad (A8.29)$$

In restricted number of gases, Equation A8.26 is modified as (Raju, 2006)

$$\frac{\alpha}{N} = C\exp\left(-\frac{D}{(E/N)^{1/2}}\right)\text{m}^2 \qquad (A8.30)$$

Then $C = F$ and $D = \sqrt{3.1}$ B or $\sqrt{2.82}$ B, according to whether $T = 300$ K or 273 K.

REFERENCES

Barszczewska, W., J. Kopyra, J. Wnorowska and I. Szamrej, *J. Phys. Chem. A*, 107, 11427, 2003.

Christodoulides, A. A., L. G. Christophorou, R. Y. Pai, and C. M. Tung, *J. Chem. Phys.*, 70, 1156, 1979 (see page 1159 of the article).

Chu, S. C. and P. D. Burrow, *Chem. Phys. Lett.*, 172, 17, 1990.

Eisberg, R. and R. Resnick, *Quantum Physics of Atoms, Molecules, Solids, Nuclei and Particles*, II edn, John Wiley and Sons, New York, NY, 1985, p. 56.

Raju, G. G., *Gaseous Electronics: Theory and Practice*, CRC Press, Taylor & Francis, Boca Raton, FL, 2006, p. 87.

Sauers, I., L. G. Christophorou, and J. G. Carter, *J. Chem. Phys.*, 71, 3016, 1979.

Skalny, J. D., S. Matejcik, T. Mikoviny, J. Vencko, G. Senn, A. Stamatovic, and T. D. Märk, *Int. J. Mass Spectrom*, 205, 77, 2001.

Spanel, P., D. Smith, S. Matejcik, A. Kiendler, and T. D. Märk, *Int. J. Mass Spectrom. Ion Proc.*, 167/168, 1, 1977.

Wilson, J. F., F. J. Davies, D. R. Nelson, and R. N. Compton, *J. Chem. Phys.*, 62, 4204, 1975.

Appendix 9: Quadrupole Moments of Target Particles

CONTENTS

The quadrupole moment of a molecule (μ_Q) is expressed in units of DÅ (Debye Ångstrom) with a value of 3.3351×10^{-30} (Coulomb meter) $\times 10^{-10}$ (meter) $= 3.3351 \times 10^{-40}$ C m^2. The earlier unit is

$$ea_0^2 = 1.602 \times 10^{-19} \text{ (Coulomb)} \times (5.2917 \times 10^{-11})^2 \text{ (meter}^2) = 4.4859 \times 10^{-40} \text{ Cm}^2$$

Molecule		μ_Q	
Name	Formula	($\times 10^{-41}$ C m^2)	Reference
Ammonia	NH_3	−77.40	
Benzene	C_6H_6	−304	Ritchie and Watson (2000)
Bromine	Br_2	171.8	
Bromine monofluoride	BrF	30.32	
Carbon dioxide	CO_2	−142.7	Jain and Baluja (1992)
Carbon disulfide	CS_2	60.3	Sohn et al. (1987)
Carbon monosulfide	CS	3.36	
Carbon monoxide	CO	−94.71	
Carbon tetrachloride	CCl_4	0	
Carbonyl sulfide	COS	103.8	Sohn et al. (1987)
Chlorine	Cl_2	89.45	
Chlorine monofluoride	ClF	44.70	
Chlorocyanogen	ClCN	−130.1	
Difluorine monoxide	F_2O	70.06	
Difluoromethane	CH_2F_2	73.39	
Ethylene	C_2H_4	130.8	
Fluorine	F_2	28.35	
Formaldehyde	CH_2O	−2.0162	
Hexafluorobenzene	C_6F_6	283	Ritchie and Watson (2000)
Hydrogen	H_2	27.80	Engelhardt and Phelps (1963)
		−27.80	Jain and Baluja (1992)
Hydrogen chloride	HCl	128.2	Jain and Baluja (1992)
Hydrogen cyanide	HCN	79.58	Jain and Baluja (1992)
Hydrogen fluoride	HF	78.03	Jain and Baluja (1992)
Hydrogen sulfide	H_2S	31.85	Jain and Baluja (1992)
Lithium	Li_2	462.8	Jain and Baluja (1992)
Methane	CH_4	0	Jain and Baluja (1992)
Methyl bromide	CH_3Br	118.4	
Nitrogen	N_2	46.65	Engelhardt et al. (1964)
Nitric oxide	NO	121.8	
Oxygen	O_2	11.43	Jain and Baluja (1992)
Ozone	O_3	263.5	
Phosphine	PH_3	37.1	Jain and Baluja (1992)
Phosphorous trifluoride	PF_3	804	
Propyne	C_3H_4	160.8	
Silane	SiH_4	0	Jain and Baluja (1992)
Sodium fluoride	NaF	−65.69	
Sulfur dioxide	SO_2	−176.8	
Sulfur hexafluoride	SF_6	0	
Sulfur tetrafluoride	SF_4	336.9	
Trifluoroborane	BF_3	126.0	
Water	H_2O	−4.337	Jain and Baluja (1992)

The conversion factor is

$$\text{DÅ} = 0.749 \times ea_0^2$$

REFERENCES

Engelhardt, A. G. and A. V. Phelps, *Phys. Rev.*, 131, 2115, 1963.

Engelhatdt, A. G., A. V. Phelps, and C. G. Risk, *Phys. Rev.*, 135, A1566, 1964.

Jain, A. and K. L. Baluja, *Phys. Rev. A*, 45,202, 1992.

Ritchie, G. L. D. and J. N. Watson, *Chem. Phys. Lett.*, 322, 143, 2000.

Sohn, W., K.-H. Kochem, K. M. Scheuerlein, K. Jung, and H. Ehrhardt, *J. Phys. B: At. Mol. Phys.*, 20, 3217, 1987.

Appendix 10: Relative Dielectric Strength of Gases

CONTENTS

TABLE A10.1
Relative Electric Strengths (RES) at 0.1 MPa and Molecular Weights of Dielectric Gases

Number	Gas: Formula	Gas: Name	Molecular Weight	RES	Chapter
1	He	Helium	4	0.15	3
2	Ne	Neon	20	0.25	6
3	H_2	Hydrogen	2	0.50	18
4	CH_3NH_2	Methyl amine	31	0.80	88
5	CO_2	Carbon dioxide	44	0.88	27
6	N_2	Nitrogen	28	1.0	25
7	C_2H_5Cl	Chloroethane	64	1.00	95
8	CH_4	Methane	16	1.00	64
9	CH_3CHO	Ethanal	44	1.01	87
10	CF_4	Tetrafluoromethane	88	1.01	70
11	CO	Carbon monoxide	28	1.02	11
12	S_2Cl_2	Sulfur chloride	135	1.02	
13	CH_2ClF	Chlorofluoromethane	68	1.03	A4
14	$(CH_3)_2NH$	Dimethylamine	45	1.04	
15	$C_2H_5NH_2$	Ethylamine	45	1.06	
16	C_2H_2	Acetylene	26	1.10	38
17	SO_2	Sulfur dioxide	64	1.3	36
18	$CHCl_2F$	Dichlorofluoromethane	103	1.33	A4, A5
19	$CBrF_3$	Bromotrifluoromethane	149	1.35	49
20	$CHClF_2$	Chlorodifluoromethane	86	1.40	A4, A5
21	$CClF_3$	Chlorotrifluoromethane	104	1.43	53
22	Cl_2	Chlorine	71	1.55	12
23	C_2F_6	Hexafluoroethane	138	1.82	103
24	C_3F_8	Perfluoropropane	188	2.19	115
25	CH_3I	Iodomethane	142	2.20	63
26	C_2ClF_5	Chloropentafluoroethane	154	2.3	A5
27	CCl_2F_2	Dichlorofluoromethane	121	2.42	57
28	SF_6	Sulfur hexafluoride	146	2.5	90
29	SOF_2	Thionyl fluoride	86	2.50	
30	$C_2Cl_2F_4$	Dichlorotetrafluoroethane	171	2.52	A5
31	$ClFO_3$	Perchlorylfluoride	102	2.73	
32	C_4F_8	Perfluoro-2-butene	200	2.8	118
33	C_4F_{10}	Perfluorobutane	238	3.08	139
34	CF_3CN	Trifluoroacetonitrile	95	3.5	A7
35	CCl_3F	Trichlorofluoromethane	137	3.50	73

continued

TABLE A10.1 (continued)
Relative Electric Strengths (RES) at 0.1 MPa and Molecular Weights of Dielectric Gases

Number	Gas: Formula	Gas: Name	Molecular Weight	RES	Chapter
36	$CHCl_2F$	Dichlorofluoromethane	119	4.2	A4, A5
37	C_3F_5N	Pentafluoronitropropene	145	4.5	
38	C_4F_7N	Heptafluoronitrobutene	195	5.5	
39	C_5F_8	Pentafluorobutadiene	212	5.5	
40	CCl_4	Carbon tetrachloride	154	6.33	50

Sources: Data from Raju, G. G., *Gaseous Electronics*, Taylor & Francis, Boca Raton, FL, 2005; Vijh, A. K., *IEEE Trans. Elec. Insul.*, EI-12, 313, © (1977) IEEE.

Note: Numbers in Figure A10.1 correspond to the first column. Rearranged by the author in order of increasing RES. RES 1 = 30.54 kV for air at 0.1 MPa.

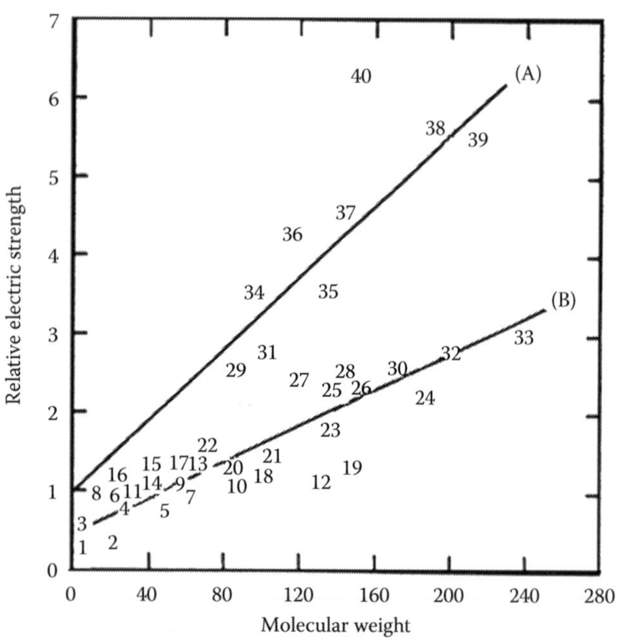

FIGURE A10.1 Relative electric strength as a function of molecular weight in gases. The numbers on the plot correspond to Table A10.1. Class A gases have a higher dielectric strength than class B gases. Class A gases have a –CN group or a –SO fragment and do not contain a hydrogen atom.

REFERENCES

Raju, G. G., *Gaseous Electronics*, Taylor & Francis, Boca Raton, FL, 2005.

Vijh, A. K., *IEEE Trans. Elec. Insul.*, EI-12, 313, 1977.

Index

B

C

D

F

G

H

I

Q

R

S

T

U

V

W

X

Z

Printed and bound by CPI Group (UK) Ltd, Croydon, CR0 4YY
24/10/2024
01778991-0001